QUANTUM THEORY

Quantum Theory
A Wide Spectrum

by

E.B. MANOUKIAN

*Suranaree University of Technology,
Nakhon Ratchasima,
Thailand*

 Springer

A C.I.P. Catalogue record for this book is available from the Library of Congress.

ISBN-10 1-4020-4189-6 (HB)
ISBN-13 978-1-4020-4189-1 (HB)
ISBN-10 1-4020-4190-X (e-book)
ISBN-13 978-1-4020-4190-7 (e-book)

Published by Springer,
P.O. Box 17, 3300 AA Dordrecht, The Netherlands.

www.springer.com

Printed on acid-free paper

All Rights Reserved
© 2006 Springer
No part of this work may be reproduced, stored in a retrieval system, or transmitted
in any form or by any means, electronic, mechanical, photocopying, microfilming, recording
or otherwise, without written permission from the Publisher, with the exception
of any material supplied specifically for the purpose of being entered
and executed on a computer system, for exclusive use by the purchaser of the work.

Contents

Acknowledgments XV

Preface XVII

1 **Fundamentals** 1
   1.1 Selective Measurements 2
   1.2 A, B, C to Probabilities 8
   1.3 Expectation Values and Matrix Representations 10
      1.3.1 Probabilities and Expectation Values 10
      1.3.2 Representations of Simple Machines 13
   1.4 Generation of States, Inner-Product Spaces, Hermitian
      Operators and the Eigenvalue Problem 15
      1.4.1 Generation of States and Vector Spaces 16
      1.4.2 Transformation Functions and Wavefunctions in
             Different Descriptions 18
      1.4.3 An Illustration 19
      1.4.4 Generation of Inner Product Spaces 23
      1.4.5 Hermitian Operators and the Eigenvalue Problem 24
   1.5 Pure Ensembles and Mixtures 25
   1.6 Polarization of Light: An Interlude 29
   1.7 The Hilbert Space; Rigged Hilbert Space 33
   1.8 Self-Adjoint Operators and Their Spectra 39
   1.9 Wigner's Theorem on Symmetry Transformations 55
   1.10 Probability, Conditional Probability and Measurement 65
      1.10.1 Correlation of a Physical System and an Apparatus ... 66
      1.10.2 Probability and Conditional Probability 68
      1.10.3 An Exactly Solvable Model 70
   Problems 79

2 Symmetries and Transformations ... 81
- 2.1 Galilean Space-Time Coordinate Transformations ... 81
- 2.2 Successive Galilean Transformations and the Closed Path ... 86
- 2.3 Quantum Galilean Transformations and Their Generators ... 89
- 2.4 The Transformation Function $\langle \mathbf{x}|\mathbf{p}\rangle$ ... 98
- 2.5 Quantum Dynamics and Construction of Hamiltonians ... 100
  - 2.5.1 The Time Evolution: Schrödinger Equation ... 100
  - 2.5.2 Time as an Operator? ... 101
  - 2.5.3 Construction of Hamiltonians ... 102
  - 2.5.4 Multi-Particle Hamiltonians ... 104
  - 2.5.5 Two-Particle Systems and Relative Motion ... 104
  - 2.5.6 Multi-Electron Atoms with Positions of the Electrons Defined Relative to the Nucleus ... 105
  - 2.5.7 Decompositions into Clusters of Particles ... 106
- Appendix to §2.5: Time-Evolution for Time-Dependent Hamiltonians ... 109
- 2.6 Discrete Transformations: Parity and Time Reversal ... 112
- 2.7 Orbital Angular Momentum and Spin ... 116
- 2.8 Spinors and Arbitrary Spins ... 121
  - 2.8.1 Spinors and Generation of Arbitrary Spins ... 121
  - 2.8.2 Rotation of a Spinor by 2π Radians ... 129
  - 2.8.3 Time Reversal and Parity Transformation ... 130
  - 2.8.4 Kramers Degeneracy ... 132
- Appendix to §2.8: Transformation Rule of a Spinor of Rank One Under a Coordinate Rotation ... 133
- 2.9 Supersymmetry ... 136
- Problems ... 139

3 Uncertainties, Localization, Stability and Decay of Quantum Systems ... 143
- 3.1 Uncertainties, Localization and Stability ... 143
  - 3.1.1 A Basic Inequality ... 143
  - 3.1.2 Uncertainties ... 144
  - 3.1.3 Localization and Stability ... 145
  - 3.1.4 Localization, Stability and Multi-Particle Systems ... 148
- 3.2 Boundedness of the Spectra of Hamiltonians From Below ... 151
- 3.3 Boundedness of Hamiltonians From Below: General Classes of Interactions ... 152
- 3.4 Boundedness of Hamiltonian From Below: Multi-Particle Systems ... 163
  - 3.4.1 Multi-Particle Systems with Two-Body Potentials ... 164
  - 3.4.2 Multi-Particle Systems and Other Potentials ... 166
  - 3.4.3 Multi-Particle Systems with Coulomb Interactions ... 167
- 3.5 Decay of Quantum Systems ... 168
- Appendix to §3.5: The Paley-Wiener Theorem ... 174

Problems 178

4 Spectra of Hamiltonians 181
 4.1 Hamiltonians with Potentials Vanishing at Infinity 182
 4.2 On Bound-States 187
  4.2.1 A Potential Well 187
  4.2.2 Limit of the Potential Well 190
  4.2.3 The Dirac Delta Potential 190
  4.2.4 Sufficiency Conditions for the Existence of a Bound-State for $\nu = 1$ 192
  4.2.5 Sufficiency Conditions for the Existence of a Bound-State for $\nu = 2$ 194
  4.2.6 Sufficiency Conditions for the Existence of a Bound-State for $\nu = 3$ 195
  4.2.7 No-Binding Theorems 197
 4.3 Hamiltonians with Potentials Approaching Finite Constants at Infinity 199
 4.4 Hamiltonians with Potentials Increasing with No Bound at Infinity 200
 4.5 Counting the Number of Eigenvalues 203
  4.5.1 General Treatment of the Problem 203
  4.5.2 Counting the Number of Eigenvalues 206
  4.5.3 The Sum of the Negative Eigenvalues 216
 Appendix to §4.5: Evaluation of Certain Integrals 219
 4.6 Lower Bounds to the Expectation Value of the Kinetic Energy: An Application of Counting Eigenvalues 220
  4.6.1 One-Particle Systems 220
  4.6.2 Multi-Particle States: Fermions 222
  4.6.3 Multi-Particle States: Bosons 224
 4.7 The Eigenvalue Problem and Supersymmetry 224
  4.7.1 General Aspects 224
  4.7.2 Construction of Supersymmetric Hamiltonians 226
  4.7.3 The Eigenvalue Problem 230
 Problems 244

5 Angular Momentum Gymnastics 249
 5.1 The Eigenvalue Problem 251
 5.2 Matrix Elements of Finite Rotations 254
 5.3 Orbital Angular Momentum 258
  5.3.1 Transformation Theory 258
  5.3.2 Half-Odd Integral Values? 259
  5.3.3 The Spherical Harmonics 262
  5.3.4 Addition Theorem of Spherical Harmonics 267
 5.4 Spin 269
  5.4.1 General Structure 269

		5.4.2 Spin 1/2 270
		5.4.3 Spin 1 272
		5.4.4 Arbitrary Spins 274
	5.5	Addition of Angular Momenta 275
	5.6	Explicit Expression for the Clebsch-Gordan Coefficients 284
	5.7	Vector Operators 290
	5.8	Tensor Operators 296
	5.9	Combining Several Angular Momenta: 6-j and 9-j Symbols .. 304
	5.10	Particle States and Angular Momentum; Helicity States 307
		5.10.1 Single Particle States 307
		5.10.2 Two Particle States 317
	Problems 324

6 Intricacies of Harmonic Oscillators 329
   6.1 The Harmonic Oscillator 329
   6.2 Transition to and Between Excited States in the Presence of a Time-Dependent Disturbance 335
   6.3 The Harmonic Oscillator in the Presence of a Disturbance at Finite Temperature 340
   6.4 The Fermi Oscillator 343
   6.5 Bose-Fermi Oscillators and Supersymmetric Bose-Fermi Transformations 346
   6.6 Coherent State of the Harmonic Oscillator 349
   Problems 356

7 Intricacies of the Hydrogen Atom 359
   7.1 Stability of the Hydrogen Atom 360
   7.2 The Eigenvalue Problem 363
   7.3 The Eigenstates 366
   7.4 The Hydrogen Atom Including Spin and Relativistic Corrections 370
   Appendix to §7.4: Normalization of the Wavefunction Including Spin and Relativistic Corrections 378
   7.5 The Fine-Structure of the Hydrogen Atom 379
   Appendix to §7.5: Combining Spin and Angular Momentum in the Atom 383
   7.6 The Hyperfine-Structure of the Hydrogen Atom 384
   7.7 The Non-Relativistic Lamb Shift 391
      7.7.1 The Radiation Field 391
      7.7.2 Expression for the Energy Shifts 394
      7.7.3 The Lamb Shift and Renormalization 398
   Appendix to §7.7: Counter-Terms and Mass Renormalization 401
   7.8 Decay of Excited States 403
   7.9 The Hydrogen Atom in External Electromagnetic Fields..... 406
      7.9.1 The Atom in an External Magnetic Field 406

	7.9.2 The Atom in an External Electric Field 412

Problems 414

8 Quantum Physics of Spin 1/2 and Two-Level Systems; Quantum Predictions Using Such Systems 419

8.1 General Properties of Spin 1/2 and Two-Level Systems 420
    8.1.1 General Aspects of Spin 1/2 420
    8.1.2 Spin 1/2 in External Magnetic Fields 423
    8.1.3 Two-Level Systems; Exponential Decay 427
8.2 The Pauli Hamiltonian; Supersymmetry 432
    8.2.1 The Pauli Hamiltonian 432
    8.2.2 Supersymmetry 434
8.3 Landau Levels; Expression for the g-Factor 436
    8.3.1 Landau Levels 436
    8.3.2 Expression for the g-Factor 440
8.4 Spin Precession and Radiation Losses 441
8.5 Anomalous Magnetic Moment of the Electron 444
    8.5.1 Observational Aspect of the Anomalous Magnetic Moment 445
    8.5.2 Computation of the Anomalous Magnetic Moment 446
8.6 Density Operators and Spin 453
    8.6.1 Spin in a General Time-Dependent Magnetic Field ... 453
    8.6.2 Scattering of Spin 1/2 Particle off a Spin 0 Target 454
    8.6.3 Scattering of Spin 1/2 Particles off a Spin 1/2 Target . 459
8.7 Quantum Interference and Measurement; The Role of the Environment 462
    8.7.1 Interaction with an Apparatus and Unitary Evolution Operator 463
    8.7.2 Interaction with a Harmonic Oscillator in a Coherent State 467
    8.7.3 The Role of the Environment 469
8.8 Ramsey Oscillatory Fields Method and Spin Flip; Monitoring the Spin 473
    8.8.1 Ramsey Apparatus and Interference; Spin Flip 473
    8.8.2 Monitoring the Spin 478
8.9 Schrödinger's Cat and Quantum Decoherence 482
8.10 Bell's Test 486
    8.10.1 Bell's Test 486
    8.10.2 Basic Processes 490
Appendix to §8.10. Entangled States; The C-H Inequality 499
8.11 Quantum Teleportation and Quantum Cryptography 501
    8.11.1 Quantum Teleportation 501
    8.11.2 Quantum Cryptography 503

	8.12	Rotation of a Spinor	508
	8.13	Geometric Phases	513
		8.13.1 The Berry Phase and the Adiabatic Regime	513
		8.13.2 Degeneracy	518
		8.13.3 Aharonov-Anandan (AA) Phase	520
		8.13.4 Samuel-Bhandari (SB) Phase	529
	8.14	Quantum Dynamics of the Stern-Gerlach Effect	531
		8.14.1 The Quantum Dynamics	531
		8.14.2 The Intensity Distribution	535
	Append	dix to §8.14: Time Evolution and Intensity Distribution	540
	Proble	ems	544

9 Green Functions 547
  9.1 The Free Green Functions 548
  9.2 Linear and Quadratic Potentials 555
  9.3 The Dirac Delta Potential 558
  9.4 Time-Dependent Forced Dynamics 561
  9.5 The Law of Reflection and Reconciliation with the Classical Law ... 565
  9.6 Two-Dimensional Green Function in Polar Coordinates: Application to the Aharonov-Bohm Effect ... 570
  9.7 General Properties of the Full Green Functions and Applications ... 580
    9.7.1 A Matrix Notation 580
    9.7.2 Applications 582
    9.7.3 An Integral Expression for the (Homogeneous) Green Function ... 586
  9.8 The Thomas-Fermi Approximation and Deviations Thereof 587
  9.9 The Coulomb Green Function: The Full Spectrum 590
    9.9.1 An Integral Equation 590
    9.9.2 The Negative Spectrum $p^0 < 0, \lambda < 0$ 594
    9.9.3 The Positive Spectrum $p^0 > 0$ 596
  Problems 598

10 Path Integrals 601
  10.1 The Free Particle 602
  10.2 Particle in a Given Potential 604
  10.3 Charged Particle in External Electromagnetic Fields: Velocity Dependent Potentials ... 608
  10.4 Constrained Dynamics 614
    10.4.1 Classical Notions 614
    10.4.2 Constrained Path Integrals 623
    10.4.3 Second Class Constraints and the Dirac Bracket 627
  10.5 Bose Excitations 628

    10.6 Grassmann Variables: Fermi Excitations 633
        10.6.1 Real Grassmann Variables 633
        10.6.2 Complex Grassmann Variables 637
        10.6.3 Fermi Excitations 640
   Problems 645

11 The Quantum Dynamical Principle 649
    11.1 The Quantum Dynamical Principle 650
    11.2 Expressions for Transformations Functions 656
    11.3 Trace Functionals 665
    11.4 From the Quantum Dynamical Principle to Path Integrals ... 669
    11.5 Bose/Fermi Excitations 672
    11.6 Closed-Time Path and Expectation-Value Formalism 675
   Problems 681

12 Approximating Quantum Systems 683
    12.1 Non-Degenerate Perturbation Theory 684
    12.2 Degenerate Perturbation Theory 688
    12.3 Variational Methods 690
    12.4 High-Order Perturbations, Divergent Series; Padé
        Approximants 695
    12.5 WKB Approximation 703
        12.5.1 General Theory 703
        12.5.2 Barrier Penetration 709
        12.5.3 WKB Quantization Rules 712
        12.5.4 The Radial Equation 715
    12.6 Time-Dependence; Sudden Approximation and the Adiabatic
        Theorem 716
        12.6.1 Weak Perturbations 717
        12.6.2 Sudden Approximation 720
        12.6.3 The Adiabatic Theorem 724
    12.7 Master Equation; Exponential Law, Coupling to the
        Environment 727
        12.7.1 Master Equation 728
        12.7.2 Exponential Law 733
        12.7.3 Coupling to the Environment 734
   Problems 736

13 Multi-Electron Atoms: Beyond the Thomas-Fermi Atom .. 739
    13.1 The Thomas-Fermi Atom 740
    Appendix A To §13.1: The TF Energy Gives the Leading
        Contribution to $E(Z)$ for Large Z 746
    Appendix B to §13.1: The TF Density Actually Gives the Smallest
        Value for the Energy Density Functional in (13.1.6) 752
    13.2 Correction due to Electrons Bound Near the Nucleus 753

- 13.3 The Exchange Term 756
- 13.4 Quantum Correction 759
- 13.5 Adding Up the Various Contributions: Estimation of $E(Z)$ ... 762
- Problems 762

14 Quantum Physics and the Stability of Matter 765
- 14.1 Lower Bound to the Multi-Particle Repulsive Coulomb Potential Energy 767
- Appendix to §14.1: A Thomas-Fermi-Like Energy Functional and No Binding 769
- 14.2 Lower and Upper Bounds for the Ground-State Energy and the Stability of Matter 774
  - 14.2.1 A Lower Bound 774
  - 14.2.2 Upper Bounds 777
- 14.3 Investigation of the High-Density Limit for Matter and Its Stability... 780
  - 14.3.1 Upper Bound of the Average Kinetic Energy of Electrons in Matter 780
  - 14.3.2 Inflation of Matter................................. 781
- 14.4 The Collapse of "Bosonic Matter"......................... 783
  - 14.4.1 A Lower Bound 784
  - 14.4.2 An Upper Bound.................................... 786
- Appendix to §14.4: Upper Bounds for $\langle H_1 \rangle$ in (14.4.47) 793
- Problems 796

15 Quantum Scattering 799
- 15.1 Interacting States and Asymptotic Boundary Conditions 800
- 15.2 Particle Detection and Connection between Configuration and Momentum Spaces in Scattering 807
- Appendix to §15.2: Some Properties of $F(u,v)$ 812
- 15.3 Differential Cross Sections 814
  - 15.3.1 Expression for the Differential Cross Section 814
  - 15.3.2 Sufficiency Conditions for the Validity of the Born Expansion 816
  - 15.3.3 Two-Particle Scattering 818
- 15.4 The Optical Theorem and Its Interpretation; Phase Shifts ... 821
  - 15.4.1 The Optical Theorem 821
  - 15.4.2 Phase Shifts Analysis 825
- 15.5 Coulomb Scattering...................................... 830
  - 15.5.1 Asymptotically "Free" Coulomb Green Functions 830
  - 15.5.2 Asymptotic Time Development of a Charged Particle State 832
  - 15.5.3 The Full Green Function G_+ Near the Energy Shell... 833
  - 15.5.4 The Scattering Amplitude via Evolution Operators ... 834

Contents XIII

    15.6 Functional Treatment of Scattering Theory 838
    15.7 Scattering at Small Deflection Angles at High Energies:
         Eikonal Approximation................................... 842
         15.7.1 Eikonal Approximation 842
         15.7.2 Determination of Asymptotic "Free" Green Function
              of the Coulomb Interaction 845
    15.8 Multi-Channel Scatterings of Clusters and Bound Systems ... 846
         15.8.1 Channels and Channel Hamiltonians 847
         15.8.2 Interacting States Corresponding to Preparatory
              Channels 851
         15.8.3 Transition Probabilities and the Optical Theorem 853
         15.8.4 Basic Processes 854
         15.8.5 Born Approximation, Connectedness and Faddeev
              Equations 858
         15.8.6 Phase Shifts Analysis 865
    15.9 Passage of Particles through Media; Neutron Interferometer .. 867
         15.9.1 Passage of Charged Particles through Hydrogen 867
         15.9.2 Neutron Interferometer........................... 871
    Problems 877

16 Quantum Description of Relativistic Particles 881
    16.1 The Dirac Equation and Pauli's Fundamental Theorem...... 884
    Appendix to §16.1: Pauli's Fundamental Theorem 889
    16.2 Lorentz Covariance, Boosts and Spatial Rotations 892
         16.2.1 Lorentz Transformations 892
         16.2.2 Lorentz Covariance, Boosts and Spatial Rotations 894
         16.2.3 Lorentz Invariant Scalar Products of Spinors, Lorentz
              Scalars and Lorentz Vectors 898
    16.3 Spin, Helicity and $\mathcal{P}, \mathcal{C}, \mathcal{T}$ Transformations 900
         16.3.1 Spin & Helicity 900
         16.3.2 $\mathcal{P}, \mathcal{C}, \mathcal{T}$ Transformations 902
    16.4 General Solution of the Dirac Equation..................... 903
    16.5 Massless Dirac Particles 912
    16.6 Physical Interpretation, Localization and Particle Content ... 916
         16.6.1 Probability, Probability Current and the Initial Value
              Problem... 917
         16.6.2 Diagonalization of the Hamiltonian and Definitions
              of Position Operators 919
         16.6.3 Origin of Relativistic Corrections in the Hydrogen
              Atom 926
         16.6.4 The Positron and Emergence of a Many-Particle
              Theory.. 931
    Appendix to §16.6: Exact Treatment of the Dirac Equation in the
         Bound Coulomb Problem 933

16.7 The Klein-Gordon Equation 937
  16.7.1 Setting Up Spin 0 Equations 937
  16.7.2 A Continuity Equation 940
  16.7.3 General Solution of the Free Feshbach-Villars Equation 941
  16.7.4 Diagonalization of the Hamiltonian and Definition of Position Operators 942
  16.7.5 The External Field Problem 944
16.8 Relativistic Wave Equations for Any Mass and Any Spin 947
  16.8.1 $M > 0$: 947
  16.8.2 $M = 0$: 950
16.9 Spin & Statistics 953
  16.9.1 Quantum Fields 954
  16.9.2 Lagrangian for Spin 0 Particles 955
  16.9.3 Lagrangian for Spin 1/2 Particles 957
  16.9.4 Schwinger's Constructive Approach 958
  16.9.5 The Spin and Statistics Connection 962
Appendix to §16.9: The Action Integral 965
Problems 968

Mathematical Appendices 971

I Variations of the Baker-Campbell-Hausdorff Formula 973
  1. Integral Expression for the Product of the Exponentials of Operators 973
  2. Derivative of the Exponential of Operator-Valued Functions . 973
  3. The Classic Baker-Campbell-Hausdorff Formula 975
  4. A Modification of the Baker-Campbell-Hausdorff Formula ... 975

II Convexity and Basic Inequalities 977
  1. General Convexity Theorem 977
  2. Minkowski's Inequality for Integrals 978
  3. Hölder's Inequality for Integrals 979
  4. Young's Inequality for Integrals 980

III The Poisson Equation in 4D 981
  1. The Poisson Equation 982
  2. Generating Function 983
  3. Expansion Theorem 984
  4. Generalized Orthogonality Relation 985

References 987

Index 999

Acknowledgments

It is my pleasure to thank several colleagues who have contributed directly to my learning of the subject and of related ones. These include, Prof. T. F. Morris, Prof. E. Prugovečki, Prof. P. R. Wallace, Prof. W. R. Raudorf, Prof. C. S. Lam, Prof. S. Morris, Prof. R. Sharma, Prof. B. Frank, Prof. Y. Takahashi, Prof. A. Z. Capri, Prof. A. N. Kamal, Prof. S. D. Jog, Prof. E. Jeżak, Prof. A. Ungkitchanukit and Prof. C.-H. Eab. I am also indebted to many of my graduate students, notably to Dr. C. Muthaporn, Mr. S. Siranan, Mr. N. Yongram, Mr. S. Sirininlakul, Mr. S. Sukkhasena, Ms. K. Limboonsong, Mr. P. Viriyasrisuwattana, Mr. A. Rotjanakusol, Ms. D. Charuchittipan, Ms. J. Osaklung and Ms. N. Jearnkulprasert, who through their many questions, several discussions and collaborations, have been very helpful in my way of analyzing the subject. I want to give my special thanks to Prof. W. E. Thirring for clarifying some aspects related to Chapter 14. Over the years, I have benefited greatly from the writings of Prof. J. Schwinger and from few of his lectures I was fortunate to attend. He had one of the greatest minds in physics of our time.

Most of my graduate students have participated in typing this rather difficult manuscript, as well as, Ms. P. Pechmai, who has contributed to the typing at the early stages of the project. I am grateful to all of them. In this regard my special thanks and appreciation go to Dr. C. Muthaporn and Mr. S. Siranan for bringing the manuscript to its final form and for the long hours they have spent to achieve this. A large portion of the book was written at the Suranaree University of Technology, and I could have found no more congenial atmosphere for completing this pleasant task than the one existing here. This atmosphere was essentially created by my colleague and the Rector of our university, Prof. P. Suebka. I want to express my gratitude to him for giving me the privileged opportunity to lecture on the subject matter of this book, and for his constant encouragement and optimism. This project would never have been completed without the patience and understanding of Mrs. Tuenjai Nokyod. I dedicate this book to her.

Preface

This book is based on lectures given in quantum theory over the years at various levels culminating into graduate level courses given to the students in physics. It is a modern self-contained textbook and covers most aspects of the theory and important recent developments with fairly detailed presentations. In addition to traditional or so-called standard topics, it emphasizes on modern ones and on theoretical techniques which have become indispensable in the theory. I have included topics which I believe every serious graduate student in physics should know. This volume is also a useful source of information and provides background for research in this discipline and related ones as well. As such, the book should be valuable to the graduate student, the instructor, the researcher and to all those concerned with the intricacies of this subject. To make this work accessible to a wider audience, some of the technical details occurring in the presentations have been relegated to appendices. A glance at the Contents will reveal that although the book is fairly advanced, it develops the entire formalism afresh. As for prerequisites, a familiarity with general concepts and methods of quantum physics as well as with basic mathematical techniques which most students entering graduate school seem to have is, however, required. The evident interest of my students in my quantum theory courses has led me quite often to expand and refine my notes that eventually became the book. I often witness many of my earlier students, who have already taken my courses, coming back to sit in my lectures and continue to do so. Some of these learners are A-students. In developing the formalism, at the very early stages, and of the rules for computations, I have followed a method based on Schwinger's (1970, 1991, 2001) elegant and incisive approach of direct analyses of selective measurements, rather than of the historical one, as well as in the introduction of quantum generators and the development of transformation theory. The selective measurement approach has its roots in Dirac's abstract presentation in terms of projection operators and provides tremendous insight into the physics behind the formalism. Other authors who have also shown interest in this approach include Kaempffer (1965), Gottfried (1989) and Sakurai (1994).

Some of the highlights of the book are: 1) Selective measurements. Direct analyses of such measurements and extensions thereof lead to the development of the underlying rules of the theory in the most natural and elegant way. 2) Wigner's Theorem on symmetry transformations. This theorem is of central importance in quantum theory and provides the nature of symmetry transformations and is the starting point on how to implement them. 3) Continuous transformations as well as supersymmetry and discrete transformations. 4) Hilbert space concepts and self-adjoint operators. 5) General study of the spectra of Hamiltonians. 6) Localizability, uncertainties and stability of quantum systems, such as of the H-atom, and their relations to boundedness of the corresponding Hamiltonians from below. 7) Decay of quantum systems and the Paley-Wiener Theorem. 8) Harmonic oscillators at finite temperatures, with external sources and coherent states. Bose-Fermi oscillators. 9) Hyperfine splitting of the H-atom for arbitrary angular momentum states. 10) The non-relativistic Lamb shift. 11) The anomalous magnetic moment of the electron. 12) Measurement, interference and the role of the environment. 13) Schrödinger's cat and quantum decoherence. 14) Bell's test. 15) Quantum teleportation and quantum cryptography. 16) Geometric phases under non-adiabatic and non-cyclic conditions. The AB effect. Rotation of a spinor by 2π radians. Neutron interferometry. 17) Analytical quantum dynamical treatment of the Stern-Gerlach effect. 18) Ramsey oscillatory fields method and applications. 19) Green functions, and how they provide information on different aspects of the theory in a unified manner. 20) Path integrals and constrained dynamics. 21) The quantum dynamical principle as a powerful, simple and most elegant way of doing quantum physics. This approach has not yet been sufficiently stressed in the literature and it is expected to play a very special role in the near future not only as a practical way for computations but also as a technically rigorous method. 22) The stability of matter in this monumental theory. This problem is undoubtedly one of the most important and serious problems that quantum physics has resolved. The Pauli exclusion principle is not only sufficient for stability but it is also necessary. 23) The intriguing problem of "bosonic matter" and the collapse of matter if the Pauli exclusion principle were abolished with the energy released upon the collapse of two such macroscopic objects in contact being comparable to that of an atomic bomb. 24) Systematics of quantum scattering including a detailed treatment of the Coulomb problem. Emphasis is also put on the connection between configuration and momentum spaces in a scattering process. 25) Spinors, quantum description of relativistic particles, helicity and relativistic equations for any mass and any spin. As the energy and momentum of a particle become large enough, the Schrödinger equation, with a non-relativistic kinetic energy, becomes inapplicable. One is then confronted with the development of a formalism to describe quantum particles in the relativistic regime. The chapter in question emerging from this endeavor provides the precursor of relativistic quantum theory of fields. 26) Spin & Statistics, as probably one of the most important results not only in physics but in all of the sciences, in general. The spin and statistics connection is responsible for the stability of matter, without it the universe would collapse. 27) Detailed mathematical appendices, with proofs, tailored to our needs which may be otherwise not easy to read in the mathematics literature.

The above are some of the topics covered in addition to the more standard ones. I have made much effort in providing a pedagogical approach to some of the more difficult ones just mentioned. These relatively involved topics are treated in a more simplified manner than that in a technical journal without, however, sacrificing rigor, thus making them more accessible to a wider audience and not only to the mathematically inclined reader. The problems given at the end of each chapter form an integral part of the book and should be attempted by every serious reader. Some of these problems are research oriented. With the rapid progress in quantum physics, I hope that this work will fill a gap, which I feel does exist, and will be useful, and also provides a challenge, to all those concerned with our quantum world.

July, 2005 E. B. M.

1
Fundamentals

This chapter begins with the development of the formalism of quantum theory being sought. The first three sections deal with the early stages of the formalism, with setting up the language and the preliminary rules of computations. The procedure follows a method based on Schwinger's elegant and incisive manner of direct analyses of selective measurements and extensions thereof, and has its roots in Dirac's abstract presentation in terms of projection operators. Several examples of selective measurements will be given. The method developed provides tremendous insight into the physics behind the formalism and it leads naturally to the notion of probability associated with observations, to the generation of states, of wavefunctions in different descriptions and to various basic operations occurring in quantum mechanics, as well as to the emergence of Hermitian operators and inner-product spaces (§1.4). Preparation of pure ensembles of systems and mixtures is the subject matter of §1.5. In §1.6, the transmission of photons with given polarizations through polarizers is used to provide an illustration of rules developed earlier. The physical spaces in which *computations* are carried out are, in general, the Hilbert space, as an extension of the (finite) inner-product spaces encountered before, and the Rigged Hilbert space which are introduced in §1.7. Self-adjoint operators, representing observables, and their associated spectra are studied in §1.8. We will see that symmetry operations are implemented by either so-called unitary or anti-unitary operators and is the content of a famous theorem due E. P. Wigner (Wigner's Theorem on symmetry transformations) which is proved in §1.9. This theorem is of central importance and a *key* one in quantum physics and deserves the special attention given here. The concept of probability and measurement with detailed illustrations are given in §1.10, emphasizing, in the process, the physical significance of a conditional probability associated with a measurement. This section also deals with non-ideal apparatuses that may disturb the physical system under consideration. Additional pertinent material related to this section will be given in §8.7–§8.9.

1.1 Selective Measurements

From the possible values that a physical quantity, under consideration, may take on, one may select, through a filtering process, a special range of its values or select some of its particular values for further investigations by a process referred to as a *selective measurement*. Some examples of selective measurements are given in Figure 1.1. Such selective measurements may be considered, for example, as a preparatory stage for a system before undergoing a subsequent analysis. By a selective measurement, for example, one may prepare the momentum of a particle within a given range before it participates in a collision process with other particles. As a final selective measurement, in a typical experiment, one may be interested in counting the number of particles with spin emerging, in turn, from a given physical process, with the component of spin, along some, specified direction.

We consider first physical quantities which may take on only a finite number of discrete values. An example of such a physical quantity is the component of spin of a particle of spin s, along a given axis, which may take on $(2s+1)$ values. Generalizations to physical quantities which may take on an infinite number of possible discrete values and/or may take on values from a continuous set of values will be dealt with later.

Suppose that the measurement of a physical quantity A (also called an observable), as a physical attribute of a system, can lead to a certain finite set of discrete real values $\{a, a', a'', \ldots\}$. In general, the measurement of another physical quantity B may destroy the assigned value in a previous measurement of the physical quantity A, and both quantities cannot be measured simultaneously. In such a case A and B are said to be *incompatible*. Otherwise they are said to be *compatible* observables.

To obtain the optimum information about a system one needs to introduce a complete set of compatible observables, say, $\{A_1, \ldots, A_k\}$. By this it is meant that any observable not belonging to this set and which is not a function of these observables is incompatible with at least one of them. To simplify the notation, we will denote such a complete set $\{A_1, \ldots, A_k\}$ of compatible observables simply by A. Each of the values in $\{a, a', a'', \ldots\}$ given above will then, in general, stand for k-tuplet of real numbers.

Through a filtering process, as in a Stern-Gerlach experiment (see Figure 1.1 (d)), an ensemble of identical systems each having a definite value, say a, for A may be prepared. Each one of such prepared systems is said to be in the state a. If these prepared systems are fed, in turn, into another filtering machine which selects and transmits systems having only the value a' for A, then 100% of these systems will be transmitted if $a' = a$ and none will be transmitted if $a' \neq a$.

With a filtering process, we introduce the symbol

$$\Lambda(a) = |a\rangle\langle a| \tag{1.1.1}$$

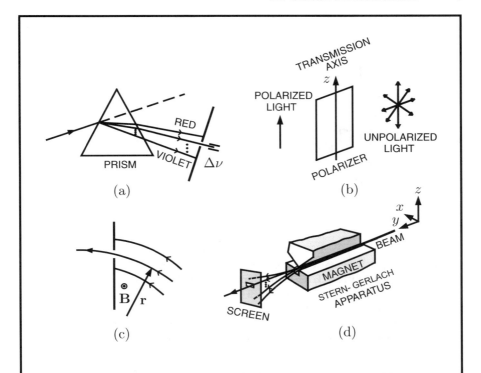

Fig. 1.1. Some examples of idealized selective measurements. (a) Selection of a frequency range $\Delta\nu$ for light. (b) Selection of linearly polarized light. (c) Selection of a momentum range $\Delta p = q\Delta r B/c$ for a charged particle of charge q initially in a uniform magnetic field B. (d) Selection of a particular component of spin by blocking systems with other orientations through a filtered beam by a Stern-Gerlach apparatus. A particle of magnetic moment $\boldsymbol{\mu}$ experiences, classically, a force $\mathbf{F} = \nabla(\boldsymbol{\mu} \cdot \mathbf{B})$ in a non-uniform magnetic field $\mathbf{B}$.

to denote the operation which selects and transmits only those systems in state a. From the description given in the previous paragraph, we may consider the successive operations to be defined through (see Figure 1.2 for an example):

$$\Lambda(a')\Lambda(a) = \Lambda(a') \langle a' | a \rangle \tag{1.1.2}$$

where $\langle a' | a \rangle = \delta(a', a)$ is the *numerical* factor

$$\delta(a', a) = \begin{cases} 1, & \text{for } a' = a \\ 0, & \text{for } a' \neq a \end{cases} \tag{1.1.3}$$

with $\Lambda(a)\,0 = \mathbf{0}$ standing for the operation which accepts no system whatsoever. For $a' = a$, the second selective measurement, symbolized by $\Lambda(a')$, simply accepts and transmits 100% $(\delta(a', a) = 1)$ of the systems prepared by

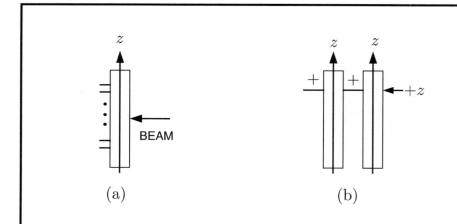

Fig. 1.2. (a) Schematic representation of the Stern-Gerlach apparatus of Figure 1.1 (d) with $\partial \mathbf{B}/\partial z \neq \mathbf{0}$, showing the splitting of a beam of spin s particles into $(2s+1)$ components. (b) Spin 1/2 particles initially prepared with component of spin in the $+z$ direction. In an obvious notation here, $\delta(+z, +z) = 1$ and $\delta(-z, +z) = 0$ for the corresponding numerical factors.

the first selective measurement, symbolized by $\Lambda(a)$. One is naturally led to introduce the identity operation

$$\mathbf{1} = \sum_a |a\rangle\langle a| \qquad (1.1.4)$$

which simply accepts and transmits all systems with no discrimination in any of the states a corresponding to all the values taken by the physical quality A (i.e., by the complete set of compatible observables $\{A_1, \ldots, A_k\}$.)

If A and B are incompatible, we may still consider the selective measurement $\Lambda(a) = |a\rangle\langle a|$ followed by the selective measurement $\Lambda(b) = |b\rangle\langle b|$: $\Lambda(b)\Lambda(a)$. This is a $|b\rangle\langle a|$-type of an operation which initially prepares systems in state a and then, through another filtering process, transmits a sub-ensemble of systems in state b. Since only a fraction of the systems in state a are expected to be finally transmitted through the B-filter, the operation $\Lambda(b)\Lambda(a)$, in analogy to (1.1.2), may be defined (see Figure 1.3, for an example) through:

$$\Lambda(b)\Lambda(a) = |b\rangle\langle a| \left(\langle b|a\rangle\right) \qquad (1.1.5)$$

reflecting the fact that it is a $|b\rangle\langle a|$-type selective operation and also providing a *numerical* factor $\langle b|a\rangle$ as a measure of the fraction of the systems initially prepared, by the A-filter, in state a, to be finally transmitted through the B-filter and found in state b.

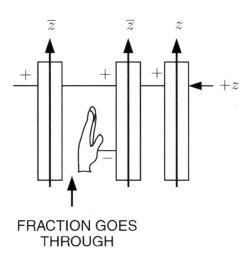

Fig. 1.3. Spin 1/2 particles initially prepared with the component of spin in the $+z$ axis direction. This experimental set up is represented, in an obvious notation, by $|+1/2, \bar{z}\rangle\langle+1/2, z| (\langle+1/2, \bar{z}|+1/2, z\rangle)$ with $\langle+1/2, \bar{z}|+1/2, z\rangle$ being a numerical factor providing a measure of the fraction of particles going through the middle apparatus with the $-\bar{z}$ component of spin blocked (as shown by the appearance of the hand). The rules for the computation of numerical factors such as $\langle+1/2, \bar{z}|+1/2, z\rangle$ will emerge naturally later. Here for the numericals $\overline{m}$, m in $\langle \overline{m}, \bar{z} | m, z \rangle$ we have, $\overline{m} = +1/2$, $m = +1/2$, for example, corresponding to spin components along the $+\bar{z}$, $+z$ directions, respectively.

Clearly more elaborate successive selective measurements (see, for example, Figure 1.4) may be considered and one may establish, in the process, the following associative law of the measurement symbols:

$$\Lambda(c)\big[\Lambda(b)\Lambda(a)\big] = \Lambda(c) |b\rangle\langle a| \left(\langle b|a\rangle \right)$$

$$= |c\rangle\langle a| \left(\langle c|b\rangle \langle b|a\rangle \right)$$

$$= \big[\Lambda(c)\Lambda(b)\big]\Lambda(a). \tag{1.1.6}$$

At this stage, it is worth *re*-examining the role of *incompatible* observables. Corresponding to an observable A, we note first from (1.1.2) that

$$\Lambda(a')\Lambda(a) = \mathbf{0}, \quad a' \neq a. \tag{1.1.7}$$

We recall that what the latter means is that the first filtering operation, via $\Lambda(a)$, has prepared systems in the state a, and for $a \neq a'$, the second

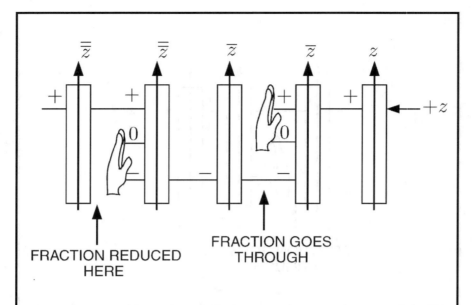

Fig. 1.4. Spin 1 (massive) particles prepared with the component of spin in the $+z$ direction. This experimental set up is represented, in an obvious notation, by successive measurement symbols $(|+1,\overline{\overline{z}}\rangle\langle+1,\overline{\overline{z}}|)\,(|-1,\overline{z}\rangle\langle-1,\overline{z}|)$ $\times\,(|+1,z\rangle\langle+1,z|) = (|+1,\overline{\overline{z}}\rangle\langle+1,z|)\,(\langle+1,\overline{\overline{z}}|-1,\overline{z}\rangle\langle-1,\overline{z}|+1,z\rangle)$. The numerical factor $\langle+1,\overline{\overline{z}}|-1,\overline{z}\rangle\langle-1,\overline{z}|+1,z\rangle$ is a measure of the fraction of the number of particles going through the five apparatuses. [The 0 in $|0,\overline{z}\rangle$ corresponds to a spin component perpendicular to the $\overline{z}$-axis.]

operation, via $\Lambda(a')$, rejects all such systems. That is, *no* systems appear in the final stage after the application of the two filtering processes. What is quite remarkable, is that if we insert a B-filter, via the application of $\Lambda(b)$, *between* the two successive operations in (1.1.7) we obtain

$$\Lambda(a')\Lambda(b)\Lambda(a) = |a'\rangle\langle a|\,(\,\langle a'|b\rangle\langle b|a\rangle\,) \qquad (1.1.8)$$

and for two *incompatible* observables A and B, $\langle a'|b\rangle\langle b|a\rangle$ is not necessarily equal to zero. In such cases

$$\Lambda(a')\Lambda(b)\Lambda(a) \neq \mathbf{0}. \qquad (1.1.9)$$

That is, by making the selective B-measurement, via $\Lambda(b)$, *after* the selective A-measurement, via $\Lambda(a)$, followed finally by a selective A-measurement, via $\Lambda(a')$, some systems may emerge in the state a' even if $a' \neq a$ (!), although this would not happen if the B-filter were absent, thus increasing the fraction of systems finally transmitted from zero to a possible non-zero value.

Re-iterating the above remarks, is that although the first selective measurement via $\Lambda(a)$ makes sure that no systems are transmitted through it in

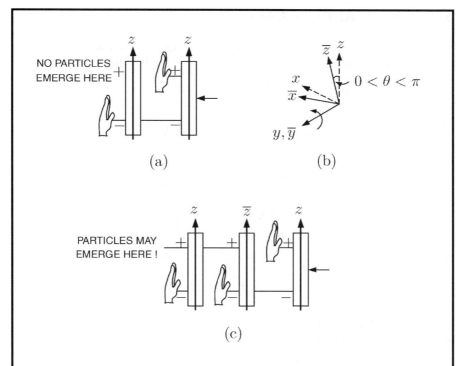

Fig. 1.5. (a) Experimental set up involving spin 1/2 particles showing that no particles finally emerge with component of spin along the $+z$ direction. (b) Orientation of the $\bar{z}$-axis relative to the z-axis. (c) The insertion of the middle filtering (Stern-Gerlach) apparatus may allow some particles with component of spin along the $+z$ direction to appear in the final stage knowing that the first filter has rejected particles in such a state! This is because $\langle +1/2, z | +1/2, \bar{z} \rangle \langle +1/2, \bar{z} | -1/2, z \rangle \neq 0$. This provides an illustration of the fact that the observables associated with measuring of components of spin along different orientations, as shown in (b), are incompatible. The rules for the computation of numerical factors such as $\langle +1/2, z | +1/2, \bar{z} \rangle \langle +1/2, \bar{z} | -1/2, z \rangle$ will be worked out later.

a state $a' \neq a$, the B-filter allows such systems in a state $a' \neq a$ be finally transmitted after the selective measurement, via $\Lambda(a')$, is carried out.

An example of the operations in (1.1.7) and (1.1.9) is given in Figure 1.5 illustrating the above remarks. Another example worked out in some details dealing with polarization of light will be given in §1.6.

The filtering Stern-Gerlach devices considered above are ideal and allow simple deductions to be made without going into the subtleties of their performance. A fairly detailed account of the Stern-Gerlach effect will be given in §8.14.

1.2 A, B, C to Probabilities

We obtain a useful identity involving the numerical factors such as $\langle c|b\rangle$. To this end we insert the identity operation in (1.1.4) as shown below

$$\Lambda(c)\Lambda(b) = \Lambda(c)\,\mathbf{1}\,\Lambda(b)$$

$$= |c\rangle\langle c| \left(\sum_a |a\rangle\langle a|\right) (|b\rangle\langle b|)$$

$$= |c\rangle\langle b| \left(\langle c|a\rangle \langle a|b\rangle + \langle c|a'\rangle \langle a'|b\rangle + \ldots \right)$$

$$= |c\rangle\langle b| \left(\sum_a \langle c|a\rangle \langle a|b\rangle\right) \quad (1.2.1)$$

and infer from (1.1.5), in a corresponding notation, that

$$\langle c|b\rangle = \sum_a \langle c|a\rangle \langle a|b\rangle . \quad (1.2.2)$$

In particular, if C is chosen to be the observable B, with c replaced by, say, b', we have from (1.1.3) and (1.2.2)

$$\sum_a \langle b'|a\rangle \langle a|b\rangle = \delta(b', b) \quad (1.2.3)$$

and for $b' = b$

$$\sum_a \langle b|a\rangle \langle a|b\rangle = 1. \quad (1.2.4)$$

We note that under the arbitrary scale transformations:

$$|a\rangle\langle b| \longrightarrow |a\rangle\langle b|\left(\lambda(b)/\lambda(a)\right) \quad (1.2.5)$$

and

$$\langle a|b\rangle \longrightarrow \langle a|b\rangle\left(\lambda(a)/\lambda(b)\right) \quad (1.2.6)$$

all the equations involving the selective measurements and successive selective measurements, (1.1.1)–(1.1.9), together with the identities (1.2.2)–(1.2.4) for the numerical factors, remain *invariant*. Because of this arbitrariness under such scale transformations, a numerical factor $\langle a|b\rangle$, although of physical interest as discussed in §1.1, cannot have a physical meaning by *itself*. The combination $\langle a|b\rangle \langle b|a\rangle$, however, from (1.2.6) remains invariant. For the subsequent analysis, we introduce the notation

$$\langle a|b\rangle \langle b|a\rangle = p_a(b). \quad (1.2.7)$$

At this stage the following basic points should be noted which are relevant to the numerical factor $p_a(b)$:

1.2 A, B, C to Probabilities

(i)
$$p_a(a) = 1. \tag{1.2.8}$$

(ii) As mentioned above, $p_a(b)$ is invariant under the scale transformations in (1.2.6).

(iii) As already noted in §1.1, the factor $\langle b|a \rangle$ in it, for example, is a measure of the *fraction* of systems all in state a that will be found in state b after the corresponding selective measurement.

(iv) $p_a(b)$ satisfies the *normalization* condition

$$\sum_b p_a(b) = 1 \tag{1.2.9}$$

as follows from (1.2.4).

Accordingly, $p_a(b)$ is qualified to represent the probability of observing a system in state b knowing that it was in state a prior to the B-measurement. A probability, however, has to be a *real* and *non-negative* number $p_a(b) \geqslant 0$. One may satisfy both of these conditions if one requires that

$$\langle a|b \rangle = \langle b|a \rangle^*. \tag{1.2.10}$$

Where $*$ denotes complex conjugation. The probability $p_a(b)$ will be then given from (1.2.7) to be

$$p_a(b) = |\langle b|a \rangle|^2 \geqslant 0, \tag{1.2.11}$$

where all the initial systems prior to a B-measurement were in state a. The numerical factor $\langle b|a \rangle$ which, in general, is a complex number is referred to as the *amplitude* of obtaining the value b for a B-measurement on a system initially known to be *in* a state a.

The scale factors $\lambda(a)$, $\lambda(b)$, in (1.2.5), (1.2.6) must then obey the rule

$$\bigl(\lambda(a)\bigr)^* = 1/\lambda(a). \tag{1.2.12}$$

That is, they are necessarily phase factors:

$$\lambda(a) = e^{i\phi(a)} \tag{1.2.13}$$

i.e., with $\phi(a)$ denoting a real number.

As an application, consider feeding systems, *all* prepared in the state a, into a B-filtering apparatus via the application of the $\Lambda(b)$-operation, and then into a C-filtering apparatus, via the application of the $\Lambda(c)$-operation. This sequence of measurements, including the preparatory one, is represented by

$$\Lambda(c)\Lambda(b)\Lambda(a) = |c\rangle\langle a| \bigl(\langle c|b \rangle \langle b|a \rangle \bigr) \tag{1.2.14}$$

(see Figure 1.4, for an explicit situation). The probability of obtaining the value b for a B-measurement, then the value c for a C-measurement, on a system initially *in* the state a is given by

1 Fundamentals

$$\text{Prob}_a\left[b \text{ then } c\right] = \left|\langle c|b\rangle\langle b|a\rangle\right|^2$$

$$= \left|\langle c|b\rangle\right|^2 \left|\langle b|a\rangle\right|^2$$

$$= p_b(c)\, p_a(b). \tag{1.2.15}$$

It is convenient to introduce the concept of the trace operation. For any number α, we define

$$\text{Tr}\left[\alpha\, |b\rangle\langle a|\right] = \alpha\, \langle a|b\rangle. \tag{1.2.16}$$

Hence $p_a(b)$ may be rewritten as

$$p_a(b) = \text{Tr}\left[\Lambda(b)\Lambda(a)\right]. \tag{1.2.17}$$

We observe from (1.2.11), that the probability of obtaining the value b of a measurement of the observable B on a system in the state a is the same as the probability of obtaining the value a of a measurement of the observable A on a system in the state b. This suggests to consider the reversal of a sequence of selective measurements, called the adjoint, defined as follows

$$\left(|b\rangle\langle a|\right)^\dagger = |a\rangle\langle b| \tag{1.2.18}$$

$$\left(\left(|c\rangle\langle d|\right)|b\rangle\langle a|\right)^\dagger = \left(|a\rangle\langle b|\right)\left(|d\rangle\langle c|\right). \tag{1.2.19}$$

Upon expanding the left-hand and right-hand sides of (1.2.19) we may infer from (1.2.10) that

$$\langle d|b\rangle^\dagger = \langle d|b\rangle^*. \tag{1.2.20}$$

That is, the adjoint transformation introduces the complex conjugation of numerical factors.

1.3 Expectation Values and Matrix Representations

1.3.1 Probabilities and Expectation Values

Consider systems *all* in the state a being fed into a B-filtering machine (see, e.g., Figure 1.6) transmitting systems in states b, b', ... The probability that a transmitted system is found in state b is given by $p_a(b)$ in (1.2.11). Accordingly, the expectation value of the observable B of systems in state a is

$$\sum_b b\, p_a(b) \equiv \langle B\rangle_a \tag{1.3.1}$$

which we have conveniently denoted by $\langle B\rangle_a$. The latter may be rewritten as

$$\langle B\rangle_a = \sum_b b\, \text{Tr}\left[\Lambda(b)\Lambda(a)\right]$$

1.3 Expectation Values and Matrix Representations   11

$$= \sum_b \text{Tr}\left[b\Lambda(b)\Lambda(a)\right] \qquad (1.3.2)$$

(see (1.2.16), (1.2.17). This suggests to introduce the object

$$B = \sum_b b\,\Lambda(b) \equiv \sum_b b\,|b\rangle\langle b| \qquad (1.3.3)$$

as a linear combination of B-selective measurement symbols which has far reaching consequences. For simplicity of the notation, we have used the same symbol B for this object as the physical quantity it represents. In particular we may write

$$\langle B\rangle_a = \text{Tr}\left[B\Lambda(a)\right]. \qquad (1.3.4)$$

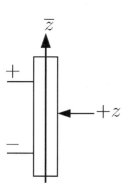

Fig. 1.6. A beam of spin 1/2 particles all with component of spin along the $+z$ direction is fed into a Stern-Gerlach apparatus and is split into two beams with components of spin prepared along the $+\bar{z}$ and $-\bar{z}$ directions, respectively. The initial beam is referred to as being completely polarized.

The expression in (1.3.3) in turn suggests to introduce more general objects like

$$M = \sum_{b,b'} \langle b|M|b'\rangle\,|b\rangle\langle b'| \qquad (1.3.5)$$

where $\langle b|M|b'\rangle$ are numerical factors. Immediate consequences of this definition are

$$M_1 M_2 = \sum_{b,b'} \left(\sum_{b''} \langle b|M_1|b''\rangle\,\langle b''|M_2|b'\rangle\right)|b\rangle\langle b'| \qquad (1.3.6)$$

$$\text{Tr}\,[M] = \sum_b \langle b|M|b\rangle \tag{1.3.7}$$

$$M^\dagger = \sum_{b,b'} \langle b|M|b'\rangle^* \left(|b\rangle\langle b'|\right)^\dagger$$

or

$$M^\dagger = \sum_{b,b'} \langle b|M|b'\rangle^* |b'\rangle\langle b|\,. \tag{1.3.8}$$

From (1.3.6)–(1.3.8), we may infer that the numerical factors $\langle b|M|b'\rangle$ may be interpreted as the matrix elements of a matrix labelled by the possible values of the observable B. In particular, we note, according to the definition (1.3.5) that

$$M_1 M_2 = \sum_{b,b'} \langle b|M_1 M_2|b'\rangle\,|b\rangle\langle b'| \tag{1.3.9}$$

and upon comparison with (1.3.6) we obtain the expected result of matrix multiplication definition that

$$\sum_{b''} \langle b|M_1|b''\rangle\,\langle b''|M_2|b'\rangle = \langle b|M_1 M_2|b'\rangle\,. \tag{1.3.10}$$

One may rewrite the expression in (1.3.5) directly for the object $M^\dagger$ and compare it with (1.3.8) to conclude that

$$\langle b|M^\dagger|b'\rangle = \langle b'|M|b\rangle^*\,. \tag{1.3.11}$$

Upon multiplying M in (1.3.5) by $|b''\rangle\langle b|$ and taking the trace gives

$$\text{Tr}\,[M\,|b''\rangle\langle b|\,] = \langle b|M|b''\rangle\,. \tag{1.3.12}$$

For the identity operation the latter gives

$$\langle b|\mathbf{1}|b''\rangle = \text{Tr}\,[\mathbf{1}\,|b''\rangle\langle b|\,] \tag{1.3.13}$$

and from (1.1.4) and (1.2.3) that

$$\langle b|\mathbf{1}|b''\rangle = \delta(b,b'') \tag{1.3.14}$$

which from (1.3.5) leads finally to

$$\mathbf{1} = \sum_b |b\rangle\langle b| \tag{1.3.15}$$

emphasizing the fact that the identity operation accepts all systems, without discrimination, whether it is written in the A-description or the B-description.

M in (1.3.5) may be also rewritten in a mixed-description as

1.3 Expectation Values and Matrix Representations

$$M = \sum_{a,b} \langle b|M|a\rangle \, |b\rangle\langle a| \,. \tag{1.3.16}$$

To show the equivalence of (1.3.16) and (1.3.5) we note from the former that

$$\langle b|M|a\rangle = \text{Tr}\left[M\,|a\rangle\langle b|\right]. \tag{1.3.17}$$

On the other hand, upon multiplying M in (1.3.5) by $|a\rangle\langle b|$ and taking the trace yields

$$\text{Tr}\left[M\,|a\rangle\langle b|\right] = \sum_{b'} \langle b|M|b'\rangle \, \langle b'|a\rangle \,. \tag{1.3.18}$$

Finally upon multiplying M in (1.3.5) by the identity in the A-description we have

$$M = \sum_{a,b,b'} \left(\langle b|M|b'\rangle \, \langle b'|a\rangle\right) |b\rangle\langle a| \,. \tag{1.3.19}$$

Upon comparison of (1.3.18)/(1.3.19) with (1.3.16)/(1.3.17) establishes the equivalence. Another equivalent expression for M is

$$M = \sum_{a,b} \langle a|M|b\rangle \, |a\rangle\langle b| \,. \tag{1.3.20}$$

The following rules then easily follow:

$$M_1 M_2 = \sum_{a,b,c} \langle b|M_1|a\rangle \, \langle a|M_2|c\rangle \, |b\rangle\langle c| \tag{1.3.21}$$

$$\langle a|M^\dagger|b\rangle = \langle b|M|a\rangle^* \,. \tag{1.3.22}$$

The equivalence of the descriptions of M given in (1.3.5), (1.3.16) and (1.3.20) may be then summarized by multiplying M from the right and left by the identity written in any description and noting, in particular, the identity

$$\sum_b \langle a|M|b\rangle \, \langle b|c\rangle = \langle a|M|c\rangle \,. \tag{1.3.23}$$

1.3.2 Representations of Simple Machines

For a concrete example of an object of the form in (1.3.5), consider the following machine (apparatus). It consists of two parts. The first part is a Stern-Gerlach apparatus (a filter) which transmits a beam of spin 1/2 particles with their spin components prepared in the $+z$ direction of the z-axis, while the beam of particles with spin components in the $-z$ direction is blocked. The resulting beam is then fed into a *Ramsey apparatus*,[1] consisting of the second part of the machine. In its simplest description, this apparatus consists of an oscillatory time-dependent magnetic field $\mathbf{B}(t)$ switched on for some time τ, followed by a uniform time-independent magnetic field $\mathbf{B}_0$ for some time T, and then finally the oscillatory magnetic field $\mathbf{B}(t)$ is switched on again for an additional time τ.

---

[1] Ramsey (1990) based on the 1989 Nobel Prize in Physics Lectures. The underlying theory will be discussed in some detail in §8.8.

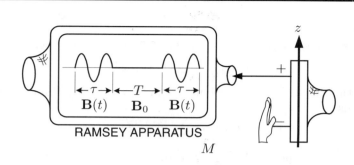

Fig. 1.7. A machine (apparatus) M consisting of a Stern-Gerlach apparatus (a filter) transmitting a beam of spin 1/2 particles with spin components along the $+z$ direction, with the beam then fed into a Ramsey apparatus as described in the text. The machine M may be conveniently represented by the expression in (1.3.24), (1.3.25) and is of the form in (1.3.5).

The above machine is depicted in Figure 1.7, and may be represented in the form

$$M = \sum_{m,m'=\pm 1/2} |m, z\rangle \langle m, z|M|m', z\rangle \langle m', z| \qquad (1.3.24)$$

with

$$\langle m, z|M|m', z\rangle = \delta(m', +1/2) \langle m, z|M|+1/2, z\rangle \qquad (1.3.25)$$

where the Ramsey apparatus with an initial beam of particles with spin component in the $+z$ direction fed into it is represented by the quantity $\langle m, z|M|+1/2, z\rangle$.

Other examples of objects of the form in (1.3.5), (1.3.16), with increasing complexity, are given in Figure 1.8.

From (1.3.5), (1.3.16), the machines M_1, M_2, M_3, M_4 in Figure 1.8 may be then represented in the simple forms:

$$M_1 = \Lambda(a) \qquad (1.3.26)$$

$$M_2 = \sum_b |b\rangle \langle b|a\rangle \langle a| \qquad (1.3.27)$$

$$M_3 = \sum_{b \in \Delta} |b\rangle \langle b|a\rangle \langle a| \qquad (1.3.28)$$

$$M_4 = \sum_{b,a'} |b\rangle \langle b|M_4|a'\rangle \langle a'| \qquad (1.3.29)$$

with

1.4 Generation of States, Inner-Product Spaces, Hermitian Operators ...

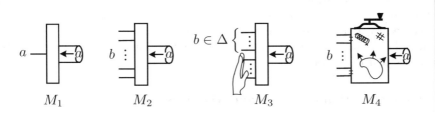

Fig. 1.8. Examples of four machines which may be represented in the form (1.3.5), (1.3.16). These machines select only systems for which a measurement of a physical quantity A, characteristic of the systems, yields a given fixed value a and reject all other systems. Machine M_1 is the simplest one represented by the measurement symbol $\Lambda(a)$ in (1.1.1). M_3 is a filtering machine which transmits only systems with given values of a physical quantity B, characteristic of the systems, obtained through a filtering process, as in a S-G apparatus, within some specified range Δ of b values. M_2 is identical to M_3 except that it transmits all the systems with any b values without discrimination. M_4 is a generalization of the machine M in Figure 1.7, where after the selection of systems with a given a value, the systems may, in general, go through complicated processes, such as the collisions of the underlying particles, absorptions, and so on, and the machine, then through a filtering process, transmits the emerging systems having b values of a physical quantity B characteristic of the systems.

$$\langle b|M_4|a'\rangle = \delta(a',a)\langle b|M_4|a\rangle \qquad (1.3.30)$$

and the $\langle b|M_4|a\rangle$ are some complex quantities. [See also the representation of the machine M of Figure 1.7 in (1.3.24), (1.3.25).]

The adjoint operation in (1.3.22) for a machine introduces, formally, to a machine operating in reverse, and from (1.3.16),

$$M^\dagger = \sum_{a,b} |a\rangle \langle b| \langle b|M|a\rangle^* . \qquad (1.3.31)$$

The successive operations of two machines M_2 followed by M_1 is given in (1.3.21). The significance of the trace operation in (1.3.18) will be considered in (1.4.18).

1.4 Generation of States, Inner-Product Spaces, Hermitian Operators and the Eigenvalue Problem

One may regard the significance of a selective measurement $|b\rangle\langle a|$, after the selection of all systems in a state specified by a value a of a physical quantity A, characteristic of the systems, as a two-stage process. The first being the annihilation of the selected systems in state specified by a and subsequently, as the second stage, the production of systems in a final state specified by a value b of a physical quantity B, characteristic of the systems, with the $\langle a|$ and $|b\rangle$ symbols being associated with the two stages of the process just discussed.

1.4.1 Generation of States and Vector Spaces

The symbols $|b\rangle$, for example, acquire a significance mathematically as they may be represented as vectors which generate a vector space of dimensionality directly obtained from the associated observable B (see (1.3.3)) representing a complete set of compatible observables, say, $B = \{B_1, \ldots, B_k\}$ with $b = \{b_1, \ldots, b_k\}$. The dimensionality of the generated vector space coincides with the number of different vectors that one may define as $b_1, \ldots, b_k$ take on consistently their allowed real physical values. It is often convenient, but not always so, to use the notation $|b\rangle$ as well for the corresponding vector representation. The **0** vector, in this vector space, is associated with the measurement of producing no systems at all. The state of a system characterized by a given fixed values taken by the k-tuplet $\{b_1, \ldots, b_k\}$ corresponding to the complete set of observables $\{B_1, \ldots, B_k\}$ is also often denotes by $|b\rangle$.

To see how such a vector space, as mentioned above, arises, consider, for example, the function of machine M_4 in Figure 1.8 which is represented in the form (see (1.3.29), (1.3.30))

$$M_4 = \sum_b |b\rangle \langle b|M_4|a\rangle \langle a| \qquad (1.4.1)$$

with the $\langle b|M_4|a\rangle$ denoting some complex quantities. This machine may be considered to operate, effectively, in two general stages. In the first stage, it annihilates all the systems selected in state specified by a, and finally creates systems in some new state $|\psi\rangle$ given by

$$|\psi\rangle = \sum_b |b\rangle \langle b|M_4|a\rangle \qquad (1.4.2)$$

for *a priori* given fixed value a of a physical quantity A characteristic of the systems. From (1.4.1), the machine M_4 may be then also represented in the compact form

$$M_4 = |\psi\rangle \langle a| \qquad (1.4.3)$$

for some fixed value a.

That is, starting from a system *initially prepared in a state* $|a\rangle$ and fed into the machine M_4, the latter produces a system in some final state which may be also denoted by $|\psi\rangle$. From (1.1.3), this operation procedure of the machine M_4 may be then defined by

$$M_4 |a\rangle = |\psi\rangle. \qquad (1.4.4)$$

1.4 Generation of States, Inner-Product Spaces, Hermitian Operators ...

With $\langle b|M_4|a\rangle$ as some complex quantities, the state $|\psi\rangle$ in (1.4.2) is written as a linear combination of the states $|b\rangle$. This is very much as having a vector space with the $|b\rangle$, corresponding to a complete set of observables characteristic of the system into consideration, providing a basis for such a vector space. To define a vector space, however, we have to consider the *addition* of states such as $|\psi\rangle$ and also *define* a $\mathbf{0}$ vector in it. This is done below.

To the above end, for each given fixed value c of some physical quantity C, characteristic of the systems in consideration, consider the state $|\Phi_c\rangle$ produced by a B-filtering machine

$$M_0 = \sum_{b \in \Delta} |b\rangle\langle b| \qquad (1.4.5)$$

with a given fixed range Δ of b values, from an initially prepared state $|c\rangle$:

$$|\Phi_c\rangle = \sum_{b \in \Delta} |b\rangle\langle b|c\rangle \qquad (1.4.6)$$

written as a linear combination of $|b\rangle$ states. On the other hand, for any given fixed value a, we may produce a state $|\chi\rangle$ from the successive operations of a machine M_4' followed by that of machine M_0 from an initially prepared state $|a\rangle$, where

$$M_4' = \sum_c |c\rangle\langle c|M_4'|a\rangle\langle a| \qquad (1.4.7)$$

which is of the same form as M_4 in Figure 1.8 with observable B replaced by an observable C. That is,

$$|\chi\rangle = \sum_{b \in \Delta} \left(\sum_c |b\rangle\langle b|c\rangle\langle c|M_4'|a\rangle \right)$$

$$= \sum_c \langle c|M_4'|a\rangle \sum_{b \in \Delta} |b\rangle\langle b|c\rangle$$

$$= \sum_{b \in \Delta} |b\rangle \sum_c \langle b|c\rangle\langle c|M_4'|a\rangle \qquad (1.4.8)$$

again written as a linear combination of $|b\rangle$ states.

Upon comparison of the second equality in (1.4.8) with (1.4.6), we obtain

$$|\chi\rangle = \sum_c |\Phi_c\rangle\langle c|M_4'|a\rangle \qquad (1.4.9)$$

with, in general, $\langle c|M_4'|a\rangle$ denoting complex quantities, the vectors $|\chi\rangle$, $|\Phi_c\rangle$ are written as linear combination of $|b\rangle$ states. Equation (1.4.9) provides a *linear superposition* of vectors $|\Phi_c\rangle$ in the underlying vector space leading to a vector $|\chi\rangle$ also in the same vector space.

The **0** vector in the underlying vector space may be simply defined by carrying out a selective measurement via the symbol $\Lambda(b')$ with $b' \notin \Delta$, on the state $|\Phi_c\rangle$ in (1.4.6)

$$\Lambda(b')|\Phi_c\rangle = \mathbf{0}. \tag{1.4.10}$$

In analogy to (1.2.17), (1.1.5), we consider the successive measurement symbols

$$\Lambda(b)|\psi\rangle\langle\psi| = |b\rangle\langle\psi|\left(\langle b|\psi\rangle\right) \tag{1.4.11}$$

with $\langle b|\psi\rangle$ as a measure of the fraction of systems found in the state $|b\rangle$, and from (1.2.17),

$$p_\psi(b) = |\langle b|\psi\rangle|^2 \tag{1.4.12}$$

is interpreted as the probability that the physical quantity B, characteristic of the system, takes the value b if the system is in the state $|\psi\rangle$, provided

$$\sum_b |\langle b|\psi\rangle|^2 = 1 \tag{1.4.13}$$

giving a normalization condition for the generally complex quantities $\langle b|\psi\rangle$.

Now we use the expression for $|\psi\rangle$ in (1.4.2), and consider the application of the selective measurement provided by $\Lambda(b)$ of a system in the state $|\psi\rangle$ giving

$$|b\rangle\langle b|M_4|a\rangle = |b\rangle\langle b|\psi\rangle \tag{1.4.14}$$

(see also (1.3.23)) leading to the identification

$$\langle b|M_4|a\rangle = \langle b|\psi\rangle \equiv \psi(b) \tag{1.4.15}$$

where with $\langle b|\psi\rangle$, in general, a complex quantity, we have denoted it by $\psi(b)$. Thus we may rewrite (1.4.2) as

$$|\psi\rangle = \sum_b |b\rangle\,\psi(b). \tag{1.4.16}$$

Conversely, with a a priori given and fixed, and $\langle b|M_4|a\rangle$ denoting, in general, some complex quantity which may be denoted, say, by $\psi(b)$ in (1.4.1), equations (1.4.2), (1.4.14) lead to the identification $\langle b|\psi\rangle = \psi(b)$. Finally note that the application of the selective measurement, denoted by $\Lambda(b)$, on the state $|\psi\rangle$ in (1.4.16) confirms this notation.

1.4.2 Transformation Functions and Wavefunctions in Different Descriptions

Equation (1.4.16) emphasizes again the expansion of the state $|\psi\rangle$ in terms of the $|b\rangle$ states, with the b values corresponding to a complete set of compatible observables characteristic of the system into consideration. $|\psi\rangle$ is referred to as a *state vector*, and $\psi(b)$ as the *wavefunction* in the B-description.

1.4 Generation of States, Inner-Product Spaces, Hermitian Operators ...

Upon the application of a selective measurement $\Lambda(a)$, of a physical quantity A characteristic of the system in state $|\psi\rangle$, one obtains

$$\psi(a) = \sum_b \langle a|b\rangle \, \psi(b) \qquad (1.4.17)$$

providing the transformation law of a wavefunction from the B-description to the A-description, with the amplitude $\langle a|b\rangle$, as is referred to below (1.2.11), is also called the *transformation function* from the B- to A-descriptions.

The trace operation

$$\operatorname{Tr}\left[\,|a\rangle\langle b|\, M_4\right] = \langle b|M_4|a\rangle = \psi(b) \qquad (1.4.18)$$

corresponding to the machine M_4 in (1.4.1), as also introduced in (1.3.29), gives the wavefunction $\psi(b)$.

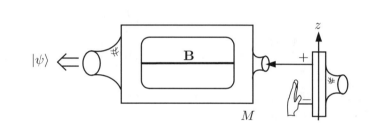

Fig. 1.9. A machine (apparatus) M consisting of two parts, a Stern-Gerlach apparatus (a filter) transmitting a beam of spin 1/2 particles with spin components along the $+z$ direction, with the beam then fed into a region of constant magnetic field. As far as a particle is concerned the machine M may be represented as in (1.4.24). If the initial state of a particle is different from $|+1/2, z\rangle$, such as being in a state $|+1/2, x\rangle$, the interesting situation of *absorption* by the machine arises.

1.4.3 An Illustration

As an illustration, consider the simple machine (apparatus) M in Figure 1.9. It consists of two parts. The fist part is a filter which transmits particles of spin 1/2 with spin components along the $+z$ direction while those with components along the $-z$ direction are not transmitted. If we represent the state $|+1/2, z\rangle$ by

$$|+1/2, z\rangle = \begin{pmatrix} 1 \\ 0 \end{pmatrix} \qquad (1.4.19)$$

and its adjoint by

1 Fundamentals

$$\langle z, +1/2 | = \begin{pmatrix} 1 \\ 0 \end{pmatrix}^\dagger = \begin{pmatrix} 1 & 0 \end{pmatrix}. \tag{1.4.20}$$

Then the filter may be represented by the selective measurement symbol

$$\Lambda_z(+1/2) = \begin{pmatrix} 1 \\ 0 \end{pmatrix} \begin{pmatrix} 1 & 0 \end{pmatrix} = \begin{pmatrix} 1 & 0 \\ 0 & 0 \end{pmatrix}. \tag{1.4.21}$$

The second part consists of a region of constant magnetic field

$$\mathbf{B} = (B, 0, 0). \tag{1.4.22}$$

Let μ denote the magnetic dipole moment of a particle. If t_0 is the time spent by a particle in the magnetic field $\mathbf{B}$, then as we will see later when studying the physics of spin 1/2 in Chapter 8 (see (8.1.28) later), that as far as a particle is concerned, the second part of the machine may be represented by the 2×2 (non-Hermitian) matrix

$$M(\mathbf{B}) = \begin{pmatrix} \cos\dfrac{\mu B t_0}{\hbar} & i\sin\dfrac{\mu B t_0}{\hbar} \\ i\sin\dfrac{\mu B t_0}{\hbar} & \cos\dfrac{\mu B t_0}{\hbar} \end{pmatrix}. \tag{1.4.23}$$

From (1.4.21) and (1.4.23), the combined machine M in Figure 1.9 may be then represented simply by

$$M = \begin{pmatrix} \cos\dfrac{\mu B t_0}{\hbar} & 0 \\ i\sin\dfrac{\mu B t_0}{\hbar} & 0 \end{pmatrix}. \tag{1.4.24}$$

Hence, if the particles are initially prepared in the state $|+1/2, z\rangle$, so that 100% of them are transmitted through the first stage, via the filtering machine in (1.4.21), the machine M in Figure 1.9, from (1.4.4), produces particles each in the state

$$|\psi\rangle = M |+1/2, z\rangle$$

$$= \begin{pmatrix} \cos\dfrac{\mu B t_0}{\hbar} & 0 \\ i\sin\dfrac{\mu B t_0}{\hbar} & 0 \end{pmatrix} \begin{pmatrix} 1 \\ 0 \end{pmatrix} \tag{1.4.25}$$

or

$$|\psi\rangle = \begin{pmatrix} 1 \\ 0 \end{pmatrix} \cos\dfrac{\mu B}{\hbar} t_0 + i \begin{pmatrix} 0 \\ 1 \end{pmatrix} \sin\dfrac{\mu B}{\hbar} t_0. \tag{1.4.26}$$

We may rewrite (1.4.26) as

1.4 Generation of States, Inner-Product Spaces, Hermitian Operators ...

$$|\psi\rangle = \psi(\{+1/2, z\})\,|+1/2, z\rangle + \psi(\{-1/2, z\})\,|-1/2, z\rangle \qquad (1.4.27)$$

with

$$\psi(\{+1/2, z\}) = \cos\frac{\mu B}{\hbar}t_0, \quad \psi(\{-1/2, z\}) = i\sin\frac{\mu B}{\hbar}t_0. \qquad (1.4.28)$$

By comparing (1.4.28)/(1.4.27) with (1.4.15)/(1.4.16), we may infer that

$$\langle +1/2, z|\psi\rangle = \cos\frac{\mu B}{\hbar}t_0, \quad \langle -1/2, z|\psi\rangle = i\sin\frac{\mu B}{\hbar}t_0 \qquad (1.4.29)$$

and from (1.4.12), (1.4.13) that

$$|\langle +1/2, z|\psi\rangle|^2 + |\langle -1/2, z|\psi\rangle|^2 = \cos^2\frac{\mu B}{\hbar}t_0 + \sin^2\frac{\mu B}{\hbar}t_0$$

$$= 1 \qquad (1.4.30)$$

as expected.

As will be seen later (§8.1), that particles in the states $|\pm 1/2, x\rangle$ may be represented as

$$|+1/2, x\rangle = \frac{1}{\sqrt{2}}\begin{pmatrix}1\\1\end{pmatrix}, \quad |-1/2, x\rangle = \frac{1}{\sqrt{2}}\begin{pmatrix}1\\-1\end{pmatrix}. \qquad (1.4.31)$$

For the selective measurement symbol $\Lambda_x(+1/2)$, we then have the representation

$$\Lambda_x(+1/2) = \frac{1}{2}\begin{pmatrix}1&1\\1&1\end{pmatrix}. \qquad (1.4.32)$$

Upon making a selective measurement via $\Lambda_x(+1/2)$ of a system in the state $|\psi\rangle$ in (1.4.26), we obtain, in analogy to (1.4.17),

$$\psi(\{+1/2, x\}) = \frac{1}{\sqrt{2}}\psi(\{+1/2, z\}) + \frac{1}{\sqrt{2}}\psi(\{-1/2, z\}) \qquad (1.4.33)$$

where we have used the normalizability of the state $|+1/2, x\rangle$ as given in (1.4.31), and the identifications in (1.4.28).

Upon the comparison of (1.4.33) with (1.4.17) we obtain

$$\langle +1/2, x|+1/2, z\rangle = \frac{1}{\sqrt{2}} = \langle +1/2, x|-1/2, z\rangle. \qquad (1.4.34)$$

Similarly, one derives that

$$\langle -1/2, x|+1/2, z\rangle = \frac{1}{\sqrt{2}} \qquad (1.4.35)$$

$$\langle -1/2, x|-1/2, z\rangle = -\frac{1}{\sqrt{2}} \qquad (1.4.36)$$

thus obtaining the transformation function $\langle m', x | m, z \rangle$ for, $m', m = \pm 1/2$, and we have developed the transformation from a description of spin along the z-axis to one along the x-axis, with wavefunctions $\psi(\{m, z\})$, $\psi(\{m', x\})$ in these descriptions, respectively.

With particles in an initial state $|+1/2, z\rangle$ fed into the machine M in Figure 1.9, the presence of the filter as part of the machine seems redundant since 100% of the particles in this initial state will be transmitted through the filter. The interesting situation then arises if the particles are initially in a different state, say, in the state $|+1/2, x\rangle$, which we now consider.

According to (1.4.34), only 50% of the particles will go through the filter, i.e., we will have absorption, and the machine M will produce a particular state $|\Phi\rangle_{\text{ABS}}$ from $|+1/2, x\rangle$ reflecting this fact. In detail

$$|\Phi\rangle_{\text{ABS}} = M\,|+1/2, x\rangle$$

$$= \begin{pmatrix} \cos \dfrac{\mu B t_0}{\hbar} & 0 \\ i\sin \dfrac{\mu B t_0}{\hbar} & 0 \end{pmatrix} \frac{1}{\sqrt{2}} \begin{pmatrix} 1 \\ 1 \end{pmatrix}$$

$$= \frac{1}{\sqrt{2}} \cos \frac{\mu B}{\hbar} t_0 \begin{pmatrix} 1 \\ 0 \end{pmatrix} + \frac{i}{\sqrt{2}} \sin \frac{\mu B}{\hbar} t_0 \begin{pmatrix} 0 \\ 1 \end{pmatrix} \qquad (1.4.37)$$

and as expected

$$\left| \frac{1}{\sqrt{2}} \cos \frac{\mu B}{\hbar} t_0 \right|^2 + \left| \frac{i}{\sqrt{2}} \sin \frac{\mu B}{\hbar} t_0 \right|^2 = \frac{1}{2} \qquad (1.4.38)$$

showing 50% absorption.

In general, consider an *initial* state $|c\rangle$, with c a priori fixed value taken by a different physical quantity C characteristic of the system into consideration, then from (1.4.3), machine M_4 in Figure 1.8, will produce, from the initial state $|c\rangle$ fed into it, a state

$$|\Phi\rangle_{\text{ABS}} = M_4 |c\rangle$$

$$= |\psi\rangle \langle a | c \rangle \qquad (1.4.39)$$

or from (1.4.16), (1.4.15)

$$|\Phi\rangle_{\text{ABS}} = \sum_b |b\rangle \langle b | \psi \rangle \langle a | c \rangle . \qquad (1.4.40)$$

From (1.4.12), (1.4.40) we then have

$$\left[1 - |\langle a | c \rangle|^2 \right] \times 100\%$$

absorption by the machine M_4.

1.4.4 Generation of Inner Product Spaces

Finally, we are led to consider the measurement symbol $|\psi\rangle\langle\phi|$, with $|\psi\rangle$, $|\phi\rangle$ two states, written as linear combinations of $|b\rangle$ states. The trace operation (see (1.2.16)) in the following

$$\mathrm{Tr}\left[|\psi\rangle\langle\phi|\right] = \langle\phi|\psi\rangle \tag{1.4.41}$$

leads to the definition of an *inner product*, which from (1.4.16) may be written as

$$\langle\phi|\psi\rangle = \sum_b \phi^*(b)\psi(b) \tag{1.4.42}$$

where we have used the property that the adjoint transformation takes the complex conjugation of numericals,

$$|\phi\rangle^\dagger = \sum_b \phi^*(b)\langle b| \equiv \langle\phi| \tag{1.4.43}$$

denoting the right-hand side of the above by $\langle\phi|$ as an expansion in terms of the adjoints $\langle b|$. The vector space generated by the adjoints $\langle b|$ is called the *dual* vector space to the vector space generated the $|b\rangle$ vectors having similar properties as the initial vector space itself. A vector $|b\rangle$ and its adjoint $\langle b|$ are often referred to as a *ket* and as a *bra*, respectively.

From (1.2.17), we also have,

$$p_\psi(|\phi\rangle) = |\langle\phi|\psi\rangle|^2 \tag{1.4.44}$$

representing the probability that the system is found in the state $|\phi\rangle$ if it is initially in the state $|\psi\rangle$, and the trace operation in (1.4.41) gives the corresponding amplitude $\langle\phi|\psi\rangle$.

A vector space on which an inner product is defined is called an *inner product space*. Thus with the inner product given in (1.4.42), we have thus introduced such an inner product space from the vector space generated by the $|b\rangle$ vectors.

It is easily seen that (1.4.42) actually provides a definition of an inner product by explicitly verifying the following properties,

(i)
$$\langle\psi|\psi\rangle = \sum_b |\psi(b)|^2 \geqslant 0 \tag{1.4.45}$$

(ii)
$$\langle\phi|\psi\rangle^* = \langle\psi|\phi\rangle \tag{1.4.46}$$

(iii)
$$\langle\alpha\phi|\psi\rangle = \alpha^*\langle\phi|\psi\rangle, \qquad \langle\phi|\alpha\psi\rangle = \alpha\langle\phi|\psi\rangle \tag{1.4.47}$$

(iv)
$$\langle\phi|\psi_1+\psi_2\rangle = \langle\phi|\big(|\psi_1\rangle+|\psi_2\rangle\big)$$
$$= \langle\phi|\psi_1\rangle + \langle\phi|\psi_2\rangle \qquad (1.4.48)$$

for any complex number α. In regard to $|\psi_1+\psi_2\rangle$ see (1.4.9).

The length or the norm of a vector $|\psi\rangle$ is defined by

$$\|\psi\| = \sqrt{\langle\psi|\psi\rangle} \qquad (1.4.49)$$

and hence the normalization condition (1.4.13) reads

$$\|\psi\|^2 = \sum_b |\psi(b)|^2 = 1. \qquad (1.4.50)$$

From (1.4.17), we note that

$$\sum_b \phi^*(b)\psi(b) = \sum_a \phi^*(a)\psi(a) \qquad (1.4.51)$$

establishing the description independence of an inner product corresponding to any two physical quantities B and A, characteristics of the systems in question.

The following basic inequalities follow from the definitions of the inner product and the norm of vectors given above

$$|\langle\phi|\psi\rangle| \leqslant \|\phi\|\,\|\psi\| \qquad (1.4.52)$$

referred to as the *Cauchy-Schwarz inequality*, and

$$\|\phi+\psi\| \leqslant \|\phi\| + \|\psi\| \qquad (1.4.53)$$

referred to as the *triangular inequality*.

1.4.5 Hermitian Operators and the Eigenvalue Problem

The definition in (1.3.3), together with (1.1.2) and (1.1.3), lead to the eigenvalue problem of a *Hermitian* operator

$$B|b\rangle = b|b\rangle \qquad (1.4.54)$$

or

$$B_i|b\rangle = b_i|b\rangle, \qquad i = 1,\ldots,k \qquad (1.4.55)$$

where $B = \{B_1,\ldots,B_k\}$, $|b\rangle = \{b_1,\ldots,b_k\}$. The Hermiticity condition reads

$$B^\dagger = B \qquad (1.4.56)$$

and follows from the reality of the b values and the definition of the adjoint through (1.2.18)–(1.2.20).

Clearly, for any given value b_i in (1.4.55), there will be, in general, more than one vector $|b_1, \ldots, b_i, \ldots, b_k\rangle$, satisfying (1.4.55), as the other b_j ($j \neq i$) take on, consistently, their allowed values. The number of such distinct vectors $|b_1, \ldots, b_k\rangle$ defines the *degree of degeneracy* of the eigenvalue b_i of B_i. Once all the eigenvalues $|b_1, \ldots, b_k\rangle$ are specified, then there will be only one eigenvector $|b_1, \ldots, b_k\rangle$ corresponding to the eigenvalue b in question. That is, by definition, the eigenvalue b of a *complete set of compatible* observables $\{B_1, \ldots, B_k\} \equiv B$ is *non-degenerate*. The set of values $\{b\}$ of B is called its spectrum, and the set of vectors $|b\rangle$ provides a basis for the generated vector space, with

$$\langle b|b'\rangle = \delta(b,b') \equiv \prod_i \delta(b_i, b'_i) \tag{1.4.57}$$

providing the orthonormality condition of the vectors $|b\rangle$, as follows from (1.1.3).

In the case when the eigenvalues b take on an infinite number of discrete values and/or continuous values, one is faced with convergence problems. These problems are dealt with by defining the concept of a Hilbert space (§1.7), as an underlying vector space, and introduce in turn the concept of self-adjoint operators (§1.7), representing observables, operating on vectors in such a vector space.

1.5 Pure Ensembles and Mixtures

In the elementary selective measurements experiments discussed in §1.1–§1.3, all the initial systems were prepared in some definite state a. Such a collection of systems all prepared in the *same* state is referred to as an *ensemble* or more appropriately as a *pure ensemble*. More generally, one may design an apparatus which may prepare an ensemble of systems, for further experimentation, such that every system in the ensemble may be represented by a creation symbol $|\psi\rangle$ in the form

$$|\psi\rangle = \sum_b |b\rangle \psi(b) \tag{1.5.1}$$

satisfying the normalization in (1.4.50).

The expectation value of an observable B for systems in the state $|\psi\rangle$ is then

$$|B\rangle_\psi = \text{Tr}\left[|\psi\rangle\langle\psi|B\right]$$

$$= \sum_b b|\psi(b)|^2 = \langle\psi|B|\psi\rangle \tag{1.5.2}$$

26   1 Fundamentals

(see (1.3.1)–(1.3.4), (1.3.17), (1.4.18)).

The operator
$$\rho = |\psi\rangle\langle\psi| \qquad (1.5.3)$$
is called the *density* or *statistical* operator for the ensemble of systems all in the state $|\psi\rangle$. Immediate properties of ρ, as defined in (1.5.3), are
$$\mathrm{Tr}\,[\rho] = 1 \qquad (1.5.4)$$
$$\mathrm{Tr}\,[\rho^2] = 1. \qquad (1.5.5)$$

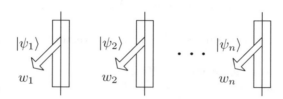

Fig. 1.10. A figure displaying n machines preparing *independently* n ensemble of systems in states $|\psi_1\rangle, |\psi_2\rangle, \ldots, |\psi_n\rangle$, respectively, constituting $w_1\%, w_2\%, \ldots, w_n\%$ of the total number of prepared systems.

One may also consider more general preparatory procedures such as described in Figure 1.10, where we have, say, n machines which produce (prepare) *independently* n ensemble of systems, respectively, in states $|\psi_1\rangle, |\psi_2\rangle, \ldots, |\psi_n\rangle$ with $w_1\%$ of systems in state $|\psi_1\rangle$, $w_2\%$ of systems in state $|\psi_2\rangle$, $\ldots$, $w_n\%$ of systems in state $|\psi_n\rangle$, respectively, with
$$\sum_{i=1}^{n} w_i = 1. \qquad (1.5.6)$$

A collection of such ensembles is called a *mixture*. The expectation values of an observable B in these different states are $\langle\psi_1|B|\psi_1\rangle, \langle\psi_2|B|\psi_2\rangle, \ldots, \langle\psi_n|B|\psi_n\rangle$ and hence the average over all the systems produced is given by
$$\langle B \rangle = \sum_{i=1}^{n} w_i \langle\psi_i|B|\psi_i\rangle. \qquad (1.5.7)$$

In this case the statistical operator associated with the mixture is given by
$$\rho = \sum_{i=1}^{n} w_i |\psi_i\rangle\langle\psi_i| \qquad (1.5.8)$$

1.5 Pure Ensembles and Mixtures

and it is easily checked that the latter may *not* be rewritten in the form $|\psi\rangle\langle\psi|$ for any state $|\psi\rangle$. In terms of ρ, (1.5.7) reduces to

$$\langle B \rangle = \text{Tr}\left[\rho B\right]. \tag{1.5.9}$$

In particular,

$$\text{Tr}\left[\rho\right] = \sum_{i=1}^{n} w_i = 1. \tag{1.5.10}$$

Unlike the pure ensemble case in (1.5.5), for a mixture we have

$$\text{Tr}\left[\rho^2\right] = \sum_{i=1}^{n} w_i^2 < \sum_{i=1}^{n} w_i \tag{1.5.11}$$

that is,

$$\text{Tr}\left[\rho^2\right] < 1 \tag{1.5.12}$$

since for at least two of the $w_1, w_2, \ldots, w_n$ non-zero, say, w_i and w_j, $w_i^2 < w_i$ (and $w_j^2 < w_j$).

In an extreme case when all the states $|\psi_1\rangle, \ldots, |\psi_n\rangle$ occur equally likely, $w_i = 1/n$ and we have, what is called, a completely *random* mixture.

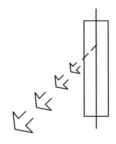

Fig. 1.11. A machine which because of its very nature (such as malfunctioning or for any other reason) produces systems in a statistically fluctuating manner in states $|\Phi_1\rangle, |\Phi_2\rangle, \ldots, |\Phi_n\rangle$, described by an overall state $|\Phi^\alpha\rangle = \sum_{i=1}^{n} \alpha_i |\Phi_i\rangle$ with statistically fluctuating coefficients $(\alpha_1, \ldots, \alpha_n)$.

The density operator in the form given in (1.5.8) also holds if we have a machine which because of its very nature (such as malfunctioning or for any other reason) produces systems in a statistically fluctuating manner in states $|\Phi_1\rangle, |\Phi_2\rangle, \ldots, |\Phi_n\rangle$ as summarized in Figure 1.11. This situation and the one spelled out in Figure 1.10 may be described by the same underlying theory. To this end we may define a state

$$|\Phi^\alpha\rangle = \sum_{i=1}^{n} \alpha_i |\Phi_i\rangle \tag{1.5.13}$$

where because of the experimental set up as given in Figure 1.10 or because of the fluctuating nature of a machine (Figure 1.11) or both, the coefficients $(\alpha_1, \ldots, \alpha_n) = \alpha$ may statistically fluctuate. Since the coefficients α_i are, in general, complex numbers, we may write

$$\alpha_j = r_j\, e^{i\delta_j} \tag{1.5.14}$$

where r_j, δ_j are real.

For a fixed value of α, the expectation value of an observable B in the state (1.5.13) is given by

$$\langle B\rangle\Big|_\alpha = \sum_{i,j=1}^{n} r_i\, r_j\, e^{-i\gamma_{ij}} \langle \Phi_i |B|\Phi_j\rangle \tag{1.5.15}$$

where

$$\gamma_{ij} = \delta_i - \delta_j \tag{1.5.16}$$

and

$$\sum_{i=1}^{n} r_i^2 = 1. \tag{1.5.17}$$

Let

$$f(r_1, \ldots, r_n, \gamma_{12}, \gamma_{13}, \ldots, \gamma_{23}, \ldots, \gamma_{n-1,n}) \tag{1.5.18}$$

denote the probability density describing the statistical distribution of $r_1, \ldots, r_n, \gamma_{12}, \ldots, \gamma_{n-1,n}$, whose explicit knowledge is not essential. To obtain the expectation value of the observable B over all fluctuations of the coefficients $r_i r_j \exp(-i\gamma_{ij})$ in (1.5.15), one has to average the latter over the density in (1.5.18), subject to the constraint (1.5.17). Suppose w_{ij} denotes the latter average. Accordingly, the overall expectation of the observable B is given by

$$\langle B \rangle = \sum_{i,j=1}^{n} w_{ij} \langle \Phi_i |B|\Phi_j\rangle. \tag{1.5.19}$$

We note, in particular, that

$$w_{ij}^* = w_{ji}. \tag{1.5.20}$$

That is, w_{ij} denotes the matrix elements of a Hermitian $n \times n$ matrix w. From elementary matrix algebra, we know that we may then find a unitary matrix U, i.e., $U^\dagger = U^{-1}$ such that

$$U^\dagger w U = \text{diag}\left[w_1, \ldots, w_n\right] \tag{1.5.21}$$

where, from (1.5.17), $w_i \geqslant 0$ and satisfy (1.5.10). Accordingly, let

$$U_{ij}^* \Phi_i = \psi_j \tag{1.5.22}$$

where
$$U_{ij}^* U_{kj} = \delta_{ik} \tag{1.5.23}$$
$$U_{im}^* w_{ij} U_{jk} = w_m \delta_{mk} \tag{1.5.24}$$

and a summation over repeated indices in (1.5.22)–(1.5.24) is understood. Hence
$$\langle B \rangle = \sum_{i=1}^{n} w_i \langle \psi_i | B | \psi_i \rangle \tag{1.5.25}$$

coinciding with the formula in (1.5.7).

For a completely randomized system where the probability density f in (1.5.18) has a *constant* value over its domain of definition, the unitary operator U above becomes the identity operator with $w_{ii} = 1/n$ (no sum over i here). To see this, note that for the phase averages ($i < j$) in this case we have
$$\frac{1}{2\pi} \int_0^{2\pi} \mathrm{d}\gamma_{ij}\, e^{-i\gamma_{ij}} = \delta_{ij} \tag{1.5.26}$$

and by symmetry or by explicit calculation (see Problem 1.4)
$$\langle r_1^2 \rangle = \cdots = \langle r_n^2 \rangle = \frac{1}{n} \tag{1.5.27}$$

under the constraint in (1.5.17). Accordingly,
$$\langle B \rangle = \frac{1}{n} \sum_{i=1}^{n} \langle \psi_i | B | \psi_i \rangle \tag{1.5.28}$$

for a completely random mixture.

1.6 Polarization of Light: An Interlude

An illustration of most of the developments provided so far is readily given by examining polarization aspects of light. The relative simplicity of dealing, at this stage, with a photon, as opposed to, say, a spin 1/2 particle, is that the *vector* character of light, being of spin 1, allows one readily to decompose the polarization states along different directions by using the same elementary geometry as one uses in decomposing the three dimensional position vector of a particle along different directions in Euclidean space. The situation dealing with the decomposition of spin 1/2 states along different axes is different and will be dealt with later. One should be careful, however, that for a photon, like any other massless particle, the direction of polarization is perpendicular to the direction of its propagation.

30   1 Fundamentals

It is convenient to denote the polarization state of light polarized along a unit vector making an angle ϕ with the x-axis simply by $|\phi\rangle$. An x-polarized state may be then written as $|0\rangle$ and a y-polarized state as $|\pi/2\rangle$.

From the vector nature of light we may decompose $|\phi\rangle$ as (see Figure 1.12)

$$|\phi\rangle = |0\rangle \cos\phi + |\pi/2\rangle \sin\phi. \tag{1.6.1}$$

The states $|0\rangle$, $|\pi/2\rangle$ may be represented by the column vectors:

$$|0\rangle = \begin{pmatrix} 1 \\ 0 \\ 0 \end{pmatrix}, \qquad (\langle 0| = \begin{pmatrix} 1 & 0 & 0 \end{pmatrix}) \tag{1.6.2}$$

$$|\pi/2\rangle = \begin{pmatrix} 0 \\ 1 \\ 0 \end{pmatrix}, \qquad (\langle \pi/2| = \begin{pmatrix} 0 & 1 & 0 \end{pmatrix}) \tag{1.6.3}$$

and hence

$$|\phi\rangle = \begin{pmatrix} \cos\phi \\ \sin\phi \\ 0 \end{pmatrix}, \qquad (\langle \phi| = \begin{pmatrix} \cos\phi & \sin\phi & 0 \end{pmatrix}). \tag{1.6.4}$$

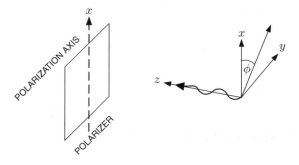

Fig. 1.12. Light polarized along a unit vector in the xy-plane making an angle ϕ with the x-axis on its way to an x-polarizer.

The selective measurement symbols may be then explicitly represented as follows:

$$\Lambda(0) = \begin{pmatrix} 1 & 0 & 0 \\ 0 & 0 & 0 \\ 0 & 0 & 0 \end{pmatrix} \tag{1.6.5}$$

1.6 Polarization of Light: An Interlude

$$\Lambda(\pi/2) = \begin{pmatrix} 0 & 0 & 0 \\ 0 & 1 & 0 \\ 0 & 0 & 0 \end{pmatrix} \tag{1.6.6}$$

$$\Lambda(\phi) = \begin{pmatrix} \cos^2\phi & \sin\phi\cos\phi & 0 \\ \sin\phi\cos\phi & \sin^2\phi & 0 \\ 0 & 0 & 0 \end{pmatrix}. \tag{1.6.7}$$

In particular for successive selective measurements,

$$\Lambda(0)\Lambda(\phi) = \begin{pmatrix} \cos^2\phi & \sin\phi\cos\phi & 0 \\ 0 & 0 & 0 \\ 0 & 0 & 0 \end{pmatrix} \tag{1.6.8}$$

$$\Lambda(\phi)\Lambda(0) = \begin{pmatrix} \cos^2\phi & 0 & 0 \\ \sin\phi\cos\phi & 0 & 0 \\ 0 & 0 & 0 \end{pmatrix} \tag{1.6.9}$$

and for $0 < \phi < \pi/2$

$$\Lambda(0)\Lambda(\phi) \neq \Lambda(\phi)\Lambda(0) \tag{1.6.10}$$

establishing the *non-commutativity* of measurements of polarization along the two different orientations.

To make contact with formula (1.1.5), (1.6.8) may be written as

$$\Lambda(0)\Lambda(\phi) = |0\rangle\langle\phi|\cos\phi \tag{1.6.11}$$

with

$$\langle 0|\phi\rangle = \cos\phi. \tag{1.6.12}$$

The latter is also directly obtained from (1.6.1). By referring to Figure 1.12 we may infer that the probability that a photon goes through the x-polarizer is

$$|\langle 0|\phi\rangle|^2 = \cos^2\phi \tag{1.6.13}$$

which is the famous *Malus formula*.

The density operator corresponding to the state $|\phi\rangle$ in Figure 1.12 is

$$\rho = |\phi\rangle\langle\phi| = \begin{pmatrix} \cos^2\phi & \sin\phi\cos\phi & 0 \\ \sin\phi\cos\phi & \sin^2\phi & 0 \\ 0 & 0 & 0 \end{pmatrix} \tag{1.6.14}$$

as given in (1.6.7). The probability that a photon goes through the x-polarizer shown in Figure 1.12 is equivalently given by

$$\operatorname{Tr}[\Lambda(0)\rho] = \cos^2\phi. \tag{1.6.15}$$

For an initial *mixture*, the density operator ρ may be written as

$$\rho = w_1\Lambda(0) + w_2\Lambda(\pi/2) \tag{1.6.16}$$

for light propagating along the z-axis. For such a mixture going to a ϑ-polarizer, that is a polarizer with polarization axis making an angle ϑ with the x-axis, the probability of transmission is given by

$$\mathrm{Tr}\left[\Lambda(\vartheta)\rho\right] = w_1\cos^2\vartheta + w_2\sin^2\vartheta \tag{1.6.17}$$

as is easily worked out.

For *unpolarized* light, that is light with a complete random polarization, $w_1 = w_2 = 1/2$, (1.6.17) gives

$$\mathrm{Tr}\left[\Lambda(\vartheta)\rho_{\mathrm{unpol}}\right] = \frac{1}{2} \tag{1.6.18}$$

independently of the orientation of the polarization axis, specified by the angle ϑ, of the polarizer.

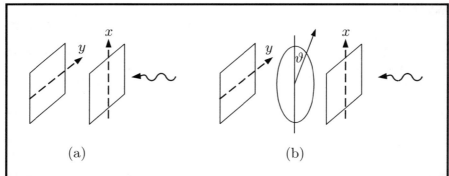

Fig. 1.13. (a) An x-polarizer followed by a y-polarizer corresponding to successive selective measurement symbols. No light may go through the combined polarization system. (b) A ϑ-polarizer is inserted between the x- and y-polarizers in (a). The corresponding successive selective measurements is, for $0 < \vartheta < \pi/2$, for example, not the **0**-operation. Depending on the initial state for light, a photon may go through the arrangement with the three polarizers.

Finally we consider the successive selective measurements provided by the polarizers in Figure 1.13. For the successive measurements corresponding to Figure 1.13 (a)

$$\Lambda(\pi/2)\Lambda(0) = \mathbf{0} \tag{1.6.19}$$

and *no* photon may go through the two polarizers. On the other hand if we insert a ϑ-polarizer between these two polarizers as shown in Figure 1.13 (b), we obtain

$$\Lambda(\pi/2)\Lambda(\vartheta)\Lambda(0) = \begin{pmatrix} 0 & 0 & 0 \\ \sin\vartheta\cos\vartheta & 0 & 0 \\ 0 & 0 & 0 \end{pmatrix} \neq \mathbf{0} \qquad (1.6.20)$$

with $0 < \vartheta < \pi/2$, for example, and a photon may *go through* the new arrangement with the three polarizers.

What is remarkable about the arrangement in Figure 1.13 (b) is that although the x-polarizer makes sure to eliminate light polarized along the y-axis to go through it, the insertion of the ϑ-polarizer, depending on the initial state, may allow y-polarized light to be finally transmitted through the y-polarizer. The naïve impression that the insertion of a polarizer between two polarizers will reduce the final intensity of light transmitted through the system is not necessarily true. To this end, suppose that light is initially polarized along an axis making an angle ϕ, with the x-axis given in (1.6.4). Consider, using in the process (1.6.20), the successive selective measurements

$$\Lambda(\pi/2)\Lambda(\vartheta)\Lambda(0)\Lambda(\phi) = |\pi/2\rangle\langle\phi|\sin\vartheta\cos\vartheta\cos\phi. \qquad (1.6.21)$$

Thus the insertion of the ϑ-polarizer, as in Figure 1.13 (b), increases the intensity of light transmitted from 0 to $100 \times \left(\sin^2\vartheta\cos^2\vartheta\cos^2\phi\right)$ %.

The arrangements given in Figure 1.13 together with the quantitative details given above provide an illustration of the theory developed with incompatible observables at the end of §1.1. The system of the three polarizers in Figure 1.13 (b) is represented by the matrix

$$M = \Lambda(\pi/2)\Lambda(\vartheta)\Lambda(0) = \begin{pmatrix} 0 & 0 & 0 \\ \sin\vartheta\cos\vartheta & 0 & 0 \\ 0 & 0 & 0 \end{pmatrix} \qquad (1.6.22)$$

in the notation of §1.3, §1.4, as introduced in (1.6.20).

Although aspects of the polarization of light are easily treated, as shown above, and many authors use light to illustrate concepts in quantum mechanics, a detailed quantum description of light is far from straightforward.

1.7 The Hilbert Space; Rigged Hilbert Space

In quantum physics, one generally deals not only with finite dimensional inner-product spaces but also with infinite dimensional ones. The Hilbert space concept is such a generalization which applies to both cases. In view of applications, we consider only separable Hilbert spaces, the property of which is spelled out in point (iii) below.

Definition of a Hilbert Space

A set of vectors is called a Hilbert space, denoted by $\mathcal{H}$, if

(i) $|f_1\rangle$, $|f_2\rangle$ are in $\mathcal{H}$, then so is $\alpha_1|f_1\rangle + \alpha_2|f_2\rangle$ for all complex numbers α_1, α_2. For the $\mathbf{0}$ vector in $\mathcal{H}$, $\mathbf{0} + |f\rangle = |f\rangle$ for all $|f\rangle$ in $\mathcal{H}$. This provides just the definition of a *vector space*.

(ii) It is equipped with an inner product $\langle f_1|f_2\rangle$ for all $|f_1\rangle$, $|f_2\rangle$ in $\mathcal{H}$, which is in general a complex number such that

$$\langle f_1|f_2 + f_3\rangle = \langle f_1|f_2\rangle + \langle f_1|f_3\rangle$$

$$\langle \alpha f_1|f_2\rangle = \alpha^* \langle f_1|f_2\rangle$$

$$\langle f_1|f_2\rangle^* = \langle f_2|f_1\rangle$$

$$\langle f|f\rangle \geq 0$$

and $\langle f|f\rangle = 0$ if and only if $|f\rangle$ is the zero vector. Property (ii) provides the definition of an *inner product space*. The norm of a vector $|f\rangle$, denoted by $\|f\|$, is defined by

$$\|f\| = \sqrt{\langle f|f\rangle}.$$

(iii) In $\mathcal{H}$ there exists a sequence of orthonormal vectors $\{|f_1\rangle, |f_2\rangle, \ldots\}$, called a basis, i.e., $\langle f_i|f_j\rangle = \delta_{ij}$, and for any f in $\mathcal{H}$, we may find constants $c_k^{(N)}$ such that

$$\left\| f - \sum_{k=1}^N c_k^{(N)} f_k \right\| \to 0 \quad \text{as } N \to \infty.$$

Property (iii) constitutes of what is called its *separability* condition of $\mathcal{H}$.

(iv) A sequence of vectors $\{|f^1\rangle, |f^2\rangle, \ldots\}$ in $\mathcal{H}$ is called a Cauchy sequence, if given any $\varepsilon > 0$, we may find a positive integer N such that

$$\|f^{(n)} - f^{(m)}\| < \varepsilon$$

whenever $n > N$ and $m > N$. Then every Cauchy sequence in $\mathcal{H}$ converges to a vector $|f\rangle$ in $\mathcal{H}$. That is,

$$\|f - f^{(n)}\| \longrightarrow 0$$

for $n \to \infty$. This is called the *completeness* property of $\mathcal{H}$.

Immediate consequences of the above definitions are the following inequalities. For any $|f\rangle$, $|g\rangle$ in $\mathcal{H}$

$$|\langle f|g\rangle| \leq \|f\| \|g\| \tag{1.7.1}$$

referred to as the *Cauchy-Schwarz* inequality (see also (1.4.52)),

$$\|f + g\| \leq \|f\| + \|g\| \tag{1.7.2}$$

1.7 The Hilbert Space; Rigged Hilbert Space

referred to as the *triangular* inequality (see also (1.4.53)). Finally let $\{|f_1\rangle, |f_2\rangle, \ldots\}$ be an orthonormal (i.e., $\langle f_i | f_j \rangle = \delta_{ij}$) basis in $\mathcal{H}$. For any $|f\rangle$ in $\mathcal{H}$ define

$$\left|f^{(n)}\right\rangle = \sum_{k=1}^{n} |f_k\rangle \langle f_k | f \rangle. \tag{1.7.3}$$

Then

$$\|f^{(n)}\| \leqslant \|f\| \tag{1.7.4}$$

and is referred to as *Bessel's* inequality.

The proofs of these inequalities are elementary and follow. If at least one of $|f\rangle$, $|g\rangle$ is the zero vector then (1.7.1) is obvious. If two non-zero vectors $|f\rangle$, $|g\rangle$ are orthogonal, that is $\langle f | g \rangle = 0$, then (1.7.1) is again obvious. If $\langle f | g \rangle \neq 0$, define

$$a = \alpha \frac{|\langle g | f \rangle|}{\langle f | g \rangle} \tag{1.7.5}$$

where α is a real number. Then

$$0 \leqslant \|f + ag\|^2 = \|f\|^2 + 2\alpha |\langle g | f \rangle| + \alpha^2 \|g\|^2 \tag{1.7.6}$$

which upon minimizing over α gives (1.7.1).

For the triangular inequality, we have

$$\|f + g\|^2 = \|f\|^2 + 2\operatorname{Re}\langle f | g \rangle + \|g\|^2. \tag{1.7.7}$$

But by making use of (1.7.1),

$$\operatorname{Re}\langle f | g \rangle \leqslant |\langle f | g \rangle| \leqslant \|f\| \|g\| \tag{1.7.8}$$

that is

$$\|f + g\|^2 \leqslant (\|f\| + \|g\|)^2 \tag{1.7.9}$$

which is equivalent to (1.7.2).

For Bessel's inequality we note that

$$\left\langle f - f^{(n)} \,\middle|\, f^{(n)} \right\rangle = 0. \tag{1.7.10}$$

Hence

$$\|f\|^2 = \left\|f - f^{(n)} + f^{(n)}\right\|^2 = \left\|f - f^{(n)}\right\|^2 + \left\|f^{(n)}\right\|^2$$

$$\geqslant \left\|f^{(n)}\right\|^2 \tag{1.7.11}$$

giving (1.7.4).

A typical example of a Hilbert space is the set, denoted by $\ell^2(\infty)$, of all vectors $|a\rangle = (a_1, a_2, \ldots)$, involving components as complex numbers such that

$$\|a\|^2 = \sum_{k=1}^{\infty} |a_k|^2 < \infty \tag{1.7.12}$$

with the addition law defined by

$$\alpha \, |a\rangle + \beta \, |b\rangle = (\alpha a_1 + \beta b_1, \alpha a_2 + \beta b_2, \ldots) \tag{1.7.13}$$

and the inner-product given by

$$\langle a \, | \, b \rangle = \sum_{k=1}^{\infty} a_k^* b_k. \tag{1.7.14}$$

[The proof of the separability and completeness of this space is relegated to Problem 1.7.]

A particularly important example of a Hilbert space is the space of square-integrable functions, denoted by $L^2(\mathbb{R}^3)$, associated with a particle of spin 0. For any $f(\mathbf{x})$, $g(\mathbf{x})$ in $L^2(\mathbb{R}^3)$, $\mathbf{x}$ in $\mathbb{R}^3$,

$$\langle f \, | \, g \rangle = \int d^3\mathbf{x} \, f^*(\mathbf{x}) \, g(\mathbf{x}) \tag{1.7.15}$$

and

$$\|f\|^2 = \int d^3\mathbf{x} \, |f(\mathbf{x})|^2 < \infty. \tag{1.7.16}$$

For a (massive) particle of spin s, having $(2s+1)$ components along a given direction, we may introduce the space $L^2(\mathbb{R}^3, \mathbb{C}^{2s+1})$ (where $\mathbb{C}$ stands for complex), and define

$$\langle f \, | \, g \rangle = \sum_{\sigma} \int d^3\mathbf{x} \, f^*(\mathbf{x}, \sigma) \, g(\mathbf{x}, \sigma) \tag{1.7.17}$$

where the sum is over all $\sigma = -s, -s+1, \ldots, s$.

For n *distinguishable* spin 0 particles, for example, one may introduce the space $L^2(\mathbb{R}^{3n})$ with inner product

$$\langle f \, | \, g \rangle = \int d^3\mathbf{x}_1 \cdots d^3\mathbf{x}_n \, f^*(\mathbf{x}_1, \ldots, \mathbf{x}_n) \, g(\mathbf{x}_1, \ldots, \mathbf{x}_n). \tag{1.7.18}$$

Of particular interest is the case dealing with n *indistinguishable* particles of spin s, such as n electrons ($s = 1/2$), n neutral pions ($s = 0$), etc. In such cases we have to invoke the *spin and statistics* connection and restrict to subclasses of square-integrable functions $\psi(\mathbf{x}_1, \sigma_1; \ldots; \mathbf{x}_n, \sigma_n)$ with definite symmetries as defined below for the interchange of any two particles:

$$\psi(\ldots; \mathbf{x}_i, \sigma_i; \ldots; \mathbf{x}_j, \sigma_j; \ldots) = -\psi(\ldots; \mathbf{x}_j, \sigma_j; \ldots; \mathbf{x}_i, \sigma_i; \ldots) \tag{1.7.19}$$

for half-odd integer spin s, with such particles referred to as *fermions*, and

1.7 The Hilbert Space; Rigged Hilbert Space

$$\psi(\ldots;\mathbf{x}_i,\sigma_i;\ldots;\mathbf{x}_j,\sigma_j;\ldots) = +\psi(\ldots;\mathbf{x}_j,\sigma_j;\ldots;\mathbf{x}_i,\sigma_i;\ldots) \qquad (1.7.20)$$

for integer spin s, and such particles are referred to as *bosons*.

The state of a system is described by a unit vector $|\psi\rangle$ in some Hilbert space $\mathcal{H}$. Since, as we have already seen in §1.2, §1.3, §1.5 and we have ample opportunity to see this further later on, physical quantities (such as probabilities, expectation values, ...), to be compared with experiments, involve the combination $|\psi\rangle\langle\psi|$, and the states $|\psi\rangle$ and $|\psi\rangle e^{i\phi}$, defined up to arbitrary *phase* factors, are equivalent. The set $\{|\psi\rangle e^{i\phi}\}$, also denoted by $|\psi\rangle$ for convenience, with ϕ varying over all real numbers, is called a unit *ray*. Accordingly, the *states of a physical system are represented by unit rays*.

Given a Hilbert space $\mathcal{H}$, not every unit ray $|\psi\rangle$ in $\mathcal{H}$, however, is necessarily a physically realizable state. From the very definition of a Hilbert space, for example, a linear combination of two states $|\psi_1\rangle$ and $|\psi_2\rangle$ with different charges Q_1, Q_2 is also in $\mathcal{H}$. But such a vector is not physically realizable. Similarly the superposition of two vectors $|\psi_1\rangle$, $|\psi_2\rangle$ with integer and half-odd integer angular momentum states, respectively, resulting a vector $|\psi\rangle$, is not physically realizable, if the principle of rotational invariance is invoked. The reason is that under a rotation by an angle 2π, for example, about the quantization z-axis, as we shall see later, $|\psi_1\rangle$ and $|\psi_2\rangle$ transform in different ways and the projection operator $|\psi\rangle\langle\psi|$ does not remain invariant. In detail, if

$$|\psi\rangle = \alpha_1 |\psi_1\rangle + \alpha_2 |\psi_2\rangle \qquad (1.7.21)$$

then under the above specified rotation with

$$|\psi\rangle\langle\psi| \longrightarrow |\psi'\rangle\langle\psi'| \qquad (1.7.22)$$

$$|\psi'\rangle = \alpha_1 |\psi_1\rangle - \alpha_2 |\psi_2\rangle. \qquad (1.7.23)$$

Accordingly, not every unit ray in a Hilbert space is necessarily physically realizable and any rule which singles out such rays is referred to as a *superselection* rule.

Any given vector $|f\rangle$ in a Hilbert space $\mathcal{H}$ may be also defined as an *anti-linear functional*, through the inner product as follows. For $|g\rangle$ a vector in $\mathcal{H}$, and with an inner product $\langle g|f\rangle$, one may consider $|f\rangle$ as a functional in $|g\rangle$, with the property

$$\langle \alpha_1\psi_1 + \alpha_2\psi_2 | f \rangle = \alpha_1^* \langle \psi_1 | f \rangle + \alpha_2^* \langle \psi_2 | f \rangle \qquad (1.7.24)$$

for any two complex numbers α_1, α_2. The "linearity" condition in (1.7.24) is implicit. "Anti" in anti-linear refers to the fact that the coefficients α_1, α_2 in (1.7.24) are complex conjugated which follows from the definition of the inner product $\langle g|f\rangle$.

The above definition is useful in studying properties of observables with corresponding operators having a continuous spectrum. Such operators are studied in §1.8. But for the moment, consider, for example, the position operator in one dimension. This may be written as

$$X = \int_{-\infty}^{\infty} \mathrm{d}x \, x \, |x\rangle\langle x| \qquad (1.7.25)$$

with the sharp selective measurement symbol $\Lambda(x) = |x\rangle\langle x|$ satisfying (compare with (1.1.2), (1.1.3))

$$\Lambda(x')\Lambda(x) = \Lambda(x)\langle x'|x\rangle \qquad (1.7.26)$$

$$\langle x'|x\rangle = \delta(x'-x) \qquad (1.7.27)$$

with the latter denoting the Dirac delta. The identity operator, which accepts and transmits all particles crossing anywhere the x-axis with no discrimination, is given by the expression

$$\mathbf{1} = \int_{-\infty}^{\infty} \mathrm{d}x \, |x\rangle\langle x|. \qquad (1.7.28)$$

If the system is initially prepared to be in a state $|\psi\rangle$, then one may write

$$|\psi\rangle = \mathbf{1}|\psi\rangle = \int_{-\infty}^{\infty} \mathrm{d}x \, |x\rangle\langle x|\psi\rangle \qquad (1.7.29)$$

and the probability density that a particle crosses the x-axis at point x is

$$\mathrm{Tr}\left[\Lambda(x)|\psi\rangle\langle\psi|\right] = |\langle\psi|x\rangle|^2. \qquad (1.7.30)$$

Although $|x\rangle$ is obviously not a vector in the underlying Hilbert space $\mathcal{H}$, it is nevertheless rigorously defined as an anti-linear functional on vectors $|\psi\rangle$ in $\mathcal{H}$ through

$$\langle \alpha_1\psi_1 + \alpha_2\psi_2 | x \rangle = \alpha_1^* \langle \psi_1 | x \rangle + \alpha_2^* \langle \psi_2 | x \rangle. \qquad (1.7.31)$$

The set of all anti-linear functionals consists then not only of vectors $|f\rangle$, as discussed above, but also generalized ket vectors (such as $|x\rangle$) and is obviously larger than the Hilbert space $\mathcal{H}$ itself. The triplet consisting of $\mathcal{H}$, the set of all anti-linear functionals and the domain, depending on the problem in hand, consisting of those vectors in $\mathcal{H}$ on which the anti-linear functionals are defined, is referred to as a *Rigged* Hilbert space.

It is interesting to dwell further on the anti-linear functional $|\mathbf{x}\rangle$ as a functional of functions belonging to $L^2(\mathbb{R}^3)$. Given some function $\psi(\cdot)$ in $L^2(\mathbb{R}^3)$, the anti-linear functional $|\mathbf{x}\rangle$ may be represented by $\delta^3(\mathbf{x} - \mathbf{x}')$, in a space larger than $L^2(\mathbb{R}^3)$, as defined through

$$\langle \psi | \mathbf{x} \rangle = \langle \psi(\cdot) | \delta^3(\mathbf{x} - \cdot) \rangle$$

$$= \int d^3x' \, \psi^*(x') \, \delta^3(x - x') = \psi^*(x). \tag{1.7.32}$$

A sharp momentum state $|p\rangle$ of a particle, as we shall see later, may be written as

$$|p\rangle = \int d^3x \, e^{ip \cdot x/\hbar} \, |x\rangle \tag{1.7.33}$$

where $\hbar$ is the Planck constant h divided by 2π. As an anti-linear functional on functions $\psi(\cdot)$ in $L^2(\mathbb{R}^3)$, $|p\rangle$ may be represented by $e^{ip \cdot x/\hbar}$ in, obviously, a larger space than $L^2(\mathbb{R}^3)$, due to its lack of square-integrability, defined through

$$\langle \psi | p \rangle = \left\langle \psi(\cdot) \middle| e^{ip \cdot (\cdot)/\hbar} \right\rangle$$

$$= \int d^3x \, \psi^*(x) \, e^{ip \cdot x/\hbar} \tag{1.7.34}$$

(a Fourier transform). Obviously, the *domain* of $|p\rangle$ consists of functions $\psi(x)$ in $L^2(\mathbb{R}^3)$ for which the integral in (1.7.34) exists.

Finally consider the anti-linear functional

$$|\Phi\rangle = \int d^3x \, |x\rangle \, \Phi(x) \tag{1.7.35}$$

where $\Phi(x)$ is not a square-integrable function, and hence necessarily belongs to a space larger than $L^2(\mathbb{R}^3)$. Its domain consists of those functions $\psi(x)$ in $L^2(\mathbb{R}^3)$ such that

$$\langle \psi | \Phi \rangle = \int d^3x \, \psi^*(x) \, \Phi(x) \tag{1.7.36}$$

exists.

The moral of the Rigged Hilbert space formalism is that physics dictates, in general, to introduce in addition to a Hilbert space $\mathcal{H}$, a space, say, D^+, larger than $\mathcal{H}$ of anti-linear functionals, including generalized ket vectors, with a domain of definition for the anti-linear functionals, say, D, contained in $\mathcal{H}$. The Rigged Hilbert space is then written as the triplet $D \subset \mathcal{H} \subset D^+$. The Rigged Hilbert formalism has clarified some of the ambiguous manipulations carried out in earlier days of quantum physics.

1.8 Self-Adjoint Operators and Their Spectra

An observable B which may take on a finite number of real discrete values, $\{b, b', \ldots\}$, is represented (see §1.3) by a Hermitian operator

$$B = \sum_b b \, |b\rangle\langle b| \tag{1.8.1}$$

$$B^\dagger = B$$

(using the same notation for the latter as the observable it represents) in a finite dimensional inner-product space generated by the eigenvectors $|b\rangle$. That is, any vector $|f\rangle$ in the generated inner-product space may be written as (see (1.4.2), (1.4.16))

$$|f\rangle = \sum_b |b\rangle f(b) \tag{1.8.2}$$

where

$$f(b) = \langle b|f\rangle \tag{1.8.3}$$

are complex numbers. The operator B maps any vector $|f\rangle$, in the generated inner product space, into another vector

$$B|f\rangle = \sum_b |b\rangle g(b) \equiv |g\rangle \tag{1.8.4}$$

where

$$g(b) = b f(b) \tag{1.8.5}$$

in the same inner-product space. In particular, the Hermiticity of B implies that

$$\langle f_2|B|f_1\rangle = \sum_b b f_2^*(b) f_1(b)$$

$$= \langle B f_2|f_1\rangle \tag{1.8.6}$$

for any two vectors in the generated inner-product space.

The selective-measurement symbol

$$\Lambda(b) = |b\rangle\langle b| \tag{1.8.7}$$

defines a projection operator, i.e.,

$$\Lambda(b)\Lambda(b) = \Lambda(b) \tag{1.8.8}$$

(see (1.1.2)). One may also define the following projection operator

$$P_B(b') = \sum_{b \leqslant b'} \Lambda(b) \tag{1.8.9}$$

by summing over the projection operators over all the eigenvalues of B, from the lowest, up to any given eigenvalue b' as indicated in the summation sign in (1.8.9). Properties of the projection operators in (1.8.9) are easily established and will be spelled out for the more general cases to be discussed below.

In the general case when an observable, say, A may take on infinite number of real discrete values and/or continuous real values, one is faced with

1.8 Self-Adjoint Operators and Their Spectra

convergence problems and one carries out physical computations, more generally, in a Hilbert space as defined in §1.7. The object of physical interest representing an observable, such as A, is a *self-adjoint* operator, as defined below, and coincides with the definition above, for a Hermitian operator, in the simpler case discussed above with a finite number of real discrete values, operating in a finite dimensional inner-product space. Hermitian operators may, however, be defined in the general case as well, but in quantum physics it is the more general concept of a self-adjoint operator that is relevant as representing a given observable. We do not wish to get too technical about the distinction between these two types of operators and we provide the bare minimum in this respect.

An operator A maps, in general, a vector $|f\rangle$ in the underlying Hilbert space $\mathcal{H}$ into some other vector $A|f\rangle$ in $\mathcal{H}$. The totality of all vectors $\{|f\rangle\}$ in $\mathcal{H}$ such that for each $|f\rangle$ in this set, $A|f\rangle$ is also in $\mathcal{H}$, i.e., $\|Af\| < \infty$, is called the domain of A. We are interested in operators with domains rich enough such that, as a generalization of (1.8.2), any vector in $\mathcal{H}$ may be well approximated by a vector in the domain of the operator in question. [Technically, this means that for any vector $|h\rangle$ in $\mathcal{H}$, and any given $\varepsilon > 0$, we may find a vector $|h_\varepsilon\rangle$ in the domain of A such that $\|h - h_\varepsilon\| < \varepsilon$.] For *such* an operator, the adjoint $A^\dagger$ may be defined through

$$\langle g|A|f\rangle = \langle A^\dagger g|f\rangle \tag{1.8.10}$$

for all $|f\rangle$ in the domain of A.

If the domains of A and $A^\dagger$ coincide and

$$\langle g|A|f\rangle = \langle Ag|f\rangle \tag{1.8.11}$$

(see (1.8.6) for the special case) then A is said to be self-adjoint and one writes $A^\dagger = A$. [In the case (1.8.11) holds for all $|f\rangle$ and $|g\rangle$ in the domain of A, but the domains of A and $A^\dagger$ do not coincide, A is said to be Hermitian.]

The remaining part of this section deals with the *spectra* of self-adjoint operators and the meaning of the eigenvalue problem. This topic is of central importance in quantum physics. We restrict, however, the presentation to those aspects which are only relevant to the rest of the volume.

Given a self-adjoint operator A we may write

$$A = \int_{-\infty}^{\infty} d\lambda\, \lambda\, \delta(\lambda - A) \tag{1.8.12}$$

which has obviously a meaning whenever a measurement of the observable with which the operator A is associated takes a given real value, say, λ_0.

Define

$$P_A(\lambda) = \Theta(\lambda - A) = \int_{-\infty}^{\lambda} d\lambda'\, \delta(\lambda' - A) \tag{1.8.13}$$

(see (1.8.9) for comparison), where $\Theta(\lambda)$ is the step function, $\Theta(\lambda) = 1$ for $\lambda > 0$ and $\Theta(\lambda) = 0$ otherwise. Then formally

$$\mathrm{d}P_A(\lambda) = \delta(\lambda - A)\,\mathrm{d}\lambda. \tag{1.8.14}$$

We may then write from (1.8.12)

$$A = \int_{-\infty}^{\infty} \lambda\,\mathrm{d}P_A(\lambda). \tag{1.8.15}$$

The latter is called the spectral decomposition of A.

The $P_A(\lambda)$ are projection operators whose properties are readily established from (1.8.13). They are:

(i) Since a step-function is bounded by 1,

$$\|P_A(\lambda)\,f\| \leqslant \|f\| \tag{1.8.16}$$

for *all* $|f\rangle$ in the underlying Hilbert space $\mathcal{H}$, i.e., (1.8.16) is defined for all vectors $|f\rangle$ in $\mathcal{H}$. In particular, $P_A(\lambda)$ is self-adjoint.

(ii)
$$P_A(-\infty) = \mathbf{0}. \tag{1.8.17}$$

(iii)
$$P_A(+\infty) = \mathbf{1} = \int_{-\infty}^{\infty} \mathrm{d}P_A(\lambda) \tag{1.8.18}$$

giving the identity operator (compare with (1.1.4), see also (1.8.1), (1.8.9)). The equality on the right-hand side of (1.8.18) is called the resolution of the identity operator.

(iv)
$$\langle f|P_A(\lambda_1)|f\rangle \leqslant \langle f|P_A(\lambda_2)|f\rangle \tag{1.8.19}$$

for $\lambda_1 < \lambda_2$.

(v)
$$P_A(\lambda_1)\,P_A(\lambda_2) = P_A(\lambda_0) \tag{1.8.20}$$

where
$$\lambda_0 = \min(\lambda_1, \lambda_2). \tag{1.8.21}$$

(vi) From (1.8.18),

$$\|f\|^2 = \langle f|f\rangle = \int_{-\infty}^{\infty} \mathrm{d}\,\langle f|P_A(\lambda)|f\rangle = \int_{-\infty}^{\infty} \mathrm{d}\|P_A(\lambda)f\|^2 \tag{1.8.22}$$

where in writing the last equality we have used (1.8.20).

To study the spectrum of A and hence infer about the possible values that may be obtained by a measurement of the observable with which A is associated, we consider the operator

$$[P_A(\lambda_0 + \varepsilon) - P_A(\lambda_0 - \varepsilon)] \equiv \int_{\lambda_0 - \varepsilon}^{\lambda_0 + \varepsilon} \mathrm{d}P_A(\lambda). \tag{1.8.23}$$

1.8 Self-Adjoint Operators and Their Spectra   43

Suppose λ_0 is an isolated point of the spectrum of A. That is, for *some* $\varepsilon > 0$, A takes no values in $(\lambda_0 - \varepsilon, \lambda_0 + \varepsilon)$ except the value λ_0. The integrand in (1.8.23) then makes a finite "jump" from zero when A takes on this value and the subspace

$$[P_A(\lambda_0 + \varepsilon) - P_A(\lambda_0 - \varepsilon)]\mathcal{H} \qquad (1.8.24)$$

including the case with the limit $\varepsilon \to +0$, is not empty. Hence for $\varepsilon \to +0$, the latter contains at least one *vector*, say, $|f\rangle$, and

$$[P_A(\lambda_0 + 0) - P_A(\lambda_0 - 0)]|f\rangle = |f\rangle. \qquad (1.8.25)$$

Accordingly, from (1.8.15), property (iii) in (1.8.18) and property (iv) in (1.8.22)

$$\begin{aligned}\|(A - \lambda_0)f\|^2 &= \int_{-\infty}^{\infty} (\lambda - \lambda_0)^2 \, \mathrm{d}\|P_A(\lambda)f\|^2 \\ &= \int_{\lambda_0 - 0}^{\lambda_0 + 0} (\lambda - \lambda_0)^2 \, \mathrm{d}\|P_A(\lambda)f\|^2 = 0 \end{aligned} \qquad (1.8.26)$$

where we have also used (1.8.23), (1.8.25). Hence $(A - \lambda_0)|f\rangle$ is the *zero* vector, i.e.,

$$A|f\rangle = \lambda_0|f\rangle \qquad (1.8.27)$$

which is the familiar eigenvalue equation for a (discrete) eigenvalue λ_0. This occurs whenever the integrand in (1.8.23) makes a "jump" from zero when A takes on the value $\lambda = \lambda_0$ and the point λ_0 is isolated in $(\lambda_0 - \varepsilon, \lambda_0 + \varepsilon)$ with no other values occurring in this interval for some $\varepsilon > 0$, and (1.8.24) is non-empty for $\varepsilon \to +0$.

On the other hand, suppose that for some value λ_0, that the observable in question may take, and for *all* $\varepsilon > 0$, no matter how small, one may find a point $\lambda_1 \neq \lambda_0$ in $(\lambda_0 - \varepsilon, \lambda_0 + \varepsilon)$ in the spectrum of A, and

$$[P_A(\lambda_0 + \varepsilon) - P_A(\lambda_0 - \varepsilon)] \qquad (1.8.28)$$

becomes the zero operator for $\varepsilon \to +0$. For such a point

$$[P_A(\lambda_0 + 0) - P_A(\lambda_0 - 0)]\mathcal{H} \qquad (1.8.29)$$

is *empty* and hence contains no *non-zero* vectors. As we shall see, this refers to the fact that there are no *eigenvectors* in $\mathcal{H}$ corresponding to the continuous spectrum of A.

For $\varepsilon > 0$, however, any non-zero *vector* in the subspace

$$[P_A(\lambda_0 + \varepsilon) - P_A(\lambda_0 - \varepsilon)]\mathcal{H} \qquad (1.8.30)$$

of $\mathcal{H}$ for which (1.8.23) is not the zero operator, necessarily depends on ε. Let $|f(\varepsilon)\rangle$ be a non-zero in this non-empty subspace of $\mathcal{H}$. That is,

$$[P_A(\lambda_0 + \varepsilon) - P_A(\lambda_0 - \varepsilon)] |f(\varepsilon)\rangle = |f(\varepsilon)\rangle. \tag{1.8.31}$$

For $\varepsilon > 0$, we may consider $|f(\varepsilon)\rangle$ to be normalized, i.e., $\||f(\varepsilon)\|\| = 1$. Hence

$$\|(A - \lambda_0) f(\varepsilon)\|^2 = \int_{\lambda_0 - \varepsilon}^{\lambda_0 + \varepsilon} (\lambda - \lambda_0)^2 \, d\|P_A(\lambda) f(\varepsilon)\|^2 \tag{1.8.32}$$

where we have used (1.8.31), (1.8.23).

The right-hand side of (1.8.32) is bounded above by

$$\varepsilon^2 \int_{\lambda_0 - \varepsilon}^{\lambda_0 + \varepsilon} d\|P_A(\lambda) f(\varepsilon)\|^2 = \varepsilon^2 \int_{-\infty}^{\infty} d\|P_A(\lambda) f(\varepsilon)\|^2 = \varepsilon^2 \tag{1.8.33}$$

where we have used (1.8.22), and the normalizability of $f(\varepsilon)$ for $\varepsilon > 0$.

That is, for any $\varepsilon > 0$, for which (1.8.30) is not empty (and is otherwise empty for $\varepsilon \to +0$, we may find a vector $|f(\varepsilon)\rangle$, depending on ε, in $\mathcal{H}$ such that

$$\|(A - \lambda_0) f(\varepsilon)\| \leqslant \varepsilon \tag{1.8.34}$$

and $(A - \lambda_0) f(\varepsilon)$ can be made closer and closer to the zero vector by making ε smaller and smaller. Equation (1.8.34) covers the earlier case in (1.8.27) for the eigenvalue equation as well by taking the limit $\varepsilon \to +0$.

One encounters this latter case quite often in elementary quantum physics courses and a classic example of this is the one dealing with the position observable and is given below.

Try to set up an eigenvalue equation for the position of a particle, via the equation

$$\mathbf{x} f_{\mathbf{x}_0}(\mathbf{x}) = \mathbf{x}_0 f_{\mathbf{x}_0}(\mathbf{x}). \tag{1.8.35}$$

We may rewrite the latter as

$$(\mathbf{x} - \mathbf{x}_0) f_{\mathbf{x}_0}(\mathbf{x}) = \mathbf{0} \tag{1.8.36}$$

whose solution, up to a multiplicative constant, is

$$f_{\mathbf{x}_0}(\mathbf{x}) = \delta^3(\mathbf{x} - \mathbf{x}_0). \tag{1.8.37}$$

[Note that any function of x multiplied by (1.8.37) is evaluated at $\mathbf{x} = \mathbf{x}_0$ and hence reduces to a constant.] The solution (1.8.37) is obviously not square-integrable. What the above analysis, however, shows, and this is in conformity with experimental limitations, that is given any $\varepsilon > 0$, as small as one wishes, one may find a square-integrable function $\delta_\varepsilon(\mathbf{x}, \mathbf{x}_0)$, depending on ε, such that

$$\|(\mathbf{x} - \mathbf{x}_0) \delta_\varepsilon(\mathbf{x}, \mathbf{x}_0)\| = \left(\int d^3 x \, |\mathbf{x} - \mathbf{x}_0|^2 |\delta_\varepsilon(\mathbf{x}, \mathbf{x}_0)|^2 \right)^{1/2} \leqslant \varepsilon. \tag{1.8.38}$$

Such an explicit function is given by

1.8 Self-Adjoint Operators and Their Spectra

$$\delta_\varepsilon(\mathbf{x}, \mathbf{x}_0) = \left(\frac{2}{\pi}\right)^{3/4} \frac{1}{\varepsilon^{3/2}} \exp\left[-\frac{(\mathbf{x}-\mathbf{x}_0)^2}{\varepsilon^2}\right]. \tag{1.8.39}$$

This is obviously normalized ($\varepsilon > 0$), and the left-hand side of (1.8.38) is equal to $\varepsilon/2 < \varepsilon$.

A similar analysis may be given for the momentum operator of a particle (see Problem 1.8).

The above analysis leads to the definition of the spectrum of a self-adjoint operator A as consisting of all (real) λ_0 such that

$$[P_A(\lambda_0 + \varepsilon) - P_A(\lambda_0 - \varepsilon)] \neq \mathbf{0} \tag{1.8.40}$$

for *all* $\varepsilon > 0$.

For the eigenvalue equation, $(A-\lambda_0)$ annihilates an eigenvector, i.e., some vector in $\mathcal{H}$, hence the inverse operator $(A - \lambda_0)^{-1}$ does not exist in all $\mathcal{H}$.

On the other hand for $\varepsilon > 0$ for which (1.8.40) is true, that is the subspace in (1.8.30) is not empty, and, is otherwise empty for $\varepsilon \to +0$,

$$(A - \lambda_0)|f(\varepsilon)\rangle \equiv |g(\varepsilon)\rangle \tag{1.8.41}$$

as well as $|f(\varepsilon)\rangle$, are non-zero vectors in $\mathcal{H}$. In such cases $(A - \lambda_0)^{-1}$ may be defined in $\mathcal{H}$ since there are no vectors $|f\rangle$ in $\mathcal{H}$ that make $(A - \lambda_0)|f\rangle$ zero. Using the normalizability condition for $|f(\varepsilon)\rangle$, we have from (1.8.34), (1.8.41)

$$\|g(\varepsilon)\| \leq \varepsilon = \varepsilon \|(A - \lambda_0)^{-1} g(\varepsilon)\| \tag{1.8.42}$$

which leads to

$$\frac{\|(A - \lambda_0)^{-1} g(\varepsilon)\|}{\|g(\varepsilon)\|} \geq \frac{1}{\varepsilon}. \tag{1.8.43}$$

The latter says that although the inverse $(A - \lambda_0)^{-1}$ may be defined in such cases it is an unbounded operator as the right-hand side of (1.8.43) may be made larger and larger as ε is chosen smaller and smaller.

Consider any two distinct points λ_1 and λ_2 in the spectrum of A. For any $\varepsilon_1 > 0$, $\varepsilon_2 > 0$, and any normalized vectors $|f_1(\varepsilon_1)\rangle$, $|f_2(\varepsilon_2)\rangle$ such that (see (1.8.34))

$$\|(A - \lambda_1) f_1(\varepsilon_1)\| \leq \varepsilon_1 \tag{1.8.44}$$

$$\|(A - \lambda_2) f_2(\varepsilon_2)\| \leq \varepsilon_2 \tag{1.8.45}$$

the orthogonality relation

$$\lim_{\varepsilon_1 \to 0} \lim_{\varepsilon_2 \to 0} \langle f_1(\varepsilon_1) | f_2(\varepsilon_2)\rangle = \lim_{\varepsilon_2 \to 0} \lim_{\varepsilon_1 \to 0} \langle f_1(\varepsilon_1) | f_2(\varepsilon_2)\rangle = 0 \tag{1.8.46}$$

easily follows. To this end,

$$(\lambda_1 - \lambda_2) \langle f_1(\varepsilon_1) | f_2(\varepsilon_2)\rangle = \langle f_1(\varepsilon_1) | (A - \lambda_2) | f_2(\varepsilon_2)\rangle$$

$$- \langle f_1(\varepsilon_1)|(A - \lambda_1)|f_2(\varepsilon_2)\rangle. \tag{1.8.47}$$

Hence from the Cauchy-Schwarz inequality (1.7.1), (1.8.44) and (1.8.45),

$$|\langle f_1(\varepsilon_1)|f_2(\varepsilon_2)\rangle| \leqslant \frac{(\varepsilon_1 + \varepsilon_2)}{|\lambda_1 - \lambda_2|} \tag{1.8.48}$$

leading to the result in (1.8.46) for $\lambda_1 \neq \lambda_2$. [Note that (1.8.48) is an *inequality* and its left-hand side may indeed *vanish* for some ε_1, ε_2.]

The results in (1.8.46) establish the familiar statement of orthogonality not only of two eigenvectors, with eigenvalue equations defined as in (1.8.27), but also in a limiting sense of the orthogonality of two eigenfunctions with λ_1 and λ_2 in the continuous spectrum *and also* the orthogonality relation with λ_1, an eigenvalue (i.e., belonging to the discrete spectrum) and λ_2 belonging to the continuous spectrum in a unified manner. In particular, for the very last situations with λ_1 in the discrete spectrum and λ_2 in the continuous one we may write

$$A|f_1\rangle = \lambda_1 |f_1\rangle \tag{1.8.49}$$

$$\|(A - \lambda_2)f_2(\varepsilon_2)\| \leqslant \varepsilon_2 \tag{1.8.50}$$

and

$$\lim_{\varepsilon_2 \to 0} \langle f_1 | f_2(\varepsilon_2)\rangle = 0 \tag{1.8.51}$$

whose importance cannot be overemphasized.

Since the inverse $(A - \xi)^{-1}$, where now ξ, in general, is taken to be some complex number is important in studying the spectrum of A, we consider further some of its properties. $(A - \xi)^{-1}$ is called the resolvent of A. To this end, we may formally write

$$(A - \xi)^{-1} = \int_{-\infty}^{\infty} \frac{d\lambda}{(\lambda - \xi)} \delta(\lambda - A)$$

$$= \int_{-\infty}^{\infty} \frac{1}{(\lambda - \xi)} dP_A(\lambda) \tag{1.8.52}$$

where in the last equality we have used (1.8.14).

If, as we will shortly investigate, (1.8.52) exists, we have

$$\|(A - \xi)^{-1}f\|^2 = \int_{-\infty}^{\infty} \frac{1}{|\lambda - \xi|^2} d\|P_A(\lambda)f\|^2. \tag{1.8.53}$$

For $\operatorname{Im}\xi \neq 0$, this integral is obviously meaningful since $|\lambda - \xi|^2 \geqslant |\operatorname{Im}\xi|^2 > 0$,

$$\|(A - \xi)^{-1}f\|^2 \leqslant \frac{1}{|\operatorname{Im}\xi|^2} \int_{-\infty}^{\infty} d\|P_A(\lambda)f\|^2 = \frac{\|f\|^2}{|\operatorname{Im}\xi|^2} \tag{1.8.54}$$

for all $|f\rangle$.

If for ξ real and some $\varepsilon > 0$

$$[P_A(\xi + \varepsilon) - P_A(\xi - \varepsilon)] = \mathbf{0} \tag{1.8.55}$$

then

$$\left\|(A - \xi)^{-1} f\right\|^2 = \int_{-\infty}^{\xi - \varepsilon} \frac{1}{|\lambda - \xi|^2} \, \mathrm{d}\|P_A(\lambda) f\|^2$$

$$+ \int_{\xi + \varepsilon}^{\infty} \frac{1}{|\lambda - \xi|^2} \, \mathrm{d}\|P_A(\lambda) f\|^2 \leqslant \frac{\|f\|^2}{\varepsilon^2} \tag{1.8.56}$$

and $(A - \xi)^{-1}$ exists. All complex ξ (i.e., $\mathrm{Im}\,\xi \neq 0$), and all real ξ such that (1.8.55) is true, constitute of what is called the *resolvent* set of A.

Finally, for $\xi = \lambda_0$ real, with λ_0 an isolated point of the spectrum of A, i.e., in particular,

$$[P_A(\lambda_0 + 0) - P_A(\lambda_0 - 0)] |f\rangle \neq \mathbf{0} \tag{1.8.57}$$

obviously (1.8.53) does not exist as we already know. On the other hand for $\xi = \lambda_0$ real such that

$$[P_A(\lambda_0 + \varepsilon) - P_A(\lambda_0 - \varepsilon)] \neq \mathbf{0} \tag{1.8.58}$$

for *all* $\varepsilon > 0$, and is otherwise the zero operator for $\varepsilon \to +0$,

$$\left\|(A - \xi)^{-1} f\right\|^2 \geqslant \int_{\lambda_0 - \varepsilon}^{\lambda_0 + \varepsilon} \frac{1}{|\lambda - \lambda_0|^2} \, \mathrm{d}\|P_A(\lambda) f\|^2 \geqslant \frac{\|f\|^2}{\varepsilon^2} \tag{1.8.59}$$

(1.8.53) exists but $(A - \xi)^{-1}$ is an unbounded operator as the right-hand side of the above inequality may be made larger and larger by making ε smaller and smaller, which we also already knew. These last two cases corresponding to values of ξ as given through (1.8.57)–(1.8.59) constitute the spectrum of A.

Suppose that for a given self-adjoint operator A, a real number λ_0 belongs to its continuous spectrum or is an eigenvalue of infinite degeneracy, then, by definition

$$\dim\left([P_A(\lambda_0 + \varepsilon) - P_A(\lambda_0 - \varepsilon)]\mathcal{H}\right) = \infty \tag{1.8.60}$$

for *all* $\varepsilon > 0$. Because of this property, eigenvalues of infinite degeneracy and the continuous spectrum are grouped together and constitute of what is called the *essential spectrum* of a self-adjoint operator.

On the other hand, eigenvalues of at most *finite* degeneracy (i.e., which are non-degenerate or of finite degree of degeneracy) constitute of what is called the *discrete* spectrum of a self-adjoint operator. In this case, suppose that some real number λ_0 belongs to the discrete spectrum of a given self-adjoint operator A. Then we can always find *some* $\delta > 0$ such that

$$0 \neq \dim\left([P_A(\lambda_0 + \delta) - P_A(\lambda_0 - \delta)]\mathcal{H}\right) < \infty \qquad (1.8.61)$$

including the limiting case $\delta \to +0$, i.e.,

$$0 \neq \dim\left([P_A(\lambda_0 + 0) - P_A(\lambda_0 - 0)]\mathcal{H}\right) < \infty. \qquad (1.8.62)$$

Two powerful propositions establish results concerning the nature of essential and discrete spectra of self-adjoint operators.

Proposition 1.8.1
For a real number λ_0 to belong to the essential spectrum of a given self-adjoint operator A it is necessary and sufficient that there exists an infinite sequence $\{|f_n\rangle\}$ of orthonormal vectors such that

$$\|(A - \lambda_0)f_n\| \longrightarrow 0 \qquad \text{for} \quad n \to \infty. \qquad (1.8.63)$$

To establish[2] this, suppose first that λ_0 belongs to the essential spectrum of A. Then let ε_0 be any positive number such that

$$\varepsilon_0 > \frac{|\lambda_0|}{2}. \qquad (1.8.64)$$

Choose a number $\lambda_1 \neq \lambda_0$ and a corresponding ε_1 such that

$$\varepsilon_1 = |\lambda_1 - \lambda_0| < \varepsilon_0. \qquad (1.8.65)$$

Similarly, choose a number $\lambda_2 \neq \lambda_0$ and a corresponding ε_2 such that

$$|\lambda_2 - \lambda_0| = \varepsilon_2 < \varepsilon_1 \qquad (1.8.66)$$

and so on (see Figure 1.14).

We have thus generated a sequence $\{\varepsilon_n\}$ such that

$$\varepsilon_0 > \varepsilon_1 > \varepsilon_2 > \ldots > \varepsilon_n > \ldots \qquad (1.8.67)$$

and (by definition of λ_0 as belonging to the essential spectrum) correspond to infinite dimensional spaces:

$$[P_A(\lambda_0 + \varepsilon_0) - P_A(\lambda_0 - \varepsilon_0)]\mathcal{H} \supset [P_A(\lambda_0 + \varepsilon_1) - P_A(\lambda_0 - \varepsilon_1)]\mathcal{H}$$

$$\supset \ldots$$

$$\supset [P_A(\lambda_0 + \varepsilon_n) - P_A(\lambda_0 - \varepsilon_n)]\mathcal{H}$$

---

[2] The proofs of Propositions 1.8.1 and 1.8.2 may be omitted at a first reading.

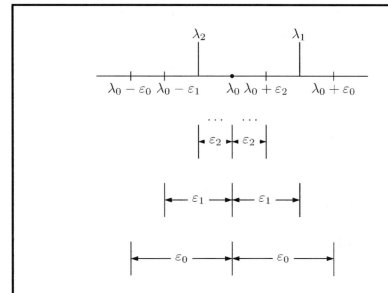

Fig. 1.14. The figure shows the process of generating the sequence $\{\varepsilon_n\}$ for forming intervals about λ_0 in the essential spectrum of a self-adjoint operator.

$$\supset \ldots . \tag{1.8.68}$$

We may, therefore, select an infinite sequence $\{|f_n\rangle\}$ of orthonormal vectors such that

$$\left[P_A(\lambda_0 + \varepsilon_n) - P_A(\lambda_0 - \varepsilon_n)\right]|f_n\rangle = |f_n\rangle. \tag{1.8.69}$$

Hence

$$\|(A - \lambda_0)f_n\|^2 = \int_{\lambda_0 - \varepsilon_n}^{\lambda_0 + \varepsilon_n} (\lambda - \lambda_0)^2 \, d\|P_A(\lambda)f_n\|^2 \leqslant \varepsilon_n^2. \tag{1.8.70}$$

Since in the process of construction we may arrange such that $\varepsilon_n^2 \to 0$ for $n \to \infty$, it follows that

$$\|(A - \lambda_0)f_n\| \longrightarrow 0 \quad \text{for} \quad n \to \infty. \tag{1.8.71}$$

Conversely, suppose that there exists an infinite sequence $\{|f_n\rangle\}$ of orthonormal vectors such that

$$\|(A - \lambda_0)f_n\| \longrightarrow 0 \quad \text{for} \quad n \to \infty. \tag{1.8.72}$$

We then have to show that λ_0 belongs to the essential spectrum of A.

To the above end, for any $\varepsilon > 0$

$$\|(A-\lambda_0)f_n\|^2 \geqslant \int_{-\infty}^{\lambda_0-\varepsilon} (\lambda-\lambda_0)^2 \, d\|P_A(\lambda)f_n\|^2$$
$$+ \int_{\lambda_0+\varepsilon}^{\infty} (\lambda-\lambda_0)^2 \, d\|P_A(\lambda)f_n\|^2. \quad (1.8.73)$$

On the other hand

$$\int_{-\infty}^{\lambda_0-\varepsilon} (\lambda-\lambda_0)^2 \, d\|P_A(\lambda)f_n\|^2 \geqslant \varepsilon^2 \|P_A(\lambda_0-\varepsilon)f_n\|^2 \quad (1.8.74)$$

and

$$\int_{\lambda_0+\varepsilon}^{\infty} (\lambda-\lambda_0)^2 \, d\|P_A(\lambda)f_n\|^2 \geqslant \varepsilon^2 \|[\mathbf{1}-P_A(\lambda_0+\varepsilon)]f_n\|^2 \quad (1.8.75)$$

where we have used (1.8.17), (1.8.18), (1.8.20).

Hence

$$\|(A-\lambda_0)f_n\|^2 \geqslant \varepsilon^2 \|P_A(\lambda_0-\varepsilon)f_n\|^2 + \varepsilon^2 \|[\mathbf{1}-P_A(\lambda_0+\varepsilon)]f_n\|^2 \quad (1.8.76)$$

and from this and (1.8.72) we conclude that

$$\lim_{n\to\infty} \|P_A(\lambda_0-\varepsilon)f_n\|^2 = 0 \quad (1.8.77)$$

$$\lim_{n\to\infty} \|[\mathbf{1}-P_A(\lambda_0+\varepsilon)]f_n\|^2 = 0 \quad (1.8.78)$$

for *any* $\varepsilon > 0$. These results may be equivalently rewritten as

$$\lim_{n\to\infty} \langle f_n|P_A(\lambda_0-\varepsilon)|f_n\rangle = 0 \quad (1.8.79)$$

$$\lim_{n\to\infty} \langle f_n|P_A(\lambda_0+\varepsilon)|f_n\rangle = 1 \quad (1.8.80)$$

or, by combining the two, we may write

$$\lim_{n\to\infty} \langle f_n|\left[P_A(\lambda_0+\varepsilon)-P_A(\lambda_0-\varepsilon)\right]|f_n\rangle = 1 \quad (1.8.81)$$

for *any* $\varepsilon > 0$.
Therefore

$$[P_A(\lambda_0+\varepsilon)-P_A(\lambda_0-\varepsilon)]\mathcal{H} \quad (1.8.82)$$

is not empty and it remains to show that the latter is infinite dimensional. Suppose that this is not true. That is, it is a finite dimensional space. Select an orthonormal set $\{g_1,\ldots,g_k\}$, $k < \infty$ in it. Then

$$\langle f_n|\left[P_A(\lambda_0+\varepsilon)-P_A(\lambda_0-\varepsilon)\right]|f_n\rangle = \sum_{m=1}^{k} |\langle g_m|f_n\rangle|^2 \leqslant 1 \quad (1.8.83)$$

1.8 Self-Adjoint Operators and Their Spectra

for any $\varepsilon > 0$ and all n.

Since $\{|f_n\rangle\}$ constitutes an orthonormal set, Bessel's inequality (1.7.4) reads

$$\sum_{\ell=1}^{n} |\langle g_m | f_\ell \rangle|^2 \leqslant 1 \tag{1.8.84}$$

and the convergence of this series for $n \to \infty$ in particular for all $m = 1, \ldots, k$, implies that

$$\lim_{n \to \infty} |\langle g_m | f_n \rangle|^2 = 0 \tag{1.8.85}$$

for $m = 1, \ldots, k$. Due to the finite number of terms in the sum in (1.8.83), we may take the limit $n \to \infty$ inside the summation sign in (1.8.83) to infer that

$$\langle f_n | [P_A(\lambda_0 + \varepsilon) - P_A(\lambda_0 - \varepsilon)] | f_n \rangle \longrightarrow 0 \tag{1.8.86}$$

for $n \to \infty$ in contradiction with (1.8.81). That is, for any $\varepsilon > 0$, the space (1.8.82) must be infinite dimensional. This establishes the statement made in the proposition.

Before stating and establishing the next important proposition regarding now discrete spectra of self-adjoint operators, we need some preliminary results.

To the above end, we first recall the definitions of the infimum and supremum of a set of real numbers. A real number $\underline{\lambda}$ is a lower bound of a set of real numbers if $\underline{\lambda}$ is less than or equal to every element of the set. If in addition, no lower bound of the set is greater than $\underline{\lambda}$, then $\underline{\lambda}$ is referred to as the infimum of the set. Similarly, $\overline{\lambda}$ is an upper bound of a set of real numbers if $\overline{\lambda}$ is greater than or equal to every element of the set. If in addition, no upper bound of the set is less than $\overline{\lambda}$ then $\overline{\lambda}$ is referred to as the supremum of the set.

We define[3] the following real numbers associated with a self-adjoint operator A for which a lower bound for its spectrum (§3.2, §3.3) exists:

$$\lambda_n(A) = \sup_{|f_1\rangle,\ldots,|f_{n-1}\rangle} \left\{ \inf_{|\psi\rangle \in [|f_1\rangle,\ldots,|f_{n-1}\rangle]^\perp} \langle \psi | A | \psi \rangle \right\} \tag{1.8.87}$$

where $|f_1\rangle, \ldots, |f_{n-1}\rangle, |\psi\rangle$ are normalized vectors, $|f_1\rangle, \ldots, |f_{n-1}\rangle$ are not necessarily independent, and $|\psi\rangle \in [|f_1\rangle, \ldots, |f_{n-1}\rangle]^\perp$ means that $\langle \psi | f_i \rangle = 0$ for $i = 0, \ldots, n-1$, i.e., the $|\psi\rangle$ are orthogonal to $|f_1\rangle, \ldots, |f_{n-1}\rangle$.

That is, we define by $[|f_1\rangle, \ldots, |f_{n-1}\rangle]^\perp$ the space generated by the vectors (in the domain of A) orthogonal to $|f_1\rangle, \ldots, |f_{n-1}\rangle$.

Since,

$$[|f_1\rangle, \ldots, |f_{n-1}\rangle]^\perp \supset [|f_1\rangle, \ldots, |f_{n-1}\rangle, |f_n\rangle]^\perp \tag{1.8.88}$$

with the set on the left-hand side involving, in general, of more vectors than the set on the right-hand side, we may infer that

---

[3] This treatment follows that of Reed and Simon (1978).

$$\inf_{|\psi\rangle \in [|f_1\rangle,\ldots,|f_{n-1}\rangle]^\perp} \langle\psi|A|\psi\rangle \leqslant \inf_{|\phi\rangle \in [|f_1\rangle,\ldots,|f_{n-1}\rangle,|f_n\rangle]^\perp} \langle\phi|A|\phi\rangle \qquad (1.8.89)$$

and then considering the operation of taking the supremum over all $f_1, \ldots, f_{n-1}, f_n$ we obtain

$$\lambda_n(A) \leqslant \lambda_{n+1}(A). \qquad (1.8.90)$$

We also need the following results. Let a and b be any two real numbers such that
$$a < \lambda_n(A) < b. \qquad (1.8.91)$$

Then

(i)
$$\dim\left([P_A(a)]\mathcal{H}\right) < n \qquad (1.8.92)$$

(ii)
$$n \leqslant \dim\left([P_A(b)]\mathcal{H}\right). \qquad (1.8.93)$$

To establish (i), suppose that $\dim\left([P_A(a)]\mathcal{H}\right) = k \geqslant n$. That is, there exist $k \geqslant n$ independent vectors in $[P_A(a)]\mathcal{H}$. Given any $k - 1$ vectors $|f_1\rangle, \ldots, |f_{k-1}\rangle$, not necessarily independent then

$$[|f_1\rangle, \ldots, |f_{k-1}\rangle]^\perp \cap [P_A(a)]\mathcal{H} \qquad (1.8.94)$$

is not empty. Let ψ be a (normalized) vector (in the domain of A) belonging to the latter space. Then

$$\langle\psi|A|\psi\rangle = \int_{-\infty}^{a} \lambda \, d\|P_A(\lambda)\psi\|^2 \leqslant a, \qquad (1.8.95)$$

i.e., $\lambda_k(A) \leqslant a$. Since by hypothesis $k \geqslant n$, (1.8.90) implies that $\lambda_n(A) \leqslant a$ in contradiction with the fact that $a < \lambda_n(A)$. This establishes (i) in (1.8.92).

To establish (ii), suppose that $\dim\left([P_A(b)]\mathcal{H}\right) = k \leqslant n - 1$, and that $|g_1\rangle, \ldots, |g_k\rangle$ are k independent vectors that generate this space. Let $|\psi\rangle$ be any (normalized) vector (in the domain of A) such that $|\psi\rangle \in [|g_1\rangle, \ldots, |g_k\rangle]^\perp \equiv [\mathbf{1} - P_A(b)]\mathcal{H}$. Then

$$\langle\psi|A|\psi\rangle = \int_{b}^{\infty} \lambda \, d\|P_A(\lambda)\psi\|^2 \geqslant b, \qquad (1.8.96)$$

i.e., $\lambda_{k+1}(A) \geqslant b$, and hence $\lambda_n(A) \geqslant b$, since by hypothesis $n \geqslant k + 1$. This contradicts the fact that $\lambda_n(A) < b$ and establishes (ii) in (1.8.93).

Now we are ready to state and establish the following proposition.

Proposition 1.8.2

Either there are n eigenvalues (counting degeneracy) $\lambda_1(A) \leq \ldots \leq \lambda_n(A)$ *below the bottom of the essential spectrum of A, where* $\lambda_n(A)$ *is the* n^{th} *eigenvalue, or*

$$\lambda_n(A) = \inf \{\lambda, \text{ with } \lambda \text{ belonging to the essential spectrum}\}$$

with $\lambda_n(A) = \lambda_{n+1}(A) = \ldots$, *and there may be at most* $(n-1)$ *eigenvalues (counting degeneracy) below* $\lambda_n(A)$.

To establish the above statements, we first note that for all $\varepsilon > 0$, we have shown earlier in (1.8.92), (1.8.93) that

$$\dim\left([P_A(\lambda_n - \varepsilon)]\mathcal{H}\right) < n \tag{1.8.97}$$

$$n \leq \dim\left([P_A(\lambda_n + \varepsilon)]\mathcal{H}\right) \tag{1.8.98}$$

respectively, where we have simply written λ_n for $\lambda_n(A)$.

We consider the two possibilities:

(i)
$$\dim\left([P_A(\lambda_n + \varepsilon_0)]\mathcal{H}\right) < \infty \tag{1.8.99}$$

for *some* $\varepsilon_0 > 0$, or the possibility that

(ii)
$$\dim\left([P_A(\lambda_n + \varepsilon_0)]\mathcal{H}\right) = \infty \tag{1.8.100}$$

for *all* $\varepsilon_0 > 0$.

For the case (i), (1.8.97)–(1.8.99), imply that

$$1 \leq \dim\left([P_A(\lambda_n + \varepsilon_0) - P_A(\lambda_n - \varepsilon_0)]\mathcal{H}\right) < \infty \tag{1.8.101}$$

for some $\varepsilon_0 > 0$. That is, λ_n is an eigenvalue (of at most finite degeneracy). Hence (1.8.98), (1.8.101) imply that we may find a $\delta_0 > 0$ such that

$$n \leq \dim\left([P_A(\lambda_n)]\mathcal{H}\right) = \dim\left([P_A(\lambda_n + \delta_0)]\mathcal{H}\right). \tag{1.8.102}$$

This means that there are exactly $(n-1)$ eigenvalues strictly below λ_n. This is because if there are, say, $n+k$ eigenvalues less or equal to λ_n, i.e., $\lambda'_1 \leq \ldots \leq \lambda'_{n+k} \leq \lambda_n$, then (1.8.97) implies that $\dim\left([P_A(\lambda'_{n+k})]\mathcal{H}\right) \leq n+k-1$. But $\lambda'_{n+k} \leq \lambda_n$ means that $n+k-1 \leq n-1$, which is true only if $k=0$, and $\lambda'_n \equiv \lambda_n$. We also note from (1.8.97) that $\dim\left([P_A(\lambda_1 - \varepsilon)]\mathcal{H}\right) = 0$ for all $\varepsilon > 0$ and hence the spectrum set is empty below λ_1. This settles the first part of the proposition.

For the case (ii), given in (1.8.100), (1.8.97) then necessarily implies that

$$\dim\left(\left[P_A(\lambda_n+\varepsilon)-P_A(\lambda_n-\varepsilon)\right]\mathcal{H}\right)=\infty \tag{1.8.103}$$

for *all* $\varepsilon > 0$. Hence λ_n, in this case, belongs to the essential spectrum of A. Let a be any real number such that $a < \lambda_n - \varepsilon$, i.e., $a + \varepsilon < \lambda_n$. Hence from (1.8.92),

$$\dim\left(\left[P_A(a+\varepsilon)\right]\mathcal{H}\right) \leqslant n-1. \tag{1.8.104}$$

Also

$$\dim\left(\left[P_A(a-\varepsilon)\right]\mathcal{H}\right) \leqslant \dim\left(\left[P_A(a+\varepsilon)\right]\mathcal{H}\right), \tag{1.8.105}$$

i.e.,

$$\dim\left(\left[P_A(a+\varepsilon)-P_A(a-\varepsilon)\right]\mathcal{H}\right) < \infty \tag{1.8.106}$$

for *all* $\varepsilon > 0$. That is, a cannot belong to the essential spectrum and this is for all $\varepsilon > 0$. This in turn means that λ_n is the bottom of the essential spectrum as stated in the second part of the proposition. Now suppose $\lambda_{n+1} > \lambda_n$, and note that

$$\lambda_{n+1} - \frac{(\lambda_{n+1}-\lambda_n)}{2} = \lambda_n + \frac{(\lambda_{n+1}-\lambda_n)}{2} = \frac{(\lambda_{n+1}+\lambda_n)}{2} \tag{1.8.107}$$

which from (1.8.92) and (1.8.93) imply the contradictory statements that

$$\dim\left(\left[P_A\!\left(\frac{\lambda_{n+1}+\lambda_n}{2}\right)\right]\mathcal{H}\right) \tag{1.8.108}$$

is $\leqslant n+1$ and $= \infty$, respectively. That is, we must have $\lambda_{n+1} = \lambda_n$.

Also for $a < \lambda_n - \varepsilon$, we have from (1.8.104), (1.8.105)

$$\dim\left(\left[P_A(a-\varepsilon)\right]\mathcal{H}\right) \leqslant n-1 \tag{1.8.109}$$

for all $\varepsilon > 0$, hence there may be at most $(n-1)$ eigenvalues below λ_n. This establishes the second part of the proposition.

For future developments, we establish an order relationship between the eigenvalues of two self-adjacent operators A and B, whose spectra are bounded from below, such that for all vectors $|\psi\rangle$ in their domains

$$\langle\psi|A|\psi\rangle \geqslant \langle\psi|B|\psi\rangle. \tag{1.8.110}$$

From the very definitions of the infimum and supremum of a set of real numbers it is not difficult to see, as shown below, that

$$\lambda_n(A) \geqslant \lambda_n(B). \tag{1.8.111}$$

To this end note that if, relative to a given space $[|f_1\rangle,\ldots,|f_{n-1}\rangle]^\perp$ (see (1.8.87)), $\underline{c}'_B$ provides the infimum corresponding to the operator B, then from (1.8.110), $\underline{c}'_B$ also gives a lower bound to $\langle\psi|A|\psi\rangle$ for all $|\psi\rangle \in$

$[|f_1\rangle, \ldots, |f_{n-1}\rangle]^\perp$. Since, by definition, no such a lower bound can be greater than the corresponding $\underline{c}'_A$, we conclude that $\underline{c}'_B \leqslant \underline{c}'_A$. The result then follows by considering the supremum of all such $\underline{c}'_B$ and $\underline{c}'_A$ as we consider all vectors $|f_1\rangle, \ldots, |f_{n-1}\rangle$.

A special function of a self-adjoint operator A is the *unitary* operator defined by

$$U(t) = e^{itA} \tag{1.8.112}$$

depending on a real parameter t, which may be rewritten as

$$U(t) = \int_{-\infty}^{\infty} d\lambda \, e^{it\lambda} \, \delta(\lambda - A) \tag{1.8.113}$$

and from (1.8.14) as

$$U(t) = \int_{-\infty}^{\infty} e^{it\lambda} \, dP_A(\lambda) \tag{1.8.114}$$

providing its spectral decomposition. It satisfies the basic property

$$U^{-1}(t) = U^\dagger(t), \tag{1.8.115}$$

i.e.,

$$U^\dagger(t) U(t) = \mathbf{1} = U(t) U^\dagger(t) \tag{1.8.116}$$

and also the group property

$$U(t_1) U(t_2) = U(t_1 + t_2). \tag{1.8.117}$$

Particular attention will be given later to the study of some basic properties of an important self-adjoint operator — the Hamiltonian for various physical systems.

1.9 Wigner's Theorem on Symmetry Transformations

Invariance of physical laws under some given transformations lead to conservation laws and the underlying transformations are referred to as *symmetry* transformations. For example, invariance of a physical law under the rotation of one's coordinate system in describing the underlying theory leads to the conservation of angular momentum, and invariance under time translation (by setting, for example, one's clocks back by a certain amount) and under space translation (by shifting the origin of one's coordinate system) lead, respectively, to energy and momentum conservations. Other transformations, for example, involve space reflection (also known as parity transformation), time reversal, and charge conjugation, where in the latter every particle in the physical process under consideration is replaced by its anti-particle. [It is remarkable that by combining relativity with quantum mechanics, leading to what is called quantum field theory, one concurs that the *simultaneous* transformation of charge conjugation, parity transformation and time reversal is a symmetry transformation based on such a merger.]

Invoking invariance properties in developing a dynamical theory, conveniently, narrows down one's choices in providing the final stages of the theory. Not all transformations are, obviously, symmetry transformations of a given physical system. But invoking the invariance of a system under such transformations may provide the starting point in describing the underlying dynamical theory, and then one may consistently modify the interaction in the theory to take into account any symmetry breaking.

A celebrated theorem due to Wigner, in the thirties originating on symmetry due to rotation in space, spells out the nature of the transformations implemented on elements of a Hilbert space under symmetry transformations.

To see how these implemented transformations on elements of a Hilbert space occur, we reconsider the general physical question arising in quantum physics (§1.2–§1.6). One prepares a system in a state $|\psi\rangle$. The question then arises as to what is the probability of finding the system is a state $|\phi\rangle$ if $|\psi\rangle$ is what we initially have? The latter is given by

$$\mathrm{Tr}\left[\,|\phi\rangle\langle\phi|\,|\psi\rangle\langle\psi|\,\right] = |\langle\phi|\psi\rangle|^2. \tag{1.9.1}$$

More generally, one may have a mixture described by a density operator (§1.5)

$$\rho = \sum_i w_i\,|\psi_i\rangle\langle\psi_i|. \tag{1.9.2}$$

The probability of finding a system in state $|\phi\rangle$ if ρ, in (1.9.2), is what we initially have is

$$\mathrm{Tr}\left[\,|\phi\rangle\langle\phi|\,\rho\right] = \sum_i w_i\,|\langle\phi|\psi_i\rangle|^2. \tag{1.9.3}$$

That is, in both cases one is confronted with the problem of computing $|\langle\phi|\psi\rangle|^2$ for given vectors $|\psi\rangle$, $|\phi\rangle$.

If $|\psi'\rangle$, $|\phi'\rangle$ denote the vectors $|\psi\rangle$, $|\phi\rangle$ resulting under a symmetry transformation, then the invariance of the corresponding probabilities may be stated by the equality

$$|\langle\psi'|\phi'\rangle|^2 = |\langle\psi|\phi\rangle|^2. \tag{1.9.4}$$

That is, under a symmetry transformation, $\{|\psi'\rangle,|\phi'\rangle\}$ give an equivalent physical description as $\{|\psi\rangle,|\phi\rangle\}$.

Each of the vectors $|\psi'\rangle$, $|\phi'\rangle$, $|\psi\rangle$, $|\phi\rangle$ may be scaled by *arbitrary* phase factors without changing the physically relevant probabilities given in (1.9.4). Accordingly, one needs to consider only unit *rays* (see §1.7) generated by such vectors. Although such overall phase factors are not important, the *relative* phases occurring when adding two or more vectors are physically relevant with far reaching consequences. Such details will be dealt with later such as

1.9 Wigner's Theorem on Symmetry Transformations

in §1.10, §8.7, §8.8 and Chapter 8, in general. For the time being we note that in computing an expression like

$$|\langle \alpha_1\psi_1 + \alpha_2\psi_2 | \beta_1\phi_1 + \beta_2\phi_2 \rangle|^2 \tag{1.9.5}$$

overall phase factors of the vectors $(\alpha_1 |\psi_1\rangle + \alpha_2 |\psi_2\rangle)$, $(\beta_1 |\phi_1\rangle + \beta_2 |\phi_2\rangle)$, for complex numbers α_1, α_2, β_1, β_2 are unimportant, and one is dealing with rays generated by the two vectors. There are, however, in general important relative phase factors which arise, when expanding the expression in (1.9.5), in terms of scalar products of $|\psi_1\rangle$, $|\psi_2\rangle$ with $|\phi_1\rangle$, $|\phi_2\rangle$.

To see how rays $|\psi\rangle$, $|\psi'\rangle$, as occurring in (1.9.4), are related and hence, in the process, spell out Wigner's Theorem of Symmetry Transformations, we provide and re-iterate the following definitions.

Definition 1. *An operator L is called linear, or else anti-linear, if given any vectors $|\psi\rangle$, $|\phi\rangle$, $(\alpha |\psi\rangle + \beta |\phi\rangle)$,*

$$L(\alpha |\psi\rangle + \beta |\phi\rangle) = \alpha L |\psi\rangle + \beta L |\phi\rangle, \tag{1.9.6}$$

or else

$$L(\alpha |\psi\rangle + \beta |\phi\rangle) = \alpha^* L |\psi\rangle + \beta^* L |\phi\rangle, \tag{1.9.7}$$

for the corresponding rays generated, respectively, by the vectors $L|\psi\rangle$, $L|\phi\rangle$, $L(\alpha |\psi\rangle + \beta |\phi\rangle)$. [Note that these equalities do not necessarily hold for the vectors themselves but only for the corresponding rays just described.]

Definition 2. *A linear or else anti-linear operator U, as given as given in Definition 1, is called unitary or else anti-unitary if*

$$\langle U\psi | U\phi \rangle = \langle \psi | \phi \rangle, \tag{1.9.8}$$

or else

$$\langle U\psi | U\phi \rangle = \langle \psi | \phi \rangle^*, \tag{1.9.9}$$

for the corresponding rays generated by vectors $|\psi\rangle$, $|\phi\rangle$.

Wigner's Theorem:

Under a symmetry transformation, these exists a unitary, or else, an anti-unitary, operator U such that (1.9.6), (1.9.8) or else (1.9.7), (1.9.9), hold with

$$|\psi'\rangle = U |\psi\rangle \tag{1.9.10}$$

$$|\phi'\rangle = U |\phi\rangle \tag{1.9.11}$$

$$(\alpha |\psi\rangle + \beta |\phi\rangle)' = U(\alpha |\psi\rangle + \beta |\phi\rangle) \tag{1.9.12}$$

58   1 Fundamentals

as applied to rays. The latter means that $|\psi'\rangle$, $|\phi'\rangle$, $(\alpha|\psi\rangle+\beta|\phi\rangle)'$ are defined up to overall phase factors.

The proof of the theorem is not difficult but long if one spells out all the details.

To establish the validity of the theorem, we proceed through various progressive steps.

1. Let $\{|f_1\rangle, |f_2\rangle, \ldots\}$ be an orthonormal basis in $\mathcal{H}$. Any vector $|\psi\rangle$ in $\mathcal{H}$ may written as
$$|\psi\rangle = \sum_k \alpha_k |f_k\rangle. \qquad (1.9.13)$$

By definition, under a symmetry transformation,
$$|\langle f'_k | f'_\ell \rangle| = |\langle f_k | f_\ell \rangle| = \delta_{k\ell},$$
i.e.,
$$\langle f'_k | f'_\ell \rangle = \delta_{k\ell}. \qquad (1.9.14)$$

Also
$$|\langle f'_k | \psi' \rangle| = |\langle f_k | \psi \rangle| = |\alpha_k|.$$

Hence let
$$a_k = \langle f'_k | \psi' \rangle \qquad (1.9.15)$$

then we have
$$|a_k| = |\alpha_k|. \qquad (1.9.16)$$

2. Consider the vector
$$|\phi\rangle = |\psi'\rangle - \sum_k a_k |f'_k\rangle$$

with a_k defined in (1.9.15). Then
$$\|\phi\|^2 = \|\psi'\|^2 - \sum_k |a_k|^2$$
$$= \|\psi\|^2 - \sum_k |\alpha_k|^2 = 0.$$

That is, $|\phi\rangle$ is the zero vector, and we may write
$$|\psi'\rangle = \sum_k a_k |f'_k\rangle \qquad (1.9.17)$$

with the a_k satisfying (1.9.16). Thus $|f'_1\rangle, |f'_2\rangle, \ldots$ provide an orthonormal basis for expanding the transformed state $|\psi'\rangle$.

1.9 Wigner's Theorem on Symmetry Transformations

3. Consider the three vectors ($j \neq k$, $j \neq 1$, $k \neq 1$)

$$|\psi_{1j}\rangle = |f_1\rangle + |f_j\rangle \tag{1.9.18}$$

$$|\psi_{1k}\rangle = |f_1\rangle + |f_k\rangle \tag{1.9.19}$$

$$|\psi_{1jk}\rangle = |f_1\rangle + |f_j\rangle + |f_k\rangle. \tag{1.9.20}$$

From (1.9.18), (1.9.16),

$$|\psi'_{1j}\rangle = e^{i\delta_{1j}}|f'_1\rangle + e^{i\delta_{j1}}|f'_j\rangle \tag{1.9.21}$$

which may be rewritten as

$$|\psi'_{1j}\rangle = e^{i\delta_{1j}}\big[|f'_1\rangle + e^{i\eta_{j1}}|f'_j\rangle\big] \tag{1.9.22}$$

where the subscripts $j1$ in η_{j1} indicate that the corresponding phase factor may, in general, depend on $|f_j\rangle$ *as well* as on $|f_1\rangle$.
Similarly,

$$|\psi'_{1k}\rangle = e^{i\delta_{1k}}\big[|f'_1\rangle + e^{i\eta_{k1}}|f'_k\rangle\big] \tag{1.9.23}$$

$$|\psi'_{1jk}\rangle = e^{i\delta_{1jk}}\big[|f'_1\rangle + e^{i\eta_{jk1}}|f'_j\rangle + e^{i\eta_{k1j}}|f'_k\rangle\big]. \tag{1.9.24}$$

By definition

$$|\langle\psi'_{1j}|\psi'_{1jk}\rangle| = |\langle\psi_{1j}|\psi_{1jk}\rangle| \tag{1.9.25}$$

from which we obtain

$$\left|1 + e^{i(\eta_{jk1} - \eta_{j1})}\right| = 2. \tag{1.9.26}$$

That is,

$$e^{i\eta_{jk1}} = e^{i\eta_{j1}} \tag{1.9.27}$$

and the phase factor multiplying $|f'_j\rangle$ within the square brackets in (1.9.24) is *independent* of the vector $|f_k\rangle$.
Similarly,

$$|\langle\psi'_{1k}|\psi'_{1jk}\rangle| = |\langle\psi_{1k}|\psi_{1jk}\rangle| \tag{1.9.28}$$

implies that

$$e^{i\eta_{k1j}} = e^{i\eta_{k1}} \tag{1.9.29}$$

and the phase factor multiplying $|f'_k\rangle$ within the square brackets in (1.9.24) is *independent* of the vector $|f_j\rangle$.
Accordingly, under a symmetry transformation, $|\psi_{1j}\rangle$, $|\psi_{1k}\rangle$, $|\psi_{1jk}\rangle$ in (1.9.18), (1.9.19), (1.9.20), respectively, transform to $|\psi'_{1j}\rangle$, $|\psi'_{1k}\rangle$ as given in (1.9.22), (1.9.23), *and*

$$|\psi'_{1jk}\rangle = e^{i\delta_{1jk}}\big[|f'_1\rangle + e^{i\eta_{j1}}|f'_j\rangle + e^{i\eta_{k1}}|f'_k\rangle\big] \tag{1.9.30}$$

and the phase factors multiplying $|f'_j\rangle$, and $|f'_k\rangle$ within these square brackets are the same as the corresponding ones within the square brackets in (1.9.22) and (1.9.23).

Since the phase factors $e^{i\eta_{j1}}$, $e^{i\eta_{k1}}$ are understood to be defined relative to, and may depend on, $|f'_1\rangle$, we will suppress in the sequel the subscript 1 in η_{j1}, η_{k1}.

4. Suppose first that $\alpha_1 \neq 0$. Then by definition

$$|\langle \psi'_{1k}|\psi'\rangle| = |\langle \psi_{1k}|\psi\rangle| \tag{1.9.31}$$

which implies that

$$\left|1 + e^{-i\eta_k}\frac{a_k}{a_1}\right| = \left|1 + \frac{\alpha_k}{\alpha_1}\right|. \tag{1.9.32}$$

That is,

$$\operatorname{Re}\left(e^{-i\eta_k}\frac{a_k}{a_1}\right) = \operatorname{Re}\left(\frac{\alpha_k}{\alpha_1}\right). \tag{1.9.33}$$

Finally from the constraints $|a_k| = |\alpha_k|$ in (1.9.16), this implies that

$$\left[\operatorname{Im}\left(e^{-i\eta_k}\frac{a_k}{a_1}\right)\right]^2 = \left[\operatorname{Im}\left(\frac{\alpha_k}{\alpha_1}\right)\right]^2. \tag{1.9.34}$$

5. By definition,

$$|\langle \psi'_{1jk}|\psi'\rangle| = |\langle \psi_{1jk}|\psi\rangle| \tag{1.9.35}$$

which implies that

$$\left|1 + e^{-i\eta_j}\frac{a_j}{a_1} + e^{-i\eta_k}\frac{a_k}{a_1}\right| = \left|1 + \frac{\alpha_j}{\alpha_1} + \frac{\alpha_k}{\alpha_1}\right|. \tag{1.9.36}$$

This equality together (1.9.33) and the constraints in (1.9.16) give

$$\left[\operatorname{Im}\left(e^{-i\eta_j}\frac{a_j}{a_1}\right)\right]\left[\operatorname{Im}\left(e^{-i\eta_k}\frac{a_k}{a_1}\right)\right] = \left[\operatorname{Im}\left(\frac{\alpha_j}{\alpha_1}\right)\right]\left[\operatorname{Im}\left(\frac{\alpha_k}{\alpha_1}\right)\right]. \tag{1.9.37}$$

We use this equality in conjunction with (1.9.34). If from the latter

$$\operatorname{Im}\left(e^{-i\eta_k}\frac{a_k}{a_1}\right) = +\operatorname{Im}\left(\frac{\alpha_k}{\alpha_1}\right) \tag{1.9.38}$$

then (1.9.37) implies that *simultaneously*

$$\operatorname{Im}\left(e^{-i\eta_j}\frac{a_j}{a_1}\right) = +\operatorname{Im}\left(\frac{\alpha_j}{\alpha_1}\right) \tag{1.9.39}$$

and vice versa, occurring with the same signs for all $j \neq k$ not equal to 1.

1.9 Wigner's Theorem on Symmetry Transformations

Similarly, if

$$\text{Im}\left(e^{-i\eta_k}\frac{a_k}{a_1}\right) = -\text{Im}\left(\frac{\alpha_k}{\alpha_1}\right) \tag{1.9.40}$$

then

$$\text{Im}\left(e^{-i\eta_j}\frac{a_j}{a_1}\right) = -\text{Im}\left(\frac{\alpha_j}{\alpha_1}\right) \tag{1.9.41}$$

simultaneously, and vice versa, occurring with the *same* signs for all $j \neq k$ not equal to 1.

From (1.9.33) and (1.9.38), (1.9.39), or (1.9.40), (1.9.41), we conclude that we have two alternatives

$$a_k = a_1 e^{i\eta_k}\left(\frac{\alpha_k}{\alpha_1}\right) \tag{1.9.42}$$

for all $k \neq 1$ uniformly for the expansion coefficients of $|\psi'\rangle$ in (1.9.17) *or else*

$$a_k = a_1 e^{i\eta_k}\left(\frac{\alpha_k}{\alpha_1}\right)^* \tag{1.9.43}$$

again for all $k \neq 1$ uniformly for the expansion coefficients of $|\psi'\rangle$.

That is, under a symmetry transformation, $|\psi\rangle$ transforms to

$$|\psi'\rangle = \frac{a_1}{\alpha_1^{\#}} \sum_k \alpha_k^{\#} e^{i\eta_k} |f'_k\rangle \tag{1.9.44}$$

with $e^{i\eta_1} = 1$ and $\alpha^{\#}$ denotes *either* α or else α^*. The same $\#$ rule for the transformation applies to *every* coefficient in $|\psi\rangle$ multiplying the vectors $|f_k\rangle$.

The case $\alpha_1 = 0$ is easily treated. To this end, we introduce the vector $|\chi\rangle$

$$|\chi\rangle = |f_1\rangle + \sum_{k\geqslant 2} \alpha_k |f_k\rangle. \tag{1.9.45}$$

Since

$$|\langle f'_k | \psi'\rangle| = |\langle f_k | \psi\rangle| = |\langle f_k | \chi\rangle| \tag{1.9.46}$$

we obtain

$$a_k = e^{i\eta_k}\alpha_k^{\#} a_1, \quad k \geqslant 2 \tag{1.9.47}$$

where a_1 is some phase factor.

That is, in all cases, we have either

$$|\psi'\rangle = \sum_k \alpha_k e^{i\eta_k} |f'_k\rangle \tag{1.9.48}$$

or else

$$|\psi'\rangle = \sum_k \alpha_k^* e^{i\eta_k} |f'_k\rangle \tag{1.9.49}$$

up to overall phase factors, thus defining the corresponding ray transformations. The phases $\mathrm{e}^{\mathrm{i}\eta_k}$ are independent of the expansion coefficients, and for each k, $\mathrm{e}^{\mathrm{i}\eta_k}$ is defined in terms of the pair of vectors $|f_k\rangle, |f_1\rangle$ only. Here it is also worth recalling that the vectors $|f_1\rangle, |f_2\rangle, \ldots$ are pairwise orthogonal and hence the relative phases $\exp \mathrm{i}(\eta_j - \eta_k)$, for $j \neq k$, never occur.

6. It remains to establish, that under a symmetry transformation, if one vector transforms under the # rule of complex conjugation (respectively unaltered) rule, then every other vector transforms under the same # rule.[4] To establish this, we assume otherwise and, in turn, run into a contradiction.

Accordingly, suppose that for two given vectors $|\psi\rangle$, $|\phi\rangle$,

$$|\psi\rangle = \sum_k \alpha_k |f_k\rangle \tag{1.9.50}$$

$$|\phi\rangle = \sum_k \beta_k |f_k\rangle \tag{1.9.51}$$

we have

$$|\psi'\rangle = \sum_k \alpha_k \mathrm{e}^{\mathrm{i}\eta_k} |f'_k\rangle \tag{1.9.52}$$

$$|\phi'\rangle = \sum_k \beta_k^* \mathrm{e}^{\mathrm{i}\eta_k} |f'_k\rangle . \tag{1.9.53}$$

Then

$$|\langle \psi' | \phi' \rangle| = |\langle \psi | \phi \rangle| \tag{1.9.54}$$

implies that

$$\sum_{j,k} \mathrm{Im}\,(\alpha_j \alpha_k^*)\,\mathrm{Im}\,(\beta_j \beta_k^*) = 0. \tag{1.9.55}$$

We consider the various possible cases regarding (1.9.55).

[I] There is at least one pair (j,k) with $j \neq k$, such that

$$\mathrm{Im}\,(\alpha_j \alpha_k^*) \neq 0 \tag{1.9.56}$$

$$\mathrm{Im}\,(\beta_j \beta_k^*) \neq 0. \tag{1.9.57}$$

We may then introduce the following vector

$$|\chi\rangle = \frac{1}{\sqrt{2}}\big[\,|f_j\rangle - \mathrm{i}\,|f_k\rangle\,\big] \equiv \sum_i \gamma_i |f_i\rangle . \tag{1.9.58}$$

---

[4] The importance of considering this step was particularly and rightly, emphasized by Weinberg (1995).

1.9 Wigner's Theorem on Symmetry Transformations

If this vector transforms like the vector $|\phi\rangle$ we obtain

$$\text{Im}\left(\alpha_j \alpha_k^*\right) \text{Im}\left(\gamma_j \gamma_k^*\right) = 0. \tag{1.9.59}$$

But since $\text{Im}\left(\gamma_j \gamma_k^*\right) = 1/2$, the above equality contradicts (1.9.56). Similarly, if $|\chi\rangle$ transforms like $|\psi\rangle$ we obtain a contradiction with (1.9.57). That is, $|\psi\rangle$ and $|\phi\rangle$ must follow the same # rule.

[II] No such a pair as in [I] exists, but there exists a triplet (j, k, ℓ), all unequal, such that

$$\begin{aligned} \text{Im}\left(\alpha_j \alpha_k^*\right) &\neq 0, & \text{Im}\left(\alpha_j \alpha_\ell^*\right) &= 0 \\ \text{Im}\left(\beta_j \beta_k^*\right) &= 0, & \text{Im}\left(\beta_j \beta_\ell^*\right) &\neq 0. \end{aligned} \tag{1.9.60}$$

Note that if either (or both) of $\text{Im}\left(\alpha_j \alpha_\ell^*\right)$, $\text{Im}\left(\beta_j \beta_k^*\right)$ are not equal to zero we are back to case [I]. We introduce the vector

$$|\xi\rangle = \frac{1}{\sqrt{3}}\left[|f_j\rangle - \mathrm{i}|f_k\rangle - \mathrm{i}|f_\ell\rangle\right] \equiv \sum_i \gamma_i |f_i\rangle \tag{1.9.61}$$

where $\text{Im}\left(\gamma_j \gamma_k^*\right) = \text{Im}\left(\gamma_j \gamma_\ell^*\right) = 1/3$, $\text{Im}\left(\gamma_k \gamma_\ell^*\right) = 0$. Again we run into a contradiction with $\text{Im}\left(\alpha_j \alpha_k^*\right) \neq 0$, $\text{Im}\left(\beta_j \beta_\ell^*\right) \neq 0$ if $|\xi\rangle$ transforms either as $|\phi\rangle$ or $|\psi\rangle$. That is, $|\psi\rangle$ and $|\phi\rangle$ must follow the same # rule.

[III] No such a triplet may be found as in [II], but we may find a quadruplet (j, k, ℓ, m), all unequal, such that

$$\begin{aligned} \text{Im}\left(\alpha_j \alpha_k^*\right) &\neq 0, & \text{Im}\left(\alpha_j \alpha_m^*\right) &= 0 \\ \text{Im}\left(\alpha_k \alpha_\ell^*\right) &= 0, & \text{Im}\left(\alpha_\ell \alpha_m^*\right) &= 0 \end{aligned} \tag{1.9.62}$$

and

$$\begin{aligned} \text{Im}\left(\beta_j \beta_k^*\right) &= 0, & \text{Im}\left(\beta_j \beta_m^*\right) &= 0 \\ \text{Im}\left(\beta_k \beta_\ell^*\right) &= 0, & \text{Im}\left(\beta_\ell \beta_m^*\right) &\neq 0. \end{aligned} \tag{1.9.63}$$

We may then introduce the vector

$$|\eta\rangle = \frac{1}{\sqrt{4}}\left[|f_j\rangle - \mathrm{i}|f_k\rangle + |f_\ell\rangle - \mathrm{i}|f_m\rangle\right] \equiv \sum_i \gamma_i |f_i\rangle \tag{1.9.64}$$

and note that

$$\text{Im}\left(\gamma_j \gamma_k^*\right) = \text{Im}\left(\gamma_j \gamma_m^*\right)$$

$$= \text{Im}\left(\gamma_\ell \gamma_m^*\right) = \text{Im}\left(\gamma_\ell \gamma_k^*\right)$$

$$= \frac{1}{4}, \tag{1.9.65}$$

$$\text{Im}\left(\gamma_j \gamma_\ell^*\right) = \text{Im}\left(\gamma_k \gamma_m^*\right) = 0. \tag{1.9.66}$$

We then run into a contradiction and conclude of the necessity that the same transformation rule holds for $|\psi\rangle$ and $|\phi\rangle$.

[IV] For all pairs (j,k), $j \neq k$, $\text{Im}\,(\alpha_j \alpha_k^*) = 0$. This in turn leads to two possibilities. Suppose $\text{Im}\,\alpha_k = 0$ for *all* k. Then $\alpha_k = \alpha_k^*$ and $|\psi\rangle$ transforms the same way as $|\phi\rangle$, up to an overall phase. Suppose at least for one j, $\text{Im}\,\alpha_j \neq 0$. Then $\text{Im}\,(\alpha_j \alpha_k^*) = 0$ gives

$$\alpha_k = \alpha_k^* \left(\frac{\alpha_j}{\alpha_j^*} \right) \tag{1.9.67}$$

since α_j/α_j^* is just a phase factor, we again reach the same conclusion regarding $|\psi\rangle$ and $|\phi\rangle$.

[V] For all pairs (j,k), $j \neq k$, $\text{Im}\,(\beta_j \beta_k^*) = 0$. This case is treated in the same way as case [IV] by reversing the roles of $|\phi\rangle$ and $|\psi\rangle$.

Therefore we have reached the conclusion that we may introduce an operator U such that

$$e^{i\eta_k} |f_k'\rangle = U |f_k\rangle \tag{1.9.68}$$

and

$$U \left(\sum_k \alpha_k |f_k\rangle \right) = \sum_k \alpha_k U |f_k\rangle \tag{1.9.69}$$

or else

$$U \left(\sum_k \alpha_k |f_k\rangle \right) = \sum_k \alpha_k^* U |f_k\rangle \tag{1.9.70}$$

and therefore

$$|\psi'\rangle = U |\psi\rangle. \tag{1.9.71}$$

It is worth recalling that any vector $|\psi\rangle$ may be expanded in terms of the orthonormal basis $\{|f_k\rangle\}$, and the transformed vector $|\psi'\rangle$ is then expanded in terms of the orthonormal basis $\{|f_k'\rangle\}$. Also due to the orthogonality of the vectors $|f_k'\rangle$, the relative phases $\exp i(\eta_j - \eta_k)$, for $j \neq k$, never occur.

Hence for any two rays $|\psi\rangle$, $|\phi\rangle$ one has

$$\langle U\psi | U\phi \rangle = \sum_k \alpha_k^* \beta_k = \langle \psi | \phi \rangle \tag{1.9.72}$$

or else

$$\langle U\psi | U\phi \rangle = \sum_k \alpha_k \beta_k^* = \langle \psi | \phi \rangle^*$$

$$= \langle \phi | \psi \rangle. \tag{1.9.73}$$

Finally consider any linear combination such as

$$\alpha\,|\psi\rangle + \beta\,|\phi\rangle = \sum_k \left(\alpha\alpha_k + \beta\beta_k\right)|f_k\rangle. \tag{1.9.74}$$

Under a symmetry transformation, we have shown that

$$\left(\alpha\,|\psi\rangle + \beta\,|\phi\rangle\right)' = \sum_k \left(\alpha\alpha_k + \beta\beta_k\right)\mathrm{e}^{\mathrm{i}\eta_k}|f'_k\rangle \tag{1.9.75}$$

or else

$$\left(\alpha\,|\psi\rangle + \beta\,|\phi\rangle\right)' = \sum_k \left(\alpha^*\alpha_k^* + \beta^*\beta_k^*\right)\mathrm{e}^{\mathrm{i}\eta_k}|f'_k\rangle \tag{1.9.76}$$

up to overall phase factors for the combination $\left(\alpha\,|\psi\rangle + \beta\,|\phi\rangle\right)'$ which do not necessarily coincide with those of the separate vectors $|\psi'\rangle$, $|\phi'\rangle$. According to (1.9.68), the right-hand of (1.9.75), or else (1.9.76), is given by

$$\alpha U\,|\psi\rangle + \beta U\,|\phi\rangle \tag{1.9.77}$$

or else by

$$\alpha^* U\,|\psi\rangle + \beta^* U\,|\phi\rangle \tag{1.9.78}$$

which establish the linearity, or else the anti-linearity, of U upon identifying $\left(\alpha\,|\psi\rangle + \beta\,|\phi\rangle\right)'$ with $U\left(\alpha\,|\psi\rangle + \beta\,|\phi\rangle\right)$, for the corresponding rays.

1.10 Probability, Conditional Probability and Measurement

As a physical attribute of a given system, consider the measurement of some quantity B that may take on values from a discrete set of real numbers $\{b, b', \ldots\}$. For simplicity of the notation, let B also denote the self-adjoint operator associated with this physical quantity satisfying the eigenvalue equation

$$B\,|b\rangle = b\,|b\rangle \tag{1.10.1}$$

$$\langle b'\,|\,b\rangle = \delta(b', b). \tag{1.10.2}$$

The states of the apparatus, which registers in which state it has found the physical system, will be denoted by $|a', \nu'\rangle$, where a' corresponds to a "needle" registering-value. The latter will be denoted by a_b if the "needle" registers the value b for the physical system. We consider the situation in which the apparatus may, in general, disturb the physical system causing the latter to make a transition from a state specified by the value b, as registered by the apparatus to a state specified by some other value, say, b' afterwards. In $|a', \nu'\rangle$, ν' denotes the collection of all other quantum numbers needed to specify the state of the apparatus. After a value for B is registered by the apparatus, ν' may also change and take on some other set of values, say, ν'' specifying together with the "needle" registering value a_b, the new state of the apparatus. In this section, the states $|a', \nu'\rangle$ of the apparatus will be taken to satisfy the orthonormality condition[5]

$$\langle a'', \nu'' | a', \nu' \rangle = \delta(a'', a') \delta(\nu'', \nu'). \tag{1.10.3}$$

1.10.1 Correlation of a Physical System and an Apparatus

Initially, the state of the apparatus before it is switched on, or before it interacts with the physical system, will be denoted by $|a, \nu\rangle$. The initial state of the combined system consisting of the physical system under consideration and the apparatus will be taken to be of the form

$$|\psi_0\rangle = \left(\sum_b c_b |b\rangle\right) |a, \nu\rangle, \quad \sum_b |c_b|^2 = 1. \tag{1.10.4}$$

The situation with the physical system and the registering apparatus is depicted pictorially in Figure 1.15 for a given b value.

During the registration process by the apparatus, the combined system will evolve and finally after the registration has been completed, will be described by a state having the structure

$$|\psi\rangle = \sum_{b,b',\nu'} c_b |b'\rangle |a_b, \nu'\rangle C(b', a_b, \nu'; b, a, \nu) \tag{1.10.5}$$

showing a correlation has occurred between the apparatus and the physical system and also incorporating a general disturbing transition as a result of the interaction between these two sub-systems.

We introduce the class of unitary operators which lead from the state $|\psi_0\rangle$ to the possible states $|\psi\rangle$ in (1.10.5). [For an apparatus that is not switched on, the identity operator is the corresponding unitary operator.] In general, such a unitary operator has the structure

$$U = \sum_{\substack{b',a',\nu' \\ b'',a'',\nu''}} |b'\rangle |a', \nu'\rangle U(b', a', \nu'; b'', a'', \nu'') \langle b''| \langle a'', \nu''| \tag{1.10.6}$$

leading to

$$U |\psi_0\rangle = \sum_{b,b',a',\nu'} c_b |b'\rangle |a', \nu'\rangle U(b', a', \nu'; b, a, \nu) \tag{1.10.7}$$

which, in particular, requires from (1.10.5) that

---

[5] For greater generality in applications, such orthogonality conditions will be relaxed in Chapter 8 (cf. §8.7, §8.9) allowing less restrictive apparatuses.

1.10 Probability, Conditional Probability and Measurement   67

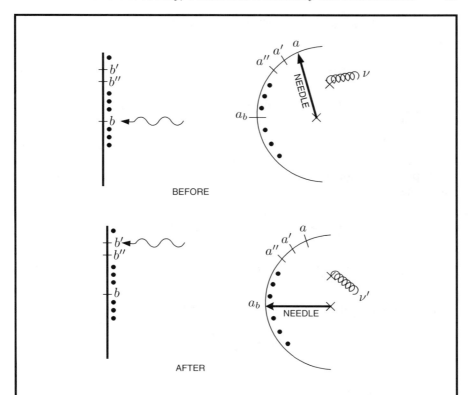

Fig. 1.15. Distinguishable states of the physical system in question are shown on the left-hand side of the figure. The registering apparatus may, in general, disturb the physical system to jump from a state $|b\rangle$ to a state, say, $|b'\rangle$ following the registration process by interacting with this system. The "needle" registering-value a' together with a collection of other possible quantum numbers ν' specify the state of the apparatus.

$$U(b', a', \nu'; b, a, \nu) = \delta(a', a_b) C(b', a_b, \nu'; b, a, \nu) \qquad (1.10.8)$$

implying that the apparatus has registered the value "b" for the physical system under consideration, and that

$$U|\psi_0\rangle = |\psi\rangle \qquad (1.10.9)$$

with $|\psi\rangle$ defined in (1.10.5).

Unitarity of U means that

$$U^\dagger U = \mathbf{1}, \qquad UU^\dagger = \mathbf{1} \qquad (1.10.10)$$

giving, respectively, the general constraints

$$\sum_{b'',a'',\nu''} U^*(b'',a'',\nu'';b''',a''',\nu''')U(b'',a'',\nu'';b',a',\nu')$$

$$= \delta(b''',b')\delta(a''',a')\delta(\nu''',\nu') \quad (1.10.11)$$

$$\sum_{b'',a'',\nu''} U(b',a',\nu';b'',a'',\nu'')U^*(b''',a''',\nu''';b'',a'',\nu'')$$

$$= \delta(b',b''')\delta(a',a''')\delta(\nu',\nu'''). \quad (1.10.12)$$

In particular, for $b''' = b'$, $a''' = a'$, $\nu''' = \nu'$ we have the normalization conditions:

$$\sum_{b'',a'',\nu''} |U(b'',a'',\nu'';b',a',\nu')|^2 = 1 \quad (1.10.13)$$

$$\sum_{b'',a'',\nu''} |U(b',a',\nu';b'',a'',\nu'')|^2 = 1. \quad (1.10.14)$$

For the sequel, it is convenient to rewrite (1.10.5) as

$$|\psi\rangle = \sum_b c_b \left|\Phi^{(b)}\right\rangle \quad (1.10.15)$$

where

$$\left|\Phi^{(b)}\right\rangle = \sum_{b',\nu'} |b'\rangle |a_b, \nu'\rangle C(b', a_b, \nu'; b, a, \nu) \quad (1.10.16)$$

and (1.10.8), (1.10.13) imply that

$$\sum_{b',\nu'} |C(b', a_b, \nu'; b, a, \nu)|^2 = 1 \quad (1.10.17)$$

and the inequality

$$\sum_{\nu'} |C(b', a_b, \nu'; b, a, \nu)|^2 \leq 1. \quad (1.10.18)$$

Equations (1.10.2), (1.10.3), (1.10.17) give the normalization condition

$$\left\langle \Phi^{(b')} \middle| \Phi^{(b)} \right\rangle = \delta(b', b). \quad (1.10.19)$$

1.10.2 Probability and Conditional Probability

We may define the density operator

$$\rho = |\psi\rangle\langle\psi| \quad (1.10.20)$$

and ask pertinent physical questions.

1.10 Probability, Conditional Probability and Measurement

After the experiment has been completed on the physical system in question, what is the probability that the "experimentalist" will find (read) the registered value a_b on the apparatus regardless in which state the physical system has made a transition to? From (1.10.15), (1.10.19), (1.10.20) this probability is given by

$$\text{Tr}\left[\left|\Phi^{(b)}\right\rangle\!\left\langle\Phi^{(b)}\right|\rho\right] = |c_b|^2 \tag{1.10.21}$$

as expected.

After the registration process of the apparatus has been completed, what is the probability that a B-filter will find the system in the state $|b\rangle$ and a reading on the registering apparatus to be a_b? This may be obtained directly from (1.10.5) to be

$$|c_b|^2 \sum_{\nu'} \left|C(b, a_b, \nu'; b, a, \nu)\right|^2 \leqslant |c_b|^2 \tag{1.10.22}$$

where we have used the inequality (1.10.18), implying, in general, that the probability in question is reduced over the value $|c_b|^2$ due to the possibility of the apparatus causing the physical system to make a transition.

The importance of the expressions for the probability on the left-hand side of (1.10.22) and the one on the right-hand side of (1.10.21) cannot be overemphasized. They lead to the following, almost tautological, question. First we note that in the initial state $|\psi_0\rangle$ in (1.10.4), there is a summation over all b. If a measurement is made on the system and the apparatus reading is a_b, then one may ask the question: what is the probability that the system immediately afterwards is in the state $|b\rangle$? Such a probability is what probabilists call a *conditional probability*:

"*Given* that the apparatus yielded the value a_b, *what* is the probability that the system is in the state $|b\rangle$?"

This probability, written as,

$$\text{Prob}\left[\text{system in state } |b\rangle \ / \ \text{apparatus yielded value } a_b\right] \tag{1.10.23}$$

is given by the ratio:

$$\frac{\text{Prob}\left[\text{system in state } |b\rangle \text{ and apparatus yielded } a_b\right]}{\text{Prob}\left[\text{apparatus yielded } a_b \text{ regardless of the state of the system}\right]}.$$
$$\tag{1.10.24}$$

From the left-hand side of (1.10.22), and (1.10.21), this probability works out to be

$$\frac{|c_b|^2 \sum_{\nu'} \left|C(b, a_b, \nu'; b, a, \nu)\right|^2}{|c_b|^2} = \sum_{\nu'} \left|C(b, a_b, \nu'; b, a, \nu)\right|^2 \tag{1.10.25}$$

70   1 Fundamentals

and the $|c_b|^2$ term (for $|c_b|^2 \neq 0$) *cancels out* in the final expression for the probability. This is "as if" the measurement carried out on the system, giving the read value a_b on the apparatus, has *forced* all the expansion coefficients in (1.10.4) to be *zero* with the exception of c_b, for the particular b value in question, and replaced this latter coefficient by *one*!

For an idealistic, rather simplistic description of an apparatus, the final state $|\psi\rangle$ for the combined physical system and the apparatus may be taken to be of the simple form

$$|\psi\rangle = \sum_{b'} c_{b'} |b'\rangle |a_{b'}\rangle \qquad (1.10.26)$$

and the probability in (1.10.23) then reduces from (1.10.24) to

$$\frac{|c_b|^2}{|c_b|^2} = 1 \qquad (1.10.27)$$

giving the idealistic confirmation of measurement that *if* (i.e., *given* that) the apparatus yields the value a_b, then the system is found in the state $|b\rangle$ with probability one, and as if all the coefficients $c_{b'}$ in (1.10.26) have been replaced by zero with the exception of c_b which has been effectively replaced by 1, as mentioned earlier.

1.10.3 An Exactly Solvable Model

As a further illustration, consider a two-level physical system such as for a particle of spin 1/2 which will be considered below. The states of such a system will be denoted by $|\pm\rangle$. If the system is in the state $|+\rangle$, then an apparatus, detecting the system, will be found in some corresponding state denoted, say, by $|+\rangle$. Similarly, if the system is in the state $|-\rangle$, the apparatus will be found in some corresponding state $|-\rangle$.

The states of the apparatus are shown in Figure 1.16, where the initial state of the apparatus, corresponding to a neutral position, is taken to be $|\mathbf{0}\rangle$ with the "needle" of the apparatus initially pointing at $\mathbf{0}$.

As we will see later, the state $|\mathbf{0}\rangle$ may be expanded as follows

$$|\mathbf{0}\rangle = \frac{1}{\sqrt{2}} \left(e^{-i\pi/4} |+\rangle + e^{i\pi/4} |-\rangle \right). \qquad (1.10.28)$$

The origin of the phase factors $\exp(\mp i\pi/4)$ in the combination of the states $|\pm\rangle$ in (1.10.28) will become clear later and have to do with the needle spin state along the y-axis and $1/\sqrt{2}$ is a normalization factor.

The state of the physical system plus the apparatus before their mutual interaction will be taken to be

$$|\psi_0\rangle = \left(c_+ |+\rangle + c_- |-\rangle \right) |\mathbf{0}\rangle \qquad (1.10.29)$$

1.10 Probability, Conditional Probability and Measurement

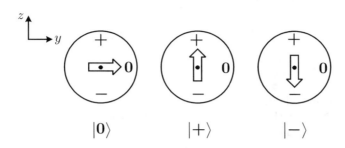

Fig. 1.16. The figure denotes the states of the apparatus with the first one on the left-hand side denoting its neutral state before it interacts with the physical system.

$|c_+|^2 + |c_-|^2 = 1$, and for the final state

$$|\psi\rangle = \left\{ \left(c_+ + \frac{1}{2}[c_+(\cos\kappa - 1) + ic_-\sin\kappa] \right) |+\rangle \right.$$

$$\left. + \frac{1}{2}[ic_+\sin\kappa + c_-(\cos\kappa - 1)] |-\rangle \right\} |+\rangle$$

$$+ \left\{ \left(c_- + \frac{1}{2}[ic_+\sin\kappa + c_-(\cos\kappa - 1)] \right) |-\rangle \right.$$

$$\left. + \frac{1}{2}[c_+(\cos\kappa - 1) + ic_-\sin\kappa] |+\rangle \right\} |-\rangle \quad (1.10.30)$$

showing that a correlation has occurred between the apparatus and the physical system, and also incorporating a general disturbing transition specified by the "angle" κ. For an almost ideal apparatus, κ is arbitrarily small — see (1.10.40)–(1.10.43). For an ideal apparatus, with $\kappa \to 0$, (1.10.30) becomes

$$|\psi\rangle_{\text{ideal}} = c_+ |+\rangle |+\rangle + c_- |-\rangle |-\rangle. \quad (1.10.31)$$

This state defines the perfect correlation between the apparatus and the physical system with the "needle" pointing in the same direction as the spin of the particle, respectively, in each case. For such an ideal apparatus, (1.10.31) is of the form in (1.10.26), and, for example, (for $c_+ \neq 0$),

$$\text{Prob}\left[\text{system in state } |+\rangle \text{ and apparatus in state } |+\rangle\right] = |c_+|^2, \quad (1.10.32)$$

$$\text{Prob}\left[\text{apparatus in state } |+\rangle\right] = |c_+|^2 \quad (1.10.33)$$

and hence from (1.10.24)

$$\text{Prob}\left[\text{system in state } |+\rangle \,/\, \textit{if} \text{ apparatus in state } |+\rangle\right] = \frac{|c_+|^2}{|c_+|^2} = 1 \tag{1.10.34}$$

as in (1.10.27).

If $c_+ \neq 0$, $c_- \neq 0$, in (1.10.31), neither the physical system nor the apparatus is in a definite state.[6]

More generally, the state $|\psi_0\rangle$ in (1.10.29) evolves to state $|\psi\rangle$ in (1.10.30) via the unitary operator

$$U = \mathbf{1} + A + B \tag{1.10.35}$$

where

$$A = -\Big[(1-i)\,|+\rangle\langle+| + (1+i)\,|-\rangle\langle-|\Big]$$

$$\times \frac{1}{2}\Big[|+\rangle\langle+| + |-\rangle\langle-| - |-\rangle\langle+| - |+\rangle\langle-|\Big] \tag{1.10.36}$$

$$B = \Big[(\cos\kappa - 1)\big(|+\rangle\langle+| + |-\rangle\langle-|\big) + i\sin\kappa\big(|-\rangle\langle+| + |+\rangle\langle-|\big)\Big]$$

$$\times \frac{1}{2}\Big[|+\rangle\langle+| + |-\rangle\langle-| + |-\rangle\langle+| + |+\rangle\langle-|\Big]. \tag{1.10.37}$$

That is (see Problem 1.10),

$$U|\psi_0\rangle = |\psi\rangle \tag{1.10.38}$$

with $|\psi_0\rangle$, $|\psi\rangle$ given, respectively, in (1.10.29), (1.10.30) and

$$U^\dagger U = UU^\dagger = \mathbf{1}. \tag{1.10.39}$$

Later (§8.7) we will see how the unitary operator U in (1.10.35)–(1.10.37) actually arises as an elementary interaction between the variables of the physical system and the apparatus. The system consisting of the apparatus and the physical system based on (1.10.30) will be analyzed in detail below.

For the interpretation of κ in (1.10.30), (1.10.36), (1.10.37), suppose that a system is initially in the state $|+\rangle$, i.e., $c_+ = 1$, $c_- = 0$. After the interaction of the apparatus with the system, the probability that the apparatus will be in the state $|+\rangle$ is from (1.10.30) given by

$$\frac{1}{4}\Big[(1+\cos\kappa)^2 + \sin^2\kappa\Big] = \frac{1}{2}(1+\cos\kappa). \tag{1.10.40}$$

Now given that the apparatus is in the state $|+\rangle$, the (conditional) probability that the physical state has made a transition to the state $|-\rangle$ is from (1.10.30), (1.10.40) with $c_+ = 1$, $c_- = 0$,

---

[6] Such a state is called an entangled state. The properties of entangled states will be studied in detail in Chapter 8, in particular, in §8.10.

$$\frac{\frac{1}{4}\sin^2\kappa}{\frac{1}{2}(1+\cos\kappa)} = \frac{1}{2}\frac{\sin^2\kappa}{(1+\cos\kappa)} \qquad (1.10.41)$$

which for an ideal apparatus, $\kappa \to 0$, goes to zero.

Conversely, the probability that the system remains in the state $|+\rangle$ after the interaction with the apparatus irrespective of the state of the apparatus is, from (1.10.30) (with $c_+ = 1$, $c_- = 0$), given by

$$\frac{1}{4}\left[(\cos\kappa+1)^2 + (\cos\kappa-1)^2\right] = \frac{1}{2}(1+\cos^2\kappa). \qquad (1.10.42)$$

Given that the system remains in the state $|+\rangle$, the (conditional) probability that the apparatus is in the state $|-\rangle$ is then from (1.10.30), (1.10.42).

$$\frac{1}{2}\frac{(1-\cos\kappa)^2}{(1+\cos^2\kappa)} \qquad (1.10.43)$$

which would again vanish for an ideal apparatus, as expected.

Finally, the probability that the system makes a transition to the state $|-\rangle$ *and* the apparatus is found in the state $|-\rangle$ is $(\sin^2\kappa)/4$, again for $c_+ = 1$, $c_- = 0$.

Figure 1.17 depicts experiments, where a spin 1/2 particle prepared in the state $|+1/2, z\rangle$ goes through a Stern-Gerlach set up. In the subsequent analysis, we make the identifications $|\pm\rangle \to |\pm 1/2, \bar{z}\rangle$ in (1.10.29), (1.10.30). In part (d) of the figure, an operating apparatus is inserted of the type just described. As in Figure 1.3 in §1.1, the numerical factor m in $|m, z\rangle$, with $m = \pm 1/2$, correspond, respectively, to spin components along the $\pm z$ directions. The coefficients $c_\pm$ in (1.10.29), (1.10.30) are now given by

$$c_\pm = \langle \pm 1/2, \bar{z} | +1/2, z\rangle. \qquad (1.10.44)$$

In part (a) of the Figure, the probability that the particle emerges and a spin flip occurs, i.e., with the spin component being along the $-z$ direction, is obtained from the successive measurements symbols

$$\Lambda_z(-1/2)\Lambda_{\bar{z}}(-1/2)\Lambda_z(+1/2) = |-1/2, z\rangle\langle +1/2, z|$$
$$\times \left(\langle -1/2, z|-1/2, \bar{z}\rangle \langle -1/2, \bar{z}|+1/2, z\rangle\right) \qquad (1.10.45)$$

relative to the z-, $\bar{z}$-, z-axes, respectively, to be given by

$$|\langle -1/2, z|-1/2, \bar{z}\rangle|^2 |\langle -1/2, \bar{z}|+1/2, z\rangle|^2. \qquad (1.10.46)$$

Similarly for the probability of a non-flip of spin, we have

$$|\langle +1/2, z|-1/2, \bar{z}\rangle|^4 \qquad (1.10.47)$$

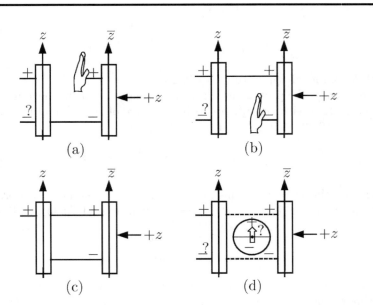

Fig. 1.17. A spin 1/2 particle initially prepared in the state $|+1/2, z\rangle$ goes through the Stern-Gerlach set up illustrated, with a non-zero $(0 < \theta < \pi)$ angle of orientation of the $\bar{z}$-axis relative to the z-axis. In part (a), the probability that the particle emerges and has a spin flip, i.e., is in the $-z$ direction, is $|\langle -1/2, z| -1/2, \bar{z}\rangle|^2 |\langle -1/2, \bar{z}| +1/2, z\rangle|^2$. In part (b), the corresponding probability is $|\langle -1/2, z| +1/2, \bar{z}\rangle|^2 |\langle +1/2, \bar{z}| +1/2, z\rangle|^2$. In (c), this probability is zero due to destructive *interference*. In part (d), an operating apparatus is inserted to determine the component of the spin, in the intermediate stage, is along the $+\bar{z}$ or the $-\bar{z}$ directions but the "experimentalist" does not take a reading and hence does not know this result. The probability in this case that the particle emerges with a spin-flip is, up to an additional term of the order $(\sin^2 \kappa)/2$, the sum of the corresponding probabilities in parts (a) and (b). For κ arbitrarily small, the interference term becomes small. This idealized experimental set-up mimics the famous "double-slit" experiment.

in part (a).

For part (b), the probabilities that the particle emerges and with a spin flip or with a non-flip of spin are, respectively,

$$|\langle -1/2, z| +1/2, \bar{z}\rangle|^2 |\langle +1/2, \bar{z}| +1/2, z\rangle|^2 \qquad (1.10.48)$$

$$|\langle +1/2, z| +1/2, \bar{z}\rangle|^4 . \qquad (1.10.49)$$

To consider the situation in parts (c) and (d), we first define corresponding density operators

1.10 Probability, Conditional Probability and Measurement

$$\rho_0 = |\psi_0\rangle\langle\psi_0| \tag{1.10.50}$$

$$\rho = |\psi\rangle\langle\psi| \tag{1.10.51}$$

where the states $|\psi_0\rangle$, $|\psi\rangle$ are given, respectively, in (1.10.29), (1.10.30).

In part (c), the probabilities that the particle emerges with a spin flip or with a non-flip of spin are, respectively,

$$0 = |\langle -1/2, z | +1/2, z\rangle|^2 = \text{Tr}\left[\Lambda_z(-1/2)\rho_0\right]$$

$$= |c_+|^2 |\langle +1/2, \bar{z} | -1/2, z\rangle|^2 + |c_-|^2 |\langle -1/2, \bar{z} | -1/2, z\rangle|^2$$

$$+ \left[c_+^* c_- \langle +1/2, \bar{z} | -1/2, z\rangle \langle -1/2, z | -1/2, \bar{z}\rangle\right.$$

$$\left. + c_-^* c_+ \langle -1/2, \bar{z} | -1/2, z\rangle \langle -1/2, z | +1/2, \bar{z}\rangle\right] \tag{1.10.52}$$

or

$$1 = |\langle +1/2, z | +1/2, z\rangle|^2 = \text{Tr}\left[\Lambda_z(+1/2)\rho_0\right]$$

$$= |c_+|^2 |\langle +1/2, \bar{z} | +1/2, z\rangle|^2 + |c_-|^2 |\langle -1/2, \bar{z} | +1/2, z\rangle|^2$$

$$+ \left[c_+^* c_- \langle +1/2, \bar{z} | +1/2, z\rangle \langle +1/2, z | -1/2, \bar{z}\rangle\right.$$

$$\left. + c_-^* c_+ \langle -1/2, \bar{z} | +1/2, z\rangle \langle +1/2, z | +1/2, \bar{z}\rangle\right]. \tag{1.10.53}$$

The numerical values 0 or 1 on the extreme left-hand sides of (1.10.52), (1.10.53) are obtained from the consideration of the successive measurements symbols

$$\Lambda_z(\mp 1/2)\Lambda_z(+1/2) = |\mp 1/2, z\rangle\langle +1/2, z| \left(\langle \mp 1/2, z | +1/2, z\rangle\right)$$

$$= |\mp 1/2, z\rangle\langle +1/2, z| \delta(\mp z, +z). \tag{1.10.54}$$

The third terms in the square brackets on the extreme right-hand sides of (1.10.52), (1.10.53) are referred to as *interference* terms, showing that destructive and constructive interferences, have, respectively, occurred in these experiments.

Now we come to the interesting situation depicted in part (d), where the apparatus has been inserted. The corresponding probabilities are given by

$$\text{Tr}\left[\Lambda_z(\mp 1/2)\rho\right] \tag{1.10.55}$$

where the operating apparatus is inserted but no reading of the apparatus is undertaken.

From (1.10.30), (1.10.51), (1.10.55) a lengthy but straightforward calculation (see Problem 1.13) shows that the probabilities in (1.10.55) work out to be equal to

$$|c_+|^2 |\langle +1/2, \bar{z}|\mp 1/2, z\rangle|^2 + |c_-|^2 |\langle -1/2, \bar{z}|\mp 1/2, z\rangle|^2$$

$$+ (\sin \kappa) \Big[\big(|c_-|^2 - |c_+|^2\big) \sin \kappa + i\big(c_- c_+^* - c_-^* c_+\big) \cos \kappa \Big]$$

$$\times \frac{1}{2}\Big(|\langle \mp 1/2, z|+1/2, \bar{z}\rangle|^2 - |\langle \mp 1/2, z|-1/2, \bar{z}\rangle|^2 \Big)$$

$$+ (\sin \kappa) \Big[\big(|c_-|^2 - |c_+|^2\big) \cos \kappa + i\big(c_- c_+^* - c_-^* c_+\big) \sin \kappa \Big]$$

$$\times \frac{i}{2}\Big(\langle -1/2, \bar{z}|\mp 1/2, z\rangle \langle \mp 1/2, z|+1/2, \bar{z}\rangle$$

$$- \langle +1/2, \bar{z}|\mp 1/2, z\rangle \langle \mp 1/2, z|-1/2, \bar{z}\rangle \Big) \quad (1.10.56)$$

which is to be compared with the extreme right-hand sides of (1.10.52), (1.10.53) given, in part (c) of Figure 1.7 in the absence of the apparatus.

For a negligibly disturbing apparatus, specified by a very small κ, the *interference* terms in the square brackets on the right-hand sides of (1.10.52), (1.10.53) essentially disappear, as seen from (1.10.56) for $\kappa \simeq 0$, by the mere insertion of an unread apparatus! and completely disappear for $\kappa \to 0$.

Let us be more quantitative. To this end, if the angle between the $\bar{z}$- and z-axes is β, then as we will see later

$$\langle +1/2, z|+1/2, \bar{z}\rangle = \cos\frac{\beta}{2} = \langle -1/2, z|-1/2, \bar{z}\rangle \quad (1.10.57)$$

$$\langle -1/2, z|+1/2, \bar{z}\rangle = -\sin\frac{\beta}{2}, \quad \langle +1/2, z|-1/2, \bar{z}\rangle = \sin\frac{\beta}{2}. \quad (1.10.58)$$

Upon setting

$$\frac{1}{2}\sin^2\kappa = \varepsilon^2 \quad (1.10.59)$$

the probabilities corresponding to parts (a), (b), (c), (d) in Figure 1.17 are directly obtained from (1.10.46)–(1.10.49), (1.10.52), (1.10.53), (1.10.56), by using (1.10.57)–(1.10.59), and are spelled out in Figure 1.18, where $c_\pm$ are defined in (1.10.44).

In particular, we note from Figure 1.17 (a), (b) that blocking one of the "$\bar{z}$-outlets", for a spin projection, increases the probability that a particle emerges with a spin-flip from a 0 value (versus part (c)) to a non-vanishing value of $(\sin^2 \beta)/4$, where in part (c) both "$\bar{z}$-outlets" are kept open.

Unlike the situations in part (c) and (d), the blocking of the "$\bar{z}$-outlet" shown in part (a) (and similarly in part (b)), the particle does not necessarily

1.10 Probability, Conditional Probability and Measurement   77

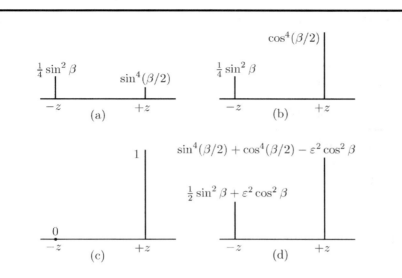

Fig. 1.18. In parts (a), (b), the probabilities are given for the particle emerging has a spin flip (spin non-flip), i.e., is in the state $|-1/2, \bar{z}\rangle$ ($|+1/2, \bar{z}\rangle$). In parts (c), (d), the probabilities are given for the particle emerging with a spin flip or with a non-flip of spin, where both of the "$\bar{z}$-outlets" are kept "open". In part (d), the apparatus is inserted in the intermediate stage. For the corresponding experiments refer to Figure 1.17. The angle between the $\bar{z}$- and z-axes is equal to β. The probability spikes are not drawn to scale for any specific value of β.

emerge from the left-hand side of the experimental set-up and hence the *sum* of the probabilities $(\sin^2 \beta)/4$, for spin-flip, and $\sin^4(\beta/2)$, for a spin non-flip, shown by the spikes in Figure 1.18 (a), is less than one.

When the apparatus is inserted in part (d), but unread by the "experimentalist", the probabilities, up to the $\pm\varepsilon^2 \cos^2 \beta$ terms, are the sum of the corresponding ones in parts (a) and (b), respectively. That is, for κ arbitrarily small, and hence small ε, the mere insertion of the apparatus essentially washes away interferences as occurring in part (c), and they completely disappear for $\kappa \to 0$.

In part (c), the particle emerges with a non-flip of spin. The probabilities of 0 (for a spin-flip) and 1 (for a spin non-flip) are *not* simply the sum of the corresponding probabilities in parts (a) and (b). There are *interference* terms

$$-\frac{(\sin^2 \beta)}{2} \tag{1.10.60}$$

and

$$+\frac{(\sin^2 \beta)}{2} = \left(1 - \sin^4 \frac{\beta}{2} - \cos^4 \frac{\beta}{2}\right) \tag{1.10.61}$$

within the square brackets on the right-hand sides of (1.10.52), (1.10.53), respectively, leading finally to the value 0 and 1 for the probabilities mentioned above.

That is, the fate of a single particle, in the experimental set-up in part (c) of Figure 1.17 is inferred from the intensity distributions computed in part (c) of Figure 1.18 and these probabilities are not obtained by simply summing the corresponding probabilities in parts (a) and (b) which would exclude the interference terms discussed above. This has led to such a statement, attributed to Paul Dirac, referring to a photon, that a particle interferes with itself! The interference, however, occurs between different amplitudes when adding them up to obtain a given probability of occurrence as in (1.10.52), (1.10.53). It is misleading to think of a particle in the quantum world as a particle in a classical sense, otherwise one would run into a typical classical argument on how a single particle may simultaneously go through both "$\pm \bar{z}$-outlets" leading finally to the built up of the interference pattern observed in several experiments, with very low particle densities used at each given time, but are run for long periods of time so designed to avoid arguments based on particle-particle interactions as necessarily the source of interference.

Finally we note in reference to part (d) in Figure 1.17, if the apparatus is *read* and found to be in the state $|+\rangle$, then one may enquire about the probability that the particle is in state $|+1/2, \bar{z}\rangle$ before it enters the second Stern-Gerlach apparatus. As in (1.10.23), one is then dealing with a *conditional probability* stated as follows: *given* that the state of the apparatus is $|+\rangle$, then *what* is the probability that the state of the spin of the particle is given by $|+1/2, \bar{z}\rangle$? From (1.10.30), (1.10.44), (1.10.56)–(1.10.58), (1.10.24), this conditional probability works out to be (see Problem 1.15)

$$\left(1 + \frac{\cos^2 \frac{\beta}{2} \sin^2 \kappa + \sin^2 \frac{\beta}{2} (1 - \cos \kappa)^2}{\cos^2 \frac{\beta}{2} (1 + \cos \kappa)^2 + \sin^2 \frac{\beta}{2} \sin^2 \kappa}\right)^{-1}. \quad (1.10.62)$$

As in (1.10.25), we see that "as if" the measurement carried out by the apparatus, and found to be in the state $|+\rangle$ has, for $\kappa \to 0$, forced the coefficient $c_+ = \cos \beta/2$ in (1.10.29) (see (1.10.44), (1.10.57), (1.10.58)) to be replaced by one and $c_- = \sin \beta/2$ to be replaced by zero.

The experimental set-up in Figure 1.17 mimics the so-called "double-slit" experiment (see also §8.7–§8.9, §9.1, §9.5) elaborated upon repeatedly in the literature and notably by R. P. Feynman.

Later on in §8.7, we will see how the unitary operator U in (1.10.35)–(1.10.38) arises as an elementary interaction between the variables of the physical system and the apparatus.

There are many views, descriptions and interpretations on what measurement theory in quantum physics is and how it may be achieved. Each view involves internal consistency problems that have to be checked and further generalizations may be often needed based on such consistency checks.

Additional details on measurement theory and related aspects will be given in Chapter 8. In particular situations will be encountered where the orthogonality conditions in (1.10.3) are necessarily relaxed. We will see, as one may argue, that taking into account of the *environment* coupled to the meter (apparatus) variables and consisting of everything else monitoring the observables being measured, provides the different alternative readings of a meter being sought. That is, after a measurement, the system is found in one of its alternative states rather than in a superposition of them. The information thus obtained on the system can be then described in usually perceived classical terms. The destruction of such quantum superpositions, resulting from the interaction with the environment, is referred to as *quantum decoherence*.

Problems

1.1. Derive the expression for the measure of fraction of the number of particles transmitted in Figure 1.4 and Figure 1.5 (c).

1.2. Show that the inner product in (1.4.42) does indeed satisfy the properties in (1.4.46)–(1.4.48), and establish the description independence of the inner product as given in (1.4.51).

1.3. Establish the inequalities in (1.4.52), (1.4.53) using (1.4.42), (1.4.49).

1.4. For a sphere of unit radius in the n-dimensional Euclidean space, a point on the surface of the sphere may be defined in terms of the n variables

$$x_1 = \cos \vartheta_1$$
$$x_k = \sin \vartheta_1 \cdots \sin \vartheta_{k-1} \cos \vartheta_k, \qquad k = 2, \ldots, n-1$$
$$x_n = \sin \vartheta_1 \cdots \sin \vartheta_{n-2} \sin \vartheta_{n-1}$$

with $0 \leqslant \vartheta_i \leqslant \pi$ $(i = 1, \ldots, n-2)$, $0 \leqslant \vartheta_{n-1} \leqslant 2\pi$, $\sum_{i=1}^{n} x_i^2 = 1$.

(i) Show that the surface element on the sphere is given by

$$d\Omega = \left(\sin^{n-2} \vartheta_1 d\vartheta_1\right)\left(\sin^{n-3} \vartheta_2 d\vartheta_2\right) \cdots \left(\sin^1 \vartheta_{n-2} d\vartheta_{n-2}\right)\left(d\vartheta_{n-1}\right).$$

(ii) Using the definition $\langle x_j^2 \rangle = \int x_j^2 \, d\Omega / \int d\Omega$, show explicitly that

$$\langle x_1^2 \rangle = \cdots = \langle x_n^2 \rangle = \frac{1}{n}$$

as in (1.5.27).

1.5. Verify the property (1.6.10) which is relevant to the incompatibility of measurements of polarizations along two different directions.

1.6. Derive (1.6.21) corresponding to the transmission of light through the arrangement of polarizers in Figure 1.13 (b).

1.7. Establish the separability and completeness of the space $\ell^2(\infty)$ defined in (1.7.12)–(1.7.14).

1.8. Extend the analysis given for the position operator through (1.8.35)–(1.8.39) to the momentum operator $\mathbf{p} = -i\hbar\boldsymbol{\nabla}$ (§2.3) in the x-description.

[Hint. For any $\varepsilon > 0$, define the normalized function

$$F^\varepsilon_{\boldsymbol{\beta}_0}(\mathbf{x}) = \left(\frac{\varepsilon^2}{\pi\hbar^2}\right)^{3/4} \exp\left(\frac{i}{\hbar}\mathbf{p}_0 \cdot \mathbf{x}\right) \exp\left(-\frac{\mathbf{x}^2\varepsilon^2}{2\hbar^2}\right)$$

and show that $\left\|(-i\hbar\boldsymbol{\nabla} - \mathbf{p}_0)F^\varepsilon_{\mathbf{p}_0}\right\| < \varepsilon.$]

1.9. Establish the properties in (1.8.16)–(1.8.22) of the projection operator $P_A(\lambda)$ associated with a self-adjoint operator A.

1.10. Show that the unitary operator in (1.10.35)–(1.10.37) leads from the state $|\psi_0\rangle$ in (1.10.29) to the state $|\psi\rangle$ in (1.10.30)/(1.10.38) and verify explicitly the unitarity condition in (1.10.39).

1.11. Work out the expressions for the probabilities in (1.10.40)–(1.10.43).

1.12. Derive the equalities on the extreme right-hand sides of (1.10.52), (1.10.53).

1.13. Show that the probabilities in (1.10.55) are given in detail in (1.10.56).

1.14. From the probabilities given in (1.10.57), (1.10.58), establish the probability spikes given in Figure 1.18.

1.15. Work out the conditional probability given in (1.10.62).

2

Symmetries and Transformations

This chapter is an extension of the first dealing with the formalism of quantum theory being sought. The present one is based on symmetries and deals with their implementations and the mechanics of the transformations of the underlying variables in the theory. The main symmetries and the corresponding transformations in non-relativistic quantum mechanics are the Galilean ones. Different labellings of an event by two Galilean frames (§2.1, §2.2) are *continuously* related and the corresponding transformations, which relate the two labellings, reduce to the identity one in the limit that the two frames coincide. Accordingly, we may infer from Wigner's Theorem on symmetry transformations (§1.9), that such symmetries must be implemented, in the underlying Hilbert space of the theory, by *unitary*, rather than by anti-unitary, operators as they must continuously reduce to the identity in the limit of coincident Galilean frames — the identity element being, of course, a unitary operator. Group properties of these transformations are derived and their corresponding generators are introduced. These group relations give rise to basic commutation rules to be satisfied by the generators and their *physical* meanings emerge naturally. One of such generators is the Hamiltonian generating time translations and describing the dynamics of systems. Explicit Galilean invariant Hamiltonians will be constructed for several physical systems of interest. We also introduce discrete symmetries, and consider further properties of the generator of spatial rotations. Special attention will be given to so-called spinors which allow the description of particles of any spin. Finally supersymmetry is introduced in analogy to the Galilean ones and basic properties are established which will find several applications in later chapters.

2.1 Galilean Space-Time Coordinate Transformations

Consider a Cartesian coordinate system $\overline{F}$ resulting from a given one F by carrying a c.c.w. rotation of the coordinate system by an angle φ about

2 Symmetries and Transformations

a unit vector **n**, translating its origin by a given amount and setting the corresponding frame in uniform motion relative to that of F (see Figure 2.1). The corresponding structure of the rotation matrix with matrix elements R^{ij} is worked out in Figure 2.2.

A point P with coordinate vector label **x** in F will be assigned the coordinate vector label $\overline{\mathbf{x}}$ in $\overline{F}$. The clocks in $\overline{F}$ will be set back relative to those in F by a given amount.

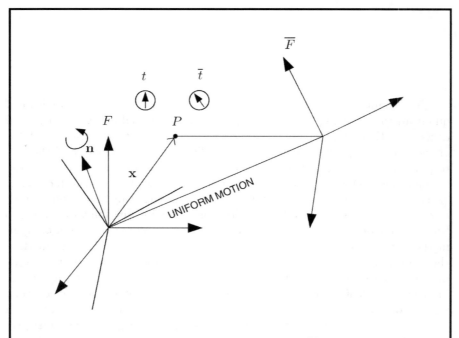

Fig. 2.1. A space-time point P is labelled, in F and $\overline{F}$, respectively, by $(\mathbf{x}, t)$ and $(\overline{\mathbf{x}}, \overline{t})$. These labellings are related through (2.1.1), (2.1.2).

Accordingly, the labelling of point P: $(\mathbf{x}, t)$ and $(\overline{\mathbf{x}}, \overline{t})$ in the frames F and $\overline{F}$, respectively, are related by the transformations

$$\overline{x}^i = R^{ij} x^j - a^i - v^i t \tag{2.1.1}$$

$$\overline{t} = t - \tau \tag{2.1.2}$$

(with a sum over j in (2.1.1) understood), where a^i, v^i are independent of time t, $-\mathbf{v}$ denotes the velocity of frame F relative to $\overline{F}$, and the rotation matrix R with matrix elements R^{ij} satisfies

$$R^{ij} R^{ik} = \delta^{jk} \tag{2.1.3}$$

2.1 Galilean Space-Time Coordinate Transformations

(with a sum over i) and from Figure 2.2, it is explicitly given by

$$R^{ik} = \delta^{ik} - \varepsilon^{ijk} n^j \sin\varphi + \left(\delta^{ik} - n^i n^k\right)(\cos\varphi - 1)$$

$$\equiv R^{ik}(\varphi, \mathbf{n}) \tag{2.1.4}$$

where

$$\varepsilon^{ijk} = \begin{cases} +1, & \text{if } (i,j,k) \text{ is an even permutation of } (1,2,3) \\ -1, & \text{if } (i,j,k) \text{ is an odd permutation of } (1,2,3) \\ 0, & \text{if two or three of the indices are the same.} \end{cases} \tag{2.1.5}$$

The transformations (2.1.1), (2.1.2) define the Galilean space-time coordinate transformations. We note that $\bar{x}^i$ in (2.1.1) defines the space coordinate labelling in $\bar{F}$ at a given fixed time in terms of x and t.

In particular, we consider an infinitesimal rotation by an angle $\delta\varphi$ and introduce the vector

$$\delta\boldsymbol{\omega} = \mathbf{n}\delta\varphi \tag{2.1.6}$$

To second order in $\delta\varphi$, the rotation matrix R may be written as

$$R = \mathbf{1} + A + B \tag{2.1.7}$$

where (⊤ for transpose)

$$A^\top = -A \tag{2.1.8}$$

$$B^\top = B = \frac{A^2}{2} \tag{2.1.9}$$

and

$$A^{ik} = -\varepsilon^{ijk}\delta\omega^j. \tag{2.1.10}$$

Also up to second order in $\delta\varphi$

$$R^{-1} = \mathbf{1} - A + B. \tag{2.1.11}$$

In establishing (2.1.9), the following identity

$$\varepsilon^{ijk}\varepsilon^{ilm} = \left(\delta^{jl}\delta^{km} - \delta^{jm}\delta^{kl}\right) \tag{2.1.12}$$

is quite useful.

For infinitesimal first order transformations with parameters $\delta\mathbf{a}$, $\delta\mathbf{v}$, $\delta\boldsymbol{\omega}$, $\delta\tau$ the Galilean transformations (2.1.1), (2.1.2) reduce to

$$\bar{\mathbf{x}} = \mathbf{x} - \delta\mathbf{a} - \delta\mathbf{v}t - \delta\boldsymbol{\omega} \times \mathbf{x} \tag{2.1.13}$$

$$\bar{t} = t - \delta\tau \tag{2.1.14}$$

84   2 Symmetries and Transformations

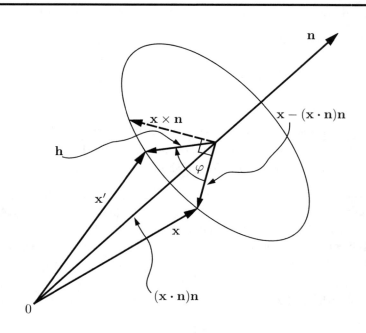

Fig. 2.2. Rotation of a vector **x** c.w. by an angle φ about a unit vector **n**. This is equivalent to the rotation of the *coordinate* system, instead, c.c.w. about **n** and keep the direction of **x** fixed. From the figure

$$\mathbf{x}' = (\mathbf{x} \cdot \mathbf{n})\mathbf{n} + \mathbf{h}$$

where

$$\mathbf{h} = [\mathbf{x} - (\mathbf{x} \cdot \mathbf{n})\mathbf{n}]\cos\varphi + \mathbf{x} \times \mathbf{n}\sin\varphi$$

or

$$\mathbf{x}' = \mathbf{x} - \mathbf{n} \times \mathbf{x} \sin\varphi + [\mathbf{x} - (\mathbf{x} \cdot \mathbf{n})\mathbf{n}](\cos\varphi - 1)$$

with $\overline{\mathbf{x}}$ in (2.1.13) given at a given fixed time t. We set

$$\delta \mathbf{x} = \mathbf{x} - \overline{\mathbf{x}} = \delta \mathbf{a} + \delta \mathbf{v} t + \delta \boldsymbol{\omega} \times \mathbf{x} \tag{2.1.15}$$

$$\delta t = t - \overline{t} = \delta \tau. \tag{2.1.16}$$

We provide an exponential representation of the rotation matrix $R = (R^{ik})$ in (2.1.24).

To this end, we define the matrices M^j, $j = 1, 2, 3$, with matrix elements

$$\left[M^j\right]^{ik} = \mathrm{i}\varepsilon^{ijk} \tag{2.1.17}$$

$i, k = 1, 2, 3$. In particular,

2.1 Galilean Space-Time Coordinate Transformations

$$[\mathbf{n} \cdot \mathbf{M}]^{ik} = i\varepsilon^{ijk} n^j \tag{2.1.18}$$

$$[(\mathbf{n} \cdot \mathbf{M})^2]^{ik} = \delta^{ik} - n^i n^k \tag{2.1.19}$$

$$(\mathbf{n} \cdot \mathbf{M})^3 = \mathbf{n} \cdot \mathbf{M} \tag{2.1.20}$$

and we may rewrite the matrix R as

$$R = 1 + i\mathbf{n} \cdot \mathbf{M} \sin\varphi + (\mathbf{n} \cdot \mathbf{M})^2 (\cos\varphi - 1). \tag{2.1.21}$$

Upon differentiation the latter with respect to φ and using (2.1.19), (2.1.20) we obtain

$$\frac{dR}{d\varphi} = (i\mathbf{n} \cdot \mathbf{M}) R. \tag{2.1.22}$$

The integration of this equation, using the boundary condition

$$R\big|_{\varphi=0} = 1, \tag{2.1.23}$$

leads to the expression

$$R = \exp[i\varphi \, \mathbf{n} \cdot \mathbf{M}] \tag{2.1.24}$$

or using the standard notation

$$\boldsymbol{\omega} = \varphi \mathbf{n} \tag{2.1.25}$$

(see also (2.1.6)), one has

$$R = \exp[i\boldsymbol{\omega} \cdot \mathbf{M}]. \tag{2.1.26}$$

The matrices M^j, satisfy the commutation relations

$$[M^i, M^j] = i\varepsilon^{ijk} M^k \tag{2.1.27}$$

as is easily obtained upon using the identity in (2.1.12). These matrices are explicitly given by

$$M^1 = i\begin{pmatrix} 0 & 0 & 0 \\ 0 & 0 & -1 \\ 0 & 1 & 0 \end{pmatrix} \tag{2.1.28}$$

$$M^2 = i\begin{pmatrix} 0 & 0 & 1 \\ 0 & 0 & 0 \\ -1 & 0 & 0 \end{pmatrix} \tag{2.1.29}$$

$$M^3 = i\begin{pmatrix} 0 & -1 & 0 \\ 1 & 0 & 0 \\ 0 & 0 & 0 \end{pmatrix} \tag{2.1.30}$$

Also (with ⊤ for transpose, Tr for trace, det for determinant)

$$\left(M^j\right)^\dagger = M^j, \quad \left(M^j\right)^\top = -M^j \tag{2.1.31}$$

$$\mathbf{M}^2 = 2\mathbf{1} \tag{2.1.32}$$

$$\text{Tr}\,[\mathbf{n}\cdot\mathbf{M}] = 0 \tag{2.1.33}$$

$$R^\top = R^\dagger = R^{-1} \tag{2.1.34}$$

$$RR^\top = R^\top R = \mathbf{1} \tag{2.1.35}$$

$$\det R = 1 \tag{2.1.36}$$

$$\text{Tr}\,(R) = 2\cos\varphi + 1. \tag{2.1.37}$$

Finally, we derive a useful identity involving the rotation matrix. To this end, from the expression of $\mathbf{x}'$ given in Figure 2.2, we have

$$\frac{d}{d\varphi}\mathbf{x}' = -\mathbf{n}\times\mathbf{x}'. \tag{2.1.38}$$

Let

$$\mathbf{A}(\varphi) = \exp\left[-\varphi\mathbf{n}\cdot(\mathbf{x}\times\boldsymbol{\nabla})\right]\mathbf{x}\exp\left[\varphi\mathbf{n}\cdot(\mathbf{x}\times\boldsymbol{\nabla})\right] \tag{2.1.39}$$

which upon using the commutator

$$\left[\nabla^i, x^j\right] = \delta^{ij} \tag{2.1.40}$$

leads to

$$\frac{d}{d\varphi}\mathbf{A}(\varphi) = -\mathbf{n}\times\mathbf{A}(\varphi). \tag{2.1.41}$$

Using the boundary condition

$$\mathbf{A}(\varphi)\big|_{\varphi=0} = \mathbf{x} \tag{2.1.42}$$

and (2.1.38), together with the expressions (2.1.24), (2.1.26), one may infer the identity

$$\exp\left[-\boldsymbol{\omega}\cdot(\mathbf{x}\times\boldsymbol{\nabla})\right]\mathbf{x}\exp\left[\boldsymbol{\omega}\cdot(\mathbf{x}\times\boldsymbol{\nabla})\right] = \exp\left[i\boldsymbol{\omega}\cdot\mathbf{M}\right]\mathbf{x} \tag{2.1.43}$$

where $(\boldsymbol{\omega}\cdot\mathbf{M}\mathbf{x})^l = i\omega^j \varepsilon^{ljk} x^k$ in the sense of matrix multiplication.

2.2 Successive Galilean Transformations and the Closed Path

Consider two successive Galilean transformations written in matrix form:

2.2 Successive Galilean Transformations and the Closed Path

$$\bar{x} = Rx - a - vt \tag{2.2.1}$$

$$\bar{t} = t - \tau \tag{2.2.2}$$

$$\bar{\bar{x}} = \overline{R}\bar{x} - \bar{a} - \bar{v}\bar{t} \tag{2.2.3}$$

$$\bar{\bar{t}} = \bar{t} - \bar{\tau}. \tag{2.2.4}$$

These allow us to relate the final configuration to the initial one through

$$\bar{\bar{x}} = (\overline{R}R)x - (\bar{a} + \overline{R}a - \bar{v}\tau) - (\bar{v} + \overline{R}v)t \tag{2.2.5}$$

$$\bar{\bar{t}} = t - (\tau + \bar{\tau}). \tag{2.2.6}$$

A Galilean transformation is specified by the following quadruplet

$$(\tau, a, v, R). \tag{2.2.7}$$

From (2.2.5), (2.2.6), the following group properties easily follow:

(1)
$$(\bar{\tau}, \bar{a}, \bar{v}, \overline{R})(\tau, a, v, R) = (\tau + \bar{\tau}, \bar{a} + \overline{R}a - \bar{v}\tau, \bar{v} + \overline{R}v, \overline{R}R) \tag{2.2.8}$$

(2) The identity element I is given by

$$I = (0, 0, 0, \mathbf{1}) \tag{2.2.9}$$

(3) The inverse is worked out to be

$$(\tau, a, v, R)^{-1} = (-\tau, -R^{-1}(a + v\tau), -R^{-1}v, R^{-1}) \tag{2.2.10}$$

as one easily checks that

$$(\tau, a, v, R)^{-1}(\tau, a, v, R) = (\tau, a, v, R)(\tau, a, v, R)^{-1}$$

$$= (0, 0, 0, \mathbf{1}) \tag{2.2.11}$$

(4) Finally, we have the associativity rule

$$(\tau_3, a_3, v_3, R_3)[(\tau_2, a_2, v_2, R_2)(\tau_1, a_1, v_1, R_1)]$$

$$= [(\tau_3, a_3, v_3, R_3)(\tau_2, a_2, v_2, R_2)](\tau_1, a_1, v_1, R_1) \tag{2.2.12}$$

Of utmost importance for the subsequent analysis are the following successive transformations forming a closed path given by

$$(\tau, a, v, R) = (\tau_2, a_2, v_2, R_2)^{-1}(\tau_1, a_1, v_1, R_1)^{-1}$$

$$\times (\tau_2, a_2, v_2, R_2)(\tau_1, a_1, v_1, R_1) \tag{2.2.13}$$

represented pictorially by the box diagram

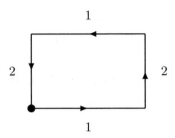

emphasizing the reversal of the transformations in the third and the fourth segments of the path.

The resulting elements in (τ, a, v, R) arising from the closed path transformation given in (2.2.13) are readily worked out from the group properties spelled out above. They are given by

$$\tau = 0 \tag{2.2.14}$$

$$a = -R_2^{-1}\left[a_2 + v_2\tau_2 + R_1^{-1}(a_1 - a_2 - R_2 a_1 + v_2\tau_1 - v_1\tau_2) - v_2\tau_2\right] \tag{2.2.15}$$

$$v = -R_2^{-1}\left[v_2 + R_1^{-1}(v_1 - v_2 - R_2 v_1)\right] \tag{2.2.16}$$

$$R = R_2^{-1} R_1^{-1} R_2 R_1. \tag{2.2.17}$$

The infinitesimal transformation corresponding to (τ, a, v, R) in (2.2.13) follows from (2.2.14)–(2.2.17). The corresponding infinitesimal parameters are readily obtained from the latter four equations and are given by

$$\delta\tau = 0 \tag{2.2.18}$$

$$\delta a = A_2 \delta a_1 - A_1 \delta a_2 + \delta v_1 \delta\tau_2 - \delta v_2 \delta\tau_1 \tag{2.2.19}$$

$$\delta v = A_2 \delta v_1 - A_1 \delta v_2 \tag{2.2.20}$$

$$\delta(R-1) = A_2 A_1 - A_1 A_2 \equiv [A_2, A_1] \tag{2.2.21}$$

where A is defined in (2.1.10) (see also (2.1.9)) with a corresponding infinitesimal $\delta\boldsymbol{\omega}$.

In detail (2.2.19)–(2.2.21) are given by

$$\delta\mathbf{a} = -(\delta\boldsymbol{\omega}_2 \times \delta\mathbf{a}_1 - \delta\boldsymbol{\omega}_1 \times \delta\mathbf{a}_2) + \delta\mathbf{v}_1 \delta\tau_2 - \delta\mathbf{v}_2 \delta\tau_1 \tag{2.2.22}$$

$$\delta\mathbf{v} = (\delta\boldsymbol{\omega}_1 \times \delta\mathbf{v}_2 - \delta\boldsymbol{\omega}_2 \times \delta\mathbf{v}_1) \tag{2.2.23}$$

$$\delta(R-1) = -\varepsilon^{ijk}\delta\omega^j \tag{2.2.24}$$

and

$$\delta\boldsymbol{\omega} = \delta\boldsymbol{\omega}_1 \times \delta\boldsymbol{\omega}_2. \tag{2.2.25}$$

Hence under the closed path transformation (2.2.13), we have the transformations $\mathbf{x} \to \mathbf{x}_f$, $t \to t_f = t$, leading to the infinitesimal expressions

$$\delta\mathbf{x} = \mathbf{x} - \mathbf{x}_f = \delta\mathbf{a} + \delta\mathbf{v}t + \delta\boldsymbol{\omega} \times \mathbf{x} \tag{2.2.26}$$

$$\delta t = t - t_f = 0 \tag{2.2.27}$$

(compare with (2.1.15), (2.1.16)), and the parameters $\delta\mathbf{a}$, $\delta\mathbf{v}$, $\delta\boldsymbol{\omega}$ are given, respectively, in (2.2.22), (2.2.23), (2.2.25).

2.3 Quantum Galilean Transformations and Their Generators

The Galilean transformations include the trivial identity transformation I in (2.2.9) which may be obtained from an arbitrary Galilean transformation, specified by the quadruplet (τ, a, v, R), when the parameters in τ, a, v, $(R-1)$ are led continuously go to zero. This continuity property of the transformations approaching the identity in the just discussed limit implies from the celebrated Wigner's Theorem (§1.9), that Galilean symmetry transformations on the states in quantum physics must be represented by *unitary* rather than by anti-unitary operators with the identity being obviously a (trivial) unitary operator. The transformations are then continuously connected with the identity.

Accordingly, invariance under a Galilean transformation dictates that a state $|\psi\rangle$ obeys the transformation rule $|\psi\rangle \to |\overline{\psi}\rangle$:

$$|\overline{\psi}\rangle = U|\psi\rangle \tag{2.3.1}$$

with

$$U^\dagger = U^{-1} \tag{2.3.2}$$

guaranteeing that a transition amplitude $\langle\phi|\psi\rangle$ transforms as a Galilean *scalar*, i.e.,

$$\langle\overline{\phi}|\overline{\psi}\rangle = \langle\phi|\psi\rangle. \tag{2.3.3}$$

For systems involving non-zero spins, $|\psi\rangle$ is a multi-component object (cf. (2.7.20), §2.8, (2.8.24)) and (2.3.3) involves a sum over such components, i.e., in detail it is of the form

90   2 Symmetries and Transformations

$$\sum_a \langle \overline{\phi}^a | \overline{\psi}^a \rangle = \sum_b \langle \phi^b | \psi^b \rangle. \qquad (2.3.4)$$

In the coordinate description, invariance under space translations and boosts, providing a uniform motion of a frame relative to another, but not involving rotations, means that

$$\langle \overline{\mathbf{x}} | \overline{\psi} \rangle = \langle \mathbf{x} | \psi \rangle, \quad \langle \overline{\mathbf{x}} | = \langle \mathbf{x} | U^\dagger \qquad (2.3.5)$$

and the former equality may be rewritten in the more familiar form

$$\overline{\psi}(\overline{\mathbf{x}}) = \psi(\mathbf{x}). \qquad (2.3.6)$$

By definition of a scalar, (2.3.6) holds under rotations as well for spin 0. On the other hand for non-zero spins, the transformation law in (2.3.6) cannot hold true under rotations. This is most apparent for spin 1, with $\psi(\mathbf{x})$ as a vector (field), since under rotations, a component $\overline{\psi}_j(\overline{\mathbf{x}})$ will be a linear combination of the components $\psi_i(\mathbf{x})$ (see (2.7.18)). This is fully exploited in §2.8 for arbitrary spins and the transformation law (2.3.6), under rotations, in the general case is simply modified to an expression of the form

$$\overline{\psi}^a(\overline{\mathbf{x}}) = A^{ab} \psi^b(\mathbf{x}) \qquad (2.3.7)$$

where the (A^{ab}) are some matrices which depend on spin.

As discussed above in the beginning of this section, since the Galilean transformations are continuously connected with the identity, then for infinitesimal transformations, one may write a unitary operator U as

$$U = \mathbf{1} + iG \qquad (2.3.8)$$

where G has the property that it vanishes when $\delta\tau$, δa, δv, $\delta \omega$ all approach zero (see, in particular, (2.1.15), (2.1.16)), and the i factor is chosen so that G is a self-adjoint operator $G^\dagger = G$.

The operator G in (2.3.8) is dimensionless, and the operator coefficients of $\delta\tau$, δa, δv, $\delta \omega$, respectively[1] in G have, what is called in quantum physics, *natural* units. These operators, as we will see later, have important physical meanings and are, in general, counterparts of classical quantities which, however, are defined in terms of different units. In order to make the comparison with these classical standards, an *overall conversion factor* is introduced by dividing G, in (2.3.8), by this conversion factor which necessarily has the dimensions of *action*. The operator coefficients of $\delta\tau$, δa, δv, $\delta \omega$ are then all defined in the same units as their classical standards. The unit of *action* in quantum physics found empirically is that provided by $\hbar$ (Planck's constant

---

[1] See (2.3.13) for details concerning these operator coefficients whose physical interpretations will then follow.

2.3 Quantum Galilean Transformations and Their Generators

divided by 2π).[2] By introducing this conversion factor we may rewrite (2.3.8) as

$$U = \mathbf{1} + \frac{i}{\hbar} G \qquad (2.3.9)$$

in conformity with the classical standards. For example, as we will see below, the coefficient of δa will be associated with momentum and hence by dividing by the unit of action $\hbar$ in (2.3.9) it will have the same units as in classical physics.

We consider successive Galilean transformations corresponding to the overall closed path in (2.2.13). That is, we consider the unitary operator

$$U = U_2^\dagger U_1^\dagger U_2 U_1 \qquad (2.3.10)$$

For infinitesimal transformations, the right-hand side of (2.3.10) is given by

$$\left(1 - \frac{i}{\hbar} G_2 + \ldots\right)\left(1 - \frac{i}{\hbar} G_1 + \ldots\right)\left(1 + \frac{i}{\hbar} G_2 + \ldots\right)\left(1 + \frac{i}{\hbar} G_1 + \ldots\right)$$

$$= \mathbf{1} + \frac{1}{\hbar^2}[G_1, G_2] + \ldots \qquad (2.3.11)$$

where we note that G_1 (respectively G_2) are of first order in the infinitesimal parameters $\delta\tau_1$, δa_1, δv_1, $\delta\omega_1$ (respectively $\delta\tau_2$, δa_2, δv_2, $\delta\omega_2$) and second order terms in these parameters, e.g. $(\delta\tau_1)^2$, $(\delta\tau_1 \delta a_1)$, ..., (respectively $(\delta\tau_2)^2$, $(\delta\tau_2 \delta a_2)$, ...), cancel out. The latter fact is emphasized by writing dots ... in each of the factors on the left-hand side of (2.3.11). Upon writing U on the left-hand of side of (2.3.10) as in (2.3.9) we obtain

$$G = \frac{1}{i\hbar}[G_1, G_2] \qquad (2.3.12)$$

The most general structure of G in (2.3.10) is

$$G = \delta\mathbf{a} \cdot \mathbf{P} + \delta\mathbf{v} \cdot \mathbf{N} - \delta\tau H + \delta\boldsymbol{\omega} \cdot \mathbf{J} + \delta\phi \mathbf{1} \qquad (2.3.13)$$

guaranteeing the fact that G goes to zero when $\delta\mathbf{a}$, $\delta\mathbf{v}$, $\delta\tau$, $\delta\boldsymbol{\omega}$ all go to zero provided $\delta\phi$, contributing to a phase factor, vanishes as well in these limits. The operators $\mathbf{P}$, $\mathbf{N}$, H, $\mathbf{J}$ are called *generators* of the Galilean transformations. $\mathbf{P}$ generates space translations, $\mathbf{N}$, which is sometimes referred as the "booster", generates uniform motion, H generates time translations, and $\mathbf{J}$ generates space rotations.

Similarly, for G_1, G_2 we may write

---

[2] The question arises as to what happens if one chooses another constant with units of action, such as a multiple of $\hbar$: $\kappa\hbar$ for some $\kappa > 0$, as a conversion factor. The energy levels of the hydrogen atom (§7.2), for example, would come out to be $-me^4/2\kappa^2\hbar^2 n^2$. By confronting theory with one or a finite number of experiments then fixes the value of this conversion factor.

2 Symmetries and Transformations

$$G_j = \delta\mathbf{a}_j \cdot \mathbf{P} + \delta\mathbf{v}_j \cdot \mathbf{N} - \delta\tau_j H + \delta\boldsymbol{\omega}_j \cdot \mathbf{J} \tag{2.3.14}$$

$j = 1, 2$. Clearly, the addition of any multiple of the identity in (2.3.14) cannot contribute to the commutator $[G_1, G_2]$ in (2.3.12).

We note that in (2.3.14) we have written $-\delta\tau_j H$ rather than $\delta\tau_j H$ as this provides, in particular, the correct overall sign for H corresponding to the observable it represents. [This is spelled out in §2.5.] G in (2.3.12) is anti-symmetric in the interchange of the indices 1 and 2 in $[G_1, G_2]$. This property is explicitly verified in the expressions for $\delta\mathbf{v}$, $\delta\tau$, $\delta\boldsymbol{\omega}$ respectively, given in (2.2.22), (2.2.23), (2.2.25). [Also we note from (2.2.18) that $\delta\tau = 0$] Accordingly, the coefficient $\delta\phi$ of the identity **1** in (2.3.13) must be also anti-symmetric in the indices 1 and 2 and is to be constructed out of dot products of $\delta\mathbf{a}_1$, $\delta\mathbf{v}_1$, $\delta\boldsymbol{\omega}_1$, $\delta\mathbf{a}_2$, $\delta\mathbf{v}_2$, $\delta\boldsymbol{\omega}_2$. The most general structure of $\delta\phi$ is then

$$\delta\phi = M\left(\delta\mathbf{a}_1 \cdot \delta\mathbf{v}_2 - \delta\mathbf{a}_2 \cdot \delta\mathbf{v}_1\right) + B\left(\delta\boldsymbol{\omega}_1 \cdot \delta\mathbf{v}_2 - \delta\boldsymbol{\omega}_2 \cdot \delta\mathbf{v}_1\right)$$
$$+ E\left(\delta\boldsymbol{\omega}_1 \cdot \delta\mathbf{a}_2 - \delta\boldsymbol{\omega}_2 \cdot \delta\mathbf{a}_1\right) \tag{2.3.15}$$

The coefficients E, B will turn out to be zero.

All told, we explicitly carry out the commutation relation in (2.3.12) by using the expressions in (2.3.14). Upon the comparison of the coefficients with those arising in (2.3.13), (2.3.15) and by using, in the process, the equalities (2.2.18)–(2.2.21), (2.2.22)–(2.2.25), we are led to Table 2.1.

From the commutation relation 6 in Table 2.1, it is easily derived that

$$E = -\frac{1}{6\hbar^2}\varepsilon^{ijk}\left[J^k, \left[P^i, J^j\right]\right] \tag{2.3.16}$$

which is readily worked out to be zero (see Problem 2.4). Similarly, from the commutator 7 in the Table, it is easily shown that $B = 0$.

Instead of transforming a state $|\psi\rangle$, under a Galilean transformation, one may keep $|\psi\rangle$ fixed and transform instead the observables $A \to \overline{A}$ such that expectation values

$$\langle\overline{\psi}|A|\overline{\psi}\rangle = \langle\psi|\overline{A}|\psi\rangle \tag{2.3.17}$$

are unaltered. From (2.3.1), this requires that

$$\overline{A} = U^\dagger A U. \tag{2.3.18}$$

[This may be generally referred to as an observable in a Heisenberg-like picture.] For infinitesimal transformations given in (2.3.9) we then have

$$\delta A = A - \overline{A} = \frac{1}{i\hbar}[A, G]. \tag{2.3.19}$$

Under a Galilean transformation we may then write from (2.3.13), (2.3.14)

$$i\hbar\delta A = \delta a^j\left[A, P^j\right] + \delta v^j\left[A, N^j\right]$$

2.3 Quantum Galilean Transformations and Their Generators

$$-\delta\tau\left[A,H\right]+\delta\omega^{j}\left[A,J^{j}\right]. \tag{2.3.20}$$

As an application of (2.3.20), we consider the operator

$$\mathbf{Q}=\frac{1}{M}\left(t\mathbf{P}-\mathbf{N}\right) \tag{2.3.21}$$

for $\delta\tau=0$. From (2.3.20) and Table 2.1

$$\delta\mathbf{Q}=\delta\mathbf{a}+\delta\mathbf{v}t+\delta\boldsymbol{\omega}\times\mathbf{Q} \tag{2.3.22}$$

which upon comparison with the transformation law in (2.1.15), one identifies $\mathbf{Q}$ with the position operator which, in the sequel, is denoted by $\mathbf{X}$.

Table 2.1 and/or (2.3.22) then lead to the following commutation relations involving $\mathbf{X}$:

$$\left[X^{i},X^{j}\right]=0 \tag{2.3.23}$$

$$\left[X^{i},J^{j}\right]=i\hbar\varepsilon^{ijk}X^{k} \tag{2.3.24}$$

$$\left[X^{i},P^{j}\right]=i\hbar\delta^{ij} \tag{2.3.25}$$

Table 2.1. Commutation relations of the generators of the Galilean transformations as obtained from (2.3.12) upon the comparison of the appropriate coefficient-parameters defining the transformations.

COEFFICIENT	RESULTING COMMUTATOR
1. $\delta a_1^i \delta a_2^j$	$\left[P^i, P^j\right] = 0$
2. $(\delta a_1^i \delta\tau_2 - \delta a_2^i \delta\tau_1)$	$\left[H, P^i\right] = 0$
3. $(\delta v_1^i \delta\tau_2 - \delta v_2^i \delta\tau_1)$	$\left[H, N^i\right] = i\hbar P^i$
4. $\delta\omega_1^i \delta\omega_2^j$	$\left[J^i, J^j\right] = i\hbar\varepsilon^{ijk} J^k$
5. $(\delta a_1^i \delta v_2^j - \delta a_2^i \delta v_1^j)$	$\left[P^i, N^j\right] = i\hbar M\delta^{ij}$
6. $(\delta a_1^i \delta\omega_2^j - \delta a_2^i \delta\omega_1^j)$	$\left[P^i, J^j\right] = i\hbar\left(\varepsilon^{ijk} P^k - E\delta^{ij}\right)$
7. $(\delta\omega_1^i \delta v_2^j - \delta\omega_2^i \delta v_1^j)$	$\left[N^i, J^j\right] = i\hbar\left(\varepsilon^{ijk} N^k - B\delta^{ij}\right)$
8. $(\delta\omega_1^i \delta\tau_2 - \delta\omega_2^i \delta\tau_1)$	$\left[H, J^i\right] = 0$
9. $\delta v_1^i \delta v_2^j$	$\left[N^i, N^j\right] = 0$

$$[H, X^i] = -\frac{i\hbar P^i}{M}. \tag{2.3.26}$$

On the other hand for $\delta \mathbf{a} = \mathbf{0}$, $\delta \mathbf{v} = \mathbf{0}$, $\delta \boldsymbol{\omega} = \mathbf{0}$, (2.3.20) gives for $\delta\tau \neq 0$

$$i\hbar\,\delta X^i = -\delta\tau[X^i, H] = -i\hbar\,\delta\tau\frac{P^i}{M} \tag{2.3.27}$$

where we have also used (2.3.26). Hence from $\delta X^i(t) = X^i(t) - \overline{X}^i(t) = X^i(t) - \overline{X}^i(\bar{t} + \delta t)$, (2.3.27) gives

$$\dot{\mathbf{X}} = \frac{\mathbf{P}}{M} \tag{2.3.28}$$

where we have used the fact that $\overline{\mathbf{X}}(\bar{t}) = \mathbf{X}(t)$, for $\delta\mathbf{a} = \mathbf{0}$, $\delta\mathbf{v} = \mathbf{0}$, $\delta\boldsymbol{\omega} = \mathbf{0}$.

Finally, we note from Table 2.1 again that

$$\left[\frac{\mathbf{P}^2}{2M}, X^i\right] = -i\hbar\frac{P^i}{M} = [H, X^i]. \tag{2.3.29}$$

Here in writing the last equality in (2.3.29) we have used (2.3.26). From (2.3.29) and the commutators 1 and 2 in Table 2.1 and (2.3.25) one is led to the following general expression for H

$$H = \frac{\mathbf{P}^2}{2M} + H_\mathrm{I} \tag{2.3.30}$$

where

$$[H_\mathrm{I}, P^i] = 0 \tag{2.3.31}$$

and

$$[H_\mathrm{I}, X^i] = 0. \tag{2.3.32}$$

The expressions (2.3.28), (2.3.30) lead inescapably one to identify $\mathbf{P}$ with the momentum operator associated with the system in question, M with its mass, and H with the total Hamiltonian operator.

The momentum operator $\mathbf{P}$ generates space translations. More specifically, for an infinitesimal numerical quantity $\delta\mathbf{a}$, the unitary operator

$$U(\delta\mathbf{a}\cdot\mathbf{P}) = 1 + \frac{i}{\hbar}\delta\mathbf{a}\cdot\mathbf{P} \tag{2.3.33}$$

leads from (2.3.25) (see also (2.3.18)) to

$$U^\dagger(\delta\mathbf{a}\cdot\mathbf{P})\mathbf{X}U(\delta\mathbf{a}\cdot\mathbf{P}) = \mathbf{X} - \delta\mathbf{a}. \tag{2.3.34}$$

Similarly,

$$U(-\delta\mathbf{p}\cdot\mathbf{X}) = 1 - \frac{i}{\hbar}\delta\mathbf{p}\cdot\mathbf{X} \tag{2.3.35}$$

2.3 Quantum Galilean Transformations and Their Generators

for an infinitesimal numerical $\delta \mathbf{p}$, leads from

$$U^\dagger(-\delta \mathbf{p} \cdot \mathbf{X})\mathbf{P}U(-\delta \mathbf{p} \cdot \mathbf{X}) = \mathbf{P} - \delta \mathbf{p} \qquad (2.3.36)$$

to a shift of the momentum operator $\mathbf{P}$ by the numerical quantity $-\delta \mathbf{p}$.

For a rotation by an infinitesimal angle $\delta\phi$ about a unit vector $\mathbf{n}$ ($\delta\boldsymbol{\omega} = \mathbf{n}\delta\phi$) we have also seen that

$$U(\delta\boldsymbol{\omega} \cdot \mathbf{J}) = 1 + \frac{i}{\hbar}\delta\boldsymbol{\omega} \cdot \mathbf{J} \qquad (2.3.37)$$

which leads to

$$U^\dagger(\delta\boldsymbol{\omega} \cdot \mathbf{J})\mathbf{X}U(\delta\boldsymbol{\omega} \cdot \mathbf{J}) = \mathbf{X} - \delta\boldsymbol{\omega} \times \mathbf{X} \qquad (2.3.38)$$

where we will see in §2.7, that $\mathbf{J}$ is identified with the angular momentum operator.

A finite space translation by a numerical quantity $\mathbf{a}$ may be obtained by writing $\delta\mathbf{a} = \mathbf{a}/N$, and consider successive N infinitesimal translations each by an amount $\mathbf{a}/N$ for $N \to \infty$:

$$\left(U\left(\frac{\mathbf{a}}{N} \cdot \mathbf{P}\right)\right)^N \to \exp\left(\frac{i}{\hbar}\mathbf{a} \cdot \mathbf{P}\right) \qquad (2.3.39)$$

(see also Problem 2.5).

Similarly upon writing $\delta\mathbf{p} = \mathbf{p}/N$, $\delta\boldsymbol{\omega} = \boldsymbol{\omega}/N$ we have for finite momentum shifts and finite rotations:

$$\left(U\left(-\frac{\mathbf{p}}{N} \cdot \mathbf{X}\right)\right)^N \to \exp\left(-\frac{i}{\hbar}\mathbf{p} \cdot \mathbf{X}\right) \qquad (2.3.40)$$

$$\left(U\left(\frac{\delta\boldsymbol{\omega}}{N} \cdot \mathbf{J}\right)\right)^N \to \exp\left(\frac{i}{\hbar}\boldsymbol{\omega} \cdot \mathbf{J}\right) \qquad (2.3.41)$$

for $N \to \infty$, respectively, where $\boldsymbol{\omega} = \mathbf{n}\phi$, for numericals $\mathbf{p}$.

That is, under a shift of a coordinate system by an amount $\mathbf{a}$, a state vector $|\psi\rangle$ changes in the following way

$$|\psi\rangle \to \exp\left(\frac{i}{\hbar}\mathbf{a} \cdot \mathbf{P}\right)|\psi\rangle . \qquad (2.3.42)$$

Similarly, when a coordinate system is rotated c.c.w. through an angle ϕ about a unit vector $\mathbf{n}$, a state vector $|\psi\rangle$ changes by the rule

$$|\psi\rangle \to \exp\left(\frac{i}{\hbar}\varphi\mathbf{n} \cdot \mathbf{J}\right)|\psi\rangle \equiv |\psi'\rangle . \qquad (2.3.43)$$

We now provide explicit representations of the generators $\mathbf{P}$, $\mathbf{X}$.

In the **X**-description, under an infinitesimal space translation (see (2.3.5))

$$\langle \overline{\mathbf{x}} | \overline{\psi} \rangle = \langle \mathbf{x} | \psi \rangle \tag{2.3.44}$$

where
$$|\overline{\psi}\rangle = U |\psi\rangle \tag{2.3.45}$$

with
$$\langle \overline{\mathbf{x}} | = \langle \mathbf{x} - \delta\mathbf{a} | \tag{2.3.46}$$

giving
$$\langle \mathbf{x} - \delta\mathbf{a} | U\psi \rangle = \langle \mathbf{x} | \psi \rangle \tag{2.3.47}$$

or
$$(U\psi)(\mathbf{x}) = \psi(\mathbf{x} + \delta\mathbf{a}). \tag{2.3.48}$$

A Taylor expansion of the right-hand side expression leads to

$$(U\psi)(\mathbf{x}) = \psi(\mathbf{x}) + \delta\mathbf{a} \cdot \boldsymbol{\nabla}\psi(\mathbf{x}) \tag{2.3.49}$$

from which we may infer that

$$U = 1 + \delta\mathbf{a} \cdot \boldsymbol{\nabla} \tag{2.3.50}$$

leading to the following representation for the momentum operator

$$\mathbf{P} = -i\hbar \boldsymbol{\nabla}. \tag{2.3.51}$$

Similarly, in the **P**-description, under a momentum shift (translation),

$$\langle \overline{\mathbf{p}} | \overline{\psi} \rangle = \langle \mathbf{p} | \psi \rangle, \qquad \overline{\mathbf{p}} = \mathbf{p} - \delta\mathbf{p} \tag{2.3.52}$$

leading to
$$(U\psi)(\mathbf{p}) = \psi(\mathbf{p} + \delta\mathbf{p}) \tag{2.3.53}$$

and from a Taylor expansion, we may infer that

$$U = 1 + \delta\mathbf{p} \cdot \boldsymbol{\nabla}_{\mathbf{p}} \tag{2.3.54}$$

thus giving the following representation for the position operator (see (2.3.35))

$$\mathbf{X} = i\hbar \boldsymbol{\nabla}_{\mathbf{p}}. \tag{2.3.55}$$

Generation of time translations and coordinate rotations will be dealt with in detail later in §2.5 and §2.7, respectively.

Space translations are provided by the unitary operator (see (2.3.42))

$$U = \exp\left(\frac{i}{\hbar} \mathbf{a} \cdot \mathbf{P}\right). \tag{2.3.56}$$

Invariance under space translations implies the vanishing of the commutator of **P** with H (see entry 2 in Table 2.1) and the transformation provided by

2.3 Quantum Galilean Transformations and Their Generators

the unitary operator in (2.3.56) is referred to as a *symmetry* transformation. The latter, in particular means, that the momentum operator satisfies the time independent property

$$\mathbf{P}(t) = e^{iH/\hbar}\, \mathbf{P}\, e^{-iH/\hbar} = \mathbf{P} \qquad (2.3.57)$$

and implies the *conservation* of momentum. In particular, (2.3.57) establishes (see also (2.3.17)) the time independence of the expectation value

$$\langle \psi | \mathbf{P}(t) | \psi \rangle = \langle \psi | \mathbf{P} | \psi \rangle \qquad (2.3.58)$$

in a state $|\psi\rangle$.

Similarly, rotation of a coordinate system, is implemented by the unitary operator (see (2.3.43))

$$U = \exp\left(\frac{i}{\hbar} \varphi \mathbf{n} \cdot \mathbf{J}\right) \qquad (2.3.59)$$

in the underlying Hilbert space. As a symmetry operation, rotational invariance, leading to the vanishing of the commutator of $\mathbf{J}$ with H (entry 8 in Table 2.1), implies the time *independence* of $\mathbf{J}(t)$, similarly defined as $\mathbf{P}(t)$ in (2.3.57), and gives rise to the *conservation* of angular momentum, and, in particular, to the time-independence of expectation values $\langle \psi | \mathbf{J}(t) | \psi \rangle$.

Of particular interest in using (2.3.20) is to derive the equations satisfied by $\mathbf{X}(t)$ and $\mathbf{P}(t)$, under pure time translations. These may be obtained from (2.3.20) by setting $\delta \mathbf{a} = \mathbf{0}$, $\delta \mathbf{v} = \mathbf{0}$, $\delta \boldsymbol{\omega} = \mathbf{0}$. For example,

$$\frac{i}{\hbar}[\mathbf{X}(t), H]\delta\tau = \delta \mathbf{X}(t) = \mathbf{X}(t) - \overline{\mathbf{X}}(t)$$

$$= \mathbf{X}(t) - \overline{\mathbf{X}}(\bar{t} + \delta\tau) \qquad (2.3.60)$$

or

$$i\hbar \dot{\mathbf{X}}(t) = [\mathbf{X}(t), H] \qquad (2.3.61)$$

since $\mathbf{X}(t) = \overline{\mathbf{X}}(\bar{t})$. Similarly, we have

$$i\hbar \dot{\mathbf{P}}(t) = [\mathbf{P}(t), H]. \qquad (2.3.62)$$

Equations (2.3.61), (2.3.62) are referred to as *Heisenberg's Equations of Motion*.

Finally, consider a system of n particles of masses $m_1, \ldots, m_n$ with associated position operators $\mathbf{X}_1, \ldots, \mathbf{X}_n$ and define the center of mass position operator

$$\mathbf{X} = \sum_\alpha \frac{m_\alpha}{M} \mathbf{X}_\alpha \qquad (2.3.63)$$

where $M = \sum_\alpha m_\alpha$ is the sum of the masses of the particles. Invariance under an infinitesimal coordinate translation means as before

$$\langle \overline{\mathbf{x}}_1, \ldots, \overline{\mathbf{x}}_n | \overline{\psi} \rangle = \langle \mathbf{x}_1, \ldots, \mathbf{x}_n | \psi \rangle. \tag{2.3.64}$$

This leads to the following chain of equalities

$$\langle \overline{\mathbf{x}}_1 - \delta \mathbf{a}, \ldots, \overline{\mathbf{x}}_n - \delta \mathbf{a} | \overline{\psi} \rangle = \langle \mathbf{x}_1, \ldots, \mathbf{x}_n | \psi \rangle \tag{2.3.65}$$

$$\langle \mathbf{x}_1, \ldots, \mathbf{x}_n | U \psi \rangle = \langle \mathbf{x}_1 + \delta \mathbf{a}, \ldots, \mathbf{x}_n + \delta \mathbf{a} | \psi \rangle \tag{2.3.66}$$

$$(U\psi)(\mathbf{x}_1, \ldots, \mathbf{x}_n) = \psi(\mathbf{x}_1 + \delta \mathbf{a}, \ldots, \mathbf{x}_n + \delta \mathbf{a})$$

$$= \left(1 + \delta \mathbf{a} \cdot \sum_\alpha \boldsymbol{\nabla}_\alpha\right) \psi(\mathbf{x}_1, \ldots, \mathbf{x}_n). \tag{2.3.67}$$

From (2.3.50), (2.3.51), we may introduce the total momentum operator

$$\mathbf{P} = \sum_\alpha \mathbf{P}_\alpha \tag{2.3.68}$$

with

$$\mathbf{P}_\alpha = -i\hbar \boldsymbol{\nabla}_\alpha \tag{2.3.69}$$

denoting the momentum operator associated with the αth particle as expected. A property consistent with (2.3.23), (2.3.63), is that the operators $\mathbf{X}_\alpha$ associated with different particles commute as well,

$$\left[X_\alpha^i, X_\beta^j\right] = 0. \tag{2.3.70}$$

From (2.3.69), one also has

$$\left[X_\alpha^i, P_\beta^j\right] = i\hbar \delta^{ij} \delta_{\alpha\beta} \tag{2.3.71}$$

and

$$\left[P_\alpha^i, P_\beta^j\right] = 0. \tag{2.3.72}$$

2.4 The Transformation Function $\langle \mathbf{x} | \mathbf{p} \rangle$

In this short section we derive the explicit expression for the transformation function $\langle \mathbf{x} | \mathbf{p} \rangle$ from the momentum: **p**-description to the position: **x**-description.

The resolution of the identity in the **p**-description is written as

$$1 = \int \frac{d^3 \mathbf{p}}{(2\pi\hbar)^3} |\mathbf{p}\rangle \langle \mathbf{p}|. \tag{2.4.1}$$

The transformation function $\langle \mathbf{x} | \mathbf{p} \rangle$ from the **p**-description to the **x**-description arises through (see also (1.4.17)) the relation

2.4 The Transformation Function $\langle \mathbf{x}|\mathbf{p}\rangle$

$$\langle \mathbf{x}|\psi\rangle = \int \frac{d^3 \mathbf{p}}{(2\pi\hbar)^3} \langle \mathbf{x}|\mathbf{p}\rangle \langle \mathbf{p}|\psi\rangle. \tag{2.4.2}$$

We consider infinitesimal shifts by numerical factors $\delta \mathbf{x}$, $\delta \mathbf{p}$: $\mathbf{x} \to \mathbf{x} - \delta \mathbf{x}$, $\mathbf{p} \to \mathbf{p} - \delta \mathbf{p}$ provided through unitary operators (see §2.3, (2.3.5), (2.3.46), (2.3.39)) explicitly given by

$$\langle \mathbf{x} - \delta \mathbf{x}| = \langle \mathbf{x}| \left(1 - \frac{i}{\hbar} \delta \mathbf{x} \cdot \mathbf{P}\right) \tag{2.4.3}$$

and (see (2.3.51), (2.3.52), (2.3.54))

$$|\mathbf{p} - \delta \mathbf{p}\rangle = \left(1 - \frac{i}{\hbar} \delta \mathbf{p} \cdot \mathbf{X}\right) |\mathbf{p}\rangle. \tag{2.4.4}$$

The formal identifications

$$\mathbf{P}|\mathbf{p}\rangle = \mathbf{p}|\mathbf{p}\rangle \tag{2.4.5}$$

$$\langle \mathbf{x}|\mathbf{X} = \mathbf{x}\langle \mathbf{x}| \tag{2.4.6}$$

then lead for infinitesimal numericals $\delta \mathbf{x}$, $\delta \mathbf{p}$ to

$$\delta \langle \mathbf{x}|\mathbf{p}\rangle = \langle \mathbf{x}|\mathbf{p}\rangle - \langle \mathbf{x} - \delta \mathbf{x}|\mathbf{p} - \delta \mathbf{p}\rangle$$

$$= \frac{i}{\hbar} \langle \mathbf{x}|(\delta \mathbf{x} \cdot \mathbf{P} + \delta \mathbf{p} \cdot \mathbf{X})|\mathbf{p}\rangle$$

$$= \frac{i}{\hbar} (\delta \mathbf{x} \cdot \mathbf{p} + \delta \mathbf{p} \cdot \mathbf{x}) \langle \mathbf{x}|\mathbf{p}\rangle$$

$$= \frac{i}{\hbar} \langle \mathbf{x}|\mathbf{p}\rangle \delta(\mathbf{x} \cdot \mathbf{p}) \tag{2.4.7}$$

and upon integration to

$$\langle \mathbf{x}|\mathbf{p}\rangle = \exp\left(\frac{i}{\hbar} \mathbf{x} \cdot \mathbf{p}\right) \tag{2.4.8}$$

where the integration constant has been set equal to one by adopting the normalization condition

$$\int \frac{d^3 \mathbf{p}}{(2\pi\hbar)^3} \langle \mathbf{x}|\mathbf{p}\rangle \langle \mathbf{p}|\mathbf{x}'\rangle = \delta^3(\mathbf{x} - \mathbf{x}'). \tag{2.4.9}$$

The importance of the expression (2.4.8) for the transformation function $\langle \mathbf{x}|\mathbf{p}\rangle$ cannot be overemphasized. From (2.4.2) it leads to the Fourier transform

$$\psi(\mathbf{x}) = \int \frac{d^3 \mathbf{p}}{(2\pi\hbar)^3} e^{i\mathbf{x}\cdot\mathbf{p}/\hbar} \psi(\mathbf{p}) \tag{2.4.10}$$

as a physical transformation from the **p**-description to the **x**-description.

The inverse transform from the **x**-description to the **p**-description similarly follows to be given by

$$\psi(\mathbf{p}) = \int d^3\mathbf{x}\, e^{-i\mathbf{x}\cdot\mathbf{p}/\hbar}\, \psi(\mathbf{x}) \tag{2.4.11}$$

with

$$\langle \mathbf{p}|\mathbf{x}\rangle = \exp\left(-\frac{i}{\hbar}\mathbf{x}\cdot\mathbf{p}\right). \tag{2.4.12}$$

Finally, the resolution of the identity in the **x**-description may be written as

$$\mathbf{1} = \int d^3\mathbf{x}\, |\mathbf{x}\rangle\,\langle\mathbf{x}| \tag{2.4.13}$$

with the normalization condition (2.4.9):

$$\langle \mathbf{x}'|\mathbf{x}\rangle = \delta^3(\mathbf{x}'-\mathbf{x}). \tag{2.4.14}$$

From (2.4.8), (2.4.13), we then have the following normalization condition for momenta

$$\langle \mathbf{p}'|\mathbf{p}\rangle = (2\pi\hbar)^3 \delta^3(\mathbf{p}'-\mathbf{p}). \tag{2.4.15}$$

2.5 Quantum Dynamics and Construction of Hamiltonians

2.5.1 The Time Evolution: Schrödinger Equation

Had we written $+\delta\tau_j H$ instead of $-\delta\tau_j H$ in (2.3.14), the commutator $[H, N^j]$ in Table 2.1 would have been equal to $-i\hbar P^i$ instead of $+i\hbar P^i$. This in turn would have led H in (2.3.30) to be of the form $-\left[\mathbf{P}^2/2M + H'_I\right]$, where H'_I is some operator which satisfies (2.3.31), (2.3.32) [as before, $\mathbf{P}^2$ does *not* commute with $\mathbf{X}$], with the wrong sign for the kinetic energy term. That is, for an infinitesimal transformation $t \to t - \delta\tau$ (see (2.1.16), (2.1.2)) the corresponding unitary transformation must be given from (2.3.14) to be

$$U(-\delta\tau H) = \mathbf{1} - \frac{i}{\hbar}\delta\tau H \tag{2.5.1}$$

with H denoting the Hamiltonian of the system having the general form (see (2.3.30)–(2.3.32))

$$H = \frac{\mathbf{P}^2}{2M} + H_I \tag{2.5.2}$$

where H_I is referred to as the internal energy and satisfies

$$[H_I, \mathbf{P}] = 0 \tag{2.5.3}$$

2.5 Quantum Dynamics and Construction of Hamiltonians

$$[H_I, \mathbf{X}] = \mathbf{0} \tag{2.5.4}$$

for a Galilean *invariant* theory.

To obtain the unitary operator for finite time translations we write $\delta\tau = \tau/N$, as in §2.3, (see, e.g., (2.3.39)) and consider successive infinitesimal transformations for $N \to \infty$:

$$\left(U\left(-\frac{\tau}{N}H\right)\right)^N \to \exp\left(-\frac{i}{\hbar}\tau H\right). \tag{2.5.5}$$

Let $|\psi_t\rangle$ denote a state determined at time t in the frame F (see §2.1–§2.3). For infinitesimal $\delta\tau$ this state is given by $|\overline{\psi}_t\rangle$ in $\overline{F}$:

$$|\overline{\psi}_t\rangle = \left(1 - \frac{i}{\hbar}\delta\tau H\right)|\psi_t\rangle. \tag{2.5.6}$$

For pure time translations, we recall from §2.1, that the frames F and $\overline{F}$ are the *same*, except that time readings are carried out, at the *same* instant of time, by two clocks with one set simply $-\delta\tau$ units of time back relative to the other. That is, $|\overline{\psi}_t\rangle$ denotes the state $|\psi_{t+\delta\tau}\rangle$. Hence from (2.5.6)

$$|\psi_{t+\delta\tau}\rangle = \left(1 - \frac{i}{\hbar}\delta\tau H\right)|\psi_t\rangle \tag{2.5.7}$$

or

$$|\psi_{t+\delta\tau}\rangle - |\psi_t\rangle = -\frac{i}{\hbar}\delta\tau H |\psi_t\rangle$$

which by taking the limit $\delta\tau \to 0$ gives the familiar *Schrödinger* equation

$$i\hbar \frac{\partial}{\partial t}|\psi_t\rangle = H|\psi_t\rangle. \tag{2.5.8}$$

In particular, by setting $|\psi_0\rangle \equiv |\psi\rangle$ for $t = 0$, (2.5.8) may be integrated to give

$$|\psi_t\rangle = e^{-itH/\hbar}|\psi\rangle \tag{2.5.9}$$

for the time evolution of a state $|\psi\rangle$.

2.5.2 Time as an Operator?

It is important to emphasize that time is a parameter and is not promoted to an operator on physical grounds. To see this, suppose that t_{op} stands for the corresponding "operator". Then for $\delta\mathbf{a} = \mathbf{0}$, $\delta\mathbf{v} = \mathbf{0}$, $\delta\boldsymbol{\omega} = \mathbf{0}$, and numerical $\delta\tau \neq 0$, (2.3.20) implies that

$$i\hbar\, \delta\tau\, t_{op} = -\delta\tau [t_{op}, H] \tag{2.5.10}$$

102   2 Symmetries and Transformations

which upon comparison with (2.1.16) leads to[3]

$$[t_{\mathrm{op}}, H] = -i\hbar. \tag{2.5.11}$$

This then implies that for any real numerical E

$$\exp\left(\frac{i}{\hbar}Et_{\mathrm{op}}\right) H \exp\left(-\frac{i}{\hbar}Et_{\mathrm{op}}\right) = H + E. \tag{2.5.12}$$

Now let λ_0 be in the spectrum of H. Whether λ_0 is in the discrete or in the continuous spectrum, given any $\varepsilon > 0$ (see §1.8, (1.8.34)), we can find a vector $|\psi_\varepsilon\rangle$ such that

$$\|(H - \lambda_0)\psi_\varepsilon\| < \varepsilon. \tag{2.5.13}$$

Upon setting $|\phi_\varepsilon\rangle = \exp\left(\frac{i}{\hbar}Et_{\mathrm{op}}\right)|\psi_\varepsilon\rangle$, (2.5.13) in turn implies that

$$\left\|\left[\exp\left(\frac{i}{\hbar}Et_{\mathrm{op}}\right) H \exp\left(-\frac{i}{\hbar}Et_{\mathrm{op}}\right) - \lambda_0\right]\phi_\varepsilon\right\| < \varepsilon \tag{2.5.14}$$

or from (2.5.12) that

$$\|[H - (\lambda_0 - E)]\phi_\varepsilon\| < \varepsilon. \tag{2.5.15}$$

That is, $\lambda_0 - E$ is also in the spectrum of H for all E. One may choose E, in particular, an arbitrarily large positive number implying that H is unbounded from below — a physically unacceptable property of a Hamiltonian and for the stability of the system in question. That is, time is to be treated as a parameter and not as an operator. [Some authors have, nevertheless, suggested promoting time, under some circumstances, to the status of an operator, but we will no go into this here.] Special attention will be given later in §3.3, §3.4 to the boundedness of Hamiltonians from below and the corresponding stability problem.

2.5.3 Construction of Hamiltonians

We now proceed to construct Hamiltonians consistent with (2.5.2)–(2.5.4).

To this end consider a system of particles with masses m_α, associated position and momentum operators $\mathbf{X}_\alpha$ and $\mathbf{P}_\alpha$, respectively, where $\alpha = 1, \ldots, n$ (see (2.3.63)–(2.3.72)). The operators $\mathbf{X}_\alpha$, $\mathbf{P}_\alpha$ associated with different particles commute.

With $\mathbf{X}$ representing the position of the whole system, that is of the center of mass position operator, $\mathbf{P}$ its total momentum and M the sum of the masses, we here record the definitions

---

[3] A "relativistic version": $[P^\mu, X^\nu] = -i\hbar g^{\mu\nu}$ for which $(g^{\mu\nu}) = \mathrm{diag}[-1, 1, 1, 1]$, or $[P^\mu, X^\nu] = i\hbar g^{\mu\nu}$ for which $(g^{\mu\nu}) = \mathrm{diag}[1, -1, -1, -1]$, also gives (2.5.11) with $P^0 = H/c$, $X^0 = ct_{\mathrm{op}}$ and c denoting the speed of light. One, however, does *not* need to use this to obtain (2.5.11) and our result follows *directly* from (2.3.20) and the identification in (2.1.16).

2.5 Quantum Dynamics and Construction of Hamiltonians

$$\mathbf{X} = \sum_\alpha \frac{m_\alpha}{M} \mathbf{X}_\alpha \tag{2.5.16}$$

$$\mathbf{P} = \sum_\alpha \mathbf{P}_\alpha \tag{2.5.17}$$

$$M = \sum_\alpha m_\alpha. \tag{2.5.18}$$

In Table 2.2, we consider commutation relations of various combinations of the particles' operators with P^j and X^j.

Table 2.2. Commutation relations of various combinations of the particles' operators with P^j and X^j.

1. $\left[\left(X_\alpha^i - X_\beta^i\right), P^j\right] = \sum_\rho \left(\left[X_\alpha^i, P_\rho^j\right] - \left[X_\beta^i, P_\rho^j\right]\right)$
   $= i\hbar\delta^{ij} \sum_\rho (\delta_{\alpha\rho} - \delta_{\beta\rho}) = 0$

2. $\left[\left(X_\alpha^i - X_\beta^i\right), X^j\right] = \sum_\rho \frac{m_\rho}{M} \left[\left(X_\alpha^i - X_\beta^i\right), X_\rho^j\right] = 0$

3. $\left[\left(X_\alpha^i - X^i\right), P^j\right] = \sum_\rho \left[X_\alpha^i, P_\rho^j\right] - i\hbar\delta^{ij} = 0$

4. $\left[\left(X_\alpha^i - X^i\right), X^j\right] = 0$

5. $\left[\left(P_\alpha^i - P_\beta^i\right), P^j\right] = 0$

6. $\left[\left(P_\alpha^i - P_\beta^i\right), X^j\right] = \sum_\rho \frac{m_\rho}{M} \left(\left[P_\alpha^i, X_\rho^j\right] - \left[P_\beta^i, X_\rho^j\right]\right)$
   $= -\frac{i\hbar}{M} (m_\alpha - m_\beta) \delta^{ij}$

7. $\left[\left(P_\alpha^i - \frac{m_\alpha}{M} P^i\right), P^j\right] = 0$

8. $\left[\left(P_\alpha^i - \frac{m_\alpha}{M} P^i\right), X^j\right] = \left(-i\hbar\frac{m_\alpha}{M} + i\hbar\frac{m_\alpha}{M}\right) \delta^{ij} = 0$

From the Table, we see that H_{I} in (2.5.2), *consistent* with (2.5.3) and (2.5.4), may be chosen, in particular, a function of: $(\mathbf{X}_\alpha - \mathbf{X}_\beta)$, $(\mathbf{X}_\alpha - \mathbf{X})$, $\mathbf{P}_\alpha - (m_\alpha/M)\mathbf{P}$.

That is, for a very large class of Hamiltonians *consistent* with (2.5.2)–(2.5.4), we may choose

$$H = \frac{\mathbf{P}^2}{2M} + \sum_\alpha^n \frac{\left(\mathbf{P}_\alpha - \frac{m_\alpha}{M}\mathbf{P}\right)^2}{2m_\alpha} + V \tag{2.5.19}$$

where
$$V = \sum_{\alpha \neq \beta}^{n} V_{\alpha\beta}(\mathbf{X}_\alpha - \mathbf{X}_\beta) + \sum_{\alpha}^{n} V_\alpha(\mathbf{X}_\alpha - \mathbf{X}). \tag{2.5.20}$$

Additional conditions may be spelled out for V by invoking, for example, rotational invariance as given in entry 8 in Table 2.1.

In the remaining part of this section, we work out some details of the class of Hamiltonians in (2.5.19) of physical interest.

2.5.4 Multi-Particle Hamiltonians

In detail, we may write

$$\sum_{\alpha}^{n} \frac{(\mathbf{P}_\alpha - \frac{m_\alpha}{M}\mathbf{P})^2}{2m_\alpha} = \sum_{\alpha}^{n} \frac{\mathbf{P}_\alpha^2}{2m_\alpha} - \frac{1}{M}\mathbf{P} \cdot \sum_{\alpha}^{n} \mathbf{P}_\alpha + \frac{\mathbf{P}^2}{2M}$$

$$= \sum_{\alpha}^{n} \frac{\mathbf{P}_\alpha^2}{2m_\alpha} - \frac{\mathbf{P}^2}{2M}. \tag{2.5.21}$$

That is, the expression on the left-hand side amounts automatically in removing the center of mass motion, and (2.5.19) leads to a familiar expression

$$H = \sum_{\alpha}^{n} \frac{\mathbf{P}_\alpha^2}{2m_\alpha} + V \tag{2.5.22}$$

where V has the structure in (2.5.20).

For example, for a system of charged particles of charges q_α and masses m_α, $\alpha = 1, \ldots, n$, with Coulomb interactions, in the coordinate description, one has

$$H = \sum_{\alpha}^{n} \frac{\mathbf{P}_\alpha^2}{2m_\alpha} + \frac{1}{2} \sum_{\alpha \neq \beta}^{n} \frac{q_\alpha q_\beta}{|\mathbf{x}_\alpha - \mathbf{x}_\beta|}. \tag{2.5.23}$$

It is important to realize that the Hamiltonian $(H' \equiv H_{\text{I}})$:

$$H' = \sum_{\alpha}^{n} \frac{(\mathbf{P}_\alpha - \frac{m_\alpha}{M}\mathbf{P})^2}{2m_\alpha} + V \tag{2.5.24}$$

provides, and is appropriately referred to as, the *internal* energy as it has, as seen in (2.5.21), the center of mass motion *removed*.

2.5.5 Two-Particle Systems and Relative Motion

We consider the structure given (2.5.19) for $n = 2$, in a coordinate description, and consider the motion of one particle relative to the other. To this end, we write

2.5 Quantum Dynamics and Construction of Hamiltonians

$$\mathbf{r} = \mathbf{x}_1 - \mathbf{x}_2 \tag{2.5.25}$$

$$\mathbf{R} = \frac{1}{M}(m_1\mathbf{x}_1 + m_2\mathbf{x}_2) \tag{2.5.26}$$

$$M = m_1 + m_2. \tag{2.5.27}$$

It is easily verified that

$$\nabla_1 - \frac{m_1}{M}\nabla_\mathbf{R} = \nabla_\mathbf{r} \tag{2.5.28}$$

$$\nabla_2 - \frac{m_2}{M}\nabla_\mathbf{R} = -\nabla_\mathbf{r} \tag{2.5.29}$$

which directly lead from (2.5.19) to the expression

$$H = -\frac{\hbar^2}{2M}\nabla_\mathbf{R}^2 - \frac{\hbar^2}{2\mu}\nabla_\mathbf{r}^2 + V(\mathbf{r}) \tag{2.5.30}$$

with

$$\frac{1}{\mu} = \frac{1}{m_1} + \frac{1}{m_2} \tag{2.5.31}$$

and μ defining the so-called reduced mass of the system of two particles. Here we note that since $\mathbf{x}_1 - \mathbf{R} = m_2\mathbf{r}/M$, $\mathbf{x}_2 - \mathbf{R} = -m_1\mathbf{r}/M$, $V(\mathbf{r})$ is some function of $\mathbf{r}$. The first term in (2.5.30) describes the center of mass free motion. The remaining part describes the relative motion of the two particles.

2.5.6 Multi-Electron Atoms with Positions of the Electrons Defined Relative to the Nucleus

Consider a system of N electrons with

$$m_\alpha \equiv m, \quad \alpha = 1, \ldots, N \tag{2.5.32}$$

and a nucleus of mass $m_{N+1} \equiv m_0$.

That is, $M = Nm + m_0$. We consider a coordinate description of the system. To this end we define the variables

$$\mathbf{r}_\alpha = \mathbf{x}_\alpha - \mathbf{x}_{N+1}, \quad \alpha = 1, \ldots, N \tag{2.5.33}$$

$$\mathbf{R} = \frac{1}{M}\left(m(\mathbf{x}_1 + \ldots + \mathbf{x}_N) + m_0\mathbf{x}_{N+1}\right). \tag{2.5.34}$$

It is readily obtained that

$$\nabla_{\mathbf{x}_\alpha} - \frac{m}{M}\nabla_\mathbf{R} = \nabla_{\mathbf{r}_\alpha}, \quad \alpha = 1, \ldots, N \tag{2.5.35}$$

$$\nabla_{\mathbf{x}_{N+1}} - \frac{m_0}{M}\nabla_{\mathbf{R}} = -\sum_{\alpha=1}^{N}\nabla_{\mathbf{r}_\alpha}. \tag{2.5.36}$$

The expression in (2.5.19) then gives for the multi-electron atom in the coordinate description, the Hamiltonian

$$H = \frac{\mathbf{P}^2}{2M} + \sum_{\alpha=1}^{N}\frac{\mathbf{P}_\alpha^2}{2m} + \frac{1}{2m_0}\left(\sum_{\alpha=1}^{N}\mathbf{P}_\alpha\right)^2$$

$$+ \frac{1}{2}\sum_{\alpha\neq\beta}^{N}\frac{e^2}{|\mathbf{r}_\alpha - \mathbf{r}_\beta|} - \sum_{\alpha=1}^{N}\frac{Ze^2}{|\mathbf{r}_\alpha|} \tag{2.5.37}$$

where $Z|e|$ is the charge of the nucleus. For a neutral atom $N = Z$. Here $\mathbf{P}_\alpha = -i\hbar\nabla_{\mathbf{r}_\alpha}$. Since $m_0, M \gg m$, one, in practical applications, may neglect the first and the third terms on the right-hand side of (2.5.37). In any case, since $\mathbf{P}^2$ and $\left(\sum_\alpha^N \mathbf{P}_\alpha\right)^2$ are positive operators, that is, their expectation values are non-negative, we have the following lower bound for atoms

$$\langle\psi|H|\psi\rangle \geqslant \langle\psi|H_{\mathrm{AT}}|\psi\rangle \tag{2.5.38}$$

for the expectation values, where

$$H_{\mathrm{AT}} = \sum_{\alpha=1}^{N}\frac{\mathbf{P}_\alpha^2}{2m} + \frac{1}{2}\sum_{\alpha\neq\beta}^{N}\frac{e^2}{|\mathbf{r}_\alpha - \mathbf{r}_\beta|} - \sum_{\alpha=1}^{N}\frac{Ze^2}{|\mathbf{r}_\alpha|}. \tag{2.5.39}$$

2.5.7 Decompositions into Clusters of Particles

As a final application of the expression (2.5.19) for a Hamiltonian, we consider the class of two-body particle-particle interactions given by the first term in (2.5.20). The interest here consists in grouping the n particles into k disjoint subsets of particles, which we refer to as clusters, with the first containing n_1 particles,..., and with the kth containing n_k particles.

The above grouping is most convenient if one is interested in studying the properties of the different clusters as separate entities. In a scattering experiment, for example, one may have, as a final state of the process, particles emerging (within experimental limitations) into localized clusters (with non-vanishing intra-clusteral interactions) which are widely separated one from the other (with negligible inter-clusteral interactions). [Intra-clusteral interactions refer to interactions occurring between particles within the same cluster, while inter-clusteral referring to interactions between *different* clusters, due to particle-particle interactions with the particles belonging to the different clusters.] That is, these clusters of particles emerge as separate entities and the properties of some or all of them may explored.

2.5 Quantum Dynamics and Construction of Hamiltonians

The positions of the particles in the clusters are conveniently labelled as

$$\{\mathbf{x}_{11}, \ldots, \mathbf{x}_{1n_1}\}, \ldots, \{\mathbf{x}_{k1}, \ldots, \mathbf{x}_{kn_k}\} \qquad (2.5.40)$$

and their masses as

$$\{m_{11}, \ldots, m_{1n_1}\}, \ldots, \{m_{k1}, \ldots, m_{kn_k}\}. \qquad (2.5.41)$$

The center of mass positions of the clusters are then defined by

$$\mathbf{R}_\alpha = \frac{1}{M_\alpha} \sum_{\beta=1}^{n_\alpha} m_{\alpha\beta} \mathbf{x}_{\alpha\beta} \qquad (2.5.42)$$

with

$$M_\alpha = \sum_{\beta=1}^{n_\alpha} m_{\alpha\beta} \qquad (2.5.43)$$

$\alpha = 1, \ldots, k$, and M_α denoting the sum of the masses of the particles in the αth cluster. For the center of mass position of the k clusters we have

$$\mathbf{R} = \frac{1}{M} \sum_{\alpha=1}^{k} M_\alpha \mathbf{R}_\alpha \qquad (2.5.44)$$

with

$$M = \sum_{\alpha=1}^{k} M_\alpha. \qquad (2.5.45)$$

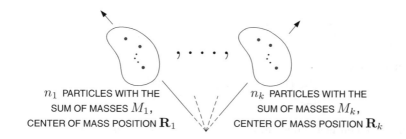

Fig. 2.3. Grouping of n particles into k clusters of particles. Such a grouping is convenient for the study of properties of these clusters as separate entities.

Upon defining the αth cluster total momentum operator

$$\mathbf{P}_{\alpha \cdot} = \sum_{\beta=1}^{n_\alpha} \mathbf{P}_{\alpha\beta} \qquad (2.5.46)$$

2 Symmetries and Transformations

and using the identity

$$\sum_{\beta=1}^{n_\alpha} \left(\mathbf{P}_{\alpha\beta} - \frac{m_{\alpha\beta}}{M_\alpha} \mathbf{P}_{\alpha\cdot} \right) = \mathbf{0} \tag{2.5.47}$$

one may write the Hamiltonian in question as

$$H = \frac{\mathbf{P}^2}{2M} + \sum_{\alpha=1}^{k} \sum_{\beta=1}^{n_\alpha} \frac{\left(\mathbf{P}_{\alpha\beta} - \frac{m_{\alpha\beta}}{M_\alpha} \mathbf{P}_{\alpha\cdot} \right)^2}{2m_{\alpha\beta}} + \sum_{\alpha=1}^{k} \frac{\left(\mathbf{P}_{\alpha\cdot} - \frac{M_\alpha}{M} \mathbf{P} \right)^2}{2M_\alpha}$$

$$+ V_E + V_A \tag{2.5.48}$$

where

$$V_A = \sum_{\alpha=1}^{k} \sum_{\beta \neq \gamma}^{n_\alpha} V_{\alpha\beta\gamma} \left(\mathbf{x}_{\alpha\beta} - \mathbf{x}_{\alpha\gamma} \right) \tag{2.5.49}$$

$$V_E = \sum_{\substack{\alpha=1 \\ (\alpha \neq \beta)}}^{k} \sum_{\beta=1}^{k} \sum_{\gamma=1}^{n_\alpha} \sum_{\rho=1}^{n_\beta} V_{\alpha\beta\gamma\rho} \left(\mathbf{x}_{\alpha\gamma} - \mathbf{x}_{\beta\rho} \right). \tag{2.5.50}$$

The potential energies V_E and V_A are, respectively, responsible for the *inter*-clusteral and the *intra*-clusteral interactions.

From (2.5.21) we note that, in reference to (2.5.48), one has the center of mass motion of an αth cluster conveniently removed in

$$\sum_{\beta=1}^{n_\alpha} \frac{\left(\mathbf{P}_{\alpha\beta} - \frac{m_{\alpha\beta}}{M_\alpha} \mathbf{P}_{\alpha\cdot} \right)^2}{2m_{\alpha\beta}} = \sum_{\beta=1}^{n_\alpha} \frac{\mathbf{P}_{\alpha\beta}^2}{2m_{\alpha\beta}} - \frac{\mathbf{P}_{\alpha\cdot}^2}{2M_\alpha}. \tag{2.5.51}$$

Similarly, in

$$\sum_{\alpha=1}^{k} \frac{\left(\mathbf{P}_{\alpha\cdot} - \frac{M_\alpha}{M} \mathbf{P} \right)^2}{2M_\alpha} = \sum_{\alpha=1}^{k} \frac{\mathbf{P}_{\alpha\cdot}^2}{2M_\alpha} - \frac{\mathbf{P}^2}{2M} \tag{2.5.52}$$

one has the center of mass motion of the k clusters removed.

At this stage, if one wishes, one may introduce variables as in (2.5.33), (2.5.34) in reference to *each* cluster, and also introduce such variables for the system of the k clusters. That is, one may set

$$\mathbf{r}_{\alpha\beta} = \mathbf{x}_{\alpha\beta} - \mathbf{x}_{\alpha n_\alpha}, \quad \beta = 1, \ldots, n_\alpha - 1 \tag{2.5.53}$$

for $\alpha = 1, \ldots, k$ and $\mathbf{R}_\alpha$ as already given in (2.5.42). On the other hand for the system of clusters, one may set

$$\boldsymbol{\eta}_\alpha = \mathbf{R}_\alpha - \mathbf{R}_k, \quad \alpha = 1, \ldots, k-1 \tag{2.5.54}$$

2.5 Quantum Dynamics and Construction of Hamiltonians

and $\mathbf{R}$ as defined in (2.5.44). Such variables may not always be the most convenient ones and different variables may be introduced instead.

As an example, consider two clusters ($k = 2$) with the first consisting of one particle ($n_1 = 1$) and the second consisting of two particles ($n_2 = 2$). This may correspond, for example, to a situation in a scattering process involving a positron e^+ and a hydrogen atom with the latter consisting of an electron-proton bound state. The variables (2.5.53), (2.5.54) in question are

$$\mathbf{r}_{11} = \mathbf{x}_{11} \equiv \mathbf{R}_1 \tag{2.5.55}$$

$$\mathbf{r}_{21} = \mathbf{x}_{21} - \mathbf{x}_{22} \equiv \mathbf{r} \tag{2.5.56}$$

$$\boldsymbol{\eta} = \mathbf{R}_1 - \mathbf{R}_2 \tag{2.5.57}$$

and the following Hamiltonian

$$H = -\frac{\hbar^2}{2M}\nabla_\mathbf{R}^2 - \frac{\hbar^2}{2}\left(\frac{1}{M_1} + \frac{1}{M_2}\right)\nabla_\eta^2 - \frac{\hbar^2}{2}\left(\frac{1}{m_{21}} + \frac{1}{m_{22}}\right)\nabla_\mathbf{r}^2$$

$$+ V_A(\mathbf{r}) + V_1\left(\boldsymbol{\eta} + \frac{m_{21}}{M_2}\mathbf{r}\right) + V_2\left(\boldsymbol{\eta} - \frac{m_{22}}{M_2}\mathbf{r}\right) \tag{2.5.58}$$

is of the form in (2.5.48), where $V_A(\mathbf{r})$ is responsible for the intra-clusteral interaction of the second cluster, and $V_E = V_1 + V_2$ is a potential energy responsible for the inter-clusteral interaction, that is, the interaction between the two clusters. Here $M_1 = m_{11}$, $M_2 = m_{21} + m_{22}$. The decomposition in (2.5.58) may, for example, be convenient if V_E vanishes sufficiently rapidly for $|\boldsymbol{\eta}| \to \infty$ and, initially and/or finally in some scattering process, one is dealing with two non-interacting clusters.

Other examples of clusters decompositions may be also carried out (see Problem 2.7) in a similar manner.

In the appendix to this section, the time evolution of states given in (2.5.9) will be generalized to time-dependent Hamiltonians.

Appendix to §2.5: Time-Evolution for Time-Dependent Hamiltonians

The purpose of this appendix is to describe the time evolution of a state for time-dependent Hamiltonians. A state $|\psi_{t+\delta\tau}\rangle$ approaches the state $|\psi_t\rangle$ for $\delta t \to 0$, and must coincide with (2.5.7) for a time-independent Hamiltonian. Quite generally we may then write

$$|\psi_{t+\delta\tau}\rangle = \left(1 - \frac{i}{\hbar}\delta\tau H(t)\right)|\psi_t\rangle \tag{A-2.5.1}$$

to first order in $\delta\tau$, where the Hamiltonian $H(t)$ is assumed to be, in general, time-dependent. As in (2.5.8), (A-2.5.1) leads, by taking the limit $\delta\tau \to 0$, to the Schrödinger equation

$$i\hbar\frac{\partial}{\partial t}\ket{\psi_t} = H(t)\ket{\psi_t}. \tag{A-2.5.2}$$

To integrate (A-2.5.2), we denote the initial state $\ket{\psi_0}$ by $\ket{\psi}$. To this end, we have by an elementary iterative formal procedure

$$\ket{\psi_t} \simeq \ket{\psi}$$

$$\ket{\psi_t} \simeq \ket{\psi} + \left(\frac{-i}{\hbar}\right)\int_0^t dt_1\, H(t_1)\ket{\psi}$$

$$\ket{\psi_t} \simeq \ket{\psi} + \left(\frac{-i}{\hbar}\right)\int_0^t dt_1\, H(t_1)\ket{\psi} + \left(\frac{-i}{\hbar}\right)^2 \int_0^t dt_2 \int_0^{t_2} dt_1\, H(t_2)H(t_1)\ket{\psi}$$

$$\vdots$$

and finally obtain

$$\ket{\psi_t} = \ket{\psi} + \sum_{n\geqslant 1}\left(\frac{-i}{\hbar}\right)^n \int_0^t dt_n \int_0^{t_n} dt_{n-1} \ldots \int_0^{t_2} dt_1\, H(t_n)H(t_{n-1})\ldots H(t_1)\ket{\psi} \tag{A-2.5.3}$$

where we note that

$$t_n \geqslant t_{n-1} \geqslant \ldots \geqslant t_2 \geqslant t_1 \tag{A-2.5.4}$$

and the important *ordering* of the operators

$$H(t_n)H(t_{n-1})\ldots H(t_2)H(t_1) \tag{A-2.5.5}$$

in (A-2.5.3) from right to left *corresponding* to the ordering in (A-2.5.4).

To rewrite (A-2.5.3) in a more manageable form, we introduce the *chronological time ordering* operation defined by

$$(H(t)H(t'))_+ = (H(t')H(t))_+ = H(t)H(t') \tag{A-2.5.6}$$

if $t \geqslant t'$, and more generally

$$(H(t_1)\ldots H(t_n))_+ = (H(t_{i_1})\ldots H(t_{i_n}))_+$$
$$= H(t_n)\ldots H(t_1) \tag{A-2.5.7}$$

if $t_n \geqslant t_{n-1} \geqslant \ldots \geqslant t_1$, and $\{t_{i_1},\ldots,t_{i_n}\}$ is any permutation of $\{t_1,\ldots,t_n\}$.

From the definition in (A-2.5.6), we have the following equality

2.5 Quantum Dynamics and Construction of Hamiltonians

$$\int_0^t dt_2 \int_0^t dt_1 \, (H(t_2)H(t_1))_+ = \int_0^t dt_2 \int_0^{t_2} dt_1 \, H(t_2)H(t_1)$$

$$+ \int_0^t dt_1 \int_0^{t_1} dt_2 \, H(t_1)H(t_2)$$

$$= 2 \int_0^t dt_2 \int_0^{t_2} dt_1 \, H(t_2)H(t_1) \qquad \text{(A-2.5.8)}$$

and more generally from (A-2.5.7)

$$\int_0^t dt_n \int_0^t dt_{n-1} \ldots \int_0^t dt_1 \, (H(t_n) \ldots H(t_1))_+$$

$$= n! \int_0^t dt_n \int_0^{t_n} dt_{n-1} \ldots \int_0^{t_2} dt_1 \, H(t_n) \ldots H(t_1). \qquad \text{(A-2.5.9)}$$

Equation (A-2.5.9) allows us to rewrite the solution in (A-2.5.3) as

$$|\psi_t\rangle = \sum_{n \geqslant 0} \frac{(-i/\hbar)^n}{n!} \int_0^t dt_n \int_0^t dt_{n-1} \ldots \int_0^t dt_1 \, (H(t_n) \ldots H(t_1))_+ |\psi\rangle$$

$$= \sum_{n \geqslant 0} \frac{(-i/\hbar)^n}{n!} \left(\int_0^t dt' \, H(t') \right)^n_+ \qquad \text{(A-2.5.10)}$$

or

$$|\psi_t\rangle = \left(\exp -\frac{i}{\hbar} \int_0^t dt' \, H(t') \right)_+ |\psi\rangle \qquad \text{(A-2.5.11)}$$

generalizing the solution in (2.5.9) for time-dependent Hamiltonians.

We note that the adjoint transformation of (A-2.5.7) leads to

$$\left[(H(t_1) \ldots H(t_n))_+ \right]^\dagger = \left[(H(t_{i_1}) \ldots H(t_{i_n}))_+ \right]^\dagger$$

$$= H(t_1) \ldots H(t_{n-1}) H(t_n) \qquad \text{(A-2.5.12)}$$

if $t_1 \leqslant \ldots \leqslant t_{n-1} \leqslant t_n$. Thus we may introduce the chronological time anti-ordering operation defined by

$$(H(t_1) \ldots H(t_n))_- = (H(t_{i_1}) \ldots H(t_{i_n}))_-$$

$$= H(t_1) \ldots H(t_{n-1}) H(t_n) \qquad \text{(A-2.5.13)}$$

if $t_1 \leqslant \ldots \leqslant t_{n-1} \leqslant t_n$, and $\{t_{i_1}, \ldots, t_{i_n}\}$ is any permutation of $\{t_i, \ldots, t_n\}$.

From (A-2.5.11)–(A-2.5.13), we then have for the adjoint of the time evolution operator

$$\left[\left(\exp -\frac{i}{\hbar} \int_0^t dt' \, H(t') \right)_+ \right]^\dagger = \left(\exp \frac{i}{\hbar} \int_0^t dt' \, H(t') \right)_-. \qquad \text{(A-2.5.14)}$$

2.6 Discrete Transformations: Parity and Time Reversal

In this section, we consider two discrete transformations. The first consists of the operation of reflecting the sense of every spatial coordinate axis. This amounts to the transformation $\mathbf{x} \to \mathbf{x}'$:

$$\mathbf{x}' \to -\mathbf{x} \tag{2.6.1}$$

as a reflection through the origin of the space coordinate system. The other is the reversal of the direction of time flow. We will see that on the physical states, a space reflection is implemented by a *unitary* operator $\mathcal{P}$, referred to as the parity transformation, while time reversal is implemented by an *anti-unitary* operator which we denote by $\mathcal{T}$. The problem encountered in trying to represent $\mathcal{P}$ by an anti-unitary operator rather than a unitary one will be discussed explicitly below. The analysis of the problem encountered in representing $\mathcal{T}$ by a unitary operator rather than by an anti-unitary one is similarly carried out and is left as an exercise to the reader (see Problem 2.8). A third discrete transformation, referred to as charge conjugation, denoted by $\mathcal{C}$, will be discussed later in Chapter 16. As mentioned earlier, if one combines special relativity with quantum mechanics, leading to what is called quantum field theory, then one may establish that the combined transformation $\mathcal{CPT}$ is a symmetry transformation based on such a merger.

A point Q in a given (right-handed) coordinate system labelled by $\mathbf{x}$, will be labelled by $\mathbf{x}'$ in a (left-handed) coordinate system arising from the initial one by reflecting the sense of every spatial coordinate axis (see Figure 2.3).

For infinitesimal Galilean transformations, we recall that (see (2.1.15), (2.1.16))

$$\delta\mathbf{x} = \delta\mathbf{a} + \delta\mathbf{v}t + \delta\boldsymbol{\omega} \times \mathbf{x} \tag{2.6.2}$$

$$\delta t = \delta\tau \tag{2.6.3}$$

where $\delta\boldsymbol{\omega}$ is defined in (2.1.6). For the combined Galilean transformation and space reflection, for example, we have

$$(\delta\mathbf{x})' = -\delta\mathbf{a} - \delta\mathbf{v}t - \delta\boldsymbol{\omega} \times \mathbf{x} \tag{2.6.4}$$

$$(\delta t)' = \delta\tau \tag{2.6.5}$$

which amount to the replacements, $\delta\mathbf{a} \to -\delta\mathbf{a}$, $\delta\mathbf{v} \to -\delta\mathbf{v}$, $\delta\boldsymbol{\omega} \to \delta\boldsymbol{\omega}$, $\delta\tau \to \delta\tau$.

With the change of description given in (2.6.4), we associate an operator $\mathcal{P}$, operating on physical states, whose nature will be soon established.

In reference to Wigner's Theorem (§1.9), to establish whether $\mathcal{P}$ should be anti-unitary or unitary, we consider the case in (2.6.4) with $\delta\mathbf{a} = \mathbf{0}$, $\delta\mathbf{v} = \mathbf{0}$,

2.6 Discrete Transformations: Parity and Time Reversal   113

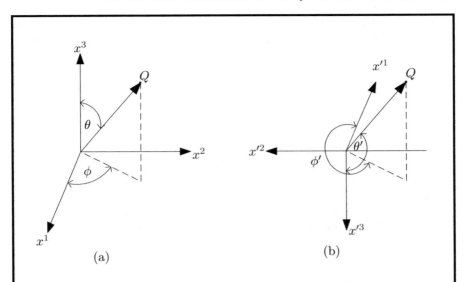

Fig. 2.4. A point Q labelled by **x** in the right-handed coordinate system in (a) and is labelled by **x'** in the left-handed coordinate system in (b) obtained from the initial one by reflecting the sense of each of its coordinate axes. We note that $\theta = \pi - \theta'$, $\phi = \phi' - \pi$, $\mathbf{x} = -\mathbf{x}'$, or equivalently $\theta' = \pi - \theta$, $\phi' = \phi + \pi$, $\mathbf{x}' = -\mathbf{x}$.

$\delta\boldsymbol{\omega} = \mathbf{0}$ first. Since the operation consisting of a space reflection followed by a time shift $\delta\tau$ in one's clocks readings and the operation of a time shift $\delta\tau$ in one's clocks readings followed by a space reflection are equivalent we may infer that

$$\left(1 - \frac{\mathrm{i}}{\hbar}\delta\tau H\right)\mathcal{P} = \mathcal{P}\left(1 - \frac{\mathrm{i}}{\hbar}\delta\tau(H + c\mathbf{1})\right) \qquad (2.6.6)$$

up to a phase factor specified by an additional structure $(1/\hbar)\delta\varphi\mathbf{1}$, with $\delta\varphi = -c\delta\tau$, as appearing on the right-hand of (2.6.6), with c some real constant. This equality implies that

$$\mathrm{i}H\mathcal{P} = \mathcal{P}\mathrm{i}(H + c\mathbf{1}). \qquad (2.6.7)$$

If $\mathcal{P}$ is anti-unitary, that is, in particular, it complex conjugates numerical factors, we obtain after re-arrangement of terms

$$\mathcal{P}H\mathcal{P}^{-1} = -(H + c\mathbf{1}). \qquad (2.6.8)$$

A Hamiltonian, in general, is unbounded from *above*. This is true even for a free particle. Equation (2.6.8) says that for every eigenstate $|\psi\rangle$ of H with a given non-negative energy E arbitrarily *large*, H has an eigenstate $\mathcal{P}|\psi\rangle$ with corresponding energy $-(E + c)$. With E arbitrarily large, this in turn implies that the Hamiltonian is unbounded from *below* and is physically unacceptable. The latter argument may be made rigorous by stating that for any $\varepsilon > 0$, however small,

$$\|(H - E)\psi_\varepsilon\| < \varepsilon \tag{2.6.9}$$

if and only if

$$\|[H + (E + c)]\mathcal{P}\psi_\varepsilon\| < \varepsilon \tag{2.6.10}$$

implying the unboundedness of the spectrum of H from below for arbitrarily large non-negative E. That is, $\mathcal{P}$ must be *unitary*.

With $\mathcal{P}$ as a unitary operator, we consider the more general cases corresponding to (2.6.4), (2.6.5).

Since the operation consisting of a space reflection, followed by an infinitesimal Galilean transformation, specified by the parameters $\delta\mathbf{a}$, $\delta\mathbf{v}$, $\delta\tau$, $\delta\boldsymbol{\omega}$, is equivalent to first performing the infinitesimal changes $-\delta\mathbf{a}$, $-\delta\mathbf{v}$, $\delta\tau$, $\delta\boldsymbol{\omega}$, and then a space reflection, we may infer that

$$\left[1 + \frac{i}{\hbar}(\delta\mathbf{a}\cdot\mathbf{P} + \delta\mathbf{v}\cdot\mathbf{N} - \delta\tau H + \delta\boldsymbol{\omega}\cdot\mathbf{J})\right]\mathcal{P}$$

$$= \mathcal{P}\left[1 + \frac{i}{\hbar}(-\delta\mathbf{a}\cdot\mathbf{P} - \delta\mathbf{v}\cdot\mathbf{N} - \delta\tau H + \delta\boldsymbol{\omega}\cdot\mathbf{J})\right] \tag{2.6.11}$$

where we have dispensed with a phase factor specified by an additional structure $(1/\hbar)\delta\varphi\mathbf{1}$ within the square brackets on the right-hand side (see Problem 2.9).

Unitarity of $\mathcal{P}$ then implies from (2.6.11) that

$$H\mathcal{P} - \mathcal{P}H = 0 \tag{2.6.12}$$

$$\mathbf{J}\mathcal{P} - \mathcal{P}\mathbf{J} = \mathbf{0} \tag{2.6.13}$$

$$\mathbf{P}\mathcal{P} + \mathcal{P}\mathbf{P} = \mathbf{0} \tag{2.6.14}$$

$$\mathbf{N}\mathcal{P} + \mathcal{P}\mathbf{N} = \mathbf{0}. \tag{2.6.15}$$

From (2.6.14), (2.6.15) and (2.3.21) we may also infer that

$$\mathbf{X}\mathcal{P} + \mathcal{P}\mathbf{X} = \mathbf{0} \tag{2.6.16}$$

as expected.

If parity is not a symmetry transformation, such as in beta decay (a weak interaction process), the Hamiltonian is so constructed to reflect this fact by ensuring, in particular, that $H\mathcal{P} - \mathcal{P}H \neq 0$.

A very similar analysis to the one given through (2.6.6)–(2.6.10) implies that time reversal is implemented by an *anti-unitary* operator rather than by a unitary operator (see Problem 2.8).

2.6 Discrete Transformations: Parity and Time Reversal

The infinitesimal Galilean transformations in (2.6.2), (2.6.3) followed by the reversal of the time translation $t - \delta\tau \to -t + \delta\tau$ give

$$(\delta\mathbf{x})' = \delta\mathbf{a} + \delta\mathbf{v}t + \delta\boldsymbol{\omega} \times \mathbf{x} \tag{2.6.17}$$

$$(\delta t)' = -\delta\tau \tag{2.6.18}$$

where $\delta\mathbf{v} \to -\delta\mathbf{v}$, $\delta\boldsymbol{\omega} \to \delta\boldsymbol{\omega}$ under the reversal of the direction of time flow. We note the invariance of the product $\delta\mathbf{v}t$.

By using the identification of the position operator in (2.3.21), we spell out an infinitesimal Galilean transformation, at a given time t, as

$$1 + \frac{i}{\hbar}\left[(\delta\mathbf{a} + t\delta\mathbf{v}) \cdot \mathbf{P} - M\delta\mathbf{v} \cdot \mathbf{X} - \delta\tau H + \delta\boldsymbol{\omega} \cdot \mathbf{J}\right]. \tag{2.6.19}$$

Since the operation of time reversal followed by an infinitesimal Galilean transformation specified by $\delta\mathbf{a}$, $\delta\mathbf{v}$, $\delta\tau$, $\delta\boldsymbol{\omega}$, at a given time t as the latter is shown in (2.6.19), is equivalent to an infinitesimal Galilean transformation specified by $\delta\mathbf{a}$, $-\delta\mathbf{v}$, $-\delta\tau$, $\delta\boldsymbol{\omega}$, at time $-t$, followed by a time reversal, we obtain

$$\left(1 + \frac{i}{\hbar}\left[(\delta\mathbf{a} + t\delta\mathbf{v}) \cdot \mathbf{P} - M\delta\mathbf{v} \cdot \mathbf{X} - \delta\tau H + \delta\boldsymbol{\omega} \cdot \mathbf{J}\right]\right)\mathcal{T}$$

$$= \mathcal{T}\left(1 + \frac{i}{\hbar}\left[(\delta\mathbf{a} + t\delta\mathbf{v}) \cdot \mathbf{P} + M\delta\mathbf{v} \cdot \mathbf{X} + \delta\tau H + \delta\boldsymbol{\omega} \cdot \mathbf{J}\right]\right) \tag{2.6.20}$$

where as for the case of the parity transformation in (2.6.11), we have dispensed with a phase factor contribution in (2.6.20) (see Problem 2.9). It is worth recalling the complex conjugation nature of numerical factors of the operator $\mathcal{T}$. Accordingly, (2.6.20) leads to

$$H\mathcal{T} - \mathcal{T}H = 0 \tag{2.6.21}$$

$$\mathbf{J}\mathcal{T} + \mathcal{T}\mathbf{J} = \mathbf{0} \tag{2.6.22}$$

$$\mathbf{P}\mathcal{T} + \mathcal{T}\mathbf{P} = \mathbf{0} \tag{2.6.23}$$

and

$$(t\mathbf{P} - M\mathbf{X})\mathcal{T} + \mathcal{T}(t\mathbf{P} + M\mathbf{X}) = \mathbf{0}. \tag{2.6.24}$$

The letter two equations then give

$$\mathbf{X}\mathcal{T} - \mathcal{T}\mathbf{X} = \mathbf{0}. \tag{2.6.25}$$

We note that the operator $\mathcal{T}$ does *not* operate on the real parameter t in (2.6.24).

2.7 Orbital Angular Momentum and Spin

In this section we investigate the nature of the generator $\mathbf{J}$ (see (2.3.13), (2.3.43)) of space coordinate *rotations*. Important commutation relations involving $\mathbf{J}$ are given in Table 2.1 below (2.3.15) and in (2.3.24), where we recall that $E = 0$, $B = 0$ in the Table. From these commutation relations, and from (2.3.21) and (2.3.20) it is not difficult to derive that for infinitesimal Galilean transformations

$$\delta \mathbf{J} = (\delta \mathbf{a} + t \delta \mathbf{v}) \times \mathbf{P} - M \delta \mathbf{v} \times \mathbf{X} + \delta \boldsymbol{\omega} \times \mathbf{J}. \tag{2.7.1}$$

On the other hand from (2.3.22), (2.3.27)

$$\delta \mathbf{X} = (\delta \mathbf{a} + t \delta \mathbf{v}) + \delta \boldsymbol{\omega} \times \mathbf{X} - \delta \tau \frac{\mathbf{P}}{M} \tag{2.7.2}$$

and from (2.3.20), Table 2.1,

$$\delta \mathbf{P} = M \delta \mathbf{v} + \delta \boldsymbol{\omega} \times \mathbf{P}. \tag{2.7.3}$$

From the latter two equations or directly from (2.3.20) (see Problem 2.10), one has

$$\delta (\mathbf{X} \times \mathbf{P}) = (\delta \mathbf{a} + t \delta \mathbf{v}) \times \mathbf{P} - M \delta \mathbf{v} \times \mathbf{X} + \delta \boldsymbol{\omega} \times (\mathbf{X} \times \mathbf{P}) \tag{2.7.4}$$

where we note, in particular, that

$$[H, \mathbf{X} \times \mathbf{P}] = 0. \tag{2.7.5}$$

As a matter of fact $\mathbf{X} \times \mathbf{P}$ satisfies the same commutation relations in Table 2.1 and in (2.3.24) as $\mathbf{J}$. Hence upon the comparison of (2.7.1) with (2.7.4) we conclude that $\mathbf{J}$ has the very general form

$$\mathbf{J} = (\mathbf{X} \times \mathbf{P}) + \mathbf{S} \tag{2.7.6}$$

where the operator $\mathbf{S}$ necessarily, as is easily established, satisfies the commutation relations

$$[P^i, S^j] = 0 \tag{2.7.7}$$

$$[X^i, S^j] = 0 \tag{2.7.8}$$

$$[H, S^i] = 0 \tag{2.7.9}$$

$$[S^i, S^j] = \mathrm{i} \varepsilon^{ijk} S^k. \tag{2.7.10}$$

The operator

$$\mathbf{X} \times \mathbf{P} \equiv \mathbf{L} \tag{2.7.11}$$

denotes the familiar orbital angular momentum. On the other hand **S** in (2.7.6) is a *translational independent* contribution to **J** and is referred to as the spin or the internal angular momentum. We note, in particular, that (2.7.7), (2.7.8) imply that

$$[L^i, S^j] = 0. \tag{2.7.12}$$

To obtain further insight into the nature of the spin operator **S**, we consider the transformations of a Galilean scalar (field) and a Galilean vector (field) under a rotation of the coordinate system.

A Galilean scalar (field) $\psi(\mathbf{x})$, under a *rotation* of the coordinate system, is defined by (see (2.1.1)) the transformation law

$$\psi'(R\mathbf{x}) = \psi(\mathbf{x}) \tag{2.7.13}$$

or by

$$\psi'(\mathbf{x}) = \psi(R^{-1}\mathbf{x}). \tag{2.7.14}$$

Upon using (2.3.43), and (2.1.26), (2.1.34), (2.1.43) we obtain from (2.7.14)

$$\left(\left(\exp \frac{i}{\hbar}\boldsymbol{\omega} \cdot \mathbf{J}\right)\psi\right)(\mathbf{x}) = \exp\left[\boldsymbol{\omega} \cdot (\mathbf{x} \times \boldsymbol{\nabla})\right]\psi(\mathbf{x}) \tag{2.7.15}$$

thus leading to the representation

$$\mathbf{J} = \mathbf{x} \times (-i\hbar \boldsymbol{\nabla}) \tag{2.7.16}$$

and

$$\mathbf{S} = 0 \tag{2.7.17}$$

reflecting the spin 0 character of a scalar (field).

A Galilean vector (field) $\boldsymbol{\Phi}(\mathbf{x})$ under the rotation of the coordinate system, is defined by the transformation law[4]

$$\Phi'^i(R\mathbf{x}) = R^{ij}\Phi^j(\mathbf{x}) \tag{2.7.18}$$

or by

$$\Phi'^i(\mathbf{x}) = R^{ij}\Phi^j(R^{-1}\mathbf{x}) \tag{2.7.19}$$

where, in the same way as the vector **x** itself (see (2.1.1)), the presence of the matrix (R^{ij}) on the right-hand sides of (2.7.18), (2.7.19) reflects the vector character of $\boldsymbol{\Phi}$. Of course, we have to check that the spin associated with $\boldsymbol{\Phi}$ is indeed equal to one which is the spin content of a vector (field). The presence of the matrix (R^{ij}), in turn, implies the invariance of an amplitude under a rotation of the coordinate system:

---

[4] Vector fields also arise as a particular case within a spinor analysis context as given in §2.8.

$$\sum_{i=1}^{3} \langle \Phi'^i | \psi'^i \rangle = \sum_{i=1}^{3} \langle \Phi^i | \psi^i \rangle \tag{2.7.20}$$

where we have used (2.1.3), and (2.1.36) for the evaluation of the Jacobian of the transformation which is equal to one.

From (2.1.26), (2.1.34), (2.1.43), we have for (2.7.19) the explicit expression

$$\Phi'(\mathbf{x}) = \exp\left[i\frac{\boldsymbol{\omega}}{\hbar} \cdot (\mathbf{x} \times (-i\hbar\boldsymbol{\nabla}) + \hbar\mathbf{M})\right] \Phi(\mathbf{x}) \tag{2.7.21}$$

leading from (2.3.43) to the representation

$$\mathbf{J} = \mathbf{x} \times (-i\hbar\boldsymbol{\nabla}) + \hbar\mathbf{M} \tag{2.7.22}$$

and hence to

$$\mathbf{S} = \hbar\mathbf{M} \tag{2.7.23}$$

where the matrices M^j are defined in (2.1.17), (2.1.28)–(2.1.30), (2.1.32). In particular,

$$\mathbf{S}^2 = 2\hbar\mathbf{1} \tag{2.7.24}$$

which upon using the definition that for a particle of spin s (see §5.1, §5.4) we must have $\mathbf{S}^2 = \hbar^2 s(s+1)\mathbf{1}$, (2.7.24) then gives $s = 1$, thus verifying the spin content of Φ.

Now we turn to a system of n particles. Using the definitions given in (2.5.16)–(2.5.18), we rewrite (2.7.6) as

$$\mathbf{J} = \sum_{\alpha=1}^{n} (\mathbf{X}_\alpha \times \mathbf{P}_\alpha + \mathbf{S}_\alpha) \tag{2.7.25}$$

with $\mathbf{X}$ and $\mathbf{P}$ given in (2.5.16), and (2.5.17), respectively, and $\mathbf{S}$ is to be determined. $\mathbf{J}$ may be rewritten in the form

$$\mathbf{J} = \mathbf{X} \times \mathbf{P} + \sum_{\alpha=1}^{n} [(\mathbf{X}_\alpha - \mathbf{X}) \times \mathbf{P}_\alpha + \mathbf{S}_\alpha] \tag{2.7.26}$$

On the other hand since

$$\sum_{\alpha=1}^{n} \frac{m_\alpha}{M} (\mathbf{X}_\alpha - \mathbf{X}) = 0 \tag{2.7.27}$$

and hence

$$\sum_{\alpha=1}^{n} \frac{m_\alpha}{M} (\mathbf{X}_\alpha - \mathbf{X}) \times \mathbf{P} = 0 \tag{2.7.28}$$

one may finally rewrite (2.7.25) as

$$\mathbf{J} = \mathbf{X} \times \mathbf{P} + \sum_{\alpha=1}^{n} \left[(\mathbf{X}_\alpha - \mathbf{X}) \times \left(\mathbf{P}_\alpha - \frac{m_\alpha}{M}\mathbf{P}\right) + \mathbf{S}_\alpha\right]. \tag{2.7.29}$$

2.7 Orbital Angular Momentum and Spin

This leads to the following expression for the total internal angular momentum **S** of the system of particles as one composite object made up of the n particles:

$$\mathbf{S} = \sum_{\alpha=1}^{n} \left[(\mathbf{X}_\alpha - \mathbf{X}) \times \left(\mathbf{P}_\alpha - \frac{m_\alpha}{M} \mathbf{P} \right) + \mathbf{S}_\alpha \right] \quad (2.7.30)$$

with $(\mathbf{X}_\alpha - \mathbf{X})$ denoting the operator for the position of the αth particle relative to the center of mass, and $(\mathbf{P}_\alpha - m_\alpha \mathbf{P}/M)$ denoting the operator of its momentum as determined in the center of mass. The first term in the summand in (2.7.30) represents the orbital angular momentum operator of the αth particle in the center of mass system. The internal angular momentum **S** in (2.7.30) is the total angular momentum of the system of particles relative to the center of mass. It is easily checked from (2.7.30) that

$$[P^i, S^j] = 0, \quad [X^i, S^j] = 0 \quad (2.7.31)$$

as they should be for a proper definition of spin (see (2.7.7), (2.7.8)).

Consider the decomposition of **J** in (2.7.25) into $k \neq 1$ clusters (see §2.5), rather than one, as in (2.7.29), (2.7.30), composed, respectively, of $n_1, \ldots, n_k$ particles. To this end, it is more convenient to introduce the operators $\mathbf{X}_{\alpha\beta}$, $\mathbf{P}_{\alpha\beta}$, $\mathbf{X}_{\alpha\cdot}$, $\mathbf{X}$, $\mathbf{P}_{\alpha\cdot}$, $\mathbf{P}$ where

$$\mathbf{X}_{\alpha\cdot} = \sum_{\beta=1}^{n_\alpha} \frac{m_{\alpha\beta}}{M_\alpha} \mathbf{X}_{\alpha\beta} \quad (2.7.32)$$

$$\mathbf{X} = \sum_{\alpha=1}^{k} \frac{M_\alpha}{M} \mathbf{X}_{\alpha\cdot} \quad (2.7.33)$$

denoting, respectively, the position operator associated with the center of mass of the αth cluster, and the position operator of the center of mass of the k clusters with the latter coinciding with the center of mass of the $\sum_{\alpha=1}^{k} n_\alpha$ particles. The operators $\mathbf{P}_{\alpha\beta}$, $\mathbf{P}_{\alpha\cdot}$, $\mathbf{P}$ and the masses $m_{\alpha\beta}$, M_α, M are defined in §2.5. In particular, M_α denotes the sum of the masses of the particles in the αth cluster.

Suppose $\mathbf{S}_{\alpha\beta}$ denotes the spin of the βth particle in the αth cluster. Then

$$\mathbf{J} = \sum_{\alpha=1}^{k} \sum_{\beta=1}^{n_\alpha} (\mathbf{X}_{\alpha\beta} \times \mathbf{P}_{\alpha\beta} + \mathbf{S}_{\alpha\beta}) \quad (2.7.34)$$

may be rewritten as

$$\mathbf{J} = \mathbf{X} \times \mathbf{P} + \sum_{\alpha=1}^{k} \left[(\mathbf{X}_{\alpha\cdot} - \mathbf{X}) \times \left(\mathbf{P}_{\alpha\cdot} - \frac{M_\alpha}{M} \mathbf{P} \right) + \mathbf{S}_{(\alpha)} \right] \quad (2.7.35)$$

where

$$S_{(\alpha)} = \sum_{\beta=1}^{n_\alpha} \left[(X_{\alpha\beta} - X_{\alpha\cdot}) \times \left(P_{\alpha\beta} - \frac{m_{\alpha\beta}}{M} P_{\alpha\cdot} \right) + S_{\alpha\beta} \right] \quad (2.7.36)$$

is the total internal angular momentum of the αth cluster as a composite object made up of n_α particles. The following commutation relations again hold true

$$\left[P^i_{\alpha\cdot}, S^j_{(\alpha)} \right] = 0, \quad \left[X^i_{\alpha\cdot}, S^j_{(\alpha)} \right] = 0. \quad (2.7.37)$$

Also

$$\left[P^i, S^j_{(\alpha)} \right] = 0, \quad \left[X^i, S^j_{(\alpha)} \right] = 0 \quad (2.7.38)$$

for all $i, j = 1, 2, 3$.

For a two cluster decomposition ($k = 2$), consisting, respectively, of n_1 and n_2 particles, $\mathbf{J}$ in (2.7.35) may be rewritten as

$$\mathbf{J} = \mathbf{X} \times \mathbf{P} + \sum_{\alpha=1}^{2} \left[(X_{\alpha\cdot} - X) \times \left(P_{\alpha\cdot} - \frac{M_\alpha}{M} P \right) + S_{(\alpha)} \right]$$

$$= \mathbf{X} \times \mathbf{P} + \mathbf{L}_r + \mathbf{S}_{(1)} + \mathbf{S}_{(2)} \quad (2.7.39)$$

where $\mathbf{L}_r$ is the total orbital angular momentum of the two clusters residing in their center of mass given by

$$\mathbf{L}_r = \sum_{\alpha=1}^{2} (X_{\alpha\cdot} - X) \times \left(P_{\alpha\cdot} - \frac{M_\alpha}{M} P \right)$$

$$= (X_{1\cdot} - X_{2\cdot}) \times \left(\frac{M_2 P_{1\cdot} - M_1 P_{2\cdot}}{M_1 + M_2} \right) \quad (2.7.40)$$

and $\mathbf{L}_r + \mathbf{S}_{(1)} + \mathbf{S}_{(2)}$ is the internal angular momentum of the two clusters.

Pertinent properties of the angular momentum operator, in general, as well as of the orbital angular momentum operator and spin, in particular, will be studied in detail later in Chapter 5. Arbitrary spins are also studied in §2.8.

Finally we note from the transformation law (2.3.43) of a state $|\psi\rangle$ under a coordinate rotation

$$|\psi'\rangle = \exp\left(\frac{i}{\hbar} \varphi \mathbf{n} \cdot \mathbf{J} \right) |\psi\rangle, \quad (2.7.41)$$

in a coordinate description, we have

$$\psi'(\mathbf{x}) = \left(\exp\left(\frac{i}{\hbar} \varphi \mathbf{n} \cdot \mathbf{J} \right) \psi \right)(\mathbf{x}) \quad (2.7.42)$$

and from (2.7.6), (2.7.11), (2.7.12) and the identity (2.1.43)

$$\psi'(\mathbf{x}') = \left(\exp\left(\frac{i}{\hbar}\varphi \mathbf{n}\cdot\mathbf{S}\right)\psi\right)(\mathbf{x}) \tag{2.7.43}$$

where $\mathbf{x}' = R\mathbf{x}$, with R defined in (2.1.26). Note that in (2.7.42), (2.7.43) the argument of ψ' is $\mathbf{x}$ when the right-hand side of the equation of its transformation law involves $\mathbf{J}$, while it is $\mathbf{x}'$ when it involves only the spin $\mathbf{S}$, respectively.

2.8 Spinors and Arbitrary Spins

In generalizing the concept of spin 0 and spin 1 studied in §2.7, we introduce the concept of a spinor. Spinors are essential objects needed to describe spin 1/2 and, more generally, half-odd integral spins. One may even describe integer spins by using these curious objects as we will see below.

2.8.1 Spinors and Generation of Arbitrary Spins

We begin by defining on how a spinor transforms under a coordinate rotation. Spinors of so-called of rank 1 and their transformation rule under a coordinate rotation are introduced from first principles in the appendix to this chapter. At a first reading, one may skip the reading of this appendix. The burden of this section is to see how these mathematical entities together with their underlying definitions describe arbitrary spins.

A spinor (field) of rank k, $\psi^{a_1\ldots a_k}(\mathbf{x})$, depending on k indices $a_1\ldots a_k$ each taking the possible values 1 and 2, has the following transformation law under a coordinate rotation by an angle φ about a unit vector $\mathbf{n}$:

$$\psi'^{a_1\ldots a_k}(\mathbf{x}') = \left(\exp\left[\frac{i}{2}\varphi\mathbf{n}\cdot\boldsymbol{\sigma}\right]\right)^{a_1 b_1}\cdots\left(\exp\left[\frac{i}{2}\varphi\mathbf{n}\cdot\boldsymbol{\sigma}\right]\right)^{a_k b_k}\psi^{b_1\ldots b_k}(\mathbf{x}) \tag{2.8.1}$$

where a *summation* over the repeated indices $b_1,\ldots,b_k$ is understood. In the notation of §2.1, $\mathbf{x}' = R\mathbf{x}$. Here $\boldsymbol{\sigma} = (\sigma_1, \sigma_2, \sigma_3)$, where $\sigma_1, \sigma_2, \sigma_3$ are 2×2 matrices introduced by Pauli

$$\sigma_1 = \begin{pmatrix} 0 & 1 \\ 1 & 0 \end{pmatrix}, \quad \sigma_2 = \begin{pmatrix} 0 & -i \\ i & 0 \end{pmatrix}, \quad \sigma_3 = \begin{pmatrix} 1 & 0 \\ 0 & -1 \end{pmatrix} \tag{2.8.2}$$

which together the unit matrix

$$\mathbf{1} = \begin{pmatrix} 1 & 0 \\ 0 & 1 \end{pmatrix} \tag{2.8.3}$$

constitute a complete set of matrices in the vector space consisting of complex 2×2 matrices. That is, any element in such a vector space may be written as a linear combination of the matrices $\mathbf{1}, \sigma_1, \sigma_2, \sigma_3$. $[\cdot]^{ab}$ denotes the matrix elements of the matrix in question.

In the sense of §2.7, spin 0 (see (2.7.17)), is described by a spinor of rank 0. To describe arbitrary spins, we are interested in *symmetric* spinors of rank k. That is, spinors $\psi^{a_1 \cdots a_k}(\mathbf{x})$ which are completely symmetric in the indices $a_1 \ldots a_k$.

The matrix $\exp(i\varphi\, \mathbf{n} \cdot \boldsymbol{\sigma}/2)$ in (2.8.1) may be rewritten in the simple form

$$\mathbf{1} \cos \frac{\varphi}{2} + i(\mathbf{n} \cdot \boldsymbol{\sigma}) \sin \frac{\varphi}{2}. \tag{2.8.4}$$

To see this note that

$$\frac{d}{d\varphi}\left[\mathbf{1} \cos \frac{\varphi}{2} + i(\mathbf{n} \cdot \boldsymbol{\sigma}) \sin \frac{\varphi}{2}\right] = \frac{i}{2}\mathbf{n} \cdot \boldsymbol{\sigma} \left[\mathbf{1} \cos \frac{\varphi}{2} + i(\mathbf{n} \cdot \boldsymbol{\sigma}) \sin \frac{\varphi}{2}\right] \tag{2.8.5}$$

where we have used the useful identity

$$\sigma_i \sigma_j = \delta^{ij} + i\varepsilon^{ijk}\sigma_k. \tag{2.8.6}$$

Upon integration of (2.8.5), we obtain

$$\exp\left(i\frac{\varphi}{2}\mathbf{n} \cdot \boldsymbol{\sigma}\right) = \mathbf{1} \cos \frac{\varphi}{2} + i(\mathbf{n} \cdot \boldsymbol{\sigma}) \sin \frac{\varphi}{2} \tag{2.8.7}$$

with the boundary condition

$$\exp\left(i\frac{\varphi}{2}\mathbf{n} \cdot \boldsymbol{\sigma}\right)\bigg|_{\varphi=0} = \mathbf{1}. \tag{2.8.8}$$

[Another way of deriving (2.8.7) is given in Problem 2.13.]

As the indices $a_1, \ldots, a_k$, take on the values 1 or 2, a completely *symmetric* spinor $\psi^{a_1 \cdots a_k}(\mathbf{x})$ define $(k+1)$ functions which we denote by $\psi_0(\mathbf{x}), \ldots, \psi_k(\mathbf{x})$ and, in particular, we may choose

$$\psi^{11 \cdots 11}(\mathbf{x}) \equiv \psi_k(\mathbf{x}) \tag{2.8.9}$$

$$\psi^{22 \cdots 22}(\mathbf{x}) \equiv \psi_0(\mathbf{x}). \tag{2.8.10}$$

For arbitrary $a_1, \ldots, a_k$ taking the values 1 or 2, we may expand $\psi^{a_1 \cdots a_k}(\mathbf{x})$ in terms of these functions. To this end, let i of the indices $a_1, \ldots, a_k$ take on the value 1 and the $(k-i)$ remaining of them take on the value 2. We introduce the orthonormal set consisting of the objects

$$C^{a_1 \cdots a_k}(i) = \frac{1}{\sqrt{k!\, i!\, (k-i)!}} \left(\delta^{a_1 1} \cdots \delta^{a_i 1} \delta^{a_{i+1} 2} \cdots \delta^{a_k 2}\right)_{\text{permut.}} \tag{2.8.11}$$

with $i = 0, 1, \ldots, k$. Here permut. stands for a summation over *all* permutations of $\{a_1, \ldots, a_k\}$. The orthonormality property reads:

2.8 Spinors and Arbitrary Spins

$$\sum_{a_1,\ldots,a_k=1,2} C^{a_1\ldots a_k}(i) C^{a_1\ldots a_k}(i') = \delta_{ii'}. \tag{2.8.12}$$

The orthogonality property for $i \neq i'$ is obvious since the number of ones (and hence also the number of twos) will be different in this case. The overall normalization factor in (2.8.11) arises in the following manner. Out of the k permutations of $\{a_1,\ldots,a_k\}$, $i!\,(k-i)!$ are redundant. Hence we may write

$$\left(\delta^{a_1 1}\ldots\delta^{a_i 1}\delta^{a_{i+1} 2}\ldots\delta^{a_k 2}\right)_{\text{permut.}}$$

$$= i!(k-i)!\left(\delta^{a_1 1}\ldots\delta^{a_i 1}\delta^{a_{i+1} 2}\ldots\delta^{a_k 2}\right)'_{\text{permut.}} \tag{2.8.13}$$

where $\left(\delta^{a_1 1}\ldots\delta^{a_i 1}\delta^{a_{i+1} 2}\ldots\delta^{a_k 2}\right)'_{\text{permut.}}$ stands for a summation over all non-redundant permutations. The latter will consist of exactly $C_i^k \equiv k!/i!(k-i)!$ orthonormal terms. That is,

$$\sum_{a_1,\ldots,a_k=1,2} \left(\delta^{a_1 1}\ldots\delta^{a_i 1}\delta^{a_{i+1} 2}\ldots\delta^{a_k 2}\right)_{\text{permut.}}$$

$$\times \left(\delta^{a_1 1}\ldots\delta^{a_i 1}\delta^{a_{i+1} 2}\ldots\delta^{a_k 2}\right)_{\text{permut.}}$$

$$= (i!(k-i)!)^2 \, C_i^k = k!\,i!\,(k-i)! \tag{2.8.14}$$

leading to the expression in (2.8.11).

Hence we may rewrite

$$\psi^{a_1\ldots a_k}(\mathbf{x}) = \sum_{i=0}^{k} C^{a_1\ldots a_k}(i)\psi_i(\mathbf{x}) \tag{2.8.15}$$

with, in particular,

$$C^{a_1\ldots a_k}(k) = \delta^{a_1 1}\ldots\delta^{a_k 1} \tag{2.8.16}$$

$$\psi^{a_1\ldots a_k}(0) = \delta^{a_1 2}\ldots\delta^{a_k 2} \tag{2.8.17}$$

to be compared, respectively, with (2.8.9), (2.8.10).

The structures $C^{a_1\ldots a_k}(k)$ in (2.8.11) may be rewritten in a more convenient form as follows. We introduce two independent variables g_1, g_2. Then it is not difficult to see that (2.8.11) may be rewritten in the equivalent form

$$C^{a_1\ldots a_k}(i) = \frac{1}{\sqrt{k!\,i!\,(k-i)!}} \frac{\partial}{\partial g_{a_1}}\ldots\frac{\partial}{\partial g_{a_k}}(g_1)^i(g_2)^{k-i}. \tag{2.8.18}$$

At this stage, it is more convenient to introduce the parameters s and m defined by[5]

---

[5] To simplify the notation in what follows, we use the symbol m rather than the more common one m_s in (2.8.20).

$$k = 2s, \quad k = 1, 2, \ldots \tag{2.8.19}$$

$$m = i - s, \quad i = 0, 1, \ldots, k. \tag{2.8.20}$$

Hence s and m take the possible values

$$s = \frac{1}{2}, 1, \frac{3}{2}, 2, \ldots \tag{2.8.21}$$

$$m = -s, -s+1, \ldots, s-1, s. \tag{2.8.22}$$

We may then rewrite (2.8.15) as

$$\psi^{a_1 \ldots a_{2s}}(\mathbf{x}) = \sum_{m=-s}^{s} C^{a_1 \ldots a_{2s}}(s+m)\psi(\mathbf{x}, m) \tag{2.8.23}$$

where $\psi(\mathbf{x}, m) = \psi_{s+m}(\mathbf{x})$ in a coordinate description.

From the orthonormality condition (2.8.12), we have the unitarity of the transformation in (2.8.23)

$$\sum_{a_1,\ldots,a_{2s}=1,2} \int d^3\mathbf{x}\, (\psi^{a_1 \ldots a_{2s}}(\mathbf{x}))^\dagger \phi^{a_1 \ldots a_{2s}}(\mathbf{x})$$

$$= \sum_{m=-s}^{s} \int d^3\mathbf{x}\, \psi^*(\mathbf{x}, m)\phi(\mathbf{x}, m) \tag{2.8.24}$$

and, in particular,

$$1 = \sum_{a_1,\ldots,a_{2s}=1,2} \|\psi^{a_1 \ldots a_{2s}}\|^2 = \sum_{m=-s}^{s} \int d^3\mathbf{x}\, |\psi(\mathbf{x}, m)|^2 \tag{2.8.25}$$

when normalized to one.

More generally, we may write instead of (2.8.23) in general

$$\psi^{a_1 \ldots a_{2s}} = \sum_{m=-s}^{s} C^{a_1 \ldots a_{2s}}(s+m)\psi(m). \tag{2.8.26}$$

The matrix in (2.8.7) is easily worked out, by using in the process (2.8.2), to be

$$\exp\left(\frac{i}{2}\varphi\, \mathbf{n}\cdot\boldsymbol{\sigma}\right) = \begin{pmatrix} \cos\frac{\varphi}{2} + in^3 \sin\frac{\varphi}{2} & i(n^1 - in^2)\sin\frac{\varphi}{2} \\ i(n^1 + in^2)\sin\frac{\varphi}{2} & \cos\frac{\varphi}{2} - in^3 \sin\frac{\varphi}{2} \end{pmatrix}$$

$$\equiv A \tag{2.8.27}$$

where $\mathbf{n} = (n^1, n^2, n^3)$. The inverse of the matrix A is obtained by replacing φ by $-\varphi$ in (2.8.27):

$$A^{-1} = \begin{pmatrix} \cos\frac{\varphi}{2} - in^3 \sin\frac{\varphi}{2} & -i\left(n^1 - in^2\right) \sin\frac{\varphi}{2} \\ -i\left(n^1 + in^2\right) \sin\frac{\varphi}{2} & \cos\frac{\varphi}{2} + in^3 \sin\frac{\varphi}{2} \end{pmatrix}. \tag{2.8.28}$$

The transformation

$$A^{a_1 b_1} \ldots A^{a_{2s} b_{2s}} \psi^{b_1 \ldots b_{2s}} \tag{2.8.29}$$

corresponding to the right-hand side of (2.8.26) may be explicitly carried out. To this end, we introduce two new variables h_1, h_2 defined by

$$h_a = g_b \left[A^{-1}\right]^{ba} \tag{2.8.30}$$

leading to

$$[A]^{ab} \frac{\partial}{\partial g_b} = \frac{\partial}{\partial h_a} \tag{2.8.31}$$

$$g_a = h_b [A]^{ba}. \tag{2.8.32}$$

Accordingly, from (2.8.1), (2.8.18)–(2.8.20), (2.8.26), we have

$$[A]^{a_1 b_1} \ldots [A]^{a_{2s} b_{2s}} \psi^{b_1 \ldots b_{2s}} = \sum_{m=-s}^{s} \left(\frac{1}{\sqrt{(2s)!(s+m)!(s-m)!}} \right.$$

$$\left. \times \frac{\partial}{\partial h_{a_1}} \ldots \frac{\partial}{\partial h_{a_{2s}}} \left(h_b [A]^{b1}\right)^{s+m} \left(h_c [A]^{c2}\right)^{s-m} \psi(m) \right). \tag{2.8.33}$$

Upon using the elementary binomial expansion

$$(a+b)^n = \sum_{q=0}^{n} \frac{n!}{q!(n-q)!} (a)^q (b)^{n-q} \tag{2.8.34}$$

for any non-negative integer n, as applied to

$$\left(h_b [A]^{b1}\right)^{s+m} \equiv \left(h_1 [A]^{11} + h_2 [A]^{21}\right)^{s+m} \tag{2.8.35}$$

$$\left(h_c [A]^{c2}\right)^{s-m} \equiv \left(h_1 [A]^{12} + h_2 [A]^{22}\right)^{s-m} \tag{2.8.36}$$

the right-hand side of (2.8.33) may be rewritten as

$$\sum_{m=-s}^{s} C^{a_1 \ldots a_{2s}}(s+m) \widetilde{\psi}(m) \tag{2.8.37}$$

where

$$\widetilde{\psi}(m) = \sum_{q,m'} \frac{\sqrt{(s+m)!(s-m)!(s+m')!(s-m')!}}{(s+m-q)!(m'-m+q)!q'!(s-m'-q)!}$$

$$\times \left([A]^{11}\right)^{s+m-q} \left([A]^{21}\right)^{m'-m+q} \left([A]^{12}\right)^{q} \left([A]^{22}\right)^{s-m'-q} \psi(m') \quad (2.8.38)$$

giving the transformation law for $\psi(m)$, and the q-sum is over all non-negative integers q for which the arguments of the factorials involving q are non-negative integers.

Upon writing
$$\psi(m) = \langle s, m | \psi \rangle \quad (2.8.39)$$
the spin operator $\mathbf{S}$ is defined through (see, in particular, (2.7.43), (2.8.23), (2.8.37), (2.8.1)),

$$\left\langle s, m \middle| \widetilde{\psi} \right\rangle = \langle s, m | \exp\left(\frac{i}{\hbar} \varphi \mathbf{n} \cdot \mathbf{S}\right) |\psi\rangle \quad (2.8.40)$$

which from (2.8.38) gives

$$\langle s, m | \exp\left(\frac{i}{\hbar} \varphi \mathbf{n} \cdot \mathbf{S}\right) = \sum_{q,m'} \frac{\sqrt{(s+m)!(s-m)!(s+m')!(s-m')!}}{(s+m-q)!(m'-m+q)!q'!(s-m'-q)!}$$

$$\times \left([A]^{11}\right)^{s+m-q} \left([A]^{21}\right)^{m'-m+q} \left([A]^{12}\right)^{q} \left([A]^{22}\right)^{s-m'-q} \langle s, m' |. \quad (2.8.41)$$

By differentiating the latter with respect to φ, using the expression for the matrix elements $[A]^{ab}$ in (2.8.27), and setting $\varphi = 0$, one easily obtains

$$\langle s, m | \mathbf{n} \cdot \mathbf{S} = (n^1 + in^2)\frac{\hbar}{2}\sqrt{(s-m)(s+m+1)} \langle s, m+1 |$$

$$+ (n^1 - in^2)\frac{\hbar}{2}\sqrt{(s+m)(s-m+1)} \langle s, m-1 |$$

$$+ n^3 \hbar m \langle s, m |. \quad (2.8.42)$$

Hence, in particular,

$$S^1 |s, m\rangle = \frac{\hbar}{2}\sqrt{(s-m)(s+m+1)} |s, m+1\rangle$$

$$+ \frac{\hbar}{2}\sqrt{(s+m)(s-m+1)} |s, m-1\rangle \quad (2.8.43)$$

$$S^2 |s, m\rangle = -i\frac{\hbar}{2}\sqrt{(s-m)(s+m+1)} |s, m+1\rangle$$

$$+ i\frac{\hbar}{2}\sqrt{(s+m)(s-m+1)}\,|s, m-1\rangle \qquad (2.8.44)$$

$$S^3\,|s,m\rangle = \hbar m\,|s,m\rangle. \qquad (2.8.45)$$

Upon applying the operators S^1, S^2, S^3 to (2.8.43)–(2.8.45) and using the latter equations all over again, one immediately obtains

$$\mathbf{S}^2\,|s,m\rangle = \hbar^2 s(s+1)\,|s,m\rangle \qquad (2.8.46)$$

$$[S^i, S^j]\,|s,m\rangle = i\hbar\varepsilon^{ijk} S^k\,|s,m\rangle. \qquad (2.8.47)$$

Using the definition that a particle is of spin s if $\mathbf{S}^2 = \hbar^2 s(s+1)\mathbf{1}$ (see §5.1, §5.4), (2.8.46) establishes the fact that the symmetric spinor $\psi^{a_1\cdots a_{2s}}$ is associated with spin s, and that the components of the spin along the x^3-axis, as dictated by (2.8.45), are given by $-\hbar s$, $\hbar(-s+1),\ldots,\hbar(s-1)$, $\hbar s$. The spin matrix $[\langle s', m'|\mathbf{S}|s,m\rangle]$ for any spin may be readily constructed from (2.8.43)–(2.8.45).

Equations (2.8.45), (2.8.46) also lead, respectively, to

$$\hbar(m-m')\,\langle s',m'|s,m\rangle = 0 \qquad (2.8.48)$$

$$\hbar^2\,[s(s+1) - s'(s'+1)]\,\langle s',m'|s,m\rangle = 0 \qquad (2.8.49)$$

establishing the orthogonality (orthonormality) of the eigenstates $|s,m\rangle$

$$\langle s',m'|s,m\rangle = \delta_{s',s}\delta_{m',m}. \qquad (2.8.50)$$

With the normalization condition in (2.8.25), $\psi(\mathbf{x}, m)$ denotes the wavefunction of a particle of spin s. In a matrix notation $\psi(\mathbf{x}) = (\psi(\mathbf{x}, s), \ldots, \psi(\mathbf{x}, -s))^\top$, one has, from (2.8.1), (2.8.23), (2.8.38), (2.8.40), the transformation law in (2.7.43) with the matrix elements of $\exp(i\varphi\,\mathbf{n}\cdot\mathbf{S}/\hbar)$ given below in (2.8.51), as directly obtained from (2.8.41), that

$$\langle s,m|\exp\left(\frac{i}{\hbar}\varphi\,\mathbf{n}\cdot\mathbf{S}\right)|s',m'\rangle$$

$$= \delta_{ss'}\sum_q \frac{\sqrt{(s+m)!(s-m)!(s+m')!(s-m')!}}{(s+m-q)!(m'-m+q)!q!(s-m'-q)!}$$

$$\times\,([A]^{11})^{s+m-q}\,([A]^{21})^{m'-m+q}\,([A]^{12})^{q}\,([A]^{22})^{s-m'-q} \qquad (2.8.51)$$

where the sum is over all non-negative integers q such that the arguments of the factorials are non-negative integers.

For example, for $\mathbf{n} = (0, 1, 0)$, we have from (2.8.27), (2.8.51)

$$\langle s,m|\exp\left(\frac{i}{\hbar}\varphi S^2\right)|s',m'\rangle$$

$$= \delta_{ss'} \sum_q \frac{\sqrt{(s+m)!(s-m)!(s+m')!(s-m')!}}{(s+m-q)!(m'-m+q)!q'!(s-m'-q)!}$$

$$\times (-1)^{m'-m+q} \left(\cos\frac{\varphi}{2}\right)^{2s-2q+m-m'} \left(\sin\frac{\varphi}{2}\right)^{2q+m'-m}. \qquad (2.8.52)$$

For spin $s = 1/2$,

$$\psi^a(\mathbf{x}) = \sum_{m=-s}^{s} C^a\left(\frac{1}{2}+m\right)\psi(\mathbf{x},m), \quad a = 1, 2 \qquad (2.8.53)$$

$$C^a(1) = \delta^{a1}, \quad C^a(0) = \delta^{a2} \qquad (2.8.54)$$

and the spin matrix is from (2.8.43)–(2.8.45),

$$\mathbf{S} = \frac{\hbar}{2}\boldsymbol{\sigma} \qquad (2.8.55)$$

and, for example, from (2.8.52)

$$\left[\exp\frac{i\varphi}{\hbar}S^2\right] = \begin{pmatrix} \cos\frac{\varphi}{2} & \sin\frac{\varphi}{2} \\ -\sin\frac{\varphi}{2} & \cos\frac{\varphi}{2} \end{pmatrix}. \qquad (2.8.56)$$

From (2.8.53), (2.8.54), the spinor may be written in a column matrix form

$$\psi(\mathbf{x}) = \begin{pmatrix} \psi(\mathbf{x},+1/2) \\ \psi(\mathbf{x},-1/2) \end{pmatrix} = \begin{pmatrix} 1 \\ 0 \end{pmatrix}\psi(\mathbf{x},+1/2) + \begin{pmatrix} 0 \\ 1 \end{pmatrix}\psi(\mathbf{x},-1/2). \qquad (2.8.57)$$

For spin $s = 1$,

$$\psi^{a_1 a_2}(\mathbf{x}) = \sum_{m=-1,0,1}^{s} C^{a_1 a_2}(1+m)\psi(\mathbf{x},m) \qquad (2.8.58)$$

and from (2.8.18) or (2.8.11):

$$C^{a_1 a_2}(1+1) = \delta^{a_1 1}\delta^{a_2 1} \qquad (2.8.59)$$

$$C^{a_1 a_2}(1+0) = \frac{1}{\sqrt{2}}\left(\delta^{a_1 1}\delta^{a_2 2} + \delta^{a_1 2}\delta^{a_2 1}\right) \qquad (2.8.60)$$

$$C^{a_1 a_2}(1-1) = \delta^{a_1 2}\delta^{a_2 2} \qquad (2.8.61)$$

and the spin $\mathbf{S} = (S^1, S^2, S^3)$ has from (2.8.43)–(2.8.45) the components

$$S^1 = \frac{\hbar}{\sqrt{2}}\begin{pmatrix} 0 & 1 & 0 \\ 1 & 0 & 1 \\ 0 & 1 & 0 \end{pmatrix} \qquad (2.8.62)$$

$$S^2 = i\frac{\hbar}{\sqrt{2}} \begin{pmatrix} 0 & -1 & 0 \\ 1 & 0 & -1 \\ 0 & 1 & 0 \end{pmatrix} \qquad (2.8.63)$$

$$S^3 = \hbar \begin{pmatrix} 1 & 0 & 0 \\ 0 & 0 & 0 \\ 0 & 0 & -1 \end{pmatrix} \qquad (2.8.64)$$

in the diagonal representation for S^3. For (2.8.52), we have

$$\left[\exp\frac{i\varphi}{\hbar}S^2\right] = \begin{pmatrix} (1+\cos\varphi)/2 & (\sin\varphi)/\sqrt{2} & (1-\cos\varphi)/2 \\ -(\sin\varphi)/\sqrt{2} & \cos\varphi & (\sin\varphi)/\sqrt{2} \\ (1-\cos\varphi)/2 & -(\sin\varphi)/\sqrt{2} & (1+\cos\varphi)/2 \end{pmatrix}. \qquad (2.8.65)$$

We will recover the structures (2.8.59)–(2.8.61) in §5.5 when adding two spin 1/2's. It is remarkable that the system (2.8.18)–(2.8.20) provides the general solution for describing a spin s out of $2s$ spin 1/2's and generalizes the expressions (2.8.59)–(2.8.61).

As mentioned before spin 0 may be described by a spinor of rank 0. That is, it transforms as

$$\psi'(\mathbf{x}') = \psi(\mathbf{x}) \qquad (2.8.66)$$

(see also (2.7.13)). It may be also described by a second rank *anti*-symmetric spinor $\Phi^{a_1 a_2}(\mathbf{x})$ (see Problem 2.15).

2.8.2 Rotation of a Spinor by 2π Radians

An intriguing result which distinguishes half-odd integral spins ($s = 1/2, 3/2, \ldots$) from integral spins ($s = 0, 1, 2, \ldots$) is that under a rotation of a coordinate system by 2π radians, about any axis, thus bringing us back to the same initial situation, a half-odd integral spin spinor, is read to acquire an overall *minus* sign multiplied by the initial spinor. This relative phase change for such spin values is physically observable and will be discussed later in §8.12. Particles with half-odd integer spins are referred to as fermions, while particles with integer spins are referred to as bosons.

To see how this phase change occurs put $\varphi = 2\pi$ in (2.8.7) to obtain

$$\exp\left(i\frac{2\pi}{2}\mathbf{n}\cdot\boldsymbol{\sigma}\right) = \mathbf{1}\cos\pi + i(\mathbf{n}\cdot\boldsymbol{\sigma})\sin\pi = -\mathbf{1}. \qquad (2.8.67)$$

Hence (2.8.1), with $k = 2s$, leads to

$$\psi'^{a_1 \ldots a_{2s}}(\mathbf{x}) = (-1)^{2s}\psi^{a_1 \ldots a_{2s}}(\mathbf{x}) \qquad (2.8.68)$$

establishing the relative phase change which results under a 2π rotation for $s = 1/2, 3/2, \ldots$. Only through a rotation by 4π (or by integer multiples of 4π) a $+1$ phase is obtained for half-odd integral spins!

For the matrix element (2.8.51), we also have

$$\langle s, m| \exp\left(i\frac{2\pi}{\hbar}\mathbf{n}\cdot\mathbf{S}\right)|s', m'\rangle = (-1)^{2s}\delta_{ss'}\delta_{mm'} \tag{2.8.69}$$

and hence also for the wavefunctions (see (2.8.23), (2.8.38))

$$\psi'(\mathbf{x}, m) = (-1)^{2s}\psi(\mathbf{x}, m). \tag{2.8.70}$$

2.8.3 Time Reversal and Parity Transformation

We next consider the time reversal operation for arbitrary spins. To this end, consider first the situation for spin $1/2$.

In a vector space generated by 2×2 matrices, the most general expression for the time reversal operator $\mathcal{T}$ for spin $1/2$ is

$$\mathcal{T} = (\mathbf{1}a + ib\mathbf{n}\cdot\boldsymbol{\sigma})K \tag{2.8.71}$$

expanded in terms of the complete set of 2×2 matrices $\mathbf{1}$, σ_i, $i = 1, 2, 3$, where a and b are, in general, complex numbers, $\mathbf{n}$ is a unit vector, and K is the *complex conjugation* operation to ensure, in particular, the anti-unitary nature of $\mathcal{T}$ (see §2.6). The latter must satisfy the anti-commutation relation in (2.6.22) with the spin operator $\mathbf{S}$ in (2.8.55). That is, we must have

$$(\mathbf{1}a + ib\mathbf{n}\cdot\boldsymbol{\sigma})\sigma^{j*} + \sigma^j(\mathbf{1}a + ib\mathbf{n}\cdot\boldsymbol{\sigma}) = 0 \tag{2.8.72}$$

for $j = 1, 2, 3$. It is easily checked that the solution consistent with (2.8.72) is

$$a = 0, \quad \mathbf{n} = (0, 1, 0) \tag{2.8.73}$$

i.e.,

$$\mathcal{T} = ib\sigma_2 K. \tag{2.8.74}$$

The anti-unitarity nature of $\mathcal{T}$ then implies that $|b|^2 = 1$, where we note, in particular, the *reality* of the matrix $(i\sigma_2)$. That is, up to an overall phase factor, b, which we define for a basic spin $1/2$ state to be one, (2.8.74) gives

$$\mathcal{T} = (i\sigma_2)K. \tag{2.8.75}$$

For two successive time reversals, we obtain

$$\mathcal{T}^2 = -1. \tag{2.8.76}$$

A spinor then transforms as

2.8 Spinors and Arbitrary Spins

$$\psi'^a(\mathbf{x}) = K[i\sigma_2]^{ab}\psi^b(\mathbf{x}) \tag{2.8.77}$$

under time reversal. Using the identity

$$[i\sigma_2]^{ab} = \delta^{a1}\delta^{b2} - \delta^{a2}\delta^{b1} \tag{2.8.78}$$

we obtain, from (2.8.53), the following transformation law of a spin 1/2 wavefunction

$$\psi(\mathbf{x},+1/2) \to \psi^*(\mathbf{x},-1/2) \tag{2.8.79}$$

$$\psi(\mathbf{x},-1/2) \to -\psi^*(\mathbf{x},+1/2) \tag{2.8.80}$$

causing, in particular, a spin flip.

For an arbitrary spin s, we define from (2.8.77), the following transformation law under time reversal

$$\psi'^{a_1\ldots a_{2s}}(\mathbf{x}) = K[i\sigma_2]^{a_1 b_1}\ldots[i\sigma_2]^{a_{2s} b_{2s}}\psi^{b_1\ldots b_{2s}}(\mathbf{x}) \tag{2.8.81}$$

where an overall complex conjugation operation is retained to ensure the anti-unitarity nature of the time reversal operation.

To see how a wavefunction $\psi(\mathbf{x},m)$ (see (2.8.23), (2.8.25)) transforms under time reversal, it is easily proved from (2.8.18), (2.8.78), that

$$[i\sigma_2]^{a_1 b_1}\ldots[i\sigma_2]^{a_{2s} b_{2s}}C^{b_1\ldots b_{2s}}(s+m)$$

$$= (-1)^{s+m}C^{a_1\ldots a_{2s}}(s-m). \tag{2.8.82}$$

Hence

$$\psi'^{a_1\ldots a_{2s}}(\mathbf{x}) = \sum_{m=-s}^{s}(-1)^{s+m}C^{a_1\ldots a_{2s}}(s-m)\psi^*(\mathbf{x},m)$$

$$= \sum_{m=-s}^{s}C^{a_1\ldots a_{2s}}(s+m)\psi'(\mathbf{x},m) \tag{2.8.83}$$

where

$$\psi'(\mathbf{x},m) = (-1)^{s-m}\psi^*(\mathbf{x},-m) \tag{2.8.84}$$

giving the transformation law $\psi(\mathbf{x},m) \to \psi'(\mathbf{x},m)$. For $s = 1/2$ we recover the expressions in (2.8.79), (2.8.80).

Again we have a distinction between half-odd integral spins and integral spins when the time reversal operation is applied twice in succession. From (2.8.81), (2.8.78), we have

$$T^2 = (-1)^{2s}\mathbf{1}. \tag{2.8.85}$$

The parity operator is similarly treated, using now the commutation relation in (2.6.13), and one may infer that (see Problem 2.17)

$$\psi'^a(\mathbf{x}) = \xi \psi^a(-\mathbf{x}) \qquad (2.8.86)$$

where ξ is a phase factor, and a similar expression for a spinor of rank k then also follows.

An important consequence of time reversal as a *symmetry* operation arises from the following consideration.

Upon multiplying the Schrödinger equation

$$i\hbar \frac{\partial}{\partial t} |\psi\rangle = H |\psi\rangle \qquad (2.8.87)$$

from the left by $\mathcal{T}$, we may rewrite the resulting equation as

$$-i\hbar \frac{\partial}{\partial t}(\mathcal{T}|\psi\rangle) = H_\mathcal{T}(\mathcal{T}|\psi\rangle) \qquad (2.8.88)$$

where

$$H_\mathcal{T} = \mathcal{T} H \mathcal{T}^{-1} \qquad (2.8.89)$$

and we have used the complex conjugation nature of $\mathcal{T}$ in (2.8.81) replacing i by $-i$ in (2.8.88). For time reversal as a symmetry operation, (2.6.21) implies that

$$\mathcal{T} H \mathcal{T}^{-1} = H. \qquad (2.8.90)$$

A theory which is invariant under time reversal, that is satisfying (2.8.90) gives rise to the following degeneracy problem of energy levels discussed below.

2.8.4 Kramers Degeneracy

For time reversal as a *symmetry* operation for which $\mathcal{T}^2 = -1$, each eigenvalue of H is at least two-fold degenerate. This result easily follows by noting that

$$H|\psi\rangle = E|\psi\rangle \qquad (2.8.91)$$

implies, from (2.8.90), that $\mathcal{T}|\psi\rangle$ is an eigenstate of H with the same eigenvalue E. Also from the chain of equalities

$$\langle \mathcal{T}\psi|\psi\rangle = \langle \mathcal{T}\mathcal{T}\psi|\mathcal{T}\psi\rangle^* = -\langle \psi|\mathcal{T}\psi\rangle^* = -\langle \mathcal{T}\psi|\psi\rangle, \qquad (2.8.92)$$

we note that $\mathcal{T}|\psi\rangle$ and $|\psi\rangle$ are *orthogonal* and are independent eigenstates of H. This type of degeneracy is referred to as Kramers degeneracy. For example, for the simple hydrogen atom, each energy level is at least two-fold degenerate due to the electron spin.

Appendix to §2.8: Transformation Rule of a Spinor of Rank One Under a Coordinate Rotation

We are interested in the transformation rule of spinors under a coordinate rotation. To introduce spinors with this in mind, we define for each point labelled by $\mathbf{x}$ in $\mathbb{R}^3$, a complex 2×2 matrix

$$X = \mathbf{x} \cdot \boldsymbol{\sigma} \equiv \begin{pmatrix} x^3 & x^1 - ix^2 \\ x^1 + ix^2 & -x^3 \end{pmatrix} \quad \text{(A-2.8.1)}$$

where $\boldsymbol{\sigma} = (\sigma_1, \sigma_2, \sigma_3)$, with σ_1, σ_2, σ_3 denoting the Pauli matrices (see (2.8.2)).

Some properties of the Pauli matrices are collected here for convenience and are given by

$$\sigma_i \sigma_j = \delta_{ij} + i\varepsilon_{ijk}\sigma_k \quad \text{(A-2.8.2)}$$

$$\sigma_i \sigma_j \sigma_k = \sigma_i \delta_{jk} - \sigma_j \delta_{ik} + \sigma_k \delta_{ij} + i\varepsilon_{ijk} \quad \text{(A-2.8.3)}$$

$$\operatorname{Tr}(\sigma_i) = 0, \quad \frac{1}{2}\operatorname{Tr}(\sigma_i \sigma_j) = \delta_{ij}, \quad \frac{1}{2}\operatorname{Tr}(\sigma_i \sigma_j \sigma_k) = i\varepsilon_{ijk} \quad \text{(A-2.8.4)}$$

and $\varepsilon_{ijk} \equiv \varepsilon^{ijk}$ is defined in (2.1.5).

From the second equality in (A-2.8.4), we may solve for $\mathbf{x}$ in terms of X

$$\mathbf{x} = \frac{1}{2}\operatorname{Tr}(\boldsymbol{\sigma} X). \quad \text{(A-2.8.5)}$$

We consider the transformations $X \to X'$ defined by

$$X' = AXA^\dagger = \mathbf{x}' \cdot \boldsymbol{\sigma} \quad \text{(A-2.8.6)}$$

where $\mathbf{x}'$ (see (2.1.4)) is given by

$$x'^i = R^{ij}(\varphi, \mathbf{n}) x^j \quad \text{(A-2.8.7)}$$

for all 2×2 complex unitary matrices $A \equiv A(\varphi, \mathbf{n})$, corresponding to coordinate rotations by angles $\{\phi\}$ about unit vectors $\{\mathbf{n}\}$.

We note that

$$\det X = -\mathbf{x} \cdot \mathbf{x} \quad \text{(A-2.8.8)}$$

implying from (A-2.8.6) and the fact that

$$\mathbf{x}' \cdot \mathbf{x}' = \mathbf{x} \cdot \mathbf{x} \quad \text{(A-2.8.9)}$$

the invariance property

$$\det X = \det X' \quad \text{(A-2.8.10)}$$

under a coordinate rotation. Equation (A-2.8.10) also follows directly from the unitarity condition of the matrices A in (A-2.8.6).

For a subsequent coordinate rotation

$$x'''^i = R^{ij}(\varphi', \mathbf{n}')x'^j \quad \text{(A-2.8.11)}$$

$$X'' = A'X'A'^\dagger = (A'A)X(A'A)^\dagger \equiv \mathbf{x}'' \cdot \boldsymbol{\sigma} \quad \text{(A-2.8.12)}$$

Since the product of unitary operators defines a unitary operator, the set of all such transformations in (A-2.8.6) forms a group.

The most general structure of a 2×2 complex matrix A is given by

$$A = \mathbf{1}a + i\mathbf{b} \cdot \boldsymbol{\sigma} \quad \text{(A-2.8.13)}$$

giving

$$A^\dagger A = |a|^2 + \mathbf{b} \cdot \mathbf{b}^* + i(b^{*i}b^j \varepsilon^{ijk} - b^{*k}a + a^* b^k)\sigma_k. \quad \text{(A-2.8.14)}$$

Unitarity then implies that

$$|a|^2 + \mathbf{b} \cdot \mathbf{b}^* = 1 \quad \text{(A-2.8.15)}$$

$$b^{*i}b^j \varepsilon^{ijk} - b^{*k}a + a^* b^k = 0. \quad \text{(A-2.8.16)}$$

On the other hand (A-2.8.6), (A-2.8.13) lead to the equality

$$\left(|a|^2 - \mathbf{b} \cdot \mathbf{b}^*\right)\sigma_j + \left(b^{j*}\mathbf{b} \cdot \boldsymbol{\sigma} + b^j \mathbf{b}^* \cdot \boldsymbol{\sigma}\right) - (ab^{*i} + a^* b^i)\varepsilon^{ijk}\sigma_k$$

$$= \cos\varphi \, \sigma_j + n^j \mathbf{n} \cdot \boldsymbol{\sigma}(1 - \cos\varphi) - n^i \sin\varphi \, \varepsilon^{ijk}\sigma_k \quad \text{(A-2.8.17)}$$

where we have made use of (A-2.8.2)–(A-2.8.4).

Upon multiplying (A-2.8.17) by σ_j, we obtain

$$3|a|^2 - \mathbf{b} \cdot \mathbf{b}^* = 2\cos\varphi + 1 \quad \text{(A-2.8.18)}$$

$$a\mathbf{b}^* \cdot \boldsymbol{\sigma} + a^* \mathbf{b} \cdot \boldsymbol{\sigma} = \mathbf{n} \cdot \boldsymbol{\sigma} \sin\varphi. \quad \text{(A-2.8.19)}$$

Equations (A-2.8.15), (2.8.18) then imply that

$$a = e^{i\delta} \cos(\varphi/2) \quad \text{(A-2.8.20)}$$

$$b^j = e^{i\delta_j} n^j \sin(\varphi/2) \quad \text{(A-2.8.21)}$$

where δ, δ_j are real. On the other hand (A-2.8.19) is equivalent to the conditions

$$ab^{*j} + a^* b^j = n^j \sin\varphi \quad \text{(A-2.8.22)}$$

which from (A-2.8.20), (A-2.8.21) imply that $\cos(\delta - \delta_j) = 1$ for $j = 1, 2, 3$, and hence that

$$b^j = e^{i\delta} n^j \sin(\varphi/2). \quad \text{(A-2.8.23)}$$

Accordingly, a rotation by an angle φ about a unit vector $\mathbf{n}$ induces the transformation $X \to X'$ where X, X' are, respectively, defined in (A-2.8.1), (A-2.8.6), with the matrix A, up to an overall phase factor which by convention we set equal to one, is given by

$$A = \mathbf{1} \cos \varphi/2 + \mathrm{i} \mathbf{n} \cdot \boldsymbol{\sigma} \sin \varphi/2. \qquad \text{(A-2.8.24)}$$

It is readily checked that
$$\det A = 1. \qquad \text{(A-2.8.25)}$$

Conversely, it is easily shown (see Problem 2.18) that the set of all transformations $X \to X'$ with

$$X' = BXB^\dagger \qquad \text{(A-2.8.26)}$$

for any 2×2 complex unitary matrices B of determinant one each one induces a coordinate rotation. With B having the general structure as given on the right-hand side of (A-2.8.13), $\det B = 1$ implies that

$$a^2 + \mathbf{b} \cdot \mathbf{b} = 1. \qquad \text{(A-2.8.27)}$$

This together with the unitarity condition in (A-2.8.15) imply that a and $\mathbf{b}$ are real and may be parameterized as in (A-2.8.20), (A-2.8.21) with $e^{\mathrm{i}\delta} \equiv 1$. Equation (A-2.8.18) is then automatically satisfied.

The matrices A in (A-2.8.24) operate on two dimensional objects

$$\psi = \begin{pmatrix} \psi^1 \\ \psi^2 \end{pmatrix} \qquad \text{(A-2.8.28)}$$

referred to as spinors or more precisely as spinors of rank 1, guaranteeing the invariance property of the inner product

$$\sum_{a=1,2} \psi'^{*a} \psi'^a = \sum_{a=1,2} \psi^{*a} \psi^a \qquad \text{(A-2.8.29)}$$

under a coordinate rotation, where

$$\psi'^a = [A]^{ab} \psi^b \qquad \text{(A-2.8.30)}$$

and $[\cdot]^{ab}$ denotes a matrix element of the matrix in question.

Equation (A-2.8.30), with A given in (A-2.8.24), provides the rule of transformation of a spinor of rank 1 under a coordinate rotation by an angle φ about a unit vector $\mathbf{n}$. The transformation law for a spinor of rank k under a coordinate rotation is defined in (2.8.1). Equation (2.8.7), shows the equivalent exponential form of the operator A in (A-2.8.24):

$$A = \exp\left(\mathrm{i}\frac{\varphi}{2} \mathbf{n} \cdot \boldsymbol{\sigma}\right). \qquad \text{(A-2.8.31)}$$

2.9 Supersymmetry

In quantum field theory, supersymmetric transformations give rise to transformations between bosons and fermions. As a unifying principle, supersymmetric Lagrangians in field theory have this appealing invariance property built in. Such a symmetry then puts constraints on physical processes involving the underlying particles. Although nature is not necessarily supersymmetric, the imposition of such a symmetry on a field theory has turned out to have some advantages over its non-symmetric counterpart. In particular, it has bean shown, in general, that the number of divergences (encountered in defining coupling parameters, masses and related physical quantities) are consequently reduced. Also the imposition of a symmetry narrows down the class of interactions that one may write down. With the hope of recovering more realistic theories, one may then break such a symmetry guided by some physical principles.

Supersymmetric transformations may be also defined in quantum mechanics, in general, as an abstract notion which turn out to have far reaching consequences and yield to several physical applications. For example, in the elementary Bose/Fermi oscillator, as a prototype of field theory models, the explicit transformations between Bose and Fermi states may be readily demonstrated. This and other examples will be given later (see, e.g., §6.5, §4.7).

In this section, we define supersymmetric transformations which parallel the treatment of the Galilean transformations carried out in §2.2, §2.3. The closed path transformations lead to basic commutation/anti-commutation relations of the supersymmetry generators and of the Hamiltonian in a way similar to the earlier study. Consequences of these relations and physical applications will be given later as mentioned above.

Together with the time variable t, as a c-number, we introduce anti-commuting c-number variables θ, θ^* having the following transformations:

$$t \to t - \tau + i\xi^*\theta - i\theta^*\xi = \bar{t} \tag{2.9.1}$$

$$\theta \to \theta + \xi = \bar{\theta}, \quad \bar{\theta}^* = \theta^* + \xi^* \tag{2.9.2}$$

where by anti-commuting c-numbers θ, θ^* one means

$$\{\theta, \theta\} = 0, \quad \{\theta^*, \theta^*\} = 0, \quad \{\theta, \theta^*\} = 0 \tag{2.9.3}$$

where

$$\{A, B\} \equiv AB + BA. \tag{2.9.4}$$

Equations (2.9.3), in particular, imply that

$$(\theta)^2 = 0, \quad (\theta^*)^2 = 0. \tag{2.9.5}$$

The variables θ, $\bar{\theta}$ are referred to as *Grassmann* variables.[6]

As before, τ is a c-number. On the other hand, θ, θ^*, ξ, ξ^*, $\bar{\xi}$, $\bar{\xi}^*$,..., anti-commute with each other but commute with t, τ, $\bar{\tau}$,....

For a subsequent transformation to (2.9.1), (2.9.2), we have

$$\bar{\bar{t}} = \bar{t} - \bar{\tau} + i\bar{\xi}^*\bar{\theta} - i\bar{\theta}^*\bar{\xi} \qquad (2.9.6)$$

$$\bar{\bar{\theta}} = \bar{\theta} + \bar{\xi} \qquad (2.9.7)$$

and similarly for $\bar{\bar{\theta}}^*$, or

$$\bar{\bar{t}} = t - \left(\tau + \bar{\tau} - i\bar{\xi}^*\xi + i\xi^*\bar{\xi}\right) + i\left(\xi^* + \bar{\xi}^*\right)\theta - i\theta^*\left(\xi + \bar{\xi}\right) \qquad (2.9.8)$$

$$\bar{\bar{\theta}} = \theta + \left(\xi + \bar{\xi}\right). \qquad (2.9.9)$$

The transformations in (2.9.1), (2.9.2) may be specified by the pair

$$(\tau, \xi) \qquad (2.9.10)$$

which for simplicity of notation we have suppressed the entry ξ^*.

From (2.9.1), (2.9.2), (2.9.8), (2.9.9) we have the following group properties:

(1)
$$(\tau_2, \xi_2)(\tau_1, \xi_1) = (\tau_1 + \tau_2 - i\xi_2^*\xi_1 + i\xi_1^*\xi_2, \xi_1 + \xi_2). \qquad (2.9.11)$$

(2) For the identity element I
$$I = (0, 0). \qquad (2.9.12)$$

(3) The inverse is given by
$$(\tau, \xi)^{-1} = (-\tau, -\xi) \qquad (2.9.13)$$

as is easily checked that

$$(-\tau, -\xi)(\tau, \xi) = (0, 0). \qquad (2.9.14)$$

(4) Finally, we have the associativity rule

$$(\tau_3, \xi_3)\left[(\tau_2, \xi_2)(\tau_1, \xi_1)\right] = \left[(\tau_3, \xi_3)(\tau_2, \xi_2)\right](\tau_1, \xi_1). \qquad (2.9.15)$$

We consider successive transformations following a closed path given by

$$(\tau_2, \xi_2)^{-1}(\tau_1, \xi_1)^{-1}(\tau_2, \xi_2)(\tau_1, \xi_1) \qquad (2.9.16)$$

represented pictorially by the box diagram

---

[6] A fairly detailed treatment of Grassmann variables will be given in §10.6.

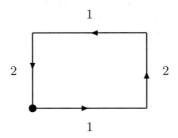

emphasizing the reversal of the transformations in the third and the fourth segments of the path.

The group properties (1)–(4) above then lead to the following rule associated with the closed path:

$$(\tau_2, \xi_2)^{-1} (\tau_1, \xi_1)^{-1} (\tau_2, \xi_2) (\tau_1, \xi_1) = (\tau, \xi) \qquad (2.9.17)$$

where

$$\tau = -2\,\mathrm{i}\,\xi_2^* \xi_1 + 2\,\mathrm{i}\,\xi_1^* \xi_2 \qquad (2.9.18)$$

$$\xi = 0. \qquad (2.9.19)$$

Since the above transformations are connected with the identity, the corresponding transformations in the vector space of physical states is represented by unitary transformations. For infinitesimal transformations, we may write for the overall transformation

$$U = 1 + \frac{\mathrm{i}}{\hbar} G + \ldots \qquad (2.9.20)$$

where

$$G = -\delta\tau H + \mathrm{i}\,\delta\xi^* Q - \mathrm{i}\,Q^\dagger\,\delta\xi \qquad (2.9.21)$$

and

$$U_j = 1 + \frac{\mathrm{i}}{\hbar} G_j + \ldots \qquad (2.9.22)$$

$$G_j = -\delta\tau_j H + \mathrm{i}\,\delta\xi_j^* Q - \mathrm{i}\,Q^\dagger\,\delta\xi_j \qquad (2.9.23)$$

$j = 1, 2$, and Q, $Q^\dagger$ are generators of supersymmetric transformations.

Upon writing

$$U = U_2^{-1} U_1^{-1} U_2 U_1 \qquad (2.9.24)$$

we have the identity

$$G = \frac{1}{\mathrm{i}\hbar}[G_1, G_2] \qquad (2.9.25)$$

and, from (2.9.18), (2.9.19),

$$\delta\tau = -2\,\mathrm{i}\,\delta\xi_2^*\,\delta\xi_1 + 2\,\mathrm{i}\,\delta\xi_1^*\,\delta\xi_2 \qquad (2.9.26)$$

$$\delta\xi = 0 \qquad (2.9.27)$$

$\delta\xi_1$, $\delta\xi_1^*$, $\delta\xi_2$, $\delta\xi_2^*$ anti-commute with Q, $Q^\dagger$ and anti-commute with each other. $\delta\xi_1$, $\delta\xi_1^*$, $\delta\xi_2$, $\delta\xi_2^*$ however, commute with $\delta\tau_1$, $\delta\tau_2$, H.

The commutator on the right-hand side of (2.9.25) is easily worked out to

$$[G_1, G_2] = (\mathrm{i}\delta\tau_2\,\delta\xi_1^* - \mathrm{i}\delta\tau_1\,\delta\xi_2^*)\,[H, Q]$$
$$+ (\mathrm{i}\delta\tau_1\,\delta\xi_2 - \mathrm{i}\delta\tau_2\,\delta\xi_1)\,[Q^\dagger, H]$$
$$+ (\delta\xi_1^*\,\delta\xi_2^*)\,\{Q, Q\}$$
$$+ (\delta\xi_1\,\delta\xi_2)\,\{Q^\dagger, Q^\dagger\}$$
$$+ (\delta\xi_1^*\,\delta\xi_2 - \delta\xi_2^*\,\delta\xi_1)\,\{Q, Q^\dagger\} \qquad (2.9.28)$$

which, from (2.9.21), (2.9.25)–(2.9.27), and upon the comparison of (2.9.28) with $\mathrm{i}\hbar G$ in (2.9.21) yields to

$$[H, Q] = 0 \qquad (2.9.29)$$

$$[H, Q^\dagger] = 0 \qquad (2.9.30)$$

$$\{Q, Q\} = 0 \quad \text{or} \quad (Q)^2 = 0 \qquad (2.9.31)$$

$$\{Q^\dagger, Q^\dagger\} = 0 \quad \text{or} \quad (Q^\dagger)^2 = 0 \qquad (2.9.32)$$

and

$$H = \frac{1}{2\hbar}\{Q, Q^\dagger\}. \qquad (2.9.33)$$

Applications of the results obtained above will be given later (e.g., in §4.7, §6.5 also in §8.2).

Problems

2.1. Establish the properties (2.1.18)–(2.1.20), (2.1.27)–(2.1.30) for the matrices M_1, M_2, M_3 involved in the expression for the rotation matrix R in (2.1.21)

2.2. Work out in detail the group properties (2.2.8)–(2.2.12) of the Galilean transformation and also the infinitesimal changes in the coordinate labels given in (2.2.26), (2.2.27) together with (2.2.22)–(2.2.25) under the closed path transformation (2.2.13).

140   2 Symmetries and Transformations

2.3. Derive the entries in Table 2.1 below (2.3.15) by using the commutation relation in (2.3.12) and the expressions for the G and G_j in (2.3.13), and (2.3.14)/(2.3.15), respectively.

2.4. Obtain the expression for the coefficient E in (2.3.16) and show explicitly it is equal to zero. Carry out a similar analysis for the coefficient B in entry 7 of Table 2.1, to show finally that it is zero as well.

2.5. Provide another derivation of (2.3.39) for finite translations without building it up from infinitesimal translations.

2.6. Using the position variables defined in (2.5.55)–(2.5.57), show that the Hamiltonian in (2.5.58) is of the form (2.5.48) for the two cluster problem with one cluster consisting of one particle and the other involving two particles.

2.7. Extend the analysis of the two cluster case given in Problem 2.6, to the case where each cluster consists of two particles as in the interaction of two hydrogen atoms. Choose your variables conveniently.

2.8. Repeat the analysis in §2.6, where it was shown that the parity transformation is to be implemented by a unitary operation, to show that the time reversal transformation is to be implemented by an anti-unitary one.

2.9. Show that consistently that no phase terms need to be introduced on the right-hand sides of (2.6.11), (2.6.20).

2.10. Show that (2.7.2), (2.7.3) lead to (2.7.4) for $\delta \mathbf{L}$, where $\mathbf{L}$ denotes the orbital angular momentum operator, in (2.7.11).

2.11. Establish the commutation relations for spin in (2.7.7)–(2.7.10) as they follow from the earlier commutation relations involving $\mathbf{J}$, $\mathbf{X}$, H, $\mathbf{P}$.

2.12. Show that the spin of a system of particles, as one composite object, obtained in (2.7.30) satisfies the commutation relations in (2.7.31).

2.13. Since
$$\exp\left(\mathrm{i}\frac{\varphi}{2}\mathbf{n}\cdot\boldsymbol{\sigma}\right) = \cos\left(\frac{\varphi}{2}\mathbf{n}\cdot\boldsymbol{\sigma}\right) + \mathrm{i}\sin\left(\frac{\varphi}{2}\mathbf{n}\cdot\boldsymbol{\sigma}\right)$$
use the fact that $(\mathbf{n}\cdot\boldsymbol{\sigma})^2 = 1$ and the facts that the cosine and sine functions are, respectively, even and odd functions, to obtain another derivation of (2.8.7).

2.14. Show that the expression in (2.8.11) may be rewritten as in (2.8.18) in terms of the two variables g_1, g_2.

2.15. Prove that spin 0 may be described by a second rank anti-symmetric spinor and verify explicitly, in the process, the spin 0 content of such a description.

2.16. Establish the identity in (2.8.82).

2.17. Repeat the analysis given through (2.8.71)–(2.8.77), (2.8.81) for time reversal, to the parity operation to derive, in the process, (2.8.86).

2.18. Provide the details to show that the set of all transformation $X \to X'$, with X' given in (A-2.8.26), induce a coordinate rotation.

2.19. Work out the closed path exact supersymmetric transformation law given in (2.9.17)–(2.9.19).

2.20. Prove the commutation rule $[G_1, G_2]$ given in (2.9.28) involving the supersymmetry generators and the Hamiltonian, and finally derive (2.9.29)–(2.9.33).

3

Uncertainties, Localization, Stability and Decay of Quantum Systems

Uncertainties, localization and stability are significant features concerning quantum systems. These are aspects which distinguish the latter systems from classical ones. This chapter is involved with such key points including related ones as the boundedness of the spectra of Hamiltonians of physical systems from below. This property of boundedness from below is important as otherwise a system would collapse to such a level as to release, in the process, an infinite amount of energy which is physically meaningless and the system, in question, would be unstable. The first section deals with analyses related to investigating the probability of the "fall" of a particle into another in a bound state as well as in determining a lower bound for the average spatial extension of such a system and in deriving an upper bound for the average kinetic energy into consideration in addition to other details. This includes investigating the nature of the resistance of a system with a large number of Fermi particles to the increase in its density. The boundedness of the spectra of Hamiltonians from below is the subject of §3.2–§3.4 for several classes of interactions and §3.4 is involved with multi-particle systems. The final section §3.5 is concerned with the decay of quantum systems and special emphasis is put on the celebrated Paley-Wiener Theorem for describing a physically consistent theory of quantum decay in which the underlying Hamiltonians are bounded from below.

3.1 Uncertainties, Localization and Stability

3.1.1 A Basic Inequality

Our starting inequality to study uncertainties and localization in quantum physics is the following one:

$$\int d^\nu x \left| [(\boldsymbol{\nabla} - i\mathbf{a}) + \alpha (\mathbf{x} - \mathbf{b}) g(\mathbf{x})] \psi(\mathbf{x}) \right|^2 \geqslant 0 \qquad (3.1.1)$$

where **a**, **b** are real constant vectors, α is a real parameter and $g(\mathbf{x})$ is a real function. For greater generality, the space dimension, denoted by the natural number ν, is left arbitrary.

Upon expanding the above inequality and multiplying the latter by $\hbar^2$, we obtain

$$\left\langle (\mathbf{p} - \mathbf{a}\hbar)^2 \right\rangle - \alpha \hbar^2 \left\langle \boldsymbol{\nabla} \cdot [(\mathbf{x} - \mathbf{b})g(\mathbf{x})] \right\rangle + \alpha^2 \hbar^2 \left\langle (\mathbf{x} - \mathbf{b})^2 g^2(\mathbf{x}) \right\rangle \geqslant 0. \quad (3.1.2)$$

Here we have used the notation $\langle A \rangle$ for $\langle \psi | A | \psi \rangle$. Minimizing the left-hand side of the above over α gives the useful bound

$$\left\langle (\mathbf{p} - \mathbf{a}\hbar)^2 \right\rangle \geqslant \frac{\hbar^2}{4} \frac{\langle \boldsymbol{\nabla} \cdot (\mathbf{x} - \mathbf{b})g(\mathbf{x}) \rangle^2}{\langle (\mathbf{x} - \mathbf{b})^2 g^2(\mathbf{x}) \rangle}. \quad (3.1.3)$$

3.1.2 Uncertainties

For example, for $g(\mathbf{x}) = 1$, (3.1.3) leads to the familiar Heisenberg uncertainty principle inequality

$$\left\langle (\mathbf{x} - \mathbf{b})^2 \right\rangle \left\langle (\mathbf{p} - \mathbf{a}\hbar)^2 \right\rangle \geqslant \frac{\nu^2 \hbar^2}{4}. \quad (3.1.4)$$

Equation (3.1.4) puts a lower limit on the product of the expectations of the deviations squared of the position and momentum about the constant vectors **b** and $\mathbf{a}\hbar$, respectively. For $\mathbf{b} = \langle \mathbf{x} \rangle$, $\mathbf{a}\hbar = \langle \mathbf{p} \rangle$, these lead to the definitions of the variances or the standard deviations squared.

An uncertainty relation of the form in (3.1.4) is also obtained for other two non-commuting self-adjoint operators **A**, **B** with

$$\mathbf{A} \cdot \mathbf{B} - \mathbf{B} \cdot \mathbf{A} = iC \quad (3.1.5)$$

giving rise, formally, to a self-adjoint operator C.

Upon using the Cauchy-Schwarz inequality

$$|\langle \mathbf{A}\psi | \cdot \mathbf{B}\psi \rangle - \langle \mathbf{B}\psi | \cdot \mathbf{A}\psi \rangle| \leqslant 2\sqrt{\langle \mathbf{A}^2 \rangle} \sqrt{\langle \mathbf{B}^2 \rangle} \quad (3.1.6)$$

where

$$\langle \mathbf{A}\psi | \cdot \mathbf{B}\psi \rangle = \sum_i \langle A_i \psi | B_i \psi \rangle$$

we obtain

$$|\langle C \rangle| \leqslant 2\sqrt{\langle \mathbf{A}^2 \rangle} \sqrt{\langle \mathbf{B}^2 \rangle}. \quad (3.1.7)$$

For $\mathbf{A} = \mathbf{x} - \mathbf{b}$, $\mathbf{B} = \mathbf{p} - \mathbf{a}\hbar$, $C = \nu \hbar$, (3.1.7) leads to (3.1.4).

Another useful inequality which follows from (3.1.2) is the following

$$-\frac{\mu}{2\hbar^2} \left\langle \mathbf{x}^2 g^2(\mathbf{x}) \right\rangle \leqslant \left\langle \frac{\mathbf{p}^2}{2\mu} - \frac{\boldsymbol{\nabla} \cdot (\mathbf{x} g(\mathbf{x}))}{2} \right\rangle \quad (3.1.8)$$

obtained by setting $\mathbf{a} = 0$, $\mathbf{b} = 0$ and choosing $\alpha = \mu/\hbar^2$.

3.1.3 Localization and Stability

An interesting application follows from (3.1.3) by setting $g(\mathbf{x}) = 1/|\mathbf{x} - \mathbf{b}|$ and $\mathbf{a} = \mathbf{0}$. This gives the important lower bound for the expectation value of the kinetic operator for $\nu = 3$:

$$\left\langle \frac{\mathbf{p}^2}{2M} \right\rangle \geqslant \frac{\hbar^2}{2M} \left\langle \frac{1}{|\mathbf{x} - \mathbf{b}|} \right\rangle^2 \tag{3.1.9}$$

for a given mass M.

On the other hand, we may invoke the Cauchy-Schwarz inequality

$$1 = \int d^3\mathbf{x} \, |\psi(\mathbf{x})|^2 = \int d^3\mathbf{x} \frac{|\psi(\mathbf{x})|}{|\mathbf{x} - \mathbf{b}|^{1/2}} |\mathbf{x} - \mathbf{b}|^{1/2} |\psi(\mathbf{x})|$$

$$\leqslant \left\langle \frac{1}{|\mathbf{x} - \mathbf{b}|} \right\rangle^{1/2} \left\langle |\mathbf{x} - \mathbf{b}| \right\rangle^{1/2} \tag{3.1.10}$$

for a normalized state, to obtain from (3.1.9), (3.1.10)

$$\left\langle \frac{\mathbf{p}^2}{2M} \right\rangle \left\langle |\mathbf{x} - \mathbf{b}| \right\rangle^2 \geqslant \frac{\hbar^2}{2M}. \tag{3.1.11}$$

By using the symbol $\mathbf{X}$ for the random variable associated with the position, we have for the probability of occurrence of the event $\{|\mathbf{X} - \mathbf{b}| \leqslant \delta\}$ for any $\delta > 0$:

$$\text{Prob}\left[|\mathbf{X} - \mathbf{b}| \leqslant \delta\right] = \int d^3\mathbf{x} \, |\psi(\mathbf{x})|^2 \, \Theta(\delta - |\mathbf{x} - \mathbf{b}|) \tag{3.1.12}$$

where $\Theta(c)$ is the step function, i.e., equal to 1 for $c > 0$, and zero for $c < 0$. Since in the integral we have the constraint

$$1 \leqslant \delta/|\mathbf{x} - \mathbf{b}| \tag{3.1.13}$$

and $\Theta(c) \leqslant 1$, we have from (3.1.12) the bound

$$\text{Prob}\left[|\mathbf{X} - \mathbf{b}| \leqslant \delta\right] \leqslant \left\langle \frac{1}{|\mathbf{x} - \mathbf{b}|} \right\rangle \delta \tag{3.1.14}$$

which from (3.1.9) leads to

$$\frac{\text{Prob}\left[|\mathbf{X} - \mathbf{b}| \leqslant \delta\right]}{\delta} \leqslant \sqrt{\frac{2M}{\hbar^2}} \left\langle \frac{\mathbf{p}^2}{2M} \right\rangle^{1/2}. \tag{3.1.15}$$

The physical interpretation of this result is clear. *If* for arbitrary small δ, $\text{Prob}\left[|\mathbf{X} - \mathbf{b}| \leqslant \delta\right]$, is *non*-vanishing, then the expectation value of the kinetic energy is necessarily arbitrarily large. That is, for a particle which has a *non*-vanishing probability of being found within a sphere of radius δ about a fixed point, specified by the vector **b**, its average kinetic energy is necessarily large for an arbitrarily small sphere of localization.

In collision theory, for example, we may identify M with the reduced mass μ of two particles, set $\mathbf{b} = \mathbf{0}$ in (3.1.15), with $|\mathbf{x}|$ denoting the distance between the two particles. Accordingly, in order to bring the "colliding" particles arbitrarily close to each other, there must exist an arbitrarily large lower bound to the average kinetic energy of the relative motion.

For a negative-energy state, the average kinetic energy cannot be arbitrarily large and (3.1.15) may be used in the following manner. Consider the Hamiltonian of relative motion of a two-particle system (§2.5),

$$H = \frac{\mathbf{p}^2}{2\mu} + V \qquad (3.1.16)$$

where $\mu = m_1 m_2/(m_1 + m_2)$ is the reduced mass.

Suppose that $V(\mathbf{x})$ remains invariant under the scalings $m_1 \to \kappa m_1$, $m_2 \to \kappa m_2$, by an arbitrary positive parameter κ, and the scalings

$$\lambda_1 \longrightarrow (\kappa)^{\delta_1} \lambda_1, \quad \lambda_2 \longrightarrow (\kappa)^{\delta_2} \lambda_2, \ldots \qquad (3.1.17)$$

where $\lambda_1, \lambda_2, \ldots$ are coupling parameters, and $\delta_1, \delta_2, \ldots$ are some real numbers. For the Coulomb interaction, for example, V is independent of m_1, m_2, and $\lambda_1 = q_1 q_2$, $\delta_1 = 0$ where q_1, q_2 denote the charges of the two particles. For the Newtonian gravitational interaction, with λ_1 chosen to coincide with the gravitational coupling constant G, $\delta_1 = -2$.

In Chapter 4, the nature of the spectra of Hamiltonians is studied under some general sufficiency conditions satisfied by the potentials. In general, we consider a strictly negative energy-state of a Hamiltonian, if there exists one, $|\psi(m_1, m_2, \lambda_i)\rangle$ satisfying, by definition,

$$\langle \psi(m_1, m_2, \lambda_i) | H | \psi(m_1, m_2, \lambda_i) \rangle < 0 \qquad (3.1.18)$$

with the understanding that energy is required to "break up" the system by separating the two particles and make the energy of the system non-negative. Here we have labelled the state $|\psi(m_1, m_2, \lambda_i)\rangle$ by m_1, m_2 and by λ_i, standing for $\lambda_1, \lambda_2, \ldots$, for a reason that will become clear below.

A physical system is one that has the spectrum of its corresponding Hamiltonian bounded from below (§3.3), that is, its spectrum does not go down to $-\infty$. Otherwise, the system would collapse to such a level as to release an infinite (!) amount of energy which is physically meaningless and the system would be unstable.

Suppose that the ground-state energy $-E[m_1, m_2, \lambda_i]$ of the Hamiltonian (3.1.16), that is, corresponding to the lowest point of its spectrum (§3.3), is finite and is strictly negative. Then for the state $|\psi(m_1, m_2, \lambda_i)\rangle$ satisfying (3.1.18) we may write

3.1 Uncertainties, Localization and Stability

$$-E[m_1, m_2, \lambda_i] \leqslant \langle \psi(m_1, m_2, \lambda_i)|H|\psi(m_1, m_2, \lambda_i)\rangle < 0. \quad (3.1.19)$$

Let $\left|\psi\left(\frac{m_1}{2}, \frac{m_2}{2}, \left(\frac{1}{2}\right)^{\delta_i}\lambda_i\right)\right\rangle$ denote the state $|\psi(m_1, m_2, \lambda_i)\rangle$ with m_1, m_2, λ_i replaced, respectively, by $\frac{m_1}{2}$, $\frac{m_2}{2}$, $\left(\frac{1}{2}\right)^{\delta_i}\lambda_i$.

Clearly, the state $\left|\psi\left(\frac{m_1}{2}, \frac{m_2}{2}, \left(\frac{1}{2}\right)^{\delta_i}\lambda_i\right)\right\rangle$ cannot lead for the expectation value

$$\left\langle \psi\left(\frac{m_1}{2}, \frac{m_2}{2}, \left(\frac{1}{2}\right)^{\delta_i}\lambda_i\right) \middle| H \middle| \psi\left(\frac{m_1}{2}, \frac{m_2}{2}, \left(\frac{1}{2}\right)^{\delta_i}\lambda_i\right)\right\rangle \quad (3.1.20)$$

where H is given in (3.1.16), a numerical value lower than $-E[m_1, m_2, \lambda_i]$, for the given H, otherwise this would contradict the fact that $-E[m_1, m_2, \lambda_i]$ is the ground-state energy of H. That is,

$$-E[m_1, m_2, \lambda_i]$$
$$\leqslant \left\langle \psi\left(\frac{m_1}{2}, \frac{m_2}{2}, \left(\frac{1}{2}\right)^{\delta_i}\lambda_i\right) \middle| \left(\frac{\mathbf{p}^2}{2\mu} + V\right) \middle| \psi\left(\frac{m_1}{2}, \frac{m_2}{2}, \left(\frac{1}{2}\right)^{\delta_i}\lambda_i\right)\right\rangle \quad (3.1.21)$$

or

$$-E[2m_1, 2m_2, (2)^{\delta_i}\lambda_i]$$
$$\leqslant \left\langle \psi(m_1, m_2, \lambda_i) \middle| \left(\frac{\mathbf{p}^2}{4\mu} + V\right) \middle| \psi(m_1, m_2, \lambda_i)\right\rangle. \quad (3.1.22)$$

On the other hand, the inequality (3.1.18) implies, for the Hamiltonian (3.1.16), that for the expectation value of $\mathbf{p}^2/4\mu$:

$$\left\langle \psi(m_1, m_2, \lambda_i) \middle| \frac{\mathbf{p}^2}{4\mu} \middle| \psi(m_1, m_2, \lambda_i)\right\rangle$$
$$< -\left\langle \psi(m_1, m_2, \lambda_i) \middle| \left(\frac{\mathbf{p}^2}{4\mu} + V\right) \middle| \psi(m_1, m_2, \lambda_i)\right\rangle$$
$$\leqslant E[2m_1, 2m_2, (2)^{\delta_i}\lambda_i] \quad (3.1.23)$$

where in writing the last inequality we have used (3.1.22).

Equation (3.1.23) gives the following upper bound for the expectation value of the kinetic energy operator for such a system

$$\left\langle \frac{\mathbf{p}^2}{2\mu}\right\rangle < 2E[2m_1, 2m_2, (2)^{\delta_i}\lambda_i]. \quad (3.1.24)$$

From (3.1.11), with $\mathbf{b} = \mathbf{0}$, we have the following *non*-vanishing lower bound for the expectation value $\langle |\mathbf{x}| \rangle$ of the separation distance between the two particles

$$\langle |\mathbf{x}| \rangle > \sqrt{\frac{\hbar^2}{2\mu}} \left(2E[2m_1, 2m_2, (2)^{\delta_i}\lambda_i] \right)^{-1/2} \qquad (3.1.25)$$

as follows from (3.1.24).

From (3.1.15), (3.1.24) we also obtain, with $\mathbf{b} = \mathbf{0}$, $M \to \mu$,

$$\text{Prob}\left[|\mathbf{X}| \leqslant \delta\right] < 2\delta \left(\frac{\mu}{\hbar^2} E\left[2m_1, 2m_2, (2)^{\delta_i}\lambda_i\right] \right)^{1/2}. \qquad (3.1.26)$$

That is, for a physical system for which $0 < E[2m_1, 2m_2, (2)^{\delta_i}\lambda_i] < \infty$, (3.1.26) gives the satisfactory result of a vanishingly small probability, for a vanishingly small δ, and rigorously vanishes for the "fall" of one particle into the other.

A similar inequality to (3.1.26) will be also derived for multi-fermion systems in (3.1.37), and also later on, in Chapter 14, in our study of the important problem of the *stability of matter* with special attention given to the number of particles involved and the Fermi character, that is of the underlying Pauli exclusion principle, of the electrons. The stability of the hydrogen *atom* is studied in §7.1.

3.1.4 Localization, Stability and Multi-Particle Systems

Here we are interested in the localization problem of identical Fermi particles, such as electrons. Consider the (anti-symmetric) wave function $\Psi(\mathbf{x}_1\sigma_1, \ldots, \mathbf{x}_N\sigma_N)$ of such N particles in the coordinate description, where $\sigma_1, \ldots, \sigma_N$ are spin indices. A single particle probability density, normalized to one, may be defined by

$$h(\mathbf{x}) = \sum_{\sigma_1, \ldots, \sigma_N} \int d^3\mathbf{x}_2 \ldots d^3\mathbf{x}_N \, |\psi(\mathbf{x}_1\sigma_1, \ldots, \mathbf{x}_N\sigma_N)|^2. \qquad (3.1.27)$$

The probability that any one of the particles is found within a sphere of radices δ, about some point in space specified by a vector $\mathbf{b}$ is given by

$$\text{Prob}\left[|\mathbf{X}_1 - \mathbf{b}| \leqslant \delta\right] = \int d^3\mathbf{x} \, h(\mathbf{x}) \, \Theta(\delta - |\mathbf{x} - \mathbf{b}|) \qquad (3.1.28)$$

where $\Theta(a)$ is the step function, i.e., $\Theta(a) = 1$ for $a > 0$ and $\Theta(a) = 0$ for $a < 0$.

By Hölder's inequality in Appendix II, we may bound

$$\int d^3\mathbf{x} \, h(\mathbf{x}) \, \Theta(\delta - |\mathbf{x} - \mathbf{b}|) \leqslant \left(\int d^3\mathbf{x} \, h(\mathbf{x})^{5/3} \right)^{3/5}$$

3.1 Uncertainties, Localization and Stability

$$\times \left(\int d^3\mathbf{x}\, \Theta(\delta - |\mathbf{x} - \mathbf{b}|) \right)^{2/5} \qquad (3.1.29)$$

where we have used the fact that $(\Theta(a))^{5/2} = \Theta(a)$.

Hence from (3.1.28), (3.1.29) we have

$$\text{Prob}\left[|\mathbf{X}_1 - \mathbf{b}| \leqslant \delta\right] \left(\frac{1}{v}\right)^{2/5} \leqslant \left(\int d^3\mathbf{x}\, (h(\mathbf{x}))^{5/3} \right)^{3/5} \qquad (3.1.30)$$

where

$$v = \frac{4\pi}{3}\delta^3 \qquad (3.1.31)$$

is the volume in which a particle is confined.

In the sequel, we use the notation,

$$\left\langle \psi \left| \sum_{i=1}^{N} \frac{\mathbf{p}_i^2}{2m} \right| \psi \right\rangle \equiv T \qquad (3.1.32)$$

for the expectation value of the kinetic energy of the N particles, where m denotes the mass of a particle.

In §4.6, (see (4.6.24), (4.6.16), (4.6.17)), we derive the following bound on the average kinetic energy T, where the exclusion principle plays a key role,

$$\left(\int d^3\mathbf{x}\, (h(\mathbf{x}))^{5/3} \right)^{3/5} \leqslant \underline{c}\, \frac{T^{3/5}}{N} \qquad (3.1.33)$$

with

$$\underline{c} = \left(\frac{10m}{3\hbar^2}\right)^{3/5} \left(\frac{2(2s+1)}{3\pi}\right)^{2/5} \qquad (3.1.34)$$

and s is the spin of a fermion.

Since

$$\text{Prob}\left[|\mathbf{X}_1 - \mathbf{b}| \leqslant \delta, \ldots, |\mathbf{X}_N - \mathbf{b}| \leqslant \delta\right] \leqslant \text{Prob}\left[|\mathbf{X}_1 - \mathbf{b}| \leqslant \delta\right] \qquad (3.1.35)$$

(3.1.30), (3.1.33) give

$$\text{Prob}\left[|\mathbf{X}_1 - \mathbf{b}| \leqslant \delta, \ldots, |\mathbf{X}_N - \mathbf{b}| \leqslant \delta\right] \left(\frac{N}{v}\right)^{2/5} \leqslant \underline{c}\left(\frac{T}{N}\right)^{3/5}. \qquad (3.1.36)$$

As an application of this bound, consider a system of fermions, which may be interacting, and with a non-vanishing probability of occurrence of the event $\{|\mathbf{X}_1 - \mathbf{b}| \leqslant \delta, \ldots, |\mathbf{X}_N - \mathbf{b}| \leqslant \delta\}$, i.e., of being localized as indicated. As N becomes larger and larger, i.e., we increase the particle density, the left-hand side of (3.1.36) goes to infinity for $N \to \infty$. This can be true only if T goes to infinity as well, and not slower than N. This shows a kind of "resistance" of Fermi particles to the increase in density, by the increase of their average kinetic energy.

As another application, consider a negative-energy state of the fermions. For a stable system, the ground-state energy $E_N, -E_N$ cannot grow faster than N. For example, $-E_N$ cannot grow like N^α with $\alpha > 1$. The reason is that otherwise the formation of a single system consisting of $2N$ particles will be favored over two separate systems brought into contact, each consisting of N particles, and the energy released of such a contact, being proportional to $[(2N)^\alpha - 2(N)^\alpha]$, will be overwhelmingly large for large N, e.g., $N \sim 10^{23}$ (see also Chapter 14).

Accordingly, with a ground-state energy E_N, with typical bounds $-\underline{a}N \leqslant E_N < -\bar{a}N$, where $\underline{a}, \bar{a}$ are some positive constants, almost an identical reasoning as the one lading to (3.1.24) for the average kinetic energy, with similar scaling properties of the interactions, for example with Coulombic ones, shows that $T \leqslant AN$,[1] with A denoting some positive finite constant independent of N.

For such a negative-energy state, we may then further bound the right-hand side of (3.3.36), giving

$$\text{Prob}\,[|\mathbf{X}_1 - \mathbf{b}| \leqslant \delta, \ldots, |\mathbf{X}_N - \mathbf{b}| \leqslant \delta]\left(\frac{N}{v}\right)^{2/5} \leqslant \underline{c}A^{3/5} \qquad (3.1.37)$$

where the right-hand side is some finite constant. For a non-vanishing probability of having the fermions within a sphere of radius δ as indicated, in order that the left-hand side of (3.1.7) remains finite, in conformity with its right-hand side, as N grows without bound, it is necessary that the volume v grows at least as fast as N. That is, in particular, the radius of spatial extension of the fermionic system grows not any slower than $N^{1/3}$ for $N \to \infty$. These properties will be quantitatively analyzed in Chapter 14 in considering the problem of the stability of matter, where the exclusion principle, plays a key role and is based on Manoukian and Sirininlakul (2005).

A lower bound to the expectation value of $\sum_i |\mathbf{x}_i|/N$, as a measure of the extension of the above system is also readily obtained.

To the above end,

$$\left\langle \sum_{i=1}^N \frac{|\mathbf{x}_i|}{N} \right\rangle = \sum_{\sigma_1,\ldots,\sigma_N} \int d^3\mathbf{x}_1 \ldots d^3\mathbf{x}_N \left(\sum_{i=1}^N \frac{|\mathbf{x}_i|}{N}\right) |\psi(\mathbf{x}_1\sigma_1,\ldots,\mathbf{x}_N\sigma_N)|^2$$

$$= \int d^3\mathbf{x}\,|\mathbf{x}|\,h(\mathbf{x}). \qquad (3.1.38)$$

But for any $\delta > 0$,

---

[1] Actually just the bound $-\underline{a}N \leqslant E_N < 0$ is sufficient to establish this, in the process, that $T \leqslant AN$ (see (3.1.19)).

$$\int d^3\mathbf{x}\, |\mathbf{x}|\, h(\mathbf{x}) \geqslant \int_{|\mathbf{x}|>\delta} d^3\mathbf{x}\, |\mathbf{x}|\, h(\mathbf{x}) > \delta \int_{|\mathbf{x}|>\delta} d^3\mathbf{x}\, h(\mathbf{x}) \qquad (3.1.39)$$

and

$$\int_{|\mathbf{x}|>\delta} d^3\mathbf{x}\, h(\mathbf{x}) = \text{Prob}\left[|\mathbf{X}| > \delta\right] = 1 - \text{Prob}\left[|\mathbf{X}| \leqslant \delta\right]. \qquad (3.1.40)$$

Accordingly, from (3.1.35), (3.1.37)–(3.1.40),

$$\left\langle \sum_{i=1}^{N} \frac{|\mathbf{x}_i|}{N} \right\rangle > \delta \left[1 - \left(\frac{v}{N}\right)^{2/5} \underline{c}\, A^{3/5}\right] = \delta \left[1 - \left(\frac{4\pi}{3N}\delta^3\right)^{2/5} \underline{c}\, A^{3/5}\right] \qquad (3.1.41)$$

where in writing the last equality, we have used (3.1.31). Upon optimizing the right-hand side of (3.1.41) over δ, gives

$$\delta = \left(\frac{3N}{4\pi}\right)^{1/3} \left(\frac{5}{11\underline{c}}\right)^{5/6} \frac{1}{A^{1/2}} \qquad (3.1.42)$$

which from (3.1.41) leads to

$$\left\langle \sum_{i=1}^{N} \frac{|\mathbf{x}_i|}{N} \right\rangle > \frac{6}{11}\left(\frac{5}{11\underline{c}}\right)^{5/6}\left(\frac{3N}{4\pi}\right)^{1/3}\frac{1}{A^{1/2}} \qquad (3.1.43)$$

giving a lower bound proportional to $N^{1/3}$, where $\underline{c}$ is defined in (3.1.34).

Bosonic systems behave differently, and so-called "bosonic matter" will be analyzed in detail in Chapter 14 (see also Problem 14.10).

3.2 Boundedness of the Spectra of Hamiltonians From Below

If the spectrum of a self-adjoint operator A is bounded from below then its spectral decomposition (§1.8, (1.8.15)) may be explicitly written as

$$A = \int_{L_A}^{\infty} \lambda\, dP_A(\lambda) \qquad (3.2.1)$$

where $|L_A| < \infty$. For any vector $|f\rangle$ in the domain of A, we then have the following lower bound for its expectation value

$$\langle f|A|f\rangle = \int_{L_A}^{\infty} \lambda\, d\langle f|P_A(\lambda)|f\rangle \geqslant L_A \int_{L_A}^{\infty} d\langle f|P_A(\lambda)|f\rangle. \qquad (3.2.2)$$

Hence for all $|f\rangle$ in the domain of A, the resolution of the identity (see (1.8.18)) gives the bound

$$\langle f|A|f\rangle \geqslant L_A \|f\|^2. \tag{3.2.4}$$

Conversely, suppose that for some self-adjoint operator A, and for all vectors $|f\rangle$ in its domain, the bound (3.2.4) holds true. Then we will show that for all λ in the spectrum of A, $\lambda \geqslant L_A$. To do this we assume otherwise and run into a contradiction. That is, suppose that for some λ_0 in the spectrum of $A : \lambda_0 < L_A$, and let

$$\varepsilon = \frac{L_A - \lambda_0}{2} > 0. \tag{3.2.5}$$

Then (§1.8),

$$[P_A(\lambda_0 + \varepsilon) - P_A(\lambda_0 - \varepsilon)]\mathcal{H} \tag{3.2.6}$$

is not empty. Let $|f_0\rangle$, in the domain of A, be a vector belonging to (3.2.6), i.e.,

$$[P_A(\lambda_0 + \varepsilon) - P_A(\lambda_0 - \varepsilon)]|f_0\rangle = |f_0\rangle. \tag{3.2.7}$$

Hence

$$\langle f_0|A|f_0\rangle = \int_{\lambda_0-\varepsilon}^{\lambda_0+\varepsilon} \lambda \, \mathrm{d}\,\|P_A(\lambda)f_0\|^2 \leqslant (\lambda_0+\varepsilon) \int_{\lambda_0-\varepsilon}^{\lambda_0+\varepsilon} \mathrm{d}\,\|P_A(\lambda)f_0\|^2 \tag{3.2.8}$$

or using, in the process, the equality in (3.2.5), we have

$$\langle f_0|A|f_0\rangle \leqslant (\lambda_0+\varepsilon)\|f_0\|^2 = (L_A - \varepsilon)\|f_0\|^2 < L_A \|f_0\|^2 \tag{3.2.9}$$

which is in contradiction with (3.2.4).

That is, if for all $|f\rangle$ in the domain of A for which (3.2.4) is true, any λ in the spectrum of A, is such that $\lambda \geqslant L_A$.

One way of obtaining a lower bound of a Hamiltonian H is through the examination of its resolvent $(H - \xi)^{-1}$. According to the treatment given in §1.8, if ξ is a real parameter belonging to the spectrum of H, then either $(H - \xi)^{-1}$ does not exist or if it exists then it is an unbounded operator. Thus, if one may find a real number ξ_0 such that for *all* $\xi < \xi_0$, $(H - \xi)^{-1}$ exists and is a bounded operator, then clearly ξ_0 provides a lower bound to the spectrum of H. Such a method will be applied in some of the subsequent investigations carried out in examining the boundedness of Hamiltonians from below.

3.3 Boundedness of Hamiltonians From Below: General Classes of Interactions

This section is entirely involved with the investigation of the boundedness of Hamiltonians from below (§3.2) for large classes of interactions. These

3.3 Boundedness of Hamiltonians From Below

classes will be dealt with in various parts [A] to [G] given below. Needless to say, an interaction, specified by a given potential, may belong to more than one of the classes considered. Lower bounds of Hamiltonians for multi-particle systems will be considered in the next section.

The lower bounds derived are not necessarily optimal ones but they establish the important property of boundedness of Hamiltonians from below given that some sufficiency conditions are satisfied by the potentials.

[A] The simplest potential $V(\mathbf{x})$ is one that is *bounded* everywhere. That is,
$$|V(\mathbf{x})| \leqslant C \ (< \infty) \tag{3.3.1}$$
for all $\mathbf{x}$. Accordingly (for normalized $|\psi\rangle$),
$$|\langle\psi|V|\psi\rangle| \leqslant C \|\psi\|^2 = C \tag{3.3.2}$$
and the positivity of the kinetic energy operator implies that $(\mathbf{p}^2 = -\hbar^2 \boldsymbol{\nabla}^2)$
$$\left\langle \psi \left| \left(\frac{\mathbf{p}^2}{2\mu} + V \right) \right| \psi \right\rangle \geqslant -C. \tag{3.3.3}$$

An example of such a given potential is
$$V(\mathbf{x}) = \lambda_1 \Theta(R - |\mathbf{x}|) + \lambda_2 e^{-\beta|\mathbf{x}|} \tag{3.3.4}$$
with $0 < R < \infty$, $\beta > 0$, and one may take
$$|V(\mathbf{x})| \leqslant |\lambda_1| + |\lambda_2| \equiv C \tag{3.3.5}$$

[B] We may use the inequality in (3.1.8) to obtain lower in bounds for Hamiltonians with a class of potentials $V(\mathbf{x})$ related to a real function $g(\mathbf{x})$ given by
$$V(\mathbf{x}) = -\frac{1}{2} \boldsymbol{\nabla} \cdot (\mathbf{x} g(\mathbf{x})). \tag{3.3.6}$$
From (3.1.8), this leads to[2]
$$-\frac{\mu}{2\hbar^2} \langle \mathbf{x}^2 g^2(\mathbf{x}) \rangle \leqslant \left\langle \frac{\mathbf{p}^2}{2\mu} + V(\mathbf{x}) \right\rangle. \tag{3.3.7}$$

The applications of (3.3.6), (3.3.7) are endless. These couple of equations will be used to obtain lower bounds for Hamiltonians for several specific interactions later on, notably in §4.2, §6.1, §7.1.

As another application of (3.1.8), where the potential, in this case, is not chosen exactly in the form in (3.3.6), consider the following general class of potentials, ($|\mathbf{x}| = r$, space dimension $\nu = 3$),

---

[2] Note that parts or all of $\mathbf{x}^2 g^2(\mathbf{x})$ may be also reconsidered as part of the potential energy itself.

154   3 Uncertainties, Localization, Stability and Decay of Quantum Systems

$$V(\mathbf{x}) = \frac{\lambda_1 e^{-\beta_1 r}}{r} + \frac{\lambda_2}{r} + \frac{\lambda_3}{r^2} \qquad (3.3.8)$$

where

$$\beta_1 > 0, \lambda_3 > -\frac{\hbar^2}{8\mu} \qquad (3.3.9)$$

which are obviously bounded from below as follows

$$V(\mathbf{x}) \geqslant -\frac{|\lambda_1|}{r} + \frac{\lambda_2}{r} + \frac{\lambda_3}{r^2}. \qquad (3.3.10)$$

Upon choosing $g(\mathbf{x})$ in (3.1.8) to be the function

$$g(\mathbf{x}) = \frac{\gamma_1}{r} + \frac{\gamma_2}{r^2} \qquad (3.3.11)$$

with

$$\gamma_1 = \frac{2(|\lambda_1| - \lambda_2)}{\left(1 + \sqrt{1 + 8\mu \frac{\lambda_3}{\hbar^2}}\right)} \qquad (3.3.12)$$

$$\gamma_2 = \frac{\hbar^2}{2\mu}\left(1 - \sqrt{1 + 8\mu \frac{\lambda_3}{\hbar^2}}\right) \qquad (3.3.13)$$

we obtain from (3.1.8) by a re-arrangement of terms,

$$-\frac{\mu}{2\hbar^2}\gamma_1^2 \leqslant \left\langle \frac{\mathbf{p}^2}{2\mu} - \frac{(|\lambda_1| - \lambda_2)}{r} + \frac{\lambda_3}{r^2} \right\rangle \qquad (3.3.14)$$

which from (3.3.8), (3.3.10) gives

$$-\frac{\mu}{2\hbar^2}\gamma_1^2 < \left\langle \frac{\mathbf{p}^2}{2\mu} + V(\mathbf{x}) \right\rangle. \qquad (3.3.15)$$

[C] We now consider another class of potentials. To this end we first write the potential in

$$H = \frac{\mathbf{p}^2}{2\mu} + V \qquad (3.3.16)$$

as

$$V = V\Theta(V) + V\Theta(-V) \qquad (3.3.17)$$

where $\Theta(a)$ is the step function and hence $\Theta(a) + \Theta(-a) = 1$. Since $V\Theta(V) \geqslant 0$ we have

$$V \geqslant V\Theta(-V) \equiv -v \qquad (3.3.18)$$

where $v \geqslant 0$. The step function $\Theta(-V)$ picks up the points $\mathbf{x}$ for which the potential V is non-positive. Another standard notation for $V\Theta(-V)$ is

$$V\Theta(-V) = -|V|_- \qquad (3.3.19)$$

3.3 Boundedness of Hamiltonians From Below

where $|V|_- \geqslant 0$.

In order to obtain a lower bound for H in (3.3.16) it is sufficient, from (3.3.18), to consider instead the Hamiltonian

$$H' = \frac{\mathbf{p}^2}{2\mu} - v \qquad (3.3.20)$$

since

$$\langle H \rangle \geqslant \langle H' \rangle. \qquad (3.3.21)$$

To the above end, set

$$v^2 = u. \qquad (3.3.22)$$

We consider a class of potentials for which

$$\int d^3\mathbf{x}\, d^3\mathbf{y}\, u(\mathbf{x}) \frac{1}{|\mathbf{x}-\mathbf{y}|^2} u(\mathbf{y}) < \infty. \qquad (3.3.23)$$

The latter integral may be also rewritten as

$$\frac{1}{4\pi\hbar^2} \int \frac{d^3\mathbf{p}}{|\mathbf{p}|} |\tilde{u}(\mathbf{p})|^2 \quad (<\infty) \qquad (3.3.24)$$

by using, in the process of the derivation, the integral expression

$$\frac{1}{|\mathbf{p}|} = \frac{4\pi\hbar^2}{(2\pi\hbar)^3} \int d^3\mathbf{x} \frac{e^{-i\mathbf{p}\cdot\mathbf{x}/\hbar}}{|\mathbf{x}|^2} \qquad (3.3.25)$$

where $\tilde{u}(\mathbf{p})$ is the Fourier transform of $u(\mathbf{x})$:

$$\tilde{u}(\mathbf{p}) = \int d^3\mathbf{x}\, u(\mathbf{x}) e^{-i\mathbf{p}\cdot\mathbf{x}/\hbar}. \qquad (3.3.26)$$

From the identity

$$\langle \psi | u | \psi \rangle = \int d^3\mathbf{x}\, d^3\mathbf{y} \left\langle \psi \left| \sqrt{-\boldsymbol{\nabla}^2} \right| \mathbf{x} \right\rangle K(\mathbf{x},\mathbf{y}) \left\langle \mathbf{y} \left| \sqrt{-\boldsymbol{\nabla}^2} \right| \psi \right\rangle \qquad (3.3.27)$$

where

$$K(\mathbf{x},\mathbf{y}) = \left\langle \mathbf{x} \left| \frac{1}{\sqrt{-\boldsymbol{\nabla}^2}} u \frac{1}{\sqrt{-\boldsymbol{\nabla}^2}} \right| \mathbf{y} \right\rangle \qquad (3.3.28)$$

and an application of the Cauchy-Schwarz inequality in $\mathbb{R}^6$, we obtain

$$\langle \psi | u | \psi \rangle \leqslant \langle \psi | (-\boldsymbol{\nabla}^2) | \psi \rangle \left(\int d^3\mathbf{x}\, d^3\mathbf{y}\, |K(\mathbf{x},\mathbf{y})|^2 \right)^{1/2} \qquad (3.3.29)$$

where

$$\int d^3\mathbf{x}\, d^3\mathbf{y}\, |K(\mathbf{x},\mathbf{y})|^2 = \int d^3\mathbf{y} \left\langle \mathbf{y} \left| \frac{1}{\sqrt{-\boldsymbol{\nabla}^2}} u \frac{1}{(-\boldsymbol{\nabla}^2)} u \frac{1}{\sqrt{-\boldsymbol{\nabla}^2}} \right| \mathbf{y} \right\rangle$$

$$= \frac{\hbar}{8} \int \frac{d^3\mathbf{p}}{(2\pi\hbar)^3} \frac{|\tilde{u}(\mathbf{p})|^2}{|\mathbf{p}|} \tag{3.3.30}$$

or

$$\langle \psi | u | \psi \rangle \leqslant aT, \quad u = v^2 \tag{3.3.31}$$

with

$$T = \left\langle \frac{\mathbf{p}^2}{2\mu} \right\rangle \tag{3.3.32}$$

$$a = \frac{\mu}{4\hbar^3 \pi^{3/2}} \left(\int \frac{d^3\mathbf{p}}{|\mathbf{p}|} |\tilde{u}(\mathbf{p})|^2 \right)^{1/2}. \tag{3.3.33}$$

We use (3.3.31) to derive a lower bound for H'.
Upon dividing the following inequality by $4T$

$$4T\sqrt{\langle v^2 \rangle} \leqslant 4T^2 + \langle v^2 \rangle \tag{3.3.34}$$

and using the bound (3.3.31) we arrive at

$$\sqrt{\langle v^2 \rangle} \leqslant T + \frac{a}{4} \tag{3.3.35}$$

or equivalently at

$$-\frac{a}{4} \leqslant T - \sqrt{\langle v^2 \rangle}. \tag{3.3.36}$$

Finally, the application of the Cauchy-Schwarz inequality to $(\psi^*\psi = |\psi|^2 = |\psi||\psi|)$

$$\langle v \rangle = \int d^3\mathbf{x} \, (v|\psi|) \, |\psi| \tag{3.3.37}$$

gives

$$\langle v \rangle \leqslant \sqrt{\langle v^2 \rangle}. \tag{3.3.38}$$

This together with (3.3.36) yield, from (3.3.16), (3.3.20), (3.3.21), the following lower bound for the Hamiltonian

$$-\frac{a}{4} \leqslant \left\langle \frac{\mathbf{p}^2}{2\mu} + V \right\rangle. \tag{3.3.39}$$

As an application consider the potential

$$V = -v = -\lambda \frac{e^{-\beta r}}{\sqrt{r}}, \quad \lambda > 0, \beta > 0 \tag{3.3.40}$$

which with $u = v^2$, we have for $\tilde{u}(\mathbf{p})$ in (3.3.26), the expression

$$\tilde{u}(\mathbf{p}) = \frac{4\pi \lambda^2 \hbar^2}{(\mathbf{p}^2 + 4\beta^2 \hbar^2)} \tag{3.3.41}$$

giving for a in (3.3.33)

3.3 Boundedness of Hamiltonians From Below

$$a = \frac{\sqrt{2}\mu\lambda^2}{2\beta\hbar^2} \tag{3.3.42}$$

thus obtaining the bound

$$-\frac{\sqrt{2}}{8}\frac{\mu\lambda^2}{\beta\hbar^2} \leqslant \left\langle \frac{\mathbf{p}^2}{2\mu} - \frac{\lambda e^{-\beta r}}{\sqrt{r}} \right\rangle. \tag{3.3.43}$$

[D] The lower bound derived in this subsection is important in that it provides a sufficient condition for a potential so that the Hamiltonian is positive thus giving a *No-Binding Theorem*.

The method of development is similar to the one given in part [C]. To this end, we use the bound in (3.3.18), and obtain directly from (3.3.29)–(3.3.31), now working with $v(\mathbf{x})$ rather than with $v^2(\mathbf{x})$, the bound

$$\langle \psi | v | \psi \rangle \leqslant b \left\langle \frac{\mathbf{p}^2}{2\mu} \right\rangle \tag{3.3.44}$$

where

$$b = \frac{\mu}{4\hbar^3 \pi^{3/2}} \left(\int \frac{d^3\mathbf{p}}{|\mathbf{p}|} |\tilde{v}(\mathbf{p})|^2 \right)^{1/2}. \tag{3.3.45}$$

Therefore

$$\left\langle \frac{\mathbf{p}^2}{2\mu} + V \right\rangle \geqslant \left\langle \frac{\mathbf{p}^2}{2\mu} - v \right\rangle \geqslant (1-b) \left\langle \frac{\mathbf{p}^2}{2\mu} \right\rangle. \tag{3.3.46}$$

Since $\langle \mathbf{p}^2/2\mu \rangle \geqslant 0$, the Hamiltonian in question is positive for

$$b < 1 \tag{3.3.47}$$

providing a sufficient condition for "no-binding".

For the Yukawa potential, for example, $r = |\mathbf{x}|$,

$$V(\mathbf{x}) = -\lambda \frac{e^{-\beta r}}{r} = -v(\mathbf{x}) \tag{3.3.48}$$

with $\lambda > 0$, $\beta > 0$,

$$\tilde{v}(\mathbf{p}) = \frac{4\pi\hbar^2\lambda}{\mathbf{p}^2 + \beta^2\hbar^2} \tag{3.3.49}$$

giving

$$b = \frac{\lambda\mu\sqrt{2}}{\hbar^2 \beta} \tag{3.3.50}$$

and implying the positivity of the Hamiltonian

$$H = \frac{\mathbf{p}^2}{2\mu} - \frac{\lambda e^{-\beta r}}{r} \tag{3.3.51}$$

for
$$\lambda < \frac{\hbar^2 \beta}{\sqrt{2}\mu} \quad (3.3.52)$$

or equivalently for
$$\frac{\sqrt{2}\mu\lambda}{\hbar^2} < \beta \quad (3.3.53)$$

as a sufficiency condition to be satisfied by β.

[E] The analysis given in this part deals, in general, in $\nu = 3, 2, 1$ dimensions of space. We consider the following class of potentials. A potential V in this class is defined as any potential which is square-integrable over a finite region about the origin and is bounded beyond this region. That is, we may find finite constants $R_\nu > 0$, $C_\nu > 0$ such that

$$\int_{|\mathbf{x}|<R_\nu} d^\nu \mathbf{x} \, |V(\mathbf{x})|^2 \leqslant a_\nu < \infty \quad (3.3.54)$$

and
$$|V(\mathbf{x})| \leqslant C_\nu, \quad \text{for} \quad |\mathbf{x}| \geqslant R_\nu. \quad (3.3.55)$$

Below we will derive the following inequality
$$\|V\psi\| \leqslant A_\nu \|H_0\psi\| + B_\nu \quad (3.3.56)$$

where
$$H_0 = \frac{\mathbf{p}^2}{2\mu} \quad (3.3.57)$$

and A_ν, B_ν are positive constants, depending on ν, and $0 < A_\nu < 1$. Such an inequality is usually referred to as a Kato bound.[3]

We first show how inequality (3.3.56) leads to a lower bound to the Hamiltonian $H = H_0 + V$. We recall, (see §1.8, §3.2) that if a real parameter ξ belongs to the spectrum of H, then either $(H-\xi)^{-1}$ does not exist or if it exists then it is an unbounded operator. We will see that for all reals

$$\xi < -B/(1-A) \quad (3.3.58)$$

$(H-\xi)^{-1}$ exists and is a bounded operator, where, for simplicity of notation here, we have suppressed the index ν of space dimension in A_ν, B_ν. That is, $-B/(1-A)$ provides a lower bound to the spectrum of H.

To reach the above conclusion, we note that, for complex ξ, for example, one may formally write

$$\frac{1}{(H-\xi)} = \frac{1}{(H_0-\xi)} + \frac{1}{(H-\xi)}[(H_0-\xi)-(H-\xi)]\frac{1}{(H_0-\xi)}$$

---
[3] Cf. Kato (1966, 1967) and an earlier classic paper: Kato (1951a).

3.3 Boundedness of Hamiltonians From Below

$$= \frac{1}{(H_0 - \xi)} + \frac{1}{(H - \xi)}(-V)\frac{1}{(H_0 - \xi)} \tag{3.3.59}$$

leading to the formal expansion

$$\frac{1}{(H - \xi)} = \frac{1}{(H_0 - \xi)}\sum_{k \geq 0}(-1)^k \left[V\frac{1}{(H_0 - \xi)}\right]^k. \tag{3.3.60}$$

For a strictly *negative* ξ, we may use (3.3.56) to obtain

$$\left\|V\frac{1}{(H_0 - \xi)}f\right\| = A\left\|H_0(H_0 - \xi)^{-1}f\right\| + \frac{B}{|\xi|}\|f\|$$

$$\leqslant \left(A + \frac{B}{|\xi|}\right)\|f\| \tag{3.3.61}$$

where we have used the fact that H_0 is a positive operator. For $\xi < -B/(1 - A)$, $A + B/|\xi| < 1$, we then have

$$\left\|\frac{1}{(H - \xi)}f\right\| \leqslant \frac{1}{|\xi|}\sum_{k \geq 0}\left(A + \frac{B}{|\xi|}\right)^k\|f\|$$

$$= \frac{1}{(|\xi|(1 - A) - B)}\|f\|$$

$$= \frac{1}{(-\xi(1 - A) - B)}\|f\| < \infty \tag{3.3.62}$$

for all $|f\rangle$, and ξ satisfying (3.3.58).

Hence we may conclude that for $\nu = 3, 2, 1$:

$$\left\langle \frac{\mathbf{p}^2}{2\mu} + V \right\rangle \geqslant -B_\nu/(1 - A_\nu) \tag{3.3.63}$$

with $0 < A_\nu < 1$, $0 < B_\nu$, as defined through (3.3.56). Therefore it remains to find such constants A_ν and B_ν in deriving (3.3.56), in general, for $\nu = 3, 2, 1$. The Coulomb potential $q_1 q_2/|\mathbf{x}|$, for example, belongs to the class of potentials considered for $\nu = 3$, i.e., it satisfies (3.3.54) and (3.3.55).

The derivation of (3.3.56) follows. To this end we use the definition of the Fourier-transform in ν dimensions:

$$\psi(\mathbf{x}) = \int \frac{d^\nu \mathbf{p}}{(2\pi\hbar)^\nu} e^{i\mathbf{p}\cdot\mathbf{x}/\hbar}\widetilde{\psi}(\mathbf{p}) \tag{3.3.64}$$

and derive the following chain of inequalities:

$$|\psi(\mathbf{x})| \leqslant \int \frac{d^\nu \mathbf{p}}{(2\pi\hbar)^\nu}\left|\widetilde{\psi}(\mathbf{p})\right|$$

$$= \int d^\nu \mathbf{p} \frac{|\widetilde{\psi}(\mathbf{p})| \left[1 + b_\nu \left(\frac{\mathbf{p}^2}{2\mu}\right)^2\right]^{1/2}}{(2\pi\hbar)^{\nu/2}} \frac{\left[1 + b_\nu \left(\frac{\mathbf{p}^2}{2\mu}\right)^2\right]^{-1/2}}{(2\pi\hbar)^{\nu/2}} \quad (3.3.65)$$

or

$$|\psi(\mathbf{x})|^2 \leqslant I_\nu \left[\int \frac{d^\nu \mathbf{p}}{(2\pi\hbar)^\nu} |\widetilde{\psi}(\mathbf{p})|^2 + b_\nu \int \frac{d^\nu \mathbf{p}}{(2\pi\hbar)^\nu} \left(\frac{\mathbf{p}^2}{2\mu}\right)^2 |\widetilde{\psi}(\mathbf{p})|^2\right] \quad (3.3.66)$$

where in (3.3.65) we have multiplied and divided the integrand by $\left[1 + b_\nu \left(\mathbf{p}^2/2\mu\right)^2\right]^{1/2}$, and in writing (3.3.66) we have used the Cauchy-Schwarz inequality, and

$$I_\nu = \int \frac{d^\nu \mathbf{p}}{(2\pi\hbar)^\nu} \frac{1}{\left[1 + b_\nu \left(\frac{\mathbf{p}^2}{2\mu}\right)^2\right]} = \frac{\mu^{\nu/2}}{2\hbar^\nu b_\nu^{\nu/4}} \frac{1}{[\nu + (\pi - 3)\delta_{\nu,3}]}. \quad (3.3.67)$$

For each ν, b_ν is so far an arbitrary strictly positive constant, $\nu = 1, 2, 3$.

Accordingly, from (3.3.66), (3.3.67), we have

$$|\psi(\mathbf{x})|^2 \leqslant I_\nu \left\{1 + b_\nu \|H_0 \psi\|^2\right\}. \quad (3.3.68)$$

We will use this upper bound for $|\psi(\mathbf{x})|^2$ for $|\mathbf{x}| < R_\nu$ only. Quite generally,

$$\int d^\nu \mathbf{x} |V(\mathbf{x})|^2 |\psi(\mathbf{x})|^2 = \int_{|\mathbf{x}| < R_\nu} d^\nu \mathbf{x} |V(\mathbf{x})|^2 |\psi(\mathbf{x})|^2$$

$$+ \int_{|\mathbf{x}| \geqslant R_\nu} d^\nu \mathbf{x} |V(\mathbf{x})|^2 |\psi(\mathbf{x})|^2. \quad (3.3.69)$$

For the second integral we have from (3.3.55)

$$\int_{|\mathbf{x}| \geqslant R} d^\nu \mathbf{x} |V(\mathbf{x})|^2 |\psi(\mathbf{x})|^2 \leqslant C_\nu^2 \|\psi\|^2 = C_\nu^2. \quad (3.3.70)$$

On the other hand for the first integral on the right-hand side of (3.3.69), we use the bound (3.3.68). All told, we finally obtain from (3.3.54)

$$\|V\psi\| \leqslant A_\nu \|H_0 \psi\| + B_\nu \quad (3.3.71)$$

where

$$A_\nu = (a_\nu b_\nu I_\nu)^{1/2} \quad (3.3.72)$$

$$B_\nu = \left(C_\nu^2 + a_\nu I_\nu\right)^{1/2} \quad (3.3.73)$$

3.3 Boundedness of Hamiltonians From Below

and we have used the inequality $\sqrt{c_1^2 + c_2^2} \leq c_1 + c_2$ for real and positive c_1 and c_2.

Now for each $\nu = 3, 2, 1$, we choose the positive constant b_ν such that

$$0 < A_\nu < 1 \tag{3.3.74}$$

thus obtaining the lower bound for the Hamiltonian as given in (3.3.63).

For each $\nu = 3, 2, 1$, we spell out a convenient expression for the lower bound just derived for the corresponding Hamiltonian.

To the above end, for $\nu = 3$, let

$$\gamma_3 = \left(\frac{a_3 \mu^{3/2}}{2\pi \hbar^3 C_3^2} \right)^{2/3} \tag{3.3.75}$$

$$\lambda = \left(\frac{a_3 \mu^{3/2}}{2\pi \hbar^3 C_3^2 b_3^{3/4}} \right)^{1/6} \tag{3.3.76}$$

to obtain $A_3 = \gamma_3/\lambda$, and for (3.3.63):

$$\left\langle \frac{\mathbf{p}^2}{2\mu} + V \right\rangle \geq -C_3 \frac{\lambda \sqrt{1 + \lambda^6}}{\lambda - \gamma_3}, \quad \nu = 3 \tag{3.3.77}$$

where now λ is and arbitrary positive parameter such that $\lambda > \gamma_3$. For a given constant γ_3, as defined in (3.3.75), λ may be then fixed by optimization.

For $\nu = 2$, let

$$\gamma_2 = \frac{a_2 \mu}{4\hbar^2 C_2} \tag{3.3.78}$$

$$\eta = \left(\frac{a_2 \mu}{4\hbar^2 C_2^2 b_2^{1/2}} \right)^{1/2} \tag{3.3.79}$$

to obtain $A_2 = \gamma_2/\eta$, we then have the lower bound

$$\left\langle \frac{\mathbf{p}^2}{2\mu} + V \right\rangle \geq -C_2 \frac{\eta \sqrt{1 + \eta^2}}{\eta - \gamma_2}, \quad \nu = 2 \tag{3.3.80}$$

with η an arbitrary positive parameter, such that $\eta > \gamma_2$, and may be fixed by optimization.

For $\nu = 1$, let

$$\gamma_1 = \left(\frac{a_1 \mu^{1/2}}{2\hbar C_1^{3/2}} \right)^2 \tag{3.3.81}$$

$$\rho = \left(\frac{a_1 \mu^{1/2}}{2\hbar C_1^2 b_1^{1/4}} \right)^{3/2} \tag{3.3.82}$$

giving $A_1 = \gamma_1/\rho$. This finally leads to the lower bound

162   3 Uncertainties, Localization, Stability and Decay of Quantum Systems

$$\left\langle \frac{\mathbf{p}^2}{2\mu} + V \right\rangle \geq -C_1 \frac{\rho\sqrt{1+\rho^{2/3}}}{\rho - \gamma_1}, \quad \nu = 1 \qquad (3.3.83)$$

with ρ an arbitrary positive parameter, such that $\rho > \gamma_1$, and may be fixed by optimization.

[F] We consider a class of *one* dimensional potentials. To this end, we bound the potential V in the Hamiltonian

$$H = \frac{p^2}{2\mu} + V \qquad (3.3.84)$$

as

$$V \geq -v \qquad (3.3.85)$$

with $v \geq 0$ as defined in (3.3.18). The class consists of any $v(x) \geq 0$ such that

$$0 < \int_{|x|<R} dx\, v(x) \leq a < \infty \qquad (3.3.86)$$

where $0 < R < \infty$, and

$$v(x) \leq C \quad \text{for} \quad |x| \geq R \qquad (3.3.87)$$

where $0 \leq C < \infty$.

For $b > 0$

$$|\psi(x)| \leq \int_{-\infty}^{\infty} \frac{dp}{2\pi\hbar} \frac{1}{\left[1+b\frac{p^2}{2\mu}\right]^{1/2}} \left[1+b\frac{p^2}{2\mu}\right]^{1/2} |\tilde{\psi}(p)| \qquad (3.3.88)$$

or

$$|\psi(x)|^2 \leq \sqrt{\frac{\mu}{2b\hbar^2}} \left[1 + b\left\langle \frac{p^2}{2\mu} \right\rangle\right]. \qquad (3.3.89)$$

We use this inequality for $|x| < R$ only.

Therefore from (3.3.86)–(3.3.88),

$$\langle v \rangle \leq a\sqrt{\frac{\mu b}{2\hbar^2}} \left\langle \frac{p^2}{2\mu} \right\rangle + \left(C + a\sqrt{\frac{\mu}{2b\hbar^2}}\right). \qquad (3.3.90)$$

We may choose

$$b = \frac{2\hbar^2}{\mu a^2} \qquad (3.3.91)$$

thus obtaining from (3.3.90)

$$-\left(C + \frac{\mu a^2}{2\hbar^2}\right) \leq \left\langle \frac{p^2}{2\mu} - v \right\rangle \qquad (3.3.92)$$

and hence finally leading to the lower bound

3.4 Boundedness of Hamiltonian From Below: Multi-Particle Systems

$$-\left(C + \frac{\mu a^2}{2\hbar^2}\right) \leq \left\langle \frac{p^2}{2\mu} + V \right\rangle. \qquad (3.3.93)$$

[G] As a by-product of counting the number of the bound-states of given potentials, the theory of which is developed in §4.5 for Hamiltonians in $\nu = 3, 2, 1$ space dimensions we have the following bounds. The proofs are given in that section.[4] As before, we let $v(\mathbf{x})$ be defined as in (3.3.18).

For $\nu = 3$, if

$$\int d^3\mathbf{x}\,(v(\mathbf{x}))^{5/2} < \infty \qquad (3.3.94)$$

then (see (4.5.92))

$$\left\langle \frac{\mathbf{p}^2}{2\mu} + V \right\rangle \geq -\frac{4}{15\pi}\left(\frac{2\mu}{\hbar^2}\right)^{3/2} \int d^3\mathbf{x}\,(v(\mathbf{x}))^{5/2}. \qquad (3.3.95)$$

For $\nu = 2$, if

$$\int d^2\mathbf{x}\,(v(\mathbf{x}))^2 < \infty \qquad (3.3.96)$$

then (see (4.5.97))

$$\left\langle \frac{\mathbf{p}^2}{2\mu} + V \right\rangle \geq -\frac{3\mu}{4\hbar^2}\int d^2\mathbf{x}\,(v(\mathbf{x}))^2. \qquad (3.3.97)$$

Finally, for $\nu = 1$, if

$$\int_{-\infty}^{\infty} dx\,(v(x))^{3/2} < \infty \qquad (3.3.98)$$

then (see (4.5.101))

$$\left\langle \frac{p^2}{2\mu} + V \right\rangle \geq -\frac{4}{3}\frac{\sqrt{2\mu}}{\hbar}\int_{-\infty}^{\infty} dx\,(v(x))^{3/2}. \qquad (3.3.99)$$

In the next section, we provide lower bounds to Hamiltonians for multi-particle systems.

3.4 Boundedness of Hamiltonian From Below: Multi-Particle Systems

In the present section, we derive lower bounds to Hamiltonians for multi-particle systems. First we treat systems of interacting particles without imposing any statistics on the particles. This analysis is then followed with some estimates involving Coulomb interactions of fermions then of bosons taking their appropriate statistics into account.

---
[4] For definiteness, here one may assume that on the negative real axis, the corresponding Hamiltonians have, at most, only eigenvalues.

3.4.1 Multi-Particle Systems with Two-Body Potentials

We consider a Hamiltonian in the form

$$H = \sum_{i=1}^{N} \frac{\mathbf{p}_i^2}{2m_i} + \sum_{i<j}^{N} V_{ij}(\mathbf{x}_i - \mathbf{x}_j) \qquad (3.4.1)$$

where necessarily the interaction part involves negative contributions, otherwise H is a positive operator.

The two-body potentials V_{ij} are chosen to satisfy the sufficiency conditions

$$\int_{|\mathbf{x}| \leq R_{ij}} d^3\mathbf{x}\, |V_{ij}(\mathbf{x})|^2 \leq a_{ij} < \infty \qquad (3.4.2)$$

where $0 < R_{ij} < \infty$ and

$$|V_{ij}(\mathbf{x})| \leq C_{ij} < \infty \qquad \text{for} \qquad |\mathbf{x}| \geq R_{ij}. \qquad (3.4.3)$$

We note that upon the change of variables

$$\mathbf{x}_1 - \mathbf{x}_2 = \mathbf{X}_1, \qquad \mathbf{x}_1 + \mathbf{x}_2 = \mathbf{X}_2 \qquad (3.4.4)$$

in

$$\|V_{12}\psi\|^2 = \int d^3\mathbf{x}_1 \cdots d^3\mathbf{x}_N\, |V_{12}(\mathbf{x}_1 - \mathbf{x}_2)|^2\, |\psi(\mathbf{x}_1, \ldots, \mathbf{x}_N)|^2 \qquad (3.4.5)$$

for example, we may write

$$\|V_{12}\psi\|^2 = \frac{1}{8} \int d^3\mathbf{X}_1\, d^3\mathbf{X}_2\, d^3\mathbf{x}_3 \cdots d^3\mathbf{x}_N\, |V_{12}(\mathbf{X}_1)|^2$$

$$\times \left|\psi\left(\frac{\mathbf{X}_2 + \mathbf{X}_1}{2}, \frac{\mathbf{X}_2 - \mathbf{X}_1}{2}, \mathbf{x}_3, \ldots, \mathbf{x}_N\right)\right|^2. \qquad (3.4.6)$$

This suggests to set

$$\psi\left(\frac{\mathbf{X}_2 + \mathbf{X}_1}{2}, \frac{\mathbf{X}_2 - \mathbf{X}_1}{2}, \mathbf{x}_3, \ldots, \mathbf{x}_N\right) \equiv \Phi(\mathbf{X}_1, \mathbf{X}_2, \mathbf{x}_3, \ldots, \mathbf{x}_N) \qquad (3.4.7)$$

then it is easy to show that the Fourier transforms are related by

$$\tilde{\psi}(\mathbf{p}_1, \mathbf{p}_2, \mathbf{p}_3, \ldots, \mathbf{p}_N) = \frac{1}{8} \tilde{\Phi}\left(\frac{\mathbf{p}_1 - \mathbf{p}_2}{2}, \frac{\mathbf{p}_1 + \mathbf{p}_2}{2}, \mathbf{p}_3, \ldots, \mathbf{p}_N\right). \qquad (3.4.8)$$

Consider the integral

$$I_{12}(\mathbf{X}_1) = \int d^3\mathbf{X}_2\, d^3\mathbf{x}_3 \cdots d^3\mathbf{x}_N\, |\Phi(\mathbf{X}_1, \mathbf{X}_2, \mathbf{x}_3, \ldots, \mathbf{x}_N)|^2 \qquad (3.4.9)$$

3.4 Boundedness of Hamiltonian From Below: Multi-Particle Systems

$$= \int \frac{d^3q_2}{(2\pi\hbar)^3} \frac{d^3p_3}{(2\pi\hbar)^3} \cdots \frac{d^3p_N}{(2\pi\hbar)^3} \left| \int \frac{d^3q_1}{(2\pi\hbar)^3} e^{i q_1 \cdot X_1/\hbar} \tilde{\Phi}(q_1, q_2, p_3, \ldots, p_N) \right|^2. \tag{3.4.10}$$

We now use the Cauchy-Schwarz inequality with $b > 0$ in the following

$$\left| \int \frac{d^3 q_1}{(2\pi\hbar)^3} \left[1 + b\left(q_1^2\right)^2 \right]^{1/2} \left| \tilde{\Phi}(q_1, q_2, p_3, \ldots, p_N) \right| e^{i q_1 \cdot X_1/\hbar} \left[1 + b\left(q_1^2\right)^2 \right]^{-1/2} \right|^2$$

$$\leqslant I \int \frac{d^3 q_1}{(2\pi\hbar)^3} \left[1 + b\left(q_1^2\right)^2 \right] \left| \tilde{\Phi}(q_1, q_2, p_3, \ldots, p_N) \right|^2 \tag{3.4.11}$$

where

$$I = \int \frac{d^3 q}{(2\pi\hbar)^3} \frac{1}{\left[1 + b\left(q^2\right)^2 \right]}$$

$$= \frac{1}{4\sqrt{2}\,\pi\hbar^3 b^{3/4}} \tag{3.4.12}$$

to obtain for $I_{12}(X_1)$ in (3.4.10) the X_1-independent bound

$$I_{12}(X_1) \leqslant I \int \frac{d^3 q_1}{(2\pi\hbar)^3} \frac{d^3 q_2}{(2\pi\hbar)^3} \frac{d^3 p_3}{(2\pi\hbar)^3} \cdots \frac{d^3 p_N}{(2\pi\hbar)^3}$$

$$\times \left[1 + b\left(q_1^2\right)^2 \right] \left| \tilde{\Phi}(q_1, q_2, p_3, \ldots, p_N) \right|^2$$

$$= \frac{I}{8} \int \frac{d^3 p_1}{(2\pi\hbar)^3} \frac{d^3 p_2}{(2\pi\hbar)^3} \cdots \frac{d^3 p_N}{(2\pi\hbar)^3}$$

$$\times \left[1 + b\left(\frac{(p_1 - p_2)^2}{4}\right)^2 \right] \left| \tilde{\Phi}\left(\frac{p_1 - p_2}{2}, \frac{p_1 + p_2}{2}, p_3, \ldots, p_N\right) \right|^2$$

$$= 8I \int \frac{d^3 p_1}{(2\pi\hbar)^3} \cdots \frac{d^3 p_N}{(2\pi\hbar)^3} \left[1 + b\left(\frac{(p_1 - p_2)^2}{4}\right)^2 \right] \left| \tilde{\psi}(p_1, p_2, p_3, \ldots, p_N) \right|^2 \tag{3.4.13}$$

where we have finally used (3.4.8).
Since

$$\frac{(p_1 - p_2)^2}{4} \leqslant \frac{p_1^2 + p_2^2}{2} \leqslant (m_1 + m_2)\left(\frac{p_1^2}{2m_1} + \frac{p_2^2}{2m_2}\right)$$

$$\leqslant (m_1 + m_2) \sum_{i=1}^{N} \frac{\mathbf{p}_i^2}{2m_i} \qquad (3.4.14)$$

we obtain the bound

$$I_{12}(\mathbf{X}_1) \leqslant 8I \left\{ 1 + b(m_1 + m_2)^2 \|H_0\psi\|^2 \right\}. \qquad (3.4.15)$$

We use this inequality in (3.4.6) for $|\mathbf{X}_1| < R_{12}$ only.
From the definition (3.4.7), (3.4.9), (3.4.10) and (3.4.2), (3.4.3) we have

$$\|V_{12}\psi\|^2 \leqslant Ia_{12}\left\{1 + b(m_1+m_2)^2 \|H_0\psi\|^2\right\} + C_{12}^2 \qquad (3.4.16)$$

or

$$\|V_{12}\psi\| \leqslant \sqrt{Ia_{12}b}\,(m_1+m_2)\|H_0\psi\| + \sqrt{C_{12}^2 + Ia_{12}}. \qquad (3.4.17)$$

Accordingly

$$\left\| \sum_{i<j}^{N} V_{ij}\psi \right\| \leqslant \sum_{i<j}^{N} \|V_{ij}\psi\|$$

$$\leqslant A\|H_0\psi\| + B \qquad (3.4.18)$$

giving a Kato bound (see also (3.3.71)), where

$$A = \sum_{i<j}^{N} (m_i + m_j)\sqrt{\frac{a_{ij}b^{1/4}}{4\sqrt{2}\,\pi\hbar^3}} \qquad (3.4.19)$$

$$B = \sum_{i<j}^{N} \sqrt{C_{ij}^2 + \frac{a_{ij}}{4\sqrt{2}\,\pi\hbar^3 b^{3/4}}} \qquad (3.4.20)$$

and we have used the value of the integral I in (3.4.12).
As in (3.3.63), the above then gives the lower bound

$$\left\langle \sum_{i=1}^{N} \frac{\mathbf{p}_i^2}{2m_i} + \sum_{i<j}^{N} V_{ij}(\mathbf{x}_i - \mathbf{x}_j) \right\rangle \geqslant -\frac{B}{1-A} \qquad (3.4.21)$$

with A, B defined in (3.4.19), (3.4.20) and the relevant constants in (3.4.2), (3.4.3). The positive constant b in (3.4.19) is so *chosen* to make $A < 1$.

3.4.2 Multi-Particle Systems and Other Potentials

A straightforward extension of the basic inequality in (3.1.8) for multi-particle states which may include many-body potentials is obtained by introducing real vector fields $F_j(\mathbf{x}_1, \ldots, \mathbf{x}_N), j = 1, \ldots, N$, of the position vectors of N particles, and define the potential energy

3.4 Boundedness of Hamiltonian From Below: Multi-Particle Systems

$$V(\mathbf{x_1}, \ldots, \mathbf{x_N}) = -\sum_{j=1}^{N} \nabla_j \cdot \mathbf{F}_j(\mathbf{x_1}, \ldots, \mathbf{x_N}). \quad (3.4.22)$$

Positivity implies that

$$\sum_{j=1}^{N} \left\| \left(\frac{\hbar \nabla_j}{\sqrt{2m_j}} + \frac{\sqrt{2m_j}}{\hbar} \mathbf{F}_j \right) \psi \right\|^2 \geq 0 \quad (3.4.23)$$

which upon integration by parts yields the elementary bound

$$\left\langle \sum_{j=1}^{N} \frac{\mathbf{p}_j^2}{2m_j} + V(\mathbf{x_1}, \ldots, \mathbf{x_N}) \right\rangle \geq -\sum_{j=1}^{N} \frac{2m_j}{\hbar^2} \langle \mathbf{F}_j^2 \rangle \quad (3.4.24)$$

Note that the lower bound on the right-hand side of (3.4.24) applies also to other potentials $U(\mathbf{x_1}, \ldots, \mathbf{x_N})$ which are bounded below by $V(\mathbf{x_1}, \ldots, \mathbf{x_N})$, i.e., for which $U(\mathbf{x_1}, \ldots, \mathbf{x_N}) \geq V(\mathbf{x_1}, \ldots, \mathbf{x_N})$.

Although the estimate in (3.4.24) is, in general, far from being optimal, it is nevertheless useful in establishing boundedness from below for some specific interactions. For an application of (3.4.22), (3.4.24), see Problem 3.5, and see also (3.4.29)–(3.4.32) below.

3.4.3 Multi-Particle Systems with Coulomb Interactions

We consider the multi-particle systems with Coulomb interactions described by Hamiltonian

$$H = \sum_{i=1}^{N} \frac{\mathbf{p}_i^2}{2m} + \sum_{i<j}^{N} \frac{e^2}{|\mathbf{x}_i - \mathbf{x}_j|} - \sum_{i=1}^{N} \sum_{j=1}^{k} \frac{Z_j e^2}{|\mathbf{x}_i - \mathbf{R}_j|} + \sum_{i<j}^{k} \frac{Z_i Z_j e^2}{|\mathbf{R}_i - \mathbf{R}_j|} \quad (3.4.25)$$

consisting of N negatively charged particles of charges e, masses m, and $k \geq 2$ positively charged particles of charges $Z_1 |e|, \ldots, Z_k |e|$, such that

$$\sum_{i=1}^{k} Z_i = N \quad (3.4.26)$$

i.e., we consider neutral systems. Here the positive charges are considered to be fixed. The Hamiltonian of an atom with atomic number Z is obtain from (3.4.25) by deleting the last term in it and setting $k = 1$, $Z_1 = Z$.

The derivation of the lower bounds for systems of fermions and bosons require special tools and will be given in detail in Chapter 14. Here we simply record the bounds obtained there.

For identical spin 1/2 (fermions) negatively charged particles, such as electrons, we have for $k \geq 2$ the bound[5]

---

[5] The numerical values 8.310 and 5.235 in (3.4.27) and (3.4.28), respectively, may be further improved, i.e., decreased but we will not attempt to do so.

$$\langle H \rangle_F \geq -8.310 \frac{me^4}{2\hbar^2} N \left[1 + \left(\frac{1}{N} \sum_{i=1}^{k} Z_i^{7/3}\right)^{1/2}\right]^2. \tag{3.4.27}$$

On the other hand, for identical spin 0 (bosons) negatively charged particles, we have[5] for $k \geq 2$

$$\langle H \rangle_B \geq -5.235 \frac{me^4}{2\hbar^2} N^{5/3} \left[1 + \left(\frac{1}{N} \sum_{i=1}^{k} Z_i^{7/3}\right)^{1/2}\right]^2. \tag{3.4.28}$$

An expression for the ground-state energy of an atom, as a function of the atomic number Z, will be obtained in Chapter 13 based on physical grounds as an extension of the so-called Thomas-Fermi atom in which the *latter* corresponds to the large Z limit. We here derive a conservative lower bound to the ground-state energy for atoms. To do this we use the simple bound

$$H\big|_{k=1, N=Z} \geq \sum_{i=1}^{Z} \left(\frac{\mathbf{p}_i^2}{2m} - \frac{Ze^2}{|\mathbf{x}_i|}\right) \tag{3.4.29}$$

where we have put the nucleus at the origin, i.e., $\mathbf{R} = \mathbf{0}$ and used the positivity of the second term in (3.4.25). A conservative lower bound for an atom is $-(me^4/2\hbar^2)Z^3$. This is easily obtained from the right-hand side of the inequality in (3.4.29) by knowing the ground state energy of a hydrogenic atom or, for example directly from (3.4.29) by defining

$$\mathbf{F}_i = \frac{Ze^2}{2} \frac{\mathbf{x}_j}{|\mathbf{x}_j|} \tag{3.4.30}$$

which gives

$$-\sum_{j=1}^{N} \boldsymbol{\nabla}_j \cdot \mathbf{F}_j = -\sum_{j=1}^{N} \frac{Ze^2}{|\mathbf{x}_j|} \tag{3.4.31}$$

and

$$-\sum_{j=1}^{N} \frac{2m^2}{\hbar^2} \mathbf{F}_j^2 = -\frac{me^4}{2\hbar^2} Z^3 \tag{3.4.32}$$

with $Z = N$, leading from (3.4.24) to the rough lower bound stated above. One may also derive improved bounds (cf., Problem 13.14). A fairly detailed investigation of the ground-state energy of atoms, as a function of Z, will be carried out in Chapter 13.

3.5 Decay of Quantum Systems

Consider a Hamiltonian H which is bounded from below by a *finite* number L_H,

3.5 Decay of Quantum Systems

$$\langle \psi | H | \psi \rangle \geqslant L_H \tag{3.5.1}$$

for normalized $|\psi\rangle$ (§3.2), where L_H is often negative. One may then conveniently introduce the Hamiltonian $H' = H - L_H$ and write the inner product $\langle \psi | \psi \rangle$ as (see also (§3.2))

$$\langle \psi | \psi \rangle = \int_0^\infty d\lambda \, \langle \psi | \delta(\lambda - H') | \psi \rangle = 1 \tag{3.5.2}$$

where

$$\langle \psi | \delta(\lambda - H') | \psi \rangle \equiv \frac{F(\lambda)}{2\pi\hbar} \tag{3.5.3}$$

denotes the probability density, per unit energy, of finding the energy of the system in the state $|\psi\rangle$ around the value λ in the energy scale translated by the amount $-L_H$.

In particular, if $\chi_\Delta(\lambda)$ denotes the characteristic function of a set Δ, i.e., $\chi_\Delta(\lambda) = 1$ if $\lambda \in \Delta$, and $\chi_\Delta(\lambda) = 0$ if $\lambda \notin \Delta$, then

$$\int_0^\infty \frac{d\lambda}{2\pi\hbar} \chi_\Delta(\lambda) F(\lambda) \tag{3.5.4}$$

represents the probability of finding the energy of the system in the state $|\psi\rangle$ to have values in the set Δ in the shifted energy scale.

Given that a system in the state $|\psi\rangle \equiv |\psi(0)\rangle$ has developed in time to the state

$$|\psi(t)\rangle = e^{-itH/\hbar} |\psi\rangle \tag{3.5.5}$$

a quantity of physical interest is the probability of finding the system in the *same* state $|\psi\rangle$ at time t. This is given by (§1.2–§1.5)

$$\text{Tr}\left[|\psi\rangle\langle\psi||\psi(t)\rangle\langle\psi(t)|\right] = |\langle\psi|\psi(t)\rangle|^2. \tag{3.5.6}$$

According to (3.5.2), (3.5.3), the corresponding amplitude to the above probability may be written as

$$\langle \psi | \psi(t) \rangle = e^{-itL_H/\hbar} A(t) \tag{3.5.7}$$

where

$$A(t) = \int_{-\infty}^\infty \frac{d\lambda}{2\pi\hbar} F(\lambda) e^{-it\lambda/\hbar} \tag{3.5.8}$$

as a Fourier transform with the constraint

$$F(\lambda) = 0 \quad \text{for} \quad \lambda < 0. \tag{3.5.9}$$

Since $|A(t)|^2$ denotes the probability that the system at time t is found in its initial state,[6] it is also referred to as the survival probability of the system. Guided by this interpretation, one may formally define

---

[6] $|A(t)|^2$ equivalently represents the following probability. Given that the system is in the state $|\psi\rangle$, it represents the probability that it was in the state $|\psi(t)\rangle$ at time $t < 0$.

$$\frac{1}{2}\int_{-\infty}^{\infty} dt\, |A(t)|^2 \equiv T \tag{3.5.10}$$

as a measure of the lifetime of the system in the state $|\psi\rangle$.

An interesting application which follows from (3.5.10) is the following one. Suppose that the energy of the system is confined to some interval Δ of length $\Delta(\lambda)$. That is, $F(\lambda) = 0$ for $\lambda \notin \Delta$, and from (3.5.2), (3.5.3)

$$\int_0^\infty \frac{d\lambda}{2\pi\hbar} \chi_\Delta(\lambda) F(\lambda) = 1 \tag{3.5.11}$$

in the notation of (3.5.4). An elementary application of the Cauchy-Schwarz inequality then gives

$$1 = \left(\int_{-\infty}^\infty \frac{d\lambda}{2\pi\hbar} \chi_\Delta(\lambda) F(\lambda)\right)^2 \leq \left(\int_{-\infty}^\infty \frac{d\lambda}{2\pi\hbar} \chi_\Delta(\lambda)\right)\left(\int_{-\infty}^\infty \frac{d\lambda}{2\pi\hbar} (F(\lambda))^2\right)$$

$$= \frac{\Delta(\lambda)}{2\pi\hbar} 2T \tag{3.5.12}$$

where we have used the property

$$\int_{-\infty}^\infty \frac{d\lambda}{2\pi\hbar} |F(\lambda)|^2 = \int_{-\infty}^\infty dt\, |A(t)|^2. \tag{3.5.13}$$

From (3.5.12) one then obtains an *energy-time uncertainty principle*,

$$\frac{h}{2} \leq \Delta(\lambda) T. \tag{3.5.14}$$

That is, the shorter the energy "width" $\Delta(\lambda)$ of the state $|\psi\rangle$ is, the longer is its lifetime against decay. In the limiting case of zero "width" $\Delta(\lambda) \to 0$, one obtains a non-decaying system of infinite lifetime! The mere fact that atoms in excited states decay from one energy level to a lower one, is an indication of the finite "widths" of such energy levels.

Some properties of $A(t)$ which follow directly from its definition in (3.5.8) are

$$A(0) = 1 \tag{3.5.15}$$

where we have used (3.5.2), (3.5.3),

$$A^*(t) = A(-t) \tag{3.5.16}$$

and formally

$$i\hbar A'(0) = \int_0^\infty \lambda\, d\lambda\, F(\lambda) \tag{3.5.17}$$

$$\hbar|A'(t)| \leq \int_0^\infty \lambda\, d\lambda\, F(\lambda) \tag{3.5.18}$$

where we have used (3.5.9). The integral on the right-hand sides of (3.5.17), (3.5.18) denotes the mean energy in the state $|\psi\rangle$, in the shifted energy scale.

Upon using the notation

$$A^*(t)A(t) = P(t) \tag{3.5.19}$$

for the survival probability in (3.5.6), we note that

$$\frac{d}{dt}P(t) = -A'(-t)A(t) + A(-t)A'(t) \tag{3.5.20}$$

as follows from (3.5.16). This leads to

$$\left.\frac{d}{dt}P(t)\right|_{t=0} = 0 \tag{3.5.21}$$

for the derivative of the survival probability at the origin.

In particular, one learns from (3.5.21) that the familiar decay law $\exp(-\Gamma t/\hbar)$, for $t > 0$, *cannot* hold near the *origin* $t \to +0$. A concrete example of a physical situation, where (3.5.21) may be explicitly verified is readily given. Consider a particle of spin 1/2 of magnetic dipole moment μ in a uniform time-independent magnetic field $\mathbf{B} = (B,0,0)$. Then as a special case of an analysis that will be given later in §8.8, leading to (8.8.47), the survival probability $P(t)$ with the initial state of the spin in the state $|+1/2, z\rangle$ is given by

$$P(t) = \cos^2\left(\frac{\mu B t}{\hbar}\right) \tag{3.5.22}$$

and in the neighborhood of the origin, this probability has the behavior

$$P(t) \simeq 1 - \frac{\mu^2 B^2 t^2}{\hbar^2} \tag{3.5.23}$$

consistent with (3.5.21) at $t = 0$.[7]

The exponential decay law with the amplitude $A(t)$ given by

$$A(t) = e^{-i\lambda_0 t/\hbar} e^{-\Gamma|t|/2\hbar} \tag{3.5.24}$$

for some λ_0, *cannot* certainly hold true for *all* t. In particular it cannot hold true for $t \to \infty$. The reason is that it yields a density $F(\lambda)$ which is non-zero for $\lambda < 0$, for arbitrary large $|\lambda|$ which is inconsistent with the boundedness of a Hamiltonian from below.[8] This is seen by the explicit evaluation of the integral

---

[7] This same behavior in (3.5.23) follows more generally in an oscillating magnetic field $\mathbf{B} = (B\cos\omega t, B\sin\omega t, B_0)$ as will be seen by examining (8.8.45) later on.

[8] The expression in (3.5.24), as it stands, is likewise inconsistent with analyticity properties when extended to the complex time domain $t \to t + i\tau$.

$$F(\lambda) = \int_{-\infty}^{\infty} dt\, e^{-i\lambda_0 t/\hbar} e^{-\Gamma|t|/2\hbar} e^{it\lambda/\hbar} \qquad (3.5.25)$$

which gives the familiar Breit-Weisskopf-Wigner expression

$$\frac{F(\lambda)}{2\pi\hbar} = \frac{1}{\pi} \frac{\Gamma/2}{(\lambda - \lambda_0)^2 + \Gamma^2/4} \qquad (3.5.26)$$

which, as mentioned above, does not vanish for $\lambda < 0$, for $|\lambda|$ arbitrarily large. Thus the exponential decay law cannot be true for all t.

On the other hand, if one chooses the density in the form in (3.5.26) to be true only for $\lambda \geq 0$, then one has to normalize it first, obtaining the density

$$\frac{F_0(\lambda)}{2\pi\hbar} = C \begin{cases} \Gamma/2 \left[(\lambda - \lambda_0)^2 + \Gamma^2/4\right]^{-1}, & \lambda \geq 0 \\ 0, & \lambda < 0 \end{cases} \qquad (3.5.27)$$

where

$$C = \frac{\pi}{2} + \tan^{-1}\left(\frac{2\lambda_0}{\Gamma}\right). \qquad (3.5.28)$$

A contour integration in the complex λ-plane, enclosing the pole at $\lambda = \lambda_0 - i\Gamma/2$, $\lambda_0 > 0$, for $t > 0$ gives (see Problem 3.6)

$$A(t)/C = \pi e^{-i\lambda_0 t/\hbar} e^{-\Gamma t/2\hbar} + R(t) \qquad (3.5.29)$$

where for $t \to \infty$

$$|R(t)| = \mathcal{O}\left(\frac{1}{t}\right). \qquad (3.5.30)$$

That is, for $t \to \infty$, $|A(t)|$ would vanish *slower* than an exponential law for $t \to \infty$.

The exponential decay law, however, is not ruled out, however, for t not close to the origin and for t not in the truly asymptotic region $|t| \to \infty$. An interesting example of an exponential decay will be worked out in detail in §8.1 (see (8.1.94)) for a two-level system interacting with an infinite number of harmonic oscillators.

The above analyses have shown the following typical behaviors $P(t) = 1 - \mathcal{O}(t^2)$ for $t \simeq 0$, $P(t) = \mathcal{O}(1/t)$ for $t \to \infty$ for the survival probability, and that the classic exponential law for intermediate t is not ruled out.

In the appendix to this section, we prove a theorem due to Paley and Wiener[9] tailored to our physical problem at hand. Given that the Hamiltonian of the physical system is *bounded from below* and $|A(t)|$ is *square-integrable* then we will see that

$$\int_{-\infty}^{\infty} dt \frac{|\ln|A(t)||}{1 + t^2} < \infty. \qquad (3.5.31)$$

---

[9] Paley and Wiener (1934).

3.5 Decay of Quantum Systems

To gain further insight into the finiteness property in (3.5.31), note the following. For any real $c > 0$,

$$|\ln c| = 2\Theta(c-1)\ln c - \ln c \tag{3.5.32}$$

where $\Theta(x)$ is the step function, i.e., $\Theta(x) = 1$ for $x > 1$, $\Theta(x) = 0$ for $x < 1$. Also

$$\Theta(c-1)\ln c \leqslant c^2 \tag{3.5.33}$$

$$|\ln c| \leqslant 2c^2 - \ln c \tag{3.5.34}$$

and with $c = |A(t)|$, we obtain from (3.5.34) upon integration over t

$$\int_{-\infty}^{\infty} dt \frac{|\ln|A(t)||}{1+t^2} \leqslant 2\int_{-\infty}^{\infty} dt \frac{|A(t)|^2}{1+t^2} - \int_{-\infty}^{\infty} dt \frac{\ln|A(t)|}{1+t^2}. \tag{3.5.35}$$

Since

$$\int_{-\infty}^{\infty} dt \frac{|A(t)|^2}{1+t^2} \leqslant \int_{-\infty}^{\infty} dt \, |A(t)|^2 \tag{3.5.36}$$

then in order to establish (3.5.31), with the square-integrability condition of $|A(t)|$, satisfied, it remains to obtain a *finite* lower bound (number) for the integral

$$\int_{-\infty}^{\infty} dt \frac{\ln|A(t)|}{1+t^2} \tag{3.5.37}$$

in (3.5.35).

A question that arises in reference to (3.5.31), (3.5.35), (3.5.36) is the following: "If we have square-integrability of $|A(t)|$, why bother with the lower bound on the left-hand side of (3.5.35) — that is with the expression in (3.5.31)?" The answer is that the condition in (3.5.31), via (3.5.35), follows if it is given, in particular, that the Hamiltonian of the system is bounded from below. The exponential law in (3.5.24) here comes to the "rescue" as a counter example for which the spectrum, as seen in (3.5.26), is unbounded from below. It is easily checked that $|A(t)|$, with $A(t)$ in (3.5.24), is square-integrable, while

$$\frac{|\ln|A(t)||}{1+t^2} = \frac{|t|}{1+t^2} \frac{\Gamma}{2\hbar} \tag{3.5.38}$$

is, clearly, not integrable. The integral on the left-hand side of (3.5.31) provides a rigorous constraint on $A(t)$, and hence on the survival probability $P(t)$, to be satisfied[10] for a physically consistent theory of quantum decay in which the Hamiltonian is bounded from below.

---

[10] The importance of the Paley-Wiener theorem was emphasized by Khalfin (1957). A general lucid overview treatment by Fonda et al. (1978) should be also noted.

Appendix to §3.5: The Paley-Wiener Theorem

To prove[11] the finiteness condition in (3.5.31), it is more convenient to start from the complex conjugate

$$A^*(t) = \int_{-\infty}^{\infty} \frac{d\lambda}{2\pi\hbar} F(\lambda) e^{it\lambda/\hbar} \qquad \text{(A-3.5.1)}$$

instead of $A(t)$.

Consider the following transform

$$I(t', \tau') = \frac{1}{\pi} \int_{-\infty}^{\infty} dt \frac{A^*(t)\tau'}{(t-t')^2 + \tau'^2} \qquad \text{(A-3.5.2)}$$

the so-called Poisson integral formula for a half-plane, where t', τ' are reals. The part of the integrand multiplying $A^*(t)$ in (A-3.5.2) has the following properties:

$$\frac{1}{\pi} \frac{\tau'}{(t-t')^2 + \tau'^2} \to \delta(t-t') \qquad \text{for} \qquad \tau' \to +0 \qquad \text{(A-3.5.3)}$$

formally, and

$$\frac{1}{\pi} \int_{-\infty}^{\infty} dt \frac{\tau'}{(t-t')^2 + \tau'^2} = 1 \qquad \text{(A-3.5.4)}$$

$$\frac{1}{\pi} \int_{-\infty}^{\infty} dt \frac{\tau' e^{iat}}{(t-t')^2 + \tau'^2} = e^{ia(t'+i\tau')} \qquad \text{(A-3.5.5)}$$

for $a > 0$.

Accordingly, from (A-3.5.1), (A-3.5.2), (A-3.5.5),

$$I(t', \tau') = \int_{-\infty}^{\infty} \frac{d\lambda}{2\pi\hbar} F(\lambda) \int_{-\infty}^{\infty} \frac{dt}{\pi} \frac{\tau' e^{i\lambda t/\hbar}}{(t-t')^2 + \tau'^2}$$

$$= \int_{-\infty}^{\infty} \frac{d\lambda}{2\pi\hbar} F(\lambda) e^{i\lambda(t'+i\tau')/\hbar}$$

$$\equiv B(t' + i\tau') \qquad \text{(A-3.5.6)}$$

and the transform in (A-3.5.2) has the remarkable property of extending the Fourier transform of $A^*(t)$ to the complex domain thus introducing the function of the complex variable $t + i\tau$: $B(t + i\tau)$.

Since $F(\lambda) = 0$ for $\lambda < 0$, however, we have *for $\tau \geq 0$*

$$\int_{-\infty}^{\infty} dt \, |B(t + i\tau)|^2 = \int_{0}^{\infty} \frac{d\lambda}{2\pi\hbar} (F(\lambda))^2 e^{-2\tau\lambda/\hbar}$$

---

[11] The proof of this theorem may be omitted at a first reading.

$$\leqslant \int_0^\infty \frac{d\lambda}{2\pi\hbar} |F(\lambda)|^2 = \int_{-\infty}^\infty dt\, |A(t)|^2 \qquad (\text{A-3.5.7})$$

and (A-3.5.6) defines the Fourier transform in the complex-time *upper* plane $\tau \geqslant 0$ and is analytic in this region.[12]

From (A-3.5.6), (A-3.5.2), we rewrite

$$B(t' + i\tau') = \frac{1}{\pi} \int_{-\infty}^\infty dt\, \frac{A^*(t)\tau'}{(t-t')^2 + \tau'^2} \qquad (\text{A-3.5.8})$$

and consider the transformation of the upper complex plane into the unit circle

$$\frac{z' - i}{z' + i} = re^{i\phi} \qquad (\text{A-3.5.9})$$

where

$$z' = t' + i\tau' \qquad (\text{A-3.5.10})$$

and $r < 1$. The boundary of the unit circle is given by $r = 1$. In reference to the real variable t in (A-3.5.8), we transform the integration along the real axis to an angular one along a unit circle defined by the transformation

$$\frac{t-i}{t+i} = e^{i\theta}. \qquad (\text{A-3.5.11})$$

The following are easily established

$$d\theta = 2\frac{dt}{1+t^2}, \quad t = -\frac{\sin\theta}{1 - \cos\theta} \qquad (\text{A-3.5.12})$$

$$t' = -\frac{2r\sin\phi}{1 + r^2 - 2r\cos\phi}, \quad \tau' = \frac{1 - r^2}{1 + r^2 - 2r\cos\phi} \qquad (\text{A-3.5.13})$$

$$(t-t')^2 + \tau'^2 = |t - z'|^2 = 2\frac{1 + r^2 - 2r\cos(\phi - \theta)}{(1 + r^2 - 2r\cos\phi)(1 - \cos\theta)} \qquad (\text{A-3.5.14})$$

and hence

$$\frac{\tau'(1+t^2)}{(t-t')^2 + \tau'^2} = \frac{1 - r^2}{(1 + r^2 - 2r\cos(\phi - \theta))}. \qquad (\text{A-3.5.15})$$

Therefore denoting $A^*(t)$, in terms of the new variable $e^{i\theta}$, by $a(e^{i\theta})$, and $B(t' + i\tau')$ in terms of $re^{i\phi}$ by $b(re^{i\phi})$, we have from (A-3.5.15), (A-3.5.12) and (A-3.5.8)

---

[12] A classic reference on the connection between analyticity and the Fourier transform is Titchmarsh (1937). For a relatively modern treatment, see, Rudin (1966).

176   3 Uncertainties, Localization, Stability and Decay of Quantum Systems

$$b(re^{i\phi}) = \frac{1}{2\pi} \int_0^{2\pi} d\theta \, a(e^{i\theta}) \frac{1-r^2}{(1+r^2 - 2r\cos(\phi-\theta))} \qquad \text{(A-3.5.16)}$$

showing that $b(re^{i\theta})$ is the so-called Poisson integral (for a circle) of $a(e^{i\theta})$, where the former is obtained for $r < 1$, i.e., inside the unit circle, from an integral of $a(e^{i\theta})$ along its circumference.

Using the integral

$$\frac{1}{2\pi} \int_0^{2\pi} d\theta \frac{1-r^2}{(1+r^2 - 2r\cos(\phi-\theta))} = 1 \qquad \text{(A-3.5.17)}$$

it is easily shown (see Problem 3.9), that

$$\int_0^{2\pi} d\phi \, |b(re^{i\phi})|^2 \leqslant \int_0^{2\pi} d\theta \, |a(re^{i\theta})|^2. \qquad \text{(A-3.5.18)}$$

To establish the finiteness condition in (3.5.31), we consider the integral

$$\frac{1}{2\pi} \int_0^{2\pi} d\phi \ln |b(re^{i\phi})|. \qquad \text{(A-3.5.19)}$$

Using the notation $\rho e^{i\phi} = w$, $\rho \leqslant r$ we note that although $b(w)$ is analytic in the circle of radius r, it may have zeros, and in (A-3.5.19), we are interested in $\ln|b(w)|$. Accordingly, suppose that $b(w)$ has zeros at:

$$w = 0 \quad \text{of order} \quad m,$$

$w = \alpha_1, \ldots, \alpha_n$ of orders $c_1, \ldots, c_n$ such that $|\alpha_1| < r, \ldots, |\alpha_n| < r$,

$w = \alpha_{n+1}, \ldots, \alpha_N$ of orders $c_{n+1}, \ldots, c_N$ such that $|\alpha_{n+1}| \to r, \ldots, |c_N| \to r$

and conveniently define[13]

$$h(w) = \frac{b(w)}{w^m} \left(\prod_{j=1}^n \left(\frac{\alpha_j^* w - r^2}{r(w - \alpha_j)} \right)^{c_j} \right) \left(\prod_{i=n+1}^N \left(\frac{\alpha_i}{w - \alpha_i} \right)^{c_i} \right) \qquad \text{(A-3.5.20)}$$

thus removing the zeros from $b(w)$.

Upon setting $\alpha_j = |\alpha_j| \exp(i\delta_j)$, we note that for the expression in the j-product in (A-3.5.20)

$$\left| \frac{\alpha_j^* w - r^2}{r(w - \alpha_j)} \right|_{\rho=r} = 1. \qquad \text{(A-3.5.21)}$$

Directly from (A-3.5.20),

---

[13] Cf. Rudin (1966), pp. 299–300, except here we are allowing zeros at the origin as well.

$$|h(0)| = \left(\left|\frac{b(w)}{w^m}\right|_{w=0}\right) \prod_{j=1}^{n} \left|\frac{r}{\alpha_j}\right|^{c_j} \qquad (A\text{-}3.5.22)$$

and for $\rho = r$, $w = r\exp i\phi$,

$$\ln\left|h\left(re^{i\phi}\right)\right| = \ln\left|b\left(re^{i\phi}\right)\right| - m\ln r - \sum_{i=n+1}^{N} c_i \ln\left|1 - e^{i(\phi - \delta_i)}\right| \qquad (A\text{-}3.5.23)$$

where in writing the latter we have used (A-3.5.21).

Since in (A-3.5.20), we have removed all the zeros of $b(w)$, we have from Cauchy's theorem

$$\frac{1}{2\pi}\int_0^{2\pi} d\phi \ln\left|h\left(re^{i\phi}\right)\right| = \frac{1}{2\pi i}\oint dw \frac{\ln|h(w)|}{w} = \ln|h(0)|. \qquad (A\text{-}3.5.24)$$

Using the integral[14]

$$\int_0^{2\pi} d\phi \ln\left|1 - e^{i(\phi - \delta)}\right| = 0 \qquad (A\text{-}3.5.25)$$

we have from (A-3.5.22)–(A-3.5.25),

$$\frac{1}{2\pi}\int_0^{2\pi} d\phi \ln\left|b\left(re^{i\phi}\right)\right| = \ln\left|\frac{b(w)}{w^m}\right|_{w=0} + \sum_{i=1}^{n} c_i \ln\left|\frac{r}{\alpha_i}\right| + m\ln r. \qquad (A\text{-}3.5.26)$$

This is known as *Jensen's formula*.

Since $c_i \geq 1$, $|r/\alpha_i| > 1$ in (A-3.5.26), we have the following bound

$$\frac{1}{2\pi}\int_0^{2\pi} d\phi \ln\left|b\left(re^{i\phi}\right)\right| \geq \ln\left|\frac{b(w)}{w^m}\right|_{w=0} + m\ln r. \qquad (A\text{-}3.5.27)$$

Using (3.5.34) in conjunction with (A-3.5.27), with $c = \left|b\left(re^{i\phi}\right)\right|$ gives,

$$\frac{1}{2\pi}\int_0^{2\pi} d\phi \left|\ln\left|b\left(re^{i\phi}\right)\right|\right| \leq \frac{1}{\pi}\int_0^{2\pi} d\phi \left|b\left(re^{i\phi}\right)\right|^2$$

$$- \ln\left|\frac{b(w)}{w^m}\right|_{w=0} - m\ln r. \qquad (A\text{-}3.5.28)$$

On the other hand, we may use (A-3.5.18) to obtain from (A-3.5.28)

$$\frac{1}{2\pi}\int_0^{2\pi} d\phi \left|\ln\left|b\left(re^{i\phi}\right)\right|\right| \leq \frac{1}{\pi}\int_0^{2\pi} d\theta \left|a\left(e^{i\theta}\right)\right|^2$$

---

[14] Cf. Rudin (1966), pp. 299,300, and Problem 3.10.

$$-\ln\left|\frac{b(w)}{w^m}\right|_{w=0} - m\ln r. \qquad (A\text{-}3.5.29)$$

Upon taking the limit $r \to 1$, and using the t variable in (A-3.5.11), (A-3.5.12) we have from (A-3.5.29)

$$\int_{-\infty}^{\infty} dt \frac{|\ln|A(t)||}{1+t^2} \leqslant 2 \int_{-\infty}^{\infty} dt \frac{|A(t)|^2}{1+t^2} - \pi\ln\left|\frac{b(w)}{w^m}\right|_{w=0} \qquad (A\text{-}3.5.30)$$

and since

$$\int_{-\infty}^{\infty} dt \frac{|A(t)|^2}{1+t^2} \leqslant \int_{-\infty}^{\infty} dt\, |A(t)|^2 \qquad (A\text{-}3.5.31)$$

the square-integrability of $|A(t)|$ implies the finiteness of the integral in (3.5.31) by noting finally that from (A-3.5.27) we have the lower bound

$$\int_{-\infty}^{\infty} dt \frac{\ln|A(t)|}{1+t^2} \geqslant \pi\ln\left|\frac{b(w)}{w^m}\right|_{w=0} \qquad (A\text{-}3.5.32)$$

is finite as a zero of order m of $b(w)$ at $w = 0$, if there is one, has been removed.

The *key* assumption in the above analysis is the *boundedness* of the Hamiltonian from below, i.e., that $|L_H|$ is finite. As a matter of fact, if H is not bounded from below, then we cannot define the Hamiltonian H', as done in the beginning of §3.5. In the latter case, if we define

$$\langle\psi|\psi(t)\rangle = \int_{-\infty}^{\infty} \frac{d\lambda}{2\pi\hbar} F(\lambda) e^{-it\lambda/\hbar} \qquad (A\text{-}3.5.33)$$

with $F(\lambda)$ not zero for $\lambda < 0$, then instead of (A-3.5.7), we obtain

$$\int_{-\infty}^{\infty} dt\, |B(t+i\tau)|^2 = \int_{-\infty}^{\infty} \frac{d\lambda}{2\pi\hbar} (F(\lambda))^2 e^{-2\tau\lambda/\hbar}$$

$$\geqslant e^{b2\tau/\hbar} \int_{-a}^{-b} \frac{d\lambda}{2\pi\hbar} |F(\lambda)|^2 \qquad (A\text{-}3.5.34)$$

where $0 < b < a$ for which $F(\lambda) \neq 0$ in the interval $(-a, -b)$. Clearly for τ positive and arbitrary large, the right-hand side of the above inequality (A-3.5.34) increases without bound, destroying the square-integrality of $|B(t+i\tau)|$ in t as opposed to the case in (A-3.5.7).

Problems

3.1. Show that the integrals (3.3.23) and (3.3.24) are equivalent.
3.2. Derive the expressions in (3.3.25) and (3.3.30).

3.3. Apply the formula (3.3.77) to find lower bounds for a Hamiltonian with potential $V(\mathbf{x})$ in (3.3.8) for $\lambda_3 \equiv 0$, $\lambda_1 < 0$, $\lambda_2 < 0$, $\beta_1 > 0$ in $\nu = 3$ dimensions, and compare your bound with the one in (3.3.15).

3.4. Use the general bound in (3.4.21) to find a lower bound to the Hamiltonian in (3.4.1) with

$$V_{ij}(\mathbf{x}_i - \mathbf{x}_j) = -\frac{Gm_im_j}{|\mathbf{x}_i - \mathbf{x}_j|}$$

for the Newtonian gravitational potential.

3.5. Introduce a vector fields $\mathbf{F}_j(\mathbf{x}_1, \ldots, \mathbf{x}_N; \mathbf{R}_1, \ldots, \mathbf{R}_k)$, for $j = 1, \ldots, N$, $\mathbf{R}_1, \ldots, \mathbf{R}_k$ fixed, to generate the potential energy for matter in (3.4.25) by applying the definition of the potential in (3.4.22). Using your expression for $\mathbf{F}_j$ obtain a lower bound for the ground-state energy by the application of (3.4.24). This estimate is rather rough in comparison to the ones in (3.4.27), (3.4.28). Is it possible to choose the position vectors $\mathbf{R}_1, \ldots, \mathbf{R}_k$ of the nuclei optimally to get an improved estimate?

3.6. Verify the normalization of the density in (3.5.27)/(3.5.28). By a contour integration in the complex λ-plane enclosing the pole at $\lambda = \lambda_0 - i\Gamma/2$, $\lambda_0 > 0$, for $t > 0$, show that $A(t)$ in (3.5.8) is given as in (3.5.29) and obtain an expression for $|R(t)|$. What is the next order behavior to $1/t$ for $t \to \infty$?

3.7. Suppose that the density in (3.5.26) is non-zero only for $0 \leqslant \lambda \leqslant c$, where c is a finite constant. Does the amplitude $A(t)$, in this case, involve an exponentially damping term?

3.8. Derive the expressions for the integrals given in (A-3.5.4), (A-3.5.5).

3.9. Prove the inequality in (A-3.5.18). [Hint: Note, in the process of the demonstration, that

$$a^*(e^{i\theta})\, a(e^{i\theta'}) + a(e^{i\theta})\, a^*(e^{i\theta'}) \leqslant \left|a(e^{i\theta})\right|^2 + \left|a(e^{i\theta'})\right|^2$$

and use (A-3.5.17).]

3.10. It may be amusing to explicitly evaluate the integral in (A-3.5.25) and show that it is equal to zero without using contour integration.

4

Spectra of Hamiltonians

The chapter is involved with several aspects concerning the spectra of Hamiltonians. In §4.1, §4.3, §4.4, the nature of the spectra of some general classes of Hamiltonian is determined under sufficiency conditions satisfied by the underlying potentials. These sufficiency conditions are readily verified and if satisfied, the nature of the spectra may be inferred. This is obviously very important if the exact solution of the problem in question is not known. Even if the solution may be explicitly obtained, the knowledge of the nature of the spectrum of a Hamiltonian prior to its determination is quite useful. Bound-states are studied in §4.2. In this section sufficiency conditions are given for their existence as well as for their absence for various physical systems. In §4.5, we carry out an analysis for counting the eigenvalues (if any) of given Hamiltonians. This investigation is then used in §4.6 to derive lower bounds to the expectation values of kinetic energy operators for single- and many-particle systems. The results obtained in these two sections will find important applications to multi-electron atoms in Chapter 13 and to the problem of the stability of matter in Chapter 14. The final section (§4.7) deals with the role of supersymmetry in solving the eigenvalue problem and in the construction of supersymmetric Hamiltonians. In this section, we make use of the general properties of supersymmetry transformations obtained in §2.9.

Before getting into the details of this chapter we note the following. Suppose that for some vector $|\psi\rangle$, in the domain of a given Hamiltonian H,

$$\langle\psi|H|\psi\rangle < 0. \tag{4.1}$$

In §4.1, for example, we give sufficiency conditions to be satisfied by the potential in question such that *if* the negative spectrum of H is not empty, then the latter consists of eigenvalues of finite degeneracy (the discrete spectrum). The condition in (4.1) would then imply that the negative spectrum of such a H is not empty consisting of a discrete one, as part of the spectrum. This property of the spectrum follows by noting that by the spectral decomposition of H (§1.8, (1.8.15)), (4.1) may be rewritten as

$$\int_{-\infty}^{\infty} \lambda \, \mathrm{d}\|P_H(\lambda)\psi\|^2 < 0 \tag{4.2}$$

thus admitting negative ($\lambda < 0$) eigenvalues for H.

4.1 Hamiltonians with Potentials Vanishing at Infinity

In this section we are concerned with Hamiltonians

$$H = \frac{\mathbf{p}^2}{2\mu} + V(\mathbf{x}) \tag{4.1.1}$$

in $\nu = 3, 2, 1$ dimensional spaces, with potentials satisfying the following sufficiency conditions. For any $0 < R < \infty$,

$$\int_{|\mathbf{x}|<R} \mathrm{d}^\nu\mathbf{x} \, |V(\mathbf{x})|^2 < \infty \tag{4.1.2}$$

and

$$|V(\mathbf{x})| \xrightarrow[|\mathbf{x}|\to\infty]{} 0. \tag{4.1.3}$$

Such potentials are said to be locally square-integrable with vanishing property at infinity. For $\nu = 3$, the Coulomb potential $\pm e^2/|\mathbf{x}|$, for example, lies in this category.

Since condition (4.1.3), in particular, implies that for a sufficiently large R, we may find a finite positive constant C such that $|V(\mathbf{x})| \leqslant C$ for $|\mathbf{x}| \geqslant R$, the analysis given in class [E] of §3.3 shows that such Hamiltonians are necessarily *bounded from below*. We will denote the lower bound of such a Hamiltonian by $-\ell_H$ ($= L_H$ in our earlier notation in §3.2), emphasizing its strict negativity if it arises.

For a two-particle system, for example, with a reduced mass μ, when the two particles are widely separated, condition (4.1.3) means that the interaction between them goes to zero, while the relative kinetic energy may take on any value from zero up to arbitrarily large positive values. Accordingly, the spectrum would include a continuous one on the positive real axis $[0, \infty)$.

Of particular interest is the situation when $-\ell_H$ is strictly negative. We would then have, in addition to the continuous spectrum discussed above, negative eigenvalues falling in the interval $[-\ell_H, 0)$. What is interesting, as the following theorem shows, is that these negative eigenvalues are at *most finitely* degenerate and there is no continuous spectrum on the negative real axis. In the introduction to this chapter, we have seen that if one can find any vector $|\psi\rangle$ such that $\langle\psi|H|\psi\rangle < 0$, then $-\ell_H$ is strictly negative. Needless to say, as mentioned earlier, in order for the Hamiltonian to have part of its spectrum consisting of negative values it is necessary that the negative part $V\Theta(-V)$, as defined in (3.3.18), of the potential V is non-vanishing, otherwise the Hamiltonian is a positive operator.

4.1 Hamiltonians with Potentials Vanishing at Infinity

Theorem 4.1.1
A Hamiltonian H in (4.1.1) with a potential $V(\mathbf{x})$ satisfying (4.1.2), (4.1.3) is bounded from below. It necessarily has a continuous spectrum on the positive real axis $[0, \infty)$ with no continuous spectrum on the negative real axis. It has also no eigenvalues of infinite degeneracy. In addition, if the lower bound $-\ell_H$ is strictly negative, it has also negative eigenvalues (a discrete spectrum) of at most finite degeneracy. [By finite it is meant to be non-infinite thus including the possibility of non-degeneracy.]

The importance of this result in quantum physics is obvious and cannot be overemphasized.

To establish[1] this result we write

$$\frac{1}{H - \xi_0} = \frac{1}{H_0 - \xi_0} + \frac{1}{H - \xi_0}(-V)\frac{1}{H_0 - \xi_0} \qquad (4.1.4)$$

where ξ_0 is any fixed real negative number such that

$$\xi_0 < \min[0, -\ell_H]. \qquad (4.1.5)$$

[Actually Proposition 4.1.2 below, alone, implies that $-\ell_H \leqslant 0$, and hence we may take $\xi_0 < -\ell_H$.] As before $H_0 = \mathbf{p}^2/2\mu$.

We first need the following two propositions.

Proposition 4.1.1
For any infinite sequence $\{|f_n\rangle\}$ of orthonormal vectors,

$$\left\| \frac{1}{[H + |\xi_0|]} V \frac{1}{[H_0 + |\xi_0|]} f_n \right\| \to 0 \quad \text{for} \quad n \to \infty \qquad (4.1.6)$$

where ξ_0 as given in (4.1.5).

To prove this we define

$$V_L(\mathbf{x}) = \begin{cases} V(\mathbf{x}), & |\mathbf{x}| \leqslant L \\ 0, & |\mathbf{x}| > L. \end{cases} \qquad (4.1.7)$$

We note that we may explicitly write

$$\left\langle \mathbf{x} \left| V_L \frac{1}{[H_0 + |\xi_0|]} f_n \right\rangle = V_L(\mathbf{x}) \langle \Phi_{\mathbf{x}} | f_n \rangle \qquad (4.1.8)$$

where we conveniently set

---
[1] The proofs of this theorem and of the following two propositions may be omitted at a first reading.

$$\Phi_{\mathbf{x}}(\mathbf{y}) = \int \frac{d^\nu \mathbf{p}}{(2\pi\hbar)^\nu} \frac{e^{-i\mathbf{p}\cdot(\mathbf{x}-\mathbf{y})/\hbar}}{\left[\frac{\mathbf{p}^2}{2\mu} + |\xi_0|\right]} \tag{4.1.9}$$

and hence

$$\langle \Phi_{\mathbf{x}} | f_n \rangle = \int d^\nu \mathbf{y}\, \Phi_{\mathbf{x}}^*(\mathbf{y}) f_n(\mathbf{y}). \tag{4.1.10}$$

Also

$$\|\Phi_{\mathbf{x}}\|^2 = \int d^\nu \mathbf{y}\, |\Phi_{\mathbf{x}}(\mathbf{y})|^2 = \int \frac{d^\nu \mathbf{p}}{(2\pi\hbar)^\nu} \frac{1}{\left[\frac{\mathbf{p}^2}{2\mu} + |\xi_0|\right]^2}$$

$$\equiv I^{(\nu)} < \infty \tag{4.1.11}$$

for $\nu = 3, 2, 1$, and the latter norm is independent of $\mathbf{x}$. Accordingly

$$|\langle \Phi_{\mathbf{x}} | f_n \rangle|^2 \leqslant \|\Phi_{\mathbf{x}}\|^2 \|f_n\|^2 = I^{(\nu)} < \infty \tag{4.1.12}$$

and is *bounded* independently of *all* $\mathbf{x}$ and n.

Also we note from Bessel's inequality (1.7.4) that

$$\sum_{j=1}^n |\langle \Phi_{\mathbf{x}} | f_j \rangle|^2 \leqslant \|\Phi_{\mathbf{x}}\|^2 \equiv I^{(\nu)} < \infty. \tag{4.1.13}$$

The finiteness of the right-hand side of this inequality for all $\mathbf{x}$ implies that

$$\lim_{k \to \infty} |\langle \Phi_{\mathbf{x}} | f_k \rangle|^2 = 0 \quad \text{for all} \quad \mathbf{x} \tag{4.1.14}$$

for the converges of the series on the left-hand side of (4.1.13) for $n \to \infty$.

The boundedness of $|\langle \Phi_{\mathbf{x}} | f_n \rangle|^2$ independently of $\mathbf{x}$ and n, as given in (4.1.12), and the obvious square-integrability of V_L (see (4.1.2), (4.1.7)), allow us to take the limit $n \to \infty$ inside the integral:

$$\lim_{n \to \infty} \int d^\nu \mathbf{x}\, |V_L(\mathbf{x})|^2 |\langle \Phi_{\mathbf{x}} | f_n \rangle|^2 = \int d^\nu \mathbf{x}\, |V_L(\mathbf{x})|^2 \left(\lim_{n \to \infty} |\langle \Phi_{\mathbf{x}} | f_n \rangle|^2\right) = 0 \tag{4.1.15}$$

where we have used (4.1.14).

Finally we consider the bound

$$\left\|\frac{1}{[H + |\xi_0|]} V \frac{1}{[H_0 + |\xi_0|]} f_n\right\| \leqslant \frac{1}{[|\xi_0| - \ell_H]} \left\|(V - V_L) \frac{1}{[H_0 + |\xi_0|]} f_n\right\|$$

$$+ \frac{1}{[|\xi_0| - \ell_H]} \left\|V_L \frac{1}{[H_0 + |\xi_0|]} f_n\right\| \tag{4.1.16}$$

and the bounds

4.1 Hamiltonians with Potentials Vanishing at Infinity

$$\left\|(V - V_L) \frac{1}{[H_0 + |\xi_0|]} f_n\right\| \leq \frac{1}{|\xi_0|} \max_{|\mathbf{x}|>L} |V(\mathbf{x})|, \qquad (4.1.17)$$

$$\left\|V_L \frac{1}{[H_0 + |\xi_0|]} f_n\right\| \leq \left(\int d^\nu \mathbf{x}\, |V_L(\mathbf{x})|^2 \,|\langle \Phi_\mathbf{x}|f_n\rangle|^2\right)^{1/2} \qquad (4.1.18)$$

to obtain from (4.1.15), upon taking the limit $n \to \infty$, that

$$\lim_{n\to\infty} \left\|\frac{1}{[H + |\xi_0|]} V \frac{1}{[H_0 + |\xi_0|]} f_n \right\| \leq \frac{1}{[|\xi_0|(|\xi_0| - \ell_H)]} \max_{|\mathbf{x}|>L} |V(\mathbf{x})|. \qquad (4.1.19)$$

The left-hand side of this inequality is independent of L. Hence finally upon taking the limit $L \to \infty$ and using (4.1.3) the result given in (4.1.6) follows.

Proposition 4.1.2
The spectrum of the operator $(H - \xi_0)^{-1}$ includes a continuous spectrum consisting of the interval $[0, 1/|\xi_0|]$ and does not contain eigenvalues of infinite degeneracy. In particular, the spectrum of H includes a continuous spectrum consisting of the positive axis $[0, \infty)$.

To establish the validity of the proposition, we use the notations

$$A = \frac{1}{H - \xi_0}(-V)\frac{1}{H_0 - \xi_0} \qquad (4.1.20)$$

$$B = \frac{1}{H_0 - \xi_0}. \qquad (4.1.21)$$

Clearly, the operator B has only a *continuous* spectrum consisting of the interval $[0, 1/|\xi_0|]$.

As a hypothesis, suppose that some real number λ_0 belongs to the essential spectrum, if not empty, of the operator $(H - \xi_0)^{-1}$. [That is, λ_0 belongs to its continuous spectrum or is an eigenvalue of infinite degeneracy.] According, to Proposition 1.8.1, there must then exists an infinite sequence $\{|f_n\rangle\}$ of orthonormal vectors such that

$$\|(A + B - \lambda_0)f_n\| \to 0 \quad \text{for} \quad n \to \infty \qquad (4.1.22)$$

where are have used (4.1.4), (4.1.20), (4.1.21).
Hence from the inequality

$$\|(B - \lambda_0)f_n\| \leq \|(A + B - \lambda_0)f_n\| + \|Af_n\| \qquad (4.1.23)$$

together with (4.1.6), (4.1.22), we conclude that

$$\|(B - \lambda_0)f_n\| \to 0 \quad \text{for} \quad n \to \infty. \qquad (4.1.24)$$

That is, we may infer from Proposition 1.8.1, and (4.1.22), (4.1.24), in particular, that any real number λ_0 which belongs to the continuous spectrum of $(A+B)$ necessarily belongs to the continuous spectrum of B as well, recalling that B has *no*, so-called, eigenvalues of infinite degeneracy. Because of the latter property of the spectrum of B we also conclude that $(A+B)$ cannot have such eigenvalues of infinite degeneracy.

Conversely, suppose that some real number λ_0 belongs to the (continuous) spectrum of B. That is, there exists an infinite sequence $\{|f_n\rangle\}$ of orthonormal vectors such that

$$\|(B-\lambda_0)f_n\| \to 0 \quad \text{for} \quad n \to \infty. \tag{4.1.25}$$

Hence from the inequality

$$\|(A+B-\lambda_0)f_n\| \leq \|Af_n\| + \|(B-\lambda_0)f_n\| \tag{4.1.26}$$

together with (4.1.6), (4.1.25) we may infer that

$$\|(A+B-\lambda_0)f_n\| \to 0 \quad \text{for} \quad n \to \infty. \tag{4.1.27}$$

That is, any real number λ_0 in the (continuous) spectrum of B necessarily belongs to the continuous spectrum of $(A+B)$ and from this, together with the conclusion following (4.1.24), we may infer that the continuous spectrum of $(A+B)$ consists of the interval $[0, 1/|\xi_0|]$. Also that $(A+B)$ has no eigenvalues of infinite degeneracy. This completes the proof of the proposition by finally noting that the continuous spectrum of H then consists of the positive axis $[0, \infty)$ alone.

To complete the proof of the theorem, it is instructive to reconsider the expression for the resolvent of H. To this end let

$$R(\xi_0) = \frac{1}{H - \xi_0} \tag{4.1.28}$$

with ξ_0 defined to be fixed as before.

Suppose λ is some complex number, then the following expressions are easily derived:

$$R(\xi_0) \frac{(-1/\lambda)}{[R(\xi_0) - \frac{1}{\lambda}]} = \frac{1}{[H - (\xi_0 + \lambda)]} \tag{4.1.29}$$

and

$$\frac{1}{[R(\xi_0) - \frac{1}{\lambda}]} = -\lambda - \lambda^2 \frac{1}{[H - (\xi_0 + \lambda)]}. \tag{4.1.30}$$

The equality in (4.1.29) leads to the bound:

$$\left\| \frac{1}{[H - (\xi_0 + \lambda)]} f \right\| \leq \frac{1}{|\lambda|} \frac{1}{[|\xi_0| - \ell_H]} \left\| \frac{1}{[R(\xi_0) - \frac{1}{\lambda}]} f \right\|. \tag{4.1.31}$$

On the other hand, (4.1.30) gives the bound:

$$\left\| \frac{1}{[R(\xi_0) - \frac{1}{\lambda}]} f \right\| \leq |\lambda| + |\lambda^2| \left\| \frac{1}{[H - (\xi_0 + \lambda)]} f \right\|. \qquad (4.1.32)$$

From (4.1.31), (4.1.32), we may then conclude, with λ now specialized to be real and not equal to zero, that $[H - (\xi_0 + \lambda)]^{-1}$ does not exist if and only if $[R(\xi_0) - 1/\lambda]^{-1}$ does not exist. Also that $[H - (\xi_0 + \lambda)]^{-1}$ exists as an unbounded operator if and only if $[R(\xi_0) - 1/\lambda]^{-1}$ exists as an unbounded operator.

In particular, the above states that if a real number $1/\lambda$ belongs to the discrete spectrum of $R(\xi_0)$ then the real value $(\lambda - |\xi_0|)$ necessarily belongs to the discrete spectrum of H and *vice versa*. Since the spectrum of H is empty in the region $(-\infty, -\ell_H)$, it is necessary that $\lambda > 0$. On the other hand, the statement in Proposition 4.1.2 does not rule out the possibility of having eigenvalues (of at most finite degeneracy) for $R(\xi_0)$ for $1/\lambda < 0$ or for $1/\lambda > 1/|\xi_0|$. These facts together with the necessary condition $\lambda > 0$, just given, show that if $-\ell_H$ is strictly negative, i.e., the spectrum of H in $[-\ell_H, 0)$ is *not empty*, then H has also a discrete spectrum (eigenvalues of at most finite degeneracy) falling in the interval $[-\ell_H, 0)$. This completes the proof of the theorem.

4.2 On Bound-States

4.2.1 A Potential Well

An elementary though important example of a potential falling in the category of Theorem 4.1.1 is the one-dimensional potential well problem defined by:

$$H = \frac{-\hbar^2}{2m} \frac{d^2}{dx^2} + V(x) \qquad (4.2.1)$$

where

$$V(x) = \begin{cases} -U_0, & |x| \leq L \\ 0, & |x| > L \end{cases} \qquad (4.2.2)$$

$U_0 > 0$.

According to the theorem, this Hamiltonian has part of its spectrum a continuous one on the positive real axis $[0, \infty)$. It is instructive to consider the discrete spectrum in the light of Proposition 1.8.2.

To the above end, we set

$$\frac{2m}{\hbar^2}(U_0 - |E|) = K^2, \quad \frac{2m}{\hbar^2}|E| = k^2 \qquad (4.2.3)$$

in the Schrödinger equation

$$\left[\frac{-\hbar^2}{2m}\frac{d^2}{dx^2} + V(x)\right]\psi(x) = E\psi(x) \tag{4.2.4}$$

with $E < 0$, corresponding to a bound-state, to obtain

$$\left[\frac{d^2}{dx^2} + K^2\right]\psi(x) = 0, \quad |x| \leqslant L \tag{4.2.5}$$

$$\left[\frac{d^2}{dx^2} - k^2\right]\psi(x) = 0, \quad |x| > L \tag{4.2.6}$$

having, respectively, even and odd solutions

$$\left.\begin{array}{l}\psi(x) = A\cos(Kx), \, |x| \leqslant L \\ \psi(x) = Be^{-k|x|}, \quad |x| > L\end{array}\right\} \tag{4.2.7}$$

and

$$\left.\begin{array}{l}\psi(x) = C\sin(Kx), \, |x| \leqslant L \\ \psi(x) = De^{-k|x|}, \quad x > L \\ \psi(x) = -De^{-k|x|}, \, x < -L.\end{array}\right\} \tag{4.2.8}$$

The boundary conditions implied by the continuity of $\psi'(x)/\psi(x)$ at $|x| = L$ lead immediately to the equations:

$$\tan\xi = \frac{\sqrt{a^2 - \xi^2}}{\xi} \tag{4.2.9}$$

and

$$\tan\xi = \frac{-\xi}{\sqrt{a^2 - \xi^2}} \tag{4.2.10}$$

corresponding, respectively, to the even and odd solutions, where we have set

$$\xi = KL, \quad a^2 = \frac{2mU_0 L^2}{\hbar^2}. \tag{4.2.11}$$

The energy levels may be thus written from (4.2.3), (4.2.11) as

$$E = -U_0\left[1 - \frac{\xi^2}{a^2}\right]. \tag{4.2.12}$$

We may quite generally write

$$\frac{a}{\pi} = N + \varepsilon \tag{4.2.13}$$

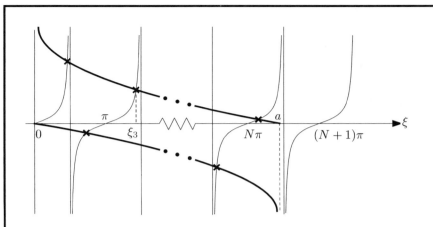

Fig. 4.1. The upper and lower curves are given, respectively, by $f(\xi) = \sqrt{a^2 - \xi^2}/\xi$, $g(\xi) = -\xi/\sqrt{a^2 - \xi^2}$, while the eigenvalues are given by $E^{(c)} = -U_0(1 - \xi_c^2/a^2)$, with ξ_c denoting critical values corresponding to the intersections of the former curves (denoted by crosses on the graphs) with the $\tan \xi$ curves. One may generally write $a/\pi = N + \varepsilon$, where N is a non-negative integer and $0 \leqslant \varepsilon < 1$. For $0 < \varepsilon < 1$, the number of eigenvalues corresponding to the upper curve is $(N+1)$. For $0 < \varepsilon \leqslant 0.5$, the number of eigenvalues corresponding to the lower curve is N, while for $0.5 < \varepsilon < 1$, it is $(N+1)$. For $\varepsilon = 0$, $a = N\pi$, $N \geqslant 1$, the largest E value is at the bottom of the continuous spectrum, with $2N$ eigenvalues falling below it. The Figure is of a qualitative nature only and is not based on actual numerical values.

where N is a non-negative integer and $0 \leqslant \varepsilon < 1$. The bound-state energies E are obtained as described in Figure 4.1.

For $0 < \varepsilon \leqslant 0.5$, the total number of bound-states are $n_0 = (2N+1)$, while for $0.5 < \varepsilon < 1$, the number of bound-states are $n_0 = (2N+2)$ — See Figure 4.1.

From (4.2.11), (4.2.12) we also have the following expression for the bound-states

$$E_k = -U_0 + \frac{\hbar^2}{2mL^2}\xi_k^2, \quad k = 1, \ldots, n_0 \qquad (4.2.14)$$

where $\xi_1, \ldots, \xi_{n_0}$ are the ξ values corresponding to the crosses (intersections) in Figure 4.1. The eigenvalues $E_1, \ldots, E_{n_0}$ are arranged in a non-decreasing order. In the notation of Proposition 1.8.2, $E_k = \lambda_k(H)$, with E_{n_0} falling below the bottom of the continuous spectrum of H.

For $\varepsilon = 0$, $\xi_{n_0} = a = N\pi$ (see the Figure) and $E_{n_0} = 0$ is at the bottom of the continuous spectrum of H with $n_0 - 1$ eigenvalues falling below E_{n_0} in conformity with Proposition 1.8.2.

The limit $a \to 0$ will be considered below.

Quite generally, we note that for $N \geqslant 1$, for example,

$$a - \frac{\pi}{2} \leqslant \xi_{n_0}. \tag{4.2.15}$$

The above example will be useful in studying the spectrum of Hamiltonians with potentials increasing with no bound at infinity in §4.4.

4.2.2 Limit of the Potential Well

In reference to (4.2.13), (4.2.11) if

$$\frac{a}{\pi} = \varepsilon, \quad 0 < \varepsilon \leqslant 0.5 \tag{4.2.16}$$

the analysis given in the figure caption of Figure 4.2, shows that there is only one bound-state corresponding to an *even* solution.

We consider the limit $\varepsilon \to +0$, i.e., $a \to +0$, with $U_0 \to \infty$, $L \to 0$ such that

$$2U_0 L = \lambda \tag{4.2.17}$$

is a finite non-vanishing positive constant.

From (4.2.3), (4.2.11) and (4.2.9) we then have from

$$\tan\left(\sqrt{\frac{2m}{\hbar^2}(U_0 - |E|)}\, L\right) = \sqrt{\frac{|E|}{U_0 - |E|}} \tag{4.2.18}$$

that is in the above limit of U_0 large and L small

$$\frac{\sqrt{\frac{2m}{\hbar^2}}\,\lambda}{2\sqrt{U_0}} = \sqrt{\frac{|E|}{U_0}} \tag{4.2.19}$$

or that

$$E = -\frac{m\lambda^2}{2\hbar^2}. \tag{4.2.20}$$

The eigenstate corresponding to (4.2.20) is from (4.2.7)

$$\psi(x) = \sqrt{k}\, e^{-k|x|} \tag{4.2.21}$$

where $k = \sqrt{2m|E|}/\hbar$.

4.2.3 The Dirac Delta Potential

It is instructive to compare the limiting solution (4.2.20), (4.2.21), for $U_0 \to \infty$, $L \to 0$ and λ a finite non-zero positive constant as defined in (4.2.17), with the one obtained directly from the Dirac delta potential

$$V(x) = -\lambda\, \delta(x), \quad \lambda > 0. \tag{4.2.22}$$

First in reference to the inequality (3.1.8) with $\nu = 1$, we set

$$g(x) = \frac{\lambda}{|x|} \tag{4.2.23}$$

hence

$$\frac{\partial}{\partial x} x\, g(x) = \lambda \frac{\partial}{\partial x} \frac{x}{|x|}$$

$$= \lambda \frac{\partial}{\partial x} [\Theta(x) - \Theta(-x)]$$

$$= 2\lambda \delta(x) \tag{4.2.24}$$

where $\Theta(x)$ is the step function.

Hence from (3.1.8), with $x^2 g^2(x) = \lambda^2$, we have

$$-\frac{m\lambda^2}{2\hbar^2} \leqslant \left\langle \frac{p^2}{2m} - \lambda \delta(x) \right\rangle \tag{4.2.25}$$

giving the lower bound $-m\lambda^2/2\hbar^2$ for the spectrum.

Now we solve for the bound-state problem corresponding to the equation

$$\left[-\frac{\hbar^2}{2m} \frac{d^2}{dx^2} - \lambda \delta(x) \right] \psi(x) = E\, \psi(x) \tag{4.2.26}$$

with $E < 0$. For $|x| > 0$, this gives the properly normalized solution

$$\psi(x) = \sqrt{k}\, e^{-k|x|} \tag{4.2.27}$$

where $k = \sqrt{2m|E|}/\hbar$. We note that

$$\lim_{x \to +0} \psi(x) = \lim_{x \to -0} \psi(x) = \sqrt{k}. \tag{4.2.28}$$

On the other hand

$$\psi'(x) = \sqrt{k}\, e^{-k|x|} \begin{cases} -k, & x > 0 \\ +k, & x < 0 \end{cases} \tag{4.2.29}$$

and hence

$$\lim_{x \to +0} \psi'(x) - \lim_{x \to -0} \psi'(x) = -2k\sqrt{k}. \tag{4.2.30}$$

Upon integration of (4.2.26) over x from $-\varepsilon$ to ε we obtain

$$-\frac{\hbar^2}{2m}(\psi'(\varepsilon) - \psi'(-\varepsilon)) - \lambda\psi(0) = E \int_{-\varepsilon}^{\varepsilon} dx\, \psi(x) \tag{4.2.31}$$

which because of the continuity of $\psi(x)$ in (4.2.28) and the discontinuity of $\psi'(x)$ as given in (4.2.30) for $\varepsilon \to 0$ gives

$$-\frac{\hbar^2}{2m}\left(-2k\sqrt{k}\right) - \lambda\sqrt{k} = 0, \tag{4.2.32}$$

$k = \lambda m/\hbar^2$, or

$$E = -\frac{m\lambda^2}{2\hbar^2} \tag{4.2.33}$$

consistent with (4.2.20) and coincides with the lower bound obtained in (4.2.25), with E in (4.2.33) as the only bound-state energy with corresponding eigenstate given in (4.2.27).

In the following theorems of this section, we assume, for every Hamiltonian into consideration, that, on the *negative* real axis, it may have at most eigenvalues. In Theorem 4.1.1, for example, we have seen that a Hamiltonian with a potential satisfying the sufficiency conditions (4.1.2), (4.1.3) will have such a property.

According to the introduction to this chapter we may then conclude that, in each case, if we can find *any* vector $|\psi\rangle$ such that

$$\langle \psi | H | \psi \rangle < 0 \tag{4.2.34}$$

then the corresponding Hamiltonian

$$H = \frac{\mathbf{p}^2}{2\mu} + V(\mathbf{x}) \tag{4.2.35}$$

has at least one bound-state.

4.2.4 Sufficiency Conditions for the Existence of a Bound-State for $\nu = 1$

Theorem 4.2.1
If

$$-\infty < \int_{-\infty}^{\infty} dx\, V(x) < 0, \quad \int_{-\infty}^{\infty} dx\, |x|\, |V(x)| < \infty \tag{4.2.36}$$

then the Hamiltonian in (4.2.35) admits at least one bound-state.

To establish the validity of theorem, we use the properly normalized trial function

$$\psi(x) = C\, \exp(-\alpha |x|/2), \quad \alpha > 0 \tag{4.2.37}$$

where

$$C = \sqrt{\frac{\alpha}{2}}. \tag{4.2.38}$$

It is easily verified that

$$\left\langle \psi \left| \frac{p^2}{2\mu} \right| \psi \right\rangle = \frac{\hbar^2 \alpha^2}{8\mu} \tag{4.2.39}$$

and hence
$$\langle \psi | H | \psi \rangle = \frac{\hbar^2 \alpha^2}{8\mu} + \frac{\alpha}{2} \int_{-\infty}^{\infty} dx \, e^{-\alpha|x|} V(x) \qquad (4.2.40)$$

which may be rewritten as
$$\langle \psi | H | \psi \rangle = \frac{\hbar^2 \alpha^2}{8\mu} + \frac{\alpha}{2} \int_{-\infty}^{\infty} dx \, V(x) + \frac{\alpha}{2} \int_{-\infty}^{\infty} dx \, \left(e^{-\alpha|x|} - 1 \right) V(x). \qquad (4.2.41)$$

We use the elementary bound
$$\left| e^{-\alpha|x|} - 1 \right| \leqslant \alpha |x| \qquad (4.2.42)$$

for all $|x|$, to obtain from (4.2.41)
$$\langle \psi | H | \psi \rangle \leqslant \frac{\alpha}{2} \left[\alpha \left(\frac{\hbar^2}{4\mu} + \int_{-\infty}^{\infty} dx \, |x| \, |V(x)| \right) + \int_{-\infty}^{\infty} dx \, V(x) \right] \qquad (4.2.43)$$

and the latter is strictly negative for
$$(0 <) \, \alpha < \left[-\int_{-\infty}^{\infty} dx \, V(x) \right] \Big/ \left[\frac{\hbar^2}{4\mu} + \int_{-\infty}^{\infty} dx \, |x| \, |V(x)| \right] \qquad (4.2.44)$$

remembering the conditions stated in the theorem, proving the existence of a vector $|\psi\rangle$ satisfying (4.2.34).

The elementary bound (4.2.42) follows by noting that for $y \geqslant 0$
$$1 - e^{-y} = \int_0^y dy' \, e^{-y'} \qquad (4.2.45)$$

and
$$\left| 1 - e^{-y} \right| = \int_0^y dy' \, e^{-y'} \leqslant \int_0^y dy' = y. \qquad (4.2.46)$$

Potentials satisfying the conditions of Theorem 4.1.1 and (4.2.36) are numerous. Such an example is given by the potential
$$V(x) = U_0 \Theta (L - |x|) + \lambda_1 e^{-\alpha|x|} \qquad (4.2.47)$$

where $\alpha > 0$, $L > 0$ and
$$2 L U_0 + \frac{2 \lambda_1}{\alpha} < 0 \qquad (4.2.48)$$

to ensure that $\int_{-\infty}^{\infty} dx \, V(x) < 0$.

4.2.5 Sufficiency Conditions for the Existence of a Bound-State for $\nu = 2$

Theorem 4.2.2

If

$$-\infty < \int d^2\mathbf{x}\, V(\mathbf{x}) < -\frac{\hbar^2 \pi}{4\mu} \qquad (4.2.49)$$

and

$$0 < \int d^2\mathbf{x}\, |\mathbf{x}|\, |V(\mathbf{x})| < \infty \qquad (4.2.50)$$

then the Hamiltonian (4.2.35) admits at least one bound-state.

To establish this result, we choose the trial function

$$\psi(\mathbf{x}) = \frac{\alpha}{\sqrt{2\pi}} e^{-\alpha|\mathbf{x}|/2}, \quad \alpha > 0 \qquad (4.2.51)$$

to obtain

$$\left\langle \psi \left| \frac{\mathbf{p}^2}{2\mu} \right| \psi \right\rangle = \frac{\hbar^2 \alpha^2}{8\mu} \qquad (4.2.52)$$

$$\langle \psi | H | \psi \rangle = \frac{\hbar^2 \alpha^2}{8\mu} + \frac{\alpha^2}{2\pi} \int d^2\mathbf{x}\, e^{-\alpha|\mathbf{x}|} V(\mathbf{x})$$

$$= \frac{\hbar^2 \alpha^2}{8\mu} + \frac{\alpha^2}{2\pi} \int d^2\mathbf{x}\, V(\mathbf{x}) + \frac{\alpha^2}{2\pi} \int d^2\mathbf{x}\, \left(e^{-\alpha|\mathbf{x}|} - 1 \right) V(\mathbf{x}). \qquad (4.2.53)$$

Hence

$$\langle \psi | H | \psi \rangle \leqslant \frac{\alpha^2}{2\pi}\left[\frac{\hbar^2 \pi}{4\mu} + \int d^2\mathbf{x}\, V(\mathbf{x}) + \alpha \int d^2\mathbf{x}\, |\mathbf{x}|\, |V(\mathbf{x})| \right] \qquad (4.2.54)$$

and the right-hand side is strictly negative if one chooses

$$(0 <)\, \alpha < -\frac{\left[\frac{\hbar^2 \pi}{4\mu} + \int d^2\mathbf{x}\, V(\mathbf{x}) \right]}{\int d^2\mathbf{x}\, |\mathbf{x}|\, |V(\mathbf{x})|} \qquad (4.2.55)$$

thus establishing the validity of the theorem, remembering the conditions (4.2.49), (4.2.50) stated (and hence, in particular, the strict positivity of the right-hand side of the inequality in (4.2.55)).

In the light of this theorem, we may quite generally define a (space) scale parameter

$$R = \frac{3}{2} \frac{\int d^2\mathbf{x}\, |\mathbf{x}|\, |V(\mathbf{x})|}{|\int d^2\mathbf{x}\, V(\mathbf{x})|} \qquad (4.2.56)$$

and an energy scale parameter

$$U_0 = \frac{3}{2\pi R^3} \int d^2\mathbf{x}\, |\mathbf{x}|\, |V(\mathbf{x})|. \qquad (4.2.57)$$

The sufficiency condition (4.2.49), in particular, for the existence of a bound-state may be then rewritten as

$$\frac{4\mu}{\hbar^2} R^2 U_0 > 1 \qquad (4.2.58)$$

with $\int d^2\mathbf{x}\, V(\mathbf{x}) < 0$.

The parameters R and U_0 conveniently coincide with the parameters of a two dimensional symmetrical well

$$V(\mathbf{x}) = \begin{cases} -U_0, & |\mathbf{x}| \leqslant R \\ 0, & |\mathbf{x}| > R \end{cases} \qquad (4.2.59)$$

where $U_0 > 0$.

4.2.6 Sufficiency Conditions for the Existence of a Bound-State for $\nu = 3$

Theorem 4.2.3

If

$$-\infty < \int d^3\mathbf{x}\, V(\mathbf{x}) < 0 \qquad (4.2.60)$$

and

$$0 < \frac{4\pi\hbar^2}{\mu} \int d^3\mathbf{x}\, |\mathbf{x}|\, |V(\mathbf{x})| < \left(\int d^3\mathbf{x}\, V(\mathbf{x})\right)^2 \qquad (4.2.61)$$

then the Hamiltonian H in (4.2.35) admits at least one bound-state.

To establish this result we choose the normalized trial function

$$\psi(\mathbf{x}) = \sqrt{\frac{\alpha^3}{8\pi}}\, e^{-\alpha|\mathbf{x}|/2}, \quad \alpha > 0 \qquad (4.2.62)$$

giving

$$\left\langle \psi \left| \frac{\mathbf{p}^2}{2\mu} \right| \psi \right\rangle = \frac{\hbar^2 \alpha^2}{8\mu}$$

and

$$\langle \psi | H | \psi \rangle = \frac{\hbar^2 \alpha^2}{8\mu} + \frac{\alpha^3}{8\pi} \int d^3\mathbf{x}\, V(\mathbf{x}) + \frac{\alpha^3}{8\pi} \int d^3\mathbf{x} \left(e^{-\alpha|\mathbf{x}|} - 1 \right) V(\mathbf{x}) \quad (4.2.63)$$

and hence

$$\langle \psi | H | \psi \rangle \leqslant \frac{\hbar^2 \alpha^2}{8\mu} \left[\alpha^2 A - \alpha B + 1 \right] \qquad (4.2.64)$$

where
$$A = \frac{\mu}{\hbar^2 \pi} \int d^3x \, |\mathbf{x}| \, |V(\mathbf{x})| \qquad (4.2.65)$$

$$B = \frac{\mu}{\hbar^2 \pi} \left| \int d^3x \, V(\mathbf{x}) \right|. \qquad (4.2.66)$$

The expression in the square brackets on the right-hand side of (4.2.64) may be rewritten as
$$A \left(\alpha - \alpha_+ \right) \left(\alpha - \alpha_- \right) \qquad (4.2.67)$$
where
$$\alpha_\pm = \frac{B}{2A} \left[1 \pm \left(1 - \frac{4A}{B^2} \right)^{1/2} \right] \qquad (4.2.68)$$

and the condition (4.2.61) ensures that $4A/B^2 < 1$.

Accordingly, we may choose $\alpha = B/2A$, thus making (4.2.67), or the right-hand side of (4.2.64), strictly negative. This completes the proof of the theorem.

As in the two-dimensional case, we may, in the light of the above theorem, define a space scale parameter by
$$R = \frac{4}{3} \frac{\int d^3x \, |\mathbf{x}| \, |V(\mathbf{x})|}{|\int d^3x \, V(\mathbf{x})|} \qquad (4.2.69)$$

and an energy scale parameter
$$U_0 = \frac{1}{\pi R^4} \int d^3x \, |\mathbf{x}| \, |V(\mathbf{x})|. \qquad (4.2.70)$$

The sufficiency condition in (4.2.61), in particular, for the existence of a bound-state then reads
$$1 < \frac{4}{9} \frac{\mu \, U_0 R^2}{\hbar^2} \qquad (4.2.71)$$
with (4.2.60) holding true.

The parameters R and U_0 above conveniently coincide, with those of a spherical potential well:
$$V(\mathbf{x}) = \begin{cases} -U_0, & |\mathbf{x}| \leqslant R \\ 0, & |\mathbf{x}| > R \end{cases} \qquad (4.2.72)$$

with $U_0 > 0$.

As another illustration, consider the Yukawa potential
$$V(\mathbf{x}) = -\lambda \frac{e^{-\beta|\mathbf{x}|}}{|\mathbf{x}|}, \quad \beta > 0, \, \lambda > 0. \qquad (4.2.73)$$

This gives

$$\int d^3x \, |x| \, |V(x)| = \frac{8\pi\lambda}{\beta^3} \qquad (4.2.74)$$

and

$$\int d^3x \, V(x) = \frac{-4\pi\lambda}{\beta^2}. \qquad (4.2.75)$$

The sufficiency condition (4.2.61) for the existence of a bound-state then reads

$$1 < \frac{\mu\lambda}{2\beta\hbar^2} \qquad (4.2.76)$$

or equivalently reads

$$0 < \beta < \frac{\mu\lambda}{2\hbar^2}. \qquad (4.2.77)$$

4.2.7 No-Binding Theorems

As a by-product of establishing the boundedness of Hamiltonians for a special class of potentials studied in class [D] of §3.3, we here first recall the following No-Binding Theorem established there.

Theorem 4.2.4
Let $-v(\mathbf{x})$ be the negative part of the potential $V(\mathbf{x})$ as defined in (3.3.18), i.e.,

$$v(\mathbf{x}) = -V(\mathbf{x})\,\Theta(-V) \qquad (4.2.78)$$

and define the Fourier-transform

$$\widetilde{v}(\mathbf{p}) = \int d^3x \, v(\mathbf{x}) \, e^{-i\mathbf{p}\cdot\mathbf{x}/\hbar}. \qquad (4.2.79)$$

Then if

$$b \equiv \frac{\mu}{4\hbar^3 \pi^{3/2}} \left(\int \frac{d^3p}{|\mathbf{p}|} \, |\widetilde{v}(\mathbf{p})|^2 \right)^{1/2} \qquad (4.2.80)$$

as a sufficiency condition, is such that

$$b < 1 \qquad (4.2.81)$$

then the Hamiltonian (4.2.35) admits no bound-states as it is strictly bounded below by zero (see (3.3.46), (3.3.47)).

By using the integral expression (3.3.25) for $1/|\mathbf{p}|$, the constant b may be also rewritten as

$$b = \frac{\mu}{2\pi\hbar^2} \left(\int d^3x \, d^3x' \, v(\mathbf{x}) \frac{1}{|\mathbf{x}-\mathbf{x}'|^2} v(\mathbf{x}') \right)^{1/2}. \qquad (4.2.82)$$

For the Yukawa potential $V(\mathbf{x}) = -\lambda e^{-\beta|\mathbf{x}|}/|\mathbf{x}|$ the condition (4.2.81) implies the sufficiency condition $\sqrt{2}\mu\lambda/\hbar^2 < \beta$ for the absence of bound-states (see bound-states see (3.3.48)–(3.3.53)).

As a second theorem of no-binding for $\nu = 3, 2, 1$, we make direct use of the inequality (3.1.8) to obtain:

Theorem 4.2.5

Given any real function $g(\mathbf{x})$ such that the potential defined by

$$V(\mathbf{x}) = \frac{\mu}{2\hbar^2} \mathbf{x}^2 g^2(\mathbf{x}) - \frac{1}{2} \nabla \cdot (\mathbf{x}\, g(\mathbf{x})) \tag{4.2.83}$$

satisfies (4.1.2), (4.1.3), then directly from Theorem 4.1.1, (3.1.8), we have

$$0 \leqslant \left\langle \frac{\mathbf{p}^2}{2\mu} + V(\mathbf{x}) \right\rangle \tag{4.2.84}$$

and the Hamiltonian in question has only a continuous spectrum (see Theorem 4.1.1) consisting of the positive real axis as $-\ell_H$ is zero.

The number of potentials that may be constructed from (4.2.83) by appropriately choosing real functions $g(\mathbf{x})$ are endless (see, e.g., §3.3 class [B]. See also §6.1, §7.1).

Finally, we provide a theorem of no-binding for the radial part of a Hamiltonian, for a given spherically symmetric potential $V(r)$, $r = |\mathbf{x}|$, specified by an arbitrary orbital quantum number (§5.1) $\ell = 0, 1, \ldots$.

The radial part of the Hamiltonian, of a spherically symmetric potential, of a given ℓ is given by (see, e.g., §7.2)

$$H_\ell = -\frac{\hbar^2}{2\mu} \frac{1}{r^2} \left(\frac{\partial}{\partial r} r^2 \frac{\partial}{\partial r} \right) + \frac{\hbar^2}{2\mu\, r^2} \ell(\ell+1) + V(r). \tag{4.2.85}$$

As in (4.2.78) define the negative part of the potential $V(r)$

$$V(r)\, \Theta(-V(r)) = -v(r). \tag{4.2.86}$$

We then have the following theorem.

Theorem 4.2.6

For

$$\frac{2\mu}{\hbar^2} \int_0^\infty r\, dr\, v(r) < 1 \tag{4.2.87}$$

the radial Hamiltonian (4.2.85) admits no bound-states for all $\ell = 0, 1, \ldots$.

The proof of this theorem is given at the end of §4.5.

For the Yukawa potential $-v(r) = -\lambda e^{-\beta r}/r$, $r = |\mathbf{x}|$, $\lambda > 0$, (4.2.87) gives the sufficiency condition

$$\frac{2\mu\lambda}{\hbar^2} < \beta \tag{4.2.88}$$

for no binding for any $\ell = 0, 1, \ldots$.

4.3 Hamiltonians with Potentials Approaching Finite Constants at Infinity

A further interesting application of Theorem 4.2.6, is the following one which includes the spherical well potential $-v(r) = -U_0 \Theta(R-r)$, $U_0 > 0$, as a special case. For any given (space) scale parameter $R > 0$, define the energy scale parameter

$$U_0 = \frac{2}{R^2} \int_0^\infty r \, dr \, v(r) \tag{4.2.89}$$

then according to (4.2.87) the radial Hamiltonian (4.2.85) admits no bound-states (for all $\ell = 0, 1, \ldots$) for

$$\mu \frac{R^2 U_0}{\hbar^2} < 1 \tag{4.2.90}$$

as a sufficiency condition.

4.3 Hamiltonians with Potentials Approaching Finite Constants at Infinity

In this section we are interested in a special class of potentials such that for any $0 < R < \infty$, $\nu = 1, 2, 3$,

$$\int_{|\mathbf{x}|<R} d^\nu \mathbf{x} \, |V(\mathbf{x})|^2 < \infty \tag{4.3.1}$$

and there exists a real finite constant C such that

$$|V(\mathbf{x}) - C| \xrightarrow[|\mathbf{x}| \to \infty]{} 0 \tag{4.3.2}$$

in reference to a Hamiltonian as given in (4.1.1).

A specific example of a potential belonging to this class is the potential

$$V(\mathbf{x}) = \gamma \tanh^2(\beta|\mathbf{x}|) + \alpha e^{-\rho|\mathbf{x}|} \tag{4.3.3}$$

with γ, β, α, ρ real constants and $\rho > 0$. In this case $C = \gamma$.

We note that quite generally for any $V(\mathbf{x})$

$$|V(\mathbf{x}) - C|^2 \leq 2 \left(|V(\mathbf{x})|^2 + C^2 \right) \tag{4.3.4}$$

and hence $V(\mathbf{x}) - C$ is, from (4.3.1), locally square-integrable, i.e., for any $0 < R < \infty$,

$$\int_{|\mathbf{x}|<R} d^\nu \mathbf{x} \, |V(\mathbf{x}) - C|^2 < \infty. \tag{4.3.5}$$

We may rewrite the Hamiltonian in question as

$$H = H' + C \tag{4.3.6}$$

where
$$H' = \frac{\mathbf{p}^2}{2\mu} + (V(\mathbf{x}) - C). \tag{4.3.7}$$

According to the analysis given in subsection [E] of §3.3, H' is bounded form below, with a lower bound of the spectrum given by, say, $-\ell_{H'}$. Hence from (4.3.6), we have for $|\psi\rangle$ normalized,
$$\langle\psi|H|\psi\rangle = C + \langle\psi|H'|\psi\rangle \geqslant (C - \ell_{H'})$$

i.e., a lower bound of the spectrum of H is given by
$$-\ell_H = C - \ell_{H'}. \tag{4.3.8}$$

Finally, with H_0 replaced by
$$H'_0 = \frac{\mathbf{p}^2}{2\mu} + C \tag{4.3.9}$$

and $V(\mathbf{x})$ replaced by $V(\mathbf{x}) - C$, we may refer to the analysis already given in §4.1. Instead of (4.1.5), we choose ξ_0 to be any fixed real number such that
$$\xi_0 < \min[C, -\ell_H]. \tag{4.3.10}$$

From Proposition 4.1.2, in particular, we may infer that H has a continuous spectrum consisting of the interval (C, ∞) and has no eigenvalues of infinite degeneracy, and in particular $-\ell_H \leqslant C$. Also if $-\ell_H < C$, then H has also a discrete spectrum (of at most finite degeneracy) falling in the interval $[-\ell_H, C)$.

We may summarize the above by stating the following.

Theorem 4.3.1
A Hamiltonian with the potential $V(\mathbf{x})$ satisfying (4.3.1), (4.3.2), with $|C| < \infty$, is bounded from below. It has a continuous spectrum consisting of the interval $[C, \infty)$ and has no eigenvalues of infinite degeneracy. Also if the lower bound $-\ell_H$, of H is such that $-\ell_H < C$, then H has also eigenvalues (of at most finite degeneracy) falling in the interval $[-\ell_H, C)$.

4.4 Hamiltonians with Potentials Increasing with No Bound at Infinity

In this section we are concerned with the spectrum of Hamiltonians with potentials $V(\mathbf{x})$ belonging to the following important class of potentials:
$$V(\mathbf{x}) \geqslant 0, \quad V(\mathbf{x}) \xrightarrow[|\mathbf{x}|\to\infty]{} +\infty \tag{4.4.1}$$

4.4 Hamiltonians with Potentials Increasing with No Bound at Infinity

and $V(\mathbf{x})$ is locally square-integrable, $\nu = 1, 2, 3$.

The classic example in this category includes the harmonic oscillator potential

$$V(\mathbf{x}) = \frac{1}{2}\mu\omega^2 |\mathbf{x}|^2. \tag{4.4.2}$$

The underlying Hamiltonian

$$H = \frac{\mathbf{p}^2}{2\mu} + V(\mathbf{x}) \tag{4.4.3}$$

is obviously positive, and we have the following theorem:

Theorem 4.4.1
H has only a discrete spectrum.

To establish this theorem, we first note that for *any* positive constant C, we may find a real positive constant R large enough such that for $|\mathbf{x}| > R$

$$V(\mathbf{x}) \geq C \tag{4.4.4}$$

and for $|\mathbf{x}| \leq R$, we may obviously simply write

$$V(\mathbf{x}) \geq 0. \tag{4.4.5}$$

Given any such a positive constant C, we define the function $u(x)$ of one-variable by

$$u(x) = \begin{cases} -C/\nu, & |x| \leq R \\ 0, & |x| > R \end{cases} \tag{4.4.6}$$

where $\nu = 3, 2, 1$.

Let $\mathbf{x} = (x_1, \ldots, x_\nu)$. Then if

$$|\mathbf{x}| = \left(\sum_{i=1}^{\nu} x_i^2\right)^{1/2} \leq R \tag{4.4.7}$$

one necessarily has $|x_i| \leq R$ for all i. From (4.4.6) we may then infer that

$$C + \left(\sum_{i=1}^{\nu} u(x_i)\right) = 0, \quad \text{for} \quad |\mathbf{x}| \leq R. \tag{4.4.8}$$

On the other hand for $|\mathbf{x}| > R$, it is easily seen that

$$C + \sum_{i=1}^{\nu} u(x_i) \leq C. \tag{4.4.9}$$

That is, in all cases, we have from (4.4.4), (4.4.5), (4.4.8), (4.4.9):

$$V(\mathbf{x}) \geqslant C + \sum_{i=1}^{\nu} u(x_i) \qquad (4.4.10)$$

for all $\mathbf{x}$, i.e. $V(\mathbf{x})$ cannot be smaller than $C + \sum_{i=1}^{\nu} u(x_i)$.

We define the Hamiltonian

$$H' = \frac{\mathbf{p}^2}{2\mu} + \sum_{i=1}^{\nu} u(x_i) \qquad (4.4.11)$$

then

$$\langle \psi | H | \psi \rangle \geqslant C + \langle \psi | H' | \psi \rangle \qquad (4.4.12)$$

for normalized $|\psi\rangle$. Also from (4.4.3), (4.4.10), (4.4.11), (1.8.87), (1.8.110) and (1.8.111) we have[2]

$$\lambda_n(H) \geqslant \lambda_n(H') + C. \qquad (4.4.13)$$

The eigenvalues of the operator H' may be determined from the one-dimensional problem considered in §4.2, with the corresponding Hamiltonian defined in (4.2.1), (4.2.2). In the notation of the latter we set $U_0 = C/\nu$, $L \equiv R$. Let $(\mathbf{k})$ stand for ν-tuplet of numbers $(k_1, \ldots, k_\nu)$ corresponding to the eigenvalues in (4.2.14):

$$E_{k_i} = -\frac{C}{\nu} + \frac{\hbar^2}{2mR^2} \xi_{k_i}^2. \qquad (4.4.14)$$

In reference to the operator H' one has

$$E_{(\mathbf{k})} = -C + \frac{\hbar^2}{2mR^2} \sum_{i=1}^{\nu} \xi_{k_i}^2. \qquad (4.4.15)$$

Let n denote the total number of $E_{(\mathbf{k})}$ values in (4.4.15) which are arranged in a non-decreasing order $E_1 \leqslant \ldots \leqslant E_n$, with $E_k \equiv \lambda_k(H')$. In particular

$$\lambda_n(H') = -C + \frac{\hbar^2}{2mR^2} \nu \xi_{n_0}^2 \qquad (4.4.16)$$

where n_0 corresponds to n for the one-dimensional case as defined in (4.2.14). Equation (4.2.15) reads (see also (4.2.11), and note that $U_0 = C/\nu$ here)

$$\left(\sqrt{\frac{2mC}{\nu}} \frac{R}{\hbar} - \frac{\pi}{2} \right) \leqslant \xi_{n_0} \qquad (4.4.17)$$

---

[2] The proof for the spectrum of H follows the treatment by Reed and Simon (1978).

where obviously C and, correspondingly, R, may be chosen large enough so that the left-hand side of this inequality is strictly positive, hence from (4.4.16)

$$\lambda_n(H') + C \geq \frac{\hbar^2 \nu}{2m} \left(\sqrt{\frac{2mC}{\nu}} \frac{1}{\hbar} - \frac{\pi}{2R} \right)^2. \quad (4.4.18)$$

Also (4.2.13) reads

$$\frac{1}{\pi} \sqrt{\frac{2mC}{\nu}} \frac{R}{\hbar} = N + \varepsilon. \quad (4.4.19)$$

From Figure 4.1, we know that we always have $n \geq N$. Accordingly, the limit $C \to \infty$, $R \to \infty$ corresponds from (4.4.19) to $N \to \infty$ and hence also to $n \to \infty$. From (4.4.18) and (4.4.13) we then have

$$\lim_{n \to \infty} \lambda_n(H) = +\infty. \quad (4.4.20)$$

Using Proposition 1.8.2 we may then conclude that the positive operator H has only a *discrete* spectrum.

4.5 Counting the Number of Eigenvalues

4.5.1 General Treatment of the Problem

We introduce the spectral decomposition of a Hamiltonian (§1.8, (1.8.15))

$$H = \int_{-\infty}^{\infty} \lambda \, \mathrm{d}P_H(\lambda). \quad (4.5.1)$$

Suppose that for some given real and a specified number ξ, H may have at most eigenvalues $\leq \xi$.[3] Then we may write

$$P_H(\xi) = \sum_{\substack{\lambda \leq \xi \\ \nu(\lambda)}} \int_{-\infty}^{\xi} \mathrm{d}\lambda' \, \delta(\lambda' - H) \, |\lambda, \nu(\lambda)\rangle \langle \lambda, \nu(\lambda)|$$

$$= \sum_{\substack{\lambda \leq \xi \\ \nu(\lambda)}} \int_{-\infty}^{\xi} \mathrm{d}\lambda' \, \delta(\lambda' - \lambda) \, |\lambda, \nu(\lambda)\rangle \langle \lambda, \nu(\lambda)|$$

$$= \sum_{\lambda, \nu(\lambda)} |\lambda, \nu(\lambda)\rangle \langle \lambda, \nu(\lambda)| \, \Theta(\xi - \lambda) \quad (4.5.2)$$

---

[3] In most of this section but not all, we consider negative ξ values, and, and when convenient, we replace ξ by $-\xi$, with $\xi > 0$ in the latter.

where $\nu(\lambda)$ specifies the degree of degeneracy of the eigenvalue λ, $|\lambda, \nu(\lambda)\rangle$ are corresponding eigenvectors, and $\Theta(\xi - \lambda)$ is defined here as $\Theta(x) = 1$ for $x \geqslant 1$, $\Theta(x) = 0$ for $x < 0$.

From (4.5.2), we may introduce, a density of states

$$\langle \mathbf{x}| P_H(\xi) |\mathbf{x}\rangle = \sum_{\lambda, \nu(\lambda)} |\psi_{\lambda, \nu(\lambda)}(\mathbf{x})|^2 \, \Theta(\xi - \lambda) \qquad (4.5.3)$$

where

$$\psi_{\lambda, \nu(\lambda)}(\mathbf{x}) = \langle \mathbf{x}| \lambda, \nu(\lambda) \rangle$$

are the eigenstates in the $\mathbf{x}$-description.

Upon integration of (4.5.3) over $\mathbf{x}$, we obtain

$$\int d^\nu \mathbf{x} \, \langle \mathbf{x}| P_H(\xi) |\mathbf{x}\rangle = \sum_{\lambda, \nu(\lambda)} \Theta(\xi - \lambda)$$

$$\equiv N(H, \xi) \qquad (4.5.4)$$

where

$$N(H, \xi) = \text{Number of states with eigenvalues } \leqslant \xi$$
$$= \text{Number of eigenvalues (counting the degree of degeneracy)}$$
$$\leqslant \xi. \qquad (4.5.5)$$

One may also introduce a non-local density of states defined by

$$\langle \mathbf{x}| P_H(\xi) |\mathbf{x}'\rangle = \sum_{\lambda, \nu(\lambda)} \psi_{\lambda, \nu(\lambda)}(\mathbf{x}) \, \psi^*_{\lambda, \nu(\lambda)}(\mathbf{x}') \, \Theta(\xi - \lambda) \qquad (4.5.6)$$

which will be useful later on.

Equation (4.5.4) also allows one to obtain an expression for the degree of degeneracy of an eigenvalue, say, λ_0. Let λ_1, λ_2 be eigenvalues of H such that $\lambda_1 < \lambda_0 < \lambda_2$. Define

$$\varepsilon = \min(\lambda_2 - \lambda_0, \lambda_0 - \lambda_1) \qquad (4.5.7)$$

and

$$\bar{\lambda}_0 = \lambda_0 + \varepsilon, \quad \underline{\lambda}_0 = \lambda_0 - \varepsilon \qquad (4.5.8)$$

then from (4.5.4)

$$\int d^\nu \mathbf{x} \, \left[\langle \mathbf{x}| P_H(\bar{\lambda}_0) |\mathbf{x}\rangle - \langle \mathbf{x}| P_H(\underline{\lambda}_0) |\mathbf{x}\rangle\right] = \sum_{\lambda, \nu(\lambda)} \left[\Theta(\bar{\lambda}_0 - \lambda) - \Theta(\underline{\lambda}_0 - \lambda)\right]$$

$$= \sum_{\nu(\lambda)} 1 \qquad (4.5.9)$$

for the degree of degeneracy of the eigenvalue of λ_0. Later in Chapter 9, we will see how the left-hand sides of (4.5.3), (4.5.4), (4.5.9), may be evaluated in terms of so-called Green functions.

Another useful expression that may be obtained from (4.5.4) is the following. Upon integration (4.5.4) over ξ as follows from $-\infty$ to 0, we have

$$\int_{-\infty}^{0} d\xi \int d^\nu x \, \langle x| P_H(\xi) |x\rangle = \sum_{\lambda,\nu(\lambda)} \int_{-\infty}^{0} d\xi \, \Theta(\xi - \lambda)$$

$$= \int_{-\infty}^{0} d\xi \, N(H,\xi) \qquad (4.5.10)$$

providing a useful expression for the negative of the *sum* of the negative eigenvalues (if any) of H.

As another application, we consider for $\xi > 0$

$$\langle x| (H-V) P_H(\xi) |x\rangle = \sum_{\lambda,\nu(\lambda)} \langle x| (H-V) |\lambda,\nu(\lambda)\rangle \langle \lambda,\nu(\lambda) |x\rangle \, \Theta(\xi - \lambda) \qquad (4.5.11)$$

which from the equality

$$\langle x| V = \int d^\nu x' \, \langle x| V |x'\rangle \langle x'|$$

$$= \int d^\nu x' \, \delta^\nu(x-x') V(x') \langle x'|$$

$$= V(x) \langle x| \qquad (4.5.12)$$

for a local potential:

$$\langle x| V |x'\rangle = V(x') \, \delta^\nu(x-x') \qquad (4.5.13)$$

leads to

$$\langle x| (H-V) P_H(\xi) |x\rangle = \sum_{\lambda,\nu(\lambda)} \langle x| H - V(x) |\lambda,\nu(\lambda)\rangle \langle \lambda,\nu(\lambda) |x\rangle \, \Theta(\xi - \lambda)$$

$$= \sum_{\lambda,\nu(\lambda)} \psi^*_{\lambda,\nu(\lambda)}(x) \, [\lambda - V(x)] \, \psi_{\lambda,\nu(\lambda)}(x) \, \Theta(\xi - \lambda). \qquad (4.5.14)$$

From the first equality in (4.5.14), if $H - V(x) = -\hbar^2 \nabla^2/2\mu$ defines the kinetic energy, then an integration of (4.5.14) over x gives ($\xi > 0$)

$$\int d^\nu x \, \langle x| (H-V) P_H(\xi) |x\rangle = \sum_{\lambda,\nu(\lambda)} \langle \lambda,\nu(\lambda)| \frac{p^2}{2\mu} |\lambda,\nu(\lambda)\rangle \, \Theta(\xi-\lambda). \qquad (4.5.15)$$

4 Spectra of Hamiltonians

As we will see later, this expression is important for multi-electron systems as used in Chapter 13. For example, if only one particle occupies a given state and the states, up to the maximum possible energy $\leqslant \xi$, are filled then (4.5.15) gives the sum of the average kinetic energy of all the particles with the most energetic having the maximum possible energy allowed $\leqslant \xi$.

The purpose of the remainder of this section is to find an upper bound for the number $N(H, \xi)$, in (4.5.4), of the eigenvalues for $\xi < 0$, and the negative of the sum of the eigenvalues in (4.5.10), by considering in the process the $\xi \to 0_-$ limit, for a given Hamiltonian H. Lower-bound expressions will be also obtained for the ground-state energy.

4.5.2 Counting the Number of Eigenvalues

We rewrite the Hamiltonian $H(g) = H_0 + V$, where H_0 is the free Hamiltonian $\mathbf{p}^2/2\mu$, in the form

$$H(g) = H_0 + gV \tag{4.5.16}$$

by introducing a variable coupling parameter $g \geqslant 0$, with $g = 1$ corresponding to the Hamiltonian in question.

As in (3.3.18), (3.3.19), we introduce the negative part of the potential

$$V \Theta(-V) \equiv -v \equiv -|V|_- . \tag{4.5.17}$$

Accordingly,

$$H(g) \geqslant H_0 - gv. \tag{4.5.18}$$

Let $N(H(g), -\xi)$ denote (see (4.5.4)) the number of eigenvalues of $H(g) \leqslant -\xi$, with $\xi > 0$. The inequality (4.5.18), together with those in (1.8.110), (1.8.111) imply that

$$\lambda_n(H(g)) \geqslant \lambda_n(H_0 - g\,v). \tag{4.5.19}$$

That is, the number of bound-states of $H_0 - g\,v$ cannot be less than that of $H(g)$:

$$N(H_0 - g\,v, -\xi) \geqslant N(H_0 + gV, -\xi). \tag{4.5.20}$$

Similarly for $0 < g' < g$,

$$H_0 - g'v \geqslant H_0 - g\,v \tag{4.5.21}$$

and hence

$$N(H_0 - g\,v, -\xi) \geqslant N(H_0 - g'v, -\xi). \tag{4.5.22}$$

Finally for $\xi_1 > \xi_2 > 0$,

$$N(H_0 - g\,v, -\xi_2) \geqslant N(H_0 - g\,v, -\xi_1) \tag{4.5.23}$$

obviously holds true recalling the definition in (4.5.5).

Let $\lambda(g_0)$ denote an eigenvalue of $H_0 - g_0 v$ for some g_0. $\lambda(g)$, being negative, is a decreasing function of g (see (4.5.19) and Proposition 1.8.2) as g increases beyond the value g_0 (see Figure 4.2), the curve traced by $\lambda(g)$ necessarily cuts every horizontal axis, such as $g = g'_0$, $g = g''_0, \ldots$, with $g_0 < g'_0 < g''_0 < \ldots$, in the manner shown in Figure 4.2, otherwise the Hamiltonian $H_0 - g_0 v$ will have more eigenvalues than, say, $H_0 - g'_0 v$ in contradiction with (4.5.22).

Also every eigenvalue $\lambda(g)$ of $H_0 - g v$, for which $\lambda(g) < -\xi$, for a given ξ, falls on a curve which cuts the vertical axis $\lambda = -\xi$ (see Figure 4.2) at some g-value, say, equal to g' corresponds to the eigenvalue $\lambda = -\xi$ of the Hamiltonian $H_0 - g' v$. Finally, note that the order of eigenvalues (see (1.8.90)) is preserved as g increases.

All told we have the important relation:

$$N(H_0 - g v, -\xi) = [\text{Number of } g'\text{'s in } 0 < g' \leqslant g \text{ for which } H_0 - g' v$$

$$\text{has the eigenvalue } \lambda = -\xi]. \quad (4.5.24)$$

According to (4.5.24), we are led to consider the eigenvalue problem

$$\left(\frac{\mathbf{p}^2}{2\mu} - g' v\right) |\psi\rangle = -\xi |\psi\rangle, \quad \|\psi\| = 1 \quad (4.5.25)$$

which may be rewritten in the form

$$A |\phi\rangle = \frac{1}{g'} |\phi\rangle \quad (4.5.26)$$

where A is the positive operator

$$A = \sqrt{v} \, \frac{1}{\left(\frac{\mathbf{p}^2}{2\mu} + \xi\right)} \, \sqrt{v} \quad (4.5.27)$$

and

$$|\phi\rangle = \sqrt{v} \, |\psi\rangle. \quad (4.5.28)$$

In Figure 4.2, the eigenvalues, say, $\{\lambda(g''_0)\}$ of $H_0 - g''_0 v$ are defined by the intersections of curves $\{\lambda(g)\}$, as g varies, with the horizontal axis $g = g''_0$. These same curves $\{\lambda(g)\}$ cut the vertical axis $\lambda = -\xi$ at corresponding points $\{g'\}$, where the $1/g'$ are the eigenvalues of the positive operator A, in (4.5.27), with the eigenvalue equation with the eigenvalue equation given in (4.5.26).

We introduce the spectral representation (1.8.15) of A:

$$A = \int_0^\infty \gamma \, dP_A(\gamma). \quad (4.5.29)$$

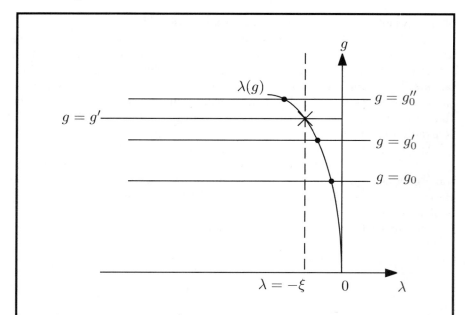

Fig. 4.2. $\lambda(g)$, an eigenvalue of $H_0 - gv$, being negative, is a decreasing function of g and necessarily cuts all the horizontal axes $g = g'_0, g = g''_0, \ldots,$ as g increases beyond a given value g_0. Also every eigenvalue $\lambda(g) < -\xi$, of $H_0 - gv$, falling on a curve which cuts the vertical axis $\lambda = -\xi$, say, at $g = g'$, corresponds to the eigenvalue $\lambda = -\xi$ of the Hamiltonian $H_0 - g'v$.

We will have occasion to use powers of the positive operator A

$$A^\rho = \int_0^\infty \gamma^\rho \, \delta(\gamma - A) \, \mathrm{d}\gamma$$

$$= \int_0^\infty \gamma^\rho \, \mathrm{d}P_A(\gamma) \qquad (4.5.30)$$

(see (1.8.14), (1.8.15)), where ρ is not necessarily an integer. Explicit representations of such powers of positive operators are formally given. For example

$$A^{-1/2} = \frac{2}{\sqrt{\pi}} \int_0^\infty \mathrm{d}x \, \exp\left(-x^2 A\right) \qquad (4.5.31)$$

and for $\rho \leqslant -1/2$,

$$A^\rho = \frac{1}{C_\rho} \int_0^\infty \frac{\mathrm{d}x}{x^{2\rho+1}} \, \exp\left(-x^2 A\right) \qquad (4.5.32)$$

where

4.5 Counting the Number of Eigenvalues

$$C_\rho = \int_0^\infty \frac{dx}{x^{2\rho+1}} \exp(-x^2). \tag{4.5.33}$$

Positive powers of A may be obtained from (4.5.32), e.g., by multiplying the latter by appropriate powers of A:

$$A^{1/2} = \frac{2}{\sqrt{\pi}} \int_0^\infty dx \, \exp(-x^2 A) \, A \tag{4.5.34}$$

$$A^{3/2} = \frac{2}{\sqrt{\pi}} \int_0^\infty dx \, \exp(-x^2 A) \, A^2 \tag{4.5.35}$$

and more generally

$$A^\rho = \frac{1}{C_\sigma} \int_0^\infty \frac{dx}{x^{2\sigma+1}} \exp(-x^2 A) \, A^m, \quad \sigma \leqslant -1/2 \tag{4.5.36}$$

with $\rho \equiv \sigma + m$, $m = 0, 1, \ldots$, in a convenient notation.

Other representations may be also given. For example,

$$A^\rho = C'_\sigma \int_0^\infty dx \, x^{\sigma-1} \frac{A^{m+1}}{(A+x)}, \quad 0 < \sigma < 1 \tag{4.5.37}$$

with $\rho \equiv \sigma + m$, $m = 0, 1, \ldots$, and

$$C'_\sigma = \frac{\sin(\pi\sigma)}{\pi}. \tag{4.5.38}$$

Later on, in (4.5.93), we will encounter the case $\rho = 3/2$.
From (4.5.30), for $\rho \geqslant 0$, in particular,

$$\langle \mathbf{x}| \, A^\rho \, |\mathbf{x}\rangle \geqslant \int_{1/g}^{\gamma_0} \gamma^\rho \, d \langle \mathbf{x}| \, P_A(\gamma) \, |\mathbf{x}\rangle$$

$$\geqslant \frac{1}{g^\rho} \left[\langle \mathbf{x}| \, P_A(\gamma_0) \, |\mathbf{x}\rangle - \left\langle \mathbf{x} \left| P_A \left(\frac{1}{g} \right) \right| \mathbf{x} \right\rangle \right] \tag{4.5.39}$$

recalling that $P_A(\gamma)$ is self-adjoint, thus (4.5.39) involves real numbers only, where $\gamma_0 > 1/g$, and is otherwise arbitrary. Upon integration over $\mathbf{x}$ we obtain (see (4.5.4), (4.5.24))

$$\int d^\nu \mathbf{x} \, \langle \mathbf{x}| A^\rho | \mathbf{x}\rangle \geqslant \frac{1}{g^\rho} \times [\text{number of all } g'\text{'s, counting degeneracy, as eigenvalues}$$

of A in (4.5.26), such that $1/\gamma_0 < g' \leqslant g$]. (4.5.40)

Since the point $\gamma_0 > 1/g$ is arbitrary, we may take the limit $\gamma_0 \to \infty$ in (4.5.39) and obtain from (4.5.24), (4.5.40)

$$N(H_0 - gv, -\xi) \leqslant g^\rho \int d^\nu \mathbf{x} \, \langle \mathbf{x}| A^\rho | \mathbf{x}\rangle \tag{4.5.41}$$

and from (4.5.20)

$$N(H_0 + V, -\xi) \leq \int d^\nu x \, \langle x | A^\rho | x \rangle \qquad (4.5.42)$$

with $g = 1$.

In reference to the definition (4.5.17), the following inequality will be also needed:

$$V \, \Theta(-V) + \xi/2 \geq (V + \xi/2) \, \Theta(-V - \xi/2) \qquad (4.5.43)$$

where $\xi > 0$, whose validity is easily established by considering, in turn, the three cases: $V > 0$, $-\xi/2 < V < 0$, $V < -\xi/2$.

From (4.5.43), we obtain

$$H_0 + V \, \Theta(-V) \geq H_0 + \left(V + \frac{\xi}{2} \right) \Theta\left(-V - \frac{\xi}{2} \right) - \frac{\xi}{2} \qquad (4.5.44)$$

and hence

$$N(H_0 + V \, \Theta(-V), -\xi) \leq N\left(H_0 + \left(V + \frac{\xi}{2} \right) \Theta\left(-V - \frac{\xi}{2} \right), -\frac{\xi}{2} \right). \qquad (4.5.45)$$

In (4.5.2), we are interested in Hamiltonians which are bounded from below (§§3.2–3.4, §4.1). Accordingly, an important situation arises for which the spectrum of a Hamiltonian is empty for energies $\leq -\xi_0$ for some specific value ξ_0 of ξ, i.e., for which $N(H_0 + V, -\xi_0) = 0$. Clearly, such a value $-\xi_0$ would then provide a lower bound to the spectrum of the Hamiltonian $H_0 + V$.

In what follows, we investigate the nature of $N(H_0 + V, -\xi)$ in space dimensions $\nu = 3, 2, 1$. The radial part of the Hamiltonian of a spherically symmetric potential is also considered. Subsequently, we examine the expression in (4.5.10) for the negative of the sum of the negative eigenvalues of a Hamiltonian. In all of these investigations, one encounters existence of integrals of the form $\int d^\nu x \, v^\beta(x) < \infty$ of the potentials in (4.5.17), for some positive powers β, which are implicitly assumed to be satisfied.

We will show below that the number of eigenvalues of $H = H_0 + V$, counting degeneracy, with energies $\leq -\xi$, for $\xi > 0$, satisfies, in a compact form, the inequality

$$N(H, -\xi) \leq a_\nu (\xi)^{(\nu^2 - 4\nu + 2)/2} \int d^\nu x \, (v(x))^{(5\nu - \nu^2 - 2)/2} \qquad (4.5.46)$$

for $\nu = 3, 2, 1$, where

$$a_\nu = \left(\frac{\mu}{2\hbar^2} \right)^{\nu/2} (\pi)^{(\nu^2 - 5\nu + 4)/2}. \qquad (4.5.47)$$

4.5 Counting the Number of Eigenvalues

The three-dimensional case ($\nu = 3$):

In three dimensions ($\nu = 3$), we choose $\rho = 2$ on the right-hand side of (4.5.42). Thus with the definition of A in (4.5.27), we obtain for the right-hand side of (4.5.42):

$$\int d^3x\, d^3x'\, v(\mathbf{x})\, v(\mathbf{x'}) \left| \left\langle \mathbf{x} \left| \frac{1}{\left[\frac{\mathbf{p}^2}{2\mu} + \xi\right]} \right| \mathbf{x'} \right\rangle \right|^2 \qquad (4.5.48)$$

where (see the Appendix to this section), with $\xi > 0$,

$$\left\langle \mathbf{x} \left| \frac{1}{\left[\frac{\mathbf{p}^2}{2\mu} + \xi\right]} \right| \mathbf{x'} \right\rangle = \frac{\mu}{2\pi\hbar^2} \frac{1}{|\mathbf{x}-\mathbf{x'}|} \exp\left(-\frac{|\mathbf{x}-\mathbf{x'}|}{\hbar}\sqrt{2\mu\xi}\right) \qquad (4.5.49)$$

giving from (4.5.42) the so-called Schwinger inequality:[4]

$$N(H_0 + V, -\xi) \leqslant \left(\frac{\mu}{2\pi\hbar^2}\right)^2 \int d^3x\, d^3x'\, v(\mathbf{x}) \frac{e^{-2|\mathbf{x}-\mathbf{x'}|\sqrt{2\mu\xi}/\hbar}}{|\mathbf{x}-\mathbf{x'}|^2} v(\mathbf{x'}). \qquad (4.5.50)$$

Since the integrand in (4.5.50) is positive, and the exponential factor is bounded above by one, we may further bound the right-hand side of (4.5.50) from above by

$$\left(\frac{\mu}{2\pi\hbar^2}\right)^2 \int d^3x\, d^3x'\, v(\mathbf{x}) \frac{1}{|\mathbf{x}-\mathbf{x'}|^2} v(\mathbf{x'}). \qquad (4.5.51)$$

This expression is nothing but the constant b^2 in (4.2.82), (4.2.80) of Theorem 4.2.4. In particular, for $b < 1$, the latter theorem states that the Hamiltonian admits no bound-states (see also (3.3.46), (3.3.47)). This is consistent with (4.5.50), showing that $N(H_0 + V, -\xi) = 0$ for any $\xi > 0$, however small, in case $b < 1$.

Now we use (4.5.50) to derive a lower bound for the spectrum of the Hamiltonian in question as well. To this end, we use Young's inequality (see Appendix II), with $p = q = 2$, to obtain

$$\int d^3x\, d^3x'\, v(\mathbf{x}) \frac{e^{-2|\mathbf{x}-\mathbf{x'}|\sqrt{2\mu\xi}/\hbar}}{|\mathbf{x}-\mathbf{x'}|^2} v(\mathbf{x'})$$

$$\leqslant \left(\int d^3x\, (v(\mathbf{x}))^2\right) \left(\int d^3x\, \frac{e^{-2|\mathbf{x}|\sqrt{2\mu\xi}/\hbar}}{|\mathbf{x}|^2}\right) \qquad (4.5.52)$$

which from (4.5.50) gives

---
[4] Schwinger (1961b).

$$N(H_0 + V, -\xi) \leq \left(\frac{\mu}{2\hbar^2}\right)^{3/2} \frac{1}{\pi\sqrt{\xi}} \int d^3x \, v^2(\mathbf{x}) \qquad (4.5.53)$$

which coincides with (4.5.46)/(4.5.47) for $\nu = 3$.

Clearly, if for any $\delta > 0$, we choose

$$-\xi = -\frac{(1+\delta)}{\pi^2} \left(\frac{\mu}{2\hbar^2}\right)^3 \left(\int d^3x \, (v(\mathbf{x}))^2\right)^2 \qquad (4.5.54)$$

then the right-hand side of (4.5.53) is necessarily less than one, implying that $N(H_0 + V, -\xi) = 0$ for such a $-\xi$, and the spectrum of the Hamiltonian is empty for energies $\leq -\xi$. That is, (4.5.54) gives the following lower bound for the ground-state energy of the Hamiltonian,

$$-\frac{(1+\delta)}{\pi^2} \left(\frac{\mu}{2\hbar^2}\right)^3 \left(\int d^3x \, (v(\mathbf{x}))^2\right)^2 \qquad (4.5.55)$$

for any $\delta > 0$.

The two-dimensional case ($\nu = 2$):

For $\nu = 2$, $\rho = 2$, the right-hand side of (4.5.42) is given by

$$\int d^2x \, \langle \mathbf{x} | A^2 | \mathbf{x} \rangle = \int d^2x \, d^2x' \, v(\mathbf{x}) \left| \left\langle \mathbf{x} \left| \frac{1}{\left[\frac{\mathbf{p}^2}{2\mu} + \xi\right]} \right| \mathbf{x}' \right\rangle \right|^2 v(\mathbf{x}') \qquad (4.5.56)$$

where (see the appendix to this section)

$$\left\langle \mathbf{x} \left| \frac{1}{\left[\frac{\mathbf{p}^2}{2\mu} + \xi\right]} \right| \mathbf{x}' \right\rangle = \frac{\mu}{\pi\hbar^2} K_0\left(\frac{|\mathbf{x} - \mathbf{x}'|}{\hbar}\sqrt{2\mu\xi}\right) \qquad (4.5.57)$$

and $K_0(x)$ is a modified Bessel function with asymptotics

$$K_0(x) \xrightarrow[x \to 0]{} -\ln x \qquad (4.5.58)$$

$$K_0(x) \xrightarrow[|x| \to \infty]{} \sqrt{\frac{\pi}{2|x|}} e^{-|x|}. \qquad (4.5.59)$$

From (4.5.42), we obtain

$$N(H_0 + V, -\xi) \leq \left(\frac{\mu}{\pi\hbar^2}\right)^2 \int d^3x \, d^3x' \, v(\mathbf{x}) \left(K_0\left(\frac{|\mathbf{x} - \mathbf{x}'|}{\hbar}\sqrt{2\mu\xi}\right)\right)^2 v(\mathbf{x}'). \qquad (4.5.60)$$

Using the integral,

4.5 Counting the Number of Eigenvalues

$$\int d^2\mathbf{x} \left(K_0 \left(\frac{|\mathbf{x}|}{\hbar} \sqrt{2\mu\xi} \right) \right)^2 = \frac{\pi\hbar^2}{2\mu\xi} \tag{4.5.61}$$

and Young's inequality, as done in (4.5.52), we obtain from (4.5.60), (4.5.61),

$$N(H_0 + V, -\xi) \leqslant \left(\frac{\mu}{2\hbar^2} \right) \frac{1}{\pi\xi} \int d^2\mathbf{x}\, v^2(\mathbf{x}) \tag{4.5.62}$$

which coincides with (4.5.46)/(4.5.47) for $\nu = 2$.
Accordingly, if for any $\delta > 0$, we choose

$$-\xi = -\frac{(1+\delta)}{\pi} \left(\frac{\mu}{2\hbar^2} \right) \int d^2\mathbf{x}\, v^2(\mathbf{x}) \tag{4.5.63}$$

then the spectrum of the Hamiltonian is empty for energies $\leqslant -\xi$, as the right-hand side of (4.5.62) will be less than one. This gives the following lower bound for the Hamiltonian

$$-\frac{(1+\delta)}{\pi} \left(\frac{\mu}{2\hbar^2} \right) \int d^2\mathbf{x}\, v^2(\mathbf{x}) \tag{4.5.64}$$

for any $\delta > 0$.

The one-dimensional case ($\nu = 1$):

For the one dimensional case we choose $\rho = 1$ in (4.5.42), to obtain

$$\int_{-\infty}^{\infty} dx\, \langle x\,|A|\,x \rangle = \int_{-\infty}^{\infty} dx\, v(x) \left\langle x \left| \frac{1}{\left[\frac{p^2}{2\mu} + \xi \right]} \right| x \right\rangle \tag{4.5.65}$$

where

$$\left\langle x \left| \frac{1}{\left[\frac{p^2}{2\mu} + \xi \right]} \right| x \right\rangle = \int_{-\infty}^{\infty} \frac{dp}{2\pi\hbar} \frac{1}{\left[\frac{p^2}{2\mu} + \xi \right]}$$

$$= \sqrt{\frac{\mu}{2\xi}} \frac{1}{\hbar} \tag{4.5.66}$$

and hence

$$N(H_0 + V, -\xi) \leqslant \sqrt{\frac{\mu}{2\xi}} \frac{1}{\hbar} \int_{-\infty}^{\infty} dx\, v(x) \tag{4.5.67}$$

which coincides with (4.5.46)/(4.5.47) for $\nu = 1$.
Accordingly, by choosing

$$-\xi = -(1+\delta) \frac{\mu}{2\hbar^2} \left(\int_{-\infty}^{\infty} dx\, v(x) \right)^2 \tag{4.5.68}$$

for any $\delta > 0$, we have the following lower bound for the ground-state energy

$$-(1+\delta)\frac{\mu}{2\hbar^2}\left(\int_{-\infty}^{\infty} dx\ v(x)\right)^2. \tag{4.5.69}$$

The radial part for a spherically symmetric potential:

The radial part of the Hamiltonian of a spherically symmetric potential energy $V(r)$, where $r = |\mathbf{x}|$, $\nu = 3$, is given by (see, e.g., §7.2)

$$H_\ell = -\frac{\hbar^2}{2\mu}\frac{1}{r^2}\left(\frac{\partial}{\partial r}r^2\frac{\partial}{\partial r}\right) + \frac{\hbar^2}{2\mu}\frac{\ell(\ell+1)}{r^2} + V(r) \tag{4.5.70}$$

where $\ell = 0, 1, \ldots$ define so-called orbital angular momentum quantum numbers (§5.1).

The Dirac delta in spherical coordinates is given by

$$\langle \mathbf{x} | \mathbf{x}' \rangle = \delta^3(\mathbf{x} - \mathbf{x}') = \frac{\delta(r - r')}{r^2}\frac{\delta(\theta - \theta')}{\sin\theta}\delta(\phi - \phi') \tag{4.5.71}$$

so that

$$\int_0^\infty r^2\,dr \int_0^\pi \sin\theta\,d\theta \int_0^{2\pi} d\phi\ \delta^3(\mathbf{x} - \mathbf{x}') = 1. \tag{4.5.72}$$

Accordingly, for the radial part only, we may write

$$\langle r | r' \rangle = \frac{\delta(r - r')}{r^2}. \tag{4.5.73}$$

Upon defining the "free part" of (4.5.70) by

$$H_{0\ell} = -\frac{\hbar^2}{2\mu}\frac{1}{r^2}\left(\frac{\partial}{\partial r}r^2\frac{\partial}{\partial r}\right) + \frac{\hbar^2}{2\mu}\frac{\ell(\ell+1)}{r^2} \tag{4.5.74}$$

and choosing $\rho = 1$, $\xi = 0$ in (4.5.42) we obtain

$$N_\ell \equiv N(H_{0\ell} + V, 0) \leq \int_0^\infty r^2\,dr\ \langle r|A|r\rangle \tag{4.5.75}$$

where

$$A = \sqrt{v(r)}\,\frac{1}{H_{0\ell}}\,\sqrt{v(r)} \tag{4.5.76}$$

$$v(r) = -V(r)\,\Theta(-V(r)). \tag{4.5.77}$$

We explicitly have

$$\langle r|A|r\rangle = v(r)\left\langle r\left|\frac{1}{H_{0\ell}}\right|r\right\rangle$$

$$= \frac{v(r)}{2} \int_0^\infty dr' \left[\delta\left(r' - (r+0)\right) + \delta\left(r' - (r-0)\right) \right] \frac{1}{H_{0\ell}} \frac{\delta(r-r')}{r^2} \quad (4.5.78)$$

with a symmetric average taken over r in performing the trace operation, and where we have used (4.5.73).

It is easily verified that

$$\left(\frac{2\mu}{\hbar^2}\right) H_{0\ell} \frac{1}{(2\ell+1)} \left[\frac{1}{r} \left(\frac{r'}{r}\right)^\ell \Theta(r-r') + \frac{1}{r'} \left(\frac{r}{r'}\right)^\ell \Theta(r'-r) \right]$$

$$= \frac{\delta(r-r')}{r^2}. \quad (4.5.79)$$

Since

$$\frac{\partial}{\partial r} \Theta(r-r') = \delta(r-r') \quad (4.5.80)$$

this gives from (4.5.78)

$$\langle r | A | r \rangle = \frac{2\mu}{\hbar^2} \frac{v(r)}{r(2\ell+1)} \quad (4.5.81)$$

which leads finally to the so-called Bargmann inequality[5]

$$N_\ell \leqslant \frac{2\mu}{\hbar^2} \frac{1}{(2\ell+1)} \int_0^\infty r\, dr\, v(r) \quad (4.5.82)$$

as an upper bound for the number of bound-states corresponding to an orbital quantum number ℓ.

An application of (4.5.82) was given at the end of §4.2, here it is spelled out in more details and includes the spherical well potential $-v(r) = -U_0 \Theta(R-r)$ as a special case. For $v(r)$ any function of r such that[6]

$$0 < \int_0^\infty dr\, v(r) < \infty, \quad 0 < \int_0^\infty r\, dr\, v(r) < \infty. \quad (4.5.83)$$

These allow us to introduce a (space) scale parameter

$$R = 2 \frac{\int_0^\infty r\, dr\, v(r)}{\int_0^\infty dr\, v(r)} \quad (4.5.84)$$

and an energy scale parameter

$$U_0 = \frac{2}{R^2} \int_0^\infty r\, dr\, v(r). \quad (4.5.85)$$

---
[5] Bargmann (1952).
[6] Note that the finiteness ($< \infty$) of any one of the integrals in (4.5.83) does not necessarily imply the finiteness ($< \infty$) of the other.

4 Spectra of Hamiltonians

Upon substitution of (4.5.85) in (4.5.82) we obtain

$$N_\ell \leqslant \frac{\mu R^2 U_0}{\hbar^2 (2\ell + 1)}. \tag{4.5.86}$$

Hence for

$$\frac{\mu R^2 U_0}{\hbar^2} < 1 \tag{4.5.87}$$

H_ℓ admits no bound-states for any given ℓ.

4.5.3 The Sum of the Negative Eigenvalues

Now we use the expression on the right-hand side of (4.5.10) to derive upper bounds for the negative of the sum of the negative eigenvalues (if any), counting degeneracy, of a Hamiltonian H. These will find useful applications, for example, to multi-particle systems in §4.6 and in Chapter 14.

The expression for the negative of the sum of the negative energies of H is obtained by integrating $N(H, \xi)$, in (4.5.4), (4.5.5), over ξ from $-\infty$ to 0, as shown in (4.5.10). With the substitution $\xi \to -\xi$, it will be shown below that this sum, in $\nu = 3, 2, 1$ dimensions, satisfies, in a compact form, the inequality,[7]

$$\int_0^\infty d\xi \, N(H, -\xi) \leqslant C_\nu \int d^\nu \mathbf{x} \, (v(\mathbf{x}))^{(\nu+2)/2} \tag{4.5.88}$$

where $-v(\mathbf{x})$, as before, is the negative part of the potential V in $H = H_0 + V$ defined in (4.5.17), and

$$C_\nu = \left(\frac{2\mu}{\hbar^2}\right)^{\nu/2} \frac{4}{\nu(\nu+2)} \left(\frac{3}{4}\right)^{(3-\nu)(\nu-1)} \left(\frac{1}{\pi}\right)^{(\nu-1)(\nu-2)/2} \tag{4.5.89}$$

$\nu = 3, 2, 1$.

The inequality in (4.5.88) is established below. For the clarity of the presentations, each of the respective dimensions are treated separately.

The three-dimensional case:

The expression in (4.5.53), is not suitable for the integration to be carried out over ξ when applying (4.5.10). We may, however, use the upper bound expression of the inequality in (4.5.45) as long as we replace $-v(\mathbf{x})$ by $-v(\mathbf{x}) + \xi/2$ and ξ by $\xi/2$ in (4.5.53). Due to the step function restriction on the right-hand side of the inequality (4.5.45) we now have the constraint $0 < \xi/2 < v(\mathbf{x})$. Accordingly,

$$\int_0^\infty d\xi \, N(H_0 + V, -\xi) \leqslant \left(\frac{\mu}{2\hbar^2}\right)^{3/2} \frac{\sqrt{2}}{\pi} \int d^3\mathbf{x} \int_0^{2v(\mathbf{x})} \left(v(\mathbf{x}) - \frac{\xi}{2}\right)^2 \frac{d\xi}{\sqrt{\xi}} \tag{4.5.90}$$

---

[7] See also Lieb and Thirring (1976).

or

$$\int_0^\infty d\xi\, N(H_0 + V, -\xi) \leq \frac{4}{15\pi} \left(\frac{2\mu}{\hbar^2}\right)^{3/2} \int_0^\infty d^3\mathbf{x}\, (v(\mathbf{x}))^{5/2} \qquad (4.5.91)$$

which coincides with the inequality in (4.5.88) for $\nu = 3$, and is referred to as a Lieb-Thirring bound,[8] providing an upper bound for the negative of the sum of the negative eigenvalues (if any), counting degeneracy, of the Hamiltonian in question.

Needless to say, since the ground-state energy cannot be less than the sum of the negative eigenvalues, (4.5.91) gives the following lower bound for the ground-state energy

$$-\frac{4}{15\pi} \left(\frac{2\mu}{\hbar^2}\right)^{3/2} \int d^3\mathbf{x}\, (v(\mathbf{x}))^{5/2} \qquad (4.5.92)$$

which is to be compared with (4.5.55). Which one provides a more optimal bound depends on the potential v.

The two-dimensional case ($\nu = 2$):

This lower dimensional cases is a bit more difficult to handle. For $\rho = 1$, the trace on the right-hand side of (4.5.42) is *infinite* (see Problem 4.4) and hence not useful. For $\rho = 2$, the right-hand side of (4.5.62) is not integrable over ξ. On the other hand, for $\rho = 3/2$ all the relevant integrals are convergent.

To the above end, we use the inequality (see Problem 4.5)

$$\int d^2\mathbf{x}\, \left\langle \mathbf{x} \left| A^{3/2} \right| \mathbf{x} \right\rangle \leq \int d^2\mathbf{x}\, \left\langle \mathbf{x} \left| v^{3/2} \frac{1}{\left[\frac{\mathbf{p}^2}{2\mu} + \xi\right]^{3/2}} \right| \mathbf{x} \right\rangle \qquad (4.5.93)$$

where A is defined in (4.5.27).

Upon using the integral

$$\left\langle \mathbf{x} \left| \frac{1}{\left[\frac{\mathbf{p}^2}{2\mu} + \xi\right]^{3/2}} \right| \mathbf{x} \right\rangle = \int \frac{d^2\mathbf{p}}{(2\pi\hbar)^2} \frac{1}{\left[\frac{\mathbf{p}^2}{2\mu} + \xi\right]^{3/2}}$$

$$= \frac{\mu}{\pi\hbar^2} \frac{1}{\sqrt{\xi}} \qquad (4.5.94)$$

we obtain from (4.5.93) the bound

---
[8] Lieb and Thirring (1975).

$$N(H_0 + V, -\xi) \leq \frac{\mu}{\pi\hbar^2} \frac{1}{\sqrt{\xi}} \int d^2\mathbf{x} \, v^{3/2}(\mathbf{x}). \tag{4.5.95}$$

Finally using the upper bound in (4.5.45), as done in writing (4.5.90), together with (4.5.20), the above inequality leads to

$$\int_0^\infty d\xi \, N(H_0 + V, -\xi) \leq \frac{\sqrt{2\mu}}{\pi\hbar^2} \int d^2\mathbf{x} \int_0^{2v(\mathbf{x})} \frac{d\xi}{\sqrt{\xi}} \left(v(\mathbf{x}) - \frac{\xi}{2}\right)^{3/2}$$

$$= \frac{2\mu}{\pi\hbar^2} \frac{\Gamma(1/2) \, \Gamma(5/2)}{\Gamma(3)} \int d^2\mathbf{x} \, (v(\mathbf{x}))^2 \tag{4.5.96}$$

or the inequality which coincides with the one in (4.5.88) for $\nu = 2$:

$$\int_0^\infty d\xi \, N(H_0 + V, -\xi) \leq \frac{3\mu}{4\hbar^2} \int d^2\mathbf{x} \, (v(\mathbf{x}))^2. \tag{4.5.97}$$

Needless to say, minus times the expression on the right-hand side of (4.5.97) provides a lower bound to the ground-state energy. The expression obtained in (4.5.64), however, gives a more optimal one for sufficiently small $\delta > 0$. Finally, (4.5.95) also leads to another lower bound for the ground-state energy given by

$$-(1+\delta)\frac{\mu^2}{\pi^2\hbar^4} \left(\int d^2\mathbf{x} \, (v(\mathbf{x}))^{3/2}\right)^2 \tag{4.5.98}$$

for any $\delta > 0$. Which one of the expressions in (4.5.64) or (4.5.98) provides a better bound, depends on the potential $v(\mathbf{x})$.

The one-dimensional case ($\nu = 1$):

Upon using (4.5.45), as done in (4.5.90), (4.5.96) leads to

$$\int_0^\infty d\xi \, N(H_0 + V, -\xi) \leq \frac{\sqrt{\mu}}{\hbar} \int_{-\infty}^\infty dx \int_0^{2v(x)} \frac{d\xi}{\sqrt{\xi}} \left(v(x) - \frac{\xi}{2}\right) \tag{4.5.99}$$

or

$$\int_0^\infty d\xi \, N(H_0 + V, -\xi) \leq \frac{4}{3} \frac{\sqrt{2\mu}}{\hbar} \int_{-\infty}^\infty dx \, (v(x))^{3/2} \tag{4.5.100}$$

which coincides with the inequality in (4.5.88) for $\nu = 1$.

This provides the following lower bound for the ground-state energy

$$-\frac{4}{3} \frac{\sqrt{2\mu}}{\hbar} \int_{-\infty}^\infty dx \, (v(x))^{3/2}. \tag{4.5.101}$$

As before which one of (4.5.69) or (4.5.101) is better depends on the potential v.

We rewrite the lower bound for a ground-state energy for a potential, whose negative part is denoted by $-v$ as defined in (4.5.17), for $\nu = 3, 2, 1$, and derived in (4.5.55), (4.5.64), (4.5.69), in a compact form

$$E_\nu \geq -a_\nu \left(\int d^\nu \mathbf{x}\, (v(\mathbf{x}))^{q(\nu)+\nu/2} \right)^{1/q(\nu)} \tag{4.5.102}$$

where

$$q(\nu) = \frac{4\nu - \nu^2 - 2}{2} = \begin{cases} 1/2, & \nu = 3 \\ 1, & \nu = 2 \\ 1/2, & \nu = 1 \end{cases} \tag{4.5.103}$$

$$a_\nu = (1+\delta) \left(\frac{1}{\pi}\right)^{\nu-1} \left(\frac{\mu}{2\hbar^2}\right)^{2(\nu/2 - q(\nu)) + 1} \tag{4.5.104}$$

for any $\delta > 0$.

Similarly, for the lower bound of the ground-state energy obtained from the sum of the negative eigenvalues, as derived in (4.5.91), (4.5.97), (4.5.100), we have in a compact form,[9]

$$E_\nu \geq -b_\nu \int d^\nu \mathbf{x}\, (v(\mathbf{x}))^{1+\nu/2} \tag{4.5.105}$$

where

$$b_\nu = \left(\frac{2\mu}{\hbar^2}\right)^{\nu/2} \frac{4}{\nu(\nu+2)} \left(\frac{3}{4}\right)^{(3-\nu)(\nu-1)} \left(\frac{1}{\pi}\right)^{(\nu-1)(\nu-2)/2}. \tag{4.5.106}$$

Needless to say, for $\nu = 1, 2, 3$, the existence of the integrals in (4.5.102), (4.5.105) is assumed.

Appendix to §4.5: Evaluation of Certain Integrals

We evaluate the expression on the left-hand side of (4.5.49). To this end we are led to consider the integral

$$\left\langle \mathbf{x} \left| \frac{1}{\left[\frac{\mathbf{p}^2}{2\mu} + \xi\right]} \right| \mathbf{x}' \right\rangle = \int \frac{d^3 \mathbf{p}}{(2\pi\hbar)^3} \frac{e^{i\mathbf{p}\cdot(\mathbf{x}-\mathbf{x}')/\hbar}}{\left[\frac{\mathbf{p}^2}{2\mu} + \xi\right]}.$$

The angular integration is readily evaluated yielding for the latter the integral

$$\frac{\mu}{2\pi^2 \hbar^2} \frac{1}{i\eta} \int_{-\infty}^{\infty} p\, dp\, \frac{e^{i\eta p/\hbar}}{p^2 + 2\mu\xi}$$

---

[9] See also Lieb (2000), for similar and other inequalities and for improvements of coefficients such as a_ν, b_ν.

integrating symmetrically over p, where $\eta \equiv |\mathbf{x} - \mathbf{x}'|$. In the complex p-plane, the integrand has simple poles at $p \pm i\sqrt{2\mu\xi}$. We may close the contour in the upper complex p-plane since the infinite semi-circle contour in the upper plane will not contribute. This gives immediately by the residue theorem at the pole $p = i\sqrt{2\mu\xi}$ the result given in (4.5.49).

For the corresponding two dimensional case given in (4.5.57), we note that for the integral

$$\int \frac{d^2\mathbf{p}}{(2\pi\hbar)^2} \frac{e^{i\mathbf{p}\cdot(\mathbf{x}-\mathbf{x}')/\hbar}}{\left[\frac{\mathbf{p}^2}{2\mu} + \xi\right]}$$

the angular part is given by

$$\int_0^{2\pi} d\theta\, e^{ip\eta \cos\theta/\hbar} = 2\pi J_0(p\eta/\hbar)$$

where $J_0(x)$ is the Bessel function of order zero. On the other hand,

$$\int_0^\infty \frac{x\, dx}{(x^2 + a^2)} J_0(bx) = K_0(ab) \qquad \text{(A-4.5.1)}$$

is a well known modified Bessel function of order zero, thus obtaining (4.5.57). [See also (4.5.58), (4.5.59).]

4.6 Lower Bounds to the Expectation Value of the Kinetic Energy: An Application of Counting Eigenvalues

The results obtained in §4.5 will be used to obtain lower bounds for the expectation value of the kinetic energy operator as a useful application. The latter will be important, in particular, for studying stability problems of multi-particle systems as given in Chapter 14.

4.6.1 One-Particle Systems

We first consider a particle in three dimensions, introduce the probability density

$$\rho(\mathbf{x}) = |\psi(\mathbf{x})|^2 \qquad (4.6.1)$$

and define the positive function

$$f(\mathbf{x}) = \gamma \frac{\rho^\alpha(\mathbf{x})}{\int d^3\mathbf{x}\, \rho^{\alpha+1}(\mathbf{x})} T \qquad (4.6.2)$$

where

$$T = \left\langle \psi \left| \frac{\mathbf{p}^2}{2\mu} \right| \psi \right\rangle \qquad (4.6.3)$$

4.6 Lower Bounds to the Expectation Value of the Kinetic Energy

and the parameters are positive and will be defined shortly.

We note that explicitly

$$\left\langle \psi \left| \frac{\mathbf{p}^2}{2\mu} - f \right| \psi \right\rangle = -(\gamma - 1)T. \tag{4.6.4}$$

In reference to the bound in (4.5.92), for example, one formally[10] has

$$\left\langle \psi \left| \frac{\mathbf{p}^2}{2\mu} - f \right| \psi \right\rangle \geqslant -\frac{4}{15\pi} \left(\frac{2\mu}{\hbar^2} \right)^{3/2} \int d^3\mathbf{x} \, (f(\mathbf{x}))^{5/2}$$

$$= -\frac{4}{15\pi} \left(\frac{2\mu}{\hbar^2} \right)^{3/2} \gamma^{5/2} T^{5/2} \frac{\int d^3\mathbf{x} \, (\rho(\mathbf{x}))^{5\alpha/2}}{\left(\int d^3\mathbf{x} \, \rho^{\alpha+1} \right)^{5/2}} \tag{4.6.5}$$

where we have used (4.6.2) in writing the last equality. This suggests to choose $5\alpha/2 = \alpha + 1$, or $\alpha = 2/3$. Hence with $\gamma > 1$, (4.6.4), (4.6.5) give

$$\left(\frac{\gamma - 1}{\gamma^{5/2}} \right)^{2/3} \left(\frac{15\pi}{4} \right)^{2/3} \frac{\hbar^2}{2\mu} \int d^3\mathbf{x} \, (\rho(\mathbf{x}))^{5/3} \leqslant T. \tag{4.6.6}$$

Optimizing over γ, gives $\gamma = 5/3$, or[11]

$$\frac{3}{5} \left(\frac{3\pi}{2} \right)^{2/3} \frac{\hbar^2}{2\mu} \int d^3\mathbf{x} \, (\rho(\mathbf{x}))^{5/3} \leqslant T. \tag{4.6.7}$$

Needless to say, $f(\mathbf{x})$ in (4.6.2) is not the potential energy for any given physical Hamiltonian. It is just introduced in order to be able to obtain a lower bound for T. Also all the integrals in (4.6.2)–(4.6.7) are implicitly assumed to be finite.

Similarly, by choosing $\alpha = 2/\nu$, $\gamma = (2 + \nu)/\nu$, we have, in reference to the bound in (4.5.105),

$$A_\nu \frac{\hbar^2}{2\mu} \int d^\nu \mathbf{x} \, (\rho(\mathbf{x}))^{(2+\nu)/\nu} \leqslant T \tag{4.6.8}$$

with

$$A_\nu = \left(\frac{\nu}{2} \right)^{2/\nu} \left(\frac{\nu}{2+\nu} \right) \left(\frac{4}{3} \right)^{2(3-\nu)(\nu-1)/\nu} (\pi)^{(\nu-1)(\nu-2)/\nu} \tag{4.6.9}$$

for $\nu = 1, 2, 3$ (see Problem 4.6). It is easily checked that (4.6.8) coincides with (4.6.7) for $\nu = 3$.

---

[10] For related technical details and subtleties see also, Lieb and Thirring (1976), p. 273.

[11] This inequality is referred to as a Lieb-Thirring kinetic energy inequality: Lieb and Thirring (1975).

On the other hand, by choosing $\alpha = 1$, using the normalization condition,

$$\int d^\nu x\, \rho(\mathbf{x}) = 1 \qquad (4.6.10)$$

and optimizing over γ, as in (4.6.6), we obtain in reference to the bound (4.5.102), the following inequality for the expectation value of the kinetic energy,

$$\frac{\nu}{1+\varepsilon} \left(\frac{\pi}{2}\right)^{(\nu-1)(4-\nu)/\nu} \frac{\hbar^2}{2\mu} \left(\int d^\nu x\, \rho^2(\mathbf{x})\right)^{2/\nu} \leqslant T \qquad (4.6.11)$$

for $\nu = 1, 2, 3$ (see Problem 4.7), for *any* $\varepsilon > 0$.

In particular, for $\nu = 3$, (4.6.11) gives

$$\frac{3}{1+\varepsilon} \left(\frac{\pi}{2}\right)^{2/3} \frac{\hbar^2}{2\mu} \left(\int d^3 x\, \rho^2(\mathbf{x})\right)^{2/3} \leqslant T \qquad (4.6.12)$$

for *any* $\varepsilon > 0$, which is to be compared with (4.6.7).

To the above end, we may write

$$\rho^{5/3} = \rho^{4/3}\rho^{1/3} \qquad (4.6.13)$$

and use Hölder's inequality (see Appendix II), with $p = 3/2$, $q = 3$ to obtain

$$\int d^3 x\, \rho^{5/3}(\mathbf{x}) \leqslant \left(\int d^3 x\, \rho^2(\mathbf{x})\right)^{2/3} \qquad (4.6.14)$$

where we have used the normalization condition in (4.6.10), with $\nu = 3$.

Accordingly, the left-hand side of the inequality (4.6.12) is bounded *below* by

$$\frac{(3)^{1/3}}{1+\varepsilon} \left(\frac{3\pi}{2}\right)^{2/3} \frac{\hbar^2}{2\mu} \int d^3 x\, \rho^{5/3}(\mathbf{x}). \qquad (4.6.15)$$

Since for sufficiently small $\varepsilon > 0$, $3^{1/3}/(1+\varepsilon) > 3/5$, (4.6.12) provides a better bound than the one in (4.6.7) for $\varepsilon > 0$ small enough. We leave it as an exercise to the reader for the comparisons of (4.6.8), (4.6.11) for $\nu = 2, 1$, in a similar fashion.

4.6.2 Multi-Particle States: Fermions

We consider N identical fermions, each of mass m, and introduce the particle number density in three dimensions:

$$\rho(\mathbf{x}) = N \sum_{\sigma_1,\ldots,\sigma_N} \int d^3 x_2\, d^3 x_3 \ldots d^3 x_N\, |\psi(\mathbf{x}\sigma_1, \mathbf{x}_2\sigma_2, \ldots, \mathbf{x}_N\sigma_N)|^2 \qquad (4.6.16)$$

4.6 Lower Bounds to the Expectation Value of the Kinetic Energy

where $\sigma_1, \ldots, \sigma_N$ specify spin projection values taking each $(2s+1)$ values for a particle of spin s (§5.4).

The total number of particles N is obtained from the normalization condition

$$\int d^3\mathbf{x}\, \rho(\mathbf{x}) = N. \tag{4.6.17}$$

The (normalized) wavefunctions $\psi(\mathbf{x}_1\sigma_1, \ldots, \mathbf{x}_N\sigma_N)$ are assumed to satisfy the appropriate statistics which in this case are anti-symmetric in the exchange of any two particles which amounts to the interchange of the position-spin labellings: $(\mathbf{x}_i\sigma_i) \Leftrightarrow (\mathbf{x}_j\sigma_j)$.

As in (4.6.2), we introduce the positive function

$$f(\mathbf{x}) = \frac{5}{3} \frac{\rho^{2/3}(\mathbf{x})}{\int d^3\mathbf{x}\, \rho^{5/3}(\mathbf{x})} T \tag{4.6.18}$$

where

$$T = \left\langle \psi \left| \sum_{i=1}^{N} \frac{\mathbf{p}_i^2}{2m} \right| \psi \right\rangle \tag{4.6.19}$$

then it is easily verified that

$$\left\langle \psi \left| \sum_{i=1}^{N} f(\mathbf{x}_i) \right| \psi \right\rangle = \frac{5}{3} T. \tag{4.6.20}$$

We consider the operator

$$\sum_{i=1}^{N} \left[\frac{\mathbf{p}_i^2}{2m} - f(\mathbf{x}_i) \right] \tag{4.6.21}$$

defining a hypothetical Hamiltonian of N non-interacting fermions which, however, interact with the external "potential" $-f(\mathbf{x})$.

To obtain a lower bound to the spectrum of the "Hamiltonian" in (4.6.21), we note that, allowing for multiplicity and spin degeneracy, we can put the N fermions in the lowest energy of levels of the "Hamiltonian" in conformity with Pauli's exclusion principle, if $N \leqslant$ number of such levels. If N is larger than this number of levels, the remaining free fermions may be chosen to have arbitrary small ($\to 0$) kinetic energies, and be infinitely separated, to define the lowest energy of the Hamiltonian in (4.6.21). That is, in all cases, the Hamiltonian (4.6.21) is bounded below by $(2s+1)$ times the *sum* of the negative energy levels of the Hamiltonian $[\mathbf{p}^2/m - f(\mathbf{x})]$, allowing, in the sum, for multiplicity but not for spin degeneracy. A bound to the latter sum has been already determined in (4.5.91), and with $v(\mathbf{x}) \to f(\mathbf{x})$, $\mu \to m$, we obtain

$$\left\langle \psi \left| \sum_{i=1}^{N} \left[\frac{\mathbf{p}_i^2}{2m} - f(\mathbf{x}_i) \right] \right| \psi \right\rangle \geqslant -(2s+1) \frac{4}{15\pi} \left(\frac{2m}{\hbar^2} \right)^{3/2} \int d^3\mathbf{x}\, (f(\mathbf{x}))^{5/2}. \tag{4.6.22}$$

224   4 Spectra of Hamiltonians

Hence from (4.6.18)–(4.6.22), we have

$$-\frac{2}{3}T \geqslant -(2s+1)\frac{4}{15\pi}\left(\frac{2m}{\hbar^2}\right)^{3/2}\left(\frac{5}{3}\right)^{5/2}T^{5/2}\left(\int d^3x\,\rho^{5/3}(x)\right)^{-3/2}. \tag{4.6.23}$$

or[12]

$$\frac{3}{5}\left(\frac{3\pi}{2(2s+1)}\right)^{2/3}\frac{\hbar^2}{2m}\int d^3x\,\rho^{5/3}(x) \leqslant T. \tag{4.6.24}$$

Again, we note that $f(x)$ is just introduced in order to derive a lower bound for T and is not the potential for any given physical Hamiltonian.

We leave it as an exercise to the reader to consider, formally, the $\nu = 2, 1$ cases (see Problem 4.8).

4.6.3 Multi-Particle States: Bosons

For simplicity of the notation, we consider (identical) bosons of spin 0. We introduce the particle number density:

$$\rho(\mathbf{x}) = N\int d^3x_2 \ldots d^3x_N\, |\psi(\mathbf{x}, \mathbf{x}_2, \ldots, \mathbf{x}_N)|^2. \tag{4.6.25}$$

A very conservative lower bound for the bosonic case may be directly obtained from that of the fermionic one in (4.6.24) without any further work by simply replacing $(2s+1)$ in the latter by N since in the present case one can put the N bosons in the lowest energy level of $[\mathbf{p}^2/2m - f(\mathbf{x})]$, where $f(\mathbf{x})$ is similarly defined as in (4.6.18). This leads to the inequality

$$\frac{3}{5}\left(\frac{3\pi}{2N}\right)^{2/3}\frac{\hbar^2}{2m}\int d^3x\,\rho^{5/3}(\mathbf{x}) \leqslant T \tag{4.6.26}$$

for bosons, where the $N^{-2/3}$ factor on the left-hand side should be noted. An improvement (i.e., a larger value) to the numerical $(3/5)(3\pi/2)^{2/3}$ may be obtained but we will not go into it here.

A similar analysis may be carried out in $\nu = 2, 1$ dimensions (see Problem 4.8).

4.7 The Eigenvalue Problem and Supersymmetry

4.7.1 General Aspects

The purpose of this section is to construct supersymmetric Hamiltonians, and show how one may study the eigenvalue problem of a given class of Hamiltonians using supersymmetry as a tool for doing so.

---

[12] This inequality is referred to as a Lieb-Thirring kinetic energy inequality: Lieb and Thirring (1975).

4.7 The Eigenvalue Problem and Supersymmetry

In §2.9, we have defined supersymmetric transformations and, by considering successive infinitesimal transformations in a closed path, have derived the following commutation/anti-commutation relations involving the supersymmetry generators Q, $Q^\dagger$ and the Hamiltonian H:

$$[H, Q] = 0, \quad [H, Q^\dagger] = 0 \tag{4.7.1}$$

$$\{Q, Q\} = 0, \quad \{Q^\dagger, Q^\dagger\} = 0 \tag{4.7.2}$$

and, in turn, obtained the explicit expression for H in terms of Q, $Q^\dagger$:

$$H = \frac{1}{2\hbar} \{Q, Q^\dagger\}. \tag{4.7.3}$$

The unitary operator for infinitesimal supersymmetry transformations, specified by infinitesimal anti-commuting c-numbers $\delta\xi$, $\delta\xi^*$ (see (2.9.2), (2.9.3)) is given by (see (2.9.21), (2.9.22), (2.9.20))

$$U = 1 + \frac{i}{\hbar} \left(i\, \delta\xi^* \, Q - i\, Q^\dagger \, \delta\xi \right). \tag{4.7.4}$$

For any state $|\phi\rangle$ in the domain of Q, (4.7.3) implies that

$$\langle \phi | H | \phi \rangle = \frac{1}{2\hbar} \langle \phi | Q Q^\dagger | \phi \rangle + \frac{1}{2\hbar} \langle \phi | Q^\dagger Q | \phi \rangle$$

$$= \frac{1}{2\hbar} \|Q^\dagger \phi\|^2 + \frac{1}{2\hbar} \|Q \phi\|^2 \geq 0. \tag{4.7.5}$$

That is, the spectrum of H satisfying (4.7.3) is necessarily non-negative.

A state for which

$$Q |\psi\rangle = 0 \tag{4.7.6}$$

and

$$Q^\dagger |\psi\rangle = 0 \tag{4.7.7}$$

is called a supersymmetric state as it remains invariant under the transformation implemented by the transformation in (4.7.4). Given generators Q, $Q^\dagger$, the actual construction of supersymmetric states will be spelled out below by solving the equations (4.7.6), (4.7.7). Clearly, from (4.7.5), a supersymmetric state defines the ground-state of the Hamiltonian H corresponding to the lowest point of its spectrum of zero energy.

A theory for which the supersymmetry generators commute with H, as given in (4.7.1), and the ground-state is supersymmetric, is said to be supersymmetric. A special class of theories is for which the commutation relations in (4.7.1) hold true but for a ground-state

$$Q |\psi\rangle \neq 0 \tag{4.7.8}$$

$$Q^\dagger |\psi\rangle = 0 \tag{4.7.9}$$

(or $Q^\dagger |\psi\rangle \neq 0$, $Q|\psi\rangle = 0$). These theories are said to be spontaneously broken theories. For such a theory the ground-state energy is strictly positive as follows directly from (4.7.5). The latter is also necessarily degenerate. This is easily shown by noting that the states $Q|\psi\rangle$, $|\psi\rangle$, are orthogonal

$$\langle \psi |Q| \psi\rangle = \langle Q^\dagger \psi |\psi\rangle = 0 \tag{4.7.10}$$

where we have used the equality in (4.7.9), signaling the importance of this equality, and

$$H|\psi\rangle = E|\psi\rangle \tag{4.7.11}$$

implies from (4.7.1), upon multiplying (4.7.11) by Q, that

$$H(Q|\psi\rangle) = E(Q|\psi\rangle). \tag{4.7.12}$$

Here we are interested, however, in the role of supersymmetry in studying the eigenvalue problem of quantum physics.

We first develop a method to construct supersymmetric Hamiltonians. This is then followed by showing how supersymmetry, seemingly unrelated to the problem at hand, may be used, as mentioned above, to study the eigenvalue problem in quantum physics for a large class of Hamiltonians. Supersymmetry will be also applied later in various chapters, notably in Chapter 6 in studying the Bose-Fermi oscillator, in Chapter 8 in reference to the Pauli Hamiltonian for a charged spin-1/2 particle in an external electromagnetic field, also discussed here. Several examples are also given in the present chapter.

4.7.2 Construction of Supersymmetric Hamiltonians

We first consider the one-dimensional case and define the following operators:

$$Q = \sqrt{\frac{\hbar}{m}}\,[p - i\,w(x)]\,\Psi^\dagger \tag{4.7.13}$$

$$Q^\dagger = \sqrt{\frac{\hbar}{m}}\,[p + i\,w(x)]\,\Psi \tag{4.7.14}$$

where $[x, p] = i\hbar$,

$$\{\Psi, \Psi^\dagger\} = 1, \quad \{\Psi, \Psi\} = 0, \quad \{\Psi^\dagger, \Psi^\dagger\} = 0 \tag{4.7.15}$$

and $w(x)$ is a real function of x. The operators Q, $Q^\dagger$ were chosen to be linear in p to ensure, in particular, that H as given in (4.7.3) includes a kinetic energy term $p^2/2\mu$. So-called Fermi operators satisfying anti-commutation

4.7 The Eigenvalue Problem and Supersymmetry

relations in (4.7.15), were introduced to satisfy the conditions in (4.7.2). The operators Ψ, $\Psi^\dagger$ are assumed to commute with p and $w(x)$.

We specify the range of x to be $a < x < b$. Implicit in the definition of $Q^\dagger$, given in (4.7.14) as the adjoint of Q, is that p is self-adjoint. That is, it is necessary that

$$\langle g | pf \rangle = \langle pg | f \rangle. \tag{4.7.16}$$

In detail, (4.7.16) implies from

$$\langle g | pf \rangle = \langle pg | f \rangle - i\hbar \left[g^*(b) f(b) - g^*(a) f(a) \right] \tag{4.7.17}$$

that is, we must have

$$g^*(b) f(b) - g^*(a) f(a) = 0. \tag{4.7.18}$$

Thus in defining a self-adjoint operator associated with the momentum of a particle, one has to impose restrictions on the functions f, g satisfying (4.7.18) and, in turn, define the associated domain of the self-adjoint operator in question introduced.

Of particular interest in applications are the cases where $(a \to -\infty, b \to \infty)$, $(a = 0, b \to \infty)$. With $r = |\mathbf{x}|$, for example, the latter case may be applied to the radial part of the Schrödinger equation (assuming that it may separated into a radial part) with the usual boundary conditions of $f(r)$ vanishing at $r = 0$ and $r \to \infty$. In many cases encountered in practice are those with vanishing boundary conditions at a and b.

From (4.7.13)–(4.7.15), we readily obtain the explicit expression for H to be

$$H = \left(\frac{p^2}{2m} + \frac{w^2(x)}{2m} \right) \mathbf{1} + \frac{\hbar}{2m} w'(x) \left[\Psi^\dagger, \Psi \right] \tag{4.7.19}$$

where we have used (4.7.3).

The anti-commutation relations in (4.7.15) for the fermi operators Ψ, $\Psi^\dagger$ allow us to introduce the convenient representations for the latter

$$\Psi^\dagger = \sigma^+ = \begin{pmatrix} 0 & 1 \\ 0 & 0 \end{pmatrix} \tag{4.7.20}$$

$$\Psi = \sigma^- = \begin{pmatrix} 0 & 0 \\ 1 & 0 \end{pmatrix} \tag{4.7.21}$$

as obtained from the Pauli-matrices, with

$$\left[\Psi^\dagger, \Psi \right] = \sigma_3 \tag{4.7.22}$$

leading to the supersymmetric Hamiltonian[13]

---

[13] Throughout, it is understood that the first term in (4.7.23) is multiplied by the identity matrix.

$$H = \frac{1}{2m}(p^2 + w^2(x)) + \frac{\hbar}{2m}w'(x)\sigma_3. \tag{4.7.23}$$

For a supersymmetric theory, the ground-state $|\psi_0\rangle$ is defined by the conditions

$$Q|\psi_0\rangle = 0, \quad Q^\dagger|\psi_0\rangle = 0 \tag{4.7.24}$$

(see (4.7.6), (4.7.7)). That is, it is the solution of the differential equations

$$\left[-i\hbar\frac{d}{dx} - iw(x)\right]\sigma^+\psi_0(x) = 0 \tag{4.7.25}$$

$$\left[-i\hbar\frac{d}{dx} + iw(x)\right]\sigma^-\psi_0(x) = 0. \tag{4.7.26}$$

Upon multiplying (4.7.25) by $-\sigma^-$, and (4.7.26) by $+\sigma^+$ and adding lead to

$$\frac{d}{dx}\psi_0(x) = \frac{w(x)}{\hbar}\sigma_3\psi_0(x). \tag{4.7.27}$$

The general solution of (4.7.27) is given by

$$\psi_0(x) = \alpha\begin{pmatrix}1\\0\end{pmatrix}F(x) + \beta\begin{pmatrix}0\\1\end{pmatrix}\frac{1}{F(x)} \tag{4.7.28}$$

where α, β are some constants and

$$F(x) = \exp\left(\frac{1}{\hbar}\int^x dx\, w(x)\right). \tag{4.7.29}$$

The following elementary consideration shows that for an infinite interval, i.e., $a \to -\infty$ and/or $b \to \infty$, if $F(x)$ is square-integrable then $1/F(x)$ is not and vice versa. To see this, note that for a function $f(x)$ such that

$$0 < \int_a^b dx\, |f(x)|^2 < \infty \tag{4.7.30}$$

we may use the Cauchy-Schwarz inequality to write

$$(b-a)^2 = \left|\int_a^b dx\, f(x)\frac{1}{f(x)}\right|^2 \leq \left(\int_a^b dx\, |f(x)|^2\right)\left|\int_a^b dx\, \frac{1}{|f(x)|^2}\right| \tag{4.7.31}$$

and derive the bound

$$\frac{(b-a)^2}{\int_a^b dx\, |f(x)|^2} \leq \int_a^b dx\, \frac{1}{|f(x)|^2}. \tag{4.7.32}$$

The latter shows that if (4.7.30) is true for $a \to -\infty$ and/or $b \to \infty$, then, upon taking the corresponding limits in (4.7.32), that $1/f(x)$ is not square-integrable on such infinite or semi-infinite intervals. On the other hand, for

4.7 The Eigenvalue Problem and Supersymmetry

finite intervals ($a < b$, $|a| < \infty$, $|b| < \infty$) the square-integrability of a function $f(x)$ and of $1/f(x)$ are not ruled out.

We may use the inequality in (4.7.32) to conclude that for an infinite interval if $F(x)$ is square-integrable, then for the square integrability of $\psi_0(x)$ we have to choose $\beta = 0$ (and vice-versa, vis-à-vis $1/F(x)$, to choose $\alpha = 0$).

We consider the construction of supersymmetric Hamiltonians in higher dimensions (see also Problem 4.14). To this end we provide examples of the interaction of a charged particle of charge e with an external magnetic field $\mathbf{B} = (0, 0, B)$. Upon setting $\boldsymbol{\pi} = \mathbf{p} - e\,\mathbf{A}/c$, with $\mathbf{B} = \nabla \times \mathbf{A}$, we have the following commutation relations

$$[\pi_j, \pi_k] = \frac{i e \hbar}{c} \varepsilon_{jkn} B_n \tag{4.7.33}$$

as is readily checked. In two dimensions we may define the generators

$$Q = \sqrt{\frac{\hbar}{m}}(\pi_2 + i\pi_1)\sigma^+, \quad Q^\dagger = \sqrt{\frac{\hbar}{m}}(\pi_2 - i\pi_1)\sigma^- \tag{4.7.34}$$

which verify the relations $Q^2 = 0$, $(Q^\dagger)^2 = 0$, and from the relations $\{\sigma^+, \sigma^-\} = 1$, $[\sigma^+, \sigma^-] = \sigma_3$, and (4.7.33) for $j = 1$, $k = 2$, we immediately obtain the supersymmetric Hamiltonian

$$H = \frac{1}{2\hbar}\{Q, Q^\dagger\} = \frac{\pi_1^2 + \pi_2^2}{2m} - \frac{e\hbar}{4mc} gB\sigma_3 \tag{4.7.35}$$

with $g = 2$, which is the celebrated Pauli Hamiltonian of spin $-1/2$ studied in Chapter 8, with the so-called g-factor *restricted* to the value of 2.[14,15] The value of $g = 2$ is only approximate for the electron (see §8.5) and such a departure may be formally interpreted as a breaking of supersymmetry.

In three dimensions, we introduce Fermi operators Ψ_1, Ψ_2 as 4×4 matrices:

$$\left.\begin{array}{c} \Psi_1 = \begin{pmatrix} \sigma^- & 0 \\ 0 & -\sigma^- \end{pmatrix}, \quad \Psi_2 = \begin{pmatrix} 0 & 0 \\ 1 & 0 \end{pmatrix}, \quad [\Psi_1^\dagger, \Psi_1] = \begin{pmatrix} \sigma_3 & 0 \\ 0 & \sigma_3 \end{pmatrix} \\[6pt] \{\Psi_1, \Psi_2\} = 0, \quad \{\Psi_1^\dagger, \Psi_2^\dagger\} = 0, \quad \{\Psi_i, \Psi_j^\dagger\} = 1\delta_{ij} \end{array}\right\} \tag{4.7.36}$$

with obvious dimensionalities of the unit matrices.

We define the generators Q, $Q^\dagger$

$$\sqrt{\frac{m}{\hbar}}\, Q = (\pi_2 + i\pi_1)\Psi_1^\dagger + \pi_3 \Psi_2^\dagger, \quad \sqrt{\frac{m}{\hbar}}\, Q^\dagger = (\pi_2 - i\pi_1)\Psi_1 + \pi_3 \Psi_2 \tag{4.7.37}$$

---

[14] Here and in (4.7.38), the so-called scalar potential U is zero.
[15] See also Khare and Maharana (1984), Cooper *et al.* (1995) and de Crombrugghe and Rittenberg (1983).

230   4 Spectra of Hamiltonians

where recalling that $\mathbf{B} = (0, 0, B)$, give from (4.7.33) that $[\pi_2, \pi_3] = 0$, $[\pi_1, \pi_3] = 0$, i.e., π_3 commutes with π_1, π_2. Thus we immediately verify that $Q^2 = 0$, $(Q^\dagger)^2 = 0$. The supersymmetric Hamiltonian is then readily worked out from (4.7.36), (4.7.37) and from the latter commutativity property of π_3 with π_1, π_2 to be

$$H = \frac{1}{2m}\left(\pi_1^2 + \pi_2^2 + \pi_3^2\right) - \frac{e\hbar g}{4mc}B\sigma_3 \qquad (4.7.38)$$

arising from two identical copies, each as a 2×2 matrix. This is again the Pauli-Hamiltonian, now in three dimensions, with g-factor equal to two, and a magnetic field $\mathbf{B}$ *along* the x_3-axis.

4.7.3 The Eigenvalue Problem

We now use supersymmetry as a tool to study the eigenvalue problem of the *discrete* spectrum of a large class of Hamiltonians (to be defined below), i.e., in particular, corresponding to normalizable eigenvectors (see §1.8) in the underlying Hilbert space. We restrict the study to one-dimensional cases, and as we will see below, through examples, that these cases are rich enough in applications. The higher dimensional cases in (4.7.35), (4.7.38) will be dealt with in Chapter 8.

In detail, we may rewrite the supersymmetric Hamiltonian H in (4.7.23) as

$$H = \begin{pmatrix} H_+ & 0 \\ 0 & H_- \end{pmatrix} \qquad (4.7.39)$$

where

$$H_+ = \frac{p^2}{2m} + V_+(x) \qquad (4.7.40)$$

$$V_+(x) = \frac{1}{2m}\left(w^2(x) + \hbar w'(x)\right) \qquad (4.7.41)$$

and

$$H_- = \frac{p^2}{2m} + V_-(x) \qquad (4.7.42)$$

$$V_-(x) = \frac{1}{2m}\left(w^2(x) - \hbar w'(x)\right). \qquad (4.7.43)$$

Also we note that

$$H_- = H_+ - \frac{\hbar}{m}w'(x). \qquad (4.7.44)$$

It is convenient to introduce the operators

$$A = \frac{1}{\sqrt{2m}}(p + iw(x)) \qquad (4.7.45)$$

4.7 The Eigenvalue Problem and Supersymmetry

$$A^\dagger = \frac{1}{\sqrt{2m}}(p - iw(x)) \qquad (4.7.46)$$

which allow us to rewrite formally

$$H_+ = A^\dagger A \qquad (4.7.47)$$

$$H_- = AA^\dagger \qquad (4.7.48)$$

and are referred to as supersymmetric partner Hamiltonians, and V_+, V_- as supersymmetric partner potentials. The function $w(x)$ is generally referred to as the superpotential.

We assume that $F(x)$, in (4.7.29), (4.7.28), is square-integrable on its domain of definition. From (4.7.26), (4.7.28), with $\alpha \neq 0$, (4.7.45) (4.7.47) imply that

$$A\,F(x) = 0 \qquad (4.7.49)$$

$$H_+ F(x) = 0. \qquad (4.7.50)$$

That is, $F(x)$ corresponds to an eigenstate of H_+ with zero eigenvalue. We are interested in the discrete spectrum of H_+ belonging to a special class of Hamiltonians to be defined below. From (4.7.47), such a discrete spectrum (assumed non-empty) would necessarily fall on the positive real axis with zero as the lowest point of this spectrum.

For simplicity of the notation only, we consider dimensionless variables in the remaining part of this section, and also divide, in the process, the Hamiltonians by suitable conversion energy scales thus defining their dimensionless counterparts.

Before treating the method in the remaining part of this section on how supersymmetry may be used to study the eigenvalue problem of a given class of Hamiltonians, we consider the following preparatory examples.

1. Suppose that

$$w(z) = -(z - \xi) \equiv w(z, \xi), \qquad -\infty < z < \infty \qquad (4.7.51)$$

where ξ is an arbitrary parameter. This leads to the Hamiltonians:

$$H_+(\xi) = -\frac{1}{2}\frac{d^2}{dz^2} + V_+(z, \xi) \qquad (4.7.52)$$

with

$$V_+(z, \xi) = \frac{1}{2}(z - \xi)^2 - \frac{1}{2} \qquad (4.7.53)$$

and

$$H_-(\xi) = -\frac{1}{2}\frac{d^2}{dz^2} + V_-(z, \xi) \qquad (4.7.54)$$

4 Spectra of Hamiltonians

with
$$V_-(z,\xi) = V_+(z,\xi) + 1 \qquad (4.7.55)$$

where we have made the dependence on the parameter ξ evident. In particular (see (4.7.29)), up to a proportionality constant,

$$F(z) = \exp\left(-\frac{1}{2}(z-\xi)^2\right). \qquad (4.7.56)$$

The simplicity of the relationship between the potentials $V_+(z,\xi)$, $V_-(z,\xi)$, in (4.7.55) is to be noted.

2. As another example, consider the function
$$w(z) = \frac{(\ell+1)}{z} - \frac{1}{(\ell+1)} \equiv w(z,\ell), \quad 0 < z < \infty \qquad (4.7.57)$$

where $\ell = 0, 1, 2, \ldots$. Here it is readily shown that

$$V_+(z,\ell) = \frac{\ell(\ell+1)}{2z^2} - \frac{1}{z} + \frac{1}{2(\ell+1)^2} \qquad (4.7.58)$$

and
$$V_-(z,\ell) = V_+(z,\ell+1) + \frac{(2\ell+3)}{2(\ell+1)^2(\ell+2)^2}. \qquad (4.7.59)$$

Also,
$$F(z) = (z)^{\ell+1} \exp\left(-z/(\ell+1)\right), \quad 0 < z < \infty \qquad (4.7.60)$$

with vanishing boundary conditions at $z \to 0$, $z \to \infty$. Again the interesting relationship between the potentials $V_\pm(z,\ell)$ is to be noted.

3. Finally, consider the function
$$w(z) = -\xi \tanh z \equiv w(z,\xi), \quad -\infty < z < \infty \qquad (4.7.61)$$

where the parameter $\xi > 0$. This leads to the potentials

$$V_+(z,\xi) = \frac{1}{2}\left(\xi^2 - \xi(\xi+1)\operatorname{sech}^2 z\right) \qquad (4.7.62)$$

$$V_-(z,\xi) = V_+(z,\xi-1) + \left(\xi - \frac{1}{2}\right) \qquad (4.7.63)$$

An interesting and simple relationship between the potentials $V_+(z,\xi)$ emerges again.

For $F(z)$ we have,
$$F(z) = \exp\left(-\xi \int^z dz\ \tanh z\right)$$

4.7 The Eigenvalue Problem and Supersymmetry

$$= \exp\left(\xi \ \ln(\mathrm{sech}\, z)\right)$$

or

$$F(z) = (\mathrm{sech}\, z)^\xi \qquad (4.7.64)$$

and for $\xi > 0$, $F(z)$ is square-integrable over $-\infty < z < \infty$.

The last three examples given above and, in particular, equations (4.7.55), (4.7.59), (4.7.63), for the supersymmetric partner potentials, suggest to consider the following class of supersymmetric partners $H_+(\xi), H_-(\xi)$ depending on some parameter ξ such that

$$V_-(z,\xi) = V_+(z,\xi_1) + R(\xi_1) \qquad (4.7.65)$$

where $R(\xi_1)$ is independent of z,

$$R(\xi_1) \neq 0 \qquad (4.7.66)$$

and there exists some function f such that

$$f(\xi) = \xi_1. \qquad (4.7.67)$$

For the class of so-called supersymmetric partner potentials $V_\pm(z,\xi)$ which satisfy equations of the structure in (4.7.65), we note that apart from the additive z-independent term $R(\xi_1)$, such pairs of potentials have similar shapes[16] when parameterized with, in general, different parameters.

For example, in (4.7.55), we have the identity transformation

$$f(\xi) = \xi \qquad (4.7.68)$$

and

$$R(\xi) = 1. \qquad (4.7.69)$$

In (4.7.59), we have with $\ell \equiv \xi$,

$$f(\xi) = \xi + 1 \qquad (4.7.70)$$

and

$$R(\xi + 1) = \frac{(2\xi + 3)}{2(\xi + 1)^2(\xi + 2)^2}. \qquad (4.7.71)$$

Finally in (4.7.63)

$$f(\xi) = \xi - 1 \qquad (4.7.72)$$

and

$$R(\xi - 1) = \xi - \frac{1}{2}. \qquad (4.7.73)$$

---

[16] Because of this property, partner potentials $V_\pm(z,\xi)$ satisfying (4.7.65) have been referred to as being shape invariant in Gendenshteïn (1983).

4 Spectra of Hamiltonians

For the class of supersymmetric partner potentials satisfying (4.7.65), we may write for the corresponding Hamiltonians

$$H_-(\xi) = H_+(\xi_1) + R(\xi_1). \tag{4.7.74}$$

Also by setting

$$f(\xi_k) = \xi_{k+1}, \quad k = 0, 1, \ldots \tag{4.7.75}$$

with

$$\xi_0 \equiv \xi \tag{4.7.76}$$

we may also write a corresponding expression to (4.7.74) the following one by replacing $\xi \to \xi_k$, $\xi_1 \to \xi_{k+1}$:

$$H_-(\xi_k) = H_+(\xi_{k+1}) + R(\xi_{k+1}). \tag{4.7.77}$$

By definition of the $H_+(\xi)$ considered, (4.7.49), (4.7.50) admit a square-integrable solution on its domain of definition. That is $\psi_0(z,\xi) = C(\xi)\,F(z,\xi)$ satisfies

$$A(\xi)\psi_0(z,\xi) = 0, \quad \|\psi_0(\cdot,\xi)\| = 1 \tag{4.7.78}$$

$$H_+(\xi)\psi_0(z,\xi) = 0. \tag{4.7.79}$$

For the present analysis, we suppose, as an induction hypothesis, that for all $k = 1, \ldots, K$ for some K, that

$$\phi_k(z,\xi) = A^\dagger(\xi)A^\dagger(\xi_1)\ldots A^\dagger(\xi_{k-1})\psi_0(z,\xi_k) \tag{4.7.80}$$

are square-integrable, on the domain of definition of the variable z, and when normalized, satisfy the boundary conditions imposed in the problem at hand, and then generalize the result by induction.

Consider the state $A^\dagger |\psi_0(\xi_1)\rangle$, and hence from (4.7.47), (4.7.48) we may write

$$H_+(\xi)A^\dagger(\xi)|\psi_0(\xi_1)\rangle = A^\dagger(\xi)H_-(\xi)|\psi_0(\xi_1)\rangle$$

$$= A^\dagger(\xi)\left[H_+(\xi_1) + R(\xi_1)\right]|\psi_0(\xi_1)\rangle$$

$$= R(\xi_1)A^\dagger(\xi)|\psi_0(\xi_1)\rangle \tag{4.7.81}$$

where we have used (4.7.77) and (4.7.79) with $\xi \to \xi_1$ in the latter.

Accordingly we have the properly normalized state

$$|\psi_1(\xi)\rangle = \frac{(\mathrm{i})}{\sqrt{E_1(\xi)}}\, A^\dagger(\xi)|\psi_0(\xi_1)\rangle \tag{4.7.82}$$

where the (i) factor is chosen for convenience, and

4.7 The Eigenvalue Problem and Supersymmetry

$$E_1(\xi) = R(\xi_1) \neq 0 \tag{4.7.83}$$

(see (4.7.66)) implicit in the normalization that has to be done of the $\phi_k(z,\xi)$ in (4.7.80),

$$H_+(\xi)|\psi_1(\xi)\rangle = E_1(\xi)|\psi_1(\xi)\rangle. \tag{4.7.84}$$

Similarly, from (4.7.77), (4.7.84)

$$\begin{aligned} H_+(\xi)A^\dagger(\xi)|\psi_1(\xi_1)\rangle &= A^\dagger(\xi)H_-(\xi)|\psi_1(\xi_1)\rangle \\ &= A^\dagger(\xi)\left[H_+(\xi_1) + R(\xi_1)\right]|\psi_1(\xi_1)\rangle \\ &= [E_1(\xi_1) + R(\xi_1)]\,A^\dagger(\xi)|\psi_1(\xi_1)\rangle \\ &= [R(\xi_2) + R(\xi_1)]\,A^\dagger(\xi)|\psi_1(\xi_1)\rangle \end{aligned} \tag{4.7.85}$$

where we have finally used (4.7.75) with $k = 1$. Thus we may introduce the properly normalized state

$$\begin{aligned} |\psi_2(\xi)\rangle &= \frac{(\mathrm{i})}{\sqrt{E_2(\xi)}} A^\dagger(\xi)|\psi_1(\xi_1)\rangle \\ &= \frac{(\mathrm{i})^2}{\sqrt{E_2(\xi)E_1(\xi_1)}} A^\dagger(\xi)A^\dagger(\xi_1)|\psi_0(\xi_2)\rangle \end{aligned} \tag{4.7.86}$$

where

$$E_2(\xi) = R(\xi_1) + R(\xi_2) \tag{4.7.87}$$

and in writing (4.7.86) we have invoked the definition (4.7.82) with $\xi \to \xi_1$, and hence also with $|\psi_0(\xi_1)\rangle \to |\psi_0(\xi_2)\rangle$. By hypothesis of normalizability, $E_2(\xi)E_1(\xi_1) \neq 0$.

By induction, we obviously obtain from (4.7.82), (4.7.83), (4.7.86), (4.7.87) the state

$$|\psi_k(\xi)\rangle = C_k(\xi) A^\dagger(\xi) A^\dagger(\xi_1) \ldots A^\dagger(\xi_{k-1})|\psi_0(\xi_k)\rangle = C_k(\xi)|\phi_k(\xi)\rangle \tag{4.7.88}$$

(see (4.7.80)), where the normalization constant $C_k(\xi)$ is given by

$$C_k(\xi) = (\mathrm{i})^k \left(E_k(\xi)E_{k-1}(\xi_1)\ldots E_1(\xi_{k-1})\right)^{-1/2} \tag{4.7.89}$$

and

$$\left.\begin{aligned} H_+|\psi_k(\xi)\rangle &= E_k(\xi)|\psi_k(\xi)\rangle \\ H_+|\psi_0(\xi)\rangle &= 0 \end{aligned}\right\} \tag{4.7.90}$$

with

$$E_k(\xi) = \sum_{i=1}^{k} R(\xi_i). \tag{4.7.91}$$

4 Spectra of Hamiltonians

The expression of the normalization constant $C_k(\xi)$ may be readily expressed as a function of ξ as follows.

For different k, we have from (4.7.91):

$$E_{k-m+1}(\xi) = R(\xi_1) + \sum_{j=2}^{k-m+1} R(\xi_j) \qquad (4.7.92)$$

$$E_{k-m}(\xi_1) = \sum_{j=2}^{k-m+1} R(\xi_j) \qquad (4.7.93)$$

where in writing the latter equation we have made use of (4.7.76). Hence

$$E_{k-m}(\xi_1) = E_{k-(m-1)}(\xi) - R(\xi_1). \qquad (4.7.94)$$

Upon carrying out the transformation $\xi \to \xi_{m-1}$, and hence $\xi_1 \to \xi_m$, (4.7.94) reads

$$\begin{aligned} E_{k-m}(\xi_m) &= E_{k-(m-1)}(\xi_{m-1}) - R(\xi_m) \\ &= E_{k-(m-2)}(\xi_{m-2}) - R(\xi_{m-1}) - R(\xi_m) \\ &\vdots \\ &= E_k(\xi) - [R(\xi_1) + \cdots + R(\xi_m)] \\ &= E_k(\xi) - E_m(\xi). \end{aligned} \qquad (4.7.95)$$

By choosing, in turn, $m = 1, 2, \ldots, k-1$, in this equality, we note that the normalization constant $C_k(\xi)$ in (4.7.89), (4.7.88) may be rewritten as

$$C_k(\xi) = (\mathrm{i})^k \left(E_k(\xi) \left[E_k(\xi) - E_1(\xi) \right] \cdots \left[E_k(\xi) - E_{k-1}(\xi) \right] \right)^{-1/2}. \qquad (4.7.96)$$

We may summarize the main result of this subsection thus far, as obtained from supersymmetry theory, as follows.

▶ *Summary*: "Suppose we are given a Hamiltonian of interest $H_+(\xi)$, depending on some parameter ξ, defined by

$$H_+(\xi) = -\frac{1}{2} \frac{\mathrm{d}^2}{\mathrm{d}z^2} + V_+(z, \xi) \qquad (4.7.97)$$

where the potential is defined in terms of a function $w(z, \xi)$ as follows:

$$V_+(z, \xi) = \frac{1}{2} \left(w^2(z, \xi) + w'(z, \xi) \right) \qquad (4.7.98)$$

with $w'(z, \xi) = \partial w(z, \xi)/\partial z$.

4.7 The Eigenvalue Problem and Supersymmetry

Suppose that the following Hamiltonian $H_-(\xi)$:

$$H_-(\xi) = -\frac{1}{2}\frac{d^2}{dz^2} + V_-(z,\xi) \qquad (4.7.99)$$

with

$$V_-(z,\xi) = \frac{1}{2}\left(w^2(z,\xi) - w'(z,\xi)\right) \qquad (4.7.100)$$

is *such that*

$$V_-(z,\xi) = V_+(z,\xi_1) + R(\xi_1) \qquad (4.7.101)$$

i.e., the shape invariant property of the supersymmetric partner potentials holds true, where $R(\xi_1)$ is independent of z, $R(\xi_1) \neq 0$, and there exists some function f such that

$$f(\xi) = \xi_1 \qquad (4.7.102)$$

and in general

$$f(\xi_k) = \xi_{k+1} \qquad (4.7.103)$$

for $k = 0, 1, \ldots, K$ (for some K, which may be finite or infinite), $\xi_0 = \xi$. Then with the square-integrability of the $\psi_k(z,\xi)$ in (4.7.88), (4.7.89), (4.7.96) (up to some K) implicit, on their domain of definition, and consistent with the boundary conditions imposed in the problem at hand, the $\psi_k(z,\xi)$ satisfy the eigenvalue equations in (4.7.90) with eigenvalues given in (4.7.91) for the Hamiltonian H_+. Also implicit is that $\psi_0(z,\xi)$ is square-integrable (normalized) as a solution of the equation

$$A(z,\xi)\psi_0(z,\xi) = 0 \qquad (4.7.104)$$

consistent with the boundary conditions in the problem, and defines the ground-state, where

$$A(z,\xi) = \frac{1}{\sqrt{2}}\left[-i\frac{d}{dz} + i w(z,\xi)\right]. \qquad (4.7.105)$$

The function $\psi_0(z,\xi_k)$ in (4.7.88) is defined by making the formal replacement $\xi \to \xi_k$." ◀

It is important to point out that the above construction for eigenvalues and eigenvectors just uses supersymmetry as a method of application to a given Hamiltonian $H_+(\xi)$ of interest satisfying the properties spelled out above.

We apply the method just developed to study the eigenvalue problem of two of the three examples mentioned through (4.7.51)–(4.7.64). The last one is left as an exercise to the reader (see Problem 4.15).

1. For $w(z,\xi)$ defined in (4.7.51), $-\infty < z < \infty$

$$H_-(\xi) = H_+(\xi) + 1, \quad f(\xi) = \xi, \quad R(\xi) = 1 \qquad (4.7.106)$$

(see (4.7.68), (4.7.69)). Hence from (4.7.90), (4.7.91), the eigenvalues of

$$H_+(\xi) = -\frac{1}{2}\frac{d^2}{dz^2} + \frac{1}{2}(z-\xi)^2 - \frac{1}{2}, \qquad (4.7.107)$$

are 0, and

$$E_k = \sum_{i=1}^{k}(1) = k, \quad k = 1, 2, \ldots. \qquad (4.7.108)$$

From (4.7.104), the ground-state is given by

$$\psi_0(z,\xi) = \left(\frac{1}{\pi}\right)^{1/4} \exp\left(-(z-\xi)^2\right)/2 \qquad (4.7.109)$$

and from (4.7.88), (4.7.96), the eigenvectors corresponding to the eigenvalues $\xi_k = \xi$ are

$$\psi_k(z,\xi) = \frac{C_k(\xi)}{\sqrt{2^k}}\left[-i\frac{d}{dz} + i(z-\xi)\right]^k \psi_0(z,\xi) \qquad (4.7.110)$$

for $k = 1, 2, \ldots$, respectively, since $\xi_k = \xi$,

$$C_k(\xi) = (i)^k (k!)^{-1/2} \qquad (4.7.111)$$

and is independent of ξ. Due to the exponential factor in (4.7.109), $\psi_k(z,\xi)$ is obviously square-integrable for all $k = 0, 1, \ldots$. The connection of $H_+(\xi)$, for $\xi = 0$, with the elementary harmonic oscillator problem is obvious (§6.1). The i^k factor in $C_k(0)$ leads to the conventional normalization of the harmonic oscillator wavefunctions (see §6.1) up to an overall minus sign.

2. For $w(z,\ell)$ defined in (4.7.57), $0 < z < \infty$,

$$H_-(\ell) = H_+(\ell+1) + \frac{(2\ell+3)}{2(\ell+1)^2(\ell+2)^2}$$

$$= H_+(\ell+1) + \frac{1}{2}\left[\frac{1}{(\ell+1)^2} - \frac{1}{(\ell+2)^2}\right], \qquad (4.7.112)$$

(see (4.7.71)),

$$R(\ell+1) = \frac{1}{2}\left[\frac{1}{(\ell+1)^2} - \frac{1}{(\ell+2)^2}\right] \qquad (4.7.113)$$

$$\xi = \ell, \quad \xi_k = \ell + k, \quad k = 1, 2, \ldots. \qquad (4.7.114)$$

Therefore the eigenvalues of

$$H_+(\ell) = -\frac{1}{2}\frac{d^2}{dz^2} - \frac{1}{z} + \frac{\ell(\ell+1)}{2z^2} + \frac{1}{2(\ell+1)^2} \qquad (4.7.115)$$

are from (4.7.91),

$$E_k(\ell) = \frac{1}{2}\sum_{j=1}^{k}\left[\frac{1}{(\ell+j)^2} - \frac{1}{(\ell+j+1)^2}\right]$$

$$= \frac{1}{2}\left[\frac{1}{(\ell+1)^2} - \frac{1}{(\ell+k+1)^2}\right] \quad (4.7.116)$$

and for $k \to 0$, this reduces to the eigenvalue 0 for the ground-state. We note from (4.7.104), (4.7.105), that

$$A^\dagger(\xi_j) = -\frac{i}{\sqrt{2}}\left[\frac{d}{dz} + \frac{\xi_j+1}{z} - \frac{1}{\xi_j+1}\right], \quad \xi_j = \ell+j \quad (4.7.117)$$

$$\psi_0(z,\ell) = c_\ell(z)^{\ell+1}\exp\left(-z/(\ell+1)\right) \quad (4.7.118)$$

and more generally

$$\psi_0(z,\xi_j) = c_{\xi_j}(z)^{\xi_j+1}\exp\left(-z/(\xi_j+1)\right) \quad (4.7.119)$$

with the eigenvectors $|\psi_k(\ell)\rangle$, for $k = 1, 2, \ldots$ defined in (4.7.88), (4.7.96).

The connection of $H_+(\ell)$ to the (one-dimensional) radial part of the Hamiltonian of the hydrogen atom (Chapter 7)

$$H(\ell) = -\frac{1}{2}\frac{d^2}{dz^2} - \frac{1}{z} + \frac{\ell(\ell+1)}{2z^2} \quad (4.7.120)$$

with ℓ as non-negative integers, should be noted. Here $H(\ell)$ is expressed in terms of the dimensionless variable

$$z = r\mu\, e^2/\hbar^2 \quad (4.7.121)$$

where μ is the reduced mass of the atom, and in energy conversion units of $\mu e^4/\hbar^2$. It follows from (4.7.115), (4.7.116) that the eigenvalues of $H(\ell)$ are

$$-\frac{1}{2}\frac{1}{(\ell+k+1)^2}. \quad (4.7.122)$$

Since $\ell = 0, 1, 2, \ldots$, one may introduce a natural number n by setting

$$\ell + k + 1 = n. \quad (4.7.123)$$

Therefore for a fixed n value, $k = 0, 1, \ldots, n-\ell-1$. Thus upon eliminating k in the states $|\psi_k(\ell)\rangle$ in favor of n and ℓ, the allowed values of ℓ become: $\ell = 0, 1, \ldots, n-1$ for a fixed n, where n specifies the eigenvalues in (4.7.122). In detail, we then have from (4.7.88), (4.7.96), (4.7.119) with

$$\xi_k = \ell + k = n - 1 \quad (4.7.124)$$

that

$$\psi_{n-\ell-1}(z,\ell) = c_{n-1}C_{n-\ell-1}(\ell)A^\dagger(\ell)A^\dagger(\ell+1)\cdots A^\dagger(n-2)z^n e^{-z/n}$$
$$\equiv \phi_{n,\ell}(z) \quad \text{for} \quad \ell = 0,\ldots,n-2; \quad n = 2,3,\ldots \quad (4.7.125)$$

and from (4.7.118)

$$\psi_0(z, n-1) = c_{n-1}z^n e^{-z/n}$$
$$\equiv \phi_{n,n-1}(z), \quad \text{for} \quad \ell = n-1; \quad n = 1,2,\ldots. \quad (4.7.126)$$

The normalization constant factor $C_{n-\ell-1}(\ell)$ in (4.7.125) is defined in (4.7.96), with (see (4.7.91), (4.7.113))

$$E_j(\ell) = \frac{1}{2}\left[\frac{1}{(\ell+1)^2} - \frac{1}{(\ell+j+1)^2}\right] \quad (4.7.127)$$

where now $j = 1,\ldots,n-\ell-1$.

The wavefunctions $\phi_{n,\ell}(z)$, for $\ell = 0,\ldots,n-1; n = 1,2,\ldots$, in (4.7.125), (4.7.126), are square-integrable on $0 \leq z < \infty$, as one-dimensional integrals, and vanish at $z = 0, z \to \infty$.

For example, for $n = 2$, $c_1 = 1/\sqrt{24} = (1/2)^{3/2}/\sqrt{3}$, and for $\ell = 1$, (4.7.126) gives

$$\phi_{2,1}(z) = \left(\frac{1}{2}\right)^{3/2}\frac{z^2}{\sqrt{3}}e^{-z/2}. \quad (4.7.128)$$

For $\ell = 0$, (i.e., $k = 1$), $E_1(0) = 3/8$ in (4.7.127), hence

$$C_1(0) = i\sqrt{\frac{8}{3}} \quad (4.7.129)$$

and

$$\phi_{2,0}(z) = \frac{i}{\sqrt{24}}\sqrt{\frac{8}{3}}\frac{(-i)}{\sqrt{2}}\left[\frac{d}{dz} + \frac{1}{z} - 1\right]z^2 e^{-z/2}$$
$$= \left(\frac{1}{2}\right)^{3/2} z(2-z)e^{-z/2}. \quad (4.7.130)$$

The conversion factor $z \to r$, is given in (4.7.121).

If the property of shape invariance of a pair of supersymmetric partner potentials, as defined in (4.7.65), holds, that is

$$H_-(\xi) = H_+(\xi_1) + R(\xi_1) \quad (4.7.131)$$

with (4.7.66), (4.7.67) satisfied, we may infer from (4.7.91) that

4.7 The Eigenvalue Problem and Supersymmetry

$$\sum_{i=1}^{k} R(\xi_{i+1}) + R(\xi_1) = E_{k+1}(\xi) \qquad (4.7.132)$$

with $k = 0, 1, \ldots$ (up to some K which may be infinite) are eigenvalues of $H_-(\xi)$ and note, in the process, that the eigenvalue 0 of $H_+(\xi_1)$ is missing for $H_-(\xi)$ with $E_1(\xi)$ representing the ground-state of the latter. Such a shift of the eigenvalues for a supersymmetric partner Hamiltonians is typical. The eigenvectors of $H_-(\xi)$ in (4.7.131) corresponding to the eigenvalues $E_{k+1}(\xi)$ in (4.7.132) are from (4.7.88) given by

$$|\chi_{k+1}(\xi)\rangle = C_k(\xi_1) \, A^\dagger(\xi_1) \, A^\dagger(\xi_2) \cdots A^\dagger(\xi_k) \, |\psi_0(\xi_{k+1})\rangle \qquad (4.7.133)$$

for $k = 1, 2, \ldots$ (up to some K), corresponding to eigenvalues $E_2(\xi)$, $E_3(\xi), \ldots$, respectively, and

$$|\chi_1(\xi)\rangle = |\psi_0(\xi_1)\rangle \qquad (4.7.134)$$

corresponding to $E_1(\xi)$. Implicit in this is that $|\psi_0(\xi)\rangle$ and the states in (4.7.88) (up to some K) are square-integrable and consistent with the boundary conditions imposed in the problem at hand. Implicit also, is that the latter hold true for $\xi \to \xi_1$.

Irrespective of any shape invariance properties of a given pair of supersymmetric potential partners, the following question arises. Does an eigenvalue E, say, of a supersymmetric partner H_+, i.e., corresponding, in particular, to a square-integrable wavefunction, is also an eigenvalue of the partner H_-? The situation is different whether $E = 0$ (i.e., the ground-state energy of H_+, assuming it exists) or $E \neq 0$.

Suppose $E = 0$ is the ground-state energy of H_+. That is, F in (4.7.29) is square-integrable, as the solution of (see (4.7.49))

$$A\,F = 0 \qquad (4.7.135)$$

and satisfies

$$A^\dagger \, A \, F = 0. \qquad (4.7.136)$$

From (4.7.25), (4.7.28), (4.7.29), (4.7.46), (4.7.48), we are formally led to consider the pair of equations

$$A^\dagger \left(\frac{1}{F}\right) = 0 \qquad (4.7.137)$$

$$A \, A^\dagger \left(\frac{1}{F}\right) = 0 \qquad (4.7.138)$$

and at least for problems defined on infinite or semi-infinite intervals, we may be conclude from (4.7.30)–(4.7.32) that $1/F$ is not square-integrable and hence cannot correspond to any eigenvalue (discrete spectrum).

On the other hand if $E \neq 0$ is an eigenvalue of H_+, then

$$H_+ |\psi\rangle = A^\dagger A |\psi\rangle = E |\psi\rangle \tag{4.7.139}$$

where $\|\psi\| < \infty$. Upon multiplying the above by A from the left, we obtain

$$A A^\dagger (A |\psi\rangle) = H_- (A |\psi\rangle) = E (A |\psi\rangle) \tag{4.7.140}$$

where we have used (4.7.48), and rather formally infer that E also belongs to the discrete spectrum of H_-.

In the light of the above discussion, we consider a couple of examples. Consider the superpotential

$$w(z) = -\tanh z, \quad -\infty < z < \infty \tag{4.7.141}$$

corresponding to $\xi = 1$ in (4.7.61). Then

$$H_+ = -\frac{1}{2} \frac{d^2}{dz^2} + \frac{1}{2} \left(1 - 2 \operatorname{sech}^2 z\right) \tag{4.7.142}$$

and

$$H_- = -\frac{1}{2} \frac{d^2}{dz^2} + \frac{1}{2} \tag{4.7.143}$$

(see (4.7.62), (4.7.63)). Clearly 0 cannot be an eigenvalue of H_-.

The ground-state wavefunction $F(z)$ of H_+ is given from (4.7.64) to be[17]

$$F(z) = \operatorname{sech} z \tag{4.7.144}$$

(up to a normalization constant) which is square-integrable and vanishes for $|z| \to \infty$. On the other hand, $1/F(z) = \cosh z$, apart from its notable properties for $|z| \to \infty$, is not square-integrable. A second formal solution to this, as obtained from Problem 4.9 (iii), gives the function $\sinh z$ and hence is not acceptable either. That is, 0 is not an eigenvalue of H_-, with the latter having an empty discrete spectrum, as expected.

As another example, consider the superpotential

$$w(z) = \pi \cot \pi z, \quad 0 < z < 1. \tag{4.7.145}$$

Hence

$$H_+ = -\frac{1}{2} \frac{d^2}{dz^2} - \frac{\pi^2}{2} \tag{4.7.146}$$

$$H_- = -\frac{1}{2} \frac{d^2}{dz^2} + \pi^2 \left(\cot^2 \pi z + \frac{1}{2}\right) \tag{4.7.147}$$

---

[17] Note that a second formal solution of, $H_+ g = 0$ as obtained from Problem 4.9 (iii) is given by $g(z) = [\sinh z + z \operatorname{sech} z]$ and is not acceptable because it is not square-integrable and its related bad behavior for $|z| \to \infty$.

where we note that $V_-(z) \to \infty$ for $z \to 0, 1$.

The ground-state of H_+ is given by

$$F(z) = \exp\left(\int^z \pi \mathrm{d}z \, \cot \pi z\right)$$

$$= \sin \pi z \qquad (4.7.148)$$

(up to a normalization constant), which is square-integrable on $[0, 1]$ and vanishes for $z \to 0, 1$. The function $1/F(z) = \csc \pi z$, apart from its notable properties for $z \to 0, 1$, is not square-integrable, neither is a second formal solution of $H_+ g = 0$, given by

$$g(z) = \frac{1}{\sin \pi z}\left[z - \frac{\sin 2\pi z}{2\pi}\right] \qquad (4.7.149)$$

as obtained from Problem 4.9 (iii). [Note that $g(z)$ in (4.7.149) was chosen such that $g(0) = 0$.] That is 0 is not an eigenvalue (i.e., not in the discrete spectrum) of H_-.

Now consider the eigenvalue $E = 3\pi^2/2$ of H_+ in (4.7.146) with eigenvector represented by $\psi(z) = \sin 2\pi z$ (up to a normalization factor). From (4.7.140), (4.7.145), we are then led to consider the function

$$A\psi(z) = -\frac{i}{\sqrt{2}}\left[\frac{\mathrm{d}}{\mathrm{d}z} - \pi \cot \pi z\right] \sin 2\pi z$$

$$= i\pi\sqrt{2} \sin^2 \pi z \qquad (4.7.150)$$

(vanishing at $z = 0, 1$) which is square-integrable, hence the non-zero eigenvalue $E = 3\pi^2/2$ is also in the discrete spectrum of H_-.

The formal rules spelled out above about a zero eigenvalue versus a non-zero $E \neq 0$ eigenvalue are, in general, true modulo some mathematical subtleties. A careful analysis of this involving all subtleties, with the boundary conditions at hand, for all general cases is quite tedious, and it is relatively far easier to examine a pair of supersymmetric Hamiltonian partners at a time. The moral of the formal analysis at the end of this section given through (4.7.135)–(4.7.140) is this. By studying the eigenvalue problem of a simple Hamiltonian, as in (4.7.146), one obtains information about the eigenvalues of a relatively much more complicated Hamiltonian, as in (4.7.147), with little extra work!

Much work has been done on the role of supersymmetry in quantum physics and not just on the eigenvalue problem and we refer the reader to the rapidly developing literature on this subject in physics journals. We will have ample opportunities to apply notions of supersymmetry in various other chapters as well.

Problems

4.1. Consider the spherically symmetric potential $V(r) = -\lambda e^{-r/R}$, $\lambda > 0$, $R > 0$ providing an elementary and rather formal model for the deuteron, as a loosely bound-state of a proton and a neutron with a binding energy $E \simeq -2.23$ MeV, with $\ell = 0$.
   (i) By choosing $N_0 = 1$ for $\ell = 0$, in (4.5.86) providing exactly one bound-state, obtain a lower bound for λ as a function of $\hbar^2/\mu R^2$.
   (ii) Find the exact solution for the ground-state ($\ell = 0$).
   (iii) Taking, as input data, $R = 2.180$ fm (fm $= 10^{-15}$ m), $E = -2.23$ MeV with the reduced mass of the deuteron $\mu = m_p m_n/(m_p + m_n)$, obtain a value for λ (and the potential energy "depth" U_0 in (4.2.89)) admitting only one bound state. Is your estimate consistent with the lower bound in (i)?

4.2. For the Yukawa potential $V(r) = -\lambda e^{-r/R}/(r/R)$, in three dimensions, with $\lambda > 0$, $R > 0$, choose $2\mu R^2/\hbar^2 = 3/\lambda$. Using a radial trial wavefunction $\psi(r) = A e^{-r/R_0}$ for $\ell = 0$, with respect to the measure $r^2 dr$, derive an upper bound for the ground-state energy ($\ell = 0$) as a function of λ, by optimizing over the parameter R/R_0. What can you say about the maximum number of bound-states this potential can ever have for $\ell = 0$?

4.3. For the Hamiltonian

$$H = -\frac{1}{2}\frac{d^2}{dz^2} + \pi^2 \cot^2(\pi z), \quad 0 < z < 1$$

in one dimension, with a confining potential, expressed in terms of a dimensionless variable z, and in some suitable energy unit, find the eigenvalues and the eigenvectors for this Hamiltonian. [The reader will be asked to reconsider this Hamiltonian in Problem 4.16 in the light of supersymmetry in reference to the pair of Hamiltonians in (4.7.146), (4.7.147) for direct comparison.] You may express the solutions for the eigenvectors in terms of hypergeometric functions (see, e.g., Arfken and Weber (1995) for the properties of these functions.)

4.4. Show that in two dimensions, the trace $\int d^2 x \, \langle x | A^\rho | x \rangle$ in (4.5.42) for $\rho = 1$, where A is given in (4.5.27), is infinite which necessitates another choice for ρ, e.g., $\rho = 3/2$ as in (4.5.93).

4.5. Prove the inequality in (4.5.93), where A is given in (4.5.27). [This is a hard nut to crack.]

4.6. Derive the general expression in (4.6.8) for the expectation value of the kinetic energy T which is also valid for $\nu = 2, 1$.

4.7. As in Problem 4.6, derive (4.6.11) which is valid for $\nu = 3, 2, 1$.

4.8. Extend the analyses in deriving a lower bound to T in §4.6 for multi-particle systems in $\nu = 2, 1$ dimensions as done there for $\nu = 3$. Can you extend the analysis also to higher dimensions?

4.9. Consider two possible functions $f(z)$, $g(z)$, satisfying formally the time-independent Schrödinger equation in one-dimension

$$\left[-\frac{1}{2}\frac{d^2}{dz^2} + V(z)\right] y(z) = E y(z)$$

expressed in terms of a dimensionless variable, where E is a real number (the eigenvalue).

(i) Show that this equation leads to the constraint

$$g'(z) f(z) - f'(z) g(z) = \alpha$$

where α is some constant which depends on the boundary conditions to be imposed on the functions. [Note that this property is consistent with the Hermiticity condition of the Hamiltonian in question.]

(ii) Show that for $f'(z)/f(z) = -w(z)$, where $w(z)$ is some given function, $g(z)$ satisfies the differential equation

$$g'(z) + w(z)g(z) = \frac{\alpha}{f(z)}$$

and hence the homogeneous solution of this equation coincides with the one satisfied by $f(z)$.

(iii) Show that the general solution for $g(z)$ in part (ii) is given by

$$g(z) = \alpha f(z) \int^z \frac{dz}{f^2(z)} + \beta f(z)$$

where β is a constant. That is, given a formal solution $f(z)$ of the Schrödinger equation above, then any other formal solution is related to $f(z)$ as given above. Also for the boundary conditions which specify that $\alpha \equiv 0$, the above shows that the formal solution $f(z)$ is unique.

4.10. Show that any Hamiltonian $H = \mathbf{p}^2/2m + V(\mathbf{x})$, which has a finite ground-state energy E_0 with ground-state wavefunction $\psi_0(\mathbf{x})$, may be rewritten as $H = \mathbf{A}^\dagger \cdot \mathbf{A} + E_0$, where

$$\mathbf{A} = \frac{1}{\sqrt{2m}}\left[\mathbf{p} + \frac{i\hbar(\boldsymbol{\nabla}\psi_0)}{\psi_0}\right]$$

[Note that, in one-dimension $\hbar(\partial\psi_0/\partial x)/\psi_0 = w(x)$ in (4.7.45), for $\psi_0(x)$ square-integrable.]

4.11. The *Virial Theorem.* Consider the Hamiltonian of a single particle $H = H_0 + V(\mathbf{x})$, where $H_0 = \mathbf{p}^2/2m$, and the eigenvalue equation $H\psi = E\psi$. Suppose that $\psi(\mathbf{x})$ is non-vanishing in a region $\mathcal{R}$, of volume v, only. $\mathcal{R}$ may be of finite or of infinite extension. For example, for a particle confined to a rigid sphere of finite radius, $\mathcal{R}$ is of finite extension.

(i) Show first that for a given function $\phi(\mathbf{x})$

$$\int_v d^3\mathbf{x}\, \psi^* H_0 \phi = \int_v d^3\mathbf{x}\, (H_0\psi^*)\phi$$
$$- \frac{\hbar^2}{2m} \oint_S d\mathbf{S} \cdot [\psi^*\boldsymbol{\nabla}\phi - (\boldsymbol{\nabla}\psi^*)\phi]$$

where the volume v is bounded by the closed surface S, and that

$$(\mathbf{x}\cdot\boldsymbol{\nabla}\, H_0\, \psi) = -2\, H_0\psi + H_0(\mathbf{x}\cdot\boldsymbol{\nabla}\psi).$$

(ii) Upon multiplying the eigenvalue equation $[H_0 + V(\mathbf{x})]\psi(\mathbf{x}) = E\psi(\mathbf{x})$ from the left by $\psi^*(\mathbf{x}\cdot\boldsymbol{\nabla})$, and using the results in (i) show that

$$2\langle H_0\rangle_v - \langle \mathbf{x}\cdot\boldsymbol{\nabla}V\rangle_v = \frac{\hbar^2}{2m} \oint_S d\mathbf{S}\cdot[(\boldsymbol{\nabla}\psi^*)(\mathbf{x}\cdot\boldsymbol{\nabla}\,\psi) - \psi^*\boldsymbol{\nabla}(\mathbf{x}\cdot\boldsymbol{\nabla}\,\psi)]$$

where $\langle A\rangle_v = \int_v d^3\mathbf{x}\, \psi^* A\, \psi$.

(iii) For a multi-particle system of N particles, with Hamiltonian

$$H = \sum_{i=1}^N \mathbf{p}_i^2/2m_i + V(\mathbf{x_1},\ldots,\mathbf{x_N}),$$

and eigenvalue equation

$$H\psi(\mathbf{x}_1\varepsilon_1,\ldots,\mathbf{x}_N\varepsilon_N) = E\psi(\mathbf{x}_1\varepsilon_1,\ldots,\mathbf{x}_N\varepsilon_N),$$

where $(\varepsilon_1,\ldots,\varepsilon_N) = \varepsilon$ are, in general, extra labellings needed to specify the particles, such as spin, charge, ..., show that the virial theorem in (ii) is readily generalized to

$$2\left\langle \sum_{j=1}^N \mathbf{p}_j^2/2m_j \right\rangle - \left\langle \sum_{j=1}^N (\mathbf{x}_j\cdot\boldsymbol{\nabla}_j V) \right\rangle$$
$$= \sum_\varepsilon \sum_{k=1}^N \frac{\hbar^2}{2m_k} \int d^3\mathbf{x}_1\ldots d^3\mathbf{x}_{k-1}\, d^3\mathbf{x}_{k+1}\ldots d^3\mathbf{x}_N$$
$$\times \oint d\mathbf{S}_k \cdot \left[(\boldsymbol{\nabla}_k\,\psi^*)\left(\sum_{j=1}^N \mathbf{x}_j\cdot\boldsymbol{\nabla}_j\psi\right) - \psi^*\boldsymbol{\nabla}_k\left(\sum_{j=1}^N \mathbf{x}_j\cdot\boldsymbol{\nabla}_j\psi\right)\right]$$

where, in an obvious notation, the integrations are over the configurational coordinates.

[Ref: Marc and McMillan (1985).]

4.12. Verify the commutation relation (4.7.33), and show that the generators in (4.7.34) lead from (4.7.3) to the Hamiltonian in (4.7.35).

4.13. For the Fermi operators in (4.7.36), verify the underlying anti-commutation relationships given there. Then show that the generators in (4.7.37) lead to the Hamiltonian in (4.7.38).

4.14. Show that, as direct generalization of the two Fermi operators defined in (4.7.36) as 4×4 matrices, one may introduce three Fermi operators, as 8×8 matrices,

$$\chi_1 = \begin{pmatrix} \Psi_1 & 0 \\ 0 & -\Psi_1 \end{pmatrix}, \quad \chi_2 = \begin{pmatrix} \Psi_2 & 0 \\ 0 & -\Psi_2 \end{pmatrix}, \quad \chi_3 = \begin{pmatrix} 0 & 0 \\ 1 & 0 \end{pmatrix}$$

where Ψ_1, Ψ_2 are given in (4.7.36), and satisfy the anti-commutation relations

$$\{\chi_i, \chi_j\} = \mathbf{0}, \quad \{\chi_i^\dagger, \chi_j^\dagger\} = \mathbf{0}, \quad \{\chi_i, \chi_j^\dagger\} = \mathbf{1}\delta_{ij}.$$

Use these operators to generate supersymmetric Hamiltonians in higher dimensions.

4.15. Study the eigenvalue problem for the Hamiltonian $H_+(\xi)$ with superpotential $w(z,\xi)$ in (4.7.61) with shape invariant partner potentials satisfying (4.7.63). [Determine, in particular, the number of the eigenvalues.]

4.16. From the elementary eigenvalue problem associated with the Hamiltonian H_+ in (4.7.146) for a particle confined to a one dimensional box with constant potential $-\pi^2/2$, use supersymmetry to generate the eigenvalues and eigenvectors for the far more complicated Hamiltonian H_- in (4.7.147). Compare your results with the ones obtained in Problem 4.3.

4.17. Consider the superpotential

$$w(z) = \frac{1}{z} - z, \quad 0 < z < \infty.$$

(i) Derive the expressions for the supersymmetric partner potentials V_+, V_-.

(ii) Find the ground-state of H_+, corresponding to the eigenvalue 0. Is the value 0 also the ground-state energy of H_-? Show your analysis in detail.

(iii) Refer to Chapter 6, if needed, to find the eigenvalues and eigenvectors of H_+ with the latter vanishing for $z \to 0, z \to \infty$. Do all the eigenvalues $E \neq 0$ of H_+ also belong to H_-?

5
Angular Momentum Gymnastics

Under a coordinate rotation (§2.1) by an angle φ about a unit vector $\mathbf{n}$, a state $|\psi\rangle$ obeys the transformation law (see (2.3.1), (2.3.43))

$$|\psi'\rangle = \exp\left(\frac{\mathrm{i}}{\hbar}\varphi\,\mathbf{n}\cdot\mathbf{J}\right)|\psi\rangle \tag{5.1}$$

where $\mathbf{J}$ is the angular momentum operator (§2.7), satisfying the commutation relations

$$\left[J^i, J^j\right] = \mathrm{i}\hbar\varepsilon^{ijk}J^k \tag{5.2}$$

$i, j = 1, 2, 3$.

By defining

$$\mathbf{J}^2 = (J^1)^2 + (J^2)^2 + (J^3)^2 \tag{5.3}$$

we note that

$$\left[\mathbf{J}^2, J^i\right] = 0, \quad i = 1, 2, 3. \tag{5.4}$$

Since the different components J^i do not commute, but they all commute with $\mathbf{J}^2$, we may find simultaneous eigenstates of $\mathbf{J}^2$ and one, and only one, of the components J^i which is traditionally taken to be J^3.

We denote the simultaneous eigenstates of $\mathbf{J}^2$, J^3 by $|\alpha,\beta\rangle$:

$$\mathbf{J}^2|\alpha,\beta\rangle = \alpha|\alpha,\beta\rangle \tag{5.5}$$

$$J^3|\alpha,\beta\rangle = \beta|\alpha,\beta\rangle. \tag{5.6}$$

In a given physical problem, there may be, in general, in addition to $\mathbf{J}^2$, J^3, more observables constituting together a *complete* set of compatible observables. In such cases, the simultaneous eigenstates of the corresponding complete set of commuting operators, will depend on additional parameters to α and β, which for simplicity of notation will be suppressed. We suppose that these additional operators, in the complete set of operators, each commutes with the angular momentum operator $\mathbf{J}$. The latter means that if we denote the simultaneous eigenstates of the complete set of observables, including $\mathbf{J}^2$, J^3, by $|\alpha, \beta, \kappa\rangle$, with κ denoting collectively the additional quantum numbers needed to specify the eigenstates, then

$$\langle \alpha', \beta', \kappa'|\mathbf{J}|\alpha, \beta, \kappa\rangle = 0 \quad \text{unless } \alpha' = \alpha \text{ and } \kappa' = \kappa \tag{5.7}$$

where we recall that α is the eigenvalue of $\mathbf{J}^2$ (see (5.5)).

To study the eigenvalue problem (5.5), (5.6), the introduction of the so-called ladder operators

$$J_\pm = J^1 \pm iJ^2 \tag{5.8}$$

turns out to be very useful. They satisfy, in particular, the following relations

$$[\mathbf{J}^2, J_\pm] = 0 \tag{5.9}$$

$$[J^3, J_\pm] = \pm\hbar J_\pm \tag{5.10}$$

$$[J_+, J_-] = 2\hbar J^3 \tag{5.11}$$

$$J_\mp J_\pm = \mathbf{J}^2 - (J^3)^2 \mp \hbar J^3 \tag{5.12}$$

$$J_- = J_+^\dagger$$

and from (5.7)

$$\langle \alpha', \beta', \kappa'|J_\pm|\alpha, \beta, \kappa\rangle = 0 \quad \text{unless } \alpha' = \alpha, \ \kappa' = \kappa. \tag{5.13}$$

Finally we summarize pertinent properties established in §2.7, §2.8 for arbitrary spins.

The general structure of the angular momentum $\mathbf{J}$ is given in (2.7.6) to be

$$\mathbf{J} = \mathbf{X} \times \mathbf{P} + \mathbf{S} \tag{5.14}$$

where $\mathbf{X}$ denotes the position operator associated with a particle, or of the center of mass position of a system of particles, and, the spin, $\mathbf{S}$ is the translational independent contribution to $\mathbf{J}$,

$$[P^i, S^j] = 0 \tag{5.15}$$

and

$$[X^i, S^j] = 0 \tag{5.16}$$

$$[S^i, S^j] = i\hbar \varepsilon^{ijk} S^k \tag{5.17}$$

$$[\mathbf{S}^2, S^i] = 0. \tag{5.18}$$

Arbitrary spins $s \geqslant 1/2$, may be described by a completely symmetric spinor $\psi^{a_1\cdots a_{2s}}$ (§2.8) and, under a coordinate rotation by an angle φ about a unit vector $\mathbf{n}$, it transforms as given in (2.8.1). For $s = 0$, we have

$$\psi'(\mathbf{x}') = \psi(\mathbf{x}) \tag{5.19}$$

where $\mathbf{x}' = R\mathbf{x}$.

Under a coordinate rotation by an angle 2π, about any unit vector $\mathbf{n}$, one has the profound relation (2.8.68):

$$\psi'^{a_1\cdots a_{2s}}(\mathbf{x}) = (-1)^{2s}\psi^{a_1\cdots a_{2s}}(\mathbf{x}), \tag{5.20}$$

for all $s = 0, 1/2, 1, 3/2, \ldots$. This relation emphasizes the odd nature character of half-odd integral spins. Only a rotation by 4π (or integer multiples of it) restores a $(+1)$ value for the corresponding transformed spinors for such spin values. The phase change of (-1) occurring for half-odd spin values is observable and will be discussed in §8.12.

The unitarity relation (2.8.24), (2.8.25), between spinors and wavefunctions should be noted. The matrix elements of the operator $\exp\left(\frac{i}{\hbar}\varphi\,\mathbf{n}\cdot\mathbf{S}\right)$ with respect to the simultaneous eigenstates of S^3, $\mathbf{S}^2$ (see (2.8.45), (2.8.46)) are given in (2.8.51).

This chapter covers in fairly details most of the aspects dealing with the intricacies of angular momentum and is quite complete. The eigenvalue problem (5.5), (5.6) is discussed in §5.1 followed by determining the matrix elements of finite arbitrary rotations in §5.2. Orbital angular momentum and spin are covered, respectively, in §5.3 and §5.4. The vectorial addition of two independent angular momenta and the respective rules are worked out in §5.5, §5.6. So called vector and, more generally, tensor operators are treated in §5.7, §5.8 which lie in the heart of the theory of angular momentum. In §5.9, we consider the problem of combining several independent angular momenta. The final section, §5.10, deals with the definition and construction of single and two particle states of arbitrary spins in terms of angular momentum components which have become quite useful over the years.

The association of the word "gymnastics" with a topic such as angular momentum involving "infinite" details is due to A. S. Wightman and we find it quite appropriate to include it in the heading of this chapter. This in no way, however, is meant to diminish the importance of the subject of angular momentum.

5.1 The Eigenvalue Problem

We consider the eigenvalue problem

$$\mathbf{J}^2\,|\alpha,\beta\rangle = \alpha\,|\alpha,\beta\rangle \tag{5.1.1}$$

$$J^3 |\alpha, \beta\rangle = \beta |\alpha, \beta\rangle \tag{5.1.2}$$

which on account of $\mathbf{J}^2 \geqslant (J^3)^2$, imply that

$$\alpha \geqslant \beta^2. \tag{5.1.3}$$

Since $\mathbf{J}^2$, J^i satisfy (see (5.2), (5.4)) exactly the *same* commutation relations as $\mathbf{S}^2$, S^i (see (2.8.46), (2.8.47)), the solution of the eigenvalue problem in (5.1.1), (5.1.2) may be formally inferred from (2.8.46), (2.8.47). The following independent and general derivation is, however, illuminating.

Let $\overline{\beta}$ denote the largest possible value for β satisfying (5.1.3). Then from (5.10)

$$J^3 J_+ |\alpha, \overline{\beta}\rangle = (\overline{\beta} + \hbar) J_+ |\alpha, \overline{\beta}\rangle \tag{5.1.4}$$

which is consistent with the fact that $\overline{\beta}$ is the largest possible value for β satisfying (5.1.3) only if $J_+ |\alpha, \overline{\beta}\rangle$ is the *zero* vector. The latter, in particular, implies that

$$J_- J_+ |\alpha, \overline{\beta}\rangle = 0 \tag{5.1.5}$$

which from (5.11), (5.1.1), (5.1.2) gives

$$\left(\alpha - \overline{\beta}^2 - \hbar \overline{\beta}\right) |\alpha, \overline{\beta}\rangle = 0 \tag{5.1.6}$$

and hence

$$\alpha = \overline{\beta} \left(\overline{\beta} + \hbar\right) \tag{5.1.7}$$

since $|\alpha, \overline{\beta}\rangle$ is *not* the zero vector.

Let $\underline{\beta}$ denote the smallest possible value for β satisfying (5.1.3). Then from (5.10),

$$J^3 J_- |\alpha, \underline{\beta}\rangle = (\underline{\beta} - \hbar) J_- |\alpha, \underline{\beta}\rangle \tag{5.1.8}$$

leading, in analogy to (5.1.7), to

$$\alpha = \underline{\beta} \left(\underline{\beta} - \hbar\right). \tag{5.1.9}$$

Upon adding $\underline{\beta} \overline{\beta}$ to both sides of (5.1.9), (5.1.7) and comparing these equations, we obtain

$$(\overline{\beta} + \underline{\beta})(\overline{\beta} - \underline{\beta} + \hbar) = 0. \tag{5.1.10}$$

Since $\underline{\beta} \leqslant \overline{\beta}$, this gives the unique solution

$$\underline{\beta} = -\overline{\beta}. \tag{5.1.11}$$

By repeated applications of (5.10), with the upper sign, k times lead to

$$J^3 (J_+)^k |\alpha, \underline{\beta}\rangle = (\underline{\beta} + \hbar k)(J_+)^k |\alpha, \underline{\beta}\rangle \tag{5.1.12}$$

which imply that there exists a largest non-negative integer $\overline{k}$ such that

5.1 The Eigenvalue Problem

$$\overline{\beta} = \underline{\beta} + \hbar \overline{k} \tag{5.1.13}$$

or from (5.1.11) to

$$\overline{\beta} = \frac{\hbar \overline{k}}{2} \equiv \hbar j, \qquad \underline{\beta} = -\hbar j \tag{5.1.14}$$

where we have denoted the latter by $\hbar j$. Accordingly, from (5.1.7), we may write

$$\alpha = \hbar^2 j (j+1) \tag{5.1.15}$$

and due to the very nature of $\overline{k}$ as a non-negative integer, the possible values of j are restricted to

$$j = 0, \frac{1}{2}, 1, \frac{3}{2}, \ldots. \tag{5.1.16}$$

On the other hand, from (5.1.12), (5.1.13) we may write

$$\beta = \underline{\beta} + \hbar k = -\hbar j + \hbar k \tag{5.1.17}$$

and since $k = 0, 1, \ldots, \overline{k} = 2j$, we have for the possible values of β:

$$-\hbar j, -\hbar j + \hbar, \ldots, \hbar j - \hbar, \hbar j. \tag{5.1.18}$$

Upon setting $\beta = \hbar m$, and using the standard labelling for the eigenstates of $\mathbf{J}^2$, J^3 by $|j, m\rangle$, instead of $|\alpha, \beta\rangle$, we may summarize the eigenvalue problem through

$$\mathbf{J}^2 |j, m\rangle = \hbar^2 j (j+1) |j, m\rangle \tag{5.1.19}$$

$$J^3 |j, m\rangle = \hbar m |j, m\rangle \tag{5.1.20}$$

$$\langle j', m' | j, m \rangle = \delta_{j'j} \delta_{m'm} \tag{5.1.21}$$

where the possible values of j are given in (5.1.16), and, from (5.1.18), the allowed m values are

$$m = -j, -j + 1, \ldots, j - 1, j. \tag{5.1.22}$$

With the assumption that the additional operators, to $\mathbf{J}^2$, J^3, in the complete set of operators, commute with the angular momentum operator $\mathbf{J}$, and hence also with $J_\pm$, we may identify, by using in the process (5.7),

$$J_\pm |j, m\rangle = C_\pm (j, m) |j, m \pm 1\rangle \tag{5.1.23}$$

and obtain from (5.11), (5.12),

$$\langle j, m | J_\mp J_\pm | j, m \rangle = \hbar^2 (j \mp m)(j \pm m + 1) \tag{5.1.24}$$

leading, with a standard phase convention, to

$$J_\pm |j,m\rangle = \hbar\sqrt{(j \mp m)(j \pm m + 1)}\, |j, m \pm 1\rangle. \tag{5.1.25}$$

By repeated application of J_- to $|j,m\rangle$ k times, we obtain the useful expression

$$(J_-)^k |j,m\rangle = \begin{cases} (\hbar)^k \sqrt{\dfrac{(j+m)!\,(j-m+k)!}{(j-m)!\,(j+m-k)!}}\, |j, m-k\rangle, & k = 0, 1, \ldots, j+m \\ 0, & k = j+m+1, j+m+2, \ldots. \end{cases} \tag{5.1.26}$$

That is, k is restricted to non-negative integers such that the arguments of the factorials are non-negative.

Similarly,

$$(J_+)^k |j,m\rangle = \begin{cases} (\hbar)^k \sqrt{\dfrac{(j-m)!\,(j+m+k)!}{(j+m)!\,(j-m-k)!}}\, |j, m+k\rangle, & k = 0, 1, \ldots, j-m \\ 0, & k = j-m+1, j-m+2, \ldots. \end{cases} \tag{5.1.27}$$

For further reference, we also note from the definitions in (5.8) and the relations in (5.1.25) that

$$J^1 |j,m\rangle = \frac{\hbar}{2}\sqrt{(j-m)(j+m+1)}\, |j, m+1\rangle$$
$$+ \frac{\hbar}{2}\sqrt{(j+m)(j-m+1)}\, |j, m-1\rangle \tag{5.1.28}$$

$$J^2 |j,m\rangle = -\mathrm{i}\frac{\hbar}{2}\sqrt{(j-m)(j+m+1)}\, |j, m+1\rangle$$
$$+ \mathrm{i}\frac{\hbar}{2}\sqrt{(j+m)(j-m+1)}\, |j, m-1\rangle. \tag{5.1.29}$$

5.2 Matrix Elements of Finite Rotations

Since the angular momentum components J^1, J^2, J^3 satisfy the same commutation relations as the spin components S^1, S^2, S^3, and the rules for the *action* of the operators J^i on the states $|j,m\rangle$ (see (5.1.20), (5.1.28), (5.1.29)) are *identical* to the action of the operators S^i on the states $|s, m_s\rangle$ (see (2.8.43)–(2.8.45), (2.8.22)), we may conclude directly from (2.8.51), that the explicit expressions for the matrix elements of the operator $\exp\left(\dfrac{\mathrm{i}\varphi}{\hbar}\mathbf{n}\cdot\mathbf{J}\right)$, with respect to the states $|j,m\rangle$, is given by

5.2 Matrix Elements of Finite Rotations

$$\left\langle j, m \left| \exp\left(\frac{i\varphi}{\hbar} \mathbf{n} \cdot \mathbf{J}\right) \right| j', m' \right\rangle$$

$$= \delta_{j'j} \sum_q \frac{\sqrt{(j+m)!(j-m)!(j+m')!(j-m')!}}{(j+m-q)!(m'-m+q)!q!(j-m'-q)!}$$

$$\times \left([A]^{11}\right)^{j+m-q} \left([A]^{21}\right)^{m'-m+q} \left([A]^{12}\right)^q \left([A]^{22}\right)^{j-m'-q} \tag{5.2.1}$$

where the sum is over all non-negative integers q such that the arguments of the factorials are non-negative integers. Here

$$A = \begin{pmatrix} \cos\frac{\varphi}{2} + in^3 \sin\frac{\varphi}{2} & i\left(n^1 - in^2\right)\sin\frac{\varphi}{2} \\ i\left(n^1 + in^2\right)\sin\frac{\varphi}{2} & \cos\frac{\varphi}{2} - in^3 \sin\frac{\varphi}{2} \end{pmatrix} \tag{5.2.2}$$

and $\mathbf{n} = \left(n^1, n^2, n^3\right)$.

We rewrite the matrix elements of finite rotations in (5.2.1) in the equivalent description in terms of the so-called *Euler* angles.

To the above end, we define consecutive rotations specified by the Euler angles α, β, γ in the following manner:

- A counter-clockwise rotation of the coordinate system $x\,y\,z$ about the z-axis by an angle α, defining the coordinate system $x'\,y'\,z'$,
- the above is followed by a counter-clockwise rotation of the coordinate system $x'\,y'\,z'$ about the y'-axis by an angle β, defining the coordinate system $x''\,y''\,z''$,
- finally, this is followed by a counter-clockwise rotation of the coordinate system $x''\,y''\,z''$ about the z''-axis by an angle γ, defining the final coordinate system $\bar{x}\,\bar{y}\,\bar{z}$.

The above consecutive rotations are illustrated in Figure 5.1 below.

A vector $\mathbf{x}$ with components $(x, y, z) \left(= \left(x^1, x^2, x^3\right)\right)$ in the coordinate system x, y, z, is described to have the components $\bar{x}, \bar{y}, \bar{z}$ in the final coordinate system $\bar{x}\,\bar{y}\,\bar{z}$, as follows:

$$\begin{pmatrix} \bar{x} \\ \bar{y} \\ \bar{z} \end{pmatrix} = R(\alpha, \beta, \gamma) \begin{pmatrix} x \\ y \\ z \end{pmatrix} \tag{5.2.3}$$

and

$$R(\alpha, \beta, \gamma) = \begin{pmatrix} \cos\gamma & \sin\gamma & 0 \\ -\sin\gamma & \cos\gamma & 0 \\ 0 & 0 & 1 \end{pmatrix} \begin{pmatrix} \cos\beta & 0 & -\sin\beta \\ 0 & 1 & 0 \\ \sin\beta & 0 & \cos\beta \end{pmatrix} \begin{pmatrix} \cos\alpha & \sin\alpha & 0 \\ -\sin\alpha & \cos\alpha & 0 \\ 0 & 0 & 1 \end{pmatrix}$$

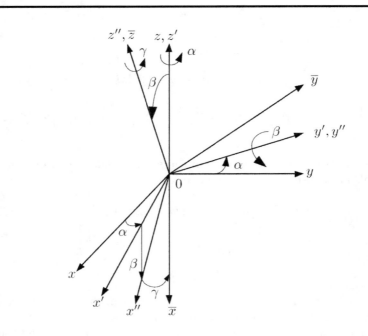

Fig. 5.1. Definition of the Euler angles (α, β, γ). The coordinate systems resulting from the *rotations* are as follows: $xyz \xrightarrow{\alpha} x'y'z' \xrightarrow{\beta} x''y''z'' \xrightarrow{\gamma} \bar{x}\bar{y}\bar{z}$, with a rotation by an angle α about the z-axis, followed by a rotation by an angle β about the y'-axis, followed by a rotation by an angle γ about the z''-axis, respectively.

$$= \begin{pmatrix} \cos\beta\cos\alpha\cos\gamma - \sin\alpha\sin\gamma & \cos\beta\sin\alpha\cos\gamma + \cos\alpha\sin\gamma & -\sin\beta\cos\gamma \\ -\cos\beta\cos\alpha\sin\gamma - \sin\alpha\cos\gamma & -\cos\beta\sin\alpha\sin\gamma + \cos\alpha\cos\gamma & \sin\beta\sin\gamma \\ \sin\beta\cos\alpha & \sin\beta\sin\alpha & \cos\beta \end{pmatrix}$$

(5.2.4)

Upon the comparison of the matrix elements of $R(\alpha, \beta, \gamma)$, given in (5.2.4), with those given in (2.1.4), corresponding to a counter-clockwise rotation by an angle φ about a unit vector

$$\mathbf{n} = (\cos\phi\sin\theta, \sin\phi\sin\theta, \cos\theta), \qquad (5.2.5)$$

one obtains the equalities

$$\sin\frac{\beta}{2} = \sin\theta \sin\frac{\varphi}{2} \qquad (5.2.6)$$

$$i e^{i(\alpha-\gamma)/2} = e^{i\phi} \qquad (5.2.7)$$

$$\cos\frac{\beta}{2} e^{i(\alpha+\gamma)/2} = \left(\cos\frac{\varphi}{2} + i\cos\theta \sin\frac{\varphi}{2}\right). \tag{5.2.8}$$

Accordingly, the matrix A in (5.2.2) may be rewritten in terms of the Euler angles in the form

$$A = \begin{pmatrix} \cos\frac{\beta}{2} e^{i(\alpha+\gamma)/2} & \sin\frac{\beta}{2} e^{-i(\alpha-\gamma)/2} \\ -\sin\frac{\beta}{2} e^{i(\alpha-\gamma)/2} & \cos\frac{\beta}{2} e^{-i(\alpha+\gamma)/2} \end{pmatrix} \tag{5.2.9}$$

from which (5.2.1) is spelled out to be

$$\left\langle j, m \left| \exp\left(\frac{i\varphi}{\hbar} \mathbf{n} \cdot \mathbf{J}\right) \right| j', m' \right\rangle$$

$$= \delta_{j'j} \sum_q \frac{\sqrt{(j+m)!(j-m)!(j+m')!(j-m')!}}{(j+m-q)!(m'-m+q)!q!(j-m'-q)!}$$

$$\times (-1)^{m'-m+q} \left(\cos\frac{\beta}{2}\right)^{2j+m-m'-2q} \left(\sin\frac{\beta}{2}\right)^{m'-m+2q} e^{im'\alpha} e^{im\gamma} \tag{5.2.10}$$

as expressed completely in terms of the Euler angles α, β, γ, where the sum is over all non-negative integers q such that the arguments of the factorials are non-negative integers.

Upon the comparison of (5.2.10) with (2.8.52) and using, in the process, (5.1.20), (5.2.7), we also have the identity

$$\left\langle j, m \left| \exp\left(\frac{i\varphi}{\hbar} \mathbf{n} \cdot \mathbf{J}\right) \right| j, m' \right\rangle = \left\langle j, m \left| \left(\exp\frac{i\gamma}{\hbar} J^3\right) \left(\exp\frac{i\beta}{\hbar} J^2\right) \right. \right.$$

$$\left. \times \left(\exp\frac{i\alpha}{\hbar} J^3\right) \right| j, m' \right\rangle \tag{5.2.11}$$

expressing the relationships between the two equivalent schemes, where all the rotations specified on the right-hand side of (5.2.11) are about the *original* axes $0z, 0y, 0z$, respectively.

We use the notation

$$\left\langle j, m \left| \left(\exp\frac{i\gamma}{\hbar} J^3\right) \left(\exp\frac{i\beta}{\hbar} J^2\right) \left(\exp\frac{i\alpha}{\hbar} J^3\right) \right| j, m' \right\rangle \equiv D^{(j)}_{mm'}(\alpha, \beta, \gamma) \tag{5.2.12}$$

and

$$\left\langle j, m \left| \left(\exp\frac{i\beta}{\hbar} J^2\right) \right| j, m' \right\rangle \equiv D^{(j)}_{mm'}(0, \beta, 0) \equiv d^{(j)}_{mm'}(\beta). \tag{5.2.13}$$

The $D^{(j)}_{mm'}(\alpha,\beta,\gamma)$ have the following property upon complex conjugation,

$$\left(D^{(j)}_{mm'}(\alpha,\beta,\gamma)\right)^* = (-1)^{m'-m} D^{(j)}_{-m,-m'}(\alpha,\beta,\gamma) \qquad (5.2.14)$$

as follows from (5.2.10).

Useful orthogonality relations of the $D^{(j)}_{mm'}(\alpha,\beta,\gamma)$ will be derived in §5.5 when combining two commuting angular momenta.

The following relations for the matrix elements $d^{(j)}_{mm'}(\beta)$ in (5.2.13) easily follow from the explicit expression given in (5.2.10)

$$d^{(j)}_{mm'}(\pi - \beta) = (-1)^{j+m'} d^{(j)}_{-m,m'}(\beta) \qquad (5.2.15)$$

$$d^{(j)}_{mm'}(\pi) = (-1)^{j-m} \delta_{m,-m'} \qquad (5.2.16)$$

$$d^{(j)}_{mm'}(-\pi) = (-1)^{j+m} \delta_{m,-m'} \qquad (5.2.17)$$

$$d^{(j)}_{mm'}(2\pi) = (-1)^{2j} \delta_{m,m'}. \qquad (5.2.18)$$

For j as obtained from the addition of an orbital angular momentum and spin, with quantum numbers ℓ and s, as studied in §5.5, $j = |\ell - s|, |\ell - s| + 1, \ldots, \ell + s$. Then for $\ell \geqslant s$, $j = \ell - s + k$, $k = 0, 1, \ldots, 2s$ and $(-1)^{2j} = (-1)^{-2s} = (-1)^{2s}$. For $\ell < s$, $j = s - \ell + k$, $k = 0, 1, \ldots, 2\ell$, $(-1)^{2j} = (-1)^{2s}$. That is

$$d^{(j)}_{mm'}(2\pi) = (-1)^{2s} \delta_{mm'} \qquad (5.2.19)$$

in conformity with (5.20).

5.3 Orbital Angular Momentum

5.3.1 Transformation Theory

From §5.2, the orbital angular momentum $\mathbf{L}$ (see (2.7.6), (2.7.11)) is defined by

$$\mathbf{L} = \mathbf{X} \times \mathbf{P}. \qquad (5.3.1)$$

Conversely, with (2.7.43) providing the definition of the transformation of a state, in the coordinate description, under a coordinate rotation, we have

$$\psi'(\mathbf{x}') = \left(\exp\left(\frac{i}{\hbar}\boldsymbol{\omega}\cdot\mathbf{S}\right)\psi\right)(\mathbf{x}) \qquad (5.3.2)$$

with $\mathbf{x}' = R\mathbf{x}$, $\boldsymbol{\omega} = \varphi\mathbf{n}$ (§2.1), or

$$\psi'(\mathbf{x}) = \left(\exp\left(\frac{i}{\hbar}\boldsymbol{\omega}\cdot\mathbf{S}\right)\psi\right)(R^{-1}\mathbf{x}). \qquad (5.3.3)$$

For example, for infinitesimal $\delta\varphi$

$$R^{-1}\mathbf{x} = \mathbf{x} + \delta\boldsymbol{\omega} \times \mathbf{x}$$

(see (2.1.13) with $\delta\boldsymbol{\omega} \to -\delta\boldsymbol{\omega}$), and

$$\psi(\mathbf{x} + \delta\boldsymbol{\omega} \times \mathbf{x}) = \psi(\mathbf{x}) + \frac{i}{\hbar}\delta\boldsymbol{\omega} \cdot \mathbf{L}\psi(\mathbf{x}) \tag{5.3.4}$$

by carrying out a Taylor expansion, where

$$\mathbf{L} = \mathbf{x} \times (-i\hbar\boldsymbol{\nabla}) \tag{5.3.5}$$

provides the coordinate description for $\mathbf{L}$. From (5.3.3), (5.3.4) give for infinitesimal $\delta\varphi$

$$\psi'(\mathbf{x}) = \left(\left[1 + \frac{i}{\hbar}\delta\boldsymbol{\omega} \cdot (\mathbf{S} + \mathbf{L})\right]\psi\right)(\mathbf{x}) \tag{5.3.6}$$

which upon comparison with the transformation law of a state in (5.1), (2.3.43), for infinitesimal $\delta\varphi$, provides the definition of $\mathbf{J}$ in terms of $\mathbf{L}$ and $\mathbf{S}$ in (2.7.6).

In spherical coordinates (r, θ, ϕ), we have the following representations

$$L^3 = -i\hbar\frac{\partial}{\partial\phi} \tag{5.3.7}$$

$$\mathbf{L}^2 = -\hbar^2\left[\frac{1}{\sin\theta}\frac{\partial}{\partial\theta}\sin\theta\frac{\partial}{\partial\theta} + \frac{1}{\sin^2\theta}\frac{\partial^2}{\partial\phi^2}\right] \tag{5.3.8}$$

and for the ladder operators (5.8),

$$L_\pm = \hbar e^{\pm i\phi}\left[i\cot\theta\frac{\partial}{\partial\phi} \pm \frac{\partial}{\partial\theta}\right]. \tag{5.3.9}$$

5.3.2 Half-Odd Integral Values?

We first formally show that according to the eigenvalue scheme for angular momentum developed in §5.1, the *orbital* angular momentum does not admit half-odd integral j values.[1] There is a long history associated with this problem beginning with some work by Pauli which, however, we will not go into it here.

With $m = j$, $k = 1$, (5.1.27) gives from (5.3.9),

$$0 = L_+ \langle\theta,\phi|j,j\rangle = \hbar e^{i\phi}\left[i\cot\theta\frac{\partial}{\partial\phi} + \frac{\partial}{\partial\theta}\right]\langle\theta,\phi|j,j\rangle \tag{5.3.10}$$

---

[1] We will eventually use the standard notation ℓ for the quantum number associated with the orbital angular momentum once it is seen below that half-odd integer values are not acceptable for the latter.

or
$$\left[i\cot\theta\frac{\partial}{\partial\phi}+\frac{\partial}{\partial\theta}\right]\langle\theta,\phi|j,j\rangle=0 \tag{5.3.11}$$

whose solution is
$$\langle\theta,\phi|j,j\rangle = A_j e^{ij\phi}(\sin\theta)^j \tag{5.3.12}$$

where
$$A_j = \sqrt{\frac{1}{2\pi}\frac{\Gamma(j+3/2)}{\Gamma(j+1)\Gamma(1/2)}} \tag{5.3.13}$$

and we have used the integral
$$\int_0^\pi d\theta\,(\sin\theta)^{2j+1} = \sqrt{\pi}\,\frac{\Gamma(j+1)}{\Gamma(j+3/2)} \tag{5.3.14}$$

for normalizing $\langle\theta,\phi|j,j\rangle$.

From (5.3.9), (5.3.12), we explicitly have
$$L_-\langle\theta,\phi|j,j\rangle = \hbar A_j e^{i(j-1)\phi}(\sin\theta)^{j-1}(-2)j[\cos\theta] \tag{5.3.15}$$

$$(L_-)^2\langle\theta,\phi|j,j\rangle = \hbar^2 A_j e^{i(j-2)\phi}(\sin\theta)^{j-2}(-2)^2$$
$$\times j\left[(j-1)\cos^2\theta - \frac{1}{2}\sin^2\theta\right]. \tag{5.3.16}$$

In particular, for $j=1/2$, (5.3.16) gives
$$(L_-)^2\langle\theta,\phi|1/2,1/2\rangle = -\frac{\hbar^2}{\pi}e^{-3i\phi/2}(\sin\theta)^{-3/2} \tag{5.3.17}$$

while (5.1.26), with $j=m=1/2$, $k=2$, states that (5.3.17) should be *zero* and hence, in particular, $L_-\langle\theta,\phi|j,j\rangle$ should be in the domain of definition of L_-. We will see later that such a contradiction does not arise for integral values of j for the orbital angular momentum. [The expression in (5.3.17) is not only different from zero but is also not even square-integrable over θ, i.e., $L_-\langle\theta,\phi|1/2,1/2\rangle$ is taken out of the space of square-integrable functions over θ, ϕ by the action of L_-.]

To proceed with other half-odd integral j values we consider the following. By repeated applications of L_- to $\langle\theta,\phi|j,j\rangle$, as in (5.3.15), (5.3.16), we have the following elementary limits

$$L_-\langle\theta,\phi|j,j\rangle \xrightarrow[\theta\to 0]{} \hbar A_j e^{i(j-1)\phi}(-2)j\theta^{j-1}$$

$$(L_-)^2\langle\theta,\phi|j,j\rangle \xrightarrow[\theta\to 0]{} \hbar^2 A_j e^{i(j-2)\phi}(-2)^2 j(j-1)\theta^{j-2}$$

$$\vdots$$

$$(L_-)^{2j}\langle\theta,\phi|j,j\rangle \xrightarrow[\theta\to 0]{} (\hbar)^{2j} A_j e^{-ij\phi}(-2)^{2j} j(j-1)\cdots(-j+1)(-j)\,\theta^{-j}.$$

That is, explicitly,

$$\lim_{\theta\to 0}\theta^j (L_-)^{2j}\langle\theta,\phi|j,j\rangle$$

$$= \begin{cases} (\hbar)^{2j} A_j\, e^{-ij\phi}(-2)^{2j}(-1)^{j+1/2}\Gamma^2(j+1)\pi^{-1}; & j=1/2, 3/2,\ldots \\ 0, & j=0,1,\ldots \end{cases}$$
(5.3.18)

where we have used the fact that $-j(-j+1)\ldots(j-1)j$ necessarily includes 0 as a factor for integral values for j.

More generally, as an induction hypothesis, suppose, by invoking (5.3.15), (5.3.16), that for some $k \geqslant 1$

$$(L_-)^k\langle\theta,\phi|j,j\rangle = (\hbar)^k A_j e^{i(j-k)\phi}(\sin\theta)^{j-k} P_k(\cos\theta, j) \qquad (5.3.19)$$

where $P_k(\cos\theta, j)$ is some polynomial of degree k in $\cos\theta$ whose coefficients depend on j.

By using the elementary property

$$\frac{\partial}{\partial\theta}P_k(\cos\theta, j) = -\sin\theta\frac{\partial}{\partial\cos\theta}P_k(\cos\theta, j) \qquad (5.3.20)$$

we obtain directly by the application of L_- to (5.3.19), that $(L_-)^{k+1}\langle\theta,\phi|j,j\rangle$ has the same structure as in (5.3.19) with k replaced by $(k+1)$ on its right-hand side.

Equation (5.3.18) gives for $\theta \to 0$:

$$P_{2j}(\cos\theta = 1, j) = (-2)^{2j}(-1)^{j+1/2}\Gamma^2(j+1)\pi^{-1}; \quad j=1/2, 3/2,\ldots.$$
(5.3.21)

That is, for such j values, $\cos\theta = 1$ is *not* a root of $P_{2j}(\cos\theta, j)$. Now since $P_{2j}(\cos\theta, j)$ is a polynomial in $\cos\theta$, and hence is continuous, the polynomial cannot abruptly vanish as we move away from $\cos\theta = 1$ to $\cos\theta < 1$ from the non-vanishing value in (5.3.21) for $\cos\theta = 1$. Hence there exists a non-empty interval $0 \leqslant \theta < \alpha_j$, where α_j depends on j and is *strictly* positive, such that $P_{2j}(\cos\theta, j) \neq 0$ for all θ in this interval. The latter in turn means that there exists a strictly positive constant C_j which depends on j such that $|P_{2j}(\cos\theta, j)| \geqslant C_j > 0$, for $j=1/2, 3/2,\ldots$, for $0 \leqslant \theta < \alpha_j$.

Finally, from (5.1.26), (5.3.19), with $k = 2j$, $m = j$, we may formally write

$$\langle\theta,\phi|j,-j\rangle = \frac{A_j}{(2j)!}e^{-ij\phi}(\sin\theta)^{-j}P_{2j}(\cos\theta, j) \qquad (5.3.22)$$

and for half-odd integral j, one has the following lower bound

$$\int_0^{2\pi} d\phi \int_0^\pi d\theta \, \sin\theta \, |\langle \theta, \phi | j, -j\rangle|^2 \geq 2\pi \int_0^{\alpha_j} d\theta \, \sin\theta \, |\langle \theta, \phi | j, -j\rangle|^2$$

$$\geq 2\pi \frac{|A_j|^2}{((2j)!)^2} C_j^2 \int_0^{\alpha_j} d\theta \, (\sin\theta)^{1-2j} \quad (5.3.23)$$

where α_j is strictly positive, and the integral on the extreme right-hand side of (5.3.23), clearly, does not exist (due to the divergence at the origin) for $j = 3/2, 5/2, \ldots$ which is in contradiction with the normalizability of the eigenstates in the eigenvalue problem.

For the non-admittance of $j = 1/2$ for orbital angular momentum, see (5.3.17), (5.3.42) and the discussions immediately below these equations.

With half-odd integral values of j not admitted for the orbital angular momentum, we will denote the latter parameter by ℓ, where now

$$\ell = 0, 1, 2, \ldots \quad (5.3.24)$$

$$m = -\ell, -\ell+1, \ldots, 0, 1, \ldots, \ell-1, \ell. \quad (5.3.25)$$

We also use the notation

$$\langle \theta, \phi | \ell, m \rangle \equiv Y_{\ell m}(\theta, \phi) \quad (5.3.26)$$

defining the so-called spherical harmonics treated next.

5.3.3 The Spherical Harmonics

With an adopted *sign convention*, (5.3.12) gives the properly normalized expressions

$$Y_{\ell\ell}(\theta, \phi) = (-1)^\ell \sqrt{\frac{(2\ell+1)}{4\pi(2\ell)!}} e^{i\ell\phi} \left[\frac{(2\ell)!}{2^\ell \ell!} (\sin\theta)^\ell\right] \quad (5.3.27)$$

where we have written the normalization constant A_l in (5.3.13) as

$$\sqrt{\frac{\Gamma(\ell+3/2)}{2\pi \Gamma(\ell+1)\Gamma(1/2)}} = \sqrt{\frac{(2\ell+1)}{4\pi(2\ell)!} \frac{(2\ell)!}{2^\ell \ell!}}. \quad (5.3.28)$$

Equation (5.1.26) leads to the following recursive relation

$$Y_{\ell\, k-1}(\theta, \phi) = \frac{1}{\hbar} \frac{1}{\sqrt{(\ell+k)(\ell-k+1)}} L_- Y_{\ell k}. \quad (5.3.29)$$

As an induction hypothesis, suppose that for all $m = \ell, \ell-1, \ldots, k$, for some k,

5.3 Orbital Angular Momentum   263

$$Y_{\ell m}(\theta, \phi) = (-1)^{(m+|m|)/2} e^{im\phi} \sqrt{\frac{(2\ell+1)}{4\pi} \frac{(\ell-|m|)!}{(\ell+|m|)!}} P_\ell^{|m|}(\cos\theta) \quad (5.3.30)$$

where

$$P_\ell^{|m|}(x) = \frac{(1-x^2)^{|m|/2}}{2^\ell \ell!} \left(\frac{d}{dx}\right)^{\ell+|m|} (x^2-1)^\ell. \quad (5.3.31)$$

For $m = \ell$, we check that $Y_{\ell\ell}(\theta, \phi)$ in (5.3.27) satisfies (5.3.30) by noting that

$$\left(\frac{d}{dx}\right)^{2\ell} (x^2-1)^\ell = (2\ell)!. \quad (5.3.32)$$

We will show that (5.3.30) also holds true with k replaced by $k-1$ as well:

1. Suppose $k > 0$. Then (5.3.29) leads to

$$Y_{\ell\, k-1}(\theta, \phi) = (-1)^{k-1} e^{i(k-1)\phi} \sqrt{\frac{(2\ell+1)}{4\pi} \frac{(\ell-k+1)!}{(\ell+k-1)!}} \frac{(\sin\theta)^{k-1}}{2^\ell \ell!} T_k(\cos\theta)$$

$$\quad (5.3.33)$$

where

$$T_k(x) = \frac{1}{(\ell+k)(\ell-k+1)} \left[2kx \left(\frac{d}{dx}\right)^{\ell+k} - (1-x^2) \right.$$

$$\left. \times \left(\frac{d}{dx}\right)^{\ell+k+1} \right] (x^2-1)^\ell. \quad (5.3.34)$$

To simplify the expression for $T_k(x)$, we use the following identities:

$$\left(\frac{d}{dx}\right)^{\ell+k-1} \left[x \frac{d}{dx}\right] = x \left(\frac{d}{dx}\right)^{\ell+k} + (\ell+k-1)\left(\frac{d}{dx}\right)^{\ell+k-1} \quad (5.3.35)$$

from which

$$x \left(\frac{d}{dx}\right)^{\ell+k} = \left(\frac{d}{dx}\right)^{\ell+k-1} \left[x \frac{d}{dx} - (\ell+k-1)\right]. \quad (5.3.36)$$

Also

$$\left(\frac{d}{dx}\right)^{\ell+k-1} \left[(1-x^2)\frac{d}{dx}\right] = -(\ell+k-1)(\ell+k-2)\left(\frac{d}{dx}\right)^{\ell+k-1}$$

$$- 2(\ell+k-1) x \left(\frac{d}{dx}\right)^{\ell+k} + (1-x^2)\left(\frac{d}{dx}\right)^{\ell+k+1} \quad (5.3.37)$$

which in conjunction with the identity (5.3.36) gives

$$(1-x^2)\left(\frac{d}{dx}\right)^{\ell+k+1}$$
$$=\left(\frac{d}{dx}\right)^{\ell+k-1}\left[(1-x^2)\frac{d^2}{dx^2}+2(\ell+k-1)x\frac{d}{dx}-(\ell+k)(\ell+k-1)\right].$$
(5.3.38)

The two identities (5.3.36), (5.3.38) immediately lead for $T_k(x)$, defined in (5.3.34), the simple expression

$$T_k(x)=\left(\frac{d}{dx}\right)^{\ell+k-1}(x^2-1)^\ell \qquad (5.3.39)$$

which from (5.3.33) establishes the correctness of (5.3.30) for $m=k-1$ as well. Here note that $(-1)^{k-1}=(-1)^{(k-1+|k-1|)/2}$.

2. On the other hand suppose $k\leqslant 0$. Then (5.3.29) immediately gives from (5.3.30)

$$Y_{\ell\,k-1}(\theta,\phi)=e^{i(k-1)\phi}\sqrt{\frac{(2\ell+1)}{4\pi}\frac{(\ell+k-1)!}{(\ell-k+1)!}}$$
$$\times\frac{(1-x^2)^{(1-k)/2}}{2^\ell\ell!}\left(\frac{d}{dx}\right)^{\ell-k+1}(x^2-1)^\ell \qquad (5.3.40)$$

which again verifies the expression (5.3.30) for $m=k-1$ as well. Here we note that $(-1)^{(k-1-(k-1))/2}=1$.

This completes the derivation of the expression for the spherical harmonics $Y_{\ell m}(\theta,\phi)$ as given in (5.3.30). The polynomials $P_\ell^{|m|}(x)$ are the so-called associated Legendre polynomials, and for $m=0$, $P_\ell^0(x)\equiv P_\ell(x)$ are referred to as the Legendre polynomials.

For $m=-\ell$, we have from (5.3.30)

$$Y_{\ell-\ell}(\theta,\phi)=e^{-i\ell\phi}\sqrt{\frac{(2\ell+1)!}{4\pi}}\frac{(\sin\theta)^\ell}{2^\ell\ell!} \qquad (5.3.41)$$

and is easily checked that unlike for the half-odd integral values (see (5.3.17))

$$L_- Y_{\ell-\ell}(\theta,\phi)=0 \qquad (5.3.42)$$

as it should be. That is, the algebraic relations (5.1.26), (5.1.27), resulting from the eigenvalue problem for angular momentum (§5.1), are internally consistent for *orbital* angular momentum *only if* ℓ has integral values given in (5.3.24). [In the earlier literature, single valuedness, for example, was a priori, often imposed under a rotation over ϕ by 2π to dismiss with half-odd integral values! The literature on this subject is too vast and many different approaches have been taken on this problem to dwell upon them here.]

Under space reflection (see Figure 2.4), $\theta \to \pi - \theta, \phi \to \phi + \pi$, hence $P_\ell^{|m|}(\cos\theta) \to (-1)^{\ell+|m|} P_\ell^{|m|}(\cos\theta)$ in (5.3.31), $(-1)^{m+|m|} = 1$, and

$$Y_{\ell\,m}(\theta,\phi) \to (-1)^\ell Y_{\ell\,m}(\theta,\phi). \tag{5.3.43}$$

Complex conjugation, on the other hand, gives

$$Y_{\ell m}^*(\theta,\phi) = (-1)^m Y_{\ell,-m}(\theta,\phi). \tag{5.3.44}$$

The spherical harmonics for some special values of ℓ and m are recorded here for convenience:

$$Y_{00}(\theta,\phi) = 1/\sqrt{4\pi} \tag{5.3.45}$$

$$Y_{1,\pm 1}(\theta,\phi) = \mp\sqrt{\frac{3}{8\pi}} e^{\pm i\phi} \sin\theta, \quad Y_{1,0}(\theta,\phi) = \sqrt{\frac{3}{4\pi}} \cos\theta \tag{5.3.46}$$

and

$$Y_{2,\pm 2}(\theta,\phi) = \sqrt{\frac{15}{32\pi}} e^{\pm 2i\phi} \sin^2\theta \tag{5.3.47}$$

$$Y_{2,\pm 1}(\theta,\phi) = \mp\sqrt{\frac{15}{32\pi}} e^{\pm i\phi} \sin 2\theta \tag{5.3.48}$$

$$Y_{2,0}(\theta,\phi) = \sqrt{\frac{5}{16\pi}} \left(3\cos^2\theta - 1\right). \tag{5.3.49}$$

For $\theta = 0, \phi = 0$,

$$Y_{\ell\,m}(0,0) = \sqrt{\frac{2\ell+1}{4\pi}} \delta_{m0}. \tag{5.3.50}$$

The orthonormality condition (5.1.21) reads

$$\int_0^{2\pi} d\phi \int_0^\pi d\theta \sin\theta\, Y_{\ell'm'}^*(\theta,\phi)\, Y_{\ell\,m}(\theta,\phi) = \delta_{\ell'\ell}\,\delta_{m'm}. \tag{5.3.51}$$

The Cartesian components $(x^1, x^2, x^3) (= (x,y,z))$ may be written in terms of spherical harmonics as

$$x^1 = \sqrt{\frac{2\pi}{3}}\, r\, [Y_{1,-1}(\theta,\phi) - Y_{1,1}(\theta,\phi)] \tag{5.3.52}$$

$$x^2 = \sqrt{\frac{2\pi}{3}}\, ir\, [Y_{1,-1}(\theta,\phi) + Y_{1,1}(\theta,\phi)] \tag{5.3.53}$$

$$x^3 = \sqrt{\frac{4\pi}{3}}\, rY_{1,0}(\theta,\phi) \tag{5.3.54}$$

with all corresponding to $\ell = 1$, emphasizing the vector character of **x**.

Selection rules for the matrix elements $\langle \ell', m' | x^i | \ell, m \rangle$ are readily obtained from the following considerations.

The elementary integral

$$\int_0^{2\pi} d\phi\, e^{i(m-m'')\phi} = 2\pi \delta_{m,m''} \tag{5.3.55}$$

shows that

$$\langle \ell', m' | x^1 | \ell, m \rangle = 0 \quad \text{for} \quad m' \neq m \pm 1 \tag{5.3.56}$$

$$\langle \ell', m' | x^2 | \ell, m \rangle = 0 \quad \text{for} \quad m' \neq m \pm 1 \tag{5.3.57}$$

$$\langle \ell', m' | x^3 | \ell, m \rangle = 0 \quad \text{for} \quad m' \neq m. \tag{5.3.58}$$

On the other hand, the addition rule of angular momenta studied in §5.5 (see also (5.8.35), (5.8.36)) with x^i corresponding to $\ell = 1$ (see (5.3.52)–(5.3.54)) shows that

$$\langle \ell', m' | x^i | \ell, m \rangle = 0 \tag{5.3.59}$$

if the value of ℓ' is not in the set $\{|\ell-1|, \ldots, \ell+1\}$.

The spherical harmonics in (5.3.26) may be readily related to the matrix elements of the $D^{(\ell)}$ matrix, with matrix elements $D^{(\ell)}_{mm'}$ given in (5.2.12), as follows.

To the above end, let

$$\mathbf{n} = (\sin\phi, -\cos\phi, 0) \tag{5.3.60}$$

$$\mathbf{N}_0 = (0, 0, 1). \tag{5.3.61}$$

Then according to (2.3.5)

$$\left\langle \mathbf{N}_0 \left| \left[\exp\left(\frac{i}{\hbar}\theta\mathbf{n}\cdot\mathbf{J}\right)\right]^\dagger \right| \ell, m \right\rangle = \langle \mathbf{N} | \ell, m \rangle \tag{5.3.62}$$

where (§2.1) in an obvious notation

$$\mathbf{N} = R(\theta, \mathbf{n})\mathbf{N}_0$$
$$= (\sin\theta\cos\phi, \sin\theta\sin\phi, \cos\theta), \tag{5.3.63}$$

and (5.3.62) may be rewritten as

$$\left\langle \theta = 0, \phi = 0 \left| \exp\left(-\frac{i}{\hbar}\theta\mathbf{n}\cdot\mathbf{J}\right) \right| \ell, m \right\rangle = Y_{\ell\, m}(\theta, \phi). \tag{5.3.64}$$

The left-hand side of (5.3.64) may be expanded as follows

$$\sum_{m'=-\ell}^{\ell} \langle \theta=0, \phi=0 | \ell, m' \rangle \left\langle \ell, m' \left| \exp\left(-\frac{i}{\hbar}\theta \mathbf{n} \cdot \mathbf{J}\right) \right| \ell, m \right\rangle$$

$$= \sqrt{\frac{2\ell+1}{4\pi}} \left\langle \ell, 0 \left| \exp\left(-\frac{i}{\hbar}\theta \mathbf{n} \cdot \mathbf{J}\right) \right| \ell, m \right\rangle \quad (5.3.65)$$

where we have used (5.3.50), to give

$$Y_{\ell m}(\theta, \phi) = \sqrt{\frac{2\ell+1}{4\pi}} \left\langle \ell, 0 \left| \exp\left(-\frac{i}{\hbar}\theta \mathbf{n} \cdot \mathbf{J}\right) \right| \ell, m \right\rangle$$

$$= \sqrt{\frac{2\ell+1}{4\pi}} \left\langle \ell, 0 \left| \exp\left(-\frac{i}{\hbar}\phi J^3\right) \exp\left(\frac{i}{\hbar}\theta J^2\right) \exp\left(\frac{i}{\hbar}\phi J^3\right) \right| \ell, m \right\rangle$$

$$= \sqrt{\frac{2\ell+1}{4\pi}} D^{(\ell)}_{0m}(\phi, \theta, 0) \equiv \sqrt{\frac{2\ell+1}{4\pi}} D^{(\ell)}_{0m}(\phi, \theta, \gamma) \quad (5.3.66)$$

where in writing the second equality we have used the identity (5.2.11), and in the last equality we have used the definition (5.2.12). We note that for $m = 0$ in $\langle \ell, 0 |$, the angle γ does not contribute in (5.3.66), i.e., the angle γ may be taken to have any value.

5.3.4 Addition Theorem of Spherical Harmonics

Finally, we establish the following theorem for the addition of the product of two spherical harmonics:

$$\frac{4\pi}{2\ell+1} \sum_{m=-\ell}^{\ell} Y^*_{\ell\, m}(\theta_1, \phi_1) Y_{\ell\, m}(\theta_2, \phi_2) = P_\ell(\cos\theta) \quad (5.3.67)$$

where (see Figure 5.2)

$$\cos\theta = \mathbf{N}_1 \cdot \mathbf{N}_2, \quad (5.3.68)$$

$$\mathbf{N}_i = (\sin\theta_i \cos\phi_i, \sin\theta_i \sin\phi_i, \cos\theta_i), \quad i = 1, 2 \quad (5.3.69)$$

and in detail

$$\cos\theta = \cos\theta_1 \cos\theta_2 + \sin\theta_1 \sin\theta_2 \cos(\phi_1 - \phi_2). \quad (5.3.70)$$

We note that due to the reality of $P_\ell(\cos\theta)$, the complex conjugation sign $*$ may be put on either $Y_{\ell m}(\theta_1, \phi_1)$ or $Y_{\ell m}(\theta_2, \phi_2)$ in (5.3.67).

To establish (5.3.67), we note that according to (2.3.5)

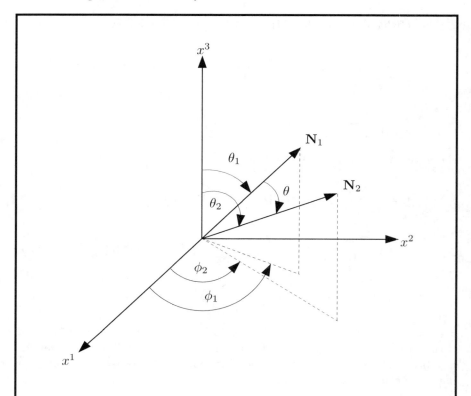

Fig. 5.2. Figure showing the angle θ as the angle between the two unit vectors $\mathbf{N}_1$, $\mathbf{N}_2$ and hence $\mathbf{N}_1 \cdot \mathbf{N}_2 = \cos\theta$.

$$\left\langle \mathbf{N}_2 \left| \left[\exp\left(\frac{i}{\hbar}\theta_1 J^2\right) \exp\left(\frac{i}{\hbar}\phi_1 J^3\right) \right]^\dagger \right| \ell, 0 \right\rangle = \langle \overline{\mathbf{N}} | \ell, 0 \rangle \quad (5.3.71)$$

where (§2.1)
$$\overline{\mathbf{N}} = R\left(\theta_1, \hat{\mathbf{x}}^2\right) R\left(\phi_1, \hat{\mathbf{x}}^3\right) \mathbf{N}_2 \quad (5.3.72)$$

where $\hat{\mathbf{x}}^2$, $\hat{\mathbf{x}}^3$ are unit vectors along x^2 and x^3 axes, respectively. In particular, $\overline{N}^3$ works out to be equal to $\cos\theta$. The corresponding ϕ-angle does not contribute to $\langle \overline{\mathbf{N}} | \ell, 0 \rangle$ since $m = 0$ in $|\ell, 0\rangle$ (see (5.3.30)). Accordingly, (5.3.71) may be rewritten as

$$\left\langle \theta_2, \phi_2 \left| \left[\exp\left(\frac{i}{\hbar}\theta_1 J^2\right) \exp\left(\frac{i}{\hbar}\phi_1 J^3\right) \right]^\dagger \right| \ell, 0 \right\rangle = Y_{\ell\,0}(\theta, 0)$$

$$\equiv \sqrt{\frac{2\ell+1}{4\pi}} P_\ell(\cos\theta). \tag{5.3.73}$$

The extreme left-hand side of this equation may be expanded in the form

$$\sum_{m=-\ell}^{\ell} \langle \theta_2, \phi_2 | \ell, m \rangle \left\langle \ell, m \left| \left[\exp\left(\frac{i}{\hbar}\theta_1 J^2\right) \exp\left(\frac{i}{\hbar}\phi_1 J^3\right) \right]^\dagger \right| \ell, 0 \right\rangle$$

$$\equiv \sum_{m=-\ell}^{\ell} Y_{\ell m}(\theta_2, \phi_2) \left[D_{0m}^{(\ell)}(\phi_1, \theta_1, 0) \right]^* \tag{5.3.74}$$

which from (5.3.66), (5.3.74) gives the identity in (5.3.67).

Additional properties of the spherical harmonics are given in §5.8.

5.4 Spin

5.4.1 General Structure

Under a coordinate rotation, spin is described by the transformation property (5.3.2) of a wavefunction

$$\psi'(\mathbf{x}') = \left(\exp\left(\frac{i}{\hbar} \varphi \mathbf{n} \cdot \mathbf{S}\right) \psi \right)(\mathbf{x}) \tag{5.4.1}$$

where $\mathbf{x}' = R\mathbf{x}$. A wavefunction $\psi(\mathbf{x})$ may be written as a $(2s+1)$ component object (see (2.8.23), (2.8.25))

$$\psi(\mathbf{x}) = \begin{pmatrix} \psi(\mathbf{x}, s) \\ \psi(\mathbf{x}, s-1) \\ \vdots \\ \psi(\mathbf{x}, -s+1) \\ \psi(\mathbf{x}, -s) \end{pmatrix} \equiv \sum_{m=-s}^{s} X^s(m) \psi(\mathbf{x}, m) \tag{5.4.2}$$

where

$$X^s(s) = \begin{pmatrix} 1 \\ 0 \\ 0 \\ \vdots \\ 0 \end{pmatrix}, X^s(s-1) = \begin{pmatrix} 0 \\ 1 \\ 0 \\ \vdots \\ 0 \end{pmatrix}, \ldots, X^s(-s) = \begin{pmatrix} 0 \\ 0 \\ \vdots \\ 0 \\ 1 \end{pmatrix} \tag{5.4.3}$$

with the normalization condition (2.8.25) for the wave functions $\psi(\mathbf{x}, m)$

$$\sum_{m=-s}^{s}\int d^3x\,|\psi(\mathbf{x},m)|^2 = 1 \qquad (5.4.4)$$

satisfied.

In detail, (5.4.1) may be rewritten in terms of components as

$$\psi'(\mathbf{x}',m') = \sum_{m=-s}^{s}\left\langle s,m'\left|\exp\left(\frac{i}{\hbar}\varphi\mathbf{n}\cdot\mathbf{S}\right)\right|s,m\right\rangle\psi(\mathbf{x},m) \qquad (5.4.5)$$

where the matrix elements $\left\langle s,m'\left|\exp\left(\frac{1}{\hbar}\varphi\mathbf{n}\cdot\mathbf{S}\right)\right|s,m\right\rangle$ are given in (2.8.51).

We may use (5.2.10), (5.2.11), (2.8.41) to find the eigenstates of the projection of the spin $\mathbf{N}\cdot\mathbf{S}$ along an arbitrary unit vector

$$\mathbf{N} = (\sin\beta\cos\alpha, \sin\beta\sin\alpha, \cos\beta) \qquad (5.4.6)$$

making, in particular an angle β with the z-axis. We consider specific elementary cases first which are easily and directly worked out before dealing with arbitrary spins.

5.4.2 Spin 1/2

For spin 1/2, $\mathbf{S} = \hbar\boldsymbol{\sigma}/2$,

$$\mathbf{N}\cdot\mathbf{S} = \frac{\hbar}{2}\begin{pmatrix}\cos\beta & \sin\beta\,e^{-i\alpha}\\ \sin\beta\,e^{i\alpha} & -\cos\beta\end{pmatrix} \qquad (5.4.7)$$

and the eigenstates $|\lambda,\mathbf{N}\rangle$ of $\mathbf{N}\cdot\mathbf{S}$ satisfying

$$\mathbf{N}\cdot\mathbf{S}\,|\lambda,\mathbf{N}\rangle = \hbar\lambda\,|\lambda,\mathbf{N}\rangle, \qquad \lambda = \pm 1/2 \qquad (5.4.8)$$

are easily worked out to be

$$|1/2,\mathbf{N}\rangle = \begin{pmatrix}\cos\dfrac{\beta}{2}\,e^{-i\alpha/2}\\[4pt] \sin\dfrac{\beta}{2}\,e^{i\alpha/2}\end{pmatrix} \qquad (5.4.9)$$

$$|-1/2,\mathbf{N}\rangle = \begin{pmatrix}-\sin\dfrac{\beta}{2}\,e^{-i\alpha/2}\\[4pt] \cos\dfrac{\beta}{2}\,e^{i\alpha/2}\end{pmatrix}. \qquad (5.4.10)$$

For example, the amplitude for a spin measurement along a $\overline{z}$-axis, making an angle β with the z-axis with $\alpha \equiv 0$, to be along the $+\overline{z}$ direction if it is initially prepared to be in state $|-1/2, z\rangle$ is given from (5.4.9), (5.4.10) to be

$$\langle +1/2, \bar{z} | -1/2, z \rangle = \left(\cos \frac{\beta}{2} \ \sin \frac{\beta}{2} \right) \begin{pmatrix} 0 \\ 1 \end{pmatrix} = \sin \frac{\beta}{2}. \tag{5.4.11}$$

In reference to Figure 1.5, we also have

$$\langle +1/2, z | +1/2, \bar{z} \rangle \langle +1/2, \bar{z} | -1/2, z \rangle = \sin \frac{\beta}{2} \cos \frac{\beta}{2}. \tag{5.4.12}$$

We recall (§2.7) that spin $\mathbf{S}$ and the momentum operator $\mathbf{P}$ commute, and if $\mathbf{p}$ denotes the momentum of a particle, then the spin $\mathbf{S}$ along $\mathbf{p}/p$ is called the *helicity* of the particle with eigenvalues $\hbar\lambda$, $\lambda = \pm 1/2$ in (5.4.8).

We note the following elementary properties:

$$\text{Tr}\left[S^i\right] = 0, \quad \text{Tr}\left[S^i S^j\right] = \frac{\hbar^2}{2}\delta^{ij} \tag{5.4.13}$$

and

$$(\mathbf{N} \cdot \mathbf{S})^2 = \frac{\hbar^2}{4}\mathbf{1}. \tag{5.4.14}$$

The spin density operator ρ associated with a mixture (§1.5) of spin components in reference to the vector $\mathbf{N}$ may be written as

$$\rho = \sum_{\lambda = \pm 1/2} w(\lambda) |\lambda, \mathbf{N}\rangle \langle \lambda, \mathbf{N}| \tag{5.4.15}$$

with

$$\sum_{\lambda = \pm 1/2} w(\lambda) = 1 \tag{5.4.16}$$

and is easily work out from (5.4.9), (5.4.10) to be given by

$$\rho = \frac{1}{2}\left[\mathbf{1} + (w(+1/2) - w(-1/2))\mathbf{N} \cdot \boldsymbol{\sigma}\right] \tag{5.4.17}$$

which has the general expected structure

$$\rho = \frac{1}{2}[\mathbf{1} + \mathbf{b} \cdot \boldsymbol{\sigma}] \tag{5.4.18}$$

satisfying the normalization condition

$$\text{Tr}[\rho] = 1. \tag{5.4.19}$$

The expectation value of the spin $\mathbf{S}$ (§1.5) is then given by

$$\text{Tr}[\rho \mathbf{S}] = \frac{\hbar}{2}(w(+1/2) - w(-1/2))\mathbf{N} \tag{5.4.20}$$

where we have used (5.4.13).

The vector $\mathbf{b}$

$$\mathbf{b} = (w(+1/2) - w(-1/2))\mathbf{N} \tag{5.4.21}$$

is often referred to as the spin-polarization vector and satisfies

$$0 \leqslant |\mathbf{b}| \leqslant 1. \tag{5.4.22}$$

For example, for $w(+1/2) = 0$, $w(-1/2) = 1$, that is, $\mathbf{b} = -\mathbf{N}$, one is dealing with a beam with spin polarized completely in the $-\mathbf{N}$ direction. For $w(+1/2) = w(-1/2) = 1/2$, that is, $\mathbf{b} = 0$, one has a completely unpolarized beam.

Detailed treatments, with various applications, of density operators for spin 1/2 will be given in §8.6.

5.4.3 Spin 1

For spin 1, we have from (2.8.62)–(2.8.64),

$$\mathbf{N} \cdot \mathbf{S} = \hbar \begin{pmatrix} \cos\beta & \dfrac{\sin\beta}{\sqrt{2}} e^{-i\alpha} & 0 \\ \dfrac{\sin\beta}{\sqrt{2}} e^{i\alpha} & 0 & \dfrac{\sin\beta}{\sqrt{2}} e^{-i\alpha} \\ 0 & \dfrac{\sin\beta}{\sqrt{2}} e^{i\alpha} & -\cos\beta \end{pmatrix} \tag{5.4.23}$$

and the eigenstates $|\lambda, \mathbf{N}\rangle$ are readily worked out to be

$$|+1, \mathbf{N}\rangle = \begin{pmatrix} \dfrac{1+\cos\beta}{2} e^{-i\alpha} \\ \dfrac{\sin\beta}{\sqrt{2}} \\ \dfrac{1-\cos\beta}{2} e^{i\alpha} \end{pmatrix} \tag{5.4.24}$$

$$|0, \mathbf{N}\rangle = \begin{pmatrix} -\dfrac{\sin\beta}{\sqrt{2}} e^{-i\alpha} \\ \cos\beta \\ \dfrac{\sin\beta}{\sqrt{2}} e^{i\alpha} \end{pmatrix} \tag{5.4.25}$$

$$|-1, \mathbf{N}\rangle = \begin{pmatrix} \dfrac{1-\cos\beta}{2} e^{-i\alpha} \\ -\dfrac{\sin\beta}{\sqrt{2}} \\ \dfrac{1+\cos\beta}{2} e^{i\alpha} \end{pmatrix}. \tag{5.4.26}$$

We note the following relations

$$(\mathbf{N}\cdot\mathbf{S})^2 = \hbar^2 \begin{pmatrix} \dfrac{1+\cos^2\beta}{2} & \dfrac{\sin\beta\cos\beta}{\sqrt{2}} e^{-i\alpha} & \dfrac{\sin^2\beta}{2} e^{-2i\alpha} \\ \dfrac{\sin\beta\cos\beta}{\sqrt{2}} e^{i\alpha} & \sin^2\beta & -\dfrac{\sin\beta\cos\beta}{\sqrt{2}} e^{-i\alpha} \\ \dfrac{\sin^2\beta}{2} e^{2i\alpha} & -\dfrac{\sin\beta\cos\beta}{\sqrt{2}} e^{i\alpha} & \dfrac{1+\cos^2\beta}{2} \end{pmatrix} \tag{5.4.27}$$

$$\left(\dfrac{\mathbf{N}\cdot\mathbf{S}}{\hbar}\right)^3 = \dfrac{\mathbf{N}\cdot\mathbf{S}}{\hbar} \tag{5.4.28}$$

$$\text{Tr}\left[S^k\right] = 0, \qquad \text{Tr}\left[\dfrac{S^k S^\ell}{\hbar^2}\right] = 2\delta^{k\ell} \tag{5.4.29}$$

$$\text{Tr}\left[\dfrac{S^k S^\ell S^m}{\hbar^3}\right] = i\varepsilon^{k\ell m}. \tag{5.4.30}$$

The spin density operator associated with a mixture of spin components in reference to the vector $\mathbf{N}$ may be written as

$$\rho = \sum_{\lambda=0,\pm 1} w(\lambda) |\lambda, \mathbf{N}\rangle\langle\lambda, \mathbf{N}| \tag{5.4.31}$$

with

$$\sum_{\lambda=0,\pm 1} w(\lambda) = 1 \tag{5.4.32}$$

and is readily worked out from (5.4.24)–(5.4.27), (5.4.23) to be given by

$$\rho = w(0)\mathbf{1} + \dfrac{(w(+1) - w(-1))}{2} \dfrac{\mathbf{N}\cdot\mathbf{S}}{\hbar}$$
$$+ \dfrac{(w(+1) + w(-1) - 2w(0))}{2} \left(\dfrac{\mathbf{N}\cdot\mathbf{S}}{\hbar}\right)^2 \tag{5.4.33}$$

verifying, in particular, the normalization condition

$$\text{Tr}\,[\rho] = 3w\,(0) + (w\,(+1) + w\,(-1) - 2w\,(0)) = 1 \tag{5.4.34}$$

where we have used (5.4.29), (5.4.32).

Upon using (5.4.29), (5.4.30), we have for the expectation value of **S**

$$\text{Tr}\,[\rho\,\mathbf{S}] = \hbar\,(w\,(+1) - w\,(-1))\,\mathbf{N} \tag{5.4.35}$$

(See also Problem 5.7).

5.4.4 Arbitrary Spins

From (2.8.42), (5.4.6), the matrix elements of the $(2s+1) \times (2s+1)$ matrix $\mathbf{N} \cdot \mathbf{S}$ are given by

$$\langle s, m | \mathbf{N} \cdot \mathbf{S} | s, m' \rangle = \frac{\hbar}{2} \sin\beta e^{i\alpha} \sqrt{(s-m)(s+m+1)}\,\delta_{m',m+1}$$

$$+ \frac{\hbar}{2} \sin\beta\, e^{-i\alpha} \sqrt{(s+m)(s-m+1)}\,\delta_{m',m-1}$$

$$+ \hbar \cos\beta\, m\, \delta_{m',m} \tag{5.4.36}$$

and from (2.8.41), (2.8.51), (5.2.11), (5.2.12), the eigenstates $|s; \lambda, \mathbf{N}\rangle$ satisfying

$$\mathbf{N} \cdot \mathbf{S}\,|s; \lambda, \mathbf{N}\rangle = \hbar\lambda\,|s; \lambda, \mathbf{N}\rangle\,, \tag{5.4.37}$$

$$\lambda = -s, -s+1, \ldots, s-1, s \tag{5.4.38}$$

are explicitly given by

$$|s; \lambda, \mathbf{N}\rangle = \begin{pmatrix} D^{(s)*}_{\lambda\,s}(\alpha,\beta,0) \\ D^{(s)*}_{\lambda\,s-1}(\alpha,\beta,0) \\ \vdots \\ D^{(s)*}_{\lambda\,-s}(\alpha,\beta,0) \end{pmatrix} \tag{5.4.39}$$

where $D^{(s)*}_{\lambda\lambda'}(\alpha,\beta,0)$, in (5.2.12), is given by the expression on the right-hand side of (5.2.10) with $j' = j = s$, $m = \lambda$, $m' = \lambda'$, $\gamma = 0$.

The spin density operator ρ associated with a mixture of spin components in reference to the vector $\mathbf{N}$ may be then written as

$$\rho = \sum_{\lambda=-s}^{s} w\,(\lambda)\,|s; \lambda, \mathbf{N}\rangle\,\langle s; \lambda, \mathbf{N}|\,. \tag{5.4.40}$$

For $\mathbf{N} = \mathbf{p}/p$, where $\mathbf{p}$ is the momentum of a particle, $\mathbf{N} \cdot \mathbf{S}$ is referred to as the *helicity*, having eigenvalues $\hbar\lambda$, $\lambda = -s, \ldots, s$.

5.5 Addition of Angular Momenta

In this section we deal with the angular momentum eigenvalue problem resulting from the addition of two commuting angular momenta. That is, we consider the angular momentum operator

$$\mathbf{J} = \mathbf{J}_1 + \mathbf{J}_2 \tag{5.5.1}$$

where $\mathbf{J}_1, \mathbf{J}_2$ are assumed to commute. A typical example, is the addition of the orbital angular momentum to the spin of the electron in the hydrogen atom discussed later in §7.4, §7.5.

For the $(\mathbf{J}_1, \mathbf{J}_2)$ system, $\mathbf{J}_1^2, J_1^3, \mathbf{J}_2^2, J_2^3$ system are commuting operators, and for a given pair (j_1, j_2), we may generate a $(2j_1+1)(2j_1+1)$ dimensional vector space $V(j_1, j_2)$ spanned by the orthonormal set consisting of the vectors

$$|j_1, m_1\rangle |j_2, m_2\rangle \equiv |j_1, m_1; j_2, m_2\rangle \equiv |m_1, m_2\rangle \tag{5.5.2}$$

where $m_1 = -j_1, -j_1+1, \ldots, j_1-1, j_1;\ m_2 = -j_2, -j_2+1, \ldots, j_2-1, j_2$.

For the combined system, $\mathbf{J}^2, J^3, \mathbf{J}_1^2, \mathbf{J}_2^2$ commute and we denote their simultaneous eigenstates by

$$|j, m; j_1, j_2\rangle \equiv |j, m\rangle. \tag{5.5.3}$$

For a given fixed pair (j_1, j_2) we may expand the eigenstates $|j, m\rangle$ in (5.5.3) in the terms of the basis $\{|m_1, m_2\rangle\}$ in (5.5.2):

$$|j, m\rangle = \sum_{m_1, m_2} |m_1, m_2\rangle \langle m_1, m_2 | j, m\rangle. \tag{5.5.4}$$

The expansion coefficients $\langle m_1, m_2 | j, m\rangle$ are referred to as Clebsch-Gordan coefficients. Unfortunately, there are many different notations used for these coefficients in the literature.

The application of $J^3 = J_1^3 + J_2^3$ to (5.5.4) yields

$$\hbar m |j, m\rangle = \sum_{m_1, m_2} \hbar |m_1 + m_2\rangle |m_1, m_2\rangle \langle m_1, m_2 | j, m\rangle$$

implying that the sum in (5.5.4) over m_1, m_2 is restricted to

$$m_1 + m_2 = m. \tag{5.5.5}$$

Given a fixed pair (j_1, j_2), one may readily find the allowed j values corresponding to $\mathbf{J}^2$ for the combined angular momentum.

To the above end, we note that m takes its maximum (minimum) value j $(-j)$ when m_1, m_2 take on their maximum (minimum) values j_1, j_2 $(-j_1, -j_2)$. For $m = \pm(j_1 + j_2)$ one necessarily has $j = j_1 + j_2$ and only one vector $|\pm j_1, \pm j_2\rangle$ contributes to the sum in (5.5.4) for each sign. Accordingly, we may identify

$$|j_1 + j_2, \pm (j_1 + j_2)\rangle \equiv |\pm j_1, \pm j_2\rangle \qquad (5.5.6)$$

and set

$$\langle \pm j_1, \pm j_2 | j_1 + j_2, \pm (j_1 + j_2)\rangle = 1 \qquad (5.5.7)$$

for the corresponding Clebsch-Gordan coefficients.

Since $m_1 = -j_1, \ldots, j_1$; $m_2 = -j_2, \ldots, j_2$, the next possible value to $m = j_1 + j_2$ is

$$m = j_1 + j_2 - 1. \qquad (5.5.8)$$

In this case we have

$$m_2 = j_2, \quad m_1 = j_1 - 1$$

or

$$m_2 = j_2 - 1, \quad m_1 = j_1. \qquad (5.5.9)$$

We then have exactly two vectors, $|j_1 - 1, j_2\rangle$, $|j_1, j_2 - 1\rangle$ contributing to the sum in (5.5.4) for the given m value in (5.5.8).

More generally, we may write for the possible values of m

$$m = j_1 + j_2 - k = m_1 + m_2 \qquad (5.5.10)$$

where

$$k = 0, 1, 2, \ldots, 2(j_1 + j_2) \qquad (5.5.11)$$

and for $k = 2(j_1 + j_2)$, $m = -(j_1 + j_2)$.

For the subsequent analysis, suppose, without loss of any generality, that $j_1 \geqslant j_2$. The reversed situation is similarly handled.

To satisfy (5.5.10), we may find the *smallest* possible non-negative integers a and b such that

$$\left.\begin{array}{c} m_2 = -j_2 + a, \quad m_1 = 2j_2 + j_1 - k - a \\ \text{or} \\ \vdots \qquad \vdots \qquad \vdots \\ \text{or} \\ m_2 = j_2 - b, \quad m_1 = j_1 - k + b \end{array}\right\}. \qquad (5.5.12)$$

Since, in particular, $-j_1 \leqslant m_1 \leqslant j_2$, these possible solutions imply that

$$-j_1 \leqslant 2j_2 + j_1 - k - a \leqslant j_1 \qquad (5.5.13)$$

$$-j_1 \leqslant j_1 - k + b \leqslant j_1 \qquad (5.5.14)$$

or that

5.5 Addition of Angular Momenta

$$2j_2 - k \leq a \leq 2(j_1 + j_2) - k \tag{5.5.15}$$

$$k - 2j_1 \leq b \leq k. \tag{5.5.16}$$

Clearly the *number* $N(k)$ of $|m_1, m_2\rangle$ states, contributing to the sum in (5.5.4), giving rise to the *same* m value given in (5.5.10) is, using in the process (5.5.12), given by

$$N(k) = m_1(\max) - m_1(\min) + 1$$

$$= m_2(\max) - m_2(\min) + 1. \tag{5.5.17}$$

Depending on the value of k, i.e., whether $0 \leq k < 2j_2$ or $2j_2 \leq k \leq 2j_1$ or $2j_1 < k$, the values of a, b and $N(k)$ are worked out in Table 5.1 for a given k, all corresponding to the same value $m = j_1 + j_2 - k$.

Table 5.1. For $m = m_1 + m_2 - k$, $j_1 \geq j_2$, depending on the value of $k = 0, 1, \ldots, 2(j_1 + j_2)$, the possible values of the pair (m_1, m_2) are given in (5.5.12) with a and b as given in this table. $N(k)$ denotes number of states $|m_1, m_2\rangle$ that contribute to the sum in (5.5.4) leading all to the same $m = j_1 + j_2 - k$ value for a given k.

	a	b	$N(k)$
$0 \leq k < 2j_2$	$2j_2 - k$	0^*	$k+1$
$2j_2 \leq k \leq 2j_1$	0	0	$2j_2 + 1$
$2j_1 < k \leq 2(j_1 + j_2)$	0	$k - 2j_1$	$2(j_1 + j_2) - k + 1$

* This result follows because we have taken $j_1 \geq j_2$.

Finally, we refer to Table 5.2, to infer, by using in the process $m = -j, -j+1, \ldots, j-1, j$, that for fixed $j_1 \geq j_2$, the possible values of j are $j = j_1 - j_2, j_1 - j_2 + 1, \ldots, j_1 + j_2$. For $j_1 \leq j_2$ we may simply interchange the indices 1, 2. That is, for a fixed pair of values (j_1, j_2), the possible values of j are

$$j = |j_1 - j_2|, |j_1 - j_2| + 1, \ldots, j_1 + j_2. \tag{5.5.18}$$

The above constraint $|j_1 - j_2| \leq j \leq j_1 + j_2$ is referred to as the triangular condition.

For a given pair (j_1, j_2) the number of independent vectors $|j, m\rangle$ in (5.5.3) is

$$\sum_{j=|j_1-j_2|}^{(j_1+j_2)} (2j+1) = (2j_1+1)(2j_2+1) \tag{5.5.19}$$

278   5 Angular Momentum Gymnastics

Table 5.2. for $m = j_1 + j_2 - k$, $j_1 \geq j_2$ the table gives, for each possible k, the corresponding m values. For a given k, the value of m is repeated $N(k)$ times corresponding to the totality of all possible pairs (m_1, m_2) giving rise to the same m value. That is, $N(k)$ denotes the number of $|m_1, m_2\rangle$ states contributing to the sum in (5.5.4) leading all to the same m value for the given k. Reading across horizontally in the table, one may readily infer the possible values of j, given a fixed pair $j_1 \geq j_2$, by using the relation that $m = -j, -j+1, \ldots, j-1, j$. This is given in (5.5.18) for any fixed pair (j_1, j_2). Note that consistency requires that, in each row, the m values increase as we go from left to right. Accordingly, some of the columns and the rows will obviously not contribute for particular pairs $j_1 \geq j_2$ and are easily figured out as we increase m by integer steps.

k	$2(j_1+j_2)$	$2(j_1+j_2)-1$	$\ldots$	$(2j_1+1)$	$2j_1$	$\ldots$	$2j_2$	$(2j_2-1)$	$\ldots$	1	0	j
	$-(j_1+j_2)$	$-(j_1+j_2)+1$	$\ldots$	$-(j_1-j_2)-1$	$-(j_1-j_2)$	$\ldots$	(j_1-j_2)	$(j_1-j_2)+1$	$\ldots$	j_1+j_2-1	j_1+j_2	j_1+j_2
		$-(j_1+j_2)+1$	$\ldots$	$-(j_1-j_2)-1$	$-(j_1-j_2)$	$\ldots$	(j_1-j_2)	$(j_1-j_2)+1$	$\ldots$	j_1+j_2-1		j_1+j_2-1
				$-(j_1-j_2)-1$	$-(j_1-j_2)$	$\ldots$	(j_1-j_2)	$(j_1-j_2)+1$				$\vdots$
					$-(j_1-j_2)$	$\ldots$	(j_1-j_2)					j_1-j_2+1
					$-(j_1-j_2)$	$\ldots$	(j_1-j_2)					j_1-j_2
$N(k)$	1	2	$\ldots$	$2j_2$	$2j_2+1$	$\ldots$	$2j_2+1$	$2j_2$	$\ldots$	2	1	

since m takes $(2j+1)$ values.

We may invert the expansion in (5.5.4), and expand $|m_1, m_2\rangle$ in (5.5.2) in terms of the kets $|j, m\rangle$,

$$|m_1, m_2\rangle = \sum_{j=|j_1-j_2|}^{j_1+j_2} |j, m\rangle \langle j, m | m_1, m_2\rangle \qquad (5.5.20)$$

where $m = m_1 + m_2$.

From the orthonormality relations

$$\langle j', m' | j, m\rangle = \delta_{j'j}\delta_{m',m} \qquad (5.5.21)$$

$$\langle m'_1, m'_2 | m_1, m_2\rangle = \delta_{m'_1 m_1}\delta_{m'_2 m_2} \qquad (5.5.22)$$

for a given pair (j_1, j_2) the following completeness relations follow from (5.5.20) and (5.5.4),

$$\sum_{j=|j_1-j_2|}^{j_1+j_2} \langle m'_1, m'_2 | j, m\rangle \langle j, m | m_1, m_2\rangle = \delta_{m'_1 m_1}\delta_{m'_2 m_2} \qquad (5.5.23)$$

with $m = m_1 + m_2 = m'_1 + m'_2$,

$$\sum_{m_1, m_2} \langle j', m' | m_1, m_2\rangle \langle m_1, m_2 | j, m\rangle = \delta_{j'j}\delta_{m'm} \qquad (5.5.24)$$

where $m = m_1 + m_2 = m'_1 + m'_2$, $m_1 = -j_1, -j_1+1, \ldots, j_1-1, j_1$, $m_2 = -j_2, -j_2+1, \ldots, j_2-1, j_2$. Equation (5.5.24) is valid for j taking any of the values in (5.5.18).

Upon applying

$$J_\pm = J_{1\pm} + J_{2\pm} \qquad (5.5.25)$$

to (5.5.4) and using (5.1.25), we obtain

$$\hbar\sqrt{(j \mp m)(j \pm m + 1)}\,|j, m \pm 1\rangle$$

$$= \sum_{m_1, m_2} \hbar\sqrt{(j_1 \mp m_1)(j_1 \pm m_1 + 1)}\,|m_1 \pm 1, m_2\rangle \langle m_1, m_2 | j, m\rangle$$

$$+ \sum_{m_1, m_2} \hbar\sqrt{(j_2 \mp m_2)(j_2 \pm m_2 + 1)}\,|m_1, m_2 \pm 1\rangle \langle m_1, m_2 | j, m\rangle$$

$$(5.5.26)$$

which upon multiplying by $\langle m'_1, m'_2|$ and using (5.5.22) gives the following useful relationship relating Clebsch-Gordan coefficients:

$$\sqrt{(j \mp m)(j \pm m + 1)}\,\langle m_1, m_2 | j, m \pm 1\rangle$$

$$= \sqrt{(j_1 \pm m_1)(j_1 \mp m_1 + 1)} \langle m_1 \mp 1, m_2 | j, m \rangle$$
$$+ \sqrt{(j_2 \pm m_2)(j_2 \mp m_2 + 1)} \langle m_1, m_2 \mp 1 | j, m \rangle. \qquad (5.5.27)$$

The following relation for the Clebsch-Gordan coefficients as following from (5.5.4) or (5.5.20) should be also noted

$$\langle m_1, m_2 | j, m' \rangle = \delta_{m', m_1 + m_2} \langle m_1, m_2 | j, m_1 + m_2 \rangle. \qquad (5.5.28)$$

Of particular interest is the addition of an arbitrary non-zero angular momentum $j_1 \geqslant 1/2$ to a spin $j_2 = 1/2$ which is easily handled. In this case $j = j_1 + 1/2$ or $j = j_1 - 1/2$.

For $j = j_1 + 1/2$, $j_2 = 1/2$, $m_2 = +1/2$, and taking the lower sign in (5.5.27) gives a recurrence relation leading to the following chain of equalities:

$$\langle m - 1/2, 1/2 | j, m \rangle = \sqrt{\frac{j+m}{j+m+1}} \langle m + 1/2, 1/2 | j, m + 1 \rangle$$

$$\langle m + 1/2, 1/2 | j, m + 1 \rangle = \sqrt{\frac{j+m+1}{j+m+2}} \langle m + 3/2, 1/2 | j, m + 2 \rangle$$

$$\vdots$$

$$\langle j - 3/2, 1/2 | j, j - 1 \rangle = \sqrt{\frac{2j-1}{2j}} \langle j - 1/2, 1/2 | j, j \rangle$$

or

$$\langle m - 1/2, 1/2 | j_1 + 1/2, m \rangle = \sqrt{\frac{j_1 + m + 1/2}{2j_1 + 1}} \langle j_1, 1/2 | j_1 + 1/2, j_1 + 1/2 \rangle \qquad (5.5.29)$$

and from (5.5.7), with the + sign, to

$$\langle m - 1/2, 1/2 | j_1 + 1/2, m \rangle = \sqrt{\frac{j_1 + m + 1/2}{2j_1 + 1}}. \qquad (5.5.30)$$

Similarly for $j = j_1 + 1/2$, $m_2 = -1/2$, and now taking the upper sign in (5.5.27) leads to

$$\langle m + 1/2, -1/2 | j_1 + 1/2, m \rangle$$

$$= \sqrt{\frac{j_1 - m + 1/2}{2j_1 + 1}} \langle -j_1, -1/2 | j_1 + 1/2, -(j_1 + 1/2) \rangle \qquad (5.5.31)$$

and from (5.5.7), with the $-$ sign, leads to

$$\langle m+1/2,-1/2|j_1+1/2,m\rangle = \sqrt{\frac{j_1-m+1/2}{2j_1+1}}. \tag{5.5.32}$$

Upon using (5.5.23) with $m'_1 = m_1$, $m'_2 = m_2 = -1/2$, we have from (5.5.32)

$$\frac{j_1-m+1/2}{2j_1+1} + |\langle m+1/2,-1/2|j_1-1/2,m\rangle|^2 = 1 \tag{5.5.33}$$

and with a definite choice of phase to

$$\langle m+1/2,-1/2|j_1-1/2,m\rangle = \sqrt{\frac{j_1+m+1/2}{2j_1+1}}. \tag{5.5.34}$$

With this choice of phase, the phase of $\langle m-1/2,1/2|j_1-1/2,m\rangle$, is uniquely determined from (5.5.24) and one obtains

$$\langle m-1/2,1/2|j_1-1/2,m\rangle = -\sqrt{\frac{j_1-m+1/2}{2j_1+1}}. \tag{5.5.35}$$

These Clebsch-Gordan coefficients are tabulated in Table 5.3.

Table 5.3. The expression for the Clebsch-Gordan coefficients $\langle j_1,m_1;1/2,m_2|j,m\rangle$ for the addition of an arbitrary $j_1 \geqslant 1/2$ angular momentum to a $j_2 = 1/2$ one.

m_2 \ j	$j_1+1/2$	$j_1-1/2$
$1/2$	$\sqrt{\dfrac{j_1+m+1/2}{2j_1+1}}$	$-\sqrt{\dfrac{j_1-m+1/2}{2j_1+1}}$
$-1/2$	$\sqrt{\dfrac{j_1-m+1/2}{2j_1+1}}$	$\sqrt{\dfrac{j_1+m+1/2}{2j_1+1}}$

For the addition of two spin 1/2's, the above Table gives

$$|1,+1\rangle = |1/2,1/2\rangle \tag{5.5.36}$$

$$|1,0\rangle = \frac{1}{\sqrt{2}}\left(|1/2,-1/2\rangle + |-1/2,1/2\rangle\right) \tag{5.5.37}$$

$$|1,-1\rangle = |-1/2,-1/2\rangle \tag{5.5.38}$$

a triplet associated with spin 1, and a singlet

$$|0,0\rangle = \frac{1}{\sqrt{2}} \left(|1/2, -1/2\rangle - |-1/2, 1/2\rangle\right) \tag{5.5.39}$$

associated with spin 0. The triplet should be compared with the corresponding expressions in (2.8.59)–(2.8.61).

As another illustration, we consider the problem of constructing simultaneous eigenstates of the commuting operators (§2.3, §2.7) $\mathbf{J}^2$, J^3, $\mathbf{L}^2$, $\mathbf{S}^2$ which we denote by $|j, m, \ell, s\rangle$. In particular, for $s = 1/2$ we may use Table 5.3 directly to construct such eigenstates. These will be important in our study of the hydrogen atom when we include spin. To this end, using the notation $|j, m, \ell, 1/2\rangle \equiv |j, m, \ell\rangle$, and setting $j_1 = \ell$, $m_2 = m'$ in the Table, we have from (for $\ell \neq 0$)

$$|j, m, \ell\rangle = \sum_{m_\ell + m' = m} |\ell, m_\ell; 1/2, m'\rangle \langle \ell, m_\ell; 1/2, m' | j, m\rangle \tag{5.5.40}$$

that

$$|j, m, \ell\rangle = -|\ell, m-1/2; 1/2, 1/2\rangle \sqrt{\frac{\ell - m + 1/2}{2\ell + 1}} + |\ell, m+1/2; 1/2, -1/2\rangle$$

$$\times \sqrt{\frac{\ell + m + 1/2}{2\ell + 1}} \tag{5.5.41}$$

for $j = \ell - 1/2$, and

$$|j, m, \ell\rangle = |\ell, m-1/2; 1/2, 1/2\rangle \sqrt{\frac{\ell + m + 1/2}{2\ell + 1}} + |\ell, m+1/2; 1/2, -1/2\rangle$$

$$\times \sqrt{\frac{\ell - m + 1/2}{2\ell + 1}} \tag{5.5.42}$$

for $j = \ell + 1/2$.

Finally, we derive a useful *orthogonality* relation of the $D^{(j)}_{mm'}(\alpha, \beta, \gamma)$ functions given in (5.2.12) by combining two independent, i.e., commuting, angular momenta $\mathbf{J}_1$, $\mathbf{J}_2$ as defined in (5.5.1).

For the above purpose, we note that

$$\left\langle j_1, m_1; j_2, m_2 \left| \exp\left(\frac{i}{\hbar} \varphi \mathbf{n} \cdot \mathbf{J}\right) \right| j_1, m'_1; j_2, m'_2 \right\rangle$$

$$= \left\langle j_1, m_1 \left| \exp\left(\frac{i}{\hbar} \varphi \mathbf{n} \cdot \mathbf{J}_1\right) \right| j_1, m'_1 \right\rangle \left\langle j_2, m_2 \left| \exp\left(\frac{i}{\hbar} \varphi \mathbf{n} \cdot \mathbf{J}_2\right) \right| j_2, m'_2 \right\rangle$$

$$\equiv D^{(j_1)}_{m_1 m'_1}(\alpha, \beta, \gamma) D^{(j_2)}_{m_2 m'_2}(\alpha, \beta, \gamma). \tag{5.5.43}$$

5.5 Addition of Angular Momenta

Upon using the expansion (5.5.20), the above may be rewritten as

$$\sum_{j,m,m'} \langle j_1, m_1; j_2, m_2 | j, m \rangle D^{(j)}_{mm'}(\alpha, \beta, \gamma) \langle j, m' | j_1, m'_1; j_2, m'_2 \rangle$$

$$= D^{(j_1)}_{m_1 m'_1}(\alpha, \beta, \gamma) D^{(j_2)}_{m_2 m'_2}(\alpha, \beta, \gamma). \quad (5.5.44)$$

We will consider those cases in which j_1, j_2 are either both integers or are both half-odd integers. Then j necessarily takes on integer values. [For other cases see Problem 5.12].

Upon multiplying (5.5.44) by $\sin \beta$ and integrating over $0 \leqslant \alpha \leqslant 2\pi, 0 \leqslant \beta \leqslant \pi, 0 \leqslant \gamma \leqslant 2\pi$, we obtain

$$\int_0^{2\pi} d\alpha \int_0^{\pi} d\beta \sin \beta \int_0^{2\pi} d\gamma\, D^{(j_1)}_{m_1 m'_1}(\alpha, \beta, \gamma) D^{(j_2)}_{m_2 m'_2}(\alpha, \beta, \gamma)$$

$$= 4\pi^2 \sum_j \langle j_1, m_1; j_2, m_2 | j, 0 \rangle \int_0^{\pi} d\beta \sin \beta\, d^{(j)}_{00}(\beta) \langle j, 0 | j_1, m'_1; j_2, m'_2 \rangle \quad (5.5.45)$$

where we have used (5.2.10) and (5.2.13). For $j = 0$, (5.2.10) gives $d^{(j)}_{00}(\beta) = 1$. For $j \neq 0$, (5.2.10) leads explicitly to

$$I^{(j)} \equiv \int_0^{\pi} d\beta \sin \beta\, d^{(j)}_{00}(\beta)$$

$$= 4 \sum_{q=0}^{j} \frac{(j!)^2 (-1)^q}{[(j-q)!\, q!]^2} \int_0^{\pi/2} dx\, (\sin x)^{2q+1} (\cos x)^{2j+1-2q} \quad (5.5.46)$$

which upon using the integral

$$\int_0^{\pi/2} dx\, (\sin x)^{2a-1} (\cos x)^{2b-1} = \frac{1}{2} \frac{\Gamma(a)\, \Gamma(b)}{\Gamma(a+b)} \quad (5.5.47)$$

for non-negative integers a, b, gives

$$I^{(j)} \equiv \frac{2}{(j+1)} \sum_{q=0}^{j} (-1)^q \frac{j!}{(j-1)!\, q!} = \frac{2}{(j+1)} (1-1)^j = 0. \quad (5.5.48)$$

That is,

$$\int_0^{\pi} d\beta \sin \beta\, d^{(j)}_{00}(\beta) = 2\delta_{j0}. \quad (5.5.49)$$

The reader will recognize (5.5.49) as a special case of the orthogonality condition of the Legendre polynomials $P_\ell(\cos \theta) = P_\ell^{|m|}(\cos \theta)$ for $m = 0$ (see (5.3.30), (5.3.65), (5.3.51), (5.2.13)), with $d^{(j)}_{00}(\beta) = P_j(\cos \beta)$.

284   5 Angular Momentum Gymnastics

Hence the right-hand side of (5.5.45) becomes equal to

$$8\pi^2 \langle j_1, m_1; j_2, m_2 | 0, 0 \rangle \langle 0, 0 | j_1, m'_1; j_2, m'_2 \rangle$$

$$= 8\pi^2 \delta_{j_1 j_2} \delta_{m_1, -m_2} \delta_{m'_1, -m'_2} (2j_1 + 1)^{-1} (-1)^{2j_1 - m_1 - m'_1}$$

$$= 8\pi^2 \delta_{j_1 j_2} \delta_{m_1, -m_2} \delta_{m'_1, -m'_2} (2j_1 + 1)^{-1} (-1)^{m'_2 - m_2} (-1)^{2(j_2 + m_2)} \quad (5.5.50)$$

as follows from the general expression of the Clebsch-Gordan coefficients that will be obtained later in (5.6.14). Finally, we use the complex conjugate property of the $D_{mm'}^{(j)}(\alpha, \beta, \gamma)$ function in (5.2.14) to obtain from (5.5.45)

$$\int_0^{2\pi} d\alpha \int_0^{2\pi} d\gamma \int_0^{\pi} d\beta \, \sin\beta \, \left(D_{m_1 m'_1}^{(j_1)}(\alpha, \beta, \gamma) \right)^* D_{m_2 m'_2}^{(j_2)}(\alpha, \beta, \gamma)$$

$$= \frac{8\pi^2 \delta_{j_1 j_2} \delta_{m_1 m_2} \delta_{m'_1 m'_2}}{2j_1 + 1} \quad (5.5.51)$$

since $(-1)^{2(j_2 + m_2)} = 1$, where j_1, j_2 are either both (non-negative) integers or are both half-odd integers (see also Problem 5.12).

Clearly, the same procedure as above may be used to obtain the integral of the product of more than two $D_{mm'}^{(j)}$ functions.

Another useful orthogonality relation is the following one

$$\int_0^{2\pi} d\alpha \int_0^{\pi} d\beta \, \sin\beta \, \left(D_{mm'}^{(j_1)}(\alpha, \beta, -\alpha) \right)^* D_{mm''}^{(j_2)}(\alpha, \beta, -\alpha) = \frac{4\pi \delta_{j_1 j_2} \delta_{m'm''}}{(2j_1 + 1)} \quad (5.5.52)$$

where, again, j_1, j_2 are both either integers or are both half-odd integers.

The proof of (5.5.52) follows by noting that the integration over α, on its left-hand side, imposes the restriction that $m' = m''$. Accordingly from (5.2.10), (5.2.12) we may rewrite the left-hand side of (5.5.52) as

$$\frac{1}{2\pi} \delta_{m'm''} \int_0^{2\pi} d\gamma \int_0^{2\pi} d\alpha \int_0^{\pi} d\beta \, \sin\beta \, \left(D_{mm'}^{(j_1)}(\alpha, \beta, \gamma) \right)^* D_{mm'}^{(j_2)}(\alpha, \beta, \gamma) \quad (5.5.53)$$

which from (5.5.51) leads to (5.5.52).

5.6 Explicit Expression for the Clebsch-Gordan Coefficients

To obtain the explicit expression for the Clebsch-Gordan coefficients for the addition of any two independent angular momenta we may proceed as follows. Taking the upper sign in the recurrence relation (5.5.27) with $m = j$, $m_2 = j + 1 - m_1$, its left-hand side is then zero and leads to

5.6 Explicit Expression for the Clebsch-Gordan Coefficients

$$\sqrt{(j_1+m_1)(j_1-m_1+1)}\,\langle m_1-1,m_2|j,j\rangle$$
$$= -\sqrt{(j_2+m_2)(j_2-m_2+1)}\,\langle m_1,m_2-1|j,j\rangle. \quad (5.6.1)$$

By successive replacements, $m_1 \to m_1+1 \to m_1+2,\ldots,j_1$ in this equation, we get the following chain of equalities

$$\langle m_1, j-m_1|j,j\rangle = -\sqrt{\frac{(j_2+j-m_1)(j_2-j+m_1+1)}{(j_1+m_1+1)(j_1-m_1)}}$$
$$\times \langle m_1+1, j-m_1-1|j,j\rangle$$

$$\langle m_1+1, j-m_1-1|j,j\rangle = -\sqrt{\frac{(j_2+j-m_1-1)(j_2-j+m_1+2)}{(j_1+m_1+2)(j_1-m_1-1)}}$$
$$\times \langle m_1+2, j-m_1-2|j,j\rangle$$

$$\vdots \qquad (5.6.2)$$

$$\langle j_1-1, j-j_1+1|j,j\rangle = -\sqrt{\frac{(j_2+j-j_1+1)(j_2-j+j_1)}{2j_1}}$$
$$\times \langle j_1, j-j_1|j,j\rangle.$$

Upon taking the product of these equalities we obtain

$$\langle m_1, j-m_1|j,j\rangle$$
$$= \frac{(-1)^{j_1-m_1}}{\sqrt{(2j_1)!}}\sqrt{\frac{(j_2-j+j_1)!}{(j_2+j-j_1)!}}\sqrt{\frac{(j_1+m_1)!(j_2+j-m_1)!}{(j_1-m_1)!(j_2-j+m_1)!}}\,\langle j_1, j-j_1|j,j\rangle \quad (5.6.3)$$

[It is easily checked that the arguments of all the factorials are non-negative integers.] To obtain the expression for $\langle j_1, j-j_1|j,j\rangle$, we use the unitarity condition

$$\sum_{m_1+m_2=j}|\langle m_1, j-m_1|j,j\rangle|^2 = 1 \quad (5.6.4)$$

as obtained from (5.5.24), and the sum

$$\sum_{m_1+m_2=j}\frac{(j_1+m_1)!(j_2+m_2)!}{(j_1-m_1)!(j_2-m_2)!}$$
$$= \frac{(j+j_1+j_2+1)!(j_2-j_1+j)!(j_1-j_2+j)!}{(2j+1)!(j_1+j_2-j)!} \quad (5.6.5)$$

with $-j_1 \leqslant m_1 \leqslant j_1$, $-j_2 \leqslant m_2 \leqslant j_2$, established below. With the phase convention, generally referred to as the Condon and Shortley convention, defined by

$$\langle j_1, j - j_1 | j, j \rangle \equiv |\langle j_1, j - j_1 | j, j \rangle| \qquad (5.6.6)$$

equations (5.6.3)–(5.6.6) immediately lead to

$$\langle m_1, j - m_1 | j, j \rangle$$

$$= (-1)^{j_1 - m_1} \sqrt{\frac{(2j+1)!\,(j_1 + j_2 - j)!}{(j + j_1 + j_2 + 1)!\,(j_1 - j_2 + j)!\,(j_2 - j_1 + j)!}}$$

$$\times \sqrt{\frac{(j_1 + m_1)!\,(j_2 + j - m_1)!}{(j_1 - m_1)!\,(j_2 - j + m_1)!}}. \qquad (5.6.7)$$

To obtain the expression for the general coefficient $\langle m_1, m_2 | j, m \rangle$, we note from (5.1.26) that

$$|j, m\rangle = \sqrt{\frac{(j+m)!}{(2j)!\,(j-m)!}} \frac{1}{(\hbar)^{j-m}} (J_-)^{j-m} |j, j\rangle \qquad (5.6.8)$$

where

$$|j, j\rangle = \sum_{m_1 = -j_1}^{j_1} |m_1, j - m_1\rangle \langle m_1, j - m_1 | j, j \rangle \qquad (5.6.9)$$

and hence amounts to evaluating

$$\sqrt{\frac{(j+m)!}{(2j)!\,(j-m)!}} \frac{1}{(\hbar)^{j-m}} \left\langle m_1, j - m_1 \left| (J_-)^{j-m} \right| j, j \right\rangle \equiv \langle m_1, j - m_1 | j, m \rangle. \qquad (5.6.10)$$

The evaluation of the expression on the left-hand side of the above is straightforward. To this end we use the binomial expansion

$$(J_-)^{j-m} = \sum_{k=0}^{j-m} \binom{j-m}{k} (J_{1-})^k (J_{2-})^{j-m-k} \qquad (5.6.11)$$

where

$$\binom{j-m}{k} = \frac{(j-m)!}{k!\,(j-m-k)!} \qquad (5.6.12)$$

since J_{1-} and J_{2-} commute, to obtain from (5.1.26), (5.6.9), (5.6.10)

$$\frac{(J_-)^{j-m}}{(\hbar)^{j-m}} |j, j\rangle$$

5.6 Explicit Expression for the Clebsch-Gordan Coefficients

$$= \sum_{m_1+m_2=j} \sqrt{\frac{(j_1+m_1)!\,(j_2+m_2)!}{(j_1-m_1)!\,(j_2-m_2)!}} \sum_k |m_1-k\rangle\,|m+m_2-j+k\rangle$$

$$\times \frac{(j-m)!}{k!\,(j-m-k)!} \sqrt{\frac{(j_1-m_1+k)!\,(j_2-m_2+j-m-k)!}{(j_1+m_1-k)!\,(j_2+m_2-j+m+k)!}} \,\langle m_1,m_2|j,j\rangle \tag{5.6.13}$$

with $-j_1 \leqslant m_1 \leqslant j_1$, $-j_2 \leqslant m_2 \leqslant j_2$, and the sum over k is restricted to all k non-negative integers such that the arguments of the factorials are non-negative.

From the expression for $\langle m_1,m_2|j,j\rangle \equiv \langle m_1,j-m_1|j,j\rangle$ in (5.6.3) together with (5.6.10) and (5.1.21), we finally obtain, the so-called Racah expression of the Clebsch-Gordan coefficients:

$$\langle m_1,m_2|j,m\rangle \equiv \langle j_1,m_1;j_2,m_2|j,m\rangle$$

$$= \sqrt{2j+1}$$

$$\times \sqrt{\frac{(j_1+j_2-j)!\,(j-m)!\,(j+m)!\,(j_1-m_1)!\,(j_2-m_2)!}{(j+j_1+j_2+1)!\,(j_1-j_2+j)!\,(j_2-j_1+j)!\,(j_1+m_1)!\,(j_2+m_2)!}}$$

$$\times \sum_k \frac{(-1)^{j_1-m_1-k}}{k!\,(j-m-k)!} \frac{(j_1+m_1+k)!\,(j_2+j-m_1-k)!}{(j_1-m_1-k)!\,(j_2-j+m_1+k)!} \tag{5.6.14}$$

where now $m_1+m_2=m$, and k is over all non-negative integers such that the arguments of the factorials are non-negative. The *reality* condition of the Clebsch-Gordan coefficients as given in (5.6.14) is to be noted.

It remains to establish the expression for the sum as given in (5.6.5). To this end, for any two strictly positive integers a and b, the formal expansion

$$(x+y)^{-a} = \sum_{k_1=0}^{\infty} (-1)^{k_1} \frac{(y)^{k_1}\,x^{-a-k_1}}{k_1!} \frac{(a+k_1-1)!}{(a-1)!} \tag{5.6.15}$$

gives immediately, upon the comparison of the product of the expansions of $(x+y)^{-a} \cdot (x+y)^{-b}$ with the expansion of $(x+y)^{-(a+b)}$,

$$\sum_{k_1+k_2=k} \frac{(a+k_1-1)!}{(a-1)!} \frac{(b+k_2-1)!}{(b-1)!\,k_1!\,k_2!} = \frac{(a+b+k-1)!}{(a+b-1)!\,k!} \tag{5.6.16}$$

with k_1, k_2 non-negative integers. Upon setting

$$k_1 = j_1 - m_1, \qquad k_2 = j_2 - m_2 \tag{5.6.17}$$

$$a-1 = j_2 - j_1 + j, \qquad b-1 = j_1 - j_2 + j \tag{5.6.18}$$

and hence with $m_1 + m_2 = j$,

$$k = k_1 + k_2 = j_1 + j_2 - j \tag{5.6.19}$$

(5.6.16) leads to (5.6.5).

In §5.5, we have added a spin 1/2 to an arbitrary non-zero spin with the Clebsch-Gordan coefficients given in Table 5.3. As an application of the general formula (5.6.14), one may consider the addition of a spin 1 to an arbitrary spin $\geqslant 1$. The results for the Clebsch-Gordan coefficients are summarized in Table 5.4.

Table 5.4. The expression for the Clebsch-Gordan coefficients $\langle j_1, m_1; 1, m_2 | j, m \rangle$ for the addition of an arbitrary $j_1 \geqslant 1$ angular momentum to a $j_2 = 1$ one.

$m_2 \backslash j$	$j_1 - 1$	j_1	$j_1 + 1$
1	$\sqrt{\dfrac{(j_1 - m)(j_1 - m + 1)}{2j_1(2j_1 + 1)}}$	$-\sqrt{\dfrac{(j_1 + m)(j_1 - m + 1)}{2j_1(j_1 + 1)}}$	$\sqrt{\dfrac{(j_1 + m)(j_1 + m + 1)}{(2j_1 + 1)(2j_1 + 2)}}$
0	$-\sqrt{\dfrac{(j_1 - m)(j_1 + m)}{j_1(2j_1 + 1)}}$	$\dfrac{m}{\sqrt{j_1(j_1 + 1)}}$	$\sqrt{\dfrac{(j_1 - m + 1)(j_1 + m + 1)}{(2j_1 + 1)(j_1 + 1)}}$
-1	$\sqrt{\dfrac{(j_1 + m + 1)(j_1 + m)}{2j_1(2j_1 + 1)}}$	$\sqrt{\dfrac{(j_1 - m)(j_1 + m + 1)}{2j_1(j_1 + 1)}}$	$\sqrt{\dfrac{(j_1 - m)(j_1 - m + 1)}{(2j_1 + 1)(2j_1 + 2)}}$

For future reference, we record the following particular Clebsch-Gordan coefficient in a unified manner $(m_1 + m_2 = m)$ *for* $j = j_1$:

$$\langle j, m_1; 1, m_2 | j, m \rangle$$

$$= \frac{1}{\sqrt{j(j+1)}} \left[m_1 \delta_{m_2, 0} \mp \frac{1}{\sqrt{2}} \sqrt{(j \mp m_1)(j \pm m_1 + 1)} \delta_{m_2, \pm 1} \right]$$

$$\equiv \langle j, m | j, m_1; 1, m_2 \rangle \tag{5.6.20}$$

as follows from the above Table.

Some symmetry properties of the Clebsch-Gordan coefficients are

$$\langle j_1, m_1; j_2, m_2 | j, m \rangle = (-1)^{j_1 + j_2 - j} \langle j_2, m_2; j_1, m_1 | j, m \rangle \tag{5.6.21}$$

$$\langle j_1, m_1; j_2, m_2 | j, m \rangle = (-1)^{j_1 + j_2 - j} \langle j_1, -m_1; j_2, -m_2 | j, -m \rangle \tag{5.6.22}$$

$$\langle j_1, m_1; j_2, m_2 | j, m \rangle = (-1)^{j_2 + m_2} \sqrt{\frac{2j + 1}{2j_2 + 1}} \langle j_2, -m_2; j, m | j_1, m_1 \rangle . \tag{5.6.23}$$

5.6 Explicit Expression for the Clebsch-Gordan Coefficients

To display such symmetry properties, it is more convenient to introduce the so-called 3-j symbols, due to Wigner, defined in terms of the Clebsch-Gordan coefficients as follows:

$$\begin{pmatrix} j_1 & j_2 & j \\ m_1 & m_2 & m \end{pmatrix} = (-1)^{j_1-j_2-m} \frac{1}{\sqrt{2j+1}} \langle j_1, m_1; j_2, m_2 | j, -m \rangle \quad (5.6.24)$$

where we note that $m_1 + m_2 + m = 0$, and the 3-j symbols are zero otherwise.

Some symmetry properties of the latter symbols are

$$\begin{pmatrix} j_1 & j_2 & j_3 \\ m_1 & m_2 & m_3 \end{pmatrix} = \begin{pmatrix} j_3 & j_1 & j_2 \\ m_3 & m_1 & m_2 \end{pmatrix} = \begin{pmatrix} j_2 & j_3 & j_1 \\ m_2 & m_3 & m_1 \end{pmatrix} \quad (5.6.25)$$

$$\begin{pmatrix} j_1 & j_2 & j_3 \\ m_1 & m_2 & m_3 \end{pmatrix} = (-1)^{j_1+j_2+j_3} \begin{pmatrix} j_2 & j_1 & j_3 \\ m_2 & m_1 & m_3 \end{pmatrix} \quad (5.6.26)$$

$$\begin{pmatrix} j_1 & j_2 & j_3 \\ m_1 & m_2 & m_3 \end{pmatrix} = (-1)^{j_1+j_2+j_3} \begin{pmatrix} j_1 & j_2 & j_3 \\ -m_1 & -m_2 & -m_3 \end{pmatrix}. \quad (5.6.27)$$

The following particular values of the 3-j symbols are to be noted

$$\begin{pmatrix} j_1 & j_2 & j_3 \\ 0 & 0 & 0 \end{pmatrix} = 0, \quad \text{if} \quad J \equiv j_1 + j_2 + j_3 \text{ is } odd \quad (5.6.28)$$

as follows directly from (5.6.27), and

$$\begin{pmatrix} j_1 & j_2 & j_3 \\ 0 & 0 & 0 \end{pmatrix} = (-1)^{J/2} \sqrt{\frac{(J-2j_1)!\,(J-2j_2)!\,(J-2j_3)!}{(J+1)!}}$$

$$\times \frac{(\frac{1}{2}J)!}{(\frac{1}{2}J-j_1)!\,(\frac{1}{2}J-j_2)!\,(\frac{1}{2}J-j_3)!} \quad (5.6.29)$$

if $J \equiv j_1 + j_2 + j_3$ is $even$.

From (5.6.24), (5.5.23), we also have the orthogonality property

$$\sum_j (2j+1) \begin{pmatrix} j_1 & j_2 & j \\ m_1 & m_2 & m \end{pmatrix} \begin{pmatrix} j_1 & j_2 & j \\ m'_1 & m'_2 & m \end{pmatrix} = \delta_{m_1 m'_1} \delta_{m_2 m'_2} \quad (5.6.30)$$

where $m = -m_1 - m_2 = -m'_1 - m'_2$ and (5.5.24) leads to

$$\sum_{m_1, m_2} \begin{pmatrix} j_1 & j_2 & j \\ m_1 & m_2 & m \end{pmatrix} \begin{pmatrix} j_1 & j_2 & j' \\ m_1 & m_2 & m' \end{pmatrix} = \frac{1}{(2j+1)} \delta_{jj'} \delta_{mm'} \quad (5.6.31)$$

and the triangular condition on j, j_1, j_2 is understood.

5.7 Vector Operators

The operators $\mathbf{X}, \mathbf{P}, \mathbf{N}, \mathbf{L}, \mathbf{S}$ and the angular momentum operator $\mathbf{J}$ itself all satisfy the same commutation relations with $\mathbf{J}$ (see §2.3):

$$[V^i, J^j] = i\hbar \varepsilon^{ijk} V^k \tag{5.7.1}$$

where $\mathbf{V}$ denotes any one of the operators mentioned above, reflecting the vector character of these operators. Any such operator satisfying the commutation relations (5.7.1) is referred to as a *vector operator*. The commutation relations $[V^i, V^j]$, however, may be different for different vector operators $\mathbf{V}$. For example, the different components of $\mathbf{P}$ commute while the different components of $\mathbf{S}$ do not.

We are interested in evaluating matrix elements of the form $\langle j, m'| \mathbf{V} |j, m\rangle$. To this end, it is more convenient to define the *spherical vector* components:

$$V(m) = \delta_{m,0} V^3 - \frac{1}{\sqrt{2}} \delta_{m,1} V_+ + \frac{1}{\sqrt{2}} \delta_{m,-1} V_- \tag{5.7.2}$$

$m = -1, 0, +1$, where

$$V_\pm = V^1 \pm i V^2 \tag{5.7.3}$$

as opposed to the Cartesian components V^1, V^2, V^3.

It is straightforward to show from (5.7.1) that

$$[J(m_1), V(m_2)] = \hbar \left[m_2 \delta_{m_1,0} \mp \frac{1}{\sqrt{2}} \sqrt{(1 \mp m_2)(1 \pm m_2 + 1)} \delta_{m_1, \pm 1} \right]$$

$$\times V(m_1 + m_2). \tag{5.7.4}$$

In particular, for $m_1 = m_2$, $[J(m_1), V(m_1)] \equiv 0$, and there are no ambiguities associated with the notations $V(\pm 2)$, arising on the right-hand side of (5.7.4), since their coefficients are always identically equal to zero.

We also have

$$J(M) |j, m\rangle = \hbar \left[m \delta_{M,0} \mp \frac{1}{\sqrt{2}} \sqrt{(j \mp m)(j \pm m + 1)} \delta_{M, \pm 1} \right] |j, m + M\rangle$$

$$\equiv |j, m + M\rangle \langle j, m + M | J(M) | j, m\rangle \tag{5.7.5}$$

and

$$\langle j, m' | J(M) = \hbar \left[m' \delta_{M,0} \mp \frac{1}{\sqrt{2}} \sqrt{(j \pm m')(j \mp m' + 1)} \delta_{M, \pm 1} \right] \langle j, m' - M |$$

$$\equiv \langle j, m' | J(M) | j, m' - M \rangle \langle j, m' - M | \tag{5.7.6}$$

where $M = -1, 0, +1$.

5.7 Vector Operators

One may rewrite the commutations relations in (5.7.4) as

$$[J(m_1), V(m_2)] = \langle 1, m_1 + m_2 | J(m_1) | 1, m_2 \rangle V(m_1 + m_2). \tag{5.7.7}$$

By taking the matrix elements of (5.7.7) between $\langle j, m' |$ and $|j, m\rangle$, and using (5.7.5), (5.7.6) we obtain

$$\langle j, m' | J(m_1) | j, m' - m_1 \rangle \langle j, m' - m_1 | V(m_2) | j, m \rangle$$
$$- \langle j, m' | V(m_2) | j, m + m_1 \rangle \langle j, m + m_1 | J(m_1) | j, m \rangle$$
$$= \langle j, m' | V(m_1 + m_2) | j, m \rangle \langle 1, m_1 + m_2 | J(m_1) | 1, m_2 \rangle. \tag{5.7.8}$$

In particular we note that

$$\langle j, m' | J(0) | j, m \rangle = \hbar m \, \delta_{m', m}. \tag{5.7.9}$$

For $m_1 = 0, m_2 \to M$, (5.7.8) then gives

$$\hbar (m' - m - M) \langle j, m' | V(M) | j, m \rangle = 0 \tag{5.7.10}$$

for the spherical vector components of a vector operator. That is, $\langle j, m' | V(M) | j, m \rangle$ is necessarily zero unless

$$m' = m + M. \tag{5.7.11}$$

From (5.7.5), we note that

$$\langle 1, 2M | J(M) | 1, M \rangle \equiv 0 \tag{5.7.12}$$

accordingly, for $m_1 = m_2 \equiv M$, (5.7.8) gives, upon using (5.7.11), that

$$\frac{\langle j, m + M | V(M) | j, m \rangle}{\langle j, m + M | J(M) | j, m \rangle} = \frac{\langle j, m + 2M | V(M) | j, m + M \rangle}{\langle j, m + 2M | J(M) | j, m + M \rangle}. \tag{5.7.13}$$

That is, in particular,

$$\frac{\langle j, m \pm 1 | V(\pm 1) | j, m \rangle}{\langle j, m \pm 1 | J(\pm 1) | j, m \rangle} = \frac{\langle j, m \pm 2 | V(\pm 1) | j, m \pm 1 \rangle}{\langle j, m \pm 2 | J(\pm 1) | j, m \pm 1 \rangle}. \tag{5.7.14}$$

Similarly, for $m_1 = \mp 1, m_2 = \pm 1, m' = m$, (5.7.8) leads to

$$\langle j, m | J(\mp 1) | j, m \pm 1 \rangle \langle j, m \pm 1 | V(\pm 1) | j, m \rangle$$
$$- \langle j, m | V(\pm 1) | j, m \mp 1 \rangle \langle j, m \mp 1 | J(\mp 1) | j, m \rangle$$
$$= \langle j, m | V(0) | j, m \rangle \langle 1, 0 | J(\mp 1) | 1, \pm 1 \rangle$$
$$= \pm \hbar \langle j, m | V(0) | j, m \rangle \tag{5.7.15}$$

where we have used (5.7.2) for $J(\mp 1)$ in writing the last equality. Upon using (5.7.14), we may rewrite (5.7.15) as

$$\left\{ |\langle j,m|J(\mp 1)|j,m\pm 1\rangle|^2 - |\langle j,m\mp 1|J(\mp 1)|j,m\rangle|^2 \right\} \frac{\langle j,m\pm 1|V(\pm 1)|j,m\rangle}{\langle j,m\pm 1|J(\pm 1)|j,m\rangle}$$

$$= \pm\hbar \langle j,m|V(0)|j,m\rangle . \qquad (5.7.16)$$

For $V(\pm 1) \equiv J(\pm 1)$, the expression in the curly brackets is $\pm\hbar \langle j,m|J(0)|j,m\rangle$. Accordingly, (5.7.16) simplifies to

$$\langle j,m|V(0)|j,m\rangle = \langle j,m|J(0)|j,m\rangle \frac{\langle j,m\pm 1|V(\pm 1)|j,m\rangle}{\langle j,m\pm 1|J(\pm 1)|j,m\rangle}. \qquad (5.7.17)$$

From equations (5.7.14), (5.7.17), we then conclude that for $M = -1, 0, 1$,

$$\langle j,m'|V(M)|j,m\rangle = \frac{1}{\hbar}\frac{\langle j,m'|J(M)|j,m\rangle}{\sqrt{j(j+1)}} C(\mathbf{V},j) \qquad (5.7.18)$$

where $C(\mathbf{V},j)$, a proportionality factor, characteristic of the vector operator $\mathbf{V}$, which may depend on j, but is necessarily *independent* of m, m', M. It should be noted that (5.7.18) is valid for *all* $m, m' = -j, -j+1, \ldots, j$. The $1/\sqrt{j(j+1)}$ factor is inserted for convenience. Equations (5.7.5) and (5.6.20) allow one to rewrite

$$\frac{\langle j,m'|J(M)|j,m\rangle}{\hbar\sqrt{j(j+1)}} = \langle j,m'|j,m;1,M\rangle \qquad (5.7.19)$$

appearing in (5.7.18), in terms of a Clebsch-Gordan coefficient as given in (5.6.20).

It has become customary to denote the proportionality factor $C(\mathbf{V},j)$ in (5.7.18) as

$$C(\mathbf{V},j) \equiv \langle j\|\mathbf{V}\|j\rangle . \qquad (5.7.20)$$

Don't let the notation scare you. Due to its independence of m, m', it may be formally evaluated and defined by

$$\langle j\|\mathbf{V}\|j\rangle = \frac{\langle j,m'|V(m)|j,m\rangle}{\langle j,m'|j,m;1,M\rangle} \qquad (5.7.21)$$

for any allowed and conveniently chosen values for m', m, for which the evaluation of the expression on the right-hand side of (5.7.21) is unambiguous. This will be done explicitly later for various cases.

From the definition of $V(M)$ in (5.7.2), (5.7.3), we may rewrite (5.7.18) as

$$\langle j,m'|\mathbf{V}|j,m\rangle = \frac{\langle j,m'|\mathbf{J}|j,m\rangle}{\hbar\sqrt{j(j+1)}} \langle j\|\mathbf{V}\|j\rangle . \qquad (5.7.22)$$

This is a particular case of the so-called Wigner-Eckart theorem that will be established quite generally in the next section.

For $\mathbf{V} = \mathbf{J}$, (5.7.22) gives

$$\langle j||\mathbf{J}||j\rangle = \hbar\sqrt{j(j+1)}. \tag{5.7.23}$$

The factor $\langle j||\mathbf{V}||j\rangle$ is referred to as *a reduced matrix element* of $\mathbf{V}$. [Some authors provide a different definition of a reduced matrix element by multiplying the latter by a given function of j. In any case such an additional normalization factor may be absorbed in the definition of a reduced matrix element.]

As shown below, the reduced matrix element $\langle j||\mathbf{V}||j\rangle$ element may be explicitly written as

$$\langle j||\mathbf{V}||j\rangle = \frac{\langle j,m|\mathbf{V}\cdot\mathbf{J}|j,m\rangle}{\hbar\sqrt{j(j+1)}} \tag{5.7.24}$$

(with $j \neq 0$) and due to its independence of m, it may be evaluated for any of its values.

To show (5.7.24), note that

$$\mathbf{V}\cdot\mathbf{J} = \sum_{N=0,\pm 1} (-1)^N V(-N) J(N) \tag{5.7.25}$$

and hence

$\langle j,m|\mathbf{V}\cdot\mathbf{J}|j,m\rangle$

$= \sum_{N=0,\pm 1} (-1)^N \langle j,m|V(-N)|j,m+N\rangle \langle j,m+N|J(N)|j,m\rangle$

$= \sum_{N=0,\pm 1} (-1)^N \langle j,m|J(-N)|j,m+N\rangle \langle j,m+N|J(N)|j,m\rangle$

$\times \dfrac{\langle j||\mathbf{V}||j\rangle}{\hbar\sqrt{j(j+1)}}$

$= \sum_{N=0,\pm 1} (-1)^N \langle j,m|J(-N)J(N)|j,m\rangle \langle j||\mathbf{V}||j\rangle$

$$= \frac{\langle j,m|\mathbf{J}^2|j,m\rangle \langle j||\mathbf{V}||j\rangle}{\hbar\sqrt{j(j+1)}} \tag{5.7.26}$$

which upon using the fact that $\mathbf{J}^2|j,m\rangle = \hbar^2 j(j+1)|j,m\rangle$ establishes (5.7.24). In writing the second equality in (5.7.26) use has been made of (5.7.22).

We may then rewrite (5.7.22), in the convenient form

$$\langle j, m' | \mathbf{V} | j, m \rangle = \langle j, m' | \mathbf{J} | j, m \rangle \frac{\langle j, m | \mathbf{V} \cdot \mathbf{J} | j, m \rangle}{\hbar^2 j(j+1)} \qquad (5.7.27)$$

for the matrix elements of a vector operator $\mathbf{V}$ we were seeking.

As an application of (5.7.27), we evaluate the matrix element $\langle j, m' | (J^3 + S^3) | j, m \rangle$. This latter will occur in our study of the Zeeman effect in §7.9. To this end, (5.7.27) leads to

$$\langle j, m' | (J^3 + S^3) | j, m \rangle = \langle j, m' | J^3 | j, m \rangle \frac{\langle j, m | (\mathbf{J}^2 + \mathbf{S} \cdot \mathbf{J}) | j, m \rangle}{\hbar^2 j(j+1)}$$

$$= \hbar m \, \delta_{m'm} \left\{ 1 + \frac{\langle j, m | \mathbf{S} \cdot \mathbf{J} | j, m \rangle}{\hbar^2 j(j+1)} \right\}. \qquad (5.7.28)$$

Upon writing

$$\mathbf{S} \cdot \mathbf{J} = \frac{1}{2} \left(\mathbf{J}^2 - \mathbf{L}^2 + \mathbf{S}^2 \right) \qquad (5.7.29)$$

and carrying out the expansion

$$|j, m\rangle = \sum_{m_\ell + m_s = m} |\ell, m_\ell; s, m_s\rangle \langle \ell, m_\ell; s, m_s | j, m \rangle \qquad (5.7.30)$$

where $j = |\ell - s|, \ldots, \ell + s$, we obtain for the expression in the early brackets in (5.7.28)

$$\left\{ 1 + \frac{j(j+1) - \ell(\ell+1) + s(s+1)}{2j(j+1)} \right\}. \qquad (5.7.31)$$

The states in (5.7.30) are simultaneous eigenstates of the commuting operators $\mathbf{J}^2, \mathbf{S}^2, \mathbf{L}^2, J^3$. [Here we note that $[\mathbf{J}^2, S^3] \neq 0$, $[\mathbf{J}^2, L^3] \neq 0$.]

In particular for $s = 1/2$, $\ell = 1, 2, \ldots$, (5.7.31) gives the famous Landé g-factor

$$g = \frac{j + 1/2}{\ell + 1/2} \qquad (5.7.32)$$

where $j = \ell \pm 1/2$. For $\ell = 0$, $s = 1/2$, (5.7.32) gives $g = 2$.

Another expression for (5.7.27) may be also provided by noting that

$$[\mathbf{V} \cdot \mathbf{J}, \mathbf{J}] = 0 \qquad (5.7.33)$$

and hence with $\mathbf{V}$ replaced by $\mathbf{V} \cdot \mathbf{J} \mathbf{J}$ in (5.7.27), we obtain

$$\langle j, m' | \mathbf{V} \cdot \mathbf{J} \mathbf{J} | j, m \rangle = \langle j, m' | \mathbf{J} | j, m \rangle \langle j, m' | \mathbf{V} \cdot \mathbf{J} | j, m \rangle. \qquad (5.7.34)$$

This then allows us to rewrite (5.7.27) simply as

$$\langle j, m' | \mathbf{V} | j, m \rangle = \frac{\langle j, m' | \mathbf{V} \cdot \mathbf{J} \mathbf{J} | j, m \rangle}{\hbar^2 j(j+1)} \qquad (5.7.35)$$

$$\equiv \left\langle j, m' \left| \frac{\mathbf{V} \cdot \mathbf{J} \mathbf{J}}{\mathbf{J}^2} \right| j, m \right\rangle \tag{5.7.36}$$

(for $j \neq 0$). This formally shows that only the component of $\mathbf{V}$ along the angular momentum may contribute to the matrix element $\langle j, m' | \mathbf{V} | j, m \rangle$.

The transformation of a vector operator $\mathbf{V}$ under arbitrary finite rotations may be worked out from (5.7.1). To this end let (see (2.3.18), (2.3.43))

$$\exp\left(\frac{-i}{\hbar}\varphi\mathbf{n}\cdot\mathbf{J}\right) V^i \exp\left(\frac{i}{\hbar}\varphi\mathbf{n}\cdot\mathbf{J}\right) = F^i(\varphi) \tag{5.7.37}$$

for a given fixed unit vector $\mathbf{n}$. Then (5.7.1) gives

$$\frac{d}{d\varphi}\mathbf{F}(\varphi) = -\mathbf{n}\times\mathbf{F}(\varphi). \tag{5.7.38}$$

On the other hand, let (§2.1)

$$G^i(\varphi) = V^k R^{ik}(\varphi, \mathbf{n}) \tag{5.7.39}$$

then

$$\frac{d}{d\varphi}\mathbf{G}(\varphi) = -\mathbf{n}\times\mathbf{G}(\varphi). \tag{5.7.40}$$

With the boundary conditions:

$$\mathbf{G}(0) = \mathbf{V} = \mathbf{F}(0) \tag{5.7.41}$$

we may infer that

$$\exp\left(\frac{-i}{\hbar}\varphi\mathbf{n}\cdot\mathbf{J}\right) V^i \exp\left(\frac{i}{\hbar}\varphi\mathbf{n}\cdot\mathbf{J}\right) = V^k R^{ik}(\varphi, \mathbf{n}). \tag{5.7.42}$$

As an application of (5.7.42), consider

$$e^{-i\phi J^3/\hbar} J^2 e^{i\phi J^3/\hbar} = J^j R^{2j}(\phi, \hat{\mathbf{x}}^3) \tag{5.7.43}$$

where $\hat{\mathbf{x}}^3$ is a unit vector along the x^3 axis, and (see (2.1.4))

$$R^{2j}(\phi, \hat{\mathbf{x}}^3) = \delta^{2j}\cos\phi - \delta^{1j}\sin\phi \tag{5.7.44}$$

thus giving

$$e^{-i\phi J^3/\hbar} J^2 e^{i\phi J^3/\hbar} = -\mathbf{n}\cdot\mathbf{J} \tag{5.7.45}$$

with $\mathbf{n}$ now given by

$$\mathbf{n} = (\sin\phi, -\cos\phi, 0). \tag{5.7.46}$$

In particular (5.7.45) implies that

$$e^{-i\phi J^3/\hbar} e^{i\theta J^2/\hbar} e^{i\phi J^3/\hbar} = \exp\left[\frac{i}{\hbar}\theta\left(e^{-i\phi J^3/\hbar} J^2 e^{i\phi J^3/\hbar}\right)\right]$$

$$= \exp\left[\frac{-i}{\hbar} \theta \mathbf{n} \cdot \mathbf{J}\right]. \tag{5.7.47}$$

We will make use of this identity in §5.10.

Finally we note that for two vector operators $\mathbf{V}_1$, $\mathbf{V}_2$, the product $\mathbf{V}_1 \cdot \mathbf{V}_2$ commutes with $\mathbf{J}$:

$$[\mathbf{J}, \mathbf{V}_1 \cdot \mathbf{V}_2] = 0 \tag{5.7.48}$$

as expected and is readily checked from (5.7.1).

5.8 Tensor Operators

From the previous section, we recall that a vector operator is one such that its spherical components satisfy the commutation relations (see (5.7.4)) ($m = -1, 0, 1;\ M = -1, 0, 1$):

$$[J(m), V(M)] = \hbar \left[M\delta_{m,0} \mp \frac{1}{\sqrt{2}} \sqrt{(J \mp M)(J \pm M + 1)} \delta_{m,\pm 1} \right]$$
$$\times V(m+M) \tag{5.8.1}$$

with $J = 1$, reflecting the vector character of the operator in question.

Equation (5.8.1) naturally leads to define a tensor operator T_M^K of rank K with components specified by $M = -K, -K+1, \ldots, K$, as one satisfying the commutation solutions

$$[J(m), T_M^K] = \hbar \left[M\delta_{m,0} \mp \frac{1}{\sqrt{2}} \sqrt{(K \mp M)(K \pm M + 1)} \delta_{m,\pm 1} \right] T_{m+M}^K \tag{5.8.2}$$

by simply changing the $J = 1$ value to general $K = 1/2, 1, \ldots$ for the possible values of an angular momentum. The spherical components $J(0)$, $J(\pm 1)$ of the angular momentum operator $\mathbf{J}$ are defined in (5.7.2).

As a generalization of the relation in (5.7.18) (see also (5.7.19),(5.7.20)), we establish the following one

$$\langle j', m' | T_M^K | j, m \rangle = \langle j', m' | j, m; K, M \rangle \langle j' || T^K || j \rangle \tag{5.8.3}$$

where $\langle j' || T^K || j \rangle$ is a reduced matrix element, which may depend on j', j, K, but is *independent* of m', m, M. Here

$$m' = M + m \tag{5.8.4}$$

$$j' = |j - K|, |j - K| + 1, \ldots, j + K. \tag{5.8.5}$$

Different normalization are used by different authors to define a reduced matrix element $\langle j' || T^K || j \rangle$. This is done by multiplying (5.8.3) by a given function of j. We have chosen such a factor to be one to make direct comparison with the vector operator case defined earlier (see (5.7.18), (5.7.19)). In any case, as we have done here, such an additional normalization factor may be absorbed in the definition of a reduced matrix element.

The result embodied in (5.8.3) is referred to as the *Wigner-Eckart Theorem*. As a Clebsch-Gordan coefficient is a geometrical factor, the physical properties of T_M^K enter *only* in the reduced matrix element. In (5.8.3), we have suppressed other quantum numbers which may be characteristics of the physical system at hand.

To establish (5.8.3), we define the ket vector

$$|\psi(j',m';j,K)\rangle = \sum_{m+M=m'} T_M^K |j,m\rangle \langle j,m;K,M|j',m'\rangle \qquad (5.8.6)$$

where $|j-K| \leqslant j' \leqslant j+K$ in integer steps.

We multiply (5.8.6), by $\langle j',m'|j,m'';K,M''\rangle$, sum over j', and use the completeness relation (5.5.23), to obtain

$$T_M^K |j,m\rangle = \sum_{j'=|j-K|}^{j+K} |\psi(j',m';j,K)\rangle \langle j',m'|j,m;K,M\rangle. \qquad (5.8.7)$$

Upon applying the operator $J(M')$, to (5.8.6) where $M' = 0, \pm 1$, writing

$$J(M') T_M^K = [J(M'), T_M^K] + T_M^K J(M') \qquad (5.8.8)$$

using (5.7.5) for $J(M')|j,m\rangle$, and the explicit commutation relations (5.8.2), we obtain by making a change of the summation variables m, M in (5.8.6), and finally using (5.5.27) that:

$$J(M') |\psi(j',m';j,K)\rangle = \hbar \left\{ m'\delta_{M',0} \mp \frac{1}{\sqrt{2}} \sqrt{(j' \mp m')(j' \pm m' + 1)} \delta_{M',\pm 1} \right\}$$

$$\times |\psi(j',m'+M';j,K)\rangle. \qquad (5.8.9)$$

Equation (5.8.9), in particular, implies from (5.7.2) that

$$J^3 |\psi(j',m';j,K)\rangle = \hbar m' |\psi(j',m';j,K)\rangle \qquad (5.8.10)$$

$$J_\pm |\psi(j',m';j,K)\rangle = \hbar \sqrt{(j' \mp m')(j' \pm m' + 1)} |\psi(j',m' \pm 1;j,K)\rangle \qquad (5.8.11)$$

and from (5.11) or (5.12)

$$\mathbf{J}^2 |\psi(j',m';j,K)\rangle = \hbar^2 j'(j'+1) |\psi(j',m';j,K)\rangle. \qquad (5.8.12)$$

With the underlying assumption that we have at hand, together with $\mathbf{J}^2$, J^3 a complete set of commuting operators, with the $|j',m'\rangle$, as before,[2]

---

[2] For simplicity of the notation, we suppress additional quantum numbers that these states may depend on.

denoting their simultaneous eigenstates, we conclude that $|\psi(j',m';j,K)\rangle$ is proportional to $|j',m'\rangle$. That is

$$|\psi(j',m';j,K)\rangle = |j',m'\rangle\,\alpha(j',m';j,K). \qquad (5.8.13)$$

It is easy to show, however, that $\alpha(j',m';j,K)$ is independent of m'. To this end (5.1.23) reads

$$J_\pm |j',m'\rangle = C_\pm(j',m')|j',m'\pm 1\rangle \qquad (5.8.14)$$

where $C_\pm(j',m')$ are the coefficients defined in (5.8.11). On the other hand (5.8.11), (5.8.13) together (5.8.14) then give

$$J_\pm |\psi(j',m';j,K)\rangle = |j',m'\pm 1\rangle\, C_\pm(j',m')\,\alpha(j',m';j,K)$$

$$= |j',m'\pm 1\rangle\, C_\pm(j',m')\,\alpha(j',m'\pm 1;j,K) \qquad (5.8.15)$$

implying the independence of the factor $\alpha(j',m';j,K)$ of m'. Hence, we may rewrite (5.8.13) as

$$|\psi(j',m';j,K)\rangle = |j',m'\rangle\,\alpha(j';j,K). \qquad (5.8.16)$$

Multiplying (5.8.7) by $\langle j'',m''|$, using (5.8.16), immediately leads to

$$\langle j',m'|T_M^K|j,m\rangle = \langle j',m'|j,m;K,M\rangle\,\alpha(j';j,K) \qquad (5.8.17)$$

which is the desired result quoted in (5.8.3) upon appropriately identifying a reduced matrix element, characteristic of the operator T_M^K, and as shown above it is necessarily *independent* of m, m', M, where $m+M=m'$.

In particular, for $K=1$, we have $T_M^1 \equiv V(m)$, and for $j'=j$, (5.8.3), reduces to the expression for the vector operator obtained earlier in (5.7.18)–(5.7.20). For j' not necessarily equal to j, (5.8.3) generalizes (5.7.18) to

$$\langle j',m'|V(M)|j,m\rangle = \langle j',m'|j,m;1,M\rangle\,\langle j'||\mathbf{V}||j\rangle \qquad (5.8.18)$$

where $m'=m+M$, $j'=|j-1|,\ldots,j+1$.

From (5.8.16), (5.8.7) we also have the following expression for the action of the operator T_M^K on the angular momentum states $|j,m\rangle$:

$$T_M^K|j,m\rangle = \sum_{j'=|j-K|}^{j+K} |j',m'\rangle\,\langle j'||T^K||j\rangle\,\langle j',m'|j,m;K,M\rangle \qquad (5.8.19)$$

where $m'=m+M$.

Out of two rank K_1, K_2 tensor operators $T_{M_1}^{K_1}$, $T_{M_2}^{K_2}$ one may construct a K rank tensor operator T_M^K, where $M=M_1+M_2$, $K=|K_1-K_2|,\ldots,K_1+K_2$ as follows

$$F(K, K_1, K_2) T_M^K = \sum_{M_1+M_2=M} T_{M_1}^{K_1} T_{M_2}^{K_2} \langle K_1, M_1; K_2, M_2 | K, M \rangle \quad (5.8.20)$$

as is readily checked, where $F(K, K_1, K_2)$ is, in general, a function of K, K_1, K_2 and is necessarily *independent* of M.

We may also invert (5.8.20) and use the completeness relation (5.5.23) to obtain

$$T_{M_1}^{K_1} T_{M_2}^{K_2} = \sum_{K=|K_1-K_2|}^{K_1+K_2} T_M^K F(K, K_1, K_2) \langle K, M | K_1, M_1; K_2, M_2 \rangle. \quad (5.8.21)$$

An important application of tensor operators is to the spherical harmonics studied in §5.3.

To the above end, for any non-negative integer L, we define a tensor operator Y_M^L of rank L, with components specified by $M = -L, -L+1, \ldots, L-1, L$, as follows. The spherical harmonics $Y_{LM}(\Omega)$ (§5.3), $\Omega = (\theta, \phi)$, are defined by

$$\langle \Omega | Y_M^L | \Omega' \rangle = \delta(\Omega - \Omega') Y_{LM}(\Omega) \quad (5.8.22)$$

where

$$\delta(\Omega - \Omega') = \frac{\delta(\theta - \theta')}{\sin \theta} \delta(\phi - \phi') \quad (5.8.23)$$

with Y_M^L satisfying the usual commutation relations in (5.8.2), with the orbital angular momentum states,

$$[J(m), Y_M^L] = \hbar \left[M \delta_{m,0} \mp \frac{1}{\sqrt{2}} \sqrt{(L \mp M)(L \pm M + 1)} \delta_{m, \pm 1} \right] Y_{M+1}^L \quad (5.8.24)$$

and the orbital angular momentum states given by

$$|L, M\rangle = \int d\Omega' \, Y_M^L |\Omega'\rangle \quad (5.8.25)$$

where

$$d\Omega = \sin \theta \, d\theta \, d\phi.$$

From (5.8.25), we then have

$$\langle \Omega | L, M \rangle = Y_{LM}(\Omega) \quad (5.8.26)$$

as expected.

Also

$$J(m) |L, M\rangle = \int d\Omega' \, [J(m), Y_M^L] |\Omega'\rangle + \int d\Omega' \, Y_M^L J(m) |\Omega'\rangle. \quad (5.8.27)$$

The second term on the right-hand side of this equation may be rewritten as

$$\sum_{\ell',m'} \int d\Omega' \, Y_M^L J(m) \, |\ell', m'\rangle \, Y_{\ell',m'}^* |\Omega'\rangle$$

$$= \sqrt{4\pi} \, Y_M^L \, J(m) \, |0, 0\rangle = 0 \tag{5.8.28}$$

using the orthonormality of the spherical harmonics, with $Y_{00}(\Omega) = 1/\sqrt{4\pi}$, and then using (5.7.5).

From (5.8.24), (5.8.25), (5.8.27) we then obtain

$$J(m) \, |L, M\rangle = \hbar \left[M \delta_{m,0} \mp \frac{1}{\sqrt{2}} \sqrt{(L \mp M)(L \pm M + 1)} \, \delta_{m,\pm 1} \right] |L, M+m\rangle \tag{5.8.29}$$

yielding in particular to

$$J^3 \, |L, M\rangle = \hbar M \, |L, M\rangle \tag{5.8.30}$$

$$\mathbf{J}^2 \, |L, M\rangle = \hbar^2 L(L+1) \, |L, M\rangle \tag{5.8.31}$$

as expected.

Now we use the general expansion in (5.8.20) for the tensor operator Y_m^ℓ. Upon taking the matrix element $\langle \Omega | \cdot | \Omega' \rangle$, of (5.8.21), integrating over Ω', and using the property

$$\int d\Omega' \, \langle \Omega | Y_{m_1}^{\ell_1} Y_{m_2}^{\ell_2} | \Omega' \rangle = \int d\Omega' \, d\Omega'' \, \langle \Omega | Y_{m_1}^{\ell_1} | \Omega'' \rangle \langle \Omega'' | Y_{m_2}^{\ell_2} | \Omega' \rangle$$

$$= Y_{\ell_1 m_1}(\Omega) \, Y_{\ell_2 m_2}(\Omega) \tag{5.8.32}$$

we obtain

$$F(\ell, \ell_1, \ell_2) \, Y_{\ell m}(\Omega) = \sum_{m_1+m_2=m} Y_{\ell_1 m_1}(\Omega) \, Y_{\ell_2 m_2}(\Omega) \, \langle \ell_1, m_1; \ell_2, m_2 | \ell, m \rangle. \tag{5.8.33}$$

To evaluate the factor $F(\ell, \ell_1, \ell_2)$ explicitly we set $\theta = 0$, $\phi = 0$, use the relation (5.3.50) to obtain from (5.8.33)

$$F(\ell, \ell_1, \ell_2) = \sqrt{\frac{(2\ell_1+1)(2\ell_2+1)}{4\pi(2\ell+1)}} \, \langle \ell_1, 0; \ell_2, 0 | \ell, 0 \rangle. \tag{5.8.34}$$

From the inverse relation (5.8.21), we have from (5.8.33), (5.8.34),

$$Y_{\ell_1 m_1}(\Omega) \, Y_{\ell_2 m_2}(\Omega) = \sum_{\ell=|\ell_1-\ell_2|}^{\ell_1+\ell_2} Y_{\ell\,m}(\Omega) \sqrt{\frac{(2\ell_1+1)(2\ell_2+1)}{4\pi(2\ell+1)}}$$

$$\times \langle \ell_1, 0; \ell_2, 0 | \ell, 0 \rangle \, \langle \ell, m | \ell_1, m_1; \ell_2, m_2 \rangle. \tag{5.8.35}$$

5.8 Tensor Operators

Upon multiplying this equation by $Y_{\ell_3 m_3}(\Omega)$, integrating over Ω, and using the orthonormality of the spherical harmonics, we obtain the following useful integral involving three spherical harmonics

$$\int d\Omega\, Y_{\ell_1 m_1}(\Omega)\, Y_{\ell_2 m_2}(\Omega)\, Y_{\ell_3 m_3}(\Omega)$$

$$= (-1)^{m_3} \sqrt{\frac{(2\ell_1+1)(2\ell_2+1)}{4\pi(2\ell_3+1)}} \langle \ell_1,0; \ell_2,0 | \ell_3,0 \rangle \langle \ell_3, m_3 | \ell_1, m_1; \ell_2, m_2 \rangle$$

$$\equiv \sqrt{\frac{(2\ell_1+1)(2\ell_2+1)(2\ell_3+1)}{4\pi}} \begin{pmatrix} \ell_1 & \ell_2 & \ell_3 \\ 0 & 0 & 0 \end{pmatrix} \begin{pmatrix} \ell_1 & \ell_2 & \ell_3 \\ m_1 & m_2 & -m_3 \end{pmatrix} \quad (5.8.36)$$

where $m_3 = m_1 + m_2$, $\ell_3 = |\ell_1 - \ell_2|, |\ell_1 - \ell_2|+1, \ldots, \ell_1 + \ell_2$, and otherwise the integral (5.8.36) is zero. In using the orthogonality relation between $Y_{\ell m}(\Omega)$ and $Y_{\ell_3 m_3}(\Omega)$, we have used the fact that (see (5.3.44))

$$Y_{\ell_3\, -m_3}(\Omega) = (-1)^{m_3} Y^*_{\ell_3 m_3}(\Omega). \quad (5.8.37)$$

In writing the last equality in (5.8.36), we have used the definition of the 3-j symbols in (5.6.24) and the reality of the Clebsch-Gordan coefficients.

As a concrete non-trivial application of the integral (5.8.36), consider the case $\ell_2 = 2$, $m_2 = 1$,

$$Y_{21}(\Omega) = -\sqrt{\frac{15}{8\pi}} \sin\theta \cos\theta\, e^{i\phi} \quad (5.8.38)$$

then for $\ell_1 = \ell_3 \equiv \ell$, $m_3 = m'$, $m_1 = m$, (5.8.36) leads to the following matrix element

$$\langle \ell, m' | \sin\theta \cos\theta\, e^{i\phi} | \ell, m \rangle = -\sqrt{(\ell-m)(\ell+m+1)}$$

$$\times \frac{(2m+1)}{(2\ell-1)(2\ell+3)} \delta_{m', m+1}. \quad (5.8.39)$$

Equation (5.8.35) also leads to useful recurrence relations for the spherical harmonics. For example, for $\ell_1 = 1$, $m_1 = 0$,

$$Y_{1\,0}(\Omega) = \sqrt{\frac{3}{4\pi}} \cos\theta \quad (5.8.40)$$

$m_2 = m$, and (5.8.35) leads to

$$\cos\theta\, Y_{\ell\, m}(\Omega) = \sqrt{\frac{(\ell-m+1)(\ell+m+1)}{(2\ell+1)(2\ell+3)}} Y_{\ell+1\, m}(\Omega)$$

$$+ \sqrt{\frac{\ell^2 - m^2}{4\ell^2 - 1}} \, Y_{\ell-1\,m}(\Omega). \qquad (5.8.41)$$

For $\ell = 0$, the second term should be set equal to zero, and $m = 0$.

Equation (5.8.41) gives directly the matrix element

$$\langle \ell, m' | \cos\theta | \ell, m \rangle = 0 \quad \text{for all } m', m \qquad (5.8.42)$$

and all ℓ.

On the other hand, upon multiplying (5.8.41) by $\cos\theta$, and using the recurrence relation (5.8.41) one more time leads to the following matrix elements

$$\langle \ell, m' | \cos^2\theta | \ell, m \rangle = \frac{[2\ell^2 + 2\ell - 1 - 2m^2]}{(2\ell - 1)(2\ell + 3)} \delta_{m',m}. \qquad (5.8.43)$$

The expression in (5.8.43) may be generalized as follows.

Let $\mathbf{n} = (\cos\phi\sin\theta, \sin\phi\sin\theta, \cos\theta)$, then quite generally, by using the symmetry of the product $n^i n^j$, we may write

$$\langle \ell, m' | n^i n^j | \ell, m \rangle = \left\langle \ell, m' \left| A_\ell \, \delta^{ij} + B_\ell \frac{(L^i L^j + L^j L^i)}{\hbar^2} \right| \ell, m \right\rangle \qquad (5.8.44)$$

where A_ℓ, B_ℓ are to be determined. Since $n^i n^i = 1$, $L^i L^i |\ell, m\rangle = \mathbf{L}^2 |\ell, m\rangle = \hbar^2 \ell(\ell+1)|\ell, m\rangle$, we obtain

$$3A_\ell = [1 - 2B_\ell \, \ell(\ell+1)]. \qquad (5.8.45)$$

On the other hand for $i = j = 3$, we have from (5.8.44), the fact that $(L^3)^2 |\ell, m\rangle = \hbar^2 m^2 |\ell, m\rangle$, and (5.8.45) that

$$A_\ell = \frac{(2\ell^2 + 2\ell - 1)}{(2\ell - 1)(2\ell + 3)}, \quad B_\ell = -\frac{1}{(2\ell - 1)(2\ell + 3)}. \qquad (5.8.46)$$

That is,

$$\langle \ell, m' | n^i n^j | \ell, m \rangle = \frac{1}{(2\ell - 1)(2\ell + 3)} \langle \ell, m' | (2\ell^2 + 2\ell - 1)\delta^{ij}$$

$$- \frac{(L^i L^j + L^j L^i)}{\hbar^2} | \ell, m \rangle. \qquad (5.8.47)$$

This equation will be useful in our treatment of the hyperfine structure of the hydrogen atom for any ℓ in §7.6.

Finally we use the integral (5.8.36) to evaluate the reduced matrix element $\langle \ell' \| Y_L \| \ell \rangle$ for the spherical-harmonic-tensor operator, and also provide an important application of the integral in (5.8.36) involving three spherical harmonics.

To the above end, the integral on the left-hand side of (5.8.36) may be rewritten as
$$(-1)^{m_1} \langle \ell_1, m_1 | Y_{m_2}^{\ell_2} | \ell_3, m_3 \rangle \qquad (5.8.48)$$
which for $m_1 = 0$, $m_2 = 0$, $m_3 = 0$ leads to
$$\langle \ell_1, 0 | Y_0^{\ell_2} | \ell_3, 0 \rangle = \sqrt{\frac{(2\ell_1 + 1)(2\ell_2 + 1)(2\ell_3 + 1)}{4\pi}} \begin{pmatrix} \ell_1 & \ell_2 & \ell_3 \\ 0 & 0 & 0 \end{pmatrix}^2 \qquad (5.8.49)$$
which upon comparison with (5.8.3) we obtain
$$\langle \ell_1 || Y^{\ell_2} || \ell_3 \rangle = (-1)^{\ell_2 - \ell_3} \sqrt{\frac{(2\ell_2 + 1)(2\ell_3 + 1)}{4\pi}} \begin{pmatrix} \ell_1 & \ell_2 & \ell_3 \\ 0 & 0 & 0 \end{pmatrix} \qquad (5.8.50)$$
where we have used (5.6.24), and the right-hand side of (5.8.50) is zero if $\ell_1 + \ell_2 + \ell_3$ is odd (see (5.6.28), (5.6.29)).

We close this section by giving an important application of (5.8.36), involving the integral of the product of three spherical harmonics, in evaluating the matrix element $\langle \ell_1, m_1 | V(|\mathbf{r}_1 - \mathbf{r}_2|) | \ell_2, m_2 \rangle$ of a potential depending on the distance
$$|\mathbf{r}_1 - \mathbf{r}_2| = \left(r_1^2 - 2r_1 r_2 \cos\theta + r_2^2 \right)^{1/2} \qquad (5.8.51)$$
between two particles. To do this, one may expand
$$V(|\mathbf{r}_1 - \mathbf{r}_2|) = \sum_{\ell=0}^{\infty} V_\ell(r_1, r_2) P_\ell(\cos\theta) \qquad (5.8.52)$$
in terms of Legendre polynomials, where[3]
$$V_\ell(r_1, r_2) = \frac{(2\ell + 1)}{2} \int_0^\pi \cos\theta \, d\theta \, V(|\mathbf{r}_1 - \mathbf{r}_2|) P_\ell(\cos\theta). \qquad (5.8.53)$$
Hence upon using the identity (see (5.3.30))
$$P_\ell(\cos\theta) = \sqrt{\frac{4\pi}{(2\ell + 1)}} Y_{\ell\, 0}(\theta, \phi) \qquad (5.8.54)$$
the above matrix element $\langle \ell_1, m_1 | V(|\mathbf{r}_1 - \mathbf{r}_2|) | \ell_2, m_2 \rangle$ becomes
$$\sum_{\ell=0}^{\infty} V_\ell(r_1, r_2) \sqrt{\frac{4\pi}{2\ell + 1}} \langle \ell_1, m_1 | Y_0^\ell | \ell_2, m_2 \rangle. \qquad (5.8.55)$$

---

[3] For example for the Coulomb potential $\lambda/|\mathbf{r}_1 - \mathbf{r}_2|$, $V_\ell(r_1, r_2) = \lambda (r_</r_>)^\ell / r_>$, $r_< = \min(r_1, r_2)$, $r_> = \max(r_1, r_2)$.

5.9 Combining Several Angular Momenta: 6-j and 9-j Symbols

In this section, we consider the problem of combining several independent angular momenta. More specifically, the combination of only three or four angular momenta are considered.

One can combine three angular momenta with quantum numbers j_1, j_2, j_3 in more than one way. We may, for example, combine j_1, j_2 to form a quantum number j_{12} and then combine the latter with j_3 to form a final quantum number j. On the other hand, we may, for example, combine j_2, j_3 to form j_{23} and then combine the latter with j_1 to form a final j value.

Accordingly, one may define (see §5.5, (5.6.24)), the following states:

$$|j, m; j_{12}(j_1 j_2), j_3\rangle$$

$$= \sqrt{2j+1}\sqrt{2j_{12}+1} \sum_{m_1,m_2,m_3,m_{12}} (-1)^{j_3-j_{12}-m+j_2-j_1-m_{12}}$$

$$\times \begin{pmatrix} j_1 & j_2 & j_{12} \\ m_1 & m_2 & -m_{12} \end{pmatrix} \begin{pmatrix} j_{12} & j_3 & j \\ m_{12} & m_3 & -m \end{pmatrix} |j_1, m_1\rangle |j_2, m_2\rangle |j_3, m_3\rangle \quad (5.9.1)$$

with obvious constraints on the summation variables m_1, m_2, m_3, m_{12} understood, and

$$|j, m; j_1, j_{23}(j_2 j_3)\rangle$$

$$= \sqrt{2j+1}\sqrt{2j_{23}+1} \sum_{m_1,m_2,m_3,m_{23}} (-1)^{j_{23}-j_1-m+j_3-j_2-m_{23}}$$

$$\times \begin{pmatrix} j_1 & j_{23} & j \\ m_1 & m_{23} & -m \end{pmatrix} \begin{pmatrix} j_2 & j_3 & j_{23} \\ m_2 & m_3 & -m_{23} \end{pmatrix} |j_1, m_1\rangle |j_2, m_2\rangle |j_3, m_3\rangle. \quad (5.9.2)$$

The two states on the left-hand sides of (5.9.1), (5.9.2) are related by

$$|j, m; j_1, j_{23}(j_2 j_3)\rangle = \sqrt{2j_{23}+1}\,(-1)^{j_1+j_2+j_3+j} \sum_{j_{12}} \sqrt{2j_{12}+1}$$

$$\times |j, m; j_{12}(j_1 j_2), j_3\rangle \begin{Bmatrix} j_3 & j_{12} & j \\ j_1 & j_{23} & j_2 \end{Bmatrix} \quad (5.9.3)$$

where the orthogonality relations (5.6.30) give

$$\begin{Bmatrix} j_1 & j_2 & j_3 \\ \ell_1 & \ell_2 & \ell_3 \end{Bmatrix} = (2j_3+1) \sum_{m_i,n_i} (-1)^{\Sigma_i j_i + \Sigma_i \ell_i + \Sigma_i n_i} \begin{pmatrix} j_1 & j_2 & j_3 \\ -m_1 & -m_2 & -m_3 \end{pmatrix}$$

$$\times \begin{pmatrix} j_1 & \ell_2 & \ell_3 \\ m_1 & n_2 & -n_3 \end{pmatrix} \begin{pmatrix} \ell_1 & j_2 & \ell_3 \\ -n_1 & m_2 & n_3 \end{pmatrix} \begin{pmatrix} \ell_1 & \ell_2 & j_3 \\ n_1 & -n_2 & m_3 \end{pmatrix}. \quad (5.9.4)$$

5.9 Combining Several Angular Momenta: 6-j and 9-j Symbols

The objects $\begin{Bmatrix} j_1 & j_2 & j_3 \\ \ell_1 & \ell_2 & \ell_3 \end{Bmatrix}$ are referred to as 6-j symbols. Some properties of the latter are the following. They are left invariant by *any* permutations of the columns, e.g.,

$$\begin{Bmatrix} j_1 & j_2 & j_3 \\ \ell_1 & \ell_2 & \ell_3 \end{Bmatrix} = \begin{Bmatrix} j_3 & j_2 & j_1 \\ \ell_3 & \ell_2 & \ell_1 \end{Bmatrix} = \begin{Bmatrix} j_2 & j_1 & j_3 \\ \ell_2 & \ell_1 & \ell_3 \end{Bmatrix}. \tag{5.9.5}$$

A 6-j symbol remains also invariant under the interchange of upper and lower arguments in each of *any two* columns, e.g.,

$$\begin{Bmatrix} j_1 & j_2 & j_3 \\ \ell_1 & \ell_2 & \ell_3 \end{Bmatrix} = \begin{Bmatrix} \ell_1 & j_2 & \ell_3 \\ j_1 & \ell_2 & j_3 \end{Bmatrix} = \begin{Bmatrix} \ell_1 & \ell_2 & j_3 \\ j_1 & j_2 & \ell_3 \end{Bmatrix}. \tag{5.9.6}$$

In particular, they satisfy the orthogonality relation

$$\sum_j (2j+1) \begin{Bmatrix} j_1 & j_2 & j \\ \ell_1 & \ell_2 & \ell \end{Bmatrix} \begin{Bmatrix} j_1 & j_2 & j \\ \ell_1 & \ell_2 & \ell' \end{Bmatrix} = (2\ell+1)^{-1} \delta_{\ell,\ell'} \tag{5.9.7}$$

and the following sum rule

$$\sum_j (-1)^{j+\ell+\ell'} (2j+1) \begin{Bmatrix} j_1 & j_2 & j \\ \ell_1 & \ell_2 & \ell \end{Bmatrix} \begin{Bmatrix} j_1 & j_2 & j \\ \ell_2 & \ell_1 & \ell' \end{Bmatrix} = \begin{Bmatrix} j_1 & \ell_1 & \ell' \\ j_2 & \ell_2 & \ell \end{Bmatrix}. \tag{5.9.8}$$

As an example, consider a particle composite of two particles of spins $\mathbf{S}_1$, $\mathbf{S}_2$ and relative angular momentum (see §2.7, (2.7.40)) $\mathbf{L}_r$ residing in their center of mass. Then the total internal angular momentum of the composite particle is given by (see (2.7.39), (2.7.40))

$$\mathbf{S} = \mathbf{S}_1 + \mathbf{S}_2 + \mathbf{L}_r \equiv \mathbf{J}_1 + \mathbf{J}_2 + \mathbf{J}_3. \tag{5.9.9}$$

For definiteness, suppose $\mathbf{S}_1$ denotes the spin of a proton: $s_1 = 1/2$, and $\mathbf{S}_2$ denotes the spin of a deuteron taken to be $s_2 = 1$. Then $j_{12} = 1/2$ or $3/2$. Consider $\ell = 1$, and that the total spin of the composite particle corresponds to $s = 3/2$ and $j_{23} = 2$.

According to (5.9.3),

$$|s=3/2, m; s_1=1/2, j_{23}=2\rangle = \sqrt{5} \sum_{j_{12}=1/2,3/2} |s=3/2, m; j_{12}(s_1 s_2), j_3=1\rangle$$

$$\times \sqrt{(2j_{12}+1)} \begin{Bmatrix} 2 & 1 & 1 \\ j_{12} & 3/2 & 1/2 \end{Bmatrix} \tag{5.9.10}$$

where we have used (5.9.5), (5.9.6), and the numerical values of the 6-j symbols may be evaluated from (5.9.4) to be

$$\begin{Bmatrix} 2 & 1 & 1 \\ 1/2 & 3/2 & 1/2 \end{Bmatrix} = \frac{1}{2\sqrt{3}}, \quad \begin{Bmatrix} 2 & 1 & 1 \\ 3/2 & 3/2 & 1/2 \end{Bmatrix} = \frac{1}{2\sqrt{30}}. \tag{5.9.11}$$

That is,

$$|s=3/2, m; s_1=1/2, j_{23}=2\rangle = \sqrt{\frac{5}{6}} |s=3/2, m; j_{12}(s_1 s_2)=1/2, j_3=1\rangle$$

$$+ \sqrt{\frac{1}{6}} |s=3/2, m; j_{12}(s_1 s_2)=3/2, j_3=1\rangle \qquad (5.9.12)$$

and the probabilities that in the state $|s=3/2, m; s_1=1/2, j_{23}=2\rangle$, the proton spin and the deuteron spin combine to give the values $j_{12}=1/2, 3/2$, respectively, are 5/6 and 1/6. These also check out the correctness of the normalization condition.

For completeness, we note that the 6-j symbols may be also rewritten from (5.9.3) in terms of amplitudes as

$$\langle j, m; j_{12}(j_1 j_2), j_3 | j, m; j_1, j_{23}(j_2 j_3) \rangle$$

$$= (-1)^{j_1+j_2+j_3+j} \sqrt{(2j_{12}+1)(2j_{23}+1)} \begin{Bmatrix} j_{12} & j_3 & j \\ j_{23} & j_1 & j_2 \end{Bmatrix} \qquad (5.9.13)$$

where for the numerical example above, the corresponding amplitudes are equal to $\sqrt{5/6}$ or $\sqrt{1/6}$ for $j_{12}=1/2$ or $j_{12}=3/2$, respectively.

Finally, we consider combining four angular momenta with quantum numbers j_1, j_2, j_3, j_4. As before, there are many ways of combining these angular momenta first in pairs. These lead to the definition of so-called 9-j symbols, generalizing further the expression (5.9.13) for 6-j symbols, to four angular momenta. Specifically,

$$\langle j, m; j_{12}(j_1 j_2), j_{34}(j_3 j_4) | j, m; j_{13}(j_1 j_3), j_{24}(j_2 j_4) \rangle$$

$$= \sqrt{(2j_{12}+1)(2j_{34}+1)(2j_{13}+1)(2j_{24}+1)} \begin{Bmatrix} j_1 & j_2 & j_{12} \\ j_3 & j_4 & j_{34} \\ j_{13} & j_{24} & j \end{Bmatrix} \qquad (5.9.14)$$

where we note that for any row $(a\,b\,c)$ or column $(a\,b\,c)^T$ of the 9-j symbols, the addition rule of angular momenta $|a-b| \leqslant c \leqslant (a+b)$ is satisfied.

The 9-j symbols may be evaluated in terms of 6-j symbols as follows:

$$\begin{Bmatrix} j_{11} & j_{12} & j_{13} \\ j_{21} & j_{22} & j_{23} \\ j_{31} & j_{32} & j_{33} \end{Bmatrix} = \sum_j (-1)^{2j} (2j+1)$$

$$\times \begin{Bmatrix} j_{11} & j_{21} & j_{31} \\ j_{31} & j_{33} & j \end{Bmatrix} \begin{Bmatrix} j_{12} & j_{22} & j_{32} \\ j_{21} & j & j_{23} \end{Bmatrix} \begin{Bmatrix} j_{13} & j_{23} & j_{33} \\ j & j_{11} & j_{12} \end{Bmatrix}. \qquad (5.9.15)$$

The 9-j symbols satisfy particularly the following symmetry relations. A 9-j symbol remains *invariant* under each of the following transformations:

cyclic permutations of its columns, cyclic permutations of its rows, a transposition as for a 3 × 3 matrix.

Orthonormality of states leads to the rule

$$\sum_{j_{13},j_{23}} (2j_{13}+1)(2j_{23}+1) \begin{Bmatrix} j_{11} & j_{12} & j_{13} \\ j_{21} & j_{22} & j_{23} \\ j_{31} & j_{32} & j_{33} \end{Bmatrix} \begin{Bmatrix} j_{11} & j_{12} & j_{13} \\ j_{21} & j_{22} & j_{23} \\ j'_{31} & j'_{32} & j'_{33} \end{Bmatrix}$$

$$= \frac{\delta(j_{31},j'_{31})\,\delta(j_{32},j'_{32})}{(2j_{31}+1)(2j_{32}+1)}. \qquad (5.9.16)$$

Ultimately, the 6-j and 9-j symbols are evaluated in terms of Clebsch-Gordan coefficients or equivalently in terms of 3-j symbols as defined in (5.6.24). The explicit expressions are given, respectively, in (5.9.4) and (5.9.15).

5.10 Particle States and Angular Momentum; Helicity States

In this section, we use some of the details worked out on angular momentum to define and construct one- and two-particle states with or without spin. These particles may be composite of several particles as discussed in §2.7. For example, a given particle may be the deuteron of spin 1, composite of a proton and a neutron. For greater generality, we also consider so-called helicity states (see also §5.4) for which the projection of the spin of a particle is taken along the direction of its momentum instead of the traditionally taken z-axis. These latter states are also important for a relativistic treatment and are essential for describing zero-mass particles. As we have already developed the whole machinery to construct helicity states as well with no difficulty, they are worked out here for completeness. The reader may wish to skip over the construction of helicity states at a first reading.

5.10.1 Single Particle States

Spin 0

Consider a particle with momentum along the z-axis:

$$\mathbf{p_0} = p(0,0,1) \qquad (5.10.1)$$

and a corresponding state $\langle \mathbf{p_0} |$ satisfying

$$\langle \mathbf{p_0} | \, \mathbf{P} = \mathbf{p_0} \, \langle \mathbf{p_0} | \qquad (5.10.2)$$

where $\mathbf{P}$ is the momentum operator.

To define a state $\langle \mathbf{p}|$ of arbitrary momentum

$$\mathbf{p} = p(\sin\theta\cos\phi, \sin\theta\sin\phi, \cos\theta) \tag{5.10.3}$$

as obtained from the state $\langle \mathbf{p}_0|$, we use the definition (2.3.5)/(2.3.43) which amounts to rotating the coordinate system appropriately instead. To this end, we consider the bra

$$\langle \mathbf{p}_0| \left[\exp\left(\frac{i}{\hbar}\theta\mathbf{n}\cdot\mathbf{J}\right) \right]^\dagger \tag{5.10.4}$$

where

$$\mathbf{n} = (\sin\phi, -\cos\phi, 0). \tag{5.10.5}$$

Since we are considering spin 0, $\mathbf{J} = \mathbf{L}$. The state in (5.10.4) is actually the state $\langle \mathbf{p}|$ of arbitrary $\mathbf{p}$ as given in (5.10.3). To see this, apply the momentum operator $\mathbf{P}$ to it:

$$\langle \mathbf{p}_0| \exp\left(-\frac{i}{\hbar}\theta\mathbf{n}\cdot\mathbf{J}\right) P^i = \langle \mathbf{p}_0| P'^i \exp\left(-\frac{i}{\hbar}\theta\mathbf{n}\cdot\mathbf{J}\right) \tag{5.10.6}$$

where

$$P'^i = \exp\left(-\frac{i}{\hbar}\theta\mathbf{n}\cdot\mathbf{J}\right) P^i \exp\left(\frac{i}{\hbar}\theta\mathbf{n}\cdot\mathbf{J}\right)$$

$$= P^k R^{ik} \exp(\theta, \mathbf{n}) \tag{5.10.7}$$

and in writing the equality, we have used the vector property of $\mathbf{P}$ as given in (5.7.42). Hence

$$\langle \mathbf{p}_0| \exp\left(-\frac{i}{\hbar}\mathbf{n}\cdot\mathbf{J}\right) P^i = p_0^k R^{ik}(\theta, \mathbf{n}) \langle \mathbf{p}_0| \exp\left(-\frac{i}{\hbar}\mathbf{n}\cdot\mathbf{J}\right) \tag{5.10.8}$$

and from (2.1.4) we explicitly have

$$R^{i3}(\theta, \mathbf{n}) = \delta^{i3}\cos\theta - \varepsilon^{ij3}n^j \sin\theta \tag{5.10.9}$$

$$p_0^k R^{ik}(\theta, \mathbf{n}) = p^i \tag{5.10.10}$$

with p^i as given in (5.10.3). That is,

$$\langle \mathbf{p}_0| \exp\left(-\frac{i}{\hbar}\theta\mathbf{n}\cdot\mathbf{J}\right) \mathbf{P} = \mathbf{p} \langle \mathbf{p}_0| \exp\left(-\frac{i}{\hbar}\theta\mathbf{n}\cdot\mathbf{J}\right) \tag{5.10.11}$$

and

$$\langle \mathbf{p}| = \langle \mathbf{p}_0| \left[\exp\left(\frac{i}{\hbar}\theta\mathbf{n}\cdot\mathbf{J}\right) \right]^\dagger. \tag{5.10.12}$$

5.10 Particle States and Angular Momentum; Helicity States

We note that

$$\langle \mathbf{p}_0 | L^3 = \langle \mathbf{p}_0 | \left(X^1 P^2 - X^2 P^1 \right) = 0 \quad (5.10.13)$$

since $\left[X^1, P^2 \right] = 0$, $\left[X^2, P^1 \right] = 0$, and $\mathbf{p}_0$ has zero components along the x^2 and x^1 axes. That is, $\langle \mathbf{p}_0 |$ is an eigenstate of L^3 with $m = 0$.

We introduce simultaneous eigenstates of the commuting operators $\mathbf{P}^2, \mathbf{L}^2, L^3$ denoted by $\langle p, \ell, m |$:

$$\langle p, \ell, m | \, \mathbf{P}^2 = p^2 \, \langle p, \ell, m | \quad (5.10.14)$$

$$\langle p, \ell, m | \, \mathbf{L}^2 = \hbar^2 \ell(\ell+1) \, \langle p, \ell, m | \quad (5.10.15)$$

$$\langle p, \ell, m | \, L^3 = \hbar m \, \langle p, \ell, m | \quad (5.10.16)$$

normalized as follows:

$$\langle p, \ell, m | p', \ell', m' \rangle = (2\pi\hbar)^3 \frac{\delta(p - p')}{p^2} \delta_{\ell\ell'} \delta_{mm'}. \quad (5.10.17)$$

The following properties of the amplitude $\langle \mathbf{p}_0 | p', \ell, m \rangle$ are obvious

$$0 = \langle \mathbf{p}_0 | L^3 | p', \ell, m \rangle = \hbar m \, \langle \mathbf{p}_0 | p', \ell, m \rangle \quad (5.10.18)$$

$$0 = \langle \mathbf{p}_0 | \left(\mathbf{P}^2 - p^2 \right) | p', \ell, m \rangle = (p'^2 - p^2) \, \langle \mathbf{p}_0 | p', \ell, m \rangle. \quad (5.10.19)$$

Accordingly, we may write quite generally

$$\langle \mathbf{p}_0 | p', \ell, m \rangle = (2\pi\hbar)^3 \frac{\delta(p - p')}{p^2} \delta_{m0} C_\ell(p) \quad (5.10.20)$$

where $C_\ell(p)$ is to be determined.

We expand the state $\langle \mathbf{p} |$ in terms of the states $\langle p, \ell, m |$ and hence introduce in the process the identity operator

$$\mathbf{1} = \sum_{\ell, m} \int_0^\infty \frac{p'^2 \, dp'}{(2\pi\hbar)^3} |p', \ell, m\rangle \langle p', \ell, m |. \quad (5.10.21)$$

From (5.10.12), (5.10.20), (5.10.21) we then obtain

$$\langle \mathbf{p} | = \sum_{\ell, m} C_\ell(p) \left\langle \ell, 0 \left| \left[\exp\left(\frac{i}{\hbar} \theta \mathbf{n} \cdot \mathbf{J}\right) \right]^\dagger \right| \ell, m \right\rangle \langle p, \ell, m | \quad (5.10.22)$$

where we have used the normalization

$$\langle p', \ell', 0 | \left[\exp\left(\frac{i}{\hbar} \theta \mathbf{n} \cdot \mathbf{J}\right) \right]^\dagger | p, \ell, m \rangle$$

$$= (2\pi\hbar)^3 \frac{\delta(p-p')}{p^2} \delta_{\ell'\ell} \left\langle \ell, 0 \left| \left[\exp\left(\frac{i}{\hbar}\theta\mathbf{n}\cdot\mathbf{J}\right) \right]^\dagger \right| \ell, m \right\rangle \quad (5.10.23)$$

since $\mathbf{J}$ commutes with both $\mathbf{L}^2$ and $\mathbf{P}^2$.

From the identity (5.7.47) and (5.3.65), (5.3.66), we have

$$\left\langle \ell, 0 \left| \left[\exp\left(\frac{i}{\hbar}\theta\mathbf{n}\cdot\mathbf{J}\right) \right]^\dagger \right| \ell, m \right\rangle$$

$$= \left\langle \ell, 0 \left| \exp\left(-\frac{i}{\hbar}\phi J^3\right) \exp\left(\frac{i}{\hbar}\theta J^2\right) \exp\left(\frac{i}{\hbar}\phi J^3\right) \right| \ell, m \right\rangle$$

$$\equiv D^{(\ell)}_{0m}(\phi, \theta, -\phi)$$

$$\equiv D^{(\ell)}_{0m}(\phi, \theta, 0)$$

$$= \sqrt{\frac{4\pi}{2\ell+1}} Y_{\ell m}(\theta, \phi). \quad (5.10.24)$$

Equation (5.10.22) then becomes

$$\langle \mathbf{p} | = \sum_{\ell, m} C_\ell(p) \sqrt{\frac{4\pi}{2\ell+1}} Y_{\ell\,m}(\theta, \phi) \langle p, \ell, m| \quad (5.10.25)$$

and

$$\langle \mathbf{p} | p', \ell, m \rangle = (2\pi\hbar)^3 C_\ell(p) \sqrt{\frac{4\pi}{2\ell+1}} \frac{\delta(p-p')}{p^2} Y_{\ell m}(\theta, \phi). \quad (5.10.26)$$

Finally, from the equality

$$(2\pi\hbar)^3 \delta_{\ell''\ell'}\delta_{m''m'} \frac{\delta(p''-p')}{p'^2} = \int \langle p'', \ell'', m'' | \mathbf{p} \rangle \frac{d^3\mathbf{p}}{(2\pi\hbar)^3} \langle \mathbf{p} | p', \ell', m' \rangle, \quad (5.10.27)$$

it is readily verified from (5.10.25), that we may set

$$C_\ell(p) = C_\ell = \sqrt{\frac{2\ell+1}{4\pi}} \quad (5.10.28)$$

thus obtaining

$$\langle \mathbf{p} | = \sum_{\ell=0}^\infty \sum_{m=-\ell}^\ell Y_{\ell m}(\widehat{\mathbf{p}}) \langle p, \ell, m| \quad (5.10.29)$$

where $\widehat{\mathbf{p}}$ is the unit vector $(\sin\theta\cos\phi, \sin\theta\sin\phi, \cos\theta)$ along $\mathbf{p}$.

From (5.10.29), we also obtain

5.10 Particle States and Angular Momentum; Helicity States

$$\langle p, \ell, m| = \int d\Omega \, Y^*_{\ell m}(\theta, \phi) \, \langle \mathbf{p}| \tag{5.10.30}$$

and note from (5.10.17), the correct normalization in (2.4.15) adopted for the states $\langle \mathbf{p}|$.

Before treating arbitrary spins we also consider a coordinate description of the above. To this end, as in (5.10.4)

$$\langle \mathbf{x}| = \langle z_0| \left[\exp\left(\frac{i}{\hbar}\theta' \mathbf{n}' \cdot \mathbf{J}\right)\right]^\dagger, \quad |\mathbf{x}| = |z_0| \equiv r, \quad \mathbf{x} = r\hat{\mathbf{x}} \tag{5.10.31}$$

$\mathbf{n}' = (\sin\phi', -\cos\phi', 0)$, and from (5.10.29), (5.10.21), (5.10.24) we readily obtain

$$\langle \mathbf{x}|\mathbf{p}\rangle = \sum_{\ell,m} Y_{\ell m}(\hat{\mathbf{x}}) \, Y^*_{\ell m}(\hat{\mathbf{p}}) \, \langle z_0|p,\ell,0\rangle. \tag{5.10.32}$$

The addition theorem of spherical harmonics in (5.3.67) and the expression for the transformation function $\langle \mathbf{x}|\mathbf{p}\rangle$ in (2.4.8), allow us to write

$$\exp\left(\frac{i}{\hbar}\mathbf{x} \cdot \mathbf{p}\right) = \sum_{\ell=0}^{\infty} \frac{(2\ell+1)}{4\pi} P_\ell(\hat{\mathbf{x}} \cdot \hat{\mathbf{p}}) \, \langle z_0|p,\ell,0\rangle. \tag{5.10.33}$$

On the other hand, the orthogonality relation (5.3.51), and the definition (5.3.30): $Y_{\ell 0}(\theta, 0) = \sqrt{(2\ell+1)/4\pi} P_\ell(\cos\theta)$ give

$$\langle z_0|p,\ell,0\rangle = 2\pi \int_{-1}^{1} d(\cos\theta) \, P_\ell(\cos\theta) \, e^{irp\cos\theta/\hbar}. \tag{5.10.34}$$

Thus we are led to introduce a function of $rp/\hbar$ which defines the amplitude $\langle z_0|p,\ell,0\rangle$. This function is referred to as a spherical Bessel function of order ℓ and may be defined as an integral over a Legendre polynomial:

$$j_\ell\left(\frac{pr}{\hbar}\right) = \frac{1}{2i^\ell} \int_{-1}^{1} d(\cos\theta) \, P_\ell(\cos\theta) \, e^{irp\cos\theta/\hbar}. \tag{5.10.35}$$

From (5.10.34) and (5.10.35), (5.10.33) becomes

$$\exp\left(\frac{i}{\hbar}\mathbf{x} \cdot \mathbf{p}\right) = \sum_{\ell=0}^{\infty} (2\ell+1) \, i^\ell \, j_\ell(pr/\hbar) \, P_\ell(\hat{\mathbf{x}} \cdot \hat{\mathbf{p}}). \tag{5.10.36}$$

The identity

$$\langle \mathbf{x}| = \int \langle \mathbf{x}|\mathbf{p}\rangle \frac{d^3\mathbf{p}}{(2\pi\hbar)^3} \langle \mathbf{p}| \tag{5.10.37}$$

allows one to write

$$\langle \mathbf{x}| = \sum_{\ell=0}^{\infty} (2\ell+1) \, i^\ell \int \frac{d^3\mathbf{p}}{(2\pi\hbar)^3} \, j_\ell(pr/\hbar) \, P_\ell(\hat{\mathbf{x}} \cdot \hat{\mathbf{p}}) \langle \mathbf{p}| \tag{5.10.38}$$

and from (5.10.30), (5.3.66)

$$\langle \mathbf{x} | p, \ell, m \rangle = 4\pi \, i^\ell \, j_\ell \, (pr/\hbar) \, Y_{\ell \, m} \, (\hat{\mathbf{x}}). \tag{5.10.39}$$

As an equivalent expression to (5.10.38), (5.10.39) gives

$$\langle \mathbf{x} | = \sum_{\ell,m} \int_0^\infty \frac{p^2 dp}{(2\pi\hbar)^3} \, 4\pi \, i^\ell \, j_\ell \, (pr/\hbar) \, Y_{\ell \, m} \, (\hat{\mathbf{x}}) \, \langle p, \ell, m |. \tag{5.10.40}$$

The normalization conditions $\langle \mathbf{x}' | \mathbf{x} \rangle = \delta^3 (\mathbf{x} - \mathbf{x}')$ (see (2.4.14)) and (5.10.17), lead from (5.10.40) to infer the closure relation

$$\int_0^\infty \frac{p^2 dp}{\hbar^3} j_\ell \, (pr/\hbar) \, j_\ell \, (pr'/\hbar) = \frac{\pi}{2} \frac{\delta (r - r')}{r^2}. \tag{5.10.41}$$

Arbitrary Spins

Consider a particle of spin s, with projection of spin along the z-axis given by $\hbar\sigma$, and momentum $\mathbf{p}_0$ (see (5.10.1)) along the z-axis. We denote such a state by $\langle \mathbf{p}_0, \sigma |$. That is,

$$\langle \mathbf{p}_0, \sigma | \mathbf{P} = \mathbf{p}_0 \, \langle \mathbf{p}_0, \sigma | \tag{5.10.42}$$

$$\langle \mathbf{p}_0, \sigma | S^3 = \hbar\sigma \, \langle \mathbf{p}_0, \sigma |. \tag{5.10.43}$$

As in (5.10.4), the state

$$\langle \mathbf{p}_0, \sigma | \left[\exp\left(\frac{i}{\hbar} \theta \mathbf{n} \cdot \mathbf{L}\right) \right]^\dagger \tag{5.10.44}$$

where $\mathbf{L}$ is the *orbital* angular momentum, describes a state of momentum $\mathbf{p}$, as given in (5.10.3), and projection of spin *along* the z-axis equal to $\hbar\sigma$. To see this, note that S^3 commutes with $\mathbf{L}$:

$$\langle \mathbf{p}_0, \sigma | \left[\exp\left(\frac{i}{\hbar} \theta \mathbf{n} \cdot \mathbf{L}\right) \right]^\dagger S^3 = \langle \mathbf{p}_0, \sigma | S^3 \left[\exp\left(\frac{i}{\hbar} \theta \mathbf{n} \cdot \mathbf{L}\right) \right]^\dagger$$

$$= \hbar\sigma \, \langle \mathbf{p}_0, \sigma | \left[\exp\left(\frac{i}{\hbar} \theta \mathbf{n} \cdot \mathbf{L}\right) \right]^\dagger. \tag{5.10.45}$$

Also (5.10.6) holds with $\mathbf{J}$ replaced by $\mathbf{L}$ since $\mathbf{P}$ and $\mathbf{S}$ commute. Hence we conclude that

$$\langle \mathbf{p}_0, \sigma | \left[\exp\left(\frac{i}{\hbar} \theta \mathbf{n} \cdot \mathbf{L}\right) \right]^\dagger = \langle \mathbf{p}, \sigma |. \tag{5.10.46}$$

As in (5.10.13)

$$\langle \mathbf{p}_0, \sigma | L^3 = 0. \tag{5.10.47}$$

5.10 Particle States and Angular Momentum; Helicity States

We introduce simultaneous eigenstates of the commuting operators $\mathbf{P}^2$, $\mathbf{L}^2$, L^3, $\mathbf{S}^2$, S^3 which we denote by $\langle p, \ell, m, s, \sigma |$, with the identity defined by

$$1 = \sum_{\ell, m, \sigma} \int \frac{p^2\, dp}{(2\pi\hbar)^3} \, |p, \ell, m, s, \sigma\rangle \langle p, \ell, m, s, \sigma|. \tag{5.10.48}$$

From (5.10.46)–(5.10.48), we readily obtain in a similar way as in (5.10.25), (5.10.29), (5.10.30),

$$\langle \mathbf{p}, \sigma| = \sum_{\ell, m} \sqrt{\frac{2\ell+1}{4\pi}}\, D^{(\ell)}_{0\,m}(\phi, \theta, -\phi)\, \langle p, \ell, m, \sigma|$$

$$\equiv \sum_{\ell, m} Y_{\ell\,m}(\theta, \phi)\, \langle p, \ell, m, \sigma| \tag{5.10.49}$$

and

$$\langle p, \ell, m, \sigma| = \int d\Omega\, Y^*_{\ell\,m}(\theta, \phi)\, \langle \mathbf{p}, \sigma|. \tag{5.10.50}$$

One may also combine the orbital angular momentum and spin, using Clebsch-Gordan coefficients, to rewrite (5.10.49) as

$$\langle \mathbf{p}, \sigma| = \sum_{\ell, m, J} Y_{\ell\,m}(\theta, \phi)\, \langle \ell, m; s, \sigma | J, M = m+\sigma\rangle\, \langle p, J, M, \ell, s| \tag{5.10.51}$$

where due to the commutativity of $\mathbf{L}^2$, $\mathbf{S}^2$, $\mathbf{P}^2$ with $\mathbf{J}^2$, J^3, the states $\langle p, J, M, \ell, s|$ are labelled by p, ℓ, and s as well.

Arbitrary Spins — Helicity States

The projection of the spin along the momentum direction of a particle is referred to as the helicity. Since

$$\mathbf{J} \cdot \mathbf{P} = \mathbf{S} \cdot \mathbf{P} \tag{5.10.52}$$

as follows from (2.7.6), helicity may be equivalently defined as the projection of the angular momentum along the direction of momentum.

From (5.7.48), we also note the commutativity properties:

$$[\mathbf{J}, \mathbf{J} \cdot \mathbf{P}] = 0 \tag{5.10.53}$$

$$[\mathbf{J}, \mathbf{P}^2] = 0 \tag{5.10.54}$$

and

$$[\mathbf{J} \cdot \mathbf{P}, \mathbf{P}^2] = 0 \tag{5.10.55}$$

314   5 Angular Momentum Gymnastics

$$[\mathbf{P}, \mathbf{J} \cdot \mathbf{P}] = 0. \tag{5.10.56}$$

According to (5.10.55), we may define a state $\langle \mathbf{p}_0, \lambda |$ labelled by $\mathbf{p}_0$ and the helicity λ such that

$$\langle \mathbf{p}_0, \lambda | \mathbf{P} = \mathbf{p}_0 \langle \mathbf{p}_0, \lambda | \tag{5.10.57}$$

(see (5.10.1), $p = |\mathbf{p}|$)

$$\langle \mathbf{p}_0, \lambda | \frac{\mathbf{J} \cdot \mathbf{P}}{p} = \hbar \lambda \langle \mathbf{p}_0, \lambda |. \tag{5.10.58}$$

On the other hand, the left-hand side of (5.10.58) is equal to $\langle \mathbf{p}_0, \lambda | J^3$. That is

$$\langle \mathbf{p}_0, \lambda | J^3 = \hbar \lambda \langle \mathbf{p}_0, \lambda |. \tag{5.10.59}$$

As in (5.10.4), (5.10.5), we consider the state

$$\langle \mathbf{p}_0, \lambda | \left[\exp\left(\frac{i}{\hbar} \theta \mathbf{n} \cdot \mathbf{J}\right) \right]^{\dagger} \tag{5.10.60}$$

for which (see (5.10.8)), (5.10.3))

$$\langle \mathbf{p}_0, \lambda | \exp\left(-\frac{i}{\hbar} \theta \mathbf{n} \cdot \mathbf{J}\right) \mathbf{P} = \mathbf{p} \langle \mathbf{p}_0, \lambda | \exp\left(-\frac{i}{\hbar} \theta \mathbf{n} \cdot \mathbf{J}\right). \tag{5.10.61}$$

Also from (5.10.53), (5.10.58)

$$\langle \mathbf{p}_0, \lambda | \exp\left(-\frac{i}{\hbar} \theta \mathbf{n} \cdot \mathbf{J}\right) \frac{\mathbf{J} \cdot \mathbf{P}}{p} = \hbar \lambda \langle \mathbf{p}_0, \lambda | \exp\left(-\frac{i}{\hbar} \theta \mathbf{n} \cdot \mathbf{J}\right). \tag{5.10.62}$$

That is,

$$\langle \mathbf{p}_0, \lambda | \left[\exp\left(\frac{i}{\hbar} \theta \mathbf{n} \cdot \mathbf{J}\right) \right]^{\dagger} = \langle \mathbf{p}, \lambda |. \tag{5.10.63}$$

We introduce simultaneous eigenstates of the commuting operators $\mathbf{P}^2$, $\mathbf{J}^2$, J^3, $\mathbf{J} \cdot \mathbf{P}$

$$\langle p, J, M, \lambda | \mathbf{P}^2 = p^2 \langle p, J, M, \lambda | \tag{5.10.64}$$

$$\langle p, J, M, \lambda | \mathbf{J}^2 = \hbar^2 J(J+1) \langle p, J, M, \lambda | \tag{5.10.65}$$

$$\langle p, J, M, \lambda | J^3 = \hbar M \langle p, J, M, \lambda | \tag{5.10.66}$$

$$\langle p, J, M, \lambda | \frac{\mathbf{J} \cdot \mathbf{P}}{p} = \hbar \lambda \langle p, J, M, \lambda | \tag{5.10.67}$$

and

5.10 Particle States and Angular Momentum; Helicity States   315

$$1 = \sum_{J,M,\lambda} \int_0^\infty \frac{p^2 dp}{(2\pi\hbar)^3} |p, J, M, \lambda\rangle \langle p, J, M, \lambda|$$
(5.10.68)

$$\langle p, J, M, \lambda | p', J', M', \lambda'\rangle = (2\pi\hbar)^3 \frac{\delta(p-p')}{p^2} \delta_{JJ'}\delta_{MM'}\delta_{\lambda\lambda'}.$$
(5.10.69)

Following a procedure as the one leading to (5.10.49), we obtain

$$\langle \mathbf{p}, \lambda | = \sum_{J,M} \sqrt{\frac{2J+1}{4\pi}} D_{\lambda M}^{(J)}(\phi, \theta, -\phi) \langle p, J, M, \lambda|$$
(5.10.70)

and upon multiplying this equation by $\left(D_{\lambda M'}^{(J')}(\phi, \theta, -\phi)\right)^*$, integrating over θ and ϕ, and using (5.5.52), it is easily worked out that

$$\langle p, J, M, \lambda | = \sqrt{\frac{2J+1}{4\pi}} \int d\Omega \left(D_{\lambda M}^{(J)}(\phi, \theta, -\phi)\right)^* \langle \mathbf{p}, \lambda |.$$
(5.10.71)

Helicity Versus Standard Spin States

Consider the spin state $\langle \mathbf{p}_0, \sigma|$ in (5.10.42), (5.10.43) and the helicity state $\langle \mathbf{p}_0, \lambda|$ in (5.10.57), (5.10.58). Clearly, from the just mentioned equations,

$$\langle \mathbf{p}_0, \sigma| \Big|_{\sigma=\lambda} = \langle \mathbf{p}_0, \lambda|$$
(5.10.72)

relating the spin (on the left) and helicity states for a particle with momentum $\mathbf{p}_0$ along the z-axis in (5.10.1).

We apply the operator $\exp(-i\theta \mathbf{n}\cdot\mathbf{J}/\hbar)$ in (5.10.4)/(5.10.5) to both sides of (5.10.72), and use the commutativity of $\mathbf{L}$ and $\mathbf{S}$ to obtain from (5.10.46), (5.10.63)

$$\langle \mathbf{p}, \sigma| \Big|_{\sigma=\lambda} \exp\left(-\frac{i}{\hbar}\theta \mathbf{n}\cdot\mathbf{S}\right) = \langle \mathbf{p}, \lambda|$$
(5.10.73)

for an arbitrary momentum $\mathbf{p}$ as defined in (5.10.3).

From (5.10.49), the left-hand side of (5.10.73) is equal to

$$\sum_{\ell,m} \sqrt{\frac{2\ell+1}{4\pi}} \left(D_{0m}^\ell(\phi, \theta, -\phi)\right) \langle p, \ell, m, \sigma| \Big|_{\sigma=\lambda} \exp\left(-\frac{i}{\hbar}\theta\mathbf{n}\cdot\mathbf{S}\right)$$

$$= \sum_{\ell,m,\sigma} \sqrt{\frac{2\ell+1}{4\pi}} D_{0m}^{(\ell)}(\phi, \theta, -\phi) D_{\lambda\sigma}^{(s)}(\phi, \theta, -\phi) \langle p, \ell, m, \sigma|$$

$$= \sum_{\ell,m,\sigma,J} \sqrt{\frac{2\ell+1}{4\pi}} D_{0m}^{(\ell)}(\phi, \theta, -\phi) D_{\lambda\sigma}^{(s)}(\phi, \theta, -\phi)$$

$$\times \langle \ell, m; s, \sigma | J, M = m + \sigma \rangle \langle p, J, M, \ell, s | \quad (5.10.74)$$

where $\langle \ell, m; s, \sigma | J, M = m + \sigma \rangle$ is a Clebsch-Gordan coefficient. Due to the commutativity of $\mathbf{L}^2$, $\mathbf{S}^2$, $\mathbf{P}^2$ with $\mathbf{J}^2$, J^3 the states $\langle p, J, M, \ell, s |$ are labelled by p, ℓ and s as well.

From (5.10.73), (5.10.70), (5.10.74)

$$\sum_{\ell,m,\sigma,J} \sqrt{\frac{2\ell+1}{4\pi}} D^{(\ell)}_{0m}(\phi, \theta, -\phi) D^{(s)}_{\lambda\sigma}(\phi, \theta, -\phi)$$

$$\times \langle \ell, m; s, \sigma | J, M = m + \sigma \rangle \langle p, J, M, \ell, s |$$

$$= \sum_{J,M} \sqrt{\frac{2J+1}{4\pi}} D^{(J)}_{\lambda M}(\phi, \theta, -\phi) \langle p, J, M, \lambda |. \quad (5.10.75)$$

We finally make use of the identity in (5.5.44) as specialized to the present case to read

$$D^{(\ell)}_{0m}(\phi, \theta, -\phi) D^{(s)}_{\lambda\sigma}(\phi, \theta, -\phi)$$

$$= \sum_{J,M} \langle \ell, 0; s, \lambda | J, \lambda \rangle D^{(J)}_{\lambda M}(\phi, \theta, -\phi) \langle J, M | \ell, m; s, \sigma \rangle. \quad (5.10.76)$$

Upon substituting (5.10.76) in the left-hand expression in (5.10.75), and using the completeness relation

$$\sum_{m,\sigma} \langle J', M' | \ell, m; s, \sigma \rangle \langle \ell, m; s, \sigma | J, M = m + \sigma \rangle = \delta_{J'J} \delta_{M'M}$$

we obtain from (5.10.75)

$$\sum_{J,M,\ell} \sqrt{\frac{2\ell+1}{4\pi}} D^{(J)}_{\lambda M}(\phi, \theta, -\phi) \langle \ell, 0; s, \lambda | J, \lambda \rangle \langle p, J, M, \ell, s |$$

$$= \sum_{J,M} \sqrt{\frac{2J+1}{4\pi}} D^{(J)}_{\lambda M}(\phi, \theta, -\phi) \langle p, J, M, \lambda | = \langle \mathbf{p}, \lambda |. \quad (5.10.77)$$

The latter leads to

$$\langle J, M, \lambda | J, M, \ell, s \rangle = \langle \ell, 0; s, \lambda | J, \lambda \rangle \sqrt{\frac{2\ell+1}{2J+1}} \quad (5.10.78)$$

where $\langle \ell, 0; s, \lambda | J, \lambda \rangle$ is just a Clebsch-Gordan coefficient with $M = \lambda$.

5.10.2 Two Particle States

Consider two particles $\mathcal{P}_1$, $\mathcal{P}_2$. Each of the particles may in turn be a cluster of particles (§2.3, §2.7), consisting, say, of n_1 and n_2 particles, respectively. The total angular momentum $\mathbf{J}$ of $\mathcal{P}_1$, $\mathcal{P}_2$ may be written as (see (2.7.29))

$$\mathbf{J} = \mathbf{X} \times \mathbf{P} + \mathbf{J}_{(1)} + \mathbf{J}_{(2)} \qquad (5.10.79)$$

where

$$\mathbf{J}_{(\alpha)} \equiv (\mathbf{X}_\alpha - \mathbf{X}) \times \left(\mathbf{P}_\alpha - \frac{M_\alpha}{M}\mathbf{P}\right) + \mathbf{S}_{(\alpha)}, \quad \alpha = 1, 2 \qquad (5.10.80)$$

and $\mathbf{S}_{(1)}$, $\mathbf{S}_{(2)}$ denote the spins of $\mathcal{P}_1$, $\mathcal{P}_2$. That is, $\mathbf{J}_{(\alpha)}$ is the total angular momentum of the αth cluster in the center of mass system of n_α the particles. Here $\mathbf{P}_1$, $\mathbf{P}_2$ are the linear momentum operators associated with the particles $\mathcal{P}_1$, $\mathcal{P}_2$.

The momentum operator

$$\mathbf{P}_r \equiv \frac{M_2 \mathbf{P}_1 - M_1 \mathbf{P}_2}{M}, \quad M = M_1 + M_2 \qquad (5.10.81)$$

is the relative momentum of $\mathcal{P}_1$, $\mathcal{P}_2$, as appearing in (2.7.40), where M_1, M_2 respectively, denote the sum of the masses of the n_1, n_2 particles, and may be also rewritten as

$$\mathbf{P}_r \equiv \frac{M_2}{M}\left(\mathbf{P}_1 - \frac{M_1}{M}\mathbf{P}\right) - \frac{M_1}{M}\left(\mathbf{P}_2 - \frac{M_2}{M}\mathbf{P}\right), \qquad (5.10.82)$$

where $\mathbf{P}_\alpha - (M_\alpha/M)\mathbf{P}$ are the momenta relative to the center of mass, with

$$\mathbf{P} = \mathbf{P}_1 + \mathbf{P}_2. \qquad (5.10.83)$$

In a coordinate description, the total angular momentum $\mathbf{J}$ of particles may be rewritten as (§2.3, §2.7)

$$\mathbf{J} = \mathbf{X} \times \mathbf{P} + \boldsymbol{\eta} \times (-i\hbar \boldsymbol{\nabla}_\eta) + \mathbf{S}_{(1)} + \mathbf{S}_{(2)} \qquad (5.10.84)$$

where

$$\mathbf{X} = \frac{M_1 \mathbf{X}_1 + M_2 \mathbf{X}_2}{M} \qquad (5.10.85)$$

with $\mathbf{X}_1$, $\mathbf{X}_2$ denoting the center of mass positions of $\mathcal{P}_1$ and $\mathcal{P}_2$

$$\boldsymbol{\eta} = \mathbf{X}_1 - \mathbf{X}_2. \qquad (5.10.86)$$

We consider the problem of defining states for two free particles $\mathcal{P}_1$, $\mathcal{P}_2$. The n_1 particles making up $\mathcal{P}_1$ may, however, be interacting with arbitrary interactions. Similarly, the n_2 particles, making up $\mathcal{P}_2$ may have arbitrary interactions. This is typical in a scattering process (see Figure 5.3), where

318   5 Angular Momentum Gymnastics

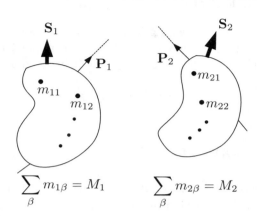

Fig. 5.3. Two clusters of particles are initially widely separated and non-interacting before they merge together in a scattering process leading finally to a variety of possible outcomes. The intra-clusteral interactions (§2.5) are not necessarily vanishing.

two clusters of particles are, for example, initially widely separated and non-interacting before they merge together, interact and produce finally a variety of possible outcomes.

Let $(\mathbf{p}_1, \sigma_1)$, $(\mathbf{p}_2, \sigma_2)$ denote the momenta and spin projections (in units of $\hbar$) along the z-axis of the particles $\mathcal{P}_1$, $\mathcal{P}_2$. We use the normalization

$$\langle \mathbf{p}'_1, \sigma'_1; \mathbf{p}'_2, \sigma'_2 | \mathbf{p}_1, \sigma_1; \mathbf{p}_2, \sigma_2 \rangle = (2\pi\hbar)^6 \delta^3\left(\mathbf{p}'_1 - \mathbf{p}_1\right) \delta^3\left(\mathbf{p}'_2 - \mathbf{p}_2\right) \delta_{\sigma'_1 \sigma_1} \delta_{\sigma'_2 \sigma_2}$$
(5.10.87)

for the two particle states $\langle \mathbf{p}_1, \sigma_1; \mathbf{p}_2, \sigma_2 |$.

From the formal definition of the Dirac deltas, it is easily verified, by a change of variables, that

$$\delta^3(\mathbf{p}'_1 - \mathbf{p}_1)\delta^3(\mathbf{p}'_2 - \mathbf{p}_2) = \delta^3(\mathbf{p}'_T - \mathbf{p}_T)\delta^3(\mathbf{p}' - \mathbf{p}) \qquad (5.10.88)$$

where

$$\mathbf{p}_T = \mathbf{p}_1 + \mathbf{p}_2 \qquad (5.10.89)$$

is the total momentum of the two particles (i.e., the center of mass momentum), and

$$\mathbf{p} = \frac{M_2 \mathbf{p}_1 - M_1 \mathbf{p}_2}{M}$$

$$= \frac{M_2}{M}\left(\mathbf{p}_1 - \frac{M_1}{M}\mathbf{p}_T\right) - \frac{M_1}{M}\left(\mathbf{p}_2 - \frac{M_2}{M}\mathbf{p}_T\right). \qquad (5.10.90)$$

5.10 Particle States and Angular Momentum; Helicity States

From (5.10.87), (5.10.88), we may re-parameterize the two particle states as $\langle \mathbf{p}_T, \mathbf{p}, \sigma_1, \sigma_2 |$ with the normalization

$$\langle \mathbf{p}'_T, \mathbf{p}', \sigma'_1, \sigma'_2 | \mathbf{p}_T, \mathbf{p}, \sigma_1, \sigma_2 \rangle = (2\pi\hbar)^6 \, \delta^3 (\mathbf{p}'_T - \mathbf{p}_T) \, \delta^3 (\mathbf{p}' - \mathbf{p}) \, \delta_{\sigma'_1 \sigma_1} \delta_{\sigma'_2 \sigma_2}. \tag{5.10.91}$$

Finally, we may introduce the states in the center of mass of the two particles $\mathcal{P}_1$, $\mathcal{P}_2$

$$\langle \mathbf{p}, \sigma_1, \sigma_2 | = \langle \mathbf{p}_T, \mathbf{p}, \sigma_1, \sigma_2 \rangle \big|_{\mathbf{p}_T = \mathbf{0}}. \tag{5.10.92}$$

Two Spin 0 Particles

In a center of mass frame, $\mathbf{p}_T = \mathbf{0}$, $\mathbf{p}_1 = -\mathbf{p}_2$ and

$$\mathbf{p} = \mathbf{p}_1 \tag{5.10.93}$$

for the relative momentum. The two-particle state $\langle \mathbf{p} |$ in the center of mass frame is obtained from a state $\langle \mathbf{p}_0 |$, with relative momentum along the z-axis, by

$$\langle \mathbf{p} | = \langle \mathbf{p}_0 | \left[\exp\left(-\frac{i}{\hbar} \theta \mathbf{n} \cdot \mathbf{L}_r \right) \right]^\dagger \tag{5.10.94}$$

where $\mathbf{n}$ is defined in (5.10.5), leading as in (5.10.29), (5.10.30) to

$$\langle \mathbf{p} | = \sum_{\ell, m} Y_{\ell m}(\hat{\mathbf{p}}) \langle p, \ell, m | \tag{5.10.95}$$

$$\langle p, \ell, m | = \int d\Omega \, Y^*_{\ell m}(\hat{\mathbf{p}}) \langle \mathbf{p} |. \tag{5.10.96}$$

The normalization conditions are as before as given in (2.4.15), (5.10.17).

Two Particles of Arbitrary Spins

Consider two particles of spins s_1, s_2 and momenta $\mathbf{p}_1$, $\mathbf{p}_2$ respectively. In the center of mass frame (5.10.93), (5.10.94), and due to the commutativity of $\mathbf{S}_1$, $\mathbf{S}_2$ with $\mathbf{L}_r$, we have

$$\langle \mathbf{p}, \sigma_1, \sigma_2 | \left(S^3_{(1)} + S^3_{(2)} \right) = \langle \mathbf{p}_0, \sigma_1, \sigma_2 | \left(S^3_{(1)} + S^3_{(2)} \right) \left[\exp\left(\frac{i}{\hbar} \theta \mathbf{n} \cdot \mathbf{L}_r \right) \right]^\dagger$$

$$= \hbar(\sigma_1 + \sigma_2) \langle \mathbf{p}_0, \sigma_1, \sigma_2 | \left[\exp\left(\frac{i}{\hbar} \theta \mathbf{n} \cdot \mathbf{L}_r \right) \right]^\dagger$$

$$= \hbar(\sigma_1 + \sigma_2) \langle \mathbf{p}, \sigma_1, \sigma_2 |. \tag{5.10.97}$$

Also
$$\langle \mathbf{p}_0, \sigma_1, \sigma_2 | L_r^3 = 0. \quad (5.10.98)$$

Equations (5.10.97), (5.10.98) then lead, as for (5.10.49), (5.10.50), to

$$\langle \mathbf{p}, \sigma_1, \sigma_2 | = \sum_{\ell,m} \sqrt{\frac{2\ell+1}{4\pi}} D_{0m}^{(\ell)}(\phi, \theta, -\phi) \langle p, \ell, m, \sigma_1, \sigma_2 |$$

$$\equiv \sum_{\ell,m} Y_{\ell,m}(\widehat{\mathbf{p}}) \langle p, \ell, m, \sigma_1, \sigma_2 | \quad (5.10.99)$$

and

$$\langle p, \ell, m, \sigma_1, \sigma_2 | = \int d\Omega \, Y_{\ell m}^*(\widehat{\mathbf{p}}) \langle \mathbf{p}, \sigma_1, \sigma_2 |. \quad (5.10.100)$$

Here $\langle p, \ell, m, \sigma_1, \sigma_2 |$ are simultaneous eigenstates of the commuting operators $\mathbf{P}^2, \mathbf{L}^2, L^3, S_{(1)}^3, S_{(2)}^3$. The normalization conditions are

$$\langle \mathbf{p}', \sigma_1', \sigma_2' | \mathbf{p}, \sigma_1, \sigma_2 \rangle = (2\pi\hbar)^3 \delta^3(\mathbf{p}' - \mathbf{p}) \delta_{\sigma_1'\sigma_1} \delta_{\sigma_2'\sigma_2} \quad (5.10.101)$$

$$\langle p', \ell', m', \sigma_1', \sigma_2' | p, \ell, m, \sigma_1, \sigma_2 \rangle = \frac{\delta(p'-p)}{p^2} \delta_{\ell'\ell} \delta_{m'm} \delta_{\sigma_1'\sigma_1} \delta_{\sigma_2'\sigma_2}. \quad (5.10.102)$$

For example, for the proton: $\mathcal{P}$, pion: π system, we may write for (5.10.99),

$$\langle \mathcal{P}(\mathbf{p}, \sigma), \pi(-\mathbf{p}) | = \sum_{\ell,m} Y_{\ell m}(\widehat{\mathbf{p}}) \langle p, \ell, m, \sigma |$$

$$= \sum_{\ell,m,J,M} Y_{\ell m}(\widehat{\mathbf{p}}) \langle \ell, m; 1/2, \sigma | J, M \rangle \langle p, J, M | \quad (5.10.103)$$

where $M = m + \sigma$ and

$$J = \frac{1}{2}, \frac{3}{2}, \ldots \quad (5.10.104)$$

$$\ell = J \pm \frac{1}{2}. \quad (5.10.105)$$

For *identical* particles, we have to consider in our description the interchange of particles $\mathcal{P}_1, \mathcal{P}_2$ where now, in particular, $s_1 = s_2 \equiv s$.

To the above end, in the center of mass frame, if $\langle \mathbf{p}, \sigma_1 |$ corresponds to one particle, then $\langle -\mathbf{p}, \sigma_2 |$ corresponds to the other particle.

From (5.10.49), (5.3.43)

$$\langle -\mathbf{p}, \sigma | = \sum_{\ell,m} (-1)^\ell Y_{\ell m}(\theta, \phi) \langle p, \ell, m, \sigma | \quad (5.10.106)$$

and, in particular, from (5.3.50)

5.10 Particle States and Angular Momentum; Helicity States

$$\langle -\mathbf{p}_0, \sigma | = \sum_\ell (-1)^\ell \sqrt{\frac{2\ell+1}{4\pi}} \langle p, \ell, 0, \sigma |. \qquad (5.10.107)$$

On the other hand, we also explicitly have (see (5.2.13), (5.2.17))

$$\langle \mathbf{p}_0, \sigma | \exp\left(-\frac{i}{\hbar}\pi L_r^2\right) = \sum_{\ell,m} \sqrt{\frac{2\ell+1}{4\pi}} d^{(\ell)}_{0m}(-\pi) \langle p, \ell, m, \sigma |$$

$$= \sum_\ell \sqrt{\frac{2\ell+1}{4\pi}} (-1)^\ell \langle p, \ell, 0, \sigma |.$$

That is,

$$\langle -\mathbf{p}_0, \sigma | = \langle \mathbf{p}_0, \sigma | \exp\left(-\frac{i}{\hbar}\pi L_r^2\right). \qquad (5.10.108)$$

Similarly, (5.10.107), leads to

$$\langle \mathbf{p}_0, \sigma | = \langle -\mathbf{p}_0, \sigma | \exp\left(-\frac{i}{\hbar}\pi L_r^2\right). \qquad (5.10.109)$$

Hence for the interchange of the two particles, we have

$$\langle \mathbf{p}_0, \sigma_1 | \langle -\mathbf{p}_0, \sigma_2 | \to \langle -\mathbf{p}_0, \sigma_2 | \langle \mathbf{p}_0, \sigma_1 |$$

$$= \langle \mathbf{p}_0, \sigma_2 | \langle -\mathbf{p}_0, \sigma_1 | \exp\left(-\frac{i}{\hbar}\pi L_r^2\right) \qquad (5.10.110)$$

where *now* L_r^2 denotes the *total* orbital angular momentum, in the x^2-direction, of the two particles in the center of mass system.

According to the spin and statistics connection to be studied later in §16.9, we have to symmetrize for integral spins: $s =$ integers, (bosons) and, anti-symmetrize for half-odd integer spins (fermions): $s = (2k+1)/2$, $k = 0, 1, \ldots$.

We are thus led to the states

$$\left[\langle \mathbf{p}_0, \sigma_1, \sigma_2 | + (-1)^{2s} \langle -\mathbf{p}_0, \sigma_2, \sigma_1 |\right]$$

$$= \left[\langle \mathbf{p}_0, \sigma_1, \sigma_2 | + (-1)^{2s} \langle \mathbf{p}_0, \sigma_2, \sigma_1 | \exp\left(-\frac{i\pi}{\hbar} L_r^2\right)\right]. \qquad (5.10.111)$$

To obtain motion in an arbitrary direction $\hat{\mathbf{p}}$, we apply the operator in (5.10.4) to (5.10.111) leading in a straightforward manner from (5.10.99), (5.3.43) to

$$\left[\langle \mathbf{p}, \sigma_1, \sigma_2 | + (-1)^{2s} \langle -\mathbf{p}, \sigma_2, \sigma_1 |\right]$$

$$= \sum_{\ell,m} Y_{\ell,m}(\hat{\mathbf{p}}) \left[\langle p, \ell, m, \sigma_1, \sigma_2 | + (-1)^{2s+\ell} \langle p, \ell, m, \sigma_2, \sigma_1 |\right]. \qquad (5.10.112)$$

Two Particles of Arbitrary Spins — Helicity States

The two particle helicity states are obtained as for the single particle states in (5.10.70) by noting that

$$\langle \mathbf{p}_0, \lambda_1, \lambda_2 | J^3 = \hbar(\lambda_1 - \lambda_2) \langle \mathbf{p}_0, \lambda_1, \lambda_2 | \tag{5.10.113}$$

where λ_1 is the spin projection of the first particle along the $+z$ direction, corresponding to the state $\langle \mathbf{p}_0, \lambda_1 |$, and λ_2 is the spin projection of the other particle along the $-z$ direction, corresponding to the state $\langle -\mathbf{p}_0, \lambda_2 |$.

From (5.10.113), (5.10.70), by using in the process (see (5.10.63)),

$$\langle \mathbf{p}, \lambda_1, \lambda_2 | = \langle \mathbf{p}_0, \lambda_1, \lambda_2 | \exp\left(-\frac{i}{\hbar}\theta \mathbf{n} \cdot \mathbf{J}\right) \tag{5.10.114}$$

we then obtain

$$\langle \mathbf{p}, \lambda_1, \lambda_2 | = \sum_{J,M} \sqrt{\frac{2J+1}{4\pi}} D^{(J)}_{\lambda m}(\phi, \theta, -\phi) \langle p, J, M, \lambda_1, \lambda_2 | \tag{5.10.115}$$

where

$$\lambda = \lambda_1 - \lambda_2. \tag{5.10.116}$$

We also note from (5.10.7), (2.1.4),

$$\exp\left(-\frac{i}{\hbar}\theta \mathbf{n} \cdot \mathbf{J}\right) J^i \exp\left(\frac{i}{\hbar}\theta \mathbf{n} \cdot \mathbf{J}\right) = J^k R^{ik}(\theta, \mathbf{n}) \tag{5.10.117}$$

and consequently

$$J^k R^{ik}(\theta, \mathbf{n}) \widehat{p}^i = J^3 \tag{5.10.118}$$

giving from (5.10.114)

$$\langle \mathbf{p}, \lambda_1, \lambda_2 | \mathbf{J} \cdot \widehat{\mathbf{p}} = \hbar(\lambda_1 - \lambda_2) \langle \mathbf{p}, \lambda_1, \lambda_2 |. \tag{5.10.119}$$

To treat identical particles, we first apply the operator $\exp(-i\pi S^2/\hbar)$ to (5.10.108) and use (5.10.107). The expression on the right-hand side of (5.10.108) then becomes

$$\langle \mathbf{p}_0, \sigma | \exp\left(-\frac{i\pi}{\hbar} J^2\right) = \sum_{\ell, m, \sigma'} \sqrt{\frac{2\ell+1}{4\pi}} d^{(\ell)}_{0m}(-\pi) d^{(s)}_{\sigma\sigma'}(-\pi) \langle p, \ell, m, \sigma' |$$

$$= (-1)^{s+\sigma} \sum_{\ell} (-1)^\ell \sqrt{\frac{2\ell+1}{4\pi}} \langle p, \ell, 0, -\sigma |$$

$$= (-1)^{s+\sigma} \langle -\mathbf{p}_0, -\sigma |$$

5.10 Particle States and Angular Momentum; Helicity States

$$\equiv \langle -\mathbf{p}_0, \sigma | \exp\left(-\frac{i\pi}{\hbar} S^2\right) \qquad (5.10.120)$$

leading to the equality

$$(-1)^{s+\sigma} \langle -\mathbf{p}_0, -\sigma | = \langle \mathbf{p}_0, \sigma | \exp\left(-\frac{i\pi}{\hbar} J^2\right). \qquad (5.10.121)$$

Upon applying the operator $\exp\left(-i\pi J^2/\hbar\right)$ one more time to (5.10.121) and using (5.20) leads to

$$\langle \mathbf{p}_0, \sigma | = (-1)^{-s+\sigma} \langle -\mathbf{p}_0, -\sigma | \exp\left(-\frac{i\pi}{\hbar} J^2\right). \qquad (5.10.122)$$

By definition of helicity states,

$$\langle \mathbf{p}_0, \lambda | J^3 = \hbar \lambda \langle \mathbf{p}_0, \lambda | \qquad (5.10.123)$$

$$\langle -\mathbf{p}_0, \lambda | (-J^3) = \hbar \lambda \langle -\mathbf{p}_0, \lambda |. \qquad (5.10.124)$$

For a vector operator $\mathbf{V}$, it is also easily checked from (5.7.42) that

$$\exp\left(-\frac{i}{\hbar}\pi J^2\right) V^k \exp\left(\frac{i}{\hbar}\pi J^2\right) = -V^k + 2\delta^{k2} V^2. \qquad (5.10.125)$$

Hence from (5.10.121), (5.10.123)–(5.10.125), we have

$$(-1)^{s-\lambda} \langle -\mathbf{p}_0, \lambda | = \langle \mathbf{p}_0, \lambda | \exp\left(-\frac{i\pi}{\hbar} J^2\right) \qquad (5.10.126)$$

for the corresponding helicity states.
Similarly, from (5.10.122)–(5.10.125), for the corresponding helicity states in (5.10.122) we have

$$\langle \mathbf{p}_0, \lambda | = (-1)^{-s-\lambda} \langle -\mathbf{p}_0, \lambda | \exp\left(-\frac{i\pi}{\hbar} J^2\right). \qquad (5.10.127)$$

Accordingly, for the interchange of the two identical particles with $s_1, s_2 \equiv s$,

$$\langle \mathbf{p}_0, \lambda_1, \lambda_2 | \to \langle -\mathbf{p}_0, \lambda_2, \lambda_1 |$$

$$= (-1)^{-2s+\lambda_2-\lambda_1} \langle \mathbf{p}_0, \lambda_2, \lambda_1 | \exp\left(-\frac{i\pi}{\hbar} J^2\right) \qquad (5.10.128)$$

where now J^2 is the total angular momentum of the two particles in the center of mass system. Hence

$$\left[\langle \mathbf{p}_0, \lambda_1, \lambda_2 | + (-1)^{2s} \langle -\mathbf{p}_0, \lambda_2, \lambda_1 |\right]$$

$$= \left[\langle \mathbf{p}_0, \lambda_1, \lambda_2 | + (-1)^{-\lambda} \langle \mathbf{p}_0, \lambda_2, \lambda_1 | \exp\left(-\frac{i\pi}{\hbar} J^2\right) \right] \quad (5.10.129)$$

$\lambda = \lambda_1 - \lambda_2$. Upon using, the expression $d^{(J)}_{-\lambda, M}(-\pi) = (-1)^{J-\lambda} \delta_{M, \lambda}$ as given in (5.2.17), we obtain from (5.10.129), (5.10.115)

$$\left[\langle \mathbf{p}, \lambda_1, \lambda_2 | + (-1)^{2s} \langle -\mathbf{p}, \lambda_2, \lambda_1 | \right]$$

$$= \sum_{J,M} \sqrt{\frac{2J+1}{4\pi}} D^{(J)}_{\lambda m}(\phi, \theta, -\phi) \left[\langle p, J, M, \lambda_1, \lambda_2 | + (-1)^J \langle p, J, M, \lambda_2, \lambda_1 | \right] \quad (5.10.130)$$

where we have used the fact that $\lambda \equiv \lambda_1 - \lambda_2$ is always an integer for both integer as well as for half-odd integer spins s ($\equiv s_1 = s_2$), and hence $(-1)^{-2\lambda} = 1$.

Helicity Versus Standard Spin States

We follow the procedure developed for single particle states. The corresponding equation to (5.10.72)

$$\langle \mathbf{p}_0, \sigma_1, \sigma_2 | \bigg|_{\substack{\sigma_1 = \lambda_1 \\ \sigma_2 = -\lambda_2}} = \langle \mathbf{p}_0, \lambda_1, \lambda_2 | \quad (5.10.131)$$

relates spin (on the left-hand side) to helicity states. Upon applying the operator $\exp(-i\theta \mathbf{n} \cdot \mathbf{J}/\hbar)$ in (5.10.4) to (5.10.131), the corresponding equation to (5.10.77) becomes

$$\sum_{J,M,\ell} \sqrt{\frac{2\ell+1}{4\pi}} D^{(J)}_{\lambda m}(\phi, \theta, -\phi) \langle s_1, \lambda_1; s_2, -\lambda_2 | s, \lambda \rangle \langle \ell, 0; s, \lambda | J, \lambda \rangle \langle p, J, M, \ell, s |$$

$$= \sum_{J,M} \sqrt{\frac{2J+1}{4\pi}} D^{(J)}_{\lambda m}(\phi, \theta, -\phi) \langle p, J, M, \lambda_1, \lambda_2 | \quad (5.10.132)$$

from which we finally obtain

$$\langle J, M, \lambda_1, \lambda_2 | J, M, \ell, s \rangle = \sqrt{\frac{2\ell+1}{2J+1}} \langle s_1, \lambda_1; s_2, -\lambda_2 | s, \lambda \rangle \langle \ell, 0; s, \lambda | J, \lambda \rangle \quad (5.10.133)$$

where $\langle s_1, \lambda_1; s_2, -\lambda_2 | s, \lambda \rangle$, $\langle \ell, 0; s, \lambda | J, \lambda \rangle$ are just Clebsch-Gordan coefficients.

Problems

5.1. Derive the relations in (5.1.26), (5.1.27) by the repeated applications of the operators J_-, J_+ as well as the relations in (5.1.28), (5.1.29).

5.2. (i) Obtain the expressions of L^3, $\mathbf{L}^2$, $L_\pm$ in spherical coordinates as given in (5.3.7)–(5.3.9).
   (ii) Evaluate the commutators: $[\mathbf{L}^2, \mathbf{x}]$, $[\mathbf{L}^2, \mathbf{p}]$, $[\mathbf{L}^2, f(|\mathbf{x}|)]$ where $f(|\mathbf{x}|)$ is a function of $|\mathbf{x}|$. Also show that $\mathbf{x} \cdot \mathbf{L} = 0$, $\mathbf{L} \cdot \mathbf{x} = 0$.
   (iii) Obtain a lower bound to the product

$$\left\langle \ell, m \left| (L^1)^2 \right| \ell, m \right\rangle \left\langle \ell, m \left| (L^2)^2 \right| \ell, m \right\rangle$$

and interpret this result.

5.3. Upon the comparison of the matrix elements of rotation by the Euler angles in (5.2.4) with that in (2.1.4) obtain the relationships spelled out in (5.2.6)–(5.2.8).

5.4. Use the definition of $d^{(j)}_{mm'}(\beta)$ in (5.2.13)/(5.2.10), for the matrix elements of finite rotation by the angles β about the x^2 axis, to prove (5.2.15)–(5.2.18).

5.5. Verify explicitly the transformation rules of the spherical harmonics in (5.3.43), (5.3.44) under space reflection and complex conjugation.

5.6. Use the transformation law of a vector $\mathbf{x} = (x^1, x^2, x^3)$, $(x^1 = x, x^2 = y, x^3 = z)$, in (5.2.3) under a rotation of a coordinate system by the Euler angles, to find the corresponding transformation law for the transformation of the spherical harmonics $Y_{1,1}, Y_{1,0}, Y_{1,-1}$. [Hint: Use the identities in (5.3.52)–(5.3.54).]

5.7. For the density operator ρ in (5.4.33) for spin 1, derive (5.4.35), and also obtain the expression for $\text{Tr}\left[\rho S^i S^j\right]$.

5.8. Derive the equalities (5.5.26), (5.5.27) relating the Clebsch-Gordan coefficients.

5.9. Consider two particles of spins $\mathbf{S}_{(1)}$, $\mathbf{S}_{(2)}$ of equal masses.
   (i) Show that the orbital angular momenta of the two particles relative to their center of mass are equal, i.e.,

$$(\mathbf{X}_1 - \mathbf{X}) \times \left(\mathbf{P}_1 - \frac{\mathbf{P}}{2} \right) = \left(\mathbf{X}_2 - \frac{\mathbf{P}}{2} \right) \times \left(\mathbf{P}_2 - \frac{\mathbf{P}}{2} \right)$$

by referring to (2.7.39), (2.7.40), where

$$\mathbf{X} = (\mathbf{X}_1 + \mathbf{X}_2)/2, \qquad \mathbf{P} = \mathbf{P}_1 + \mathbf{P}_2$$

and hence each particle carries angular momentum $\mathbf{L}_r/2$ in the notation of (2.7.40), where $\mathbf{L}_r$ is the total orbital angular momentum in the center of mass.

(ii) Consider the deuteron (D) as a bound-state of a proton (p) and a neutron (n) both of spin 1/2, with the masses taken approximately equal. Upon defining the respective magnetic moment vector operators

$$\mu_p = \frac{g_p \mu_N}{\hbar} \mathbf{S}_p, \quad \mu_n = \frac{g_n \mu_N}{\hbar} \mathbf{S}_n, \quad \mu_D = \frac{g_D \mu_N}{\hbar} \mathbf{J}$$

where $\mathbf{J}$ is the internal angular momentum of the deuteron (see (2.7.39), (2.7.40))
$$\mathbf{J} = \mathbf{L}_r + \mathbf{S}_p + \mathbf{S}_n$$
show that
$$\mu_D = \frac{\mu_N}{\hbar}\left(\frac{\mathbf{L}_r}{2} + g_p \mathbf{S}_p + g_n \mathbf{S}_n\right).$$
Here $\mu_N = |e|\hbar/2m_p c$ is the nuclear magneton, and $g_p = 5.58$, $g_n = -3.83$ are the g-factors of the proton and neutron, respectively.

(iii) Consider the deuteron in the state $j = 1$ and $(\mathbf{S}_p + \mathbf{S}_n)$ with corresponding s value equal to 1. Use the Wigner-Eckart Theorem in the form given in (5.7.35) to evaluate approximately the magnetic moment $\mu_D = g_D \mu_N$ of the deuteron corresponding to $\ell = 0$ and $\ell = 2$. [Experimentally, $g_D \simeq 1.72$ and the deuteron is predominantly, about 98%, in the $\ell = 0$ state and about 2% in the $\ell = 2$ state.]

5.10. The Hamiltonian of the dipole-dipole interaction of two spin 1/2 particles each of magnetic moment μ separated by a vector $\mathbf{a}$ is
$$H = \frac{\mu^2}{a^3}\left[\boldsymbol{\sigma}_1 \cdot \boldsymbol{\sigma}_2 - 3\frac{\boldsymbol{\sigma}_1 \cdot \mathbf{a}\, \boldsymbol{\sigma}_2 \cdot \mathbf{a}}{a^2}\right].$$
Suppose that $\mathbf{a} = a\,(0, 0, 1)$. Show that the singlet and triplet states in spin space are eigenvectors of H and find their corresponding eigenvalues.

5.11. Suppose that the initial spins of the particles in Problem 5.10, at time $t = 0$, are along the x-axis, i.e., they are in the state $(1\ 1)_1^T\, (1\ 1)_2^T/2$ (see (5.4.9)).
 (i) Find the state of the system at any time $t > 0$. Consider $\mathbf{a}$ to be fixed.
 (ii) What is the probability that the system is found in its initial state at time t?

5.12. Find an orthogonality relation for the D-functions in (5.5.51) for the cases when j is a half-odd integer.

5.13. Establish the symmetry relations in (5.6.25)–(5.6.27), and the equalities in (5.6.28), (5.6.29) of the 3-j symbols.

5.14. Introduce the spherical vector components $\boldsymbol{\nabla}(m)$ of the gradient $\boldsymbol{\nabla}$, and evaluate explicitly the commutator $[L(m_1), \boldsymbol{\nabla}(m_2)]$, where $L(m)$ are the spherical vector components of the orbital angular momentum.

5.15. Evaluate the reduced matrix element $\langle \ell_1 ||\hat{\mathbf{r}}|| \ell_2 \rangle$ of the unit vector $\hat{\mathbf{r}}$, for the position vector $\mathbf{x} = r\hat{\mathbf{r}}$. [Hint: Use (5.3.52)–(5.3.54), together with (5.8.50) and (5.6.28), (5.6.29).]

5.16. Deviation from spherical symmetry of a charge distribution, such as of the nucleus, is determined from its quadrupole moment. Quantum mechanically, the corresponding operator may be defined by $Q_0^2 = 3z^2 - r^2$.

(i) Show that $Q_0^2 = \sqrt{16\pi/5}\, r^2\, Y_{2,0}(\theta, \phi)$. Thus, up to the factor $\sqrt{16\pi/5}\, r^2$, you recognize this operator as the zeroth component of a second rank tensor and hence the notation Q_0^2.

(ii) Write the matrix element $\langle j, m | Q_0^2 | j, m \rangle$ in terms of the reduced matrix element $\langle j || Q^2 || j \rangle$.

(iii) Show that for $j \geqslant 1$

$$\langle j, m | Q_0^2 | j, m \rangle = \left(3 \frac{m^2}{j(j+1)} - 1 \right) \frac{(j+1)}{(2j-1)} \langle j, j | Q_0^2 | j, j \rangle$$

introducing $(2j + 1)$ quadrupole moments associated with the $(2j + 1)$ possible orientations of the internal angular momentum **J** of the body written in terms of the one oriented along the z-axis. The angles θ_m of the orientations are given by $\cos\theta_m = m/\sqrt{j(j+1)}$. For j large, one has almost a continuous distribution of orientations. From the result in (iii) you may then infer that for j large, one obtains its classical counterpart

$$Q(\cos\vartheta_c) = \frac{1}{2}\left(3\cos^2\vartheta_c - 1\right) Q(1)$$

where ϑ_c defines the angle between the symmetry axis of the body and the z-axis.

5.17. Evaluate the matrix element on the left-hand side of (5.8.39) directly from the general formula (5.8.47).

5.18. Verify the symmetry relations of the 9-j symbols stated below (5.9.15).

5.19. Use (5.10.30), (5.3.66) to establish the equality in (5.10.39) for $\langle \mathbf{x} | p, \ell, m \rangle$.

5.20. Follow a procedure similar to the one giving (5.10.49) to establish the equality in (5.10.70) for $\langle \mathbf{p}, \lambda |$ expressed in terms of the helicity.

6

Intricacies of Harmonic Oscillators

This chapter deals with several problems and intricate details associated with the harmonic oscillator potential. After solving the eigenvalue problem in §6.1, we study transitions that may be caused between its energy levels in the presence of a time-dependent external, i.e., classical, force coupled linearly to the position observable at zero and finite temperatures in §6.2 and §6.3, respectively. §6.4 deals with the construction of the so-called Fermi oscillator in analogy to the Bose one in §6.2. In §6.5, we combine the Bose oscillator to the Fermi one to construct supersymmetric theories and consider underlying supersymmetric transformations in the light of the analysis carried out in §2.9. In the final section (§6.6), the coherent state of the harmonic oscillator is constructed and relevant details are developed which allow one to compare the quantum mechanical problem with the corresponding classical one in the most natural way.

6.1 The Harmonic Oscillator

The Hamiltonian of the ν-dimensional harmonic oscillator is defined by

$$H = -\frac{\hbar^2}{2m}\nabla^2 + \frac{1}{2}m\omega^2\mathbf{x}^2. \tag{6.1.1}$$

Such a quadratic interaction may be considered as a perturbation, about an equilibrium point $\mathbf{x} = \mathbf{0}$ $\left(\nabla V(\mathbf{x})\big|_0 = 0\right)$ of a spherically symmetric potential $V(\mathbf{x})$ in $\mathbb{R}^\nu$, up to an additive constant. The harmonic oscillator Hamiltonian is, however, in its own right, a very useful one for various investigations and is the prototype of quantum field theories involving an arbitrary large numbers of so-called degrees of freedom.

A lower bound to the spectrum of H in (6.1.1) is readily found by choosing $g(\mathbf{x}) = \omega\hbar$, with $\mu \to m$, in (3.1.8) giving after re-arrangement of terms,

$$\frac{\nu\hbar\omega}{2} \leqslant \left\langle -\frac{\hbar^2}{2m}\nabla^2 + \frac{1}{2}m\omega^2\mathbf{x}^2 \right\rangle. \tag{6.1.2}$$

6 Intricacies of Harmonic Oscillators

Actually, the lower bound $\nu\hbar\omega/2$ coincides with the exact ground-state energy of H. To see this, introduce the trial normalized wave function

$$\Phi(\mathbf{x}) = \left(\frac{\alpha}{\pi}\right)^{\nu/4} \exp\left(-\frac{\alpha\mathbf{x}^2}{2}\right) \quad (6.1.3)$$

and note, by the very definition of the ground-state energy E_0, that the expectation value $\langle\Phi|H|\Phi\rangle$ cannot be less than E_0. That is, from (6.1.2)

$$\frac{\nu\hbar\omega}{2} \leqslant E_0 \leqslant \langle\Phi|H|\Phi\rangle. \quad (6.1.4)$$

We explicitly have

$$(H\Phi)(\mathbf{x}) = \left(-\frac{\hbar^2}{2m}\alpha^2\mathbf{x}^2 + \frac{m\omega^2}{2}\mathbf{x}^2 + \frac{\hbar^2}{2m}\nu\alpha\right)\Phi(\mathbf{x}) \quad (6.1.5)$$

and upon choosing $\alpha = m\omega/\hbar$, we obtain

$$(H\Phi)(\mathbf{x}) = \frac{\nu\hbar\omega}{2}\Phi(\mathbf{x}) \quad (6.1.6)$$

and from (6.1.4) that

$$E_0 = \frac{\nu\hbar\omega}{2}. \quad (6.1.7)$$

From Theorem 4.1.1, we may also infer that H has only a discrete spectrum.

To study the eigenvalue problem of the Hamiltonian (6.1.1), it is convenient to introduce the operators

$$\mathbf{X} = \sqrt{\frac{m\omega}{\hbar}}\,\mathbf{x}, \qquad \mathbf{P} = \frac{1}{\sqrt{m\omega\hbar}}(-i\hbar\nabla) \quad (6.1.8)$$

and rewrite H as

$$H = \frac{\hbar\omega}{2}\left(\mathbf{P}^2 + \mathbf{X}^2\right). \quad (6.1.9)$$

We also define the operator

$$\mathbf{a} = \frac{1}{\sqrt{2}}(\mathbf{X} + i\mathbf{P}). \quad (6.1.10)$$

The following are easily established (see Problem 6.1):

$$[X^i, P^j] = i\delta^{ij} \quad (6.1.11)$$

$$H = \hbar\omega\left(\mathbf{a}^\dagger \cdot \mathbf{a} + \frac{\nu}{2}\right) \quad (6.1.12)$$

$$\left[a^i, a^{j\dagger}\right] = \delta^{ij}, \qquad [a^i, a^j] = 0 \quad (6.1.13)$$

6.1 The Harmonic Oscillator

$$[H, a^i] = -\hbar\omega a^i \tag{6.1.14}$$

$$[H, a^{i\dagger}] = +\hbar\omega a^{i\dagger}. \tag{6.1.15}$$

Since for a normalized state $|\psi\rangle$, $\langle\psi|H|\psi\rangle = \hbar\omega \|\mathbf{a}\psi\|^2 + \hbar\omega\nu/2 \geqslant \hbar\omega\nu/2$, (6.1.12) alone shows that H is bounded from below by $\hbar\omega\nu/2$ as expected.

To solve the eigenvalue problem, we consider first the one-dimensional case $\nu = 1$. The several dimensional case may be directly inferred from the one-dimensional one.

To the above end, the eigenvalue problem reads

$$H|\alpha\rangle = \alpha|\alpha\rangle \tag{6.1.16}$$

and from property (6.1.14),

$$H(a)^k|\alpha\rangle = (\alpha - \hbar\omega k)(a)^k|\alpha\rangle \tag{6.1.17}$$

for $k = 0, 1, 2, \ldots$.

The boundedness of H from below implies that for a given α, there must exist a non-negative integer k and a $\alpha_{\min}$ such that

$$\alpha_{\min} = \alpha - \hbar\omega k \tag{6.1.18}$$

and

$$a|\alpha_{\min}\rangle = 0. \tag{6.1.19}$$

The latter equation implies that

$$a^\dagger a|\alpha_{\min}\rangle = 0 \tag{6.1.20}$$

and hence

$$H|\alpha_{\min}\rangle = \frac{\hbar\omega}{2}|\alpha_{\min}\rangle \tag{6.1.21}$$

where we have used (6.1.12) with $\nu = 1$. That is,

$$\alpha_{\min} = \frac{\hbar\omega}{2} \tag{6.1.22}$$

and the eigenvalues α in (6.1.16) are given from (6.1.18) to be

$$\alpha = \hbar\omega\left(n + \frac{1}{2}\right), \qquad n = 0, 1, \ldots \tag{6.1.23}$$

for $\nu = 1$.

For the subsequent analysis, we label eigenstates $|\alpha\rangle$ by the non-negative integer n instead, i.e., we have

$$H|n\rangle = \hbar\omega\left(n + \frac{1}{2}\right)|n\rangle. \tag{6.1.24}$$

From the property (6.1.15),

$$H\left(a^\dagger |n\rangle\right) = \hbar\omega\left(n + 1 + \frac{1}{2}\right)\left(a^\dagger |n\rangle\right) \quad (6.1.25)$$

and infer from (6.1.24) that

$$a^\dagger |n\rangle = c_n |n+1\rangle \quad (6.1.26)$$

where the constant c_n, depending on n, is to be determined.

To the above end, we note from (6.1.13), (6.1.12), with $\nu = 1$, that

$$a\, a^\dagger = a^\dagger a + 1 = \frac{H}{\hbar\omega} + \frac{1}{2} \quad (6.1.27)$$

and hence upon taking the norm squared of (6.1.26),

$$|c_n|^2 = \langle n| \left(\frac{H}{\hbar\omega} + \frac{1}{2}\right) |n\rangle \quad (6.1.28)$$

giving

$$c_n = \sqrt{n+1} \quad (6.1.29)$$

with a phase convention, where we have used (6.1.24) and the normalizability of the states $|n\rangle$.

Equations (6.1.26), (6.1.29) give the recurrence relation

$$|n+1\rangle = \frac{a^\dagger}{\sqrt{n+1}} |n\rangle \quad (6.1.30)$$

providing the solution

$$|n\rangle = \frac{(a^\dagger)^n}{\sqrt{n!}} |0\rangle. \quad (6.1.31)$$

On the other hand (6.1.19), with (6.1.22), imply from $a|0\rangle = 0$ that

$$\left(\frac{d}{dx} + \frac{m\omega}{\hbar}x\right)\psi_0(x) = 0 \quad (6.1.32)$$

where $\psi_0(x) = \langle x|0\rangle$, giving the (normalized) solution

$$\psi_0(x) = \left(\frac{m\omega}{\pi\hbar}\right)^{1/4} \exp\left(-\frac{m\omega x^2}{2\hbar}\right) \quad (6.1.33)$$

coinciding with the one in (6.1.3) for $\nu = 1$, with $\alpha = m\omega/\hbar$.

For the first excited state $\psi_1(x) = \langle x|1\rangle$, we obtain from (6.1.30), with $n = 0$, and (6.1.33)

$$\psi_1(x) = \left(\frac{m\omega}{\pi\hbar}\right)^{1/4} \sqrt{\frac{\hbar}{2m\omega}} \left(\frac{m\omega}{\hbar}x - \frac{\partial}{\partial x}\right) \exp\left(-\frac{m\omega x^2}{2\hbar}\right) \quad (6.1.34)$$

6.1 The Harmonic Oscillator

or

$$\psi_1(x) = \left(\frac{m\omega}{\pi\hbar}\right)^{1/4} \sqrt{2}\sqrt{\frac{m\omega}{\hbar}}\, x \exp\left(-\frac{m\omega x^2}{2\hbar}\right). \tag{6.1.35}$$

We introduce the dimensionless variable ($\nu = 1$), using rather a conventional notation

$$\rho = \sqrt{\frac{m\omega}{\hbar}}\, x \tag{6.1.36}$$

to rewrite (6.1.33), (6.1.35) as

$$\psi_0(x) = \left(\frac{m\omega}{\pi\hbar}\right)^{1/4} \exp\left(-\frac{\rho^2}{2}\right) \tag{6.1.37}$$

$$\psi_1(x) = \left(\frac{m\omega}{\pi\hbar}\right)^{1/4} \exp\left(-\frac{\rho^2}{2}\right) \sqrt{2}\,\rho$$

$$\equiv \left(\frac{m\omega}{\pi\hbar}\right)^{1/4} \exp\left(-\frac{\rho^2}{2}\right) \frac{(-1)}{\sqrt{2}} \left[e^{\rho^2}\left(\frac{d}{d\rho}\right) e^{-\rho^2}\right]. \tag{6.1.38}$$

To obtain the solution for all n, we proceed by induction. As an induction hypothesis, suppose that for some k, and $n = 0, \ldots, k$,

$$\psi_n(x) = \left(\frac{m\omega}{\pi\hbar}\right)^{1/4} e^{-\rho^2/2} \frac{(-1)^n}{\sqrt{2^n n!}} \left[e^{\rho^2}\left(\frac{d}{d\rho}\right)^n e^{-\rho^2}\right] \tag{6.1.39}$$

which obviously agree for $k = 1$, i.e., for $n = 0, 1$, with the expressions in (6.1.37), (6.1.38), respectively.

Hence from (6.1.30), (6.1.10),

$$\psi_{k+1}(x) = \left(\frac{m\omega}{\pi\hbar}\right)^{1/4} \frac{(-1)^k}{\sqrt{2^k k!}} \frac{1}{\sqrt{2(k+1)}}$$

$$\times \left(\rho - \frac{\partial}{\partial \rho}\right) e^{-\rho^2/2} \left[e^{\rho^2}\left(\frac{d}{d\rho}\right)^k e^{-\rho^2}\right]$$

$$= \left(\frac{m\omega}{\pi\hbar}\right)^{1/4} e^{-\rho^2/2} \frac{(-1)^{k+1}}{\sqrt{2^{k+1}(k+1)!}} \left[e^{\rho^2}\left(\frac{d}{d\rho}\right)^{k+1} e^{-\rho^2}\right] \tag{6.1.40}$$

thus establishing the validity of (6.1.39) for all $n = 0, 1, 2, \ldots$.

The expression within the square brackets in (6.1.39), multiplied by $(-1)^n$, defines the so-called Hermite polynomials:

$$H_n(\rho) = (-1)^n e^{\rho^2} \left(\frac{d}{d\rho}\right)^n e^{-\rho^2} \tag{6.1.41}$$

$$H_0(\rho) = 1, \quad H_1(\rho) = 2\rho, \quad H_2(\rho) = 4\rho^2 - 2, \ldots. \tag{6.1.42}$$

6 Intricacies of Harmonic Oscillators

The orthonormality condition

$$\delta_{nn'} = \langle n|n'\rangle = \int_{-\infty}^{\infty} dx\, \psi_n(x)\,\psi_{n'}(x) \tag{6.1.43}$$

as follows from the eigenvalue equation (6.1.16) and the normalizability of the states $|n\rangle$ in (6.1.30), (6.1.31), *imply* the orthogonality property of the Hermite polynomials (6.1.41) from (6.1.39) and the definition (6.1.41):

$$\int_{-\infty}^{\infty} d\rho\, e^{-\rho^2} H_n(\rho)\, H_{n'}(\rho) = \delta_{nn'} \sqrt{\pi}\, 2^n n! \tag{6.1.44}$$

as written in terms of the dimensionless variable ρ in (6.1.36).

The following matrix elements should be noted

$$\langle n'|x|n\rangle = \sqrt{\frac{\hbar}{2m\omega}}\left[\sqrt{n+1}\,\delta_{n',n+1} + \sqrt{n}\,\delta_{n',n-1}\right] \tag{6.1.45}$$

$$\langle n|x^2|n\rangle = \frac{\hbar}{m\omega}\left(n + \frac{1}{2}\right) \tag{6.1.46}$$

$$\langle n'|p|n\rangle = i\sqrt{\frac{\hbar m\omega}{2}}\left[\sqrt{n+1}\,\delta_{n',n+1} - \sqrt{n}\,\delta_{n',n-1}\right] \tag{6.1.47}$$

$$\langle n|p^2|n\rangle = \hbar m\omega\left(n + \frac{1}{2}\right). \tag{6.1.48}$$

These equations, in particular, imply that

$$\sigma_x^2(n)\,\sigma_p^2(n) = \hbar^2\left(n + \frac{1}{2}\right)^2 \geq \frac{\hbar^2}{4} \tag{6.1.49}$$

where

$$\sigma_x^2(n) \equiv \langle n|\left(x - \langle n|x|n\rangle\right)^2|n\rangle = \langle n|x^2|n\rangle \tag{6.1.50}$$

and similarly defined for $\sigma_p^2(n)$. We note that the presence of the so-called zero point energy $\hbar\omega/2$ in (6.1.24), as reflected on the right-hand of the inequality (6.1.49) is consistent with the Heisenberg uncertainty principle.

For the ν dimensional case, with H defined in (6.1.1), we may immediately infer from the additive nature of $\nabla^2 = \sum_{i=1}^{n}(\partial/\partial x^i)^2$, $\mathbf{x}^2 = \sum_{i=1}^{n}(x^i)^2$ and from (6.1.24), (6.1.39), (6.1.41), that

$$H|n_1,\ldots,n_\nu\rangle = \hbar\omega\left(n_1 + \ldots + n_\nu + \frac{\nu}{2}\right)|n_1,\ldots,n_\nu\rangle \tag{6.1.51}$$

and with $\langle \mathbf{x}|\psi_{n_1,\ldots,n_\nu}\rangle = \psi_{n_1,\ldots,n_\nu}(\mathbf{x})$,

$$\psi_{n_1,\ldots,n_\nu}(\mathbf{x}) = \left(\frac{m\omega}{\pi\hbar}\right)^{\nu/4} e^{-\rho^2/2} \prod_{j=1}^{\nu} \frac{1}{\sqrt{2^{n_j} n_j!}} H_{n_j}(\rho_j) \tag{6.1.52}$$

where $\boldsymbol{\rho} = (\rho_1,\ldots,\rho_\nu)$, $\boldsymbol{\rho} = \sqrt{m\omega/\hbar}\,\mathbf{x}$.

6.2 Transition to and Between Excited States in the Presence of a Time-Dependent Disturbance

We add a linear coupling of the position variable x to a time-dependent, c-function, $F(t)$ in the harmonic oscillator problem in one dimension, and define the Hamiltonian

$$H(t) = \frac{p^2}{2m} + \frac{1}{2}m\omega^2 x^2 - \sqrt{\frac{2m\omega}{\hbar}}\, xF(t) \qquad (6.2.1)$$

where the factor $(2m\omega/\hbar)^{1/2}$ has been introduced for convenience, and up to this proportionality factor, $F(t)$ represents an external force which is also referred to as an external source.

In terms of the annihilation and creation operators a, $a^\dagger$, (6.2.1) takes the form

$$H(t) = \hbar\omega\left(a^\dagger a + \frac{1}{2}\right) - F(t)\left(a + a^\dagger\right). \qquad (6.2.2)$$

The Hamiltonian in (6.2.2) is a prototype of field theories in "zero" dimension of space, in the presence of an external source $F(t)$, and where $\left(a + a^\dagger\right)$ is the "field" at time $t = 0$.

In the present section, we investigate the problem of transitions from the ground-state to excited states and transitions between different states, in general, due to the disturbance provided by $F(t)$, not barring, however the possibility that the system may stay in its initial state.

We choose $F(t)$ to vanish for $t \leqslant T_1$, $t \geqslant T_2$ for some $T_1, T_2, T_2 > T_1$.[1] Before $F(t)$ is switched on, i.e., for $t < T_1$, we consider the system to be in the ground-state, and choose for the initial state

$$|\psi(T_1)\rangle = |0\rangle \equiv |0_-\rangle \qquad (6.2.3)$$

borrowing a notation $|0_-\rangle$ often used in field theory.

After $F(t)$ is switched on, and then later on when it ceases to operate after the time T_2, the system may, or may not, be found in some excited state.

To investigate such possible transitions, we solve the Schrödinger equation

$$i\hbar \frac{\partial}{\partial t}|\psi(t)\rangle = H(t)|\psi(t)\rangle \qquad (6.2.4)$$

for $|\psi(T_2)\rangle$, with the initial condition in (6.2.3).

Equation (6.2.4) is readily solved by making the ansatz

$$|\psi(t)\rangle = \exp\left[-\frac{i}{\hbar}\left(\beta(t)a^\dagger + \gamma(t)\right)\right]|0_-\rangle \qquad (6.2.5)$$

---

[1] Actually, one may introduce such a function as $F(t)$ which together with its derivative are continuous, and vanishes at some given points, see, for example, §8.7, (8.7.21), §12.6, (12.6.33).

336   6 Intricacies of Harmonic Oscillators

where $\beta(t)$, $\gamma(t)$ are c-functions satisfying

$$\beta(T_1) = 0, \qquad \gamma(T_1) = 0. \tag{6.2.6}$$

Upon substitution of (6.2.5) in (6.2.4), using the fact that $a\,|0_-\rangle \equiv a\,|0\rangle = 0$ and the identity in Problem 6.6, we obtain

$$\left[\dot{\beta}(t)a^\dagger + \dot{\gamma}(t)\right]|0_-\rangle$$

$$= \left[\hbar\omega\left(\frac{1}{2} - \frac{i}{\hbar}\beta(t)a^\dagger\right) - F(t)\left(-\frac{i}{\hbar}\beta(t) + a^\dagger\right)\right]|0_-\rangle \tag{6.2.7}$$

which with the initial conditions in (6.2.6) gives the solutions

$$\beta(t) = -e^{-i\omega t}\int_{-\infty}^{t} dt'\, e^{i\omega t'}F(t') \tag{6.2.8}$$

$$\gamma(t) = \frac{\hbar\omega}{2}(t - T_1)$$

$$-\frac{i}{\hbar}\int_{-\infty}^{\infty}dt'\int_{-\infty}^{\infty}dt''\, e^{-i\omega(t''-t')}F(t'')\,\Theta(t - t'')\,\Theta(t'' - t')\,F(t') \tag{6.2.9}$$

where we have extended the integrations in (6.2.8), (6.2.9) beyond the range $T_1 < t < T_2$ since $F(t)$ vanishes there, i.e., it ceases to operate. Note that the second term on the right-hand side of (6.2.9) vanishes for $t = T_1$ in addition to the first term, and recall also the property of the step function

$$\frac{d}{dt}\Theta(t - t'') = \delta(t - t''). \tag{6.2.10}$$

Hence from (6.2.5)–(6.2.9), the ground-state persistence amplitude, normalized to one for $F(t) = 0$, is given by

$$\langle 0_+|0_-\rangle_F = \frac{\langle 0|\psi(T_2)\rangle}{\langle 0|\psi(T_2)\rangle\big|_{F=0}}$$

$$= \frac{\langle 0|\exp[-i\gamma(T_2)/\hbar]|0\rangle}{\langle 0|\exp[-i\gamma(T_2)/\hbar]|0\rangle\big|_{F=0}} \tag{6.2.11}$$

using the fact that $\langle 0|\,a^\dagger = 0$, and we have

$$\langle 0_+|0_-\rangle_F = \exp\left[-\frac{1}{\hbar^2}\int_{-\infty}^{\infty}dt''\int_{-\infty}^{\infty}dt'\, e^{-i\omega(t''-t')}F(t'')\,\Theta(t''-t')\,F(t')\right] \tag{6.2.12}$$

where we have set $\Theta(T_2 - t'') = 1$, since $F(t'') = 0$, for $t'' \geq T_2$ and have used the convenient notation $\langle 0_+|0_-\rangle_F$ for the amplitude in question. The denominator in (6.2.11) simply cancels out the phase $\exp[-i\omega(T_2 - T_1)/2]$ due to the zero point energy (see (6.2.9)).

Given that the system is initially in the ground-state prior to the switching on of $F(t)$, (6.2.12) leads for the *probability that the system will remain in the ground-state after $F(t)$ is switched on and then after it ceases to operate*, the expression

$$|\langle 0_+|0_-\rangle_F|^2 = \exp\left[-\frac{2}{\hbar^2}\int_{-\infty}^{\infty} dt''\right.$$
$$\left.\times \int_{-\infty}^{\infty} dt'\, F(t'')\, F(t')\cos\omega(t''-t')\,\Theta(t''-t')\right]. \tag{6.2.13}$$

The persistence probability in (6.2.13) may be rewritten in the more convenient form

$$|\langle 0_+|0_-\rangle_F|^2 = \exp\left(-\frac{1}{\hbar^2}\left|\int_{-\infty}^{\infty} dt\, e^{-i\omega t} F(t)\right|^2\right) \tag{6.2.14}$$

which upon introducing the Fourier transform

$$F(\omega) = \int_{-\infty}^{\infty} dt\, e^{i\omega t} F(t) \tag{6.2.15}$$

gives the simple expression

$$|\langle 0_+|0_-\rangle_F|^2 = \exp\left(-\frac{|F(\omega)|^2}{\hbar^2}\right). \tag{6.2.16}$$

From (6.2.16) we also obtain for the probability of having *any* excitation from the ground-state due to the disturbance provided by the source $F(t)$ to be given by

$$\left[1 - \exp\left(-\frac{|F(\omega)|^2}{\hbar^2}\right)\right]. \tag{6.2.17}$$

To obtain the transition amplitude $\langle n_+|0_-\rangle$ from the ground-state to an excited state $|n_+\rangle$ after a disturbing source is switched on and then off, we proceed as follows.

We write a given source $F(t)$ as a sum of two sources:[2]

$$F(t) = F_1(t) + F_2(t) \tag{6.2.18}$$

where the source $F_2(t)$ is switched on *after* the source $F_1(t)$ is switched off. Hence directly from (6.2.12), we obtain

---

[2] Such an approach is used in particle production by external sources, see, for example, Schwinger (1970).

338   6 Intricacies of Harmonic Oscillators

$$\langle 0_+|0_-\rangle_F = \langle 0_+|0_-\rangle_{F_2}\langle 0_+|0_-\rangle_{F_1}$$

$$\times \exp\left[-\frac{1}{\hbar^2}\int_{-\infty}^{\infty}dt''\int_{-\infty}^{\infty}dt'\, e^{-i\omega(t''-t')}F_2(t'')F_1(t')\right] \quad (6.2.19)$$

where we have used the fact that $F_1(t'') = 0$ when $F_2(t'') \neq 0$.

The amplitude in (6.2.19) may be rewritten in the more convenient form as

$$\langle 0_+|0_-\rangle_F = \langle 0_+|0_-\rangle_{F_2}\langle 0_+|0_-\rangle_{F_1}\exp\left[\frac{iF_2^*(\omega)}{\hbar}\frac{iF_1(\omega)}{\hbar}\right]$$

$$= \sum_{n=0}^{\infty}\langle 0_+|0_-\rangle_{F_2}\frac{[iF_2^*(\omega)/\hbar]^n}{\sqrt{n!}}\frac{[iF_1(\omega)/\hbar]^n}{\sqrt{n!}}\langle 0_+|0_-\rangle_{F_1}. \quad (6.2.20)$$

We compare this with the expression obtained from a completeness relation, referred to as a unitarity sum, i.e.,[3]

$$\langle 0_+|0_-\rangle_F = \sum_{n,m}\langle 0_+|m_-\rangle_{F_2}\langle m|n\rangle_0'\langle n_+|0_-\rangle_{F_1} \quad (6.2.21)$$

where after F_1 is switched on, the system may (or may not) make a transition to a state $|n_+\rangle$. After the source $F_1(t)$ is switched off the system will remain in the same state until $F_2(t)$ is switched on. The amplitude $\langle m|n\rangle_0'$ arises in a *force-free* interval, and if t_1 is the time F_1 is switched off and t_2 is the time F_2 is switched on, then $\langle m|n\rangle_0'$, developing in time with a free Hamiltonian, is given by $\langle m|n\rangle_0' = \exp[-i\omega n(t_2 - t_1)]\delta_{mn}$.

From (6.2.20), (6.2.21), we may infer that for given external forces $F(t)$, $F'(t)$:

$$\langle n_+|0_-\rangle_F = \frac{[ie^{-i\omega T_2}F(\omega)/\hbar]^n}{\sqrt{n!}}\langle 0_+|0_-\rangle_F \quad (6.2.22)$$

$$\langle 0_+|n_-\rangle_{F'} = \langle 0_+|0_-\rangle_{F'}\frac{[ie^{i\omega T_1'}F'^*(\omega)/\hbar]^n}{\sqrt{n!}} \quad (6.2.23)$$

where T_2, T_1' denote the times that sources F, F' are switched off and on, respectively.

That is, the probability of having a transition from the ground-state to an excited state $|n_+\rangle$ after an intervening force is switched on then off is given by

---

[3] Note that the $\pm$ signs attached to n_+, m_- in $\langle n_+|0_-\rangle_{F_1}$, $\langle 0_+|m_-\rangle_{F_2}$ refer to the stages after the source $F_1(t)$ ceases to operate and before the source $F_2(t)$ is switched on. These amplitudes *a priori* not known are obtained by comparing (6.2.21) with (6.2.20).

$$|\langle n_+|0_-\rangle_F|^2 = \frac{[|F(\omega)|^2/\hbar^2]^n}{n!}\exp\left(-\frac{|F(\omega)|^2}{\hbar^2}\right) \qquad (6.2.24)$$

where we have also used (6.2.16), and on the average the system, initially in the ground-state, will be found in the state $\langle n\rangle \equiv \overline{n} = |F(\omega)|^2/\hbar^2$ after the intervening source is switched off.

To obtain the amplitude for the transition from a state $|n_-\rangle$ to a state $|n'_+\rangle$ after a given intervening source is switched on and then off, we write

$$F(t) = F_1(t) + F_2(t) + F_3(t) \qquad (6.2.25)$$

where the source $F_2(t)$ is switched on after the source $F_1(t)$ is switched off, and $F_3(t)$ is switched on after the source $F_2(t)$ is switched off.

From (6.2.12), we have

$$\langle 0_+|0_-\rangle_F = \langle 0_+|0_-\rangle_{F_3}\langle 0_+|0_-\rangle_{F_2}\langle 0_+|0_-\rangle_{F_1}\exp\left[\frac{iF_3^*(\omega)}{\hbar}\frac{iF_2(\omega)}{\hbar}\right]$$

$$\times \exp\left[\frac{iF_3^*(\omega)}{\hbar}\frac{iF_1(\omega)}{\hbar}\right]\exp\left[\frac{iF_2^*(\omega)}{\hbar}\frac{iF_1(\omega)}{\hbar}\right]$$

$$= \sum_{n,n'}\langle 0_+|n'_-\rangle_{F_3}\langle n'_+|n_-\rangle_{F_2}\langle n_+|0_-\rangle_{F_1} \qquad (6.2.26)$$

where in the last step, we have written $\langle 0_+|0_-\rangle_F$ in terms of a unitarity sum, as in (6.2.21). In (6.2.26), $\langle n_+|0_-\rangle_{F_1}$ denotes the amplitude that we have a transition from the ground-state to a state $|n_+\rangle$ after the source $F_1(t)$ is switched off. The latter state may then make a transition to a state $|n'_+\rangle$ after an intervening source $F_2(t)$ is switched on and then off. We note that the $\mp$ signs in $\langle n'_+|n_-\rangle_{F_2}$ refer, respectively, to the stage before $F_2(t)$ is switched on and after it is switched off. The amplitude $\langle 0_+|n'_-\rangle_{F_3}$ is similarly defined.

The amplitude $\langle n'_+|n_-\rangle_{F_2}$, a priori, not known will be then obtained upon the comparison of the second equality with the first one in (6.2.26) and using, in the process, the expressions in (6.2.22), (6.2.23) for the corresponding sources.

To the above end, we note the identity

$$\exp\left[\frac{iF_3^*(\omega)}{\hbar}\frac{iF_1(\omega)}{\hbar}\right]\exp\left[\frac{iF_3^*(\omega)}{\hbar}\frac{iF_2(\omega)}{\hbar}\right]\exp\left[\frac{iF_2^*(\omega)}{\hbar}\frac{iF_1(\omega)}{\hbar}\right]$$

$$= \sum_{L,M,N}\frac{[iF_3^*(\omega)/\hbar]^{L+M}}{L!}\frac{[iF_2(\omega)/\hbar]^M[iF_2^*(\omega)/\hbar]^N}{M!}\frac{[iF_1(\omega)/\hbar]^{L+N}}{N!}.$$

$$(6.2.27)$$

Upon setting

$$L + M = n', \qquad L + N = n \qquad (6.2.28)$$

we may rewrite the summand in (6.2.27) as[4]

$$\frac{[iF_3^*(\omega)/\hbar]^{n'}}{\sqrt{n'!}} \sqrt{n'!\,n!} \, \frac{[iF_2(\omega)/\hbar]^{n'-L}\,[iF_2^*(\omega)/\hbar]^{n-L}}{L!\,(n'-L)!\,(n-L)!} \, \frac{[iF_1(\omega)/\hbar]^n}{\sqrt{n!}} \quad (6.2.29)$$

which from (6.2.26) and the identifications in (6.2.22), (6.2.23) for the corresponding sources, gives for the amplitude $\langle n'_+ | n_- \rangle_F$ for a given source $F(t)$ operating within an interval from T_1 to T_2,[5]

$$\langle n'_+ | n_- \rangle_F = \langle 0_+ | 0_- \rangle_F \sqrt{n'!\,n!}$$

$$\times \sum_{L=0}^{\min(n',n)} \frac{[iF(\omega)/\hbar]^{n'-L}\, e^{-i\omega n' T_2}\, e^{i\omega n T_1}\, [iF^*(\omega)/\hbar]^{n-L}}{(n'-L)!\,L!\,(n-L)!}. \quad (6.2.30)$$

In particular the amplitude of making no transition in the presence of a disturbing source F is

$$\langle n_+ | n_- \rangle_F = \langle 0_+ | 0_- \rangle_F \, n! \sum_{\ell=0}^{n} \frac{\left[-|F(\omega)|^2/\hbar^2\right]^\ell}{(\ell!)^2\,(n-\ell)!} \quad (6.2.31)$$

where we have divided by $\exp\left[-i\omega n(T_2 - T_1)\right]$ for proper *normalization* of $\langle n_+ | n_- \rangle_F \to \langle n | n \rangle = 1$ for $F = 0$, and where $\langle 0_+ | 0_- \rangle_F$ is given in (6.2.12).

6.3 The Harmonic Oscillator in the Presence of a Disturbance at Finite Temperature

We consider the system described by the Hamiltonian given in (6.2.1) at non-zero temperature $T \neq 0$, where T is not to be confused with the time limits $T_{1,2}$.

Temperature dependence is introduced by averaging the expression for the persistence amplitudes $\langle n_+ | n_- \rangle_F$ for *all* $n = 0, 1, 2, \ldots$ with the familiar Boltzmann factor $\exp\left[-\hbar\omega(n+1/2)/kT\right]$, where k is the Boltzmann constant. This defines the *thermal average*

$$\langle G_+ | G_- \rangle_F^T = C \sum_{n=0}^{\infty} \exp\left[-\frac{\hbar\omega}{kT}\left(n + \frac{1}{2}\right)\right] \langle n_+ | n_- \rangle_F \quad (6.3.1)$$

---

[4] This treatment parallels similar, but more involved, methods used in field theory: Manoukian (1986b).
[5] Such processes are referred to as stimulated excitations where an initial state is already in some excited state, prior to the switching on of the intervening sources.

involving all possible excitations. Here C is a normalization constant determined from the normalization condition

$$C \sum_{n=0}^{\infty} \exp\left[-\frac{\hbar\omega}{kT}\left(n + \frac{1}{2}\right)\right] = 1. \qquad (6.3.2)$$

The latter gives

$$C = 2\sinh\left(\frac{\hbar\omega}{2kT}\right). \qquad (6.3.3)$$

The expression for $\langle n_+|n_-\rangle_F$ in (6.3.1) was obtained in (6.2.31). Upon using the notation

$$-\frac{|F(\omega)|^2}{\hbar^2} \equiv \lambda \qquad (6.3.4)$$

we may rewrite (6.3.1) as[6]

$$\langle G_+|G_-\rangle_F^T = \langle 0_+|0_-\rangle_F \left[1 - e^{-\hbar\omega/kT}\right] \sum_{n=0}^{\infty} \sum_{\ell=0}^{n} \frac{n!\,e^{-\hbar\omega n/kT}}{(\ell!)^2\,(n-\ell)!} (\lambda)^{\ell}. \qquad (6.3.5)$$

The double sum in (6.3.5) may be conveniently rewritten as

$$\sum_{n=0}^{\infty} \left[e^{-\hbar\omega/kT}\left(\frac{\partial}{\partial\lambda} + 1\right)\right]^n \frac{(\lambda)^n}{n!}. \qquad (6.3.6)$$

Upon using the integral

$$(\lambda)^n = \int_{-\infty}^{\infty} dz\, \delta(z - \lambda)\, (z)^n$$

$$= \int_{-\infty}^{\infty} dz \int_{-\infty}^{\infty} \frac{dy}{(2\pi)} e^{iy(z-\lambda)} (z)^n \qquad (6.3.7)$$

we obtain for the sum in (6.3.6)

$$\sum_{n=0}^{\infty} \int_{-\infty}^{\infty} dz \int_{-\infty}^{\infty} \frac{dy}{(2\pi)} \frac{[ze^{-\hbar\omega/kT}(-iy+1)]^n}{n!} e^{iy(z-\lambda)}$$

$$= \int_{-\infty}^{\infty} dz \int_{-\infty}^{\infty} \frac{dy}{(2\pi)} \exp\left[iy\left(z - ze^{-\hbar\omega/kT} - \lambda\right)\right] \exp\left[ze^{-\hbar\omega/kT}\right]$$

$$= \int_{-\infty}^{\infty} dz\, \delta\left(z\left[1 - e^{-\hbar\omega/kT}\right] - \lambda\right) \exp\left[ze^{-\hbar\omega/kT}\right]$$

---

[6] This treatment parallels similar, but more involved, methods used in field theory: Manoukian (1990).

$$= \frac{1}{[1 - e^{-\hbar\omega/kT}]} \exp\left[\frac{\lambda}{(e^{\hbar\omega/kT} - 1)}\right]. \tag{6.3.8}$$

Hence for the thermal average in (6.3.5) we have the expression

$$\langle G_+ | G_- \rangle_F^T = \langle 0_+ | 0_- \rangle_F \exp\left[-\frac{|F(\omega)|^2}{\hbar^2} \frac{1}{e^{\hbar\omega/kT} - 1}\right]. \tag{6.3.9}$$

Here we recognize the *Bose-Einstein* $\left(e^{\hbar\omega/kT} - 1\right)^{-1}$ factor occurring in the amplitude.

From (6.3.9), (6.2.16), we obtain for the probability that the system stays in thermal equilibrium, the expression

$$\left|\langle G_+ | G_- \rangle_F^T\right|^2 = \exp\left[-\frac{|F(\omega)|^2}{\hbar^2} \coth\left(\frac{\hbar\omega}{2kT}\right)\right]. \tag{6.3.10}$$

We note that

$$\coth\left(\frac{\hbar\omega}{2kT}\right) \xrightarrow[T \to \infty]{} \left(\frac{2kT}{\hbar\omega}\right) \tag{6.3.11}$$

$$\coth\left(\frac{\hbar\omega}{2kT}\right) \xrightarrow[T \to 0]{} 1. \tag{6.3.12}$$

Hence, in particular, for the persistence probability in (6.3.10) at high temperatures, one has the exponentially damping expression

$$\left|\langle G_+ | G_- \rangle_F^T\right|^2 \xrightarrow[T \to \infty]{} \exp\left[-\frac{2kT}{\hbar\omega} \frac{|F(\omega)|^2}{\hbar^2}\right]. \tag{6.3.13}$$

The amplitude in (6.3.9) may be rewritten in a more convenient form by introducing, in the process, an integral representation for the step function

$$\Theta(t) = \frac{i}{2\pi} \int_{-\infty}^{\infty} d\omega' \frac{e^{-i\omega' t}}{\omega' + i\varepsilon}, \qquad \varepsilon \to +0 \tag{6.3.14}$$

encountered in the expression for $\langle 0_+ | 0_- \rangle_F$ in (6.2.12), to obtain

$$\langle G_+ | G_- \rangle_F^T = \exp\left[\frac{i}{\hbar^2} \int_{-\infty}^{\infty} dt' \int_{-\infty}^{\infty} dt'' \, F(t'') \, \Delta_+(t'' - t'; T) \, F(t')\right] \tag{6.3.15}$$

where

$$\Delta_+(t; T) = \int_{-\infty}^{\infty} \frac{d\omega'}{2\pi} e^{-i\omega' t} \left[\frac{(-1)}{\omega' - (\omega - i\varepsilon)} + \frac{2\pi i \, \delta(\omega' - \omega)}{e^{\hbar\omega'/kT} - 1}\right] \tag{6.3.16}$$

and T denotes the temperature.

6.4 The Fermi Oscillator

In analogy to the Bose oscillator in (6.1.12), for $\nu = 1$, where $[a, a^\dagger] = 1$, defining the Bose-Einstein statistics, we define a Fermi-oscillator with Hamiltonian

$$H = \hbar\omega \left(a_F^\dagger a_F - \frac{1}{2} \right) \tag{6.4.1}$$

where the annihilation, creation operators a_F, $a_F^\dagger$ satisfy anti-commutation relations:

$$\{a_F, a_F\} = 0, \qquad \{a_F^\dagger, a_F^\dagger\} = 0 \tag{6.4.2}$$

and

$$\{a_F, a_F^\dagger\} = 1. \tag{6.4.3}$$

We note that the zero point energy in (6.4.1) was chosen to be $-\hbar\omega/2$ rather than $+\hbar\omega/2$ in contrast to bosons. We will see in the next section that when we invoke supersymmetry, alone, implying a symmetry of the Hamiltonian for the supersymmetric version of the Bose-oscillator Hamiltonian $\hbar\omega \left(a^\dagger a + 1/2 \right)$, under the boson-fermion exchange, consistency leads to a zero point energy of $-\hbar\omega/2$ for the Fermi-oscillator.

The ground-state $|0\rangle$ is defined by

$$a_F |0\rangle = 0 \tag{6.4.4}$$

and the single particle-state by

$$|1\rangle = a_F^\dagger |0\rangle, \qquad a_F |1\rangle = |0\rangle \tag{6.4.5}$$

with

$$H |0\rangle = -\frac{\hbar\omega}{2} |0\rangle \tag{6.4.6}$$

$$H |1\rangle = +\frac{\hbar\omega}{2} |1\rangle. \tag{6.4.7}$$

We note that since $\left(a_F^\dagger\right)^2 = 0$, there are no two-particle states.

A representation of the annihilation and creation operators is given by

$$a_F = \frac{\sigma_1 - i\sigma_2}{2} = \begin{pmatrix} 0 & 0 \\ 1 & 0 \end{pmatrix} \tag{6.4.8}$$

$$a_F^\dagger = \frac{\sigma_1 + i\sigma_2}{2} = \begin{pmatrix} 0 & 1 \\ 0 & 0 \end{pmatrix}. \tag{6.4.9}$$

where σ_1, σ_2 are Pauli matrices. The ground- and single-particle states may be then represented by

$$|0\rangle = \begin{pmatrix} 0 \\ 1 \end{pmatrix}, \qquad |1\rangle = \begin{pmatrix} 1 \\ 0 \end{pmatrix}. \tag{6.4.10}$$

Now we couple the Fermi-oscillator to external *anti-commuting* sources[7] $\eta^*(t)$, $\eta(t)$ as follows

$$H(t) = \hbar\omega\left(a_F^\dagger a_F - \frac{1}{2}\right) - \eta^*(t) a_F - a_F^\dagger \eta(t) \tag{6.4.11}$$

where $\eta^*(t)$, $\eta(t)$ anti-commute with a_F, $a_F^\dagger$ and

$$\{\eta(t), \eta(t')\} = 0, \qquad \{\eta(t), \eta^*(t')\} = 0. \tag{6.4.12}$$

To solve the dynamics involved with $H(t)$ in (6.4.11), we follow the procedure given in §6.2 for bosons, subject to the constraints (6.4.2), (6.4.3), (6.4.12) and the fact that η, η^* anti-commute with both a_F and $a_F^\dagger$. As for the Bose-oscillator, we choose $\eta(t)$, $\eta^*(t)$ to vanish for $t \leqslant T_1$, $t \geqslant T_2$ for some T_1, T_2 ($T_2 > T_1$).

To the above end, the Schrödinger equation reads

$$i\hbar\frac{\partial}{\partial t}|\psi(t)\rangle = H(t)|\psi(t)\rangle. \tag{6.4.13}$$

By setting $|\psi(t)\rangle = |\chi(t)\rangle e^{i\omega t/2}$, we obtain

$$i\hbar\frac{\partial}{\partial t}|\chi(t)\rangle = \left(\hbar\omega a_F^\dagger a_F - \eta^* a_F - a_F^\dagger \eta\right)|\chi(t)\rangle. \tag{6.4.14}$$

By making the ansatz

$$|\chi(t)\rangle = e^{-i\phi(t)/\hbar} \exp\left(-\frac{i}{\hbar}\rho(t) a_F^\dagger\right)|0\rangle \tag{6.4.15}$$

where $\phi(t)$ is a c-function, $\rho(t)$ is an *anti-commuting* c-function, substituting in (6.4.13) and using the constraints (6.4.2), (6.4.3), (6.4.12), we obtain as in (6.2.8), (6.2.9),

$$\phi(t) = \int_{T_1}^{t} dt'\, e^{-i\omega(t-t')}\eta(t') \tag{6.4.16}$$

$$\rho(t) = -\frac{i}{\hbar}\int_{-\infty}^{\infty} dt'' \int_{-\infty}^{\infty} dt'\, \eta^*(t'')\,\Theta(t-t'')\,\Theta(t''-t')\, e^{-i\omega(t''-t')}\eta(t') \tag{6.4.17}$$

with the boundary conditions that $\phi(T_1) = 0$, $\rho(T_1) = 0$.

The ground-state persistence amplitude is then given by (see (6.2.11)),

---

[7] These objects are referred to as Grassmann variables as also noted in §2.9. Such variables will be studied in great detail in §10.6.

$$\langle 0_+ | 0_- \rangle_{\eta,\eta^*} = \exp\left[-\frac{1}{\hbar^2} \int_{-\infty}^{\infty} dt''\right.$$

$$\left. \times \int_{-\infty}^{\infty} dt'\, e^{-i\omega(t''-t')} \eta^*(t'')\, \Theta(t''-t')\, \eta(t')\right] \quad (6.4.18)$$

which is normalized to unity for $\eta = 0$, $\eta^* = 0$, with a persistence probability

$$\left|\langle 0_+ | 0_- \rangle_{\eta,\eta^*}\right|^2 = \exp\left[-\frac{\eta^*(\omega)\eta(\omega)}{\hbar^2}\right]$$

$$= 1 - \frac{\eta^*(\omega)\eta(\omega)}{\hbar^2} \quad (6.4.19)$$

written in terms of the Fourier-transforms of the sources. In the last step in (6.4.19) we have used the anti-commutativity property of η, η^*.

By a similar procedure as in obtaining (6.2.22), (6.2.23), we have

$$\langle n_+ | 0_- \rangle_{\eta,\eta^*} = \left[i\eta(\omega)\, e^{-i\omega T_2}/\hbar\right]^n \langle 0_+ | 0_- \rangle_{\eta,\eta^*} \quad (6.4.20)$$

$$\langle 0_+ | n_- \rangle_{\eta',\eta'^*} = \langle 0_+ | 0_- \rangle_{\eta',\eta'^*} \left[i\eta'^*(\omega)\, e^{i\omega T_1'}/\hbar\right]^n \quad (6.4.21)$$

where $n = 0, 1$, and hence we have dismissed with the factorial factor $(n!)^{-1/2}$.

In analogy to (6.2.31), we also obtain

$$\langle n_+ | n_- \rangle_{\eta,\eta^*} = \left[1 + \frac{\eta^*(\omega)\eta(\omega)}{\hbar^2}\right]^n \langle 0_+ | 0_- \rangle_{\eta,\eta^*} \quad (6.4.22)$$

for $\langle n_+ | n_- \rangle_{\eta,\eta^*}$ properly *normalized* to one for $\eta, \eta^* \to 0$, where $n = 0, 1$.
Equations (6.4.18), (6.4.20)–(6.4.22) give all the relevant amplitudes.[8]

---

[8] For completeness we note that although the thermal average in the present case with only two states is quite formal, the corresponding amplitude is given by

$$\langle G_+ | G_- \rangle_{\eta,\eta^*}^T = \exp\left[\frac{i}{\hbar^2} \int_{-\infty}^{\infty} dt' \int_{-\infty}^{\infty} dt''\, \eta^*(t'')\, \widetilde{\Delta}_+(t''-t';T)\, \eta(t')\right]$$

where

$$\widetilde{\Delta}_+(t;T) = \int_{-\infty}^{\infty} \frac{d\omega'}{2\pi}\, e^{-i\omega' t}\left[\frac{(-1)}{\omega'-(\omega-i\varepsilon)} - \frac{2\pi i\, \delta(\omega'-\omega)}{e^{\hbar\omega'/kT}+1}\right]$$

in analogy to the Bose case in (6.3.15), (6.3.16), the derivation of which is left as an exercise to the reader. Here we recognize the *Fermi-Dirac* factor $\left[\exp(\hbar\omega/kT)+1\right]^{-1}$ in the amplitude.

6.5 Bose-Fermi Oscillators and Supersymmetric Bose-Fermi Transformations

To construct a supersymmetric Bose-Fermi oscillator which combines the Bose-oscillator (§6.1) and the Fermi-oscillator (§6.4), we define the supersymmetric generators (§2.9, §4.7)

$$Q = \hbar\sqrt{2\omega}\, a_B^\dagger\, a_F \tag{6.5.1}$$

$$Q^\dagger = \hbar\sqrt{2\omega}\, a_B\, a_F^\dagger \tag{6.5.2}$$

where, together with their adjoints,

$$[a_B, a_F] = 0, \qquad \left[a_B, a_F^\dagger\right] = 0. \tag{6.5.3}$$

For convenience, in this section, the Bose operators have been denoted by a_B, $a_B^\dagger$.

The generators Q, $Q^\dagger$ lead for the Hamiltonian (§2.9) given by

$$H = \frac{1}{2\hbar}\{Q, Q^\dagger\} \tag{6.5.4}$$

the expression

$$H = \hbar\omega\left(a_B^\dagger\, a_B + a_F^\dagger\, a_F\right) \tag{6.5.5}$$

where we note, in particular, that

$$Q\,Q^\dagger = 2\hbar^2\omega\, a_B^\dagger\, a_B\, a_F\, a_F^\dagger$$

$$= 2\hbar^2\omega\, a_B^\dagger\, a_B\left(1 - a_F^\dagger\, a_F\right). \tag{6.5.6}$$

The Hamiltonian in (6.5.5) may be rewritten as

$$H = \hbar\omega\left(a_B^\dagger\, a_B + \frac{1}{2}\right) + \hbar\omega\left(a_F^\dagger\, a_F - \frac{1}{2}\right) \tag{6.5.7}$$

We recognize the first term as the Hamiltonian of the Bose-oscillator and infer a zero point energy of $-\hbar\omega/2$, rather than of $+\hbar\omega/2$, for the Fermi-oscillator.

The cancellation between the zero point energies of the Bose and Fermi oscillators is a very special and attractive feature of supersymmetry. This is quite significant when one is dealing with an infinite degrees of freedom ($\nu \to \infty$, see (6.1.12)) and similar cancellations occur in field theory which would otherwise lead to ambiguities.

The ground-state of the Bose-Fermi oscillator $|0,0\rangle$ is defined by

$$a_B|0,0\rangle = 0, \qquad a_F|0,0\rangle = 0 \tag{6.5.8}$$

6.5 B-F Oscillators and SUSY B-F Transformations

and the state with n boson excitations and m fermion ones, with $m = 0, 1$, are given by (see also (6.1.31))

$$|n, m\rangle = \frac{(a_B^\dagger)^n}{\sqrt{n!}} (a_F^\dagger)^m |0, 0\rangle. \qquad (6.5.9)$$

By using the property

$$a_B (a_B^\dagger)^n = n(a_B^\dagger)^{n-1} + (a_B^\dagger)^n a_B \qquad (6.5.10)$$

one obtains

$$H |n, m\rangle = \hbar\omega(n + m) |n, m\rangle \qquad (6.5.11)$$

as expected.

The generators Q, $Q^\dagger$ carry out the Bose $\Leftrightarrow$ Fermi transformations as follows. We explicitly obtain

$$Q^\dagger |1, 0\rangle = \hbar\sqrt{2\omega}\, a_B\, a_F^\dagger\, a_B^\dagger |0, 0\rangle \qquad (6.5.12)$$

or

$$Q^\dagger |1, 0\rangle = \hbar\sqrt{2\omega}\, |0, 1\rangle. \qquad (6.5.13)$$

Similarly we have

$$Q |0, 1\rangle = \hbar\sqrt{2\omega}\, |1, 0\rangle. \qquad (6.5.14)$$

More generally one has

$$Q^\dagger |n, 0\rangle = \hbar\sqrt{2\omega}\sqrt{n}\, |n - 1, 1\rangle \qquad (6.5.15)$$

$$Q |n, 1\rangle = \hbar\sqrt{2\omega}\sqrt{n + 1}\, |n + 1, 0\rangle \qquad (6.5.16)$$

and from (6.5.11), we note that the states $|n - 1, 1\rangle$ and $|n, 0\rangle$ are degenerate corresponding to the energy $\hbar\omega n$.

One may introduce interaction terms in the elementary Bose-Fermi oscillator Hamiltonian in (6.5.5) by adding to it, for example, a Yukawa term, describing a direct interaction between bosons and fermions, and a cubic term, describing a direct self-coupling of the bosons. To this end, we introduce a Bose "field" at time $t = 0$ defined by

$$\phi = a_B + a_B^\dagger \qquad (6.5.17)$$

and consider the Hamiltonian H defined by

$$\frac{H}{\hbar\omega} = \left(a_B^\dagger a_B + a_F^\dagger a_F\right) + \lambda_0\, a_F^\dagger a_F\, \phi + \lambda \phi^3. \qquad (6.5.18)$$

In relativistic quantum field theory, as quantum mechanics with an *infinite* degrees of freedom, a generalization of (6.5.18) shows, in particular, that the anharmonic cubic term in the Bose field gives rise to a Hamiltonian which is unbounded from below, that is, its spectrum goes down to $-\infty$. Here suffice it to say, as will be shown below, the simple Hamiltonian in (6.5.18) with the Bose "field" of one degree of freedom, leads to a spectrum which has a strictly negative part for all $\lambda \neq 0$, and emphasize that an immediate supersymmetric generalization of (6.5.18) "shifts" the spectrum to a non-negative value.

To the above end, we note the following expectation values, in the state $|0, 0\rangle$, involving the anharmonic cubic term ϕ^3:

$$\langle 0, 0 | \phi^3 | 0, 0 \rangle = 0 \tag{6.5.19}$$

$$\langle 0, 0 | \phi^3 \phi^3 | 0, 0 \rangle = 15 \tag{6.5.20}$$

$$\langle 0, 0 | \phi^3 \phi^3 \phi^3 | 0, 0 \rangle = 0 \tag{6.5.21}$$

$$\langle 0, 0 | \phi^3 a_B^\dagger a_B \phi^3 | 0, 0 \rangle = 27 \tag{6.5.22}$$

as obtained from the commutation relations involving a_B, $a_B^\dagger$.

As a trial state, we choose the normalized state

$$|\psi\rangle = \frac{[1 - \lambda \phi^3]}{\sqrt{15}\,|\lambda|} |0, 0\rangle \tag{6.5.23}$$

for $\lambda \neq 0$.

By definition of the lowest point E of the spectrum (see also Problem 6.18),

$$E \leqslant \langle \psi | H | \psi \rangle. \tag{6.5.24}$$

To obtain the expression on the right-hand side of this inequality, we use the ones for the expectation values given in (6.5.19)–(6.5.22) and note that the terms in the Hamiltonian depending on the Fermi operators a_F, $a_F^\dagger$, more specifically $a_F^\dagger a_F$, does not contribute to it. This is because a_F, $a_F^\dagger$ commute with the Bose operators, and $a_F |0, 0\rangle = 0$.

Accordingly,

$$\frac{E}{\hbar \omega} \leqslant \frac{1}{15 \lambda^2} \lambda^2 \langle 0, 0 | \phi^3 a_B^\dagger a_B \phi^3 | 0, 0 \rangle - 2 \frac{\lambda^2}{15 \lambda^2} \langle 0, 0 | \phi^3 \phi^3 | 0, 0 \rangle \tag{6.5.25}$$

or

$$E \leqslant -\frac{1}{5} \hbar \omega \tag{6.5.26}$$

for *all* $\lambda \neq 0$.

A supersymmetric version of the Hamiltonian in (6.5.18) which includes correct terms to cancel out, in particular, the negative contribution of the anharmonic term $\lambda \phi^3$ is immediate. To obtain *precisely* this cubic term in the supersymmetric Hamiltonian, we introduce the supersymmetry generators

$$Q = \hbar \sqrt{2\omega} \left(a_B^\dagger + \lambda \phi^2 \right) a_F \tag{6.5.27}$$

$$Q^\dagger = \hbar\sqrt{2\omega}\left(a_B + \lambda\phi^2\right)a_F^\dagger \qquad (6.5.28)$$

as a generalization of the non-interacting case in (6.5.1), (6.5.2).

It is readily verified, that these generators give rise from

$$H_S = \frac{1}{2\hbar}\{Q,Q^\dagger\} \qquad (6.5.29)$$

to the Hamiltonian H_S,

$$\frac{H_S}{\hbar\omega} = a_B^\dagger a_B + a_F^\dagger a_F + 4\lambda\left(a_F^\dagger a_F - \frac{1}{2}\right)\phi + \lambda\phi^3 + \lambda^2\phi^4. \qquad (6.5.30)$$

The non-negativity of the spectrum of H_S follows, as before (see §4.7), by noting that for a state $|\chi\rangle$

$$\langle\chi|H_S|\chi\rangle = \frac{1}{2\hbar}\left(\|Q^\dagger\chi\|^2 + \|Q\chi\|^2\right) \geq 0. \qquad (6.5.31)$$

It is remarkable that a supersymmetric version of a Hamiltonian provides a Hamiltonian with a non-negative spectrum especially if the former would have an unbounded spectrum from below. It is also interesting to note that supersymmetry provides constraints on the couplings (such as $\lambda_0 = 4\lambda$ in (6.5.18)/(6.5.30)), thus reducing much of the arbitrariness in choosing interactions. Such facts have been quite useful in field theory.

6.6 Coherent State of the Harmonic Oscillator

A coherent state is a very special linear combination of the states $|n\rangle$, given in §6.1, which brings the quantum mechanical treatment into very close proximity with its classical counterpart description in the most natural way and a contact with classical notions is most suitable to the experimentalist. Such a state arises naturally in the following way.

The matrix elements of the position and momentum operators $x(t)$, $p(t)$ at any time t, and the Hamiltonian $H = p^2/2m + m\omega^2 x^2/2$, with respect to the states $|n\rangle$ may be read from (6.1.45), (6.1.47), (6.1.24) to be given by

$$\langle n'|x(t)|n\rangle = \left(\frac{\hbar}{2m\omega}\right)^{1/2}\left[\sqrt{n+1}\,\delta_{n',n+1}\,e^{i\omega t} + \sqrt{n}\,\delta_{n',n-1}\,e^{-i\omega t}\right] \qquad (6.6.1)$$

$$\langle n'|p(t)|n\rangle = i\left(\frac{\hbar m\omega}{2}\right)^{1/2}\left[\sqrt{n+1}\,\delta_{n',n+1}\,e^{i\omega t} - \sqrt{n}\,\delta_{n',n-1}\,e^{-i\omega t}\right] \qquad (6.6.2)$$

$$\langle n'|H|n\rangle = \delta_{n',n}\,\hbar\omega\left(n + \frac{1}{2}\right). \qquad (6.6.3)$$

The expressions in (6.6.1), (6.6.2) are not only non-diagonal but also any resemblance to the classical solutions

$$x_c(t) = |A|\cos(\omega t - \delta) \tag{6.6.4}$$

$$p_c(t) = -|A|\, m\omega \sin(\omega t - \delta) \tag{6.6.5}$$

$$H_c = |A|^2 \frac{m\omega^2}{2} \tag{6.6.6}$$

is quite remote.

For given ω, m these classical expressions are parameterized by the amplitude $|A|$ and the phase δ, or equivalently by the *complex* number $A = |A|\, e^{i\delta}$.

We will construct a state, referred to as a coherent state, as a linear combination of the states $|n\rangle$ and conveniently parametrized by the complex number A, denoted by $|A\rangle$, such that the expectation values of the operators $x(t)$, $p(t)$, with respect to $|A\rangle$ coincide with the classical counterparts (6.6.4), (6.6.5), and such that the expectation value of H in $|A\rangle$ comes as close as possible to the classical expression in (6.6.6). As we will see below, the reason why $\langle A|H|A\rangle$ cannot exactly coincide with H_c in (6.6.6) is due to the presence of the zero point energy $\hbar\omega/2$ in the spectrum of H in (6.1.24) and due to a positivity constraint.

That is, we introduce a state

$$|A\rangle = \sum_{n=0}^{\infty} C_n(A) |n\rangle \tag{6.6.7}$$

$$\langle A|A\rangle = \sum_{n=0}^{\infty} |C_n(A)|^2 = 1 \tag{6.6.8}$$

such that

$$\langle A|x(t)|A\rangle = |A|\cos(\omega t - \delta) \equiv x_c(t) \tag{6.6.9}$$

$$\langle A|p(t)|A\rangle = -|A|\, m\omega \sin(\omega t - \delta) \equiv p_c(t) \tag{6.6.10}$$

and, as we will see the closest we can come to H_c for the expectation value of H, is given by

$$\langle A|H|A\rangle = |A|^2 \frac{m\omega^2}{2} + \frac{\hbar\omega}{2} \equiv H_c + \frac{\hbar\omega}{2}. \tag{6.6.11}$$

The coefficients squared $|C_n(A)|^2$ have the usual interpretation of the probabilities of the system, described by the state $|A\rangle$, to be found in the states $|n\rangle$.

From (6.6.1), (6.6.7), we explicitly have

6.6 Coherent State of the Harmonic Oscillator

$$\langle A|x(t)|A\rangle = \sum_{n=0}^{\infty} \left(\frac{\hbar}{2m\omega}\right)^{1/2} (C_{n+1}(A))^* C_n(A)\sqrt{n+1}\,e^{i\omega t} + \text{c.c.} \quad (6.6.12)$$

which from (6.6.9) requires that

$$\sum_{n=0}^{\infty} \left(\frac{\hbar}{2m\omega}\right)^{1/2} (C_{n+1}(A))^* C_n(A)\sqrt{n+1} = \frac{A^*}{2}. \quad (6.6.13)$$

where we recall that $A = |A|\exp i\delta$.

Equation (6.6.10), for the expectation value of $p(t)$ in the state $|A\rangle$ gives the same condition as in (6.6.13).

On the other hand, from (6.6.3), we have

$$\langle A|H|A\rangle = \hbar\omega \sum_{n=0}^{\infty} n|C_n(A)|^2 + \frac{\hbar\omega}{2} \quad (6.6.14)$$

and the non-negativity of $\sum_n n|C_n(A)|^2$ in this equation should be noted.

Upon defining the convenient complex number

$$z = A\left(\frac{m\omega}{2\hbar}\right)^{1/2} \quad (6.6.15)$$

the conditions (6.6.8), (6.6.13), (6.6.11), (6.6.14) give

$$\sum_{n=0}^{\infty} C_n^* C_n = 1 \quad (6.6.16)$$

$$\sum_{n=0}^{\infty} \frac{C_{n+1}^*}{z^*} C_n \sqrt{n+1} = 1 \quad (6.6.17)$$

$$\sum_{n=0}^{\infty} \frac{C_{n+1}^*}{z^*} \frac{C_{n+1}}{z}(n+1) = 1 \quad (6.6.18)$$

where we have assumed $n|C_n|^2 = 0$, for $n = 0$ in writing the last sum, and suppressed the A (or z) dependence of the C_n.

We note the summands in (6.6.16), (6.6.17) have the C_n factor in common, while (6.6.17), (6.6.18) have the C_{n+1}^* factor in common. It is easy to see that the following recurrence relation obtained from the comparison of the summands in (6.6.17) and (6.6.18),

$$C_{n+1} = \frac{z}{\sqrt{n+1}} C_n, \quad n = 0, 1, \ldots \quad (6.6.19)$$

satisfies all of the sums in (6.6.16)–(6.6.18) (see also Problem 6.20, (6.6.26)).

The solution of (6.6.19) is elementary and is given by

$$C_n = \frac{(z)^n}{\sqrt{n!}} e^{-|z|^2/2} \tag{6.6.20}$$

where we have finally used the normalization condition (6.6.16) again to solve for C_0.

That is, apart from the expected zero point energy $\hbar\omega/2$ in (6.6.11), the state, now parametrized by the complex number z in (6.6.15),

$$|z\rangle = e^{-|z|^2/2} \sum_{n=0}^{\infty} \frac{(z)^n}{\sqrt{n!}} |n\rangle \tag{6.6.21}$$

$$\langle z | z \rangle = 1 \tag{6.6.22}$$

gives rise to expectation values for $x(t)$, $p(t)$, coinciding with the classical solutions, and for H coming close to its classical counterpart.

Let us recapitulate the physical meanings of the complex number z, with which the coherent state in (6.6.21) is labelled. To this end, upon writing

$$z = |z| e^{i\delta} \tag{6.6.23}$$

$|z|$, up to a scaling factor, denotes the amplitude of the corresponding classical solution (6.6.4), and δ provides its phase, and the privileged state (6.6.21) brings the quantum oscillator in the light of classical notions well suited to the experimentalist. Many additional properties of the coherent state $|z\rangle$ will be obtained as we go along. From (6.6.18), (6.6.20), we also have

$$\sum_{n=0}^{\infty} n |C_n|^2 = \sum_{n=0}^{\infty} n \frac{(|z|^2)^n}{n!} e^{-|z|^2} = |z|^2 \equiv \overline{n}. \tag{6.6.24}$$

That is, $|z|^2$ denotes the mean excitation quantum number n in the state $|z\rangle$. In this respect the probability of the system, described by the state $|z\rangle$, to be found in the state $|n\rangle$, is then given from (6.6.21), (6.6.23) to be

$$|\langle n | z \rangle|^2 \equiv P_z(n) = \frac{(\overline{n})^n}{n!} e^{-\overline{n}}, \qquad n = 0, 1, \ldots \tag{6.6.25}$$

which is the celebrated Poisson probability mass function.[9]

A property of central importance of a coherent state $|z\rangle$ is that with respect to it, the annihilation operator a (§6.1) becomes a *multiplicative*[10] one, i.e.,

---

[9] For detailed properties of a very wide range of probability distributions, see: Manoukian (1986c).

[10] Many authors introduce coherent states by beginning from (6.6.26) as a defining equation for the state $|z\rangle$.

6.6 Coherent State of the Harmonic Oscillator

$$a\,|z\rangle = z\,|z\rangle \tag{6.6.26}$$

as is easily checked, and although a is not a self-adjoint operator, (6.6.26) signals its importance in making contact with classical solutions.

The unitary operator $\exp(-\mathrm{i}tH/\hbar)$ applied to $|z\rangle$ gives

$$\mathrm{e}^{-\mathrm{i}tH/\hbar}\,|z\rangle \equiv |z;t\rangle = \mathrm{e}^{-|z|^2/2}\,\mathrm{e}^{-\mathrm{i}\omega t/2} \sum_{n=0}^{\infty} \frac{(\mathrm{e}^{-\mathrm{i}\omega t}z)^n}{\sqrt{n!}}\,|n\rangle \tag{6.6.27}$$

as follows from (6.6.21) and (6.1.24). That is, we may write

$$|z;t\rangle = |z(t)\rangle\,\mathrm{e}^{-\mathrm{i}\omega t/2} \tag{6.6.28}$$

where

$$z(t) = z\,\mathrm{e}^{-\mathrm{i}\omega t} = |z|\,\mathrm{e}^{-\mathrm{i}(\omega t - \delta)}. \tag{6.6.29}$$

The coherent states (6.6.21), for all z, satisfy at *all* times t, the very optimal *minimum* uncertainty principle criterion:

$$\left\langle z\left|[x(t) - \langle x(t)\rangle]^2\right|z\right\rangle \left\langle z\left|[p(t) - \langle p(t)\rangle]^2\right|z\right\rangle = \frac{\hbar^2}{4} \tag{6.6.30}$$

where $\langle x(t)\rangle \equiv \langle z|x(t)|z\rangle$. The equality in (6.6.30) follows by noting that

$$\langle z|x^2(t)|z\rangle = \frac{2\hbar}{m\omega}|z|^2\cos^2(\omega t - \delta) + \frac{\hbar}{2m\omega} \tag{6.6.31}$$

$$\langle z|p^2(t)|z\rangle = 2\hbar m\omega|z|^2\sin^2(\omega t - \delta) + \frac{m\omega\hbar}{2} \tag{6.6.32}$$

and using (6.6.9), (6.6.10), (6.6.15).

It is expected that the coherent states $|z\rangle$, for *all* complex z, providing the totality of all possible amplitudes and phases, admit a (completeness) resolution of the identity. This is indeed the case and it reads:

$$\int \frac{\mathrm{d}(\mathrm{Re}\,z)\,\mathrm{d}(\mathrm{Im}\,z)}{\pi}\,|z\rangle\langle z| = \mathbf{1}. \tag{6.6.33}$$

This is easily established by working, for example in polar coordinates and setting $z = r\mathrm{e}^{\mathrm{i}\theta}$ in (6.6.21) and on the left-hand side of (6.6.33) (see Problem 6.24) giving

$$\frac{1}{\pi}\int_0^{\infty} r\,\mathrm{d}r \int_0^{2\pi} \mathrm{d}\theta\,|r\mathrm{e}^{\mathrm{i}\theta}\rangle\langle r\mathrm{e}^{\mathrm{i}\theta}| = \sum_{n=0}^{\infty} |n\rangle\langle n| = \mathbf{1}. \tag{6.6.34}$$

This resolution of the identity is not to be confused with the corresponding one for a self-adjoint operator (§1.8), as the z are not some eigenvalues of a self-adjoint operator and the states $|z\rangle$ do not satisfy any orthogonality conditions. As a matter of fact one explicitly has, directly from (6.6.21), and (6.1.31), that

$$\langle z'|z\rangle = \langle 0|\exp(z'^*a)\exp(za^\dagger)|0\rangle \exp\left[-\frac{(|z|^2+|z'|^2)}{2}\right] \quad (6.6.35)$$

and since

$$\left[a,\left[a,a^\dagger\right]\right] = 0 \quad (6.6.36)$$

$$\left[a^\dagger,\left[a,a^\dagger\right]\right] = 0 \quad (6.6.37)$$

an application of the Baker-Campbell-Hausdorff formula (see Appendix I) gives

$$\langle z'|z\rangle = \exp\left(z'^*z - \frac{|z|^2}{2} - \frac{|z'|^2}{2}\right). \quad (6.6.38)$$

In particular, one has

$$|\langle z'|z\rangle|^2 = \exp\left(-|z'-z|^2\right). \quad (6.6.39)$$

The coherent states in (6.6.21) also satisfy the continuity[11] condition:

$$\| |z'\rangle - |z\rangle \| \xrightarrow[z'\to z]{} 0 \quad (6.6.40)$$

which is an important property in the mathematical definition of coherent states, and follows from (6.6.22) and (6.6.38).

In the light of the minimum uncertainty principle in (6.6.30), it is worth investigating the time development of the x-space coherent wavepacket as well as of the p-description. We work out the details for the x-description only; the p-description is left as an exercise (see Problem 6.26).

From (6.6.28), (6.6.21), (6.6.29) and (6.1.31), we may write

$$|z;t\rangle = e^{-|z|^2/2}\, e^{-i\omega t/2}\, e^{z(t)a^\dagger}|0\rangle \quad (6.6.41)$$

and from the definition (6.1.10) we have

$$\langle x|z;t\rangle = e^{-|z|^2/2}\, e^{-i\omega t/2}\, \langle x|\exp\left[\frac{z(t)}{\sqrt{2}}(X-iP)\right]|0\rangle. \quad (6.6.42)$$

Using the commutation relation $[X,P] = i$ in (6.1.11) for $\nu = 1$ and the Baker-Campbell-Hausdorff formula, this gives

---

[11] More precisely this is referred to as strong continuity, while weak continuity is referred to the weaker condition $\langle \psi|z'-z\rangle \to 0$, for normalizable states $|\psi\rangle$, which follows from strong continuity by a direct application of the Cauchy-Schwarz inequality.

$$\langle x \,|\, z; t\rangle = \mathrm{e}^{-|z|^2/2}\, \mathrm{e}^{-\mathrm{i}\omega t/2}\, \mathrm{e}^{z(t)X/\sqrt{2}}\, \mathrm{e}^{-z^2(t)/4}\, \langle x|\exp\left[-\mathrm{i}\frac{z(t)}{\sqrt{2}}P\right]|0\rangle. \quad (6.6.43)$$

Upon rewriting the ground-state wavefunction in (6.1.33) in terms of X:

$$\psi_0(x) = \left(\frac{m\omega}{\pi\hbar}\right)^{1/4}\exp\left(-\frac{X^2}{2}\right) \quad (6.6.44)$$

we obtain

$$\langle x \,|\, z; t\rangle = \left(\frac{m\omega}{\pi\hbar}\right)^{1/4} \mathrm{e}^{-\mathrm{i}\omega t/2}\exp\left[-\frac{|z|^2}{2} + \frac{z(t)X}{\sqrt{2}} - \frac{z^2(t)}{4}\right]$$

$$\times \exp\left[-\frac{1}{2}\left(X - \frac{z(t)}{\sqrt{2}}\right)^2\right]. \quad (6.6.45)$$

Finally using the expression for $z(t)$ in (6.6.29), the definition (6.6.15), the fact that $X = (m\omega/\hbar)^{1/2}x$ and (6.6.4), we get

$$\langle x \,|\, z; t\rangle = \left(\frac{m\omega}{\pi\hbar}\right)^{1/4}\exp\left[-\frac{m\omega}{2\hbar}(x - x_c(t))^2\right]\exp\left[\mathrm{i}\phi(x,t)\right] \quad (6.6.46)$$

where $\exp(\mathrm{i}\phi(x,t))$ is a phase factor with

$$\phi(x,t) = -\frac{\omega t}{2} + \frac{|z|^2}{2}\sin[2(\omega t - \delta)] - \left(\frac{2m\omega}{\hbar}\right)^{1/2}|z|\,x\,\sin(\omega t - \delta). \quad (6.6.47)$$

This gives a Gaussian probability density

$$|\langle x \,|\, z; t\rangle|^2 = \left(\frac{m\omega}{\pi\hbar}\right)^{1/2}\exp\left[-\frac{m\omega}{\hbar}(x - x_c(t))^2\right] \quad (6.6.48)$$

centered about the classical solution (6.6.4), and most importantly it is *non-spreading* in time as the variance $\sigma_x^2(t)$ is time-independent given by

$$\sigma_x^2(t) = \sigma_x^2(0) = \frac{\hbar}{2m\omega}. \quad (6.6.49)$$

A similar analysis may be carried out in the p-description.

As a function of the dimensionless variable X in (6.1.8), the probability density corresponding to (6.6.48) may be rewritten as

$$|\langle X \,|\, z; t\rangle|^2 = \frac{1}{\sqrt{2\pi}}\frac{1}{(1/\sqrt{2})}\exp\left[-\frac{1}{2}\frac{(X - \sqrt{2}\,\mathrm{Re}[z(t)])^2}{1/2}\right] \quad (6.6.50)$$

where we have used (6.6.29), (6.6.15), (6.6.4), corresponding to a standard deviation equal to $1/\sqrt{2}$. In particular we note that for $z(t)$ pure imaginary, the Gaussian distribution in (6.6.50) is centered at the origin. The density in (6.6.50) is normalized with respect to the measure $\mathrm{d}X$.

Problems

6.1. Derive the properties given through (6.1.11)–(6.1.15).

6.2. By writing the Schrödinger equation in the momentum description and using the solutions given in (6.1.39), (6.1.41), in the position description for $\nu = 1$, obtain the eigenstates $\langle p|\psi\rangle \equiv \psi(p)$ in the momentum description satisfying the normalization condition

$$\int_{-\infty}^{\infty} \frac{dp}{2\pi\hbar} |\psi(p)|^2 = 1.$$

6.3. Derive the expressions for the matrix elements given through (6.1.45)–(6.1.48).

6.4. A particle is moving in a one-dimensional harmonic oscillator potential and is in the state $n = 1$.
   (i) Derive an expression for the probability of finding the particle in the range $x_1 < x < x_2$.
   (ii) Derive an expression for the probability of finding the particle's momentum in the range $p_1 < p < p_2$.

6.5. A uniform electric field E is applied to a charged particle moving in a one-dimensional harmonic oscillator potential. Find the eigenvalues and the eigenvectors of the resulting Hamiltonian.

6.6. For an arbitrary number β, show that

$$a\, e^{\beta a^\dagger} = e^{\beta a^\dagger}(a + \beta)$$

where a, $a^\dagger$ are the Bose annihilation and creation operators.

6.7. Verify that $\beta(t)$, $\gamma(t)$ in (6.2.8), (6.2.9), are, respectively, the solutions of (6.2.7) satisfying the initial conditions in (6.2.6).

6.8. Solve the equation (6.2.4) for the Hamiltonian (6.2.2) by the method developed in Appendix to §2.5.

6.9. Show that the expression in (6.2.14) is equivalent to the one in (6.2.13).

6.10. Derive the expression for the normalization constant C in (6.3.3) for the Boltzmann factor.

6.11. If the disturbing source $F(t)$ in (6.2.1) is given by

$$F(t) = \begin{cases} \hbar\omega \cos\left(\dfrac{\pi t}{2T}\right), & \text{for } |t| < T \\ 0, & \text{for } |t| \geqslant T \end{cases}$$

find the probability of having a transition from an initial state $n = 2$ to a final state $n = 3$.

6.12. Show that the double sum in (6.3.5) may be rewritten as the expression in (6.3.6).

6.13. Obtain the solutions in (6.4.16), (6.4.17).

6.14. Derive the equations (6.4.20)–(6.4.22) for the Fermi-oscillator.

Problems 357

6.15. Verify the Bose $\Leftrightarrow$ Fermi transformation rules given in (6.5.15), (6.5.16).
6.16. Obtain the values for the matrix elements in (6.5.19)–(6.5.22).
6.17. Show that the supersymmetric generators in (6.5.27), (6.5.28) lead to the Hamiltonian H_S given in (6.5.30).
6.18. Investigate further the nature of the lowest point E of the spectrum of the Hamiltonian in (6.5.18), i.e., whether $E = -\infty$ or $E > -\infty$, by constructing states, in analogy to the one in (6.5.23), containing an arbitrary large number of boson excitations.
6.19. Show by invoking, in the process, the normalization condition (6.6.16), that the recurrence relation in (6.6.19) leads to the solution given in (6.6.20).
6.20. Can you find another solution for the coefficients $C_n(A)$ in (6.6.7) consistent with the equations (6.6.16)–(6.6.18)? Comment on this.
6.21. Show directly from the expression of the probability density

$$P_z(n) = \frac{(|z|^2)^n \, e^{-|z|^2}}{n!}$$

that

$$\sum_{n=0}^{\infty} n P_z(n) = |z|^2.$$

6.22. Derive explicitly the multiplicative nature of a as given in (6.6.26) for coherent states.
6.23. Establish the equalities in (6.6.31), (6.6.32) for the expectation values of $x^2(t)$, $p^2(t)$, in the coherent state $|z\rangle$.
6.24. Derive the resolution of the identity resulting from the coherent states $|z\rangle$ as given in (6.6.34).
6.25. Prove the non-orthogonality relation of the states $|z\rangle$ as given in (6.6.38), (6.6.39) and establish the continuity property spelled out in (6.6.40).
6.26. In analogy to (6.6.46), (6.6.47) derive the expression of the time-dependent coherent states in the momentum description $\langle p | z; t \rangle$.
6.27. Use the generator $G = i\delta\xi^* Q - iQ^\dagger \delta\xi$, for infinitesimal transformations introduced in §2.9, to derive the infinitesimal transformations δa_B, $\delta a_B^\dagger$, δa_F, $\delta a_F^\dagger$ corresponding to the supersymmetric generators Q, $Q^\dagger$ in (6.5.27), (6.5.28).

7

Intricacies of the Hydrogen Atom

The purpose of this chapter is to carry out a detailed study of the hydrogen atom. The first section deals with its stability whose importance cannot be overemphasized and is based on the earlier analysis in §3.1 on uncertainties, localization and stability. In the process of this analysis, we will derive an upper bound for the average kinetic energy of the electron in the atom, and obtain a non-vanishing lower bound to its average radial extension as well as determine its exact ground-state energy without even considering first the eigenvalue problem. In particular, the boundedness of the spectrum of the corresponding Hamiltonian of the (bound) atom from below implies the vanishing of the probability of the "fall" of the electron on the proton. The eigenvalue problem is studied in §7.2, §7.3 and the inclusion of the spin of the electron and relativistic corrections is the subject of §7.4–§7.6. In §7.4, we will see how a direct generalization of the Schrödinger equation may be carried out to include such effects[1] as the fine-structure and hyperfine-structure of the atom for all orbital angular momenta. The so-called non-relativistic Lamb shift is presented in detail in §7.7 where the electron, treated non-relativistically[2] in the atom, interacts with radiation. This provides a splitting of some of the energy levels which would otherwise be degenerate by an analysis based on the Schrödinger equation, as well as on the Dirac equation, with a Coulomb interaction alone, if radiation is not included in the treatment. Here due to the ever presence of radiation accompanying the electron we will encounter the concept of mass renormalization and the analysis will provide a glimpse of the fascinating world of quantum filed theory, namely that of quantum electrodynamics. This section is followed by one dealing with

---

[1] The quantum description of relativistic particles will be studied in more detail in Chapter 16.

[2] In the atom, the speed of the electron is roughly suppressed by a factor α relative to the speed of light, where $\alpha \sim 1/137$ is the fine-structure constant (see (7.1.16), (7.4.34)).

the decay and determination of the mean lifetimes of excited states. Finally in §7.9, we consider the hydrogen atom in external electromagnetic fields.

7.1 Stability of the Hydrogen Atom

Due to the central importance of the subject matter of this section in quantum physics, we recapitulate some of the details given in §3.1 as specialized to the hydrogen atom.

First we derive a finite lower bound to the spectrum and prove that it actually coincides with the exact ground-state energy of the atom. The boundedness of the spectrum from below is important, otherwise the electron would fall into such an unbounded level releasing an unlimited amount of energy and is physically meaningless.[3]

We then use the *finite negative* property of the lower bound of the spectrum derived to investigate the localization of the electron in the atom and study further the stability of the atom.

The Hamiltonian of the relative motion of the electron to the proton is given by (§2.5, (2.5.30))

$$H = \widehat{T} - \frac{e^2}{|\mathbf{x}|}. \tag{7.1.1}$$

where $\widehat{T} = \mathbf{p}^2/2\mu$, $\mathbf{p} = -i\hbar\nabla$, μ is the reduced mass of the electron-proton system. By choosing $g(\mathbf{x}) = -e^2/|\mathbf{x}|$ directly in (3.1.8) and using the fact that $\nabla \cdot (\mathbf{x}/|\mathbf{x}|) = 2/|\mathbf{x}|$, we obtain from (3.1.8)

$$-\frac{\mu e^4}{2\hbar^2} \leqslant \left\langle \widehat{T} - \frac{e^2}{|\mathbf{x}|} \right\rangle \tag{7.1.2}$$

providing a lower bound for the ground-state energy $\underline{E}$ of H in (7.1.1). Actually, the bound $-\mu e^4/2\hbar^2$ provides the exact value for $\underline{E}$. To see this, note that for any normalized trial function $\Phi(\mathbf{x})$, we may write (see the introduction to Chapter 4)

$$-\frac{\mu e^4}{2\hbar^2} \leqslant \underline{E} \leqslant \left\langle \Phi \left| \left(\widehat{T} - \frac{e^2}{|\mathbf{x}|} \right) \right| \Phi \right\rangle. \tag{7.1.3}$$

In particular, for

$$\Phi(\mathbf{x}) = \frac{1}{\sqrt{\pi}} \beta^{3/2} e^{-\beta r}, \qquad |\mathbf{x}| = r, \quad \beta > 0 \tag{7.1.4}$$

---

[3] From the analysis given in §4.1 alone, we already know that the spectrum is bounded from below since the Coulomb potential $-e^2/|\mathbf{x}|$ satisfies the sufficiency conditions (4.1.2), (4.1.3) for $\nu = 3$, that is, it is locally square-integrable and it vanishes for $|\mathbf{x}| \to \infty$.

with $\nabla^2 \to \frac{1}{r^2}\frac{\partial}{\partial r}r^2\frac{\partial}{\partial r}$, we obtain

$$\left(\widehat{T} - \frac{e^2}{|\mathbf{x}|}\right)\Phi(\mathbf{x}) = \left[\left(\frac{2}{r} - \beta\right)\frac{\hbar^2 \beta}{2\mu} - \frac{e^2}{r}\right]\Phi(\mathbf{x}). \qquad (7.1.5)$$

Hence by choosing $\beta = \mu e^2/\hbar^2$, the right-hand side of (7.1.5) becomes $-\left(\mu e^4/2\hbar^2\right)\Phi(\mathbf{x})$, and for (7.1.3) we have

$$-\frac{\mu e^4}{2\hbar^2} \leqslant \underline{E} \leqslant -\frac{\mu e^4}{2\hbar^2} \qquad (7.1.6)$$

thus establishing that the exact ground-state energy is given by

$$\underline{E} = -\frac{\mu e^4}{2\hbar^2}. \qquad (7.1.7)$$

Incidentally, $1/\beta = \hbar^2/\mu e^2$ is the celebrated Bohr radius:

$$a_0 = \frac{\hbar^2}{\mu e^2} \qquad (7.1.8)$$

which gives an order of magnitude of the average of the radial extension of the atom in its ground-state, since

$$\langle \Phi|r|\Phi\rangle = 3a_0/2 \qquad (7.1.9)$$

where $|\Phi\rangle$ is the ground-state as defined in (7.1.4) with $\beta = 1/a_0$.

Quite *generally*, we may derive a Heisenberg uncertainty-like inequality between the average radial extension of the atom and the average kinetic energy of the electron in the following manner. By choosing $g(\mathbf{x}) = 1/|\mathbf{x}|$, $\mathbf{a} = 0, \mathbf{b} = 0$ in (3.1.3), we obtain for the electron in a normalized state $|\psi\rangle$

$$\langle \psi|\widehat{T}|\psi\rangle \geqslant \frac{\hbar^2}{2\mu}\left\langle\psi\left|\frac{1}{r}\right|\psi\right\rangle^2. \qquad (7.1.10)$$

On the other hand, we may invoke the Cauchy-Schwarz inequality

$$1 = \int d^3\mathbf{x}\,|\psi(\mathbf{x})|^2 = \int d^3\mathbf{x}\,\frac{|\psi(\mathbf{x})|}{r^{1/2}}\cdot r^{1/2}\,|\psi(\mathbf{x})|$$

$$\leqslant \left\langle\psi\left|\frac{1}{r}\right|\psi\right\rangle^{1/2}\langle\psi|r|\psi\rangle^{1/2} \qquad (7.1.11)$$

giving

$$\left\langle\psi\left|\frac{1}{r}\right|\psi\right\rangle \geqslant \frac{1}{\langle\psi|r|\psi\rangle} \qquad (7.1.12)$$

to infer from (7.1.10) that

$$\left\langle \psi \left| \widehat{T} \right| \psi \right\rangle \left\langle \psi |r|\psi \right\rangle^2 \geq \frac{\hbar^2}{2\mu}. \tag{7.1.13}$$

An interpretation of this inequality is that the smaller the average extension of the atom, the larger is the average kinetic energy of the electron. The complete balance between the increase of the average kinetic energy of the electron, as implied by localization, and any energy loss from the atom provides the final stable state of the atom.

For the hydrogen atom, as a bound-state, the average kinetic energy cannot increase by an unlimited amount and a *finite* upper bound to it may be derived as shown below. In the same way, the average radial extension of the atom cannot become arbitrarily small and cannot vanish as is also shown below. To establish these properties, we note that for a bound-state

$$\left\langle \widehat{T} - \frac{e^2}{|\mathbf{x}|} \right\rangle < 0 \tag{7.1.14}$$

or

$$\left\langle \frac{\widehat{T}}{2} \right\rangle < - \left\langle \frac{\widehat{T}}{2} - \frac{e^2}{|\mathbf{x}|} \right\rangle \tag{7.1.15}$$

where $\widehat{T}/2 - e^2/|\mathbf{x}|$ is the Hamiltonian of the hydrogen atom when μ is replaced by 2μ (see also (3.1.18), (3.1.24)). Hence for a bound-state, we have from (7.1.15), (7.1.2),

$$\left\langle \widehat{T} \right\rangle < \frac{2\mu e^4}{\hbar^2}. \tag{7.1.16}$$

This gives an *upper* bound for the average kinetic energy of the electron bound in the atom.

The inequality (7.1.16) in turn, provides from (7.1.13) the non-vanishing *lower* bound

$$\langle r \rangle > \frac{a_0}{2} \tag{7.1.17}$$

for the average radial extension of the atom, where we have used the definition of a_0 given in (7.1.8).

It is important to realize how quantum physics has built-in inequalities as in (7.1.17). The boundedness of the spectrum from below by the finite negative number in (7.1.2), (7.1.7), leading, in particular to the bound in (7.1.16), provides also a rigorous bound on the *probability* distribution of the radial extension of the atom not only for its average value as is shown below.

To the above end, (7.1.10) leads to the following chain of inequalities:

$$\sqrt{\frac{2\mu}{\hbar^2}} \left\langle \psi \left| \widehat{T} \right| \psi \right\rangle^{1/2} \geq \int d^3\mathbf{x} \frac{|\psi(\mathbf{x})|^2}{r} \geq \int_{r_0 \geq r} d^3\mathbf{x} \frac{|\psi(\mathbf{x})|^2}{r} \tag{7.1.18}$$

for *any* $0 < r_0 < \infty$, where we have used the positivity of the integrand $|\psi(\mathbf{x})|^2/r$. Since in the last integral in (7.1.18), we have the constraint $1/r \geq 1/r_0$, we obtain

$$\sqrt{\frac{2\mu}{\hbar^2}} \langle \psi | \widehat{T} | \psi \rangle^{1/2} \geq \frac{1}{r_0} \int_{r_0 \geq r} d^3\mathbf{x} \, |\psi(\mathbf{x})|^2 \qquad (7.1.19)$$

or equivalently to

$$\text{Prob}\,[r \leq r_0] \leq r_0 \sqrt{\frac{2\mu}{\hbar^2}} \langle \widehat{T} \rangle^{1/2} \qquad (7.1.20)$$

with $|\psi\rangle$ normalized. For a bound-state, (7.1.16) then leads from (7.1.20) to

$$\text{Prob}\,[r \leq r_0] \leq 2\frac{r_0}{a_0} \qquad (7.1.21)$$

giving the consistent and satisfactory result of a vanishingly small probability for a vanishingly small r_0 and rigorously vanishing for $r_0 \to 0$, i.e., for the "fall" of the electron on the proton.

7.2 The Eigenvalue Problem

From the analysis carried out in §4.1, one may readily infer the nature of the spectrum of the Hamiltonian with a Coulomb potential. The latter potential is locally square-integrable and vanishes at infinity, i.e., it satisfies conditions (4.1.2), (4.1.3). Also we have shown in §7.1, that the lower bound of the attractive Coulomb Hamiltonian is strictly negative (see (7.1.2), (7.1.7)). Hence from Theorem 4.1.1 alone, we may immediately conclude that this Hamiltonian has a discrete spectrum on the negative real axis of at most *finite* degeneracy, and a continuous spectrum on the positive real axis $[0, \infty)$. [For a repulsive Coulomb potential, the corresponding Hamiltonian is positive and hence has only a continuous spectrum on the positive real axis.]

To study the spectrum of the hydrogen atom, we rewrite the Hamiltonian (7.1.1) in detail as

$$H = \frac{p_r^2}{2\mu} + \frac{\mathbf{L}^2}{2\mu r^2} - \frac{e^2}{r}, \qquad |\mathbf{x}| = r \qquad (7.2.1)$$

where

$$p_r = -i\hbar \left(\frac{\partial}{\partial r} + \frac{1}{r} \right) \qquad (7.2.2)$$

and $\mathbf{L}^2$ is the square of the orbital angular momentum, which when written in spherical coordinates is given by

$$\mathbf{L}^2 = -\hbar^2 \left[\frac{1}{\sin\theta} \frac{\partial}{\partial \theta} \sin\theta \frac{\partial}{\partial \theta} + \frac{1}{\sin^2\theta} \frac{\partial^2}{\partial \phi^2} \right]. \qquad (7.2.3)$$

It is easily verified that $H, \mathbf{L}^2, L_z$ commute and hence we may find simultaneous eigenstates for them which we conveniently denote by $|E, \ell, m\rangle$ and set

$$H\,|E,\ell,m\rangle = E\,|E,\ell,m\rangle \tag{7.2.4}$$

$$\mathbf{L}^2\,|E,\ell,m\rangle = \hbar^2 \ell(\ell+1)\,|E,\ell,m\rangle \tag{7.2.5}$$

$$L_z\,|E,\ell,m\rangle = \hbar m\,|E,\ell,m\rangle \tag{7.2.6}$$

where ℓ takes on non-negative integer values (§5.3) and $m = -\ell, -\ell+1, \ldots, 0, 1, \ldots, \ell$. By definition the states $|E,\ell,m\rangle$ are in the domain of definition of the operators $H, \mathbf{L}^2, L_z$.

From (7.2.4), (7.2.5), we may also write

$$H\,|E,\ell,m\rangle = H_\ell\,|E,\ell,m\rangle = E\,|E,\ell,m\rangle \tag{7.2.7}$$

where

$$H_\ell = \frac{p_r^2}{2\mu} + \frac{\hbar^2 \ell(\ell+1)}{2\mu r^2} - \frac{e^2}{r} \tag{7.2.8}$$

and clearly $H_\ell, \mathbf{L}^2, L_z$ commute. We define the operators

$$D_\ell = p_r - i\left(\frac{(\ell+1)\hbar}{r} - \frac{\mu e^2}{(\ell+1)\hbar}\right). \tag{7.2.9}$$

To define the adjoint $D_\ell^\dagger$ of D_ℓ, we must impose appropriate boundary conditions, in the process, in defining the operator p_r. With a measure of integration $d^3\mathbf{x}$, we consider functions $f(\mathbf{x}), g(\mathbf{x})$ such that $rf(\mathbf{x}), rg(\mathbf{x})$ vanish not only for $r \to \infty$ but also for $r \to 0_+$. The adjoint $D_\ell^\dagger$ of D_ℓ, satisfying

$$\langle g|D_\ell|f\rangle = \left\langle D_\ell^\dagger\, g\,\middle|\, f\right\rangle \tag{7.2.10}$$

obtained by integrating by parts, is then given by

$$D_\ell^\dagger = p_r + i\left(\frac{(\ell+1)\hbar}{r} - \frac{\mu e^2}{(\ell+1)\hbar}\right) \tag{7.2.11}$$

(see Problem 7.4).

The following relations are easily established for $\ell = 0, 1, \ldots$

(i) $\quad H_{\ell+1}\, D_\ell^\dagger = D_\ell^\dagger\, H_\ell$ \hfill (7.2.12)

(ii) $\quad H_\ell\, D_\ell = D_\ell\, H_{\ell+1}$ \hfill (7.2.13)

and that

(iii) $\quad D_\ell\, D_\ell^\dagger = 2\mu\left(H_\ell + \frac{\mu e^4}{2(\ell+1)^2 \hbar^2}\right)$ \hfill (7.2.14)

(iv) $\quad D_\ell^\dagger\, D_\ell = 2\mu\left(H_{\ell+1} + \frac{\mu e^4}{2(\ell+1)^2 \hbar^2}\right).$

$$\tag{7.2.15}$$

Since $\mu e^4/2(\ell+1)^2\hbar^2$ in (7.2.14) is decreasing in ℓ, the positivity of $D_\ell D_\ell^\dagger$ implies from (7.2.14), (7.2.7) that *for the ground-state energy* $\underline{E} = -\mu e^4/2\hbar^2$, obtained in (7.1.7), $\ell = 0$; for larger ℓ this positivity is *violated*.

Quite generally, (7.2.12) and the right-hand side equality in (7.2.7) imply that
$$H_{\ell+1}\left(D_\ell^\dagger |E,\ell,m\rangle\right) = E\left(D_\ell^\dagger |E,\ell,m\rangle\right). \tag{7.2.16}$$
with the same eigenvalue E.

That is, $D_\ell^\dagger |E,\ell,m\rangle$ is an eigenstate of $H_{\ell+1}$, and we may write[4]
$$D_\ell^\dagger |E,\ell,m\rangle = c_\ell^+ |E,\ell+1,m\rangle. \tag{7.2.17}$$
From (7.2.14), one may infer that,
$$c_\ell^+ = i\left[2\mu\left(E + \frac{\mu e^4}{2(\ell+1)^2\hbar^2}\right)\right]^{1/2} \tag{7.2.18}$$
with a chosen overall phase convention.

Hence from the *positivity* of $D_\ell D_\ell^\dagger$ the sequence $D_\ell^\dagger |E,\ell,m\rangle$, $D_{\ell+1}^\dagger D_\ell^\dagger |E,\ell,m\rangle,\ldots$ must terminate for some $\ell = \ell_{\max}$ for an eigenvalue E, $\underline{E} \leqslant E < 0$. As a matter of fact if $D_\ell^\dagger |E,\ell,m\rangle$ does not vanish for such an ℓ, then $D_{\ell+1}^\dagger D_\ell^\dagger |E,\ell,m\rangle$, if not the zero vector, would violate the positivity of $D_{\ell+1} D_{\ell+1}^\dagger$, in (7.2.14) for $\ell \to \ell+1$, since $E + \mu e^4/2(\ell+1)^2\hbar^2$ would be negative. That is, there exists an $\ell = \ell_{\max}$, for an eigenvalue $E < 0$ such that $D_{\ell_{\max}}^\dagger |E,\ell_{\max},m\rangle$ is the zero vector. As we have seen above $\ell_{\max} = 0$ for the ground-state energy $\underline{E}$. Since $\ell_{\max}$ is a non-negative integer, we may set $\ell_{\max} = n-1, n = 1, 2, \ldots$, with $n = 1$ corresponding to the ground-state energy $\underline{E}$.

The vanishing of $c_{\ell_{\max}}^+ = c_{n-1}^+$, then imply that the eigenvalues E may be parameterized by n, and, from the vanishing of c_{n-1}^+ in (7.2.18)/(7.2.17), take the values:
$$E_n = -\frac{\mu e^4}{2\hbar^2 n^2}, \qquad n = 1, 2, \ldots \tag{7.2.19}$$
with $E_1 = \underline{E}$ as seen in §7.1.

For the subsequent analysis, we use the labelling $|n,\ell,m\rangle$ for $|E,\ell,m\rangle$, and we recall the condition
$$D_{n-1}^\dagger |n, n-1, m\rangle = 0. \tag{7.2.20}$$

Since for a given ℓ, m may take on $(2\ell+1)$ values, the degree of degeneracy of an energy level E_n in (7.2.19), taking into account the spin of the electron, is given by
$$2\sum_{\ell=1}^{n-1}(2\ell+1) = 2n^2. \tag{7.2.21}$$

---
[4] Here it is sufficient to assume that any operator which commutes with $H_{\ell+1}$, $\mathbf{L}^2$, L_z and which is *not* a function of the latter operators also commutes with $D_\ell^\dagger$.

7.3 The Eigenstates

The **x**-description state wavefunctions may be represented as follows:

$$\langle r, \theta, \phi | n, \ell, m \rangle \equiv R_{n\ell}(r) Y_{\ell m}(\theta, \phi) \equiv \psi_{n\ell m}(r, \theta, \phi) \qquad (7.3.1)$$

where the $Y_{\ell m}(\theta, \phi)$ are the spherical harmonics defined in (5.3.26), (5.3.30), (5.3.51), satisfying (5.1.19), (5.1.20):

$$\mathbf{L}^2 Y_{\ell m}(\theta, \phi) \equiv \hbar^2 \ell(\ell+1) Y_{\ell m}(\theta, \phi) \qquad (7.3.2)$$

$$L_z Y_{\ell m}(\theta, \phi) \equiv \hbar \, m Y_{\ell m}(\theta, \phi) \qquad (7.3.3)$$

and from (7.2.20), (7.2.11) and (7.2.2), ($\langle r | n, \ell \rangle = R_{n\ell}(r)$):

$$\left(\frac{\partial}{\partial r} - \frac{(n-1)}{r} + \frac{\mu e^2}{n\hbar^2} \right) R_{n,n-1}(r) = 0 \qquad (7.3.4)$$

whose normalized

$$\int_0^\infty r^2 \, dr \, R_{n,n-1}^2(r) = 1 \qquad (7.3.5)$$

solution is

$$R_{n,n-1}(r) = \left(\frac{2}{na_0} \right)^{3/2} \frac{\rho^{n-1}}{\sqrt{(2n)!}} \exp(-\rho/2) \qquad (7.3.6)$$

where

$$\rho = \left(\frac{2}{na_0} \right) r \qquad (7.3.7)$$

and a_0 is the Bohr radius in (7.1.8).

In particular, (7.3.6), (5.3.45) imply from (7.3.1) that

$$R_{10}(r) = \left(\frac{1}{a_0} \right)^{3/2} 2 \, e^{-r/a_0} \qquad (7.3.8)$$

and for the ground-state

$$\psi_{100}(r, \theta, \phi) = \frac{1}{\sqrt{\pi}} \left(\frac{1}{a_0} \right)^{3/2} e^{-r/a_0} \qquad (7.3.9)$$

which coincides with the expression in (7.1.4), with $\beta = \mu e^2/\hbar^2 = 1/a_0$.

A similar analysis as carried out in (7.2.16)–(7.2.18), in reference to (7.2.12), as now applied to (7.2.13), (7.2.15) with $\ell \to \ell - 1$, leads to

$$D_{\ell-1} | n, \ell, m \rangle \equiv -i \frac{\mu e^2 \sqrt{n^2 - \ell^2}}{\hbar n \ell} | n, \ell - 1, m \rangle \qquad (7.3.10)$$

for $\ell \neq 0$ (see Problem 7.7). The $-i$ factor in (7.3.10) is chosen for convenience (see, (7.2.18) and (7.3.12)). Equation (7.3.10) then gives

$$R_{n,\ell-1}(r) = \frac{i\hbar\ell n}{\mu e^2 \sqrt{n^2 - \ell^2}} D_{\ell-1} R_{n,\ell}(r). \tag{7.3.11}$$

Equation (7.3.11) provides the differential equation

$$R_{n,\ell-1}(r) = \frac{2\ell}{\sqrt{n^2 - \ell^2}} \left[\frac{d}{d\rho} + \frac{(\ell+1)}{\rho} - \frac{n}{2\ell} \right] R_{n,\ell}(r) \tag{7.3.12}$$

for $\ell \neq 0$, and $R_{n,\ell-1}$ are necessarily normalized for normalized $R_{n,\ell}$.
In particular, for $\ell = n-1$, (7.3.12) gives from (7.3.6)

$$R_{n,n-2}(r) = \left(\frac{2}{na_0}\right)^{3/2} \frac{1}{\sqrt{(2n-2)!2n}} \rho^{n-2} e^{-\rho/2} [-\rho + 2(n-1)] \tag{7.3.13}$$

for $n \geqslant 2$.

Equations (7.3.6), (7.3.12), (7.3.13) suggest developing a proof by induction in ℓ to find the explicit expression for $R_{n\ell}$. To this end, as an induction hypothesis, suppose that for all $\ell = k, k+1, \ldots, n-1$, for some $k: 0 < k \leqslant n-1$

$$R_{n,\ell}(r) = \left(\frac{2}{na_0}\right)^{3/2} \sqrt{\frac{(n-\ell-1)!}{(n+\ell)!2n}} \rho^\ell e^{-\rho/2} L_{n-\ell-1}^{2\ell+1}(\rho) \tag{7.3.14}$$

where

$$L_N^M(\rho) = \rho^{-M} \frac{e^\rho}{N!} \left(\frac{d}{d\rho}\right)^N \left[\rho^{M+N} e^{-\rho}\right]. \tag{7.3.15}$$

Equation (7.3.14) is easily checked to coincide with (7.3.6), (7.3.13) for $\ell = n-1, \ell = n-2$, respectively. We then show that (7.3.14) holds also true for ℓ replaced by $\ell - 1$.

The $L_N^M(\rho)$ are so-called associated Laguerre polynomials, having the series expansion

$$L_N^M(\rho) = \sum_{t=0}^{N} (-1)^t \frac{(M+N)!}{(N-t)!(M+t)!} \frac{(\rho)^t}{t!} \tag{7.3.16}$$

with the normalization in (7.3.15) adopted,[5] and satisfy for $M \geqslant 2$, the recurrence relation

$$\rho \frac{d}{d\rho} L_N^M(\rho) + \left[M - \frac{N+M}{M-1}\rho\right] L_N^M(\rho) - \frac{(N+1)(N+M)}{M-1} L_{N+1}^{M-2}(\rho) = 0. \tag{7.3.17}$$

Upon substitution of the expression (7.3.14) in the right-hand side of (7.3.12), it is readily checked by using the recurrence relation (7.3.17) with

---

[5] Note that different normalizations are sometimes adopted for the associated Laguerre polynomials by different authors.

$M = 2\ell + 1, \ell = k, \ldots, n-1$, for some $k > 0$, that $R_{n,\,\ell-1}(\rho)$ is also given by (7.3.14) with ℓ replaced by $\ell - 1$. This establishes the validity of (7.3.14) for all $\ell = n-1, n-2, \ldots, 1, 0$ by induction.

All told, the normalized wavefunctions in (7.3.1) are given by

$$\psi_{n\ell m}(r, \theta, \phi) = \left(\frac{2}{na_0}\right)^{3/2} \sqrt{\frac{(n-\ell-1)!}{(n+\ell)!\,2n}}\, \rho^\ell\, e^{-\rho/2}\, L^{2\ell+1}_{n-\ell-1}(\rho) Y_{\ell m}(\theta, \phi) \tag{7.3.18}$$

where ρ is defined in (7.3.7) and (7.1.8). Note that ρ depends on n. The associated Laguerre polynomials are given in (7.3.15), (7.3.16). Here $n = 1, 2, \ldots$; $\ell = 0, 1, \ldots, n-1$; $m = -\ell, -\ell+1, \ldots, -1, 0, 1, \ldots, \ell-1, \ell$. Also

$$\int_0^\infty r^2\, dr \int_0^\pi \sin\theta\, d\theta \int_0^{2\pi} d\phi\, |\psi_{n\ell m}(r, \theta, \phi)|^2 = 1. \tag{7.3.19}$$

To study the orthogonality of the wavefunctions in (7.3.18), we first recall from (5.3.51) that

$$\int_0^\pi \sin\theta\, d\theta \int_0^{2\pi} d\phi\, Y^*_{\ell' m'}(\theta, \phi) Y_{\ell m}(\theta, \phi) = \delta_{\ell'\ell}\delta_{m'm}. \tag{7.3.20}$$

Hence it remains to establish the orthogonality of $R_{n\ell}(r), R_{n'\ell}(r)$ for $n' \neq n$. Due to the n dependence, in particular, of ρ, the argument of the Laguerre polynomials in (7.3.18), establishing the orthogonality of $R_{n\ell}(r), R_{n'\ell}(r)$ for $n' \neq n$ based on any orthogonality properties of (weighted) Laguerre polynomials is not obvious. One may, however, establish this orthogonality property directly and rather easily from the Schrödinger equation satisfied by $R_{n\ell}(r)$, which by using (7.3.2), (7.2.19) and (7.2.1), is given by

$$\left[\frac{-\hbar^2}{2\mu r^2}\left(\frac{d}{dr} r^2 \frac{d}{dr}\right) - \frac{e^2}{r} + \frac{\mu e^4}{2n^2\hbar^2} + \frac{\hbar^2 \ell(\ell+1)}{2\mu r^2}\right] R_{n\ell}(r) = 0. \tag{7.3.21}$$

Upon setting

$$\eta = r/a_0 \tag{7.3.22}$$

and

$$u_{n\ell}(\eta) = \eta\, R_{n\ell}(r) \tag{7.3.23}$$

(7.3.21) simplifies to

$$\left[\frac{d^2}{d\eta^2} + \frac{2}{\eta} - \frac{1}{n^2} + \frac{\ell(\ell+1)}{\eta^2}\right] u_{n\ell}(\eta) = 0. \tag{7.3.24}$$

Finally by writing the corresponding equation to (7.3.24) for $u_{n'\ell}(\eta)$, and using the boundary condition

$$\left[u_{n'\ell}(\eta)\frac{d}{d\eta} u_{n\ell}(\eta) - u_{n\ell}\frac{d}{d\eta} u_{n'\ell}(\eta)\right]\bigg|_0^\infty = 0 \tag{7.3.25}$$

one readily derives (see Problem 7.10) the orthogonality relation

$$0 = \left(\frac{1}{n'^2} - \frac{1}{n^2}\right) \int_0^\infty d\eta \, u_{n'\ell}(\eta) \, u_{n\ell}(\eta). \tag{7.3.26}$$

Hence from (7.3.23), we have for $n' \neq n$

$$\int_0^\infty r^2 \, dr \, R_{n'\ell}(r) R_{n\ell}(r) = 0. \tag{7.3.27}$$

If the Coulomb potential in (7.1.1) is replaced by

$$V(\mathbf{x}) = -Ze^2/|\mathbf{x}| \tag{7.3.28}$$

corresponding to a nucleus of charge $Z|e|$, then (7.2.19) simply becomes

$$E_n = -\frac{\mu Z^2 e^4}{2n^2 \hbar^2} \tag{7.3.29}$$

and $1/a_0$ as appearing in (7.3.18) (also in ρ) is to be replaced by Z/a_0.

The radial wavefunctions (7.3.14) are non-vanishing at the origin only for $\ell = 0$. To see this note that

$$L_{n-1}^1(\rho) \xrightarrow[\rho \to 0]{} n \tag{7.3.30}$$

yielding

$$R_{n\ell}^2(0) = \frac{4}{n^3} \left(\frac{Z}{a_0}\right)^3 \delta_{\ell 0}. \tag{7.3.31}$$

Upon using the fact that $Y_{00}(\Omega) = 1/\sqrt{4\pi}$, as given in (5.3.45), we have for the wavefunctions in (7.3.18), ($m = 0$),

$$|\psi_{n\ell m}(0)|^2 = \frac{1}{\pi n^3} \left(\frac{Z}{a_0}\right)^3 \delta_{\ell 0}. \tag{7.3.32}$$

With the definition,

$$\langle f(r) \rangle = \int d^3\mathbf{x} \, f(r) |\psi_{n\ell m}(\mathbf{x})|^2$$

$$= \int_0^\infty r^2 \, dr \, f(r) R_{n\ell}^2(r) \tag{7.3.33}$$

where for simplicity of the notation we have suppressed the dependence of $\langle f(r) \rangle$ on n and ℓ, the following expectation values are useful:

$$\langle r \rangle = [3n^2 - \ell(\ell+1)] \left(\frac{a_0}{2Z}\right) \tag{7.3.34}$$

$$\langle r^2 \rangle = [5n^2 + 1 - 3\ell(\ell+1)] \left(\frac{a_0^2 n^2}{2Z^2} \right) \tag{7.3.35}$$

$$\left\langle \frac{1}{r} \right\rangle = \frac{Z}{a_0} \frac{1}{n^2} \tag{7.3.36}$$

$$\left\langle \frac{1}{r^2} \right\rangle = \left(\frac{Z}{a_0} \right)^2 \frac{1}{(\ell+1/2)} \frac{1}{n^3} \tag{7.3.37}$$

$$\left\langle \frac{1}{r^3} \right\rangle = \left(\frac{Z}{a_0} \right)^3 \frac{1}{\ell(\ell+1/2)(\ell+1)} \frac{1}{n^3}, \quad \ell \neq 0. \tag{7.3.38}$$

7.4 The Hydrogen Atom Including Spin and Relativistic Corrections

In the present section, we extend the work in §7.2 to include spin and relativistic corrections. This may be done from Dirac's equation studied later on dealing with the quantum description of the relativistic electron. We find it, however, more suitable to treat this topic right in this chapter for continuity, especially since the Schrödinger equation may be readily generalized to a relativistic treatment as we shall see below. Needless to say, we will, in the process, discover the Dirac equation which will be treated afresh in Chapter 16.

The time-independent Schrödinger equation for a free particle of mass M may be written as

$$\left(\frac{\mathbf{p}^2}{2M} - E \right) \psi = 0 \tag{7.4.1}$$

providing the constraint that is imposed between the energy and the momentum. Its interaction with an external time-independent[6] electromagnetic potential $(U, \mathbf{A})$ may be then obtained from (7.4.1) by the substitutions $\mathbf{p} \to \mathbf{p} - \frac{e}{c}\mathbf{A}$, $E \to E - U$.

For the relativistic free electron, we may write

$$\left[\sqrt{\mathbf{p}^2 c^2 + M^2 c^4} - E \right] \Psi = 0 \tag{7.4.2}$$

providing the relativistic constraint between the energy and the momentum,[7] where c denotes the speed of light. We multiply (7.4.2) by the operator $\sqrt{\mathbf{p}^2 c^2 + M^2 c^4} + E$ to obtain the more manageable equation

---

[6] The interaction with an external time-dependent electromagnetic potential, in general, is easily obtained by considering the time-dependent Dirac equation — see for example, the appendix to this section.

[7] The square-root operator in (7.4.2) of the positive operator $\mathbf{p}^2 c^2 + M^2 c^4$ is a well defined operator, for example, by Fourier transform theory. See also (4.5.34), and §16.1.

7.4 The Hydrogen Atom Including Spin and Relativistic Corrections

$$\left(\mathbf{p}^2 c^2 + M^2 c^4 - E^2\right) \Psi = 0. \tag{7.4.3}$$

Using the identity

$$\sigma_i \sigma_j = \delta_{ij} + i\varepsilon_{ijk}\sigma_k \tag{7.4.4}$$

for the Pauli matrices, we may rewrite (7.4.3) as

$$\left[(\boldsymbol{\sigma} \cdot \mathbf{p})(\boldsymbol{\sigma} \cdot \mathbf{p}) - \frac{(E - Mc^2)}{c}\frac{(E + Mc^2)}{c}\right] \Psi = 0. \tag{7.4.5}$$

Upon setting

$$\boldsymbol{\sigma} \cdot \mathbf{p}\, \Psi = \frac{(E + Mc^2)}{c} \Phi \tag{7.4.6}$$

$$\boldsymbol{\sigma} \cdot \mathbf{p}\, \Phi = \frac{(E - Mc^2)}{c} \Psi. \tag{7.4.7}$$

Equation (7.4.5) may be rewritten in terms of the following two equations

$$\boldsymbol{\sigma} \cdot \mathbf{p}\, \Phi - \frac{E}{c} \Psi + Mc\Psi = 0 \tag{7.4.8}$$

$$-\boldsymbol{\sigma} \cdot \mathbf{p}\, \Psi + \frac{E}{c} \Phi + Mc\Phi = 0. \tag{7.4.9}$$

The Pauli matrices are 2×2 matrices, and with Φ, Ψ each as a two-row, one-column matrix, we may combine (7.4.8), (7.4.9) into the equation

$$\left[\begin{pmatrix} 0 & \boldsymbol{\sigma} \\ -\boldsymbol{\sigma} & 0 \end{pmatrix} \cdot \mathbf{p} - \begin{pmatrix} I & 0 \\ 0 & -I \end{pmatrix}\frac{E}{c} + Mc \begin{pmatrix} I & 0 \\ 0 & I \end{pmatrix}\right] \begin{bmatrix} \Psi \\ \Phi \end{bmatrix} = 0 \tag{7.4.10}$$

which is the famous Dirac equation to be studied later in Chapter 16 and we will not consider it here further, as it stands, except in the appendix to this section dealing with normalization problems. We are here interested in (7.4.8), (7.4.9) to study, approximately, the bound-state problem of the hydrogen atom including spin and relativistic corrections.

We isolate the rest energy Mc^2 in E

$$E = \varepsilon + Mc^2 \tag{7.4.11}$$

and rewrite (7.4.8), (7.4.9), respectively, as

$$\varepsilon \Psi = c\boldsymbol{\sigma} \cdot \mathbf{p}\, \Phi \tag{7.4.12}$$

$$\left(\varepsilon + 2Mc^2\right) \Phi = c\boldsymbol{\sigma} \cdot \mathbf{p}\, \Psi. \tag{7.4.13}$$

The interaction with an external time-independent electromagnetic potential may be then obtained by the substitutions $\mathbf{p} \to \mathbf{p} - \frac{e}{c}\mathbf{A}(\mathbf{x})$, $\varepsilon \to \varepsilon - U(\mathbf{x})$, giving rise to the equations

$$(\varepsilon - U)\Phi = c\boldsymbol{\sigma} \cdot \left(\mathbf{p} - \frac{e}{c}\mathbf{A}\right)\Phi \qquad (7.4.14)$$

$$(2Mc^2 + \varepsilon - U)\Phi = c\boldsymbol{\sigma} \cdot \left(\mathbf{p} - \frac{e}{c}\mathbf{A}\right)\Psi. \qquad (7.4.15)$$

We will see below, how the familiar term $-\boldsymbol{\mu} \cdot \mathbf{B}$, where $\boldsymbol{\mu}$ is the magnetic dipole moment of the electron and $\mathbf{B} = \nabla \times \mathbf{A}$, naturally arises in the modified Hamiltonian as obtained from (7.4.14), (7.4.15).

We use the identity

$$\frac{1}{2Mc^2 + \varepsilon - U} = \frac{1}{2Mc^2}\left(1 + \frac{U - \varepsilon}{2Mc^2 + \varepsilon - U}\right) \qquad (7.4.16)$$

and note that for $-U \geqslant 0$ and $|\varepsilon| \ll 2Mc^2$, the denominator $2Mc^2 + \varepsilon - U$ does not vanish. Here we have in mind the attractive Coulomb potential

$$U(\mathbf{x}) = -\frac{e^2}{|\mathbf{x}|} \qquad (7.4.17)$$

and where $|\varepsilon| \ll 2Mc^2$.

Using the identity (7.4.16), we may write for Φ the exact equation

$$\Phi = \frac{1}{2Mc^2}\left[1 + \frac{U - \varepsilon}{(2Mc^2 + \varepsilon - U)}\right] c\boldsymbol{\sigma} \cdot \left(\mathbf{p} - \frac{e}{c}\mathbf{A}\right)\Psi. \qquad (7.4.18)$$

Upon substituting (7.4.18) in (7.4.14) we obtain for the latter

$$(\varepsilon - U)\Psi = \left[\frac{\mathbf{p}^2}{2M} + \frac{1}{2Mc^2}\boldsymbol{\sigma}\cdot\left(\mathbf{p} - \frac{e}{c}\mathbf{A}\right)\frac{(U-\varepsilon)}{(2Mc^2+\varepsilon-U)}\boldsymbol{\sigma}\cdot\left(\mathbf{p} - \frac{e}{c}\mathbf{A}\right)\right]\Psi$$

$$+ \frac{1}{2M}\left[\boldsymbol{\sigma}\cdot\left(\mathbf{p} - \frac{e}{c}\mathbf{A}\right)\boldsymbol{\sigma}\cdot\left(\mathbf{p} - \frac{e}{c}\mathbf{A}\right) - \mathbf{p}^2\right]\Psi \qquad (7.4.19)$$

where we have added and subtracted the term $\mathbf{p}^2/2M$ on the right-hand side of (7.4.19).

As an order of magnitude estimate, we note that $(\varepsilon - U) \sim Me^4/2\hbar^2 = Mc^2\alpha^2/2$, with the latter (the Rydberg) providing the energy scale of the hydrogen atom. Here $\alpha = e^2/\hbar c \sim (137)^{-1}$ is the fine-structure constant. Hence we neglect the $(\varepsilon - U)$ term in comparison to $2Mc^2$ in the denominator on the right-hand side of (7.4.19) as it is suppressed by a factor α^2. We will keep track of terms up to the order $1/c^2$. The vector potential $\mathbf{A}$ considered is due to the magnetic dipole moment (or the spin) of the proton and involves a factor $1/c$ (see (7.4.39), (7.4.40)). For the time being we will treat $\mathbf{A}$ in (7.4.19) as arbitrary, as it is readily identified, and assess its contribution later. This will save us some work in the sequel.

With the order of magnitude sought in mind, we may rewrite (7.4.19) as

7.4 The Hydrogen Atom Including Spin and Relativistic Corrections

$$(\varepsilon - U)\Psi = \left[\frac{\mathbf{p}^2}{2M} + \frac{1}{4M^2c^2}\boldsymbol{\sigma}\cdot\mathbf{p}(U-\varepsilon)\boldsymbol{\sigma}\cdot\mathbf{p}\right]\Psi$$
$$+ \frac{1}{2M}\left[\boldsymbol{\sigma}\cdot\left(\mathbf{p}-\frac{e}{c}\mathbf{A}\right)\boldsymbol{\sigma}\cdot\left(\mathbf{p}-\frac{e}{c}\mathbf{A}\right) - \mathbf{p}^2\right]\Psi. \quad (7.4.20)$$

In the appendix to this section, it is shown that the correct normalization of Ψ is given by

$$\int d^3x\, \Psi^\dagger(\mathbf{x})\left[1 + \frac{\mathbf{p}^2}{4M^2c^2}\right]\Psi(\mathbf{x}) = 1 \quad (7.4.21)$$

for the accuracy required. Accordingly, we may set

$$\Psi = \left[1 - \frac{\mathbf{p}^2}{8M^2c^2}\right]\chi \quad (7.4.22)$$

where now

$$\int d^3x\, \chi^\dagger(\mathbf{x})\chi(\mathbf{x}) = 1 \quad (7.4.23)$$

to the accuracy needed.

In terms of χ, (7.4.20) reduces to

$$(\varepsilon - U)\chi - (\varepsilon - U)\frac{\mathbf{p}^2}{8M^2c^2}\chi$$
$$= F\chi + \left[\frac{\mathbf{p}^2}{2M} - \frac{\mathbf{p}^4}{16M^3c^2} + \frac{1}{4M^2c^2}\boldsymbol{\sigma}\cdot\mathbf{p}(U-\varepsilon)\boldsymbol{\sigma}\cdot\mathbf{p}\right]\chi \quad (7.4.24)$$

where

$$F\chi = \frac{1}{2M}\left[\boldsymbol{\sigma}\cdot\left(\mathbf{p}-\frac{e}{c}\mathbf{A}\right)\boldsymbol{\sigma}\cdot\left(\mathbf{p}-\frac{e}{c}\mathbf{A}\right) - \mathbf{p}^2\right]\chi. \quad (7.4.25)$$

With $\mathbf{A}$ given in terms of the magnetic dipole moment of the proton (see (7.4.39), (7.4.40)), F involves a *factor* $1/c^2$ (see (7.4.31)). In (7.4.24), we have used the notation $\mathbf{p}^4$ for $(\mathbf{p}^2)^2$.

The following equalities are easily established (see Problem 7.12)

$$\boldsymbol{\sigma}\cdot\mathbf{p}(U-\varepsilon)\boldsymbol{\sigma}\cdot\mathbf{p}\chi = \mathbf{p}^2(U-\varepsilon)\chi + \hbar^2(\boldsymbol{\nabla}^2 U)\chi$$
$$+ i\hbar(\boldsymbol{\nabla}U)\cdot\mathbf{p}\chi + \hbar\boldsymbol{\sigma}\cdot[(\boldsymbol{\nabla}U)\times\mathbf{p}]\chi \quad (7.4.26)$$

$$(\varepsilon - U)\mathbf{p}^2\chi = \mathbf{p}^2(\varepsilon - U)\chi + \hbar^2(\boldsymbol{\nabla}^2(\varepsilon - U))\chi$$
$$+ 2i\hbar(\boldsymbol{\nabla}(\varepsilon - U))\cdot\mathbf{p}\chi \quad (7.4.27)$$

where $\mathbf{p} = -i\hbar\boldsymbol{\nabla}$, and note the positions of the various brackets in (7.4.26), (7.4.27).

7 Intricacies of the Hydrogen Atom

Upon multiplying the equality (7.4.27) by 1/2 and adding the resulting expression to the one in (7.4.26), we obtain, to the accuracy needed, for (7.4.24)

$$(\varepsilon - U)\chi = \left[\frac{\mathbf{p}^2}{2M} - \frac{\mathbf{p}^4}{16M^3c^2} + \frac{\mathbf{p}^2(U-\varepsilon)}{8M^2c^2} + \frac{\hbar^2}{8M^2c^2}(\nabla^2 U)\right.$$

$$\left. + \frac{1}{2M^2c^2}\left(\frac{1}{r}\frac{d}{dr}U\right)\mathbf{S}\cdot\mathbf{L} + F\right]\chi. \qquad (7.4.28)$$

Here $\mathbf{S} = \hbar\boldsymbol{\sigma}/2$, and we have used the relations

$$\mathbf{S}\cdot(\nabla U \times \mathbf{p}) = \left(\frac{1}{r}\frac{d}{dr}U\right)\mathbf{S}\cdot(\mathbf{x}\times\mathbf{p})$$

$$= \left(\frac{1}{r}\frac{d}{dr}U\right)\mathbf{S}\cdot\mathbf{L} \qquad (7.4.29)$$

with $\mathbf{L}$ denoting the orbital angular momentum. For an s-state for which the orbital angular momentum is zero, i.e., $\ell = 0$, it is advisable to use the expression on the extreme left-hand side of (7.4.29) and not of the first one on the right-hand side in the evaluations of the energy corrections.

Now we combine the second and the third terms on the right-hand side of (7.4.28) as follows:

$$-\frac{\mathbf{p}^2}{8M^2c^2}\left[\frac{\mathbf{p}^2}{2M} + (\varepsilon - U)\right]\chi = -\frac{\mathbf{p}^2}{8M^2c^2}\left[\frac{\mathbf{p}^2}{2M} + \frac{\mathbf{p}^2}{2M} + \mathcal{O}\left(\frac{1}{c^2}\right)\right]\chi \qquad (7.4.30)$$

where we have used, in the process, the equality (7.4.28) for $(\varepsilon - U)\chi$. That is, for the accuracy needed, the second plus the third terms on the right-hand side within the square brackets of (7.4.28) may be replaced by $-(\mathbf{p}^4/8M^3c^2)$.

Finally we use the operator identity

$$\boldsymbol{\sigma}\cdot\left(\mathbf{p}-\frac{e}{c}\mathbf{A}\right)\boldsymbol{\sigma}\cdot\left(\mathbf{p}-\frac{e}{c}\mathbf{A}\right) = \left(\mathbf{p}-\frac{e}{c}\mathbf{A}\right)^2 - \frac{e\hbar}{c}\boldsymbol{\sigma}\cdot(\nabla\times\mathbf{A}) \qquad (7.4.31)$$

(see Problem 7.13) and the properties that $\nabla\cdot\mathbf{A} = 0$, and that $\mathbf{A}$ involves a factor $1/c$ (see (7.4.39), (7.4.40)), to simplify the expression for F in (7.4.25), and rewrite (7.4.28) as

$$\left[\frac{\mathbf{p}^2}{2M} + U - \varepsilon - \frac{\mathbf{p}^4}{8M^3c^2} + \frac{\hbar^2}{8M^2c^2}(\nabla^2 U)\right.$$

$$\left. + \frac{1}{2M^2c^2}\left(\frac{1}{r}\frac{d}{dr}U\right)\mathbf{S}\cdot\mathbf{L} - \frac{e}{Mc}\mathbf{A}\cdot\mathbf{p} - \boldsymbol{\mu}\cdot\mathbf{B}\right]\chi = 0. \qquad (7.4.32)$$

Here

7.4 The Hydrogen Atom Including Spin and Relativistic Corrections

$$\boldsymbol{\mu} = \frac{e}{Mc}\mathbf{S} \tag{7.4.33}$$

is the magnetic dipole moment of the electron, $\mathbf{B} = \nabla \times \mathbf{A}$.

The magnetic dipole moment $\boldsymbol{\mu}$ in (7.4.33) is more precisely given by $\boldsymbol{\mu} = (ge/2Mc)\mathbf{S}$, where g is the so-called g-factor of the electron given approximately[8] by $g = 2\left(1 + \frac{\alpha}{2\pi}\right)$ according to quantum electrodynamics, due to Schwinger, and

$$\alpha = e^2/\hbar c \sim 1/137 \tag{7.4.34}$$

is the *fine-structure* constant. Since $\mathbf{B} = \nabla \times \mathbf{A}$ has already a factor $1/c$ (see (7.4.39), (7.4.40)) in it, one may simply take $g \simeq 2$ as given in (7.4.33), (7.4.32).[9]

We investigate the meanings of the new terms arising in the modified Schrödinger equation (7.4.32).

Since

$$\left(\mathbf{p}^2 c^2 + M^2 c^4\right)^{1/2} - Mc^2 - \frac{\mathbf{p}^2}{2M} = -\frac{\mathbf{p}^4}{8M^3 c^2} + \cdots \tag{7.4.35}$$

the $-\left(\mathbf{p}^4/8M^3 c^2\right)$ term gives the relativistic correction to the kinetic energy.

The interaction $\left(\hbar^2/8M^2 c^2\right)\nabla^2 U$ is non-vanishing only at the origin since

$$\nabla^2 \left(\frac{1}{|\mathbf{x}|}\right) = -4\pi\delta^3(\mathbf{x}) \tag{7.4.36}$$

(see Problem 7.14) implies that

$$\nabla^2 U(\mathbf{x}) = 4\pi e^2 \delta^3(\mathbf{x}) \tag{7.4.37}$$

and contributes only to s-states. It is referred to as the Darwin term. We will see in §16.6, that the physical origin of this may be explained to arise due to fluctuations of the position of the electron (referred to as "Zitterbewegung") in the atom, and the electron encounters a smeared-out Coulomb potential giving rise to the term $\nabla^2 U(\mathbf{x})$.

The term

$$\frac{1}{2M^2 c^2}\mathbf{S} \cdot [(\nabla U) \times \mathbf{p}] = \frac{1}{2M^2 c^2}\left(\frac{1}{r}\frac{d}{dr}U\right)\mathbf{S} \cdot \mathbf{L}$$

$$= -(-\nabla U) \cdot \left[\frac{1}{2M^2 c^2}\mathbf{p} \times \mathbf{S}\right] \tag{7.4.38}$$

---

[8] See §8.5 for a computation and observation of g.
[9] One may add a term $-(\kappa e/Mc)\mathbf{S} \cdot \mathbf{B}$ to the left-hand side of (7.4.32), with the latter as obtained from the Hamiltonian of the system (7.4.14), (7.4.15), giving rise to the term $-(ge/2Mc)\mathbf{S} \cdot \mathbf{B}$ instead of $-(e/Mc)\mathbf{S} \cdot \mathbf{B}$, with $g = 2(1+\kappa)$, $\kappa = \alpha/2\pi$. For the accuracy needed here, however, this is not essential.

is referred to as the *spin-orbit (SL) coupling*.[10] It may be interpreted as the interaction of an electric dipole moment $(\mathbf{p} \times \mathbf{S}/2M^2c^2)$, set up by the motion of the electron carrying a magnetic dipole moment $\boldsymbol{\mu}$ (or spin $\mathbf{S}$), with the electric field (up to the charge) $-\nabla U$ in the proton's rest frame where we will be carrying out the computations.[11]

The vector potential $\mathbf{A}$ due to the magnetic dipole moment $\boldsymbol{\mu}_\mathrm{p}$ of the proton may be written as

$$\mathbf{A} = -\boldsymbol{\mu}_\mathrm{p} \times \nabla\left(\frac{1}{r}\right) \tag{7.4.39}$$

and

$$\boldsymbol{\mu}_\mathrm{p} = \frac{|e|\, g_\mathrm{p} \mathbf{I}}{2M_\mathrm{p} c} \tag{7.4.40}$$

where g_p is the so-called g-factor of the proton approximately equal to 5.56, M_p is its mass and $\mathbf{I}$ its spin.

We have retained the last two terms in the modified Schrödinger equation (7.4.32) because they involve the factor $1/c^2$ in them. This will obviously save us some work. One should note, however, that as far as orders of magnitudes are concerned, since $M/M_\mathrm{p} \sim 0.545 \times 10^{-3}$, we have $|\boldsymbol{\mu}_\mathrm{p}| \sim 10^{-3}|\boldsymbol{\mu}|$. Accordingly, the energy corrections due to the combination of the last two terms in (7.4.32) are suppressed by a factor of 10^{-3} relative to the other corrections in (7.4.32). For this reason, and for the subsequent analyses, we separate the corresponding potentials and set

$$V_\mathrm{F} = -\frac{\mathbf{p}^4}{8M^3c^2} + \frac{\hbar^2}{8M^2c^2}(\nabla^2 U) + \frac{1}{2M^2c^2}\left(\frac{1}{r}\frac{\mathrm{d}U}{\mathrm{d}r}\right)\mathbf{S} \cdot \mathbf{L} \tag{7.4.41}$$

$$V_\mathrm{HF} = -\frac{e}{Mc}\mathbf{A} \cdot \mathbf{p} - \boldsymbol{\mu} \cdot \mathbf{B}. \tag{7.4.42}$$

Because of the suppression of the contribution of V_HF relative to that of V_F, we first treat V_F separately, and then consider V_HF as a small perturbation to the former. Every energy level resulting by including the so-called fine-structure correction due to V_F is then split due to V_HF.

The energy levels of the hydrogen atom obtained in §7.2 are of the order $\sim Mc^2\alpha^2$. The V_F interaction provides corrections to these levels which

---

[10] The 1/2 factor in the spin-orbit coupling in (7.4.38), referred to as the Thomas factor, was automatically obtained in (7.4.32) "without tears", which is otherwise obtained, by some labor, by considering successive Lorentz transformations corresponding to the motion of the electron as it changes its velocity in infinitesimal periods of time. This in turn implies that the spin of the electron precesses with angular frequency involving a factor $1/c^2$, to the leading order $(v/c)^2$.

[11] This may be equivalently interpreted as the interaction of the magnetic dipole moment $\boldsymbol{\mu}$ of the electron with a magnetic field set up by the proton, not however due to $\boldsymbol{\mu}_\mathrm{p}$, in the electron's rest frame.

7.4 The Hydrogen Atom Including Spin and Relativistic Corrections

are suppressed by a factor α^2, i.e., are of the order $Mc^2\alpha^4$, and is said to provide the *fine-structure* of the atom as mentioned above. For low lying states $n = 1$, $n = 2$, these corrections are about $\sim 1.8 \times 10^{-4}$ eV, $\sim 5 \times 10^{-5}$ eV, respectively, depending on the total angular momentum of the electron. On the other hand, because of the further suppression by a factor (M/M_p), discussed above, the corrections due to V_HF are of the order $\sim (M/M_\mathrm{p})g_\mathrm{p} Mc^2\alpha^4$, where g_p is the g-factor of the proton, and the V_HF interaction is said to provide the *hyperfine-structure* of the atom. We note that $(M/M_\mathrm{p}) = .545 \times 10^{-3} \simeq 10.2\,\alpha^2$. For the low lying state $n = 1$, the hyperfine corrections are about $\sim 1.46 \times 10^{-6}$ eV, $\sim -4.39 \times 10^{-6}$ eV for the spins of the electron and proton aligned or opposed, respectively.

The expression for the energy levels for the hydrogen atom in (7.2.19) may be rewritten as

$$-\left(\frac{MM_\mathrm{p}}{M + M_\mathrm{p}}\right)\frac{c^2\alpha^2}{2n^2} \simeq \left(1 - \frac{M}{M_\mathrm{p}}\right)\left(-\frac{Mc^2\alpha^2}{2n^2}\right)$$

$$\simeq -\frac{Mc^2\alpha^2}{2n^2} + 10.2\frac{Mc^2\alpha^4}{2n^2}. \quad (7.4.43)$$

That is, the reduced mass gives rise to a correction to the levels, when initially determined in the infinite proton mass limit, by multiplying these levels by the corrective factor $(1 - M/M_\mathrm{p})$.

Since the interaction V_F dominates over V_HF, the question on the contribution of the reduced mass effect on the fine-structure arises. The corresponding analysis is not as straightforward as one might think as it is involved with a relativistic treatment of the electron and the proton as well. The same rule as above approximately holds again and one may multiply the fine-structure splittings by the corrective factor $(1 - M/M_\mathrm{p})$, but also an additional correction arises which is the same for a fixed value of the principal quantum number n of the order $\sim 10^{-7}$ eV or less, and will not contribute, or more precisely gives negligible contributions, to the computations of energy differences between levels with the same n. On the other hand, the correction $(M/M_\mathrm{p}) \times$fine-structure is again of the order $\sim 10^{-7}$ eV or less. Accordingly, we will not dwell on the reduced mass effect *in* the fine-structure, as well as in the hyperfine-structure in the sequel.

The interaction of the electron with radiation (the photon) produces a shift, the so-called Lamb shift, of the order $\sim Mc^2\alpha^5 \ln(\alpha)$ which for the low lying state $n = 2$, $\ell = 0$ is about $\sim 4.3 \times 10^{-6}$ eV.

The fine-structure of the hydrogen atom is studied in §7.5. Every energy level resulting by the inclusion of the fine-structure correction is also split further due V_HF, providing the hyperfine structure of the atom. This is treated in §7.6. The Lamb shift, in a non-relativistic setting for the atomic electron, is worked out in §7.7.

Appendix to §7.4: Normalization of the Wavefunction Including Spin and Relativistic Corrections

Upon introducing the 4×4 matrices[12]

$$\gamma^0 = \begin{pmatrix} I & 0 \\ 0 & -I \end{pmatrix}, \quad \gamma = \begin{pmatrix} 0 & \sigma \\ -\sigma & 0 \end{pmatrix} \tag{A-7.4.1}$$

with properties,

$$(\gamma^0)^\dagger = \gamma^0, \quad \gamma^\dagger = -\gamma, \quad \gamma^0 \gamma = -\gamma \gamma^0 \tag{A-7.4.2}$$

$$(\gamma^0)^2 = \begin{pmatrix} I & 0 \\ 0 & I \end{pmatrix} \tag{A-7.4.3}$$

writing $i\hbar \partial/\partial t$ for E, and introducing the interaction of the electron with an external electromagnetic potential $U(\mathbf{x}, t)$, $\mathbf{A}(\mathbf{x}, t)$

$$i\hbar \frac{\partial}{\partial t} \rightarrow i\hbar \frac{\partial}{\partial t} - eU(\mathbf{x}, t) \tag{A-7.4.4}$$

$$\mathbf{p} \rightarrow \mathbf{p} - \frac{e}{c} \mathbf{A}(\mathbf{x}, t) \tag{A-7.4.5}$$

providing the so-called minimal electromagnetic coupling, the time-dependent Dirac equation (7.4.10) becomes

$$\frac{\gamma^0 i\hbar}{c} \frac{\partial}{\partial t} \begin{bmatrix} \Psi \\ \Phi \end{bmatrix} = \left[\gamma \cdot \left(-i\hbar \nabla - \frac{e}{c} \mathbf{A} \right) + McI + \frac{eU}{c} \gamma^0 \right] \begin{bmatrix} \Psi \\ \Phi \end{bmatrix} \tag{A-7.4.6}$$

where I, in the latter, is the 4×4 identity matrix.

The adjoint of the matrix equation (A-7.4.6) is

$$\frac{-i\hbar}{c} \frac{\partial}{\partial t} \begin{bmatrix} \Psi^\dagger & \Phi^\dagger \end{bmatrix} \gamma^0 = \begin{bmatrix} \Psi^\dagger & \Phi^\dagger \end{bmatrix} \left[-\left(i\hbar \overleftarrow{\nabla} - \frac{e}{c} \mathbf{A} \right) \cdot \gamma + McI + \frac{eU}{c} \gamma^0 \right] \tag{A-7.4.7}$$

where we have used (A-7.4.2), and the arrow on $\overleftarrow{\nabla}$ means that the gradient operates to the left.

Upon multiplying (A-7.4.6) from the left by $-(ic/\hbar) \begin{bmatrix} \Psi^\dagger & \Phi^\dagger \end{bmatrix} \gamma^0$, and (A-7.4.7) from the right by $(ic/\hbar) \gamma^0 \begin{bmatrix} \Psi & \Phi \end{bmatrix}^T$, and adding the resulting equations, we obtain

$$\frac{\partial}{\partial t} \left(\Psi^\dagger \Psi + \Phi^\dagger \Phi \right) + \nabla \cdot \mathbf{J} = 0 \tag{A-7.4.8}$$

---

[12] The Dirac equation will be studied in detail later. Here we introduce the minimum to discuss just the normalization of the wavefunction Ψ in (7.4.19), (7.4.22), (7.4.32) to the accuracy needed there.

where
$$\mathbf{J} = c \begin{bmatrix} \Psi^\dagger & \Phi^\dagger \end{bmatrix} \gamma^0 \boldsymbol{\gamma} \begin{bmatrix} \Psi \\ \Phi \end{bmatrix} \quad \text{(A-7.4.9)}$$

is the probability current density.

The normalization condition becomes
$$\int d^3\mathbf{x} \left(\Psi^\dagger \Psi + \Phi^\dagger \Phi \right) = 1. \quad \text{(A-7.4.10)}$$

From (7.4.18),
$$\Phi = \frac{1}{2Mc} \boldsymbol{\sigma} \cdot \mathbf{p} \Psi + \mathcal{O}\left(\frac{1}{c^2}\right) \quad \text{(A-7.4.11)}$$

which from (A-7.4.10) implies, by integrating by parts, that
$$\int d^3\mathbf{x} \, \Psi^\dagger \left(1 + \frac{\mathbf{p}^2}{4M^2c^2} \right) \Psi = 1 \quad \text{(A-7.4.12)}$$

to the accuracy needed in §7.4, where we have used the facts that $\boldsymbol{\sigma}^\dagger = \boldsymbol{\sigma}$ and $(\boldsymbol{\sigma} \cdot \mathbf{p})(\boldsymbol{\sigma} \cdot \mathbf{p}) = \mathbf{p}^2$. Hence, to the accuracy sought, we may write
$$\Psi = \left(1 - \frac{\mathbf{p}^2}{8M^2c^2} \right) \chi \quad \text{(A-7.4.13)}$$

yielding to the normalization condition
$$\int d^3\mathbf{x} \, \chi^\dagger \chi = 1. \quad \text{(A-7.4.14)}$$

7.5 The Fine-Structure of the Hydrogen Atom

According to (7.4.32), (7.4.41) the fine-structure of the hydrogen atom is described by the Hamiltonian
$$H = \frac{\mathbf{p}^2}{2M} + U(\mathbf{x}) + V_{\mathrm{F}}(\mathbf{x}) \quad (7.5.1)$$

where
$$V_{\mathrm{F}}(\mathbf{x}) = -\frac{(E_n - U(\mathbf{x}))^2}{2Mc^2} + \frac{\hbar^2}{8M^2c^2} (\boldsymbol{\nabla}^2 U(\mathbf{x})) + \frac{1}{2M^2c^2}\left(\frac{1}{r}\frac{dU}{dr}\right)\mathbf{S}\cdot\mathbf{L} \quad (7.5.2)$$

with $V_{\mathrm{F}}(\mathbf{x})$ treated as a small perturbation, where, using the same reasoning as in (7.4.30), we have replaced $(-\mathbf{p}^4/8M^3c^2)$ effectively by $-(E_n - U)^2/2Mc^2$ in (7.5.2) to the leading order.

We note that H does not commute with $\mathbf{L}$ and $\mathbf{S}$. It commutes, however, with $\mathbf{J}^2, \mathbf{L}^2, \mathbf{S}^2, J_z$, where $\mathbf{J} = \mathbf{L}+\mathbf{S}$ is the angular momentum of the electron. That H commutes with $\mathbf{J}^2$ follows from the fact that

380   7 Intricacies of the Hydrogen Atom

$$\mathbf{J}^2 = \mathbf{L}^2 + \mathbf{S}^2 + 2\mathbf{S} \cdot \mathbf{L} \tag{7.5.3}$$

and that every term on the right-hand side of (7.5.3) does. Since $[\mathbf{J}^2, J_z] = 0$, we also have the easily verified property $[\mathbf{S} \cdot \mathbf{L}, J_z] = 0$ which is a statement of rotational invariance of the scalar product (about the z-axis). For a radial function $U(r)$, $\nabla^2 U(r)$ is also a radial function and hence commutes with $\mathbf{L}$. Actually, $\nabla^2 U(r)$ is proportional to $\delta^3(\mathbf{x})$ (see (7.4.36)), and $\delta^3(\mathbf{x})$ is essentially $\delta(r)/4\pi r^2$ when multiplied by a function $f(r)$ of r in an integral.

We are thus led to consider the wavefunctions

$$\psi_{j,\ell,m_j,n}(r,\theta,\phi) = \langle \theta,\phi | j,\ell,m_j \rangle R_{n\ell}(r) \tag{7.5.4}$$

where $R_{n\ell}(r)$ are the radial wavefunction in (7.3.14) with $\mu \to M$. Also $|j,\ell,m_j\rangle$ are expressed in terms of the states $|\ell,m\rangle |s,m_s\rangle$ via Clebsch-Gordan coefficients:

$$|j,\ell,m_j\rangle = \sum_{m+m_s=m_j} |\ell,m\rangle |s,m_s\rangle \langle \ell,m,s,m_s | j,\ell,m_j \rangle \tag{7.5.5}$$

and are given in the appendix to this section, the details of which, however, are not necessary here. The quantum number $s = 1/2$, and for $\ell = 0$, $j = 1/2$ and for $\ell \neq 0$, $j = \ell \pm 1/2$.

Since $|j,\ell,m_j\rangle$ is a linear combination of the states $|\ell,m\rangle$, for different m, the wavefunctions in (7.5.4) are eigenvectors of the hydrogen atom problem with eigenvalues $-Me^4/2\hbar^2 n^2$.

Undoubtedly, a reader of quantum physics presented at this level is familiar with the elements of perturbation theory[13] and that the leading correction to energy levels due to V_F is given by the expectation value of V_F in the unperturbed states in (7.5.4) i.e., by the application of first order perturbation theory.[14] We note that for a given n there are exactly $2n^2$ $|j,\ell,m_j\rangle$ states. To see this, note that for $n = 1$, we have $\ell = 0$, $j = 1/2$, $m_j = \pm 1/2$, and hence we obtain two states. For $n \neq 1$, $\ell = 0$, $j = 1/2$, $m_j = \pm 1/2$, we obtain two states which we have to *add* to the states $\ell \neq 0$, $j = \ell \pm 1/2$, $m_j = -j, \ldots, j$. The total number of the latter states is

$$\sum_{\ell=1}^{n-1}\left[2\left(\ell + \frac{1}{2}\right)+1\right] + \sum_{\ell=1}^{n-1}\left[2\left(\ell - \frac{1}{2}\right)+1\right] = 2n^2 - 2 \tag{7.5.6}$$

thus confirming the above result.

We first consider s-states, i.e., for which $\ell = 0$.

---

[13] Perturbation theory will be systematically developed later on in Chapter 12.
[14] Here we remark that since H commutes with $\mathbf{J}^2, \mathbf{L}^2, \mathbf{S}^2, J_z$, the perturbation V_F is *diagonal* in the states $|j,\ell,m_j\rangle$ and so-called degenerate perturbation theory, to the leading order, reduces to elementary first order perturbation theory.

$\ell = 0$ States:

For the $\ell = 0$ states $m_j = m_s$, and we may write for (7.5.4)

$$\psi_{1/2,0,m_s,n}(r,\theta,\phi) = \frac{1}{\sqrt{4\pi}} R_{n0}(r) |1/2, m_s\rangle. \tag{7.5.7}$$

From (7.3.29), (7.3.36), (7.3.37), with $\mu \to M$, $Z \to 1$, we have

$$-\left\langle 1/2, 0, m_s, n \left| \frac{(E_n - U)^2}{2Mc^2} \right| 1/2, 0, m_s, n \right\rangle = -\frac{\alpha^2}{n} \left(\frac{Me^4}{2\hbar^2 n^2} \right) \left[2 - \frac{3}{4n} \right] \tag{7.5.8}$$

to the leading order.

For the Darwin term $(\hbar^2/8M^2c^2)(\nabla^2 U)$ we have from (7.4.37), (7.3.32),

$$\left\langle 1/2, 0, m_s, n \left| \frac{\hbar^2}{8M^2c^2} (\nabla^2 U) \right| 1/2, 0, m_s, n \right\rangle = \frac{\alpha^2}{n} \left(\frac{Me^4}{2\hbar^2 n^2} \right). \tag{7.5.9}$$

Finally an $\ell = 0$ state corresponding to no angular momentum, for the last term in (7.5.2), we may use the explicit expression given on the left-hand side of (7.4.38), i.e.,

$$\frac{1}{2M^2c^2} \mathbf{S} \cdot [(\nabla U) \times \mathbf{p}] = \frac{-i\hbar \varepsilon^{ijk}}{2M^2c^2} S^i \left(\frac{\partial}{\partial x^j} U \right) \frac{\partial}{\partial x^k} \tag{7.5.10}$$

with a summation over repeated indices understood.

The expectation value of (7.5.10) in state $|1/2, 0, m_s, n\rangle$ may be explicitly carried out and there is no need to "guess" on what its value is as is often made in the literature. To this end, by using

$$\frac{\partial}{\partial x^j} U(r) = \frac{x^j}{r} \frac{d}{dr} U(r)$$

$$= e^2 \frac{x^j}{r^3} \tag{7.5.11}$$

$$\frac{\partial}{\partial x^k} R_{n0}(r) = \frac{x^k}{r} \frac{\partial}{\partial r} R_{n0}(r) \tag{7.5.12}$$

and that

$$\int d\Omega \, \frac{x^j x^k}{r^2} = \frac{4\pi}{3} \delta^{jk} \tag{7.5.13}$$

we obtain

$$\left\langle 1/2, 0, m_s, n \left| -i\hbar \varepsilon^{ijk} S^i \left(\frac{\partial}{\partial x^j} U \right) \frac{\partial}{\partial x^k} \right| 1/2, 0, m_s, n \right\rangle$$

382   7 Intricacies of the Hydrogen Atom

$$= -i\hbar\varepsilon^{ijk} \langle 1/2, 0|S^i|1/2, 0\rangle \frac{1}{4\pi} \int d^3\mathbf{x}\, R_{n0}(r) \left(\frac{\partial}{\partial x^j} U\right) \frac{\partial}{\partial x^k} R_{n0}(r)$$

$$= -i\hbar e^2 \varepsilon^{ijk} \langle 1/2, 0|S^i|1/2, 0\rangle \frac{\delta^{jk}}{3} \frac{1}{2} \int_0^\infty \frac{r^2\, dr}{r^2} \frac{d}{dr} R_{n0}^2(r)$$

$$= \frac{i\hbar e^2 \varepsilon^{ijk}}{6} \langle 1/2, 0|S^i|1/2, 0\rangle \delta^{jk} \frac{4}{n^3} \left(\frac{Me^2}{\hbar^2}\right)^3 = 0 \qquad (7.5.14)$$

since $\varepsilon^{ijk}\delta^{jk} = 0$ and the latter is multiplied by a *finite* constant. We have also used (7.3.31), and the fact that $R_{n0}^2(r) \to 0$ for $r \to \infty$.

Unfortunately,[15] many books have the Darwin term "missing" in the corresponding expression to (7.5.2) and then use (7.5.3) to infer that the expectation value of the last term in (7.5.2) is finite and non-zero for $\ell = 0$ so that the first plus the third terms alone in (7.5.2) give the same expression as the one provided from a leading order expansion of the exact Dirac expression (see (16.6.18), (16.6.55)). The integral in question, as seen in (7.5.14), is zero and the fine-structure for $\ell = 0$ is provided by $-\frac{(E_n-U)^2}{2Mc^2}$ plus the Darwin term $(\hbar^2/8M^2c^2)(\nabla^2 U)$. The fine-structure correction to the hydrogen atom energy levels is, for $\ell = 0$, $j = 1/2$, then given from (7.5.8) and (7.5.9) to be

$$\Delta E_{n,j,\ell} = -\frac{\alpha^2}{n}\left(\frac{\alpha^2 Mc^2}{2n^2}\right)\left[1 - \frac{3}{4n}\right], \qquad \ell = 0. \qquad (7.5.15)$$

$\ell \neq 0$ States:

For $\ell \neq 0$, the Darwin term in (7.5.2) gives a zero contribution from the property (7.3.32) and (7.4.37). On the other hand as in (7.5.8), we have from (7.3.29), (7.3.36), (7.3.37)

$$-\left\langle j, \ell, m_j, n \left| \frac{(E_n-U)^2}{2Mc^2} \right| j, \ell, m_j, n \right\rangle = -\frac{\alpha^2}{n}\left(\frac{Me^4}{2\hbar^2 n^2}\right)\left[\frac{1}{\ell+1/2} - \frac{3}{4n}\right] \qquad (7.5.16)$$

for $\mu \to M$.

Also from (7.5.3), we may solve for $\mathbf{S}\cdot\mathbf{L}$

$$\mathbf{S}\cdot\mathbf{L} = \frac{1}{2}\left(\mathbf{J}^2 - \mathbf{L}^2 - \mathbf{S}^2\right) \qquad (7.5.17)$$

use (7.3.38) for $\ell \neq 0$, to obtain

$$\left\langle j, \ell, m_j, n \left| \frac{1}{2M^2c^2}\left(\frac{1}{r}\frac{d}{dr}U\right)\mathbf{S}\cdot\mathbf{L} \right| j, \ell, m_j, n \right\rangle$$

---

[15] Fortunately, there are some exceptions to these books emphasizing that the integral in question *is* zero, e.g., Bethe and Salpeter (1977), p. 60. A regularization was, however, used in their analysis to conclude that this integral is zero.

$$= \frac{\alpha^2}{2n} \left(\frac{Me^4}{2\hbar^2 n^2} \right) \frac{[j(j+1) - \ell(\ell+1) - 3/4]}{\ell(\ell+1/2)(\ell+1)} \quad (7.5.18)$$

where $j = \ell \pm 1/2$.

From (7.5.16), (7.5.18) we have

$$\Delta E_{n,j,\ell} = -\frac{\alpha^2}{n} \left(\frac{\alpha^2 Mc^2}{2n^2} \right) \left[\frac{1}{j+1/2} - \frac{3}{4n} \right]. \quad (7.5.19)$$

Upon the comparison of (7.5.19) with (7.5.15), we note that the former coincides with the latter if we formally set $\ell = 0$, $j = 1/2$ in the former. That is, (7.5.19) holds for $j = \ell - 1/2$, $\ell = 1, \ldots, n-1$, ($n \neq 1$); $j = \ell + 1/2$ for $\ell = 0, 1, \ldots, n-1$ as well.

Using the notation $n\ell_j$, we see from (7.5.19), for example, that the levels $2P_{1/2}/2P_{3/2}$ are split as well as the levels $3P_{1/2}/3P_{3/2}$ and $3D_{3/2}/3D_{5/2}$, where P, D correspond to $\ell = 1, 2$, respectively, The shift between $2P_{3/2}$ and $2P_{1/2}$ is about 10.95 GHz in frequency unit.

On the other hand for fixed n, j the fine-structure correction (7.5.19) shows that corresponding energy levels which differ by the ℓ-value are degenerate. For example, the levels $2S_{1/2}/2P_{1/2}$, with S corresponding to $\ell = 0$, are degenerate, while experimentally it has been confirmed not to be so. This splitting is, however, quite elegantly predicted by quantum electrodynamics and is referred to as the Lamb shift. The latter is reduced by a factor of α relative to the fine-structure, and also involves α in a non-trivial way (such as on the logarithm of α). A non-relativistic derivation of the Lamb shift, i.e., with the atomic electron treated non-relativistically, will be given in §7.7.

Appendix to §7.5: Combining Spin and Angular Momentum in the Atom

In (7.5.5), we combined the spin of the electron with the angular momentum. The states $|j, \ell, m_j\rangle$ in (7.5.5) may be written down directly from Table 5.3, below (5.5.35), by making the substitutions in the latter:

$$j_1 \to \ell, \quad m_1 \to m, \quad m_2 \to m_s, \quad m \to m_j. \quad (A\text{-}7.5.1)$$

Hence the states $|j, \ell, m_j\rangle$ are explicitly given by ($\ell \neq 0$):

$$|j, \ell, m_j\rangle = \frac{1}{\sqrt{(2\ell+1)}} \begin{bmatrix} |\ell, m_j - 1/2\rangle \sqrt{\ell + 1/2 + m_j} \\ |\ell, m_j + 1/2\rangle \sqrt{\ell + 1/2 - m_j} \end{bmatrix} \quad (A\text{-}7.5.2)$$

for $j = \ell + 1/2$,

$$|j, \ell, m_j\rangle = \frac{1}{\sqrt{(2\ell+1)}} \begin{bmatrix} -|\ell, m_j - 1/2\rangle \sqrt{\ell + 1/2 - m_j} \\ |\ell, m_j + 1/2\rangle \sqrt{\ell + 1/2 + m_j} \end{bmatrix} \quad (A\text{-}7.5.3)$$

for $j = \ell - 1/2$.

Accordingly, we may write

$$\langle \theta, \phi | j, \ell, m_j \rangle = \frac{\sqrt{\ell + 1/2 + m_j}}{\sqrt{(2\ell + 1)}} Y_{\ell, m_j - 1/2}(\theta, \phi) \begin{pmatrix} 1 \\ 0 \end{pmatrix}$$

$$+ \frac{\sqrt{\ell + 1/2 - m_j}}{\sqrt{(2\ell + 1)}} Y_{\ell, m_j + 1/2}(\theta, \phi) \begin{pmatrix} 0 \\ 1 \end{pmatrix} \quad \text{(A-7.5.4)}$$

for $j = \ell + 1/2$,

$$\langle \theta, \phi | j, \ell, m_j \rangle = - \frac{\sqrt{\ell + 1/2 - m_j}}{\sqrt{(2\ell + 1)}} Y_{\ell, m_j - 1/2}(\theta, \phi) \begin{pmatrix} 1 \\ 0 \end{pmatrix}$$

$$+ \frac{\sqrt{\ell + 1/2 + m_j}}{\sqrt{(2\ell + 1)}} Y_{\ell, m_j + 1/2}(\theta, \phi) \begin{pmatrix} 0 \\ 1 \end{pmatrix} \quad \text{(A-7.5.5)}$$

for $j = \ell - 1/2$.

A relation which follows from (A-7.5.4), (A-7.5.5) and will be useful later in Appendix to §16.6 in an exact treatment of the Dirac equation in the bound Coulomb problem is the following. Let

$$\mathbf{n} = \frac{\mathbf{x}}{|\mathbf{x}|} \equiv \hat{\mathbf{x}} = (\sin\theta \cos\phi, \sin\theta \sin\phi, \cos\theta) \quad \text{(A-7.5.6)}$$

then

$$\boldsymbol{\sigma} \cdot \hat{\mathbf{x}} \, \langle \theta, \phi | j, \ell = j \mp 1/2, m_j \rangle = - \langle \theta, \phi | j, \ell = j \pm 1/2, m_j \rangle. \quad \text{(A-7.5.7)}$$

The spin $\mathbf{I}$ of the proton may also added to $\mathbf{J}$,

$$\mathbf{F} = \mathbf{I} + \mathbf{J} \quad \text{(A-7.5.8)}$$

as done in (7.6.4) in the next section and the eigenvectors $|f, j, \ell, m_f\rangle$ of the commuting operators $\mathbf{F}^2$, $\mathbf{J}^2$, $\mathbf{S}^2$, $\mathbf{I}^2$, $\mathbf{L}^2$, F_z, (see Problem 7.17), where $\hbar^2 f(f+1)$, and $\hbar m_f$ are the eigenvalues of $\mathbf{F}^2$ and F_z, respectively, are similarly constructed (see Problem 7.18, see also §5.9).

7.6 The Hyperfine-Structure of the Hydrogen Atom

The hyperfine-structure of the hydrogen atom is provided by the interaction term given in (7.4.42):

$$V_{\text{HF}} = -\frac{e}{Mc} \mathbf{A} \cdot \mathbf{p} - \boldsymbol{\mu} \cdot \mathbf{B} \quad (7.6.1)$$

where $\mathbf{A}$ is given by

7.6 The Hyperfine-Structure of the Hydrogen Atom

$$\mathbf{A} = -\boldsymbol{\mu}_\mathrm{p} \times \nabla \left(\frac{1}{r}\right) \qquad (7.6.2)$$

and $\mathbf{B} = \nabla \times \mathbf{A}$. The magnetic dipole moment $\boldsymbol{\mu}_\mathrm{p}$ of the proton is defined in (7.4.40). The first term on the right-hand side of (7.6.1) may be written in various forms:

$$-\frac{e}{Mc}\mathbf{A}\cdot\mathbf{p} = \frac{i e^2 \hbar}{2 M M_\mathrm{p} c^2} g_\mathrm{p} \varepsilon^{ijk} I^i \left(\frac{\partial}{\partial x^j}\frac{1}{r}\right)\frac{\partial}{\partial x^k}$$

$$= \frac{e^2}{2 M M_\mathrm{p} c^2} g_\mathrm{p} \frac{1}{r^3}\mathbf{I}\cdot\mathbf{L}$$

$$= \frac{|e|}{Mc}\frac{1}{r^3}\boldsymbol{\mu}_\mathrm{p}\cdot\mathbf{L} \qquad (7.6.3)$$

representing an interaction between the magnetic dipole moment $\boldsymbol{\mu}_\mathrm{p}$ of the proton and the orbital angular momentum L of the electron in the atom. The $-\boldsymbol{\mu}\cdot\mathbf{B}$ term denotes the familiar interaction between the magnetic dipole moment $\boldsymbol{\mu}$ of the electron and the magnetic field set up by the proton due to $\boldsymbol{\mu}_\mathrm{p}$.

We add the spin $\mathbf{I}$ of the proton to $\mathbf{J}$ and define

$$\mathbf{F} = \mathbf{J} + \mathbf{I} \qquad (7.6.4)$$

and note that $\mathbf{F}^2$, $\mathbf{J}^2$, $\mathbf{S}^2$, $\mathbf{I}^2$, $\mathbf{L}^2$, F_z commute in pairs (see Problem 7.17). The following operator equality is useful

$$\mathbf{I}\cdot\mathbf{J} = \frac{1}{2}\left[\mathbf{F}^2 - \mathbf{I}^2 - \mathbf{J}^2\right] \qquad (7.6.5)$$

and will be used in our analysis of the hyperfine-structure.

For a given fixed pair (n,ℓ), for $\ell \neq 0$, the splitting due to the fine-structure between successive levels $j = \ell - 1/2$, $j = \ell + 1/2$ is given by (see (7.5.19))

$$\delta\left(\Delta E_{n,j,\ell}\right) = \frac{\alpha^2}{n}\left(\frac{\alpha^2 M c^2}{2 n^2}\right)\frac{1}{\ell(\ell+1)} \qquad (7.6.6)$$

and is larger by a factor of (M_p/M) relative to the hyperfine-structure effect as discussed below (7.4.42). Also for a fixed n, $\ell = 0$, we have only one j value $= 1/2$, with the fine-structure correction given in (7.5.15). Finally the differences between the energy levels in (7.2.19) for successive n's are much larger than the energy splitting due to the fine-structure and the hyperfine-structure for the n-value in question. Accordingly, we define the Hamiltonian as obtained from (7.4.32) for a fixed triplet (n,j,ℓ):

$$H_{nj\ell} = H^0_{nj\ell} + H^1_{nj\ell} \qquad (7.6.7)$$

where

$$H^0_{nj\ell} = \sum_{m_j, m'_j} |n, j, \ell, m_j\rangle \langle n, j, \ell, m_j | H^0 | n, j, \ell, m'_j\rangle \langle n, j, \ell, m'_j| \quad (7.6.8)$$

$$H^0 = \frac{\mathbf{p}^2}{2M} - \frac{e^2}{r} + V_{\mathrm{F}} \quad (7.6.9)$$

$$\langle n, j, \ell, m_j | H^0 | n, j, \ell, m'_j\rangle = \delta_{m_j, m'_j} E_{n, j, \ell} \quad (7.6.10)$$

with $E_{n,j,\ell}$ denoting the energy levels of the hydrogen atom (with $\mu \to M$) *including* the fine-structure correction in (7.5.19). Also

$$H^1_{nj\ell} = \sum_{m_j, m'_j} |n, j, \ell, m_j\rangle \langle n, j, \ell, m_j | V_{\mathrm{HF}} | n, j, \ell, m'_j\rangle \langle n, j, \ell, m'_j| \quad (7.6.11)$$

$$\langle \mathbf{x} | n, j, \ell, m_j \rangle = \psi_{j, \ell, m_j, n}(r, \theta, \phi) \quad (7.6.12)$$

with the latter defined in (7.5.4).

One should note that $\langle n, j, \ell, m_j | V_{\mathrm{HF}} | n, j, \ell, m'_j\rangle$ is an operator depending on $\mathbf{I}$ — the spin of the proton (see (7.6.1)–(7.6.3), (7.4.40)).

Now we consider the states

$$|n, f, j, \ell, m_f\rangle = \sum_{(m_i + m_j = m_f)} |n, j, \ell, m_j\rangle |i, m_i\rangle \langle n, j, \ell, m_j, i, m_i | n, f, j, \ell, m_f\rangle$$

$$(7.6.13)$$

where

$$\langle \mathbf{x} | n, f, j, \ell, m_f\rangle = \langle \theta, \phi | f, j, \ell, m_f\rangle R_{n\ell}(r) \quad (7.6.14)$$

and $\hbar^2 f(f+1)$, $\hbar m_f$, $\hbar^2 i(i+1)$, $\hbar m_i$ are the eigenvalues of $\mathbf{F}^2$, F_z, $\mathbf{I}^2$, I_z. The states $|f, j, \ell, m_f\rangle$ are readily constructed out of the states $|j, \ell, m_j\rangle$ given the Appendix to §7.5 and of the states $|i, m_i\rangle$ (see Problem 7.18), the details of which, however, are not necessary here. The important thing to note is that they are linear combinations of the states $|j, \ell, m_j\rangle |i, m_i\rangle$, and note that the states $|n, f, j, \ell, m_f\rangle$ are eigenvectors of $H^0_{nj\ell}$ with eigenvalues $E_{n,j,\ell}$ as given in (7.6.10).

$\ell = 0$ States:

We consider the states $|n, f, 1/2, 0, m_f\rangle \equiv |n, f, m_f\rangle$ in (7.6.13) for $\ell = 0$, where $f = 1, 0$.

Using the first equality in (7.6.3), we have for the matrix elements

$$\langle n, f', m'_f | -\frac{e}{Mc} \mathbf{A} \cdot \mathbf{p} | n, f, m_f\rangle = \frac{i e^2 \hbar}{2 M M_{\mathrm{p}} c^2} g_{\mathrm{p}} \langle n, f', m'_f | I^i | n, f, m_f\rangle$$

$$\times \frac{\varepsilon^{ijk}}{4\pi} \int d^3 \mathbf{x}\, R_{n0}(r) \left(\frac{\partial}{\partial x^j} \frac{1}{r}\right) \frac{\partial}{\partial x^k} R_{n0}(r)$$

7.6 The Hyperfine-Structure of the Hydrogen Atom

$$= 0 \quad (7.6.15)$$

directly from the identical evaluation of the integral in (7.5.14).

The second term on the right-hand side of (7.6.1) may be conveniently written as

$$-\boldsymbol{\mu} \cdot \mathbf{B} = -\mu^i \varepsilon^{ijk} \frac{\partial}{\partial x^j} A^k$$

$$= \left(\mu^i \mu_{\mathrm{p}}^\ell\right) \varepsilon^{kij} \varepsilon^{k\ell m} \frac{\partial}{\partial x^j} \frac{\partial}{\partial x^m} \left(\frac{1}{r}\right)$$

$$= \boldsymbol{\mu} \cdot \boldsymbol{\mu}_{\mathrm{p}} \boldsymbol{\nabla}^2 \left(\frac{1}{r}\right) - \mu^i \mu_{\mathrm{p}}^j \left(\frac{\partial}{\partial x^i} \frac{\partial}{\partial x^j} \frac{1}{r}\right) \quad (7.6.16)$$

with a summation over repeated indices understood. Upon using the definitions (7.4.33), (7.4.40) of $\boldsymbol{\mu}$, $\boldsymbol{\mu}_{\mathrm{p}}$, respectively, we obtain ($e < 0$)

$$\langle n, f', m'_f | -\boldsymbol{\mu} \cdot \mathbf{B} | n, f, m_f \rangle = -\frac{e^2}{2MM_{\mathrm{p}}c^2} g_{\mathrm{p}} \langle n, f', m'_f | S^i I^j | n, f, m_f \rangle$$

$$\times \frac{1}{4\pi} \int d^3\mathbf{x}\, R_{n0}^2(r) \left[\delta^{ij} \boldsymbol{\nabla}^2 \left(\frac{1}{r}\right) - \frac{\partial}{\partial x^i} \frac{\partial}{\partial x^j} \left(\frac{1}{r}\right) \right]$$

$$= -\frac{e^2}{2MM_{\mathrm{p}}c^2} g_{\mathrm{p}} \langle n, f', m'_f | S^i I^j | n, f, m_f \rangle \frac{\delta^{ij}}{4\pi} \frac{2}{3} \int d^3\mathbf{x}\, R_{n0}^2(r) \boldsymbol{\nabla}^2 \left(\frac{1}{r}\right) \quad (7.6.17)$$

where we have used the spherical symmetry of $R_{n0}^2(r)$. From (7.4.36), (7.3.31), with $Z \to 1$, $\mu \to M$, we obtain

$$\langle n, f', m'_f | -\boldsymbol{\mu} \cdot \mathbf{B} | n, f, m_f \rangle = \frac{4}{3n^3} \frac{e^2}{MM_{\mathrm{p}}c^2} g_{\mathrm{p}} \left(\frac{Me^2}{\hbar^2}\right)^3$$

$$\times \langle n, f', m'_f | \mathbf{S} \cdot \mathbf{I} | n, f, m_f \rangle. \quad (7.6.18)$$

From (7.6.5), or simply by the addition of two spin 1/2's, we have

$$\langle n, f', m'_f | \mathbf{S} \cdot \mathbf{I} | n, f, m_f \rangle = \delta_{ff'} \delta_{m'_f m_f} \frac{\hbar^2}{2} \left[f(f+1) - \frac{3}{2} \right] \quad (7.6.19)$$

with $f = 1, 0$, and V_{F} is diagonal in f, m_f, i.e., $H^1_{n,1/2,0}$ in (7.6.11) is diagonal in $|n, f, m_f\rangle$ for a fixed n.

From (7.6.15), (7.6.18), (7.6.19), we then have for the hyperfine-structure corrections for s-states:

$$\Delta E^{\mathrm{HF}}_{n,f} = \frac{4}{3} \frac{\alpha^2}{n} \left(\frac{M}{M_{\mathrm{p}}}\right) g_{\mathrm{p}} \left(\frac{\alpha^2 Mc^2}{2n^2}\right) \left[f(f+1) - \frac{3}{2} \right], \qquad \ell = 0 \quad (7.6.20)$$

where $f = 1$ or 0 corresponding, respectively, to the spins of the electron and proton aligned or opposed.

The frequency of radiation ($\sim 1420\,\mathrm{MHz}$) arising from the transition from the excited state $f = 1$ (the triplet state) to the ground-state $f = 0$ (the singlet state) for $n = 1$, is one of the most precisely measured quantities in physics and corresponds to a wavelength of $21.1\,\mathrm{cm}$, well known in radio astronomy.

$\ell \neq 0$ States:

We evaluate the matrix elements

$$\langle n, f', j, \ell, m'_f | H^1_{nj\ell} | n, f, j, \ell, m_f \rangle \tag{7.6.21}$$

of the perturbation $H^1_{nj\ell}$ in (7.6.11) to $H^0_{nj\ell}$ defined in (7.6.8).

To the above end we first note that we may rewrite V_{HF} in (7.6.1) as

$$V_{\mathrm{HF}} = \frac{e^2}{2MM_{\mathrm{p}}c^2} g_{\mathrm{p}} \left[\frac{\mathbf{I} \cdot \mathbf{L}}{r^3} + \mathbf{S} \cdot \mathbf{I} \,\nabla^2 \left(\frac{1}{r}\right) - S^i I^j \frac{\partial}{\partial x^i} \frac{\partial}{\partial x^j}\left(\frac{1}{r}\right) \right] \tag{7.6.22}$$

where we recall the relation (7.4.36) for $\nabla^2(1/r)$. Since $R^2_{n\ell}(r)$, vanishes for $r = 0$ if $\ell \neq 0$, (see (7.3.31)), we may *omit* $\delta^3(\mathbf{x})$ terms arising from the last two terms on the right-hand side of (7.6.22). That is, for $\ell \neq 0$ we may effectively omit the second term in (7.6.22) and effectively write the last term as

$$-\frac{\mathbf{S} \cdot \mathbf{I}}{r^3} + 3\frac{\mathbf{I} \cdot \mathbf{x}\, \mathbf{S} \cdot \mathbf{x}}{r^5}. \tag{7.6.23}$$

That is, for $\ell \neq 0$, we may *effectively* take for V_{HF} the expression

$$V_{\mathrm{HF}} = \frac{e^2}{2MM_{\mathrm{p}}c^2} g_{\mathrm{p}} \frac{1}{r^3} \left[\mathbf{I} \cdot \mathbf{L} - \mathbf{I} \cdot \mathbf{S} + 3\, \mathbf{I} \cdot \mathbf{n}\, \mathbf{S} \cdot \mathbf{n} \right] \tag{7.6.24}$$

where $\mathbf{n} = \mathbf{x}/|\mathbf{x}| = (\cos\phi \sin\theta, \sin\phi\sin\theta, \cos\theta)$.

Now we use the important identity in (5.8.47)

$$\langle \ell, m' | n^i n^j | \ell, m \rangle = \frac{1}{(2\ell - 1)(2\ell + 3)} \langle \ell, m' | (2\ell^2 + 2\ell - 1)\,\delta^{ij}$$

$$- \frac{1}{\hbar^2}\left(L^i L^j + L^j L^i \right) | \ell, m \rangle. \tag{7.6.25}$$

These matrix elements allow us to compute the expressions $\langle n, j, \ell, m_j | V_{\mathrm{HF}} | n, j, \ell, m'_j \rangle$ in (7.6.11) by using, in the process, the expansion (7.5.5) for the states $|n, j, \ell, m_j\rangle$ involving $|\ell, m\rangle$ states for different m. We may then replace V_{HF} in (7.6.24) effectively by

$$V'_{\mathrm{HF}} = \frac{e^2}{2MM_{\mathrm{p}}c^2} g_{\mathrm{p}} \frac{1}{r^3} \left[\mathbf{I} \cdot \mathbf{L} + (3a_\ell - 1)\mathbf{I} \cdot \mathbf{S} + 3b_\ell D \right] \tag{7.6.26}$$

7.6 The Hyperfine-Structure of the Hydrogen Atom

where
$$D = \frac{1}{\hbar^2}(\mathbf{I} \cdot \mathbf{L}\ \mathbf{L} \cdot \mathbf{S} + \mathbf{L} \cdot \mathbf{S}\ \mathbf{I} \cdot \mathbf{L}) \tag{7.6.27}$$

$$a_\ell = \frac{(2\ell^2 + 2\ell - 1)}{(2\ell - 1)(2\ell + 3)} \tag{7.6.28}$$

$$b_\ell = -\frac{1}{(2\ell - 1)(2\ell + 3)}. \tag{7.6.29}$$

Upon using the operator relation for $\mathbf{S} \cdot \mathbf{L}$ in (7.5.17), we note that in computing the matrix elements in (7.6.21) we may replace $\mathbf{S} \cdot \mathbf{L}$ by its eigenvalues, and replace D in (7.6.27) by

$$D' = \left[j(j+1) - \ell(\ell+1) - \frac{3}{4} \right] \mathbf{I} \cdot \mathbf{L}. \tag{7.6.30}$$

That is, we may replace V'_{HF} in (7.6.26) by

$$V''_{\mathrm{HF}} = \frac{e^2}{2MM_{\mathrm{p}}c^2} \frac{g_{\mathrm{p}}}{r^3} \left[(3a_\ell - 1)\mathbf{I} \cdot \mathbf{S} + c_{\ell j}\mathbf{I} \cdot \mathbf{L} \right] \tag{7.6.31}$$

where
$$c_{\ell j} = 1 - 3\frac{j(j+1) - \ell(\ell+1) - 3/4}{(2\ell - 1)(2\ell + 3)}. \tag{7.6.32}$$

Now we use the operator identities (see Problem 7.20):

$$\frac{1}{\hbar^2}\left[\mathbf{J}^2, \left[\mathbf{J}^2, L^i\right]\right] = 2\left(\mathbf{J}^2 L^i + L^i \mathbf{J}^2\right) - 2J^i\left(\mathbf{J}^2 + \mathbf{L}^2 - \mathbf{S}^2\right) \tag{7.6.33}$$

$$\frac{1}{\hbar^2}\left[\mathbf{J}^2, \left[\mathbf{J}^2, S^i\right]\right] = 2\left(\mathbf{J}^2 S^i + S^i \mathbf{J}^2\right) - 2J^i\left(\mathbf{J}^2 - \mathbf{L}^2 + \mathbf{S}^2\right) \tag{7.6.34}$$

to obtain

$$\frac{1}{\hbar^2}\left[\mathbf{J}^2, \left[\mathbf{J}^2, \mathbf{I} \cdot \mathbf{L}\right]\right] = 2\left(\mathbf{J}^2\ \mathbf{I} \cdot \mathbf{L} + \mathbf{I} \cdot \mathbf{L}\ \mathbf{J}^2\right) - 2\mathbf{I} \cdot \mathbf{J}\left(\mathbf{J}^2 + \mathbf{L}^2 - \mathbf{S}^2\right) \tag{7.6.35}$$

$$\frac{1}{\hbar^2}\left[\mathbf{J}^2, \left[\mathbf{J}^2, \mathbf{I} \cdot \mathbf{S}\right]\right] = 2\left(\mathbf{J}^2\ \mathbf{I} \cdot \mathbf{S} + \mathbf{I} \cdot \mathbf{S}\ \mathbf{J}^2\right) - 2\mathbf{I} \cdot \mathbf{J}\left(\mathbf{J}^2 - \mathbf{L}^2 + \mathbf{S}^2\right). \tag{7.6.36}$$

Upon using the operator identity for $\mathbf{I} \cdot \mathbf{J}$ in (7.6.5) and taking the matrix elements of (7.6.35), (7.6.36) (see Problem 7.21) between the states $\langle f', j, \ell, m'_j |$ and $| f, j, \ell, m_f \rangle$ one obtains after some algebra[16] (suppressing n)

---
[16] The equalities (7.6.37), (7.6.38) are the contents of the Wigner-Eckart Theorem in §5.7, §5.8 (see Problem 7.22).

$$\langle f', j, \ell, m'_f | \mathbf{I} \cdot \mathbf{L} | f, j, \ell, m_f \rangle = \frac{\hbar^2}{4} \frac{[j(j+1) + \ell(\ell+1) - 3/4]}{j(j+1)}$$

$$\times [f(f+1) - j(j+1) - 3/4] \delta_{f'f} \delta_{m'_f m_f} \quad (7.6.37)$$

$$\langle f', j, \ell, m'_f | \mathbf{I} \cdot \mathbf{S} | f, j, \ell, m_f \rangle = \frac{\hbar^2}{4} \frac{[j(j+1) - \ell(\ell+1) + 3/4]}{j(j+1)}$$

$$\times [f(f+1) - j(j+1) - 3/4] \delta_{f'f} \delta_{m'_f m_f} \quad (7.6.38)$$

where we have used the identity

$$\langle f', j, \ell, m'_f | \mathbf{I} \cdot \mathbf{J} | f, j, \ell, m_f \rangle = \frac{\hbar^2}{2} [f(f+1) - j(j+1) - 3/4] \delta_{f'f} \delta_{m'_f m_f}. \quad (7.6.39)$$

That is $H^1_{nj\ell}$ is diagonal in f, m_f.

We recall the values taken by j are $j = \ell \pm 1/2$. Accordingly, for $j = \ell + 1/2$

$$j(j+1) + \ell(\ell+1) - \frac{3}{4} = \ell(2\ell+3) \quad (7.6.40)$$

$$j(j+1) - \ell(\ell+1) + \frac{3}{4} = \frac{1}{2}(2\ell+3) \quad (7.6.41)$$

$$j(j+1) - \ell(\ell+1) - \frac{3}{4} = \ell \quad (7.6.42)$$

and for $j = \ell - 1/2$,

$$j(j+1) + \ell(\ell+1) - \frac{3}{4} = (2\ell-1)(\ell+1) \quad (7.6.43)$$

$$j(j+1) - \ell(\ell+1) + \frac{3}{4} = -\frac{1}{2}(2\ell-1) \quad (7.6.44)$$

$$j(j+1) - \ell(\ell+1) - \frac{3}{4} = -(\ell+1). \quad (7.6.45)$$

Upon substituting (7.6.40)–(7.6.42), (7.6.43)–(7.6.45) in turn in (7.6.37) and (7.6.38), we obtain directly from (7.6.31), for the matrix elements (7.6.21), in reference to (7.6.11),

$$\langle n, f', j, \ell, m'_f | H^1_{nj\ell} | n, f, j, \ell, m_f \rangle$$

$$= \delta_{f'f} \delta_{m'_f m_f} \frac{e^2}{2MM_p c^2} g_p \left\langle n, \ell \left| \frac{1}{r^3} \right| n, \ell \right\rangle \ell(\ell+1)$$

$$\times \hbar^2 \frac{[f(f+1) - j(j+1) - 3/4]}{2j(j+1)} \tag{7.6.46}$$

where the matrix elements $\langle n, \ell | (1/r^3) | n, \ell \rangle$ are given in (7.3.38), with $Z \to 1$, $\mu \to M$, for $\ell \neq 0$.

All told, we obtain for the hyperfine-structure correction

$$\Delta E_{n,f,j,\ell}^{\text{HF}} = \frac{\alpha^2}{2n} \left(\frac{M}{M_{\text{p}}} \right) g_{\text{p}} \left(\frac{\alpha^2 M c^2}{2n^2} \right) \frac{[f(f+1) - j(j+1) - 3/4]}{(\ell + 1/2) j(j+1)}. \tag{7.6.47}$$

If we formally set $\ell = 0$, $j = 1/2$ in the above formula, we see that it coincides with the expression in (7.6.20) for $\ell = 0$. That is, (7.6.47) applies for all $\ell = 0, \ldots, n-1$.

7.7 The Non-Relativistic Lamb Shift

This section provides a glimpse of the fascinating world of quantum field theory, namely that of quantum electrodynamics. More precisely, this section deals with the splitting of energy levels of the hydrogen atom due to the interaction of the atomic electron with radiation, that is, with the photon.

An atomic electron has an average speed v of the order of $e^2/\hbar \equiv \alpha c$, and is thus much smaller than that of the speed of light c. We shall therefore content ourselves by treating the electron non-relativistically only and not dwell on the relativistic regime here. We will also see below that a detailed treatment of radiation is not necessary for the problem at hand and the following introduction suffices.

7.7.1 The Radiation Field

The photon may be simply described as follows. Let $|\mathbf{k}, \lambda\rangle$ denote a state of a photon with momentum $\hbar \mathbf{k}$, energy $\hbar |\mathbf{k}| c$ and polarization described by a polarization three-vector $\mathbf{e}_\lambda(\mathbf{k})$, with λ taking on two possible values, say, 1 and 2, such that

$$\mathbf{k} \cdot \mathbf{e}_\lambda(\mathbf{k}) = 0 \tag{7.7.1}$$

and

$$\mathbf{e}_\lambda(\mathbf{k}) \cdot \mathbf{e}_{\lambda'}(\mathbf{k}) = \delta_{\lambda \lambda'} \tag{7.7.2}$$

$\lambda, \lambda' = 1, 2$. Without loss of general of generality we consider real polarization vectors.

With $\mathbf{k}/|\mathbf{k}|$, $\mathbf{e}_1(\mathbf{k})$, $\mathbf{e}_2(\mathbf{k})$ representing a complete set of vectors in the three-dimensional Euclidean space, we may write a completeness relation as follows

$$\delta^{ij} = \frac{k^i k^j}{|\mathbf{k}|^2} + e_1^i e_1^j + e_2^i e_2^j \tag{7.7.3}$$

$i,j = 1,2,3$. In particular, this gives us the property

$$\sum_{\lambda=1,2} e_\lambda^i e_\lambda^j = \delta^{ij} - \frac{k^i k^j}{|\mathbf{k}|^2} \quad (7.7.4)$$

which will be useful later on.

One may introduce a completeness relation over the momenta and the two polarizations, specified by λ, for a *single* photon, by

$$\sum_\lambda \int \frac{d^3\mathbf{k}}{(2\pi)^3 2|\mathbf{k}|} |\mathbf{k},\lambda\rangle\langle\mathbf{k},\lambda| = 1 \quad (7.7.5)$$

(see also Problem 7.24). The factor $1/2|\mathbf{k}|$ comes from relativity for the ever relativistic particle — the photon. It arises as follows. With the energy-momentum constraint between energy, denoted by $\hbar k^0 c$, and momentum $\hbar \mathbf{k}$, we may instead integrate over $\mathbf{k}$ and k^0 with the measure $(k^0 > 0)$

$$\Theta\left(k^0\right) dk^0 d^3\mathbf{k}\, \delta\!\left(\left(k^0\right)^2 - |\mathbf{k}|^2\right) \quad (7.7.6)$$

which leads immediately to the factor $1/2|\mathbf{k}|$ in (7.7.5). The measure in (7.7.6) turns up to be relativistically (so-called Lorentz) *invariant*.

We introduce a radiation (real) field $\mathbf{A}_{\mathrm{RAD}}(\mathbf{x})$, as well as the Hamiltonian of the atomic electron and radiation system given by

$$H = \frac{1}{2M}\left(\mathbf{p} - \frac{e}{c}\mathbf{A}_{\mathrm{RAD}}\right)^2 + H_{0,\mathrm{RAD}} - \frac{e^2}{|\mathbf{x}|} \quad (7.7.7)$$

where $H_{0,\mathrm{RAD}}$ is the free Hamiltonian of radiation whose detailed structure, as will be seen below, is not needed for the problem at hand. For simplicity we omit the spin of the electron (see also Problem 7.27).

The Hamiltonian H in (7.7.7) may be rewritten as

$$H = H_\mathrm{C} + H_{0,\mathrm{RAD}} + H_\mathrm{I} \quad (7.7.8)$$

where H_C is the Coulomb Hamiltonian

$$H_\mathrm{C} = \frac{\mathbf{p}^2}{2M} - \frac{e^2}{|\mathbf{x}|} \quad (7.7.9)$$

$$H_\mathrm{I} = -\frac{e}{2Mc}\left(\mathbf{A}_{\mathrm{RAD}}\cdot\mathbf{p} + \mathbf{p}\cdot\mathbf{A}_{\mathrm{RAD}}\right) + \frac{e^2}{2Mc^2}\mathbf{A}_{\mathrm{RAD}}^2. \quad (7.7.10)$$

We treat H_I as a perturbation to H_C in H.

To zeroth order in H_I, the radiation field $\mathbf{A}_{\mathrm{RAD}}$ may create or destroy a single photon. Let $|0\rangle$ denote the no photon (vacuum) state. Then we may define the matrix element of a single photon state as

$$\langle 0|\mathbf{A}_{\text{RAD}}(\mathbf{x})|\mathbf{k},\lambda\rangle = \sqrt{4\pi}\sqrt{\hbar c}\, e^{i\mathbf{k}\cdot\mathbf{x}}\mathbf{e}_\lambda(\mathbf{k}). \qquad (7.7.11)$$

Due to the orthogonality of the vacuum and a photon state, the vacuum expectation value of $\mathbf{A}_{\text{RAD}}(\mathbf{x})$ evidently satisfies

$$\langle 0|\mathbf{A}_{\text{RAD}}(\mathbf{x})|0\rangle = 0. \qquad (7.7.12)$$

Also the matrix elements of $\mathbf{A}_{\text{RAD}}$ between the vacuum and states containing more than one photon are also zero.

Upon taking the complex conjugate of (7.7.11) we obtain

$$\langle \mathbf{k},\lambda|\mathbf{A}_{\text{RAD}}(\mathbf{x})|0\rangle = \sqrt{4\pi}\sqrt{\hbar c}\, e^{-i\mathbf{k}\cdot\mathbf{x}}\mathbf{e}_\lambda(\mathbf{k}) \qquad (7.7.13)$$

for real polarization vectors $\mathbf{e}_\lambda(\mathbf{k})$. Here $\mathbf{A}_{\text{RAD}}(\mathbf{x})$ creates out of the vacuum a single photon.

To all orders in H_I, and in a relativistic treatment, (7.7.13) is simply replaced by

$$\langle \mathbf{k},\lambda|\mathbf{A}_{\text{RAD}}(\mathbf{x})|0\rangle|_{\text{Full}} = \sqrt{4\pi}\sqrt{\hbar c}\sqrt{Z_3}\, e^{-i\mathbf{k}\cdot\mathbf{x}}\mathbf{e}_\lambda(\mathbf{k}) \qquad (7.7.14)$$

where $0 \leqslant Z_3 \leqslant 1$, is a constant and is referred to as the photon *wavefunction renormalization constant*. The reason why (7.7.13) is to be modified by the presence of an additional multiplicative factor $\leqslant 1$ in (7.7.14), in the general case, is easy to see. In this latter situation, $\mathbf{A}_{\text{RAD}}$ may create out of the vacuum, in addition to a photon (for $Z_3 \neq 0$), electron-positron pairs and other particles. (See also Problem 7.25). The amplitude (7.7.13) is thus, in general, reduced by a factor $(\equiv \sqrt{Z_3})$, as given in (7.7.14), such that the probability that $\mathbf{A}_{\text{RAD}}$ creates all other possible particles as well adds up to one.

Now we turn to the expression in (7.7.11). The factor $\sqrt{4\pi}$ in it is introduced because of the definition of the fine-structure constant $\alpha = e^2/\hbar c$, in (7.4.34), adopted here. If the Coulomb potential in (7.7.7) is taken as $-e^2/4\pi|\mathbf{x}|$, then (7.7.11) does not involve the $\sqrt{4\pi}$ factor and the fine-structure is defined as $e^2/4\pi\hbar c$. [In field theory, the latter definition is usually adopted.] The factor $\sqrt{\hbar c}$ arises because of dimensional reasons.

As far as the second power $\mathbf{A}_{\text{RAD}}^2(\mathbf{x})$ of $\mathbf{A}_{\text{RAD}}(\mathbf{x})$ in (7.7.10) is concerned, we will only encounter its (photon) vacuum expectation value to zeroth order in e in our leading order perturbative treatment of H_I. That is, only $\langle 0|\mathbf{A}_{\text{RAD}}^2(\mathbf{x})|0\rangle$ is encountered. The latter is formally given from (7.7.5), (7.7.11), (7.7.12) and the property that the matrix element of $\mathbf{A}_{\text{RAD}}$ between the vacuum and states containing more than one photon are zero, by the expression

$$e^2\langle 0|\mathbf{A}_{\text{RAD}}^2(\mathbf{x})|0\rangle = e^2\sum_\lambda \int \frac{d^3k}{(2\pi)^3 2|\mathbf{k}|}\, |\langle 0|\mathbf{A}(\mathbf{x})|\mathbf{k},\lambda\rangle|^2 \qquad (7.7.15)$$

to the leading order.

394   7 Intricacies of the Hydrogen Atom

This formal expression, although an infinite constant (as is easily checked) is common to all the energy levels and hence does not contribute to the computation of the energy level differences we are seeking. More precisely, after carrying out a so-called renormalization, as a physical requirement due to the ever presence of radiation accompanying a charged particle, this term gets altogether cancelled out in the process as will be seen through (7.7.37)–(7.7.40) below.

For the free Hamiltonian $H_{0,\text{RAD}}$ of radiation in (7.7.7), we have, by definition,

$$H_{0,\text{RAD}} |\mathbf{k}, \lambda\rangle = \hbar|\mathbf{k}|\, c\, |\mathbf{k}, \lambda\rangle \qquad (7.7.16)$$

corresponding to the energy of a photon of momentum $\hbar\mathbf{k}$.

We also note from the orthogonality relation of the polarization vectors $\mathbf{e}_\lambda(\mathbf{k})$ of a photon and of its momentum $\hbar\mathbf{k}$, as given in (7.7.1), the commutativity relation

$$\mathbf{p} \cdot \langle 0|\mathbf{A}_{\text{RAD}}(\mathbf{x})|\mathbf{k}, \lambda\rangle = \langle 0|\mathbf{A}_{\text{RAD}}(\mathbf{x})|\mathbf{k}, \lambda\rangle \cdot \mathbf{p} \qquad (7.7.17)$$

holds true ($\mathbf{p} = -i\hbar\boldsymbol{\nabla}$) by using (7.7.11).

7.7.2 Expression for the Energy Shifts

Let $|\psi_n\rangle$ denote formally the state of the atomic electron in the absence of radiation. That is, the $|\psi_n\rangle$ are the eigenvectors of H_C, with eigenvalue E_n, where n denotes the principal quantum number, suppressing for simplicity of the notation other quantum numbers.

In the present case, we are working in a larger space due to the presence of radiation and the state $|\psi_n\rangle$ above, corresponding to such a larger space, will be written as $|\psi_n\rangle|0\rangle \equiv |\psi_n; 0\rangle$ emphasizing, in the mean time, that it involves no photons. This state is often referred to as the *bare* state of the electron in the atom.

We assume that in the presence of the interaction H_I with radiation, the atomic electron may be described by some state $|\phi_n\rangle$, in this larger space, at least for small H_I, corresponding to the same n, with an eigenvalue denoted by ε_n.

One may then write

$$(H_C + H_{0,\text{RAD}})|\psi_n; 0\rangle = E_n |\psi_n; 0\rangle \qquad (7.7.18)$$

$$H |\phi_n) = \varepsilon_n |\phi_n) \, . \qquad (7.7.19)$$

Also for the state $|\psi_n\rangle|\mathbf{k}, \lambda\rangle \equiv |\psi_n; \mathbf{k}, \lambda\rangle$,

$$(H_C + H_{0,\text{RAD}})|\psi_n; \mathbf{k}, \lambda\rangle = (E_n + \hbar\omega)|\psi_n; \mathbf{k}, \lambda\rangle \qquad (7.7.20)$$

where

7.7 The Non-Relativistic Lamb Shift

$$\omega = |\mathbf{k}|c. \tag{7.7.21}$$

Upon multiplying (7.7.19) from the left by $(\psi_{n'}; 0|$ and using (7.7.18), we obtain

$$(\varepsilon_n - E_{n'})(\psi_{n'}; 0|\phi_n) = (\psi_{n'}; 0|H_\mathrm{I}|\phi_n). \tag{7.7.22}$$

In particular, for $n' = n$, this gives

$$\Delta E_n = \frac{(\psi_n; 0|H_\mathrm{I}|\phi_n)}{(\psi_n; 0|\phi_n)} \tag{7.7.23}$$

where

$$\Delta E_n = \varepsilon_n - E_n. \tag{7.7.24}$$

On the other hand, by multiplying (7.7.19) from the left by a single photon state $\langle \mathbf{k}, \lambda|$, we generate a state corresponding to H_C only given by[17]

$$\langle \mathbf{k}, \lambda|\phi_n\rangle = \frac{1}{(\varepsilon_n - H_\mathrm{C} - \hbar\omega)}\langle \mathbf{k}, \lambda|H_\mathrm{I}|\phi_n\rangle. \tag{7.7.25}$$

From (7.7.11), (7.7.12), the fact that the matrix element of $\mathbf{A}_\mathrm{RAD}$ between the vacuum and multi-photon slates are zero, and from (7.7.25), (7.7.5), (7.7.1), we have *up to order e^2 in H_I*, the following key matrix element plus the term in (7.7.27) to follow:

$$-\frac{e}{Mc}(\psi_{n'}; 0|\mathbf{p} \cdot \mathbf{A}_\mathrm{RAD}(\mathbf{x})|\phi_n)$$

$$= \frac{4\pi\hbar c e^2}{M^2 c^2} \sum_\lambda \int \frac{d^3\mathbf{k}}{(2\pi)^3 2|\mathbf{k}|}$$

$$\times \left\langle \psi_{n'} \left| \mathbf{p} \cdot \mathbf{e}_\lambda(\mathbf{k}) e^{i\mathbf{k}\cdot\mathbf{x}} \frac{1}{(E_n - H_\mathrm{C} - \hbar\omega)} e^{-i\mathbf{k}\cdot\mathbf{x}} \mathbf{e}_\lambda(\mathbf{k}) \cdot \mathbf{p} \right| \psi_n \right\rangle \tag{7.7.26}$$

obtained by simply inserting **1**, given in (7.7.5), between $(\mathbf{p} \cdot \mathbf{A}_\mathrm{RAD} + \mathbf{A}_\mathrm{RAD} \cdot \mathbf{p})$ and $|\phi_n\rangle$, and making use of (7.7.11), (7.7.17) and (7.7.25).

The $\mathbf{A}_\mathrm{RAD}^2$ term in (7.7.10) already involves e^2, and hence we may immediately write to this order

$$\frac{e^2}{2Mc^2}(\psi_n; 0|\mathbf{A}_\mathrm{RAD}^2(\mathbf{x})|\phi_n) = \frac{e^2}{2Mc^2}(\psi_n; 0|\mathbf{A}_\mathrm{RAD}^2(\mathbf{x})|\psi_n; 0)$$

$$= \frac{e^2}{2Mc^2}\langle 0|\mathbf{A}_\mathrm{RAD}^2(\mathbf{x})|0\rangle \tag{7.7.27}$$

---

[17] The method to follow is a slight variation of perturbation theory. The latter will be studied afresh in Chapter 12.

and is *independent* of n (and related quantum numbers). The formal expression on the right-hand side of (7.7.27), although infinite (see (7.7.15)), is common to all the levels (see (7.7.23)), and hence will not contribute to the differences in the levels, and it may be omitted in the analysis.[18] On physical grounds, however, this integral cancels out, as will be seen below, when defining the concept of a free electron, accompanied by the ever present radiation in the process of renormalization as mentioned before.

Since the expression in (7.7.26) (as well the one in (7.7.27)) is already of order e^2, as originating from H_I, we may set $(\psi_\mathbf{n};0|\phi_n)$ in the denominator on the right-hand side of (7.7.23) equal to one.

Making the change of notation $\Delta E_n \to \delta E_n$, for having omitted the term in (7.7.27), common to all levels, (and for working up to e^2 in H_I), we then have from (7.7.26) with $n' = n$, and (7.7.23)

$$\delta E_n = \frac{\alpha \hbar^2}{4\pi^2 M^2}$$

$$\sum_\lambda \int \frac{d^3\mathbf{k}}{|\mathbf{k}|} \left\langle \psi_n \left| \mathbf{p} \cdot \mathbf{e}_\lambda(\mathbf{k}) e^{i\mathbf{k}\cdot\mathbf{x}} \frac{1}{(E_n - H_\mathrm{C} - \hbar\omega)} e^{-i\mathbf{k}\cdot\mathbf{x}} \mathbf{e}_\lambda(\mathbf{k}) \cdot \mathbf{p} \right| \psi_n \right\rangle. \tag{7.7.28}$$

If we introduce the resolution of the identity of H_C (see (1.8.18)), then we may formally write

$$\frac{1}{(E_n - H_\mathrm{C} - \hbar\omega)} = \int \frac{1}{(E_n - E - \hbar\omega)} dP_{H_\mathrm{C}}(E) \tag{7.7.29}$$

and singularities, in general, develop for $\hbar\omega = E_n - E > 0$. For $E = E_1$, for example, and $n = 2$, the system may decay to the state $n = 1$, with the emission of a photon of energy $\hbar\omega$ as just given. Accordingly, we interpret the meaning of the singularities arising from the denominator[19] $E_n - H_\mathrm{C} - \hbar\omega$, appearing in (7.7.28), as follows. We add to it a $+i\varepsilon$ term with $\varepsilon \to +0$. The shift then acquires from (7.7.28) a negative imaginary part, as will be shown later by using in the process that

$$\operatorname{Im} \frac{1}{(E_n - H_\mathrm{C} - \hbar\omega + i\varepsilon)} = -\pi\delta(E_n - H_\mathrm{C} - \hbar\omega) \tag{7.7.30}$$

and one may write

$$\delta E_n = \operatorname{Re}(\delta E_n) - \frac{i}{2}\Gamma \tag{7.7.31}$$

with $\Gamma > 0$. Thus a state, specified by n and related quantum numbers, will develop in time by acquiring an exponential damping factor

---

[18] To be rigorous, one may, in intermediate steps, insert a large (so-called ultraviolet) cut-off Λ, with $\hbar|\mathbf{k}|c \leqslant \Lambda$ in the integral in (7.7.15) before considering the limit $\Lambda \to \infty$.

[19] See also the interpretation of such singularities of Green's functions in Chapter 9.

7.7 The Non-Relativistic Lamb Shift 397

$$\exp -\frac{i}{\hbar}t\left(-\frac{i}{2}\Gamma\right) = \exp -\frac{t}{\hbar}\frac{\Gamma}{2} \qquad (7.7.32)$$

in $t > 0$, leading to the decay of excited states, as described above, for a correct interpretation of the vanishing of the denominator in (7.7.28). This will be taken up in §7.8. [See also §3.5.]

The real part $\text{Re}(\delta E_n)$ in (7.7.31), corresponding to the energy shift, we are seeking, then is obtained by taking the principal value of $(E_n - H_C - \hbar\omega)^{-1}$ and this should be understood in the sequel.

Since

$$e^{i\mathbf{k}\cdot\mathbf{x}}\left(\frac{\mathbf{p}^2}{2M} + U_C(\mathbf{x})\right)e^{-i\mathbf{k}\cdot\mathbf{x}} = \frac{(\mathbf{p}-\hbar\mathbf{k})^2}{2M} + U_C(\mathbf{x}) \qquad (7.7.33)$$

and for any function of $[\mathbf{p}^2/2M + U_C(\mathbf{x})]$, where with $U_C(\mathbf{x})$, in this case, given just to be $-e^2/|\mathbf{x}|$, we have

$$e^{i\mathbf{k}\cdot\mathbf{x}}\frac{1}{E_n - H_C - \hbar\omega}e^{-i\mathbf{k}\cdot\mathbf{x}} = \frac{1}{E_n - \frac{(\mathbf{p}^2-\hbar\mathbf{k})^2}{2M} - U_C - \hbar\omega} \qquad (7.7.34)$$

in (7.7.28).

Thus we may rewrite (7.7.28) as

$$\delta E_n = \frac{\alpha\hbar^2}{4\pi^2 M^2}\sum_\lambda \int \frac{d^3 k}{|\mathbf{k}|}$$

$$\times \left\langle \psi_n \left| \mathbf{p}\cdot\mathbf{e}_\lambda(\mathbf{k}) \frac{1}{\left(E_n - H_C + \frac{\mathbf{p}^2}{2M} - \frac{(\mathbf{p}-\hbar\mathbf{k})^2}{2M} - \hbar\omega\right)} \mathbf{e}_\lambda(\mathbf{k})\cdot\mathbf{p} \right| \psi_n \right\rangle.$$

$$(7.7.35)$$

This expression is not complete by itself, and before evaluating the energy shifts, there are various physical points that will be considered first.

The corresponding expression to (7.7.35) in the absence of the Coulomb interaction, that is, for the electron in the presence of radiation only will be worked out first.

In analogy to (7.7.22), we then have

$$\left(\varepsilon(\mathbf{p}) - \frac{\mathbf{p}'^2}{2m}\right)(\mathbf{p}';0|\mathbf{p}) = (\mathbf{p}';0|H_I|\mathbf{p}) \qquad (7.7.36)$$

and by including the term in (7.7.27), we obtain, to the leading order,

$$\left(\varepsilon(\mathbf{p}) - \frac{\mathbf{p}'^2}{2m}\right)(\mathbf{p}';0|\mathbf{p})$$

$$= \frac{e^2}{2Mc^2}\langle 0|\mathbf{A}_{\text{RAD}}^2(\mathbf{x})|0\rangle(2\pi\hbar)^3\delta^3(\mathbf{p}'-\mathbf{p}) + \frac{\alpha\hbar^2}{4\pi^2 M^2}$$

$$\times \sum_\lambda \int \frac{d^3k}{|\mathbf{k}|} \left\langle \mathbf{p}' \left| \mathbf{p}' \cdot \mathbf{e}_\lambda(\mathbf{k}) \frac{1}{\left(\frac{\mathbf{p}^2}{2M} - \frac{(\mathbf{p}'-\hbar\mathbf{k})^2}{2M} - \hbar\omega \right)} \mathbf{e}_\lambda(\mathbf{k}) \cdot \mathbf{p} \right| \mathbf{p} \right\rangle.$$
(7.7.37)

In the appendix to this chapter, it is shown that for a free electron[20] for $H_I \neq 0$, i.e., in the ever presence of radiation, (7.7.37) gives rise to a shift in the energy of an electron and, in particular, its mass gets shifted from the mass parameter appearing in $\mathbf{p}^2/2M$ in the original Hamiltonian. This necessitates to *renormalize* the theory. In simplest form, this amounts to adding an operator δH_c to the original Hamiltonian such that the corresponding expression to the right-hand side of (7.7.37) is zero when computed with the modified Hamiltonian and the mass now appearing in the theory represents the physically observed mass of the electron. Related details to these are spelled out in the appendix to this section. Here it is sufficient to note that δH_c, for the problem at hand, is formally given by

$$\delta H_c = -\frac{\alpha\hbar^2}{4\pi^2 M^2} \sum_\lambda \int \frac{d^3k}{|\mathbf{k}|} \mathbf{p} \cdot \mathbf{e}_\lambda(\mathbf{k}) \frac{1}{\left(\frac{\mathbf{p}^2}{2M} - \frac{(\mathbf{p}-\hbar\mathbf{k})^2}{2M} - \hbar\omega \right)} \mathbf{e}_\lambda(\mathbf{k}) \cdot \mathbf{p}$$

$$- \frac{e^2}{2Mc^2} \langle 0 | \mathbf{A}_{\text{RAD}}^2(\mathbf{x}) | 0 \rangle \qquad (7.7.38)$$

of course working to *lowest* order in e^2 in H_I, and the modified interaction H_I' becomes

$$H_I' = H_I + \delta H_c. \qquad (7.7.39)$$

The last term in (7.7.38) is a c-number, while $\mathbf{p}$ in the first term is an operator.

With the above definition, then the electron stays stable, at least to order e^2 in H_I, *retaining* its kinetic energy $\mathbf{p}^2/2M$ as directly seen from (7.7.37), when H_I is replaced by H_I' as defined in (7.7.39), (7.7.38). By doing so, the mass M appearing now in (7.7.39) represents the *physically* observed mass taking the ever present radiation accompanying an electron into account. Also note that the c-number in (7.7.27) will automatically cancel out when working with the interaction Hamiltonian H_I'. In the literature δH_c is referred to a *counter-term* and the physically observed mass is referred to as the *renormalized mass*.

7.7.3 The Lamb Shift and Renormalization

The shift in (7.7.35) is now to be replaced by

$$\delta E_n^{\text{Ren}} = \frac{\alpha\hbar^2}{4\pi^2 M^2} \sum_\lambda \int \frac{d^3k}{|\mathbf{k}|} \langle \psi_n | \mathbf{p} \cdot \mathbf{e}_\lambda(\mathbf{k})$$

---

[20] By a free electron, of momentum $\mathbf{p}$, it is meant that its energy should be equal to $\mathbf{p}^2/2M$.

7.7 The Non-Relativistic Lamb Shift

$$\times \left[\frac{1}{\left(E_n - H_C + \frac{\mathbf{p}^2}{2M} - \frac{(\mathbf{p}-\hbar\mathbf{k})^2}{2M} - \hbar\omega\right)} - \frac{1}{\left(\frac{\mathbf{p}^2}{2M} - \frac{(\mathbf{p}-\hbar\mathbf{k})^2}{2M} - \hbar\omega\right)} \right]$$

$$\times \mathbf{e}_\lambda(\mathbf{k}) \cdot \mathbf{p} \, |\psi_n\rangle$$

$$= \frac{\alpha\hbar^2}{4\pi^2 M^2} \sum_\lambda \int \frac{d^3\mathbf{k}}{|\mathbf{k}|} \langle \psi_n | \, \mathbf{p} \cdot \mathbf{e}_\lambda(\mathbf{k}) \frac{1}{\left[\left(\frac{\mathbf{P}}{Mc}\right) \cdot \hbar\mathbf{k}c - \hbar|\mathbf{k}|c\left(1 + \frac{\hbar|\mathbf{k}|c}{2Mc^2}\right)\right]}$$

$$\times (H_C - E_n) \frac{1}{\left[E_n - H_C + \left(\frac{\mathbf{P}}{Mc}\right) \cdot \hbar\mathbf{k}c - \hbar|\mathbf{k}|c\left(1 + \frac{\hbar|\mathbf{k}|c}{2Mc^2}\right)\right]} \mathbf{e}_\lambda(\mathbf{k}) \cdot \mathbf{p} \, |\psi_n\rangle$$

(7.7.40)

where Ren in $\delta E_n^{\mathrm{Ren}}$ stands for renormalized. We are interested in the cases $n = 2$ in (7.7.40).

By counting powers of $|\mathbf{k}|$ in the integral on the extreme right-hand side of (7.7.40); for $|\mathbf{k}| \to \infty$, we note that the integral in question is convergent at high energies $\hbar|\mathbf{k}|c \to \infty$. Although this integral in (7.7.40) is convergent at high $|\mathbf{k}|$,[21] having started with a non-relativistic Hamiltonian, and hence in a non-relativistic setting, integration over high energies exceeding the (relativistic) rest energy Mc^2 of the electron is not justified. Accordingly, we restrict the integration over $\mathbf{k}$ in (7.7.40) for energies $\hbar|\mathbf{k}|c$ smaller than Mc^2 by providing an upper bound cut-off at Mc^2 as an upper limit to the integral in question and as a natural cut-off for the applicability of the non-relativistic treatment. The main values taken by the atomic electron $|\mathbf{p}|$ are of the order αMc, and hence $(|\mathbf{p}|/Mc)\,\hbar|\mathbf{k}|c$, in order of magnitude, is suppressed by a factor of α relative $\hbar|\mathbf{k}|c$.

Accordingly, one formally replaces (7.7.40) by

$$\delta E_n^{\mathrm{Ren}} = -\frac{\alpha\hbar^2}{4\pi^2 M^2} \int_{(|\mathbf{k}|<Mc/\hbar)} \frac{d^3\mathbf{k}}{|\mathbf{k}|\,(\hbar|\mathbf{k}|c)} \langle \psi_n | \, \mathbf{p} \cdot \mathbf{e}_\lambda(\mathbf{k}) \, (H_C - E_n)$$

$$\times \frac{1}{(E_n - H_C - \hbar|\mathbf{k}|c)} \mathbf{e}_\lambda(\mathbf{k}) \cdot \mathbf{p} \, |\psi_n\rangle$$

(7.7.41)

or with the integration expressed in terms of the photon energy $\hbar\omega$, we have

$$\delta E_n^{\mathrm{Ren}} = \frac{\alpha}{4\pi^2 M^2 c^2} \sum_\lambda \int_0^{Mc^2} d(\hbar\omega) \, d\Omega \, \langle \psi_n | \, \mathbf{p} \cdot \mathbf{e}_\lambda(\mathbf{k}) \, (H_C - E_n)$$

---

[21] The convergence has resulted by retaining the exponential factors in (7.7.26) which provide additional powers in $1/|\mathbf{k}|$, as seen from (7.7.34), for $|\mathbf{k}| \to \infty$ in examining the integral in (7.7.40). This was particularly emphasized by Au and Feinberg (1974), Grotch (1981).

$$\times \frac{1}{(H_C - E_n + \hbar\omega)} \mathbf{e}_\lambda(\mathbf{k}) \cdot \mathbf{p} |\psi_n\rangle \qquad (7.7.42)$$

which is known as Bethe's non-relativistic approximation.[22] We note that the values taken by $(H_C - E_n)$ may be comparable to $\hbar\omega$, in its integration range, and both terms should be kept in the denominator in (7.7.42).[23]

We may sum over λ in (7.7.42) by using (7.7.4). Also to carry out the Ω-integral, we note the following angular integration

$$\int d\Omega \left(\delta^{ab} - \frac{k^a k^b}{|\mathbf{k}|} \right) = \frac{8\pi}{3} \delta^{ab}. \qquad (7.7.43)$$

Therefore, by finally carrying out the $(\hbar\omega)$-integral, we obtain

$$\delta E_n^{\text{Ren}} = \frac{2}{3\pi} \frac{\alpha}{M^2 c^2} \left\langle \psi_n \left| [\mathbf{p}, H_C] \cdot \ln\left(\frac{Mc^2}{|H_C - E_n|}\right) \mathbf{p} \right| \psi_n \right\rangle \qquad (7.7.44)$$

where we have replaced $\mathbf{p}(H_C - E_n)$ by $[\mathbf{p}, H_C]$ in (7.7.44), since $\langle\psi_n|(H_C - E_n)\mathbf{p} = 0$, and noted the large energy scale provided by Mc^2.

The right-hand of (7.7.44) involves the Coulomb Hamiltonian H_C, the hydrogen atomic functions ψ_n and the eigenvalues E_n, which are all *known*.

We are particularly interested in the splitting of the levels $2S_{1/2}$, $2P_{1/2}$. These levels are degenerate not only in the Schrödinger theory but in the Dirac theory as well, in the absence of radiation.

The following expression, involving the above mentioned known quantities, has been evaluated numerically by several authors[24]

$$\frac{\langle 2, \ell | [\mathbf{p}, H_C] \cdot \ln\left(\frac{\text{Ry}}{|H_C - E_n|}\right) \mathbf{p} | 2, \ell \rangle}{\langle 2, 0 | [\mathbf{p}, H_C] \cdot \mathbf{p} | 2, 0 \rangle} = -2.81 \delta_{\ell,0} + .03 \delta_{\ell,1} \qquad (7.7.45)$$

where $n = 2$, Ry stands for the Rydberg energy equal to $Me^4/2\hbar^2 = Mc^2\alpha^2/2$.

Finally by using (7.4.37), it is readily shown (see Problem 7.26) that

$$\langle n, \ell, m | [\mathbf{p}, H_C] \cdot \mathbf{p} | n, \ell, m \rangle = 2\pi e^2 \hbar^2 |\psi_{n\ell m}(0)|^2$$

$$= \frac{2M^3 c^4}{n^3} \alpha^4 \delta_{\ell,0} \qquad (7.7.46)$$

where we have also used (7.3.32) for $Z = 1$.

---

[22] Bethe (1947), in which the expression in (7.7.44) was first derived.

[23] Actually, $H_C - E_n$ provides a lower bound cut-off to the $\hbar\omega$-integral signalling the fact that the Coulomb potential cannot be considered as a perturbation in a perturbation series expansion, in the analysis of the Lamb shift, as seen by the logarithmic singularity in α in (7.7.47).

[24] In particular by Bethe *et al.* (1950), Schwartz and Tiemann (1959).

Equations (7.7.44)–(7.7.46) then lead to the following expression

$$\delta E^{\text{Ren}}_{2,\ell} = \frac{4}{3\pi}\left(\frac{1}{8}\right)\alpha^5 Mc^2\left[\left(\ln\left(\frac{2}{\alpha^2}\right) - 2.81\right)\delta_{\ell,0} + .03\delta_{\ell,1}\right] \qquad (7.7.47)$$

giving for the Lamb shift

$$\frac{\delta E^{\text{Ren}}_{2,0} - \delta E^{\text{Ren}}_{2,1}}{h} = 1043.5\,\text{MHz} \qquad (7.7.48)$$

which compares favorably well with the experimental value of $\simeq 1058\,\text{MHz}$.[25]

There are many other small corrections which contribute to the shift of energy levels and a more detailed treatment will not only necessitate to use the full machinery of (relativistic) quantum electrodynamics but to dwell on some phenomenological aspects as well (see for example, Problem 7.29).

Appendix to §7.7: Counter-Terms and Mass Renormalization

By summing over λ and using (7.7.4), the integral corresponding to the second term on the right-hand side of (7.7.37) becomes

$$-\alpha\frac{\hbar^2}{4\pi^2 M^2}\int\frac{d^3k}{|\mathbf{k}|}\frac{1}{\left(\frac{(\hbar k)^2}{2M} - \frac{\mathbf{p}\cdot\hbar\mathbf{k}}{M} + \hbar\omega\right)}\left(\mathbf{p}^2 - \frac{(\mathbf{p}\cdot\mathbf{k})^2}{|\mathbf{k}|^2}\right) \qquad (\text{A-7.7.1})$$

multiplied by $(2\pi\hbar)^3\,\delta^3(\mathbf{p} - \mathbf{p}')$.

Accordingly, upon integration over $\mathbf{p}'$ in (7.7.37) we obtain to order e^2,

$$\varepsilon(\mathbf{p}) - \frac{\mathbf{p}^2}{2M} = \frac{e^2}{2Mc^2}\langle 0|\mathbf{A}^2_{\text{RAD}}(\mathbf{x})|0\rangle$$

$$-\alpha\frac{\hbar^2}{4\pi^2 M^2}\int\frac{d^3k}{|\mathbf{k}|}\frac{1}{\left(\frac{(\hbar k)^2}{2M} - \frac{\mathbf{p}\cdot\hbar\mathbf{k}}{M} + \hbar\omega\right)}\left(\mathbf{p}^2 - \frac{(\mathbf{p}\cdot\mathbf{k})^2}{|\mathbf{k}|^2}\right). \qquad (\text{A-7.7.2})$$

The latter may be conveniently rewritten as

$$\varepsilon(\mathbf{p}) = \left(1 - \frac{\delta M}{M}\right)\frac{\mathbf{p}^2}{2M} + a + b(\mathbf{p}) \qquad (\text{A-7.7.3})$$

where

$$\frac{\delta M}{M} = \frac{8\alpha}{3\pi}\int_0^\infty\frac{d(\hbar\omega)}{(\hbar\omega + 2Mc^2)} \qquad (\text{A-7.7.4})$$

whose evaluation requires an ultraviolet, i.e., a high-energy, cut-off,

---

[25] The early experiment was due to Lamb and Retherford (1947).

$$a = \frac{e^2}{2Mc^2} \langle 0 | \mathbf{A}_{\text{RAD}}^2(\mathbf{x}) | 0 \rangle \qquad (A\text{-}7.7.5)$$

$$b(\mathbf{p}) = -\frac{\alpha \hbar^3}{4\pi^2 M^3} p^a p^b p^c \int \frac{d^3 k}{|\mathbf{k}|} \frac{1}{\left(\frac{(\hbar k)^2}{2M} + \hbar \omega\right)}$$

$$\times \frac{1}{\left(\frac{(\hbar k)^2}{2M} - \frac{\mathbf{p} \cdot \hbar \mathbf{k}}{M} + \hbar \omega\right)} k^a \left(\delta^{bc} - \frac{k^b k^c}{|\mathbf{k}|^2}\right). \qquad (A\text{-}7.7.6)$$

The moral of the calculation leading to (A-7.7.3) is the following.
We may rewrite the first term on the right-hand side of (A-7.7.3) as

$$\left(1 - \frac{\delta M}{M}\right) \frac{\mathbf{p}^2}{2M} \simeq \frac{\mathbf{p}^2}{2(M + \delta M)}. \qquad (A\text{-}7.7.7)$$

That is, the mass parameter M one starts with in the Hamiltonian does not represent the physically observed mass. Due to the ever present radiation accompanying the (charged) electron, its mass gets shifted from M to $M + \delta M$, and the latter is actually the physically observed mass of the electron $M_{\text{phys}} \simeq M + \delta M$. It is also referred to as the *renormalized mass*.

Accordingly, if rewrite the Hamiltonian as

$$H = \frac{\mathbf{p}^2}{2M_0} + H_{\text{I}} + H_{0,\text{RAD}} \qquad (A\text{-}7.7.8)$$

then M_0 does not represent the physically observed mass. It is referred to as the *bare mass*. One may, however, rewrite $M_0 \simeq M_{\text{phys}} - \delta M$, to obtain

$$\frac{\mathbf{p}^2}{2M_0} \simeq \frac{\mathbf{p}^2}{2M_{\text{phys}}} + \frac{\delta M}{M_{\text{phys}}} \frac{\mathbf{p}^2}{2M_{\text{phys}}}. \qquad (A\text{-}7.7.9)$$

Hence if we express the Hamiltonian in terms of the physical mass M_{phys}, then an additional term arises in the Hamiltonian, referred to as a *counter-term*, which contributes in making the kinetic energy one calculates to be $\mathbf{p}^2/2M_{\text{phys}}$, as it should be. The situation is, however, more complicated than that, in that we have also to subtract the term $a + b(\mathbf{p})$ from H, to ensure that the kinetic energy is just $\mathbf{p}^2/2M_{\text{phys}}$.

All told, we define a new Hamiltonian

$$H' = \frac{\mathbf{p}^2}{2M} - \frac{e^2}{|\mathbf{x}|} + H_{\text{I}} + H_{0,\text{RAD}} + \delta H_c \qquad (A\text{-}7.7.10)$$

to replace (7.7.8), where the counter-term δH_c is given by

$$\delta H_c = \frac{\delta M}{M} \frac{\mathbf{p}^2}{2M} - a - b(\mathbf{p}) \qquad (A\text{-}7.7.11)$$

to *lowest* order, where $b(\mathbf{p})$ is now an operator, and M in (A-7.7.10), (A-7.7.11) now denotes the *physically observed mass*. This is what we have done in computing δE_n^{Ren} in (7.7.40) with M in it denoting the physical mass. We note that although δM in (A-7.7.4) is divergent, it does not have to be computed (as a cut-off integral) explicitly in calculating the physical quantity in (7.7.40). Also the term in (7.7.27), here denoted by a, is automatically cancelled out in the process of renormalization.

For the non-relativistic electron, considered above, it is easily shown (see Problem 7.28) that

$$b(\mathbf{p}) = \alpha \frac{\mathbf{p}^2}{2M} \times \mathcal{O}\left(\frac{|\mathbf{p}|}{Mc}\right) \quad \text{(A-7.7.12)}$$

for $|\mathbf{p}|/Mc \ll 1$, hence the last term in (A-7.7.11) is small, consistent with the fact that the kinetic energy of the electron is $\mathbf{p}^2/2M$, in the ever presence of radiation, where M is the renormalized mass.[26]

7.8 Decay of Excited States

As discussed through (7.7.29)–(7.7.32), by inserting the $+i\varepsilon$, $\varepsilon \to +0$, factor in the denominator in (7.7.28) giving $(E_n - H_C - \hbar\omega + i\varepsilon)^{-1}$, we obtain from (7.7.42) for the decay constant Γ in (7.7.31), (7.7.32),

$$\frac{\Gamma}{2} = -\operatorname{Im} \delta E_n^{\text{Ren}}\bigg|_{i\varepsilon} = \frac{\alpha}{4\pi M^2 c^2}\left(\frac{8\pi}{3}\right)\int d(\hbar\omega)$$

$$\times \langle \psi_n | [\mathbf{p}, H_C] \delta(\hbar\omega + H_C - E_n) \cdot \mathbf{p} | \psi_n \rangle, \varepsilon \to +0 \quad (7.8.1)$$

Since $\hbar\omega > 0$, the resolution of the identity of H_C, inserted just before the Dirac delta in (7.8.1) will be restricted to

$$\sum_{\substack{n' \\ (E_{n'} < E_n)}} |\psi_{n'}\rangle\langle\psi_{n'}| \quad (7.8.2)$$

giving from (7.8.1) the explicit expression

$$\Gamma = \frac{4\alpha}{3M^2c^2} \sum_{\substack{n' \\ (E_{n'} < E_n)}} |\langle \psi_{n'} | \mathbf{p} | \psi_n \rangle|^2 (E_n - E_{n'}). \quad (7.8.3)$$

The norm of a state $|\phi_n, t\rangle$ will involve the damping factor[27]

---

[26] For a mathematically rigorous treatment of renormalization theory, see: Manoukian (1983).
[27] See also §3.5.

7 Intricacies of the Hydrogen Atom

$$e^{-\Gamma t/\hbar} \tag{7.8.4}$$

for $t > 0$, and if we translate t by $\hbar/\Gamma$ the exponential factor will be reduced by the factor e^{-1}, defining $\hbar/\Gamma$ as the mean lifetime of the state specified by n and related quantum numbers.

For any Hamiltonian $H = \mathbf{p}^2/2M + U(\mathbf{x})$, it is easily shown that

$$\mathbf{p} = \frac{iM}{\hbar}[H, \mathbf{x}]. \tag{7.8.5}$$

Accordingly, we may rewrite (7.8.3) as

$$\Gamma = \frac{4\alpha}{3}\left(\frac{1}{\hbar^2 c^2}\right)\sum_{\substack{n' \\ (E_{n'} < E_n)}} (E_n - E_{n'})^3 |\langle \psi_{n'}|\mathbf{x}|\psi_n\rangle|^2. \tag{7.8.6}$$

where $e\langle \psi_{n'}|\mathbf{x}|\psi_n\rangle$ is referred to as the electric dipole moment associated with the transition from state specified by n to n'.

As an application of (7.8.6), we compute the mean lifetime of the state $2P_{1/2}$. The only state to which a transition may occur in this case is the $1S_{1/2}$ state.

Thus we are led to evaluate the matrix elements $\langle n', \ell', m'|\mathbf{x}|n, \ell, m\rangle$ with $n' = 1, \ell' = 0, m' = 0, n = 2, \ell = 1, m = 0, \pm 1$.

For the radial functions in question we have from (7.3.14)

$$R_{10}(r) = \left(\frac{1}{a_0}\right)^{3/2} 2e^{-r/a_0} \tag{7.8.7}$$

$$R_{21}(r) = \left(\frac{1}{a_0}\right)^{3/2} \frac{r}{2\sqrt{6}} e^{-r/2a_0}. \tag{7.8.8}$$

The radial integration part of $\langle 1, 0, 0|\mathbf{x}|2, 1, m\rangle$ is given from (7.8.7), (7.8.8) to be

$$\int_0^\infty dr\, r^3 R_{10}(r) R_{21}(r) = 8a_0 \left(\frac{2}{3}\right)^{9/2}. \tag{7.8.9}$$

For the angular integration part, we use the expansions of the components of the vector $\mathbf{x}$ in terms of spherical harmonics as given in (5.3.52)–(5.3.54) and use the orthogonality relation of the latter to obtain

$$\left\langle Y_{00}\left|\frac{\mathbf{x}}{r}\right|Y_{1m}\right\rangle = \frac{1}{\sqrt{6}}\left(\delta_{m,-1} - \delta_{m,1}, i(\delta_{m,-1} + \delta_{m,1}), \sqrt{2}\delta_{m,0}\right). \tag{7.8.10}$$

From (7.8.9), (7.8.10) we have

$$|\langle 1, 0, 0|\mathbf{x}|2, 1, m\rangle|^2 = \frac{2^{15}}{3^{10}} a_0^2 \tag{7.8.11}$$

which is independent of m and is a statement of rotational invariance and that there is no preferred direction in space. One may then average (7.8.11) over m, for the initial state, with arbitrary weights. [This independence of m is explicitly seen in the general case as well given in (7.8.15) below.]

Finally using the expression for the energy shift

$$E_2 - E_1 = \frac{3}{8} Mc^2 \alpha^5 \tag{7.8.12}$$

we obtain from (7.8.6), (7.8.11), (7.8.12)

$$\Gamma = \left(\frac{2}{3}\right)^8 Mc^2 \alpha^5 \tag{7.8.13}$$

corresponding to a frequency of 99.75 MHz which is much smaller than the Lamb shift between the $2S_{1/2}$ and $2P_{1/2}$ states. Γ is also referred to as the decay width of the state in question.

From (7.8.13) we obtain for the mean lifetime of the $2P_{1/2}$ state

$$\frac{\hbar}{\Gamma} \simeq 1.595 \times 10^{-9}\,\text{s}. \tag{7.8.14}$$

To evaluate Γ in (7.8.6) for an arbitrary initial state $|n, \ell, m\rangle$, we may use again the expressions for the components of the vector $\mathbf{x}$ expressed in terms of spherical harmonics, as given in (5.3.52)–(5.3.54), and the integral (5.8.36) involving the product of three spherical harmonics with $\ell_2 = 1, m_2 = 0, \pm 1, \ell_1 = \ell, m_1 = m$. A long but straightforward computation gives

$$\sum_{m'} \left|\left\langle Y_{\ell' m'} \left|\frac{\mathbf{x}}{r}\right| Y_{\ell m}\right\rangle\right|^2 = \frac{(\ell+1)}{(2\ell+1)}\delta_{\ell',\ell+1} + \frac{\ell}{(2\ell+1)}\delta_{\ell',\ell-1} \tag{7.8.15}$$

which is again independent of m and is a statement of rotational invariance and that there is no preferred direction in space.

As in (7.8.9), one is also led to evaluate the integrals

$$I_{\ell',\ell}^{n',n} = \int_0^\infty dr\, r^3 R_{n'\ell'}(r) R_{n\ell}(r) \tag{7.8.16}$$

in particular, for $\ell' = \ell + 1$, and for $\ell \neq 0$, $\ell' = \ell - 1$, as seen from (7.8.15).

We note that the expression in (7.8.6) *applies to other potentials* $U(\mathbf{x})$ *as well*, not only to the Coulomb one, where now the states $|\psi_n\rangle$ correspond to the potential $U(\mathbf{x})$, for a charged particle, of charge $\pm|e|$, interacting with radiation and this given potential. To this end note that the expression on the right-hand side of (7.8.1) follows directly from (7.7.28) by making the replacement $E_n = E_n + i\varepsilon$, and neglecting the momentum of the photon $\hbar\mathbf{k}$ in $(\mathbf{p} - \hbar\mathbf{k})^2/2M$ in (7.7.34). The latter is equivalent to the replacement of the exponential factors

$$e^{i\mathbf{k}\cdot\mathbf{x}}[\,\cdot\,]e^{-i\mathbf{k}\cdot\mathbf{x}} \tag{7.8.17}$$

in (7.7.28), (7.7.34), (7.7.35) by unity. That is, it is obtained by assuming that the wavelength of radiation is much larger than the spatial extension of the system (e.g., the atom) so that $\exp(i\mathbf{k}\cdot\mathbf{x}) \simeq 1$. This is the so-called *long wavelength approximation* associated with dipole radiation.

7.9 The Hydrogen Atom in External Electromagnetic Fields

7.9.1 The Atom in an External Magnetic Field

Quite generally, the shift of the energy levels of an atom induced by an external magnetic field is referred to as the Zeeman effect. We consider the hydrogen atom in an external uniform magnetic field $\mathbf{B}_{\text{ext}}$ such that the shift induced by the field is large in comparison to the one corresponding to the hyperfine structure but much smaller in comparison to the Rydberg — the energy unit of the elementary hydrogen atom.

In view to the above application, this amounts to replacing $\mathbf{B}$ and $\mathbf{A}$ in (7.4.32)/(7.4.42), respectively, by $\mathbf{B}_{\text{ext}}$ and $\mathbf{A}_{\text{ext}}$, and the latter will be taken as

$$\mathbf{A}_{\text{ext}} = \frac{1}{2}(\mathbf{B}_{\text{ext}} \times \mathbf{x}). \tag{7.9.1}$$

Accordingly, for the Hamiltonian, in question, we first consider the expression

$$H = H_{\text{C}} + V_{\text{F}} - \frac{e\hbar}{2Mc}\boldsymbol{\sigma}\cdot\mathbf{B}_{\text{ext}} - \frac{e}{Mc}\mathbf{A}_{\text{ext}}\cdot\mathbf{p} \tag{7.9.2}$$

where we have omitted the $\mathbf{A}_{\text{ext}}^2$ term in comparison to $\mathbf{A}_{\text{ext}}\cdot\mathbf{p}$, based on an order of magnitude estimate given below. Here V_{F} is the potential in (7.5.2) responsible for the fine-structure splitting, and H_{C} denotes the Coulomb potential with potential energy $-e^2/|\mathbf{x}|$. In reference to the $\mathbf{A}_{\text{ext}}\cdot\mathbf{p}$ term, we also note that $\boldsymbol{\nabla}\cdot\mathbf{A}_{\text{ext}} = 0$.

With the magnetic field $\mathbf{B}_{\text{ext}}$ chosen along the z- (3-) axis: $\mathbf{B}_{\text{ext}} = (0,0,B)$, and with $\mathbf{A} = B(-y,x,0)/2$, we may write

$$-\frac{e}{Mc}\mathbf{A}_{\text{ext}}\cdot\mathbf{p} = -\frac{eB}{2Mc}L_z. \tag{7.9.3}$$

For the magnetic quantum number m small, this term, as an order of magnitude estimate, is of the order $-e\hbar B/Mc$. We consider magnetic fields B such that

$$\frac{|eB|\hbar}{2Mc} \ll \text{Ry} \tag{7.9.4}$$

where Ry denotes the Rydberg unit of energy $\text{Ry} = Me^4/2\hbar^2 \equiv Mc^2\alpha^2/2$.

For the quadratic term in $\mathbf{A}_{\text{ext}}$, we have

7.9 The Hydrogen Atom in External Electromagnetic Fields

$$\frac{e^2}{2Mc^2}\mathbf{A}_{\text{ext}}^2 = \frac{e^2 B^2}{8Mc^2}(x^2+y^2)$$

$$= \left(\frac{e\hbar B}{Mc}\right)^2 \frac{M}{8\hbar^2}(x^2+y^2). \quad (7.9.5)$$

For the principal quantum number n small, we may estimate (x^2+y^2) by the Bohr radius squared $a_0^2 = (\hbar^2/Me^2)^2$. That is, as an order of magnitude estimate

$$\frac{e^2}{2Mc^2}\mathbf{A}_{\text{ext}}^2 \sim \frac{1}{16}\left(\frac{e\hbar B}{Mc}\right)^2 \frac{1}{\text{Ry}} \ll \frac{|eB|\hbar}{2Mc} \quad (7.9.6)$$

where we have used (7.9.4). Accordingly, for n small, one may formally neglect the term quadratic in $\mathbf{A}_{\text{ext}}$ in comparison to the linear term for magnetic fields B satisfying (7.9.4).[28,29]

We may then rewrite (7.9.2) as

$$H = H_C + H_I \quad (7.9.7)$$

with

$$H_I = V_F - \frac{eB}{2Mc}(L_z + 2S_z) \quad (7.9.8)$$

and $S_z = \hbar\sigma_3/2$, such that (7.9.4), (7.9.6) hold true for $|m|$, n small.

With $\mathbf{J}^2$ written as in (7.5.3), we note that since $(L_z + 2S_z)$ does not commute with $\mathbf{S}\cdot\mathbf{L}$, $\mathbf{J}^2$ does not commute with H. But H commutes with $\mathbf{L}^2$, J_z and $\mathbf{S}^2$.

We are thus led to consider a linear combination of the states $\psi_{j,\ell,m_j,n}$ in (7.5.4), for different j, with the angular parts $\langle\theta,\phi|j,\ell,m_j\rangle$ as given in (A-7.5.4), (A-7.5.5). In detail

$$\psi_{j,\ell,m_j,n}(\mathbf{x}) = \frac{R_{n\ell}(r)}{\sqrt{2\ell+1}}\begin{bmatrix}\sqrt{\ell+\tfrac{1}{2}+m_j}\,Y_{\ell,m_j-1/2}(\theta,\phi)\\ \sqrt{\ell+\tfrac{1}{2}-m_j}\,Y_{\ell,m_j+1/2}(\theta,\phi)\end{bmatrix} \quad (7.9.9)$$

for $j = \ell+1/2$, $|m_j| \leq \ell+1/2$

$$\psi_{j,\ell,m_j,n}(\mathbf{x}) = \frac{R_{n\ell}(r)}{\sqrt{2\ell+1}}\begin{bmatrix}-\sqrt{\ell+\tfrac{1}{2}-m_j}\,Y_{\ell,m_j-1/2}(\theta,\phi)\\ \sqrt{\ell+\tfrac{1}{2}+m_j}\,Y_{\ell,m_j+1/2}(\theta,\phi)\end{bmatrix} \quad (7.9.10)$$

for $j = \ell-1/2$, $|m_j| \leq \ell-1/2$.

---

[28] More precisely, we deal with principal quantum numbers n such that $|eB|\hbar/Mc$ is much smaller than the differences between successive energy levels E_n.

[29] The contribution of the $e^2\mathbf{A}^2/2Mc^2$ term will be also analyzed in Chapter 12.

For $\ell \neq 0$, we introduce linear combinations

$$\phi_{n\ell m'} = \sum_{j=\ell \pm 1/2} a_{(j)}(\ell, m')\psi_{j,\ell,m',n}(\mathbf{x}) \qquad (7.9.11)$$

for $|m'| \leq \ell - 1/2$, and the states

$$\phi_{n,\ell,m'}(\mathbf{x}) = \psi_{j,\ell,m',n}|_{j=\ell+1/2} \qquad (7.9.12)$$

for $m' = \pm(\ell + 1/2)$.

For a sufficiently *strong* magnetic field B such that V_F is small compared to the second term in (7.9.8) and such that the conditions in (7.9.4), (7.9.6) hold true, we may consider the Hamiltonian

$$H' = H_C - \frac{eB}{2Mc}(L_z + 2S_z). \qquad (7.9.13)$$

[Later we treat V_F as a perturbation to H'.]

The eigenvectors of (7.9.13) are given by

$$\psi_{n\ell m}(\mathbf{x})\begin{pmatrix}1\\0\end{pmatrix}, \qquad \psi_{n\ell m}(\mathbf{x})\begin{pmatrix}0\\1\end{pmatrix} \qquad (7.9.14)$$

where $\psi_{n\ell m}(\mathbf{x})$ are the Coulomb wavefunctions given in (7.3.18), with corresponding eigenvalues

$$E_n + \eta(m + 2m_s), \qquad m_s = \pm 1/2 \qquad (7.9.15)$$

respectively, where $E_n = -\text{Ry}/n^2$, and we have set

$$-\frac{e\hbar B}{2Mc} \equiv \eta > 0. \qquad (7.9.16)$$

The Zeeman effect for a strong magnetic field is usually referred to as the Paschen-Back effect.

To treat V_F as a perturbation to H', we consider the linear combination introduced in (7.9.11), and the states in (7.9.12). The eigenvectors in (7.9.14) may be equivalently rewritten in the form of (7.9.11), (7.9.12) as follows. For $\ell \neq 0$, the first eigenvector may be rewritten as in (7.9.11) (see Problem 7.31) with

$$a_{(\ell+1/2)} = \sqrt{\frac{\ell + 1/2 + m'}{2\ell + 1}}, \qquad a_{(\ell-1/2)} = -\sqrt{\frac{\ell + 1/2 - m'}{2\ell + 1}} \qquad (7.9.17)$$

corresponding to the eigenvalue $E_n + \eta(m' + 1/2)$ with $m' = m + 1/2$, $|m'| \leq \ell - 1/2$, while for the second eigenvector,

$$a_{(\ell+1/2)} = \sqrt{\frac{\ell + 1/2 - m'}{2\ell + 1}}, \qquad a_{(\ell-1/2)} = \sqrt{\frac{\ell + 1/2 + m'}{2\ell + 1}} \qquad (7.9.18)$$

7.9 The Hydrogen Atom in External Electromagnetic Fields

corresponding to the eigenvalue $E_n + \eta(m' - 1/2)$, with $m' = m - 1/2$, $|m'| \leq \ell - 1/2$. For $m' = \pm(\ell + 1/2)$, the eigenvectors are given in (7.9.12), corresponding respectively, to the eigenvalues $E_n + \eta(\ell + 1)$, $E_n - \eta(\ell + 1)$.

Since V_F commutes with $\mathbf{L}^2, J_z$, it is *diagonal* not only in $\psi_{j,\ell,m_j,n}$ but also in $\phi_{n\ell m'}$.

The corrections to the eigenvalues of H' are then given by

$$\Delta E_{n\ell m'} = \langle \phi_{n\ell m'} | V_F | \phi_{n\ell m'} \rangle. \tag{7.9.19}$$

The latter is easily evaluated to be

$$\Delta E^{\pm}_{n\ell' m'} = \frac{\varepsilon_+ + \varepsilon_-}{2} + \eta\left(m' \pm \frac{1}{2}\right) \tag{7.9.20}$$

for $|m'| \leq \ell - 1/2$,

$$\Delta E^{\pm}_{n\ell' m'} = \varepsilon_+ + \eta\left(\ell \pm \frac{1}{2}\right) \tag{7.9.21}$$

for $m' = \pm(\ell + 1/2)$, where,

$$\varepsilon_\pm = \Delta E_{n,j\ell}|_{j=\ell \pm 1/2} \tag{7.9.22}$$

and the $\Delta E_{n,j,\ell}$ are the fine-structure corrections in (7.5.19).

More generally, since we are interested in the splitting of the energy level E_n, for given fixed n, due to the interaction term H_I, assuming that the latter induces a small correction to E_n, we consider the matrix element of $H_C + H_I$ in (7.9.7) between the state $R_{n\ell}(r)$, thus defining an effective Hamiltonian

$$H_{\text{eff}}(n, \ell) = \int d^3\mathbf{x}\, R_{n\ell}(r)\, [H_C + H_I]\, R_{n\ell}(r). \tag{7.9.23}$$

For $\ell = 0$, we have from (7.5.15),

$$H_{\text{eff}}(n, 0) = E_n - \frac{\alpha^2}{n} E_n \left(1 - \frac{3}{4n}\right) + \frac{2\eta S_z}{\hbar} \tag{7.9.24}$$

and for $\ell \neq 0$, we have from (7.5.16), (7.5.18), in particular, that

$$H_{\text{eff}}(n, \ell) = E_n - \frac{\alpha^2}{n} E_n \left[\left(\frac{1}{\ell + 1/2} - \frac{3}{4n}\right)\right.$$

$$\left. - \frac{1}{2\hbar^2} \frac{(\mathbf{J}^2 - \hbar^2 \ell(\ell+1) - \hbar^2 3/4)}{\ell(\ell + 1/2)(\ell + 1)}\right] + \frac{\eta}{\hbar}(L_z + 2S_z). \tag{7.9.25}$$

These effective Hamiltonians, as the Hamiltonian in (7.9.2), commute with J_z. Note in particular that $H_{\text{eff}}(n, \ell)$ in (7.9.23) does not commute with $\mathbf{J}^2$.

7 Intricacies of the Hydrogen Atom

The eigenvalue problem is now formulated as follows.
The eigenvectors corresponding to (7.9.24) are given by

$$\begin{bmatrix} Y_{0,0}(\theta,\phi) \\ 0 \end{bmatrix}, \quad \begin{bmatrix} 0 \\ Y_{0,0}(\theta,\phi) \end{bmatrix} \tag{7.9.26}$$

with eigenvalues

$$E_n - \frac{\alpha^2}{n} E_n \left(1 - \frac{3}{4n}\right) \pm \eta \tag{7.9.27}$$

respectively.

From (7.9.12), for $m_j \equiv m' = \pm(\ell+1/2)$, $\mathbf{J}^2 \psi_{j,\ell,m',n} = \hbar^2 j(j+1) \psi_{j,\ell,m,n'}$, with $j = (\ell + 1/2)$. That is, for $m' = \pm(\ell + 1/2)$, the eigenvectors of (7.9.25) are given by

$$\begin{bmatrix} Y_{\ell,\ell}(\theta,\phi) \\ 0 \end{bmatrix}, \quad \begin{bmatrix} 0 \\ Y_{\ell,-\ell}(\theta,\phi) \end{bmatrix} \tag{7.9.28}$$

with respective eigenvalues

$$E_n + \Delta E_{n,j,\ell}\big|_{j=\ell+1/2} \pm \eta(\ell+1) = E_n + \varepsilon_+ \pm \eta(\ell+1) \tag{7.9.29}$$

using the notation in (7.9.22).

For $|m'| \leqslant \ell - 1/2$, the situation is more involved. To this end, we consider eigenvectors, as obtained from (7.9.11), to be

$$\chi_{\ell,m'} = \frac{u}{\sqrt{2\ell+1}} \begin{bmatrix} \sqrt{\ell + \frac{1}{2} + m'}\, Y_{\ell,m'-1/2} \\ \sqrt{\ell + \frac{1}{2} - m'}\, Y_{\ell,m'+1/2} \end{bmatrix}$$

$$+ \frac{v}{\sqrt{2\ell+1}} \begin{bmatrix} -\sqrt{\ell + \frac{1}{2} - m'}\, Y_{\ell,m'-1/2} \\ \sqrt{\ell + \frac{1}{2} + m'}\, Y_{\ell,m'+1/2} \end{bmatrix}. \tag{7.9.30}$$

The eigenvalue equation then reads

$$H_{\text{eff}}(n,\ell) \chi_{\ell,m'} = (E_n + \varepsilon_{n\ell m'}) \chi_{\ell,m'}. \tag{7.9.31}$$

Upon substitution of (7.9.30) in (7.9.31) using (7.9.25), and the definitions in (7.9.22), we obtain

$$u = v \sqrt{\frac{\ell + 1/2 - m'}{\ell + 1/2 + m'}} \left(\frac{\varepsilon - \varepsilon_- - \eta(m' + 1/2)}{\varepsilon - \varepsilon_+ - \eta(m' + 1/2)} \right) \tag{7.9.32}$$

$$u = -v \sqrt{\frac{\ell + 1/2 + m'}{\ell + 1/2 - m'}} \left(\frac{\varepsilon - \varepsilon_- - \eta(m' - 1/2)}{\varepsilon - \varepsilon_+ - \eta(m' - 1/2)} \right) \tag{7.9.33}$$

7.9 The Hydrogen Atom in External Electromagnetic Fields

where ε stands for $\varepsilon_{n\ell m'}$, and by setting $\varepsilon - \eta m' = \varepsilon'$, we obtain the quadratic equation

$$\varepsilon'^2 - \varepsilon'(\varepsilon_+ + \varepsilon_-) + \left(\varepsilon_+ \varepsilon_- - \frac{\eta^2}{4}\right) - m'\frac{(\varepsilon_+ - \varepsilon_-)}{2\ell + 1}\eta = 0. \quad (7.9.34)$$

The solutions of (7.9.32)–(7.9.34) are readily obtained, by using in the process that $|u|^2 + |v|^2 = 1$, to be

$$\varepsilon^{\pm}_{n\ell m'} = \frac{\varepsilon_+ + \varepsilon_-}{2} + \Delta\varepsilon \left[m'\left(\frac{\eta}{\Delta\varepsilon}\right) \pm \frac{1}{2}\sqrt{1 + \left(\frac{\eta}{\Delta\varepsilon}\right)^2 + \frac{4m'}{2\ell+1}\left(\frac{\eta}{\Delta\varepsilon}\right)}\right] \quad (7.9.35)$$

where

$$\Delta\varepsilon = \varepsilon_+ - \varepsilon_- \equiv \frac{\alpha}{n}E_n\frac{1}{\ell(\ell+1)} \quad (7.9.36)$$

and

$$u^{\pm} = \frac{1}{\sqrt{2}}\left(1 + \frac{1}{Z_{\pm}}\left(1 + \frac{2m'}{2\ell+1}\left(\frac{\eta}{\Delta\varepsilon}\right)\right)\right)^{1/2} \quad (7.9.37)$$

$$v^{\pm} = \mp\frac{1}{\sqrt{2}}\left(1 - \frac{1}{Z_{\pm}}\left(1 + \frac{2m'}{2\ell+1}\left(\frac{\eta}{\Delta\varepsilon}\right)\right)\right)^{1/2} \quad (7.9.38)$$

where

$$Z_{\pm} = \pm\sqrt{1 + \left(\frac{\eta}{\Delta\varepsilon}\right)^2 + \frac{4m'}{2\ell+1}\left(\frac{\eta}{\Delta\varepsilon}\right)}. \quad (7.9.39)$$

For a strong magnetic field $|\eta/\Delta\varepsilon| \gg 1$, (7.9.35) reduces to

$$\varepsilon^{\pm}_{n\ell m'} \simeq \frac{\varepsilon_+ + \varepsilon_-}{2} + \eta\left(m' \pm \frac{1}{2}\right) \quad (7.9.40)$$

which together with (7.9.29), (7.9.27), coincide with the earlier solution in (7.9.20), (7.9.21). We also note that

$$u^{\pm} \to \sqrt{\frac{\ell + 1/2 \pm m'}{2\ell + 1}}, \quad v^{\pm} \to \mp\sqrt{\frac{\ell + 1/2 \mp m'}{2\ell + 1}} \quad (7.9.41)$$

which coincide with the coefficients in (7.9.17), (7.9.18).

For a weak magnetic field $|\eta/\Delta\varepsilon| \ll 1$,

$$\varepsilon^{\pm}_{n\ell m'} \simeq \varepsilon^{\pm} + \eta m'\left(1 \pm \frac{1}{2\ell+1}\right) \quad (7.9.42)$$

which upon combining with the results in (7.9.27), (7.9.29), provide a shift induced by the magnetic field given by

$$\Delta\varepsilon_{j\ell m_j} = \eta m_j g_{j,\ell} \quad (7.9.43)$$

where $|m_j| \leqslant j$,
$$g_{j,\ell} = \left(\frac{j+1/2}{\ell+1/2}\right) \qquad (7.9.44)$$
with the latter referred to as the Landé g-factor (see also (5.7.32)). The expression in (7.9.44) follows by noting that for $m_j = \pm(\ell+1/2), j = \ell + 1/2, m_j g_{j,\ell} = \pm(\ell+1)$, while for $|m_j| \leqslant \ell - 1/2, j = \ell \pm 1/2, g_{j,\ell} = (2\ell+1 \pm 1)/(2\ell+1)$.

The splitting in (7.9.43) takes the spin of the electron into account and is referred to as the *anomalous* Zeeman effect — a rather unfortunate nomenclature.[30] When the spin is not taken into account or if one is dealing with a spinless particle, the effect is usually referred to as the normal Zeeman effect, with $m_j \to m, g_{j,\ell} \to 1$.

7.9.2 The Atom in an External Electric Field

The shift of energy levels of an atom induced by an external electric field is generally referred to as the Stark effect. We consider the hydrogen atom in an external uniform electric field $\mathcal{E}$ such as the splittings induced by the latter are much smaller than that of the fine-structure but much larger than that of the hyperfine-structure.

Accordingly, we consider the Hamiltonian
$$H = H_\mathrm{C} + V_\mathrm{F} - e\mathcal{E}z, \qquad z = r\cos\theta \qquad (7.9.45)$$
where we have taken the electric field to point in the z-direction, with $-e\mathcal{E}z$ taken formally as a perturbation.

The energy levels of $H_\mathrm{C} + V_\mathrm{F}$, with the V_F contribution taken small in comparison to the differences between the energy levels E_n provided by the Coulomb potential, is determined by the two quantum numbers n and j and are given by
$$E_{n,j} = E_n - \frac{\alpha^2}{n}E_n\left[\frac{1}{j+1/2} - \frac{3}{4n}\right]. \qquad (7.9.46)$$
as obtained in (7.5.19)

We are considering shifts in $E_{n,j}$, as induced by the electric field $\mathcal{E}$, for given fixed pairs (n,j). The perturbation $-e\mathcal{E}z$ commutes with J_z but does *not* commute with $\mathbf{L}^2$ due to its dependence on θ. Accordingly, we are led to consider linear combinations, denoted by ϕ_{n,j,m_j}, of the states in (7.9.9), (7.9.10) with $\ell = j \pm 1/2$.

We may then introduce the effective Hamiltonian
$$H_\mathrm{eff}(n_j, j, m_j) = \sum_{\ell=j\pm 1/2} \sum_{\ell'=j\pm 1/2} |n, j, \ell, m_j\rangle$$

---

[30] Historically, the theory of the normal Zeeman effect ran into problems when confronting experiments until the theory of the anomalous one was used based on the incorporation of the spin of the electron in an atom.

7.9 The Hydrogen Atom in External Electromagnetic Fields

$$\times \langle n,j,\ell,m_j|H|n,j,\ell',m_j\rangle \langle n,j,\ell',m_j|. \quad (7.9.47)$$

Now we rewrite (7.9.9), (7.9.10) in terms of j on their right-hand sides:

$$\psi_{j,\ell,m_j,n} = \frac{R_{n,j-1/2}}{\sqrt{2j}} \begin{bmatrix} \sqrt{j+m_j}\, Y_{j-1/2,m_j-1/2} \\ \sqrt{j-m_j}\, Y_{j-1/2,m_j+1/2} \end{bmatrix} \quad (7.9.48)$$

for $\ell = j - 1/2$,

$$\psi_{j,\ell,m_j,n} = \frac{R_{n,j+1/2}}{\sqrt{2(j+1)}} \begin{bmatrix} -\sqrt{j+1-m_j}\, Y_{j+1/2,m_j-1/2} \\ \sqrt{j+1+m_j}\, Y_{j+1/2,m_j+1/2} \end{bmatrix} \quad (7.9.49)$$

for $\ell = j + 1/2$.
By setting

$$\phi_{j,m_j,n} = \sum_{\ell=j\pm 1/2} a_\ell \psi_{j,\ell,m_j,n} \quad (7.9.50)$$

the eigenvalue equation then reads

$$H_{\text{eff}}(n,j,m_j)\phi_{j,m_j,n} = E_{j,m_j,n}\phi_{j,m_j,n}. \quad (7.9.51)$$

To evaluate the matrix elements, $\langle n,j,\ell,m_j|H|n,j,\ell',m_j\rangle$, we recall that the matrix elements of $(H_\text{C} + V_\text{F})$ are given by $E_{n,j}\delta_{\ell\ell'}$, with $E_{n,j}$ given in (7.9.46). To find the matrix elements of $r\cos\theta$, we use the identity in (5.8.41) to carry out the angular integration. To this ends, the latter identity leads for all ℓ, ℓ', m, m':

$$\int d\Omega\, Y_{\ell'm'}(\theta,\phi)\cos\theta\, Y_{\ell m}(\theta,\phi)$$

$$= \sqrt{\frac{(\ell+1)^2 - m^2}{(2\ell+1)(2\ell+3)}}\delta_{\ell',\ell+1}\delta_{m',m} + \sqrt{\frac{\ell^2 - m^2}{4\ell^2 - 1}}\delta_{\ell',\ell-1}\delta_{m',m}. \quad (7.9.52)$$

Thus for the radial integration only the product $r^3 R_{n,j-1/2}(r)R_{n,j+1/2}(r)$ will contribute. An application of the properties of the associated Laguerre polynomials in (7.3.16) also shows (see Problem 7.33) from (7.3.14) that

$$\int_0^\infty dr\, r^3 R_{n,j-1/2}(r)R_{n,j+1/2}(r) = -\frac{3}{4}a_0 n\sqrt{(2n)^2 - (2j+1)^2}. \quad (7.9.53)$$

All told, we obtain from (7.9.52), (7.9.53)

$$\langle n,j,j\pm 1/2,m_j|r\cos\theta|n,j,j\mp 1/2,m_j\rangle$$

$$= \frac{3}{8}a_0 n m_j \frac{\sqrt{(2n)^2 - (2j+1)^2}}{j(j+1)} \equiv M' \quad (7.9.54)$$

and zero for the diagonal element $\ell = \ell'$.[31]

Upon substitution of (7.9.54) in the 2×2 matrix equation in (7.9.51), we obtain the necessary condition (the secular equation) for the existence of non-trivial solutions,

$$\det \begin{bmatrix} E_{n,j} - E_{j,m_j,n} & -eM'\mathcal{E} \\ -eM'\mathcal{E} & E_{n,j} - E_{j,m_j,n} \end{bmatrix} = 0. \tag{7.9.55}$$

That is, the energy levels are given by

$$E^{\pm}_{j,m_j,n} = E_{n,j} \pm \frac{3}{8} e a_0 \mathcal{E} m_j n \frac{\sqrt{(2n)^2 - (2j+1)^2}}{j(j+1)}. \tag{7.9.56}$$

For an electric field $\mathcal{E}$ strong enough such that the fine-structure effect may be neglected, it amounts to taking the Hamiltonian as

$$H = H_C - e\mathcal{E}z. \tag{7.9.57}$$

The latter commutes with J_z but does not commute with $\mathbf{J}^2$ and $\mathbf{L}^2$. Accordingly, the above procedure leads to considering linear combinations of several states for n not too small and the analysis becomes laborious.

Problems

7.1. Show that for the probability distribution of the kinetic energy, in particular, one has

$$\text{Prob}\,[T \geqslant T_0] \leqslant \frac{1}{T_0} \langle T \rangle$$

for any $T_0 > 0$ and infer from (7.1.16) that for the hydrogen atom

$$\text{Prob}\,[T \geqslant T_0] < \frac{1}{T_0} \left(\frac{2\mu e^4}{\hbar^2} \right).$$

7.2. For any linear combination

$$\psi(\mathbf{x}) = \sum_{n,\ell,m} a_{n\ell m} \psi_{n\ell m}(\mathbf{x})$$

with the $\psi_{n\ell m}(\mathbf{x})$ denoting the eigenstates of the hydrogen atom, and

$$\sum_{n,\ell,m} |a_{n\ell m}|^2 = 1$$

---

[31] We note, quite generally, that the matrix elements for $\ell = \ell'$ are zero because the corresponding states have a *definite* parity (see (5.3.43)), while $z = r\cos\theta$ has odd parity.

(i) derive bounds for Prob $[r \leqslant r_0]$, Prob $[T \geqslant T_0]$ and compare them with the ones in (7.1.21), and the one in Problem 7.1, respectively.

(ii) Derive the exact expressions for Prob $[r \leqslant r_0]$, Prob $[T \geqslant T_0]$, for the hydrogen atom in its ground-state. [Hint: It may be useful to show that the ground-state in the momentum description is given by $\Phi(\mathbf{p}) = 8\sqrt{\pi} a_0^{3/2} \left[1 + \mathbf{p}^2 a_0^2/\hbar^2\right]^{-2}$ normalized as
$$\int \frac{d^3\mathbf{p}}{(2\pi\hbar)^3} |\Phi(\mathbf{p})|^2 = 1.$$

(iii) Evaluate the left-hand side of the inequality (7.1.13) for any of the eigenstates $|\psi_{n\ell m}\rangle$, and show that it is consistent with the inequality in question.

7.3. With p_r defined in (7.2.2), show that

(i) $p_r^2 = -\hbar^2 \left(\dfrac{\partial^2}{\partial r^2} + \dfrac{2}{r} \dfrac{\partial}{\partial r} \right)$

(ii) $[p_r, r] = -i\hbar$

(iii) $\left[p_r, \dfrac{1}{r}\right] = \dfrac{i\hbar}{r^2}$

(iv) $\left[p_r, \dfrac{1}{r^2}\right] = \dfrac{2i\hbar}{r^3}$

(v) $\left[p_r^2, \dfrac{1}{r}\right] = \dfrac{2\hbar}{r^2} \left(ip_r - \dfrac{\hbar}{r} \right)$.

7.4. With the boundary conditions imposed on the functions in (7.2.10) for $r \to \infty$ and $r \to 0_+$, show that the adjoint D_ℓ^+ of D_ℓ is given by the expression in (7.2.11).

7.5. Verify the relations in (7.2.12)–(7.2.15).

7.6. Show that the normalized solution of (7.3.4), according to (7.3.5), is given by (7.3.6).

7.7. Use (7.2.13), (7.2.15), with $\ell \to \ell - 1$, to establish (7.3.10), for $\ell \neq 0$.

7.8. Show that (7.3.11) leads to the differential equation in (7.3.12), and, in particular, for $\ell = n - 1$, it leads to the solution in (7.3.13).

7.9. Show that (7.3.14) coincide with $R_{n,n-1}$, $R_{n,n-2}$ given, respectively, in (7.3.6), (7.3.13) for $\ell = n - 1, n - 2$.

7.10. Use (7.3.24), (7.3.25) to establish the orthogonality relation in (7.3.26), (7.3.27). [Is there an orthogonality relation for $n' = n$, $\ell' \neq \ell$ for the $R_{n\ell}$?]

7.11. Derive the expectation values in (7.3.34)–(7.3.38).

7.12. Prove the equalities in (7.4.26), (7.4.27). [Note the positions of the various brackets in these equations.]

7.13. Derive the operator identity in (7.4.31).

7.14. Derive the fundamental Poisson equation (7.4.36).

7.15. Derive afresh the expressions for V_F, V_{HF} starting from (7.4.2) for a spin 0 charged particle interacting with a proton.

7.16. Find the fine-structure correction corresponding to the one in (7.5.19) for the spin 0 case in Problem 7.15.

7.17. Show that the operators $\mathbf{F}^2$, $\mathbf{J}^2$, $\mathbf{S}^2$, $\mathbf{I}^2$, $\mathbf{L}^2$, F_z introduced in §7.6 commute in pairs.

7.18. Find the simultaneous eigenstates of the operators in Problem 7.17 in terms of the spherical harmonics and the spin eigenstates of the electron and proton.

7.19. Find the hyperfine correction corresponding to the one in (7.6.47) for the spin 0 case in Problem 7.15. Is there such a correction for $\ell = 0$ states?

7.20. Derive the operator identities in (7.6.33)–(7.6.36). Here it is worth recalling, in particular, the operator equality in (7.5.17).

7.21. Show that the matrix elements of (7.6.35), (7.6.36) lead to the expressions as indicated in (7.6.37), (7.6.38).

7.22. Derive the equalities in (7.6.37), (7.6.38) directly from the Wigner-Eckart Theorem in §5.7, §5.8.

7.23. Extend as much as possible, the analyses given in §7.1–7.3 to the attractive $1/r$ potential in two dimensional space.

7.24. Find the coefficient of the orthogonality relation of the photon states $\langle \mathbf{k}, \lambda | \mathbf{k}', \lambda' \rangle \propto \delta^3(\mathbf{k}-\mathbf{k}')\delta_{\lambda\lambda'}$ from the completeness relation in (7.7.5).

7.25. Provide a physical explanation if the extreme case with $Z_3 = 0$ is realized in (7.7.14).

7.26. Prove the equality in (7.7.46) by using in the process (7.4.37), (7.3.32) for $Z \to 1$.

7.27. Consider the addition of the spin part $-\frac{e\hbar}{2mc}\boldsymbol{\sigma} \cdot \mathbf{B}_{\text{RAD}}$, where $\mathbf{B}_{\text{RAD}} = \nabla \times \mathbf{A}_{\text{RAD}}$ to the Hamiltonian in (7.7.7).

(i) Show that
$$\langle 0 | \mathbf{B}_{\text{RAD}} | \mathbf{k}, \lambda \rangle = i\sqrt{4\pi}\sqrt{\hbar c}\, e^{i\mathbf{k}\cdot\mathbf{x}} \mathbf{k} \times \mathbf{e}_\lambda(\mathbf{k}).$$

(ii) Show that $\mathbf{p} \cdot \mathbf{e}_\lambda [\,] \mathbf{e}_\lambda \cdot \mathbf{p}$ in (7.7.26) becomes simply replaced by $\boldsymbol{\eta}(\mathbf{p}, \mathbf{k}) \cdot \mathbf{e}_\lambda [\,] \mathbf{e}_\lambda \cdot \boldsymbol{\eta}^\dagger(\mathbf{p}, \mathbf{k})$ where
$$\boldsymbol{\eta}(\mathbf{p}, \mathbf{k}) = \left(\mathbf{p} + \frac{i\hbar}{2}\boldsymbol{\sigma} \times \mathbf{k}\right).$$

(iii) Show that for any operator Q which *commutes* with $\boldsymbol{\sigma}$ and $\mathbf{k}$, the following relation holds true
$$\sum_\lambda \boldsymbol{\eta}\cdot\mathbf{e}_\lambda\, Q \mathbf{e}_\lambda\cdot\boldsymbol{\eta}^\dagger = \left(\mathbf{p} - \frac{\mathbf{k}\cdot\mathbf{p}\,\mathbf{k}}{|\mathbf{k}|^2}\right)\cdot Q\mathbf{p} + \frac{\hbar^2}{2}\mathbf{k}^2 Q + \frac{i\hbar}{2}\boldsymbol{\sigma}\cdot(\mathbf{k} \times [Q, \mathbf{p}]).$$

(iv) Carry out the renormalization in the appendix to §7.7 by including spin as described in (i).

(v) Investigate then the contribution of spin in the calculation of §7.7.

7.28. By explicit integration of $b(\mathbf{p})$ in (A-7.7.6) derive (A-7.7.12).

7.29. Suppose that the charge distribution of the proton is given by

$$\rho(\mathbf{x}) = \frac{1}{8\pi\gamma^3}\exp(-r/\gamma), \qquad r=|\mathbf{x}|, \quad \gamma>0$$

normalized to unity, with the empirical data for the root mean square radius of the proton $\sqrt{\langle r^2\rangle}=0.81\times 10^{-15}$ m.
(i) Find the value of γ.
(ii) With the Coulomb potential $U(\mathbf{x})=-e^2/|\mathbf{x}|$ now replaced by

$$U'_C(\mathbf{x}) = -e^2\int d^2\mathbf{x}\,\frac{\rho(\mathbf{x}')}{|\mathbf{x}-\mathbf{x}'|}$$

estimate the energy level shifts, with

$$\Delta U_C(\mathbf{x}) = U'_C(\mathbf{x}) - U_C(\mathbf{x})$$

treated as a perturbation.

7.30. Provide the details of the computations leading to the results given through (7.8.9)–(7.8.11) in determining the mean-life of the $2P_{1/2}$ state.

7.31. Show that the eigenvectors in (7.9.14) may be rewritten as in (7.9.11) with coefficients given in (7.9.17), (7.9.18).

7.32. Derive the relations in (7.9.32), (7.9.33) and the quadratic equation (7.9.34) for the eigenvalues. Finally show that the solutions of these equations are as given in (7.9.35)–(7.9.39)

7.33. Use the expression for the associated Laguerre polynomials in (7.3.16)/(7.3.15) and the relation of $R_{n,\ell}(r)$ to them in (7.3.14) to prove (7.9.53). Finally show that (7.9.52), (7.9.53) lead to the result given in (7.9.54).

7.34. The Schrödinger equation, with the Coulomb potential $-e^2/|\mathbf{x}|$, can also be separated in the so-called parabolic coordinate system. The latter may be defined by introducing three variables $\xi,\eta,\phi, 0\leqslant \xi<\infty$, $0\leqslant\eta<\infty, 0\leqslant\phi\leqslant 2\pi$, with

$$x = \sqrt{\xi\eta}\cos\phi, \quad y=\sqrt{\xi\eta}\sin\phi, \quad z=\frac{1}{2}(\xi-\eta).$$

In particular note that $r=(\xi+\eta)/2$, $\xi\eta=r^2\sin^2\theta$, $\phi=\tan^{-1}(y/x)$.
(i) Show that the Laplacian in this coordinate system is given by

$$\nabla^2 = \frac{4}{(\xi+\eta)}\left[\frac{\partial}{\partial\xi}\xi\frac{\partial}{\partial\xi}+\frac{\partial}{\partial\eta}\eta\frac{\partial}{\partial\eta}\right]+\frac{1}{\xi\eta}\frac{\partial^2}{\partial\phi^2}.$$

(ii) Prove that the 3D volume element is given by $(\xi+\eta)\,d\xi\,d\eta\,d\phi/4$.
(iii) Carry out a separation of variables of the Schrödinger equation, with the Coulomb potential, to show that the eigenvectors in the (ξ,η,ϕ) representation may be written as

7 Intricacies of the Hydrogen Atom

$$\psi(\mathbf{x}) = \frac{\chi_1(\xi)}{\sqrt{\xi}} \frac{\chi_2(\eta)}{\sqrt{\eta}} \frac{e^{im\phi}}{\sqrt{2\pi}}$$

where m is the magnetic quantum number, and χ_1, χ_2 satisfy the equations

$$\left[-\frac{d^2}{d\xi^2} + \frac{(m^2-1)}{4\xi^2} - \frac{MZ_1 e^2}{\hbar^2 \xi} - \frac{ME}{2\hbar^2} \right] \chi_1(\xi) = 0$$

$$\left[-\frac{d^2}{d\eta^2} + \frac{(m^2-1)}{4\eta^2} - \frac{MZ_2 e^2}{\hbar^2 \eta} - \frac{ME}{2\hbar^2} \right] \chi_2(\eta) = 0$$

with Z_1, Z_2 as separation constants and $Z_1 + Z_2 = 1$. [For a hydrogenic atom with potential $-Ze^2/|\mathbf{x}|$, then $Z_1 + Z_2 = Z$.] You may express part of the solutions of the above equations, as factors, in terms of the associated Laguerre polynomials in (7.3.15), (7.3.16) as functions of ξ, η.

8

Quantum Physics of Spin 1/2 and Two-Level Systems; Quantum Predictions Using Such Systems

Spin 1/2 and two-level systems are simple enough that quite often they allow explicit solutions to their underlying problems and provide a wealth of information on quantum systems, in general. This chapter is devoted entirely to these structures and their intricate details. It includes important quantum predictions that are made by their direct analyses which may be tested experimentally.

General properties of the above systems are studied in §8.1, including the exponential decay law (§3.5) in two-level systems. The Pauli Hamiltonian as a generalization of the Schrödinger equation of a non-relativistic spin 1/2 particle is discussed in §8.2 and makes contact with supersymmetry introduced in §2.9 and further elaborated upon in §4.7, §6.5. The so-called Landau levels are treated in §8.3 and special emphasis is put on the g-factor of the electron. Spin precession and accompanying radiation losses is the subject matter of §8.4. A derivation of the anomalous magnetic moment of the non-relativistic electron in the ever presence of radiation accompanying a charged particle is given in §8.5. This gives a deviation of the g-factor of the electron from the value 2 which is remarkably quite accurate. Density operators (§1.5) and the scattering of spin 1/2 particles off spin 0 and spin 1/2 targets are studied in §8.6. The analysis carried out in §1.10 on probability and measurement is extended in much detail in §8.7 emphasizing the role of the environment, surrounding a physical system and a measuring device, and the so-called quantum decoherence as well as the quantum superposition law in the light of classical notions of measurements. The Ramsey oscillatory fields method, based on the Ramsey apparatus, is introduced in §8.8 which provides interesting applications to interference phenomena, spin flips, and in monitoring spin, in general, as a particle moves in different magnetic field zones. The role of the superposition law for macroscopic systems, such as Schrödinger's cat, and quasi-macroscopic (mesoscopic) systems, often referred to as kittens, and the important role of quantum decoherence, due to the environment, are studied in §8.9 as extensions of the work in §8.7. Bell's test together with the analysis of basic processes in the light of Local Hidden Variables is the subject matter of §8.10. Quantum teleportation and quantum cryptography which rely on fundamental and mysterious aspects of quantum theory are treated in §8.11. The rotation of a spinor by 2π radians, initially introduced in §2.8 ((2.8.68), (2.8.70)), is analyzed from a practical dynamical point of view in §8.12. The theoretical foundations of geometric phases are developed in §8.13. There has been much interest in recent years in geometric phases both experimentally and theoretically. Finally we provide an analytical quantum dynamical treatment of the Stern-Gerlach effect for charged as well as neutral spin 1/2 particles in §8.14.

8.1 General Properties of Spin 1/2 and Two-Level Systems

Much has been developed in the previous chapters on spin, in general, and spin 1/2, in particular.[1] Here we discuss some of general aspects of spin 1/2 which are useful in describing its quantum dynamics and treating problems associated with quantum measurement. Some pertinent aspects of two-level systems, in general, are also investigated.

8.1.1 General Aspects of Spin 1/2

Under a c.c.w. rotation of a coordinate system about a unit vector $\mathbf{n}$ by an angle φ, a wavefunction, in the coordinate description, transforms as (see (2.7.42))

$$\psi'(\mathbf{x}) = \left(\exp\left(\frac{i}{\hbar}\mathbf{n}\cdot\mathbf{J}\right)\psi\right)(\mathbf{x}) \tag{8.1.1}$$

where $\mathbf{J}$ is the total angular momentum.

The spin $\mathbf{S}$ arises (§2.7, §5.4), by rewriting (8.1.1) in terms of the coordinate labels $\mathbf{x}' = R\mathbf{x}$, with R the rotation matrix defined in (2.1.4), (2.1.24), via (see (2.7.43))

$$\psi'(\mathbf{x}') = \left(\exp\left(\frac{i}{\hbar}\mathbf{n}\cdot\mathbf{S}\right)\psi\right)(\mathbf{x}). \tag{8.1.2}$$

As an angular momentum operator, the spin components satisfy the commutation relations (see (2.7.10)),

$$\left[S^i, S^j\right] = i\hbar\varepsilon^{ijk}S^k. \tag{8.1.3}$$

An elementary way of obtaining a representation of the spin operator $\mathbf{S}$ for spin 1/2 is the following. With the spin quantization along the z-axis (§5.4), the eigenvalue equations

---
[1] Cf., §2.7, §2.8, §5.4, §5.10.

8.1 General Properties of Spin 1/2 & Two-Level Systems

$$S^3 \begin{pmatrix} 1 \\ 0 \end{pmatrix} = \frac{\hbar}{2} \begin{pmatrix} 1 \\ 0 \end{pmatrix}, \qquad S^3 \begin{pmatrix} 0 \\ 1 \end{pmatrix} = -\frac{\hbar}{2} \begin{pmatrix} 0 \\ 1 \end{pmatrix} \qquad (8.1.4)$$

for spin along or opposite the orientation of the z-axis, provides the representation

$$S^3 = \frac{\hbar}{2} \begin{pmatrix} 1 & 0 \\ 0 & -1 \end{pmatrix}. \qquad (8.1.5)$$

On the other hand, the commutation relations (see (5.10)),

$$[S^3, S_\pm] = \pm \hbar S_\pm \qquad (8.1.6)$$

where

$$S_\pm = S^1 \pm iS^2 \qquad (8.1.7)$$

give,

$$S_+ = \hbar \begin{pmatrix} 0 & 1 \\ 0 & 0 \end{pmatrix}, \qquad S_- = \hbar \begin{pmatrix} 0 & 0 \\ 1 & 0 \end{pmatrix} \qquad (8.1.8)$$

from which,

$$S^1 = \frac{\hbar}{2} \begin{pmatrix} 0 & 1 \\ 1 & 0 \end{pmatrix}, \qquad S^2 = \frac{\hbar}{2} \begin{pmatrix} 0 & -i \\ i & 0 \end{pmatrix}. \qquad (8.1.9)$$

Upon writing $\mathbf{S} = \hbar \boldsymbol{\sigma}/2$, the so-called Pauli matrices $\sigma_1, \sigma_2, \sigma_3$ as the components of $\boldsymbol{\sigma}$, satisfy the important relations,

$$\sigma_j \sigma_k = \delta_{jk} + i\varepsilon_{jkl}\sigma_l \qquad (8.1.10)$$

from which, or directly from (8.1.3),

$$[\sigma_j, \sigma_k] = 2i\varepsilon_{jkl}\sigma_l. \qquad (8.1.11)$$

The transformation rule in (8.1.2), in terms of the components ψ^a of the spinor ψ (§2.8), reads (see (2.8.1), (2.8.7))

$$\psi'^a(\mathbf{x}') = \left[\exp\left(i\frac{\varphi}{2}\mathbf{n}\cdot\boldsymbol{\sigma}\right)\right]^{ab} \psi^b(\mathbf{x})$$

$$= \left(\cos\frac{\varphi}{2}\delta^{ab} + i\sin\frac{\varphi}{2}[\mathbf{n}\cdot\boldsymbol{\sigma}]^{ab}\right)\psi^b(\mathbf{x}). \qquad (8.1.12)$$

The matrix $\exp\left(i\varphi\, \mathbf{n}\cdot\boldsymbol{\sigma}/2\right)$ is given explicitly by

$$\exp\left(i\frac{\varphi}{2}\mathbf{n}\cdot\boldsymbol{\sigma}\right) = \begin{pmatrix} \cos\frac{\varphi}{2} + in_3\sin\frac{\varphi}{2} & (in_1 + n_2)\sin\frac{\varphi}{2} \\ (in_1 - n_2)\sin\frac{\varphi}{2} & \cos\frac{\varphi}{2} - in_3\sin\frac{\varphi}{2} \end{pmatrix} \qquad (8.1.13)$$

where $\mathbf{n} = (n_1, n_2, n_3)$.

For the spin $\mathbf{S}$ along an arbitrary unit vector $\mathbf{N} = (\sin\theta\cos\phi, \sin\theta\sin\phi, \cos\theta)$,

$$\mathbf{N}\cdot\mathbf{S} \equiv S_{\mathbf{N}} = \frac{\hbar}{2}\begin{pmatrix} \cos\theta & \sin\theta\, e^{-i\phi} \\ \sin\theta\, e^{i\phi} & -\cos\theta \end{pmatrix} \tag{8.1.14}$$

with eigenstates $|1/2, \mathbf{N}\rangle$, $|-1/2, \mathbf{N}\rangle$,

$$S_{\mathbf{N}}|\pm 1/2, \mathbf{N}\rangle = \pm\frac{\hbar}{2}|\pm 1/2, \mathbf{N}\rangle \tag{8.1.15}$$

$$|1/2, \mathbf{N}\rangle = \begin{pmatrix} \cos\frac{\theta}{2}\, e^{-i\phi/2} \\ \sin\frac{\theta}{2}\, e^{i\phi/2} \end{pmatrix}, \quad |-1/2, \mathbf{N}\rangle = \begin{pmatrix} -\sin\frac{\theta}{2}\, e^{-i\phi/2} \\ \cos\frac{\theta}{2}\, e^{i\phi/2} \end{pmatrix}. \tag{8.1.16}$$

In particular,

$$|1/2, \mathbf{N}\rangle = \cos\frac{\theta}{2}\, e^{-i\phi/2}\begin{pmatrix}1\\0\end{pmatrix} + \sin\frac{\theta}{2}\, e^{i\phi/2}\begin{pmatrix}0\\1\end{pmatrix} \tag{8.1.17}$$

and with $\hat{\mathbf{z}}$ a unit vector along the z-axis,

$$|1/2, \hat{\mathbf{z}}\rangle \equiv |+z\rangle = \begin{pmatrix}1\\0\end{pmatrix}, \quad |-1/2, \hat{\mathbf{z}}\rangle \equiv |-z\rangle = \begin{pmatrix}0\\1\end{pmatrix} \tag{8.1.18}$$

one obtains the amplitudes

$$\langle +z | 1/2, \mathbf{N}\rangle = \cos\frac{\theta}{2}\, e^{-i\phi/2} \tag{8.1.19}$$

$$\langle -z | 1/2, \mathbf{N}\rangle = \sin\frac{\theta}{2}\, e^{i\phi/2}. \tag{8.1.20}$$

Similarly,

$$\langle +z | -1/2, \mathbf{N}\rangle = -\sin\frac{\theta}{2}\, e^{-i\phi/2} \tag{8.1.21}$$

$$\langle -z | -1/2, \mathbf{N}\rangle = \cos\frac{\theta}{2}\, e^{i\phi/2}. \tag{8.1.22}$$

For example, for a particle with spin initially prepared in the state $|+z\rangle$, which goes through a filtering process represented by a selective measurement $|1/2, \mathbf{N}\rangle\langle 1/2, \mathbf{N}|$ symbol, the amplitude of a spin flip, i.e., that of the spin of the particle to be found in the state $|-z\rangle$ is given by (§5.4)

$$\langle -z | 1/2, \mathbf{N}\rangle \langle 1/2, \mathbf{N} | +z\rangle = \sin\frac{\theta}{2}\cos\frac{\theta}{2}. \tag{8.1.23}$$

This gives a probability of $(\sin^2\theta)/4$ of a spin flip, with a maximum probability of 25%, for $\theta = \pi/2$.

For an apparatus which allows both components $|\pm 1/2, \mathbf{N}\rangle$ to go through, the spin flip amplitude is reduced to zero

$$\langle -z| \left[|1/2, \mathbf{N}\rangle\langle 1/2, \mathbf{N}| + |-1/2, \mathbf{N}\rangle\langle -1/2, \mathbf{N}| \right] |+z\rangle$$

$$= \langle -z|\mathbf{1}|+z\rangle = \langle -z|+z\rangle = 0. \tag{8.1.24}$$

As another example suppose a measurement of spin of a particle is carried out along a direction specified by a unit vector $\mathbf{n}_1 = (\sin\theta_1, 0, \cos\theta_1)$ followed by a measurement of spin along $\mathbf{n}_2 = (\sin\theta_2, 0, \cos\theta_2)$, where $\theta_1 \neq \theta_2$. If the spin of the particle is initially prepared in the state $|+z\rangle$, then the probability that a measurement of spin along $\mathbf{n}_1$ is found to be parallel to $\mathbf{n}_1$ followed by a measurement of spin along $\mathbf{n}_2$ is found to be, say, parallel to $\mathbf{n}_2$ as well, is given by

$$|\langle +\mathbf{n}_2|+\mathbf{n}_1\rangle \langle +\mathbf{n}_1|+z\rangle|^2 = \cos^2\left(\frac{\theta_1-\theta_2}{2}\right) \cos^2\left(\frac{\theta_1}{2}\right) \tag{8.1.25}$$

where we have used the notation $|+1/2, \mathbf{n}\rangle \equiv |+\mathbf{n}\rangle$. On the other hand for the reverse process of measurement of spin along $\mathbf{n}_2$, and found to be parallel to $\mathbf{n}_2$, followed by a measurement along $\mathbf{n}_1$, and also found to be parallel to $\mathbf{n}_1$, is given by

$$|\langle +\mathbf{n}_1|+\mathbf{n}_2\rangle \langle +\mathbf{n}_2|+z\rangle|^2 = \cos^2\left(\frac{\theta_1-\theta_2}{2}\right) \cos^2\left(\frac{\theta_2}{2}\right) \tag{8.1.26}$$

demonstrating, in particular, the inequivalence of the orders in which the measurements are carried out.

8.1.2 Spin 1/2 in External Magnetic Fields

Consider a neutral particle of spin 1/2, such as the neutron (§8.2), with magnetic moment $\mu < 0$. Restricting for simplicity to the dynamics of the spin only,[2] the Hamiltonian for the interaction of spin with an external uniform time-independent magnetic field $\mathbf{B} = |B|\mathbf{n}$ is given by

$$H = -\mu |B|\mathbf{n}\cdot\boldsymbol{\sigma} = |\mu B|\mathbf{n}\cdot\boldsymbol{\sigma}. \tag{8.1.27}$$

Hence from (8.1.13), the time evolution operator is given by

$$U(t) = \exp\left(-\frac{i}{\hbar}tH\right) = \begin{pmatrix} \cos\frac{\omega t}{2} - in_3\sin\frac{\omega t}{2} & -(in_1+n_2)\sin\frac{\omega t}{2} \\ -(in_1-n_2)\sin\frac{\omega t}{2} & \cos\frac{\omega t}{2} + in_3\sin\frac{\omega t}{2} \end{pmatrix}$$
$$\tag{8.1.28}$$

---

[2] The more complete description, where the kinetic energy of the particle is not neglected, or more precisely is not assumed to be negligible in comparison to the spin part, will be studied in detail in §8.14 in the Stern-Gerlach effect.

8 Quantum Physics of Spin 1/2 & Two-Level Systems

with $\omega = 2|\mu B|/\hbar$. For a particle initially prepared in the state $|+z\rangle$, the amplitude of a spin flip at time $t > 0$, is then

$$\langle -z|U(t)|+z\rangle = -(in_1 - n_2)\sin\frac{\omega t}{2} \tag{8.1.29}$$

giving a probability of spin flip of $(n_1^2 + n_2^2)\sin^2 \omega t/2$. For a magnetic field in the x-y plane and at time $t = \pi/\omega$, the spin flip probability reaches a 100%.

Consider the time-dependent Hamiltonian

$$H(t) = -\mu B \mathbf{R}(t) \cdot \boldsymbol{\sigma} \tag{8.1.30}$$

for the interaction of spin with a time-dependent magnetic field $B\mathbf{R}(t)$, where $\mathbf{R}(t)$ is the *unit* vector

$$\mathbf{R}(t) = \big(\sin\theta\cos\omega t, \sin\theta\sin\omega t, \cos\theta\big). \tag{8.1.31}$$

As an initial state $|\psi(0)\rangle$ of spin, we choose it to be the eigenstate of $H(0)$. That is,

$$|\psi(0)\rangle = \begin{pmatrix} \cos\dfrac{\theta}{2} \\ \sin\dfrac{\theta}{2} \end{pmatrix} \tag{8.1.32}$$

up to a phase factor.

Using the notation $-\mu B = \hbar\omega_0/2$, the Hamiltonian in (8.1.30) may be rewritten as

$$H(t) = \frac{\hbar\omega_0}{2}\begin{pmatrix} \cos\theta & \sin\theta\, e^{-i\omega t} \\ \sin\theta\, e^{i\omega t} & -\cos\theta \end{pmatrix}. \tag{8.1.33}$$

By setting

$$|\psi(t)\rangle = \begin{pmatrix} \alpha_+(t) \\ \alpha_-(t) \end{pmatrix} \tag{8.1.34}$$

the Schrödinger equation leads to the simultaneous equations

$$\dot{\alpha}_+ = -i\frac{\omega_0}{2}\Big[\cos\theta\,\alpha_+ + \sin\theta\, e^{-i\omega t}\alpha_-\Big] \tag{8.1.35}$$

$$\dot{\alpha}_- = i\frac{\omega_0}{2}\Big[\cos\theta\,\alpha_- - \sin\theta\, e^{i\omega t}\alpha_+\Big]. \tag{8.1.36}$$

With the initial condition in (8.1.32), the solutions of (8.1.35), (8.1.36) are readily obtained (see Problem 8.2) to be conveniently written in the form

$$\alpha_+(t) = \cos\frac{\theta}{2}\left[\frac{\Omega + \omega - \omega_0}{2\Omega}e^{i(\Omega-\omega)t/2} + \frac{\Omega - \omega + \omega_0}{2\Omega}e^{-i(\Omega+\omega)t/2}\right] \tag{8.1.37}$$

8.1 General Properties of Spin 1/2 & Two-Level Systems

$$\alpha_-(t) = \sin\frac{\theta}{2} e^{i\omega t} \left[\frac{\Omega - \omega - \omega_0}{2\Omega} e^{i(\Omega-\omega)t/2} + \frac{\Omega + \omega + \omega_0}{2\Omega} e^{-i(\Omega+\omega)t/2} \right] \tag{8.1.38}$$

where

$$\Omega = \sqrt{\omega^2 - 2\omega\omega_0 \cos\theta + \omega_0^2}. \tag{8.1.39}$$

This solution will find an interesting application in §8.13 on geometric phases.

As an alternate initial condition to (8.1.32), consider the following one

$$|\psi(0)\rangle = \begin{pmatrix} \cos\dfrac{\alpha}{2} \\ \sin\dfrac{\alpha}{2} \end{pmatrix} \tag{8.1.40}$$

where[3]

$$\sin\alpha = \frac{\omega_0 \sin\theta}{\Omega} \tag{8.1.41}$$

$$\cos\alpha = \frac{\omega_0 \cos\theta - \omega}{\Omega} \tag{8.1.42}$$

with Ω defined in (8.1.39). The solution then takes the particularly simple form

$$|\psi(t)\rangle = \begin{pmatrix} \cos\dfrac{\alpha}{2} e^{-i\omega t/2} \\ \sin\dfrac{\alpha}{2} e^{i\omega t/2} \end{pmatrix} e^{-it\Omega/2}. \tag{8.1.43}$$

An application of this solution will be made in §8.13 on geometric phases.

Numerous additional applications of spin 1/2 devoted to concrete situations will be given in the remaining sections of this chapter. We next consider the dynamics of spin in a general time-dependent magnetic field.

By absorbing the magnetic moment μ of the particle in question in the magnetic field, we consider the Schrödinger equation

$$i\hbar \frac{d}{dt} |\psi(t)\rangle = H(t) |\psi(t)\rangle \tag{8.1.44}$$

with time-dependent Hamiltonian

$$H(t) = \frac{\hbar}{2} \mathbf{K}(t) \cdot \boldsymbol{\sigma} \tag{8.1.45}$$

where $\mathbf{K}(t)$ is a time-dependent vector.

---

[3] Such an initial condition is also considered in Lin (2002) and references therein.

We prepare the spin of the particle in an initial state $|\psi(0)\rangle$ as an eigenstate of $\mathbf{n}_0 \cdot \boldsymbol{\sigma}$, where $\mathbf{n}_0$ is an arbitrary unit vector,

$$\mathbf{n}_0 = \left(\sin\theta_0 \cos\phi_0, \sin\theta_0 \sin\phi_0, \cos\theta_0 \right) \tag{8.1.46}$$

with

$$\mathbf{n}_0 \cdot \boldsymbol{\sigma} |\psi(0)\rangle = |\psi(0)\rangle. \tag{8.1.47}$$

Hence $|\psi(0)\rangle$ may be chosen to be

$$|\psi(0)\rangle = \begin{pmatrix} \cos\dfrac{\theta_0}{2} \, e^{-i\phi_0/2} \\ \sin\dfrac{\theta_0}{2} \, e^{i\phi_0/2} \end{pmatrix} \tag{8.1.48}$$

up to a phase factor.

Let $\mathbf{n}(t)$ be a unit vector satisfying the equation[4]

$$\dot{\mathbf{n}}(t) = \mathbf{K}(t) \times \mathbf{n}(t) \tag{8.1.49}$$

with initial condition

$$\mathbf{n}(0) = \mathbf{n}_0 \tag{8.1.50}$$

then, as is easily shown below, the solution $|\psi(t)\rangle$ of (8.1.44) is an eigenstate of $\mathbf{n}(t) \cdot \boldsymbol{\sigma}$, that is,

$$\mathbf{n}(t) \cdot \boldsymbol{\sigma} |\psi(t)\rangle = |\psi(t)\rangle. \tag{8.1.51}$$

To establish (8.1.51), we partition the interval $[0, t]$ into infinitesimally small sub-intervals. For $t = 0$, (8.1.51) coincides with (8.1.47) since $\mathbf{n}(0)$ satisfies (8.1.50). Therefore, it is sufficient to show that (8.1.51) is true for infinitesimally small Δt to complete the demonstration by induction.

To the above end, it is readily checked that

$$\frac{d}{dt}\big[\mathbf{n}(t) \cdot \boldsymbol{\sigma} |\psi(t)\rangle\big] = -\frac{i}{2}\mathbf{K}(t) \cdot \boldsymbol{\sigma} \big[\mathbf{n}(t) \cdot \boldsymbol{\sigma} |\psi(t)\rangle\big] \tag{8.1.52}$$

where we have used (8.1.49) and (8.1.44). Hence for $\Delta t \simeq 0$

$$\mathbf{n}(\Delta t) \cdot \boldsymbol{\sigma} |\psi(\Delta t)\rangle = \mathbf{n}(0) \cdot \boldsymbol{\sigma} |\psi(0)\rangle - \frac{i}{2}\mathbf{K}(0) \cdot \boldsymbol{\sigma} \big[\mathbf{n}(0) \cdot \boldsymbol{\sigma} |\psi(0)\rangle\big] \Delta t$$

$$= |\psi(0)\rangle - \frac{i}{2}\mathbf{K}(0) \cdot \boldsymbol{\sigma} |\psi(0)\rangle \Delta t$$

$$= |\psi(\Delta t)\rangle \tag{8.1.53}$$

---

[4] Cf. Lin (2002), and references therein; Wagh and Rakhecha (1993); Feynman et al. (1957).

8.1 General Properties of Spin 1/2 & Two-Level Systems

as following from (8.1.44), (8.1.47), (8.1.52). That is (8.1.51) is true for $t = \Delta t$ as well and hence for all t by induction.

Since the unit vector $\mathbf{n}(t)$, may be parameterized as

$$\mathbf{n}(t) = \big(\sin\theta(t)\cos\phi(t), \sin\theta(t)\sin\phi(t), \cos\theta(t)\big) \tag{8.1.54}$$

the solution of the Schrödinger equation (8.1.44) may be written from (8.1.51) as

$$|\psi(t)\rangle = \begin{pmatrix} \cos\dfrac{\theta(t)}{2}\, e^{-i\phi(t)/2} \\[4pt] \sin\dfrac{\theta(t)}{2}\, e^{i\phi(t)/2} \end{pmatrix} \tag{8.1.55}$$

up to a phase factor.

8.1.3 Two-Level Systems; Exponential Decay

Suppose that transitions occur between two levels, which we denote by $|0\rangle$ and $|1\rangle$ with corresponding energies E_0 and E_1, with $E_1 > E_0$. We may introduce creation $a_F^\dagger$ and annihilation a_F operators connecting these two levels

$$a_F^\dagger |0\rangle = |1\rangle \tag{8.1.56}$$

$$a_F |1\rangle = |0\rangle . \tag{8.1.57}$$

For transitions that are restricted to these two levels, we have $(a_F^\dagger)^2 = 0$, $(a_F)^2 = 0$, for example, and the operators a_F, $a_F^\dagger$ satisfy *anti*-commutation rules (cf. (6.4.2), (6.4.3)).

By representing the states $|0\rangle$, $|1\rangle$, for example, by $(0\ 1)^\top$, $(1\ 0)^\top$, we may write the Hamiltonian, in the absence of interaction, i.e., the free Hamiltonian as

$$H_0 = \begin{pmatrix} E_1 & 0 \\ 0 & E_0 \end{pmatrix} = \frac{E_0 + E_1}{2} + \frac{E_1 - E_0}{2}\sigma_3 \tag{8.1.58}$$

where σ_3 is a Pauli matrix.

For the interaction Hamiltonian causing transitions between these two levels, we consider the simple structure

$$H_I = \begin{pmatrix} 0 & V \\ V^* & 0 \end{pmatrix} \tag{8.1.59}$$

where V may be time-dependent.

Thus upon writing the solution of the time evolution problem as

$$|\psi(t)\rangle = e^{-i(E_0 + E_1)t/2\hbar} \begin{pmatrix} a(t) \\ b(t) \end{pmatrix} \tag{8.1.60}$$

8 Quantum Physics of Spin 1/2 & Two-Level Systems

the Schrödinger equation for $|\psi(t)\rangle$ leads to

$$i\hbar \frac{d}{dt}\begin{pmatrix} a(t) \\ b(t) \end{pmatrix} = \frac{\hbar\omega}{2}\sigma_3 \begin{pmatrix} a(t) \\ b(t) \end{pmatrix} + \begin{pmatrix} 0 & V \\ V^* & 0 \end{pmatrix}\begin{pmatrix} a(t) \\ b(t) \end{pmatrix} \quad (8.1.61)$$

where

$$E_1 - E_0 \equiv \hbar\omega \quad (8.1.62)$$

and we have the normalization condition

$$|a(t)|^2 + |b(t)|^2 = 1. \quad (8.1.63)$$

We may introduce the unit vector

$$\mathbf{n}(t) = \Big(a(t)b^*(t) + a^*(t)b(t), i[a(t)b^*(t) - a^*(t)b(t)], |a(t)|^2 - |b(t)|^2\Big) \quad (8.1.64)$$

and an angular frequency vector

$$\mathbf{\Omega} = \frac{1}{2}\Big(V + V^*, i(V - V^*), \hbar\omega\Big) \quad (8.1.65)$$

depending on V, and it may, in general, be time-dependent. It is then readily checked directly from (8.1.61) that it may be rewritten in the form[5] (8.1.49), i.e.,

$$\frac{d}{dt}\mathbf{n}(t) = \mathbf{\Omega} \times \mathbf{n}(t). \quad (8.1.66)$$

It is easily verified, that with the parametrization $a(t) = \cos(\theta(t)/2)\,e^{-i\phi(t)/2}$, $b(t) = \sin(\theta(t)/2)\,e^{i\phi(t)/2}$, $\mathbf{n}(t)$ in (8.1.64) coincides with the expression in (8.1.54). The system in (8.1.61), in particular, will be considered in §8.12 in analyzing the fundamental problem of the rotation of a spinor, corresponding to a two-level system, by 2π radians.

We close this section, by considering the following Hamiltonian to describe transitions in a two-level system[6]

$$H = \frac{E_0 + E_1}{2} + \frac{E_1 - E_0}{2}\sigma_3 + \sum_k \hbar\omega_k\, b_k^\dagger b_k + H_I \quad (8.1.67)$$

$$H_I = a_F^\dagger \sum_k \lambda_k b_k + a_F \sum_k \lambda_k^* b_k^\dagger \quad (8.1.68)$$

where $a_F^\dagger$, a_F are the creation, annihilation operators introduced in (8.1.56), (8.1.57) and may be represented as

$$a_F^\dagger = \begin{pmatrix} 0 & 1 \\ 0 & 0 \end{pmatrix} \equiv \sigma^+ \quad (8.1.69)$$

---

[5] Feynman et al. (1957), see also see this paper for other related details.
[6] The energy levels may be considered to denote the mid-points of the corresponding energy linewidths.

8.1 General Properties of Spin 1/2 & Two-Level Systems

$$a_F = \begin{pmatrix} 0 & 0 \\ 1 & 0 \end{pmatrix} \equiv \sigma^- \tag{8.1.70}$$

satisfying anti-commutation relations, and $b_k^\dagger$, b_k are creation, annihilation operators of a photon of energy $\hbar\omega_k$ satisfying commutation relations.

Unlike the interaction in (8.1.59), where V is a priori given external potential, the interaction in (8.1.68), which may also cause transitions between the two levels, is due to photons which are treated dynamically though in a simplified manner.

Suppose that the two-level system is initially in the state $|1\rangle$, we then investigate the nature of the survival probability (§3.5) for the two-level system to stay in the state $(1\ 0)^\top$.

To do this, we consider the Schrödinger equation

$$i\hbar \frac{d}{dt}|\psi(t)\rangle = H|\psi(t)\rangle \tag{8.1.71}$$

write

$$|\psi(t)\rangle = e^{-itH_0/\hbar}|\phi(t)\rangle \tag{8.1.72}$$

where H_0 is the free Hamiltonian part in (8.1.67), to obtain (the so-called interaction picture)

$$i\hbar \frac{d}{dt}|\phi(t)\rangle = H_I(t)|\phi(t)\rangle \tag{8.1.73}$$

where

$$H_I(t) = e^{itH_0/\hbar} H_I e^{-itH_0/\hbar}$$

$$= \sigma^+ \sum_k \lambda_k b_k e^{-i(\omega_k-\omega)t} + \sigma^- \sum_k \lambda_k^* b_k^\dagger e^{i(\omega_k-\omega)t}. \tag{8.1.74}$$

Hence upon writing

$$|\phi(t)\rangle = \begin{pmatrix} |A(t)\rangle \\ |B(t)\rangle \end{pmatrix} \tag{8.1.75}$$

in (8.1.73), we have

$$i\hbar \frac{d}{dt}|A(t)\rangle = \sum_k \lambda_k b_k e^{-i(\omega_k-\omega)t}|B(t)\rangle \tag{8.1.76}$$

$$i\hbar \frac{d}{dt}|B(t)\rangle = \sum_k \lambda_k^* b_k^\dagger e^{i(\omega_k-\omega)t}|A(t)\rangle \tag{8.1.77}$$

where we have used the expressions for the matrices in (8.1.69), (8.1.70).

Upon multiplying (8.1.77) by the operator b_k and using the commutation relations of b_k, $b_k^\dagger$, we obtain

$$i\hbar \frac{d}{dt} \langle 0|b_k|B(t)\rangle = \lambda_k^* e^{i(\omega_k - \omega)t} \langle 0|A(t)\rangle \qquad (8.1.78)$$

where we have also used the fact that $\langle 0|b_k^\dagger = 0$, and $|0\rangle$, here, denotes the 'no-photon' state.

Equations (8.1.76), (8.1.78) then yield

$$\frac{d}{dt} \langle 0|A(t)\rangle = -\frac{1}{\hbar^2} \sum_k |\lambda_k|^2 \int_0^t d\tau \, e^{-i(\omega_k - \omega)\tau} \langle 0|A(t-\tau)\rangle. \qquad (8.1.79)$$

The survival probability of the system to stay in the state $\begin{pmatrix} 1 & 0 \end{pmatrix}^\mathsf{T}$ is then

$$P(t) = |\langle 0|A(t)\rangle|^2. \qquad (8.1.80)$$

To see this note that if we write $|\psi(t)\rangle = \begin{pmatrix} \alpha(t) & \beta(t) \end{pmatrix}^\mathsf{T}$, then

$$\begin{pmatrix} \langle 0|\alpha(t)\rangle \\ \langle 0|\beta(t)\rangle \end{pmatrix} = \langle 0| e^{-itH_0/\hbar} \begin{pmatrix} |A(t)\rangle \\ |B(t)\rangle \end{pmatrix}$$

$$= \begin{pmatrix} e^{-iE_1 t/\hbar} \langle 0|A(t)\rangle \\ e^{-iE_0 t/\hbar} \langle 0|B(t)\rangle \end{pmatrix} \qquad (8.1.81)$$

and $|\langle 0|\alpha(t)\rangle|^2 = |\langle 0|A(t)\rangle|^2$, where we have used the fact that $\langle 0|b_k^\dagger = 0$ as arising from the application of the free photon Hamiltonian to $\langle 0|$.

Using a continuous variable extension of the photon energy, replacing the sum over k in (8.1.79) by an integral over ω', and setting

$$\langle 0|A(t)\rangle = F(t) \qquad (8.1.82)$$

we may rewrite (8.1.79) as

$$\frac{d}{dt} F(t) = -\frac{1}{\hbar^2} \int_0^\infty d\omega' \, |\lambda(\omega')|^2 \, n(\omega') \int_0^t d\tau \, e^{-i(\omega' - \omega)\tau} F(t-\tau) \qquad (8.1.83)$$

where $\lambda(\omega')$ is a continuous variable extension and $n(\omega')$ denotes the density of such states.

We use the relation

$$e^{-i(\omega' - \omega)\tau} = \frac{d}{d\tau}\left[\frac{e^{-i(\omega' - \omega)\tau} - 1}{-i(\omega' - \omega)} \right] \qquad (8.1.84)$$

and integrate by parts over τ in (8.1.83) to obtain

$$\frac{d}{dt} F(t) = -F(t)\left(I(t) + \frac{d}{dt} R[F; t] \right) \qquad (8.1.85)$$

8.1 General Properties of Spin 1/2 & Two-Level Systems

where

$$I(t) = \frac{1}{\hbar^2} \int_0^\infty d\omega' \, |\lambda(\omega')|^2 \, n(\omega') \left[\frac{e^{-i(\omega'-\omega)t} - 1}{-i(\omega'-\omega)} \right] \quad (8.1.86)$$

$$R[F;t] = \frac{1}{\hbar^2} \int_0^\infty d\omega' \, |\lambda(\omega')|^2 \, n(\omega')$$

$$\times \int_0^t dt' \int_0^{t'} d\tau \left[\frac{e^{-i(\omega'-\omega)\tau} - 1}{-i(\omega'-\omega)} \right] \left[\frac{1}{F(t')} \frac{d}{dt'} F(t'-\tau) \right]. \quad (8.1.87)$$

Equation (8.1.85) gives for the survival probability in (8.1.80)/(8.1.82),

$$P(t) = C(t) \exp\bigl[-G(t)\bigr],$$

$$G(t) = \int_0^t dt' \, \bigl[I(t') + I^*(t')\bigr] \quad (8.1.88)$$

and

$$C(t) = \bigl|\exp(-R[F;t])\bigr|^2. \quad (8.1.89)$$

The time-integral in (8.1.88) is easily evaluated to yield

$$G(t) = \frac{1}{\hbar^2} \int_0^\infty d\omega' \, |\lambda(\omega')|^2 \, n(\omega') \left[\frac{\sin(\omega'-\omega)t/2}{(\omega'-\omega)/2} \right]^2$$

$$= \frac{2t}{\hbar^2} \int_{-\omega t/2}^\infty dx \, \left|\lambda\!\left(\omega\left[1 + \frac{2x}{\omega t}\right]\right)\right|^2 n\!\left(\omega\left[1 + \frac{2x}{\omega t}\right]\right) \left[\frac{\sin x}{x}\right]^2. \quad (8.1.90)$$

The function $\sin x/x$ peaks at the origin, and is concentrated mainly in the region $|x| \leq \pi$. We make the Markov approximation[7] by assuming that $|\lambda(\omega')|^2 \, n(\omega')$ is a slowly varying function around the point of resonance $\omega' = \omega$ and hence for $\omega t/2 \geq \pi$, it may be taken outside the integral in (8.1.90), evaluated at $\omega' = \omega$, with increasing accuracy for $t \gg 1/\omega$, thus obtaining

$$G(t) \simeq \frac{2t}{\hbar^2} |\lambda(\omega)|^2 \, n(\omega) \int_{-\omega t/2}^\infty dx \left[\frac{\sin x}{x}\right]^2 \quad (8.1.91)$$

and for $t \gg 1/\omega$,[8]

$$G(t) \simeq \frac{2t}{\hbar^2} |\lambda(\omega)|^2 \, n(\omega) \, \pi \quad (8.1.92)$$

which allows us to set

---

[7] The Markov approximation will be discussed further in §12.7 in terms of so-called correlation functions.
[8] Note that for $\hbar\omega$ expressed in eV, $t \gg 1/\omega \simeq (1 \text{ eV}/\hbar\omega) \times 10^{-15}$ s.

432   8 Quantum Physics of Spin 1/2 & Two-Level Systems

$$\frac{2}{\hbar^2} |\lambda(\omega)|^2 \, n(\omega) \, \pi \equiv \gamma \tag{8.1.93}$$

For $1/\gamma \gg t$, but $t \gg 1/\omega$, it is not difficult to show that if we formally replace $(\mathrm{d}F(t' - \tau)/\mathrm{d}t')/F(t')$ by γ, then $C(t) \simeq \exp\left[\mathcal{O}(\gamma^2 t^2)\right]$.

Accordingly, in the above mentioned time limits, we obtain for the survival probability $P(t)$ the exponential decay law

$$P(t) \simeq e^{-\gamma t} \tag{8.1.94}$$

with γ defined in (8.1.93). For t short enough, this defines a weak coupling limit by $1/\gamma \gg t$. The time t, however, is taken large enough so that $t \gg \omega$. In particular, the latter condition implies that we must have $\gamma \ll \omega$ which is a well known property relating decay widths $\Gamma = \hbar\gamma$ and energy shifts $\hbar\omega$.

Remarks 1
1. *The exponential decay law in (8.1.94) holds for t not too small and t not too large, as discussed above and is in the light of the general analysis of decay in §3.5.*
2. *If one formally replaces $|\lambda(\omega')|^2 \, n(\omega')$ by a constant in (8.1.83) and extends the ω' limit of the corresponding integral to $-\infty$, thus obtaining a Dirac delta $\delta(\tau)$ in the integrand, then this equation may be readily integrated to yield the exponential decay law in (8.1.94). Such a procedure, although used by some authors, involves, according to the above analysis, the implicit assumption of a limit set on the time variable, and the exponential decay law (8.1.94) cannot hold true exactly for all t. Otherwise one would run into a contradiction with the general analysis of decay in §3.5.*
3. *Equation (8.1.83) may be also used self-consistently to investigate the behavior of the survival probability $P(t)$ in the truly asymptotic limits $t \to 0$, $t \to \infty$ by basing the analysis on some sufficiency convergence conditions to be satisfied and will be left as an exercise to the reader (see Problem 8.3).*

Another derivation of the exponential decay law (8.1.94) will be given in §12.7 by working directly with the density operator.

8.2 The Pauli Hamiltonian; Supersymmetry

8.2.1 The Pauli Hamiltonian

The non-relativistic quantum dynamical equation of a spin 1/2 charged particle of charge e in an external vector potential **A** and scalar potential U may be derived in the following manner.

One may start with the Schrödinger equation for a free particle

8.2 The Pauli Hamiltonian; Supersymmetry

$$\left(\frac{\mathbf{p}^2}{2M} - i\hbar\frac{\partial}{\partial t}\right)|\psi\rangle = 0 \tag{8.2.1}$$

which upon using the identity (7.4.4) leads to[9]

$$\left[\frac{(\mathbf{p}\cdot\boldsymbol{\sigma})(\mathbf{p}\cdot\boldsymbol{\sigma})}{2M} - i\hbar\frac{\partial}{\partial t}\right]|\psi\rangle = 0. \tag{8.2.2}$$

Now one makes the so-called minimal coupling substitutions

$$\mathbf{p} \longrightarrow \mathbf{p} - \frac{e}{c}\mathbf{A}, \qquad i\hbar\frac{\partial}{\partial t} \longrightarrow i\hbar\frac{\partial}{\partial t} - U \tag{8.2.3}$$

to obtain, by finally using the identity (7.4.31) (see also Problem 7.13),

$$\left[\frac{1}{2M}\left(\mathbf{p} - \frac{e}{c}\mathbf{A}\right)^2 - \frac{ge}{2Mc}\mathbf{S}\cdot\mathbf{B} + U\right]|\psi\rangle = i\hbar\frac{\partial}{\partial t}|\psi\rangle \tag{8.2.4}$$

known as the Pauli equation, where $\mathbf{S} = \hbar\boldsymbol{\sigma}/2$, $g = 2$, $\mathbf{B} = \nabla\times\mathbf{A}$.

The so-called g-factor of the electron, for example, is given approximately by $g = 2(1+\alpha/2\pi)$, rather than 2, according to the leading order correction in quantum electrodynamics, where $\alpha = e^2/\hbar c$ is the fine-structure constant.[10] A computation of g within the realm of quantum mechanics for the interaction of a non-relativistic electron with radiation will be given in §8.5. Its observational aspect will be also discussed there.

By formally replacing $ge\hbar/4Mc$ by μ in (8.2.4) and then taking the limit $e \to 0$ in the latter equation, one also obtains the Pauli equation for a neutral spin 1/2 particle interacting with the magnetic field given by $-\boldsymbol{\mu}\cdot\mathbf{B}$, and $\boldsymbol{\mu} = \mu\boldsymbol{\sigma}$ denoting the magnetic dipole moment. This will be used in the study of the spin precession of the neutron in §8.4, where the radiation loss due to the interaction of its spin with the magnetic field set up by radiation is also investigated.

An application of (8.2.4) was given in §7.9 for the interaction of spin 1/2 with a constant sufficiently strong magnetic field $\mathbf{B}$ along the z-axis[11] such that the potential V_F, responsible for the fine-structure of the hydrogen atom, is small in comparison to the magnetic field contribution, where U is the coulomb potential due to the proton. The Pauli equation in (8.2.4), as it stands, provides an approximation to the more precise treatment, including relativistic corrections given in §7.4, for the hydrogen atom.

---

[9] See also §7.4, in general.

[10] For a phenomenological treatment of the g-factor, one may formally add a term $-(\kappa e/Mc)\mathbf{S}\cdot\mathbf{B}$ to the left-hand side of (8.2.4) giving rise to $-(ge/2Mc)\mathbf{S}\cdot\mathbf{B}$, where now $g = 2(1 + \kappa/2)$, for the interaction with the magnetic field. This, however, is no substitute to a dynamical treatment of g as done in §8.5 for the electron.

[11] See (7.9.13) and below it, where the magnetic field is not too strong so that quadratic term in $\mathbf{A}$ may be neglected (see (7.9.6)).

434 8 Quantum Physics of Spin 1/2 & Two-Level Systems

For $\mathbf{B}$ a uniform time-independent magnetic field, say, along the z-axis, with $U = 0$, the eigenvalue problem corresponding to (8.2.4) will be studied in the next section, and is also discussed below to some extent, leading to the so-called Landau energy levels. Equation (8.2.4) will be used in §8.14 to study the quantum dynamics involved in the Stern-Gerlach effect.

For a purely time-dependent magnetic field $\mathbf{B}(t)$, the contribution of the spin involves an overall multiplicative factor in the solution of (8.2.4) as follows. By setting,

$$|\psi(t)\rangle = \left(\exp\left[\frac{i}{\hbar}\int_0^t dt'\, \boldsymbol{\mu}\cdot \mathbf{B}(t')\right]\right)_+ |\phi(t)\rangle \qquad (8.2.5)$$

where $(\cdot)_+$ denotes the time-ordered product (see Appendix to §2.5), then $|\phi(t)\rangle$ satisfies the simpler equation

$$\left[\frac{1}{2M}\left(\mathbf{p} - \frac{e}{c}\mathbf{A}\right)^2 + U\right]|\phi\rangle = i\hbar\frac{\partial}{\partial t}|\phi\rangle \qquad (8.2.6)$$

where $\boldsymbol{\mu} = \mu\boldsymbol{\sigma}$, $\mu = ge\hbar/4Mc$ in (8.2.5).

In particular, for a constant magnetic field $\mathbf{B} = B\mathbf{n}$, where $\mathbf{n}$ is a constant unit vector, (8.2.5) gives (see (8.1.12)),

$$|\psi\rangle = \left[\cos\left(\frac{\mu B t}{\hbar}\right)\mathbf{1} + i\sin\left(\frac{\mu B t}{\hbar}\right)\mathbf{n}\cdot\boldsymbol{\sigma}\right]|\phi\rangle. \qquad (8.2.7)$$

8.2.2 Supersymmetry

In §4.7 through (4.7.33)–(4.7.38), we have seen that for a magnetic field $\mathbf{B}$ along the z-axis, one may define supersymmetry generators Q, $Q^\dagger$ (see (4.7.34), (4.7.37)) such that in $2D$ and $3D$, the Pauli Hamiltonian in (8.2.4), with $U = 0$, has the formal structure (see §2.9) of a supersymmetric Hamiltonian

$$H = \frac{1}{2\hbar}\{Q, Q^\dagger\} \qquad (8.2.8)$$

taking the forms ($\boldsymbol{\pi} = \mathbf{p} - e\mathbf{A}/c$)

$$H = \frac{\pi_1^2 + \pi_2^2}{2M} - \frac{e\hbar}{4Mc}gB\sigma_3 \qquad (8.2.9)$$

and

$$H = \frac{\pi_1^2 + \pi_2^2 + \pi_3^2}{2M} - \frac{e\hbar}{4Mc}gB\sigma_3 \qquad (8.2.10)$$

respectively, *where* $g = 2$, $\mathbf{B} = (0, 0, B)$, for a particle of mass M.

One of the most attractive features of the supersymmetric Hamiltonian in (8.2.8) is the formal non-negativity of its spectrum (see (4.7.5)).

In particular, for a constant magnetic field $\mathbf{B} = (0, 0, B)$, $B > 0$ with vector potential in the gauge

8.2 The Pauli Hamiltonian; Supersymmetry

$$\mathbf{A} = \frac{B}{2}(-y, x, 0) \tag{8.2.11}$$

with motion restricted in 2D in the x-y plane, we will see in the next section, that the eigenstates are given by

$$\psi_{n,m,\sigma}(\mathbf{r}) = C_{n,m}\,(\rho^2)^{|m|/2}\, L_n^{|m|}(\rho^2)\,\exp\,\mathrm{i}m\phi\,\exp(-\rho^2/2)\begin{pmatrix}\delta_{\sigma,+1}\\ \delta_{\sigma,-1}\end{pmatrix} \tag{8.2.12}$$

where $n = 0, 1, \ldots;\ m = 0, \pm 1, \pm 2, \ldots$,

$$\rho^2 = \frac{M\omega}{2\hbar}r^2, \qquad \omega = \frac{|eB|}{Mc} \tag{8.2.13}$$

$$\mathbf{r} = (x, y), \qquad x = r\cos\phi, \qquad y = r\sin\phi \tag{8.2.14}$$

$$C_{n,m} = \left(\frac{M\omega}{2\pi\hbar}\right)^{1/2} \frac{\left[\left(n + \frac{|m|+m}{2}\right)!\left(n + \frac{|m|-m}{2}\right)!\right]^{1/2}}{(n+|m|)!}(-1)^{n+|m|+m} \tag{8.2.15}$$

for $e < 0$ (as for the electron), and $L_n^{|m|}(\rho^2)$ are the associated Laguerre polynomials introduced in (7.3.16) now of argument ρ^2.

The eigenstates satisfy the normalizability condition

$$\int \mathrm{d}^2\mathbf{r}\ |\psi_{n,m,\sigma}(\mathbf{r})|^2 = 1. \tag{8.2.16}$$

The eigenvalues are given by

$$E_{n,m,\sigma} = \hbar\omega\left(n + \frac{|m|+m}{2} + \frac{1}{2} + \frac{g}{4}\sigma\delta_{\sigma,\pm 1}\right) \tag{8.2.17}$$

for $e < 0$, $n = 0, 1, \ldots;\ m = 0, \pm 1, \ldots$, and we have deliberately kept g in the expression for $E_{n,m,\sigma}$ in (8.2.17).

For $g = 2$, the ground-state energy (which is infinitely degenerate) is *zero*, as it should be by supersymmetry, corresponding to $n = 0$, $m = 0, -1, -2, \ldots$. As a matter of fact a ground-state vector, corresponding to $m = 0$, or -1, or $\ldots$, and $n = 0$, is given from (8.2.12) to be

$$\psi_{0,m,-1}(\mathbf{r}) = \frac{1}{\sqrt{\pi|m|!}}\left(\frac{M\omega}{2\hbar}\right)^{(|m|+1)/2}(x-\mathrm{i}y)^{|m|}\exp\left(-\frac{M\omega}{4\hbar}(x^2+y^2)\right)\begin{pmatrix}0\\1\end{pmatrix}. \tag{8.2.18}$$

On the other hand from (4.7.34)

$$Q = \sqrt{\frac{\hbar}{m}}\left(-\mathrm{i}\left(\frac{\partial}{\partial y} + \frac{M\omega y}{2}\right) + \left(\frac{\partial}{\partial x} + \frac{M\omega x}{2}\right)\right)\begin{pmatrix}0 & 1\\ 0 & 0\end{pmatrix} \tag{8.2.19}$$

for $e < 0$, satisfy the supersymmetry relations (4.7.6), (4.7.7). Condition (4.7.7) is trivially satisfied. Condition (4.7.6) is readily checked by explicitly applying Q in (8.2.19) to (8.2.18) giving

$$Q\psi_{0,m,-1} = 0 \begin{pmatrix} 0 \\ 1 \end{pmatrix} \qquad (8.2.20)$$

as expected for $g = 2$.

8.3 Landau Levels; Expression for the g-Factor

In this section, we derive the expression for the Landau levels quoted in §8.2 through (8.2.11)–(8.2.18), and also obtain a useful relationship between the g-factor and such levels. The latter will be used, in the process in §8.5, to compute the anomalous magnetic moment of the electron.

8.3.1 Landau Levels

Consider the Hamiltonian H in (8.2.9) with the electron ($e < 0$) restricted to a plane with a transverse constant magnetic field to it and vector potential in the gauge given in (8.2.11).

The eigenvalue equation for the Pauli matrix σ_3 may be conveniently written as

$$\sigma_3 \begin{pmatrix} \delta_{\sigma,+1} \\ \delta_{\sigma,-1} \end{pmatrix} = \sigma \begin{pmatrix} \delta_{\sigma,+1} \\ \delta_{\sigma,-1} \end{pmatrix}, \qquad \sigma = \pm 1 \qquad (8.3.1)$$

where we note that σ_3 commutes with H.

Let

$$H_0 = \frac{\pi_1^2 + \pi_2^2}{2M} \qquad (8.3.2)$$

and hence $H = H_0 + |eB|\hbar g \sigma_3 / 4Mc$, where

$$\pi_1 = p_x - \frac{M\omega}{2} y, \qquad \pi_2 = p_y + \frac{M\omega}{2} x \qquad (8.3.3)$$

$$[\pi_1, \pi_2] = -\mathrm{i} M \hbar \omega \qquad (8.3.4)$$

and ω is the so-called cyclotron angular frequency given by

$$\omega = |eB|/Mc. \qquad (8.3.5)$$

Upon defining the operators (see also (4.7.34))

$$A = \frac{1}{\sqrt{2m\hbar\omega}} (\pi_2 + \mathrm{i}\pi_1), \qquad A^\dagger = \frac{1}{\sqrt{2m\hbar\omega}} (\pi_2 - \mathrm{i}\pi_1) \qquad (8.3.6)$$

we have

$$[A, A^\dagger] = 1 \qquad (8.3.7)$$

$$H_0 = \hbar\omega \left(A^\dagger A + 1/2\right) \qquad (8.3.8)$$

$$[H_0, A] = -\hbar\omega A, \qquad [H_0, A^\dagger] = +\hbar\omega A^\dagger \qquad (8.3.9)$$

and the following commutation relationships with the third component $L_z = xp_y - yp_x$ of the orbital angular momentum

$$[L_z, A] = -\hbar A, \quad [L_z, A^\dagger] = +\hbar A^\dagger \qquad (8.3.10)$$

$$[L_z, H_0] = 0. \qquad (8.3.11)$$

The unit of energy $\hbar\omega$ is sometimes referred to as the Landau-Larmor energy.

The commutation relations (8.3.7), (8.3.9) are just those of the harmonic oscillator problem. Hence, by using in the process of the commutativity of L_z with H_0 in (8.3.11), we have simultaneous eigenstates $|N, m\rangle$ of $A^\dagger A$, L_z,

$$A^\dagger A |N, m\rangle = N |N, m\rangle \qquad (8.3.12)$$

$$L_z |N, m\rangle = \hbar m |N, m\rangle \qquad (8.3.13)$$

and by using (8.3.9), (8.3.11) and (6.1.30)

$$A^\dagger |k, q\rangle = \sqrt{k+1} \, |k+1, q+1\rangle \qquad (8.3.14)$$

using a general notation of the states $|k, q\rangle$ for convenience.

For the ground state energy of $A^\dagger A$, we have

$$A^\dagger A |0, -q\rangle = 0 \qquad (8.3.15)$$

$$L_z |0, -q\rangle = -\hbar q |0, -q\rangle \qquad (8.3.16)$$

$$A |0, -q\rangle = 0 \qquad (8.3.17)$$

and, as we will see below, the square-integrability of a state $|0, -q\rangle$ requires that the integer q to be non-negative, i.e., $q = 0, 1, 2, \ldots$.

From the definition of A in (8.3.6), (8.3.17) leads to the following differential equation for $\langle x, y|0, -q\rangle \equiv \psi_{0,-q}(x, y)$

$$\left[-i\left(\frac{\partial}{\partial y} + \frac{M\omega}{2}y\right) + \left(\frac{\partial}{\partial x} + \frac{M\omega}{2}x\right) \right] \psi_{0,-q}(x, y) = 0. \qquad (8.3.18)$$

This equation together (8.3.16), then give the normalized solution

$$\psi_{0,-q}(x, y) = \frac{1}{\sqrt{\pi q!}} \left(\frac{M\omega}{2\hbar}\right)^{(q+1)/2} (x-iy)^q \exp\left(-\frac{M\omega}{4\hbar}(x^2 + y^2)\right) \qquad (8.3.19)$$

and $q = 0, 1, \ldots$ for square-integrability of $\psi_{0,-q}(\mathbf{r})$,

$$\int d^2\mathbf{r} \, |\psi_{0,-q}(\mathbf{r})|^2 = 1 \qquad (8.3.20)$$

where $\mathbf{r} = (x, y)$.

Upon setting $x = r\cos\phi$, $y = r\sin\phi$, (8.3.19) may be rewritten as

$$\psi_{0,-q}(\mathbf{r}) = \frac{\sqrt{M\omega/2\pi\hbar}}{\sqrt{q!}} \left(\frac{M\omega}{2\hbar}r^2\right)^{q/2} \exp(-iq\phi) \exp\left(-\frac{M\omega}{4\hbar}r^2\right). \quad (8.3.21)$$

For the purpose of obtaining all of the eigenvectors, it is convenient to set

$$\sqrt{\frac{M\omega}{2\hbar}}\, r = \rho. \quad (8.3.22)$$

In terms of the variables ρ and ϕ, $A^\dagger$ in (8.3.6) may be rewritten as

$$A^\dagger = \rho e^{i\phi} \left\{-\frac{\partial^2}{\partial \rho^2} + \frac{1}{2} - \frac{i}{2\rho^2}\frac{\partial}{\partial \phi}\right\}. \quad (8.3.23)$$

From (8.3.21), and (8.3.14) with $k=0$, $q \to -q$,

$$\langle \mathbf{r}|0,-q\rangle = \frac{\sqrt{M\omega/2\pi\hbar}}{\sqrt{q!0!}} (\rho^2)^{-q/2}\left[(\rho^2)^q\right]\exp(-iq\phi)\exp\left(-\rho^2/2\right) \quad (8.3.24)$$

$$\langle \mathbf{r}|1,1-q\rangle = \frac{\sqrt{M\omega/2\pi\hbar}}{\sqrt{q!1!}} (\rho^2)^{(1-q)/2}(-1)\left[(\rho^2)^{q-1}(q-\rho^2)\right]$$
$$\times \exp(i(1-q)\phi)\exp\left(-\rho^2/2\right)$$
$$= \frac{\sqrt{M\omega/2\pi\hbar}}{\sqrt{q!1!}} (\rho^2)^{(1-q)/2} \exp(i(1-q)\phi)\exp\left(-\rho^2/2\right)$$
$$\times (-1)\left[\exp\rho^2 \left(\frac{d}{d\rho^2}\right)^1 (\rho^2)^q \exp(-\rho^2)\right]. \quad (8.3.25)$$

Therefore, as an induction hypothesis, suppose that

$$\langle \mathbf{r}|k,k-q\rangle = \frac{\sqrt{M\omega/2\pi\hbar}}{\sqrt{q!k!}} (\rho^2)^{(k-q)/2} \exp(i(k-q)\phi)\exp\left(-\rho^2/2\right)$$
$$\times (-1)^k \left[e^{\rho^2}\left(\frac{d}{d\rho^2}\right)^k (\rho^2)^q \exp -\rho^2\right] \quad (8.3.26)$$

for some $k > 1$, which is obviously true for $k = 0, 1$, then we show that (8.3.26) is true for k replaced by $k+1$ as well. This *directly* follows by explicitly applying $A^\dagger$, in (8.3.23), to (8.3.26) and then using (8.3.14).

It is not difficult to verify that

$$\left[e^{\rho^2}\left(\frac{d}{d\rho^2}\right)^k (\rho^2)^q e^{-\rho^2}\right] = (-1)^{k-q} k! q! \sum_{l=0}^{q} \frac{(-1)^l}{(q-l)!(k-q+l)!} \frac{(\rho^2)^l}{l!},$$

8.3 Landau Levels; Expression for the g-Factor   439

$$\text{for} \quad k > q \tag{8.3.27}$$

and

$$= (\rho^2)^{q-k} k! q! \sum_{l=0}^{q} \frac{(-1)^l}{(k-l)!(q-k+l)!} \frac{(\rho^2)^l}{l!},$$

$$\text{for} \quad q > k \tag{8.3.28}$$

which are polynomials in ρ^2 of degree q. From (7.3.16), we here recognize the sums over l to denote $[(|k-q|+n)!]^{-1}$ times the associated Laguerre polynomial $L_n^{|k-q|}(\rho^2)$, of argument ρ^2, where

$$n = \min(k, q) \tag{8.3.29}$$

using the normalization adopted in (7.3.16).

That is,

$$\left[e^{\rho^2} \left(\frac{d}{d\rho^2} \right)^k (\rho^2)^q e^{-\rho^2} \right] = \frac{(-1)^{(|m|+m)/2} (\rho^2)^{(|m|-m)/2} k! q!}{(|m|+n)!} L_n^{|m|}(\rho^2) \tag{8.3.30}$$

with

$$m = k - q \tag{8.3.31}$$

(see also (8.3.26), (8.3.13)).

We note that (8.3.29), (8.3.31) allow one to write

$$q = n + \frac{|m| - m}{2} \tag{8.3.32}$$

$$k = n + \frac{|m| + m}{2}. \tag{8.3.33}$$

All told, we have from (8.3.1), (8.3.2), (8.3.12), (8.3.13), (8.3.26), (8.3.29)–(8.3.33) for $k = N$, that the eigenvectors $\psi_{n,m,\sigma}$ of H are given in (8.2.12) with eigenvalues in (8.2.17), where

$$E_{n,m,\sigma} = \hbar\omega \left(N + \frac{1}{2} + \frac{g}{4} \sigma \delta_{\sigma,\pm 1} \right) \tag{8.3.34}$$

with $N = n + (|m| + m)/2$ as given in (8.3.33) for $k = N$. [For a spin 0 charged particle with $e < 0$ simply replace g by 0.]

In particular, for the supersymmetric case (§8.2) $g = 2$, and the ground state energy 0 corresponds to $n = 0$, $\sigma = -1$, and $m = 0, -1, -2, \ldots$.

The results derived in this section, as they stand, are also valid for $e > 0$ if the choice $B < 0$ is made, i.e., if the magnetic field is taken to be along the negative of the z-axis.[12]

---

[12] In the literature the choice $eB > 0$ is sometimes made.

8.3.2 Expression for the g-Factor

We derive a convenient expression for the g-factor, as arising from the Landau levels, which will be used in the process of a computation of the anomalous magnetic moment of the electron in §8.5.

To the above end, it is more convenient to consider the problem in the *momentum* description, and work in the gauge specified by a vector potential ($B > 0$)

$$\mathbf{A} = B(-y, 0, 0). \quad (8.3.35)$$

In the momentum description, the Hamiltonian in (8.2.10) is given by

$$H = \frac{1}{2M}\left(\frac{\mathrm{i}|e|B}{c}\hbar\frac{\partial}{\partial p_2} + p_1\right)^2 + \frac{p_2^2}{2M} + \frac{p_3^2}{2M} + \frac{|e|\hbar}{4Mc}g\sigma_3 B \quad (8.3.36)$$

where now g is left arbitrary, and we note that p_1, p_3 are just c-numbers. We may then consider the limiting case with $p_1, p_3 \to 0$ and work in one-dimension along p_2.

Upon setting

$$\Omega = \frac{c}{|e|BM} \quad (8.3.37)$$

one is then led to solve the equation

$$\left[-\frac{\hbar^2}{2m}\frac{\partial^2}{\partial p_2^2} + \frac{M\Omega^2}{2}p_2^2\right]\Psi(p_2) = \varepsilon\Psi(p_2) \quad (8.3.38)$$

where for the energy E we have

$$E = \frac{1}{M^2\Omega^2}\varepsilon + \frac{g\sigma}{4M^2\Omega} \quad (8.3.39)$$

$\sigma = \pm 1$, and the eigenvalues ε are given below.

From the harmonic oscillator problem, the eigenvalues in (8.3.38) are given by $\varepsilon_n = \hbar\Omega(n + 1/2)$, $n = 0, 1, \ldots$, and hence we may write for E in (8.3.39)

$$E(n, \sigma) = \frac{|e|B\sigma\hbar}{4Mc}\left[(4n + 2)\sigma + g\right] \quad (8.3.40)$$

where we have used the fact that $(\sigma)^2 = +1$. This leads to the following useful relationship upon differentiation with respect to B,

$$-\frac{2Mc}{|e|\hbar}\frac{\partial}{\partial B}E(n = 0, \sigma = -1)\bigg|_{B=0, p_1, p_3=0} = \frac{g-2}{2}. \quad (8.3.41)$$

For the ground state corresponding to (8.3.38), we obviously have

$$\Psi_0(p_2) = \left(\frac{M\Omega}{\pi\hbar}\right)^{1/4}\exp\left(-\frac{M\Omega}{2\hbar}p_2^2\right) \quad (8.3.42)$$

normalized with respect to the measure dp_2. With a proper normalization, we will see in §8.5, how the physical $3D$ problem may be reduced to a computation in $1D$ for the anomalous magnetic moment of the electron.

8.4 Spin Precession and Radiation Losses

We consider the Hamiltonian of a neutron, an uncharged particle of spin 1/2, in a uniform and time-independent magnetic field

$$H = \frac{\mathbf{p}^2}{2M_\mathrm{n}} - \mu_\mathrm{n}\boldsymbol{\sigma}\cdot\mathbf{B} \tag{8.4.1}$$

where the magnetic moment μ_n is about

$$\mu_\mathrm{n} = -9.66\times 10^{-27}\ \mathrm{J/T}. \tag{8.4.2}$$

The commutator of the spin $\mathbf{S} = \hbar\boldsymbol{\sigma}/2$ and the Hamiltonian H is easily worked out to be

$$[\mathbf{S}, H] = 2\mathrm{i}\mu_\mathrm{n}\mathbf{S}\times\mathbf{B}. \tag{8.4.3}$$

From the time-development of an operator $O(t) = e^{itH/\hbar}\,O\,e^{-itH/\hbar}$, we have

$$\frac{\mathrm{d}}{\mathrm{d}t}\mathbf{S}(t) = \frac{2\mu_\mathrm{n}}{\hbar}\,\mathbf{S}(t)\times\mathbf{B}. \tag{8.4.4}$$

In particular, for components perpendicular $\perp$ or parallel $\parallel$ to $\mathbf{B}$,

$$\frac{\mathrm{d}^2}{\mathrm{d}t^2}\mathbf{S}_\perp(t) = -\frac{4\mu_\mathrm{n}^2 B^2}{\hbar^2}\mathbf{S}_\perp(t) \tag{8.4.5}$$

$$\frac{\mathrm{d}}{\mathrm{d}t}\mathbf{S}_\parallel(t) = 0. \tag{8.4.6}$$

Taking expectation values of the above yield the solutions

$$\langle\mathbf{S}_\perp(t)\rangle = \mathbf{a}\cos(\omega_\mathrm{n}t + \delta) \tag{8.4.7}$$

$$\langle\mathbf{S}_\parallel(t)\rangle = \langle\mathbf{S}_\parallel(0)\rangle \tag{8.4.8}$$

where $\mathbf{a}$ is a constant vector, showing the precession of the spin about the magnetic field $\mathbf{B}$, with angular frequency

$$\omega_\mathrm{n} = \frac{2|\mu_\mathrm{n} B|}{\hbar} \tag{8.4.9}$$

with the projection $\langle\mathbf{S}_\parallel(t)\rangle$ along $\mathbf{B}$ being constant in time.

Let $\mathbf{B} = (0,0,B)$, $B > 0$, then due to the precession of the spin, the neutron, from an excited state $\begin{pmatrix}1\\0\end{pmatrix}$, for $|\mathbf{p}|\to 0$, with energy

$$E_+ = |\mu_\mathrm{n}|B \tag{8.4.10}$$

may fall to the ground-state $(0\ 1)^\top$, for $|\mathbf{p}|\to 0$, of energy $E_- = -|\mu_\mathrm{n}|B$ by the emission of a photon of energy $\hbar kc = 2|\mu_\mathrm{n}|B$. The mean lifetime of such a transition is, however, too large for all practical values of **B**. For $B \simeq 1$ T, the mean lifetime τ is $\sim 10^{25}$ s, and hence such radiation losses are not significant, in general, unless one is dealing with huge magnetic fields of millions T encountered in astrophysics.

Due to the uncharged nature of the neutron the mean lifetime τ is easily estimated. To this end, the total Hamiltonian of the neutron in the magnetic field **B**, in the presence of radiation is given by (see also §7.7)

$$H_{\mathrm{I}} = \frac{\mathbf{p}^2}{2M_{\mathrm{n}}} + H_{0,\mathrm{RAD}} - \mu_{\mathrm{n}} \sigma_3 B - \mu_{\mathrm{n}} \boldsymbol{\sigma} \cdot (\boldsymbol{\nabla} \times \mathbf{A}_{\mathrm{RAD}}) \quad (8.4.11)$$

where $\mathbf{A}_{\mathrm{RAD}}$ is the radiation field (§7.7).

For a state $|\mathbf{p}\rangle \begin{pmatrix} 1 \\ 0 \end{pmatrix}$, we have, in a similar manner as obtaining (7.7.28), (7.7.37), to second order in the radiation field, due to the presence of radiation, the energy shift $\Delta \varepsilon$, (see also (7.7.35))

$$\Delta \varepsilon \, (2\pi)^3 \, \delta^3(\mathbf{p} - \mathbf{p}')$$

$$= \mu_{\mathrm{n}}^2 (4\pi \hbar c) \sum_\lambda \int \frac{d^3 \mathbf{k}}{(2\pi)^3 \, 2|\mathbf{k}|} \begin{pmatrix} 1 \\ 0 \end{pmatrix}^{\mathrm{T}} \langle \mathbf{p}'| \Bigg\{ (i\boldsymbol{\sigma} \times \mathbf{k}) \cdot \mathbf{e}_\lambda$$

$$\times \frac{1}{\dfrac{\mathbf{p}^2}{2M_{\mathrm{n}}} + |\mu_{\mathrm{n}}|B - \dfrac{(\mathbf{p}' - \hbar \mathbf{k})^2}{2M_{\mathrm{n}}} + |\mu_{\mathrm{n}}| \sigma_3 B - \hbar |\mathbf{k}| c + i\varepsilon}$$

$$\times (-i\boldsymbol{\sigma} \times \mathbf{k}) \cdot \mathbf{e}_\lambda \Bigg\} \begin{pmatrix} 1 \\ 0 \end{pmatrix} |\mathbf{p}\rangle \quad (8.4.12)$$

where the $+i\varepsilon$ is to account for the decay under study.

Let

$$D = \frac{\mathbf{p}'^2}{2M_{\mathrm{n}}} - \frac{\mathbf{p}^2}{2M_{\mathrm{n}}} - \frac{\mathbf{p}' \cdot \hbar \mathbf{k}}{M_{\mathrm{n}}} + \hbar |\mathbf{k}| c + \frac{\hbar^2 \mathbf{k}^2}{2M_{\mathrm{n}}} - i\varepsilon \quad (8.4.13)$$

then

$$\frac{1}{D - |\mu_{\mathrm{n}}|B - |\mu_{\mathrm{n}}|\sigma_3 B} = \begin{pmatrix} \dfrac{1}{D - 2|\mu_{\mathrm{n}}|B} & 0 \\ 0 & \dfrac{1}{D} \end{pmatrix}. \quad (8.4.14)$$

Also using (see (§7.7))

$$\sum_\lambda [(\boldsymbol{\sigma} \times \mathbf{k}) \cdot \mathbf{e}_\lambda] \, [\ldots] \, [(\boldsymbol{\sigma} \times \mathbf{k}) \cdot \mathbf{e}_\lambda] = (\boldsymbol{\sigma} \times \mathbf{k}) \cdot [\ldots] (\boldsymbol{\sigma} \times \mathbf{k})$$

$$= (\delta_{ab} \mathbf{k}^2 - k_a k_b) \sigma_a [\ldots] \sigma_b. \quad (8.4.15)$$

From (8.4.14), (8.4.15), the expression for (8.4.12) becomes

$$\Delta\varepsilon(\mathbf{p})\,(2\pi)^3\,\delta^3(\mathbf{p}-\mathbf{p}')$$

$$= -\frac{\mu_n^2 \hbar c}{4\pi^2}(2\pi)^3\,\delta^3(\mathbf{p}-\mathbf{p}')\int\frac{d^3k}{|\mathbf{k}|}\left\{\frac{2\mathbf{k}^2-k_1^2-k_2^2}{D}+\frac{\mathbf{k}^2-k_3^2}{D-2|\mu_n|B}\right\}. \tag{8.4.16}$$

Upon integration over $\mathbf{p}'$, taking the limit $\mathbf{p} \to 0$, and performing the $\mathbf{k}$-angular integration in (8.4.16), we obtain

$$\Delta\varepsilon(\mathbf{0}) = -\frac{\mu_n^2 \hbar c}{\pi}\int k^3 dk \left\{\frac{4}{3}\frac{1}{\left(\hbar k c + \frac{\hbar^2 k^2}{2M} - i\varepsilon\right)}\right.$$

$$\left. + \frac{2}{3}\frac{1}{\left(\hbar k c + \frac{\hbar^2 k^2}{2M} - 2|\mu_n|B - i\varepsilon\right)}\right\}. \tag{8.4.17}$$

The first term in the curly brackets is independent of B. For a neutron, in the *absence* of an external magnetic field $\mathbf{B}$, in the ever presence of radiation interacting with its spin, the shift in energy in (8.4.17) should be *zero*. This amounts to carrying out a renormalization (see also §7.7, §8.5) by subtracting off the corresponding expression to the one within the curly brackets in (8.4.17) evaluated at $B=0$.[13]

Accordingly,

$$\Delta\varepsilon^{\text{Ren}}(\mathbf{0}) = -\frac{2}{3}\frac{\mu_n^2 \hbar c}{\pi}\int k^3 dk \left\{\frac{1}{\left(\hbar k c + \frac{\hbar^2 k^2}{2M} - 2|\mu_n|B - i\varepsilon\right)}\right.$$

$$\left. - \frac{1}{\left(\hbar k c + \frac{\hbar^2 k^2}{2M} - i\varepsilon\right)}\right\}. \tag{8.4.18}$$

By taking the imaginary part of the above (see (7.7.30)–(7.7.32), §7.8), we obtain for the decay constant[14]

---

[13] Actually this term does not contribute to the decay rate (or the mean lifetime) we are seeking.

[14] Decays due to precession of the spin in a magnetic field was also previously considered, by different methods, cf., Stump and Pollack (1998).

$$\Gamma = \frac{4}{3}\mu_n^2 \int k^3 dk \left\{ \delta\left(k + \frac{\hbar k^2}{2Mc} - \frac{2|\mu_n|B}{\hbar c}\right) - \delta\left(k + \frac{\hbar k^2}{2Mc}\right)\right\}. \qquad (8.4.19)$$

For most practical cases (see (8.4.2)), $|\mu_n|B \ll Mc^2$, and (8.4.19) gives

$$\Gamma = \frac{32}{3} \frac{|\mu_n|^5 B^3}{\hbar^3 c^3} \qquad (8.4.20)$$

where the second term in (8.4.19) does not contribute.

For $B \sim 1$ T, the mean lifetime is $\tau = \hbar/\Gamma \sim 10^{25}$ s, which is incredibly large and not significant unless huge magnetic fields are considered within the realm of astrophysics of millions T.

The real part of (8.4.18) gives rise to a very small correction (contribution) to the magnetic moment μ_n (see Problem 8.9, §8.5).

8.5 Anomalous Magnetic Moment of the Electron

In this section, we consider the measurement as well as an explicit computation of the deviation $(g-2)$ of the gyromagnetic factor of the electron.

We first consider the Hamiltonian of an electron in interaction with a uniform and time-independent magnetic field $\mathbf{B}$,

$$H = \frac{\boldsymbol{\pi}^2}{2M} - \frac{eg}{2Mc}\mathbf{S}\cdot\mathbf{B} \qquad (8.5.1)$$

where

$$\boldsymbol{\pi} = \mathbf{p} - \frac{e}{c}\mathbf{A} \qquad (8.5.2)$$

with

$$\mathbf{B} = \nabla \times \mathbf{A} \qquad (8.5.3)$$

and

$$\left[\pi^j, \pi^k\right] = \frac{i\hbar e}{c} F^{jk} \qquad (8.5.4)$$

$$F^{jk} = \varepsilon^{jkl} B^l \qquad (8.5.5)$$

$$\left[S^j, S^k\right] = i\hbar \varepsilon^{jkl} S^l. \qquad (8.5.6)$$

The following commutation relations with the Hamiltonian are then readily derived

$$[\boldsymbol{\pi}, H] = \frac{i\hbar e}{Mc} \boldsymbol{\pi} \times \mathbf{B} \qquad (8.5.7)$$

$$[\mathbf{S}, H] = \frac{i\hbar eg}{2Mc} \mathbf{S} \times \mathbf{B}. \qquad (8.5.8)$$

From the time-development of an operator, $O(t) = e^{itH/\hbar} O e^{-itH/\hbar}$, we then obtain for expectation values

$$\frac{d}{dt} \langle \boldsymbol{\pi}(t) \rangle = \frac{e}{Mc} \langle \boldsymbol{\pi}(t) \rangle \times \mathbf{B} \tag{8.5.9}$$

$$\frac{d}{dt} \langle \mathbf{S}(t) \rangle = \frac{eg}{2Mc} \langle \mathbf{S}(t) \rangle \times \mathbf{B}. \tag{8.5.10}$$

In particular for components parallel $\parallel$ or $\perp$ perpendicular to $\mathbf{B}$

$$\frac{d}{dt} \langle \boldsymbol{\pi}_\parallel(t) \rangle = \mathbf{0}, \quad \frac{d}{dt} \langle \mathbf{S}_\parallel(t) \rangle = \mathbf{0}, \tag{8.5.11}$$

$$\frac{d^2}{dt^2} \langle \boldsymbol{\pi}_\perp(t) \rangle = -\frac{e^2 B^2}{M^2 c^2} \langle \boldsymbol{\pi}_\perp(t) \rangle \tag{8.5.12}$$

$$\frac{d^2}{dt^2} \langle \mathbf{S}_\perp(t) \rangle = -\frac{e^2 g^2 B^2}{4 M^2 c^2} \langle \mathbf{S}_\perp(t) \rangle \tag{8.5.13}$$

giving rise to the solutions ($|\mathbf{B}| = B$)

$$\langle \boldsymbol{\pi}(t) \rangle = \mathbf{b}_1 \cos\left(\frac{|eB|}{Mc} t + \delta_1\right) + \mathbf{b}_2 \tag{8.5.14}$$

$$\langle \mathbf{S}(t) \rangle = \mathbf{c}_1 \cos\left(\frac{|egB|}{2Mc} t + \delta_2\right) + \mathbf{c}_2 \tag{8.5.15}$$

where $\mathbf{b}_1, \mathbf{b}_2, \mathbf{c}_1, \mathbf{c}_2$ are constant vectors and δ_1, δ_2 are phase factors. That is, $\langle \boldsymbol{\pi}(t) \rangle$, $\langle \mathbf{S}(t) \rangle$ would precess with the same angular velocity only if $|g| = 2$.

We next consider the measurement of $(g - 2)$. Its computation will then follow this study.

8.5.1 Observational Aspect of the Anomalous Magnetic Moment

With $\boldsymbol{\pi}/M$ denoting the velocity operator, we determine the commutation relation of $\mathbf{S} \cdot \boldsymbol{\pi}$ with H. As we will see below, this commutator is proportional to $(g - 2)$ and hence would vanish if g were equal to 2.

A direct evaluation, leads to

$$[S^j \pi^k, H] = \frac{ie\hbar}{2Mc} (2\delta^{jl} \varepsilon^{kmn} - g\varepsilon^{ljn} \delta^{km}) \pi^m B^n S^l \tag{8.5.16}$$

from which we obtain ($|\mathbf{B}| = B$)

$$[\mathbf{S} \cdot \boldsymbol{\pi}, H] = -\frac{i\hbar e}{2Mc} (g - 2)(\mathbf{S} \times \boldsymbol{\pi}) \cdot \mathbf{B} \tag{8.5.17}$$

$$[(\mathbf{S} \times \boldsymbol{\pi}) \cdot \mathbf{B}, H] = \frac{i\hbar e}{2Mc} (g - 2)(B)^2 \left(\delta^{ij} - \frac{B^i B^j}{B^2}\right) S^i \pi^j \tag{8.5.18}$$

and
$$\left[[\mathbf{S}\cdot\boldsymbol{\pi},H],H\right] = \frac{(e\hbar B)^2}{4M^2c^2}(g-2)^2\left(\delta^{ij} - \frac{B^i B^j}{B^2}\right)S^i\pi^j. \tag{8.5.19}$$

We note that
$$\left(\delta^{ij} - \frac{B^i B^j}{B^2}\right)S^i\pi^j = \mathbf{S}_\perp \cdot \boldsymbol{\pi}_\perp \tag{8.5.20}$$

corresponding to components perpendicular to $\mathbf{B}$, and from (8.5.16),
$$\left[\mathbf{S}_\parallel \cdot \boldsymbol{\pi}_\parallel, H\right] = 0. \tag{8.5.21}$$

According to (8.5.19)–(8.5.21) we then have for expectation values
$$\frac{d^2}{dt^2}\langle \mathbf{S}_\perp(t) \cdot \boldsymbol{\pi}_\perp(t)\rangle = -\frac{e^2(B)^2(g-2)^2}{4M^2c^2}\langle \mathbf{S}_\perp(t) \cdot \boldsymbol{\pi}_\perp(t)\rangle \tag{8.5.22}$$

$$\frac{d}{dt}\langle \mathbf{S}_\parallel(t) \cdot \boldsymbol{\pi}_\parallel(t)\rangle = 0 \tag{8.5.23}$$

giving rise to the solution
$$\langle \mathbf{S}(t) \cdot \boldsymbol{\pi}(t)\rangle = c_0 \cos\left(\frac{|eB(g-2)|}{2Mc}t + \delta\right) + d_0 \tag{8.5.24}$$

with c_0, d_0 denoting constants.

Hence for $g \neq 2$, (8.5.24) shows a periodic behavior in time with angular velocity
$$\omega_g = \frac{|eB(g-2)|}{2Mc}$$
and period $T = 2\pi/\omega_g$. For a given B and the observed period, the anomaly $(g-2)/2$ is readily determined[15] and is consistent with the value $(g-2)/2 = \alpha/2\pi$ evaluated first by Schwinger using (relativistic) quantum electrodynamics. Actually, this anomaly has been evaluated to higher orders in the fine-structure constant and the agreement between theory and experiment is quite impressive.

8.5.2 Computation of the Anomalous Magnetic Moment

The expression for the eigenvalues in (8.3.40), allowed us to write the deviation $(g-2)/2$ in the convenient from (8.3.41):

$$-\frac{2Mc}{|e|\hbar}\frac{\partial}{\partial B}E(n=0, \sigma=-1)\bigg|_{B=0, p_1, p_3 = 0} = \frac{(g-2)}{2} \tag{8.5.25}$$

(see also 8.3.1).

---
[15] Cf. Wilkinson and Crane (1963).

8.5 Anomalous Magnetic Moment of the Electron   447

As a result of the interaction of the electron with the ever present radiation accompanying it, the value of g deviates slightly from 2 and this correction is of the order α — the fine-structure constant. A formal derivation of this is given below.

The total Hamiltonian for the interaction of an electron with a uniform magnetic field $\mathbf{B} = (0, 0, B)$, $B > 0$ and the radiation field $\mathbf{A}_{\text{RAD}}$ (see also §7.7) is given by

$$H_T = H + H_{0,\text{RAD}} + H_I \qquad (8.5.26)$$

where H is the Hamiltonian in (8.4.1) in the presence of the magnetic field $\mathbf{B}$, and

$$H_I = \frac{|e|}{2Mc}(\mathbf{A}_{\text{RAD}} \cdot \mathbf{p} + \mathbf{p} \cdot \mathbf{A}_{\text{RAD}}) + \frac{e^2}{2Mc^2}\mathbf{A}_{\text{RAD}}^2$$

$$-\frac{|e|B}{2Mc^2}x_2|e|A_{\text{RAD},1} + \frac{|e|\hbar g}{4Mc}\boldsymbol{\sigma} \cdot (\boldsymbol{\nabla} \times \mathbf{A}_{\text{RAD}}) + \delta H_c' \qquad (8.5.27)$$

taking the spin of the electron into account, and where $\delta H_c'$ is a renormalization counter-term whose nature is specified below.

Let

$$|n=0\rangle \begin{pmatrix} 0 \\ 1 \end{pmatrix} \equiv |\psi_0\rangle \qquad (8.5.28)$$

corresponding to $n = 0$, $\sigma = -1$ and suppressing for simplicity the dependence on p_1, p_3.

We work out the correction $(g-2)/2$ to the order α as arising from the presence of radiation.

To the above end, we denote

$$|\psi_0\rangle|0\rangle = |\psi_0; 0\rangle \qquad (8.5.29)$$

$$|\psi_0\rangle|\mathbf{k}\lambda\rangle = |\psi_0; \mathbf{k}\lambda\rangle \qquad (8.5.30)$$

as eigenstates of $H + H_{0,\text{RAD}}$ in (8.5.26), with $|0\rangle$, $|\mathbf{k}\lambda\rangle$ denoting, respectively, a no-photon state and a single-photon state with momentum $\hbar\mathbf{k}$ and polarization specified by λ (see §7.7).

Finally, let $|\phi_0\rangle$ denote the state corresponding to $n = 0$, $\sigma = -1$ for the electron in the magnetic field $\mathbf{B}$, in interaction with radiation,

$$H_T|\phi_0\rangle = \mathcal{E}_0|\phi_0\rangle \qquad (8.5.31)$$

such that if H_I is formally replaced by zero, then $|\phi_0\rangle = |\psi_0; 0\rangle$ involving no photons.

Following the derivation of (7.7.23), we may write

$$\Delta E = \mathcal{E}_0 - E_0 = \frac{\langle\psi_0; 0|H_I|\phi_0\rangle}{\langle\psi_0; 0|\phi_0\rangle} \qquad (8.5.32)$$

for the energy shift in the presence of radiation, where

$$E_0 = E(n=0, \sigma = -1). \tag{8.5.33}$$

To second order e^2 in radiation, we then have from (8.5.32), (8.5.27), in a similar manner as in (7.7.28),

$$\Delta E = \frac{\alpha \hbar^2}{4\pi^2 M^2} \frac{1}{\langle \psi_0 | \psi_0 \rangle}$$

$$\times \left[\sum_\lambda \int \frac{d^3 k}{|\mathbf{k}|} \langle \psi_0 | \eta_\lambda \, e^{i\mathbf{k}\cdot\mathbf{x}} \frac{1}{E_0 - H - \hbar |\mathbf{k}| c} e^{-i\mathbf{k}\cdot\mathbf{x}} \eta_\lambda^\dagger | \psi_0 \rangle \right]$$

$$+ \frac{e^2}{2Mc^2} \langle 0 | \mathbf{A}_{\text{RAD}}^2 | 0 \rangle + \frac{(\psi_0; 0 | \delta H_c' | \phi_0)}{(\psi_0; 0 | \phi_0)} \tag{8.5.34}$$

where

$$\eta_\lambda = \mathbf{p} \cdot \mathbf{e}_\lambda(\mathbf{k}) - \frac{|e|B}{2c} x_2 \, e_{\lambda 1}(\mathbf{k}) + \frac{i\hbar g}{4} (\boldsymbol{\sigma} \times \mathbf{k}) \cdot \mathbf{e}_\lambda(\mathbf{k}) \tag{8.5.35}$$

and to second order e^2 in radiation, we may set $g = 2$ in (8.5.34).

In detail,

$$e^{i\mathbf{k}\cdot\mathbf{x}} H e^{-i\mathbf{k}\cdot\mathbf{x}} = \frac{(\mathbf{p} - \hbar \mathbf{k})^2}{2M} - \frac{|e|Bx_2}{Mc}(p_1 - \hbar k_1)$$

$$+ \frac{|e|\hbar B g}{4Mc} \sigma_3 + \frac{e^2 B^2 x_2^2}{2Mc^2}$$

$$= H - \frac{\mathbf{p} \cdot \hbar \mathbf{k}}{M} + \frac{\hbar^2 \mathbf{k}^2}{2M} + \frac{|e|Bx_2}{Mc} \hbar k_1. \tag{8.5.36}$$

It is convenient to work in the momentum description, and take the limits $p_1, p_3 \to 0$. We set g equal to 2, $E_0 = 2$ in (8.5.34) (see also (8.5.25)). For the wavefunction, as a function on its dependence on p_2, we have

$$\langle p_2 | \psi_0 \rangle = \left(\frac{M\Omega}{\pi \hbar} \right)^{1/4} \exp\left(-p_2^2 \frac{M\Omega}{2\hbar} \right) \equiv \psi_0(p_2) \tag{8.5.37}$$

$$\Omega = \frac{c}{|e|BM}. \tag{8.5.38}$$

With the normalization factor $\langle \psi_0 | \psi_0 \rangle$ for the three dimensional motion, as initially appearing in (8.5.34), all reference to p_1, p_3 which are formally taken to go to zero, disappears.

To make the dependence of ΔE on B explicit, we make the change of variable $p_2 \to \xi$, with

8.5 Anomalous Magnetic Moment of the Electron

$$p_2 = \sqrt{\frac{|e|\hbar B}{c}}\,\xi. \tag{8.5.39}$$

Then for an operator $F(p_2, \partial/\partial p_2)$

$$\int_{-\infty}^{\infty} dp_2\, \psi_0(p_2)\, F\!\left(p_2, \frac{\partial}{\partial p_2}\right) \psi_0(p_2)$$

$$= \int_{-\infty}^{\infty} \frac{d\xi}{\sqrt{\pi}}\, e^{-\xi^2/2}\, F\!\left(\sqrt{\frac{|e|\hbar B}{c}}\,\xi,\ \sqrt{\frac{c}{|e|\hbar B}}\,\frac{\partial}{\partial \xi}\right) e^{-\xi^2/2} \tag{8.5.40}$$

where we note that $\psi_0(p_2)$ is properly normalized with respect to the measure dp_2.

Accordingly, (8.5.34) becomes

$$\Delta E = -\frac{\alpha \hbar^2}{4\pi^2 M^2} \sum_\lambda \int \frac{d^3\mathbf{k}}{|\mathbf{k}|} \int_{-\infty}^{\infty} \frac{d\xi}{\sqrt{\pi}}$$

$$\times \left[e^{-\xi^2/2} \begin{pmatrix} 0 \\ 1 \end{pmatrix}^T \eta'_\lambda\, \frac{1}{H' + \hbar|\mathbf{k}|c}\, \eta'^\dagger_\lambda \begin{pmatrix} 0 \\ 1 \end{pmatrix} e^{-\xi^2/2} \right]$$

$$+ \frac{e^2}{2Mc^2} \langle 0|\mathbf{A}_{\text{RAD}}^2|0\rangle + \frac{(\psi_0; 0|\delta H'_c|\phi_0)}{(\psi_0; 0|\phi_0)} \tag{8.5.41}$$

where

$$H' = -\frac{|e|\hbar B}{2Mc}\,\frac{\partial^2}{\partial \xi^2} + \frac{|e|\hbar B}{2Mc}\,\xi^2 + \frac{|e|\hbar}{2Mc}\,\sigma_3 B$$

$$- \sqrt{\frac{|e|\hbar B}{c}}\,\xi\,\frac{\hbar k_2}{M} + \frac{\hbar^2 \mathbf{k}^2}{2M} + \sqrt{\frac{|e|\hbar B}{c}}\,\frac{\hbar k_1}{M}\,i\frac{\partial}{\partial \xi} \tag{8.5.42}$$

$$\eta'_\lambda = \sqrt{\frac{|e|\hbar B}{c}}\,\xi\, e_{\lambda 2} - \frac{1}{2}\sqrt{\frac{|e|\hbar B}{c}}\, e_{\lambda 1}\, i\frac{\partial}{\partial \xi} + \frac{i\hbar}{2}\,(\boldsymbol{\sigma}\times\mathbf{k})\cdot\mathbf{e}_\lambda. \tag{8.5.43}$$

With the parameters M, e appearing in (8.5.41) taken to denote renormalized, i.e., the physically observed ones (see also §7.7), $(\psi_0; 0|\delta H'_c|\phi_0)/(\psi_0; 0|\phi_0)$ is so chosen such that there is no energy shift ($\Delta E = 0$) for $B = 0$. Physically, this is a condition of the *stability* of the electron in vacuum in the ever presence of radiation accompanying the electron. In the literature, the c-number $(\psi_0; 0|\delta H'_c|\phi_0)/(\psi_0; 0|\phi_0)$ is referred to as a *contact term*.

Hence

$$\Delta E = \Delta_1 E + \Delta_2 E + \Delta_3 E \tag{8.5.44}$$

where

$$\Delta_1 E = -\frac{\alpha \hbar^2}{4\pi^2 M^2} \frac{|e|\hbar B}{c} \sum_\lambda \int \frac{d^3 k}{|\mathbf{k}|} \int_{-\infty}^{\infty} \frac{d\xi}{\sqrt{\pi}} e^{-\xi^2/2} \begin{pmatrix} 0 \\ 1 \end{pmatrix}^{\mathrm{T}}$$

$$\times Q_\lambda \frac{1}{H' + \hbar|\mathbf{k}|c} Q_\lambda \begin{pmatrix} 0 \\ 1 \end{pmatrix} e^{-\xi^2/2} \qquad (8.5.45)$$

$$Q_\lambda = \xi\, e_{\lambda 2} - \frac{1}{2} e_{\lambda 1}\, \mathrm{i}\frac{\partial}{\partial \xi} \qquad (8.5.46)$$

$$\Delta_2 E = -\frac{\alpha \hbar^2}{4\pi^2 M^2} \sqrt{\frac{|e|\hbar B}{c}} \sum_\lambda \int \frac{d^3 k}{|\mathbf{k}|} \int_{-\infty}^{\infty} \frac{d\xi}{\sqrt{\pi}} e^{-\xi^2/2} \begin{pmatrix} 0 \\ 1 \end{pmatrix}^{\mathrm{T}}$$

$$\times \left[Q_\lambda \frac{1}{H' + \hbar|\mathbf{k}|c} \left(\frac{-\mathrm{i}\hbar}{2}\right) (\boldsymbol{\sigma} \times \mathbf{k}) \cdot \mathbf{e}_\lambda \right.$$

$$\left. + \left(\frac{\mathrm{i}\hbar}{2}\right) (\boldsymbol{\sigma} \times \mathbf{k}) \cdot \mathbf{e}_\lambda \frac{1}{H' + \hbar|\mathbf{k}|c} Q_\lambda \right] \begin{pmatrix} 0 \\ 1 \end{pmatrix} e^{-\xi^2/2} \qquad (8.5.47)$$

$$\Delta_3 E = -\frac{\alpha \hbar^2}{4\pi^2 M^2} \sum_\lambda \int \frac{d^3 k}{|\mathbf{k}|} \int_{-\infty}^{\infty} \frac{d\xi}{\sqrt{\pi}} e^{-\xi^2/2} \begin{pmatrix} 0 \\ 1 \end{pmatrix}^{\mathrm{T}}$$

$$\times \left(\frac{\mathrm{i}\hbar}{2}\right) (\boldsymbol{\sigma} \times \mathbf{k}) \cdot \mathbf{e}_\lambda \left[\frac{1}{H' + \hbar|\mathbf{k}|c} - \frac{1}{\hbar|\mathbf{k}|c + \frac{\hbar^2 \mathbf{k}^2}{2M}} \right]$$

$$\times \left(\frac{-\mathrm{i}\hbar}{2}\right) (\boldsymbol{\sigma} \times \mathbf{k}) \cdot \mathbf{e}_\lambda \begin{pmatrix} 0 \\ 1 \end{pmatrix} e^{-\xi^2/2}. \qquad (8.5.48)$$

In a non-relativistic setting, starting from a non-relativistic Hamiltonian, integration over $\hbar|\mathbf{k}|c$ beyond the natural cut-off Mc^2, corresponding to the rest mass energy of the electron, is not justifiable. Accordingly, we cut-off the $\mathbf{k}$-integrals in (8.5.45), (8.5.47), (8.5.48) by imposing the restriction $\hbar|\mathbf{k}|c < Mc^2$. The very welcome damping in $|\xi|$ by the Gaussian function in these integrals should be also noted.

Since we are interested in the dependence of ΔE only up to the first power of B, we immediately obtain, by using the completeness relation of the photon polarization vectors in (7.7.4), and the elementary averages over the angular integrations in (7.7.43),

$$\Delta_1 E = -\frac{\alpha \hbar^2}{4\pi^2 M^2} \frac{|e|\hbar B}{c} \frac{8\pi}{3} \int_0^{Mc/\hbar} \frac{k\, dk}{\left[\hbar k c + \frac{\hbar^2 k^2}{2M}\right]}$$

8.5 Anomalous Magnetic Moment of the Electron

$$\times \int_{-\infty}^{\infty} \frac{d\xi}{\sqrt{\pi}} e^{-\xi^2/2} \left(\xi^2 - \frac{1}{4} \frac{\partial^2}{\partial \xi^2} \right) e^{-\xi^2/2}$$

$$= - \left(\frac{\alpha}{2\pi} \right) \left(\frac{|e|\hbar B}{2Mc} \right) \left(\frac{10}{3} \right) \ln \left(\frac{3}{2} \right) \tag{8.5.49}$$

linear in B.

The computation of the linear dependence of $\Delta_3 E$ on B is also relatively easy. To this end, we note that

$$\sum_\lambda (\boldsymbol{\sigma} \times \mathbf{k}) \cdot \mathbf{e}_\lambda (\boldsymbol{\sigma} \times \mathbf{k}) \cdot \mathbf{e}_\lambda = (\boldsymbol{\sigma} \times \mathbf{k})^2 - \frac{(\mathbf{k} \cdot (\boldsymbol{\sigma} \times \mathbf{k}))^2}{\mathbf{k}^2} = 0 \tag{8.5.50}$$

and

$$\begin{pmatrix} 0 \\ 1 \end{pmatrix}^T \sum_\lambda (\boldsymbol{\sigma} \times \mathbf{k}) \cdot \mathbf{e}_\lambda \, \sigma_3 \, (\boldsymbol{\sigma} \times \mathbf{k}) \cdot \mathbf{e}_\lambda \begin{pmatrix} 0 \\ 1 \end{pmatrix} = 2(k_3)^2. \tag{8.5.51}$$

According to (8.5.50), (8.5.51), we may effectively make the replacement

$$\frac{1}{H' + \hbar |\mathbf{k}| c} \frac{1}{\hbar |\mathbf{k}| c + \frac{\hbar^2 \mathbf{k}^2}{2M}} \longrightarrow -\frac{\frac{|e|\hbar B}{2Mc} \sigma_3}{\left[\hbar |\mathbf{k}| c + \frac{\hbar^2 \mathbf{k}^2}{2M} \right]^2} \tag{8.5.52}$$

in (8.5.48).

From (8.5.51), (8.5.52), we then have

$$\Delta_3 E = \frac{\alpha \hbar^2}{4\pi^2 M^2} \left(\frac{|e|\hbar B}{2Mc} \right) \left(\frac{\hbar^2}{4} \right) \frac{8\pi}{3} \int_0^{Mc/\hbar} \frac{k^3 dk}{\left[\hbar k c + \frac{\hbar^2 k^2}{2M} \right]^2}$$

$$= \left(\frac{\alpha}{2\pi} \right) \left(\frac{|e|\hbar B}{2Mc} \right) \frac{4}{3} \left[\ln \left(\frac{3}{2} \right) - \frac{1}{3} \right] \tag{8.5.53}$$

for the linear part in B.

The first term within the square brackets in (8.5.47) is effectively

$$+ \left(\frac{i\hbar}{2} \right) \left(\frac{\hbar}{M} \right) \sqrt{\frac{|e|\hbar B}{c}} \left(\xi e_{\lambda 2} - \frac{1}{2} e_{\lambda 1} i \frac{\partial}{\partial \xi} \right)$$

$$\times \left(-\xi k_2 + k_1 i \frac{\partial}{\partial \xi} \right) (\boldsymbol{\sigma} \times \mathbf{k}) \cdot \mathbf{e}_\lambda \tag{8.5.54}$$

multiplied by $\left[\hbar |\mathbf{k}| c + \hbar^2 \mathbf{k}^2 / 2M \right]^{-2}$. By summing over λ, using (7.7.4), noting the fact that the $\mathbf{k}$-angular integration leads to the rule spelled out in (7.7.43), and anticipating the spin average to be performed in (8.5.47) with the simple observations that

$$\begin{pmatrix}0\\1\end{pmatrix}^T \sigma_j \begin{pmatrix}0\\1\end{pmatrix} = 0, \qquad j = 1, 2 \tag{8.5.55}$$

$$\begin{pmatrix}0\\1\end{pmatrix}^T \sigma_3 \begin{pmatrix}0\\1\end{pmatrix} = -1 \tag{8.5.56}$$

we may effectively replace (8.5.54) by

$$\left(\frac{i\hbar}{2}\right)\left(\frac{\hbar}{M}\right)\sqrt{\frac{|e|\hbar B}{c}}\frac{4\pi}{3}\mathbf{k}^2\left(-i\xi\frac{\partial}{\partial\xi} + \frac{i}{2}\frac{\partial}{\partial\xi}\xi\right) \tag{8.5.57}$$

after carrying out the **k**-angular integral.

Similarly, the second term within the square brackets in (8.5.47), after carrying out the **k**-angular integration, may be effectively replaced by

$$-\left(\frac{i\hbar}{2}\right)\left(\frac{\hbar}{M}\right)\sqrt{\frac{|e|\hbar B}{c}}\frac{4\pi}{3}\mathbf{k}^2\left(\frac{i}{2}\xi\frac{\partial}{\partial\xi} - i\frac{\partial}{\partial\xi}\xi\right) \tag{8.5.58}$$

multiplied by $\left[\hbar|\mathbf{k}|c + \hbar^2\mathbf{k}^2/2M\right]^{-2}$, giving for the expression within the square brackets in question the effective net result

$$\int d\Omega \begin{pmatrix}0\\1\end{pmatrix}^T [\cdots] \begin{pmatrix}0\\1\end{pmatrix} = \frac{-\frac{\hbar^2}{4M}\sqrt{\frac{|e|\hbar B}{c}}4\pi\mathbf{k}^2}{\left[\hbar|\mathbf{k}|c + \frac{\hbar^2\mathbf{k}^2}{2M}\right]^2} \tag{8.5.59}$$

in (8.5.47) after the angular integration.

After carrying the elementary ξ-Gaussian integral, we have from (8.5.59), the following expression for $\Delta_2 E$ in (8.5.47)

$$\Delta_2 E = \left(\frac{\alpha}{2\pi}\right)\left(\frac{|e|\hbar B}{2Mc}\right)\left(\frac{\hbar^4}{M^2}\right)\int_0^{Mc/\hbar}\frac{k^3\,dk}{\left[\hbar kc + \frac{\hbar^2 k^2}{2M}\right]^2}$$

$$= \left(\frac{\alpha}{2\pi}\right)\left(\frac{|e|\hbar B}{2Mc}\right)4\left[\ln\left(\frac{3}{2}\right) - \frac{1}{3}\right] \tag{8.5.60}$$

for the linear part in B. [We note that there is no term proportional to $B^{1/2}$.]

All told, (8.5.44), (8.5.49), (8.5.53), (8.5.60), give

$$\Delta E = -\left(\frac{\alpha}{2\pi}\right)\left(\frac{|e|\hbar B}{2Mc}\right)\left[\frac{16}{9} - 2\ln\left(\frac{3}{2}\right)\right] \tag{8.5.61}$$

for the linear part in B.

From (8.5.61), (8.5.25), we may finally infer that

$$\frac{(g-2)}{2} = \left(\frac{\alpha}{2\pi}\right)\left[\frac{16}{9} - 2\ln\left(\frac{3}{2}\right)\right] \tag{8.5.62}$$

which is quite accurate since

$$\left[\frac{16}{9} - 2\ln\left(\frac{3}{2}\right)\right] \simeq 0.97 \tag{8.5.63}$$

and compares well with the fully quantum-electrodynamic value $\alpha/2\pi$.[16]

8.6 Density Operators and Spin

Detailed accounts of density operators were given in §1.5, §1.6 and §5.4 with the latter section dealing with spin 1/2 and arbitrary spins as well. Here we study general aspects of such operators as applied to concrete physical situations for spin 1/2 particles.

8.6.1 Spin in a General Time-Dependent Magnetic Field

As a first application, we consider the spin Hamiltonian in (8.1.45) for the interaction of spin 1/2 with a general time-dependent magnetic field:

$$H(t) = \frac{\hbar}{2}\mathbf{K}(t)\cdot\boldsymbol{\sigma}. \tag{8.6.1}$$

The initial state $|\psi(0)\rangle$ at time $t=0$ is prepared to be an eigenstate of $\mathbf{n}_0\cdot\boldsymbol{\sigma}$ for some vector $\mathbf{n}_0 = (\sin\theta_0\cos\phi_0, \sin\theta_0\sin\phi_0, \cos\theta_0)$. With such an initially prepared state as given in (8.1.48), the time $t=0$, the density operator $\rho(0)$ works out to be

$$\rho(0) = |\psi(0)\rangle\langle\psi(0)| = \frac{1}{2}\left[\mathbf{1} + \mathbf{n}_0\cdot\boldsymbol{\sigma}\right] \tag{8.6.2}$$

where $\mathbf{1}$ is the 2×2 unit matrix.

The time $t > 0$ density operator $\rho(t)$ is then given by

$$\rho(t) = \frac{1}{2}\left[\mathbf{1} + \mathbf{P}(t)\cdot\boldsymbol{\sigma}\right] \tag{8.6.3}$$

where for the average of spin at time $t > 0$, divided by $\hbar/2$, we have

---

[16] There is a long history of the computation of $(g-2)/2$ within non-relativistic quantum mechanics. For other attempts, cf. Arunasalam (1969), Grotch and Kazes (1977). Arunasalam also deals with Landau states.

454   8 Quantum Physics of Spin 1/2 & Two-Level Systems

$$\text{Tr}[\boldsymbol{\sigma}\rho(t)] = \mathbf{P}(t) \equiv \mathbf{n}(t) \tag{8.6.4}$$

and we have identified the latter with the unit vector $\mathbf{n}(t)$ in (8.1.54). $\mathbf{P}(t)$ is referred as the polarization vector. For $t = 0$, $\mathbf{P}(0) = \mathbf{n}(0) = \mathbf{n}_0$.

The equation for the time development of the polarization vector $\mathbf{P}(t)$ was already given in (8.1.49). Here it is more instructive to derive this equation from the equation

$$i\hbar \frac{\partial}{\partial t}\rho(t) = [H(t), \rho(t)] \tag{8.6.5}$$

for the density operator $\rho(t)$, which from (8.6.3), (8.6.1), leads to

$$\frac{\partial}{\partial t}\mathbf{P}(t) \cdot \boldsymbol{\sigma} = \big(\mathbf{K}(t) \times \mathbf{P}(t)\big) \cdot \boldsymbol{\sigma}. \tag{8.6.6}$$

Upon multiplying the latter by $\boldsymbol{\sigma}$, and taking the trace gives

$$\frac{\partial}{\partial t}\mathbf{P}(t) = \mathbf{K}(t) \times \mathbf{P}(t) \tag{8.6.7}$$

consistent with (8.1.49) as expected.

As another application, we consider the scattering of a spin 1/2 particle off a spin 0 target. Later on, we will treat the problem of the scattering of a spin 1/2 particle off a spin 1/2 target.

8.6.2 Scattering of Spin 1/2 Particle off a Spin 0 Target

For particles initially prepared to be polarized,[17] say, along the x-axis, the initial state may be taken, up to a phase factor, to be

$$|\psi_i\rangle_x = \frac{1}{\sqrt{2}} \begin{pmatrix} 1 \\ 1 \end{pmatrix}. \tag{8.6.8}$$

Similarly, for polarizations along the y-, z-axes, we have respectively, up to phase factors,

$$|\psi_i\rangle_y = \frac{1}{2}\begin{pmatrix} 1-i \\ 1+i \end{pmatrix}, \quad |\psi_i\rangle_z = \begin{pmatrix} 1 \\ 0 \end{pmatrix}. \tag{8.6.9}$$

These states lead, respectively, to the following initial density operators

$$\rho^{(i)} : \frac{1}{2}\begin{pmatrix} 1 & 1 \\ 1 & 1 \end{pmatrix}, \quad \frac{1}{2}\begin{pmatrix} 1 & -i \\ i & 1 \end{pmatrix}, \quad \begin{pmatrix} 1 & 0 \\ 0 & 0 \end{pmatrix} \tag{8.6.10}$$

with corresponding polarization vectors

$$\mathbf{P}^{(i)} : (1,0,0), \ (0,1,0), \ (0,0,1). \tag{8.6.11}$$

---
[17] Polarized along the x-axis means, that the polarization *vector* as an *average* value has only an x-component.

8.6 Density Operators & Spin

The derivation of the expression for $\rho^{(i)}$, $\mathbf{P}^{(i)}$ for an arbitrary initial state is straightforward (see Problem 8.11).

On the other hand, for an *unpolarized* beam, the density operator $\rho^{(i)}$ may not be presented in the form as $|\psi_i\rangle\langle\psi_i|$ for some $|\psi_i\rangle$. Such a density operator defines a mixture (see also §1.5) rather than a pure state some of which are given in (8.6.10). For an initial unpolarized beam, it is easy to see that the density operator is given by

$$\rho_u^{(i)} : \frac{1}{2}\begin{pmatrix} 1 & 0 \\ 0 & 1 \end{pmatrix} \qquad (8.6.12)$$

giving as expected,

$$\mathbf{P}_u^{(i)} \equiv \text{Tr}\left[\boldsymbol{\sigma}\rho_u^{(i)}\right] = \mathbf{0} \qquad (8.6.13)$$

with the letter u standing for 'unpolarized'.

To find the general expression for the density operator ρ *after* the scattering process, suppose that the initial and final momenta of the projectile are, respectively, $\hbar\mathbf{k}'$ and $\hbar\mathbf{k}$. To this end, if $|\psi_f\rangle$ denotes the final state, we define a 2×2 matrix M such that

$$|\psi_f\rangle = M|\psi_i\rangle. \qquad (8.6.14)$$

That is, for initially polarized, and also for initially unpolarized projectile, the final density operator ρ may be obtained from

$$\rho = M\rho^{(i)}M^\dagger. \qquad (8.6.15)$$

For the problem at hand, the general structure of M is readily obtained. To this end, using the fact that $\mathbf{1}$, and the Pauli matrices $\sigma_1, \sigma_2, \sigma_3$ constitute a complete set of matrices in the vector space generated by 2×2 matrices, we may write

$$M = \alpha\mathbf{1} + \beta\mathbf{n}\cdot\boldsymbol{\sigma} \qquad (8.6.16)$$

where $\mathbf{n}$ is a unit vector, α, β, are in general, appropriate function of $\mathbf{k}'$, $\mathbf{k}$.

If we assume the invariance of the theory under space reflection, i.e., under parity transformation, then using the fact that $\boldsymbol{\sigma}$ (spin, angular momentum) is an axial vector (§2.6), $\mathbf{n}$ is to be chosen as an axial vector, constructed out of $\mathbf{k}'$, $\mathbf{k}$ and α, β are invariant under the transformation $\mathbf{k}' \to -\mathbf{k}'$, $\mathbf{k} \to -\mathbf{k}$. That is, in particular,

$$\mathbf{n} = \frac{\mathbf{k}' \times \mathbf{k}}{|\mathbf{k}' \times \mathbf{k}|} \qquad (8.6.17)$$

(for $|\mathbf{k}' \times \mathbf{k}| \neq 0$). α, β are, in general, functions of the scalar product $\mathbf{k}' \cdot \mathbf{k}$, $|\mathbf{k}'|$, $|\mathbf{k}|$.

With $\mathbf{n}$ given in (8.6.17), it is convenient to choose the coordinate system as shown in Figure 8.1, with $\mathbf{n}$ along the z-axis.

With the choice of the coordinate system in Figure 8.1, the matrix M takes the simple form

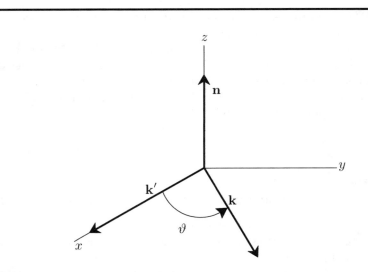

Fig. 8.1. The coordinate system for the scattering problem is conveniently chosen with $\mathbf{k'}$, $\mathbf{k}$ in the x-y plane with $\mathbf{k'}$, say along the x-axis. The unit vector $\mathbf{n}$ in (8.6.17) is then along the z-axis.

$$M = \begin{pmatrix} \alpha + \beta & 0 \\ 0 & \alpha - \beta \end{pmatrix}. \tag{8.6.18}$$

For an initially polarized beam along the y-axis for example, we have from (8.6.18), (8.6.10),

$$\rho = \frac{1}{2}\begin{pmatrix} |\alpha + \beta|^2 & -i(\alpha + \beta)(\alpha^* - \beta^*) \\ i(\alpha^* + \beta^*)(\alpha - \beta) & |\alpha - \beta|^2 \end{pmatrix} \tag{8.6.19}$$

for the final density operator. This gives the probability density $F_y(\vartheta)$, at scattering angle $\phi = \vartheta$ (see Figure 8.1),[18]

$$F_y(\vartheta) = \mathrm{Tr}[\rho] = \frac{1}{2}\left(|\alpha + \beta|^2 + |\alpha - \beta|^2\right) = |\alpha|^2 + |\beta|^2 \tag{8.6.20}$$

indicating, for simplicity of the notation, only the scattering angle ϑ of the projectile. For the final polarization we have

$$\frac{\mathrm{Tr}[\boldsymbol{\sigma}\rho]}{\mathrm{Tr}[\rho]} = \mathbf{P}_y(\vartheta) = \frac{\left(2|\alpha||\beta|\sin\delta,\ |\alpha|^2 - |\beta|^2,\ 2|\alpha||\beta|\cos\delta\right)}{|\alpha|^2 + |\beta|^2} \tag{8.6.21}$$

where we have written

---

[18] $F_y(\vartheta)$ is related to the so-called differential cross section. Differential cross sections will be studied in detail in Chapter 15.

8.6 Density Operators & Spin

$$\alpha = |\alpha|e^{i\delta_1}, \qquad \beta = |\beta|e^{i\delta_2}, \qquad e^{i\delta} = e^{i(\delta_2 - \delta_1)}. \tag{8.6.22}$$

With the definitions in (8.6.22), the matrix M in (8.6.16) may be rewritten as

$$M = e^{i\delta_1}\left(|\alpha|\mathbf{1} + |\beta|e^{i\delta}\mathbf{n}\cdot\boldsymbol{\sigma}\right) \tag{8.6.23}$$

and the knowledge of $|\alpha|$, $|\beta|$, $e^{i\delta}$ determines the scattering problem. The overall phase factor $\exp[i\delta_1]$ in (8.6.23) does not contribute to (8.6.15).

Of particular interest is the case of an initial unpolarized beam. The reason is that upon scattering, the beam becomes *polarized* along the z-axis. To this end, $\rho_{\rm u}^{(i)}$ in (8.6.12) gives, from (8.6.18), the final density operator

$$\rho_{\rm u} = \frac{1}{2}\begin{pmatrix} |\alpha + \beta|^2 & 0 \\ 0 & |\alpha - \beta|^2 \end{pmatrix} \tag{8.6.24}$$

with final probability density $F_{\rm u}(\vartheta)$ at scattering angle ϑ,

$$F_{\rm u}(\vartheta) = |\alpha|^2 + |\beta|^2 \tag{8.6.25}$$

and polarization vector

$$\mathbf{P}_{\rm u}(\vartheta) = \frac{\bigl(0, 0, 2|\alpha||\beta|\cos\delta\bigr)}{|\alpha|^2 + |\beta|^2}. \tag{8.6.26}$$

With an initially polarized beam along the z-axis, we have similarly,

$$\rho_z = \begin{pmatrix} |\alpha + \beta|^2 & 0 \\ 0 & 0 \end{pmatrix}, \qquad F_z(\vartheta) = |\alpha + \beta|^2 \tag{8.6.27}$$

$$\mathbf{P}_z(\vartheta) = (0, 0, 1). \tag{8.6.28}$$

For a given angle ϑ, probability densities such as $F_{\rm u}(\vartheta)$, $F_{\rm u}(-\vartheta)$, for specific initial conditions given by $\rho^{(i)}$, and polarization vectors, such as $\mathbf{P}_{\rm u}(\vartheta)$ in (8.6.26), the numbers $|\alpha|$, $|\beta|$, $\exp[i\delta_1]$ defining the general matrix M in (8.6.23), up to an overall phase factor, may be determined from the underlying dynamics.

Before discussing the scattering of particles of spin 1/2 off a target of spin 1/2, we recast the above problem in a more general form.

We denote the 2×2 identity matrix $\mathbf{1}$ by σ_0, and write

$$\rho^{(i)} = \sum_{\mu=0}^{3} C_\mu \sigma_\mu. \tag{8.6.29}$$

Upon multiplying the latter by σ_ν and taking the trace, one obtains

$$C_\nu = \frac{1}{2}\operatorname{Tr}\left[\sigma_\nu \rho^{(i)}\right] \equiv \frac{1}{2}\langle\sigma_\nu\rangle^{(i)} \tag{8.6.30}$$

where $\langle\sigma_0\rangle^{(i)} = 1$ with $\mathrm{Tr}\,[\rho^{(i)}]$ normalized to unity. This leads to the expression

$$\rho^{(i)} = \frac{1}{2}\sum_{\mu=0}^{3} \langle\sigma_\mu\rangle^{(i)} \sigma_\mu \tag{8.6.31}$$

and for the final density operator

$$\rho = \frac{1}{2}\sum_{\mu=0}^{3} \langle\sigma_\mu\rangle^{(i)} M\,\sigma_\mu\,M^\dagger \tag{8.6.32}$$

giving

$$\mathrm{Tr}[\sigma_\nu \rho] = \frac{1}{2}\sum_{\mu=0}^{3} \langle\sigma_\mu\rangle^{(i)} \mathrm{Tr}\,[\sigma_\nu\,M\,\sigma_\mu\,M^\dagger]. \tag{8.6.33}$$

In particular,

$$F(\vartheta) = \frac{1}{2}\sum_{\mu=0}^{3} \langle\sigma_\mu\rangle^{(i)} \mathrm{Tr}\,[M\,\sigma_\mu\,M^\dagger] \tag{8.6.34}$$

for the probability density at angle ϑ, and with

$$\frac{\mathrm{Tr}[\sigma_\nu \rho]}{\mathrm{Tr}[\rho]} = \langle\sigma_\nu\rangle \tag{8.6.35}$$

we have

$$F(\vartheta)\,\langle\sigma_\nu\rangle = \frac{1}{2}\sum_{\mu=0}^{3} \langle\sigma_\mu\rangle^{(i)} \mathrm{Tr}\,[\sigma_\nu\,M\,\sigma_\mu\,M^\dagger]. \tag{8.6.36}$$

Equations (8.6.34), (8.6.36) give the final expectation values of σ_ν in terms of the initial expectation values of the σ_μ operators.

For an initial unpolarized beam, $\langle\sigma_\mu\rangle^{(i)} = \delta_{\mu 0}$, hence (8.6.34) gives

$$F_\mathrm{u}(\vartheta) = \frac{1}{2}\mathrm{Tr}\,[M\,M^\dagger] \tag{8.6.37}$$

and (8.6.36) reduces to

$$F_\mathrm{u}(\vartheta)\,\langle\sigma_\nu\rangle_\mathrm{u} = \frac{1}{2}\mathrm{Tr}\,[\sigma_\nu\,M\,M^\dagger]. \tag{8.6.38}$$

For a beam initially polarized along the z-axis, $\langle\sigma_\mu\rangle^{(i)} = \delta_{\mu 0} + \delta_{\mu 3}$ (see (8.6.11)),

$$F_z(\vartheta) = \frac{1}{2}\mathrm{Tr}\,[M\,M^\dagger] + \frac{1}{2}\mathrm{Tr}\,[M\,\sigma_3\,M^\dagger]. \tag{8.6.39}$$

From (8.6.18) one explicitly has

$$\mathrm{Tr}\,[M\,\sigma_\nu\,M^\dagger] = \mathrm{Tr}\,[M^\dagger\,\sigma_\nu\,M] \tag{8.6.40}$$

which by using (8.6.37), (8.6.38), (8.6.39) leads to the interesting result

$$F_z(\vartheta) = F_u(\vartheta) \left[1 + \langle \sigma_3 \rangle_u \right] \tag{8.6.41}$$

showing the modification of the probability density due to the initial polarization of the beam over the density with initially unpolarized beam. We note that $F_z(\vartheta)$ is expressed in terms of expectation values for the final state of the scattering problem with the initially unpolarized beam.

The sign of $\langle \sigma_3 \rangle_u$ in (8.6.41) determines an asymmetry in the number of particles with spin along or opposite the orientation of the z-axis as arising from an initially unpolarized beam. Such an asymmetry is then reflected in the observed probability density $F_z(\vartheta)$.

8.6.3 Scattering of Spin 1/2 Particles off a Spin 1/2 Target

We introduce Pauli matrices $\boldsymbol{\sigma}^{(B)}$, $\boldsymbol{\sigma}^{(T)}$ corresponding to the beam and target, and unit matrices $\mathbf{1}^{(B)} = \sigma_0^{(B)}$, $\mathbf{1}^{(T)} = \sigma_0^{(T)}$ operating in their respective spin spaces.

The initial density operator $\rho^{(i)}$ may be quite generally expanded in terms of the sixteen components[19]

$$\sigma_0^{(B)}\sigma_0^{(T)}, \ldots, \sigma_3^{(B)}\sigma_0^{(T)}, \ldots, \sigma_3^{(B)}\sigma_3^{(T)} \tag{8.6.42}$$

$$\rho^{(i)} = \sum_{\mu=0}^{3}\sum_{\nu=0}^{3} C_{\mu\nu} \sigma_\mu^{(B)} \sigma_\nu^{(T)}. \tag{8.6.43}$$

Upon multiplying the latter by $\sigma_\lambda^{(B)} \sigma_\kappa^{(T)}$, taking the trace and using the fact that the initial polarizations of the beam and the target are independent, we obtain

$$C_{\lambda k} = \frac{1}{4} \operatorname{Tr}\left[\sigma_\lambda^{(B)} \sigma_\kappa^{(T)} \rho^{(i)}\right] = \frac{1}{4} \left\langle \sigma_\lambda^{(B)} \sigma_k^{(T)} \right\rangle^{(i)} \tag{8.6.44}$$

for a normalized density $\operatorname{Tr}\left[\rho^{(i)}\right] = 1$.

The final density then takes the form

$$\rho = \frac{1}{4}\sum_{\mu=0}^{3}\sum_{\nu=0}^{3} \left\langle \sigma_\mu^{(B)} \sigma_\nu^{(T)} \right\rangle M \sigma_\mu^{(B)} \sigma_\nu^{(T)} M^\dagger. \tag{8.6.45}$$

For initially unpolarized beam and target, (8.6.45) gives

$$F_u(\vartheta) = \frac{1}{4} \operatorname{Tr}\left[MM^\dagger\right] \tag{8.6.46}$$

and

---

[19] These sixteen matrices may be rewritten as 4×4 matrices, cf. Fernow (1976). See also this elegant paper for other details.

$$F_{\mathrm{u}}(\vartheta)\left\langle\sigma_\lambda^{(\mathrm{B})}\right\rangle_{\mathrm{u}} = \frac{1}{4}\mathrm{Tr}\left[\sigma_\lambda^{(\mathrm{B})}MM^\dagger\right] \tag{8.6.47}$$

$$F_{\mathrm{u}}(\vartheta)\left\langle\sigma_\kappa^{(\mathrm{T})}\right\rangle_{\mathrm{u}} = \frac{1}{4}\mathrm{Tr}\left[\sigma_\kappa^{(\mathrm{T})}MM^\dagger\right] \tag{8.6.48}$$

and as a correlation of the polarizations of the beam and target

$$F_{\mathrm{u}}(\vartheta)\left\langle\sigma_\lambda^{(\mathrm{B})}\sigma_\kappa^{(\mathrm{T})}\right\rangle_{\mathrm{u}} = \frac{1}{4}\mathrm{Tr}\left[\sigma_\lambda^{(\mathrm{B})}\sigma_\kappa^{(\mathrm{T})}MM^\dagger\right]. \tag{8.6.49}$$

For initially polarized beam and target along $p_{\mathrm{B}}\hat{\mathbf{z}}$, $p_{\mathrm{T}}\hat{\mathbf{z}}$, respectively, where $\hat{\mathbf{z}}$ is a unit vector along the z-axis and p_{B}, p_{T} take on the values ± 1,

$$\left\langle\sigma_\mu^{(\mathrm{B})}\sigma_\nu^{(\mathrm{T})}\right\rangle^{(\mathrm{i})} = \delta_{\mu 0}\delta_{\nu 0} + p_{\mathrm{B}}\delta_{\mu 3}\delta_{\nu 0} + p_{\mathrm{T}}\delta_{\mu 0}\delta_{\nu 3} + p_{\mathrm{B}}p_{\mathrm{T}}\delta_{\mu 3}\delta_{\nu 3} \tag{8.6.50}$$

where again we have used the fact that the initial polarizations of the beam and target are independent. Equation (8.6.50) gives the probability density

$$F(\vartheta; p_{\mathrm{B}}\hat{\mathbf{z}}, p_{\mathrm{T}}\hat{\mathbf{z}}) = \frac{1}{4}\mathrm{Tr}\left[MM^\dagger\right] + \frac{1}{4}p_{\mathrm{B}}\,\mathrm{Tr}\left[M\sigma_3^{(\mathrm{B})}M^\dagger\right]$$
$$+ \frac{1}{4}p_{\mathrm{T}}\,\mathrm{Tr}\left[M\sigma_3^{(\mathrm{T})}M^\dagger\right] + \frac{1}{4}p_{\mathrm{B}}p_{\mathrm{T}}\,\mathrm{Tr}\left[M\sigma_3^{(\mathrm{B})}\sigma_3^{(\mathrm{T})}M^\dagger\right]. \tag{8.6.51}$$

For a process which is invariant under space reflection and time reversal, the matrix $M(\hat{\mathbf{k}}', \hat{\mathbf{k}}, \boldsymbol{\sigma}^{(\mathrm{B})}, \boldsymbol{\sigma}^{(\mathrm{T})})$ expressed as a function of the unit vectors $\hat{\mathbf{k}}'$, $\hat{\mathbf{k}}$ and the Pauli matrices, satisfies

$$M(\hat{\mathbf{k}}', \hat{\mathbf{k}}, \boldsymbol{\sigma}^{(\mathrm{B})}, \boldsymbol{\sigma}^{(\mathrm{T})}) = M(-\hat{\mathbf{k}}', -\hat{\mathbf{k}}, \boldsymbol{\sigma}^{(\mathrm{B})}, \boldsymbol{\sigma}^{(\mathrm{T})}) \tag{8.6.52}$$

under space reflection and

$$M(\hat{\mathbf{k}}', \hat{\mathbf{k}}, \boldsymbol{\sigma}^{(\mathrm{B})}, \boldsymbol{\sigma}^{(\mathrm{T})}) = M(-\hat{\mathbf{k}}, -\hat{\mathbf{k}}', -\boldsymbol{\sigma}^{(\mathrm{B})}, -\boldsymbol{\sigma}^{(\mathrm{T})}) \tag{8.6.53}$$

under time reversal.

Below we will see later that under the constraints (8.6.52), (8.6.53), the following equalities follow,

Lemma 1.

(i) $\mathrm{Tr}\left[M\,\boldsymbol{\sigma}^{(\mathrm{B}),(\mathrm{T})}\,M^\dagger\right] = \mathrm{Tr}\left[M^\dagger\,\boldsymbol{\sigma}^{(\mathrm{B}),(\mathrm{T})}\,M\right]$
(ii) $\mathrm{Tr}\left[M\,\sigma_3^{(\mathrm{B})}\sigma_3^{(\mathrm{T})}\,M^\dagger\right] = \mathrm{Tr}\left[M^\dagger\,\sigma_3^{(\mathrm{B})}\sigma_3^{(\mathrm{T})}\,M\right]$

From (8.6.46)–(8.6.49) and the above two equalities give for (8.6.51)

$$F(\vartheta; p_{\mathrm{B}}\hat{\mathbf{z}}, p_{\mathrm{T}}\hat{\mathbf{z}}) = F_\mu(\vartheta)\left[1 + p_{\mathrm{B}}\left\langle\sigma_3^{(\mathrm{B})}\right\rangle_{\mathrm{u}} + p_{\mathrm{T}}\left\langle\sigma_3^{(\mathrm{T})}\right\rangle_{\mathrm{u}}\right.$$

$$+ p_B p_T \left\langle \sigma_3^{(B)} \sigma_3^{(T)} \right\rangle_u \bigg]. \tag{8.6.54}$$

This equation shows how the probability density is obtained, with the beam and target initially polarized along $p_B\hat{z}$, $p_T\hat{z}$, respectively, from that with initially unpolarized beam and target. The first term within the square brackets on the right-hand side of (8.6.54) corresponds to the case that neither the beam nor the target is polarized (from initially unpolarized ones), while the second, third and fourth terms correspond, respectively, to the cases where only the beam is polarized, only the target is polarized and finally both the beam and target are polarized.

The proof of the above lemma is straightforward, and easily follows by finding, for example, the general expression for M under the constraints in (8.6.52), (8.6.53) as done below.

The matrix M may be written as

$$M = A \mathbf{1}^{(B)} + \mathbf{B} \cdot \boldsymbol{\sigma}^{(B)} \tag{8.6.55}$$

where in addition to $\hat{\mathbf{k}}$, $\hat{\mathbf{k}}'$ the matrices A and $\mathbf{B}$ may depend on $\boldsymbol{\sigma}^{(T)}$.

Consider the three orthonormal vectors

$$\mathbf{a} = \frac{\hat{\mathbf{k}} - \hat{\mathbf{k}}'}{|\hat{\mathbf{k}} - \hat{\mathbf{k}}'|}, \quad \mathbf{b} = \frac{\hat{\mathbf{k}} + \hat{\mathbf{k}}'}{|\hat{\mathbf{k}} + \hat{\mathbf{k}}'|}, \quad \mathbf{c} = \frac{\hat{\mathbf{k}}' \times \hat{\mathbf{k}}}{|\hat{\mathbf{k}}' \times \hat{\mathbf{k}}|} \tag{8.6.56}$$

for non-zero denominators, where $\mathbf{c}$ is a unit vector along the z-axis. Since $\boldsymbol{\sigma}^{(T)}$ is an axial vector, A is of the form

$$A = \alpha_1 \mathbf{1}^{(T)} + \alpha_2 \mathbf{c} \cdot \boldsymbol{\sigma}^{(T)}. \tag{8.6.57}$$

On the other hand, the vector $\mathbf{B}$ may, in general, be written as

$$\mathbf{B} = \beta_1 \mathbf{a} + \beta_2 \mathbf{b} + \beta_3 \mathbf{c}. \tag{8.6.58}$$

Now it is a simple matter[20] to show that the constraints (8.6.52), (8.6.53) imply that

$$\beta_1 = \alpha_3 \mathbf{a} \cdot \boldsymbol{\sigma}^{(T)}, \quad \beta_2 = \alpha_4 \mathbf{b} \cdot \boldsymbol{\sigma}^{(T)} \tag{8.6.59}$$

$$\beta_3 = \alpha_0 \mathbf{1}^{(B)} + \alpha_5 \mathbf{c} \cdot \boldsymbol{\sigma}_3^{(T)} \tag{8.6.60}$$

where $\alpha_0, \ldots, \alpha_5$ are scalars.

The expression for M in (8.6.55) then gives explicitly

$$\text{Tr}\left[M\sigma_j^{(B)}M^\dagger\right] = 2\,\text{Tr}'\left[AB_j^\dagger\right] + 2\,\text{Tr}'\left[B_j A^\dagger\right] - 2\mathrm{i}\varepsilon_{jkl}\,\text{Tr}'\left[B_k B_l^\dagger\right] \tag{8.6.61}$$

---

[20] Cf. Goldberger and Watson (1964), p. 391.

462   8 Quantum Physics of Spin 1/2 & Two-Level Systems

where Tr′ corresponds to tracing over the target variables only. Using, in the process, the orthogonality of the vector in (8.6.56), and the structure of **B** in (8.6.58)–(8.6.60), this shows that the last term in (8.6.61) is equal to zero. This establishes part (i) of the lemma for $\boldsymbol{\sigma}^{(B)}$ upon the comparison of (8.6.61) with $\text{Tr}\left[M^\dagger \sigma_j^{(B)} M\right]$. The proof of the part (i) for $\boldsymbol{\sigma}^{(T)}$ is similar.

The proof of part (ii) of the lemma easily follows (see Problem 8.14 (ii)) by noting, in the process, that

$$B_3 = \beta_3 = \alpha_0 \mathbf{1}^{(T)} + \alpha_5 \sigma_3^{(T)} \tag{8.6.62}$$

which depends on the component $\sigma_3^{(T)}$ of $\boldsymbol{\sigma}^{(T)}$ only.

Incidently (8.6.55), (8.6.56)–(8.6.60) give the following general expression for M:

$$\begin{aligned} M = {} & \alpha_1 \mathbf{1}^{(B)} \mathbf{1}^{(T)} + \eta_1 \boldsymbol{\sigma}^{(B)} \cdot \mathbf{a}\, \boldsymbol{\sigma}^{(T)} \cdot \mathbf{a} + \eta_2 \boldsymbol{\sigma}^{(B)} \cdot \mathbf{b}\, \boldsymbol{\sigma}^{(T)} \cdot \mathbf{b} \\ & + \eta_3 \boldsymbol{\sigma}^{(B)} \cdot \mathbf{c}\, \boldsymbol{\sigma}^{(T)} \cdot \mathbf{c} + \eta_4 \left(\boldsymbol{\sigma}^{(B)} \mathbf{1}^{(T)} + \mathbf{1}^{(B)} \boldsymbol{\sigma}^{(T)} \right) \cdot \mathbf{c} \\ & + \eta_5 \left(\boldsymbol{\sigma}^{(B)} \mathbf{1}^{(T)} - \mathbf{1}^{(B)} \boldsymbol{\sigma}^{(T)} \right) \cdot \mathbf{c}. \end{aligned} \tag{8.6.63}$$

8.7 Quantum Interference and Measurement; The Role of the Environment

In §1.10, an example was given of the interaction (see (1.10.29)–(1.10.31), (1.10.35)–(1.10.38)) of a spin 1/2 physical system and an apparatus, also described as a spin 1/2 system, and studied the role of quantum interference (see (1.10.52), (1.10.53), (1.10.54)) in the presence of the apparatus allowing imperfection for the latter, specified by some parameter κ, in the measurement process. We have seen that up to small corrections corresponding to a small κ, how this example provides a model for the disappearance of interference (see (1.10.56)) by the mere presence of the measuring and unread apparatus. The limit $\kappa \to 0$, defines an ideal apparatus for which a perfect *correlation* occurs between the physical system and the apparatus as given in (1.10.31).

In this section, we show how the above process may be formally implemented by an interaction Hamiltonian, involving the system and meter variables, leading to the unitary operator spelled out in (1.10.35)–(1.10.39). We also provide another illustration of the interaction of the spin 1/2 system with an apparatus with the latter described by a harmonic oscillator in a coherent state. We will see how this example may be used to provide a model with an almost perfect *correlation* occurring between the system and the apparatus with a built-in imperfection in the latter naturally arising from the non-orthogonality of the coherent states.

Finally, we consider the role of the environment, surrounding a quantum system (e.g., as a response of excitations generated in some medium nearby,...), as part of the measuring process, on the meter readings. The environment, coupled to the meter variables, consists of everything else monitoring the observables being measured and provides, as one may argue, the different alternative readings of the meter being sought. This gives a natural way of producing classical correlations between the system and the detector (meter) eliminating the coherence between different states thus destroying quantum superpositions. The destruction of the superposition, referred to as *quantum decoherence*, makes sure that the system is in one of its alternative states rather than in a superposition of them. The information thus obtained on the system by the meter, "hooked up" to the environment, can be then described in usually perceived classical terms.

8.7.1 Interaction with an Apparatus and Unitary Evolution Operator

To describe the apparatus in §1.10 as a spin 1/2 system, we introduce its spin variables operating in the apparatus vector space. To this end, we introduce the spin operator $\hbar \Sigma/2$, with

$$\Sigma_1 = \begin{bmatrix} 0 & 1 \\ 1 & 0 \end{bmatrix}, \quad \Sigma_2 = \begin{bmatrix} 0 & -i \\ i & 0 \end{bmatrix}, \quad \Sigma_3 = \begin{bmatrix} 1 & 0 \\ 0 & -1 \end{bmatrix} \qquad (8.7.1)$$

which are the Pauli matrices operating in the vector space in question. To the three matrices in (8.7.1), we adjoin the unit matrix

$$\mathbf{1} = \begin{bmatrix} 1 & 0 \\ 0 & 1 \end{bmatrix}. \qquad (8.7.2)$$

The initial state of the apparatus, given in (1.10.28), is then simply

$$\frac{1}{2} \begin{bmatrix} 1 - i \\ 1 + i \end{bmatrix} \qquad (8.7.3)$$

which is the eigenvector of the component Σ_2:

$$\begin{bmatrix} 0 & -i \\ i & 0 \end{bmatrix} \frac{1}{2} \begin{bmatrix} 1 - i \\ 1 + i \end{bmatrix} = +\frac{1}{2} \begin{bmatrix} 1 - i \\ 1 + i \end{bmatrix} \qquad (8.7.4)$$

with the "needle" of the apparatus initially pointing at **0** in Figure 1.16 along the positive direction of the y-axis.

The initial state of the system and apparatus in (1.10.29) may be also simply written in matrix form as

$$|\psi_0\rangle = \begin{pmatrix} c_+ \\ c_- \end{pmatrix} \frac{1}{2} \begin{bmatrix} 1 - i \\ 1 + i \end{bmatrix} \qquad (8.7.5)$$

8 Quantum Physics of Spin 1/2 & Two-Level Systems

with the variables of the system and the apparatus operating in different vector spaces using different notations for the matrices for the two systems with square ones for the apparatus matrices.

We consider the following pair of the apparatus variables

$$\frac{1}{2}(\mathbf{1} - \Sigma_1) = \frac{1}{2}\begin{bmatrix} 1 & -1 \\ -1 & 1 \end{bmatrix} = \frac{1}{\sqrt{2}}\begin{bmatrix} 1 \\ -1 \end{bmatrix} \frac{1}{\sqrt{2}}\begin{bmatrix} 1 \\ -1 \end{bmatrix}^\mathrm{T} \tag{8.7.6}$$

$$\frac{1}{2}(\mathbf{1} + \Sigma_1) = \frac{1}{2}\begin{bmatrix} 1 & 1 \\ 1 & 1 \end{bmatrix} = \frac{1}{\sqrt{2}}\begin{bmatrix} 1 \\ 1 \end{bmatrix} \frac{1}{\sqrt{2}}\begin{bmatrix} 1 \\ 1 \end{bmatrix}^\mathrm{T}. \tag{8.7.7}$$

with $[1\ -1]^\mathrm{T}/\sqrt{2}$, $[1\ 1]^\mathrm{T}/\sqrt{2}$ denoting, respectively, the apparatus spin states along the negative/positive x-axis.

These matrices, apart being Hermitian, have the following interesting properties

$$\frac{1}{2}\begin{bmatrix} 1 & \mp 1 \\ \mp 1 & 1 \end{bmatrix} \frac{1}{2}\begin{bmatrix} 1 & \mp 1 \\ \mp 1 & 1 \end{bmatrix} = \frac{1}{2}\begin{bmatrix} 1 & \mp 1 \\ \mp 1 & 1 \end{bmatrix} \tag{8.7.8}$$

and

$$\frac{1}{2}\begin{bmatrix} 1 & -1 \\ -1 & 1 \end{bmatrix} \frac{1}{2}\begin{bmatrix} 1 & 1 \\ 1 & 1 \end{bmatrix} = \begin{bmatrix} 0 & 0 \\ 0 & 0 \end{bmatrix}. \tag{8.7.9}$$

As a Hamiltonian, in spin space, we consider the following simple one

$$H = -\lambda \sigma_3 \frac{1}{2}\begin{bmatrix} 1 & -1 \\ -1 & 1 \end{bmatrix} - \lambda q \sigma_1 \frac{1}{2}\begin{bmatrix} 1 & 1 \\ 1 & 1 \end{bmatrix} \tag{8.7.10}$$

where σ_1, σ_3 are Pauli matrices pertinent to the physical system, λ is a coupling parameter and q, as we will see below, is a measure of imperfection of the apparatus related to κ in §1.10. The coupling parameter λ is not arbitrary and is specified by the time of operation of the apparatus until the measurement is completed and the correlation between the apparatus and the physical system, of the type given in (1.10.30), required is achieved. The parameter q, however, may be controlled, taking the value 0 in the limit of an ideal apparatus, and should, in general, be small.

Due to the property in (8.7.9), the two parts of the Hamiltonian in (8.7.10) automatically commute, although σ_1 and σ_3 do not.

The construction of the time evolution operator from (8.7.10) is straightforward.

From the commutativity of the two terms in (8.7.10), the evolution operator may be written as the product of two factors

$$U(T) = \exp\left(\frac{\mathrm{i}\lambda T}{\hbar}\sigma_3 \frac{1}{2}\begin{bmatrix} 1 & -1 \\ -1 & 1 \end{bmatrix}\right) \exp\left(\frac{\mathrm{i}\lambda q T}{\hbar}\sigma_1 \frac{1}{2}\begin{bmatrix} 1 & 1 \\ 1 & 1 \end{bmatrix}\right). \tag{8.7.11}$$

Since $(\sigma_3)^2 = \mathbf{1}$, we have from (8.7.8), for the first factor in the product in (8.7.11)

$$1 + \sum_{\substack{n \geqslant 1 \\ (n=\text{odd})}} \frac{(i\lambda T/\hbar)^n}{n!} \sigma_3 \frac{1}{2}\begin{bmatrix} 1 & -1 \\ -1 & 1 \end{bmatrix} + \sum_{\substack{n \geqslant 2 \\ (n=\text{even})}} \frac{(i\lambda T/\hbar)^n}{n!} \frac{1}{2}\begin{bmatrix} 1 & -1 \\ -1 & 1 \end{bmatrix}$$

$$= 1 + \left(i \sin\left(\frac{\lambda T}{\hbar}\right) \sigma_3 + \left(\cos\left(\frac{\lambda T}{\hbar}\right) - 1\right) \mathbf{1}\right) \frac{1}{2}\begin{bmatrix} 1 & -1 \\ -1 & 1 \end{bmatrix}. \quad (8.7.12)$$

The second factor in the product in (8.7.11) is similarly handled, giving for $U(T)$ the expression

$$U(T) = 1 + \left(i \sin\left(\frac{\lambda T}{\hbar}\right) \sigma_3 + \left(\cos\left(\frac{\lambda T}{\hbar}\right) - 1\right) \mathbf{1}\right) \frac{1}{2}\begin{bmatrix} 1 & -1 \\ -1 & 1 \end{bmatrix}$$

$$+ \left(i \sin\left(\frac{\lambda q T}{\hbar}\right) \sigma_1 + \left(\cos\left(\frac{\lambda q T}{\hbar}\right) - 1\right) \mathbf{1}\right) \frac{1}{2}\begin{bmatrix} 1 & 1 \\ 1 & 1 \end{bmatrix} \quad (8.7.13)$$

where we have finally used (8.7.9) to carry out explicitly the product of the two factors in (8.7.11).

The implementation of measurement by the interaction of the apparatus with the physical system, means that a perfect correlation between these two systems occurs for $q = 0$. For an ideal apparatus, for which $q = 0$, we must have for a perfect correlation,

$$|\psi\rangle = U(T)|\psi_0\rangle = c_+ \begin{pmatrix} 1 \\ 0 \end{pmatrix}\begin{bmatrix} 1 \\ 0 \end{bmatrix} + c_- \begin{pmatrix} 0 \\ 1 \end{pmatrix}\begin{bmatrix} 0 \\ 1 \end{bmatrix} \quad (8.7.14)$$

which states that if the spin is up, then the "needle" of the apparatus is also up (see Figure 1.16), and similarly, vis-à-vis the second term in (8.7.14), if the spin is down so is the "needle" of the apparatus.[21] Hence the correlation to be reached between the apparatus and the system by the measurement process, as given in (8.7.14), dictates that $\lambda T/\hbar$ in (8.7.13) cannot be chosen to take any arbitrary value. It is easy to see that to achieve this correlation condition in (8.7.14), as implied by measurement, we may choose[22]

$$\lambda T/\hbar = \pi/2. \quad (8.7.15)$$

The unitary operator (8.7.13) then becomes

---

[21] In the state (8.7.14), neither the state of the physical system nor the state of the apparatus is well defined, and this state is referred to as an entangled one (see also §8.10). Later on below, we will see how one may argue, by taking into account of the environment, surrounding the combined system above, either the first state (coefficient of c_+) or the second state (coefficient of c_-) in (8.7.14) is selected in a measurement, by a process referred to as quantum decoherence, in conformity with one's classical notions of a measurement.

[22] Such correlations measurements restrictions are typical in dynamical investigations of measurement theory, cf. Perès (1986), Yurke and Stoler (1986). Such conditions are realized experimentally, cf. Itano et al. (1990).

$$U(T) = \mathbf{1} + \begin{pmatrix} -1+i & 0 \\ 0 & -1-i \end{pmatrix} \frac{\begin{bmatrix} 1 & -1 \\ -1 & 1 \end{bmatrix}}{2} + \begin{pmatrix} \cos\left(\frac{\pi q}{2}\right) - 1 & i\sin\left(\frac{\pi q}{2}\right) \\ i\sin\left(\frac{\pi q}{2}\right) & \cos\left(\frac{\pi q}{2}\right) - 1 \end{pmatrix} \frac{\begin{bmatrix} 1 & 1 \\ 1 & 1 \end{bmatrix}}{2}$$
(8.7.16)

which for $q = 0$ leads to (8.7.14). The unit matrix in (8.7.16) is given by

$$\mathbf{1} = \begin{pmatrix} 1 & 0 \\ 0 & 1 \end{pmatrix} \begin{bmatrix} 1 & 0 \\ 0 & 1 \end{bmatrix}. \tag{8.7.17}$$

Upon comparison of (8.7.16) with (1.10.35)–(1.10.38) it is easy to see that the parameter κ, as a measure of the imperfection of the apparatus, is related to q by the simple relation

$$\kappa = \pi q/2. \tag{8.7.18}$$

The final state, as obtained from (8.7.16), (8.7.5), quite generally, is given by

$$|\psi\rangle = \left(c_+ \begin{pmatrix} 1 \\ 0 \end{pmatrix} + \frac{1}{2} \begin{pmatrix} \left(\cos\left(\frac{\pi q}{2}\right) - 1\right) c_+ + i\sin\left(\frac{\pi q}{2}\right) c_- \\ i\sin\left(\frac{\pi q}{2}\right) c_+ + \left(\cos\left(\frac{\pi q}{2}\right) - 1\right) c_- \end{pmatrix}\right) \begin{bmatrix} 1 \\ 0 \end{bmatrix}$$
$$+ \left(c_- \begin{pmatrix} 0 \\ 1 \end{pmatrix} + \frac{1}{2} \begin{pmatrix} \left(\cos\left(\frac{\pi q}{2}\right) - 1\right) c_+ + i\sin\left(\frac{\pi q}{2}\right) c_- \\ i\sin\left(\frac{\pi q}{2}\right) c_+ + \left(\cos\left(\frac{\pi q}{2}\right) - 1\right) c_- \end{pmatrix}\right) \begin{bmatrix} 0 \\ 1 \end{bmatrix}$$
(8.7.19)

As expected, this coincides with the expression in (1.10.30).

The relevant probabilities for the above combined systems are now straightforward to compute and the reader may refer to §1.10 for these expressions. [Here $(1\ 0)^{\mathsf{T}}$, $(0\ 1)^{\mathsf{T}}$ in (8.7.19), refer, respectively, to the $\pm \bar{z}$ directions there.]

In the above illustration, T in (8.7.11) denotes the time of operation of the apparatus. To be rigorous in the treatment of the "switching on" of the apparatus at time $t = 0$ and "switching it off" at time $t = T$, we may replace the formal discontinuous expression

$$\xi_0(t) = \Theta(T-t) - \Theta(t) \tag{8.7.20}$$

reflecting this property, by a smooth function

$$\xi(t) = \begin{cases} 0, & -\infty < t < 0 \\ \dfrac{e\varepsilon}{t} \exp(-\varepsilon/t), & 0 \leqslant t < \varepsilon \\ 1, & \varepsilon \leqslant t < T - \varepsilon \\ \dfrac{e\varepsilon}{T-t} \exp(-\varepsilon/(T-t)), & T - \varepsilon \leqslant t < T \\ 0, & T \leqslant t < \infty \end{cases} \tag{8.7.21}$$

where ε is an arbitrarily chosen small number much less than T. This remarkable function is not only a continuous function of t but may be also differentiated an arbitrary number of times. It vanishes for $t < 0$, $t > T$, and $\xi(t) = 1$ for $\varepsilon \leqslant t < T - \varepsilon$ as given above.

The coupling λ is then to be replaced by $\lambda_0 \xi(t)$, where λ_0 is the coupling parameter between the system and the apparatus during the interaction period, and λ, now, is an effective coupling related to λ_0 as spelled out below.

To the above end, we integrate $\xi(t)$ over t to obtain

$$\int_{-\infty}^{\infty} dt\, \xi(t) = T - 2\varepsilon \left(1 - e \int_{1}^{\infty} \frac{d\tau}{\tau} e^{-\tau} \right)$$

$$= T - 0.8073\, \varepsilon \tag{8.7.22}$$

recognizing the exponential integral in (8.7.22). We may then set

$$\lambda_0 (T - 0.8073\, \varepsilon) = \lambda T \tag{8.7.23}$$

for a given $0 < \varepsilon \ll T$.

8.7.2 Interaction with a Harmonic Oscillator in a Coherent State

As another illustration we consider the interaction of a spin 1/2 with an apparatus described by a harmonic oscillator, in a coherent state, with Hamiltonian

$$H = -\lambda \sigma_3 a^\dagger a + \hbar \omega a^\dagger a \tag{8.7.24}$$

omitting, for simplicity the zero point energy. The initial state of the system plus the apparatus is taken to be

$$|\Phi_0\rangle = \begin{pmatrix} c_+ \\ c_- \end{pmatrix} |-i\alpha_0\rangle \tag{8.7.25}$$

where $|-i\alpha_0\rangle$ is a coherent state (§6.6, (6.6.21), (6.6.15), (6.6.28), (6.6.4)), providing a very close description of a classical state of the apparatus. For convenience, we choose α_0 to be real (and positive) so that $i\alpha_0$ is pure imaginary and hence in configuration space, the *initial* state (see (6.6.50)) corresponds to a Gaussian distribution centered at the origin. α_0 will be taken to be a large number. As a matter of fact, since $\alpha_0 = |A|(m\omega/2\hbar)^{1/2}$ (see (6.6.15), (6.6.4)), it will be quite large for macroscopic values taken by $|A|$, m.[23]

---

[23] For example, for $|A| \sim 10^{-3}$ meters, $m \sim 10^{-3}$ kg, one has the estimate $\alpha_0 \sim 10^{12}\, (\omega \cdot s)^{1/2}$, with ω in s^{-1}. During a short time a microscopic particle interacting with an apparatus, ω will be also not small due to the rapid response of the oscillator (apparatus) to the particle in the short time. An upper time limit of response of the oscillator is also set up by its decay time induced by the environment.

If $\alpha_\pm = i\alpha_0 \exp(i\delta_\pm)$, where $\exp(i\delta_\pm)$ are phase factors, i.e., $\delta_\pm$ are real, then we recall from (6.6.38) that

$$\langle \alpha_+ | \alpha_- \rangle = \exp\left(-\alpha_0^2 \left(1 - e^{i(\delta_- - \delta_+)}\right)\right) \qquad (8.7.26)$$

and the states $|\alpha_\pm\rangle$ are optimally almost orthogonal for large α_0^2, if, in particular,

$$\delta_- - \delta_+ = \pi. \qquad (8.7.27)$$

The final state Φ is easily obtained from (8.7.25), (8.7.24), (6.6.27) to be

$$\Phi = c_+ \begin{pmatrix} 1 \\ 0 \end{pmatrix} \left| -i\, e^{i(\lambda T/\hbar - \omega T)} \alpha_0 \right\rangle + c_- \begin{pmatrix} 0 \\ 1 \end{pmatrix} \left| -i\, e^{i(-\lambda T/\hbar - \omega T)} \alpha_0 \right\rangle. \qquad (8.7.28)$$

The two coherent states, on the right-hand side of (8.7.28) are not orthogonal.

The state in (8.7.28) corresponds, however, to an almost perfect correlation, as in (8.7.14), between the physical system and the apparatus as a successful measurement, for (see (8.7.27))

$$2\lambda T/\hbar = \pi. \qquad (8.7.29)$$

This gives the almost perfectly correlated state[24]

$$|\Phi\rangle = c_+ \begin{pmatrix} 1 \\ 0 \end{pmatrix} |\alpha\rangle + c_- \begin{pmatrix} 0 \\ 1 \end{pmatrix} |-\alpha\rangle \qquad (8.7.30)$$

with $\alpha = \alpha_0 \exp(-i\omega T)$, and $|-i\alpha_0\rangle$ denoting the neutral state of the apparatus before its interaction with the physical system. Equation (8.7.30) is to be compared with (8.7.14).

The density operator corresponding to (8.7.30) is given by

$$\rho = |c_+|^2 \begin{pmatrix} 1 & 0 \\ 0 & 0 \end{pmatrix} |\alpha\rangle\langle\alpha| + |c_-|^2 \begin{pmatrix} 0 & 0 \\ 0 & 1 \end{pmatrix} |-\alpha\rangle\langle-\alpha|$$

$$+ c_+ c_-^* \begin{pmatrix} 0 & 1 \\ 0 & 0 \end{pmatrix} |\alpha\rangle\langle-\alpha| + c_+^* c_- \begin{pmatrix} 0 & 0 \\ 1 & 0 \end{pmatrix} |-\alpha\rangle\langle\alpha|. \qquad (8.7.31)$$

If one is not interested in reading the apparatus, and thus the latter is unread, one may introduce the so-called *reduced* density operator ρ_{RED} by taking the trace over the coherent states in (8.7.31). This gives the remarkably simple expression

$$\rho_{\text{RED}} = |c_+|^2 \begin{pmatrix} 1 & 0 \\ 0 & 0 \end{pmatrix} + |c_-|^2 \begin{pmatrix} 0 & 0 \\ 0 & 1 \end{pmatrix}$$

---

[24] A correlation will be also achieved for other values of $\lambda T/\hbar$. The "tuning" condition in (8.7.29), however, provides an optimum one as dictated by the definition of a measurement with the (almost) perfect correlation occurring between the apparatus and the physical system as given in (8.7.30).

$$+ \left[c_+ c_-^* \begin{pmatrix} 0 & 1 \\ 0 & 0 \end{pmatrix} + c_+^* c_- \begin{pmatrix} 0 & 0 \\ 1 & 0 \end{pmatrix} \right] \exp(-2\alpha_0^2) \qquad (8.7.32)$$

and the non-diagonal terms in (8.7.32) are practically equal to zero for macroscopic large values of α_0^2 as discussed above. Thus we obtain a mixture (see §1.5) of spin states for practical purposes.

Upon writing

$$\begin{pmatrix} 1 & 0 \\ 0 & 0 \end{pmatrix} = |+1/2, \bar{z}\rangle \langle +1/2, \bar{z}|, \qquad \begin{pmatrix} 0 & 0 \\ 0 & 1 \end{pmatrix} = |-1/2, \bar{z}\rangle \langle -1/2, \bar{z}|$$

$$\begin{pmatrix} 0 & 1 \\ 0 & 0 \end{pmatrix} = |+1/2, \bar{z}\rangle \langle -1/2, \bar{z}|, \qquad \begin{pmatrix} 0 & 0 \\ 1 & 0 \end{pmatrix} = |-1/2, \bar{z}\rangle \langle +1/2, \bar{z}| \qquad (8.7.33)$$

in the notation of §1.10, we note that the apparatus has a built in imperfection for measurement (see also Figure 1.17). In particular, the probabilities of a spin flip or a non-flip, corresponding to the experiment in Figure 1.17 (d) in the presence of our new apparatus are, respectively,

$$\langle -1/2, z|\rho_{\text{RED}}|-1/2, z\rangle = \frac{\sin^2 \beta}{2} - \frac{\sin^2 \beta}{2} e^{-2\alpha_0^2} \qquad (8.7.34)$$

$$\langle +1/2, z|\rho_{\text{RED}}|+1/2, z\rangle = \sin^4(\beta/2) + \cos^4(\beta/2) + \frac{\sin^2 \beta}{2} e^{-2\alpha_0^2} \qquad (8.7.35)$$

where we have used (1.10.44), (1.10.57), (1.10.58), and the interference terms (see (1.10.52), (1.10.53)) initially present in the experiment in part (c) of Figure 1.17, up to the exponentially damped terms involving $\exp(-2\alpha_0^2)$, disappear. The expressions in (8.7.34), (8.7.35) should be compared with those in Figure 1.18 (d).

8.7.3 The Role of the Environment

Macroscopic systems, such as meters associated with the above two examples, in the real world are never in isolation from the environment. The latter, coupled to a meter's variables, consists of everything else monitoring an observable being measured and may include additional degrees of freedom associated with the apparatus itself, providing, as one may argue, the different alternative readings on a meter being sought.[25] Modellings of the environment, as applied to the above two examples, will be described and shown as to how quantum decoherence may set in producing classical correlations between the physical system into consideration and the detector (meter) in conformity with one's classical conception of measurements.

Consider the pure correlated state[26] between the physical system and the meter given in (8.7.14). Referring to such a state, it is not clear what

---
[25] Zurek (1991).
[26] Such a state, with $c_+ \neq 0$, $c_- \neq 0$ is referred to as an entangled one. Entangled states will be considered in detail in §8.10.

the different alternatives being measured. For example, for $c_\pm = \pm i$, and by using the expansions,

$$\begin{pmatrix}1\\0\end{pmatrix} = \frac{1+i}{2} \cdot \frac{1}{2}\begin{pmatrix}1-i\\1+i\end{pmatrix} + \frac{1-i}{2} \cdot \frac{1}{2}\begin{pmatrix}1+i\\1-i\end{pmatrix} \qquad (8.7.36)$$

$$\begin{pmatrix}0\\1\end{pmatrix} = \frac{1-i}{2} \cdot \frac{1}{2}\begin{pmatrix}1-i\\1+i\end{pmatrix} + \frac{1+i}{2} \cdot \frac{1}{2}\begin{pmatrix}1+i\\1-i\end{pmatrix} \qquad (8.7.37)$$

and similarly for the meter's variables, we may rewrite the same state $|\psi\rangle$ in (8.7.14), with $c_\pm = \pm i$, as

$$|\psi\rangle = -\frac{1}{2}\begin{pmatrix}1-i\\1+i\end{pmatrix} \cdot \frac{1}{2}\begin{bmatrix}1-i\\1+i\end{bmatrix} + \frac{1}{2}\begin{pmatrix}1+i\\1-i\end{pmatrix} \cdot \frac{1}{2}\begin{bmatrix}1+i\\1-i\end{bmatrix}. \qquad (8.7.38)$$

This raises the ambiguity as what spin components are actually being measured — the z-components or the y-components. On the other hand, the density operator as obtained from the state in (8.7.14) is given by

$$\rho = |c_+|^2 \begin{pmatrix}1\\0\end{pmatrix}(1\ 0)\begin{bmatrix}1\\0\end{bmatrix}[1\ 0] + |c_-|^2 \begin{pmatrix}0\\1\end{pmatrix}(0\ 1)\begin{bmatrix}0\\1\end{bmatrix}[0\ 1]$$

$$+ c_+ c_-^* \begin{pmatrix}1\\0\end{pmatrix}(0\ 1)\begin{bmatrix}1\\0\end{bmatrix}[0\ 1] + c_+^* c_- \begin{pmatrix}0\\1\end{pmatrix}(1\ 0)\begin{bmatrix}0\\1\end{bmatrix}[1\ 0]. \qquad (8.7.39)$$

The coefficients $|c_+|^2$, $|c_-|^2$ are interpreted as classical probabilities, and if the non-diagonal terms, proportional to $c_+ c_-^*$, $c_- c_+^*$, were absent, one may infer that the detector and the system are either in the spin up or spin down states, in conformity with one's classical perception of measurements, and not in superpositions of these states, and the set of alternatives to measurement of spin components along the z-axis have been selected.

Now we provide a straightforward illustration of the quantum decoherence as induced by the environment. One may consider the measurement process, involving the physical system (S), the meter (M) and the environment (E), as a two-step[27] one. In the first step of the measurement, a correlation is established between the system and the meter, as given in (8.7.14). In the second step, the environment becomes correlated with the meter, as a result of its interaction with the latter. If we develop an elementary modelling of the environment as a two-level system, one may then consider, in the process, the state $|\psi\rangle$, in (8.7.14), to be replaced by

$$|\psi\rangle \frac{1}{2}\begin{Bmatrix}1-i\\1+i\end{Bmatrix} \qquad (8.7.40)$$

leading to the final state of the system/meter/environment:

---
[27] Zurek (1991).

8.7 Quantum Interference & Measurement

$$|\psi_{\text{SME}}\rangle = c_+ \begin{pmatrix} 1 \\ 0 \end{pmatrix} \begin{bmatrix} 1 \\ 0 \end{bmatrix} \begin{Bmatrix} 1 \\ 0 \end{Bmatrix} + c_- \begin{pmatrix} 0 \\ 1 \end{pmatrix} \begin{bmatrix} 0 \\ 1 \end{bmatrix} \begin{Bmatrix} 0 \\ 1 \end{Bmatrix} \qquad (8.7.41)$$

using the notation $\{\cdot\}$ for the environment states.

Since one is interested in the state of the system and the meter reading only, we may obtain the reduced density operator ρ_{SM}, concerning the system and the meter, by tracing over the environment states obtaining

$$\rho_{\text{SM}} = \underset{\text{E}}{\text{Tr}}\left(|\psi_{\text{SME}}\rangle\langle\psi_{\text{SME}}|\right)$$

$$= |c_+|^2 \begin{pmatrix} 1 \\ 0 \end{pmatrix} (1\ 0) \begin{bmatrix} 1 \\ 0 \end{bmatrix} [1\ 0] + |c_-|^2 \begin{pmatrix} 0 \\ 1 \end{pmatrix} (0\ 1) \begin{bmatrix} 0 \\ 1 \end{bmatrix} [0\ 1] \qquad (8.7.42)$$

transforming the quantum superpositions into statistical mixtures (see also §1.5) readily interpreted in classical terms. As mentioned in the introduction to this section, this reduction of the density operator from a pure to a mixture is referred to as quantum decoherence. Also (8.7.42) implies that the interaction between the meter and the environment has actually led to the selection of the alternatives being measured — the components of spin along the z-axis. Also the basis $[1\ 0]^T$, $[0\ 1]^T$ in the meter's vector space, referred to as the pointer basis, have been selected. This removes the ambiguity as to what the alternatives being measured are.

The interaction between the meter and the environment may be implemented (in analogy to the one in (8.7.10) with $q \to 0$) by the Hamiltonian

$$H_{\text{ME}} = -\lambda_0 \begin{bmatrix} 1 & 0 \\ 0 & -1 \end{bmatrix} \frac{1}{2} \begin{Bmatrix} 1 & -1 \\ -1 & 1 \end{Bmatrix} \qquad (8.7.43)$$

which for $\lambda_0 \tau = \pi/2$, with τ referring to the interaction time, leads from the state in (8.7.40) to the state in (8.7.41) (see Problem 8.17).

A more realistic and more interesting modelling of the environment may be given, pertaining to the second example in (8.7.24), where the environment is represented by a collection of harmonic oscillators involving infinitely many degrees of freedom. As the interaction of the meter and the environment, we consider the model[28] interaction Hamiltonian

$$H_{\text{ME}} = a^\dagger \sum_k \lambda_k b_k + a \sum_k \lambda_k^* b_k^\dagger \qquad (8.7.44)$$

where b_k, $b_k^\dagger$ denote the annihilation, creation operators associated with the various degrees of freedom of the environment (also referred to as a reservoir). The environment is taken initially in its ground-state. This interaction will be considered in detail in §12.7.

---

[28] This is reminiscent of the interaction of a charged particle with the ever present electromagnetic field surrounding it with corresponding physical consequences and that of renormalization (see §7.7, §8.5).

We set
$$\gamma = \frac{2\pi}{\hbar^2}|\lambda(\omega)|^2 n(\omega) \qquad (8.7.45)$$

where $\lambda(\omega)$ is a continuous variable extension of λ_k replacing the summation over k in (8.7.44) by an integral, $n(\omega)$ denotes the density of such states and ω is introduced in (8.7.24). The explicit expression of γ is not important for a qualitative discussion of quantum decoherence. Then for weak coupling and a time t short enough such that the oscillator, representing the meter, has not changed much given by

$$\frac{\gamma t}{2} \ll 1 \qquad (8.7.46)$$

but for t much larger than the correlation time of the environment, the reduced density operator, pertinent to the system and meter only, is given by (§12.7)

$$\rho_{\mathrm{SM}}(t) \simeq |c_+|^2 \begin{pmatrix} 1 & 0 \\ 0 & 0 \end{pmatrix} \left|\alpha e^{-\gamma t/2}\right\rangle \left\langle \alpha e^{-\gamma t/2}\right|$$

$$+ |c_-|^2 \begin{pmatrix} 0 & 0 \\ 0 & 1 \end{pmatrix} \left|-\alpha e^{-\gamma t/2}\right\rangle \left\langle -\alpha e^{-\gamma t/2}\right|$$

$$+ c_+ c_-^* \begin{pmatrix} 0 & 1 \\ 0 & 0 \end{pmatrix} \left|\alpha e^{-\gamma t/2}\right\rangle \left\langle -\alpha e^{-\gamma t/2}\right| \exp(-2\alpha_0^2(1-e^{-\gamma t}))$$

$$+ c_- c_+^* \begin{pmatrix} 0 & 0 \\ 1 & 0 \end{pmatrix} \left|-\alpha e^{-\gamma t/2}\right\rangle \left\langle \alpha e^{-\gamma t/2}\right| \exp(-2\alpha_0^2(1-e^{-\gamma t}))$$

$$(8.7.47)$$

evolving from the density operator in (8.7.31), *up* to phase factors which are unimportant for discussing quantum decoherence, with α a complex number.

The condition (8.7.46) implies that

$$\exp(-2\alpha_0^2(1-e^{-\gamma t})) \simeq \exp(-2\alpha_0^2 \gamma t) \qquad (8.7.48)$$

and for α_0^2, associated with the meter (see (8.7.25)) taking on macroscopic large values such that

$$\alpha_0 \gg \frac{1}{2\gamma t} \qquad (8.7.49)$$

the non-diagonal part in (8.7.47) will be washed away relative to the diagonal one, demonstrating how quantum decoherence may arise destroying quantum superpositions. Decoherence then occurs exponentially on a decoherence time scale $\sim 1/(\gamma \alpha_0^2)$.

8.8 Ramsey Oscillatory Fields Method and Spin Flip; Monitoring the Spin

8.8.1 Ramsey Apparatus and Interference; Spin Flip

The Ramsey separated fields method,[29] in its simplest form, consists of two oscillatory magnetic fields each acting for some time τ separated for some time T with no oscillations. More precisely, the magnetic field as a function of time may be taken to be of the form

$$\mathbf{B}(t) = \begin{cases} (B\cos\omega t, B\sin\omega t, B_0), & 0 \leq t < \tau \\ (0, 0, B_0), & \tau \leq t < \tau + T \\ (B\cos\omega t, B\sin\omega t, B_0), & \tau + T \leq t < 2\tau + T \end{cases} \qquad (8.8.1)$$

where $B_0 > 0$, $B > 0$ are constants. We note that for time T, the amplitude of oscillations is reduced to zero as is shown pictorially in Figure 8.2.

It is remarkable that an interference pattern arises as a consequence of having the two *separated* field zones, as will be seen below, for the spin intensity distributions for a beam of particles, with spin, entering one Ramsey zone and finally leaving the other Ramsey zone.

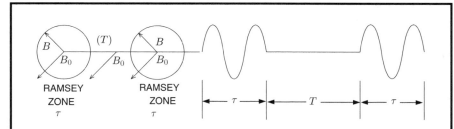

Fig. 8.2. Two separated oscillatory fields zones, each acting for a time τ, with zero amplitude for a time T in between. [More than two oscillatory fields zones may be also utilized.]

Consider an uncharged particle of spin 1/2 and magnetic moment $\boldsymbol{\mu} = \mu\boldsymbol{\sigma}$, $\mu < 0$. The interaction of the magnetic moment with the magnetic field $\mathbf{B}(t)$ is given by

$$H = |\mu|\boldsymbol{\sigma}\cdot\mathbf{B}(t) = \begin{pmatrix} \hbar\omega_0/2 & |\mu|Be^{-i\omega t} \\ |\mu|Be^{i\omega t} & -\hbar\omega_0/2 \end{pmatrix} \qquad (8.8.2)$$

---

[29] Ramsey (1990), based on the 1989 Noble Prize in Physics Lectures. For an earlier related classic paper by Ramsey, cf. Ramsey (1950).

474   8 Quantum Physics of Spin 1/2 & Two-Level Systems

where
$$\hbar\omega_0 = 2|\mu|B_0. \tag{8.8.3}$$

In (8.8.2), B is taken to be *zero* for $\tau \leqslant t < \tau + T$.

From (8.8.2), we are led to solve the system

$$i\hbar \begin{pmatrix} \dot{\alpha}_+(t) \\ \dot{\alpha}_-(t) \end{pmatrix} = \begin{pmatrix} \frac{\hbar\omega_0}{2}\alpha_+(t) + |\mu|Be^{-i\omega t}\alpha_-(t) \\ -\frac{\hbar\omega_0}{2}\alpha_-(t) + |\mu|Be^{i\omega t}\alpha_+(t) \end{pmatrix}. \tag{8.8.4}$$

Upon setting
$$\alpha_\pm(t) = \beta_\pm(t)e^{\mp i\omega_0 t/2} \tag{8.8.5}$$

we obtain the two equations

$$\ddot{\beta}_+ + i(\omega - \omega_0)\dot{\beta}_+ + \frac{\mu^2 B^2}{\hbar^2}\beta_+ = 0 \tag{8.8.6}$$

$$\ddot{\beta}_- - i(\omega - \omega_0)\dot{\beta}_- + \frac{\mu^2 B^2}{\hbar^2}\beta_- = 0. \tag{8.8.7}$$

After a lengthy but straightforward manipulations of (8.8.4)–(8.8.7), we obtain the following expression for the unitary evolution operator from some time t_1 to t within a zone

$$U(t, t_1) = \begin{pmatrix} U_{11}(t, t_1) & U_{12}(t, t_1) \\ U_{21}(t, t_1) & U_{22}(t, t_1) \end{pmatrix} \tag{8.8.8}$$

where

$$U_{11}(t, t_1) = \left[\cos a(t - t_1) + i\left(\frac{\omega - \omega_0}{2a}\right)\sin a(t - t_1)\right]e^{-i\omega(t-t_1)/2} \tag{8.8.9}$$

$$U_{12}(t, t_1) = -i\frac{|\mu|B}{\hbar a}\sin a(t - t_1)e^{-i\omega(t+t_1)/2} \tag{8.8.10}$$

$$U_{21}(t, t_1) = -i\frac{|\mu|B}{\hbar a}\sin a(t - t_1)e^{i\omega(t+t_1)/2} \tag{8.8.11}$$

$$U_{22}(t, t_1) = \left[\cos a(t - t_1) - i\left(\frac{\omega - \omega_0}{2a}\right)\sin a(t - t_1)\right]e^{i\omega(t-t_1)/2} \tag{8.8.12}$$

$$a = \left[\left(\frac{\omega - \omega_0}{2}\right)^2 + \frac{\mu^2 B^2}{\hbar^2}\right]^{1/2}. \tag{8.8.13}$$

For the intermediate time range $\tau \leqslant t < \tau + T$, the unitary time evolution operator is simply

$$U_0(\tau + T, \tau) = U_0(T) = \begin{pmatrix} e^{-i\omega_0 T/2} & 0 \\ 0 & e^{i\omega_0 T/2} \end{pmatrix}. \tag{8.8.14}$$

8.8 Ramsey Oscillatory Fields Method & Spin Flip

For the full process, from $0 \leq t < \tau + T$, the time evolution operator is given by

$$U_F(2\tau + T, 0) = U(2\tau + T, \tau + T)U_0(T)U(\tau, 0) \tag{8.8.15}$$

where U, U_0 are, respectively, given in (8.8.8), (8.8.14).

In particular, for a particle initially prepared in the state $(1\ 0)^T$, the amplitude of a spin flip after the particle has gone through the process in Figure 8.2 is given from (8.8.15) to be

$$A_{12}(\text{spin flip}) = \begin{pmatrix} 0 \\ 1 \end{pmatrix}^T U_F(2\tau + T, 0) \begin{pmatrix} 1 \\ 0 \end{pmatrix}$$

$$= \frac{-i|\mu|B}{\hbar a}\sin(\alpha\tau)e^{i\omega\tau}e^{i\omega T/2}$$

$$\times \left\{ e^{i(\omega-\omega_0)T/2}\left[\cos a\tau + i\left(\frac{\omega - \omega_0}{2a}\right)\sin a\tau\right]\right.$$

$$\left. + e^{-i(\omega-\omega_0)T/2}\left[\cos a\tau - i\left(\frac{\omega - \omega_0}{2a}\right)\sin a\tau\right]\right\}. \tag{8.8.16}$$

This gives for the probability of a spin flip

$$\text{Prob}[\text{spin flip}] = \frac{4\mu^2 B^2}{\hbar^2 a^2}\sin^2 a\tau \left[\cos\left((\omega - \omega_0)\frac{T}{2}\right)\cos a\tau\right.$$

$$\left. - \left(\frac{\omega - \omega_0}{2a}\right)\sin\left((\omega - \omega_0)\frac{T}{2}\right)\sin a\tau\right]^2 \tag{8.8.17}$$

where a is defined in (8.8.13), and ω_0 is given in (8.8.3).

At resonance $\omega = \omega_0$, $a = |\mu|B/\hbar$ and (8.8.17) reduces to

$$\text{Prob}[\text{spin flip}]_{\text{resonace}} = \sin^2\left(\frac{2|\mu|B\tau}{\hbar}\right) \tag{8.8.18}$$

and a 100% probability for a spin flip is formally attained for

$$\tau = \frac{\pi}{4}\frac{\hbar}{|\mu|B}. \tag{8.8.19}$$

To investigate the nature of the interference occurring in (8.8.16), (8.8.17), we first determine the amplitudes for a spin flip or of a non-flip of spin after the particle has gone through the first Ramsey zone before it enters the second one. These are respectively given by

$$A_1(\text{spin flip}) = \begin{pmatrix} 0 \\ 1 \end{pmatrix}^T U_0(T)U(\tau, 0)\begin{pmatrix} 1 \\ 0 \end{pmatrix}$$

$$= \frac{-\mathrm{i}|\mu|B}{\hbar a} \sin(a\tau) \mathrm{e}^{\mathrm{i}\omega_0 T/2} \mathrm{e}^{\mathrm{i}\omega\tau/2} \qquad (8.8.20)$$

$$A_1(\text{non–flip of spin}) = \begin{pmatrix}1\\0\end{pmatrix}^{\mathrm{T}} U_0(T) U(\tau, 0) \begin{pmatrix}1\\0\end{pmatrix}$$

$$= \mathrm{e}^{-\mathrm{i}\omega_0 T/2} \mathrm{e}^{-\mathrm{i}\omega\tau/2} \left[\cos a\tau + \mathrm{i}\left(\frac{\omega - \omega_0}{2a}\right)\sin a\tau\right]. \qquad (8.8.21)$$

Similarly for a particle entering the second Ramsey zone, the amplitude of a spin flip or of a non-flip of spin, and ending up in the state $(0\ 1)^{\mathrm{T}}$, at the end of the experiment, are, respectively,

$$A_2(\text{spin flip}) = \begin{pmatrix}0\\1\end{pmatrix}^{\mathrm{T}} U(2\tau + T, \tau + T) \begin{pmatrix}1\\0\end{pmatrix}$$

$$= \frac{-\mathrm{i}|\mu|B}{\hbar a} \sin(a\tau)\, \mathrm{e}^{\mathrm{i}\omega(3\tau + 2T)/2} \qquad (8.8.22)$$

$$A_2(\text{non–flip of spin}) = \begin{pmatrix}0\\1\end{pmatrix}^{\mathrm{T}} U(2\tau + T, \tau + T) \begin{pmatrix}0\\1\end{pmatrix}$$

$$= \left[\cos a\tau - \mathrm{i}\left(\frac{\omega - \omega_0}{2a}\right)\sin a\tau\right] \mathrm{e}^{\mathrm{i}\omega\tau/2}. \qquad (8.8.23)$$

From (8.8.16), (8.8.22), (8.8.23), we see that the amplitude for a spin flip for a particle going through the whole process in Figure 8.2, may be rewritten as

$$A_{21}(\text{spin flip}) = [A_2(\text{spin flip}) A_1(\text{non–flip of spin})$$

$$+ A_2(\text{non–flip of spin}) A_1(\text{spin flip})]$$

$$\equiv A(-++) + A(--+) \qquad (8.8.24)$$

We note that (8.8.24) is a statement of completeness as it follows from the insertion of the identity between $U_0(T)$ and $U(2\tau + T, \tau + T)$ in (8.8.15):

$$U(2\tau + T, \tau + T) U_0(T) U(\tau, 0)$$

$$= U(2\tau + T, \tau + T) \left\{\begin{pmatrix}1\\0\end{pmatrix}(1\ 0) + \begin{pmatrix}0\\1\end{pmatrix}(0\ 1)\right\} U_0(T) U(\tau, 0) \qquad (8.8.25)$$

leading immediately to the two terms on the right-hand side of (8.8.24) upon taking the matrix element $(0\ 1)\,(\cdot)\,(1\ 0)^{\mathrm{T}}$ of (8.8.25).

8.8 Ramsey Oscillatory Fields Method & Spin Flip   477

The probability of a spin flip in (8.8.17) may be then rewritten as

$$\text{Prob[spin flip]} = |A(-++)|^2 + |A(--+)|^2$$
$$+ (A^*(-++)A(--+) + A(-++)A^*(--+)) \quad (8.8.26)$$

exhibiting an interference term

$$[A^*(-++)A(--+) + A(-++)A^*(--+)]. \quad (8.8.27)$$

Such terms are observed and are responsible for providing narrow resonance curves about $\omega = \omega_0$ for the transition probability in corresponding experiments.

It is interesting to insert the apparatus (a meter) of §8.7 in (8.7.10), described as a spin 1/2 system, in the intermediate stage between the two Ramsey zones in Figure 8.2, the interaction Hamiltonian in that region becomes

$$H(t) = \begin{cases} \sigma_3 \left(|\mu|B_0 \begin{bmatrix} 1 & 0 \\ 0 & 1 \end{bmatrix} - \frac{\lambda}{2} \begin{bmatrix} 1 & -1 \\ -1 & 1 \end{bmatrix} \right), & \tau \leqslant t < \tau + t_0 \\ \\ \sigma_3 |\mu|B_0 \begin{bmatrix} 1 & 0 \\ 0 & 1 \end{bmatrix}, & \tau + t_0 \leqslant t < \tau + T \end{cases} \quad (8.8.28)$$

restricting for simplicity only (see also Problem 8.20) to an ideal apparatus. Here we use the notation $t_0 = \pi\hbar/2\lambda$ (see (8.7.15)) for the time of operation of the apparatus, assumed to be finite, to reach the correlated state given in (8.7.14), rather than T, with the latter reserved to denote the time of no oscillations of the magnetic field between the two Ramsey zones. T is chosen such that $T \geqslant t_0$.

The unitary time evolution operator between the two Ramsey zones in this case, then follows from (8.8.28), (8.8.14), to be given by

$$\widehat{U}_0(T) = \begin{pmatrix} e^{-i\omega_0 T/2} & 0 \\ 0 & e^{i\omega_0 T/2} \end{pmatrix} \left\{ \mathbf{1} + (i\sigma_3 - 1)\frac{1}{2}\begin{bmatrix} 1 & -1 \\ -1 & 1 \end{bmatrix} \right\} \quad (8.8.29)$$

instead of (8.8.14), and for the full process we obtain

$$U_F(2\tau + T, 0) = U(2\tau + T, \tau + T)\widehat{U}_0(T)U(\tau, 0). \quad (8.8.30)$$

Given the initial state

$$|\psi_0\rangle = \begin{pmatrix} 1 \\ 0 \end{pmatrix} \frac{1}{2} \begin{bmatrix} 1 - i \\ 1 + i \end{bmatrix} \quad (8.8.31)$$

of the particle-apparatus system, the amplitude of a spin flip of the particle is readily worked out from (8.8.28)–(8.8.31) to be

$$A(-++)\begin{bmatrix}1\\0\end{bmatrix} + A(--+)\begin{bmatrix}0\\1\end{bmatrix}. \tag{8.8.32}$$

This gives for the probability of a spin flip, the simple expression

$$|A(-++)|^2 + |A(--+)|^2 \tag{8.8.33}$$

showing the disappearance of the interference term in (8.8.27) by the mere insertion of the meter between the two Ramsey zones.

8.8.2 Monitoring the Spin

The insertion of a meter between the two Ramsey zones, as just discussed, leads us to investigate the fate of the spin of a particle as it goes through the Ramsey apparatus, as described by the magnetic fields in (8.8.1), when the spin of the particle, as it comes out of the first zone, is monitored by the meter. We prepare the particle to be initially in the state $(1\ 0)^\top$ before entering the first zone.

A machine, as a composite system consisting of a filter, which accepts a particle in the state $(1\ 0)^\top$ only, and a Ramsey apparatus, in the absence of a meter, may be represented by the matrix (see also Figure 1.7, (1.3.24), (1.3.25), (1.3.5))

$$M = U(2\tau + T, \tau + T)U_0(T)U(\tau, 0)\begin{pmatrix}1 & 0\\0 & 0\end{pmatrix} \tag{8.8.34}$$

which may be rewritten as

$$M = \begin{pmatrix}M_{11} & 0\\M_{21} & 0\end{pmatrix} \tag{8.8.35}$$

with

$$M_{11} = e^{-i\omega_0 T/2} U_{11}(2\tau + T, \tau + T)U_{11}(\tau, 0)$$
$$+ e^{i\omega_0 T/2} U_{12}(2\tau + T, \tau + T)U_{21}(\tau, 0) \tag{8.8.36}$$

$$M_{21} = e^{-i\omega_0 T/2} U_{21}(2\tau + T, \tau + T)U_{11}(\tau, 0)$$
$$+ e^{i\omega_0 T/2} U_{22}(2\tau + T, \tau + T)U_{21}(\tau, 0) \tag{8.8.37}$$

(see (8.8.8)–(8.8.12)).

The amplitude that the particle comes out of the machine with no change in its spin state is then given by

$$(1\ 0)\, M \begin{pmatrix}1\\0\end{pmatrix} = M_{11} \tag{8.8.38}$$

8.8 Ramsey Oscillatory Fields Method & Spin Flip 479

giving the probability (see also (8.8.17))

$$\text{Prob[same state]} = 1 - \frac{4\mu^2 B^2}{\hbar^2 a^2} \sin^2 a\tau \left[\cos\left((\omega - \omega_0)\frac{T}{2}\right) \cos a\tau \right.$$
$$\left. - \left(\frac{\omega - \omega_0}{2a}\right) \sin\left((\omega - \omega_0)\frac{T}{2}\right) \sin a\tau \right]^2. \quad (8.8.39)$$

In the presence of the meter, the corresponding machine may be represented by the matrix

$$M_0 = U(2\tau + T, \tau + T)\,\widehat{U}_0(T)\,U(\tau, 0) \begin{pmatrix} 1 & 0 \\ 0 & 0 \end{pmatrix} \quad (8.8.40)$$

where $\widehat{U}_0(T)$ is given in (8.8.29). For the apparatus in the initial state, appearing tin (8.8.31), we have

$$\widehat{U}_0(T)\frac{1}{2}\begin{bmatrix} 1-i \\ 1+i \end{bmatrix} = \begin{pmatrix} e^{-i\omega_0 T/2} & 0 \\ 0 & 0 \end{pmatrix} \begin{bmatrix} 1 \\ 0 \end{bmatrix} + \begin{pmatrix} 0 & 0 \\ 0 & e^{i\omega_0 T/2} \end{pmatrix} \begin{bmatrix} 0 \\ 1 \end{bmatrix} \quad (8.8.41)$$

leading from (8.8.40) to

$$\text{Prob}_0\text{[same state]} = \sum_{x=0,2} \left(\frac{\mu^2 B^2}{\hbar^2 a^2} \sin^2 a\tau\right)^x \left(1 - \frac{\mu^2 B^2}{\hbar^2 a^2} \sin^2 a\tau\right)^{2-x} \quad (8.8.42)$$

corresponding to the probability in (8.8.39) now in the *presence* of the meter, as indicated by writing Prob_0 for such a probability.

At resonance $\omega = \omega_0$, the expressions for the probabilities in (8.8.39), (8.8.42) simplify to

$$\text{Prob[same state]} = \cos^2\left(\frac{2|\mu|B\tau}{\hbar}\right)$$
$$= \cos^4\left(\frac{|\mu|B\tau}{\hbar}\right) + \sin^4\left(\frac{|\mu|B\tau}{\hbar}\right) - \frac{1}{2}\sin^2\left(\frac{2|\mu|B\tau}{\hbar}\right) \quad (8.8.43)$$

(see also (8.8.18)),

$$\text{Prob}_0\text{[same state]} = \cos^4\left(\frac{|\mu|B\tau}{\hbar}\right) + \sin^4\left(\frac{|\mu|B\tau}{\hbar}\right) \quad (8.8.44)$$

respectively.

Upon the comparison of the probabilities in (8.8.43), (8.8.44), we learn, under the above experimental situation, that monitoring the spin of the particle by the meter, as the particle comes out of the first zone, suppresses the

480   8 Quantum Physics of Spin 1/2 & Two-Level Systems

probability of change of state, i.e., *reduces the probability of decay* (unless $2|\mu|B\tau/\hbar = n\pi$, $n = 0, 1, 2, \ldots$).

It is interesting to consider the more general case of a system involving N Ramsey zones. As before a meter is inserted between each zone to monitor the spin of the particle as it comes out of a zone, with a finite time of operation of each meter $t_0 = \pi\hbar/2\lambda$, so that the correlated state with the system is reached with $T \geqslant t_0$. The situation is depicted in Figure 8.3.

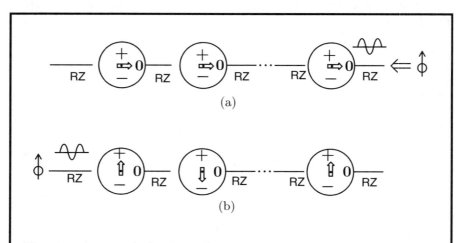

Fig. 8.3. A spin 1/2 (uncharged) particle going through N Ramsey zones (RZ). (a) An apparatus is set between each zone with the "needle" of each pointing initially towards the neutral **0** direction. The initial (a) and final (b) directions of the spin of the particle are the same in this experiment. Part (b) shows a possible configuration of the "needles" directions after the particle has gone through $N - 1$ zones.

The time evolution operator of the combined state of the jth meter and the spin of the particle coming out of the jth zone, $j = 1, \ldots, N-1$, is of the form in (8.8.29). The expression for the time evolution operator within each Ramsey zone may be read from (8.8.8)–(8.8.12) with the appropriate values for t_1 and t in them. For the first zone, for example, $t_1 = 0$, $t = \tau$, and the $(1,1)$ element of $U(\tau, 0)$ is (8.8.9) and so on.

From (8.8.41), the state of a jth meter may be either $\begin{bmatrix} 1 & 0 \end{bmatrix}_j^T$ or $\begin{bmatrix} 0 & 1 \end{bmatrix}_j^T$. In order that the final state of the spin of the particle be $(1\ 0)^T$, i.e., be in the same state as the initial one, at the end of the experiment, the particle may have only an even number $(0, 2, \ldots)$ of spin flips as the particle moves from the initial spin state to each of the meters between the Ramsey zones and to its final state. Also due to the orthogonality of the states $\begin{bmatrix} 1 & 0 \end{bmatrix}_j^T$, $\begin{bmatrix} 0 & 1 \end{bmatrix}_j^T$, for each j, the relative phases of the elements of the U operators in (8.8.8)–

8.8 Ramsey Oscillatory Fields Method & Spin Flip

(8.8.12) are unimportant. Here we note that we have N such U operators corresponding to the N Ramsey zones.

All told, (8.8.8)–(8.8.12) lead, in the presence of the meters

$$\text{Prob}_0[\text{same state}] = \sum_{\substack{x=0 \\ (x=\text{even})}}^{N} \binom{N}{x} \left(\frac{\mu^2 B^2}{\hbar^2 a^2} \sin^2 a\tau\right)^x \left(1 - \frac{\mu^2 B^2}{\hbar^2 a^2} \sin^2 a\tau\right)^{N-x}$$

$$\equiv P_0^{(N)}[\text{same state}] \tag{8.8.45}$$

which is a generalization of the expression in (8.8.42) with the latter given for $N = 2$.

At resonance $\omega = \omega_0$, (8.8.45) reduces to

$$P_0^{(N)}[\text{same state}] = \sum_{\substack{x=0 \\ (x=\text{even})}}^{N} \binom{N}{x} \left(\sin^2 \frac{\mu B}{\hbar}\tau\right)^x \left(\cos^2 \frac{\mu B}{\hbar}\tau\right)^{N-x}. \tag{8.8.46}$$

In the absence of the meters and at resonance $\omega = \omega_0$, the corresponding probability to (8.8.46) is simply given by (see Problem 8.21)

$$P^{(N)}[\text{same state}] = \cos^2\left(\frac{N\mu B}{\hbar}\tau\right). \tag{8.8.47}$$

For $N = 3$, for example,

$$P_0^{(3)}[\text{same state}] = 4\cos^6\left(\frac{\mu B}{\hbar}\tau\right) - 6\cos^4\left(\frac{\mu B}{\hbar}\tau\right) + 3\cos^2\left(\frac{\mu B}{\hbar}\tau\right) \tag{8.8.48}$$

and

$$P^{(3)}[\text{same state}] = P_0^{(3)}[\text{same state}]$$
$$+ 12\left(1 - \cos^2\left(\frac{\mu B}{\hbar}\tau\right)\right)\left(\frac{1}{2} - \cos^2\left(\frac{\mu B}{\hbar}\tau\right)\right)\cos^2\left(\frac{\mu B}{\hbar}\tau\right) \tag{8.8.49}$$

and one learns that for

$$0 < \cos^2\left(\frac{\mu B}{\hbar}\tau\right) < 1/2 \tag{8.8.50}$$

monitoring the spin, in the above experimental situation, *enhances the probability of decay* of the system (i.e., change of state), and for

$$1/2 < \cos^2\left(\frac{\mu B}{\hbar}\tau\right) < 1 \tag{8.8.51}$$

it *reduces the probability of decay*. Similar considerations may be given for $N > 3$.

The $x = 0$ term in (8.8.45), given by

$$\left[1 - \frac{\mu^2 B^2}{\hbar^2 a^2} \sin^2 a\tau\right]^N \qquad (8.8.52)$$

corresponds to the physically interesting case where each time the meter "encounters" the particle its spin is *always* found in its initial state. For τ *finite*, i.e., non-zero, and, in general, for $\sin^2 a\tau \neq 0$, the expression within the square-brackets in (8.8.52) is necessarily less than one. Accordingly for a system involving a large number of Ramsey zones (i.e., N large), the probability that the spin of the particle is found by the meter, as the particle emerges from a zone, always in its initial state becomes negligibly small.[30]

The above analyses show that measurements made on a system during the course of its time evolution lead, in general, to situations where the probability of decay of the system (change of state), at the end of the experiment, may be either reduced or enhanced. Such conclusions depend, however, very much on the experimental situation considered and on the variables involved.[31]

8.9 Schrödinger's Cat and Quantum Decoherence

The "Schrödinger cat" paradox arises when one considers the interpretation of the superposition principle of quantum physics as is extended to macroscopic systems. The classic example of this is the one dealing with Schrödinger's 1935 thought experiment consisting, in a simple description, of a vessel containing a live cat coupled by a lethal device to a radioactive

---

[30] If τ_0 denotes the total time of oscillations of the magnetic fields in the N zones, one may write $\tau = \tau_0/N$. On using the elementary inequality $\sin^2 x \leq x^2$, one obtains the following *lower* bound $[1 - \mu^2 B^2 \tau_0^2/\hbar^2 N^2]^N$ to the probability in (8.8.52). One is then tempted to infer that the latter probability approaches 1 for large N. Such a conclusion, however, cannot be true for finite τ. Also in an experiment, the total time T_0 of no oscillations is finite, and for the time t_0 of operation of a meter we have $t_0 \leq T_0/(N-1)$. The limit $N \to \infty$, would then imply the unrealistic condition of an instantaneous operation time of an apparatus.

[31] There is a long history of the role of measurements (continuous, frequent,...), made on a system during the course of its evolution, on its decay, cf. Khalfin (1990); Degasperis *et al.* (1973); Misra and Sudarshan (1977). The name "Zeno" effect, as their effect, was coined by the last two authors, and their work has led to numerous investigations. Cf. Nakazato *et al.* (1995); Koshino and Shimizu (2004) and references therein, and many other investigations by several authors with variations in its definition and different experimental situations with varied consequences following from them. In generally, one may refer to the *suppression* of the decay of a system noted as achieved by measurements made on it during the course of its evolution as a "quantum Zeno effect", while refer to the *enhancement* of its decay as a "quantum anti-Zeno effect".

substance. If a decay in it occurs, this triggers the device to release a deadly gas and the cat dies. On the other hand, if no decay in it occurs, the cat lives (see Figure 8.4). The radioactive decay law obeying the probabilistic rules of quantum physics, a decay may or may not occur within a given time specified by the "experimentalist" and it depends on the half-life of the radioactive substance.

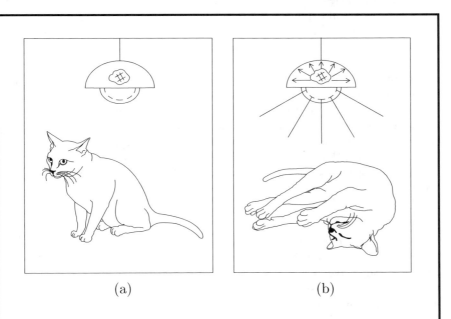

(a)   (b)

Fig. 8.4. (a) Vessel containing an alive cat in the presence of a radioactive substance in which no decay has occurred. (b) A decay has occurred which in turn has triggered a device to release a deadly gas and the cat dies. Unless one looks into the vessel, should one assume that the cat is in a superposition of cat alive/cat dead states?

According to quantum theory, the cat may be then found in a superposition state of being alive/dead correlated with the radioactive substance in which a no decay/decay has occurred. A state of this form may formally have the structure

$$|\Psi\rangle = \frac{1}{\sqrt{2}}\Big[\big|\text{no decay, cat alive}\big\rangle + \big|\text{decay, cat dead}\big\rangle\Big] \quad (8.9.1)$$

where for simplicity one may assume equal amplitudes for both configuration states. Such a state is called an entangled one as it may not be rewritten as the product of two states (see Appendix to §8.10). Also in (8.9.1), neither the substance nor the cat is in a definite state.

In our usual perception of the world, the cat is either alive or dead and not in a superposition of the two states. Such "cat" states are not meaningful in the macroscopic world and one faces the question as to what the significance of the superposition of macroscopic states, in general, is. As done at the end of §8.7, one may then argue that the environment, surrounding a macroscopic system and interacting with it, will destroy such superpositions thus inducing naturally superselection rules preventing them to be observed. Such quantum decoherences should then occur in such short times for us to perceive.

A cat is a very complex system consisting of a very large number of degrees of freedom with its different parts evolving in time in complicated ways whether the cat is dead or alive. Admittedly, it is also rather too simplistic to introduce a state for a cat either alive or dead in a meaningful way. There are, however, so-called mesoscopic states which have macroscopic and microscopic features and have been actually prepared in the laboratory.[32]

In this context, the state of the "cat" as being alive or dead is replaced by some classical notion such as simply as the position of a particle.

For example, the Monroe et. al. experiment involved in preparing an atom in a superposition of two spatially separated but localized wavepackets, thus creating a state

$$|\phi\rangle = \frac{1}{\sqrt{2}} \left[|x_1\rangle |\uparrow\rangle + |x_2\rangle |\downarrow\rangle \right] \qquad (8.9.2)$$

where $|x_1\rangle$, $|x_2\rangle$ refer to wavepacket states corresponding to separated positions of the atom, and $|\uparrow\rangle$, $|\downarrow\rangle$ refer to internal states of the atom. The extension of the wavepackets was about 7 nm, and the separation between the wavepackets was not smaller than the rather macroscopic distance of 80 nm which is large in comparison to the atomic dimension of the order 0.1 nm. In the Brune et. al. experiment, the coupling of a two-level atom with a few photon coherent field in a cavity was considered generating a Schrödinger cat-like state of radiation and the quantum decoherence in a measurement process was observed for such a mesoscopic state. The decoherence is considered as being due to dissipation corresponding to absorption by the cavity walls. Mesoscopic states are often referred to as kitten-like states.

One may model the latter experiment by considering the coupling of a spin 1/2 system and a harmonic oscillator to generate a Schrödinger cat entangled state

$$|\phi\rangle = \frac{1}{\sqrt{2}} \left[\begin{pmatrix} 1 \\ 0 \end{pmatrix} |\alpha\rangle + \begin{pmatrix} 0 \\ 1 \end{pmatrix} |-\alpha\rangle \right] \qquad (8.9.3)$$

where $(1\ 0)^T$, $(0\ 1)^T$ correspond to two states of an atom, while the coherent states $|\alpha\rangle$, $|-\alpha\rangle$ correspond to two configurations of radiation. In Schrödinger's cat thought experiment, $(1\ 0)^T / (0\ 1)^T$ correspond to the ra-

---

[32] Monroe et al. (1996); Brune et al. (1996); Brune et al. (1992); see also Gerry and Knight (1997).

dioactive substance in which no decay/a decay has occurred, while $|\alpha\rangle$ / $|-\alpha\rangle$ correspond to the cat alive, dead, respectively.

By considering an interaction of the harmonic oscillator with the environment of the form given in (8.7.44), we may infer from the analysis given through (8.7.44)–(8.7.49), that quantum coherence will disappear exponentially on a decoherence time scale $\sim 1/(\gamma|\alpha|^2)$ (see end of §8.7), where γ, for example, is defined in (8.7.45). We also recall from (6.6.24) that $|\alpha|^2 = \bar{n}$, with $\bar{n}$ denoting the mean number of the oscillator quanta. That is, the larger $\bar{n}$ is the more rapidly coherence is destroyed. For macroscopic systems, $\bar{n}$ is very large and decoherence occurs too rapidly to be observed, while for mesoscopic ones, with $\bar{n}$ not too large decoherence is expected to set in slowly.

A simple model that generates the Schrödinger entangled state in (8.9.3) may be readily given. Consider an uncharged spin 1/2 particle with magnetic moment μ. As an initial condition, we take the spin state $(1\ 0)^\top$.

As a first stage, we consider the interaction of the spin of a particle with a constant magnetic field $\mathbf{B} = (0, 1, 0)$, $B > 0$, $\mu < 0$, for a time $t = \hbar\pi/(4|\mu|B)$, with Hamiltonian $H = -\boldsymbol{\mu} \cdot \mathbf{B}$. This generates the state

$$\exp\left(-\frac{it}{\hbar}(-\boldsymbol{\mu} \cdot \mathbf{B})\right)\begin{pmatrix}1\\0\end{pmatrix} = \frac{1}{\sqrt{2}}\left[\begin{pmatrix}1\\0\end{pmatrix} + \begin{pmatrix}0\\1\end{pmatrix}\right]. \quad (8.9.4)$$

As a second stage, we consider the interaction of the spin, in the "initial" state (8.9.4), with a harmonic oscillator taken in an initial coherent state $|-i\alpha_0\rangle$. This interaction will be taken to be proportional to the simple form

$$\left[\begin{pmatrix}1\\0\end{pmatrix}(1\ 0) - \begin{pmatrix}0\\1\end{pmatrix}(0\ 1)\right]a^\dagger a \quad (8.9.5)$$

with a Hamiltonian ($\lambda > 0$)

$$H = \hbar\omega a^\dagger a - \lambda\sigma_3 a^\dagger a \quad (8.9.6)$$

for a time $t = \hbar\pi/2\lambda$. This generates the state (see (8.7.28))

$$\exp\left(-\frac{i}{\hbar}tH\right)\frac{1}{\sqrt{2}}\begin{pmatrix}1\\1\end{pmatrix}|-i\alpha_0\rangle = \frac{1}{\sqrt{2}}\left[\begin{pmatrix}1\\0\end{pmatrix}\exp\left(-\frac{it}{\hbar}(\hbar\omega - \lambda)a^\dagger a\right)|-i\alpha_0\rangle\right.$$

$$\left.+ \begin{pmatrix}0\\1\end{pmatrix}\exp\left(-\frac{it}{\hbar}(\hbar\omega + \lambda)a^\dagger a\right)|-i\alpha_0\rangle\right]$$

$$= \frac{1}{\sqrt{2}}\left[\begin{pmatrix}1\\0\end{pmatrix}|\alpha\rangle + \begin{pmatrix}0\\1\end{pmatrix}|-\alpha\rangle\right] \quad (8.9.7)$$

where

$$\alpha = e^{-i\phi}\alpha_0, \quad \phi = \frac{\hbar\omega}{\lambda}\frac{\pi}{2}. \quad (8.9.8)$$

This second stage mimics an interaction which finally establishes the correlation between the radioactive substance and the cat as discussed below (8.9.3), [or formally the interaction of an atom in a superposition of two of its levels with a coherent field in a cavity (see also Problem 8.22).]

As the third stage, one may also generate a pure superposition of the cat (kitten) states as follows. By applying an identical magnetic field as in stage 1, for the same time $t = \hbar\pi/(4|\mu|B)$, we generate the state

$$\frac{1}{2}\begin{pmatrix}1\\0\end{pmatrix}\left[\left|e^{-i\phi'}\alpha_0\right\rangle - \left|-e^{-i\phi'}\alpha_0\right\rangle\right] + \frac{1}{2}\begin{pmatrix}0\\1\end{pmatrix}\left[\left|e^{-i\phi'}\alpha_0\right\rangle + \left|-e^{-i\phi'}\alpha_0\right\rangle\right] \quad (8.9.9)$$

for a phase ϕ'.

Finally, as the fourth stage, one may perform a selective measurement, selecting the state $(0\ 1)^\top$ component of the spin, and generate, in the process, the superposition state

$$N\left[\left|e^{-i\phi'}\alpha_0\right\rangle + \left|-e^{-i\phi'}\alpha_0\right\rangle\right] \quad (8.9.10)$$

where N is a normalization factor. A coupling of the form in (8.7.44), where the harmonic oscillator in (8.9.6) interacts with the environment, represented by a collection of harmonic oscillators involving infinitely many degrees of freedom, leads to a destruction of such a superposition exponentially on a decoherence time scale $\sim 1/(\gamma|\alpha_0|^2)$, where γ is defined in (8.7.45).[33]

8.10 Bell's Test

8.10.1 Bell's Test

Consider the commutation relations for spin

$$[S_i, S_j] = i\hbar\varepsilon_{ijk}S_k. \quad (8.10.1)$$

Let $\mathbf{n}_1$, $\mathbf{n}_2$ be any two unit vectors specifying two different directions. Since the components X_i of the position operator commute, and also commute with spin (see (2.7.8)), we may multiply (8.10.1) by the components n_{1i}, n_{2j} and sum over i and j to obtain

$$[S_{\mathbf{n}_1}, S_{\mathbf{n}_2}] = i\hbar|\mathbf{n}_1 \times \mathbf{n}_2|S_{\mathbf{N}} \quad (8.10.2)$$

---

[33] As pointed out in the footnote to the interaction in (8.7.44), quantum decoherence arising from the non-isolation of a measuring (or detection) system from the environment and hence from its interaction with it, is reminiscent of the interaction of the non-isolated charged particle with the ever present electromagnetic field surrounding it with corresponding physical consequences and that of renormalization. [It is interesting to note that the electromagnetic field is also essentially represented by harmonic oscillators with infinitely many degrees of freedom.]

where $S_\mathbf{n} = \mathbf{n} \cdot \mathbf{S}$, $\mathbf{N} = (\mathbf{n}_1 \times \mathbf{n}_2)/|\mathbf{n}_1 \times \mathbf{n}_2|$, and $\mathbf{n}_1 \times \mathbf{n}_2$ is not the zero vector.

For spin 1/2, the explicit expression for $S_\mathbf{n}$ is given in (8.1.14), and for any unit vector $\mathbf{n}$, $S_\mathbf{n}$ has the eigenvalues $\pm\hbar/2$. The commutation relations (8.10.2) imply, in particular, that the components of spin along any two different directions cannot be defined simultaneously. As a matter of fact, if one argues, incorrectly, that there may be a simultaneous eigenvector for $S_{\mathbf{n}_1}$ and $S_{\mathbf{n}_2}$, for which $\mathbf{n}_1 \times \mathbf{n}_2 \neq \mathbf{0}$, then from (8.10.2), one would run into a contradiction that this eigenvector is also an eigenvector of $S_\mathbf{N}$ with zero eigenvalue!

Suppose a pair of spin 1/2 particles are prepared in a singlet state (cf. (5.5.39))

$$|\psi\rangle = \frac{1}{\sqrt{2}} \left(|\mathbf{n}\rangle_1 |-\mathbf{n}\rangle_2 - |-\mathbf{n}\rangle_1 |\mathbf{n}\rangle_2 \right) \quad (8.10.3)$$

where we have used the notation $|\pm 1/2, \mathbf{n}\rangle \equiv |\pm\mathbf{n}\rangle$ (see (8.1.16), (8.1.25)), and the two particles move freely with momenta in opposite directions and cease to interact. The quantization axis in (8.10.3) was chosen, arbitrarily, along a unit vector $\mathbf{n}$ (see also Problem 8.24).

An actual process will be discussed below giving rise to a singlet state as in (8.10.3). This state is not factorable as the product of two states and is referred to as an *entangled* state.[34] According to (8.10.3), if the measurement of spin of one of the particles is found to be along, say, $\mathbf{n}$, then one would infer rather *instantaneously*, with probability one, that the component of spin of the other particle is along $-\mathbf{n}$. This together with the fact that all the components of spin cannot be simultaneously defined, as implied by quantum physics, has led Einstein, Podolsky and Rosen (EPR) in 1935[35] to a serious criticism of quantum mechanics.[36]

The EPR argument, tailored to the problem at hand, is of the following nature.

'Devil's Advocate Argument

From the measurement of spin of one of the particles, call it particle 1, and found, say, to be along $\mathbf{n}$, one may conclude instantaneously, because of the correlation implied by (8.10.3), that the component of spin of the other particle, call it particle 2, is along $-\mathbf{n}$ without ever disturbing this latter particle. With no such disturbance, one may invoke locality, as a no-action at a distance, to infer that the value of the component of spin found indirectly for particle 2 must have existed prior to a measurement done on particle 1.

---

[34] The mathematical aspect of entangled states is given in the appendix to this section.

[35] Einstein et al. (1935).

[36] Actually their criticism was reformulated by D. Bohm in terms of spin, while the original EPR argument was based on positions and momenta of particles.

Since **n** was arbitrary, one may also infer that all the components of spin of particle 2 were known to begin with. That is, all the components of spin of a particle are definite in clear contradiction with quantum mechanics and the underlying theory of the latter is incomplete.'

The above led to the belief that perhaps quantum mechanics is a limiting case of a more complete local theory which, involves, so-called, hidden variables. Such theories are referred to as *Local Hidden Variables* (LHV) theories. In 1964, and in subsequent years, John Bell[37] has put such theories to a test. Several tests have been also proposed in the literature by various authors. We refer to all such tests as Bell-like tests. We will discuss one originating from the work of Clauser, Horne, Shimoney and Holt (CHSH).[38]

To the above end, and in view of applications to a system of two particles, as described below (8.10.3), and other similar processes, we consider the following in the light of LHV theories.

Let λ denote collectively the random variables expected to be relevant to the system under study with corresponding probability density or probability mass function $d\rho(\lambda)$ normalized as

$$\int_\Lambda d\rho(\lambda) = 1 \qquad (8.10.4)$$

summed over the set Λ of all values that λ may take on.

One is interested in determining coincidence and single counts obtained in the measurements of the spins of the particles, after emerging from the process, making angles, say, χ_1, χ_2 with some given directions.

Suppose that the system is in a state specified by λ. We may introduce the following probabilities of counts:

$$p(\chi_1, \chi_2; \lambda), \quad p(\chi_1, -; \lambda), \quad p(-, \chi_2; \lambda) \qquad (8.10.5)$$

correspondingly, respectively, to coincidence counts when measurements are made on both particles' spins, to a count when a measurement is made on only one particle (call it particle 1), and, finally, to a count when a measurement is made on particle 2 only.

In such a framework, one makes the key assumption that if the system is in any given state specified by λ, the probability count obtained from measurements performed on one particle is independent of the probability count corresponding to the other particle after they have emerged from the process. That is, the probability counts are necessarily factorable,

$$p(\chi_1, \chi_2; \lambda) = p(\chi_1, -; \lambda) p(-, \chi_2; \lambda) \qquad (8.10.6)$$

for all λ in Λ, implying their independence, with all determined in the same state λ.

---

[37] Bell's insight has been of great significance in science, in general. Many of his contributions to this problem have been collected, e.g., in: Bell (1989).

[38] Clauser and Horne (1974); Clauser and Shimony (1978); Clauser et al. (1969).

Now we use the fact that for any four numbers $0 \leqslant x_1, x_2, x'_1, x'_2 \leqslant 1$, we have the following elementary inequality

$$-1 \leqslant x_1 x_2 - x_1 x'_2 + x'_1 x_2 + x'_1 x'_2 - x'_1 - x_2 \leqslant 0 \quad (8.10.7)$$

as established in the appendix to this section.

Accordingly upon setting

$$\left.\begin{array}{l} x_1 = p(\chi_1, -; \lambda) \\ x_2 = p(-, \chi_2; \lambda) \\ x'_1 = p(\chi'_1, -; \lambda) \\ x'_2 = p(-, \chi'_2; \lambda) \end{array}\right\} \quad (8.10.8)$$

for four angles, χ_1, χ_2, χ'_1, χ'_2, and using the fact that probabilities, as in (8.10.5), necessarily must fall in the range $[0, 1]$, we have from (8.10.7) upon multiplying the latter by $d\rho(\lambda)$ and summing (integrating) over λ:

$$-1 \leqslant p(\chi_1, \chi_2) - p(\chi_1, \chi'_2) + p(\chi'_1, \chi_2) + p(\chi'_1, \chi'_2) - p(\chi'_1, -) - p(-, \chi_2) \leqslant 0 \quad (8.10.9)$$

where

$$p(\chi_1, \chi_2) = \int_\Lambda d\rho(\lambda) p(\chi_1, -; \lambda) p(-, \chi_2; \lambda)$$

$$= \int_\Lambda d\rho(\lambda) p(\chi_1, \chi_2; \lambda) \quad (8.10.10)$$

etc., where we have used, in particular, the factorization assumption in (8.10.6).

The inequality in (8.10.9) is expressed in terms of probability counts which may be determined experimentally putting LHV theories to a test. $p(\chi_1, \chi_2)$ denotes the joint probability count for measurements of both spins, while $p(\chi_1, -)$, $p(-, \chi_2)$ correspond to probability counts with measurements of only one of the spins.

In the sequel, the probabilities computed from quantum theory corresponding to $p(\chi_1, \chi_2)$, $p(\chi_1, -)$, $p(-, \chi_2)$ will be denoted, respectively, by $P(\chi_1, \chi_2)$, $P(\chi_1, -)$, $P(-, \chi_2)$ with a capital "P".

In order to obtain a violation of the inequality in (8.10.9) experimentally, it is sufficient to choose any four angles χ_1, χ_2, χ'_1, χ'_2 that do the job since, according to the LHV reasoning, (8.10.9) must be true for all angles. Experiments show violation of the inequalities and are consistent with the quantum mechanical predictions. Experiments of optical nature have been performed

490   8 Quantum Physics of Spin 1/2 & Two-Level Systems

and a classic one involving two photons with measurements made on photon polarization correlations is one due to Aspect *et. al.*[39,40]

8.10.2 Basic Processes

Now we investigate the nature of the inequality (8.10.9) with the probabilities appearing in it as computed from quantum theory for specific processes.

The Process:

$$e^- e^- \longrightarrow e^- e^-$$

We prepare a pair of electrons, with one spin up and the other spin down along the z-axis and initial momenta $\mathbf{p}$ and $-\mathbf{p}$ along the y-axis (see Figure 8.5). There is a non-zero amplitude[41] that the scattered electrons move along the x-axis, as shown in Figure 8.5. In this particular case, the probabilities $P(\chi_1, \chi_2)$, $P(\chi_1, -)$, $P(-, \chi_2)$ take quite simple forms.[42]

Prior to the computation of the above probabilities let us investigate their precise physical meanings by analyzing the possible *outcomes* of an experiment.

In the above experiment, depicted in Figure 8.5, a measurement of spin of one of the particles emerging from the process is measured along a unit vector $\mathbf{n}_1$, making an angle χ_1 with the z-axis, while the spin of the other particle is measured along a unit vector $\mathbf{n}_2$, making an angle χ_2 with the z-axis. The outcomes of such an experiment are shown in Figure 8.6. Since we are dealing with spin 1/2 particles, a spin measurement along a unit vector $\mathbf{n}$ gives only two possible answers, the spin is either *along* $\mathbf{n}$ *or* in *opposite* direction to $\mathbf{n}$. That is, the particle's spin is in the state $|+\mathbf{n}\rangle$ or $|-\mathbf{n}\rangle$ in our earlier notation in (8.10.3).

That is, there are *four* possible outcomes of the experiment, as shown in Figure 8.6, with corresponding probabilities of occurrence

$$P[+\mathbf{n}_1, +\mathbf{n}_2], \ P[-\mathbf{n}_1, +\mathbf{n}_2], \ P[+\mathbf{n}_1, -\mathbf{n}_2], \ P[-\mathbf{n}_1, -\mathbf{n}_2]. \tag{8.10.11}$$

---

[39] Aspect et al. (1982).
[40] For many other experiments of different nature, cf. Clauser and Shimony (1978); Chiao et al. (1994); Bell (1989).
[41] That there is a non-zero amplitude for the process may be shown to be true from quantum electrodynamics, for example, to the leading order in the fine-structure constant, see: Manoukian and Yongram (2004).
[42] For the scattered electrons moving along other axes than the x-axis these probabilities turn out to have complicated dependences on the initial speed of the electrons relative to the speed of light, in general, as discussed below, see: Manoukian and Yongram (2004); Yongram and Manoukian (2003), for such details.

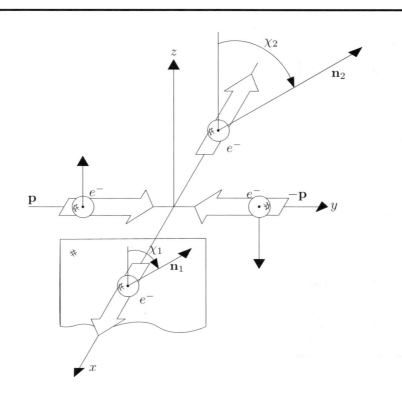

Fig. 8.5. A possible configuration of the process $e^-e^- \to e^-e^-$, where the initial spins are prepared to be one up and one down along the z-axis, and initial momenta $\mathbf{p}$ and $-\mathbf{p}$ are along the y-axis. There is a non-zero amplitude that the scattered electrons move with momenta along the x-axis as shown. This particular configuration yields to the simple probabilities given in (8.10.24), (8.10.26), (8.10.27) with spin measurements along directions specified by the unit vectors $\mathbf{n}_1$, $\mathbf{n}_2$. There are only four possible outcomes of the spin measurements of the emerging particles along the unit vectors $\mathbf{n}_1$, $\mathbf{n}_2$. These are spelled out in Figure 8.6.

Here, for example, read $P[-\mathbf{n}_1, +\mathbf{n}_2]$ as the probability that a measurement of spin of a particle, call it 1, along $\mathbf{n}_1$ is found in its opposite direction, while the spin of particle 2, measured along $\mathbf{n}_2$ is found to lie in the same direction. With the physical meanings of the probabilities in (8.10.11) made clear, we have

$$P(\chi_1, \chi_2) \equiv P[+\mathbf{n}_1, +\mathbf{n}_2] \tag{8.10.12}$$

$$P(\chi_1 + \pi, \chi_2) \equiv P[-\mathbf{n}_1, +\mathbf{n}_2] \tag{8.10.13}$$

$$P(\chi_1, \chi_2 + \pi) \equiv P[+\mathbf{n}_1, -\mathbf{n}_2] \tag{8.10.14}$$

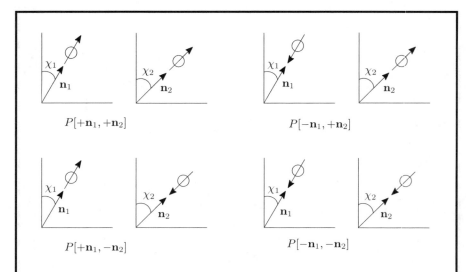

Fig. 8.6. The *four* possible outcomes of spin measurements (of the two emerging particles) along the unit vectors $\mathbf{n}_1$, $\mathbf{n}_2$ in the process depicted in Figure 8.5. The corresponding probabilities of occurrence appear under each possible case. Some of these probabilities may be zero for some $\mathbf{n}_1$, $\mathbf{n}_2$ and these may be read off from (8.10.24) in conjunction with (8.10.12)–(8.10.15).

$$P(\chi_1 + \pi, \chi_2 + \pi) \equiv P[-\mathbf{n}_1, -\mathbf{n}_2] \qquad (8.10.15)$$

corresponding to only four possible outcomes.

For *any* angles χ_1, χ_2, the *normalization* of probability reads as

$$P(\chi_1, \chi_2) + P(\chi_1 + \pi, \chi_2) + P(\chi_1, \chi_2 + \pi) + P(\chi_1 + \pi, \chi_2 + \pi) = 1. \quad (8.10.16)$$

If a measurement of spin is made only on one particle, say, particle, call it, 1, then with $P[\pm\mathbf{n}_1, -]$ denoting the probability that a measurement of its spin along $\mathbf{n}_1$ is found in the same or opposite direction to $\mathbf{n}_1$, respectively, we have

$$P(\chi_1, -) \equiv P[\mathbf{n}_1, -] \qquad (8.10.17)$$

$$P(\chi_1 + \pi, -) \equiv P[-\mathbf{n}_1, -] \qquad (8.10.18)$$

and the normalization of probability in this case reads

$$P(\chi_1, -) + P(\chi_1 + \pi, -) = 1. \qquad (8.10.19)$$

Similar expressions are given for $P(-, \chi_2)$, $P(-, \chi_2 + \pi)$.

Now we go back to the process depicted in Figure 8.5, to compute the probabilities $P(\chi_1, \chi_2)$, $P(\chi_1, -)$, $P(-, \chi_2)$ and the corresponding ones with $\chi_1, \chi_2 \to \chi_1 + \pi, \chi_2 + \pi$. From the conservation of total angular momentum,

8.10 Bell's Test

and the indistinguishability of the electrons, obeying the Fermi-Dirac statistics, that is, being described by an anti-symmetric state under the exchange of the two electrons, imply that the latter initial state, in spin-space, is a singlet-one,

$$|\Phi\rangle = \frac{1}{\sqrt{2}} \left[\begin{pmatrix} 1 \\ 0 \end{pmatrix}_1 \begin{pmatrix} 0 \\ 1 \end{pmatrix}_2 - \begin{pmatrix} 0 \\ 1 \end{pmatrix}_1 \begin{pmatrix} 1 \\ 0 \end{pmatrix}_2 \right]. \quad (8.10.20)$$

With measurements of spins along the axes

$$\mathbf{n}_1 = (0, \sin\chi_1, \cos\chi_1), \quad \mathbf{n}_2 = (0, \sin\chi_2, \cos\chi_2) \quad (8.10.21)$$

(see Figure 8.5), and corresponding final states

$$\begin{pmatrix} e^{-i\pi/4} \cos\chi_1/2 \\ e^{i\pi/4} \sin\chi_1/2 \end{pmatrix}_1, \quad \begin{pmatrix} e^{-i\pi/4} \cos\chi_2/2 \\ e^{i\pi/4} \sin\chi_2/2 \end{pmatrix}_2 \quad (8.10.22)$$

(see (8.1.16)), we obtain the amplitude

$$\left(e^{i\pi/4} \cos\chi_1/2 \quad e^{-i\pi/4} \sin\chi_1/2 \right)_1 \left(e^{i\pi/4} \cos\chi_2/2 \quad e^{-i\pi/4} \sin\chi_2/2 \right)_2 |\Phi\rangle$$

$$= -\frac{1}{\sqrt{2}} \sin\left(\frac{\chi_1 - \chi_2}{2} \right) \quad (8.10.23)$$

giving the joint probability[43]

$$P(\chi_1, \chi_2) = \frac{1}{2} \sin^2\left(\frac{\chi_1 - \chi_2}{2} \right) \equiv P[+\mathbf{n}_1, +\mathbf{n}_2]. \quad (8.10.24)$$

The probability $P(\chi_1, -)$ may be obtained from the square of the norm of the state (see (8.10.20), (8.10.22)):

$$\left(e^{i\pi/4} \cos\chi_1/2 \quad e^{-i\pi/4} \sin\chi_1/2 \right)_1 |\Phi\rangle$$

$$= \frac{1}{\sqrt{2}} \left[e^{i\pi/4} \cos\frac{\chi_1}{2} \begin{pmatrix} 0 \\ 1 \end{pmatrix}_2 + e^{-i\pi/4} \sin\frac{\chi_1}{2} \begin{pmatrix} 1 \\ 0 \end{pmatrix}_2 \right] \quad (8.10.25)$$

giving simply

$$P(\chi_1, -) = \frac{1}{2} \quad (8.10.26)$$

and similarly

$$P(-, \chi_2) = \frac{1}{2}. \quad (8.10.27)$$

Upon defining

---

[43] This coincides with the quantum electrodynamics calculation (Manoukian and Yongram (2004)), to the leading order in the fine-structure constant as obtained *for* the process depicted in Figure 8.5.

$$S = P(\chi_1, \chi_2) - P(\chi_1, \chi_2') + P(\chi_1', \chi_2) + P(\chi_1', \chi_2') - P(\chi_1', -) - P(-, \chi_2) \tag{8.10.28}$$

for four angles χ_1, χ_2, χ_1', χ_2', we note that for

$$\chi_1 = 45°, \quad \chi_2 = 90°, \quad \chi_1' = 135°, \quad \chi_2' = 180° \tag{8.10.29}$$

for example, we obtain $S = -1.207$, which violates the LHV theories inequality (8.10.9) (from below), according to the quantum mechanical computations of the underlying probabilities. Such violations are in conformity with experiments and are consistent with the quantum theory predictions.

Remarks 2
1. The joint probability of spin correlation in (8.10.24), whose physical meaning is spelled out in (8.10.12)–(8.10.15), see also Figure 8.5, is what is technically called a conditional probability. That is, given that the process in Figure 8.5 has occurred, then (8.10.24) gives the joint probability of spins correlations of the emerging particles.
2. The expression in (8.10.24), corresponding to the process in Figure 8.5, is valid even in the relativistic regime where β is not small. Here $\beta = v/c$ denotes the speed of any one of the incoming electrons and c denotes the speed of light. On the other hand for the scattered electrons moving along a different axis, than that of the x-axis, the corresponding conditional probability of the joint spin correlations has, in general, a complicated dependence on β.[44] Only when the formal limit $\beta \to 0$ is taken (the non-relativistic regime) then these conditional joint probabilities of spin correlations for these different scattering axes coincide with the one in (8.10.24) as well.
3. From (8.10.24), (8.10.26), (8.10.27) we note that, in general, $P(\chi_1, \chi_2) \neq P(\chi_1, -)P(-, \chi_2)$ showing the dependence of the two events corresponding to the spin measurements of the two particles.

The Process:

$$e^+ e^- \longrightarrow \gamma\gamma$$

A very investing process, relevant to the above analysis, is the one of positron-electron annihilation into two photons. We consider the process depicted in Figure 8.7. The electron, positron are prepared with spins up and down along the z-axis, and with momenta $\mathbf{p}$ and $-\mathbf{p}$, along the y-axis, respectively. Again there is a non-zero amplitude that such a process occurs. The created photons move with opposite momenta and, in here, along the z-axis. For this situation, the (conditional) probabilities $P(\chi_1, \chi_2)$, $P(\chi_1, -)$, $P(-, \chi_2)$ take particularly simple forms.

---

[44] Such details, based on quantum electrodynamics calculations, have been investigated in Manoukian and Yongram (2004), Yongram and Manoukian (2003), and are beyond the scope of the present analysis.

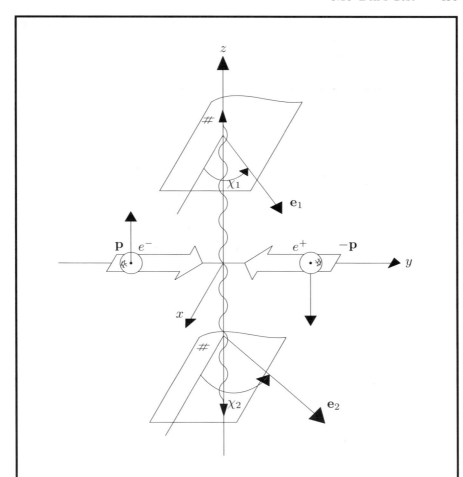

Fig. 8.7. A possible configuration of the process $e^+e^- \to \gamma\gamma$, where the spins of the electron, positron are prepared to be up and down along the z-axis, and with momenta $\mathbf{p}$ and $-\mathbf{p}$ along the y-axis, respectively. There is a non-zero amplitude that the created photons, emerging with opposite momenta, move along the z-axis as shown. Photon polarization correlations are measured with polarization vectors making angles χ_1, χ_2 with the x-axis.

To obtain the above probabilities, it is convenient to define right-handed (R-H) and left-handed (L-H) polarization vectors associated with a photon with momentum $\hbar\mathbf{k}$, say, moving *along* the z-axis in the positive direction:

$$e_+ = \frac{1}{\sqrt{2}}(e^1 + \mathrm{i}e^2), \qquad e_- = \frac{1}{\sqrt{2}}(e^1 - \mathrm{i}e^2) \qquad (8.10.30)$$

where e^1, e^2 are (linear) polarization vector components along the x- and y-axis, respectively (see also §8.5). The action of the parity operation on R-H and L-H polarizations is spelled out in Figure 8.8.

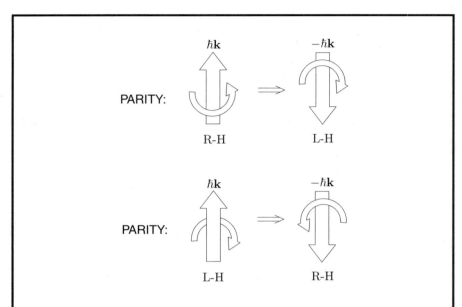

Fig. 8.8. The role of the parity operation on the momentum and right-handed (R-H) and left-handed (L-H) polarization vectors. The photon has also a so-called intrinsic parity of minus one.

To find the two-photon state, we note the following: (1) the photons satisfy the Bose-Einstein statistics, and hence such a state must be even under the interchange of the two-photons. (2) The positron and electron have opposite intrinsic parities[45] and hence the two-photon state must be odd under a parity transformation. (3) The total angular momentum along the z-axis is zero.

The R-H and L-H polarizations may be obtained from (8.10.30), (1.6.2), (1.6.3), to be represented as follows:

$$\frac{1}{\sqrt{2}}\begin{pmatrix}1\\i\\0\end{pmatrix}, \text{ as R-H } (e_{+1}) \text{ for } k_z > 0, \text{ as L-H } (e_{-2}) \text{ for } k_z < 0 \quad (8.10.31)$$

---

[45] Intrinsic parities of some particles will be investigated in Chapter 16. The vector potential **A**, for example, being coupled to the *current* **J** density via Maxwell's equations imply that the intrinsic parity of a photon is odd. Since we are considering here *two* photons, the intrinsic parity of a photon does not play an important role in this analysis.

$$\frac{1}{\sqrt{2}}\begin{pmatrix}1\\-i\\0\end{pmatrix}, \text{ as R-H } (e_{+2}) \text{ for } k_z < 0, \text{ as L-H } (e_{-1}) \text{ for } k_z > 0. \quad (8.10.32)$$

Accordingly, the two-photon polarization correlation state may be written as

$$|\Psi\rangle = \frac{i}{\sqrt{2}}(e_{+1}e_{+2} - e_{-1}e_{-2})$$

$$= \frac{i}{2\sqrt{2}}\left[\begin{pmatrix}1\\i\\0\end{pmatrix}_1 \begin{pmatrix}1\\-i\\0\end{pmatrix}_2 - \begin{pmatrix}1\\-i\\0\end{pmatrix}_1 \begin{pmatrix}1\\i\\0\end{pmatrix}_2\right] \quad (8.10.33)$$

which is consistent with the Bose character of the photons of opposite momenta. From Figure 8.8, this state has also an odd parity: $e_{+1}e_{+2} \leftrightarrow e_{-1}e_{-2}$. To show that the total angular momentum, associated with $|\Psi\rangle$, along the z-axis, is zero, we note that with the convention of the representation of e^1 and e^2 by the column vectors in (1.6.2), (1.6.3), we may write

$$S_3 = i\hbar \begin{pmatrix}0 & -1 & 0\\1 & 0 & 0\\0 & 0 & 0\end{pmatrix}_1 + i\hbar \begin{pmatrix}0 & -1 & 0\\1 & 0 & 0\\0 & 0 & 0\end{pmatrix}_2 \quad (8.10.34)$$

for the total angular momentum (spin) for the two photons along the z-axis, and note that

$$S_3|\Psi\rangle = 0. \quad (8.10.35)$$

Finally, since

$$\begin{pmatrix}1\\\pm i\\0\end{pmatrix} = \begin{pmatrix}1\\0\\0\end{pmatrix} \pm i \begin{pmatrix}0\\1\\0\end{pmatrix} \quad (8.10.36)$$

the state in (8.10.33) may be rewritten as

$$|\Psi\rangle = \frac{1}{\sqrt{2}}\left[\begin{pmatrix}1\\0\\0\end{pmatrix}_1 \begin{pmatrix}0\\1\\0\end{pmatrix}_2 - \begin{pmatrix}0\\1\\0\end{pmatrix}_1 \begin{pmatrix}1\\0\\0\end{pmatrix}_2\right] \quad (8.10.37)$$

which is an entangled state similar to the one in (8.10.20), leading to the amplitude

$$(\cos\chi_1 \; \sin\chi_1 \; 0)_1 \; (\cos\chi_2 \; \sin\chi_2 \; 0)_2 \; |\Psi\rangle = -\frac{1}{\sqrt{2}}\sin(\chi_1 - \chi_2) \quad (8.10.38)$$

(see (1.6.4)), and the conditional probability,

$$P(\chi_1, \chi_2) = \frac{1}{2}\sin^2(\chi_1 - \chi_2). \quad (8.10.39)$$

The probability $P(\chi_1, -)$ is obtained from the square of the norm of the state

$$\begin{pmatrix} \cos\chi_1 & \sin\chi_1 & 0 \end{pmatrix}_1 |\Psi\rangle = \frac{1}{\sqrt{2}} \left[\cos\chi_1 \begin{pmatrix} 0 \\ 1 \\ 0 \end{pmatrix}_2 - \sin\chi_1 \begin{pmatrix} 1 \\ 0 \\ 0 \end{pmatrix}_2 \right] \quad (8.10.40)$$

giving simply
$$P(\chi_1, -) = 1/2 \quad (8.10.41)$$

and similarly
$$P(-, \chi_2) = 1/2. \quad (8.10.42)$$

For four angles

$$\chi_1 = 0°, \quad \chi_2 = 23°, \quad \chi_1' = 45°, \quad \chi_2' = 67°, \quad (8.10.43)$$

for example, S, defined in (8.10.28) is, according to the quantum mechanical computation, equal to -1.207 violating the LHV theories inequality (8.10.9) from below. For

$$\chi_1 = 0°, \quad \chi_2 = 67°, \quad \chi_1' = 135°, \quad \chi_2' = 23° \quad (8.10.44)$$

we obtain $S = 0.207$, violating the LHV theories inequality (8.10.9) from above according to the quantum mechanical computations of the underlying probabilities. Such violations are in conformity with experiments and are consistent with the quantum theory predictions.

Remarks 3

1. *The probability in (8.10.39) is a conditional probability given that the process depicted in Figure 8.7 has occurred and may be also obtained directly from quantum electrodynamics, to the leading order, and is valid for all speeds $0 \leqslant \beta \leqslant 1$ of e^+, e^-. For the line of momenta of the pair of photons in different directions than the one in Figure 8.7, $P(\chi_1, \chi_2)$ turns out to have a complicated dependence*[46] *on β, in general.*
2. *In the formal limit $\beta \to 0$ (the non-relativistic regime) of e^+, e^-, $P(\chi_1, \chi_2)$, as given in (8.10.39), holds true for all directions of the line of momenta of the two photons.*
3. *The normalization condition for the photon (massless spin 1 particle) reads*

$$P(\chi_1, \chi_2) + P(\chi_1 + \frac{\pi}{2}, \chi_2) + P(\chi_1, \chi_2 + \frac{\pi}{2}) + P(\chi_1 + \frac{\pi}{2}, \chi_2 + \frac{\pi}{2}) = 1 \quad (8.10.45)$$

instead of (8.10.16).
4. *From (8.10.39), (8.10.41), (8.10.42), we note that, in general, $P(\chi_1, \chi_2) \neq P(\chi_1, -)P(-, \chi_2)$ showing the dependence of the two events corresponding to the polarizations measurements of the two photons.*

---

[46] Manoukian and Yongram (2004), Yongram and Manoukian (2003).

Many other processes have been discussed in the literature.[47] It is interesting to note that for specific configurations, specific collision processes lead to entangled states and speed independent expressions for the probabilities. More generally, however, such probabilities depend on the speeds of the parent particles[48] and makes a Bell-like test for testing LHV theories more challenging.

The Bell-like tests have not only put quantum theory on a pedestal, and emphasized how reliable the theory is, but also created enormous interest in the foundation of its underlying theory. Its inherit "non-locality", as some practitioners put it, remains to be a problem, while for others as a mere problem of interpretation. Operationally, however, quantum theory is in pretty good shape. It is consistent with experiments — the final verdict of a theory.

Appendix to §8.10. Entangled States; The C-H Inequality

Entangled States

Consider two sets of independent vectors $|\alpha_i\rangle$, $|\beta_j\rangle$,

$$\langle \alpha_i | \alpha_j \rangle = \delta_{ij}, \quad \langle \beta_i | \beta_j \rangle = \delta_{ij} \tag{A-8.10.1}$$

then for any vector

$$|\psi\rangle = \sum_i c_i |\alpha_i\rangle |\beta_i\rangle \tag{A-8.10.2}$$

such that at least two of the coefficients c_i are non-zero, *cannot* be rewritten as a product

$$|\psi\rangle = |\psi_1\rangle |\psi_2\rangle \tag{A-8.10.3}$$

where

$$|\psi_1\rangle = \sum_i a_i |\alpha_i\rangle \tag{A-8.10.4}$$

$$|\psi_2\rangle = \sum_i b_i |\beta_i\rangle. \tag{A-8.10.5}$$

To show this, suppose, without any loss of generality that $c_1 \neq 0$ and $c_2 \neq 0$. Upon multiplying (A-8.10.2), in turn, by $\langle \alpha_1 | \langle \beta_1 |$, $\langle \alpha_2 | \langle \beta_2 |$ and using (A-8.10.3) we obtain

$$c_1 = a_1 b_1, \quad c_2 = a_2 b_2. \tag{A-8.10.6}$$

On the other hand by multiplying (A-8.10.2) in turn by $\langle \alpha_1 | \langle \beta_2 |$, $\langle \alpha_2 | \langle \beta_1 |$ and using (A-8.10.3) we obtain

---

[47] Cf. Clauser and Shimony (1978).
[48] Manoukian and Yongram (2004), Yongram and Manoukian (2003).

$$0 = a_1 b_2, \quad 0 = a_2 b_1 \tag{A-8.10.7}$$

which upon comparison with (A-8.10.6) leads to the *contradiction* that at least one of c_1, c_2 is zero.

A state as defined in (A-8.10.2), with at least two of the coefficients c_i non-zero is called an *entangled state*.

The Clauser-Horne (C-H) Inequality

Consider four numbers $0 \leqslant x_1, x_2, x'_1, x'_2 \leqslant 1$, and set

$$U = x_1 x_2 - x_1 x'_2 + x'_1 x_2 + x'_1 x'_2 - x'_1 - x_2. \tag{A-8.10.8}$$

We first derive the upper bound $U \leqslant 0$.
For $x_1 \geqslant x'_1$, we may rewrite

$$U = (x_1 - 1)x_2 + x'_1(x_2 - 1) + x'_2(x'_1 - x_1)$$

$$\leqslant 0 \tag{A-8.10.9}$$

since every term is non-positive.
For $x_1 < x'_1$, we may rewrite

$$U = x_1(x_2 - x'_2) + (x'_1 - 1)x_2 - x'_1(1 - x'_2)$$

$$\leqslant x_1(x_2 - x'_2) + (x'_1 - 1)x_2 - x_1(1 - x'_2)$$

$$= x_1 x_2 + (x'_1 - 1)x_2 - x_1$$

$$= x_1(x_2 - 1) + (x'_1 - 1)x_2 \leqslant 0. \tag{A-8.10.10}$$

We now derive the lower bound $-1 \leqslant U$.
For $x'_1 \geqslant x_1$,

$$U + 1 = (1 - x'_1)(1 - x_2) + x_1 x_2 + x'_2(x'_1 - x_1)$$

$$\geqslant 0 \tag{A-8.10.11}$$

since every term is non-negative.
For $x_1 > x'_1$,

$$U + 1 = (1 - x'_1)(1 - x_2) - (x_1 - x'_1)(x'_2 - x_2) + x'_1 x_2$$

$$\geqslant (x_1 - x'_1)(1 - x_2) - (x_1 - x'_1)(x'_2 - x_2) + x'_1 x_2$$

$$= (x_1 - x'_1)(1 - x'_2) + x'_1 x_2 \geqslant 0. \tag{A-8.10.12}$$

All told, we have

$$-1 \leqslant U \leqslant 0. \tag{A-8.10.13}$$

8.11 Quantum Teleportation and Quantum Cryptography

Quantum teleportation, in its simplest form, is a process of transferring the quantum state of a particle onto another particle. On the other hand, quantum cryptography is a process of sending coded messages between two parties with the aim of minimizing, or even abolishing, the risk that the message be intercepted by an unwanted third party. These methods rely on such fundamental and mysterious aspects of quantum theory such as entanglement and on the general basic fact that a quantum system can be in a superposition of different states. Research in such problems, under the heading of quantum information, has quite flourished in recent years and uses, in the process, the very basics of quantum theory in important potential applications. Quantum teleportation and quantum cryptography are discussed next.

8.11.1 Quantum Teleportation

Suppose a person — traditionally called Alice — has a spin-1/2 particle (or any two-level system), call it particle 1, in a state

$$\begin{pmatrix} \alpha \\ \beta \end{pmatrix}_1 \qquad (8.11.1)$$

$|\alpha|^2 + |\beta|^2 = 1$, and she wants another person — traditionally called Bob — at a distant location, to have a particle, of spin-1/2, call it particle 3, in this state. Quantum theory provides an answer to the transfer of the state of one particle to another one as follows.

Consider two particles, each of spin 1/2, call them particles 2 and 3, where 3 denotes the particle in question above. Suppose particles 2 and 3 are in the entangled state

$$|\Psi_{23}^-\rangle = \frac{1}{\sqrt{2}} \left[\begin{pmatrix} 0 \\ 1 \end{pmatrix}_2 \begin{pmatrix} 1 \\ 0 \end{pmatrix}_3 - \begin{pmatrix} 1 \\ 0 \end{pmatrix}_2 \begin{pmatrix} 0 \\ 1 \end{pmatrix}_3 \right] \qquad (8.11.2)$$

i.e., one has no information on the states of particles 2 and 3 except that they are in opposite spin states. One of these particles, referred to as particle 2, is sent to Alice, and the other, referred to as particle 3, is sent to Bob. Alice wants to transfer the state in (8.11.1) of particle 1 with her to Bob's particle 3. The entangled state $|\Psi_{23}^-\rangle$ in (8.11.2) between particles 2 and 3 plays a key role in such a transfer. Because of this, particles 2 and 3 are referred to as entangled ancillary pair of particles.

If Alice succeeds, by a specific measurement, of putting particles 1 and 2, with her, in the entangled state $|\Psi_{12}^-\rangle$, then particle 3, with Bob — at a distant location from Alice — will be projected into the initial state of particle 1 in (8.11.1) as shown below.

To the above end, we introduce the four orthogonal entangled states:

$$|\Psi_{ab}^{\pm}\rangle = \frac{1}{\sqrt{2}}\left[\begin{pmatrix}0\\1\end{pmatrix}_a \begin{pmatrix}1\\0\end{pmatrix}_b \pm \begin{pmatrix}1\\0\end{pmatrix}_a \begin{pmatrix}0\\1\end{pmatrix}_b\right] \qquad (8.11.3)$$

$$|\Phi_{ab}^{\pm}\rangle = \frac{1}{\sqrt{2}}\left[\begin{pmatrix}0\\1\end{pmatrix}_a \begin{pmatrix}0\\1\end{pmatrix}_b \pm \begin{pmatrix}1\\0\end{pmatrix}_a \begin{pmatrix}1\\0\end{pmatrix}_b\right]. \qquad (8.11.4)$$

These four states are referred to as Bell states. Also a measurement which puts two particles in one of these states is referred to as a Bell state measurement.

Now it is straightforward to show by some algebra (see Problem 8.25) that we may write

$$\begin{pmatrix}\alpha\\\beta\end{pmatrix}_1 |\Psi_{23}^{-}\rangle = -\frac{1}{2}\left\{\begin{pmatrix}\alpha\\\beta\end{pmatrix}_3 |\Psi_{12}^{-}\rangle + \begin{pmatrix}-\alpha\\\beta\end{pmatrix}_3 |\Psi_{12}^{+}\rangle \right.$$

$$\left. - \begin{pmatrix}\beta\\\alpha\end{pmatrix}_3 |\Phi_{12}^{-}\rangle + \begin{pmatrix}-\beta\\\alpha\end{pmatrix}_3 |\Phi_{12}^{+}\rangle\right\} \qquad (8.11.5)$$

where the left-hand side represents the initial state of the three particles 1, 2, 3.

Clearly, all Alice has to do is to carry out a Bell state measurement such that to put particles 1 and 2 in the entangled state $|\Psi_{12}^{-}\rangle$. Then from (8.11.5), we see that particle 3, with Bob, will be necessarily projected onto the state

$$\begin{pmatrix}\alpha\\\beta\end{pmatrix}_3 \qquad (8.11.6)$$

as was initially for particle 1 (see also (8.11.1)).

If it is unknown onto which of the four states $|\Psi_{12}^{-}\rangle, |\Psi_{12}^{+}\rangle, |\Phi_{12}^{-}\rangle, |\Phi_{12}^{+}\rangle$ Alice's Bell state measurement of particles 1 and 2 are projected, then according to (8.11.5) there are equal probabilities of 25% that they could be found in any of the four states.

What happens if Alice's Bell state measurement yields the entangled state $|\Psi_{12}^{+}\rangle$ or $|\Phi_{12}^{-}\rangle$ or $|\Phi_{12}^{+}\rangle$ instead of $|\Psi_{12}^{-}\rangle$. In such cases, Alice would inform Bob, by classical means, such as by telephone, to apply the unitary operator (up to phase factors)

$$\sigma_3 \quad \text{or} \quad \sigma_1 \quad \text{or} \quad \sigma_2 \qquad (8.11.7)$$

respectively, on the state of particles 3, with him, since

$$\sigma_3 \begin{pmatrix}-\alpha\\\beta\end{pmatrix}_3 = -\begin{pmatrix}\alpha\\\beta\end{pmatrix}_3, \quad \sigma_1\begin{pmatrix}\beta\\\alpha\end{pmatrix}_3 = \begin{pmatrix}\alpha\\\beta\end{pmatrix}_3, \quad \sigma_2\begin{pmatrix}-\beta\\\alpha\end{pmatrix}_3 = -i\begin{pmatrix}\alpha\\\beta\end{pmatrix}_3 \qquad (8.11.8)$$

to put his particle 3 in the initial state of particle 1.

The application of any such a unitary operator is a dynamical process. For example, if μ is the magnetic dipole magnetic moment of the particle, such as a neutron, one may formally apply a magnetic fields in the directions x_3, x_1, x_2, respectively, in reference to (8.11.5), for specific periods of times, to achieve such transformations (see Problem 8.26).

In regard to the teleportation carried out above, *up* to a unitary transformation to be carried out by Bob, it should be noted that the initial state (8.11.1) of particle 1 need not be known neither to Alice nor to Bob. Also after particles 1 and 2 are entangled in a Bell state, particle 1 would not be in its original state, and hence particle 3 is not a clone.

The identity in (8.11.5) leads also to the following interesting result that if particle 1 is initially entangled with another particle, call it particle 4, a Bell state measurement by Alice which puts particles 1 and 2 in an entangled state, automatically puts particles 4 and 3, with the latter particle with Bob, in an entangled state as well. To see this, suppose that particles 1 and 4 are initially in the entangled state $|\Psi_{14}^-\rangle$, then by setting α, $\beta = 0$ or 1 in (8.11.5) and multiplying the resulting equations by

$$\begin{pmatrix} 0 \\ 1 \end{pmatrix}_4 \quad \text{or} \quad \begin{pmatrix} 1 \\ 0 \end{pmatrix}_4 \tag{8.11.9}$$

as the case be, one obtains (see Problem 8.25)

$$|\Psi_{14}^-\rangle |\Psi_{23}^-\rangle = -\frac{1}{2}\Big\{ |\Psi_{34}^-\rangle |\Psi_{12}^-\rangle + |\Psi_{34}^+\rangle |\Psi_{12}^+\rangle$$
$$+ |\Phi_{34}^-\rangle |\Phi_{12}^-\rangle - |\Phi_{34}^+\rangle |\Phi_{12}^+\rangle \Big\}. \tag{8.11.10}$$

Again a Bell state measurement which puts particles 1 and 2, with Alice, in the state $|\Psi_{12}^-\rangle$, then projects particles 3 and 4 in the entangled state $|\Psi_{34}^-\rangle$ initially shared by particles 1 and 4. Such a quantum teleportation has been referred to as an entangled swapping.[49]

For earlier investigations, see the work of Bennett *et. al.*[50] Several experiments[51] have been carried out on teleportation, with most of them involving photons, confirming this fascinating predictions of quantum theory. For the rather rapid progress in the field, one may consult the research journals.

8.11.2 Quantum Cryptography

Before going into the role of quantum theory in modern cryptography and its future, we first discuss a classic cryptographic system the so-called "Vernam cipher" or the "one-time pad scheme" introduced by Gilbert Vernam in the thirties which provides perfect secrecy of communication.

---
[49] Zukowski *et al.* (1993).
[50] Bennett *et al.* (1993).
[51] Cf. Bouwmeester *et al.* (1997); Nielsen *et al.* (1998); Miranowicz and Tamaki (2002).

Suppose Alice wants to send a message to Bob at a distant location. Such a message may be sent in code consisting of 1's and 0's called bits. Letters, numerals, a blank space, the comma and so on are represented by ordered sequences of 1's and 0's making up a message which may be then readily read. In order to avoid an unwanted outsider – traditionally called Eve – from reading the message, however, Alice also produces another string of bits, randomly chosen, called the key as long as the message. She then encodes each bit of her message using the key generated thus introducing a scrambled text by using the following simple rule.

Let n denote the number of bits in her message. She adds the kth digit of the key to the kth digit of the message, with $k = 1, 2, \ldots, n$, thus generating a scrambled message, using the rule that

$$\left. \begin{array}{ll} 0+0, & \text{is replaced by } 0 \\ 0+1, & \text{is replaced by } 1 \\ 1+0, & \text{is replaced by } 1 \\ 1+1, & \text{is replaced by } 0 \end{array} \right\}. \qquad (8.11.11)$$

For example,

$$\left. \begin{array}{rl} \text{key:} & 0\ 1\ 1\ 0\ 1\ 0\ 0 \\ \text{Message:} & 1\ 0\ 1\ 1\ 1\ 1\ 0 \\ \text{Scrambled Message:} & 1\ 1\ 0\ 1\ 0\ 1\ 0 \end{array} \right\}. \qquad (8.11.12)$$

Now she transmits the scrambled message publicly (i.e., by telephone, radio,...) to Bob. She must also provide the key to Bob in secrecy in order that he may read the actual message by using the simple rule given above involving the key and the scrambled message.

The scrambled message is of no use to Eve if she does not have the key to decode the message. [Eve may guess and make up factorials of keys leading, in general, to different messages.]

The practical difficulty of the above (classical) procedure is that to transmit a message in secrecy one has to transmit a key in secrecy. Also what happens if Alice does not know the location of Bob and the two have not even met before to share a secret key to communicate in secrecy? On the other hand, should she trust a messenger who, at an earlier time to the availability of the message, was asked to deliver her key to Bob? For example, the messenger may make a copy of the key without even "disturbing" it, i.e., without even anybody else knowing it. This is unlike the situation in quantum physics where a measurement may, in general, disturb the system that has been tampered with. Finally it is not advisable to re-use a key in subsequent transmissions of messages if secrecy is of concern and hence the nomenclature "one-time pad scheme".

Now we will see how quantum theory may be used to generate a random key that Alice and Bob may share and how they may test the presence of an intruder.

8.11 Quantum Teleportation & Quantum Cryptography

For definiteness consider the scattering process in Figure 8.5 of the pair of spin 1/2 particles, with the pair put in the singlet state (8.10.20), such that, after scattering, one particle from each pair goes to Alice while the other goes to Bob. After scattering the particles from each pair are supposed to travel (in opposite directions) along the x-axis as depicted in the figure. When a particle reaches Alice she randomly chooses the y- or z-axis, by tossing a coin for example, to measure the spin direction of the particle in question. Bob, likewise, carries out a measurement of the spin orientation of the corresponding particle by randomly choosing the y- or z-axis. They both record their results by assigning 0 if the spin of a particle is opposite in direction to the axis chosen and 1 if it in the same direction.

After all the measurements (assumed to be sufficiently large in number[52]) have been made, Bob then selects randomly a subset of his ordered measurements, again assumed to be large in number, and communicates publicly with Alice which measurements he has selected (e.g., 2nd, 5th, 8th,...) and the corresponding axes (y or z) he had chosen and the results recorded for this subset of ordered set of measurements. Alice then checks the results she has recorded for the *same* axes, say N in number (assumed to be large enough), common with Bob's in the corresponding subset of the ordered set of measurements. If her results are exactly *opposite* to those of Bob's, as imposed by the entanglement of their pairs of particles, for these common axes, then with some confidence, which will be quantified and estimated below, she will announce publicly to Bob that no intruder has spied upon them.

Alice and Bob then communicate publicly the axes (y or z) they had chosen earlier in their remaining ordered set of measurements but *not* of their corresponding results obtained. Finally they select together randomly entries, equal in number to the number of bits, say n, making up Alice's message, from the ordered set of this remaining measurements having common axes to both of them. For example, for the purpose of an illustration only, suppose that $n = 3$, and the axes chosen in the remaining set of ordered measurements are as follows

	\multicolumn{9}{c}{Order of measurements}

	1	3	4	6	7	9	10	11	15
Alice's chosen axes	y	z	z	y	z	y	z	z	y
Bob's chosen axes	y	z	y	z	z	z	y	z	y
A possible selection	y	z	–	–	z	–	–	–	–
A possible selection	–	z	–	–	z	–	–	–	y

then two possible selections of entries with axes common to Alice and Bob are given above.

Assuming that they have not been spied on, and a selection of entries have been made, as just described, Alice knows the results (0 or 1) obtained by

---

[52] This should be much larger than the number of bits of the message Alice wants to send to Bob.

Bob, corresponding to the entries, by merely examining her own (opposite) results. The string of bits of n entries appearing in this selection in Bob's measurements may be then taken to be the *key*. A scrambled message may be sent by Alice to Bob, using the key obtained above, by the "Vernam cipher" method, thus completing her task. Therefore it remains to quantify and estimate her degree of confidence of not being spied upon.

To the above end, let $\hat{\mathbf{z}}$, $\hat{\mathbf{y}}$ denote unit vectors along the z, y axes, respectively. We use the notation $P_{AB}[\varepsilon_1\hat{\mathbf{z}}, \varepsilon_2\hat{\mathbf{z}}]$, with $\varepsilon_1, \varepsilon_2 = \pm 1$, for the probability that a measurement of spin along $\hat{\mathbf{z}}$ for the respective particles, in a given pair, by Alice and Bob are found along $\varepsilon_1\hat{\mathbf{z}}$, $\varepsilon_2\hat{\mathbf{z}}$ respectively. The probability $P_{AB}[\varepsilon_1\hat{\mathbf{y}}, \varepsilon_2\hat{\mathbf{y}}]$ is similarly defined. *Given* that Alice has obtained the spin of her particle, in a pair, to be along $\varepsilon_1\hat{\mathbf{z}}$ for a spin measurement along $\hat{\mathbf{z}}$, we also introduce the *conditional probability* $P_{B/A}[\varepsilon_2\hat{\mathbf{z}}/\varepsilon_1\hat{\mathbf{z}}]$ that Bob obtains for his particle, in the pair, to be along $\varepsilon_2\hat{\mathbf{z}}$ for a spin measurement along $\hat{\mathbf{z}}$.

In the *absence* of an intruder, we have according to (8.10.24), (8.10.26),

$$P_{B/A}[\varepsilon_2\hat{\mathbf{z}}/\varepsilon_1\hat{\mathbf{z}}] \equiv \frac{P_{AB}[\varepsilon_1\hat{\mathbf{z}}, \varepsilon_2\hat{\mathbf{z}}]}{P_A[\varepsilon_1\hat{\mathbf{z}}]} = \delta(\varepsilon_1, -\varepsilon_2) \qquad (8.11.13)$$

$$P_{B/A}[\varepsilon_2\hat{\mathbf{y}}/\varepsilon_1\hat{\mathbf{y}}] \equiv \frac{P_{AB}[\varepsilon_1\hat{\mathbf{y}}, \varepsilon_2\hat{\mathbf{y}}]}{P_A[\varepsilon_1\hat{\mathbf{y}}]} = \delta(\varepsilon_1, -\varepsilon_2). \qquad (8.11.14)$$

If p_z^A denotes the probability that Alice picks the z-axis for a spin measurement, and hence a probability of $(1 - p_z^A)$ that she picks the y-axis, then in reference to a spin of observations in the N pairs of observations with common axes for Alice and Bob, the probability that Bob gets an observation for spin in opposite direction, for his particle, to that of Alice's is given by

$P_B[\text{spin opposite to that of Alice's}]$

$$= \sum_{\varepsilon_1,\varepsilon_2=\pm 1} \left(P_{AB}[\varepsilon_1\hat{\mathbf{z}}, \varepsilon_2\hat{\mathbf{z}}]p_z^A + P_{AB}[\varepsilon_1\hat{\mathbf{y}}, \varepsilon_2\hat{\mathbf{y}}]\left(1 - p_z^A\right) \right) = 1. \qquad (8.11.15)$$

In the presence of an intruder, (8.11.15) is not necessarily true. To investigate this pertinent situation, suppose that Eve, the intruder, is aware that Alice and Bob have planned to measure the spins of their respective particles along $\hat{\mathbf{z}}$ or $\hat{\mathbf{y}}$. Accordingly, before each particle reaches Bob, Eve measures its spin along the z or y-axis as well, with the latter axes chosen, for example, by tossing a coin. By doing so, she puts the particle, in question in some "initial" state before it reaches Bob.

With the above tampering by Eve, if for the N measurements recorded by Alice, mentioned earlier, corresponding to the *same* axes common with Bob's for the ordered set of pairs, in anticipation of preparing a key, Alice finds at least one result not to be opposite to the corresponding one of Bob's, then she will be certain of the presence of the intruder, and the process is discontinued. On the other hand, if for every one of the N measurements her

8.11 Quantum Teleportation & Quantum Cryptography

results are opposite to the corresponding ones of Bob, she will fail to detect Eve's presence.

Accordingly, if p denotes the probability that for any one of the N measurements, Bob and Alice obtain opposite results for spins, corresponding to either of the common axes y or z, then the probability that Alice fails to detect Eve's intrusion, after examining the N measurements, is p^N. The probability of Alice's detection of the intruder is then

$$\text{Prob[detection of intruder]} = 1 - (p)^N. \qquad (8.11.16)$$

The latter is referred to as the *power* of the test, with the test being the detection of the intruder. Such a probability close to one, which will be the case for $p < 1$ and N sufficiently large, quantifies Alice's confidence in her statement that "no intruder has spied" on Bob and her.

To determine p, let

$$\mathbf{n}(\vartheta) = (0, \sin\vartheta, \cos\vartheta) \qquad (8.11.17)$$

then, with the letter E corresponding to Eve, the following are readily obtained, by using, in particular, (8.10.24), (8.10.27), (8.1.16), if Eve carries her measurements along $\mathbf{n}(\vartheta)$:

z: Common Axis of Alice and Bob

State "Prepared" by Eve

$\left\{\begin{array}{l} P_{AE}[+\hat{\mathbf{z}},+\hat{\mathbf{n}}(\vartheta)]=\frac{1}{2}\sin^2\frac{\vartheta}{2} \\ P_{AE}[-\hat{\mathbf{z}},+\hat{\mathbf{n}}(\vartheta)]=\frac{1}{2}\cos^2\frac{\vartheta}{2} \end{array}\right\}$ $\begin{pmatrix} e^{-i\pi/4}\cos\frac{\vartheta}{2} \\ e^{+i\pi/4}\sin\frac{\vartheta}{2} \end{pmatrix}$ $\left\{\begin{array}{l} P_{B/E}[-\hat{\mathbf{z}}/+\hat{\mathbf{n}}(\vartheta)]=\sin^2\frac{\vartheta}{2} \\ P_{B/E}[+\hat{\mathbf{z}}/+\hat{\mathbf{n}}(\vartheta)]=\cos^2\frac{\vartheta}{2} \end{array}\right\}$

$\left\{\begin{array}{l} P_{AE}[+\hat{\mathbf{z}},-\hat{\mathbf{n}}(\vartheta)]=\frac{1}{2}\cos^2\frac{\vartheta}{2} \\ P_{AE}[-\hat{\mathbf{z}},-\hat{\mathbf{n}}(\vartheta)]=\frac{1}{2}\sin^2\frac{\vartheta}{2} \end{array}\right\}$ $\begin{pmatrix} -e^{-i\pi/4}\sin\frac{\vartheta}{2} \\ e^{+i\pi/4}\cos\frac{\vartheta}{2} \end{pmatrix}$ $\left\{\begin{array}{l} P_{B/E}[-\hat{\mathbf{z}}/-\hat{\mathbf{n}}(\vartheta)]=\cos^2\frac{\vartheta}{2} \\ P_{B/E}[+\hat{\mathbf{z}}/-\hat{\mathbf{n}}(\vartheta)]=\sin^2\frac{\vartheta}{2} \end{array}\right\}$

y: Common Axis of Alice and Bob

State "Prepared" by Eve

$\left\{\begin{array}{l} P_{AE}[+\hat{\mathbf{y}},+\hat{\mathbf{n}}(\vartheta)]=\frac{1}{4}(1-\sin\vartheta) \\ P_{AE}[-\hat{\mathbf{y}},+\hat{\mathbf{n}}(\vartheta)]=\frac{1}{4}(1+\sin\vartheta) \end{array}\right\}$ $\begin{pmatrix} e^{-i\pi/4}\cos\frac{\vartheta}{2} \\ e^{+i\pi/4}\sin\frac{\vartheta}{2} \end{pmatrix}$ $\left\{\begin{array}{l} P_{B/E}[-\hat{\mathbf{y}}/+\hat{\mathbf{n}}(\vartheta)]=\frac{1}{2}(1-\sin\vartheta) \\ P_{B/E}[+\hat{\mathbf{y}}/+\hat{\mathbf{n}}(\vartheta)]=\frac{1}{2}(1+\sin\vartheta) \end{array}\right\}$

$\left\{\begin{array}{l} P_{AE}[+\hat{\mathbf{y}},-\hat{\mathbf{n}}(\vartheta)]=\frac{1}{4}(1+\sin\vartheta) \\ P_{AE}[-\hat{\mathbf{y}},-\hat{\mathbf{n}}(\vartheta)]=\frac{1}{4}(1-\sin\vartheta) \end{array}\right\}$ $\begin{pmatrix} -e^{-i\pi/4}\sin\frac{\vartheta}{2} \\ e^{+i\pi/4}\cos\frac{\vartheta}{2} \end{pmatrix}$ $\left\{\begin{array}{l} P_{B/E}[-\hat{\mathbf{y}}/-\hat{\mathbf{n}}(\vartheta)]=\frac{1}{2}(1+\sin\vartheta) \\ P_{B/E}[+\hat{\mathbf{y}}/-\hat{\mathbf{n}}(\vartheta)]=\frac{1}{2}(1-\sin\vartheta) \end{array}\right\}$

Hence for $\vartheta = 0$ or $\vartheta = \pi/2$, if p_ϑ^E denotes the corresponding probability that Eve picks the direction specified by $\mathbf{n}(0)$ or $\mathbf{n}(\pi/2)$ for a spin measurement of a particle before it goes off to Bob, then

$$p = \sum_{\vartheta=0,\pi/2} \sum_{\varepsilon_1,\varepsilon_2=\pm} \left(P_{AE}[\varepsilon_1 \hat{\mathbf{z}}, \varepsilon_2 \mathbf{n}(\vartheta)] P_{B/E}[-\varepsilon_1 \hat{\mathbf{z}}/\varepsilon_2 \mathbf{n}(\vartheta)] p_z^A \right.$$

$$\left. + P_{AE}[\varepsilon_1 \hat{\mathbf{y}}, \varepsilon_2 \mathbf{n}(\vartheta)] P_{B/E}[-\varepsilon_1 \hat{\mathbf{y}}/\varepsilon_2 \mathbf{n}(\vartheta)] \left(1 - p_z^A\right) \right) p_\vartheta^E$$

$$= \left(1 - \frac{1}{2} p_z^A\right) - \left(\frac{1}{2} - p_z^A\right) p_z^E \quad (8.11.18)$$

where $p_z^E = p_\vartheta^E \big|_{\vartheta=0}$.

Clearly if $p_z^A \simeq 0$ and $p_z^E \simeq 0$ or $p_z^A \simeq 1$, $p_z^E \simeq 1$, then $p \simeq 1$ which will not be advantageous to Alice. On the other hand for $p_z^A = 1/2$, which, for example, corresponds to the case of tossing a balanced coin for a large sample, $p = 3/4$ irrespective of what the value of p_z^E is corresponding to Eve's choice of the z- or y-axis. For this rather natural 'choice' of $p_z^A = 1/2$ by Alice, we obtain from (8.11.16), (8.11.18)

$$\text{Prob[detection of intruder]} = 1 - (3/4)^N \quad (8.11.19)$$

which is already $\simeq 0.99982$ for $N = 30$ (see also Problem 8.27).

There are other schemes, improvements, and other additional details that have been and are being developed in this rapidly growing field, and the reader may consult the literature[53] on such developments including the research journals.

8.12 Rotation of a Spinor

This section addresses the important question of the observability of the overall minus sign (§2.8, (2.8.68), (2.8.70), §8.1) acquired by a spinor for spin 1/2 under the operation of rotation through 2π radians. An interesting way of investigating the nature of this phase change is by the application of a Ramsey (§8.8) like method[54] with oscillatory fields causing transitions between hyperfine energy levels of a molecule.

In §8.1, we have seen how the nature of a spinor arises for a two-level system, as part of a multi-level system. For a given time interval of length

---

[53] Cf. Bennett et al. (1992); Tittel et al. (1998).
[54] Klempt (1976).

8.12 Rotation of a Spinor

T, we consider an interaction which may cause transitions between only two such states, which we denote by $|0\rangle$, $|1\rangle$.

Here we find it more convenient to represent the states $|0\rangle$, $|1\rangle$ by $\begin{pmatrix}1 & 0\end{pmatrix}^T$, $\begin{pmatrix}0 & 1\end{pmatrix}^T$, respectively, rather than the other way around. The interaction Hamiltonian, during the time interval in question, will be taken to be of the form

$$H_I(0,1) = \begin{pmatrix} 0 & V_{01} \\ V_{01}^* & 0 \end{pmatrix}. \tag{8.12.1}$$

If E_0, E_1 denote the energies associated with the states $|0\rangle$, $|1\rangle$ then the free Hamiltonian, restricted to these two states may be written as

$$H_0(0,1) = \frac{(E_0+E_1)}{2} + \frac{E_0-E_1}{2}\sigma_3 \tag{8.12.2}$$

where σ_3 is a Pauli matrix.

We consider the system to be initially in the state $|0\rangle$, and subject it to an interaction with an alternating field described by an interaction Hamiltonian

$$H_I^1(0,2) = \begin{pmatrix} 0 & \hbar b_1 e^{i\omega t} \\ \hbar b_1^* e^{-i\omega t} & 0 \end{pmatrix}, \qquad 0 \leqslant t < \tau \tag{8.12.3}$$

acting during the time interval specified, which may cause transitions between $|0\rangle$ and some other state $|2\rangle$ only. Here b_1 is a complex number, and we denote the energy of the state $|2\rangle$ by E_2 The interaction in (8.12.3) defines the first Ramsey zone.

For a time T, following the interaction in (8.12.3), we subject the system to an interaction with Hamiltonian which may cause transitions between $|0\rangle$ and state $|1\rangle$ only

$$H_I(0,1) = \begin{pmatrix} 0 & \hbar b_0 e^{i\omega_0 t} \\ \hbar b_0 e^{-i\omega_0 t} & 0 \end{pmatrix}, \qquad \tau \leqslant t < \tau + T \tag{8.12.4}$$

with b_0 is taken to be real, and ω_0 to satisfy the *resonance* condition $\omega_0 = (E_1 - E_0)/\hbar$.

Finally, we subject the system to a second Ramsey zone with interaction Hamiltonian which may cause transitions between the states $|0\rangle$ and $|2\rangle$ only for a time interval of length τ

$$H_I^2(0,2) = \begin{pmatrix} 0 & \hbar b_2 e^{i\omega t} \\ \hbar b_2^* e^{-i\omega t} & 0 \end{pmatrix}, \qquad \tau + T \leqslant t \leqslant 2\tau + T. \tag{8.12.5}$$

We take

$$b_1 = b e^{i\phi_1}, \qquad b_2 = b e^{i\phi_2} \tag{8.12.6}$$

thus introducing a phase difference between the fields in the two Ramsey zones.[55] The interaction Hamiltonian in (8.12.4) is crucial for the subsequent

---

[55] Klempt (1976); Ramsey and Silsbee (1951); Ramsey (1990).

considerations, and introduces, during a time interval of length T, an interaction occurring within this time interval between the two Ramsey zones, specified by (8.12.3), (8.12.5), as opposed to a free one of the type treated in §8.8.

To describe the dynamics of the system, we are led to find the expression for the unitary operator $U(t, t_1; j)$ giving rise to the time development from some t_1 to t for a generic system specified by the integer j ($j = 1, 2$):

$$i\hbar \begin{pmatrix} \dot{\alpha}_0 \\ \dot{\alpha}_2 \end{pmatrix} = \begin{pmatrix} E_0 & \hbar b_j e^{i\omega t} \\ \hbar b_j^* e^{-i\omega t} & E_2 \end{pmatrix} \begin{pmatrix} \alpha_0 \\ \alpha_2 \end{pmatrix}. \tag{8.12.7}$$

The same analysis as carried out in §8.8, leads to the following explicit expression for $U(t, t_1; j)$:

$$U(t, t_1; j) = \begin{pmatrix} U_{00}(t, t_1; j) & U_{01}(t, t_1; j) \\ U_{20}(t, t_1; j) & U_{22}(t, t_1; j) \end{pmatrix} \tag{8.12.8}$$

where

$$U_{00}(t, t_1; j) = \left[\cos a(t - t_1) - i\frac{\Delta}{2a} \sin a(t - t_1) \right] e^{-i\alpha(t-t_1)} e^{i\frac{\omega}{2}(t-t_1)} \tag{8.12.9}$$

$$U_{02}(t, t_1; j) = -i\frac{|b|}{a} \sin a(t - t_1) e^{-i\alpha(t-t_1)} e^{i\left(\frac{\omega}{2}(t+t_1)+\phi_j\right)} \tag{8.12.10}$$

$$U_{20}(t, t_1; j) = -i\frac{|b|}{a} \sin a(t - t_1) e^{-i\alpha(t-t_1)} e^{-i\left(\frac{\omega}{2}(t+t_1)+\phi_j\right)} \tag{8.12.11}$$

$$U_{00}(t, t_1; j) = \left[\cos a(t - t_1) + i\frac{\Delta}{2a} \sin a(t - t_1) \right] e^{-i\alpha(t-t_1)} e^{-i\frac{\omega}{2}(t-t_1)} \tag{8.12.12}$$

$$\Delta = \omega - \frac{E_2 - E_0}{\hbar} \tag{8.12.13}$$

$$a = \left[\left(\frac{\Delta}{2}\right)^2 + |b|^2 \right]^{1/2} \tag{8.12.14}$$

$$b_j = |b| e^{i\phi_j} \tag{8.12.15}$$

$$\alpha = \frac{E_2 + E_0}{2\hbar}. \tag{8.12.16}$$

Hence the initial state

$$|\psi(0)\rangle = |0\rangle \tag{8.12.17}$$

develops in time τ to

$$|\psi(\tau)\rangle = a_0(\tau) |0\rangle + a_2(\tau) |2\rangle \tag{8.12.18}$$

by going through the *first* Ramsey zone, where

8.12 Rotation of a Spinor

$$a_0(\tau) = U_{00}(\tau, 0; 1) \tag{8.12.19}$$

$$a_2(\tau) = U_{20}(\tau, 0; 1). \tag{8.12.20}$$

It is interesting to note that the unitary operator $U(t, t_1; j)$ in (8.12.8)–(8.12.16), corresponding to the Ramsey zones, may be rewritten in the compact form (see Problem 8.28)

$$U(t, t_1; j) = \exp\left[-\frac{i}{\hbar}(t - t_1)H_0(0, 2)\right] \exp\left[i\frac{\Delta}{2}(t - t_1)\sigma_3\right]$$

$$\times \exp\left[-ia(t - t_1)\mathbf{n}_j \cdot \boldsymbol{\sigma}\right] \tag{8.12.21}$$

where $H_0(0, 2)$ is defined as in (8.12.2) *with E_1 replaced by E_2,*

$$\mathbf{n}_j = \left(\frac{|b|}{a}\cos[\phi_j + \omega(t - t_1)], -\frac{|b|}{a}\sin[\phi_j + \omega(t - t_1)], \frac{\Delta}{2a}\right). \tag{8.12.22}$$

According to the interaction (8.12.4), acting in the time interval $[\tau, \tau+T)$, the state $|\psi(\tau)\rangle$ in (8.12.18) develops in time before going through the second Ramsey zone, via the unitary operator

$$U(\tau + T, \tau) = \exp\left[-\frac{i}{\hbar}TH_0(0, 1)\right] \exp\left[-i\frac{\vartheta}{2}\mathbf{n} \cdot \boldsymbol{\sigma}\right] \tag{8.12.23}$$

as following from (8.12.9)–(8.12.16), (8.12.21), (8.12.22), at *resonance* $\omega_0 = (E_1 - E_0)/\hbar$, and hence with $\Delta \to 0$, $a \to |b_0|$, where

$$\frac{\vartheta}{2} = |b_0|T \tag{8.12.24}$$

$$\mathbf{n} = (\cos\omega_0\tau, -\sin\omega_0\tau, 0) \tag{8.12.25}$$

and we recall that (see (2.8.7), (2.8.4)),

$$\exp\left[-i\frac{\vartheta}{2}\mathbf{n} \cdot \boldsymbol{\sigma}\right] = \begin{pmatrix} \cos\frac{\vartheta}{2} & -ie^{i\omega_0\tau}\sin\frac{\vartheta}{2} \\ -ie^{-i\omega_0\tau}\sin\frac{\vartheta}{2} & \cos\frac{\vartheta}{2} \end{pmatrix}. \tag{8.12.26}$$

That is, during the time interval $\tau \leqslant t < \tau + T$, a spinor, in the presence of the interaction in (8.12.4), gets rotated by the angle ϑ, and the state $|\psi(\tau)\rangle$ in (8.12.18) develops in time, before the system enters the second Ramsey zone, to

$$|\psi(\tau + T)\rangle = e^{-iE_0T/\hbar}\cos\frac{\vartheta}{2} a_0(\tau)|0\rangle - ie^{-iE_1T/\hbar}e^{-i\omega_0\tau}\sin\frac{\vartheta}{2}|1\rangle$$

$$+ e^{-iE_2T/\hbar}a_2(\tau)|2\rangle \tag{8.12.27}$$

as follows directly from (8.12.23)–(8.12.26), and the fact that the state $|2\rangle$ develops freely in the time interval in question of length T.

By going through the second Ramsey zone, the state $|\psi(\tau+T)\rangle$ develops in time, according to the unitary operator (8.12.8)–(8.12.12), with $j = 2$, to

$$|\psi(2\tau+T)\rangle = a_0 |0\rangle + a_1 |1\rangle + a_2 |2\rangle \qquad (8.12.28)$$

with

$$a_0 = U_{00}(2\tau+T, \tau+T; 2)\, e^{-iE_0 T/\hbar} a_0(\tau) \cos\frac{\vartheta}{2}$$

$$+ U_{02}(2\tau+T, \tau+T; 2)\, e^{-iE_2 T/\hbar} a_2(\tau) \qquad (8.12.29)$$

$$a_1 = -i e^{-iE_1(\tau+T)/\hbar}\, e^{-i\omega_0 \tau} \sin\frac{\vartheta}{2} \qquad (8.12.30)$$

$$a_2 = U_{20}(2\tau+T, \tau+T; 2)\, e^{-iE_0 T/\hbar} a_0(\tau) \cos\frac{\vartheta}{2}$$

$$+ U_{22}(2\tau+T, \tau+T; 2)\, e^{-iE_2 T/\hbar} a_2(\tau) \qquad (8.12.31)$$

and $a_0(\tau)$, $a_2(\tau)$ are, respectively, defined in (8.12.19), (8.12.20).

In particular, a_2 denotes the amplitude of a transition from the state $|0\rangle$ to the state $|2\rangle$, just after going through the second Ramsey zone, if the spinor representing the two-level system, corresponding to the states $|0\rangle$, $|1\rangle$ is rotated by an angle ϑ.

The above transition probability is worked out from the expression of a_2 in (8.12.31) to be

$$P_\vartheta\big[\text{Transition } |0\rangle \to |2\rangle\,\big] = |a_2|^2$$

$$= \frac{|b|^2}{a^2} \sin^2 a\tau \bigg[F_+ \cos^2 a\tau - \frac{\Delta}{2a} F_0 \sin 2a\tau$$

$$+ \left(\frac{\Delta}{2a}\right)^2 F_- \sin^2 a\tau \bigg] \qquad (8.12.32)$$

where

$$F_\pm\left(\cos\frac{\vartheta}{2}, \phi_2 - \phi_1\right) = 1 + \cos^2\frac{\vartheta}{2} \pm 2\cos\frac{\vartheta}{2} \cos\left(T\Delta + \phi_2 - \phi_1\right) \qquad (8.12.33)$$

$$F_0\left(\cos\frac{\vartheta}{2}, \phi_2 - \phi_1\right) = \cos\frac{\vartheta}{2} \sin\left(T\Delta + \phi_2 - \phi_1\right). \qquad (8.12.34)$$

We may compare this probability, for two cases of interest corresponding to no spinor rotation $\vartheta = 0$ and to the case corresponding to a full rotation for $\vartheta = 2\pi$. These are respectively given by

$$P_\vartheta\bigg|_{\vartheta=0} = 4\frac{|b|^2}{a^2} \sin^2 a\tau \left[\cos a\tau \cos \Omega - \frac{\Delta}{2a} \sin a\tau \sin \Omega\right]^2 \quad (8.12.35)$$

$$P_\vartheta\bigg|_{\vartheta=2\pi} = 4\frac{|b|^2}{a^2} \sin^2 a\tau \left[\cos a\tau \sin \Omega + \frac{\Delta}{2a} \sin a\tau \cos \Omega\right]^2 \quad (8.12.36)$$

and are obviously different, in general, where

$$\Omega = \frac{1}{2}(T\Delta + \phi_2 - \phi_1). \quad (8.12.37)$$

In particular, one may consider the difference in the transition probability P_ϑ between two cases: one with a phase difference $\phi_2 - \phi_1 = \pi/2$, and one with the reversal of the phase difference $\phi_2 - \phi_1 = -\pi/2$. Denoting the change in the transition probability on such reversal of the phase difference by

$$\Delta P_\vartheta = P_\vartheta\bigg|_{\phi_2-\phi_1=\pi/2} - P_\vartheta\bigg|_{\phi_2-\phi_1=-\pi/2} \quad (8.12.38)$$

we have from (8.12.32) the explicit expression

$$\Delta P_\vartheta = 4\frac{|b|^2}{a^2} \sin^2 a\tau \left\{ \left[\left(\frac{\Delta}{2a}\right)^2 \sin^2 a\tau - \cos^2 a\tau\right] \sin T\Delta \right.$$
$$\left. - \frac{\Delta}{2a} \sin 2a\tau \cos T\Delta \right\} \cos \frac{\vartheta}{2} \quad (8.12.39)$$

leading to a complete reversal in sign from $\vartheta = 0$ to $\vartheta = 2\pi$, of the change in the transition probability under a reversal of the phase difference $\phi_2 - \phi_1 = \pm\pi/2$ between the fields in the two Ramsey zones. This sort of experiment has been performed[56] and is consistent with the reversal of sign of the change in the transition probability. A 4π radian rotation restores everything back to the $\vartheta = 0$ case.

8.13 Geometric Phases

8.13.1 The Berry Phase and the Adiabatic Regime

Consider a time-dependent Hamiltonian $H(t)$ which varies in time t over some interval $[0, T]$, such that for each such t, $H(t)$ has a discrete non-degenerate spectrum, with eigenvalue equations

---

[56] Klempt (1976). For some other experimental studies with different procedures, see Rauch et al. (1975); Werner et al. (1975); Byrne (1978); Klein and Opat (1976); Stoll et al. (1977).

$$H(t)|\eta_n(t)\rangle = E_n(t)|\eta_n(t)\rangle, \qquad \langle\eta_n(t)|\eta_m(t)\rangle = \delta_{mn}. \tag{8.13.1}$$

A process is said to be an adiabatic one, if the change of the Hamiltonian in time is so slow, measured by a large value of the length T of the time interval (i.e., $T \to \infty$), such that if the system is initially the state $|\eta_n(0)\rangle$ of $H(0)$, then at any later time t, in the above interval, it is found, up to a phase factor, in the state $|\eta_n(t)\rangle$ of $H(t)$.[57] That is, the state of the system evolves together with the Hamiltonian. In particular, if $H(T)$ coincides with $H(0)$, then the system will be found, approximately, in the state $|\eta_n(0)\rangle$ for T large enough, up to a possible phase factor.

Now suppose that the time-dependence of $H(t)$ is implicit and comes from the dependence of the latter on an external time-dependent k-vector $\mathbf{R}(t)$, with $k \geq 2$, i.e., $H(t) = H(\mathbf{R}(t))$, $|\eta_n(t)\rangle = |\eta_n(\mathbf{R}(t))\rangle$. In particular,

$$\frac{\mathrm{d}}{\mathrm{d}t}|\eta_n(t)\rangle = \boldsymbol{\nabla}_{\mathbf{R}}|\eta_n(\mathbf{R}(t))\rangle \cdot \dot{\mathbf{R}}(t). \tag{8.13.2}$$

Also suppose that $\mathbf{R}(t)$ traces a closed curve C in the parameter space in which it varies such that $\mathbf{R}(T) = \mathbf{R}(0)$. That is, the Hamiltonian parametrized by the vector $\mathbf{R}(t)$, returns to its initial form $H(T) = H(0)$, as $\mathbf{R}(t)$ follows a *cyclic* motion in the parameter space.

The time T is taken to be much larger than any typical quantum mechanical oscillation period in the problem.

In a remarkable paper, Berry[58] made the observation, in the adiabatic regime, that if the system is initially in the state $|\psi(0)\rangle = |\eta_n(\mathbf{R}(0))\rangle$, then $|\psi(T)\rangle$ develops in addition to the familiar phase factor

$$\exp\left(-\frac{\mathrm{i}}{\hbar}\int_0^T \mathrm{d}t'\, E_n(t')\right) \tag{8.13.3}$$

a phase factor which depends on the *geometry* determined by the path traversed by $\mathbf{R}(t)$ in the parameter space. Specifically,

$$\langle\psi(0)|\psi(T)\rangle = \exp\left(-\frac{\mathrm{i}}{\hbar}\int_0^T \mathrm{d}t'\, E_n(t')\right)\exp(\mathrm{i}\gamma_n)\,\langle\eta_n(0)|\eta_n(T)\rangle \tag{8.13.4}$$

with

$$\gamma_n = \mathrm{i}\int_0^T \mathrm{d}t'\,\langle\eta_n(t')|\dot{\eta}_n(t')\rangle. \tag{8.13.5}$$

To derive (8.13.4), note that according to the adiabatic hypothesis, we may write

---

[57] The adiabatic regime, as an approximation method, will be treated in Chapter 12. The relevant technical details of this approximation, however, are not necessary here to follow the present study.

[58] Berry (1984).

$$|\psi(t)\rangle = \exp\left(-\frac{i}{\hbar}\int_0^t dt' E_n(t')\right) \exp\left(i\gamma_n(t)\right) |\eta_n(t)\rangle + \mathcal{O}\left(\frac{1}{T}\right) \quad (8.13.6)$$

and, as indicated, up to corrections which vanish for the parameter $T \to \infty$, $\gamma_n(t)$ is to be *determined*. Upon substitution of (8.13.6) in the Schrödinger equation

$$i\hbar\frac{d}{dt}|\psi(t)\rangle = H(t)|\psi(t)\rangle \quad (8.13.7)$$

and taking the inner product of the resulting equation with $\langle\eta_n(t)|$, one obtains

$$\dot{\gamma}_n = i\langle\eta_n(t)|\dot{\eta}_n(t)\rangle \quad (8.13.8)$$

which formally leads to (8.13.4), (8.13.5).

To investigate further the nature of (8.13.4), we consider phase transformations of the eigenstates

$$|\eta_n(t)\rangle \longrightarrow |\eta_n(t)\rangle e^{i\alpha_n(t)}. \quad (8.13.9)$$

Since the eigenvalues $E_n(t)$, in (8.13.1), obviously do not change, under the transformations in (8.13.9), and

$$\gamma_n \longrightarrow \gamma_n - (\alpha_n(T) - \alpha_n(0)) \quad (8.13.10)$$

$$\langle\eta_n(0)|\eta_n(T)\rangle \longrightarrow \langle\eta_n(0)|\eta_n(T)\rangle \exp i(\alpha_n(T) - \alpha_n(0)) \quad (8.13.11)$$

we note that the expression on the right-hand side of (8.13.4) remains *invariant* under the transformations in (8.13.9). Accordingly, by recalling that $H(T) = H(0)$, and that $|\eta_n(T)\rangle$ denotes the *same* state as $|\eta_n(0)\rangle$, we may choose, by construction, the *eigenstates* $|\eta_n(t)\rangle$, by appropriate choice of phases, such that $|\eta_n(T)\rangle = |\eta_n(0)\rangle$. We may then rewrite (8.13.4) as

$$\langle\psi(0)|\psi(T)\rangle = \exp\left(-\frac{i}{\hbar}\int_0^T dt' E_n(t')\right) \exp(i\gamma_n(C)) \quad (8.13.12)$$

with

$$\gamma_n(C) = i\oint_C \langle\eta_n(\mathbf{R})|\nabla_\mathbf{R}\eta_n(\mathbf{R})\rangle \cdot d\mathbf{R} \quad (8.13.13)$$

where we have finally used (8.13.2).

The phase $\gamma_n(C)$ is referred to as the *Berry phase*, in the adiabatic regime, while the familiar factor in (8.13.3) as the *dynamical phase* factor. We note that the integral in (8.13.13) depends on the segments of the curve C and not on the durations of time taken to travel over these segments and hence the term geometric.

That $\exp(i\gamma_n(C))$ is a phase factor follows by noting that the normalization condition $\langle\eta_n(\mathbf{R}(t))|\eta_n(\mathbf{R}(t))\rangle = 1$, i.e., $\nabla_\mathbf{R}\langle\eta_n(\mathbf{R})|\eta_n(\mathbf{R})\rangle = 0$, implies that

$$\langle \eta_n(\mathbf{R})|\nabla_\mathbf{R}\eta_n(\mathbf{R})\rangle = -\langle \nabla_\mathbf{R}\eta_n(\mathbf{R})|\eta_n(\mathbf{R})\rangle$$
$$= -\langle \eta_n(\mathbf{R})|\nabla_\mathbf{R}\eta_n(\mathbf{R})\rangle^* \qquad (8.13.14)$$

which means that $i\langle \eta_n(\mathbf{R})|\nabla_\mathbf{R}\eta_n(\mathbf{R})\rangle$ in (8.13.5) is real. [Note that for $k=1$, the expression on the right-hand side of (8.13.13) vanishes, which explains the condition $k \geqslant 2$ stated earlier for the k-vector $\mathbf{R}(t)$.]

$\gamma_n(C)$ in (8.13.13) is independent of phase transformations of the eigenstates

$$|\eta_n(\mathbf{R})\rangle \longrightarrow |\eta_n(\mathbf{R})\rangle \, e^{i\beta_n(\mathbf{R})}. \qquad (8.13.15)$$

To the above end, we set

$$i\langle \eta_n(\mathbf{R})|\nabla_\mathbf{R}\eta_n(\mathbf{R})\rangle = \mathbf{A}(\mathbf{R}) \qquad (8.13.16)$$

and observe that the latter then transforms according to the rule

$$\mathbf{A}(\mathbf{R}) \longrightarrow \mathbf{A}(\mathbf{R}) - \nabla_\mathbf{R}\beta(\mathbf{R}) \qquad (8.13.17)$$

reminiscent of gauge transformations of a vector potential. The invariance of $\gamma_n(C)$ under the phase transformations (8.13.15), then follows upon the substitution of (8.13.17) in (8.13.13).

By the application of Stokes's theorem, $\gamma_n(C)$ may be also rewritten as a surface integral

$$\gamma_n(C) = \int (\nabla_\mathbf{R} \times \mathbf{A}(\mathbf{R})) \cdot d\mathbf{S} \qquad (8.13.18)$$

Since in the transformations (8.13.15), the phases are just c-numbers and hence they commute, such transformations are referred to as *abelian* transformations as opposed to the cases when degeneracy is involved in the eigenvalue problem in (8.13.1) as will be discussed later.

As an illustration consider the Hamiltonian of spin 1/2 in an external magnetic field $\mathbf{B} = B\mathbf{R}(t)$,

$$H(\mathbf{R}(t)) = -\mu B \mathbf{R}(t) \cdot \boldsymbol{\sigma} \qquad (8.13.19)$$

where $\mathbf{R}(t)$ is the unit vector,

$$\mathbf{R}(t) = (\sin\theta\cos\omega t, \, \sin\theta\sin\omega t, \, \cos\theta) \qquad (8.13.20)$$

and β, θ are time-independent. We set $-\mu B \equiv \hbar\omega_0/2$.

For $t = T = 2\pi/\omega$, $\mathbf{R}(t) = \mathbf{R}(0)$, and for any fixed t in the interval $[0,T]$ of consideration, an eigenstate of the Hamiltonian $H(\mathbf{R}(t))$ is

$$|\eta_+(t)\rangle = \begin{pmatrix} \cos\frac{\theta}{2} \\ \sin\frac{\theta}{2}e^{i\omega t} \end{pmatrix} \qquad (8.13.21)$$

satisfying the eigenvalues equation

$$H(\mathbf{R}(t))\,|\eta_+(t)\rangle = \frac{\hbar\omega_0}{2}\,|\eta_+(t)\rangle \qquad (8.13.22)$$

and, $|\eta_+(T)\rangle = |\eta_+(0)\rangle$,

$$\mathrm{i}\,\langle \eta_+(t)|\dot{\eta}_+(t)\rangle = -\frac{\omega}{2}(1-\cos\theta). \qquad (8.13.23)$$

The adiabatic regime is defined by the very slow change of the Hamiltonian, that is, for a very large T much larger than the characteristic quantum oscillation period $T_0 \sim 1/|\omega_0|$, i.e., for $|\omega_0| \gg \omega$.

From (8.13.4), we then have

$$\langle \psi(0)|\psi(T)\rangle = \exp(-\mathrm{i}\omega_0 T/2)\exp(\mathrm{i}\gamma_n(C)) \qquad (8.13.24)$$

with

$$\gamma_n(C) = -\pi(1-\cos\theta) \qquad (8.13.25)$$

where $2\pi(1-\cos\theta)$ denotes the *solid* angle, subtended at the origin of the sphere of unit radius, swept out by the unit vector $\mathbf{R}(t)$, along the magnetic field $\mathbf{B}(t)$, which traces the closed curve C on the surface of the sphere, with the latter surface defining the parameter space (see Figure 8.9).

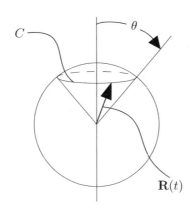

Fig. 8.9. $-2\gamma_n(C)$ in (8.13.25) of the Berry phase, denotes the solid angle, subtended at the origin of a sphere of unit radius, swept up by $\mathbf{R}(t)$ while tracing the closed curve C in the parameter space.

It is interesting to compare the expression in (8.13.24) with the exact solution given in (8.1.37), (8.1.38), (8.1.34), to see the adiabatic hypothesis at work.

To the above end, Ω in (8.1.39) is given by

$$\Omega \simeq |\omega_0| - \frac{|\omega_0|}{\omega_0}\omega\cos\theta \qquad (8.13.26)$$

for $|\omega_0| \gg \omega$, which from (8.1.34), (8.1.37), (8.1.38),

$$|\psi(t)\rangle \simeq e^{-i\omega_0 t/2} e^{-i\omega t(1-\cos\theta)/2} \begin{pmatrix} \cos\frac{\theta}{2} \\ \sin\frac{\theta}{2} e^{i\omega t} \end{pmatrix} + \mathcal{O}\left(\frac{\omega}{|\omega_0|}\right). \qquad (8.13.27)$$

With the initial state $|\psi(0)\rangle$ given in (8.1.32), this leads, in the adiabatic regime, to the result in (8.13.24).

Several experiments[59] have measured the Berry phase for particles with different spins including that for spin 1/2.

The analogue of the geometrical phase for classical systems has been also found.[60] The classic example of this is the change in the direction of swing of the so-called Foucault pendulum after a full rotation of the earth.[61]

8.13.2 Degeneracy

We briefly consider the generalization of (8.13.4) to the cases when degeneracy[62] is involved.

To the above end, consider the eigenvalues problem

$$H(t)|\eta(t,a)\rangle = E(t)|\eta(t,a)\rangle, \qquad a = 1,\ldots,N \qquad (8.13.28)$$

for an N-fold degenerate eigenvalue $E(t)$ for all t in $[0,T]$, where we have suppressed a generic quantum number n for simplicity of the notation. As before if $\mathbf{R}(t)$ in $H(t) = H(\mathbf{R}(t))$ is varied slowly over the long time interval of length T, i.e., in the adiabatic regime, and assuming that the given space of degenerate levels does not cross other levels, we may write for $T \to \infty$,

$$|\psi_a(t)\rangle = |\eta(t,b)\rangle c_{ab}(t) \exp\left(-\frac{i}{\hbar}\int_0^t dt'\, E_n(t')\right) \qquad (8.13.29)$$

as a linear combination of the degenerate states $|\eta(t,b)\rangle$ with a summation over b understood, where

$$|\psi_a(0)\rangle = |\eta(0,a)\rangle \qquad (8.13.30)$$

for a given a.

Upon substituting (8.13.29) in the Schrödinger equation gives

---

[59] Cf. Bitter and Dubbers (1987); Tycko (1987).
[60] Hannay (1985).
[61] For a demonstration of this among other things, by formulating the classical dynamical problem in complex form, cf. Manoukian and Yongram (2002).
[62] Wilczek and Zee (1984).

8.13 Geometric Phases

$$\dot{c}_{ab}(t) = i A_{bk}(t) c_{ak}(t) \qquad (8.13.31)$$

where

$$A_{bk}(t) = i \langle \eta(t,b) | \dot{\eta}(t,k) \rangle . \qquad (8.13.32)$$

Equation (8.13.31) may be readily integrated as a time-ordered product as in (A-2.5.11), (A-2.5.2) obtaining for the $N \times N$ *matrix* $\mathbf{c}(t)$ the expression:

$$[\mathbf{c}(T)]_{ab} = \left[\left(\exp\left[i \int_0^T dt' \, \mathbf{A}(t') \right] \right)_+ \right]_{ba} \qquad (8.13.33)$$

which, in the adiabatic regime, allows us to write

$$|\psi_a(T)\rangle = |\eta(T,b)\rangle \left[\left(\exp\left[i \int_0^T dt' \, \mathbf{A}(t') \right] \right)_+ \right]_{ba} \exp\left(-\frac{i}{\hbar} \int_0^T dt' \, E_n(t') \right) . \qquad (8.13.34)$$

The state $|\psi_a(T)\rangle$ evolves into a linear combination of the $|\eta(T,b)\rangle$ for any given a with initial condition given in (8.13.30).

We consider general transformations, which generalize those in (8.13.9),

$$|\eta(t,b)\rangle \longrightarrow \Omega_{bc}(t) |\eta(t,c)\rangle \qquad (8.13.35)$$

$$\langle \eta(t,b)| \longrightarrow \langle \eta(t,c)| \Omega_{cb}^{-1}(t) \qquad (8.13.36)$$

with Ω as $N \times N$ matrices whose inverses are assumed to exist. Since such matrices do not necessarily commute, in general, these transformations are referred to as *non-abelian* ones as opposed to the abelian ones in (8.13.9).

Under the transformations (8.13.35), (8.13.36), the $A_{bk}(t)$ in (8.13.32) then transforms as[63]

$$A_{bk}(t) \longrightarrow A_{cd}(t) \Omega_{cb}^{-1}(t) \Omega_{kd}(t) + i \dot{\Omega}_{kc}(t) \Omega_{cb}^{-1}(t). \qquad (8.13.37)$$

Additional details on the problem of degeneracy are worked out in Problem 8.31.

In the study of the Berry phase above in this section, the Hamiltonian was parametrized by a k-vector with cyclic motion in parameter space, and the assumption of an adiabatic regime was a key one in its development. The initial state of a system in question was also assumed to be an eigenstates of the instantaneous Hamiltonian at $t = 0$ i.e., of $H(0)$. Aharonov and Anandan[64] generalized Berry's analysis by giving up the adiabaticity assumption. They also regard the cyclic parameters as labelling the states, rather than the Hamiltonian. This is discussed next. Later on, even the cyclicity of the parametric motion will be given up in studying the generation of geometric phases.

---

[63] Such transformations are well known in non-abelian gauge theories of fundamental interactions.
[64] Aharonov and Anandan (1987).

8.13.3 Aharonov-Anandan (AA) Phase

The AA phase arising by even giving up the adiabaticity assumption is perhaps best introduced by considering first a couple of examples.

Consider the time-independent (!) Hamiltonian

$$H = \frac{\hbar \omega_0}{2} \sigma_3 \tag{8.13.38}$$

and the initial state

$$|\psi(0)\rangle = \begin{pmatrix} \cos \frac{\theta}{2} \\ \sin \frac{\theta}{2} \end{pmatrix} \tag{8.13.39}$$

which is not an eigenstates of H for general θ.

The solution of the corresponding Schrödinger equation is elementary and is given by

$$|\psi(t)\rangle = \begin{pmatrix} \cos \frac{\theta}{2} \, e^{-i\omega_0 t/2} \\ \sin \frac{\theta}{2} \, e^{i\omega_0 t/2} \end{pmatrix}. \tag{8.13.40}$$

The following expectation value is readily worked out

$$\langle \psi(t) | H | \psi(t) \rangle = \frac{\hbar \omega_0}{2} \cos \theta. \tag{8.13.41}$$

We may then introduce the dynamical phase factor by

$$\exp\left[-\frac{i}{\hbar} \int_0^t dt' \, \langle \psi(t') | H | \psi(t') \rangle \right] = \exp\left[-\frac{it}{2} \omega_0 \cos \theta \right] \tag{8.13.42}$$

and rewrite the solution in (8.13.40) explicitly as

$$|\psi(t)\rangle = e^{-(i\omega_0 t \cos \theta)/2} \, e^{-i\omega_0 t(1-\cos\theta)/2} \begin{pmatrix} \cos \frac{\theta}{2} \\ e^{it\omega_0} \sin \frac{\theta}{2} \end{pmatrix}. \tag{8.13.43}$$

Also for $2/\hbar$ times the averaged spin, we have

$$\langle \psi(t) | \boldsymbol{\sigma} | \psi(t) \rangle = \big(\sin \theta \cos \omega_0 t, \, \sin \theta \sin \omega_0 t, \, \cos \theta \big) \equiv \mathbf{n}(t). \tag{8.13.44}$$

Accordingly, for[65] $t = \tau = 2\pi/|\omega_0|$, as the parameter $\mathbf{n}(t)$ traces a *closed* curve in the parameter space, of the type shown in Figure 8.9,

---

[65] We use the notation τ rather than T in order not to confuse it with the time parameter T used in the adiabatic regime for which the latter is taken to be large.

8.13 Geometric Phases

$$|\psi(\tau)\rangle = e^{-(i\omega_0 \tau \cos\theta)/2} \, e^{-i\varepsilon\pi(1-\cos\theta)} |\psi(0)\rangle \tag{8.13.45}$$

where $\varepsilon = \text{sign}\,\omega_0$. Again in addition to the dynamical phase factor in (8.13.42), $|\psi(t)\rangle$ acquires a *geometric* phase factor, with $2\pi(1 - \cos\theta)$ denoting the solid angle, subtended at the origin of a unit sphere, swept out by $(2/\hbar)$ times the averaged spin. Here we note that there is no question of adiabaticity assumption, and the Hamiltonian in (8.13.38) is not only not parametrized by $\mathbf{n}(t)$ but is also time-independent.

As another example consider the time-dependent Hamiltonian

$$H(t) = \frac{\hbar\omega_0}{2} \mathbf{N}(t) \cdot \boldsymbol{\sigma} = \frac{\hbar\omega_0}{2} \begin{pmatrix} \cos\theta & \sin\theta\, e^{-i\omega t} \\ \sin\theta\, e^{i\omega t} & -\cos\theta \end{pmatrix} \tag{8.13.46}$$

where

$$\mathbf{N}(t) = \big(\sin\theta\cos\omega t, \sin\theta\sin\omega t, \cos\theta\big). \tag{8.13.47}$$

As in (8.1.41), (8.1.42), we define the angle α, and consider the initial state

$$|\psi(0)\rangle = \begin{pmatrix} \cos\dfrac{\alpha}{2} \\ \sin\dfrac{\alpha}{2} \end{pmatrix} \tag{8.13.48}$$

which is not an eigenstate of $H(0)$ in (8.13.46). The solution of the corresponding Schrödinger equation has been given in (8.1.43) to be

$$|\psi(t)\rangle = \begin{pmatrix} \cos\dfrac{\alpha}{2}\, e^{-i\omega t/2} \\ \sin\dfrac{\alpha}{2}\, e^{i\omega t/2} \end{pmatrix} e^{-it\Omega/2} \tag{8.13.49}$$

where $\Omega = \big(\omega^2 - 2\omega\omega_0\cos\theta + \omega_0^2\big)^{1/2}$ defined in (8.1.39).

To obtain the expression for the dynamical phase factor, we compute the expectation value

$$\langle\psi(t)|H(t)|\psi(t)\rangle = \frac{\hbar\omega_0}{2}\cos(\theta - \alpha). \tag{8.13.50}$$

Also we note that

$$\langle\psi(t)|\boldsymbol{\sigma}|\psi(t)\rangle = \big(\sin\alpha\cos\omega t, \sin\alpha\sin\omega t, \cos\alpha\big) \equiv \mathbf{n}(t). \tag{8.13.51}$$

Hence we may rewrite (8.13.49) as

$$|\psi(t)\rangle = \exp\left[-\frac{i}{2}\omega_0 t \cos(\theta-\alpha)\right] \exp\left[-\frac{i}{2}\omega t(1-\cos\alpha)\right] \begin{pmatrix} \cos\dfrac{\alpha}{2} \\ \sin\dfrac{\alpha}{2}\, e^{i\omega t} \end{pmatrix} \tag{8.13.52}$$

where we have used the fact that

$$\omega - \omega_0 \cos(\theta - \alpha) + \Omega = \omega(1 - \cos\alpha) \qquad (8.13.53)$$

(see (8.1.41), (8.1.42)).

For $t = \tau = 2\pi/\omega$, as the vector parameter $\mathbf{n}(t)$ traces a closed curve in the parameter space, we may write from (8.13.52)

$$\langle\psi(0)|\psi(\tau)\rangle = \exp\left(-\frac{i}{2}\omega_0\tau\cos(\theta-\alpha)\right)\exp\left(-i\pi\omega(1-\cos\alpha)\right) \qquad (8.13.54)$$

recognizing the geometric phase factor as the second factor on the right-hand side of (8.13.54).

More generally, suppose that a normalized state $|\psi(t)\rangle$ evolving according to the Schrödinger equation

$$i\hbar\frac{d}{dt}|\psi(t)\rangle = H(t)|\psi(t)\rangle \qquad (8.13.55)$$

is such that

$$|\psi(\tau)\rangle = e^{i\alpha(\tau)}|\psi(0)\rangle \qquad (8.13.56)$$

for some τ, where $\exp(i\alpha(\tau))$ is a phase factor. We denote the closed curve $\Pi\{|\psi(t)\rangle : 0 \leqslant t \leqslant \tau\}$ by C, where Π is the projection map which maps each (non-zero) vector to the ray on which it lies.

Needless to say, for $t \neq \tau$, in the neighborhood of τ, $|\psi(t)\rangle$ is not necessarily related to $|\psi(0)\rangle$ by just a phase factor. Accordingly, before using the Schrödinger equation to determine $\alpha(\tau)$) in (8.13.56), we carry out a phase transformation

$$|\psi(t)\rangle \longrightarrow e^{-if(t)}|\psi(t)\rangle \equiv \chi(t) \qquad (8.13.57)$$

such that

$$\langle\chi(0)|\chi(\tau)\rangle = 1 \qquad (8.13.58)$$

since

$$|\chi(0)\rangle = e^{-if(0)}|\psi(0)\rangle, \quad |\chi(\tau)\rangle \equiv e^{-if(\tau)}|\psi(\tau)\rangle \qquad (8.13.59)$$

(8.13.56) gives (up to additional integer multiples of 2π)

$$\alpha(\tau) = f(\tau) - f(0). \qquad (8.13.60)$$

Upon substitution of $|\chi(t)\rangle$, given in (8.13.57), in the Schrödinger equation (8.13.55) provides the simple expression

$$\dot{f}(t) = -\frac{1}{\hbar}\langle\psi(t)|H(t)|\psi(t)\rangle + i\langle\chi(t)|\dot{\chi}(t)\rangle \qquad (8.13.61)$$

which upon integration, and using (8.13.60), gives

$$\langle\psi(0)|\psi(\tau)\rangle = \exp\left(-\frac{i}{\hbar}\int_0^\tau dt\,\langle\psi(t)|H(t)|\psi(t)\rangle\right)\exp\left(i\int_0^\tau dt\,\langle\chi(t)|i\dot\chi(t)\rangle\right).$$
(8.13.62)

To investigate the nature of the last phase factor, in addition to the dynamical one

$$\exp\left(-\frac{i}{\hbar}\int_0^\tau dt\,\langle\psi(t)|H(t)|\psi(t)\rangle\right) \qquad (8.13.63)$$

we define the state vector

$$|\Phi(t)\rangle = \exp\left(\frac{i}{\hbar}\int_0^t dt'\,\langle\psi(t')|H(t')|\psi(t')\rangle\right)|\psi(t)\rangle \qquad (8.13.64)$$

thus removing the dynamical phase factor, leading to

$$\langle\Phi(0)|\Phi(\tau)\rangle = \exp\left(i\int_0^\tau dt\,\langle\chi(t)|i\dot\chi(t)\rangle\right) \qquad (8.13.65)$$

we note that due to the normalizability condition $\langle\chi(t)|\chi(t)\rangle = 1$, $\langle\chi(t)|i\dot\chi(t)\rangle$ is real, hence the expression of the right-hand side of (8.13.65) is a phase factor.

Upon substitution of the expression (8.13.64) in the Schrödinger equation (8.13.55), to obtain the resulting equation for $|\dot\Phi(t)\rangle$, and taking the inner product of the latter with $|\Phi(t)\rangle$, we obtain the simple result that

$$\langle\Phi(t)|\dot\Phi(t)\rangle = 0. \qquad (8.13.66)$$

This expression is of some significance.[66] It states, in particular, that for two neighboring points t, $t + \Delta t$, $\Delta t \simeq 0$

$$\langle\Phi(t)|\Phi(t+\Delta t)\rangle$$
$$= \langle\Phi(t)|\Phi(t)\rangle\left\{1 - \frac{(\Delta t)^2}{2}\left[\langle\dot\chi(t)|\dot\chi(t)\rangle - \langle\dot\chi(t)|\chi(t)\rangle\langle\chi(t)|\dot\chi(t)\rangle\right]\right\}$$
(8.13.67)

as is readily obtained from (8.13.57), (8.13.64), (8.13.61), where we have also used the normalizability of $|\Phi(t)\rangle$. We note that the second term in the curly brackets in (8.13.67) is not only of second order in Δt, due to (8.13.66), but is also real, and for $\Delta t \simeq 0$, the entire expression in (8.13.67) is real and positive.

Equation (8.13.67) means, in particular, that although, for infinitesimal $\Delta t \simeq 0$, $\langle\Phi(t)|\Phi(t+\Delta t)\rangle$ does not develop a phase factor, to first (not even to second) order in Δt, as we move from t to $t + \Delta t$, a *net* phase change arises, in general, according to (8.13.65), at the end of the journey for finite

---

[66] Simon (1983).

τ as we move from $t = 0$ to $t = \tau$. Such a system, with a net phase change, is said to be *non-holonomic*, and the phase change is known as an *anholonomy*.

To gain further insight into the nature of the phase acquired in (8.13.65), we use an idea[67] of Pancharatnam[68] to study the phase relationship of two (non-orthogonal) vectors $|\Phi_1\rangle, |\Phi_2\rangle$ by considering their interference effect by the norm squared of their superposition

$$\||\Phi_1\rangle + |\Phi_2\rangle\|^2 = \langle\Phi_1|\Phi_1\rangle + \langle\Phi_2|\Phi_2\rangle + 2|\langle\Phi_1|\Phi_2\rangle|\cos\delta_{12} \quad (8.13.68)$$

where

$$\langle\Phi_1|\Phi_2\rangle = |\langle\Phi_1|\Phi_2\rangle|\,e^{i\delta_{12}}. \quad (8.13.69)$$

The vectors $|\Phi_1\rangle, |\Phi_2\rangle$ are then said to be "in phase" when $\delta_{12} = 0$ giving a maximum for the norm squared in (8.13.68) for which is real and positive.

In the light of the above definition, we note that for two neighboring points $t, t + \Delta t, \Delta t \simeq 0$, we have from (8.13.67),

$$\||\Phi(t + \Delta t)\rangle + |\Phi(t)\rangle\|^2 = 2 + 2|\langle\Phi(t + \Delta t)|\Phi(t)\rangle| \quad (8.13.70)$$

implying that $|\Phi(t + \Delta t)\rangle, |\Phi(t)\rangle$ are "in phase", or that the vector $|\Phi(t)\rangle$ is "parallel transported" to itself within the infinitesimal intervals $[t, t + \Delta t]$ for $\Delta t \simeq 0$. In spite of this "parallel transport" in infinitesimal steps, for the full journey from $t = 0$ to $t = \tau$, however, we have from (8.13.65),

$$\||\Phi(\tau)\rangle + |\Phi(0)\rangle\|^2 = 2 + 2\cos\left(\int_0^\tau dt'\,\langle\chi(t')|i\dot\chi(t')\rangle\right)$$

$$= 2 + 2|\langle\Phi(\tau)|\Phi(0)\rangle|\cos\left(\int_0^\tau dt'\,\langle\chi(t')|i\dot\chi(t')\rangle\right) \quad (8.13.71)$$

implying a net phase change. As a matter of fact it is easy to show that if vectors $|\Phi_1\rangle, |\Phi_2\rangle$ are "in phase" and so are $|\Phi_2\rangle, |\Phi_3\rangle$, then $|\Phi_1\rangle, |\Phi_3\rangle$ are not necessarily "in phase" (see Problem 8.33).

Now suppose that the state $|\chi(t)\rangle$ may be parameterized by a vector $\mathbf{X}(t) = (X^1(t), \ldots, X^N(t)) : |\chi(t)\rangle = |\chi(\mathbf{X}(t))\rangle$. Then by using (8.13.67), we have for the distance squared between the states $|\Phi(t + dt)\rangle$ and $|\Phi(t)\rangle$

$$\||\Phi(t + dt)\rangle - |\Phi(t)\rangle\|^2 = \left(\langle\nabla_a\chi|\nabla_b\chi\rangle - \langle\nabla_a\chi|\chi\rangle\langle\chi|\nabla_b\chi\rangle\right)dX^a dX^b \quad (8.13.72)$$

with sums over $a, b = 1, \ldots, N$, and $\nabla_a = \partial/\partial X^a$. Hence we may define a metric, up to an overall scale factor, in parameter space by[69]

---

[67] Samuel and Bhandari (1988).
[68] Pancharatnam (1956). This author applied this idea in optics, and was successfully extended to quantum mechanics by the Samuel and Bhandari.
[69] See also Provost and Vallee (1980).

$$g_{ab} = \langle \nabla_a \chi | \nabla_b \chi \rangle - \langle \nabla_a \chi | \chi \rangle \langle \chi | \nabla_b \chi \rangle \tag{8.13.73}$$

signalling, in general, the *curved* nature of the parameter space.

The AA phase factor, appearing on the right-hand side of (8.13.65), denoted by $\exp(i\gamma_{AA}(C))$, may be rewritten as

$$\exp(i\gamma_{AA}(C)) = \exp\left(i \oint_C A_a \mathrm{d}X^a\right) \tag{8.13.74}$$

where

$$A_a = \langle \chi | i\nabla_a \chi \rangle. \tag{8.13.75}$$

We note that $\gamma_{AA}(C)$ is independent of how fast the different parts of the path defined by C is traversed.

Under a phase transformation

$$|\chi(\mathbf{X})\rangle \longrightarrow e^{i\alpha(\mathbf{X})} |\chi(\mathbf{X})\rangle \tag{8.13.76}$$

the A_a transform as

$$A_a \longrightarrow A_a - \nabla_a \alpha(\mathbf{X}) \tag{8.13.77}$$

and we note that[70] the metric in (8.13.73) may be rewritten as

$$g_{ab} = \langle (\nabla_a + iA_a)\chi | (\nabla_b + iA_a)\chi \rangle \tag{8.13.78}$$

and is obviously invariant under the transformations in (8.13.76).

As an illustration of the various concepts introduced above, consider the Schrödinger equation

$$i\hbar \frac{\mathrm{d}}{\mathrm{d}t} |\psi(t)\rangle = H(t) |\psi(t)\rangle, \quad H(t) = \frac{\hbar}{2} \mathbf{K}(t) \cdot \boldsymbol{\sigma} \tag{8.13.79}$$

where $\mathbf{K}(t)$ is an external time-dependent vector, $\mathbf{n}(t)$ is a unit vector satisfying the equation

$$\dot{\mathbf{n}}(t) = \mathbf{K}(t) \times \mathbf{n}(t) \tag{8.13.80}$$

with an initially prepared state as an eigenstate of $\mathbf{n}(0) \cdot \boldsymbol{\sigma}$:

$$\mathbf{n}(0) \cdot \boldsymbol{\sigma} |\psi(0)\rangle = |\psi(0)\rangle. \tag{8.13.81}$$

Then the solution of (8.13.79) satisfies the eigenvalue equation (§8.1, (8.1.44)–(8.1.51)),

$$\mathbf{n}(t) \cdot \boldsymbol{\sigma} |\psi(t)\rangle = |\psi(t)\rangle. \tag{8.13.82}$$

The unit vector may be parametrized as

$$\mathbf{n}(t) = (\sin\theta(t)\cos\phi(t), \sin\theta(t)\sin\phi(t), \cos\theta(t)). \tag{8.13.83}$$

---

[70] See also Samuel and Bhandari (1988).

526   8 Quantum Physics of Spin 1/2 & Two-Level Systems

Suppose that $\mathbf{n}(t)$ traces some closed curve on the unit sphere (see Figure 8.10) as it moves from $t = 0$ to $t = \tau$ with

$$\theta(\tau) = \theta(0), \quad \phi(\tau) = \phi(0) + 2\pi. \tag{8.13.84}$$

With the cyclic evolution just described, the state vector $|\chi(t)\rangle$ in (8.13.57), may be represented by

$$|\chi(t)\rangle = \begin{pmatrix} \cos\left(\frac{\theta(t)}{2}\right) \\ \sin\left(\frac{\theta(t)}{2}\right) e^{i\phi(t)} \end{pmatrix} \tag{8.13.85}$$

which deviously satisfies (8.13.82) *and* (8.13.58).

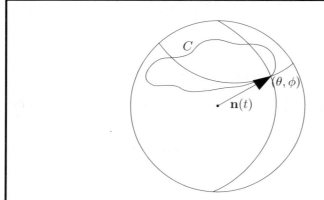

Fig. 8.10. A closed curve traced by the unit vector $\mathbf{n}$ in (8.13.83) over the unit sphere. The latter is referred to as the Poincaré sphere. [The corresponding sphere in optics involving polarization vectors is referred to as Stokes's sphere.]

From (8.13.83), (8.13.57) it is easily verified that

$$\langle \psi(t)|\boldsymbol{\sigma}|\psi(t)\rangle = \mathbf{n}(t) \tag{8.13.86}$$

and hence the dynamical phase factor in (8.13.63) for the system may be written as

$$\exp\left(-\frac{i}{2}\int_0^\tau dt\, \mathbf{K}(t)\cdot\mathbf{n}(t)\right). \tag{8.13.87}$$

On the other hand,

$$\langle \chi(t)|i\dot{\chi}(t)\rangle = -\frac{1}{2}(1 - \cos\theta(t))\dot{\phi}(t) \tag{8.13.88}$$

giving the AA phase factor

$$\exp\left(-\frac{i}{2}\int_0^T dt(1-\cos\theta(t))\dot\phi(t)\right) = \exp\left(-\frac{i}{2}\oint(1-\cos X^1)dX^2\right) \tag{8.13.89}$$

where $X^1 = \theta$, $X^2 = \phi$. This phase factor is equivalently rewritten as

$$\exp i\oint_C \mathbf{A}\cdot d\hat{\mathbf{r}} \tag{8.13.90}$$

where $\mathbf{A}$ in the latter may be taken to be

$$\mathbf{A} = -\frac{\sin\theta}{2(1+\cos\theta)}\hat{\boldsymbol{\phi}}$$

$$= \left(\frac{n_2}{2(1+n_3)}, -\frac{n_1}{2(1+n_3)}, 0\right) \tag{8.13.91}$$

and $\hat{\boldsymbol{\phi}}$ is a unit vector in the direction of increasing ϕ,

$$d\hat{\mathbf{r}} = \hat{\boldsymbol{\theta}}d\theta + \hat{\boldsymbol{\phi}}\sin\theta d\phi, \tag{8.13.92}$$

$\hat{\boldsymbol{\theta}}$ is a unit vector in the direction of increasing θ.

In reference to (8.13.72)/(8.13.73) we observe that

$$\||\Phi(t+dt)\rangle - |\Phi(t)\rangle\|^2 = \frac{1}{4}\left(\dot\theta^2 + \sin^2\theta\dot\phi^2\right)(dt)^2$$

$$= \frac{1}{4}\left((d\theta)^2 + \sin^2\theta(d\phi)^2\right) \tag{8.13.93}$$

and here we recognize the familiar metric defined on the unit sphere:[71]

---

[71] A reader who has, for example, studied relativity has undoubtedly came across the concept of the parallel transport of a vector along a curve (in curved space) which states that the covariant derivative of its components are zero. In curved space, the ordinary derivative of the component of a vector does not transform as the component of a vector, and the covariant derivative is so defined to ensure that the resulting expression does transform as a vector component. In the present case, with motion restricted to the surface of the unit sphere, with basis vectors $\hat{\boldsymbol{\theta}}$, $\hat{\boldsymbol{\phi}}\sin\theta$, spanning the tangent plane, at a point (θ,ϕ) on the sphere, the covariant derivative of the components of a vector, relative to such a basis, read

$$\frac{d}{dt}V^1 - \sin\theta\cos\theta\,\dot\phi V^2 = 0, \qquad \frac{d}{dt}V^2 + \cot\theta\left(V^2\dot\theta + V^1\dot\phi\right) = 0$$

(cf. Misner et al. (1973), p. 340). These equations may be combined as

$$\frac{d}{dt}\left(V^1 - i\sin\theta\,V^2\right) = i\cos\theta\,\dot\phi\left(V^1 - i\sin\theta\,V^2\right)$$

$$(g_{ab}) = \begin{pmatrix} 1 & 0 \\ 0 & \sin^2\theta \end{pmatrix}, \quad X^1 = \theta, \quad X^2 = \phi \tag{8.13.94}$$

up to the scale factor of $1/4$ in (8.13.93). [We note that we could have instead considered the curve traced by the vector $\mathbf{n}(t)/2$ on the surface of a sphere of radius $1/2$.]

The curl of $\mathbf{A}$ in (8.13.91) gives

$$\nabla \times \mathbf{A} = -\frac{1}{2}\hat{\mathbf{r}} = -\frac{1}{2}\frac{\hat{\mathbf{r}}}{|\hat{\mathbf{r}}|^2} \equiv \mathbf{B} \tag{8.13.95}$$

which may be interpreted as the magnetic field on the surface of the unit sphere at $\hat{\mathbf{r}}$ due to a magnetic monopole of strength $1/2$ located at the origin.[72]

The vector $\mathbf{A}$ in (8.13.91) may be rewritten as

$$\mathbf{A} = -\frac{(1-\cos\theta)}{2\sin\theta}\hat{\boldsymbol{\phi}}, \quad 0 \leqslant \theta < \pi \tag{8.13.96}$$

which develops a line of singularities for $\theta \to \pi$ corresponding to the negative z-axis. [Such a line of singularities is referred to as a Dirac string.] We could have equally chosen the vector

$$\mathbf{A}' = \frac{(1+\cos\theta)}{2\sin\theta}\hat{\boldsymbol{\phi}}, \quad 0 < \theta \leqslant \pi \tag{8.13.97}$$

in (8.13.90) which has a well defined limit for $\theta \to \pi$, but develops a line of singularities for $\theta \to 0$, corresponding to the positive z-axis. The vectors

---

giving the solution

$$\left(V^1 - i\sin\theta\, V^2\right)(\tau) = \left(V^1 - i\sin\theta\, V^2\right)(0)\,\exp\left[i\int_0^\tau \cos\theta\, \dot{\phi}\, dt\right]$$

$$= \left(V^1 - i\sin\theta\, V^2\right)(0)\,\exp\left[-i\oint(1-\cos\theta)\,d\phi\right].$$

Here we recognize (the square of) the geometric phase in (8.13.89), arising as a consequence of the curvature of the surface of the sphere, associated with the vector. The extra $1/2$ factor in the phase in (8.13.89) arises due to the spinor nature of $|\Phi\rangle$.

[72] For $\theta(t) = \pi/2$, the phase in the phase factor in (8.13.89) simply becomes $-\pi$. One may argue that by equating this phase with the one formally generated in the wavefunction of a particle of charge q satisfying Schrödinger's equation arising from the minimal coupling $\mathbf{p} \to \mathbf{p} - q\mathbf{A}/c$, due to a magnetic monopole of strength g: $-q\oint \mathbf{A}\cdot d\hat{\mathbf{r}}/\hbar c = -2\pi qg/\hbar c$, gives $q = \hbar cN/2g$, where $N = 1$ for the initial condition chosen for the wavefunction. One may then formally argue that by generalizing this to arbitrary spins, that this equality becomes $q = \hbar cN/2g$, where $N = 0, \pm 1, \pm 2, \ldots$, implying, in particular, the quantization of the electric charge (Dirac (1931)) consistent with the experimental observation. See also Aitchison (1987). The study of the geometrical phase for arbitrary spins is not straightforward, see Lin (2002).

A, **A**′ are well defined in the region $\varepsilon < \theta < \pi - \varepsilon$, for $\varepsilon > 0$ and small, and are related by a gauge transformation (see Problem 8.35). Clearly, that a singularity must exist for a vector **A** in any gauge is expected since one would otherwise obtain $\nabla \times \mathbf{A} = \mathbf{0}$ in contradiction with the presence of a monopole at the origin of the sphere in the formalism.

The AA geometric phase, as a generalization of the Berry phase, has been observed experimentally[73] for various cyclic cases.

We close this section by studying the generation of a geometric by relaxing the condition of cyclicity[74] in addition to relaxing the condition of adiabaticity already discussed.

8.13.4 Samuel-Bhandari (SB) Phase

Consider the general Hamiltonian for spin 1/2 in (8.13.79). Now suppose that $\mathbf{n}(t)$ in (8.13.83) traces an open curve C_0, instead of a closed curve as in the AA phase, starting from $t = 0$ and ending at $t = \tau_0$ (see Figure 8.11), where we let $\theta(0) = \theta_1$, $\phi(0) = \phi_1$, $\theta(\tau_0) = \theta_2$, $\phi(\tau_0) = \phi_2$.

Let $U(\tau_0, 0)$ denote the unitary operator which takes the initial state $|\Phi(0)\rangle$, as an eigenstate of $\mathbf{n}(0) \cdot \boldsymbol{\sigma}$, with the dynamical phase removed, to the state $|\Phi(\tau_0)\rangle$. As before, we are interested in the expression

$$\langle \Phi(0)|\Phi(\tau_0)\rangle = \langle \Phi(0)|U(\tau_0, \tau)|\Phi(0)\rangle \qquad (8.13.98)$$

but now involving *non-cyclic* evolution as just described, for the non-trivial case where the amplitude in (8.13.98) is *non-zero*.

The SB phase is obtained as follows. The basic idea is to rewrite[75] (8.13.98) as

$$\langle \Phi(0)|\Phi(\tau_0)\rangle = \left\langle \Phi(0)|U_G^\dagger(\tau, \tau_0)U_G(\tau, \tau_0)U(\tau_0, 0)|\Phi(0)\right\rangle \qquad (8.13.99)$$

where $U_G(\tau, \tau_0)$ is the unitary operator which is responsible for joining the end point (θ_2, ϕ_2) to the initial point (θ_1, ϕ_1) by the shortest path (i.e., the geodesic[76] on the surface of the sphere. The corresponding curve is denoted by G in Figure 8.11.

---

[73] Suter *et al.* (1988).
[74] This remarkable result was observed by Samuel and Bhandari (1988) as an extension of earlier work of Pancharatnam (1956), in optics, to quantum mechanics.
[75] Zhu *et al.* (2000).
[76] The geodesic equation passing through the points (θ_2, ϕ_2) and (θ_1, ϕ_1) is given by

$$\cot \theta = A \cos \phi + B \sin \phi$$

where

$$A = (\sin \phi_1 \cot \theta_2 - \sin \phi_2 \cot \theta_1)/ \sin(\phi_1 - \phi_2)$$
$$B = (\cos \phi_2 \cot \theta_1 - \cos \phi_1 \cot \theta_2)/ \sin(\phi_1 - \phi_2).$$

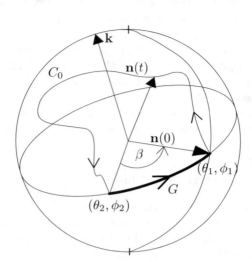

Fig. 8.11. The vector $\mathbf{n}(t)$ traces the open curve C_0 on the surface of the unit sphere. The shortest path (the geodesic) joining the point (θ_2, ϕ_2) to (θ_1, ϕ_1) is denoted by G. The unit vector $\mathbf{k}$ is perpendicular to $\mathbf{n}(0)$.

From the earlier AA phase analysis, we have

$$U_G(\tau, \tau_0) U(\tau_0, 0) |\Phi(0)\rangle = \exp\left(i\gamma_{\text{AA}}(C_0 U G)\right) |\Phi(0)\rangle \qquad (8.13.100)$$

where $\gamma_{\text{AA}}(C_0 U G)$ is now given by (see (8.13.90), (8.13.91))

$$\gamma_{\text{AA}}(C_0 U G) = \int_{C_0} \mathbf{A} \cdot d\hat{\mathbf{r}} + \int_G \mathbf{A} \cdot d\hat{\mathbf{r}}. \qquad (8.13.101)$$

That is,

$$\langle\Phi(0)|\Phi(\tau_0)\rangle = \exp\left(i\gamma_{\text{AA}}(C_0 U G)\right) \left\langle \Phi(0) | U_G^\dagger(\tau, \tau_0) | \Phi(0) \right\rangle. \qquad (8.13.102)$$

Now all one has to do in order to establish that a geometric phase arises even for the non-cyclic evolution considered is to show that

$$\left\langle \Phi(0) | U_G^\dagger(\tau, \tau_0) | \Phi(0) \right\rangle \qquad (8.13.103)$$

is real and positive. Once this is established, one may then infer that the SB phase is obtained by closing the open curve C_0 by the geodesic joining the end points of the curve C_0 and calculate it from the AA phase for the newly constructed closed curve.

By definition of the geodesic, $U_G(\tau,\tau_0)$ is given by the now familiar unitary operator of rotation by some angle β as shown in Figure 8.11. (see §2.8)[77] given by

$$U_G(\tau,\tau_0) = \cos\frac{\beta}{2} + i\mathbf{k}\cdot\boldsymbol{\sigma}\sin\frac{\beta}{2} \qquad (8.13.104)$$

where $\mathbf{k}$ is a constant unit vector perpendicular to $\mathbf{n}(0)$. But

$$\langle\Phi(0)|\boldsymbol{\sigma}|\Phi(0)\rangle = \mathbf{n}(0) \qquad (8.13.105)$$

hence the second term on the right-hand side of (8.13.104) does not contribute in computing the matrix element in (8.13.103) since $\mathbf{k}\cdot\mathbf{n}(0) = 0$. Thus we may rewrite (8.13.102) as

$$\langle\Phi(0)|\Phi(\tau_0)\rangle = \cos\frac{\beta}{2}\exp\left(i\gamma_{AA}(C_0 U G)\right) \qquad (8.13.106)$$

for the non-cyclic evolution. For the non-trivial case considered, where $\langle\Phi(0)|\Phi(\tau_0)\rangle$ is non-zero, and for the shortest path joining the end points of C_0, $|\beta| < \pi$ which completes the demonstration of the reality and positivity of the matrix element in (8.13.103).[78]

8.14 Quantum Dynamics of the Stern-Gerlach Effect

8.14.1 The Quantum Dynamics

In the present section, we provide an analytical dynamical treatment of the Stern-Gerlach (S-G) effect which is:

(1) *quantum* mechanical, as it should be, and takes into account,
(2) the *field equation* $\boldsymbol{\nabla}\cdot\mathbf{B} = 0$, where $\mathbf{B}$ is the magnetic field in the problem,
(3) the quantum counterpart of the *Lorentz force*,
(4) the *two*, rather than one, dimensional aspect of the beam hitting the observation screen,
(5) the rather non-trivial *correlations* that occur between dynamical variables, as will be seen to exist, describing the intensity distribution on the screen.

A theoretical analysis of the effect will be given below which takes into account *all* of the above five points just listed.[79] We will see that an analytical dynamical treatment to the leading order $|e|/\sqrt{\hbar c} \equiv \sqrt{\alpha}$ in for the electron, where α is the fine-structure constant, and for spin $1/2$ charged particles (e.g.,

---

[77] See also Berry (1987).

[78] For additional details on the subject of this section, one may refer to: Shapere and Wilczek (1989). Although this reprint volume was assembled way over a decade ago, it remains a useful reference source.

[79] This analysis is based on, Manoukian and Rotjanakusol (2003).

the proton), in general, leads to a unitary, i.e., positive definite, expression for the probability intensity distribution on the observation screen, where the magnetic field has a controllable uniform component along the initial average direction of propagation of the particle, in addition to a non-uniform, almost longitudinal, magnetic field lying in the plane defined by the quantization axis, in question of the spin, and the initial average direction of propagation.

With an initially prepared Gaussian wavepacket, the analysis leads to a sum of so-called bivariate normal distributions for the probability intensity distribution with *non-zero* correlations.[80] The uniform longitudinal controllable magnetic field, as will be seen, has a dual role. Although longitudinal, it reduces effectively the quantum Lorentz force contribution by reducing, in turn, the correlation between the dynamical variables describing the probability density of observation, and also provides a positive definite expression for the latter. We will also see that the analysis applies to neutral particles as well.

The importance of the consideration of the S-G effect for the electron itself is evident. To this end, it is worth recalling the statement made by Albert Einstein: "*We know, it would be sufficient to really understand the electron*".[81] The difficulty in carrying out a S-G experiment for the electron, as such a basic experiment of quantum physics, with a conventional purely (non-uniform) transverse magnetic field has been well documented in the literature.[82,83]

An obstacle in carrying out such an experiment with a standard transverse magnetic field of Stern and Gerlach, is that the Lorentz force, in the classic apparatus, causes an obvious deviation of the particle from its initial path thus leading to a blurring of the expected splitting of the beam. Because of this, the feasibility of performing the experiment with a longitudinal non-uniform magnetic field was suggested many years ago[84] and also more recently.[85] Another aspect of a transversal magnetic field is that a non-uniform magnetic field perpendicular to the non-uniform component along the quantization axis of the spin, as demanded by the field equation $\nabla \cdot \mathbf{B} = 0$, tends to cause, in general, a further splitting of the beam in a direction perpendicular to the quantization axis as well.

We consider the Pauli-Hamiltonian in (8.2.4), written the form

$$H = \frac{\left(\mathbf{p} - \frac{q}{c}\mathbf{A}\right)^2}{2M} - \boldsymbol{\mu} \cdot \mathbf{B} \qquad (8.14.1)$$

with

---

[80] Cf. Manoukian (1986c), pp. 127, 129.
[81] As quoted in: Rabi (1988).
[82] Pauli (1964). In Wheeler and Zurek (1983), p. 701. Dehmelt (1990).
[83] Batelaan *et al.* (1997); Gallup *et al.* (2001). See also Conte *et al.* (1995).
[84] Brillouin (1928).
[85] Batelaan *et al.* (1997). See also these papers for the historical development in studying the S-G effect and for their analyses of the problem.

8.14 Quantum Dynamics of the Stern-Gerlach Effect

$$\mathbf{B} = \nabla \times \mathbf{A}, \quad \boldsymbol{\mu} = \mu\boldsymbol{\sigma}, \quad \mu = \frac{q\hbar}{4Mc}g \qquad (8.14.2)$$

and g is taken to be arbitrary, where $g \simeq 2$ for the electron, and, e.g., $g = 5.59$ for the proton. We choose the vector potential $\mathbf{A}$ in the Coulomb gauge, i.e.,

$$\nabla \cdot \mathbf{A} = 0. \qquad (8.14.3)$$

In terms of the dimensionless parameter

$$\alpha_q = |q|^2/\hbar c, \quad \varepsilon(q) = \text{sign}(q) \qquad (8.14.4)$$

the interaction Hamiltonian in (8.14.1) may be written as

$$H_{\mathrm{I}} = \varepsilon(q)\sqrt{\alpha_q}\left[-\sqrt{\frac{\hbar}{c}}\mathbf{A}\cdot\mathbf{p} + \varepsilon(q)\sqrt{\alpha_q}\frac{\hbar}{2Mc}\mathbf{A}^2 - \frac{1}{4M}\sqrt{\frac{\hbar^3}{c}}g\boldsymbol{\sigma}\cdot\mathbf{B}\right]. \qquad (8.14.5)$$

For the electron $q = -|e|$, α is the fine-structure constant.

For the initial wavepacket at time $t = 0$, in the $\mathbf{x}$-description, we take the Gaussian type

$$\Psi_0(\mathbf{x}) = \frac{1}{(2\pi)^{3/4}\gamma^{3/2}}\exp\left(\frac{i}{\hbar}\mathbf{p}_0\cdot\mathbf{x}\right)\exp\left(-\frac{\mathbf{x}^2}{4\gamma^2}\right) \qquad (8.14.6)$$

where we use the notation γ^2 for the variance in order not to confuse it with the Pauli matrices, and

$$\mathbf{p}_0 = (0, p_0, 0). \qquad (8.14.7)$$

Here the x_2-axis denotes the initial average direction of propagation of the particle (see Figure 8.12).

In the absence of a magnetic field $\Psi_0(\mathbf{x})$ in (8.14.6) develops in time to

$$\Psi_0(\mathbf{x},t) = \frac{e^{i\mathbf{p}_0\cdot\mathbf{x}/\hbar}e^{-i p_0^2 t/2M\hbar}}{(2\pi)^{3/4}\gamma^{3/2}\left(1+\frac{i\hbar t}{2M\gamma^2}\right)^{3/2}}\exp\left(-\frac{\left(\mathbf{x}-\frac{\mathbf{p}_0}{M}t\right)^2}{4\gamma^2\left(1+\frac{i\hbar t}{2M\gamma^2}\right)}\right) \qquad (8.14.8)$$

and

$$|\Psi_0(\mathbf{x})|^2 = \frac{1}{(2\pi)^{3/2}\gamma^3(t)}\exp\left(-\frac{\left(\mathbf{x}-\frac{\mathbf{p}_0}{M}t\right)^2}{2\gamma^2(t)}\right) \qquad (8.14.9)$$

$$\gamma(t) = \gamma\left(1 + \frac{\hbar^2 t^2}{4M\gamma^4}\right)^{1/2}. \qquad (8.14.10)$$

We have chosen a common γ-width in all directions to simplify the grouping together of the various terms in the analysis. This is not a serious restriction.

For the magnetic field we choose the simple form

$$\mathbf{B} = (0, b - \beta x_2, \beta x_3), \quad \mathbf{A} = (bx_3, \beta x_1 x_3, \beta x_1 x_2) \qquad (8.14.11)$$

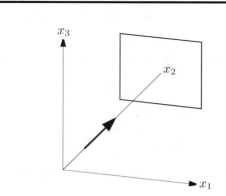

Fig. 8.12. The x_2-axis denotes the initial average direction of propagation of the particle with the observation screen being parallel to the x_1-x_3 plane. No magnetic field component is chosen along the x_1-axis.

satisfying (8.14.3), (8.14.2), where b, β are some constants. We note that the longitudinal component $(b - \beta x_2)$ is along the initial average direction of propagation specified by $\mathbf{p}_0$, with Ox_3 denoting the traditional quantization axis of the spin.

We will see that the uniform part $(0, b, 0)$ of the magnetic field, although longitudinal, may be appropriately set up to effectively reduce the quantum mechanical counterpart of the Lorentz-force contribution by reducing, in turn, the correlation that occurs between x_1, and x_3 variables on the screen, and also provide a positive definite expression for the probability distribution in question. Concerning the non-uniform part $(0, \beta x_2, \beta x_3)$, we note that since $|x_2|$, a macroscopic distance, is much larger than $|x_3|$ (providing a measure of the splitting of the beam), this non-uniform magnetic field is almost longitudinal along the direction of propagation, at the screen. In the above set up, as a working hypothesis, we treat the particles as if they are throughout in the magnetic field. Otherwise an analytical treatment is not so manageable.

The dynamics is most elegantly described in terms of the density operator, which at $t = 0$, is given by

$$\rho = w_+ \begin{pmatrix} 1 \\ 0 \end{pmatrix} |\Psi_0\rangle \langle\Psi_0| \begin{pmatrix} 1 & 0 \end{pmatrix} + w_- \begin{pmatrix} 0 \\ 1 \end{pmatrix} |\Psi_0\rangle \langle\Psi_0| \begin{pmatrix} 0 & 1 \end{pmatrix} \qquad (8.14.12)$$

where

$$w_+ + w_- = 1. \qquad (8.14.13)$$

For

$$w_+ = w_- = 1/2 \qquad (8.14.14)$$

one would be dealing with an unpolarized beam. For $t > 0$, the density operator is given by

8.14 Quantum Dynamics of the Stern-Gerlach Effect

$$\rho(t) = w_+ e^{-itH/\hbar} \begin{pmatrix} 1 \\ 0 \end{pmatrix} |\Psi_0\rangle \langle\Psi_0| \begin{pmatrix} 1 & 0 \end{pmatrix} e^{itH/\hbar}$$

$$+ w_- e^{-itH/\hbar} \begin{pmatrix} 0 \\ 1 \end{pmatrix} |\Psi_0\rangle \langle\Psi_0| \begin{pmatrix} 0 & 1 \end{pmatrix} e^{itH/\hbar}. \qquad (8.14.15)$$

The probability density is then (for $t > 0$)

$$\langle \mathbf{x}|\rho(t)|\mathbf{x}\rangle \qquad (8.14.16)$$

and for the probability density, in question, on the screen one may then, most conveniently, write it as

$$f(x_1, x_3; t) = \int_{-\infty}^{\infty} dx_2 \, \langle \mathbf{x}|\rho(t)|\mathbf{x}\rangle$$

$$= w_+ \int_{-\infty}^{\infty} dx_2 \left| \langle \mathbf{x}| e^{-itH/\hbar} \begin{pmatrix} 1 \\ 0 \end{pmatrix} |\Psi\rangle \right|^2$$

$$+ w_- \int_{-\infty}^{\infty} dx_2 \left| \langle \mathbf{x}| e^{-itH/\hbar} \begin{pmatrix} 0 \\ 1 \end{pmatrix} |\Psi\rangle \right|^2$$

$$\equiv w_+ f_+(x_1, x_3; t) + w_- f_-(x_1, x_3; t). \qquad (8.14.17)$$

8.14.2 The Intensity Distribution

With $\exp(-itH/\hbar)$ as the time-evolution operator, the following expectation values of the Heisenberg operators in the state (8.14.6), relevant to the observation screen, to the leading order in $\sqrt{\alpha_q}$, are readily obtained:

$$\langle x_1(t) \rangle = 0 \qquad (8.14.18)$$

$$\langle x_3(t) \rangle = \frac{\mu}{2M} \sigma_3 \beta t^2 \qquad (8.14.19)$$

and the important non-trivial correlation occurring between the dynamical variables $x_1(t)$, $x_3(t)$:

$$\left\langle \left(x_1(t) - \langle x_1(t)\rangle\right)\left(x_3(t) - \langle x_3(t)\rangle\right) \right\rangle = -\frac{qbt\gamma^2}{Mc} + \frac{q\beta p_0 t^2 \gamma^2}{2M^2 c} + \frac{q\beta p_0 t^4 \hbar^2}{24 M^4 \gamma^2 c}$$

$$\equiv A_{13} \qquad (8.14.20)$$

with

$$\left\langle \left(x_1(t) - \langle x_1(t)\rangle\right)^2 \right\rangle^{1/2} = \left\langle \left(x_3(t) - \langle x_3(t)\rangle\right)^2 \right\rangle^{1/2} = \gamma(t). \qquad (8.14.21)$$

Upon using the notation

$$\langle g \rangle_t^{\pm} = \int dx_1 dx_3 \, g(x_1, x_3) f_{\pm}(x_1, x_3; t) \tag{8.14.22}$$

$$\langle g \rangle_0^{\pm} = \langle g \rangle^{\pm} \tag{8.14.23}$$

where $f_{\pm}(x_1, x_2; t)$ are introduced in (8.14.17), then a straightforward but tedious evaluation of $f_{\pm}(x_1, x_3; t)$, as given in the appendix to the section, consistent with the following constraints, as dictated by the expectation values in (8.14.18)–(8.14.21), normalizability and positivity:

C.1. $\langle x_1 \rangle_t^{\pm} = 0 + \text{higher orders}$

C.2. $\langle x_3 \rangle_t^{\pm} = \dfrac{\mu \beta t^2}{2M} \langle \sigma_3 \rangle^{\pm} + \text{higher orders}$

C.3. $\sqrt{\langle x_1^2 \rangle_t^{\pm}} = \gamma(t) + \text{higher orders}$

C.4. $\left(\langle x_3^2 \rangle_t^{\pm} - \left(\langle x_3 \rangle_t^{\pm} \right)^2 \right)^{1/2} = \gamma(t) + \text{higher orders}$

C.5. $\left\langle \left(x_1 - \langle x_1 \rangle_t^{\pm} \right) \left(x_3 - \langle x_3 \rangle_t^{\pm} \right) \right\rangle_t^{\pm} = A_{13} + \text{higher orders}$

C.6. $\int dx_1 \, dx_3 \, f(x_1, x_3; t) = 1$

C.7. $f(x_1, x_3; t)$ is real and positive

where A_{13} is defined in (8.14.20), and higher orders stand relative to the parameter $\sqrt{\alpha_q}$. These lead to the following expression for the probability density in question:

$$f(x_1, x_3; t) = \dfrac{\sqrt{\det \mathbf{C}}}{2\pi} \left[w_+ \exp\left(-\dfrac{1}{2}(x_i - x_{i0}) C^{ij}(x_j - x_{j0}) \right) \right.$$

$$\left. + w_- \exp\left(-\dfrac{1}{2}(x_i + x_{i0}) C^{ij}(x_j + x_{j0}) \right) \right] \tag{8.14.24}$$

where $\mathbf{C} = [C^{ij}]$, $i,j = 1,3$, $C^{11} = C^{33} = 1/\gamma^2(t)$,

$$C^{13} = C^{31} = \dfrac{1}{\gamma^4(t)} \left[\dfrac{qbt\gamma^2}{Mc} - \dfrac{q\beta p_0 t^2 \gamma^2}{2M^2 c} - \dfrac{q\beta p_0 t^4 \hbar^2}{24 M^4 \gamma^2 c} \right] \tag{8.14.25}$$

$$x_{i0} = \dfrac{\mu \beta}{2M} t^2 \delta_{i3} \tag{8.14.26}$$

and $w_+ = w_- = 1/2$ for an unpolarized beam. It remains to check to positivity constraint C.7 (see also the appendix to this section).

The probability density in (8.14.24) is a sum of bivariate normal distributions and

$$\left[[\Sigma]^{ij} \right] = \left[[\mathbf{C}^{-1}]^{ij} \right] \tag{8.14.27}$$

8.14 Quantum Dynamics of the Stern-Gerlach Effect

is the so-called covariance matrix describing the correlation between x_1 and x_3 on the screen for $i \neq j$. Σ is a measure of dispersion in all directions in the (x_1, x_3)-plane. The multiplicative factor $\sqrt{\det \mathbf{C}}/2\pi$ is the standard normalization factor.

Finally, the constraint C.7 implies that $\det \mathbf{C} > 0$, i.e., it leads to a positivity requirement. This in turn implies that we should have

$$\frac{|q|t}{Mc}\left|b - \frac{\beta p_0 t}{2M} - \frac{\beta p_0 t^3 \hbar^2}{24 M^3 \gamma^4}\right| < 1 + \frac{\hbar^2 t^2}{4 M^2 \gamma^4}. \tag{8.14.28}$$

In reference to this inequality consider first the case with $b = 0$, i.e., the constraint

$$C < 1 + \frac{\hbar^2 t^2}{4 M^2 \gamma^4} \tag{8.14.29}$$

with

$$C = \frac{|q|\beta p_0 t^2}{2 M^2 c}\left(1 + \frac{\hbar^2 t^2}{12 M^2 \gamma^4}\right). \tag{8.14.30}$$

By setting,

$$\Delta z = \frac{|\mu|\beta t^2}{2M} \tag{8.14.31}$$

$$\frac{p_0 t}{M} = L \tag{8.14.32}$$

with the latter denoting the macroscopic distance from the particle's initial center of the wavepacket to the observation screen, we may rewrite C as

$$C = \frac{4L}{|g|}\left(\frac{M}{\hbar}\right)\frac{\Delta z}{t}\left(1 + \frac{\hbar^2 t^2}{12 M^2 \gamma^4}\right). \tag{8.14.33}$$

For the electron with $\Delta z \simeq 10^{-3}$ m, $t \simeq 10^{-6}$ s, $L \simeq 1$ m, $\gamma < 10^{-3}$ m

$$C \simeq 1.73 \times 10^7 \left(1 + \frac{1.12 \times 10^{-21}}{\gamma^4}\right) \tag{8.14.34}$$

which is a very large number and the positivity constraint (8.14.29) cannot be satisfied. On the other hand, the uniform magnetic field $(0, b, 0)$ may a priori be set at

$$b = \frac{\beta}{2}L \tag{8.14.35}$$

defined simply in terms of the non-uniform magnetic field gradient $\partial B_2/\partial x_2 = -\beta = -\partial B_3/\partial x_3$ (see (8.14.11)), and the distance to the observation screen L, independently of any of the details of the spin 1/2 charged particle considered and of the (initial) spread γ. [The uniform magnetic field component b may be, of course, chosen so that $C^{13} = 0$, but this would mean to choose a different uniform magnetic field for every different charged particle, and a different spread γ, and would not be physically as interesting.] The matrix elements in (8.14.25) then simply become

$$C^{13} = C^{31} = -\varepsilon(q)\frac{1}{3|g|}\left(\frac{\Delta z}{\gamma}\right)\left(\frac{L}{\gamma}\right)\frac{\hbar t}{M}\frac{1}{\gamma^4(t)} \tag{8.14.36}$$

and the positivity constraint

$$\frac{1}{3|g|}\frac{\Delta z}{\gamma}\frac{L}{\gamma}\frac{\hbar t}{M\gamma^2} < 1 + \frac{\hbar^2 t^2}{4M^2\gamma^4} \tag{8.14.37}$$

is readily satisfied. For example, for the electron with $\Delta z = 10^{-3}$ m, $L = 0.7$ m, $\gamma = 0.55 \times 10^{-3}$ m, $t = 4 \times 10^{-6}$ s, corresponding to an initial average speed of 1.75×10^5 m/s, a magnetic field gradient $\beta = 12.280$ T/m, and a uniform longitudinal magnetic field $b = 4.298$ T, the left-hand side of (8.14.37) is $\simeq 0.59$. In Figure 8.13, $t = 4 \times 10^{-6}$ s, and the probability density $f(x_1, x_3; t)$ for $\mathbf{B} = 0$, is plotted for $\gamma = 0.55 \times 10^{-3}$ m and the corresponding density for $\mathbf{B} \neq 0$, for the above just given parameters, is plotted in Figure 8.14, for an initially unpolarized beam, showing a clear splitting of the beam along the quantization axis. [The magnetic field b may be chosen to be even smaller. For example, for slower electrons $t = 5.93 \times 10^{-6}$ s, $L = 0.5$ m, $b = 1.4$ T consistent with (8.14.37).] The asymmetry with elongations in the second and the fourth quadrants in Figure 8.14 are easy to understand. For the electron $\varepsilon(q) = -1$, $C^{13} = C^{31} > 0$ and the probability density gets, respectively, positive amplifying contributions for $x_3 > x_{30}$, $x_1 < 0$ and $x_3 < -x_{30}$, $x_1 > 0$. This graph corresponds to a negative charged particle. The formal physical argument for this asymmetry is that it arises as a consequence of the direction of the Lorentz force, as determined by the so-called right-hand rule, on a charged particle as applied to the transverse part of the non-uniform magnetic field. For a positive charge, the elongations, as arising in opposite directions, occur in the corresponding first and third quadrants.

We note that the correlation in (8.14.20) and $C^{13} = C^{31}$ in (8.14.25) vanish for neutral particles. The analysis carried above (with $C^{13} = C^{31}$ set equal to zero), is equally valid for neutral spin 1/2 particles with magnetic moment $\boldsymbol{\mu} = \mu\boldsymbol{\sigma}$, as carried to the leading order in $M|\mu|/|g|(\hbar^3 c)^{1/2}$, and finally leads to the expression

$$f(x_1, x_3; t) = \frac{1}{2\pi\gamma^2(t)}\left\{w_+ \exp\left[-\frac{1}{2}\frac{x_1^2 + (x_3 - x_0)^2}{\gamma^2(t)}\right]\right.$$

$$\left. + w_- \exp\left[-\frac{1}{2}\frac{x_1^2 + (x_3 + x_0)^2}{\gamma^2(t)}\right]\right\} \tag{8.14.38}$$

where $x_0 = \mu\beta t^2/2M$. For an unpolarized beam, this is plotted in Figure 8.15 for $t = 4\times 10^{-6}$ s, $|x_0| = 1\times 10^{-3}$ m, $\gamma = 0.55\times 10^{-3}$ m, showing the difference of the densities for the charged and uncharged cases in the presence of an appropriately chosen longitudinal uniform magnetic field.

8.14 Quantum Dynamics of the Stern-Gerlach Effect 539

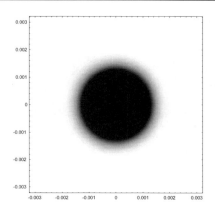

Fig. 8.13. Plot of the density $f(x_1, x_3; t)$ for $\mathbf{B} = 0$ $\gamma = 0.55 \times 10^{-3}$ m, $t = 4 \times 10^{-6}$ s.

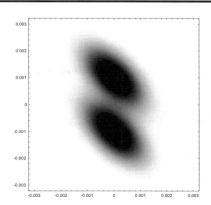

Fig. 8.14. Plot of the density $f(x_1, x_3; t)$ for the electron, based on (8.14.24), (8.14.35), (8.14.36) with $\Delta z = 10^{-3}$ m, $\gamma = 0.55 \times 10^{-3}$ m, $t = 4 \times 10^{-6}$ s, $L = 0.7$ m, corresponding to an initial average speed of 1.75×10^5 m/s, a magnetic field gradient $\beta = 12.280$ T/m, and a uniform longitudinal magnetic field $b = 4.298$ T, for an initially unpolarized beam.

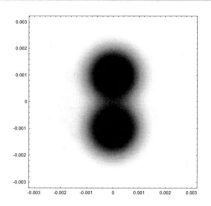

Fig. 8.15. Plot of the density $f(x_1, x_3; t)$ for uncharged particles, based on (8.14.38) for $|x_0| = 1 \times 10^{-3}$ m, $\gamma = 0.55 \times 10^{-3}$ m, $t = 4 \times 10^{-6}$ s, for an initially unpolarized beam.

Appendix to §8.14: Time Evolution and Intensity Distribution

To study the probability intensity distribution, we note the following commutation relations:

$$\left[\frac{\mathbf{p}^2}{2M}, H_\mathrm{I}\right] = \frac{i\hbar q\beta}{M^2 c} x_3 p_1 p_2 + \frac{i\hbar q}{M^2 c}(\beta x_2 + b)p_1 p_3$$
$$+ \frac{2i\hbar q\beta}{M^2 c} x_1 p_2 p_3 - \frac{i\hbar\mu\beta}{M}\sigma_2 p_2 + \frac{i\hbar\mu\beta}{M}\sigma_3 p_3 \qquad \text{(A-8.14.1)}$$

$$\left[\frac{\mathbf{p}^2}{2M}, \left[\frac{\mathbf{p}^2}{2M}, H_\mathrm{I}\right]\right] = \frac{4\hbar^2 q\beta}{M^3 c} p_1 p_2 p_3 \qquad \text{(A-8.14.2)}$$

with all the other commutators with $\mathbf{p}^2/2M$ vanish or are of higher order. We use a variation of the Baker-Campbell-Hausdorff formula (see Appendix I): if

$$[B, [A, B]] = 0 \qquad \text{(A-8.14.3)}$$

$$\left[B, [A, [A, B]]\right] = 0 \qquad \text{(A-8.14.4)}$$

8.14 Quantum Dynamics of the Stern-Gerlach Effect

$$\Big[A, [A, [A, B]]\Big] = 0 \tag{A-8.14.5}$$

for two operators A, B, then

$$e^{A+B} = \exp\left(\frac{1}{2}[A, B] + \frac{1}{6}[A, [A, B]]\right) e^B e^A. \tag{A-8.14.6}$$

We let

$$A = -\frac{it}{\hbar} \frac{\mathbf{p}^2}{2M} \tag{A-8.14.7}$$

$$B = -\frac{it}{\hbar} H_{\mathrm{I}} \tag{A-8.14.8}$$

and note that (A-8.14.3)–(A-8.14.5) hold true to the accuracy retained. Equation (A-8.14.6) then gives

$$\exp\left(-\frac{it}{\hbar}H\right) = \exp\left(-\frac{it}{2\hbar^2}\left[\frac{\mathbf{p}^2}{2M}, H_{\mathrm{I}}\right]\right) \exp\left(\frac{2it^3 q\beta}{3\hbar M^3 c} p_1 p_2 p_3\right)$$

$$\times \exp\left(-\frac{it}{\hbar} H_{\mathrm{I}}\right) \exp\left(-\frac{it}{\hbar} H_0\right) \tag{A-8.14.9}$$

where $H_0 = \mathbf{p}^2/2M$ is the free Hamiltonian and $(\exp(-itH_0/\hbar)\psi)(\mathbf{x}) = \psi_0(\mathbf{x}, t)$ is explicitly given in (8.14.8).

To carry out the time-evolution operation given in (A-8.14.9) on ψ we use, in the process, the identity

$$\exp\left[ia\frac{p}{\hbar}\right] f(x) = f(x + a). \tag{A-8.14.10}$$

The operation defined on the right-hand side of (A-8.14.9) on ψ may be then carried out. The analysis is very tedious but straightforward. Up to a normalization factor, and the phase factor $\exp(it\mathbf{p}_0 \cdot \mathbf{x}/\hbar)$, $(\exp(-iH/\hbar)\psi)(\mathbf{x})$, is given by the expression:

$$F_1\left(x_1, x_2 - \frac{p_0}{M}t, x_3; t\right) F_2\left(x_2 - \frac{p_0}{M}t; t\right) F_3(x_1, x_3; t) \tag{A-8.14.11}$$

where we have conveniently isolated the terms dependent on the variable x_2, in the two factors F_1, F_2 as we have to integrate over it as indicated in (8.14.17). With

$$x_2' = x_2 - \frac{p_0}{M}t \tag{A-8.14.12}$$

$$F = \frac{1}{4\gamma^2 \left(1 + \frac{i\hbar t}{2M\gamma^2}\right)} \equiv F(t) \tag{A-8.14.13}$$

we have

$$F_1(x_1, x_2', x_3) = \exp\left(-x_2'^2 F\right) \exp\left(-\frac{4tq\beta}{Mc} x_1 x_2' x_3 F\right)$$

$$\times \exp\left(\frac{8i\hbar q\beta t^2}{M^2 c} x_1 x_2' x_3 F^2\right) \exp\left(\frac{16q\beta\hbar^2 t^3}{3M^3 c} x_1 x_2' x_3 F^3\right) \tag{A-8.14.14}$$

and

$$F_2(x_2'; t) = \exp\left(\frac{it\mu}{\hbar} \sigma_2 \left(b - \beta \frac{p_0}{M} t\right)\right) \exp\left(-\frac{it\mu\beta\sigma_2}{\hbar} x_2'\right)$$

$$\times \exp\left(\frac{ip_0 \mu\beta t^2}{2\hbar M} \sigma_2\right) \exp\left(-\frac{\mu\beta t^2}{M} \sigma_2 x_2' F\right) \tag{A-8.14.15}$$

$$F_3(x_1, x_3; t) = \exp\left(\frac{itq\beta p_0}{\hbar Mc} x_1 x_3\right) \exp\left(\frac{it\mu\beta}{\hbar} \sigma_3 x_3\right) \exp\left(\frac{p_0 q\beta t^2}{M^2 c} x_1 x_3 F\right)$$

$$\times \exp\left(-\frac{2tqb}{Mc} x_1 x_3 F\right) \exp\left(-\frac{8ip_0 t^3 q\beta\hbar}{3M^3 c} x_1 x_3 F^2\right)$$

$$\times \exp\left(\frac{\mu\beta t^2}{M} \sigma_3 x_3 F\right) \exp\left(\frac{2i\hbar qt^2}{M^2 c} \left(\beta \frac{p_0}{M} t + b\right) x_1 x_3 F^2\right)$$

$$\times \exp\left(-(x_1^2 + x_3^2) F\right). \tag{A-8.14.16}$$

Now we have to apply the operator $(F_1 F_2 F_3)$ in (A-8.14.11) to $\begin{pmatrix} 1 \\ 0 \end{pmatrix}$ and $\begin{pmatrix} 0 \\ 1 \end{pmatrix}$, and perform the operations defined in (8.14.15)–(8.14.17). To this end, we use the identities:

$$\sigma_3 \begin{pmatrix} 1 \\ 0 \end{pmatrix} = \begin{pmatrix} 1 \\ 0 \end{pmatrix}, \quad \sigma_3 \begin{pmatrix} 0 \\ 1 \end{pmatrix} = -\begin{pmatrix} 0 \\ 1 \end{pmatrix} \tag{A-8.14.17}$$

$$\begin{pmatrix} 1 \\ 0 \end{pmatrix} = \frac{1}{2} \begin{pmatrix} 1 \\ i \end{pmatrix} + \frac{1}{2} \begin{pmatrix} 1 \\ -i \end{pmatrix} \tag{A-8.14.18}$$

$$\begin{pmatrix} 0 \\ 1 \end{pmatrix} = -\frac{i}{2} \begin{pmatrix} 1 \\ i \end{pmatrix} + \frac{i}{2} \begin{pmatrix} 1 \\ -i \end{pmatrix} \tag{A-8.14.19}$$

$$\sigma_2 \begin{pmatrix} 1 \\ i \end{pmatrix} = \begin{pmatrix} 1 \\ i \end{pmatrix}, \quad \sigma_2 \begin{pmatrix} 1 \\ -i \end{pmatrix} = -\begin{pmatrix} 1 \\ -i \end{pmatrix} \tag{A-8.14.20}$$

and the orthogonality of $(1 \ i)^\top$, $(1 \ -i)^\top$. Also we note that

$$F(t) + F^*(t) = \frac{1}{2\gamma^2(t)} \tag{A-8.14.21}$$

8.14 Quantum Dynamics of the Stern-Gerlach Effect

$$iF(t) + (iF(t))^* = \frac{1}{4\gamma^2(t)} \frac{\hbar t}{M\gamma^2} \qquad \text{(A-8.14.22)}$$

$$iF^2(t) + (iF^2(t))^* = \frac{1}{8\gamma^4(t)} \frac{\hbar t}{M\gamma^2}. \qquad \text{(A-8.14.23)}$$

From (A-8.14.11)–(A-8.14.23) we obtain, up to a normalization factor, the following expression for the x_2-integrand in (8.14.17)

$$\exp\left(-\frac{1}{2\gamma^2(t)}\left[x_2'^2 + a(t)x_1 x_2' x_3\right]\right)\left[\exp\left(\frac{t^2}{2\gamma^2(t)}\frac{\mu\beta}{M}x_2'\right)\right.$$

$$\left. + \exp\left(-\frac{t^2}{2\gamma^2(t)}\frac{\mu\beta}{M}x_2'\right)\right] f(x_1, x_3; t) \qquad \text{(A-8.14.24)}$$

where $a(t)$, of order $\sqrt{\alpha_q}$, is a function of t only, and up to a multiplicative time-dependent constant,

$$f(x_1, x_3; t) \equiv w_+ f_+(x_1, x_3; t) + w_- f_-(x_1, x_3; t)$$

$$= w_+ \exp\left(-\frac{1}{2\gamma^2(t)}\left[x_1^2 + x_3^2 - \frac{t^2}{M}\mu\beta x_3 - \frac{x_i A_{ij} x_j}{\gamma^2(t)}\right]\right)$$

$$+ w_- \exp\left(-\frac{1}{2\gamma^2(t)}\left[x_1^2 + x_3^2 + \frac{t^2}{M}\mu\beta x_3 - \frac{x_i A_{ij} x_j}{\gamma^2(t)}\right]\right).$$
$$\text{(A-8.14.25)}$$

A summation over the repeated indices $i, j = 1, 3$ in (A-8.14.25) is understood,

$$A_{13} = A_{31} = -\frac{qbt\gamma^2}{Mc} + \frac{q\beta p_0 t^2 \gamma^2}{2M^2 c} + \frac{q\beta p_0 t^4 \hbar^2}{24 M^4 \gamma^2 c}, \qquad \text{(A-8.14.26)}$$

$$A_{11} = A_{33} = 0. \qquad \text{(A-8.14.27)}$$

The expression in (A-8.14.26) is identical to the correlation of the dynamical variables $x_1(t)$, $x_3(t)$ in (8.14.20).

In reference to the x_2-integral in (8.14.17), we have, from (A-8.14.24) with

$$b(t) = \frac{\mu\beta t^2}{M} \qquad \text{(A-8.14.28)}$$

for the shifted x_2'-integral,

$$\int_{-\infty}^{\infty} dx_2' \exp\left(-\frac{1}{2\gamma^2(t)}\left(x_2'^2 + [a(t)x_1 x_3 \pm b(t)]\right)x_2'\right)$$

$$= \sqrt{2\pi}\gamma(t) \exp\left(\frac{1}{8\gamma^2(t)}[a(t)x_1 x_3 \pm b(t)]^2\right) \qquad \text{(A-8.14.29)}$$

where $[a(t)x_1x_3 \pm b(t)]^2$ is necessarily of a higher order correction in $\sqrt{\alpha_q}$.

Accordingly, for the probability density $f(x_1, x_3; t)$, we obtain the preliminary expression given in (A-8.14.25).

To satisfy, in the process, constraint C.2 in the text (see also (8.14.19)), we multiply the right-hand side of (A-8.14.25) by an overall normalizing factor $\exp\left(-\left(\mu\beta t^2/2M\right)^2/2\gamma^2(t)\right)$ giving

$$f(x_1, x_3; t) \propto w_+ \exp\left(-\frac{1}{2\gamma^2(t)}\left[x_1^2 + \left(x_3 - \frac{\mu\beta t^2}{2M}\right)^2 - \frac{x_i A_{ij} x_j}{\gamma^2(t)}\right]\right)$$

$$+ w_- \exp\left(-\frac{1}{2\gamma^2(t)}\left[x_1^2 + \left(x_3 + \frac{\mu\beta t^2}{2M}\right)^2 - \frac{x_i A_{ij} x_j}{\gamma^2(t)}\right]\right).$$

(A-8.14.30)

Consistency with the constrains C.1–C.6 in the text necessarily leads to the expression in (8.14.24) for the probability density in question below which the positivity constraint C.7 has been already analyzed in the text.

Problems

8.1. For $\mathbf{n}_1 = (\sin\theta_1, 0, \cos\theta_1)$, $\mathbf{n}_1 = (\sin\theta_1, 0, \cos\theta_1)$, show that

$$|+1/2, \mathbf{n}_1\rangle = |+1/2, \mathbf{n}_2\rangle \cos\left(\frac{\theta_1 - \theta_2}{2}\right) + |-1/2, \mathbf{n}_2\rangle \sin\left(\frac{\theta_1 - \theta_2}{2}\right).$$

8.2. (i) Solve the simultaneous equations in (8.1.35), (8.1.36) to show that the solutions with the initial conditions (8.1.32), (8.1.40) are given respectively by (8.1.37)/(8.1.38) and (8.1.43).

(ii) More generally, find the unitary operator $U(t,0)$ for the time development via the time-dependent Hamiltonian in (8.1.30).

[Hint: Find the latter by integrating (8.1.35), (8.1.36) rather than by considering the time-ordered product of the exponential of $\left(-i\int_0^t dt'\, H(t')/\hbar\right)$.]

8.3. Use (8.1.83) satisfied by the amplitude $F(t)$, self-consistently, to investigate the behavior of the survival probability in the truly asymptotic limits $t \to 0$, $t \to \infty$, and state the sufficiency conditions to be satisfied in your analysis for the validity of your results.

8.4. (i) Show that for $g = 2$ in (8.2.17), the eigenvectors corresponding to the ground state energy 0, as follow from (8.2.12), are given in (8.2.18) with $m = 0$ or -1, or ....

(ii) Verify that the supersymmetric generators Q, $Q^\dagger$ for $g = 2$, actually annihilate $\Psi_{0,m,-1}$ (see (8.2.19)).

Problems 545

8.5. Verify explicitly the expressions for the polynomials in (8.3.27), (8.3.28) to finally establish (8.3.30)–(8.3.33).
8.6. Repeat the operator analysis for the Landau levels in §8.3 in the gauge given by the vector potential in (8.3.35) rather than the one in (8.2.11). Comment on the connection between the two solutions obtained corresponding to these two gauges.
8.7. Solve the corresponding equation to the one in (8.3.38) without first putting p_1, $p_3 = 0$ in (8.3.36).
8.8. Derive the equations (8.4.3)–(8.4.6).
8.9. Compute the very small contribution, by introducing in the process an ultraviolet cut-off $Mc/\hbar$ for the k-integral, to the magnetic moment μ_n in §8.4 by considering the real part of (8.4.18). (See §8.5 for useful details).
8.10. Verify the angular integration in (8.5.59) starting from (8.5.47).
8.11. Derive the expressions for the density operators $\rho^{(i)}$ and the polarization vectors $\mathbf{P}^{(i)}$ as in (8.6.10), (8.6.11), for an arbitrary initial state.
8.12. For an arbitrary initial state $|\Psi^{(i)}\rangle = (c_+ \; c_-)^\top$, using the general expressions for $\rho^{(i)}$ and $\mathbf{P}^{(i)}$ obtained in Problem 8.11, and the expression for the matrix M in (8.6.18), to find the general forms for the final density operator ρ and the final polarization vector $\mathbf{P}$.
8.13. Refer to (8.6.27), to find the expression $F_2(\vartheta) + F_2(-\vartheta)$ for the final probability densities for an initially polarized beam along the z-axis.
8.14. (i) Show that the constraints in (8.6.52), (8.6.53) lead to the general expressions for β_1, β_2, β_3 as given in (8.6.59), (8.6.60).
   (ii) Complete the proof of part (ii) in the lemma below the constraints (8.6.52), (8.6.53).
   (iii) Finally show that (8.6.55), (8.6.56)–(8.6.60) give the structure in (8.6.63) for M.
8.15. Find the expression for the probability density for the scattering of a beam of spin $1/2$ off a spin 0 target with the resulting spin $1/2$ beam scattering off a second spin 0 target. This is referred to as double scattering.
8.16. Obtain the final state $|\Phi\rangle$ in (8.7.28) from (8.7.25), (8.7.24) and (6.6.21).
8.17. Show that the state in (8.7.40), with $|\psi\rangle$ given in (8.7.14), develops in time $\tau = \pi/2\lambda_0$ via the Hamiltonian H_{ME} in (8.7.43) to the state in (8.7.41).
8.18. Show that the system of equations in (8.8.4)–(8.8.7), lead to the time evolution operator $U(t, t_1)$ in (8.8.8) from a time t_1 to a time t.
8.19. Write down explicitly the matrix elements of the unitary matrix in (8.8.15), then obtain the probability of a spin flip as given in (8.8.17).
8.20. Insert an apparatus described by a harmonic oscillator, as introduced in §8.7, (8.7.24), instead of one described by a spin $1/2$ system, in reference to the Hamiltonian in (8.8.28), between two Ramsey zones,

546   8 Quantum Physics of Spin 1/2 & Two-Level Systems

to study the nature of interference effect in a spin flip experiment as done for the latter experiment emphasizing the differences between the two apparatuses.

8.21. Derive the expression for the probability given in (8.8.47), at resonance $\omega = \omega_0$, with no meters inserted between the Ramsey zones.

8.22. Follow the elementary model set up to generate the Schrödinger cat entangled state in (8.9.3) leading to (8.9.10), to set up a similar model to generate the Schrödinger cat state in (8.9.2).

8.23. In the spirit of the actual experiment, reformulate the model developed through (8.9.3)–(8.9.9) by considering two levels of an atom and replace, in the process, the interactions given there in the first and third stages by appropriate interactions, as done in §8.12, within Ramsey zones.

8.24. Show that the state $|\psi\rangle$ in (8.10.3) is invariant under rotations of the unit vector **n** about any axis.

8.25. Derive the expressions for the expansions of the initial states on the left-hand sides of (8.11.5) and (8.11.10) in terms of the other states as stated, respectively.

8.26. Develop simple dynamical models, as done in §8.7, to generate the unitary transformations specified in (8.11.8).

8.27. Develop the analysis in quantum cryptography in §8.11 including the evaluation of the power of the test of detecting an intruder, if an intruder is present, for the process depicted in Figure 8.7 for the production of pairs of photons one going to Alice and one to Bob from each pair.

8.28. Show that the unitary operators in (8.12.8)–(8.12.16) may be rewritten in the compact form as given in (8.12.21).

8.29. Show that $|a_2|^2$ is as given in (8.12.32) for the transition probability $P_\vartheta[\text{Transition } |0\rangle \to |2\rangle]$ as follows from (8.12.31).

8.30. Show that in the adiabatic regime, for which $|\omega_0| \gg \omega$, the exact solution in (8.1.34) goes over to the one in (8.13.27).

8.31. Find the transformation law for the matrix $\mathbf{c}(T)$ in (8.13.33) under the transformations (8.13.35), (8.13.36). [Hint: It is easier to consider (8.13.31) directly rather than the defining equation in (8.13.33).]

8.32. Reformulate the change of sign of a spinor of spin 1/2 under a rotation by 2π radians in the light of the AA phase.

8.33. Show that in the Pancharatnam definition (see (8.13.68), (8.13.69)), if vectors $|\Phi_1\rangle$, $|\Phi_2\rangle$ are "in phase" and so are $|\Phi_2\rangle$, $|\Phi_3\rangle$ then $|\Phi_1\rangle$, $|\Phi_3\rangle$ are not necessarily "in phase".

8.34. Derive (8.13.67) for the inner product $\langle \Phi(t) | \Phi(t + \Delta t) \rangle$.

8.35. Show that the vectors in **A**, **A**$'$ in (8.13.96), (8.13.97), which are well defined in the region $\varepsilon < \theta < \pi - \varepsilon$ for $\varepsilon > 0$ and small, are related by a gauge transformation.

9
Green Functions

Green functions provide information on different aspects of quantum physical systems in a unified manner and their importance cannot be overemphasized. To see how a Green function may arise, suppose that $\langle \mathbf{x}t | \psi \rangle$ denotes the state of a system at time t, in the x-description, as it has evolved from an initial state $\langle \mathbf{x}0 | \psi \rangle$. Upon the insertion of the identity operator in (2.4.13) between $\langle \mathbf{x}t|$ and $|\psi\rangle$, we may relate $\langle \mathbf{x}t|\psi\rangle$ to the initial state as follows

$$\langle \mathbf{x}t | \psi \rangle = \int d^3\mathbf{x}'\, \langle \mathbf{x}t | \mathbf{x}'0 \rangle \langle \mathbf{x}'0 | \psi \rangle. \tag{9.1}$$

This may be rewritten as

$$\psi_t(\mathbf{x}) = \int d^3\mathbf{x}'\, G(\mathbf{x}t; \mathbf{x}'0)\psi_0(\mathbf{x}') \tag{9.2}$$

where

$$G(\mathbf{x}t; \mathbf{x}'t') = \langle \mathbf{x}t | \mathbf{x}'t' \rangle \tag{9.3}$$

defines a Green function describing the evolution of the system. Unlike $\langle \mathbf{x}t|\psi\rangle$, $G(\mathbf{x}t;\mathbf{x}'t')$ is independent of the state ψ, i.e., it is not tied up to any state thus emphasizing its general character. By definition of $\langle \mathbf{x}t|\mathbf{x}'t'\rangle$, we may write

$$G(\mathbf{x}t; \mathbf{x}'t') = \langle \mathbf{x} | U(t,t') | \mathbf{x}' \rangle \tag{9.4}$$

where $U(t,t')$ is the time evolution operator from time t' to time t. For a time independent Hamiltonian $U(t,t') = \exp{-\mathrm{i}(t-t')H/\hbar}$ and one has

$$G(\mathbf{x}t; \mathbf{x}'t') = \int_{-\infty}^{\infty} d\lambda\, \langle \mathbf{x} | \delta(\lambda - H) | \mathbf{x}' \rangle \exp{-\mathrm{i}(t-t')\lambda/\hbar} \tag{9.5}$$

providing the intimate connection between a Green function and the spectrum of the underlying Hamiltonian of the system.

Also by definition of $\langle \mathbf{x}t|\mathbf{x}'t'\rangle$, we have the following completeness relation

$$G(\mathbf{x}t;\mathbf{x}'t') = \int d^3\mathbf{x}''\, G(\mathbf{x}t;\mathbf{x}''t'')G(\mathbf{x}''t'';\mathbf{x}'t') \tag{9.6}$$

for $t' < t'' < t$. With $G(\mathbf{x}t;\mathbf{x}'t')$ interpreted as the amplitude that the system at $\mathbf{x}'$ at time t' to be found later at $\mathbf{x}$ at time $t > t'$, (9.6) shows that for any given t'' ($t' < t'' < t$), the system may go through any point $\mathbf{x}''$ before ending up at $\mathbf{x}$. By repeated applications of (9.6), writing $G(\mathbf{x}t;\mathbf{x}'t')$ as integrals of the product of an arbitrary number of amplitudes, the path integral formulation will be developed in the next chapter.

In the present chapter, we carry out a detailed study of Green functions and their properties starting with those of a free particle followed by those for systems with interactions. The Green function method is also applied to study the law of reflection as well as to the celebrated quantum phenomenon known as the Aharonov-Bohm (AB) effect. Special attention will be given to the Green function for general systems, and for the Coulomb potential, in particular.

Green functions will find important applications in other chapters as well, especially, in the path integral formalism of quantum physics, in scattering theory and in the theory of multi-electron atoms.

9.1 The Free Green Functions

In this section, we consider various aspects of free Green functions mainly in three dimensions. Other dimensional cases are treated in the next few sections as well as in the problems.

Consider the general solution of the Schrödinger equation for a free particle in three dimensions

$$\psi(\mathbf{x}, t) = \left(e^{-itH_0/\hbar} \psi \right)(\mathbf{x})$$

$$= \int \frac{d^3\mathbf{p}}{(2\pi\hbar)^3}\, e^{i[\mathbf{x}\cdot\mathbf{p} - \mathbf{p}^2 t/2m]/\hbar} \psi(\mathbf{p}), \quad t > 0 \tag{9.1.1}$$

where $H_0 = -\hbar^2 \boldsymbol{\nabla}^2/2m$, with initial condition at $t = 0$

$$\psi(\mathbf{x}, 0) = \psi(\mathbf{x}) = \int \frac{d^3\mathbf{p}}{(2\pi\hbar)^3} e^{i\mathbf{x}\cdot\mathbf{p}/\hbar} \psi(\mathbf{p}). \tag{9.1.2}$$

Upon using the inverse Fourier transform of (9.1.2)

$$\psi(\mathbf{p}) = \int d^3\mathbf{x}'\, e^{-i\mathbf{x}'\cdot\mathbf{p}/\hbar} \psi(\mathbf{x}') \tag{9.1.3}$$

the general solution (9.1.1) may be rewritten as

9.1 The Free Green Functions

$$\psi(\mathbf{x},t) = \int d^3\mathbf{x}'\, G^0(\mathbf{x}t;\mathbf{x}'0)\, \psi(\mathbf{x}') \tag{9.1.4}$$

(compare with (9.2)), where

$$G^0(\mathbf{x}t;\mathbf{x}'0) = \int \frac{d^3\mathbf{p}}{(2\pi\hbar)^3} e^{i[(\mathbf{x}-\mathbf{x}')\cdot\mathbf{p} - \mathbf{p}^2 t/2m]/\hbar} \tag{9.1.5}$$

satisfying the initial condition

$$G^0(\mathbf{x}t;\mathbf{x}'0) \xrightarrow[t\to 0]{} \delta^3(\mathbf{x}-\mathbf{x}'). \tag{9.1.6}$$

By using the identity

$$[(\mathbf{x}-\mathbf{x}')\cdot\mathbf{p} - \mathbf{p}^2 t/2m] = -\frac{t}{2m}\left[\left(\mathbf{p} - m\frac{(\mathbf{x}-\mathbf{x}')}{t}\right)^2 - \frac{m^2}{t^2}|\mathbf{x}-\mathbf{x}'|^2\right]$$

changing the integration variable $\mathbf{p} \to \mathbf{p} + m(\mathbf{x}-\mathbf{x}')/t$ in (9.1.5), and carrying out the resulting Gaussian integral, we obtain

$$G^0(\mathbf{x}t;\mathbf{x}'0) = \left(\frac{m}{2\pi i\hbar t}\right)^{3/2} \exp\frac{im|\mathbf{x}-\mathbf{x}'|^2}{2\hbar t} \tag{9.1.7}$$

with $t > 0$.

Since we are considering the *case* $t > 0$ in (9.1.7), we may introduce the following Green function

$$G_+^0(\mathbf{x}t;\mathbf{x}'t') = \Theta(t-t')\left(\frac{m}{2\pi i\hbar(t-t')}\right)^{3/2} \exp\frac{im|\mathbf{x}-\mathbf{x}'|^2}{2\hbar(t-t')} \tag{9.1.8}$$

where $\Theta(t-t')$ is the step function, and will be referred to as the *retarded* Green function. It satisfies the boundary condition

$$G_+^0(\mathbf{x}t;\mathbf{x}'t') = 0 \quad \text{for} \quad t - t' < 0 \tag{9.1.9}$$

and the initial condition

$$G_+^0(\mathbf{x}t;\mathbf{x}'t') \xrightarrow[t\to t'+0]{} \delta^3(\mathbf{x}-\mathbf{x}'). \tag{9.1.10}$$

Because of the presence of the step function $\Theta(t-t')$ in the definition (9.1.8), it follows directly from (9.1.5), that $G_+^0(\mathbf{x}t;\mathbf{x}'t')$ satisfies the differential equation

$$\left[i\hbar\frac{\partial}{\partial t} - H_0\right] G_+^0(\mathbf{x}t;\mathbf{x}'t') = i\hbar\, \delta^3(\mathbf{x}-\mathbf{x}')\, \delta(t-t') \tag{9.1.11}$$

where $H_0 = -\hbar^2 \nabla^2/2m$.

We may incorporate the boundary condition, in an energy momentum description, of the retarded Green function, by rewriting the solution (9.1.8) in the form

$$G^0_+ (xt; x't') = i\hbar \int \frac{(\mathrm{d}p)}{(2\pi\hbar)^4} \frac{e^{i(x-x')\cdot p/\hbar}}{\left[p^0 - \frac{\mathbf{p}^2}{2m} + i\varepsilon \right]}, \varepsilon \to +0 \qquad (9.1.12)$$

where

$$(x - x') \cdot p = (\mathbf{x} - \mathbf{x}') \cdot \mathbf{p} - (t - t') p^0 \qquad (9.1.13)$$

and

$$(\mathrm{d}p) = \mathrm{d}p^0 \mathrm{d}^3\mathbf{p} \qquad (9.1.14)$$

with $-\infty < p^0 < \infty, -\infty < p^i < \infty$, $i = 1, 2, 3$.

To show that the expression on the right-hand side of (9.1.12) leads to the solution in (9.1.8), we go to the complex p^0-plane and note that the integrand in (9.1.12) has a pole at $p^0 = \mathbf{p}^2/2m - i\varepsilon$, i.e., in the lower complex plane. Hence for $t - t' < 0$, we may close the contour in the complex p^0-plane from above thus *avoiding* the pole, and noting that the real expression $-i(t-t')$ i Im p^0, in the exponential in (9.1.12), is negative for $t - t' < 0$ and for Im $p^0 > 0$ in the upper complex plane. Thus the infinite semi-circle contour in the upper complex p^0-plane gives *zero* contribution and we obtain (9.1.9).

On the other hand for $t - t' > 0$, we may close the contour of integration in the complex p^0-plane from below thus *enclosing* the pole. Also for $t - t' > 0$, the real quantity $-i(t - t')$ i Im p^0, in the exponential in (9.1.12), is negative for Im $p^0 < 0$ in the lower complex p^0-plane. Thus the infinite semi-circle contour in the lower complex p^0-plane gives zero (see Figure 9.1). From the residue theorem, we then obtain for $t - t' > 0$

$$G^0_+ (xt; x't) = (-2\pi i)(i\hbar) \int \frac{\mathrm{d}^3\mathbf{p}}{(2\pi\hbar)^4} \exp i \left[(\mathbf{x} - \mathbf{x}') \cdot \mathbf{p} - \mathbf{p}^2 (t - t')/2m \right]/\hbar \qquad (9.1.15)$$

which from (9.1.5) coincides with (9.1.8).

Upon rewriting $G^0_+ (xt; x't')$ as

$$G^0_+ (xt; x't') = \Theta(t - t') \int \frac{\mathrm{d}^3\mathbf{p}}{(2\pi\hbar)^3} e^{i[(\mathbf{x}-\mathbf{x}')\cdot\mathbf{p}-\mathbf{p}^2(t-t')/2m]/\hbar} \qquad (9.1.16)$$

the following *completeness* relation is readily obtained

$$\int \mathrm{d}^3\mathbf{x}'' G^0_+ (xt; x''t'') G^0_+ (x''t''; x't') = G^0_+ (xt; x't') \qquad (9.1.17)$$

where t'' is arbitrary such that $t > t'' > t'$.

9.1 The Free Green Functions

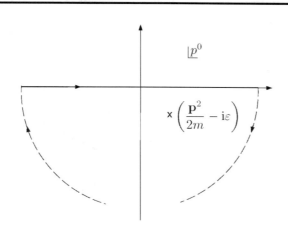

Fig. 9.1. The contour of integration chosen in the complex p^0-plane to evaluate the p^0-integral in (9.1.12) for $t - t' > 0$. The contour encloses a pole at $p^0 = \mathbf{p}^2/2m - i\varepsilon$. For $t - t' > 0$, $(t - t')\,\text{Im}\,p^0 < 0$ for $\text{Im}\,p^0 < 0$, in the lower complex p^0-plane, and the infinite semi-circle contour gives zero contribution. The p^0-integral in (9.1.12) is then evaluated by an application of the residue theorem.

The physical interpretation of (9.1.17) was given in the introduction to this chapter.

As an application, and as an idealistic toy model for the double slit experiment, suppose that particles are constrained to pass through two given points,[1] as shown in Figure 9.2, before hitting an observation screen. That is, we need the constraint

$$d^3 x'' \longrightarrow \delta(x_1'')\,\delta(x_3'')\left[\delta(x_2'' - d) + \delta(x_2'' + d)\right]d^3 x'' \qquad (9.1.18)$$

to evaluate the amplitude of finding a particle at a given point on the screen.

Using the notation $\mathbf{x} = (x_1, x_2, x_3)$, the geometry in Figure 9.2 together with the constraints in (9.1.18) give

$$|\mathbf{x} - \mathbf{x}''|^2 = \mathbf{x}^2 + d^2 \mp 2x_2 d \qquad (9.1.19)$$

for the upper and lower "slits", respectively, and

$$|\mathbf{x}' - \mathbf{x}''|^2 = L^2 + d^2. \qquad (9.1.20)$$

---

[1] For more realistic computations, allowing a spread, i.e., an extension, of the slits see Problem 9.5

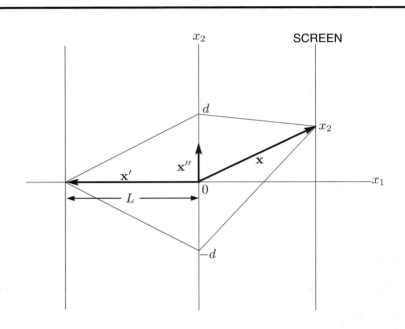

Fig. 9.2. Idealistic toy model of the double-slit experiment describing constrained motion. Here the vector $\mathbf{x}''$ is restricted as defined through the integration measure *in* (9.1.18). We are interested in determining the amplitude that a particle hits the screen at $\mathbf{x}$ given that it has "originated" from the point described by the vector $\mathbf{x}'$.

Hence the amplitude of finding a particle on the screen at $\mathbf{x} = (x_1, x_2, x_3)$ is given by

$$\left(\frac{m}{2\pi i\hbar T}\right)^{3/2} \left(\frac{m}{2\pi i\hbar T'}\right)^{3/2} \exp im\left(\frac{\mathbf{x}^2 + d^2}{2\hbar T}\right) \exp im\left(\frac{L^2 + d^2}{2\hbar T'}\right)$$

$$\times \left[\exp\left(im\frac{x_2 d}{\hbar T}\right) + \exp\left(-im\frac{x_2 d}{\hbar T}\right)\right] \quad (9.1.21)$$

where $T = t - t''$, $T' = t'' - t'$. This gives a probability density $\propto \cos^2(mx_2 d/\hbar T)$ of finding a particle at $\mathbf{x}$, with constructive interference at $x_2 = n\pi\hbar T/md$ and destructive interference at $x_2 = (2n+1)\pi\hbar T/2md$ for $n = 0, \pm 1, \pm 2, \ldots$. The non-normalizability of this probability density is due to the simplified toy model adopted.

A useful expression for the Green function $G_+^0(\mathbf{x}t; \mathbf{x}'t')$ in (9.1.12) may be also obtained by explicitly carrying out the **p**-integral to obtain a one-dimensional representation.

To the above end, we set

9.1 The Free Green Functions   553

$$\mathbf{x} - \mathbf{x}' = \boldsymbol{\rho}, \qquad t - t' = T \tag{9.1.22}$$

$$(\mathbf{x} - \mathbf{x}') \cdot \mathbf{p} = \rho p \cos\theta \tag{9.1.23}$$

$$\mathrm{d}^3\mathbf{p} = 2\pi p^2 \mathrm{d}p \sin\theta \mathrm{d}\theta \tag{9.1.24}$$

and carry out the angular integration in (9.1.12) to obtain

$$G_+^0(\mathbf{x}t;\mathbf{x}'t') = -\frac{m}{2\pi^2 \hbar \rho} \int_{-\infty}^{\infty} \frac{\mathrm{d}p^0}{2\pi\hbar} \, e^{-ip^0 T/\hbar} \int_0^{\infty} p\, \mathrm{d}p \, \frac{\left[e^{ip\rho/\hbar} - e^{-ip\rho/\hbar}\right]}{(p^2 - 2mp^0 - i\varepsilon)} \tag{9.1.25}$$

or

$$G_+^0(\mathbf{x}t;\mathbf{x}'t') = -\frac{m}{2\pi^2 \hbar \rho} \int_{-\infty}^{\infty} \frac{\mathrm{d}p^0}{2\pi\hbar} \, e^{-ip^0 T/\hbar} \int_{-\infty}^{\infty} p\, \mathrm{d}p \, \frac{e^{ip\rho/\hbar}}{(p^2 - 2mp^0 - i\varepsilon)} \tag{9.1.26}$$

where $T > 0$.

Regarding the denominator in (9.1.26), for $\varepsilon \to +0$, we may make the substitutions

$$p^2 - 2mp^0 - i\varepsilon \longrightarrow \left[p - \left(\sqrt{2mp^0} + i\varepsilon\right)\right]\left[p + \left(\sqrt{2mp^0} + i\varepsilon\right)\right] \tag{9.1.27}$$

for $p^0 > 0$, and

$$p^2 - 2mp^0 - i\varepsilon \longrightarrow \left(p - i\sqrt{2m|p^0|}\right)\left(p + i\sqrt{2m|p^0|}\right) \tag{9.1.28}$$

for $p^0 < 0$.

Since $\rho > 0$, for $\mathbf{x} \neq \mathbf{x}'$, we close the contour from *above* in the complex p-plane to obtain,

$$G_+^0(\mathbf{x}t;\mathbf{x}'t') = -\frac{mi}{2\pi\hbar\rho} \int_{-\infty}^{\infty} \frac{\mathrm{d}p^0}{2\pi\hbar} \, e^{-ip^0 T/\hbar} \left[\Theta(p^0) \, e^{i\sqrt{2mp^0}\rho/\hbar}\right.$$
$$\left. + \Theta(-p^0) \, e^{-\sqrt{2m|p^0|}\rho/\hbar}\right] \tag{9.1.29}$$

which may be rewritten as

$$G_+^0(\mathbf{x}t;\mathbf{x}'t') = \int_{-\infty}^{\infty} \frac{\mathrm{d}p^0}{2\pi\hbar} \, e^{-ip^0 T/\hbar} \widetilde{G}_+^0(\mathbf{x},\mathbf{x}';p^0), \quad T > 0 \tag{9.1.30}$$

where

$$\widetilde{G}_+^0(\mathbf{x},\mathbf{x}';p^0) = \frac{m}{2\pi i\hbar|\mathbf{x}-\mathbf{x}'|} \exp\left(\frac{i}{\hbar}\sqrt{2mp^0}\,|\mathbf{x}-\mathbf{x}'|\right). \tag{9.1.31}$$

Equation (9.1.30) is equivalently given by

$$G_+^0 (\mathbf{x}t; \mathbf{x}'t') = \frac{m}{2\pi i \hbar \rho} \int_0^\infty \frac{dp^0}{2\pi \hbar} \left[e^{-ip^0 T/\hbar} e^{i\rho\sqrt{2mp^0}/\hbar} + e^{ip^0 T/\hbar} e^{-\rho\sqrt{2mp^0}/\hbar} \right]. \tag{9.1.32}$$

One may also define the *advanced* Green function

$$G_-^0 (\mathbf{x}t; \mathbf{x}'t') = \Theta(t' - t) G^0 (\mathbf{x}t; \mathbf{x}'t') \tag{9.1.33}$$

satisfying the differential equation

$$\left[i\hbar \frac{\partial}{\partial t} - H_0 \right] G_-^0 (\mathbf{x}t; \mathbf{x}'t') = -i\hbar \delta^3 (\mathbf{x} - \mathbf{x}') \delta(t' - t). \tag{9.1.34}$$

Its integral representation corresponding to the one in (9.1.12) is given by

$$G_-^0 (\mathbf{x}t; \mathbf{x}'t') = -i\hbar \int \frac{(dp)}{(2\pi \hbar)^4} \frac{e^{i(x-x')p/\hbar}}{\left(p^0 - \frac{\mathbf{p}^2}{2m} - i\varepsilon \right)}, \quad \varepsilon \to +0. \tag{9.1.35}$$

The integrand, as a function of p^0, has a pole at $p^0 = \mathbf{p}^2/2m + i\varepsilon$ in the upper complex p^0-plane. Hence for $t - t' < 0$ (see (9.1.13)), we may close the p^0-contour from above thus obtaining the expression in (9.1.33) by an application of the residue theorem. For $t - t' > 0$, the residue theorem gives zero by closing the p^0-contour from below.

From the identity

$$\frac{1}{p^0 - \frac{\mathbf{p}^2}{2m} + i\varepsilon} - \frac{1}{p^0 - \frac{\mathbf{p}^2}{2m} - i\varepsilon} = -2\pi i \delta \left(p^0 - \frac{\mathbf{p}^2}{2m} \right) \tag{9.1.36}$$

or directly from the definition of the step function we note the relation

$$G_+^0 (\mathbf{x}t; \mathbf{x}'t') + G_-^0 (\mathbf{x}t; \mathbf{x}'t') = G^0 (\mathbf{x}t; \mathbf{x}'t'). \tag{9.1.37}$$

Almost identical derivations as the one in obtaining (9.1.7) in three dimensions, shows that for arbitrary ν dimensions ($\nu = 1, 2, \ldots$), e.g., for $t > t'$

$$G_+^0 (\mathbf{x}t; \mathbf{x}'t') = \left(\frac{m}{2\pi i \hbar (t - t')} \right)^{\nu/2} \exp \frac{im |\mathbf{x} - \mathbf{x}'|^2}{2\hbar (t - t')} \tag{9.1.38}$$

satisfying the boundary condition

$$G_+^0 (\mathbf{x}t; \mathbf{x}'t') \xrightarrow[t \to t']{} \delta^\nu (\mathbf{x} - \mathbf{x}'). \tag{9.1.39}$$

From (9.1.11), (9.1.34) and (9.1.37), we note that unlike $G_\pm^0 (\mathbf{x}t; \mathbf{x}'t')$, $G^0 (\mathbf{x}t; \mathbf{x}'t')$ satisfies a *homogeneous* differential equation

$$\left[i\hbar \frac{\partial}{\partial t} - H_0 \right] G^0 (\mathbf{x}t; \mathbf{x}'t') = 0. \tag{9.1.40}$$

It will still be referred to as a Green function. We next treat Green functions in the presence of interactions.

9.2 Linear and Quadratic Potentials

Consider the Hamiltonian

$$H = \frac{p^2}{2m} - Ex \qquad (9.2.1)$$

in one dimension. The following commutation relations are readily derived

$$[x, p^2] = 2i\hbar p \qquad (9.2.2)$$

$$[x, [x, p^2]] = -2\hbar^2 \qquad (9.2.3)$$

$$[p^2, [x, p^2]] = 0. \qquad (9.2.4)$$

Upon setting

$$A = iETx/\hbar, \qquad B = -iTp^2/2m\hbar \qquad (9.2.5)$$

and using the modified Baker-Campbell-Hausdorff formula (see Appendix I), as applied to the problem at hand

$$\exp(A+B) = \exp\left(\frac{1}{2}[A,B] + \frac{1}{6}[A,[A,B]]\right) \exp B \exp A \qquad (9.2.6)$$

we obtain

$$\langle x | \exp -\frac{iT}{\hbar} H | x' \rangle$$
$$= \exp\left(-\frac{i}{6}\frac{E^2 T^3}{\hbar m}\right) \exp\left(\frac{iTEx'}{\hbar m}\right) \left\langle x \left| \exp\left(\frac{iET^2}{2\hbar m}p - \frac{iT}{\hbar}\frac{p^2}{2m}\right) \right| x' \right\rangle \qquad (9.2.7)$$

where we have used the fact that A is a multiplicative operator when applied to $|x'\rangle$.

The last factor on the right-hand side of (9.2.7) may be rewritten as

$$\langle x | \exp\left(\frac{iET^2}{2\hbar m}p - \frac{iT}{\hbar}\frac{p^2}{2m}\right) | x' \rangle$$
$$= \int_{-\infty}^{\infty} \frac{dp}{2\pi\hbar} \langle x | p \rangle \langle p | x' \rangle \exp\left(\frac{iET^2}{2\hbar m}p\right) \exp\left(\frac{-iT}{\hbar}\frac{p^2}{2m}\right) \qquad (9.2.8)$$

where $\langle x|p \rangle = \exp(ixp/\hbar)$ (see §2.4).

Upon completing the squares in the exponentials in (9.2.8), and integrating over p, as in (9.1.5), we obtain for (9.2.8), the expression

$$\left(\frac{m}{2\pi i\hbar T}\right)^{1/2} \exp\left[\frac{iT}{8\hbar m}\left(ET + \frac{2m}{T}(x-x')\right)^2\right] \quad (9.2.9)$$

giving from (9.2.7)

$$\langle xt|x't'\rangle = \left(\frac{m}{2\pi i\hbar T}\right)^{1/2} \exp\frac{i}{\hbar}\left[\frac{m(x-x')^2}{2T} + \frac{E(x+x')T}{2} - \frac{E^2}{24m}T^3\right] \quad (9.2.10)$$

with $t - t' = T$, which obviously satisfies the boundary condition in (9.1.39).

For the quadratic Hamiltonian,

$$H = \frac{p^2}{2m} + \frac{1}{2}m\omega^2 x^2 \quad (9.2.11)$$

the method just applied to the linear potential is not the best one for this case. The reason is that the commutators of the operators p^2 and x^2 with $[x^2, p^2]$, and their commutation relations with the resulting ones and so on, go on and on unlike the situation in (9.2.2)–(9.2.4) for the linear potential. In this case, it is easier to solve the differential equation satisfied by $G(xt; x't')$ directly.

To the above end, we consider the equation with $t' = 0, t > 0$

$$\left[i\hbar\frac{\partial}{\partial t} + \frac{\hbar^2}{2m}\frac{\partial^2}{\partial x^2} - \frac{1}{2}m\omega^2 x^2\right]G(xt; x'0) = 0. \quad (9.2.12)$$

Quite generally, we may write

$$G = \exp\frac{i}{\hbar}\left[Ax^2 + Bx + C\right] \quad (9.2.13)$$

where A, B, C may, in general, be functions of x' and t, with the solution subjected to satisfy the boundary condition in (9.1.39). Obviously, the exponential in (9.2.13) cannot depend on x^3 or on higher powers of x.

Upon substituting (9.2.13) in (9.2.12), we obtain the following differential equations:

$$\dot{A} + \frac{2}{m}A^2 + \frac{1}{2}m\omega^2 = 0 \quad (9.2.14)$$

$$\dot{B} + \frac{2AB}{m} = 0 \quad (9.2.15)$$

$$\dot{C} + \frac{B^2}{2m} - \frac{i\hbar A}{m} = 0 \quad (9.2.16)$$

for the coefficients of x^2, x and the constant term, respectively, and where $\dot{A} = \partial A/\partial t$. Equation (9.2.15) gives $A = -m\dot{B}/2B$.

9.2 Linear and Quadratic Potentials

By introducing the variable $u = 1/B$, and replacing the expression $A = m\dot{u}/2u$ in (9.2.14), leads to the equation $\ddot{u} + \omega u = 0$, giving the solution

$$A = \frac{m\omega}{2}\left(\frac{-\alpha \sin\omega t + \beta \cos\omega t}{\alpha \cos\omega t + \beta \sin\omega t}\right). \tag{9.2.17}$$

To satisfy the boundary condition (9.1.39), with $A \to m/2t$, for $t \to 0$ (see (9.1.38), (9.2.13)), we set $\alpha = 0$. That is, we have

$$A = \frac{m\omega}{2}\cot\omega t, \qquad B = \frac{1}{\beta \sin\omega t} \tag{9.2.18}$$

and β, which may depend only x', will be determined.

Substituting the expressions for A and B, given in (9.2.18), in (9.2.16) and using the integrals

$$\int \frac{dz}{\sin^2 z} = -\cot z, \qquad \int dz \cot z = \ln(\sin z) \tag{9.2.19}$$

gives the general solution

$$C = \frac{\cot\omega t}{2m\omega\beta^2} + \frac{i\hbar}{2}\ln(\sin\omega t) + C_0 \tag{9.2.20}$$

where C_0 is independent of t.

For G, in (9.2.13), we then have

$$G = \frac{1}{\sqrt{\sin\omega t}}\exp\frac{i}{\hbar}\left[\frac{m\omega}{2}x^2\cot\omega t + \frac{x}{\beta\sin\omega t} + \frac{\cot\omega t}{2m\beta^2}\right]\exp\frac{i}{\hbar}C_0. \tag{9.2.21}$$

Upon comparison with (9.1.38), (9.1.39) for $t \to 0$, we may infer that

$$\frac{1}{\beta} = -m\omega x', \qquad \exp\frac{i}{\hbar}C_0 = \left(\frac{m\omega}{2\pi i\hbar}\right)^{1/2} \tag{9.2.22}$$

giving the final expression,

$$G(xt; x't') = \left(\frac{m\omega}{2\pi i\hbar \sin\omega T}\right)^{1/2}\exp\frac{im\omega}{2\hbar}\left[(x^2 + x'^2)\cot\omega T - \frac{2xx'}{\sin\omega T}\right] \tag{9.2.23}$$

where $T = t - t'$.

We apply the expression in (9.2.23) to provide an independent derivation of the eigenvalues and eigenvectors of the harmonic oscillator.

To the above end, we use the representation for the generating function of the product of two Hermite polynomials:[2]

---

[2] See, for example, Morse and Feshbach (1953), p. 786. [The overall factor $1/\sqrt{1+z^2}$ on the left-hand side of the equation in question here should read $1/\sqrt{1-z^2}$ as is easily checked by the normalizability property of $\exp(-x^2/2)H_n(x)$ by first putting $x = x'$ in (9.2.24).]

$$\frac{1}{\sqrt{1-z^2}} \exp - \left[\frac{x^2 + x'^2 - 2xx'z}{(1-z^2)} \right]$$

$$= \exp\left[-\left(x^2 + x'^2\right)\right] \sum_{n=0}^{\infty} \frac{(z)^n}{2^n n!} H_n(x) H_n(x'). \qquad (9.2.24)$$

We set $z = \exp(-i\omega T)$, giving

$$\frac{z}{1-z^2} = \frac{1}{2i \sin \omega T}, \qquad \frac{1+z^2}{1-z^2} = \frac{1}{i} \cot \omega T. \qquad (9.2.25)$$

Upon multiplying (9.2.24) by

$$\left(\frac{m\omega z}{\pi \hbar}\right)^{1/2} \exp\left(x^2 + x'^2\right)/2$$

and making the replacements $x \to (m\omega/\hbar)^{1/2} x$, $x' \to (m\omega/\hbar)^{1/2} x'$, we obtain from (9.2.24), *and* (9.2.23),

$$G(xt; x't') = \sum_{n=0}^{\infty} \exp\left(-\frac{i}{\hbar} T \left[\hbar\omega (n + 1/2)\right]\right) \left(\frac{m\omega}{\pi \hbar}\right)^{1/4} \frac{e^{-m\omega x^2/2\hbar}}{\sqrt{2^n \, n!}} H_n(x)$$

$$\times \left(\frac{m\omega}{\pi \hbar}\right)^{1/4} \frac{e^{-m\omega x'^2/2\hbar}}{\sqrt{2^n \, n!}} H_n(x') \qquad (9.2.26)$$

from which the eigenvalues and eigenvectors are directly read.

9.3 The Dirac Delta Potential

Consider a particle in a one dimensional Dirac delta potential $\lambda \delta(x)$. To solve for the Green function $G_+(xt'x't')$ satisfying the differential equation

$$\left[i\hbar \frac{\partial}{\partial t} + \frac{\hbar^2}{2m} \frac{\partial^2}{\partial x^2} - \lambda \delta(x)\right] G_+(xt; x't') = i\hbar \, \delta(x - x') \, \delta(t - t') \qquad (9.3.1)$$

we carry out a Fourier transform

$$G_+(xt; x't') = \int_{-\infty}^{\infty} \frac{dp^0}{2\pi\hbar} e^{-ip^0(t-t')/\hbar} \widetilde{G}_+\left(x, x'; p^0\right) \qquad (9.3.2)$$

where $\widetilde{G}_+\left(x, x'; p^0\right)$ satisfies the equation

$$\left[p^0 + \frac{\hbar^2}{2m} \frac{\partial^2}{\partial x^2} - \lambda \delta(x)\right] \widetilde{G}_+\left(x, x'; p^0\right) = i\hbar \, \delta(x - x'). \qquad (9.3.3)$$

The latter may be rewritten as the integral equation

$$\widetilde{G}_+\left(x,x';p^0\right) = \widetilde{G}_+^0\left(x,x';p^0\right)$$
$$-\frac{i}{\hbar}\int_{-\infty}^{\infty}dx''\,\widetilde{G}_+^0\left(x,x'';p^0\right)\lambda\,\delta\left(x''\right)\widetilde{G}_+\left(x'',x';p^0\right) \quad (9.3.4)$$

where $\widetilde{G}_+^0\left(x,x';p^0\right)$ is the free counter part of $\widetilde{G}_+\left(x,x';p^0\right)$ satisfying the equation

$$\left[p^0+\frac{\hbar^2}{2m}\frac{\partial^2}{\partial x^2}\right]\widetilde{G}_+^0\left(x,x';p^0\right) = i\hbar\,\delta\left(x-x'\right) \quad (9.3.5)$$

having the integral form (see (9.1.12))

$$\widetilde{G}_+^0\left(x,x';p^0\right) = i\hbar\int_{-\infty}^{\infty}\frac{dp}{2\pi\hbar}\frac{e^{ip\,(x-x')/\hbar}}{\left[p^0-\dfrac{p^2}{2m}+i\varepsilon\right]} \quad (9.3.6)$$

whose solution (see Problem 9.1) is

$$\widetilde{G}_+^0\left(x,x';p^0\right) = \sqrt{\frac{m}{2p^0}}\exp\left(i\sqrt{2mp^0}\,|x-x'|/\hbar\right). \quad (9.3.7)$$

Upon the application of

$$\left[p^0+\frac{\hbar^2}{2m}\frac{\partial^2}{\partial x^2}\right]$$

to (9.3.4) and using (9.3.5), it is readily verified that (9.3.3) is satisfied.

The integral in (9.3.4) may be explicitly carried out to give

$$\widetilde{G}_+\left(x,x';p^0\right) = \widetilde{G}_+^0\left(x,x';p^0\right) - \frac{i\lambda}{\hbar}\widetilde{G}_+^0\left(x,0;p^0\right)\widetilde{G}_+\left(0,x';p^0\right) \quad (9.3.8)$$

from which

$$\widetilde{G}_+\left(0,x';p^0\right) = \frac{\widetilde{G}_+^0\left(0,x';p^0\right)}{\left[1+i\dfrac{\lambda}{\hbar}\widetilde{G}_+^0\left(0,0;p^0\right)\right]} \quad (9.3.9)$$

and hence from (9.3.8) again, we obtain

$$\widetilde{G}_+\left(x,x';p^0\right) = \widetilde{G}_+^0\left(x,x';p^0\right) - i\frac{\lambda}{\hbar}\frac{\widetilde{G}_+^0\left(x,0;p^0\right)\widetilde{G}_+^0\left(0,x';p^0\right)}{\left[1+i\dfrac{\lambda}{\hbar}\widetilde{G}_+^0\left(0,0;p^0\right)\right]}. \quad (9.3.10)$$

In detail, (9.3.7), (9.3.10) give

$$\widetilde{G}_+\left(x,x';p^0\right) = \sqrt{\frac{m}{2p^0}}\,e^{i\sqrt{2mp^0}|x-x'|/\hbar} - i\frac{\lambda}{\hbar}\frac{m}{2}\frac{1}{\sqrt{p^0}}\frac{e^{i\sqrt{2mp^0}(|x|+|x'|)/\hbar}}{\left[\sqrt{p^0}+i\dfrac{\lambda}{\hbar}\sqrt{\dfrac{m}{2}}\right]}. \quad (9.3.11)$$

For $\lambda < 0$, it is convenient to write

$$\frac{-\lambda\Theta(-\lambda)}{2\sqrt{p^0}\left[\sqrt{p^0}+\frac{i\lambda}{\hbar}\sqrt{\frac{m}{2}}\right]} = \frac{|\lambda|\Theta(-\lambda)}{\left[p^0+\frac{m\lambda^2}{2\hbar^2}\right]} - \frac{|\lambda|\Theta(-\lambda)}{2\sqrt{p^0}\left[\sqrt{p^0}+\frac{i|\lambda|}{\hbar}\sqrt{\frac{m}{2}}\right]} \quad (9.3.12)$$

and for $\lambda > 0$,

$$\frac{-\lambda\Theta(\lambda)}{2\sqrt{p^0}\left[\sqrt{p^0}+\frac{i\lambda}{\hbar}\sqrt{\frac{m}{2}}\right]} = -\frac{|\lambda|\Theta(\lambda)}{2\sqrt{p^0}\left[\sqrt{p^0}+\frac{i|\lambda|}{\hbar}\sqrt{\frac{m}{2}}\right]} \quad (9.3.13)$$

to obtain for all $\lambda < 0$, $\lambda > 0$,

$$\widetilde{G}_+(x,x';p^0) = \frac{im|\lambda|}{\hbar}\Theta(-\lambda)\frac{e^{i\sqrt{2mp^0}(|x|+|x'|)/\hbar}}{\left[p^0+\frac{m\lambda^2}{2\hbar^2}+i\varepsilon\right]}$$

$$+ \sqrt{\frac{m}{2p^0}}\left\{e^{i\sqrt{2mp^0}|x-x'|/\hbar} - i\frac{|\lambda|}{\hbar}\sqrt{\frac{m}{2}}\frac{e^{i\sqrt{2mp^0}(|x|+|x'|)/\hbar}}{\left[\sqrt{p^0}+\frac{i|\lambda|}{\hbar}\sqrt{\frac{m}{2}}\right]}\right\} \quad (9.3.14)$$

where we have used the fact that $\Theta(\lambda)+\Theta(-\lambda)=1$.

For $\lambda < 0$, $\widetilde{G}_+(x,x';p^0)$ develops a pole at

$$p^0 = \frac{-m\lambda^2}{2\hbar^2} \quad (9.3.15)$$

(compare with (4.2.33)),[3] coming from the denominator of the first term in (9.3.14). By adding $+i\varepsilon$ to p^0 in the latter denominator, in conformity with (9.1.12), and closing the p^0 contour in the complex p^0-plane from below, and by an application of the residue theorem (see Problem 9.8), this pole gives rise to $G_+(xt;x't')$, in (9.3.2), the contribution

$$-\frac{2\pi i}{2\pi\hbar}\frac{im|\lambda|}{\hbar}\theta(-\lambda)\exp\left[-\frac{m|\lambda|}{\hbar^2}(|x|+|x'|)\right]\exp\left[\frac{im\lambda^2}{2\hbar^3}(t-t')\right] \quad (9.3.16)$$

leading to the bound state wavefunction

$$\psi(x) = \frac{\sqrt{m|\lambda|}}{\hbar}\exp\left[-\frac{m|\lambda|}{\hbar^2}|x|\right] \quad (9.3.17)$$

consistent with (4.2.27).

---

[3] Here we let λ carry its own sign.

The second expression in the curly brackets in (9.3.14) also contains useful information. Suppose $x' < 0$, $p^0 > 0$. Then for $x > 0$, the entire second term in (9.3.14) becomes equal to

$$\sqrt{\frac{m}{2p^0}} \, e^{i\sqrt{2mp^0}(|x|+|x'|)/\hbar} \left[\frac{\sqrt{p^0}}{\sqrt{p^0} + \frac{i|\lambda|}{\hbar}\sqrt{\frac{m}{2}}} \right] \quad (9.3.18)$$

from which we may infer that the transmission coefficient is given by

$$T = \frac{\sqrt{p^0}}{\sqrt{p^0} + \frac{i|\lambda|}{\hbar}\sqrt{\frac{m}{2}}}. \quad (9.3.19)$$

For $x < 0$, we have for the reflection coefficient, directly from the second term within the curly brackets in (9.3.14), the expression

$$R = \frac{-\frac{i|\lambda|}{\hbar}\sqrt{\frac{m}{2}}}{\sqrt{p^0} + \frac{i|\lambda|}{\hbar}\sqrt{\frac{m}{2}}}. \quad (9.3.20)$$

9.4 Time-Dependent Forced Dynamics

We consider the forced linear potential, with the Schrödinger equation given by

$$\left[i\hbar \frac{\partial}{\partial t} + \frac{\hbar^2}{2m} \frac{\partial^2}{\partial x^2} + xF(t) \right] \psi(x,t) = 0 \quad (9.4.1)$$

where $F(t)$ is a given c-function.

Let $\eta(t)$ provide the classical solution of the problem. That is, it satisfies the classical equation of motion

$$m\ddot{\eta}(t) = F(t) \quad (9.4.2)$$

with boundary conditions taken to be

$$\eta(t') = x', \, \eta(t) = x \quad (9.4.3)$$

for some $t > t'$.

We define the quantum deviation from the classical path

$$z = x - \eta \quad (9.4.4)$$

and note that in terms of the new variable z, we have to make the replacement

$$\frac{\partial}{\partial t} \longrightarrow \frac{\partial}{\partial t} + \frac{\partial z}{\partial t}\frac{\partial}{\partial z} = \frac{\partial}{\partial t} - \dot{\eta}\frac{\partial}{\partial z} \qquad (9.4.5)$$

for the time derivative in (9.4.1).

In terms of the new variable, the Schrödinger equation (9.4.1) reads

$$\left[i\hbar\frac{\partial}{\partial t} - i\hbar\,\dot{\eta}\frac{\partial}{\partial z} + \frac{\hbar^2}{2m}\frac{\partial^2}{\partial z^2} + (z + \eta(t))\,F(t)\right]\psi = 0. \qquad (9.4.6)$$

Upon setting

$$\psi = \exp\left(\frac{imz\dot{\eta}}{\hbar}\right)\Phi \qquad (9.4.7)$$

the following differential equation for Φ is obtained

$$\left[i\hbar\frac{\partial}{\partial t} + \frac{\hbar^2}{2m}\frac{\partial^2}{\partial z^2} + \left(\frac{m\dot{\eta}^2}{2} + \eta F\right)\right]\Phi = 0 \qquad (9.4.8)$$

Here we recognize, in the last term in the square brackets, the classical Lagrangian

$$L_c(t) = \frac{m\dot{\eta}^2(t)}{2} + \eta(t)\,F(t). \qquad (9.4.9)$$

This suggests to set

$$\Phi = \left[\exp\frac{i}{\hbar}\int_0^t d\tau\, L_c(\tau)\right]\chi \qquad (9.4.10)$$

to obtain the free Schrödinger equation

$$\left[i\hbar\frac{\partial}{\partial t} + \frac{\hbar^2}{2m}\frac{\partial^2}{\partial z^2}\right]\chi = 0. \qquad (9.4.11)$$

Hence from (9.4.7), (9.4.10), the solution of (9.4.1) is given by

$$\psi(x,t) = \left(\exp\frac{im}{\hbar}(x-\eta)\dot{\eta}\right)\left(\exp\frac{i}{\hbar}\int_0^t d\tau\, L_c(\tau)\right)\chi(x-\eta,t). \qquad (9.4.12)$$

To obtain the Green function, which has the advantageous of not being tied up to any wavefunctions, we note that from

$$\psi(x,t) = \int_{-\infty}^{\infty} dx'\, G(xt; x't')\,\psi(x'; t') \qquad (9.4.13)$$

we may write for χ in (9.4.12)

$$\chi(x-\eta, t) = \left(\exp-\frac{im}{\hbar}(x-\eta)\dot{\eta}\right)\left(\exp-\frac{i}{\hbar}\int_0^t d\tau\, L_c(\tau)\right)$$

$$\times \int_{-\infty}^{\infty} dx'\, G(xt; x't')\left(\exp\frac{im}{\hbar}(x'-\eta')\dot{\eta}'\right)$$

$$\times \left(\exp \frac{i}{\hbar} \int_0^{t'} d\tau\, L_c(\tau) \right) \chi(x' - \eta', t') \qquad (9.4.14)$$

where $\eta' = \eta(t')$, $\dot{\eta}' = \dot{\eta}(t')$.

But for the free Schrödinger equation in (9.4.11) we have

$$\chi(z, t) = \int_{-\infty}^{\infty} dz'\, G^0(zt; z't') \chi(z'; t'). \qquad (9.4.15)$$

Upon comparing (9.4.14) with (9.4.15), this gives

$$G(xt; x't') = \exp \frac{im}{\hbar} \left[(x - \eta)\dot{\eta} - (x' - \eta')\dot{\eta}' \right] \left(\exp \frac{i}{\hbar} \int_{t'}^{t} d\tau\, L_c(\tau) \right)$$

$$\times G^0(x - \eta, t; x' - \eta', t') \qquad (9.4.16)$$

where

$$G^0(x - \eta, t; x' - \eta', t')$$

$$= \left(\frac{m}{2\pi i \hbar (t - t')} \right)^{1/2} \exp \left[\frac{im}{2\hbar(t - t')} (x - \eta(t) - x' + \eta(t'))^2 \right]. \qquad (9.4.17)$$

With the boundary conditions in (9.4.3), we obtain

$$G(xt; x't') = \left(\frac{m}{2\pi i \hbar (t - t')} \right)^{1/2} \exp \frac{i}{\hbar} \int_{t'}^{t} d\tau\, L_c(\tau). \qquad (9.4.18)$$

On the other hand the explicit expression for $\eta(\tau)$, in terms of $F(\tau)$, is from (9.4.2), (9.4.3) given by

$$\eta(\tau) = \int_{\tau}^{t} d\tau' \frac{K(\tau')}{m} - \frac{(t - \tau)}{(t - t')} \int_{t'}^{t} d\tau' \frac{K(\tau')}{m} + x \frac{(\tau - t')}{(t - t')} + x' \frac{(t - \tau)}{(t - t')} \qquad (9.4.19)$$

where

$$K(\tau) = \int_{\tau}^{t} d\tau' F(\tau') \qquad (9.4.20)$$

suppressing the t dependence for simplicity of the notation, and note that $K(t) = 0$.

Upon substitution of (9.4.19), (9.4.20) in (9.4.18), this leads to

$$G(xt; x't') = \left(\frac{m}{2\pi i \hbar T} \right)^{1/2} \exp \frac{i}{\hbar} \left\{ \frac{m(x - x')^2}{2\,T} + \frac{(x - x')}{T} \int_{t'}^{t} d\tau\, K(\tau) \right.$$

$$+\frac{1}{2mT}\left(\int_{t'}^{t}d\tau\,K(\tau)\right)^{2}-\frac{1}{2m}\int_{t'}^{t}d\tau\,(K(\tau))^{2}+x'K(t')\bigg\}$$
(9.4.21)

where $T = t - t'$.

As another application of a time-dependent forced dynamics, we consider the interaction described by the Schrödinger equation

$$\left[i\hbar\frac{\partial}{\partial t}+\frac{\hbar^{2}}{2m}\frac{\partial^{2}}{\partial x^{2}}-\frac{1}{2}m\omega^{2}x^{2}+xF(t)\right]\psi(x,t)=0.$$
(9.4.22)

The corresponding classical solution $\eta(t)$ satisfies the equation of motion

$$m\left(\ddot{\eta}+\omega^{2}\eta\right)=F$$
(9.4.23)

with boundary conditions taken to be

$$\eta(t')=x',\,\eta(t)=x.$$
(9.4.24)

In terms of the quantum deviation $z = x - \eta$, from the classical solution, and in terms of the function Φ defined through

$$\psi=\exp\left(\frac{imz}{\hbar}\dot{\eta}\right)\Phi$$
(9.4.25)

(see (9.4.7)), (9.4.22) becomes

$$\left[i\hbar\frac{\partial}{\partial t}+\frac{\hbar^{2}}{2m}\frac{\partial^{2}}{\partial z^{2}}-\frac{m\omega^{2}z^{2}}{2}+\left(\frac{m\dot{\eta}^{2}}{2}-\frac{m\omega^{2}\eta^{2}}{2}+\eta F\right)\right]\Phi=0$$
(9.4.26)

where

$$\frac{m\dot{\eta}^{2}}{2}-\frac{m\omega^{2}\eta^{2}}{2}+\eta F=L_{c}$$
(9.4.27)

is the classical Lagrangian.

Upon setting

$$\Phi=\left[\exp\frac{i}{\hbar}\int_{0}^{t}d\tau\,L_{c}(\tau)\right]\chi$$
(9.4.28)

the equation corresponding to the one in (9.4.11) becomes

$$\left[i\hbar\frac{\partial}{\partial t}+\frac{\hbar^{2}}{2m}\frac{\partial^{2}}{\partial z^{2}}-\frac{1}{2}m\omega^{2}z^{2}\right]\chi=0.$$
(9.4.29)

The same reasoning as given in deriving (9.4.12)–(9.4.17) shows that

$G(xt;x't')$

$$=\exp\frac{im}{\hbar}\left[(x-\eta)\dot{\eta}-(x'-\eta')\dot{\eta}'\right]\left(\exp\frac{i}{\hbar}\int_{t'}^{t}d\tau\,L_{c}(\tau)\right)\left(\frac{m\omega}{2\pi i\hbar\sin\omega T}\right)^{1/2}$$

$$\times \exp \frac{i m \omega}{2\hbar} \left\{ \left[(x-\eta)^2 + (x'-\eta')^2 \right] \cot \omega T - \frac{2(x-\eta)(x'-\eta')}{\sin \omega T} \right\}$$
(9.4.30)

where we have used (9.2.23) for the harmonic oscillator Green function with variables z, z', and we recall that $\eta' = \eta(t')$, $\dot{\eta}' = \dot{\eta}(t')$, $(t-t') = T$.

With the boundary conditions in (9.4.24), the expression in (9.4.30) becomes

$$G(xt; x't') = \left(\frac{m\omega}{2\pi i\hbar \sin \omega T} \right)^{1/2} \exp \frac{i}{\hbar} \int_{t'}^{t} d\tau\, L_c(\tau). \qquad (9.4.31)$$

Upon solving the classical system in (9.4.23), (9.4.24), as done for the linear potential, and substituting in (9.4.31) gives after some labor

$$G(xt; x't') = \left(\frac{m\omega}{2\pi i\hbar \sin \omega T} \right)^{1/2} \exp \left(i \frac{m\omega}{\hbar \sin \omega T} \left\{ \frac{(x^2 + x'^2)}{2} \cos \omega T \right. \right.$$

$$- xx' + \frac{1}{m\omega} \int_{t'}^{t} d\tau\, F(\tau) [x \sin \omega (\tau - t') + x' \sin \omega (t - \tau)]$$

$$\left. \left. - \frac{1}{m^2 \omega^2} \int_{t'}^{t} d\tau\, F(\tau) \sin \omega (t-\tau) \int_{t'}^{\tau} d\tau' \sin \omega (\tau'-t') F(\tau') \right\} \right). \quad (9.4.32)$$

The factor $\exp\left(i \int_{t'}^{t} d\tau\, L_c(\tau)/\hbar\right)$ in (9.4.18), (9.4.31), with the *classical* action is a general property of quadratic interactions such as: $\left[m\omega^2(t) x^2/2 - xF(t) \right]$ (see Problem 9.23).

9.5 The Law of Reflection and Reconciliation with the Classical Law

In this section, we use the method of Green functions to carry out a simplified, but illuminating, analytical treatment of the reflection[4] of a particle off a reflecting (infinite) plane surface which is taken to be the $z = 0$ plane. To this end, we first determine the Green function in half-space $z \geq 0$, denoted by $G^0_{+>}(\mathbf{x}t; \mathbf{x}'t')$, with boundary conditions

$$G^0_{+>}(\mathbf{x}t; \mathbf{x}'t') \Big|_{z=0} = 0 \qquad G^0_{+>}(\mathbf{x}t; \mathbf{x}'t') \Big|_{z'=0} = 0 \qquad (9.5.1)$$

where we have used the notation $\mathbf{x} = (x, y, z)$, and for $z < 0$, $z' < 0$, a particle is assumed not to be able to penetrate.

---
[4] The law of reflection has received some special, rather non-technical, but nevertheless fascinating treatment by Feynman (1985). Here we are considering massive non-relativistic particles.

Since the physics of the problem occurs in the region $-\infty < x < \infty$, $-\infty < y < \infty$, $0 \leq z < \infty$, we may develop Fourier transforms in the first two variables and a Fourier-sine transform in the last variable to satisfy the boundary conditions in (9.5.1). That is, we use a representation for the delta distribution given by[5]

$$\delta^3(\mathbf{x}, \mathbf{x}') = \int_{\mathbb{R}^2} \frac{d^2\mathbf{p}}{(2\pi\hbar)^2} \int_0^\infty \frac{2}{\pi\hbar} dq \exp\left[i\mathbf{p} \cdot (\mathbf{r} - \mathbf{r}')/\hbar\right] \sin\frac{qz}{\hbar} \sin\frac{qz'}{\hbar} \quad (9.5.2)$$

where $\mathbf{x} = (\mathbf{r}, z)$, with $\mathbf{r}, \mathbf{r}'$ parallel to the (x, y)-plane, to obtain for $G^0_{+>}$ (see (9.1.12))

$$G^0_{+>}(\mathbf{x}t; \mathbf{x}'t') = i\hbar \int_{\mathbb{R}^2} \frac{d^2\mathbf{p}}{(2\pi\hbar)^2} \int_0^\infty \frac{2}{\pi\hbar} dq$$

$$\times \int_{-\infty}^\infty \frac{dp^0}{2\pi\hbar} \frac{e^{i\mathbf{p}\cdot(\mathbf{r}-\mathbf{r}')/\hbar} e^{-ip^0 T/\hbar}}{\left[p^0 - \left(\frac{\mathbf{p}^2}{2m} + \frac{q^2}{2m} - i\varepsilon\right)\right]} \sin\frac{qz}{\hbar} \sin\frac{qz'}{\hbar}. \quad (9.5.3)$$

where $T = t - t' > 0$. The latter may be integrated over p^0, to yield

$$G^0_{+>}(\mathbf{x}t; \mathbf{x}'t') = \int_{\mathbb{R}^2} \frac{d^2\mathbf{p}}{(2\pi\hbar)^2} \int_0^\infty \frac{2}{\pi\hbar} dq\, e^{i\mathbf{p}\cdot(\mathbf{r}-\mathbf{r}')/\hbar} e^{-i(\mathbf{p}^2+q^2)T/2m\hbar}$$

$$\times \sin\frac{qz}{\hbar} \sin\frac{qz'}{\hbar} \quad (9.5.4)$$

leading finally to the closed form expression

$$G^0_{+>}(\mathbf{x}t; \mathbf{x}'t') = \left(\frac{m}{2\pi i\hbar T}\right)^{3/2} \exp\frac{im|\mathbf{r} - \mathbf{r}'|^2}{2\hbar T}$$

$$\times \left[\exp\frac{im(z - z')^2}{2\hbar T} - \exp\frac{im(z + z')^2}{2\hbar T}\right]. \quad (9.5.5)$$

The Green function $G^0_{+>}$ satisfies the completeness relation

$$\int_{\mathbb{R}^2} d^2\mathbf{r}'' \int_0^\infty dz'' G^0_{+>}(\mathbf{x}t; \mathbf{x}''t'') G^0_{+>}(\mathbf{x}''t''; \mathbf{x}'t') = G^0_{+>}(\mathbf{x}t; \mathbf{x}'t') \quad (9.5.6)$$

where $t > t'' > t'$.

As a simplified description, and as a working hypothesis, a particle which reaches within the interval $0 \leq z \leq \delta$, for some given small $\delta > 0$, providing loosely speaking a "skin depth" for the reflecting body, above the $z = 0$ plane, is considered to have reached the reflecting body.

---

[5] We follow the treatment given in: Manoukian (1987c).

9.5 Law of Reflection and Reconciliation with Classical Law

Suppose that a particle initially emitted at time $t' = 0$ from a point Q at a height $z' \gg \delta$, above the $z = 0$ plane, reaches the reflecting body (location within unknown), at some time, say, T'. *Given* that this has occurred with *probability one*, we determine *partial* contributions to the full *conditional* amplitude of finding the reflected particle at any given z at time t.

To obtain the conditional amplitude of detecting a particle at some height z, given that the particle has reached the reflecting body, we consider, in the process, a Gaussian region, along the z-axis, about the point $z = \delta/2$, with standard deviation σ and integrate, for simplicity, symmetrically for the amplitude along the z-axis. We thus obtain for the latter amplitude the expression $(T = t - T')$,

$$\left(\frac{m}{2\pi i \hbar T'}\right)^{1/2} \left(\frac{m}{2\pi i \hbar T}\right)^{1/2} I_3 \tag{9.5.7}$$

where, using the z-dependent part of the Green function in (9.5.5),

$$I_3 = \int_{-\infty}^{\infty} \frac{dZ}{\sqrt{2\pi}\sigma} \left[\exp\frac{im(z-Z)^2}{2\hbar T} - \exp\frac{im(z+Z)^2}{2\hbar T}\right]$$

$$\times \left[\exp\frac{im(Z-z')^2}{2\hbar T'} - \exp\frac{im(Z+z')^2}{2\hbar T'}\right] \exp-\frac{(Z-\delta/2)^2}{2\sigma^2}. \tag{9.5.8}$$

Here we have extended the Z-integration beyond the region $0 < Z < \delta$, which is justified provided the integral

$$\frac{1}{\sqrt{2\pi}\sigma} \int_{-\infty}^{0} dZ \exp-\frac{(Z-\delta/2)^2}{2\sigma^2} = \frac{1}{\sqrt{2\pi}\sigma} \int_{\delta}^{\infty} dZ \exp-\frac{(Z-\delta/2)^2}{2\sigma^2}$$

$$= \frac{1}{\sqrt{2\pi}} \int_{\delta/2\sigma}^{\infty} dZ \exp-Z^2/2 \tag{9.5.9}$$

is small, where the first equality follows from symmetry. An upper bound to the integral in (9.5.9) is given by (see Problem 9.11)

$$\frac{1}{\sqrt{2\pi}} \int_{a}^{\infty} dZ \exp-Z^2/2 \leqslant \sqrt{\frac{2}{\pi}} \frac{\exp-a^2/2}{a}, \quad a = \delta/2\sigma > 0. \tag{9.5.10}$$

For $\delta/2\sigma = 10$, for example, the upper bound in (9.5.10) is bounded above by 16×10^{-24} in comparison to one for the normalized Gaussian distribution.

The integral in (9.5.8) may be explicitly carried out by using the integral of three Gaussian functions

$$\int_{-\infty}^{\infty} dz \, \frac{\exp-(z_1-z)^2/2\sigma_1^2}{\sqrt{2\pi}\sigma_1} \frac{\exp-(z_2-z)^2/2\sigma_2^2}{\sqrt{2\pi}\sigma_2} \frac{\exp-(z_3-z)^2/2\sigma_3^2}{\sqrt{2\pi}\sigma_3}$$

$$= \frac{1}{2\pi\sqrt{A}} \exp\left[-\frac{B}{2A}\right] \qquad (9.5.11)$$

where

$$A = \sigma_1^2\sigma_2^2 + \sigma_2^2\sigma_3^2 + \sigma_3^2\sigma_1^2 \qquad (9.5.12)$$

$$B = \sigma_3^2(z_1 - z_2)^2 + \sigma_2^2(z_1 - z_3)^2 + \sigma_1^2(z_2 - z_3)^2. \qquad (9.5.13)$$

Hence upon normalizing the amplitude in (9.5.7), we obtain for the conditional probability density of detecting the particle at z given that it had reached the reflecting body, and was emitted initially from a point of z-coordinate z', the expression $(z', z \gg \delta)$

$$P_3 = \frac{\frac{2T'}{\sqrt{\pi}}\sqrt{\frac{\sigma^2}{C}}\exp\left(-\frac{\sigma^2}{C}z^2 T'^2\right)\left[\cosh\left(\frac{2TT'\sigma^2 zz'}{C}\right) - \cos\left(\frac{2(T+T')\sigma^4 zz'}{\hbar C/m}\right)\right]}{\left[\exp\left(\frac{\sigma^2 z'^2 T^2}{C}\right) - \exp\left(-\frac{(T+T')^2\sigma^6 z'^2}{T'^2\hbar^2 C/m^2}\right)\right]} \qquad (9.5.14)$$

where

$$C = \sigma^4(T+T')^2 + T^2 T'^2 \frac{\hbar^2}{m^2}. \qquad (9.5.15)$$

The probability density is non-zero for almost all finite z. However for a large (classical) mass such that

$$\frac{m}{T'}\sigma^2 \gg \hbar \qquad (9.5.16)$$

with an order of magnitude reference set up by $\hbar$, and a macroscopic limit

$$\sigma \ll \frac{\sqrt{TT'}}{T+T'}\min(z, z'), \qquad (9.5.17)$$

P_3 simplifies to

$$P_3 \simeq \frac{1}{\sqrt{\pi}\sigma}\frac{T'}{T+T'}\exp\left[-\frac{T'^2}{\sigma^2(T+T')^2}\left(z - z'\frac{T}{T'}\right)^2\right] \qquad (9.5.18)$$

and is easily verified to be normalized over z. The density is highly peaked at the classical value for z, for small σ^2 satisfying (9.5.17), but with m sufficiently large so that (9.5.16) is also satisfied.

Now *given* that a particle had reached the reflecting body and is detected at a point (x, y, z), we determine partial contributions to the full *conditional amplitude* that the particle's detection point has (x, y)-coordinates x and y, which was initially emitted from the point (x', y', z').

Given that the above experiment has been realized, a partial contribution to the full conditional amplitude will be obtained as coming from an integration over a Gaussian region in the (x, y) plane with standard deviations

9.5 Law of Reflection and Reconciliation with Classical Law

σ_1, σ_2, respectively, about an arbitrary point $(\bar{x}, \bar{y})$. In this case, using the x, y dependent part of (9.5.5), and the product formula for three Gaussian functions in (9.5.11), one easily obtains for the partial contribution to the conditional amplitude the expression

$$\frac{i\hbar}{2\pi m} \frac{T+T'}{\sqrt{A_1 A_2}} e^{i\alpha} \exp\left[-\frac{\sigma_1^2}{C_1}(T+T')^2 \left(\frac{x'T + xT'}{T+T'} - \bar{x}\right)^2\right]$$

$$\times \exp\left[-\frac{\sigma_2^2}{C_2}(T+T')^2 \left(\frac{y'T + yT'}{T+T'} - \bar{y}\right)^2\right] \quad (9.5.19)$$

where

$$A_j = \left[\sigma_j^2 (T+T') \frac{i\hbar}{m} - \frac{TT'\hbar^2}{m^2}\right] \quad (9.5.20)$$

$$C_j = \left(\sigma_j^4 (T+T')^2 + T^2 T'^2 \frac{\hbar^2}{m^2}\right) \quad (9.5.21)$$

and α, in the phase factor $\exp(i\alpha)$ in (9.5.19), is given by

$$\alpha = \frac{\hbar}{2m}(T+T')TT' \left[\frac{1}{C_1}\left(\frac{x'T + xT'}{T+T'} - \bar{x}\right)^2 + \frac{1}{C_2}\left(\frac{y'T + yT'}{T+T'} - \bar{y}\right)^2\right] \quad (9.5.22)$$

The partial amplitude (9.5.19) is properly normalized. We obtain the full amplitude by multiplying (9.5.19) by $2\pi\sigma_1\sigma_2$ and taking the limits $\sigma_1 \to \infty$, $\sigma_2 \to \infty$, giving unity, as expected, since one is then covering the entire reflecting surface. We note that since the partial amplitudes in (9.5.19) do not vanish for different pairs $(\bar{x}, \bar{y})$, the reflections may occur from almost anywhere on the surface.

Again for a large (classical) mass m such that

$$m \frac{T+T'}{TT'} \sigma_j^2 \gg \hbar \quad (9.5.23)$$

with an order of magnitude reference set up by $\hbar$, the partial amplitudes (9.5.19) become arbitrary small for all $\bar{x}$, $\bar{y}$, that is, as we move all around the reflecting surface, and for sufficiently small σ_j, unless

$$\bar{x} - \frac{x'T + xT'}{T+T'}, \quad \bar{y} - \frac{y'T + yT'}{T+T'} \quad (9.5.24)$$

are arbitrarily small. On the other hand, with the condition (9.5.16) satisfied with (9.5.17) not violated, the probability density P_3 in (9.5.18) becomes small unless

$$z - z'\frac{T}{T'} \quad (9.5.25)$$

is arbitrary small. The conditions (9.5.16), (9.5.23) may be satisfied for a classical particle, for large m. For a sufficiently large m, which will be the case for a classical particle as just mentioned, the standard deviations $\sigma, \sigma_1, \sigma_2$ may be taken to be quite small, giving a point-like impression for the reflection region about $(\bar{x}, \bar{y})$, without violating the bounds in (9.5.16), (9.5.23), thus recovering the law of reflection, corresponding to the expressions in (9.5.24), (9.5.25) being arbitrarily small.

9.6 Two-Dimensional Green Function in Polar Coordinates: Application to the Aharonov-Bohm Effect

The two dimensional retarded free Green function is from (9.1.38) given by

$$G_+^0 (\mathbf{x}t; \mathbf{x}'t') = \frac{m}{2\pi i \hbar T} \exp \frac{im |\mathbf{x} - \mathbf{x}'|^2}{2\hbar T} \tag{9.6.1}$$

where $T = t - t' > 0$. Working in polar coordinates, $\mathbf{x} = r(\cos\phi, \sin\phi)$, we may rewrite the latter as

$$G_+^0 (\mathbf{x}t; \mathbf{x}'t') = \frac{m}{2\pi i \hbar T} \exp\left(\frac{im(r^2 + r'^2)}{2\hbar T}\right) \exp\left(\frac{mrr' \cos(\phi - \phi')}{i\hbar T}\right). \tag{9.6.2}$$

We use the generating function of the modified Bessel function[6] $I_k(z)$

$$\exp \frac{z}{2}\left(s + \frac{1}{s}\right) = \sum_{k=-\infty}^{\infty} (s)^k I_k(z) \tag{9.6.3}$$

where

$$I_{-k}(z) = I_k(z) \tag{9.6.4}$$

for integer k values.

The modified Bessel functions $I_\nu(z)$, with ν not necessarily an integer, satisfy the differential equation

$$\left[z^2 \frac{d^2}{dz^2} + z\frac{d}{dz} - (z^2 + \nu^2)\right] I_\nu(z) = 0. \tag{9.6.5}$$

Upon choosing $s = \exp i(\phi - \phi')$, we may rewrite the Green function in (9.6.2) as

$$G_+^0 (\mathbf{x}t; \mathbf{x}'t') = \sum_{k=-\infty}^{\infty} e^{ik(\phi - \phi')} F_{0k}^+ (rt; r't') \tag{9.6.6}$$

where $(T > 0)$

---

[6] The classic reference on Bessel functions is: Watson (1966).

9.6 2D Green Function in Polar Coordinates

$$F_{0k}^{+}(rt;r't') = \frac{m}{2\pi i\hbar T} \exp\left(\frac{im(r^2 + r'^2)}{2\hbar T}\right) I_{|k|}(-i\rho) \qquad (9.6.7)$$

$$\rho = \frac{mrr'}{\hbar T}. \qquad (9.6.8)$$

The expansion (9.6.6) is just a Fourier series expansion in complex form, where

$$\delta(\phi - \phi') = \sum_{k=-\infty}^{\infty} \frac{e^{ik(\phi-\phi')}}{2\pi} \qquad (9.6.9)$$

for the angular part of the Dirac delta distribution.

We may use the Poisson sum formula (see Problem 9.13),

$$\sum_{k=-\infty}^{\infty} f(k) = \sum_{k=-\infty}^{\infty} \int_{-\infty}^{\infty} d\lambda\, e^{i2\pi k\lambda} f(\lambda) \qquad (9.6.10)$$

for a given function $f(\lambda)$, to rewrite (9.6.6) as

$$G_{+}^{0}(\mathbf{x}t;\mathbf{x}'t') = \sum_{k=-\infty}^{\infty} H_{k}^{0}(\mathbf{x}t;\mathbf{x}'t') \qquad (9.6.11)$$

where

$$H_{k}^{0}(\mathbf{x}t;\mathbf{x}'t') = \int_{-\infty}^{\infty} d\lambda\, e^{i\lambda(\phi-\phi'+2\pi k)} F_{0\lambda}^{+}(rt;r't') \qquad (9.6.12)$$

and for $T > 0$, we have from (9.6.7)

$$F_{0\lambda}^{+}(rt;r't') = \frac{m}{2\pi i\hbar T} \exp\left(\frac{im(r^2 + r'^2)}{2\hbar T}\right) I_{|\lambda|}(-i\rho) \qquad (9.6.13)$$

providing an alternative representation of the two dimensional Green function G_{+}^{0} in polar coordinates. Note that $\exp(ik(\phi - \phi'))F_{0k}^{+}$ and H_{k}^{0} are *not* the same functions (see (9.6.10)).

From (9.6.11), (9.6.12), we may infer that the amplitude for a particle to go from (r', ϕ') to (r, ϕ) may be written as a sum of amplitudes, each specified by an integer k. For $k \neq 0$, $k \neq -1$, an amplitude in question corresponds to one to go from (r', ϕ') to (r, ϕ) by winding around the origin exactly $|k|$ times on the way to the point (r, ϕ). The sense of a rotation is specified by the sign of k with a c.c.w. one for $k \geq 1$ and a c.w., one for $k \leqslant -2$. For $k = 0$, $k = -1$, we have a c.c.w., c.w. rotations, respectively, not making full circles around the origin. A given integer is appropriately referred to as a winding number.

The differential equation satisfied by $G_{+}^{0}(\mathbf{x}t;\mathbf{x}'t')$ including the factor $\Theta(t - t')$, in polar coordinates is given by

$$\left[i\hbar \frac{\partial}{\partial t} + \frac{\hbar^2}{2m} \left(\frac{1}{r} \frac{\partial}{\partial r} r \frac{\partial}{\partial r} + \frac{1}{r^2} \frac{\partial^2}{\partial \phi^2} \right) \right] G_+^0 (\mathbf{x}t; \mathbf{x}'t')$$

$$= i\hbar \frac{\delta(r-r')}{r} \delta(\phi - \phi') \delta(t - t'), \quad (9.6.14)$$

where we have used the definition

$$\delta^2(\mathbf{x} - \mathbf{x}') = \frac{\delta(r - r')}{r} \delta(\phi - \phi'), \quad (9.6.15)$$

implies from (9.6.6) and (9.6.9), (9.6.13) that

$$\left[i\hbar \frac{\partial}{\partial t} + \frac{\hbar^2}{2m} \left(\frac{1}{r} \frac{\partial}{\partial r} r \frac{\partial}{\partial r} - \frac{k^2}{r^2} \right) \right] F_{0k}^+ (rt; r't') = \frac{i\hbar}{2\pi} \frac{\delta(r-r')}{r} \delta(t-t'). \quad (9.6.16)$$

We now apply the polar coordinate decompositions carried out above to study the celebrated Aharonov-Bohm effect.[7]

We consider an "infinitely" long, current carrying, solenoid of circular cross section of arbitrary small radius $r_0 \to 0$, placed all along the z-axis, and, in polar coordinates, we take for the vector potential outside the solenoid (see Problem 9.14) the expression

$$\mathbf{A} = \frac{\Phi}{2\pi r} \hat{\boldsymbol{\phi}} \quad (9.6.17)$$

where $\hat{\boldsymbol{\phi}}$ is a unit vector in the direction of increasing of the angle ϕ; Φ is the flux

$$\Phi = \oint \mathbf{B} \cdot d\mathbf{S} \quad (9.6.18)$$

and $\mathbf{B}$ is the magnetic field inside the solenoid, with the surface integral carried over the cross sectional area of the solenoid which survives for $r_0 \to 0$. The magnetic field *outside* the solenoid is *zero*, and one explicitly checks from (9.6.17) that $\nabla \times \mathbf{A} = 0$ there.

Although the magnetic field is zero outside the solenoid, the line integral over a closed path, through which passes the solenoid, is *not* zero, and is given by

$$\oint \mathbf{A} \cdot d\mathbf{x} = \Phi \quad (9.6.19)$$

with $d\mathbf{x} = \hat{\boldsymbol{\phi}} r\, d\phi + \hat{\mathbf{r}}\, dr$, or as obtained directly from (9.6.18) by the application of Stokes's theorem. Because of this non-vanishing flux, the latter may have an observable effect, outside the solenoid, even though the magnetic field is zero there. This is the Aharonov-Bohm effect.

---

[7] This was also studied in the path integral formalism in Gerry and Singh (1983); Schulman (1971).

9.6 2D Green Function in Polar Coordinates

To study this effect, we consider the Green function of a charged particle of charge q in the presence of the vector potential $\mathbf{A}$. It satisfies the two dimensional differential equation

$$\left[i\hbar\frac{\partial}{\partial t} - \frac{\left(\mathbf{p} - \frac{q}{c}\mathbf{A}\right)^2}{2m}\right] G_+\left(\mathbf{x}t;\mathbf{x}'t'\right) = i\hbar\,\delta^2\left(\mathbf{x} - \mathbf{x}'\right)\delta\left(r - r'\right) \quad (9.6.20)$$

where c is the speed of light.

For the vector potential in (9.6.17), and in polar coordinates, (9.6.20) reads

$$\left[i\hbar\frac{\partial}{\partial t} + \frac{\hbar^2}{2m}\left[\left(\frac{1}{r}\frac{\partial}{\partial r}r\frac{\partial}{\partial r}\right) + \frac{1}{r^2}\left(\frac{\partial}{\partial \phi} - i\lambda_0\right)^2\right]\right] G_+\left(\mathbf{x}t;\mathbf{x}'t'\right)$$

$$= i\hbar\,\frac{\delta\left(r - r'\right)}{r}\delta\left(\phi - \phi'\right)\delta\left(t - t'\right) \quad (9.6.21)$$

where

$$\lambda_0 \equiv \frac{q\Phi}{2\pi\hbar c} \quad (9.6.22)$$

expressed in terms of the flux Φ.

To solve for G_+, we carry out a Fourier series as in (9.6.6):

$$G_+\left(\mathbf{x}t;\mathbf{x}'t'\right) = \sum_{k=-\infty}^{\infty} e^{ik(\phi-\phi')} F_k^+\left(rt;r't'\right) \quad (9.6.23)$$

to obtain for $F_k^+\left(rt;r't'\right)$ the differential equation

$$\left(i\hbar\frac{\partial}{\partial t} + \frac{\hbar^2}{2m}\left[\frac{1}{r}\frac{\partial}{\partial r}r\frac{\partial}{\partial r} - \frac{1}{r^2}(k - \lambda_0)^2\right]\right) F_k^+\left(rt;r't'\right)$$

$$= \frac{i\hbar}{2\pi}\frac{\delta\left(r - r'\right)}{r}\delta\left(t - t'\right). \quad (9.6.24)$$

Upon making the ansatz

$$F_k^+\left(rt;r't'\right) = \frac{m}{2\pi i\hbar T}\exp\left(\frac{im\left(r^2 + r'^2\right)}{2\hbar T}\right) F^+(k,z) \quad (9.6.25)$$

where $z = -i\rho$ (see (9.6.8), (9.6.7)), we obtain from (9.6.24) the following *two* equations (see Problem 9.15):

$$i\hbar\,\delta(T)\,\frac{m}{2\pi i\hbar T}\exp\left(\frac{im\left(r^2 + r'^2\right)}{2\hbar T}\right) F_k^+(k,z) = \frac{i\hbar}{2\pi}\delta(T)\,\frac{\delta\left(r - r'\right)}{r} \quad (9.6.26)$$

and for $T > 0$ we have

$$\left(z^2 \frac{d}{dz^2} + z\frac{d}{dz} - \left[z^2 + (k-\lambda_0)^2\right]\right) F^+(k,z) = 0. \qquad (9.6.27)$$

Equation (9.6.26) provides a boundary condition constraint as we will see below, where we recall that z depends on T, while (9.6.27) provides the solution $F^+(k,z)$.

The solutions which satisfy (9.6.27) are $I_{|\nu|}(z)$, $I_{-|\nu|}(z)$, $K_{|\nu|}(z)$, where $\nu = (k-\lambda_0)$. $K_{|\nu|}(z)$ diverges for $z \to 0$, i.e., for $rr'/\hbar T \to 0$ and hence at the origin. For $\nu = n$, an integer, $I_{-|\nu|}(z) = I_{|\nu|}(z)$ (see (9.6.4)). For ν not an integer $I_{-|\nu|}(z)$ also diverges for $z \to 0$. Hence we are left with $I_{|\nu|}(z)$. Therefore it remains to verify that the asymptotic behavior of $I_{|\nu|}(z)$ for $T \to 0$ is consistent with (9.6.26).

With $z = -i\rho = -imrr'/\hbar T$, $r > 0$, $r' > 0$, and hence ρ is real and positive, the asymptotic behavior of $I_{|\nu|}(z)$ for $T \to 0$, is given by[8]

$$I_{|\nu|}(-i\rho) \xrightarrow[\rho\to\infty]{} \sqrt{\frac{i}{2\pi\rho}} \left[e^{-i\rho} + \frac{1}{i} e^{i(\rho-|\nu|\pi)}\right] \qquad (9.6.28)$$

or

$$I_{|\nu|}(-i\rho) \xrightarrow[\rho\to\infty]{} \sqrt{\frac{2}{\pi\rho}}\, e^{-i|\nu|\pi/2} \cos\left(\rho - \left(|\nu|+\frac{1}{2}\right)\frac{\pi}{2}\right). \qquad (9.6.29)$$

Multiplying the expression on the right-hand side of (9.6.28) by

$$\frac{m}{2\pi i\hbar T} \exp\left(\frac{im(r^2+r'^2)}{2\hbar T}\right)$$

for $T \to 0$, we obtain for the resulting expression the behavior

$$\frac{1}{2\pi}\frac{1}{\sqrt{rr'}}\left[\delta(r-r') + \frac{1}{i}\delta(r+r')e^{-|\nu|\pi i}\right] \qquad (9.6.30)$$

which is consistent with (9.6.26) since for $r > 0$, $r' > 0$, the second term in (9.6.30) is zero. For $\rho \to \infty$, i.e., ρ large, but $T \neq 0$, both terms in (9.6.28) contribute since the factor $(m/2\pi i\hbar T)$ in (9.6.25) would not be relevant in this case.

For $T > 0$, the Green function in (9.6.23) may be then rewritten as

$$G_+(\mathbf{x}t;\mathbf{x}'t') = \frac{m}{2\pi i\hbar T} \exp\left(\frac{im(r^2+r'^2)}{2\hbar T}\right) \sum_{k=-\infty}^{\infty} e^{ik(\phi-\phi')} I_{|k-\lambda_0|}(-i\rho). \qquad (9.6.31)$$

---

[8] See Watson (1966), p. 203; Gradshteyn and Ryzhik (1965), p. 962. $I_{|\nu|}(-i\rho)$ is related to $J_{|\nu|}(\rho)$ with the latter having the familiar cosine behavior, divided by $\sqrt{\pi\rho/2}$, for $\rho \to \infty$ (*op. cit.* p. 203, p. 952, respectively) (see Problem 9.16).

9.6 2D Green Function in Polar Coordinates

From the definition of λ_0 in (9.6.22) we note that *if*

$$\frac{q\Phi}{2\pi\hbar c} = N \tag{9.6.32}$$

is an integer, i.e., flux is quantized, then by making a change of the summation variable $k \to k + N$, $G_+(xt; \mathbf{x}'t')$, up to the overall phase factor $\exp iN(\phi - \phi')$, coincides with the free propagator in (9.6.6). Interesting situations arise when the flux is not quantized, that is when λ_0 is *not* an integer. In such cases, we may write

$$\lambda_0 = N + \delta_0 \tag{9.6.33}$$

where N is some integer,

$$0 < \delta_0 < 1 \tag{9.6.34}$$

and (9.6.31) takes the form

$$G_+(\mathbf{x}t; \mathbf{x}'t') = \frac{m}{2\pi i\hbar T} \exp\left(\frac{im(r^2 + r'^2)}{2\hbar T}\right) e^{iN(\phi - \phi')}$$

$$\times \sum_{k=-\infty}^{\infty} e^{ik(\phi-\phi')} I_{|k-\delta_0|}(-i\rho) \tag{9.6.35}$$

where now $|k - \delta_0| > 0$ for all k and $I_{|k-\delta_0|}(-i\rho)$ vanishes for $\rho \to 0$). That is, each term in the summand in (9.6.35) has the built in vanishing property for $r \to 0$ and/or $r' \to 0$, i.e., at the origin, thus making the solenoid a forbidden region, for the case in (9.6.33), (9.6.34), for each winding number.

By using the equality in (9.6.10) and making a change of the continuous variable $\lambda \to \lambda + \lambda_0$, (9.6.31) may be rewritten in the form

$$G_+(\mathbf{x}t; \mathbf{x}'t') = \frac{m}{2\pi i\hbar T} \exp\left(\frac{im(r^2 + r'^2)}{2\hbar T}\right) e^{i\lambda_0(\phi - \phi')}$$

$$\times \sum_{k=-\infty}^{\infty} \left(e^{iq\Phi/\hbar c}\right)^k \int_{-\infty}^{\infty} d\lambda \exp i\lambda(\phi - \phi' + 2\pi k) I_{|\lambda|}(-i\rho). \tag{9.6.36}$$

This expression is to be compared with the free one in (9.6.11)–(9.6.13). The term $\exp i\lambda_0(\phi - \phi')$ is nothing but the exponential of $iq/\hbar c$ times the integral

$$\int_{(r',\phi')}^{(r,\phi)} \mathbf{A} \cdot d\mathbf{x} = \Phi(\phi - \phi')/2\pi \tag{9.6.37}$$

obtained by integrating along a direct line segment joining the end points. This term is also obtained by a naive examination of (9.6.21). The first factor in the summand in (9.6.36) is the kth power of the exponential of $(iq/\hbar c)$

times the flux Φ. That is, the amplitude for the particle to arrive at (r,ϕ) is, up to the phase factor $\exp i\lambda_0 (\phi - \phi')$, the sum of amplitudes of arrival at (r,ϕ) after an arbitrary number of rotations around the solenoid acquiring *additional* phase factors $(\exp iq\Phi/\hbar c)^k$ with the sense of a rotation specified by the sign of k. For a quantized flux defined in (9.6.32), these phases disappear. On the other hand for a non-quantized flux, as given in (9.6.33), (9.6.34), the mere presence of such, in general, different phases may have an observable effect even though the magnetic field is zero outside the solenoid! This is the celebrated Aharonov-Bohm effect. The importance of the phase factor

$$\exp\left(\frac{iq}{\hbar c}\oint \mathbf{A}\cdot d\mathbf{x}\right) = \exp\left(\frac{iq}{\hbar c}\Phi\right) \tag{9.6.38}$$

cannot be overemphasized. Such a phase factor is usually referred to as a Wu-Yang[9] phase.

Now we come back to our expression in (9.6.35). In practice, one is interested, in general, to the limit $mrr'/\hbar T \to \infty$ involving large distances $r \to \infty$ and/or $r' \to \infty$, with $T \neq 0$. In such cases we may use the asymptotic behavior given in (9.6.28) to write formally for $\rho \to \infty$, with $T \neq 0$ in (9.6.35),

$$\sqrt{\frac{2\pi\rho}{i}} \sum_{k=-\infty}^{\infty} e^{ik(\phi-\phi')} I_{|k-\delta_0|}(-i\rho)$$

$$\longrightarrow e^{-i\rho}\sum_{k=-\infty}^{\infty} e^{ik(\phi-\phi')} + \frac{1}{i} e^{i\rho} \sum_{k=-\infty}^{\infty} e^{ik(\phi-\phi')} e^{-i|k-\delta_0|\pi}$$

$$= 2\pi e^{-i\rho} \delta(\phi-\phi') + \frac{1}{i} e^{i\rho} \sum_{k=-\infty}^{\infty} e^{ik(\phi-\phi')} e^{-i|k-\delta_0|\pi} \tag{9.6.39}$$

where $\rho \to \infty$ means ρ large. Here we note that since we are considering the case $T \neq 0$, the second term on the right-hand side of (9.6.39) does not give rise to a $\delta(r+r')$ term to the asymptotic behavior of (9.6.35) and gives a non-zero contribution.

As an application of (9.6.35), with the asymptotic expression in (9.6.39), we consider the case in (9.6.33), (9.6.34) with $N = 0$, $0 < \delta_0 < 1$. [The situation with $-1 < \delta_0 < 0$ may be handled similarly.]

To the above end, suppose, for the purpose of a simple but concrete illustration, one has an initial state $\psi_0(\mathbf{x}')$ which is uniform in ϕ': $-\Delta < \phi' < \Delta$ for some small angle Δ, and is zero otherwise, with $|\mathbf{x}'| = R$ a fixed large radial distance from the solenoid with the latter located at the origin 0 (see Figure 9.3). This will allow us to sum the series on the extreme right-hand side of (9.6.39) explicitly. That is, equivalently one has an extended uniform

---

[9] Wu and Yang (1975).

9.6 2D Green Function in Polar Coordinates

source, rather than a point-like one, which initially emits particles. The source is situated at a fixed large radial distance from the origin and is described by an arc subtended by a small angle 2Δ symmetrically about the angle zero, say, with $0 < \Delta \ll \pi/4$.[10]

To determine the probability of particle detection, for $T \neq 0$, we must then integrate, in the process, over ϕ', on the extreme right-hand side of (9.6.39), from $-\Delta$ to Δ. That is, one is summing over all initial configurations and hence equivalently adding up all amplitudes from every point on the source and ending up at the observation point in question for some $T > 0$.

Thus the amplitude for a particle to be found at an angle ϕ, for $|\mathbf{x}| = |\mathbf{x}'| \to \infty$, is up to an unimportant multiplicative factor independent of δ_0, with ϕ not in the range $(-\Delta, \Delta)$, is given from (9.6.39) to be

$$A = \sum_{k=-\infty}^{\infty} e^{ik\phi} \frac{\sin k\Delta}{k\Delta} e^{-i|k-\delta_0|\pi} \qquad (9.6.40)$$

We note that for ϕ not in $(-\Delta, \Delta)$, the term in (9.6.39) involving $\delta(\phi - \phi')$

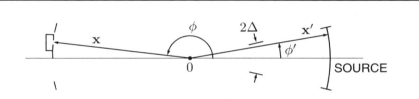

Fig. 9.3. The solenoid is placed at the origin 0, and the extended source is described by an arc subtended by a very small angle 2Δ symmetrically about the angle zero. The detector and the source are at large radial distances $|\mathbf{x}| = |\mathbf{x}'| \equiv R \to \infty$ from the origin.

does not contribute. Also note that $\sin k\Delta/k\Delta \to 1$ for $k \to 0$.

Since $0 < \delta_0 < 1$, (9.6.40) may be rewritten as

$$A = e^{-i\delta_0\pi} \sum_{k=0}^{\infty} e^{-ik(\phi-\pi)} \frac{\sin k\Delta}{k\Delta} + e^{i\delta_0\pi} \sum_{k=1}^{\infty} e^{ik(\phi-\pi)} \frac{\sin k\Delta}{k\Delta} \qquad (9.6.41)$$

where note that $e^{-ik\pi} = e^{ik\pi}$.

To verify the Aharonov-Bohm effect, it is sufficient to look at the point $\phi = \pi$ in (9.6.41) (see Figure 9.3.). To this end we use the sum[11]

---

[10] It is worth noting that a non-zero integer part N of λ_0 may contribute in specific situations.

[11] For the summation of such trigonometric functions see: Gradshteyn and Ryzhik (1965), pp. 38–41.

$$\sum_{k=1}^{\infty}\frac{\sin k\Delta}{k\Delta}=\frac{\pi-\Delta}{2\Delta} \tag{9.6.42}$$

to obtain from (9.6.41), at $\phi=\pi$,

$$|A|^2=\left(\frac{\pi}{\Delta}\right)^2\cos^2(\delta_0\pi)+\sin^2(\delta_0\pi) \tag{9.6.43}$$

giving the relative intensity, in the presence of the solenoid with $0<\delta_0<1$, the result

$$\cos^2(\delta_0\pi)+\left(\frac{\Delta}{\pi}\right)^2\sin^2(\delta_0\pi)<1 \tag{9.6.44}$$

where Δ/π is some small number. The mere fact that the intensity is altered (reduced) is a statement of the existence of this effect.

It is remarkable that the expression for the amplitude A in (9.6.41) may be summed exactly. To this end it is sufficient to consider $\pi/2<\phi<3\pi/2$.

We use the identities

$$1+2\sum_{k=1}^{\infty}\frac{\sin k\Delta}{k\Delta}\cos k(\phi-\pi)=\begin{cases}\pi/\Delta, & \pi-\Delta<\phi<\pi+\Delta \\ 0, & \pi/2<\phi<\pi-\Delta,\ \pi+\Delta<\phi<3\pi/2\end{cases} \tag{9.6.45}$$

$$\sum_{k=1}^{\infty}\frac{\sin k\Delta}{k\Delta}\sin k(\phi-\pi)=\frac{1}{4\Delta}\ln\left[\frac{1+\cos(\phi+\Delta)}{1+\cos(\phi-\Delta)}\right],\quad \pi/2<\phi<3\pi/2 \tag{9.6.46}$$

being, respectively, even and odd functions of $(\phi-\pi)$, to obtain from (9.6.41):

$$|A|^2=\left(\frac{\pi}{\Delta}\cos(\delta_0\pi)+\frac{\sin(\delta_0\pi)}{2\Delta}\ln\left[\frac{1+\cos(\phi-\Delta)}{1+\cos(\phi+\Delta)}\right]\right)^2+\sin^2(\delta_0\pi) \tag{9.6.47}$$

for $\pi-\Delta<\phi<\pi+\Delta$, which reduces to (9.6.43) for $\phi=\pi$, and

$$|A|^2=\sin^2(\delta_0\pi)\left[1+\frac{1}{4\Delta^2}\left(\ln\left[\frac{1+\cos(\phi-\Delta)}{1+\cos(\phi+\Delta)}\right]\right)^2\right] \tag{9.6.48}$$

for $\pi/2<\phi<\pi-\Delta,\ \pi+\Delta<\phi<3\pi/2$. It is interesting to see that (9.6.48) gives a non-vanishing contribution, in the regions thus defined (see Figure 9.3), due to the presence of the solenoid with $0<\delta_0<1$.

The densities in (9.6.47), (9.6.48) are plotted in Figure 9.4 for $\delta_0=+1/4$, $\Delta=\pi/100$. With a uniform initial wavefunction with sharp cut-offs at $\phi'=\pm\Delta$ adopted, these densities become arbitrarily large in the limits $\phi-\pi\to\pm\Delta$, in their respective intervals, which is the price one pays with an idealistic initial wavefunction (see also Problem 9.18). In the region $\pi-\Delta<\phi<\pi+\Delta$, just around the point $\phi=\pi$ (see also (9.6.44)), the density in question is reduced from one over the solenoid-free case (equivalently for $\delta_0\to 0$), while

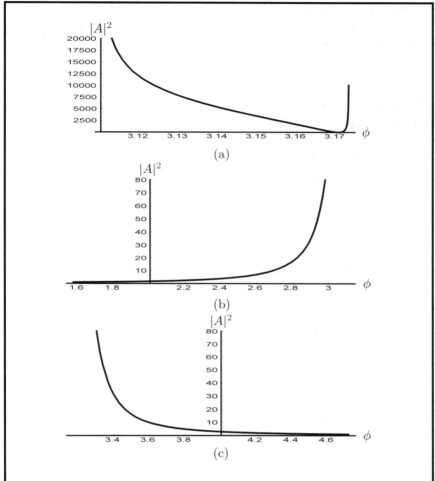

Fig. 9.4. Plots of the densities $|A|^2$ in (9.6.47), (9.6.48) for $\delta_0 = +1/4$, $\Delta = \pi/100$ in the regions $\pi - \Delta < \phi < \pi + \Delta$, $\pi/2 < \phi < \pi - \Delta$, $\pi + \Delta < \phi < 3\pi/2$, respectively, in the graphs (a), (b), (c), with ϕ expressed in radians.

outside this interval the intensity, as mentioned above, is non-zero but rapidly goes to zero to the right and left of $\pi + \Delta$, $\pi - \Delta$, respectively. The Aharonov-Bohm effect (with $\delta_0 \neq 0$) is clearly seen. A similar analysis may be carried out for $-1 < \delta < 0$.

We note that the illustration of the Aharonov-Bohm effect given above is based on the scattering of the charged particle off the origin due to the non-zero flux only, i.e., is purely of electromagnetic origin in a quantum setting. In Problem 9.17, the reader is asked to extend this analysis, to the more difficult case, when a particle may be scattered off the origin (by the solenoid), even for $\delta_0 \to 0$, by the application of appropriate boundary conditions at the origin, and study how this scattering is modified by the activation of the solenoid, i.e., for $\delta_0 > 0$, as an illustration of the Aharonov-Bohm effect.

9.7 General Properties of the Full Green Functions and Applications

9.7.1 A Matrix Notation

We use the following convenient matrix notations as

$$\langle \mathbf{x} | G_\pm(T) | \mathbf{x}' \rangle = G_\pm(\mathbf{x}\, T; \mathbf{x}'0) \tag{9.7.1}$$

$$\langle \mathbf{x} | \mathbf{1} | \mathbf{x}' \rangle = \delta^\nu(\mathbf{x} - \mathbf{x}') \tag{9.7.2}$$

with $G_\pm(T)$ satisfying the differential equation

$$\left[i\hbar \frac{\partial}{\partial T} - H \right] G_\pm(T) = \pm i\hbar\, \mathbf{1}\, \delta(T). \tag{9.7.3}$$

As before, the Green functions $G_\pm$ considered are those with built in boundary conditions as specified by the $\pm i\varepsilon$ prescription adopted and spelled out below. This equation may be integrated directly but it is more instructive to proceed as follows. We introduce the Fourier transform

$$G_\pm(T) = \pm i\hbar \int_{-\infty}^{\infty} \frac{dp^0}{2\pi\hbar} e^{-ip^0 T/\hbar} G_\pm(p^0) \tag{9.7.4}$$

where

$$\left[p^0 - H \right] G_\pm(p^0) = \mathbf{1}. \tag{9.7.5}$$

Upon introducing the resolution of the identity (see (1.8.18), (1.8.14))

$$\mathbf{1} = \int_{-\infty}^{\infty} d\lambda\, \delta(\lambda - H) \tag{9.7.6}$$

we have from (9.7.5)

$$G_\pm(p^0) = \int_{-\infty}^{\infty} \frac{1}{(p^0 - \lambda \pm i\varepsilon)} d\lambda\, \delta(\lambda - H), \tag{9.7.7}$$

$$G_\pm(T) = \pm i\hbar \int_{-\infty}^{\infty} \frac{dp^0}{2\pi\hbar} \int_{-\infty}^{\infty} \frac{d\lambda\, \delta(\lambda - H)}{(p^0 - \lambda \pm i\varepsilon)} e^{-ip^0 T/\hbar}. \tag{9.7.8}$$

9.7 General Properties of Full Green Functions & Applications

It is easily shown that the denominator in (9.7.8) (see Problem 9.20) may be written as

$$\frac{1}{(p^0 - \lambda \pm i\varepsilon)} = \mp i \int_0^\infty d\alpha \exp\left[\pm i(p^0 - \lambda \pm i\varepsilon)\alpha\right] \qquad (9.7.9)$$

from which

$$G_\pm(T) = \hbar \int_{-\infty}^\infty d\lambda\, \delta(\lambda - H) \int_0^\infty d\alpha\, \exp\left[\mp i\alpha(\lambda \mp i\varepsilon)\right]$$

$$\times \int_{-\infty}^\infty \frac{dp^0}{2\pi\hbar} \exp\left[\pm i p^0(\hbar\alpha \mp T)/\hbar\right]$$

$$= \int_{-\infty}^\infty d\lambda\, \delta(\lambda - H) \exp\left[-iT(\lambda \mp i\varepsilon)/\hbar\right] \int_0^\infty d\alpha\, \delta\left(\alpha \mp \frac{T}{\hbar}\right). \qquad (9.7.10)$$

Since α is positive, $G_+(T)$ is zero for $T < 0$, and $G_-(T)$ is zero for $T > 0$. That is, upon integration over α, (9.7.10) is equivalent to

$$G_\pm(T) = \Theta(\pm T) \int_{-\infty}^\infty d\lambda\, \delta(\lambda - H) \exp\left[-iT(\lambda \mp i\varepsilon)/\hbar\right] \qquad (9.7.11)$$

or

$$G_\pm(T) = \Theta(\pm T) \exp\left[-iTH/\hbar\right] \exp\left(-\varepsilon|T|\right). \qquad (9.7.12)$$

For $\varepsilon \to +0$, it is easily verified that $G_\pm(T)$ satisfy (9.7.3).
We may rewrite (9.7.11) as

$$G_\pm(T) = \Theta(\pm T)\, G(T) \qquad (9.7.13)$$

where

$$G(T) = \int_{-\infty}^\infty d\lambda\, \delta(\lambda - H) \exp(-iT\lambda/\hbar) \qquad (9.7.14)$$

for $\varepsilon \to +0$.
From (9.7.1), we also have

$$G_\pm(\mathbf{x}\,T; \mathbf{x}'\,0) = \Theta(\pm T)\, G(\mathbf{x}\,T; \mathbf{x}'\,0) \qquad (9.7.15)$$

and as in (9.5),

$$G(\mathbf{x}\,T; \mathbf{x}'\,0) = \int_{-\infty}^\infty d\lambda\, \langle \mathbf{x}|\delta(\lambda - H)|\mathbf{x}'\rangle \exp(-iT\lambda/\hbar). \qquad (9.7.16)$$

Complex conjugation gives the relation

$$(G(\mathbf{x}\,T; \mathbf{x}'\,0))^* = G(\mathbf{x}', -T; \mathbf{x}, 0). \qquad (9.7.17)$$

9.7.2 Applications

Let
$$T/\hbar = \tau \tag{9.7.18}$$
in (9.7.16) and consider the integral

$$I(\mathbf{x}, \mathbf{x}';\xi) = \frac{1}{2\pi i} \int_{-\infty}^{\infty} \frac{d\tau}{\tau - i\varepsilon} G(\mathbf{x}\,\hbar\tau; \mathbf{x}'\,0) e^{i\xi\tau}, \quad \varepsilon \to +0 \tag{9.7.19}$$

for real ξ.

From (9.7.16), the latter is given by

$$I(\mathbf{x},\,\mathbf{x}';\xi) = \int_{-\infty}^{\infty} d\lambda \, \langle \mathbf{x}|\delta(\lambda - H)|\mathbf{x}'\rangle \frac{1}{2\pi i} \int_{-\infty}^{\infty} \frac{d\tau}{\tau - i\varepsilon} e^{i(\xi - \lambda)\tau}. \tag{9.7.20}$$

For $\lambda > \xi$, we may close the contour of the integral in the complex τ-plane from below and obtain zero from the residue theorem since $\varepsilon \to +0$. On the other hand, for $\lambda < \xi$, we may close the τ-contour from above to obtain, from the residue theorem, that the τ-integral multiplied by $(1/2\pi i)$, gives one. The latter just provides the definition of the step function $\Theta(\xi - \lambda)$. That is,

$$I(\mathbf{x},\,\mathbf{x}';\xi) = \int_{-\infty}^{\infty} d\lambda \, \langle \mathbf{x}|\delta(\lambda - H)|\mathbf{x}'\rangle \, \Theta(\xi - \lambda)$$

$$= \int_{-\infty}^{\xi} d\lambda \, \langle \mathbf{x}|\delta(\lambda - H)|\mathbf{x}'\rangle$$

$$= \langle \mathbf{x}|P_H(\xi)|\mathbf{x}'\rangle \tag{9.7.21}$$

where in writing the last equality we have used (1.8.14).

Suppose that for $\lambda < \xi$, we are dealing with the discrete spectrum of the Hamiltonian in question, and ξ is not in its spectrum, then from (4.5.3), (9.7.21) for $\mathbf{x} = \mathbf{x}'$, we have the useful relationship

$$\frac{1}{2\pi i} \int_{-\infty}^{\infty} \frac{d\tau}{\tau - i\varepsilon} G(\mathbf{x}, \hbar\tau; \mathbf{x}, 0) e^{i\xi\tau} = \sum_{\lambda,\nu(\lambda)} |\psi_{\lambda,\nu(\lambda)}(\mathbf{x})|^2 \Theta(\xi - \lambda)$$

$$\equiv n(\mathbf{x}; \xi) \tag{9.7.22}$$

defining a density of states, where

$$\psi_{\lambda,\nu(\lambda)}(\mathbf{x}) = \langle \mathbf{x}|\lambda, \nu(\lambda)\rangle \tag{9.7.23}$$

9.7 General Properties of Full Green Functions & Applications 583

are the eigenstates in the **x**-description, and $\nu(\lambda)$, not to confuse it with the dimensionality of space ν, specifies the degenerate states corresponding to the eigenvalue λ.

Upon integrating (9.7.21) over **x**, for ξ not in the spectrum of H, we obtain

$$\int d^\nu x \, \frac{1}{2\pi i} \int_{-\infty}^{\infty} \frac{d\tau}{\tau - i\varepsilon} G(\mathbf{x}, \hbar\tau; \mathbf{x}, 0) e^{i\xi\tau} = \sum_{\lambda, \nu(\lambda)} \Theta(\xi - \lambda)$$

$$\equiv N(H; \xi). \tag{9.7.24}$$

That is, the operation defined on the left-hand side of (9.7.24) gives the number $N(H, \xi)$ of eigenvalues of H, taking into account degeneracy, less than ξ. The importance of (9.7.24) cannot be overemphasized.

To find the number of eigenvalues, taking into account degeneracy, falling between two points $\xi_1, \xi_2, \xi_2 > \xi_1$, with these two points not in the spectrum of H, may be obtained from (9.7.24) to be given by

$$\int d^\nu x \, \frac{1}{2\pi i} \int_{-\infty}^{\infty} \frac{d\tau}{\tau - i\varepsilon} G(\mathbf{x}, \hbar\tau; \mathbf{x}, 0) \left(e^{i\xi_2\tau} - e^{i\xi_1\tau} \right). \tag{9.7.25}$$

From (4.5.6), (9.7.19), (9.7.21),

$$\frac{1}{2\pi i} \int_{-\infty}^{\infty} \frac{d\tau}{\tau - i\varepsilon} G(\mathbf{x}, \hbar\tau; \mathbf{x}', 0) e^{i\xi\tau} = \sum_{\lambda, \nu(\lambda)} \psi_{\lambda, \nu(\lambda)}(\mathbf{x}) \psi^*_{\lambda, \nu(\lambda)}(\mathbf{x}') \Theta(\xi - \lambda)$$

$$\equiv n(\mathbf{x}, \mathbf{x}'; \xi) \tag{9.7.26}$$

defines a non-local density of states. This expression will be used in Chapter 13 dealing with multi-electron atoms. The following property obtained by complex conjugation is to be noted

$$(n(\mathbf{x}, \mathbf{x}'; \xi))^* = n(\mathbf{x}', \mathbf{x}; \xi) \tag{9.7.27}$$

as obtained, for example, from the expression on the left-hand side of (9.7.26) and (9.7.17).

From (9.7.16)

$$\left(i\hbar \frac{\partial}{\partial T} - H \right) G(\mathbf{x}T; \mathbf{x}' \, 0) = 0 \tag{9.7.28}$$

and with

$$H = \frac{\mathbf{p}^2}{2\mu} + V(\mathbf{x}) \tag{9.7.29}$$

we have from (9.7.16)

$$\left[i\frac{\partial}{\partial \tau} - V(\mathbf{x})\right] G(\mathbf{x}, \hbar\tau; \mathbf{x}', 0) = \int_{-\alpha}^{\infty} d\lambda \, [\lambda - V(\mathbf{x})] \langle \mathbf{x}|\delta(\lambda - H)|\mathbf{x}' \rangle \, e^{-i\tau\lambda}$$

$$= \int_{-\alpha}^{\infty} d\lambda \, \langle \mathbf{x}|(H - V(\mathbf{x}))\delta(\lambda - H)|\mathbf{x}' \rangle \, e^{-i\tau\lambda}. \tag{9.7.30}$$

From (1.8.14), (4.5.14), (4.5.15), we may use (9.7.30) to infer that

$$\int d^\nu x \, \frac{1}{2\pi i} \int_{-\infty}^{\infty} \frac{d\tau}{\tau - i\varepsilon} e^{i\xi\tau} \left[i\frac{\partial}{\partial\tau} - V\right] G(\mathbf{x}, \hbar\tau; \mathbf{x}', 0)\bigg|_{\mathbf{x}'=\mathbf{x}}$$

$$= \sum_{\lambda, \nu(\lambda)} \left\langle \lambda, \nu(\lambda) \left| \frac{\mathbf{p}^2}{2\mu} \right| \lambda, \nu(\lambda) \right\rangle \Theta(\xi - \lambda). \tag{9.7.31}$$

This expression will be also useful in studying multi-electron atoms in §13.1 (see also (4.5.15)).

A similar expression to (9.7.31) as follows from (9.7.28) is

$$\int d^\nu x \, \frac{1}{2\pi i} \int_{-\infty}^{\infty} \frac{d\tau}{\tau - i\varepsilon} e^{i\xi\tau} \left[i\frac{\partial}{\partial\tau} G(\mathbf{x}, \hbar\tau; \mathbf{x}, 0)\right] = \sum_{\lambda, \nu(\lambda)} \lambda \, \Theta(\xi - \lambda) \tag{9.7.32}$$

giving the sum of eigenvalues of H less than ξ.

Before considering specific cases, we note that, if λ_0 is an eigenvalue of H and ξ_2, ξ_1, not in its spectrum, are such that $\xi_1 < \lambda_0 < \xi_2$ and that there are no other eigenvalues between ξ_1 and ξ_2 then (9.7.25) gives the degree of degeneracy of λ_0. In detail, from (9.7.24), (9.7.25), this may be written as

$$\int d^\nu x \, \frac{1}{2\pi i} \int_{-\infty}^{\infty} \frac{d\tau}{\tau - i\varepsilon} G(\mathbf{x}, \hbar\tau; \mathbf{x}, 0)(e^{i\xi_2\tau} - e^{i\xi_1\tau}) = \sum_{\nu(\lambda_0)} 1. \tag{9.7.33}$$

An elementary, though rather formal application of (9.7.22), since the latter was written for a discrete spectrum, is to the free electron gas. In $3D$ we have

$$\left[i\frac{\partial}{\partial\tau} - \frac{\mathbf{p}^2}{2m}\right] G_{\pm\sigma,\sigma'}(\mathbf{x}, \hbar\tau; \mathbf{x}', 0) = i\delta_{\sigma,\sigma'}\delta(\tau)\delta^3(\mathbf{x} - \mathbf{x}') \tag{9.7.34}$$

where σ, σ' are spin indices. From (9.7.15), (9.7.34), we have

$$G_{\sigma,\sigma'}(\mathbf{x}, \hbar\tau; \mathbf{x}', 0) = \delta_{\sigma,\sigma'} \int \frac{d^3\mathbf{p}}{(2\pi\hbar)^3} e^{i\mathbf{p}\cdot(\mathbf{x}-\mathbf{x}')/\hbar} e^{-i\mathbf{p}^2\tau/2m}. \tag{9.7.35}$$

9.7 General Properties of Full Green Functions & Applications

Upon setting
$$G(\mathbf{x}, \hbar\tau; \mathbf{x}, 0) = \sum_\sigma G_{\sigma,\sigma}(\mathbf{x}, \hbar\tau; \mathbf{x}, 0) \qquad (9.7.36)$$

we have from (9.7.22), (9.7.35) for the density of states
$$n(\mathbf{x}; \xi) = 2 \int \frac{d^3\mathbf{p}}{(2\pi\hbar)^3} \frac{1}{2\pi i} \int_{-\infty}^{\infty} \frac{d\tau}{\tau - i\varepsilon} e^{i(\xi - p^2/2m)\tau}$$
$$= 2 \int \frac{d^3\mathbf{p}}{(2\pi\hbar)^3} \Theta\left(\xi - \frac{p^2}{2m}\right) \qquad (9.7.37)$$

leading to the familiar expression
$$n(\mathbf{x}; \xi) = \frac{1}{3\pi^2} \left(\frac{2m\xi}{\hbar^2}\right)^{3/2}. \qquad (9.7.38)$$

As an illustration of the formula (9.7.25), consider the Green function in (9.2.23) for the one-dimensional harmonic oscillator. Then from (9.2.23)
$$G(x\,T; x\,0) = \left(\frac{m\omega}{2\pi i\hbar \sin\omega T}\right)^{1/2} \exp\left(-\frac{im\omega x^2}{\hbar} \tan\left(\frac{\omega T}{2}\right)\right) \qquad (9.7.39)$$

and
$$\int_{-\infty}^{\infty} dx\, G(x\,T; x\,0) = \frac{e^{-i\omega T/2}}{1 - e^{-i\omega T}}. \qquad (9.7.40)$$

Hence
$$\int_{-\infty}^{\infty} dx \frac{1}{2\pi i} \int_{-\infty}^{\infty} \frac{d\tau}{\tau - i\varepsilon} G(x\,\hbar\tau; x\,0) \left(e^{i\xi_2\tau} - e^{i\xi_1\tau}\right)$$
$$= \frac{1}{2\pi i} \int_{-\infty}^{\infty} \frac{d\tau}{\tau - i\varepsilon} \frac{e^{-i\hbar\omega\tau/2}}{(1 - e^{-i\hbar\omega\tau/2})} \left(e^{i\xi_2\tau} - e^{i\xi_1\tau}\right) \qquad (9.7.41)$$

which upon using the expansion
$$\frac{e^{-i\hbar\omega\tau/2}}{1 - e^{-i\hbar\omega\tau/2}} = \sum_{n=0}^{\infty} e^{-i\hbar\omega(n+1/2)\tau} \qquad (9.7.42)$$

and the integral representation of the step function
$$\Theta(\xi) = \frac{1}{2\pi i} \int_{-\infty}^{\infty} \frac{d\tau}{\tau - i\varepsilon} e^{i\xi\tau} \qquad (9.7.43)$$

we obtain for (9.7.41), the explicit expression for the number of eigenvalues between ξ_1 and ξ_2:
$$\sum_{n=0}^{\infty} \left[\Theta(\xi_2 - \hbar\omega(n+1/2)) - \Theta(\xi_1 - \hbar\omega(n+1/2))\right]. \qquad (9.7.44)$$

Other properties of the Green functions will be considered in other chapters as well.

9.7.3 An Integral Expression for the (Homogeneous) Green Function

From (9.7.16), (2.4.1), we may write

$$G(\mathbf{x}\, T; \mathbf{x}'0) = \int \frac{\mathrm{d}^\nu \mathbf{p}}{(2\pi\hbar)^\nu} \langle \mathbf{p}|\mathbf{x}'\rangle \int_{-\infty}^{\infty} \mathrm{d}\lambda \, \langle \mathbf{x}|\delta(\lambda - H)|\mathbf{p}\rangle \exp(-\mathrm{i}T\lambda/\hbar) \tag{9.7.45}$$

in ν dimensions, which satisfies the homogeneous equation (9.7.28).

Using (2.4.8) and defining

$$F(\mathbf{p}, \mathbf{x}, T) = \mathrm{e}^{-\mathrm{i}\mathbf{p}\cdot\mathbf{x}/\hbar}\, \mathrm{e}^{\mathrm{i}\mathbf{p}^2 T/2m\hbar} \int_{-\infty}^{\infty} \mathrm{d}\lambda\, \langle \mathbf{x}|\delta(\lambda - H)|\mathbf{p}\rangle\, \mathrm{e}^{-\mathrm{i}T\lambda/\hbar} \tag{9.7.46}$$

with the latter being independent of $\mathbf{x}'$, we rewrite (9.7.45) as

$$G(\mathbf{x}\, \hbar\tau; \mathbf{x}'0) = \int \frac{\mathrm{d}^\nu \mathbf{p}}{(2\pi\hbar)^\nu} \exp\mathrm{i}\left[\frac{\mathbf{p}\cdot(\mathbf{x}-\mathbf{x}')}{\hbar} - \mathbf{p}^2\tau/2m\right] F(\mathbf{p}, \mathbf{x}, \hbar\tau)$$

$$\equiv f(\mathbf{x}-\mathbf{x}', \mathbf{x}, \hbar\tau). \tag{9.7.47}$$

We introduce the variables

$$\boldsymbol{\zeta} \equiv \mathbf{x} - \mathbf{x}' \tag{9.7.48}$$

$$\boldsymbol{\eta} = \mathbf{x} \tag{9.7.49}$$

noting that

$$\frac{\partial}{\partial x_i} = \frac{\partial}{\partial \zeta_i} + \frac{\partial}{\partial \eta_i} \tag{9.7.50}$$

$$\nabla_\mathbf{x}^2 = \nabla_{\boldsymbol{\zeta}}^2 + \nabla_{\boldsymbol{\eta}}^2 + 2\nabla_{\boldsymbol{\zeta}} \cdot \nabla_{\boldsymbol{\eta}} \tag{9.7.51}$$

and rewriting ($T = \hbar\tau$)

$$G(\mathbf{x}\, \hbar\tau; \mathbf{x}'0) = \int \frac{\mathrm{d}^\nu \mathbf{p}}{(2\pi\hbar)^\nu} \mathrm{e}^{\mathrm{i}[\mathbf{p}\cdot\boldsymbol{\zeta}/\hbar - \mathbf{p}^2\tau/2m]} F(\mathbf{p}, \boldsymbol{\eta}, \hbar\tau) \tag{9.7.52}$$

we have from (9.7.28), (9.7.29)

$$0 = \int \frac{\mathrm{d}^\nu \mathbf{p}'}{(2\pi\hbar)^\nu} \mathrm{e}^{\mathrm{i}[\mathbf{p}'\cdot\boldsymbol{\zeta}/\hbar - \mathbf{p}'^2\tau/2m]}$$

$$\times \left\{\mathrm{i}\frac{\partial}{\partial\tau} + \frac{\mathrm{i}\hbar}{\mu}\mathbf{p}'\cdot\nabla_{\boldsymbol{\eta}} + \frac{\hbar^2}{2\mu}\nabla_{\boldsymbol{\eta}}^2 - V(\boldsymbol{\eta})\right\} F(\mathbf{p}', \boldsymbol{\eta}, \hbar\tau). \tag{9.7.53}$$

Upon multiplying the latter by $\exp(-\mathrm{i}\mathbf{p}\cdot\boldsymbol{\zeta}/\hbar)$ and integrating over $\boldsymbol{\zeta}$ (or multiplying by $\exp(\mathrm{i}\mathbf{p}\cdot\mathbf{x}'/\hbar)$ and integrating over $\mathbf{x}'$), we obtain

$$\left\{i\frac{\partial}{\partial\tau}+\frac{i\hbar}{\mu}\mathbf{p}\cdot\boldsymbol{\nabla}_{\boldsymbol{\eta}}+\frac{\hbar^{2}}{2\mu}\boldsymbol{\nabla}_{\boldsymbol{\eta}}^{2}-V(\boldsymbol{\eta})\right\}F(\mathbf{p},\boldsymbol{\eta},\hbar\tau)=0. \qquad (9.7.54)$$

It is more convenient to have an exponential representation

$$F(\mathbf{p},\boldsymbol{\eta},\hbar\tau)=\exp\bigl(-iU(\mathbf{p},\boldsymbol{\eta},\tau,\hbar)\bigr) \qquad (9.7.55)$$

and rewrite (9.7.47), (9.7.52) as

$$G(\mathbf{x}\,\hbar\tau;\mathbf{x}'0)=\int\frac{d^{\nu}\mathbf{p}}{(2\pi\hbar)^{\nu}}\exp i\left[\frac{\mathbf{p}\cdot\boldsymbol{\zeta}}{\hbar}-\frac{\mathbf{p}^{2}}{2\mu}\tau\right]\exp\bigl(-iU(\mathbf{p},\boldsymbol{\eta},\tau,\hbar)\bigr) \qquad (9.7.56)$$

with the initial condition

$$U(\mathbf{p},\boldsymbol{\eta},\tau,\hbar)\big|_{\tau=0}=0 \qquad (9.7.57)$$

so that

$$G(\mathbf{x}\,\hbar\tau;\mathbf{x}'0)\xrightarrow[\tau\to 0]{}\delta^{\nu}(\mathbf{x}-\mathbf{x}'). \qquad (9.7.58)$$

From (9.7.54), (9.7.55), the following differential equation for $U(\mathbf{p},\boldsymbol{\eta},\tau,\hbar)$ is obtained

$$\left[\frac{\partial}{\partial\tau}U+\frac{\hbar\mathbf{p}}{\mu}\cdot\boldsymbol{\nabla}_{\boldsymbol{\eta}}U-\frac{i\hbar^{2}}{2\mu}\boldsymbol{\nabla}_{\boldsymbol{\eta}}^{2}U-\frac{\hbar^{2}}{2\mu}\boldsymbol{\nabla}_{\boldsymbol{\eta}}U\cdot\boldsymbol{\nabla}_{\boldsymbol{\eta}}U-V\right]=0 \qquad (9.7.59)$$

with U satisfying the initial condition in (9.7.57).

In (9.7.54)–(9.7.59), we have used the parameter τ needed in the applications in (9.7.22), (9.7.24)–(9.7.26), (9.7.31)–(9.7.33). Some of these applications will be considered in the next section as well as, later on, in Chapter 13 in the study of multi-electron atoms.

9.8 The Thomas-Fermi Approximation and Deviations Thereof

In view of applications of the formulae (9.7.22), (9.7.24)–(9.7.26), (9.7.31)–(9.7.33), written in terms of integrals in the integration variable τ, we wish to investigate the expressions in (9.7.56)–(9.7.59) for $G(\mathbf{x}\,\hbar\tau;\mathbf{x}'0)$ on its dependence on $\hbar$ for a given τ.[12]

To the above end, to the leading order in $\hbar$, we have from (9.7.59),

$$\frac{\partial}{\partial\tau}U-V\simeq 0 \qquad (9.8.1)$$

or

---

[12] Note that in this section, $G(\mathbf{x}\,\hbar\tau;\mathbf{x}'0)$ is expressed in terms of the parameter of interest τ and not T.

$$U \simeq U_0 \equiv V\tau \tag{9.8.2}$$

(see (9.7.57)), leading from (9.7.56) to

$$G(\mathbf{x}\hbar\tau;\mathbf{x}'0) = \int \frac{d^\nu \mathbf{p}}{(2\pi\hbar)^\nu} \exp i\left[\frac{\mathbf{p}\cdot\boldsymbol{\zeta}}{\hbar} - \frac{\mathbf{p}^2}{2\mu}\tau\right]\exp(-iV\tau)$$

$$= \left(\frac{\mu}{2\pi i\hbar^2\tau}\right)^{\nu/2} \exp i\left(\frac{\mu|\mathbf{x}-\mathbf{x}'|^2}{2\hbar^2\tau} - V(\mathbf{x})\tau\right) \tag{9.8.3}$$

which will be referred to as the Thomas-Fermi semi-classical approximation.

Upon substituting (9.8.3) for $\mathbf{x} = \mathbf{x}'$ in (9.7.22), for example, we obtain

$$n(\mathbf{x};\xi) = \int \frac{d^\nu \mathbf{p}}{(2\pi\hbar)^\nu} \Theta\left(\xi - V(\mathbf{x}) - \frac{\mathbf{p}^2}{2\mu}\right) \tag{9.8.4}$$

which for $\nu = 3$, gives

$$n(\mathbf{x};\xi) = \frac{1}{6\pi^2}\left(\frac{2\mu(\xi - V(\mathbf{x}))}{\hbar^2}\right)^{3/2}. \tag{9.8.5}$$

This semi-classical solution is usually referred to as the Thomas-Fermi approximation.

Several applications of the expression in (9.8.3) will be given in our study of multi-electron atoms in Chapter 13.

To find the deviation of $G(\mathbf{x}\hbar\tau;\mathbf{x}'0)$, for example, from the Thomas-Fermi semi-classical approximation, in view of the applications mentioned above, we write for U satisfying (9.7.57), to order $\hbar^2$ and a given τ,

$$U = U_0 + \hbar\,\delta U_1 + \hbar^2 \delta U_2 \tag{9.8.6}$$

thus obtaining from (9.7.59),

$$\frac{\partial}{\partial \tau}\delta U_1 + \frac{\mathbf{p}}{\mu}\cdot\boldsymbol{\nabla} U_0 = 0 \tag{9.8.7}$$

written in terms of the variable $\mathbf{x}$ (see (9.8.2)). That is,

$$\delta U_1 = -\frac{\tau^2}{2\mu}\,\mathbf{p}\cdot\boldsymbol{\nabla} V(\mathbf{x}). \tag{9.8.8}$$

For δU_2 we get

$$\frac{\partial}{\partial \tau}\delta U_2 = \frac{i}{2\mu}\tau \boldsymbol{\nabla}^2 V + \frac{\tau^2}{2\mu}(\boldsymbol{\nabla} V)^2 + \frac{\tau^2}{2\mu^2}(\mathbf{p}\cdot\boldsymbol{\nabla})^2 V \tag{9.8.9}$$

giving

9.8 The Thomas-Fermi Approximation and Deviations Thereof

$$\delta U_2 = \frac{i}{4\mu} \tau^2 \boldsymbol{\nabla}^2 V + \frac{\tau^3}{6\mu} (\boldsymbol{\nabla} V)^2 + \frac{\tau^3}{6\mu^2} (\mathbf{p} \cdot \boldsymbol{\nabla})^2 V. \tag{9.8.10}$$

We set

$$A^{jk} = \frac{i\tau}{2\mu} \delta^{jk} + \frac{i\tau^3}{6\mu^2} \hbar^2 \nabla^j \nabla^k V \tag{9.8.11}$$

$$B^j = \frac{-i\hbar}{2\mu} \tau^2 \nabla^j V \tag{9.8.12}$$

$$C^j = - \left(A^{-1}\right)^{jk} \frac{B^k}{2} \tag{9.8.13}$$

and note that

$$\det A = \left(\frac{i\tau}{2\mu}\right)^\nu + \left(\frac{i\tau}{2\mu}\right)^{\nu-1} \frac{i\tau^3}{6\mu^2} \hbar^2 \boldsymbol{\nabla}^2 V \tag{9.8.14}$$

and

$$\frac{1}{\sqrt{\det A}} = \left(\frac{2\mu}{i\tau}\right)^{\nu/2} \left[1 - \frac{\tau^2}{6\mu} \hbar^2 \boldsymbol{\nabla}^2 V\right] \tag{9.8.15}$$

to order $\hbar^2$, for a given τ.

From (9.8.6), (9.8.2), (9.8.8), (9.8.10)–(9.8.12), we may write for $G(\mathbf{x}\hbar\tau;\mathbf{x}0)$ in (9.7.56)

$$\exp\left(-i\left[V(\mathbf{x})\tau + \frac{i\hbar^2}{4\mu}\tau^2\boldsymbol{\nabla}^2 V + \frac{\hbar^2}{6\mu}\tau^3(\boldsymbol{\nabla} V)^2\right]\right)$$

$$\times \int \frac{d^\nu \mathbf{p}}{(2\pi\hbar)^\nu} \exp - \left(p^j A^{jk} p^k + B^j p^j\right) \tag{9.8.16}$$

with a summation over repeated indices understood.

Upon changing the integration variables in (9.8.16) as

$$p^j \to p^j + C^j \tag{9.8.17}$$

we have

$$p^j A^{jk} p^k + B^j p^j \to p^j A^{jk} p^k + \left(B^j C^j + C^j A^{jk} C^k\right) \tag{9.8.18}$$

where

$$B^j C^j + C^j A^{jk} C^k = \frac{1}{2} B^j C^j$$

$$\simeq \frac{1}{4} \frac{\hbar \tau^2}{2\mu} \nabla^j V \left(\frac{2\mu}{i\tau} \delta^{jk}\right) \frac{\hbar \tau^2}{2\mu} \nabla^k V$$

$$= \frac{-i\hbar^2 \tau^3}{8\mu} (\nabla V)^2 \qquad (9.8.19)$$

to order $\hbar^2$ for a given τ.

Finally we use the integral

$$\int d^\nu \mathbf{p}\, \exp\left(-p^j A^{jk} p^k\right) = \frac{(\pi)^{\nu/2}}{\sqrt{\det A}} \qquad (9.8.20)$$

to obtain from (9.8.16), (9.8.18), (9.8.19), (9.8.15)

$$G(\mathbf{x}\,\hbar\tau;\mathbf{x}\,0) \simeq \left(\frac{\mu}{2\pi i\hbar^2 \tau}\right)^{\nu/2} e^{-iV(\mathbf{x})\tau} \left[1 + \frac{\hbar^2 \tau^2}{12\mu} \nabla^2 V - \frac{i\hbar^2 \tau^3}{24\mu} (\nabla V)^2\right]. \qquad (9.8.21)$$

The latter may be also rewritten as

$$G(\mathbf{x}\,\hbar\tau;\mathbf{x}\,0) \simeq \left[1 + \frac{\hbar^2 \tau^2}{12\mu} (\nabla^2 V) - \frac{i\hbar^2 \tau^3}{24\mu} (\nabla V)^2\right]$$

$$\times \int \frac{d^\nu \mathbf{p}}{(2\pi\hbar)^\nu} \exp -i\left[\mathbf{p}^2/2\mu + V(\mathbf{x})\right]\tau \qquad (9.8.22)$$

giving the following deviation from the Thomas-Fermi semi-classical approximation

$$\delta G(\mathbf{x}\hbar\tau;\mathbf{x}\,0) = \frac{\hbar^2 \tau^2}{12\mu} \left[(\nabla^2 V) - \frac{i\tau}{2}(\nabla V)^2\right]$$

$$\times \int \frac{d^\nu \mathbf{p}}{(2\pi\hbar)^\nu} \exp -i\left[\mathbf{p}^2/2\mu + V(\mathbf{x})\right]\tau \qquad (9.8.23)$$

for a *given* τ.

The above expression will find an important application to multi-electron atoms when, in the process, we carry out the τ-integral in (9.7.32).

9.9 The Coulomb Green Function: The Full Spectrum

9.9.1 An Integral Equation

The Green functions $G_\pm(\mathbf{x}t;\mathbf{x}'t')$ for a given potential $V(\mathbf{x})$ satisfies the equation

$$\left[i\hbar\frac{\partial}{\partial t} + \frac{\hbar^2}{2\mu}\nabla^2 - V(\mathbf{x})\right] G_\pm(\mathbf{x}t;\mathbf{x}'t') = \pm i\hbar\, \delta^3(\mathbf{x} - \mathbf{x}')\, \delta(t - t') \qquad (9.9.1)$$

in 3D, and their free counterparts (§9.1), with corresponding boundary conditions, satisfy

9.9 The Coulomb Green Function: The Full Spectrum

$$\left[i\hbar\frac{\partial}{\partial t} + \frac{\hbar^2}{2\mu}\nabla^2\right] G^0_\pm(\mathbf{x}t; \mathbf{x}'t') = \pm i\hbar\, \delta^3(\mathbf{x} - \mathbf{x}')\, \delta(t - t'). \qquad (9.9.2)$$

The solution of (9.9.1) may be written as

$$G_\pm(\mathbf{x}t; \mathbf{x}'t') = G^0_\pm(\mathbf{x}t; \mathbf{x}'t')$$

$$\mp \frac{i}{\hbar} \int d^3\mathbf{x}'' dt''\, G^0_\pm(\mathbf{x}t; \mathbf{x}''t'')\, V(\mathbf{x}'')\, G_\pm(\mathbf{x}''t''; \mathbf{x}'t') \qquad (9.9.3)$$

as is easily verified. We introduce the Fourier transforms

$$G_\pm(\mathbf{x}t; \mathbf{x}'t') = \pm i\hbar \int_{-\infty}^{\infty} \frac{dp^0}{2\pi\hbar} \int \frac{d^3\mathbf{p}\, d^3\mathbf{p}'}{(2\pi\hbar)^3} G_\pm(\mathbf{p}, \mathbf{p}'; p^0)$$

$$\times e^{i\mathbf{p}\cdot\mathbf{x}/\hbar} e^{-i\mathbf{p}'\cdot\mathbf{x}'/\hbar} e^{-ip^0(t-t')/\hbar} \qquad (9.9.4)$$

$$G^0_\pm(\mathbf{x}t; \mathbf{x}'t') = \pm i\hbar \int_{-\infty}^{\infty} \frac{dp^0}{2\pi\hbar} \int \frac{d^3\mathbf{p}\, d^3\mathbf{p}'}{(2\pi\hbar)^3} \frac{\delta^3(\mathbf{p} - \mathbf{p}')}{\left(p^0 - \frac{\mathbf{p}^2}{2\mu} \pm i\varepsilon\right)}$$

$$\times e^{i\mathbf{p}\cdot\mathbf{x}/\hbar} e^{-i\mathbf{p}'\cdot\mathbf{x}'/\hbar} e^{-ip^0(t-t')/\hbar}. \qquad (9.9.5)$$

For the potential in question

$$V(\mathbf{x}) = \frac{\lambda}{r}, \qquad r = |\mathbf{x}| \qquad (9.9.6)$$

$$\frac{1}{r} = 4\pi\hbar^2 \int \frac{d^3\mathbf{p}}{(2\pi\hbar)^3} \frac{e^{i\mathbf{p}\cdot\mathbf{x}/\hbar}}{\mathbf{p}^2} \qquad (9.9.7)$$

we then obtain

$$G_\pm(\mathbf{p}, \mathbf{p}'; p^0) = \frac{\delta^3(\mathbf{p} - \mathbf{p}')}{\left(p^0 - \frac{\mathbf{p}^2}{2\mu} \pm i\varepsilon\right)} + \frac{4\pi\lambda}{\hbar(2\pi)^3} \int \frac{d^3\mathbf{p}''}{(\mathbf{p} - \mathbf{p}'')^2} \frac{G_\pm(\mathbf{p}'', \mathbf{p}'; p^0)}{\left(p^0 - \frac{\mathbf{p}^2}{2\mu} \pm i\varepsilon\right)}. \qquad (9.9.8)$$

It is convenient at this stage to set

$$p^0 = -\frac{p_0^2}{2\mu}. \qquad (9.9.9)$$

This procedure, as we will see later, applies to the continuous spectrum as well.[13] For simplicity of the notation we write $G_\pm(\mathbf{p}, \mathbf{p}'; p^0)$ simply as $G(\mathbf{p}, \mathbf{p}')$ up to (9.9.29).

---

[13] We use the Schwinger representation of the Coulomb Green function: Schwinger (1964).

We introduce the unit vectors in 4D-Euclidean space

$$\boldsymbol{\eta} = \left(\frac{p_0^2 - \mathbf{p}^2}{p_0^2 + \mathbf{p}^2}, \frac{2p_0 \mathbf{p}}{p_0^2 + \mathbf{p}^2} \right) \tag{9.9.10}$$

$$\boldsymbol{\eta}^2 = 1 \tag{9.9.11}$$

with the latter equality easily verified.

Let

$$\cos \chi = \frac{p_0^2 - \mathbf{p}^2}{p_0^2 + \mathbf{p}^2}, \qquad \sin \chi = \frac{2p_0 |\mathbf{p}|}{p_0^2 + \mathbf{p}^2}. \tag{9.9.12}$$

Upon differentiation of $\cos \chi$, as a function of the variable $|\mathbf{p}| = p$, we obtain

$$\sin \chi \, d\chi = \frac{4pp_0^2}{(p_0^2 + \mathbf{p}^2)^2} \, dp \tag{9.9.13}$$

which upon multiplying by $\sin \chi \sin \theta \, d\theta \, d\phi$, leads to

$$d^3 \mathbf{p} = \left(\frac{p_0^2 + \mathbf{p}^2}{2p_0} \right)^3 d\Omega \tag{9.9.14}$$

where $d\Omega$ is the solid angle element in $4D$ (see Appendix III, (III.8)),

$$d\Omega = \sin \theta \, d\theta \, d\phi \sin^2 \chi \, d\chi \tag{9.9.15}$$

where $0 \leqslant \chi \leqslant \pi$.

From the property of the Dirac deltas

$$\int d^3 \mathbf{p}' \delta^3 (\mathbf{p} - \mathbf{p}') = 1 = \int d\Omega' \delta (\Omega - \Omega') \tag{9.9.16}$$

we may infer from (9.9.14), that

$$\delta^3 (\mathbf{p} - \mathbf{p}') = \left(\frac{2p_0}{p^2 + p_0^2} \right)^3 \delta (\Omega - \Omega') . \tag{9.9.17}$$

In detail,

$$\delta (\Omega - \Omega') = \frac{\delta (\theta - \theta') \delta (\phi - \phi') \delta (\chi - \chi')}{\sin \theta \sin^2 \chi}. \tag{9.9.18}$$

If we set

$$\boldsymbol{\eta}' = \left(\frac{p_0^2 - p'^2}{p_0^2 + p'^2}, \frac{2p_0 \mathbf{p}'}{p_0^2 + p'^2} \right) \tag{9.9.19}$$

one readily obtains

$$\frac{1}{(\mathbf{p} - \mathbf{p}')^2} = \frac{4p_0^2}{(p_0^2 + \mathbf{p}^2)(p_0^2 + p'^2)} \frac{1}{(\boldsymbol{\eta} - \boldsymbol{\eta}')^2}. \tag{9.9.20}$$

9.9 The Coulomb Green Function: The Full Spectrum

The properties in (9.9.14), (9.9.17), (9.9.20) simplify the form of the integral equation in (9.9.8) and the solution will be then read off from the analysis provided on the Poisson equation in $4D$ in Appendix III.

To the above end, set

$$\Lambda(\Omega, \Omega') = -\frac{(p_0^2 + p^2)^2 G(\mathbf{p}, \mathbf{p}') (p_0^2 + p'^2)^2}{16\mu p_0^3} \tag{9.9.21}$$

then we see upon multiplying (9.9.8) by

$$-\frac{(p_0^2 + p^2)^2 (p_0^2 + p'^2)^2}{16\mu p_0^3}$$

and using (9.9.14), (9.9.17), (9.9.20), the integral equation (9.9.8) becomes replaced by

$$\Lambda(\Omega, \Omega') = \delta(\Omega - \Omega') - \frac{\lambda\mu}{2\pi^2 \hbar p_0} \int \frac{d\Omega''}{(\boldsymbol{\eta} - \boldsymbol{\eta}'')^2} \Lambda(\Omega'', \Omega'). \tag{9.9.22}$$

The above may be also rewritten as

$$\int d\Omega'' \left[\delta(\Omega - \Omega'') + \frac{\lambda\mu}{2\pi^2 \hbar p_0} \frac{1}{(\boldsymbol{\eta} - \boldsymbol{\eta}'')^2} \right] \Lambda(\Omega'', \Omega') = \delta(\Omega - \Omega'). \tag{9.9.23}$$

On the other hand, from (III.3.11), (III.3.12) and (III.3.2) in Appendix III with $\boldsymbol{\eta}^2 = 1 = \boldsymbol{\eta}''^2$:

$$\delta(\Omega - \Omega'') = \sum_{n=0}^{\infty} \frac{(n+1)}{2\pi^2} U_n(\boldsymbol{\eta} \cdot \boldsymbol{\eta}'') \tag{9.9.24}$$

$$\frac{1}{(\boldsymbol{\eta} - \boldsymbol{\eta}'')^2} = \sum_{n=0}^{\infty} U_n(\boldsymbol{\eta} \cdot \boldsymbol{\eta}'') \tag{9.9.25}$$

where the U_n are Chebyshev's polynomials of type II (see Appendix III), which formally lead for (9.9.23) the expression

$$\int d\Omega'' \sum_{n=0}^{\infty} \left[\frac{(n+1)}{2\pi^2} + \frac{\lambda\mu}{2\pi^2 \hbar p_0} \right] U_n(\boldsymbol{\eta} \cdot \boldsymbol{\eta}'') \Lambda(\Omega'', \Omega')$$

$$= \sum_{n=0}^{\infty} \frac{(n+1)}{2\pi^2} U_n(\boldsymbol{\eta} \cdot \boldsymbol{\eta}'). \tag{9.9.26}$$

Now we use the orthogonality/completeness relation over the solid angle $d\Omega''$ in (III.4.1) of Theorem IV.1 in Appendix III, to infer from (9.9.26) that

$$\Lambda\left(\Omega,\Omega'\right)=\frac{1}{2\pi^2}\sum_{n=0}^{\infty}\frac{(n+1)^2}{\left(n+1+\dfrac{\lambda\mu}{p_0\hbar}\right)}U_n\left(\boldsymbol{\eta}\cdot\boldsymbol{\eta}'\right) \qquad (9.9.27)$$

which may be rewritten as

$$\Lambda\left(\Omega,\Omega'\right)=\frac{1}{2\pi^2}\sum_{n=1}^{\infty}\frac{n^2}{\left(n+\dfrac{\lambda\mu}{p_0\hbar}\right)}U_{n-1}\left(\boldsymbol{\eta}\cdot\boldsymbol{\eta}'\right). \qquad (9.9.28)$$

9.9.2 The Negative Spectrum $p^0 < 0, \lambda < 0$

From (9.9.21), we then have for the Green function

$$G\left(\mathbf{p},\mathbf{p}'\right)=-\frac{16\mu p_0^3}{(p_0^2+p^2)^2(p_0^2+p'^2)^2}\frac{1}{2\pi^2}\sum_{n=1}^{\infty}\frac{n^2}{\left(n+\dfrac{\lambda\mu}{p_0\hbar}\right)}U_{n-1}\left(\boldsymbol{\eta}\cdot\boldsymbol{\eta}'\right). \qquad (9.9.29)$$

In terms of the variable p^0 in (9.9.9), with $p^0 \to p^0 \pm i\varepsilon$, (9.9.29) reads

$$G_{\pm}\left(\mathbf{p},\mathbf{p}';p^0\right)=\frac{1}{\mu^2\pi^2}\frac{(p^0)^2}{\left(p^0-\dfrac{\mathbf{p}^2}{2\mu}\pm i\varepsilon\right)^2\left(p^0-\dfrac{\mathbf{p}'^2}{2\mu}\pm i\varepsilon\right)^2}$$

$$\times\sum_{n=1}^{\infty}\frac{nU_{n-1}\left(\boldsymbol{\eta}\cdot\boldsymbol{\eta}'\right)}{\left(p^0+\dfrac{\mu\lambda^2}{2\hbar^2 n^2}\pm i\varepsilon\right)}\left(\sqrt{-2\mu p^0}+\frac{|\lambda|\mu}{n\hbar}\right). \qquad (9.9.30)$$

For $\mathbf{p}' = \mathbf{p}$, $\boldsymbol{\eta} = \boldsymbol{\eta}'$, $U_{n-1}(1) = n$ (see (III.2.10)) in Appendix III, and for $(\lambda < 0)$

$$p^0 \to -\frac{\mu\lambda^2}{2\hbar^2 n^2} \qquad (9.9.31)$$

for a given fixed n, (9.9.30) leads to

$$\left(p^0+\frac{\mu\lambda^2}{2\hbar^2 n^2}\right)G_{\pm}\left(\mathbf{p},\mathbf{p};p^0\right)\to n^2\left(\frac{8|\lambda|^5\mu^5}{n^5\hbar^5\pi^2\left(\mathbf{p}^2+\dfrac{\lambda^2\mu^2}{n^2\hbar^2}\right)^4}\right) \qquad (9.9.32)$$

giving the degree of degeneracy n^2 for the eigenvalues in (9.9.31), and the normalized momentum probability density in the state n:

$$|\psi_n(\mathbf{p})|^2=\frac{8|\lambda|^5\mu^5}{n^5\hbar^5\pi^2\left(\mathbf{p}^2+\dfrac{\lambda^2\mu^2}{n^2\hbar^2}\right)^4}. \qquad (9.9.33)$$

9.9 The Coulomb Green Function: The Full Spectrum

The normalizability of the latter follows from the value of the integral

$$\int \frac{d^3\mathbf{p}}{(\mathbf{p}^2 + a^2)^4} = \frac{\pi^2}{8|a|^5} \tag{9.9.34}$$

where the integration measure here is $d^3\mathbf{p}$. [Relative to the measure $d^3\mathbf{p}/(2\pi\hbar)^3$, one must multiply (9.9.33) by $(2\pi\hbar)^3$ regarding normalizability.]

The wavefunctions in the momentum description may be also obtained from (9.9.30) for $\mathbf{p} \neq \mathbf{p}'$. To this end, one may expand $U_{n-1}(\boldsymbol{\eta} \cdot \boldsymbol{\eta}')$ in terms of Gegenbauer polynomials[14] defined by

$$C_{n-\ell-1}^{\ell+1}(x) = \frac{1}{2^\ell \, \ell!} \left(\frac{d}{dx}\right)^\ell U_{n-1}(x) \tag{9.9.35}$$

in the form

$$U_{n-1}(\boldsymbol{\eta} \cdot \boldsymbol{\eta}') = \sum_{\ell=0}^{n-1} F_{\ell\,n}(\chi) F_{\ell\,n}(\chi') P_\ell(\cos\alpha) \tag{9.9.36}$$

where

$$F_{\ell\,n}(\chi) = 2^\ell \sqrt{\frac{(n-\ell-1)!}{(n+\ell)!}} \ell! \sqrt{2\ell+1} \sin^\ell \chi \, C_{n-\ell-1}^{\ell+1}(\cos\chi), \tag{9.9.37}$$

with the P_ℓ denoting the Legendre polynomials, and

$$\cos\alpha = \cos\theta \cos\theta' + \sin\theta \sin\theta' \cos(\phi - \phi'). \tag{9.9.38}$$

Upon using the expansion (see (5.3.67)),

$$P_\ell(\cos\alpha) = \frac{4\pi}{(2\ell+1)} \sum_{m=-\ell}^{\ell} Y_{\ell\,m}(\theta,\phi) Y^*_{\ell\,m}(\theta',\phi') \tag{9.9.39}$$

and setting

$$a = \frac{|\lambda|\mu}{n\hbar} \tag{9.9.40}$$

it is not difficult to show from (9.9.30), at the pole in (9.9.31), that the wavefunction, up to a phase factor, in the momentum description is given by

$$\psi_{n\ell m}(\mathbf{p}) = \sqrt{\frac{32a^5}{\pi n}} \frac{2^{2\ell} \ell! \, (ap)^\ell}{(a^2+p^2)^{2+\ell}} \sqrt{\frac{(n-\ell-1)!}{(n+\ell)!}} C_{n-\ell-1}^{\ell+1}\left(\frac{a^2-p^2}{a^2+p^2}\right) Y_{\ell m}(\theta,\phi) \tag{9.9.41}$$

---

[14] See, for example, Gradshteyn and Ryzhik (1965), pp. 1029–1031.

where we have finally used the definitions in (9.9.12). The $\psi_{n\,\ell\,m}(\mathbf{p})$, as following from (9.9.30), are automatically normalized with respect to the measure $d^3\mathbf{p}$.

It is easily verified from (9.9.36), (9.9.37), (9.9.41) that

$$\sum_{\ell=0}^{n-1}\sum_{m=-\ell}^{\ell}|\psi_{n\ell m}(\mathbf{p})|^2 = |\psi_n(\mathbf{p})|^2 \qquad (9.9.42)$$

thus coinciding with (9.9.33).

One may formally take the Fourier transform of $\psi_{n\,\ell\,m}(\mathbf{p})$ in (9.9.41) and make a transition to the configuration space description and this is left as an exercise to the reader (see Problem 9.21)

9.9.3 The Positive Spectrum $p^0 > 0$

To obtain an expression for the Green functions $G_\pm(\mathbf{p},\mathbf{p}';p^0)$, for $p^0 > 0$, we go back to (9.9.28).

To the above end, we note that for any ξ

$$(1-\xi)^2 + \xi(\boldsymbol{\eta}-\boldsymbol{\eta}')^2 = 1 + \xi^2 - 2\xi\boldsymbol{\eta}\cdot\boldsymbol{\eta}' \qquad (9.9.43)$$

since $\boldsymbol{\eta}^2 = 1 = \boldsymbol{\eta}'^2$. Hence from (III.2.1), in Appendix III, with $|\xi| < 1$, we may write

$$\frac{1}{\left[(1-\xi)^2 + \xi(\boldsymbol{\eta}-\boldsymbol{\eta}')^2\right]} = \sum_{n=1}^{\infty} \xi^{n-1} U_{n-1}(\boldsymbol{\eta}\cdot\boldsymbol{\eta}') \qquad (9.9.44)$$

or upon multiplying the above by ξ, and explicitly taking the derivative with respect to ξ on both sides we obtain

$$\frac{(1-\xi^2)}{\left[(1-\xi)^2 + \xi(\boldsymbol{\eta}-\boldsymbol{\eta}')^2\right]^2} = \sum_{n=1}^{\infty} n\,\xi^{n-1} U_{n-1}(\boldsymbol{\eta}\cdot\boldsymbol{\eta}'). \qquad (9.9.45)$$

By multiplying this equation by ξ, and taking the derivative with respect to ξ gives the useful equation

$$\frac{d}{d\xi}\frac{\xi(1-\xi^2)}{\left[(1-\xi)^2 + \xi(\boldsymbol{\eta}-\boldsymbol{\eta}')^2\right]^2} = \sum_{n=1}^{\infty} n^2\,\xi^{n-1} U_{n-1}(\boldsymbol{\eta}\cdot\boldsymbol{\eta}'). \qquad (9.9.46)$$

In reference to (9.9.28), we note from (9.9.9) that with $p^0 \to p^0 \pm i\varepsilon$, $p^0 > 0$, corresponding to $G_\pm$, respectively,

$$p_0^2 = -2\mu p^0 \mp i\varepsilon. \qquad (9.9.47)$$

Hence concerning the Green function G_+, we have
$$p_0^2 = (2\mu p^0)\, e^{-i\pi} \tag{9.9.48}$$
for $\varepsilon \to +0$, and for G_-,
$$p_0^2 = (2\mu p^0)\, e^{i\pi}. \tag{9.9.49}$$
Accordingly, we have the rule:
$$\frac{1}{p_0} = \frac{1}{\sqrt{2\mu p^0}} e^{\pm i\pi/2} = \frac{\pm i}{\sqrt{2\mu p^0}} \tag{9.9.50}$$
with $p^0 > 0$, for the Green functions $G_\pm$, in the interpretation of $1/p_0$ in the denominators in (9.9.28).

Let
$$\gamma = \frac{\lambda \mu/\hbar}{\sqrt{2\mu p^0}}, \qquad p^0 > 0 \tag{9.9.51}$$
and using the fact that in (9.9.28) we may write
$$\frac{1}{n \pm i\gamma} = \int_0^1 d\xi\, \xi^{(n-1\pm i\gamma)} \tag{9.9.52}$$
we have directly from (9.9.46) and (9.9.28) that
$$\Lambda(\Omega, \Omega') = \frac{1}{2\pi^2} \int_0^1 \xi^{\pm i\gamma}\, d\xi\, \frac{d}{d\xi} \frac{\xi(1-\xi^2)}{\left[(1-\xi)^2 + \xi(\boldsymbol{\eta}-\boldsymbol{\eta}')^2\right]^2}. \tag{9.9.53}$$

Finally we use the equalities in (9.9.20), (9.9.50) and the definition in (9.9.21) to obtain explicitly
$$G_\pm(\mathbf{p}, \mathbf{p}'; p^0) = \mp \frac{i}{\pi^2} \sqrt{\frac{\mu}{8p^0}}\, \frac{1}{\left[(\mathbf{p}-\mathbf{p}')^2\right]^2} \int_0^1 \xi^{\pm i\gamma}\, d\xi\, \frac{d}{d\xi} \frac{\xi(1-\xi^2)}{\left[\xi - \frac{1}{\rho_\pm}(1-\xi)^2\right]^2} \tag{9.9.54}$$
where
$$\rho_\pm = (\mathbf{p}-\mathbf{p}')^2 \frac{2p^0/\mu}{\left(p^0 - \frac{\mathbf{p}^2}{2\mu} \pm i\varepsilon\right)\left(p^0 - \frac{\mathbf{p}'^2}{2\mu} \pm i\varepsilon\right)}, \qquad p^0 > 0 \tag{9.9.55}$$
and γ is defined in (9.9.51).

The Green function in (9.9.54) for the positive spectrum $p^0 > 0$, and $\lambda > 0$, $\lambda < 0$, will find detailed applications to Coulomb scattering in Chapter 15, §15.5.

Problems

9.1. Show that in one-dimension the analogous expression to the one in (9.1.30) is given by

$$G_+^0(xt; x't') = \int_{-\infty}^{\infty} \frac{dp^0}{2\pi\hbar} \sqrt{\frac{m}{2p^0}} \exp\left(i\sqrt{2mp^0}\,|x - x'|/\hbar\right)$$

$$\times \exp\left(-ip^0(t - t')/\hbar\right).$$

9.2. Show that an integral representation of $G^0(\mathbf{x}t; \mathbf{x}'t')$ in two dimensions is given by

$$G^0(\mathbf{x}t; \mathbf{x}'t') = \frac{m}{2\pi\hbar} \int_0^{\infty} dz\, J_0\left(|\mathbf{x} - \mathbf{x}'|\sqrt{\frac{2mz}{\hbar}}\right) e^{-izt}$$

where J_0 is the zeroth order Bessel function.

9.3. Derive the Green function for the linear potential in §9.2 by the method used in deriving the Green function for the quadratic potential in the same section.

9.4. Carry out a study of the Green function in the one dimensional potential barrier

$$V(x) = \begin{cases} V_0, & 0 < x < a \\ 0, & \text{elsewhere} \end{cases}$$

where $V_0 > 0$. It is advisable to rewrite $V(x)$ as

$$V(x) = [\Theta(x) - \Theta(x - a)]V_0.$$

9.5. Obtain an expression for the probability density distribution on an observation screen in a single "slit" experiment where the "slit" is a sharp circular hole. Find then the probability of observing a particle outside the classical shadow of the hole on the screen. Extend your analysis to a double slit experiment involving two sharp circular holes of equal radii. These analyses are conveniently carried out in terms of so-called Lommel functions (see also the appendix to §15.2). [Ref: Manoukian (1989).]

9.6. Show that for a time independent $F(t) = E$ in (9.4.1), the expression in (9.4.21) goes over to the one in (9.2.10).

9.7. Carry out in detail the steps leading from (9.4.22) to (9.4.27).

9.8. In reference to the first term in (9.3.14) with the denominator $[p^0 + m\lambda^2/2\hbar^2 + i\varepsilon]$, show that by closing the p^0-contour in (9.3.2), in the complex p^0-plane, from below, the semi-circle of infinite radius gives zero contribution and hence an application of the residue theorem gives (9.3.16) for the contribution of the pole $p^0 = -m\lambda^2/2\hbar^2$.

9.9. Study the Green function of an oscillator for which ω in the harmonic oscillator problem is replaced by a general time-dependent one $\omega(t)$.

9.10. Show that (9.5.2) leads to (9.5.3) and finally to (9.5.5) for the Green function $G^0_{+>}$ in half-space.

9.11. By integrating by parts, derive the inequality in (9.5.10).

9.12. Derive the expression for the probability density given in (9.5.14).

9.13. Derive the Poisson sum formula in (9.6.10). [A classic reference on this is: Morse and Feshbach (1953), p. 467.]

9.14. (A review problem in electromagnetics). Show that the vector potential $\mathbf{A}$ of an infinitely long solenoid of circular cross section, at a distance r outside the solenoid is given by the expression in (9.6.17).

9.15. Using (9.6.24), (9.6.25) derive the two equations in (9.6.26), (9.6.27).

9.16. By relating the modified Bessel function $I_{|\nu|}(-i\rho)$ to $J_{|\nu|}(\rho)$: $I_{|\nu|}(-i\rho) = e^{-i|\nu|\pi/2} J_{|\nu|}(\rho)$ and using the known asymptotic behavior of $J_{|\nu|}(\rho)$ for $\rho \to \infty$, verify the corresponding one for $I_{|\nu|}(-i\rho)$ as given in (9.6.28), (9.6.29).

9.17. By the application of appropriate boundary conditions at the origin, as discussed at the end of §9.6, to allow the scattering of a particle off by the solenoid at the origin, even for $\delta_0 \to 0$, investigate how this scattering is modified by the activation of the solenoid, i.e., for $\delta_0 > 0$, as an illustration of the Aharonov-Bohm effect.

9.18. For the illustration of the Aharonov-Bohm effect at the end of §9.6, consider rather formally initial states such as $\exp(-|\phi'/\gamma|)$, $\exp(-|\phi'^2/\gamma^2|)$ with $\gamma > 0$ very small so that contributions from $|\phi'| > \gamma$ may be neglected. Investigate the nature of the observation intensities as functions of ϕ.

9.19. Extend the analysis provided in the example on the Aharonov-Bohm effect at the end of §9.6 to single arc-slit and to double arc-slits experiments. The reader may place the arcs in the most convenient way to simplify the analyses.

9.20. Verify the identity in (9.7.9).

9.21. Carry out a Fourier transform of $\psi_{n\ell m}(\mathbf{p})$ in (9.9.41) to $\mathbf{x}$-space. Here you will need properties of some special functions which will allow you to explicitly carry out the Fourier transform.

9.22. Investigate the nature of the Green function for the $1/|\mathbf{x}|$ potential in two and one dimensions.

9.23. Show that for a potential of the form $V(x,t) = m\omega^2(t)x^2/2 - xF(t)$, the Green function takes the form $G(xt;x't') = N(t,t') \times \exp\left(i\int_{t'}^{t} d\tau\, L_c(\tau)/\hbar\right)$, with $L_c(\tau)$ the classical Lagrangian, and set up an equation that would determine $N(t,t')$. [Note that $L_c(\tau)$ depends on x, x', t, t' as well.]

10

Path Integrals

The path integral formalism of quantum physics, as an alternative to the operator approach, was introduced by Feynman[1] in 1948. In recent years, it has been applied in almost all areas of physics and has become a powerful and essential tool to do quantum physics. There is an intimate connection between this formalism and its classical counterpart, and the action, as the time integral of the Lagrangian, appears naturally in it. In its simplest form, it involves in developing an expression for the amplitude $\langle \mathbf{x}t | \mathbf{x}'t' \rangle$ for a particle initially at $\mathbf{x}'$ at time t' to be found at $\mathbf{x}$ at a later time $t > t'$ as a sum over all paths beginning at $\mathbf{x}'$ and ending up at $\mathbf{x}$. Quantum physics being probabilistic, this expression involves, in general, in addition to the classical path joining $\mathbf{x}'$ to $\mathbf{x}$, an (uncountable) infinite number of possible paths joining these two points. The importance of a so-called Lagrangian formulation of quantum physics was emphasized by Dirac and fully exploited by Feynman in his classic work.

The present chapter deals with a fairly detailed treatment of path integrals. We approach the problem as a logical extension of our study of Green functions in the previous chapter. A key result in the analysis is to rewrite the amplitude $\langle \mathbf{x}t | \mathbf{x}'t' \rangle$, mentioned above, in terms of a completeness relation as (see also (9.6))

$$\langle \mathbf{x}t | \mathbf{x}'t' \rangle = \int d^\nu \mathbf{x}'' \, \langle \mathbf{x}t | \mathbf{x}''t'' \rangle \langle \mathbf{x}''t'' | \mathbf{x}'t' \rangle \qquad (10.1)$$

with t'' conveniently chosen so that $t' < t'' < t$. By repeated application of (10.1), one may rewrite $\langle \mathbf{x}t | \mathbf{x}'t' \rangle$ as integrals of the product of a large number of amplitudes each of which describes the time evolution in infinitesimally short times. This technique, usually referred to as the "time slicing method", allows one to readily obtain the path integral representation of the amplitude in question. §10.1, §10.2 deal, respectively, with a free particle and a particle in a given potential. For interactions involving velocity dependent potentials,

---

[1] Feynman (1948); Feynman and Hibbs (1965).

such as in the interaction of charged particles with external electromagnetic fields, the completeness relation in (10.1) is extended to sums over momenta as well. This leads to a phase space analysis and is the subject matter of §10.3 with emphasis put on the interaction of a charged particle with an external electromagnetic field which is of central importance in physics. §10.4 deals with a systematic analysis of path integrals with constrained dynamics.[2] This subject matter has become quite important in recent years and deserves the proper attention given here. §10.5, §10.6 deal, respectively, with the problem of Bose and Fermi excitations with special considerations for their interactions with external sources. In §10.6 the necessarily tools for integrations over so-called Grassmann variables are also developed in great detail in order to handle Fermi excitations in the path integral context.

10.1 The Free Particle

In this section, we derive the path integral expression for the amplitude that a free particle in $\mathbb{R}^\nu$, which is initially at $\mathbf{x}'$ at time t', is found at $\mathbf{x}$ at some later time t.

Our starting point is the completeness relation (10.1), as applied $(N-1)$ number of times which reads:

$$\langle \mathbf{x}\, t | \mathbf{x}'\, t' \rangle_0 = \int d^\nu \mathbf{x}_1 \ldots d^\nu \mathbf{x}_{N-1} \Big[\langle \mathbf{x}\, t | \mathbf{x}_{N-1}\, t_{N-1} \rangle_0$$

$$\times \langle \mathbf{x}_{N-1}\, t_{N-1} | \mathbf{x}_{N-2}\, t_{N-2} \rangle_0 \cdots \langle \mathbf{x}_1\, t_1 | \mathbf{x}'\, t' \rangle_0 \Big] \quad (10.1.1)$$

where (see (9.1.38))

$$\langle \mathbf{x}_{k+1}\, t_{k+1} | \mathbf{x}_k\, t_k \rangle_0 = \left(\frac{m}{2\pi i\hbar\,(t_{k+1} - t_k)} \right)^{\nu/2} \exp\left[\frac{i\,m\,(\mathbf{x}_{k+1} - \mathbf{x}_k)^2}{2\hbar\,(t_{k+1} - t_k)} \right]$$
$$(10.1.2)$$

$t > t_{N-1} > \cdots > t_1 > t'$, and we set $\mathbf{x} \equiv \mathbf{x}_N$, $\mathbf{x}' \equiv \mathbf{x}_0$, $t = t_N$, $t' = t_0$.

By choosing $t_{k+1} - t_k = (t - t')/N \equiv T/N$, we may rewrite (10.1.1) as

$$\langle \mathbf{x}\, t | \mathbf{x}'\, t' \rangle_0 = \int \left(\frac{m}{2\pi i\hbar\, T/N} \right)^{N\nu/2} \left(\prod_{j=1}^{N-1} d^\nu \mathbf{x}_j \right)$$

$$\times \exp\left[\frac{i}{\hbar} \sum_{k=0}^{N-1} \left(\frac{T}{N} \right) \frac{m}{2} \left(\frac{\mathbf{x}_{k+1} - \mathbf{x}_k}{T/N} \right)^2 \right]. \quad (10.1.3)$$

---

[2] Cf. Dirac (2001), Faddeev (1969); see also Senjanovic (1976).

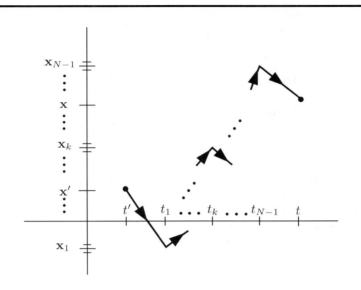

Fig. 10.1. A path contributing to the expression for $\langle \mathbf{x}t|\mathbf{x}'t'\rangle$. In the continuum limit $N \to \infty$, one has an uncountable infinite number of paths.

Upon taking the continuum limit $N \to \infty$, defining formally the measure

$$\mathscr{D}(\mathbf{x}(\cdot)) = \lim_{N \to \infty} \left(\frac{m}{2\pi i \hbar T/N}\right)^{N\nu/2} \left(\prod_{j=1}^{N-1} d^\nu \mathbf{x}_j\right) \tag{10.1.4}$$

and in this limit, converting the sum in the exponent in (10.1.3) to an integral, we have the so-called path integral expression for the amplitude $\langle \mathbf{x}t|\mathbf{x}'t'\rangle_0$:

$$\langle \mathbf{x}t|\mathbf{x}'t'\rangle_0 = \int_{\mathbf{x}(t')=\mathbf{x}'}^{\mathbf{x}(t)=\mathbf{x}} \mathscr{D}(\mathbf{x}(\cdot)) \exp\left[\frac{i}{\hbar}\int_{t'}^{t} d\tau \frac{m\dot{\mathbf{x}}^2(\tau)}{2}\right] \tag{10.1.5}$$

written as a sum over *all paths* going through the volume elements $d^\nu \mathbf{x}_1$, ..., $d^\nu \mathbf{x}_{N-1}$, respectively, about the points specified by $\mathbf{x}_1(t_1)$: $(\mathbf{x}_1, t_1)$, ..., $\mathbf{x}_{N-1}(t_{N-1})$: $(\mathbf{x}_{N-1}, t_{N-1})$, as each one of the variables $\mathbf{x}_1$, ..., $\mathbf{x}_{N-1}$ moves all over $\mathbb{R}^\nu$, i.e., as we integrate over all of these variables, *beginning at* $\mathbf{x}(t') = \mathbf{x}'$ and *ending at* $\mathbf{x}(t) = \mathbf{x}$, and take the limit $N \to \infty$.

A particular case of (10.1.5) is the amplitude for a particle to begin at the origin $\mathbf{x}' = \mathbf{0}$ and end up at the origin $\mathbf{x} = \mathbf{0}$. The expression for this amplitude may be directly read from (10.1.2), and we may write for (10.1.5) in this case:

$$\int_{\mathbf{x}(t')=\mathbf{0}}^{\mathbf{x}(t)=\mathbf{0}} \mathscr{D}(\mathbf{x}(\cdot)) \exp\left[\frac{i}{\hbar}\int_{t'}^{t} d\tau \frac{m\dot{\mathbf{x}}^2(\tau)}{2}\right] = \left(\frac{m}{2\pi i \hbar (t-t')}\right)^{\nu/2}. \tag{10.1.6}$$

This equality will be useful, as a normalization factor, in some cases for a particle interacting with a given potential as we will see in the next section.

10.2 Particle in a Given Potential

We consider a Hamiltonian of the form

$$H(t) = \frac{\mathbf{p}^2}{2m} + V(\mathbf{x},t) \equiv H_0 + V(\mathbf{x},t) \tag{10.2.1}$$

where, for greater generality, $V(\mathbf{x},t)$ may depend explicitly on time but is independent of $\mathbf{p}$. The explicit time dependence of $V(\mathbf{x},t)$ is assumed to come only from *a-priori* given c-functions of t such as, for example, in the time-dependent forced dynamics studied in §9.4. As an operator, $V(\mathbf{x},t)$ is a multiplicative one with respect to $|\mathbf{x}'\rangle$, i.e.,

$$V(\mathbf{x},t)|\mathbf{x}'\rangle = |\mathbf{x}'\rangle V(\mathbf{x}',t). \tag{10.2.2}$$

Since for a state $|\Psi(t)\rangle$, the Schrödinger equation reads

$$i\hbar \frac{\partial}{\partial t}|\Psi(t)\rangle = H(t)|\Psi(t)\rangle \tag{10.2.3}$$

and $\langle \mathbf{x}'|\Psi(t)\rangle = \langle \mathbf{x}'t|\Psi\rangle$, one has

$$i\hbar \frac{\partial}{\partial t}\langle \mathbf{x}'t| = \langle \mathbf{x}'t|H(t) \tag{10.2.4}$$

and similarly

$$-i\hbar \frac{\partial}{\partial t}|\mathbf{x}'t\rangle = H(t)|\mathbf{x}'t\rangle. \tag{10.2.5}$$

Upon setting $(t_0 + t_1)/2 = t$, $t_1 - t_0 = \varepsilon$, we may write the amplitude $\langle \mathbf{x}_1 t_1 | \mathbf{x}_0 t_0 \rangle$ as

$$\langle \mathbf{x}_1 t_1 | \mathbf{x}_0 t_0 \rangle = \left\langle \mathbf{x}_1, t + \frac{\varepsilon}{2} \middle| \mathbf{x}_0, t - \frac{\varepsilon}{2} \right\rangle. \tag{10.2.6}$$

For $\varepsilon \to 0$, then (10.2.4), (10.2.5) lead for (10.2.6)[3]

$$\langle \mathbf{x}_1 t_1 | \mathbf{x}_0 t_0 \rangle \simeq \left\langle \mathbf{x}_1 \middle| 1 - i\frac{\varepsilon}{\hbar} H(t) \middle| \mathbf{x}_0 \right\rangle \simeq \left\langle \mathbf{x}_1 \middle| \exp\left[-i\frac{\varepsilon}{\hbar} H(t)\right] \middle| \mathbf{x}_0 \right\rangle. \tag{10.2.7}$$

The exponential representation in the last equality is essential to obtain a unitary expression and satisfy the group property in time. The final verdict of the whole formalism is its consistency with the Schrödinger equation.

---

[3] We note that in (10.2.7), it is understood that we eventually take the limit $\varepsilon \to 0$, otherwise one would necessarily have the time-ordered structure given in the Appendix to §2.5.

10.2 Particle in a Given Potential

On the other hand for $\varepsilon \to 0$, the Baker-Campbell-Hausdorff formula (see Appendix I) states that

$$\exp -\frac{i\varepsilon}{\hbar} H_0 \exp -\frac{i\varepsilon}{\hbar} V(\mathbf{x}, t) \simeq \exp -\frac{i}{\hbar} \left(\varepsilon H(t) + \mathcal{O}\left(\varepsilon^2\right) \right) \qquad (10.2.8)$$

which from (10.2.7) gives

$$\langle \mathbf{x}_1 t_1 | \mathbf{x}_0 t_0 \rangle \simeq \int d^\nu \mathbf{x}' \left\langle \mathbf{x}_1 \left| e^{-i\varepsilon H_0/\hbar} \right| \mathbf{x}' \right\rangle \left\langle \mathbf{x}' \left| e^{-i\varepsilon V(\mathbf{x}, t)/\hbar} \right| \mathbf{x}_0 \right\rangle$$

$$\simeq \left(\frac{m}{2\pi i \hbar \varepsilon} \right)^{\nu/2} \exp \frac{i\varepsilon}{\hbar} \left(\frac{m}{2} \left(\frac{\mathbf{x}_1 - \mathbf{x}_0}{\varepsilon} \right)^2 - V(\mathbf{x}_0, t) \right) \qquad (10.2.9)$$

for $\varepsilon \to 0$, where we have used (10.1.2) corresponding to the free Green function.

We first check that (10.2.9) is consistent with the Schrödinger equation for a wavefunction $\langle \mathbf{x} t | \Psi \rangle$.

To the above end, we have to verify, from (10.2.9) for $\varepsilon \simeq 0$, that

$$\langle \mathbf{x}, t + \varepsilon | \Psi \rangle \simeq \left(\frac{m}{2\pi i \hbar \varepsilon} \right)^{\nu/2} \int d^\nu \mathbf{x}' \exp \frac{i\varepsilon}{\hbar} \left(\frac{m}{2} \left(\frac{\mathbf{x} - \mathbf{x}'}{\varepsilon} \right)^2 - V(\mathbf{x}', t) \right) \langle \mathbf{x}' t | \Psi \rangle \qquad (10.2.10)$$

is consistent. Upon making a change of variables $\mathbf{x}' \to \mathbf{x}' - \mathbf{x} = \mathbf{z}$, we obtain

$$\langle \mathbf{x}, t + \varepsilon | \Psi \rangle \simeq \left(\frac{m}{2\pi i \hbar \varepsilon} \right)^{\nu/2} \int d^\nu \mathbf{z} \exp \left(\frac{im}{2\hbar \varepsilon} \mathbf{z}^2 \right)$$

$$\times \left[1 - \frac{i\varepsilon}{\hbar} \left(T_\mathbf{z}(\mathbf{x}) V(\mathbf{x}, t) \right) \right] \left(T_\mathbf{z}(\mathbf{x}) \Psi(\mathbf{x}, t) \right) \qquad (10.2.11)$$

where $T_\mathbf{z}(\mathbf{x})$ is the Taylor operator

$$T_\mathbf{z}(\mathbf{x}) = \left(1 + \mathbf{z} \cdot \nabla + \frac{(\mathbf{z} \cdot \nabla)^2}{2} + \cdots \right). \qquad (10.2.12)$$

Using the integrals

$$\left(\frac{m}{2\pi i \hbar \varepsilon} \right)^{\nu/2} \int d^\nu \mathbf{z} \exp \left(\frac{im\mathbf{z}^2}{2\hbar \varepsilon} \right) = 1 \qquad (10.2.13)$$

$$\left(\frac{m}{2\pi i \hbar \varepsilon} \right)^{\nu/2} \int d^\nu \mathbf{z}\, z^j \exp \left(\frac{im\mathbf{z}^2}{2\hbar \varepsilon} \right) = 0 \qquad (10.2.14)$$

$$\left(\frac{m}{2\pi i \hbar \varepsilon} \right)^{\nu/2} \int d^\nu \mathbf{z}\, z^i z^j \exp \left(\frac{im\mathbf{z}^2}{2\hbar \varepsilon} \right) = \delta^{ij} \frac{i\hbar \varepsilon}{m} \qquad (10.2.15)$$

and so on, and keeping only terms up to order ε only in (10.2.11), we get

$$\Psi(\mathbf{x}, t + \varepsilon) - \Psi(\mathbf{x}, t) \simeq \frac{i\hbar\varepsilon}{2m} \nabla^2 \Psi(\mathbf{x}, t) - \frac{i\varepsilon}{\hbar} V(\mathbf{x}, t) \Psi(\mathbf{x}, t) \qquad (10.2.16)$$

which leads to the Schrödinger equation in the limit $\varepsilon \to 0$.

The completeness property of the Green function functions provide from (10.2.9) the limit

$$\langle \mathbf{x}\, t | \mathbf{x}'\, t' \rangle = \lim_{N \to \infty} \int \left(\frac{m}{2\pi i \hbar \varepsilon} \right)^{N\nu/2} d^\nu \mathbf{x}_1 \cdots d^\nu \mathbf{x}_{N-1}$$

$$\times \exp \frac{i}{\hbar} \sum_{k=0}^{N-1} \varepsilon \left[\frac{m}{2} \left(\frac{\mathbf{x}_{k+1} - \mathbf{x}_k}{\varepsilon} \right)^2 - V\left(\mathbf{x}_k, \hat{t}_k\right) \right] \qquad (10.2.17)$$

where $\hat{t}_k = (t_k + t_{k+1})/2 = t' + (k + 1/2)\varepsilon$, $t_k = t' + k\varepsilon$, $k = 0, 1, \ldots, N-1$, $\varepsilon = (t - t')/N = t_{k+1} - t_k$, $t_0 = t'$, $t_N = t$, $\mathbf{x}_0 = \mathbf{x}'$, $\mathbf{x}_N = \mathbf{x}$.

As in (10.1.5), we may rewrite (10.2.17) in the limit $N \to \infty$

$$\langle \mathbf{x}\, t | \mathbf{x}'\, t' \rangle = \int_{\mathbf{x}(t')=\mathbf{x}'}^{\mathbf{x}(t)=\mathbf{x}} \mathscr{D}(\mathbf{x}(\cdot)) \exp \frac{i}{\hbar} \int_{t'}^{t} d\tau \left(\frac{m}{2} \dot{\mathbf{x}}^2(\tau) - V(\mathbf{x}(\tau), \tau) \right) \qquad (10.2.18)$$

as a sum over all paths beginning at $\mathbf{x}(t') = \mathbf{x}'$ and ending at $\mathbf{x}(t) = \mathbf{x}$.

Here we recognize the Lagrangian

$$L(\tau) = \frac{m}{2} \dot{\mathbf{x}}^2(\tau) - V(\mathbf{x}(\tau), \tau) \qquad (10.2.19)$$

as a c-function, and $\int_{t'}^{t} d\tau\, L(\tau) = A$ as the action. It is not always true, however, that the Lagrangian simply appears in the integrand in the exponential in (10.2.18) in every case (see, for example, Problem 10.5).

As an application, consider the potential $V(x) = -Ex$, in one dimension (see also §9.2) and the corresponding amplitude

$$\langle x_2\, t_2 | x_1\, t_1 \rangle = \int_{x(t_1)=x_1}^{x(t_2)=x_2} \mathscr{D}(x(\cdot)) \exp \frac{i}{\hbar} \int_{t_1}^{t_2} d\tau \left[\frac{m}{2} \dot{\mathbf{x}}^2(\tau) + Ex(\tau) \right]. \qquad (10.2.20)$$

In addition to the classical path in (10.2.20), we have to consider *all* paths joining the point $x(t_1) = x_1$ to $x(t_2) = x_2$. Accordingly, any given path may be described by the function

$$x(\tau) = x_c(\tau) + y(\tau) \qquad (10.2.21)$$

where $x_c(\tau)$ is the classical solution, and $y(\tau)$ denotes the deviation of $x(\tau)$ from the classical one at time τ.

The boundary conditions are

10.2 Particle in a Given Potential

$$x(t_1) = x_1, \quad x(t_2) = x_2 \tag{10.2.22}$$

$$x_c(t_1) = x_1, \quad x_c(t_2) = x_2 \tag{10.2.23}$$

and hence

$$y(t_1) = 0 = y(t_2). \tag{10.2.24}$$

The solution of

$$\ddot{x}_c(\tau) = \frac{E}{m} \tag{10.2.25}$$

satisfying the boundary conditions in (10.2.23) is given by

$$\dot{x}_c(\tau) = \frac{E}{m}(\tau - t_1) + \left[(x_2 - x_1) - \frac{E(t_2 - t_1)^2}{m} \frac{1}{2}\right] \frac{1}{(t_2 - t_1)} \tag{10.2.26}$$

$$x_c(\tau) = \frac{E}{m} \frac{(\tau - t_1)^2}{2} + \dot{x}_c(t_1)(t_2 - t_1) + x_1. \tag{10.2.27}$$

Also note that

$$\frac{m\dot{x}^2(\tau)}{2} + Ex(\tau) = \left(\frac{m\dot{x}_c^2(\tau)}{2} + Ex_c(\tau)\right)$$

$$+ m\dot{x}_c(\tau)\dot{y}(\tau) + Ey(\tau) + \frac{m\dot{y}^2(\tau)}{2}$$

$$\equiv L_c(\tau) + \frac{m\dot{y}^2(\tau)}{2} + m\frac{d}{d\tau}(\dot{x}_c(\tau)y(\tau)) \tag{10.2.28}$$

where in writing the last equality we have identified $L_c(\tau)$ with the classical Lagrangian corresponding to the classical motion, and made use of the equation for $x_c(\tau)$ in (10.2.25).

Hence

$$\int_{t_1}^{t_2} d\tau L(\tau) = \int_{t_1}^{t_2} d\tau L_c(\tau) + \int_{t_1}^{t_2} d\tau \frac{m\dot{y}^2(\tau)}{2} + 0 \tag{10.2.29}$$

where on account of the boundary conditions in (10.2.24)

$$\int_{t_1}^{t_2} \frac{d}{d\tau}(\dot{x}_c(\tau)y(\tau)) = \dot{x}_c(t_2)y(t_2) - \dot{x}_c(t_1)y(t_1) = 0. \tag{10.2.30}$$

In detail, (10.2.26), (10.2.27) give

$$\int_{t_1}^{t_2} d\tau L_c(\tau) = \frac{m}{2}\frac{(x_2 - x_1)^2}{(t_2 - t_1)} + \frac{E(x_2 + x_1)}{2}(t_2 - t_1) - \frac{E^2}{24m}(t_2 - t_1)^3$$

$$\equiv S_c \tag{10.2.31}$$

thus defining the action for the classical path.

Finally with the change of variables $x(\tau)$ to $y(\tau)$, as given in (10.2.21) for all $t_1 \leqslant \tau \leqslant t_2$, and from the translational invariance of the measure in (10.1.4), we obtain

$$\langle x_2\, t_2 | x_1\, t_1 \rangle = \exp\left(\frac{i}{\hbar} S_c\right) \int_{y(t_1)=0}^{y(t_2)=0} \mathscr{D}(y(\cdot)) \exp \frac{i}{\hbar} \int_{t_1}^{t_2} d\tau \, \frac{m\dot{y}^2(\tau)}{2}. \tag{10.2.32}$$

From (10.1.6), (10.2.31), this leads to the expression given in (9.2.10), where now $T = t_2 - t_1$.

The derivation of $\langle x_2\, t_2 | x_1\, t_1 \rangle$ for the time-dependent potential $-xF(t)$, for a time-dependent force $F(t)$ (see §9.4) using the path integral technique is left as an exercise to the reader (see Problem 10.1) and the expression for which is given in (9.4.21).

10.3 Charged Particle in External Electromagnetic Fields: Velocity Dependent Potentials

We consider the interaction of a spin 0 charged particle, say, of charge e, with *a priori* given external electromagnetic field described by the pair of fields $\Phi(\mathbf{x}, t)$, $\mathbf{A}(\mathbf{x}, t)$. The Hamiltonian in question is given by

$$H(\mathbf{x}, \mathbf{p}, t) = \frac{(\mathbf{p} - \frac{e}{c}\mathbf{A})^2}{2m} + e\Phi$$

$$= \frac{\mathbf{p}^2}{2m} - \frac{e}{2mc}\mathbf{p} \cdot \mathbf{A} - \frac{e}{2mc}\mathbf{A} \cdot \mathbf{p} + \frac{e^2}{2mc^2}\mathbf{A}^2 + e\Phi. \tag{10.3.1}$$

It is yet not clear in what sense the path integral derivation given in the last section applies to this important system. For one thing, here we are dealing with a velocity dependent interaction.

To determine the amplitude $\langle \mathbf{x}\, t | \mathbf{x}'\, t' \rangle$, and, in the process, replace the operator $\mathbf{p}$ by a c-variable counterpart, we may use the resolution of the identity (see §2.4)

$$\mathbf{1} = \int \frac{d^\nu \mathbf{p}'}{(2\pi\hbar)^\nu} |\mathbf{p}'t\rangle \langle \mathbf{p}'t| \tag{10.3.2}$$

for any conveniently chosen time t.

To the above end, we note that we may write

$$\mathbf{A} \cdot \mathbf{p} + \mathbf{p} \cdot \mathbf{A} = (\mathbf{p} \cdot \mathbf{A} - \mathbf{A} \cdot \mathbf{p}) + 2\mathbf{A} \cdot \mathbf{p}_R$$

$$= -i\hbar(\nabla \cdot \mathbf{A}) + 2\mathbf{A} \cdot \mathbf{p}_R \tag{10.3.3}$$

where $\mathbf{p}_R$ is the operator $\mathbf{p}$ operating to its right, and

10.3 Charged Particle in External EM Fields

$$\mathbf{A} \cdot \mathbf{p} + \mathbf{p} \cdot \mathbf{A} = (\mathbf{A} \cdot \mathbf{p} - \mathbf{p} \cdot \mathbf{A}) + 2\mathbf{p}_L \cdot \mathbf{A}$$
$$= i\hbar (\nabla \cdot \mathbf{A}) + 2\mathbf{p}_L \cdot \mathbf{A} \quad (10.3.4)$$

where $\mathbf{p}_L$ operates on its left side.

Accordingly, we may rewrite the Hamiltonian H, with the operator $\mathbf{p}$ in mind, as an operator operating to its right or left as

$$H_R(\mathbf{x}, \mathbf{p}, t) = \frac{(\mathbf{p}_R)^2}{2m} + \frac{ie\hbar}{2mc}(\nabla \cdot \mathbf{A}) - \frac{e}{mc}\mathbf{A} \cdot \mathbf{p}_R + \frac{e^2}{2mc^2}\mathbf{A}^2 + e\Phi \quad (10.3.5)$$

$$H_L(\mathbf{x}, \mathbf{p}, t) = \frac{(\mathbf{p}_L)^2}{2m} - \frac{ie\hbar}{2mc}(\nabla \cdot \mathbf{A}) - \frac{e}{mc}\mathbf{p}_L \cdot \mathbf{A} + \frac{e^2}{2mc^2}\mathbf{A}^2 + e\Phi. \quad (10.3.6)$$

The expressions for the Hamiltonian as spelled out in the above two ways may be now applied in the following manner.

An amplitude $\langle \mathbf{x}_1 t_1 | \mathbf{x}_0 t_0 \rangle$, with $t = (t_0 + t_1)/2$, $\varepsilon = t_1 - t_0$, may be written as in (10.2.6)

$$\langle \mathbf{x}_1 t_1 | \mathbf{x}_0 t_0 \rangle = \left\langle \mathbf{x}_1, t + \frac{\varepsilon}{2} \middle| \mathbf{x}_0, t - \frac{\varepsilon}{2} \right\rangle. \quad (10.3.7)$$

From the dynamical equation

$$i\hbar \frac{\partial}{\partial t} \langle \mathbf{x}_1, t| = \langle \mathbf{x}_1 t | H(\mathbf{x}, \mathbf{p}, t)$$
$$= \langle \mathbf{x}_1 t | H_R(\mathbf{x}, \mathbf{p}, t)$$
$$= \langle \mathbf{x}_1 t | H_R(\mathbf{x}_1, \mathbf{p}, t) \quad (10.3.8)$$

we have for $\varepsilon \to 0$

$$\left\langle \mathbf{x}_1, t + \frac{\varepsilon}{2} \middle| \simeq \langle \mathbf{x}_1 t | \left(1 - \frac{i\varepsilon}{2\hbar} H_R(\mathbf{x}_1, \mathbf{p}, t)\right)\right.$$
$$= \int \frac{d^3 p'}{(2\pi\hbar)^3} e^{i\mathbf{x}_1 \cdot \mathbf{p}'/\hbar} \left(1 - \frac{i\varepsilon}{2\hbar} H_R(\mathbf{x}_1, \mathbf{p}', t)\right) \langle \mathbf{p}' t | \quad (10.3.9)$$

where in the last equality we have used (10.3.2), (2.4.8). We note that $H_R(\mathbf{x}_1, \mathbf{p}', t)$ is a c-function.

Similarly, from

$$-i\hbar \frac{\partial}{\partial t} |\mathbf{x}_0, t\rangle = H(\mathbf{x}, \mathbf{p}, t) |\mathbf{x}_0 t\rangle$$
$$= H_L(\mathbf{x}, \mathbf{p}, t) |\mathbf{x}_0 t\rangle$$
$$= H_L(\mathbf{x}_0, \mathbf{p}, t) |\mathbf{x}_0 t\rangle \quad (10.3.10)$$

we obtain for $\varepsilon \simeq 0$

$$\left|\mathbf{x}_0, t - \frac{\varepsilon}{2}\right\rangle \simeq \left(1 - \frac{i\varepsilon}{2\hbar} H_L(\mathbf{x}_0, \mathbf{p}, t)\right) |\mathbf{x}_0 t\rangle$$

$$= \int \frac{d^3 \mathbf{p}'}{(2\pi\hbar)^3} e^{-i\mathbf{x}_0 \cdot \mathbf{p}'/\hbar} |\mathbf{p}' t\rangle \left(1 - \frac{i\varepsilon}{2\hbar} H_L(\mathbf{x}_0, \mathbf{p}', t)\right). \quad (10.3.11)$$

Equations (10.3.9), (10.3.11), when substituted in (10.3.7) lead to

$$\langle \mathbf{x}_1 t_1 | \mathbf{x}_0 t_0 \rangle$$

$$\simeq \int \frac{d^3 \mathbf{p}'}{(2\pi\hbar)^3} e^{i(\mathbf{x}_1 - \mathbf{x}_0) \cdot \mathbf{p}'/\hbar} \left(1 - \frac{i\varepsilon}{2\hbar} \left(H_R(\mathbf{x}_1, \mathbf{p}', t) + H_L(\mathbf{x}_0, \mathbf{p}', t)\right)\right)$$

$$\simeq \int \frac{d^3 \mathbf{p}'}{(2\pi\hbar)^3} e^{i(\mathbf{x}_1 - \mathbf{x}_0) \cdot \mathbf{p}'/\hbar} \exp -\frac{i\varepsilon}{2\hbar} \left(H_R(\mathbf{x}_1, \mathbf{p}', t) + H_L(\mathbf{x}_0, \mathbf{p}', t)\right)$$

$$(10.3.12)$$

for $\varepsilon \simeq 0$.
In detail

$$H_R(\mathbf{x}_1, \mathbf{p}', t) + H_L(\mathbf{x}_0, \mathbf{p}', t)$$

$$= \frac{\mathbf{p}'^2}{m} + \frac{ie\hbar}{2mc} \left[(\boldsymbol{\nabla} \cdot \mathbf{A})_1 - (\boldsymbol{\nabla} \cdot \mathbf{A})_0\right] - \frac{e}{mc} \mathbf{p}' \cdot (\mathbf{A}_1 + \mathbf{A}_0)$$

$$+ \frac{e^2}{2mc^2} \left((\mathbf{A}_1)^2 + (\mathbf{A}_0)^2\right) + e\Phi_0 + e\Phi_1 \quad (10.3.13)$$

where the notation $f_0 = f(\mathbf{x}_0, t)$, $f_1 = f(\mathbf{x}_1, t)$ has been used.

The Gaussian $\mathbf{p}'$-integral in (10.3.12) is elementary and may be carried out to yield,

$$\langle \mathbf{x}_1 t_1 | \mathbf{x}_0 t_0 \rangle \simeq \left(\frac{m}{2\pi i\hbar\varepsilon}\right)^{3/2} \exp \frac{i\varepsilon}{\hbar} \left[\frac{m}{2} \left(\frac{\mathbf{x}_1 - \mathbf{x}_0}{\varepsilon}\right)^2 \right.$$

$$\left. + \frac{e}{c} \frac{(\mathbf{x}_1 - \mathbf{x}_0)}{\varepsilon} \cdot \frac{(\mathbf{A}_0 + \mathbf{A}_1)}{2} - e\frac{(\Phi_0 + \Phi_1)}{2} + \Delta_{01}\right]$$

$$(10.3.14)$$

for $\varepsilon \simeq 0$, where

$$\Delta_{01} = -\frac{e^2}{2mc^2} \frac{(\mathbf{A}_1 - \mathbf{A}_0)^2}{4} - \frac{ie\hbar}{4mc} \left[(\boldsymbol{\nabla} \cdot \mathbf{A})_1 - (\boldsymbol{\nabla} \cdot \mathbf{A})_0\right]. \quad (10.3.15)$$

10.3 Charged Particle in External EM Fields   611

The latter being the sum of differences, and not being divided by ε, is smaller than the first three terms in the square brackets in (10.3.14). That is for $t_1 - t_0 = \varepsilon \simeq 0$, we may take

$$\langle \mathbf{x}_1 t_1 | \mathbf{x}_0 t_0 \rangle \simeq \left(\frac{m}{2\pi i \hbar \varepsilon}\right)^{3/2} \exp \frac{i\varepsilon}{\hbar} \left[\frac{m}{2} \left(\frac{\mathbf{x}_1 - \mathbf{x}_0}{\varepsilon}\right)^2 \right.$$

$$\left. + \frac{e}{c} \frac{(\mathbf{x}_1 - \mathbf{x}_0)}{\varepsilon} \cdot \frac{(\mathbf{A}_0 + \mathbf{A}_1)}{2} - e \frac{(\Phi_0 + \Phi_1)}{2} \right]. \quad (10.3.16)$$

Now it is not difficult to verify that (10.3.16) is consistent with the Schrödinger equation to be satisfied by a wavefunction $\langle \mathbf{x} t | \Psi \rangle$. To this end, we have from (10.3.16) for $\varepsilon \simeq 0$

$$\langle \mathbf{x}, t + \varepsilon | \Psi \rangle \simeq \left(\frac{m}{2\pi i \hbar \varepsilon}\right)^{3/2} \int d^3\mathbf{x}' \exp \frac{i\varepsilon}{\hbar} \left[\frac{m}{2} \left(\frac{\mathbf{x} - \mathbf{x}'}{\varepsilon}\right)^2 \right.$$

$$\left. + \frac{e}{c} \frac{(\mathbf{x} - \mathbf{x}')}{\varepsilon} \cdot \frac{(\mathbf{A}(\mathbf{x},t) + \mathbf{A}(\mathbf{x}',t))}{2} - e \frac{(\Phi(\mathbf{x},t) + \Phi(\mathbf{x}',t))}{2} \right] \langle \mathbf{x}' t | \Psi \rangle.$$

$$(10.3.17)$$

By making a charge of integration variables $\mathbf{x}' \to \mathbf{x}' - \mathbf{x} = \mathbf{z}$, and carrying out Gaussian integrals as in (10.2.13)–(10.2.15), in the manner given in verifying (10.2.16), it is straightforward to show (see Problem 10.2), that (10.3.17) leads, upon taking the limit $\varepsilon \to 0$, to

$$i\hbar \frac{\partial}{\partial t} \Psi(\mathbf{x}, t) = H(\mathbf{x}, \mathbf{p}, t) \Psi(\mathbf{x}, t). \quad (10.3.18)$$

The expression in (10.3.16) may be then systematically used to obtain

$$\langle \mathbf{x} t | \mathbf{x}' t' \rangle = \lim_{N \to \infty} \int \left(\frac{m}{2\pi i \hbar \varepsilon}\right)^{3N/2} d^3\mathbf{x}_1 \cdots d^3\mathbf{x}_{N-1}$$

$$\times \exp \frac{i}{\hbar} \sum_{k=0}^{N-1} \varepsilon \left(m \left(\frac{\mathbf{x}_{k+1} - \mathbf{x}_k}{\varepsilon}\right)^2 \right.$$

$$+ \frac{e}{c} \frac{(\mathbf{x}_{k+1} - \mathbf{x}_k)}{\varepsilon} \cdot \frac{(\mathbf{A}(\mathbf{x}_k, \hat{t}_k) + \mathbf{A}(\mathbf{x}_{k+1}, \hat{t}_k))}{2}$$

$$\left. - e \frac{(\Phi(\mathbf{x}_k, \hat{t}_k) + \Phi(\mathbf{x}_{k+1}, \hat{t}_k))}{2} \right) \quad (10.3.19)$$

where $\hat{t}_{k+1} = (t_k + t_{k+1})/2 = t' + (k+1/2)\varepsilon$, $t_k = t' + k\varepsilon$, $k = 0, 1, \ldots, N-1$, $\varepsilon = (t - t')/N = t_{k+1} - t_k$, $t_0 = t'$, $t_N = t$, $\mathbf{x}_0 = \mathbf{x}'$, $\mathbf{x}_N = \mathbf{x}$.

In the limit $N \to \infty$, (10.3.19) may be written as

$$\langle \mathbf{x}t \,|\, \mathbf{x}'t\rangle = \int_{\mathbf{x}(t')=\mathbf{x}'}^{\mathbf{x}(t)=\mathbf{x}} \mathscr{D}(\mathbf{x}(\cdot))$$

$$\times \exp\frac{\mathrm{i}}{\hbar}\int_{t'}^{t}\mathrm{d}\tau\left(\frac{m\dot{\mathbf{x}}^2(\tau)}{2} + \frac{e}{c}\dot{\mathbf{x}}(\tau)\cdot\mathbf{A}(\mathbf{x}(\tau),\tau) - e\Phi(\mathbf{x}(\tau),\tau)\right) \quad (10.3.20)$$

where we recognize the τ-integrand as the Lagrangian of a classical charged particle in the external electromagnetic field described by $(\Phi,\mathbf{A})$.

The procedure developed above may be formally applied to Hamiltonians with velocity dependent potentials in general. Using a standard notation for such Hamiltonians $H(\mathbf{q},\mathbf{p})$, written in terms of $\mathbf{q}$'s and $\mathbf{p}$'s, we may move all the $\mathbf{p}$'s to the right of $\mathbf{q}$'s, using appropriate commutation relations, as done above, thus defining $H_\mathrm{R}(\mathbf{q},\mathbf{p})$, and similarly defining $H_\mathrm{L}(\mathbf{q},\mathbf{p})$, we may infer from (10.3.12) that

$$\langle \mathbf{q}_1 t_1 \,|\, \mathbf{q}_0 t_0\rangle \simeq \int \frac{\mathrm{d}^\nu \mathbf{p}_1}{(2\pi\hbar)^\nu}\, \mathrm{e}^{\mathrm{i}(\mathbf{q}_1-\mathbf{q}_0)\cdot\mathbf{p}_1/\hbar} \exp\frac{-\mathrm{i}\varepsilon}{2\hbar}\left(H_\mathrm{R}(\mathbf{q}_1,\mathbf{p}_1) + H_\mathrm{L}(\mathbf{q}_0,\mathbf{p}_1)\right) \quad (10.3.21)$$

where $t_1 - t_0 = \varepsilon \simeq 0$.

The expression (10.3.21) allows us to deduce formally by using the completeness of the Green functions

$$\langle \mathbf{q}t \,|\, \mathbf{q}'t'\rangle = \lim_{N\to\infty}\int \mathrm{d}^\nu \mathbf{q}_1 \cdots \mathrm{d}^\nu \mathbf{q}_{N-1}\frac{\mathrm{d}^\nu \mathbf{p}_1}{(2\pi\hbar)^\nu}\cdots\frac{\mathrm{d}^\nu \mathbf{p}_N}{(2\pi\hbar)^\nu}$$

$$\times \exp\left(\frac{\mathrm{i}}{\hbar}\sum_{k=0}^{N-1}\varepsilon\mathbf{p}_{k+1}\cdot\left(\frac{\mathbf{q}_{k+1}-\mathbf{q}_k}{\varepsilon}\right)\right)$$

$$\times \exp-\frac{\mathrm{i}}{\hbar}\left(\sum_{k=0}^{N-1}\frac{\varepsilon}{2}\left(H_\mathrm{R}(\mathbf{q}_{k+1},\mathbf{p}_{k+1}) + H_\mathrm{L}(\mathbf{q}_k,\mathbf{p}_{k+1})\right)\right) \quad (10.3.22)$$

where $\varepsilon = t_{k+1} - t_k = (t-t')/N$, $t_N = t$, $t_0 = t'$, $\mathbf{q}_N = \mathbf{q}$, $\mathbf{q}_0 = \mathbf{q}'$, as defined below (10.3.19).

Upon taking limit $N \to \infty$, we have

$$\langle \mathbf{q}t \,|\, \mathbf{q}'t'\rangle = \int_{\mathbf{q}(t')=\mathbf{q}'}^{\mathbf{q}(t)=\mathbf{q}} \mathscr{D}(\mathbf{q}(\cdot),\mathbf{p}(\cdot))$$

$$\times \exp\frac{\mathrm{i}}{\hbar}\int_{t'}^{t}\mathrm{d}\tau\left(\mathbf{p}(\tau)\cdot\dot{\mathbf{q}}(\tau) - H_c(\mathbf{q}(\tau),\mathbf{p}(\tau))\right) \quad (10.3.23)$$

10.3 Charged Particle in External EM Fields

where we have defined a classical Hamiltonian H_c by[4]

$$H_c(\mathbf{q}(\tau), \mathbf{p}(\tau)) = \frac{1}{2}\Big(H_R(\mathbf{q}(\tau), \mathbf{p}(\tau)) + H_L(\mathbf{q}(\tau), \mathbf{p}(\tau))\Big). \qquad (10.3.24)$$

The paths-measure is formally given by

$$\mathscr{D}(\mathbf{q}(\cdot), \mathbf{p}(\cdot)) = \lim_{N\to\infty}\left[\frac{d^\nu \mathbf{p}_N}{(2\pi\hbar)^\nu}\prod_{j=1}^{N-1}\left(d^\nu \mathbf{q}_j \frac{d^\nu \mathbf{p}_j}{(2\pi\hbar)^\nu}\right)\right] \qquad (10.3.25)$$

with the pre-limiting expression involving one additional ν-dimensional momentum integration, with the measure $d^\nu \mathbf{p}_N/(2\pi\hbar)^\nu$, than the $\mathbf{q}$-ones. In the limit $N \to \infty$, the number of integration variables goes to infinity and in this limit, one rather formally integrates over $\mathbf{q}(\tau)$ and $\mathbf{p}(\tau)$ in phase space for all $t' \leqslant \tau \leqslant t$ with the boundary conditions $\mathbf{q}(t') = \mathbf{q}'$, $\mathbf{q}(t) = \mathbf{q}$. The expressions in (10.3.23)–(10.3.24), for velocity dependent potentials, should, however, be taken only rather formally for general cases and are not void of ambiguities and we will not go into consistency problems of this representation here with generality. For the particular case, however, where

$$H(\mathbf{q}, \mathbf{p}) = \frac{\mathbf{p}^2}{2m} + V(\mathbf{q}). \qquad (10.3.26)$$

Equation (10.3.22) becomes

$$\langle \mathbf{q}t | \mathbf{q}'t' \rangle = \lim_{N\to\infty}\int d^\nu \mathbf{q}_1 \cdots d^\nu \mathbf{q}_{N-1} \frac{d^\nu \mathbf{p}_1}{(2\pi\hbar)^\nu} \cdots \frac{d^\nu \mathbf{p}_N}{(2\pi\hbar)^\nu}$$

$$\times \exp\left(-\frac{i}{\hbar}\sum_{k=0}^{N-1}\frac{\varepsilon}{2}\big(V(\mathbf{q}_{k+1}) + V(\mathbf{q}_k)\big)\right)$$

$$\times \exp -\frac{i}{2m\hbar}\sum_{k=0}^{N-1}\varepsilon\left((\mathbf{p}_{k+1})^2 - 2m\left(\frac{\mathbf{q}_{k+1} - \mathbf{q}_k}{\varepsilon}\right)\cdot \mathbf{p}_{k+1}\right). \qquad (10.3.27)$$

We may then explicitly integrate over the $\mathbf{p}_i$'s to obtain

$$\langle \mathbf{q}t | \mathbf{q}'t' \rangle = \lim_{N\to\infty}\left(\frac{m}{2\pi i \hbar \varepsilon}\right)^{N\nu/2}\int d^\nu \mathbf{q}_1 \cdots d^\nu \mathbf{q}_{N-1}$$

---

[4] Note that if $H_R(\mathbf{q}, \mathbf{p}) = \sum_{n,m} a_{nm} f_n(q_1, \ldots, q_\nu) h_m(p_1, \ldots, p_\nu)$, which may include a constant term, where f_n, h_m are real, then $H_L(\mathbf{q}, \mathbf{p}) = \sum_{n,m} a^*_{nm} h_m(p_1, \ldots, p_\nu) f_n(q_1, \ldots, q_\nu)$, and H_c in (10.3.24), expressed in terms of c-numbers, is real. See, for example, (10.3.5), (10.3.6), and Problem 10.5.

$$\times \exp \frac{i}{\hbar} \sum_{k=0}^{N-1} \varepsilon \left(\frac{m}{2} \left(\frac{\mathbf{q}_{k+1} - \mathbf{q}_k}{\varepsilon} \right)^2 - \frac{V(\mathbf{q}_{k+1}) + V(\mathbf{q}_k)}{2} \right)$$
(10.3.28)

which in the limit $N \to \infty$ coincides with (10.2.18), where in the latter we have, in general, an explicit time-dependence coming from a given c-function.

10.4 Constrained Dynamics

The purpose at this sections is to develop the formal theory of path integration when constraints are present in the theory. Let $\mathbf{q} = (q_1, \ldots, q_\nu)$, $\mathbf{p} = (p_1, \ldots, p_\nu)$ denote canonical conjugate variables, in general. With the path integral expression in (10.3.23) in phase space in mind, suppose because of underlying constraints in a theory such variables are not all independent and one succeeds in isolating a maximal number of independent canonical conjugate variables $(q_1^*, \ldots, q_M^*), (p_1^*, \ldots p_M^*)$, where $M < \nu$. Then according to (10.3.23) one may formally infer that

$$\langle \mathbf{q} t | \mathbf{q}' t' \rangle = \int_{\mathbf{q}(t')=\mathbf{q}'}^{\mathbf{q}(t)=\mathbf{q}} \mathscr{D}(\mathbf{q}^*(\cdot), \mathbf{p}^*(\cdot))$$

$$\times \exp \frac{i}{\hbar} \int_{t'}^{t} d\tau \left(\mathbf{p}^*(\tau) \cdot \dot{\mathbf{q}}^*(\tau) - H^*(\mathbf{q}^*(\tau), \mathbf{p}^*(\tau)) \right) \quad (10.4.1)$$

for the corresponding amplitude, with Hamiltonian H^*, expressed in terms of the independent variables, where $2(\nu - M)$ of the variables $\mathbf{q}, \mathbf{p}$ are functions of $\mathbf{q}^*, \mathbf{p}^*$, and one in turn is working in a subspace of dimension $2M < 2\nu$ of the phase space.

It is not, however, always easy to solve for such independent components and an alternative general expression is needed which deals with all the components $(\mathbf{q}, \mathbf{p})$ with the constraints imposed directly on the path integrals. This is the aim of the present section. We note that all the variables in the path integral (10.3.23) are c-variables and hence are more easily dealt with. This, as mentioned before, is the most attractive feature of path integration which solves the quantum mechanical problem in terms of c-variables and corresponding trajectories. With this in mind, we need some notions of constrained classical dynamics which are dealt with next. Due to the technical nature of this section, it may be omitted at a first reading.

10.4.1 Classical Notions

We recall Hamilton's equations

$$\dot{q}_k = \frac{\partial H}{\partial p_k}, \qquad \dot{p}_k = -\frac{\partial H}{\partial q_k} \quad (10.4.2)$$

10.4 Constrained Dynamics

where the dots denote time derivatives. The Hamiltonian H generates the time development in the sense that for $t \to t - \delta t = \bar{t}$ for a change $\delta t = t - \bar{t} = \delta \tau$ of the time variable by a parameter τ (see (2.1.15), (2.1.16)),

$$q_k(t) \to q_k(t) - \delta q_k(t) = \bar{q}_k(t) = \bar{q}_k(\bar{t} + \delta \tau)$$

$$= q_k(t) + \delta \tau \dot{q}_k(t) \tag{10.4.3}$$

where we have used the fact that $\bar{q}_k(\bar{t}) = q_k(t)$ for pure time translations,

$$p_k(t) \to p_k(t) - \delta p_k(t) = \bar{p}_k(t)$$

$$= p_k(t) + \delta \tau \dot{p}_k(t) \tag{10.4.4}$$

and

$$\delta q_k = -\frac{\partial H}{\partial p_k} \delta \tau, \qquad \delta p_k = \frac{\partial H}{\partial q_k} \delta \tau. \tag{10.4.5}$$

An important property of such a transformation is that the Jacobian is one:

$$\frac{\partial(q_k, p_k)}{\partial(\bar{q}_m, \bar{p}_m)} = 1 \tag{10.4.6}$$

and hence the measure of integration in phase space remains invariant:

$$\prod_k dq_k dp_k \longrightarrow \prod_k d\bar{q}_k d\bar{p}_k \tag{10.4.7}$$

and the transformation (10.4.3)–(10.4.5) is said to be a canonical one.

Other canonical transformations may be defined similarly by

$$\delta q_k = \frac{\partial W}{\partial p_k} \delta \xi \tag{10.4.8}$$

$$\delta p_k = -\frac{\partial W}{\partial q_k} \delta \xi \tag{10.4.9}$$

where W is referred as the corresponding generator with respect to a parameter ξ. Infinitesimal changes δt of the time variable have been defined in (2.1.14), (2.1.16) by $\delta t = \delta \tau$ and we set the parameter $\delta \xi = \delta \tau$, $W = -H$. On the other hand for pure space translations $\delta q_j = \delta a_j$, according to the definitions in (2.1.13), (2.1.15), and we must set $\delta \xi \to \delta \xi_j = \delta a_j$, $W \to p_j$. These are in conformity with the definition of the generators in (2.3.14) and consistent with (10.4.8). For a function $F(q_k, p_k)$ of the q_k's and the p_k's, (10.4.8), (10.4.9) dictates that for $F \to F - \delta F = \bar{F}$,

$$\delta F = \sum_k \left(\frac{\partial F}{\partial q_k} \delta q_k + \frac{\partial F}{\partial p_k} \delta p_k \right)$$

$$= \sum_k \left(\frac{\partial F}{\partial q_k} \frac{\partial W}{\partial p_k} - \frac{\partial F}{\partial p_k} \frac{\partial W}{\partial q_k} \right) \delta\xi. \tag{10.4.10}$$

Thus introducing the Poisson bracket defined by

$$\{F, W\}_{\text{P.B.}} = \sum_k \left(\frac{\partial F}{\partial q_k} \frac{\partial W}{\partial p_k} - \frac{\partial F}{\partial p_k} \frac{\partial W}{\partial q_k} \right), \tag{10.4.11}$$

we may rewrite (10.4.10) in the compact form[5]

$$\delta F = \{F, W\}_{\text{P.B.}} \delta\xi \tag{10.4.12}$$

Again the Jacobian of the transformation (10.4.8), (10.4.9) is one. As a consequence of this, a Poisson bracket is *invariant* under canonical transformations, i.e., in detail

$$\{A, B\}_{\text{P.B.}} = \sum_k \left(\frac{\partial A}{\partial q_k} \frac{\partial B}{\partial p_k} - \frac{\partial A}{\partial p_k} \frac{\partial B}{\partial q_k} \right) = \sum_k \left(\frac{\partial A}{\partial \bar{q}_k} \frac{\partial B}{\partial \bar{p}_k} - \frac{\partial A}{\partial \bar{p}_k} \frac{\partial B}{\partial \bar{q}_k} \right). \tag{10.4.13}$$

We recall some of the important properties of the Poisson bracket:

$$\{F_1, F_2\}_{\text{P.B.}} = -\{F_2, F_1\}_{\text{P.B.}} \tag{10.4.14}$$

$$\{\alpha_1 F_1 + \alpha_2 F_2, F_3\}_{\text{P.B.}} = \alpha_1 \{F_1, F_3\}_{\text{P.B.}} + \alpha_2 \{F_2, F_3\}_{\text{P.B.}} \tag{10.4.15}$$

for any constants α_1, α_2,

$$\{F_1 F_2, F_3\}_{\text{P.B.}} = F_1 \{F_2, F_3\}_{\text{P.B.}} + \{F_1, F_3\}_{\text{P.B.}} F_2 \tag{10.4.16}$$

and satisfy the Jacobi identity

$$\{F_1, \{F_2, F_3\}_{\text{P.B.}}\}_{\text{P.B.}} + \{F_2, \{F_3, F_1\}_{\text{P.B.}}\}_{\text{P.B.}} + \{F_3, \{F_1, F_2\}_{\text{P.B.}}\}_{\text{P.B.}} = 0. \tag{10.4.17}$$

We are, in general, interested in canonical transformations involving more than one generator defined through

$$\delta F = \sum_\alpha \{F, W_\alpha\}_{\text{P.B.}} \delta\xi_\alpha \tag{10.4.18}$$

for some parameters ξ_a, where the generators W_α satisfy the algebraic relations

$$\{W_\alpha, W_\beta\}_{\text{P.B.}} = \sum_\gamma c_{\alpha\beta}^\gamma W_\gamma \tag{10.4.19}$$

---

[5] In particular for $W\delta\xi = -H\delta\tau$, $F = F(\mathbf{q}(t), \mathbf{p}(t))$, and by definition (see (10.4.3), (10.4.4)), $\delta F = F - \bar{F} = F(\mathbf{q}(t), \mathbf{p}(t)) - F(\mathbf{q}(t) + \delta\tau\dot{\mathbf{q}}(t), \mathbf{p}(t) + \delta\tau\dot{\mathbf{p}}(t)) = -\dot{F}\delta\tau = \{F, -H\}\delta\tau$ or $\dot{F} = \{F, H\}$, where $\dot{F} = dF/dt$, as expected.

10.4 Constrained Dynamics

with the latter stating that the Poisson bracket of any two generators may be written as a (linear) combination of the generators in question, where the coefficients $c_{\alpha\beta}^{\gamma}$ are referred to as structure constants.

Transformations $\mathbf{q}, \mathbf{p} \to \bar{\mathbf{q}}, \bar{\mathbf{p}}$, $H(\mathbf{q}, \mathbf{p}) \to \bar{H}(\bar{\mathbf{q}}, \bar{\mathbf{p}})$ are canonical, leading to Hamilton's equations (10.4.2), if the difference between the Lagrangians (times dt)

$$(\mathbf{p} \cdot \dot{\mathbf{q}} - H(\mathbf{q}, \mathbf{p})) \, dt \to (\bar{\mathbf{p}} \cdot \dot{\bar{\mathbf{q}}} - \bar{H}(\bar{\mathbf{q}}, \bar{\mathbf{p}})) \, dt \qquad (10.4.20)$$

of the corresponding actions $A \to \bar{A}$ is a total differential. That is,

$$(\bar{\mathbf{p}} \cdot \dot{\bar{\mathbf{q}}} - \bar{H}(\bar{\mathbf{q}}, \bar{\mathbf{p}})) \, dt - (\mathbf{p} \cdot \dot{\mathbf{q}} - H(\mathbf{q}, \mathbf{p})) \, dt = dU \qquad (10.4.21)$$

We leave it as an exercise to the reader to establish this property for infinitesimal canonical transformations given in (10.4.12).

Now consider a Hamiltonian $H(\mathbf{q}, \mathbf{p})$ describing a system with ν_1 independent constraints, $\nu_1 < \nu$:

$$\phi_1(\mathbf{q}, \mathbf{p}) = 0, \ldots, \phi_{\nu_1}(\mathbf{q}, \mathbf{p}) = 0. \qquad (10.4.22)$$

These are referred to as *primary* constraints.

For a function $F(\mathbf{q}, \mathbf{p})$ of $\mathbf{q}, \mathbf{p}$, we shall use the notation $F|$, with a bar, when the constraints in (10.4.22) are imposed, in its definition.

The constraints in (10.4.22) may be taken into account by introducing Lagrange multipliers λ_α and by introducing the Hamiltonian

$$H_T = H + \lambda_\alpha \phi_\alpha \qquad (10.4.23)$$

with a summation over α from 1 to ν_1 understood. The resulting Hamiltonian H_T then is parameterized by the Lagrange multipliers λ_α, and the pertinent physical question and the corresponding uniqueness problem arise as to what the explicit expressions of these multipliers are.

The constraints in (10.4.22) must be satisfied at all times. That is, it is necessary to have

$$\dot{\phi}_\beta \Big| = \{\phi_\beta, H_T\}_{\text{P.B.}} \Big| = 0. \qquad (10.4.24)$$

for all β. In detail

$$\dot{\phi}_\beta = \{\phi_\beta, H\}_{\text{P.B.}} + \lambda_\alpha \{\phi_\beta, \phi_\alpha\}_{\text{P.B.}} + \{\phi_\beta, \lambda_\alpha\}_{\text{P.B.}} \phi_\alpha \qquad (10.4.25)$$

(see (10.4.16)). In particular, we note that the last term in (10.4.25) will, in general, *vanish* when the constraints in (10.4.22) are imposed.

We consider the following two cases regarding the equations in (10.4.24), if they do not provide trivial identities:

(1) Those equations that may be independent of the multipliers λ_α, and hence they cannot be used to solve for multipliers, but impose new constraints on the $\mathbf{q}$'s and the $\mathbf{p}$'s.

(2) Those that may depend at least on some of the multipliers.

In case (1), new constraints have been introduced and they are referred to as secondary constraints, defined by

$$\psi_\beta| = \dot{\phi}_\beta\Big| = 0. \tag{10.4.26}$$

We may repeat the above process, hopefully a finite number of times, by adding the newly discovered (secondary) constraints to the original set in (10.4.22) and to the sequel ones thus generated. This in conjunction with type (2) equations, which would allow us to solve for as many of Lagrange multipliers as possible, and will lead to all primary and secondary constraints and no more equations of type (1) arise. We will denote all of these constraints (primary and secondary) by $\Omega_1, \ldots, \Omega_K$.

The Hamiltonian with all these constraints may be now written as

$$H_T = H + \rho_\alpha \Omega_\alpha \tag{10.4.27}$$

with a summation over α from 1 to K. Here not all the Lagrange multipliers $\rho_1,\ldots,\rho_K$ are necessarily determined. We will see which ones are determined and which are not by considering two classes of constraints to be defined shortly below.

The constraints

$$\Omega_1(\mathbf{q},\mathbf{p}) = 0, \ldots, \Omega_K(\mathbf{q},\mathbf{p}) = 0 \tag{10.4.28}$$

hold in a K dimensional subspace of phase space which we will denote by Γ. We will still use the notation $F|$ when restricted to the subspace Γ. We assume the irreducibility of these constraints with respect to the subspace Γ, i.e., if

$$F| = 0 \tag{10.4.29}$$

then

$$F = \sum_{\alpha=1}^{K} c_\alpha \Omega_\alpha \tag{10.4.30}$$

where the coefficients c_α may depend on $\mathbf{q}, \mathbf{p}$.

A constraint is called *first* class, denoted by Φ_α, if

$$\{\Phi_\alpha, \Omega_\beta\}_{\text{P.B.}}\Big| = 0 \tag{10.4.31}$$

for *all* Ω_β. Otherwise a constraint is called of *second* class. We assume that linear combinations, with appropriate coefficients, depending, in general, on $\mathbf{q}, \mathbf{p}$, have been already carried out to bring as many of the constraints as possible to first class.

We may then write $(\Omega_\alpha) = (\Phi_\gamma, \Psi_\beta)$ where (Φ_α) constitutes of first class constraints and (Ψ_α) constitutes those of second class such that no coefficients c_α, not all zero, may be found to make $c_\alpha \Psi_\alpha$ of first class. Clearly

$$\det\{\Psi_\alpha, \Psi_\beta\}_{\text{P.B.}}| \neq 0 \tag{10.4.32}$$

since otherwise, i.e., $\det\{\Psi_\alpha, \Psi_\beta\}_{\text{P.B.}}| = 0$, would imply that we may find a vector $[c_1 \cdots c_K]^T$, with not all of the c_α equal to zero, satisfying the matrix equation

$$[\{\Psi_\alpha, \Psi_\beta\}_{\text{P.B.}}][c_\beta]| = 0. \tag{10.4.33}$$

From (10.4.16) we would then have

$$\{\Psi_\alpha, c_\beta \Psi_\beta\}_{\text{P.B.}}| = 0 \tag{10.4.34}$$

for all α since $\Psi_\beta| = 0$, which implies that $c_\beta \Psi_\beta$ is of first class thus reaching a contradiction.

With the first and second class constraints thus defined, we may rewrite (10.4.27) as

$$H_T = H + \eta_\alpha \Phi_\alpha + \rho_\beta \Psi_\beta \tag{10.4.35}$$

with the sum over α from 1 to, say, $\nu - M$, and the sum over β from 1 to, say, L. The positive integer L is necessarily *even*.[6]

All the Lagrange multipliers ρ_β, multiplying the second class constraint functions Ψ_β are determined in Γ. This follows at once by noting that

$$0 = \{H_T, \Psi_\gamma\}_{\text{P.B.}}| = \{H, \Psi_\gamma\}_{\text{P.B.}}| + \rho_\beta \{\Psi_\beta, \Psi_\gamma\}_{\text{P.B.}}| \tag{10.4.36}$$

and the fact that $\det\{\Psi_\beta, \Psi_\gamma\}_{\text{P.B.}}| \neq 0$, where we have also used (10.4.29) for the first class constraint functions Φ_α.

On the other hand,

$$0 = \{H_T, \Phi_\alpha\}_{\text{P.B.}}| = \{H, \Phi_\alpha\}_{\text{P.B.}}| + 0 \tag{10.4.37}$$

and the Lagrange multipliers η_α multiplying the *first* class constraint functions remain undetermined. For one thing we note that for a function G of $\mathbf{q}(t), \mathbf{p}(t), \dot{G}(t) = \{G, H_T\}_{\text{P.B.}}$ implies that $G(t)| = G(0)| + \dot{G}(0)|t$, in the neighborhood of $t = 0$, for a given initial condition $G(0)|$, and that $G(t)|$ is not unique since $\dot{G}(0)|$, in general, would depend on the multipliers $\eta_\alpha(0)$ and thus would be different for different values of the $\eta_\alpha(0)$ (see Problem 10.9).

Before discussing the technical details concerning the undetermined multipliers associated with the first class constraints, we provide a couple of examples.

Consider the quadratic Hamiltonian in 3D space (or in 6D phase space)

$$H = \frac{\mathbf{p}^2}{2m} + \frac{1}{2}m\omega^2 \mathbf{q}^2 \tag{10.4.38}$$

---

[6] This follows from the fact that the determinant of an anti-symmetric matrix (see (10.4.32)) of odd order is zero.

with the simple constraint function $\phi_1 = q_1 - q_2$,

$$\phi_1| = (q_1 - q_2)| = 0 \qquad (10.4.39)$$

from which one obtains

$$\{\phi_1, H\}_{\text{P.B.}} = \frac{1}{m}(p_1 - p_2). \qquad (10.4.40)$$

leading to a secondary constraint

$$\psi_1| = (p_1 - p_2)| = 0 \qquad (10.4.41)$$

and

$$\{\psi_1, H\}_{\text{P.B.}} = -m\omega^2(q_1 - q_2) \equiv -m\omega^2 \phi_1. \qquad (10.4.42)$$

Since

$$2 = \{\phi_1, \psi_1\}_{\text{P.B.}} \neq 0 \qquad (10.4.43)$$

we have two second class constraints leading to a $4D$ physical subspace Γ of the $6D$ phase space.

The following system deals with a situation involving a first class constraint only. Consider the Hamiltonian

$$H = \frac{\mathbf{p}^2}{2m} + V(\mathbf{q}^2) \qquad (10.4.44)$$

in $3D$ with constraint

$$\phi| = (q_1 p_2 - p_1 q_2)| = 0 \qquad (10.4.45)$$

implying the vanishing of the third component L_3 of the angular momentum. Due to the conservation of L_3:

$$\{H, (q_1 p_2 - p_1 q_2)\}_{\text{P.B.}} = 0 \qquad (10.4.46)$$

there is no secondary constraint.

Upon writing

$$H_T = \frac{\mathbf{p}^2}{2m} + V(\mathbf{q}^2) + \lambda(q_1 p_2 - p_1 q_2). \qquad (10.4.47)$$

On the subspace Γ of phase space of dimension $6 - 1 = 5$, specified by the constraint in (10.4.45), λ is undetermined. This ambiguity is resolved by introducing an additional constraint function χ which amounts in working in a 4 dimensional subspace $\Gamma^* \subset \Gamma$ in phase space and the Lagrange multiplier is then uniquely determined. We continue to use the notation $F|$ for a function F of $\mathbf{q}, \mathbf{p}$ when restricted to Γ^*.

We may choose

$$\chi = p_2 \cos\alpha - p_1 \sin\alpha \qquad (10.4.48)$$

where α is an arbitrary angle such that

$$\{\phi,\chi\}_{\text{P.B.}}| = -(p_2 \sin\alpha + p_1 \cos\alpha) \neq 0. \tag{10.4.49}$$

The introduction of a new constraint χ, such that $\{\phi,\chi\} \neq 0$ automatically turns the example above to one with second class constraints only and the two corresponding Lagrange multipliers are then determined.

The arbitrariness in the choice of the additional constraint χ function is often referred to as a gauge freedom — the interpretation of which will be spelled out below.

The validity of $\chi| = 0$ for all t, leads from (10.4.47), (10.4.49) to

$$-\{H,\chi\}_{\text{P.B.}}| = \lambda\{\phi,\chi\}_{\text{P.B.}}| \tag{10.4.50}$$

which allows the determination of $\lambda|$.

The interpretation of the above construction becomes clear by introducing the canonical variables:

$$Q = q_2 \cos\alpha - q_1 \sin\alpha, \quad P = \chi = p_2 \cos\alpha - p_1 \sin\alpha$$

$$q_1^* = q_1 \cos\alpha + q_2 \sin\alpha, \quad p_1^* = p_1 \cos\alpha + p_2 \sin\alpha \tag{10.4.51}$$

$$q_3^* = q_3, \quad p_3^* = p_3.$$

We leave it as an exercise to the reader (see Problem 10.10) to show that the transformation $q_1, q_2, q_3, p_1, p_2, p_3 \to q_1^*, Q, q_3^*, p_1^*, P, p_3^*$ is actually canonical consistent with Hamilton's equations. It is easily checked that the pair of constraints $\phi| = 0$, $\chi| = 0$ may be replaced by the pair $Q| = 0$, $P| = 0$. Also note that

$$\{\phi,\chi\}_{\text{P.B.}} = \frac{\partial\phi}{\partial Q} \tag{10.4.52}$$

By choosing the additional constraint $\chi| = 0$ with χ defined in (10.4.48), with a given fixed α, the dynamics, as will be seen below, may be then described in the q_1^*-q_3^* plane making an angle α with the original q_1-q_3 plane as shown in Figure 10.2 by the appropriate choice of the canonical variables in (10.4.51). The choice of the additional constraint $\chi| = 0$, fixes once and for all the choice of the plane, specified by some angle α, to describe the dynamics of the particle in question. Such a choice of an additional constraint may be referred to as fixing a gauge and the parameter α, in this context, may be called as a gauge fixing parameter.

With the canonical variables defined in (10.4.51), we may write

$$H_T \equiv \frac{(P^2 + \mathbf{p}^{*2})}{2m} + V(Q^2 + \mathbf{q}^{*2}) + \rho_1 Q + \rho_2 P \tag{10.4.53}$$

where $\mathbf{q}^* = (q_1^*, q_3^*)$, $\mathbf{p}^* = (p_1^*, p_3^*)$, with

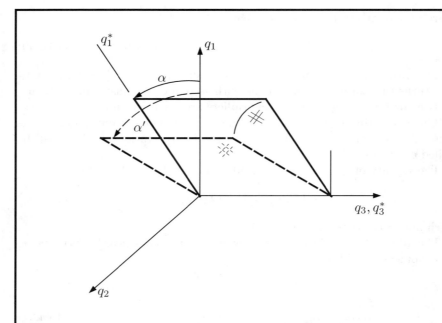

Fig. 10.2. The introduction of the additional (subsidiary) constraint χ fixes once and for all the plane in which the dynamics is described by choosing appropriately new canonical variables as defined in (10.4.51). The angle parameter α, defining the angle between the q_1^*-q_3^* and q_1-q_3 planes may be referred to as a gauge fixing parameter.

$$\left[\frac{(P^2 + \mathbf{p}^{*2})}{2m} + V(Q^2 + \mathbf{q}^{*2})\right]\bigg| = \frac{\mathbf{p}^{*2}}{2m} + V(\mathbf{q}^{*2}). \qquad (10.4.54)$$

Needless to say that $\rho_1|$, $\rho_2|$ may be determined as Q, P are second class constraints since $\{Q, P\}_{\text{P.B.}} = 1 \neq 0$.

The introduction of such additional constraints $\chi_\beta = 0$, which may be also referred to as subsidiary constraints, turn first class constraints into second class ones, which are chosen such that

$$\det\{\Phi_\alpha, \chi_\beta\}_{\text{P.B.}}\big| \neq 0 \qquad (10.4.55)$$

generalizing (10.4.49) (see also (10.4.36)) for more than one first class constraint function: $\Phi_1, \ldots, \Phi_{\nu-M}$. Clearly, then $\nu - M$ subsidiary constraint functions are needed $\chi_1, \ldots, \chi_{\nu-M}$. We will choose them such that

$$\{\chi_\alpha, \chi_\beta\}_{\text{P.B.}}\big| = 0 \qquad (10.4.56)$$

for all $\alpha, \beta = 1, \ldots, \nu - M$ in addition to the restriction in (10.4.55).

In the sequel, dealing with path integrals with constraints, we will consider only theories with first class constraints. Problems with second class constraints are, however, given in the problems section and additional comments concerning them will be made later. A generalization of the Poisson bracket, due to Dirac, will be, however, introduced to deal with second class constraints at the end of this section.

We will thus be considering $\nu - M$ first class constraints only

$$\Phi_1(\mathbf{q},\mathbf{p})| = 0, \ldots, \Phi_{\nu-M}(\mathbf{q},\mathbf{p})| = 0 \tag{10.4.57}$$

with $\nu - M$ additional subsidiary ones

$$\chi_1(\mathbf{q},\mathbf{p})| = 0, \ldots, \chi_{\nu-M}(\mathbf{q},\mathbf{p})| = 0 \tag{10.4.58}$$

with the latter satisfying (10.4.55), (10.4.56). The constraints (10.4.57), (10.4.58) lead to a $2M$ dimensional subspace Γ^* for the physical subspace of phase space, with $M < \nu$.

The constraints functions Φ_α are assumed to be irreducible in the sense that

$$\{\Phi_\alpha, \Phi_\beta\}_{\text{P.B.}} = \sum_{\gamma=1}^{\nu-M} c_{\alpha\beta}^\gamma \Phi_\gamma \tag{10.4.59}$$

and that for any function $F(\mathbf{q},\mathbf{p})$ such that $F = 0|$,

$$F(\mathbf{q},\mathbf{p}) = \sum_{\gamma=1}^{\nu-M} f^\gamma \Phi_\gamma \tag{10.4.60}$$

where the coefficients $c_{\alpha\beta}^\gamma$, f^γ may depend on $\mathbf{q}, \mathbf{p}$.

10.4.2 Constrained Path Integrals

We may choose $(\chi_\alpha, \mathbf{p}^*)$, $(Q_\alpha, \mathbf{q}^*)$, $\alpha = 1, \ldots, \nu - M$, as canonical variables, with Q_α as the canonical conjugate variable to χ_α. We may then rewrite the right-hand side of (10.4.1) as

$$\int \left(\prod_\tau \mathrm{d}\mu(\tau)\right) \exp \frac{\mathrm{i}}{\hbar} \int_{t'}^{t} \mathrm{d}\tau \left(\chi_\alpha \dot{Q}_\alpha + \mathbf{p}^* \cdot \dot{\mathbf{q}}^* - \bar{H}\right) \tag{10.4.61}$$

where

$$\mathrm{d}\mu = \left(\prod_\alpha \delta\left(Q_\alpha - Q_\alpha(\mathbf{q}^*,\mathbf{p}^*)\right) \delta(\chi_\alpha) \,\mathrm{d}Q_\alpha \mathrm{d}\chi_\alpha\right) \mathrm{d}^M \mathbf{q}^* \frac{\mathrm{d}^M \mathbf{p}^*}{(2\pi\hbar)^M} \tag{10.4.62}$$

and the Dirac deltas in (10.4.62) lead to the restriction of $\bar{H}$ in (10.4.61) to

$$\bar{H}(Q_\alpha, \chi_\alpha, \mathbf{q}^*, \mathbf{p}^*)\big|_{\chi_\alpha=0, Q_\alpha=Q_\alpha(\mathbf{q}^*,\mathbf{p}^*)} = H^*(\mathbf{q}^*, \mathbf{p}^*). \tag{10.4.63}$$

We may also make the substitution

$$\prod_\alpha \delta\left(Q_\alpha - Q_\alpha\left(\mathbf{q}^*, \mathbf{p}^*\right)\right) dQ_\alpha \rightarrow \left(\prod_\alpha \delta\left(\Phi_\alpha\right) d\Phi_\alpha\right) \left|\det\left(\frac{\partial}{\partial Q_\beta}\Phi_\gamma\right)\right| \tag{10.4.64}$$

by making use of a property of the Dirac delta distribution in one or more dimensions. Remembering our choice of (Q_α, χ_α) as pairs in the set of canonical conjugate variables, we also have

$$\frac{\partial}{\partial Q_\beta}\Phi_\gamma = \{\Phi_\gamma, \chi_\beta\}_{\text{P.B.}} \tag{10.4.65}$$

(see also (10.4.52)).

That is, we may rewrite (10.4.62) simply as

$$d\mu = \left(\prod_\alpha \delta\left(\Phi_\alpha\right) \delta\left(\chi_\alpha\right) d\Phi_\alpha d\chi_\alpha\right) \left|\det\{\Phi_\gamma, \chi_\beta\}_{\text{P.B.}}\right| d^M \mathbf{q}^* \frac{d^M \mathbf{p}^*}{(2\pi\hbar)^M}. \tag{10.4.66}$$

Now we invoke the formal property of the invariance of

$$\prod_\alpha dQ_\alpha d\chi_\alpha d^M \mathbf{q}^* d^M \mathbf{p}^* \tag{10.4.67}$$

under canonical transformations to infer that with the choice of canonical variables $(\mathbf{q}, \mathbf{p})$, we may rewrite (10.4.61) as

$$\int d\mu(\tau) \exp \frac{i}{\hbar} \int_{t'}^{t} d\tau \left(\mathbf{p} \cdot \dot{\mathbf{q}} - H + \frac{d}{d\tau}U\right) \tag{10.4.68}$$

where

$$\exp \frac{i}{\hbar} \int_{t'}^{t} d\tau \frac{d}{d\tau}U = \exp \frac{i}{\hbar}\left(U(t) - U(t')\right), \tag{10.4.69}$$

in general, is a phase factor depending, however, only on the end points evaluated at t and t' (see (10.4.21)), and the expression for U is dictated by the generators of the canonical transformation itself. Also $d\mu$ is, now given by

$$d\mu = \left(\prod_\alpha \delta\left(\Phi_\alpha\right) \delta\left(\chi_\alpha\right)\right) \left|\det\{\Phi_\gamma, \chi_\beta\}_{\text{P.B.}}\right| d^\nu \mathbf{q} \frac{d^\nu \mathbf{p}}{(2\pi\hbar)^M}. \tag{10.4.70}$$

For our application described by the system in (10.4.44), (10.4.45), $\chi| = 0$ in (10.4.48), the generator W of the canonical transformation

$$(Q, \chi, \mathbf{q}^*, \mathbf{p}^*) \rightarrow (\mathbf{q}, \mathbf{p}) \tag{10.4.71}$$

for infinitesimal transformation is given by the generator

$$W = -L_3 \delta\alpha = -(q_1 p_2 - p_1 q_2) \delta\alpha \tag{10.4.72}$$

(see (10.4.55)) in the sense that for $\alpha \simeq \delta\alpha \to 0$,

$$\delta Q = \{Q, -L_3\}_{\text{P.B.}} = -q_1 \delta\alpha \qquad (10.4.73)$$

etc., and the conservation law

$$\frac{d}{dt} L_3 = \{L_3, H\}_{\text{P.B.}} = 0 \qquad (10.4.74)$$

implies that $\dot{U} = 0$ in this case. Also

$$P\dot{Q} + \mathbf{p}^* \cdot \dot{\mathbf{q}}^* = \mathbf{p} \cdot \dot{\mathbf{q}} \qquad (10.4.75)$$

$$Q^2 + \mathbf{q}^{*2} = \mathbf{q}^2 \qquad (10.4.76)$$

$$P^2 + \mathbf{p}^{*2} = \mathbf{p}^2 \qquad (10.4.77)$$

as a consequence of rotational invariance, and note the constraints $\phi| = L_3| = 0$, $\chi| = 0$, as imposed by $\delta(\phi)\delta(\chi)$ in the measure in (10.4.70). Using the notation in (10.4.44), we have

$$H(\mathbf{q}, \mathbf{p})| = H^*(\mathbf{q}^*, \mathbf{p}^*) = \frac{\mathbf{p}^{*2}}{2m} + V(\mathbf{q}^{*2}). \qquad (10.4.78)$$

In some instances, as in the above example involving invariance under the canonical transformation, no phase factor in (10.4.68) arises. In other cases, such that when $t \to \infty$, $t' \to -\infty$ such a phase, if non-zero, may be unimportant in describing the dynamics of the problem. We shall not, however, go into such additional details here. Finite canonical transformations, not just infinitesimal ones, is the subject of Problem 10.8.[7]

Finally we show that the constrained path integral is invariant under the variation of the subsidiary constraints, i.e., when we make the replacements $\chi_\alpha \to \chi_\alpha - \delta\chi_\alpha = \chi'_\alpha$ with the χ'_α as new constraints.

Quite generally, we may write

$$\delta\chi_\alpha = \{\chi_\alpha, a_\beta \Phi_\beta\}_{\text{P.B.}} + b_\beta \Phi_\beta \qquad (10.4.79)$$

with a summation over β understood, for arbitrary b_β since $\Phi_\beta| = 0$ in Γ^*. The coefficients a_β are uniquely determined in Γ^* in terms of the $\delta\chi_\alpha$ since

$$\{\chi_\alpha, \Phi_\beta\}_{\text{P.B.}} a_\beta| = \delta\chi_\alpha| \qquad (10.4.80)$$

as a consequence of the property in (10.4.55).

With $a_\beta \Phi_\beta$ as the generator of the transformation, we also have

$$\delta\Phi_\alpha = \{\Phi_\alpha, a_\beta \Phi_\beta\}_{\text{P.B.}}$$

---

[7] For a fairly detailed treatment of canonical transformations, cf. Sudarshan and Mukunda (1974).

$$= \{\Phi_\alpha, a_\beta\}_{\text{P.B.}} \Phi_\beta + \{\Phi_\alpha, \Phi_\beta\}_{\text{P.B.}} a_\beta \qquad (10.4.81)$$

or upon using (10.4.59),

$$\delta\Phi_\alpha = M_{\alpha\gamma} \Phi_\gamma \qquad (10.4.82)$$

where

$$M_{\alpha\gamma} = \{\Phi_\alpha, a_\gamma\}_{\text{P.B.}} + c^\gamma_{\alpha\beta} a_\beta. \qquad (10.4.83)$$

The general expression of the matrix $[M_{\alpha\gamma}]$ in (10.4.83) allows an additional linear combination of the Φ_β to be present in (10.4.81). We note that the elements of this matrix are infinitesimal.

From the following

$$\{\Phi_\alpha - \delta\Phi_\alpha, \chi_\beta - \delta\chi_\beta\}_{\text{P.B.}}\big| = (\delta_{\alpha\gamma} - M_{\alpha\gamma}) \{\Phi_\gamma, \chi_\beta - \delta\chi_\beta\}_{\text{P.B.}}\big| \qquad (10.4.84)$$

we may infer that

$$\det \{\Phi_\alpha - \delta\Phi_\alpha, \chi_\beta - \delta\chi_\beta\}_{\text{P.B.}}\big| = \det(\mathbf{1} - M) \det \{\Phi_\gamma, \chi_\beta - \delta\chi_\beta\}_{\text{P.B.}}\big|. \qquad (10.4.85)$$

Also we use the property of the Dirac delta

$$\prod_\alpha \delta(\Phi_\alpha - \delta\Phi_\alpha) \equiv \prod_\alpha \delta((\delta_{\alpha\beta} - M_{\alpha\beta})\Phi_\beta)$$

$$= \frac{1}{\det(\mathbf{1} - M)} \prod_\alpha \delta(\Phi_\alpha) \qquad (10.4.86)$$

to see that for $\chi_\alpha \to \chi_\alpha - \delta\chi_\alpha = \chi'_\alpha$ we simply have the transformation

$$\left(\prod_\alpha \delta(\Phi_\alpha) \delta(\chi_\alpha)\right) \big|\det\{\Phi_\gamma, \chi_\beta\}_{\text{P.B.}}\big|$$

$$\longrightarrow \left(\prod_\alpha \delta(\Phi_\alpha) \delta(\chi'_\alpha)\right) \big|\det\{\Phi_\gamma, \chi'_\beta\}_{\text{P.B.}}\big|. \qquad (10.4.87)$$

Finally, we note that $a_\alpha \Phi_\alpha$ as the generator of the transformation

$$\frac{\mathrm{d}}{\mathrm{d}t} a_\alpha \Phi_\alpha \bigg| = \left(\frac{\mathrm{d}}{\mathrm{d}t} a_\alpha\right) \Phi_\alpha \bigg| + a_\alpha \dot{\Phi}_\alpha \bigg|$$

$$= 0. \qquad (10.4.88)$$

For our earlier application in (10.4.44), (10.4.45), (10.4.48),

$$\delta\chi = -(p_1 \cos\alpha + p_2 \sin\alpha)\delta\alpha$$

$$= a\{\chi, \phi\}_{\text{P.B.}}\big| \qquad (10.4.89)$$

hence $a = -\delta\alpha$, implying the invariance of the path integral under variations of the angle of rotation α.

10.4.3 Second Class Constraints and the Dirac Bracket

In (10.4.31), the restricted Poisson bracket $\{\Phi_\alpha, \Omega_\beta\}_{\text{P.B.}}|$ of a first class constraint Φ_α with *all* constraint functions Ω_β is zero. Otherwise a constraint function is called of second class and has been denoted by Ψ_β, $\beta = 1, \ldots, L$, where L is necessarily even.[8] It is straightforward to define, self consistently, a new bracket $\{\cdot, \cdot\}_{\text{D}}$, obtained from the Poisson bracket, such that $\{\Omega_\alpha, \Omega_\beta\}_{\text{D}}| = 0$ for *all* constraint functions. This new bracket, obtained below and referred to as the Dirac bracket, has also other desirable properties.

To the above end, note that since $\det\{\Psi_\alpha, \Psi_\beta\}_{\text{P.B}} \neq 0$ (see (10.4.32)), we may find a matrix $\mathbf{c} = [c_{\alpha\beta}]$ such that

$$\{\Psi_\alpha, \Psi_\beta\}_{\text{P.B.}} c_{\beta\gamma} = \delta_{\alpha\gamma} \tag{10.4.90}$$

On the other hand, (10.4.36) gives

$$0 = \dot{\Psi}_\alpha\Big| = \{\Psi_\alpha, H\}_{\text{P.B.}}| + \{\Psi_\alpha, \Psi_\beta\}_{\text{P.B.}} \rho_\beta| \tag{10.4.91}$$

which from (10.4.90), this leads to

$$\rho_\beta| = -c_{\beta\gamma}\{\Psi_\gamma, H\}_{\text{P.B.}}| \tag{10.4.92}$$

Accordingly, for an arbitrary function F of $\mathbf{q}(t)$, $\mathbf{p}(t)$,

$$\dot{F}\Big| = (\{F, H\}_{\text{P.B.}} - \{F, \Psi_\alpha\}_{\text{P.B.}} c_{\alpha\beta}\{\Psi_\beta, H\}_{\text{P.B.}})| + \eta_\alpha\{F, \Phi_\alpha\}_{\text{P.B.}}| \tag{10.4.93}$$

This suggests to introduce the modified bracket defined by

$$\{\cdot, \cdot\}_{\text{D}} = \{\cdot, \cdot\}_{\text{P.B.}} - \{\cdot, \Psi_\alpha\}_{\text{P.B.}} c_{\alpha\beta}\{\Psi_\beta, \cdot\}_{\text{P.B.}} \tag{10.4.94}$$

and simply write

$$\dot{F}\Big| = \{F, H\}_{\text{D}}| + \eta_\alpha\{F, \Phi_\alpha\}_{\text{P.B.}}|, \qquad \{\Omega_\alpha, H\}|_{\text{D}} = 0 \tag{10.4.95}$$

expressed in terms of the underlying Hamiltonian H of the system (see also (10.4.100)), where

$$\{F, \Phi_\alpha\}_{\text{P.B.}}| = \{F, \Phi_\alpha\}_{\text{D}}|. \tag{10.4.96}$$

From (10.4.95), we see that it is of particular interest if one has only second class constraints since in this case $\dot{F}| = \{F, H\}_{\text{D}}|$.

For two functions F, G of $\mathbf{q}(t)$, $\mathbf{p}(t)$, (10.4.94) implies that

$$\{F, G\}_{\text{D}} = \{F, G\}_{\text{P.B.}} - \{F, \Psi_\alpha\}_{\text{P.B.}} c_{\alpha\beta}\{\Psi_\beta, G\}_{\text{P.B.}} \tag{10.4.97}$$

For any two second class constraints functions Ψ_γ, $\Psi_{\gamma'}$ (10.4.94) then also leads to

---
[8] See footnote 6 below (10.4.36) for a demonstration of this.

$$\{\Psi_\gamma, \Psi_{\gamma'}\}_{\mathrm{D}}\big| = \{\Psi_\gamma, \Psi_{\gamma'}\}_{\mathrm{P.B.}}\big| - \{\Psi_\gamma, \Psi_\alpha\}_{\mathrm{P.B.}} c_{\alpha\beta} \{\Psi_\beta, \Psi_{\gamma'}\}_{\mathrm{P.B.}}\big|$$
$$= 0 \tag{10.4.98}$$

and note that
$$\{\Phi_\alpha, \Phi_\beta\}_{\mathrm{D}}\big| = 0, \qquad \{\Phi_\alpha, \Psi_\beta\}_{\mathrm{D}}\big| = 0 \tag{10.4.99}$$

Finally, for any function F of $\mathbf{q}(t)$, $\mathbf{p}(t)$,
$$\{\Psi_\alpha, F\}_{\mathrm{D}}\big| = 0, \tag{10.4.100}$$

and for *all* constraint functions Ω_β,
$$\{\Omega_\alpha, \Omega_\beta\}_{\mathrm{D}}\big| = 0. \tag{10.4.101}$$

We leave as an exercise to the reader to check the basic properties of a bracket in (10.4.14)–(10.4.17) for $\{\cdot,\cdot\}_{\mathrm{D}}$ as well.

10.5 Bose Excitations

We represent the Bose operators a_B, $a_B^\dagger$ defined in §6.1–§6.3, §6.5, as follows

$$a_B \to \frac{\mathrm{d}}{\mathrm{d}\beta^*} \tag{10.5.1}$$

$$a_B^\dagger \to \beta^*. \tag{10.5.2}$$

This representation is obviously consistent with the corresponding commutation relation in (6.1.13).

The n-particle state (see also §6.5), may be defined by

$$\langle \beta^* \mid n \rangle = \psi_n(\beta^*) = \frac{(\beta^*)^n}{\sqrt{n!}} \tag{10.5.3}$$

with the vacuum state simply given by

$$\langle \beta^* \mid 0 \rangle = \psi_0(\beta^*) = 1. \tag{10.5.4}$$

The normalization condition (§6.1)

$$\langle n \mid m \rangle = \delta_{nm} \tag{10.5.5}$$

and the explicit value of the integral

$$\int \frac{\mathrm{d}\beta^* \mathrm{d}\beta}{2\pi\mathrm{i}} \exp(-\beta^*\beta) \, (\psi_n(\beta^*))^* \, \psi_m(\beta^*) = \delta_{nm} \tag{10.5.6}$$

as obtained directly from the definition in (10.5.3), by working, for example, in polar coordinates (see Problem 10.13), allow us to introduce the identity operator

$$\mathbf{1} = \int \frac{\mathrm{d}\beta^* \mathrm{d}\beta}{2\pi\mathrm{i}} \mathrm{e}^{-\beta^*\beta} |\beta^*\rangle \langle \beta^*| \tag{10.5.7}$$

and write

$$\langle n \mid m \rangle = \langle n \mid \mathbf{1} \mid m \rangle = \delta_{nm} \tag{10.5.8}$$

consistent with (10.5.6).

We will consider, in general, any operator B, which as a function of $a_B^\dagger$, a_B, may be written in the form

$$B = \sum_{n,m} B_{nm} \left(a_B^\dagger\right)^n (a_B)^m. \tag{10.5.9}$$

From (6.1.31), (6.5.10) one easily gets

$$\langle N \mid B \mid M \rangle = \sum_{n,m} B_{nm} \frac{\sqrt{N!\, M!}}{(N-n)!} \delta(N-n, M-m) \tag{10.5.10}$$

with the restriction $n \leqslant N$, such that $N - n = M - m$, with $\delta(\cdot, \cdot)$ denoting the Kronecker delta.

In the representation (10.5.1)–(10.5.4),

$$\langle \beta^* \mid B \mid \beta'^* \rangle = \sum_{N,M} \frac{(\beta^*)^N}{\sqrt{N!}} \langle N \mid B \mid M \rangle \frac{(\beta')^M}{\sqrt{M!}}$$

$$= \sum_{N,M} \sum_{n,m} \delta(N-n, M-m) B_{nm} (\beta^*)^n (\beta')^m \frac{(\beta^*\beta')^{N-n}}{(N-n)!}$$

$$= \exp(\beta^*\beta') \sum_{n,m} B_{nm} (\beta^*)^n (\beta')^m \tag{10.5.11}$$

where in the first equality we used the completeness of the states $|N\rangle$, and in the last equality we have made a change of variable $N - n \to N$, in the process.

Upon comparison of (10.5.11) with (10.5.9) we obtain the rule

$$\langle \beta^* \mid B \mid \beta'^* \rangle = \exp(\beta^*\beta') B\big|_{a_B^\dagger \to \beta^*,\, a_B \to \beta'}. \tag{10.5.12}$$

In particular, for the identity operator, this gives

$$\langle \beta^* \mid \beta'^* \rangle = \exp(\beta^*\beta') \tag{10.5.13}$$

which is also directly verified from (10.5.3).

From (10.5.7), (10.5.13), we obtain the property

$$|\rho^*\rangle = \int \frac{\mathrm{d}\beta^*\mathrm{d}\beta}{2\pi\mathrm{i}} \mathrm{e}^{-\beta^*\beta} \mathrm{e}^{\beta^*\rho} |\beta^*\rangle. \tag{10.5.14}$$

With $\langle 0|$, denoting $\langle n = 0|$, this leads, upon multiplying (10.5.14) by $\langle 0|$, to

$$\int \frac{\mathrm{d}\beta^*\mathrm{d}\beta}{2\pi\mathrm{i}} \mathrm{e}^{-\beta^*\beta} \mathrm{e}^{\beta^*\rho} = 1. \tag{10.5.15}$$

This result may be also obtained by expanding $\exp(\beta^*\rho)$, on the left-hand side of (10.5.15) and then using (10.5.6).

Other useful relations which readily follow from (10.5.14) are

$$\int \frac{\mathrm{d}\beta^*\mathrm{d}\beta}{2\pi\mathrm{i}} \mathrm{e}^{-\beta^*\beta} \mathrm{e}^{\beta^*\rho} (\beta)^n = (\rho)^n \tag{10.5.16}$$

$$\int \frac{\mathrm{d}\beta^*\mathrm{d}\beta}{2\pi\mathrm{i}} \mathrm{e}^{-\beta^*\beta} \mathrm{e}^{\beta^*\rho} \mathrm{e}^{\beta\sigma^*} = \mathrm{e}^{\sigma^*\rho} \tag{10.5.17}$$

which in turn generalize (10.5.15).

From (10.5.7), we have the following representation for an operator B obtained by writing $B = \mathbf{1}B\mathbf{1}$,

$$B = \int \frac{\mathrm{d}\beta^*\mathrm{d}\beta}{2\pi\mathrm{i}} \frac{\mathrm{d}\beta'^*\mathrm{d}\beta'}{2\pi\mathrm{i}} \mathrm{e}^{-(\beta^*\beta+\beta'^*\beta')} |\beta^*\rangle \langle \beta^*| B |\beta'^*\rangle \langle \beta'^*|. \tag{10.5.18}$$

For the product of two operators B_1, B_2 this gives

$$B_2 B_1 = \int \frac{\mathrm{d}\beta^*\mathrm{d}\beta}{2\pi\mathrm{i}} \frac{\mathrm{d}\beta'^*\mathrm{d}\beta'}{2\pi\mathrm{i}} \mathrm{e}^{-(\beta^*\beta+\beta'^*\beta')} |\beta^*\rangle \langle \beta^*| B_{21} |\beta'^*\rangle \langle \beta'^*| \tag{10.5.19}$$

where

$$\langle \beta^*| B_{21} |\beta'^*\rangle = \int \frac{\mathrm{d}\beta_1^*\mathrm{d}\beta_1}{2\pi\mathrm{i}} \mathrm{e}^{-\beta_1^*\beta_1} \langle \beta^*| B_2 |\beta_1^*\rangle \langle \beta_1^*| B_1 |\beta'^*\rangle \tag{10.5.20}$$

with the *rule* given in (10.5.12) for computing the matrix elements on the right-hand of (10.5.19).

In the sequel we use the notation

$$\langle \beta^*| B |\beta'^*\rangle \equiv B(\beta^*, \beta'). \tag{10.5.21}$$

For a Hamiltonian $H(a_B^\dagger, a_B; t)$, the time evolution operator $U(t, t')$, $t' \to t$ is then from (10.5.12), (10.5.18), by repeated applications of (10.5.19), (10.5.20), given by

$$U(t, t') = \int \frac{\mathrm{d}\beta^*\mathrm{d}\beta}{2\pi\mathrm{i}} \frac{\mathrm{d}\beta'^*\mathrm{d}\beta'}{2\pi\mathrm{i}} \mathrm{e}^{-(\beta^*\beta+\beta'^*\beta')} |\beta^*\rangle \langle \beta'^*| U(\beta^*, \beta'; t, t') \tag{10.5.22}$$

where

$$U(\beta^*, \beta'; t, t') = \lim_{N \to \infty} \int \left(\prod_{k=1}^{N-1} \frac{d\beta_k^* d\beta_k}{2\pi i} \right)$$

$$\times \exp \left(\sum_{k=1}^{N} \left(\beta_k^* \beta_{k-1} - \frac{i}{\hbar} \varepsilon H(\beta_k^*, \beta_{k-1}; \tilde{t}_k) \right) - \sum_{k=1}^{N-1} \beta_k^* \beta_k \right) \quad (10.5.23)$$

and $\beta_N^* = \beta^*$, $\beta_0 = \beta'$, $\tilde{t}_k = t' + (k - 1/2)\varepsilon$, $k = 1, \ldots, N$, $t - t' = N\varepsilon$.

We may add and subtract the term $\beta_N^* \beta_N$ in the exponential in (10.5.23) to obtain in the limit $N \to \infty$

$$U(\beta^*, \beta'; t, t') = \int_{\beta(t')=\beta'}^{\beta^*(t)=\beta^*} \mathscr{D}(\beta^*(\cdot), \beta(\cdot)) \exp(\beta^*(t), \beta(t))$$

$$\times \exp \frac{i}{\hbar} \int_{t'}^{t} d\tau \left(i\hbar \beta^*(\tau) \dot{\beta}(\tau) - H(\beta^*(\tau), \beta(\tau); \tau) \right). \quad (10.5.24)$$

The exponentials in (10.5.24) may be combined in the form

$$\exp \frac{i}{\hbar} \int_{t'}^{t} d\tau \left(i\hbar \beta^*(\tau) \dot{\beta}(\tau) - i\hbar \beta^*(\tau) \beta(\tau) \delta(\tau - t + 0) - H(\beta^*(\tau), \beta(\tau); \tau) \right). \quad (10.5.25)$$

As an application, we consider the Hamiltonian

$$H(a_B^\dagger, a_B; t) = \hbar \omega a_B^\dagger a_B - F(t)(a_B + a_B^\dagger) \quad (10.5.26)$$

(see also (6.2.2)), where for simplicity we have not included a zero point energy.

For the Hamiltonian in (10.5.26), one is dealing in (10.5.23)/(10.5.24) with Gaussian integrals. These may be carried out directly from (10.5.23) by a method which will be explicitly worked out in detail for the corresponding case with fermions in the next section (see Problem 10.14). They may be also carried out by the method applied in §10.2 by considering the deviation of $\beta^*(\tau)$, $\beta(\tau)$ from their classical solutions with boundary conditions $\beta^*(t) = \beta^*$, $\beta(t') = \beta'$ by using, in the process, the expression in (10.5.24). These give the following explicit form

$$U(\beta^*, \beta'; t, t') = \exp \left(\beta^* \beta' e^{-i\omega(t-t')} \right) \exp \left(\frac{i}{\hbar} \beta^* \int_{t'}^{t} d\tau\, e^{-i\omega(t-\tau)} F(\tau) \right)$$

$$\times \exp \left(\frac{i}{\hbar} \beta' \int_{t'}^{t} d\tau\, e^{-i\omega(\tau-t')} F(\tau) \right)$$

$$\times \exp \left(-\frac{1}{\hbar^2} \int_{t'}^{t} d\tau \int_{t'}^{\tau} d\tau'\, F(\tau) e^{-i\omega(\tau-\tau')} F(\tau') \right). \quad (10.5.27)$$

632   10 Path Integrals

With $F(\tau)$ assumed to vanish outside the interval (t', t), we may introduce the Fourier transform

$$F(\omega) = \int_{-\infty}^{\infty} d\tau \, e^{i\omega\tau} F(\tau) \tag{10.5.28}$$

and using the notation

$$\gamma = \int_{-\infty}^{\infty} d\tau \int_{-\infty}^{\infty} d\tau' F(\tau) e^{-i\omega(\tau-\tau')} \Theta(\tau - \tau') F(\tau') \tag{10.5.29}$$

we obtain from (10.5.27), the following expression for $U(t, t')$ in (10.5.22),

$$U(t, t') = e^{-\gamma/\hbar^2} \int \frac{d\beta^* d\beta}{2\pi i} \frac{d\beta'^* d\beta'}{2\pi i} e^{-(\beta^*\beta + \beta'^*\beta')} \exp\left(\beta^*\beta' e^{-i\omega(t-t')}\right)$$

$$\times \exp\left(i\beta^* e^{-i\omega t} F(\omega)/\hbar\right) \exp\left(i\beta' e^{i\omega t'} F^*(\omega)/\hbar\right) |\beta^*\rangle \langle\beta'^*|. \tag{10.5.30}$$

In particular, this gives

$$\langle 0 | U(t, t') | 0 \rangle = \exp\left(-\frac{\gamma}{\hbar^2}\right) \tag{10.5.31}$$

(compare with (6.2.12)), where we have used (10.5.4) and the explicit integral (10.5.15) which gives unity for the integral corresponding to the measure $d\beta^* d\beta/2\pi i$ followed, again, by one for the integral corresponding to the measure $d\beta'^* d\beta'/2\pi i$.

For $\langle n | U(t, t') | 0 \rangle$, the $d\beta'^* d\beta'/2\pi i$-integral gives one, and hence

$$\langle n | U(t, t') | 0 \rangle = e^{-\gamma/\hbar^2} \int \frac{d\beta^* d\beta}{2\pi i} e^{-\beta^*\beta} \exp\left(i\beta^* e^{-i\omega t} F(\omega)/\hbar\right) \frac{(\beta)^n}{\sqrt{n!}}$$

$$= \frac{\left[iF(\omega) e^{-i\omega t}/\hbar\right]^n}{\sqrt{n!}} e^{-\gamma/\hbar^2} \tag{10.5.32}$$

where we have used (10.5.3) and (10.5.16), which is to be compared with (6.2.22).

Finally, we derive a generating function which allows the determination of all the matrix elements $\langle n | U(t, t') | m \rangle$.

To the above end, we multiply (10.5.30) form the left by $\left((c^*)^n/\sqrt{n!}\right) \langle n|$, from the right by $|m\rangle \left((c')^m/\sqrt{m!}\right)$, and sum over $n, m = 0, 1, \ldots$, to obtain

$$\sum_{n,m} \frac{(c^*)^n}{\sqrt{n!}} \langle n | U(t, t') | m \rangle \frac{(c')^m}{\sqrt{m!}} = e^{-\gamma/\hbar^2} \int \frac{d\beta^* d\beta}{2\pi i} \frac{d\beta'^* d\beta'}{2\pi i} e^{-(\beta^*\beta + \beta'^*\beta')}$$

$$\times \exp\left(\beta^*\left(\beta' e^{-i\omega(t-t')} + \frac{i}{\hbar}e^{-i\omega t}F(\omega)\right)\right)$$

$$\times e^{\beta c^*} e^{\beta'^* c'} \exp\left(\beta'\left(\frac{i}{\hbar}e^{i\omega t'}F^*(\omega)\right)\right). \tag{10.5.33}$$

The integrals may be carried out exactly thanks to the integral in (10.5.17) to yield

$$\sum_{n,m} \frac{(c^*)^n}{\sqrt{n!}} \langle n|U(t,t')|m\rangle \frac{(c')^m}{\sqrt{m!}} = e^{-\gamma/\hbar^2} \exp\left(c^*c' e^{-i\omega(t-t')}\right)$$

$$\times \exp\left(\frac{i}{\hbar}F^*(\omega)e^{+i\omega t'}c'\right)\exp\left(\frac{i}{\hbar}c^* e^{-i\omega t}F(\omega)\right). \tag{10.5.34}$$

The matrix elements $\langle n|U(t,t')|m\rangle$ are then obtained from (10.5.34) by carrying out n derivatives with respect to c^* and m derivatives with respect to c', then setting $c^* = 0$, $c' = 0$, and dividing by $\sqrt{m!n!}$. This is to be compared with the expression for $\langle nt|mt'\rangle$ in (6.2.30) in the present notation.

10.6 Grassmann Variables: Fermi Excitations

In this section, we introduce anti-commuting c-variables, referred to as Grassmann variables, and learn how to differentiate and integrate with respect to such variables. Finally Grassmann variables are used to develop path integrals to describe Fermi-particle interactions with external sources. We consider in turn real and complex Grassmann variables.

10.6.1 Real Grassmann Variables

We introduce v anti-commuting real variables $\eta_1, \ldots, \eta_v$:

$$\{\eta_j, \eta_k\} = 0 \tag{10.6.1}$$

$j, k = 1, \ldots, v$. Equation (10.6.1), in particular, implies that

$$\eta_k^2 = 0 \tag{10.6.2}$$

for $k = 1, \ldots, v$.

As a result of (10.6.2), we may expand a function of $(\eta_1, \ldots, \eta_v) \equiv \boldsymbol{\eta}$ involving at most the product of the v distinct Grassmann variables:

$$f(\boldsymbol{\eta}) = \sum_{k=0}^{v} \sum_{i_1, \ldots, i_k} f^{(k)}_{i_1 \cdots i_k} \eta_{i_1} \cdots \eta_{i_k} \tag{10.6.3}$$

where for $k = 2, \ldots, v$, the coefficients $f^{(k)}_{i_1 \cdots i_k}$ are totally anti-symmetric in $i_1, \ldots, i_k$.

The left-hand derivative with respect to a Grassmann variable η_j is defined by

$$\frac{\partial}{\partial \eta_j} \eta_{i_1} \cdots \eta_j \cdots \eta_{i_k} = (-1)^{\delta_j - 1} \eta_{i_1} \cdots \eta_{i_k} \tag{10.6.4}$$

where δ_j denotes the position of η_j in the product $\eta_{i_1} \cdots \eta_j \cdots \eta_{i_k}$ from the left, and the variable η_j is omitted on the right-hand side of (10.6.4). If the former product does not involve η_j, the right-hand side of (10.6.4) should be replaced by zero. For example, note that

$$\frac{\partial}{\partial \eta_j} \eta_k \eta_m = \delta_{jk} \eta_m - \eta_k \delta_{jm} \tag{10.6.5}$$

and

$$\left\{ \frac{\partial}{\partial \eta_j}, \eta_k \right\} = \delta_{jk}. \tag{10.6.6}$$

Similarly the right-hand derivative with respect to a η_j, denoted by $\overleftarrow{\partial/\partial \eta_j}$, is defined by

$$\eta_{i_1} \cdots \eta_j \cdots \eta_{i_k} \frac{\overleftarrow{\partial}}{\partial \eta_j} = (-1)^{\underline{\delta}_j - 1} \eta_{i_1} \cdots \eta_{i_k} \tag{10.6.7}$$

where $\underline{\delta}_j$ denotes the position of η_j in the product $\eta_{i_1} \cdots \eta_j \cdots \eta_{i_k}$ from the right, etc.

For a fixed k in $\{1, \ldots, v\}$, because of property (10.6.2), one has to investigate the meanings of only the following two integrals:

$$\int d\eta_k, \quad \int d\eta_k \, \eta_k. \tag{10.6.8}$$

Assuming translational invariance of the integrals for $\eta_k \to \eta_k + \alpha_k = \eta'_k$, where α_k is another Grassmann variable, which anti-commutes with η_k, we obtain

$$\int d\eta'_k = \int d\eta_k \tag{10.6.9}$$

$$\int d\eta'_k \, \eta'_k = \int d\eta_k \, \eta_k + \left(\int d\eta_k \right) \alpha_k$$

$$= \int d\eta_k \, \eta_k \tag{10.6.10}$$

for arbitrary α_k. Hence we conclude that

$$\int d\eta_k = 0 \tag{10.6.11}$$

10.6 Grassmann Variables: Fermi Excitations

for any k in $\{1, \ldots, v\}$.

The second integral in (10.6.8), defined as a c-number, may be normalized arbitrarily, and we choose it to be

$$\int d\eta_k \, \eta_k = \sqrt{2\pi} = -\int \eta_k \, d\eta_k \qquad (10.6.12)$$

for any k, where in writing the second equality we used the fact that the nature of η_k as a Grassmann variable requires that $d\eta_k$ also anti-commutes with the Grassmann variables to make the second integral in (10.6.8) a c-number. The latter is consistent with the definition of the differential operator $d = \sum_k d\eta_k \partial/\partial \eta_k$ satisfying the rule

$$d(\eta_i \eta_j) = d\eta_i \, \eta_j + \eta_i \, d\eta_j \qquad (10.6.13)$$

and, on the other hand, one has explicitly from (10.6.5) that

$$d(\eta_i \eta_j) = d\eta_i \, \eta_j - d\eta_j \, \eta_i \qquad (10.6.14)$$

which upon comparison with (10.6.13) implies that

$$\{d\eta_i, \eta_j\} = 0. \qquad (10.6.15)$$

From the rule given in (10.6.12), we may define the multiple integral

$$\int d\eta_1 \, \eta_1 \int d\eta_2 \, \eta_2 \int \cdots \int d\eta_v \, \eta_v = (2\pi)^{v/2} \qquad (10.6.16)$$

or

$$\int d\eta_1 \cdots \int d\eta_v \, \eta_v \cdots \eta_1 = (2\pi)^{v/2}. \qquad (10.6.17)$$

A function $f(\eta_k)$ of η_k in $\{\eta_1, \ldots, \eta_v\}$ may be written as

$$f(\eta_k) = c_0 + c \, \eta_k \qquad (10.6.18)$$

where c_0, c are c-numbers. For any other Grassmann variable α_k, which anti-commutes with η_k we have

$$\int d\eta_k \, (\eta_k - \alpha_k) f(\eta_k) = \sqrt{2\pi} \, (c_0 + c \, \alpha_k)$$

$$= \sqrt{2\pi} f(\alpha_k) \qquad (10.6.19)$$

which allows us to introduce the Dirac delta, in this context, given by

$$\delta(\eta_k - \alpha_k) = \frac{1}{\sqrt{2\pi}} (\eta_k - \alpha_k). \qquad (10.6.20)$$

636   10 Path Integrals

The reader is asked to derive some of the immediate properties of $\delta(\eta_k)$ that follow from (10.6.20) in Problem 10.15.

From the rule of integration in (10.6.12), (10.6.11), we may then introduce the representation

$$\delta(\eta_k - \alpha_k) = \int \frac{\mathrm{d}\rho_k}{2\pi\mathrm{i}} \, \mathrm{e}^{\mathrm{i}\rho_k(\eta_k - \alpha_k)} \qquad (10.6.21)$$

as is easily checked, where ρ_k is a Grassmann variable which, in particular, anti-commutes with η_k, α_k.

Since $\mathrm{d}\eta_k \, \delta(\eta_k - \alpha_k)$ commutes with all Grassmann variables, we have the useful property

$$\int \mathrm{d}\eta_1 \, \delta(\eta_1 - \alpha_1) \cdots \mathrm{d}\eta_v \, \delta(\eta_v - \alpha_v) \, (\eta_1)^{\varepsilon_1} \cdots (\eta_v)^{\varepsilon_v}$$

$$= (\alpha_1)^{\varepsilon_1} \cdots (\alpha_v)^{\varepsilon_v} \qquad (10.6.22)$$

where $\varepsilon_i = 0$ or 1, $(\cdot)^0 = 1$,

$$\int (\eta_1)^{\varepsilon_1} \cdots (\eta_v)^{\varepsilon_v} \, \mathrm{d}^v\boldsymbol{\eta} \, \delta_A^v(\boldsymbol{\eta} - \boldsymbol{\alpha}) = (\alpha_1)^{\varepsilon_1} \cdots (\alpha_v)^{\varepsilon_v} \qquad (10.6.23)$$

and

$$\mathrm{d}^v\boldsymbol{\eta} = \mathrm{d}\eta_1 \cdots \mathrm{d}\eta_{v-1}\mathrm{d}\eta_v \qquad (10.6.24)$$

$$\delta_A^v(\boldsymbol{\eta} - \boldsymbol{\alpha}) = \delta(\eta_v - \alpha_v)\delta(\eta_{v-1} - \alpha_{v-1}) \cdots \delta(\eta_1 - \alpha_1) \qquad (10.6.25)$$

and note the different orderings in $\mathrm{d}^v\boldsymbol{\eta}$ and $\delta_A^v(\boldsymbol{\eta} - \boldsymbol{\alpha})$.

By using the property that a product of two Grassmann variables $\rho_k\alpha_k$ commutes with all Grassmann variables, we may infer from (10.6.21), (10.6.25) that

$$\delta_A^v(\boldsymbol{\eta} - \boldsymbol{\alpha}) = \int \frac{\mathrm{d}_A^v\boldsymbol{\rho}}{(2\pi\mathrm{i})^v} \mathrm{e}^{\mathrm{i}\boldsymbol{\rho}\cdot(\boldsymbol{\eta} - \boldsymbol{\alpha})} \qquad (10.6.26)$$

where

$$\mathrm{d}_A^v\boldsymbol{\rho} = \mathrm{d}\rho_v\mathrm{d}\rho_{v-1}\cdots\mathrm{d}\rho_1. \qquad (10.6.27)$$

For an arbitrary function $f(\boldsymbol{\eta})$, as given in (10.6.3), we define the transform

$$\tilde{f}(\boldsymbol{\rho}) = \int f(\boldsymbol{\eta}') \, \mathrm{d}^v\boldsymbol{\eta}' \, \mathrm{e}^{\mathrm{i}\boldsymbol{\rho}\cdot\boldsymbol{\eta}'}. \qquad (10.6.28)$$

Upon multiplying (10.6.28) form the right by $(\mathrm{d}_A^v\boldsymbol{\rho}/(2\pi\mathrm{i})^v)\exp(-\mathrm{i}\boldsymbol{\rho}\cdot\boldsymbol{\eta})$ and integrating we obtain

$$\int \tilde{f}(\boldsymbol{\rho})\frac{\mathrm{d}_A^v\boldsymbol{\rho}}{(2\pi\mathrm{i})^v}\, \mathrm{e}^{-\mathrm{i}\boldsymbol{\rho}\cdot\boldsymbol{\eta}} = \int f(\boldsymbol{\eta}') \, \mathrm{d}^v\boldsymbol{\eta}' \frac{\mathrm{d}_A^v\boldsymbol{\rho}}{(2\pi\mathrm{i})^v}\, \mathrm{e}^{\mathrm{i}\boldsymbol{\rho}\cdot(\boldsymbol{\eta}' - \boldsymbol{\eta})}$$

$$= \int f(\boldsymbol{\eta}')\, d^v\boldsymbol{\eta}'\, \delta_A^v(\boldsymbol{\eta}' - \boldsymbol{\eta}) \tag{10.6.29}$$

or

$$f(\boldsymbol{\eta}) = \int \tilde{f}(\boldsymbol{\rho}) \frac{d_A^v \boldsymbol{\rho}}{(2\pi i)^v} e^{-i\boldsymbol{\rho}\cdot\boldsymbol{\eta}} \tag{10.6.30}$$

where we have used (10.6.23), (10.6.22), (10.6.26).

Finally we note that the rule for the change of integration variables in the evaluation of an integral turns out to be quite simple but surprising. Consider the change of variables $\eta_1, \ldots, \eta_v \to \alpha_1, \ldots, \alpha_v$ defined by the linear combination

$$\eta_j = C_{jk}\alpha_k \tag{10.6.31}$$

where the C_{jk} are c-numbers, and $\alpha_1, \ldots, \alpha_k$ are Grassmann variables. To the above end it is sufficient to consider the integral of the product $\eta_1, \ldots, \eta_v$. In this case, we have

$$\int d^v\boldsymbol{\eta}\, \eta_1 \cdots \eta_v = \int d^v\boldsymbol{\alpha}\, J\, C_{1k_1}\alpha_{k_1} \cdots C_{vk_v}\alpha_{k_v} \tag{10.6.32}$$

where J is the Jacobian of the transformation to be determined. Since

$$\alpha_{k_1} \cdots \alpha_{k_v} = \alpha_1 \cdots \alpha_v \varepsilon^{k_1 \cdots k_v} \tag{10.6.33}$$

where $\varepsilon^{k_1 \cdots k_v}$ is the Levi-Civita symbol equal to $+1$, -1 if $\{k_1, \ldots, k_v\}$ is and even, odd permutation of $\{1, \ldots, v\}$, respectively, and equal to zero if any two or more of the indices $k_1, \ldots, k_v$ are equal, and

$$C_{1k_1} \cdots C_{vk_v} \varepsilon^{k_1 \cdots k_v} = \det C \tag{10.6.34}$$

we immediately obtain from (10.6.32) that

$$J = (\det C)^{-1} \tag{10.6.35}$$

being the inverse of $\det C$ (!).

10.6.2 Complex Grassmann Variables

Out of two real Grassmann variables α_R, α_I satisfying $\{\alpha_R, \alpha_I\} = 0$, we may define a complex Grassmann variable $\alpha = \alpha_R + i\alpha_I$. Immediate properties which follow from this definition are

$$\{\alpha, \alpha\} = 0, \quad \{\alpha, \alpha^*\} = 0, \quad \{\alpha^*, \alpha^*\} = 0. \tag{10.6.36}$$

Also imposing a reality restriction on the product

$$\alpha^*\alpha = -2i\alpha_I\alpha_R \tag{10.6.37}$$

implies that

$$(\alpha_I \alpha_R)^* = -\alpha_I \alpha_R$$
$$= \alpha_R \alpha_I = -\alpha_I^* \alpha_R^* \qquad (10.6.38)$$

where in writing the second equality we have used the anti-commutativity of α_I, α_R, and in the last one, we have used the reality condition of these variables.

From (10.6.12), and upon writing

$$\int d\alpha_R \, \alpha_R = \sqrt{2\pi} = \left(\int d\alpha_R \, \alpha_R\right)^* = -\int (d\alpha_R)^* \alpha_R^* = -\int (d\alpha_R)^* \alpha_R \qquad (10.6.39)$$

we may infer that

$$(d\alpha_R)^* = -d\alpha_R \qquad (10.6.40)$$

and similarly

$$(d\alpha_I)^* = -(d\alpha_I). \qquad (10.6.41)$$

From the definition

$$d\alpha = d\alpha_R + i\, d\alpha_I \qquad (10.6.42)$$

it follows that

$$(d\alpha)^* = -d\alpha^*. \qquad (10.6.43)$$

Also from $\{d\alpha^*, \alpha^*\} = 0$ and

$$\left(\int d\alpha \, \alpha\right)^* = \int \alpha^* (d\alpha)^* = -\int \alpha^* d\alpha^*$$

one obtains

$$\left(\int d\alpha \, \alpha\right)^* = \int d\alpha^* \alpha^*. \qquad (10.6.44)$$

To define integrations over complex Grassmann variables, we first note that for consistency with (10.6.11), (10.6.42), (10.6.43)

$$\int d\alpha^* = 0 \qquad (10.6.45)$$

Also to obtain a consistent definition of the integral in (10.6.45), we consider the transformation of variables $\alpha_R, \alpha_I \to \alpha^*, \alpha$:

$$\begin{pmatrix} \alpha_R \\ \alpha_I \end{pmatrix} = \frac{1}{2} \begin{pmatrix} 1 & 1 \\ i & -i \end{pmatrix} \begin{pmatrix} \alpha^* \\ \alpha \end{pmatrix}, \qquad (10.6.46)$$

define the Jacobian of the transformation given in (10.6.35)

$$J = \left(\det \frac{1}{2} \begin{pmatrix} 1 & 1 \\ i & -i \end{pmatrix}\right)^{-1} = 2i \qquad (10.6.47)$$

10.6 Grassmann Variables: Fermi Excitations   639

and use the fact that $\alpha_I \alpha_R = \alpha \alpha^*/2i$ to infer that

$$2\pi = \int d\alpha_R \, d\alpha_I \, \alpha_I \, \alpha_R = \int d\alpha^* d\alpha \, (2i) \, (\alpha\alpha^*/2i) \tag{10.6.48}$$

or

$$\int \frac{d\alpha^*}{\sqrt{2\pi}} \frac{d\alpha}{\sqrt{2\pi}} \alpha\alpha^* = 1. \tag{10.6.49}$$

Hence a consistent definition with (10.6.45) is

$$\int \frac{d\alpha^*}{\sqrt{2\pi}} \alpha^* = 1 = \int \frac{d\alpha}{\sqrt{2\pi}} \alpha. \tag{10.6.50}$$

For integrations over v complex Grassmann variables $\alpha_1, \ldots, \alpha_v$, satisfying

$$\{\alpha_i, \alpha_k\} = 0, \quad \{\alpha_i, \alpha_k^*\} = 0, \quad \{\alpha_i^*, \alpha_k^*\} = 0 \tag{10.6.51}$$

consider the integral

$$I = \int \frac{d\alpha_1^* d\alpha_1}{2\pi} \cdots \frac{d\alpha_v^* d\alpha_i}{2\pi} \exp\left(-\alpha_i^* A_{ij} \alpha_j\right) \tag{10.6.52}$$

where A_{ij} are c-numbers. The commutativity of $\alpha_i^* A_{i1} \alpha_1$, for example, with all the Grassmann variables, and so on, allow us to rewrite the integral I as (since $\alpha_1^2 = 0, \ldots, \alpha_v^2 = 0$)

$$I = \int \frac{d\alpha_1^* \left(-\alpha_{j_1}^* A_{j_1 1} \alpha_1\right) d\alpha_1}{2\pi} \cdots \frac{d\alpha_v^* \left(-\alpha_{j_v}^* A_{j_v v} \alpha_v\right) d\alpha_v}{2\pi}$$

$$= \int \frac{d\alpha_1^* \, \alpha_{j_1}^* A_{j_1 1}}{\sqrt{2\pi}} \cdots \frac{d\alpha_v^* \, \alpha_{j_v}^* A_{j_v v}}{\sqrt{2\pi}}$$

$$= \int \frac{d\alpha_1^*}{\sqrt{2\pi}} \cdots \frac{d\alpha_v^*}{\sqrt{2\pi}} \alpha_{j_v}^* \cdots \alpha_{j_1}^* A_{j_v v} \cdots A_{j_v 1}. \tag{10.6.53}$$

Upon using the properties

$$\alpha_{j_v}^* \cdots \alpha_{j_1}^* = \alpha_v^* \cdots \alpha_1^* \varepsilon^{j_v \cdots j_1} \tag{10.6.54}$$

and

$$\varepsilon^{j_v \cdots j_1} A_{j_v v} \cdots A_{j_1 1} = \det A \tag{10.6.55}$$

we obtain

$$\int \frac{d\alpha_1^* d\alpha_1}{2\pi} \cdots \frac{d\alpha_v^* d\alpha_v}{2\pi} \exp\left(-\alpha_i^* A_{ij} \alpha_j\right) = \det A. \tag{10.6.56}$$

For $\det A \neq 0$, (10.6.56) gives rise to another useful integral upon making the change of variables $\alpha_j \to \alpha_j + \eta_j$, $\alpha_j^* \to \alpha_j^* + \eta_j^*$, $j = 1, \ldots, v$ where the η_j, η_j^* are Grassmann variables. This leads, from (10.6.56), to

$$\int \frac{d\alpha_1^* d\alpha_1}{2\pi} \cdots \frac{d\alpha_v^* d\alpha_v}{2\pi} \exp - (\alpha_i^* A_{ij} \alpha_j + \alpha_i^* b_i + c_i^* \alpha_i)$$

$$= (\det A) \exp \left(c_i^* A_{ij}^{-1} b_j \right) \qquad (10.6.57)$$

where
$$b_i = A_{ij} \eta_j \qquad (10.6.58)$$
$$c_i^* = \eta_j^* A_{ji} = \left(A_{ij}^\top \right) \eta_j^*. \qquad (10.6.59)$$

For A a Hermitian matrix,
$$c_i = b_i. \qquad (10.6.60)$$

10.6.3 Fermi Excitations

We represent the Fermi operators a_F, $a_F^\dagger$, defined in (6.4.1)–(6.4.3), as follows
$$a_F \longrightarrow \frac{d}{d\alpha^*} \qquad (10.6.61)$$
$$a_F^\dagger = \alpha^* \qquad (10.6.62)$$

where α^* is a Grassmann variable. The representation (10.6.61), (10.6.62) is obviously consistent with the anti-commutation relations (6.4.2), (6.4.3).

The vacuum state and the single particle state may be then defined by
$$\langle \alpha^* | 0 \rangle = \psi_0(\alpha^*) = 1 \qquad (10.6.63)$$
$$\langle \alpha^* | 1 \rangle = \psi_1(\alpha^*) = \alpha^* \qquad (10.6.64)$$

where we recall that $(\alpha^*)^2 = 0$. We may rewrite (10.6.63), (10.6.64) in a unified notation
$$\langle \alpha^* | n \rangle = \psi_n(\alpha^*) = (\alpha^*)^n, \quad n = 0, 1. \qquad (10.6.65)$$

From (10.6.56), (10.6.50), we have
$$\int \frac{d\alpha^* d\alpha}{2\pi} \exp(-\alpha^* \alpha) (\psi_n(\alpha^*))^* (\psi_m(\alpha^*)) = \delta_{nm}, \quad n = 0, 1. \qquad (10.6.66)$$

The identity operator is defined by
$$\mathbf{1} = \int \frac{d\alpha^* d\alpha}{2\pi} \exp(-\alpha^* \alpha) | \alpha^* \rangle \langle \alpha^* | \qquad (10.6.67)$$

and
$$\langle n | m \rangle = \langle n | \mathbf{1} | m \rangle = \delta_{nm} \qquad (10.6.68)$$

according to (10.6.66).

An operator F, which is a function of $a^\dagger$, a, has the general structure

10.6 Grassmann Variables: Fermi Excitations

$$F = \sum_{n,m=0,1} C_{nm} \left(a^\dagger\right)^n (a)^m \tag{10.6.69}$$

where the C_{nm} are c-numbers. It is easy to see that for $N, M = 0, 1$,

$$\langle N |F| M \rangle = {\sum}' C_{nm}\, \delta\left(N - n, M - m\right) \tag{10.6.70}$$

where the summation is over all integers n, m, $0 \leqslant n \leqslant N$, $0 \leqslant m \leqslant M$ such that $N - n = M - m$.

In the $|N\rangle$, $N = 0, 1$, basis

$$F = \sum_{N,M=0,1} |N\rangle \langle N |F| M \rangle \langle M| \tag{10.6.71}$$

with $\langle N |F| M \rangle$ given in (10.6.70). In the representation (10.6.61), (10.6.62),

$$\langle \alpha^* |F| \alpha'^* \rangle = \sum_{N,M=0,1} (\alpha^*)^N \langle N |F| M \rangle (\alpha')^M. \tag{10.6.72}$$

The latter is explicitly worked out, from (10.6.70), to be

$$\langle \alpha^* |F| \alpha'^* \rangle = C_{00} + C_{01}\alpha' + C_{10}\alpha^* + C_{11}\alpha^*\alpha' + C_{00}\alpha^*\alpha'$$

$$= (C_{00} + C_{01}\alpha' + C_{10}\alpha^* + C_{11}\alpha^*\alpha') \exp(\alpha^*\alpha'). \tag{10.6.73}$$

Comparing (10.6.73) with (10.6.69) we have the rule

$$\langle \alpha^* |F| \alpha'^* \rangle = \exp(\alpha^*\alpha')\, F \bigg|_{\substack{a^\dagger \to \alpha^* \\ a \to \alpha'}} \equiv F(\alpha^*, \alpha'). \tag{10.6.74}$$

From (10.6.67), (10.6.72), (10.6.74), we have the following convenient representation for the operator F obtained by writing $F = \mathbf{1}F\mathbf{1}$,

$$F = \int \frac{\mathrm{d}\alpha^* \mathrm{d}\alpha}{2\pi} \frac{\mathrm{d}\beta^* \mathrm{d}\beta}{2\pi} \left(\exp - (\alpha^*\alpha + \beta^*\beta)\right) |\alpha^*\rangle F(\alpha^*, \beta) \langle \beta^*|. \tag{10.6.75}$$

For F the identity, (10.6.74) gives

$$\langle \alpha^* | \alpha'^* \rangle = F(\alpha^*, \alpha') = \exp(\alpha^*\, \alpha') \tag{10.6.76}$$

coinciding with the expression obtained from the use of the completeness relation

$$\langle \alpha^* |\alpha'^*\rangle = \sum_{n=0,1} \langle \alpha^* | n \rangle \langle n | \alpha'^* \rangle$$

$$= \sum_{n=0,1} (\alpha^*)^n (\alpha')^n = 1 + \alpha^*\alpha' = \exp\alpha^*\alpha' \tag{10.6.77}$$

as expected.

For two operators F_2, F_1 which are functions of $a^\dagger$, a having the general structure in (10.6.69), we have directly from (10.6.75),

$$F_2 F_1 = \int \frac{\mathrm{d}\alpha^* \mathrm{d}\alpha}{2\pi} \frac{\mathrm{d}\beta^* \mathrm{d}\beta}{2\pi} (\exp - (\alpha^* \alpha + \beta^* \beta)) |\alpha^*\rangle F_{21}(\alpha^*, \beta) \langle \beta^*| \tag{10.6.78}$$

where, using (10.6.75), (10.6.76)

$$F_{21}(\alpha^*, \beta) = \int \frac{\mathrm{d}\alpha'^* \mathrm{d}\alpha'}{2\pi} \frac{\mathrm{d}\beta'^* \mathrm{d}\beta'}{2\pi} \left(\exp - (\alpha'^* \alpha' + \beta'^* \beta' - \beta'^* \alpha') \right)$$
$$\times F_2(\alpha^*, \beta') F_1(\alpha'^*, \beta) \tag{10.6.79}$$

or upon carrying out the elementary integrations over β'^* and α', we obtain in a convenient notation

$$F_{21}(\alpha_2^*, \beta_0) = \int \frac{\mathrm{d}\alpha_1^* \mathrm{d}\beta_1}{2\pi} (\exp - \alpha_1^* \beta_1) F_2(\alpha_2^*, \beta_1) F_1(\alpha_1^*, \beta_0). \tag{10.6.80}$$

For a Hamiltonian $H(a^\dagger, a; t)$, the time-evolution operator $U(t, t')$, $t' \to t$, is then from (10.6.74), (10.6.75), by repeated applications of (10.6.80), given by

$$U(t, t') = \int \frac{\mathrm{d}\alpha^* \mathrm{d}\alpha}{2\pi} \frac{\mathrm{d}\beta^* \mathrm{d}\beta}{2\pi} \left(\exp - (\alpha^* \alpha + \beta^* \beta) \right) |\alpha^*\rangle U(\alpha^*, \beta; t, t') \langle \beta^*| \tag{10.6.81}$$

where

$$U(\alpha^*, \beta; t, t') = \lim_{N \to \infty} \int \left(\prod_{k=1}^{N-1} \frac{\mathrm{d}\alpha_k^* \mathrm{d}\beta_k}{2\pi} \right)$$
$$\times \exp \left(\sum_{k=1}^{N} \left(\alpha_k^* \beta_{k-1} - \frac{i}{\hbar} \varepsilon H(\alpha_k^*, \beta_{k-1}; \tilde{t}_k) \right) - \sum_{k=1}^{N-1} \alpha_k^* \beta_k \right) \tag{10.6.82}$$

and $\alpha_N^* = \alpha^*$, $\beta_0 = \beta$, $\tilde{t}_k = t' + (k - 1/2)\varepsilon$, $k = 1, \ldots, N$.

Of particular interest is the vacuum expectation value $\langle 0 | U(t, t') | 0 \rangle$. Since $\langle 0 | \alpha^* \rangle = 1 = \langle \beta^* | 0 \rangle$, we may immediately carry out the α, β^* integrations. This in turn leads to the explicit evaluation of the α^*, β integrals in (10.6.81) and we may effectively set $\alpha^* = 0$, $\beta_0 = 0$ in (10.6.82) recalling that we are integrating over Grassmann variables. That is,

$$\langle 0 | U(t, t') | 0 \rangle = \lim_{N \to \infty} \int \left(\prod_{k=1}^{N-1} \frac{\mathrm{d}\alpha_k^* \mathrm{d}\beta_k}{2\pi} \right)$$

$$\times \exp \sum_{k=1}^{N} \left(\frac{1}{2} \left(\alpha_k^* - \alpha_{k-1}^* \right) \beta_{k-1} - \frac{1}{2} \alpha_k^* \left(\beta_k - \beta_{k-1} \right) - \frac{i}{\hbar} \varepsilon H \left(\alpha_k^*, \beta_k; \tilde{t}_k \right) \right) \tag{10.6.83}$$

where $\alpha_N^* = 0$, $\beta_0 = 0$.

Upon taking the limit $N \to \infty$ in (10.6.83), we obtain

$$\langle 0 | U(t, t') | 0 \rangle = \int_{\beta(t')=0}^{\alpha^*(t)=0} \mathscr{D}(\alpha^*(\cdot), \beta(\cdot)) \exp \frac{i}{\hbar} \int_{t'}^{t} d\tau \left(\frac{\hbar}{2i} \left[\dot{\alpha}^*(\tau) \beta(\tau) \right. \right.$$

$$\left. \left. - \alpha^*(\tau) \dot{\beta}(\tau) \right] - H\left(\alpha^*(\tau), \beta(\tau); \tau \right) \right). \tag{10.6.84}$$

As an application, we consider the Hamiltonian

$$H\left(a_F^\dagger, a_F; t \right) = \hbar \omega(t) a_F^\dagger a_F - \eta^*(t) a_F - a_F^\dagger \eta(t) \tag{10.6.85}$$

(see also (6.4.11)), where $\omega(t)$ is, in general, a time-dependent c-function, and $\eta^*(t)$, $\eta(t)$ are time-dependent Grassmann variables, where for simplicity we have not included a zero point energy.

The sum in the exponential in (10.6.83) or (10.6.82), with $\alpha_N^* = 0$, $\beta_0 = 0$, may be then written as

$$\sum_{k=2}^{N-1} \gamma_k \alpha_k^* \beta_{k-1} - \sum_{k=1}^{N-1} \alpha_k^* \beta_k + \frac{i}{\hbar} \varepsilon \sum_{k=1}^{N-1} \left(\eta^*(\tilde{t}_{k+1}) \beta_k + \alpha_k^* \eta(\tilde{t}_k) \right) \tag{10.6.86}$$

where

$$\gamma_k = 1 - i\varepsilon \omega(\tilde{t}_k). \tag{10.6.87}$$

The expression in (10.6.86) is conveniently rewritten as

$$- \alpha_k^* A_{kj} \beta_j - \alpha_k^* b_k - c_k^* \beta_k \tag{10.6.88}$$

where $j, k = 1, \ldots, N-1$,

$$A = [A_{ij}] = \begin{bmatrix} 1 & 0 & 0 & & & & 0 \\ -\gamma_2 & 1 & 0 & \cdot & & & \\ 0 & -\gamma_3 & 1 & & \cdot & & \\ & & & \cdot & & & \\ & & & & \cdot & & \\ & & & & & 1 & 0 \\ 0 & \cdot & \cdot \cdot & & 0 & -\gamma_{N-1} & 1 \end{bmatrix} \tag{10.6.89}$$

$\det A = 1$, and

$$b_k = -\frac{i}{\hbar}\varepsilon\eta\left(\tilde{t}_k\right), \quad c_k^* = -\frac{i}{\hbar}\varepsilon\eta^*\left(\tilde{t}_{k+1}\right). \tag{10.6.90}$$

The inverse of A is given by

$$A^{-1} = \begin{bmatrix} 1 & 0 & & 0 & 0 \\ \gamma_2 & 1 & & \ddots & \\ \vdots & \gamma_3 & & \ddots & \\ & \vdots & & 1 & 0 \\ (\gamma_2\gamma_3\cdots\gamma_{N-1}) & (\gamma_3\gamma_4\cdots\gamma_{N-1}) & \cdots & (\gamma_{N-1}) & 1 \end{bmatrix}. \tag{10.6.91}$$

From (10.6.57), we then have for $\langle 0|U(t,t')|0\rangle$ the expression

$$\lim_{N\to\infty}\exp\left(-\frac{\varepsilon^2}{\hbar^2}\sum_{k=1}^{N-1}\eta^*\left(\tilde{t}_{k+1}\right)\sum_{j=1}^{k}\gamma_{j+1}\cdots\gamma_k\eta\left(\tilde{t}_j\right)\right) \tag{10.6.92}$$

where, needless to say, $\gamma_1 \equiv 1$, and for $j = k-1$, replace $\gamma_k\gamma_k$ by γ_k, and for $j = k$, replace $\gamma_{k+1}\gamma_k$ by one.

For $\varepsilon \simeq 0$, we may rewrite

$$\gamma_{j+1}\cdots\gamma_k \simeq \exp-i\varepsilon\left(\omega\left(\tilde{t}_{j+1}\right)+\cdots+\omega\left(\tilde{t}_k\right)\right)$$
$$= \exp-i\varepsilon\left(\omega\left(\tilde{t}_2\right)+\cdots+\omega\left(\tilde{t}_k\right)\right)\exp i\varepsilon\left(\omega\left(\tilde{t}_2\right)+\cdots+\omega\left(\tilde{t}_j\right)\right). \tag{10.6.93}$$

Hence upon taking the limit $N \to \infty$ in (10.6.92) we obtain

$$\langle 0|U(t,t')|0\rangle = \exp\left(-\frac{1}{\hbar^2}\int_{t'}^{t}d\tau\int_{t'}^{t}d\tau'\eta^*(\tau)e^{-i\Omega(\tau,\tau')}\Theta(\tau-\tau')\eta(\tau')\right) \tag{10.6.94}$$

where

$$\Omega(\tau,\tau') = \int_{\tau'}^{\tau}d\tau''\omega(\tau'') \tag{10.6.95}$$

In (10.6.94), we assume that $\eta^*(\tau)$, $\eta(\tau')$ vanish outside the interval (t,t').

For $\omega(t) = \omega$, we obtain the expression in (6.4.18) in the present notation.

A very similar procedure as above (see Problem 10.16) shows, quite generally, with boundary conditions $\alpha^*(t) = \alpha^*$, $\beta(t') = \beta$, for the Hamiltonian in (10.6.85), that $U(\alpha^*,\beta;t,t')$ is given by

$$U(\alpha^*,\beta;t,t') = \exp\left(\alpha^*\beta e^{-i\Omega(t,t')}\right)\exp\left(\frac{i}{\hbar}\alpha^*\Gamma_1\right)\exp\left(\frac{i}{\hbar}\Gamma_2\beta\right)\exp\left(-\frac{1}{\hbar^2}\gamma\right) \tag{10.6.96}$$

where

$$\Gamma_1 = \int_{t'}^{t} d\tau\, \eta(\tau) e^{-i\Omega(t,\tau)} \qquad (10.6.97)$$

$$\Gamma_2 = \int_{t'}^{t} d\tau\, \eta^*(\tau) e^{-i\Omega(\tau,t')} \qquad (10.6.98)$$

and $\Omega(\tau, \tau')$ is defined in (10.6.95).

Upon using $\langle n | \alpha^* \rangle = (\alpha)^n$, $\langle \beta^* | m \rangle = (\beta^*)^m$ with $n, m = 0, 1$, in computing all the matrix elements $\langle n | U(t, t') | m \rangle$ of $U(t, t')$ in (10.6.81), and carrying out the elementary integrations over the Grassmann variables α^*, α, β^*, β we obtain from (10.6.96) (see Problem 10.17)

$$\langle n | U(t,t') | m \rangle = \left[\delta_{n0}\delta_{m0} + \delta_{n0}\delta_{m1}\frac{i}{\hbar}\Gamma_2 + \delta_{n1}\delta_{m0}\frac{i}{\hbar}\Gamma_1 \right] e^{-\gamma/\hbar^2}$$

$$+ \delta_{n1}\delta_{m1}\left[e^{-i\Omega(t,t')} + \frac{\Gamma_2\Gamma_1}{\hbar^2} \right] e^{-\gamma/\hbar^2} \qquad (10.6.99)$$

The reader is urged to show that for $\omega(t) = \omega$, (10.6.99) immediately reduces to the corresponding expressions in (6.4.20)–(6.4.22).

Problems

10.1. Follow the procedure given in §10.2 in deriving the path integral for the time-independent potential $-Fx$ to derive the corresponding expression for the time-dependent one $-F(t)x$, and compare your result with the one given in §9.4.

10.2. Verify that the integral equation in (10.3.17) leads upon taking the limit $\varepsilon \to 0$ to the Schrödinger equation in (10.3.18).

10.3. Carry out the multiple Gaussian integrals over the momenta in (10.3.27), for velocity independent potentials, to obtain the path integral in (10.3.28) in coordinate space.

10.4. Investigate the nature of the path integral of a free particle moving along a circle of fixed radius a as a non-constrained problem, i.e., by writing the Laplacian ∇^2 directly as $(1/a^2)\, \partial^2/\partial\phi^2$.

(i) Carry out a Fourier expansion of $\delta(\phi - \phi')$ and show that

$$\langle \phi t | \phi' t' \rangle = \lim_{N \to \infty} \left(\frac{ma^2}{2\pi i\hbar T/N} \right)^{N/2} \sum_{k_0=-\infty}^{\infty} \left(\prod_{j=1}^{N-1} \int_0^{2\pi} d\phi_j \sum_{k_j=-\infty}^{\infty} \right)$$

$$\times \exp\left(i\frac{ma^2}{2\hbar T/N} \sum_{j=0}^{N-1} (\phi_{j+1} - \phi_j + 2\pi k_j)^2 \right).$$

where $\phi_N = \phi$, $\phi_0 = \phi'$.

(ii) By using the Poisson sum formula, and by appropriate change of variables in part (i), show that

$$\langle \phi t | \phi' t' \rangle = a \sum_{k=-\infty}^{\infty} \langle a(\phi + 2\pi k), t | a\phi', t' \rangle_0$$

$$= a \sum_{k=-\infty}^{\infty} \int_{a\phi'}^{a(\phi+2\pi k)} \mathscr{D}(a\phi(\cdot)) \exp\left(\frac{im}{2\hbar} \int_{t'}^{t} d\tau \left(a\dot\phi(\tau) \right)^2 \right)$$

where $\langle x, t | x, t' \rangle_0$ is the free Green function. The expression above is an infinite sum of one dimensional amplitudes on the *real line* each characterized by a winding number k. Interpret this result.

10.5. Consider the following formally Hermitian operator

$$H = \frac{p^2}{2m} - \frac{i\hbar}{a^2 m} xp + \frac{1}{2a^2 m} x^2 p^2 + V(x) = H_\mathrm{R}$$

where $a^2 > 0$ is a constant. We note that the operator p stands on the right-hand side which explains the notation H_R.

(i) Show by using $[x, p] = i\hbar$, H may be rewritten with p standing on the left-hand side as

$$H = \frac{p^2}{2m} + \frac{i\hbar}{a^2 m} px + \frac{1}{2a^2 m} p^2 x^2 + V(x) = H_\mathrm{L}.$$

(ii) Carry out the path integral for this system with a velocity dependent potential as described in §10.3, to conclude that *classically* with x and p as c-numbers now

$$H_c = \frac{1}{2}(H_\mathrm{R} + H_\mathrm{L})_c = \left(1 + \frac{x^2}{a^2}\right) \frac{p^2}{2m} + V(x).$$

(iii) Integrate explicitly on the momenta of the corresponding functions to show that for the amplitude $\langle x, t | x', t' \rangle$, the resulting integrand for the corresponding one in the exponential in (10.2.18) is not simply given by the classical Lagrangian. Here you will encounter a $\delta(0)$ singularity well known in some field theories.

10.6. Verify explicitly the properties of the Poisson brackets in (10.4.14)–(10.4.17).

10.7. Investigate the nature of the transformation of the amplitude $\langle \mathbf{q}t | \mathbf{q}'t' \rangle$, with no constraints present, under a finite canonical transformation via a generator W.

10.8. Show that for a theory with first and second class constraints Φ_α, Ψ_α, respectively, that from

$$\{\Phi_\alpha, \Phi_\gamma\} = a^\sigma_{\alpha\gamma}\Phi_\sigma + b^\sigma_{\alpha\gamma}\Psi_\sigma$$

one may infer that $b^\sigma_{\alpha\gamma}| = 0$.
(Hint: Use the Jacobi identity and (10.4.32).)

10.9. For a function $G(\mathbf{q}(t), \mathbf{p}(t))$ write down explicitly the expression for $\dot{G} = \{G, H_T\}_{\text{P.B.}}$, where H_T is given in (10.4.35) and integrate the latter over t in the neighborhood of $t = 0$ for a given initial condition G at $t = 0$, to show that G depends, in general, on the undetermined Lagrange multipliers η_α at $t = 0$.

10.10. For the transformation $q_1, q_2, q_3, p_1, p_2, p_3, H \to q_1^*, Q, q_3^*, p_1^*, P, p_3^*, \bar{H}$, in (10.4.51), with α a constant,
(i) show that

$$\dot{Q} = \partial\bar{H}/\partial P, \quad \dot{P} = -\partial\bar{H}/\partial Q,$$

$$\dot{q}_i^* = \partial\bar{H}/\partial p_i^*, \quad \dot{p}_i^* = -\partial\bar{H}/\partial q_i^*, \quad i = 1, 3.$$

(ii) Show in reference to (10.4.45), (10.4.48),

$$\{\phi, \chi\}_{\text{P.B.}} = \frac{\partial\phi}{\partial Q}.$$

10.11. For a free particle moving in a circle with primary constraint $\phi_1| = (x^2 + y^2 - a^2)| = 0$, study the properties of the constraints arising from this apparently simple system.

10.12. Investigate the nature of path integrals in the presence of second class constraints and their properties under variations $\delta\chi_\alpha$ of the subsidiary constraints. [Ref.: Senjanovic (1976).]

10.13. Establish (10.5.6) by working, for example, in polar coordinates.

10.14. By a procedure similar to the one used for Fermi excitations in §10.6, carry out explicitly the multiple Gaussian integrals in (10.5.23) to obtain (10.5.27), for the Hamiltonian in (10.5.26).

10.15. Derive some properties of $\delta(\eta_k - \alpha_k)$ for Grassmann variables as follow from (10.6.20). In particular, what is $\delta(0)$ for such variables?

10.16. Use a procedure similar to the one given through (10.6.86)–(10.6.95) to derive (10.6.96) with the boundary conditions $\alpha^*(t) = \alpha^*$, $\beta(t') = \beta$.

10.17. Use the expression (10.6.96) in (10.6.82) to explicitly carry out the integrals over the Grassmann variables α^*, α, β^*, β to obtain all of the matrix elements $\langle n|U(t, t')|m\rangle$, as given in (10.6.99), with $n, m = 0, 1$. Verify that (10.6.99) reduces to the corresponding expressions in (6.4.20)–(6.4.22) for the case $\omega(t) = \omega$.

10.18. Verify that (10.5.27), (10.6.96) may be obtained directly by substituting the classical solutions $(\beta_c^*(\tau), \beta_c(\tau))$, $(\alpha_c^*(\tau), \beta_c(\tau))$ in (10.5.24), (10.6.82) for the Hamiltonians in (10.5.26), (10.6.85), respectively.

11

The Quantum Dynamical Principle

The quantum dynamical principle (QDP) provides a formalism for quantum physics which is powerful, easy to apply and is most elegant. In the form presented here, it gives rise to an expression for the variation $\delta\langle at|bt'\rangle$ of a transformation function $\langle at|bt'\rangle$ from a B-description at time t' to an A-description at time t, as arising from any changes made in the parameters of a Hamiltonian such as of the underlying *masses, coupling constants (charges,...), prescribed frequencies, external sources*, as introduced in §6.2, §6.4 and also encountered in §10.5, §10.6, and so on. Typical transformation functions considered are $\langle \mathbf{q}t|\mathbf{q}'t'\rangle$ and $\langle \mathbf{q}t|\mathbf{p}t'\rangle$ written in the $(\mathbf{q},\mathbf{p})$ language. The subsequent analysis of the expressions for $\delta\langle at|bt'\rangle$ provides then endless applications to all aspects of quantum physical problems.

One advantage of the QDP approach over the path integral one, is that the former is based on carrying out functional *differentiations* in the theory, while the latter involves in carrying out an (infinitely uncountable) multiple functional *integrals* and it is relatively easier to functionally differentiate than to deal with continual functional integrals. The formal equivalence of both formalisms is, however, established in §11.4.

The QDP is entirely due to Schwinger,[1] and remains to be an extremely powerful tool in the development of quantum physics and quantum field theory.[2]

The purpose of this chapter is to show how to do quantum physics by using variations $\delta\langle at|bt'\rangle$ of transformation functions in a systematic way. In §11.1, we derive the QDP and obtain explicit expressions for transformation functions in §11.2. So-called trace functionals are introduced in §11.3 which, in particular, provide useful information on the spectra of Hamiltonians. §11.4 deals with the connection of the path integral formalism to

---

[1] Schwinger (1951a, 1953, 1960c,b, 1962); Lam (1965); Manoukian (1985).

[2] For the reader who has some familiarity with field theory see, Manoukian (1986a, 1987b); Manoukian and Siranan (2005), on how the QDP solves the quantization problem of the present gauge theories of the fundamental interactions of physics.

the QDP one. Bose/Fermi excitations within the light of this formalism are treated in §11.5. Finally, §11.6 develops an extension of the QDP, referred to as the closed-time path formalism of Schwinger,[3] to obtain directly expectation values of observables at any time in an initially prepared state without first determining the underlying amplitudes. We will encounter applications of the QDP again later, notably, in scattering theory in Chapter 15.

It took years before the path integral formalism was widely used after it was conceived.[4] Old habits die hard, however, we expect that the QDP formalism will be also widely used in the near future not only as a practical way for computations but also as a technically rigorous method for doing quantum physics. We hope that this chapter will have some contribution to this end.

11.1 The Quantum Dynamical Principle

Consider the general Hamiltonian given by

$$H(t, \lambda) = H_1(t) + H_2(t, \lambda) \tag{11.1.1}$$

where $H_1(t)$, $H_2(t, \lambda)$ may be time-dependent but $H_2(t, \lambda)$ may, in addition, depend on some parameters denoted by λ. Here λ stands for any parameters such as *masses, coupling constants (charges,...), prescribed frequencies, external sources* (such as, e.g., in §6.2) and so on. The time dependence in $H(t, \lambda)$ is assumed to come from *a priori* given time-dependent potentials and/or external sources (see, e.g., §6.2).

The time evolution operator associated with the Hamiltonian $H(t, \lambda)$ will be denoted by $U(t, \lambda)$. From (A-2.5.2), (A-2.5.11), in the Appendix to §2.5, we have

$$i\hbar \frac{d}{dt} U(t, \lambda) = H(t, \lambda) U(t, \lambda). \tag{11.1.2}$$

For the theory given in a specific description, say, the A-description (see §1.1, (9.1)–(9.4), (10.2.3)–(10.2.5)),

$$i\hbar \frac{d}{dt} \langle at | = \langle at | H(t, \lambda). \tag{11.1.3}$$

One may also work with the Hamiltonian $H_1(t)$, introduce the corresponding time evolution operator $U_1(t)$, which are independent of λ, satisfying

$$i\hbar \frac{d}{dt} U_1(t) = H_1(t) U_1(t) \tag{11.1.4}$$

---

[3] Schwinger (1961a).

[4] This is not to mention of the much more dramatic situation involved with the Lagrange and the Hamilton formalisms for the time elapsed before they were broadly used in classical mechanics after they were formulated in turn.

11.1 The Quantum Dynamical Principle

and set,
$$i\hbar \frac{d}{dt} {}_1\langle at| = {}_1\langle at| H_1(t). \tag{11.1.5}$$

The theory, however, is described by the Hamiltonian $H(t, \lambda)$ in (11.1.1) and the time evolution operator $U(t, \lambda)$.

The physical states $\langle at|$ are clearly related to the states ${}_1\langle at|$ by
$$\langle at| = {}_1\langle at| U_1^\dagger(t) U(t, \lambda). \tag{11.1.6}$$

This suggests to introduce the unitary operator
$$V(t, \lambda) = U_1^\dagger(t) U(t, \lambda) \tag{11.1.7}$$

to rewrite (11.1.6) as
$$\langle at| = {}_1\langle at| V(t, \lambda). \tag{11.1.8}$$

We note that
$$i\hbar \frac{d}{dt} V(t, \lambda) = U_1^\dagger(t) H_2(t, \lambda) U(t, \lambda) \tag{11.1.9}$$

where we have used the definition in (11.1.7).

We are interested in studying the variations of transformation functions (cf. §1.2, §1.4) $\langle at|bt'\rangle$, with respect to the parameters λ, from, in general, one description to another.

To the above end, we use the identity

$$i\hbar \frac{d}{d\tau} \left[V(t, \lambda) V^\dagger(\tau, \lambda) V(\tau, \lambda') V^\dagger(t', \lambda') \right]$$

$$= V(t, \lambda) \left[U^\dagger(\tau, \lambda) \left(H_2(\tau, \lambda') - H_2(\tau, \lambda) \right) U(\tau, \lambda') \right] V^\dagger(t', \lambda')$$

$$= V(t, \lambda) \left[U^\dagger(\tau, \lambda) \left(H(\tau, \lambda') - H(\tau, \lambda) \right) U(\tau, \lambda') \right] V^\dagger(t', \lambda') \tag{11.1.10}$$

where we have also used (11.1.7), (11.1.9), the unitarity of $U_1(\tau)$ and finally (11.1.1). Here $\lambda' \neq \lambda$, in general.

We may integrate (11.1.10) over τ from t' to t, and use the unitarity of $V(t, \lambda)$
$$V(t, \lambda) V^\dagger(t, \lambda) = \mathbf{1}, \qquad V^\dagger(t, \lambda) V(t, \lambda) = \mathbf{1}, \tag{11.1.11}$$

evaluated at equal times and for identical parameters, to obtain

$$\left[V(t, \lambda') V^\dagger(t', \lambda') - V(t, \lambda) V^\dagger(t', \lambda) \right]$$
$$= -\frac{i}{\hbar} V(t, \lambda) \left[\int_{t'}^{t} d\tau \, U^\dagger(\tau, \lambda) \left(H(\tau, \lambda') - H(\tau, \lambda) \right) U(\tau, \lambda') \right] V^\dagger(t', \lambda'). \tag{11.1.12}$$

Now we set $\lambda' = \lambda + \delta\lambda$, to get the variational form of (11.1.12) given by

$$\delta\Big[V(t,\lambda)\,V^\dagger(t',\lambda)\Big]$$
$$= -\frac{\mathrm{i}}{\hbar} V(t,\lambda) \left[\int_{t'}^{t}\mathrm{d}\tau\, U^\dagger(\tau,\lambda)\,\delta H(\tau,\lambda)\, U(\tau,\lambda)\right] V^\dagger(t',\lambda). \quad (11.1.13)$$

Using the (q,p) language, as in Chapter 10, suppressing the indices specifying the various degrees of freedom, the Hamiltonian $H(\tau,\lambda)$ may be written as

$$H(\tau,\lambda) = H(q,p,\tau,\lambda) \quad (11.1.14)$$

$\delta H(\tau,\lambda)$ in (11.1.13) refers to the change of $H(\tau,\lambda)$, with respect to λ, with q, p (and τ) kept *fixed*.

We define a Heisenberg representation of $H(\tau,\lambda)$, at time τ, by

$$\mathbb{H}(\tau,\lambda) = U^\dagger(\tau,\lambda)\, H(q,p,\tau;\lambda)\, U(\tau,\lambda) \equiv H\big(q(\tau),p(\tau),\tau;\lambda\big). \quad (11.1.15)$$

This allows us to rewrite (11.1.13) as

$$\delta\Big[V(t,\lambda)\,V^\dagger(t',\lambda)\Big]$$
$$= -\frac{\mathrm{i}}{\hbar} V(t,\lambda) \left[\int_{t'}^{t}\mathrm{d}\tau\, \delta H\big(q(\tau),p(\tau),\tau;\lambda\big)\right] V^\dagger(t',\lambda). \quad (11.1.16)$$

provided the variation δ of $H\big(q(\tau),p(\tau),\tau;\lambda\big)$, with respect to λ, is carried out with $q(\tau)$, $p(\tau)$ kept *fixed*, in conformity with (11.1.13), since $q(\tau)$, $p(\tau)$, given by

$$q(\tau) = U^\dagger(\tau,\lambda)\, q\, U(\tau,\lambda), \qquad p(\tau) = U^\dagger(\tau,\lambda)\, p\, U(\tau,\lambda) \quad (11.1.17)$$

will, in general, depend on λ. The $q(\tau)$, $p(\tau)$ are Heisenberg representations of q, p.

Now we take the matrix elements of (11.1.16) with respect to ${}_1\langle at|$, $|bt'\rangle_1$, and use (11.1.8) to obtain

$$\delta\langle at|bt'\rangle = -\frac{\mathrm{i}}{\hbar}\int_{t'}^{t}\mathrm{d}\tau\,\big\langle at\big|\delta H\big(q(\tau),p(\tau),\tau;\lambda\big)\big|bt'\big\rangle. \quad (11.1.18)$$

This is the celebrated *Schwinger's dynamical (action) principle* or the *quantum dynamical principle*. It is expressed in terms of the physical states $|at\rangle$, $|bt'\rangle$ which depend on λ. Needless to say, q and p in (11.1.18) may carry indices corresponding to various degrees of freedom.

We recall that in (11.1.18), the variation of H, with respect to λ, is taken with $q(\tau)$, $p(\tau)$ kept fixed. Also a and b are kept fixed. After all operations associated with the variations and subsequent integrations with respect to

11.1 The Quantum Dynamical Principle

the parameters λ in (11.1.18) are carried out, these parameters may be set to have *a priori* chosen fixed values corresponding to the physical situation of the problem, thus obtaining the expression for the transformation functions $\langle at|bt'\rangle$ one is seeking.

Of particular interests, are the transformation functions $\langle qt|q't'\rangle$, $\langle qt|pt'\rangle$, $\langle pt|p't'\rangle$. For example, (11.1.18) gives

$$\delta\langle qt|q't'\rangle = -\frac{i}{\hbar}\int_{t'}^{t}d\tau\,\langle qt|\delta H(q(\tau),p(\tau),\tau;\lambda)|q't'\rangle \qquad (11.1.19)$$

and

$$\delta\langle qt|pt'\rangle = -\frac{i}{\hbar}\int_{t'}^{t}d\tau\,\langle qt|\delta H(q(\tau),p(\tau),\tau;\lambda)|pt'\rangle. \qquad (11.1.20)$$

We will make much use of these two equations in this chapter.

As an immediate application of (11.1.19), consider the Hamiltonian

$$H(q,p,\tau;F(\tau),S(\tau)) = H(q,p,\tau) - qF(\tau) + pS(\tau) \qquad (11.1.21)$$

where $F(\tau)$, $S(\tau)$ are *numerical* (i.e., c-number) functions of τ referred to as *external sources* (see also §6.2), and $H(q,p,\tau)$ is independent of them.[5] The minus sign multiplying $qF(\tau)$ is chosen for convenience.

Using the definition of the functional derivative

$$\frac{\delta}{\delta F(t)}F(\tau) = \delta(t-\tau) \qquad (11.1.22)$$

$$\frac{\delta}{\delta S(t)}S(\tau) = \delta(t-\tau) \qquad (11.1.23)$$

we obtain from (11.1.21)

$$\frac{\delta}{\delta F(t)}H(q,p,\tau;F(\tau),S(\tau)) = -q\,\delta(t-\tau) \qquad (11.1.24)$$

$$\frac{\delta}{\delta S(t)}H(q,p,\tau;F(\tau),S(\tau)) = p\,\delta(t-\tau). \qquad (11.1.25)$$

With λ replaced in turn by $F(\tau)$ and $S(\tau)$, (11.1.19) gives the important results that

$$(-i\hbar)\frac{\delta}{\delta F(\tau)}\langle qt|q't'\rangle = \langle qt|q(\tau)|q't'\rangle \qquad (11.1.26)$$

$$(i\hbar)\frac{\delta}{\delta S(\tau)}\langle qt|q't'\rangle = \langle qt|p(\tau)|q't'\rangle \qquad (11.1.27)$$

---

[5] In the Heisenberg representation, $H(q(\tau),p(\tau),\tau)$ will depend on $F(\tau)$, $S(\tau)$, and if we perform variations with respect to $F(\tau)$, $S(\tau)$, then according to (11.1.18), (11.1.19), $q(\tau)$, $p(\tau)$ should be kept fixed.

for the matrix elements of the Heisenberg operators $q(\tau)$, $p(\tau)$, for $t' < \tau < t$, where it is understood that $\langle qt|q't'\rangle$ and the matrix elements in (11.1.26), (11.1.27) depend on the external sources F, S (unless they are set equal to zero). For the simplicity of the notation only, we have suppressed this dependence on F, S in (11.1.26), (11.1.27).

For more examples of functional derivatives, consider the functional (i.e., function of function(s))

$$G[F,S] = \int_{t'}^{t} d\tau' \int_{t'}^{t} d\tau'' \, F(\tau') \, A(\tau', \tau'') \, S(\tau'') \tag{11.1.28}$$

where $A(\tau', \tau'')$ is independent of F, S. Then for $t' < \tau_1 < t$, $t' < \tau_2 < t$

$$\frac{\delta}{\delta F(\tau_1)} G[F,S] = \int_{t'}^{t} d\tau'' \, A(\tau_1, \tau'') \, S(\tau'') \tag{11.1.29}$$

$$\frac{\delta}{\delta S(\tau_2)} \frac{\delta}{\delta F(\tau_1)} G[F,S] = A(\tau_1, \tau_2) \tag{11.1.30}$$

Similarly for $t' < \tau < t$

$$\frac{\delta}{\delta F(\tau)} \int_{t'}^{t} d\tau' \int_{t'}^{t} d\tau'' \, F(\tau') \, A(\tau', \tau'') \, F(\tau'')$$

$$= \int_{t'}^{t} d\tau'' \, A(\tau, \tau'') \, F(\tau'') + \int_{t'}^{t} d\tau' \, F(\tau') \, A(\tau', \tau). \tag{11.1.31}$$

More generally, let $G[F]$ be a functional of F. Replace $F(\tau')$, wherever it appears in $G[F]$, by $F(\tau') + \varepsilon \delta(\tau - \tau')$, then the functional derivative of $G[F]$, with respect to $F(\tau)$, is formally defined by

$$\frac{\delta}{\delta F(\tau)} G[F] = \lim_{\varepsilon \to 0} \frac{G[F(\tau') + \varepsilon \delta(\tau - \tau')] - G[F(\tau')]}{\varepsilon} \tag{11.1.32}$$

keeping in the numerator terms of order ε only on the right-hand side of (11.1.32) before taking the limit $\varepsilon \to 0$.

Before considering detailed applications of (11.1.18), we generalize the latter further. To this end, consider an arbitrary function $B(q, p, \tau; \lambda)$ of the variables indicated, and define its Heisenberg representation at times τ by

$$U^\dagger(\tau, \lambda) \, B(q, p, \tau; \lambda) \, U(\tau, \lambda) = B\bigl(q(\tau), p(\tau), \tau; \lambda\bigr) \equiv \mathbb{B}(\tau, \lambda). \tag{11.1.33}$$

We note that

$$V(t, \lambda) \, \mathbb{B}(\tau, \lambda) \, V^\dagger(t', \lambda)$$

$$= V(t, \lambda) \, V^\dagger(\tau, \lambda) \, U_1^\dagger(\tau) \, B(q, p, \tau; \lambda) \, U_1(\tau) \, V(\tau, \lambda) \, V^\dagger(t', \lambda). \tag{11.1.34}$$

11.1 The Quantum Dynamical Principle   655

Accordingly, from (11.1.13), and (11.1.34) we have

$$\delta\Big[V(t,\lambda)\,\mathbb{B}(\tau,\lambda)\,V^\dagger(t',\lambda)\Big]$$

$$= -\frac{i}{\hbar}V(t,\lambda)\int_\tau^t d\tau'\,\delta H(\tau',\lambda)\,\mathbb{B}(\tau,\lambda)\,V^\dagger(t',\lambda)$$

$$+ V(t,\lambda)\,\delta\mathbb{B}(\tau,\lambda)\,V^\dagger(t',\lambda)$$

$$- \frac{i}{\hbar}V(t,\lambda)\int_{t'}^\tau d\tau'\,\mathbb{B}(\tau,\lambda)\,\delta H(\tau',\lambda)\,V^\dagger(t',\lambda) \qquad (11.1.35)$$

where, according to (11.1.34), the variation in $\delta\mathbb{B}(\tau,\lambda) = \delta B\big(q(\tau),p(\tau),\tau;\lambda\big)$, with respect to λ, is out carried by keeping $q(\tau)$ and $p(\tau)$ *fixed*.

We may use the definition of the chronological time ordering in (A-2.5.6), to combine the first and the last terms on the right-hand side of (11.1.35), thus obtaining

$$\delta\Big[V(t,\lambda)\,\mathbb{B}(\tau,\lambda)\,V^\dagger(t',\lambda)\Big]$$

$$= -\frac{i}{\hbar}V(t,\lambda)\int_{t'}^t d\tau'\,\big(\mathbb{B}(\tau,\lambda)\,\delta H(\tau',\lambda)\big)_+\,V^\dagger(t',\lambda)$$

$$+ V(t,\lambda)\,\delta\mathbb{B}(\tau,\lambda)\,V^\dagger(t',\lambda). \qquad (11.1.36)$$

Upon taking the matrix elements of the above equation with respect to ${}_1\langle at|,\,|bt'\rangle_1$ and using (11.1.8), we obtain

$$\delta\langle at\,|\mathbb{B}(\tau,\lambda)|bt'\rangle = -\frac{i}{\hbar}\int_{t'}^t d\tau'\,\langle at|\,\big(\mathbb{B}(\tau,\lambda)\,\delta\mathbb{H}(\tau',\lambda)\big)_+\,|bt'\rangle$$

$$+ \langle at|\delta\mathbb{B}(\tau,\lambda)|bt'\rangle \qquad (11.1.37)$$

with all variations taken by keeping $q(\tau)$, $p(\tau)$ for all τ in the interval from t' to t, as well as a, b, fixed.

In particular, (11.1.37), in reference to (11.1.19), implies that

$$(-i\hbar)\frac{\delta}{\delta F(\tau')}\langle qt\,|q(\tau)|q't'\rangle = (-i\hbar)\frac{\delta}{\delta F(\tau')}(-i\hbar)\frac{\delta}{\delta F(\tau)}\langle qt\,|q't'\rangle$$

$$= \langle qt|\,\big(q(\tau')\,q(\tau)\big)_+\,|q't'\rangle \qquad (11.1.38)$$

and there is no additional term, as in (11.1.37), since $q(\tau)$ in (11.1.38) should be kept fixed, when varying $F(\tau')$.

Repeated applications of (11.1.38), give

$$(-i\hbar)\frac{\delta}{\delta F(\tau_1)}\cdots(-i\hbar)\frac{\delta}{\delta F(\tau_n)}(i\hbar)\frac{\delta}{\delta S(\tau_1')}\cdots(i\hbar)\frac{\delta}{\delta S(\tau_m')}\langle qt|q't'\rangle$$

$$=\langle qt|\left(q(\tau_1)\cdots q(\tau_n)\,p(\tau_1')\cdots p(\tau_m')\right)_+|q't'\rangle \qquad (11.1.39)$$

for $t' \leqslant \tau_1, \ldots, \tau_n, \tau_1', \ldots, \tau_m' \leqslant t$. For the time ordering of several operators see (A-2.5.7). Note that all the functional derivatives operations on the left-hand side of (11.1.39) *commute*.

In particular we note that in considering the matrix elements of the product of non-commuting operators evaluated at coincident times such as in $\langle qt|p(\tau)\,q(\tau)|q't'\rangle$, the latter is obtained by functional differentiations in the following limiting way

$$(i\hbar)\frac{\delta}{\delta S(\tau+\varepsilon)}(-i\hbar)\frac{\delta}{\delta F(\tau)}\langle qt|q't'\rangle \qquad (11.1.40)$$

with $\varepsilon \to +0$. For replacing such matrix elements of the product of non-commuting operators, evaluated at coincident time τ, by functional differentiations with respect to their corresponding external sources, the time parameters in the arguments of the latter should be infinitesimally displaced, as done in (11.1.40), to preserve the initial order of the operators in question.

Thanks to relations such as in (4.5.31)–(4.5.37), the Hamiltonian does not always have to be a polynomial in the dynamical variables. For a typical Hamiltonian as in (10.3.26), the question of the ordering of non-commuting operators (at coincident times) does not arise. A matrix element such as $\langle qt|\left(q(\tau)\right)^2|q't'\rangle$ may be considered to be obtained as the limit of a symmetric average of $\langle qt|\left(q(\tau_1)\,q(\tau_2)+q(\tau_2)\,q(\tau_1)\right)/2|q't'\rangle$.

11.2 Expressions for Transformations Functions

Given a Hamiltonian $H(q,p,t)$ we derive expressions for transformation functions. As before indices corresponding to various degrees of freedom will be suppressed.

1. For a given Hamiltonian $H(q,p,t) = H$, we introduce a new Hamiltonian by multiplying H by a parameter λ and by adding to it source terms as in (11.1.21):

$$H' = \lambda H - qF(\tau) + pS(\tau). \qquad (11.2.1)$$

From (11.1.19),

$$\frac{\partial}{\partial \lambda}\langle qt|q't'\rangle_\lambda = -\frac{i}{\hbar}\int_{t'}^{t}\!\!d\tau\,\langle qt|H\bigl(q(\tau),p(\tau),\tau\bigr)|q't'\rangle_\lambda \qquad (11.2.2)$$

11.2 Expressions for Transformations Functions

where[6] $H(q, p, \tau)$ is independent of λ. Hence one may apply (11.1.39), by keeping $q(\tau)$, $p(\tau)$ fixed for all $t' \leqslant \tau < t$, to rewrite (11.2.2) as

$$\frac{\partial}{\partial \lambda} \langle qt | q't' \rangle_\lambda = -\frac{i}{\hbar} \int_{t'}^{t} d\tau \, H\left(-i\hbar \frac{\delta}{\delta F(\tau)}, i\hbar \frac{\delta}{\delta S(\tau)}, \tau\right) \langle qt | q't' \rangle_\lambda. \quad (11.2.3)$$

We may integrate this equation over λ from $\lambda = 0$ to $\lambda = 1$, to obtain

$$\langle qt | q't' \rangle = \exp\left[-\frac{i}{\hbar} \int_{t'}^{t} d\tau \, H\left(-i\hbar \frac{\delta}{\delta F(\tau)}, i\hbar \frac{\delta}{\delta S(\tau)}, \tau\right)\right] \langle qt | q't' \rangle_0 \quad (11.2.4)$$

where $\langle qt | q't' \rangle = \langle qt | q't' \rangle_{\lambda=1}$, and $\langle qt | q't' \rangle_0$, with $\lambda = 0$, is determined (see (11.2.1)) from the simple Hamiltonian

$$\widehat{H} = -qF(\tau) + pS(\tau). \quad (11.2.5)$$

The Heisenberg equations (2.3.61), (2.3.62) following from this Hamiltonian are

$$\dot{q}(\tau) = S(\tau) \quad (11.2.6)$$

$$\dot{p}(\tau) = F(\tau). \quad (11.2.7)$$

These equations may be integrated to

$$q(\tau) = q(t) - \int_{t'}^{t} d\tau' \, \Theta(\tau' - \tau) \, S(\tau') \quad (11.2.8)$$

$$p(\tau) = p(t') + \int_{t'}^{t} d\tau' \, \Theta(\tau - \tau') \, F(\tau') \quad (11.2.9)$$

and taking the matrix elements between $\langle qt |$ and $|pt' \rangle$ for $\lambda = 0$, we obtain

$$\langle qt | q(\tau) | pt' \rangle_0 = \left[q - \int_{t'}^{t} d\tau' \, \Theta(\tau' - \tau) \, S(\tau')\right] \langle qt | pt' \rangle_0 \quad (11.2.10)$$

$$\langle qt | p(\tau) | pt' \rangle_0 = \left[p + \int_{t'}^{t} d\tau' \, \Theta(\tau - \tau') \, F(\tau')\right] \langle qt | pt' \rangle_0 \quad (11.2.11)$$

where q and p within the square brackets on the right-hand sides of the above two equations are c-numbers, and we have used the relations

$$_0\langle qt | \, q(t) = q \, _0\langle qt | \quad (11.2.12)$$

$$p(t') \, | pt' \rangle_0 = p \, | pt' \rangle_0 \quad (11.2.13)$$

---

[6] $H(q(\tau), p(\tau), \tau)$ involving operators in the Heisenberg representation developing in time with the Hamiltonian H' in (11.2.1), however, depends on λ.

11 The Quantum Dynamical Principle

for $\lambda = 0$ at coincident times. Equations (11.2.10), (11.2.11) may be rewritten as

$$-i\hbar \frac{\delta}{\delta F(\tau)} \langle qt|pt'\rangle_0 = \left[q - \int_{t'}^{t} d\tau'\, \Theta(\tau' - \tau)\, S(\tau')\right] \langle qt|pt'\rangle_0 \qquad (11.2.14)$$

$$i\hbar \frac{\delta}{\delta S(\tau)} \langle qt|pt'\rangle_0 = \left[p + \int_{t'}^{t} d\tau'\, \Theta(\tau - \tau')\, F(\tau')\right] \langle qt|pt'\rangle_0. \qquad (11.2.15)$$

These equations may be integrated to yield

$$\langle qt|pt'\rangle_0 = \exp\left[\frac{i}{\hbar} q \int_{t'}^{t} d\tau\, F(\tau)\right] \exp\left[-\frac{i}{\hbar} p \int_{t'}^{t} d\tau\, S(\tau)\right] \exp\left(\frac{i}{\hbar} qp\right)$$

$$\times \exp\left[-\frac{i}{\hbar} \int_{t'}^{t} d\tau \int_{t'}^{t} d\tau'\, S(\tau)\, \Theta(\tau - \tau')\, F(\tau')\right] \qquad (11.2.16)$$

where the $\exp(iqp/\hbar)$ factor is to satisfy the boundary condition for $F = 0$, $S = 0$, i.e., for $\widehat{H} \to 0$ in (11.2.5).

Finally to obtain the expression for $\langle qt|q't'\rangle_0$, we multiply (11.2.16) by $\langle pt'|q't'\rangle = \exp(-iq'p/\hbar)$ and integrate over p, with measure $dp/2\pi\hbar$, to obtain[7]

$$\langle qt|q't'\rangle_0 = \delta\!\left(q - q' - \int_{t'}^{t} d\tau\, S(\tau)\right) \exp\left[\frac{i}{\hbar} q \int_{t'}^{t} d\tau\, F(\tau)\right]$$

$$\times \exp\left[-\frac{i}{\hbar} \int_{t'}^{t} d\tau \int_{t'}^{t} d\tau'\, S(\tau)\, \Theta(\tau - \tau')\, F(\tau')\right] \qquad (11.2.17)$$

which is to be substituted in (11.2.4). This latter expression will be, in particular, very useful when we make contact with the path integral formalism in §11.4.

2. Consider the Hamiltonian

$$H'_0 = \frac{p^2}{2m} - qF(\tau) + pS(\tau). \qquad (11.2.18)$$

Repeating the analysis given above for the transformation function $\langle qt|pt'\rangle$ we readily obtain

$$\langle qt|pt'\rangle = \exp\left[-\frac{i}{2m\hbar} \int_{t'}^{t} d\tau \left(i\hbar \frac{\delta}{\delta S(\tau)}\right)^2\right] \langle qt|pt'\rangle_0 \qquad (11.2.19)$$

where $\langle qt|pt'\rangle_0$ is given in (11.2.16). We are interested in evaluating $\langle qt|pt'\rangle$ in (11.2.19) for $S(t) = 0$.

---

[7] This equation is also valid for several degrees of freedom with an obvious notational adaptation.

11.2 Expressions for Transformations Functions

To the above end, set

$$\int_{t'}^{t} d\tau'\, \Theta(\tau - \tau')\, F(\tau') = \widehat{F}(\tau) \tag{11.2.20}$$

then the expression depending on $S(t)$ in (11.2.16) may be rewritten as

$$\exp\left[-\frac{i}{\hbar} \int_{t'}^{t} d\tau\, S(\tau)\left(p + \widehat{F}(\tau)\right)\right] \tag{11.2.21}$$

and since we will consider the limit $S(t) \to 0$, we may replace $(\delta/\delta S(\tau))^2$ in (11.2.19) when operating on the functional in (11.2.21) simply by $\left[-i(p + \widehat{F}(\tau))/\hbar\right]^2$. It is then straightforward to obtain (Problem 11.1) from (11.2.19), (11.2.16)

$$\langle qt | pt' \rangle^{(0)} = \exp\left[\frac{i}{\hbar}\left(qp - \frac{p^2}{2m}(t - t')\right)\right]$$

$$\times \exp\left[\frac{i}{\hbar} \int_{t'}^{t} d\tau\, F(\tau)\left(q - \frac{p}{m}(t - \tau)\right)\right]$$

$$\times \exp\left[-\frac{i}{2m\hbar} \int_{t'}^{t} d\tau \int_{t'}^{t} d\tau'\, F(\tau)\, (t - \tau_>)\, F(\tau')\right] \tag{11.2.22}$$

where

$$\tau_> = \max(\tau, \tau') \tag{11.2.23}$$

with (11.2.22) corresponding to the Hamiltonian

$$H_{(0)} = \frac{p^2}{2m} - qF(\tau). \tag{11.2.24}$$

To obtain $\langle qt | q't' \rangle$ for $S = 0$, we multiply (11.2.22) by $\langle pt' | q't' \rangle = \exp(-iq'p/\hbar)$ and integrate the elementary "Gaussian" integral over p with measure $dp/2\pi\hbar$ giving[8]

$$\langle qt | q't' \rangle^{(0)} = \sqrt{\frac{m}{2\pi i \hbar T}} \exp\left[\frac{im}{2\hbar T}(q - q')^2\right]$$

$$\times \exp\left[\frac{i}{\hbar} \int_{t'}^{t} d\tau\, F(\tau)\left(q' + \frac{(q - q')}{T}(\tau - t')\right)\right]$$

$$\times \exp\left[-\frac{i}{m\hbar} \int_{t'}^{t} d\tau \int_{t'}^{t} d\tau'\, F(\tau)\, \frac{(t - \tau)\, \Theta(\tau - \tau')\, (\tau' - t')}{T}\, F(\tau')\right] \tag{11.2.25}$$

---

[8] Needless to say this expression is easily rewritten when one is dealing with several degrees ν of freedom. In particular, $(m/2\pi i\hbar T)^{1/2}$ would be replaced by $(m/2\pi i\hbar T)^{\nu/2}$.

(compare with (9.4.21) — see Problem 11.2), corresponding to the Hamiltonian $H_{(0)}$ in (11.2.24), $T = t - t'$.

3. By an almost identical procedure as obtaining (11.2.4), the transformation function $\langle qt | q't' \rangle_H$ for the typical Hamiltonian

$$H = \frac{p^2}{2m} + V(q) \tag{11.2.26}$$

is worked out to be

$$\langle qt | q't' \rangle_H = \exp\left[-\frac{i}{\hbar} \int_{t'}^{t} d\tau\, V\left(-i\hbar \frac{\delta}{\delta F(\tau)}\right)\right] \langle qt | q't' \rangle^{(0)} \bigg|_{F=0} \tag{11.2.27}$$

with $\langle qt | q't' \rangle^{(0)}$ given in (11.2.25).

One may use the identity (see Problem 11.6)

$$G_1\left[-i\hbar\frac{\delta}{\delta F}\right] G_2[F]\bigg|_{F=0} = G_2\left[-i\hbar\frac{\delta}{\delta F}\right] G_1[F]\bigg|_{F=0} \tag{11.2.28}$$

for two functionals G_1, G_2, to rewrite (11.2.27) as

$$\langle qt | q't' \rangle_H = \sqrt{\frac{m}{2\pi i\hbar T}} \exp\left[\frac{im}{2\hbar T}(q-q')^2\right]$$

$$\times \exp\left[\int_{t'}^{t} d\tau \left[q' + \frac{(q-q')}{T}(\tau - t')\right] \frac{\delta}{\delta F(\tau)}\right]$$

$$\times \exp\left[\frac{i\hbar}{m}\int_{t'}^{t} d\tau \int_{t'}^{t} d\tau' \frac{(t-\tau)\,\Theta(\tau-\tau')\,(\tau'-t')}{T} \frac{\delta}{\delta F(\tau)}\frac{\delta}{\delta F(\tau')}\right]$$

$$\times \exp\left[-\frac{i}{\hbar}\int_{t'}^{t} d\tau\, V(F(\tau))\right]\bigg|_{F=0}. \tag{11.2.29}$$

Since

$$\exp\left[\int_{t'}^{t} d\tau \left[q' + \frac{(q-q')}{T}(\tau - t')\right] \frac{\delta}{\delta F(\tau)}\right] \tag{11.2.30}$$

represents the translation operator of $F(\tau)$ by $[q'+(q-q')(\tau-t')/T]$, (11.2.29) becomes

$$\langle qt | q't' \rangle_H = \sqrt{\frac{m}{2\pi i\hbar T}} \exp\left[\frac{im(q-q')^2}{2\hbar T}\right]$$

$$\times \exp\left[\frac{i}{m}\int_{t'}^{t} d\tau \int_{t'}^{t} d\tau' \frac{(t-\tau)\,\Theta(\tau-\tau')\,(\tau'-t')}{T} \frac{\delta}{\delta Q(\tau)}\frac{\delta}{\delta Q(\tau')}\right]$$

$$\times \exp\left[-\frac{i}{\hbar}\int_{t'}^{t}d\tau\, V\left(q' + \frac{(q-q')}{T}(\tau-t') + \sqrt{\hbar}\,Q(\tau)\right)\right]\Bigg|_{Q=0} \tag{11.2.31}$$

where we have made the substitution $F(\tau) \to \sqrt{\hbar}\,Q(\tau)$. We note that $\sqrt{\hbar}\,Q(\tau)$, in the argument of V in (11.2.31), takes into account the deviation of the motion of a particle from the one given by the straight line $[q' + (q-q')(\tau-t')/T]$. The effect of this deviation is obtained by the functional differentiations provided by the exponential factor involving the functional differential operators $(\delta/\delta Q(\tau))(\delta/\delta Q(\tau'))$.

4. We apply the quantum dynamical principle in (11.1.20) to the forced harmonic oscillator problem (§9.4)

$$H = \frac{p^2}{2m} + \frac{1}{2}m\omega^2 q^2 - qF(\tau) \tag{11.2.32}$$

by a method which is rich enough in that we consider variation not only of the external field $F(\tau)$ but also ω, i.e., the variation of a physical parameter in the theory.

Heisenberg's equations (2.3.61), (2.3.62) are given by

$$\dot{q}(\tau) = \frac{p(\tau)}{m} \tag{11.2.33}$$

$$\dot{p}(\tau) = -m\omega^2 q(\tau) + F(\tau). \tag{11.2.34}$$

For $t' < \tau < t$, the solutions of these equations are given by

$$q(\tau) = \left[q(t)\frac{\cos\omega(\tau-t')}{\cos\omega T} - \frac{p(t')}{m\omega}\frac{\sin\omega(t-\tau)}{\cos\omega T}\right]$$

$$+ \frac{1}{m}\int_{t'}^{t}d\tau'\, K(\tau,\tau')\, F(\tau') \tag{11.2.35}$$

$$p(\tau) = -m\omega\left[q(t)\frac{\sin\omega(\tau-t')}{\cos\omega T} - \frac{p(t')}{m\omega}\frac{\cos\omega(t-\tau)}{\cos\omega T}\right]$$

$$+ \int_{t'}^{t}d\tau'\,\frac{d}{d\tau}K(\tau,\tau')\,F(\tau') \tag{11.2.36}$$

where $T = t - t'$, and the c-function $K(\tau,\tau')$ satisfies the conditions: $(t' < \tau, \tau' < t)$

(i)
$$K(\tau,\tau')\big|_{\tau\to t} = 0 \tag{11.2.37}$$

(ii)
$$\frac{d}{d\tau}K(\tau,\tau')\bigg|_{\tau\to t'} = 0 \tag{11.2.38}$$

(iii)
$$\left(\frac{d^2}{d\tau^2} + \omega^2\right) K(\tau, \tau') = \delta(\tau - \tau') \tag{11.2.39}$$

so that $\ddot{q}(\tau) + \omega^2 q(\tau) = F(\tau)/m$, i.e.,

$$\left(\frac{d^2}{d\tau^2} + \omega^2\right) K(\tau, \tau') = 0 \quad \text{for} \quad \tau \neq \tau' \tag{11.2.40}$$

and as a function of τ

$$\left.\frac{d}{d\tau} K(\tau, \tau')\right|_{\tau'-0}^{\tau'+0} = 1. \tag{11.2.41}$$

It is not difficult to show (Problem 11.7) that ($\tau_> = \max(\tau, \tau')$, $\tau_< = \min(\tau, \tau')$)

$$K(\tau, \tau') = -\frac{\sin\omega(t - \tau_>)\cos\omega(\tau_< - t')}{\omega \cos \omega T}. \tag{11.2.42}$$

Upon taking matrix elements of (11.2.35) we obtain

$$\langle qt | q(\tau) | pt' \rangle = \left[q_c(\tau) + \frac{1}{m}\int_{t'}^{t} d\tau' \, K(\tau, \tau') \, F(\tau')\right] \langle qt | pt' \rangle \tag{11.2.43}$$

where

$$q_c(\tau) = \left[q\frac{\cos\omega(\tau - t')}{\cos\omega T} - \frac{p}{m\omega}\frac{\sin\omega(t - \tau)}{\cos\omega T}\right] \tag{11.2.44}$$

is a c-number function. From (11.1.20), we may rewrite (11.2.43) as

$$-i\hbar\frac{\delta}{\delta F(\tau)}\langle qt | pt' \rangle = \left[q_c(\tau) + \frac{1}{m}\int_{t'}^{t} d\tau' \, K(\tau, \tau') \, F(\tau')\right] \langle qt | pt' \rangle \tag{11.2.45}$$

which upon integration gives

$$\langle qt | pt' \rangle = \exp\left[\frac{i}{\hbar}\int_{t'}^{t} d\tau \, q_c(\tau) \, F(\tau)\right.$$
$$\left. + \frac{i}{2m\hbar}\int_{t'}^{t} d\tau \int_{t'}^{t} d\tau' \, F(\tau) \, K(\tau, \tau') \, F(\tau')\right] \langle qt | pt' \rangle_{F=0}. \tag{11.2.46}$$

To solve for $\langle qt | pt' \rangle_{F=0}$, we carry out another functional differentiation of (11.2.43) with respect to $F(\tau'')$, using (11.1.38), and then set $F = 0$ in the resulting expression to obtain

$$\langle qt | \left(q(\tau'') \, q(\tau)\right)_+ | pt' \rangle_{F=0} = \left[q_c(\tau) \, q_c(\tau'') - \frac{i\hbar}{m} K(\tau, \tau'')\right] \langle qt | pt' \rangle_{F=0} \tag{11.2.47}$$

11.2 Expressions for Transformations Functions

we take the symmetric limit $\tau'' \leftrightarrow \tau$, as the average of $\tau'' = \tau+0$, $\tau'' = \tau-0$, to write

$$\langle qt|q^2(\tau)|pt'\rangle_{F=0} = \left[q_c^2(\tau) - \frac{i\hbar}{m}K(\tau,\tau)\right]\langle qt|pt'\rangle_{F=0} \qquad (11.2.48)$$

and integrate over τ. This gives

$$\langle qt|\int_{t'}^{t}\!\!d\tau\, q^2(\tau)|pt'\rangle_{F=0} = \left[\int_{t'}^{t}\!\!d\tau\, q_c^2(\tau) - \frac{i\hbar}{m}\int_{t'}^{t}\!\!d\tau\, K(\tau,\tau)\right]\langle qt|pt'\rangle_{F=0}. \qquad (11.2.49)$$

Now we invoke the quantum dynamical principle (11.1.20), in regard to the $m\omega^2 q^2/2$ term in the Hamiltonian, to obtain

$$i\hbar\frac{2}{m}\frac{\partial}{\partial\omega^2}\langle qt|pt'\rangle_{F=0} = \langle qt|\int_{t'}^{t}\!\!d\tau\, q^2(\tau)|pt'\rangle_{F=0}. \qquad (11.2.50)$$

With $q_c(\tau)$ defined in (11.2.44), and $K(\tau,\tau')$ given in (11.2.42), we have

$$K(\tau,\tau) = -\frac{\sin\omega(t-\tau)\cos\omega(\tau-t')}{\omega\cos\omega T} \qquad (11.2.51)$$

leading to the integrals

$$\int_{t'}^{t}\!\!d\tau\, K(\tau,\tau) = -\frac{T}{2\omega}\tan\omega T = \frac{1}{2\omega}\frac{\partial}{\partial\omega}\ln\cos\omega T \qquad (11.2.52)$$

$$\int_{t'}^{t}\!\!d\tau\, q_c^2(\tau) = \frac{1}{m\omega}\frac{\partial}{\partial\omega}\left[q^2\frac{m\omega}{2}\tan\omega T - qp\sec\omega T + \frac{p^2}{2m}\frac{\tan\omega T}{\omega}\right]. \qquad (11.2.53)$$

From (11.2.50), (11.2.49), and using the integrals (11.2.52), (11.2.53), we obtain upon integration over: ω, and noting that $\partial/\partial\omega^2 = \partial/2\omega\partial\omega$,

$$\langle qt|pt'\rangle_{F=0} = \frac{1}{\sqrt{\cos\omega T}}\exp\left[-\frac{i}{\hbar}\left(q^2\frac{m\omega}{2}\tan\omega T\right.\right.$$

$$\left.\left.-qp\sec\omega T + \frac{p^2}{2m}\frac{\tan\omega T}{\omega}\right)\right]. \qquad (11.2.54)$$

This also satisfies the boundary condition

$$\langle qt|pt'\rangle_{F=0} \xrightarrow[\omega\to 0]{} \exp\left[\frac{i}{\hbar}\left(qp - \frac{p^2}{2m}T\right)\right]. \qquad (11.2.55)$$

The expression in (11.2.54) is to be substituted in (11.2.46). Since

$$\int_{t'}^{t}\!\!d\tau\, q_c(\tau)F(\tau) = \frac{q}{\cos\omega T}\int_{t'}^{t}\!\!d\tau\, F(\tau)\cos\omega(\tau-t')$$

$$-\frac{p}{m\omega\cos\omega T}\int_{t'}^{t}\mathrm{d}\tau\, F(\tau)\sin\omega(t-\tau) \qquad (11.2.56)$$

is linear in p, we may, after multiplying (11.2.46) by $\langle pt'|qt'\rangle_{F=0}\,\mathrm{d}p/2\pi\hbar$, integrate over p, the elementary "Gaussian" integral to obtain after rearrangement of terms the explicit expression

$$\langle qt|q't'\rangle = \sqrt{\frac{m\omega}{2\pi i\hbar\sin\omega T}}$$

$$\times \exp\left[i\frac{m\omega}{\hbar\sin\omega T}\left(\frac{(q^2+q'^2)}{2}\cos\omega T - qq'\right.\right.$$

$$+\frac{1}{m\omega}\int_{t'}^{t}\mathrm{d}\tau\, F(\tau)\left[q\sin\omega(\tau-t')+q'\sin\omega(t-\tau)\right]$$

$$-\frac{1}{m^2\omega^2}\int_{t'}^{t}\mathrm{d}\tau\int_{t'}^{t}\mathrm{d}\tau'\, F(\tau)\sin\omega(t-\tau)$$

$$\left.\left.\times\,\Theta(\tau-\tau')\sin\omega(\tau'-t')F(\tau')\right)\right] \qquad (11.2.57)$$

which coincides with the expression in (9.4.32).

5. Finally we close this section by making contact with the analysis of the dynamics given in the Appendix to §2.5. For a given Hamiltonian $H(q,p,t)$ introduce the scaled Hamiltonian

$$H'(q,p,\tau;\lambda) = \lambda H(q,p,\tau) \qquad (11.2.58)$$

then as in (11.2.2), we have

$$\frac{\partial}{\partial\lambda}\langle qt|q't'\rangle = -\frac{i}{\hbar}\int_{t'}^{t}\mathrm{d}\tau\,\langle qt|H(q(\tau),p(\tau),\tau)|q't'\rangle \qquad (11.2.59)$$

and since $H(q,p,\tau)$ is independent of λ, we have[9] by repeated application of (11.1.37),

$$\left(\frac{\partial}{\partial\lambda}\right)^n\langle qt|q't'\rangle$$

$$=\left(-\frac{i}{\hbar}\right)^n\int_{t'}^{t}\mathrm{d}\tau_1\int_{t'}^{t}\mathrm{d}\tau_2\cdots\int_{t'}^{t}\mathrm{d}\tau_n\,\langle qt|\left(\mathbb{H}(\tau_1)\cdots\mathbb{H}(\tau_n)\right)_+|q't'\rangle$$

$$(11.2.60)$$

where

---

[9] See (11.1.33)–(11.1.37).

$$\mathbb{H}(\tau) = H\big(q(\tau), p(\tau), \tau\big) \tag{11.2.61}$$

in the Heisenberg representation of $H(q,p,\tau)$, developing in time with the Hamiltonian H' in (11.2.58). For $\lambda = 0$, $\mathbb{H}(\tau) = H(q,p,\tau)$, $\langle qt| \to \langle q0|$, $|q't'\rangle \to |q'0\rangle$ on the right-hand side of (11.2.60). Accordingly $\langle qt|q't'\rangle$ for $\lambda = 1$, is obtained by a Taylor series about $\lambda = 0$ using (11.2.60), and is given by

$$\langle qt|q't'\rangle = \langle q0|\left(\exp\left[-\frac{i}{\hbar}\int_{t'}^{t}\mathrm{d}\tau\, H(q,p,\tau)\right]\right)_{+}|q'0\rangle. \tag{11.2.62}$$

In particular, for a time-independent Hamiltonian $H(q,p)$,

$$\langle qt|q't'\rangle = \langle q0|\exp\left[-\frac{i}{\hbar}(t-t')H(q,p)\right]|q'0\rangle \tag{11.2.63}$$

as expected.

11.3 Trace Functionals

Of particular interest is the transformation function $\langle qt|q't'\rangle$ for which $q = q'$, i.e., for

$$q(t) = q'(t') \tag{11.3.1}$$

as it may be used to obtain information on the spectrum of the underlying Hamiltonian.

Actually, we consider the trace of $\langle qt|qt'\rangle$ defined by

$$\int_{-L}^{L}\mathrm{d}q\,\langle qt|qt'\rangle \tag{11.3.2}$$

where L will taken to be arbitrarily large. To study the properties of the object in (11.3.2), we carry out a Fourier series analysis,

$$q(\tau) = \frac{1}{T}\sum_{n=-\infty}^{\infty} q_n \exp\left[\mathrm{i}\frac{2\pi n}{T}(\tau - t')\right] \tag{11.3.3}$$

$$F(\tau) = \frac{1}{T}\sum_{n=-\infty}^{\infty} F_n \exp\left[-\mathrm{i}\frac{2\pi n}{T}(\tau - t')\right] \tag{11.3.4}$$

$$S(\tau) = \frac{1}{T}\sum_{n=-\infty}^{\infty} S_n \exp\left[\mathrm{i}\frac{2\pi n}{T}(\tau - t')\right] \tag{11.3.5}$$

$$\frac{\delta}{\delta F(\tau)} = \sum_{n=-\infty}^{\infty} \exp\left[\mathrm{i}\frac{2\pi n}{T}(\tau - t')\right]\frac{\partial}{\partial F_n} \tag{11.3.6}$$

and note that from
$$\frac{\partial}{\partial F_n} F_m = \delta_{nm} \tag{11.3.7}$$
we have
$$\frac{\delta}{\delta F(\tau)} F(\tau') = \frac{1}{T} \sum_{n=-\infty}^{\infty} \exp\left[\frac{i2\pi n(\tau - \tau')}{T}\right] = \delta(\tau - \tau'). \tag{11.3.8}$$

Similarly, we may write
$$\frac{\delta}{\delta S(\tau)} = \sum_{n=-\infty}^{\infty} \exp\left[\frac{-i2\pi n(\tau - t')}{T}\right] \frac{\partial}{\partial S_n} \tag{11.3.9}$$

$$\frac{\partial}{\partial S_n} S_m = \delta_{nm}. \tag{11.3.10}$$

For a given time-independent Hamiltonian $H(q,p)$, we may, in particular, use (11.2.4) and (11.2.17), to write

$$\int_{-L}^{L} dq \, \langle qt | qt' \rangle = \int_{-L}^{L} dq \, \exp\left[-\frac{i}{\hbar} \int_{t'}^{t} d\tau \, H\left(-i\hbar \frac{\delta}{\delta F(\tau)}, i\hbar \frac{\delta}{\delta S(\tau)}\right)\right]$$
$$\times \delta\left(\int_{t'}^{t} d\tau \, S(\tau)\right) \exp\left[\frac{i}{\hbar} q \int_{t'}^{t} d\tau \, F(\tau)\right]$$
$$\times \exp\left[-\frac{i}{\hbar} \int_{t'}^{t} d\tau \int_{t'}^{t} d\tau' \, S(\tau') \Theta(\tau' - \tau) F(\tau)\right]. \tag{11.3.11}$$

From (11.3.4), (11.3.5)
$$\int_{t'}^{t} d\tau \, F(\tau) = F_0, \qquad \int_{t'}^{t} d\tau \, S(\tau) = S_0 \tag{11.3.12}$$

and (see Problem 11.8)
$$\int_{t'}^{t} d\tau' \, S(\tau') \int_{t'}^{\tau'} d\tau \, F(\tau) = \frac{S_0 F_0}{2} - i \sum_{n \neq 0} \frac{(S_n F_0 + S_0 F_n)}{2\pi n} + i \sum_{n \neq 0} S_n \frac{1}{2\pi n} F_n. \tag{11.3.13}$$

Therefore,
$$\int_{-\infty}^{\infty} dq \, \delta\left(\int_{t'}^{t} d\tau \, S(\tau)\right) \exp\left[\frac{i}{\hbar} q \int_{t'}^{t} d\tau \, F(\tau)\right]$$
$$\times \exp\left[-\frac{i}{\hbar} \int_{t'}^{t} d\tau \int_{t'}^{t} d\tau' \, S(\tau') \Theta(\tau' - \tau) F(\tau)\right]$$

$$= (2\pi\hbar)\,\delta(S_0)\,\delta(F_0)\,\exp\left[\frac{1}{\hbar}\sum_{n\neq 0}S_n\frac{1}{2\pi n}F_n\right]. \quad (11.3.14)$$

For the harmonic oscillator, for example, we clearly have

$$\int_{-\infty}^{\infty}dq\,\langle qt|qt'\rangle\bigg|_0 = \exp\left[i\hbar T\sum_n\left(\frac{1}{2m}\frac{\partial}{\partial S_n}\frac{\partial}{\partial S_{-n}} + \frac{m\omega^2}{2}\frac{\partial}{\partial F_n}\frac{\partial}{\partial F_{-n}}\right)\right]$$
$$\times (2\pi\hbar)\,\delta(S_0)\,\delta(F_0)\,\exp\left[\frac{1}{\hbar}\sum_{n\neq 0}S_n\frac{1}{2\pi n}F_n\right]\bigg|_0 \quad (11.3.15)$$

where $\big|_0$ means setting the external sources equal to zero after the functional differentiations are carried out.

We note that

$$\exp\left\{i\hbar T\left[\frac{1}{2m}\left(\frac{\partial}{\partial S_0}\right)^2 + \frac{m\omega^2}{2}\left(\frac{\partial}{\partial F_0}\right)^2\right]\right\}(2\pi\hbar)\,\delta(S_0)\,\delta(F_0)\bigg|_{S_0\to 0,\,F_0\to 0}$$

$$= \hbar\,\frac{1}{2\pi}\int_{-\infty}^{\infty}d\lambda_1\int_{-\infty}^{\infty}d\lambda_2\,\exp\left[-\frac{i\hbar T\lambda_1^2}{2m}\right]\exp\left[-\frac{i\hbar m\omega^2 T\lambda_2^2}{2}\right]$$

$$= \frac{(2\pi)\,\hbar}{(2\pi)}\sqrt{\frac{m}{i\hbar T}}\sqrt{\frac{1}{i\hbar m\omega^2 T}} = \frac{1}{iT\omega}. \quad (11.3.16)$$

On the other hand,

$$\exp\left[i\hbar T\sum_{n\neq 0}\left(\frac{1}{2m}\frac{\partial}{\partial S_n}\frac{\partial}{\partial S_{-n}} + \frac{m\omega^2}{2}\frac{\partial}{\partial F_n}\frac{\partial}{\partial F_{-n}}\right)\right]\exp\left[\frac{1}{\hbar}\sum_{n\neq 0}S_n\frac{1}{2\pi n}F_n\right]\bigg|_0$$

$$= \prod_{n=1}^{\infty}\exp\left(\frac{i\hbar T}{m}\frac{\partial}{\partial S_n}\frac{\partial}{\partial S_{-n}}\right)\exp\left(i\hbar T m\omega^2\frac{\partial}{\partial F_n}\frac{\partial}{\partial F_{-n}}\right)$$
$$\times \exp\left[\frac{1}{\hbar}\left(S_n\frac{1}{2\pi n}F_n - S_{-n}\frac{1}{2\pi n}F_{-n}\right)\right]\bigg|_0$$

$$= \prod_{n=1}^{\infty}\exp\left(\frac{i\hbar T}{m}\frac{\partial}{\partial S_n}\frac{\partial}{\partial S_{-n}}\right)\exp\left(-\frac{iTm\omega^2}{\hbar(2\pi n)^2}S_n S_{-n}\right)\bigg|_0. \quad (11.3.17)$$

Upon writing

$$\exp\left(-i\beta S_n S_{-n}\right) = \int_{-\infty}^{\infty}d\lambda_2\int_{-\infty}^{\infty}\frac{d\lambda_1}{2\pi}\,e^{i\lambda_1\lambda_2}\,e^{-i\lambda_2 S_{-n}}\,e^{-i\beta\lambda_1 S_n} \quad (11.3.18)$$

with $\beta = Tm\omega^2/\hbar(2\pi n)^2$, it is easily established that the right-hand side of (11.3.17) is given by

$$\prod_{n=1}^{\infty}\left[1-\left(\frac{T\omega}{2\pi n}\right)^2\right]^{-1} \tag{11.3.19}$$

which is the infinite product representation of $(T\omega/2)/\sin(T\omega/2)$, i.e.,

$$\prod_{n=1}^{\infty}\left[1-\left(\frac{T\omega}{2\pi n}\right)^2\right]^{-1} = \frac{iT\omega\,e^{-iT\omega/2}}{1-e^{-iT\omega}} = iT\omega\sum_{n=0}^{\infty}\exp\left[-i\frac{T}{\hbar}\hbar\omega\left(n+\frac{1}{2}\right)\right]. \tag{11.3.20}$$

From (11.3.15), (11.3.16), (11.3.20), we then have

$$\int_{-\infty}^{\infty} dq\,\langle qt\,|\,qt'\rangle = \sum_{n=0}^{\infty}\exp\left[-\frac{iT}{\hbar}\hbar\omega\left(n+\frac{1}{2}\right)\right] \tag{11.3.21}$$

recognizing the spectrum of the harmonic oscillator Hamiltonian.

To find the number of eigenvalues $\leqslant \xi$ of a given Hamiltonian H, we may use (9.7.24) and (11.3.11) (see also (4.5.4)) to write for the former as

$$N(H;\xi) = \frac{1}{2\pi i}\int_{-\infty}^{\infty}\frac{dT}{T-i\varepsilon}\,e^{i\xi T/\hbar}$$

$$\times \exp\left[-\frac{i}{\hbar}\int_0^T d\tau\,H\left(-i\hbar\frac{\delta}{\delta F(\tau)}, i\hbar\frac{\delta}{\delta S(\tau)}\right)\right]$$

$$\times (2\pi\hbar)\,\delta\left(\int_0^T d\tau\,S(\tau)\right)\delta\left(\int_0^T d\tau\,F(\tau)\right)$$

$$\times \exp\left[-\frac{i}{\hbar}\int_0^T d\tau\int_0^T d\tau'\,S(\tau')\,\Theta(\tau'-\tau)\,F(\tau)\right]\bigg|_0. \tag{11.3.22}$$

On the other hand, for the sum of the eigenvalues $< \xi$, with ξ not in the spectrum of H, we have from (9.7.32), (11.3.11), the expression

$$\frac{1}{2\pi i}\int_{-\infty}^{\infty}\frac{dT}{T-i\varepsilon}\,e^{i\xi T/\hbar}$$

$$\times i\hbar\frac{\partial}{\partial T}\exp\left[-\frac{i}{\hbar}\int_0^T d\tau\,H\left(-i\hbar\frac{\delta}{\delta F(\tau)}, i\hbar\frac{\delta}{\delta S(\tau)}\right)\right]$$

$$\times (2\pi\hbar)\,\delta\left(\int_0^T d\tau\,S(\tau)\right)\delta\left(\int_0^T d\tau\,F(\tau)\right)$$

$$\times \exp\left[-\frac{i}{\hbar}\int_0^T d\tau \int_0^T d\tau'\, S(\tau')\,\Theta(\tau'-\tau)\, F(\tau)\right]\bigg|_0. \qquad (11.3.23)$$

By using (11.3.14), we may rewrite the latter

$$\frac{1}{2\pi i}\int_{-\infty}^{\infty}\frac{dT}{T-i\varepsilon}\, e^{i\xi T/\hbar}\, H\!\left(-i\hbar\frac{\delta}{\delta F(T)},\, i\hbar\frac{\delta}{\delta S(T)}\right)$$

$$\times \exp\left[-\frac{i}{\hbar}\int_0^T d\tau\, H\!\left(-i\hbar\frac{\delta}{\delta F(\tau)},\, i\hbar\frac{\delta}{\delta S(\tau)}\right)\right]$$

$$\times (2\pi\hbar)\,\delta(S_0)\,\delta(F_0)\,\exp\left[\frac{1}{\hbar}\sum_{n\ne 0} S_n \frac{1}{2\pi n} F_n\right]\bigg|_0. \qquad (11.3.24)$$

For the harmonic oscillator, for example, this may be written as

$$\frac{1}{2\pi i}\int_{-\infty}^{\infty}\frac{dT}{T-i\varepsilon}\, e^{i\xi T/\hbar}\, i\hbar\frac{\partial}{\partial T}\cdot\frac{iT\omega}{iT\omega}\sum_{n=0}^{\infty}\exp\left[-iT\omega\!\left(n+\tfrac{1}{2}\right)\right]$$

$$=\sum_{n=0}^{\infty}\hbar\omega\!\left(n+\tfrac{1}{2}\right)\frac{1}{2\pi i}\int_{-\infty}^{\infty}\frac{dT}{T-i\varepsilon}\exp\left\{\frac{iT}{\hbar}\!\left[\xi-\hbar\omega\!\left(n+\tfrac{1}{2}\right)\right]\right\}$$

$$=\sum_{n=0}^{\infty}\hbar\omega\!\left(n+\tfrac{1}{2}\right)\Theta\!\left(\xi-\hbar\omega\!\left(n+\tfrac{1}{2}\right)\right) \qquad (11.3.25)$$

where we have used (11.3.17)–(11.3.20) and (11.3.16). Here ξ may be taken to fall between two consecutive eigenvalues.

11.4 From the Quantum Dynamical Principle to Path Integrals

Consider the transformation function

$$\langle qt\,|\,q't'\rangle_0 = \delta\!\left(q-q'-\int_{t'}^{t} d\tau\, S(\tau)\right)\exp\left[\frac{i}{\hbar}q\int_{t'}^{t} d\tau\, F(\tau)\right]$$

$$\times \exp\left[-\frac{i}{\hbar}\int_{t'}^{t} d\tau'\, F(\tau')\int_{\tau'}^{t} d\tau\, S(\tau)\right] \qquad (11.4.1)$$

derived in (11.2.17) for the Hamiltonian

$$\widehat{H} = -q\,F(\tau) + p\,S(\tau) \qquad (11.4.2)$$

where
$$\langle qt|\, q(t) = q \langle qt|, \qquad q(t')\,|q't'\rangle = q'\,|q't'\rangle. \tag{11.4.3}$$

We divide the time interval t' to t into N subintervals and introduce the telescopic representation

$$\int_{t'}^{t} d\tau\, S(\tau) = \sum_{k=0}^{N-1} \int_{t_k}^{t_{k+1}} d\tau\, S(\tau) = \sum_{k=0}^{N-1} \varepsilon\, S_{k+1} \tag{11.4.4}$$

with

$$t_0 = t', \qquad t_N = t \tag{11.4.5}$$

$$\varepsilon = t_{k+1} - t_k = \frac{(t-t')}{N} \tag{11.4.6}$$

and

$$S_{k+1} = \frac{1}{\varepsilon} \int_{t_k}^{t_{k+1}} d\tau\, S(\tau) \tag{11.4.7}$$

as a mean of $S(\tau)$ on the subinterval from t_k to t_{k+1}.

It is easily verified by explicit integrations over $q_1, \ldots, q_{N-1}$ that

$$\delta\left(q - q' - \int_{t'}^{t} d\tau\, S(\tau)\right)$$
$$= \int_{-\infty}^{\infty} dq_1 \int_{-\infty}^{\infty} dq_2 \cdots \int_{-\infty}^{\infty} dq_{N-1}\, \delta(q_1 - q' - \varepsilon S_1)$$
$$\times \delta(q_2 - q_1 - \varepsilon S_2) \cdots \delta(q - q_{N-1} - \varepsilon S_N). \tag{11.4.8}$$

Note that the N Dirac delta distributions imply that

$$\left.\begin{aligned}
\varepsilon(S_1 + \cdots + S_N) &= q - q' \\
\varepsilon(S_2 + \cdots + S_N) &= q - q_1 \\
&\vdots \\
\varepsilon S_N &= q - q_{N-1}.
\end{aligned}\right\} \tag{11.4.9}$$

We are interested in making the subdivisions of the time interval from t' to t finer and finer by taking the limits $\varepsilon \to 0$, $N \to \infty$.

To the above end, we effectively have

$$\int_{t'}^{t} d\tau'\, F(\tau') \int_{\tau'}^{t} d\tau\, S(\tau) \simeq \sum_{k=0}^{N-1} \varepsilon\, F_{k+1} \varepsilon\, (S_{k+2} + \ldots + S_N)$$

11.4 From QDP to Path Integrals

$$= \sum_{k=0}^{N-1} \varepsilon\, F_{k+1}(q_N - q_{k+1}) \tag{11.4.10}$$

and

$$q \int_{t'}^{t} d\tau'\, F(\tau') = q_N \sum_{k=0}^{N-1} \varepsilon\, F_{k+1} \tag{11.4.11}$$

with

$$q_N = q, \qquad q_0 = q'. \tag{11.4.12}$$

Hence the right-hand side of (11.4.1) may be formally defined as the $N \to \infty$ limit of

$$\int dq_1 \cdots dq_{N-1} \frac{dp_1}{2\pi\hbar} \cdots \frac{dp_N}{2\pi\hbar} \exp\left[\frac{i}{\hbar} \sum_{k=0}^{N-1} p_{k+1}(q_{k+1} - q_k - \varepsilon\, S_{k+1})\right]$$

$$\times \exp\left[-\frac{i}{\hbar} \sum_{k=0}^{N-1} \varepsilon\, F_{k+1}(q_N - q_{k+1})\right] \exp\left[\frac{i}{\hbar} \sum_{k=0}^{N-1} \varepsilon\, F_{k+1} q_N\right] \tag{11.4.13}$$

by introducing, in the process, the integral representation of the N Dirac delta distributions in (11.4.8), or equivalently as the $N \to \infty$ limit of

$$\int dq_1 \cdots dq_{N-1} \frac{dp_1}{2\pi\hbar} \cdots \frac{dp_N}{2\pi\hbar} \exp\left[\frac{i}{\hbar} \sum_{k=0}^{N-1} \varepsilon\, p_{k+1}\left(\frac{q_{k+1} - q_k}{\varepsilon}\right)\right]$$

$$\times \exp\left[\frac{i}{\hbar} \sum_{k=0}^{N-1} \varepsilon\, q_{k+1} F_{k+1}\right] \exp\left[-\frac{i}{\hbar} \sum_{k=0}^{N-1} \varepsilon\, p_{k+1} S_{k+1}\right]. \tag{11.4.14}$$

By taking the limit $N \to \infty$, $\langle qt | q't' \rangle_0$ may be rewritten in the form

$$\langle qt | q't' \rangle_0 = \int_{q(t')=q'}^{q(t)=q} \mathscr{D}(q(\cdot), p(\cdot))$$

$$\times \exp\left[\frac{i}{\hbar} \int_{t'}^{t} d\tau\, \left[p(\tau)\dot{q}(\tau) + q(\tau)F(\tau) - p(\tau)S(\tau)\right]\right] \tag{11.4.15}$$

where $\mathscr{D}(q(\cdot), p(\cdot))$ is defined as in (10.3.25), and, needless to say, $q(t)$, $q(t')$ here denote c-numbers.

For a given Hamiltonian $H(q, p, t)$, one obtains from (11.2.4) the path integral expression

$$\langle qt | q't' \rangle \Big|_{S=0, F=0} = \int_{q(t')=q'}^{q(t)=q} \mathscr{D}(q(\cdot), p(\cdot))$$

$$\times \exp\left[\frac{\mathrm{i}}{\hbar}\int_{t'}^{t}\mathrm{d}\tau\,\bigl[p(\tau)\dot{q}(\tau) - H\bigl(q(\tau),p(\tau),\tau\bigr)\bigr]\right]$$
(11.4.16)

thus establishing the formal connection between the quantum dynamical principle and the path integral formalisms.

As before for a Hamiltonian in the form as in (10.3.26), we may integrate over the momenta to obtain

$$\langle qt\,|\,q't'\rangle\Big|_{S=0,F=0} = \int_{q(t')=q'}^{q(t)=q}\mathscr{D}\bigl(q(\cdot)\bigr)\,\exp\left\{\frac{\mathrm{i}}{\hbar}\int_{t'}^{t}\mathrm{d}\tau\left[\frac{m\dot{q}^{2}(\tau)}{2} - V\bigl(q(\tau)\bigr)\right]\right\}$$
(11.4.17)

where $\mathscr{D}\bigl(q(\cdot)\bigr)$ is defined as in (10.1.4).

11.5 Bose/Fermi Excitations

Consider the Hamiltonian

$$H = \hbar\omega\, a_{\mathrm{B}}^{\dagger} a_{\mathrm{B}} - a_{\mathrm{B}}^{\dagger} F(\tau) - F^{*}(\tau) a_{\mathrm{B}} \qquad (11.5.1)$$

where a_{B}, $a_{\mathrm{B}}^{\dagger}$ are Bose annihilation, creation operators (§6.1), with

$$a_{\mathrm{B}}\,|0\rangle = 0,\quad a_{\mathrm{B}}\,|n\rangle = \sqrt{n}\,|n-1\rangle,\quad a_{\mathrm{B}}^{\dagger}\,|n\rangle = \sqrt{n+1}\,|n+1\rangle \qquad (11.5.2)$$

and $F(\tau)$, $F^{*}(\tau)$ are external sources (§6.2), operating within a time interval from time t' to t, and are zero at the boundaries $\tau = t'$, $\tau = t$.[10]

Let $U(\tau)$ denote the time evolution unitary operator and set

$$\langle n|\,U(\tau) = \langle n\tau| \qquad (11.5.3)$$

The Heisenberg representations of a_{B}, $a_{\mathrm{B}}^{\dagger}$ are given by

$$a_{\mathrm{B}}^{\#}(\tau) = U^{\dagger}(\tau)\,a_{\mathrm{B}}^{\#}\,U(\tau) \qquad (11.5.4)$$

where # corresponds to the operator a_{B} or its adjoint $a_{\mathrm{B}}^{\dagger}$.

The Heisenberg equations are

$$\dot{a}_{\mathrm{B}}(\tau) + \mathrm{i}\omega a_{\mathrm{B}}(\tau) = \frac{\mathrm{i}}{\hbar}\,F(\tau) \qquad (11.5.5)$$

$$\dot{a}_{\mathrm{B}}^{\dagger}(\tau) - \mathrm{i}\omega a_{\mathrm{B}}^{\dagger}(\tau) = -\frac{\mathrm{i}}{\hbar}\,F^{*}(\tau) \qquad (11.5.6)$$

with solutions

---
[10] See also (8.7.21), (12.6.33).

11.5 Bose/Fermi Excitations

$$a_B(\tau) = e^{-i\omega(\tau-t')} a_B(t') + \frac{i}{\hbar} \int_{t'}^{t} d\tau' \, e^{-i\omega(\tau-\tau')} \Theta(\tau-\tau') F(\tau') \quad (11.5.7)$$

$$a_B^\dagger(\tau) = e^{-i\omega(t-\tau)} a_B^\dagger(t) + \frac{i}{\hbar} \int_{t'}^{t} d\tau' \, e^{i\omega(\tau-\tau')} \Theta(\tau'-\tau) F^*(\tau'). \quad (11.5.8)$$

These give the matrix elements

$$\langle 0t | a_B(\tau) | 0t' \rangle = \left[\frac{i}{\hbar} \int_{t'}^{t} d\tau' \, e^{-i\omega(\tau-\tau')} \Theta(\tau-\tau') F(\tau') \right] \langle 0t | 0t' \rangle \quad (11.5.9)$$

$$\langle 0t | a_B^\dagger(\tau) | 0t' \rangle = \left[\frac{i}{\hbar} \int_{t'}^{t} d\tau' \, e^{i\omega(\tau-\tau')} F^*(\tau') \Theta(\tau'-\tau) \right] \langle 0t | 0t' \rangle \quad (11.5.10)$$

and hence from (11.5.9)

$$-i\hbar \frac{\delta}{\delta F^*(\tau)} \langle 0t | 0t' \rangle = \left[\frac{i}{\hbar} \int_{t'}^{t} d\tau' \, e^{-i\omega(\tau-\tau')} \Theta(\tau-\tau') F(\tau') \right] \langle 0t | 0t' \rangle \quad (11.5.11)$$

leading to

$$\langle 0t | 0t' \rangle = \exp\left[-\frac{1}{\hbar^2} \int_{t'}^{t} d\tau \int_{t'}^{t} d\tau' \, e^{-i\omega(\tau-\tau')} F^*(\tau) \Theta(\tau-\tau') F(\tau') \right] \quad (11.5.12)$$

to be compared with (6.2.12), where the latter was given for a real source. It is easy to see that (11.5.12) is also consistent with (11.5.10) (see Problem 11.10).

Of particular interest is the amplitude of excitations $\langle nt | 0t' \rangle$. To obtain this amplitude, note that in this case (11.5.9) is to be replaced by

$$\langle nt | a_B(\tau) | 0t' \rangle = \left[\frac{i}{\hbar} \int_{t'}^{t} d\tau' \, e^{-i\omega(\tau-\tau')} \Theta(\tau-\tau') F(\tau') \right] \langle nt | 0t' \rangle. \quad (11.5.13)$$

For $\tau = t$, we may use (11.5.2) to rewrite the above equation as

$$\sqrt{n+1} \, \langle (n+1)t | 0t' \rangle = \left[\frac{i}{\hbar} \int_{t'}^{t} d\tau' \, e^{-i\omega(t-\tau')} F(\tau') \right] \langle nt | 0t' \rangle \quad (11.5.14)$$

providing a recurrence relation in n, whose solution is elementary and is given by

$$\langle nt | 0t' \rangle = \frac{\left[\frac{i}{\hbar} \int_{t'}^{t} d\tau' \, e^{-i\omega(t-\tau')} F(\tau') \right]^n}{\sqrt{n!}} \langle 0t | 0t' \rangle. \quad (11.5.15)$$

This is to be compared with (6.2.22) and (10.5.32).

The transformation functions $\langle nt | n't' \rangle$ are obtained in a similar fashion (see Problem 11.11) and may be compared with the expression in (6.2.30) using the present notation (see also (10.5.34)). For the study of temperature dependence of Bose excitations see §6.3, §11.6.

For Fermi excitations, consider the Hamiltonian

$$H = \hbar\omega\, a_F^\dagger a_F - \eta^*(\tau)\, a_F - a_F^\dagger\, \eta(\tau). \tag{11.5.16}$$

The external sources $\eta(\tau)$, $\eta^*(\tau)$ are Grassmann variables (§10.6, §6.4), i.e.,

$$\left\{\eta(\tau), \eta^\#(\tau')\right\} = 0, \qquad \left\{\frac{\delta}{\delta\eta(\tau)}, \frac{\delta}{\delta\eta^\#(\tau')}\right\} = 0 \tag{11.5.17}$$

and

$$\left\{\frac{\delta}{\delta\eta(\tau)}, \eta^*(\tau')\right\} = 0, \qquad \left\{\frac{\delta}{\delta\eta^\#(\tau)}, \eta^\#(\tau')\right\} = \delta(\tau - \tau'). \tag{11.5.18}$$

The sources also anti-commute with a_F, $a_F^\dagger$.

Apart from these properties of the external Fermi sources, and the anti-commutation relations of the a_F, $a_F^\dagger$ (§6.4), the analysis is almost identical to the Boson case with n restricted to 0 or 1 above,

$$a_F|0\rangle = 0, \qquad a_F|1\rangle = |0\rangle, \qquad a_F^\dagger|0\rangle = |1\rangle, \tag{11.5.19}$$

and due to the anti-commutativity of the sources with a_F, $a_F^\dagger$, the commutator

$$\left[a_F(\tau), a_F^\dagger(\tau)\,\eta(\tau)\right] = \left\{a_F(\tau), a_F^\dagger(\tau)\right\}\eta(\tau) = \eta(\tau), \tag{11.5.20}$$

for example, gives rise to an anti-commutation relation as indicated.

Equation (11.5.15) is now replaced by

$$\langle nt|0t'\rangle = \left[\frac{i}{\hbar}\int_{t'}^{t} d\tau\, e^{-i\omega(t-\tau)}\,\eta(\tau)\right]^n$$

$$\times \exp\left[-\frac{1}{\hbar^2}\int_{t'}^{t} d\tau \int_{t'}^{t} d\tau'\, e^{-i\omega(\tau-\tau')}\,\eta^*(\tau)\,\Theta(\tau-\tau')\,\eta(\tau')\right] \tag{11.5.21}$$

where $n = 0$ or 1, for the fermionic case.

Needless to say interaction of bosons may be described by adding a $\lambda H_I(a^\dagger, a)$ term to the Hamiltonian (11.5.1) in the presence of the external sources followed by an application of the quantum dynamical principle. We will not, however, go into these details here. For the interaction of bosons and fermions see Problem 11.12.

11.6 Closed-Time Path and Expectation-Value Formalism

In the present section, we apply the quantum dynamical principle to determine expectation values of physical quantities for systems prepared in initially specified states.

Consider, for example, the bosonic system described by the Hamiltonian in (11.5.1) in the presence of external sources F, F^*, and suppose that one is interested in computing the expectation value $\langle 0t'|a^\dagger(\tau)|0t'\rangle$ for some $\tau > t'$. This, however, cannot be simply obtained by a functional differentiation of $\langle 0t'|0t'\rangle$ since the latter is equal to one — a constant independent of F, F^*. One may, however, use the completeness unitarity expansion

$$\langle 0t'|0t'\rangle = \sum_{n=0}^{\infty} \langle 0t'|nt\rangle \langle nt|0t'\rangle \qquad (11.6.1)$$

and note that for $t > t'$, the amplitude $\langle nt|0t'\rangle$ corresponds to time evolution in the positive sense of time, while the amplitude $\langle 0t'|nt\rangle$ corresponds to time evolution in the negative sense of time from t back to t'. Accordingly, to generate expectation values for a system which started initially, say, in the state $|0t'\rangle$, one may introduce, *a priori*, different dynamics with pairs of external sources (F_+, F_+^*) and (F_-, F_-^*) for the two segments in the positive $(t' \to t)$ and negative $(t \to t')$ senses of time, respectively. By doing so (11.6.1) becomes replaced by

$$\langle 0t'|0t'\rangle = \sum_{n=0}^{\infty} \langle 0t'|nt\rangle_- \langle nt|0t'\rangle_+ \qquad (11.6.2)$$

which is different from one for different pairs of sources, where $\pm$ refer to the dynamics in the two segments with pairs of sources $(F_\pm, F_\pm^*)$. Expectation values are then readily obtained from (11.6.2) as shown below as we have now generated two Hamiltonians H_+, H_-, arising from (11.5.1), corresponding to the above two mentioned segments.

Using the fact that $\langle 0t'|nt\rangle_-$ may be obtained from $\langle nt|0t'\rangle_+$ by complex conjugation together with the replacements (F_+, F_+^*) by (F_-, F_-^*) in the latter, it is easily seen from (11.6.2) and (11.1.18) that

$$\delta\langle 0t'|0t'\rangle\bigg| = -\frac{i}{\hbar} \langle 0t'|\int_{t'}^{t} d\tau \left[\delta\mathsf{H}_+(\tau) - \delta\mathsf{H}_-(\tau)\right]|0t'\rangle\bigg| \qquad (11.6.3)$$

and for $t' < \tau < t$,

$$-i\hbar \frac{\delta}{\delta F_+(\tau)} \langle 0t'|0t'\rangle\bigg| = \sum_{n=0}^{\infty} \langle 0t'|nt\rangle_- \langle nt|a^\dagger(\tau)|0t'\rangle_+\bigg|$$

$$= \langle 0t'|a^\dagger(\tau)|0t'\rangle\Big|$$

$$= i\hbar \frac{\delta}{\delta F_-(\tau)} \langle 0t'|0t'\rangle \Big|$$

$$= \sum_{n=0}^{\infty} \langle 0t'|a^\dagger(\tau)|nt\rangle_- \langle nt|0t'\rangle_+ \Big| \qquad (11.6.4)$$

where the bar $\Big|$ means to replace (F_+, F_-), (F_+^*, F_-^*) by common source functions F, F^*, respectively, after functional differentiations are carried out. The adjoint operation as applied to the product of operators appearing in an amplitude such as $\langle nt|0t'\rangle_-$ upon complex conjugation of the latter giving $\langle 0t'|nt\rangle_-$, reverses the order of the operator multiplication. For simplicity of the notation we have written $a^\#$ for $a_B^\#$ above.

Accordingly,

$$(i\hbar)\frac{\delta}{\delta F_-^*(\tau_1)}(i\hbar)\frac{\delta}{\delta F_-(\tau_2)}\langle 0t'|0t'\rangle = \langle 0t'|\left(a(\tau_1)\,a^\dagger(\tau_2)\right)_-|0t'\rangle \qquad (11.6.5)$$

where now $(\cdot)_-$ denotes the chronological time anti-ordering introduced in (A-2.5.13), with operators at earlier times move to the left as indicated in the latter equation.

In (11.6.2), one is dealing with two dynamical systems, one in the positive sense of time $t' \to t$, followed by one in the negative sense of time from t back to t' forming a *closed-time path*. Expectation values of physical quantities at any time $t' < \tau < t$, for a system in a specified initial state, may be then obtained by functional differentiations with respect to, say, F_+, F_+^* followed by replacing (F_+, F_-), (F_+^*, F_-^*) by common source functions F, F^*, respectively, which in turn may or may not be set equal to zero depending on the physical system into consideration. This simple, though elegant way, of computing expectation values is referred to as the *closed-time path and expectation-value formalism*.

From (11.5.7), (11.6.2), the sum over n in the latter is easily carried out, giving the expression

$$\langle 0t'|0t'\rangle = \exp\left[-\frac{1}{\hbar^2}\int_{t'}^{t}d\tau\int_{t'}^{t}d\tau'\left\{e^{-i\omega(\tau-\tau')}\left[F_+^*(\tau)\,\Theta(\tau-\tau')\,F_+(\tau')\right.\right.\right.$$

$$\left.\left.\left.+F_-^*(\tau)\,\Theta(\tau'-\tau)\,F_-(\tau')-F_-^*(\tau)\,F_+(\tau')\right]\right\}\right].$$

$$(11.6.6)$$

This satisfies the condition

$$\langle 0t'|0t'\rangle = 1 \qquad \text{for} \quad F_+ = F_- \qquad (11.6.7)$$

11.6 Closed-Time Path & Expectation Value Formalism

as expected. More conveniently, $\langle 0t'|0t'\rangle$ in (11.6.6) may be rewritten in the compact form

$$\langle 0t'|0t'\rangle = \exp\left[-\frac{i}{\hbar}\int_{t'}^{t}d\tau \int_{t'}^{t}d\tau'\, K^{\dagger}(\tau)\, G(\tau-\tau')\, K(\tau')\right] \quad (11.6.8)$$

where

$$K(\tau) = \begin{pmatrix} F_{+}(\tau) \\ F_{-}(\tau) \end{pmatrix} \quad (11.6.9)$$

and

$$G(\tau-\tau') = -\frac{i}{\hbar}e^{-i\omega(\tau-\tau')}\begin{pmatrix} \Theta(\tau-\tau') & 0 \\ -1 & \Theta(\tau'-\tau) \end{pmatrix} \quad (11.6.10)$$

As an application, consider the harmonic oscillator problem. From (6.1.8), (6.1.10), we recall that

$$q = \frac{\hbar}{\sqrt{2m\omega}}\left(a + a^{\dagger}\right). \quad (11.6.11)$$

Suppose at time $t' = 0$, the system is in the state $|0\rangle$. To compute the expectation value of any functions of the displacement operator $q(\tau)$ at any time $0 < \tau < t$ for the system initially in the state $|0\rangle$, we consider the expectation value of the operator $\exp[ikq(\tau)]$, where k is an arbitrary c-number. From (11.6.6), (11.6.11)

$$\langle 0|e^{ikq(\tau)}|0\rangle\Big|_{F_{\pm}=F} = \exp\left[\frac{k\hbar^{2}}{\sqrt{2m\omega}}\int_{0}^{t}d\tau''\,\delta(\tau''-\tau)\right.$$
$$\left.\times\left(\frac{\delta}{\delta F_{+}^{*}(\tau'')} + \frac{\delta}{\delta F_{+}(\tau'')}\right)\right]\langle 0|0\rangle\Big|_{F_{\pm}=F}. \quad (11.6.12)$$

These functional differentiations operations give rise to translation operators leading to (see Problem 11.13),

$$\langle 0|e^{ikq(\tau)}|0\rangle\Big|_{F_{\pm}=F} = \langle 0|e^{ikq(\tau)}|0\rangle\Big|_{F_{\pm}=0}\exp\left(\frac{k}{\sqrt{2m\omega}}G[F;\tau]\right) \quad (11.6.13)$$

where

$$\langle 0|e^{ikq(\tau)}|0\rangle\Big|_{F_{\pm}=0} = \exp\left[-\frac{\hbar^{2}k^{2}}{2m^{2}\omega^{2}}\int_{0}^{t}d\tau_{1}\int_{0}^{t}d\tau_{2}\right.$$
$$\left.\times\delta(\tau_{1}-\tau)\,\Theta(\tau_{1}-\tau_{2})\,\delta(\tau_{2}-\tau)\right] \quad (11.6.14)$$

$$G[F;\tau] = \int_{0}^{\tau}d\tau'\left[F^{*}(\tau')\,e^{-i\omega(\tau'-\tau)} - F(\tau')\,e^{i\omega(\tau'-\tau)}\right] \quad (11.6.15)$$

and we have used, in the process, that $1 - \Theta(\tau'-\tau) = \Theta(\tau-\tau')$ to simplify the expression in (11.6.15).

The expression in (11.6.14) may be unambiguously evaluated by noting that an elementary application of the Baker-Campbell-Hausdorff formula (Appendix I)

$$\exp(A + B) = \exp\left(\frac{1}{2}[A, B]\right) \exp(B) \exp(A) \qquad (11.6.16)$$

with $A = i\hbar ka/\sqrt{2}\,m\omega$, $B = i\hbar ka^\dagger/\sqrt{2}\,m\omega$, yields for the left-hand side of (11.6.14) $\exp\left[-\hbar^2 k^2/4m^2\omega^2\right]$, from which we here adopt the definition that $\Theta(0) = 1/2$ on the right-hand side of (11.6.14). Thus we obtain

$$\left\langle 0 \left| e^{ikq(\tau)} \right| 0 \right\rangle \bigg|_{F_\pm = F} = \exp\left(-\frac{\hbar^2 k^2}{4m^2\omega^2}\right) \exp\left(\frac{kG[F;\tau]}{\sqrt{2}\,m\omega}\right). \qquad (11.6.17)$$

In particular for the expectation value of the displacement operator at time τ, we have

$$\left\langle 0 | q(\tau) | 0 \right\rangle \bigg|_F = -\frac{i}{\sqrt{2}\,m\omega} \int_0^\tau d\tau' \left[F^*(\tau')\,e^{-i\omega(\tau'-\tau)} - F(\tau')\,e^{i\omega(\tau'-\tau)}\right]$$

$$= -\frac{i\,G[F;\tau]}{\sqrt{2}\,m\omega} \equiv \langle q(\tau) \rangle \qquad (11.6.18)$$

as a linear response to the external source F, and note that the expression on the right-hand side of (11.6.18) is real as it should be.

On the other hand, upon multiplying (11.6.17) by $\exp(-ikq)\,dk/2\pi$, where q is a c-number, and integrating over k from $-\infty$ to ∞, gives

$$\left\langle 0 | \delta(q - q(\tau)) | 0 \right\rangle \bigg|_{F=0} = \sqrt{\frac{m^2\omega^2}{\pi\hbar^2}} \exp\left[-\frac{m^2\omega^2}{\hbar^2}\left(q - \langle q(\tau) \rangle\right)^2\right] \qquad (11.6.19)$$

as the probability density for the displacement of the harmonic oscillator at time τ, which initially, at time $t' = 0$, is in its *ground-state*, where $\langle q(\tau) \rangle$ is given in (11.6.18).

The transformation function $\langle nt' | nt' \rangle$ for a closed-time path for a system which started initially in the state $|nt'\rangle$ may be derived by methods similar to the ones used in §6.4 as given below.

To the above end, we write

$$F_\pm(\tau) = F_\pm^{(1)}(\tau) + F_\pm^{(2)}(\tau) \qquad (11.6.20)$$

where $F_+^{(1)}(\tau)$ is switched on only within an interval from t'' to t', and $F_+^{(2)}(\tau)$ is switched on only within an interval from t' to t. $F_-^{(2)}(\tau)$ and $F_-^{(1)}(\tau)$ correspond to the negative sense of time from t back to t' and t' back to t'', respectively.

To obtain the expression for $\langle nt' | nt' \rangle$, we note from (11.6.6), that $\langle 0t'' | 0t'' \rangle^{F_\pm}$ may be rewritten as

11.6 Closed-Time Path & Expectation Value Formalism

$$\langle 0t''|0t''\rangle^{F\pm} = \langle 0t''|0t\rangle^{F-} \langle 0t|0t''\rangle^{F+}$$

$$\times \exp\left[\frac{1}{\hbar^2}\int_{t''}^{t}d\tau\int_{t''}^{t}d\tau'\, F_-^*(\tau)\, e^{-i\omega(\tau-\tau')} F_+(\tau')\right]. \tag{11.6.21}$$

The decompositions of the sources in (11.6.20) with their causal arrangements with $F_+^{(2)}(\tau)$, for example, switched on after $F_+^{(1)}(\tau)$ is switched off, allow us to rewrite (11.6.21) as,

$$\langle 0t''|0t''\rangle^{F\pm} = \langle 0t''|0t'\rangle^{F_-^{(1)}}\, e^{A_-^* A_+}\, \langle 0t'|0t''\rangle^{F_+^{(1)}}\, e^{A_-^* B}\, e^{-B^* A_+}\, \langle 0t'|0t''\rangle^{F_\pm^{(2)}} \tag{11.6.22}$$

where

$$A_\pm = \frac{i}{\hbar}\int_{t''}^{t'}d\tau\, e^{-i\omega(t'-\tau)} F_\pm^{(1)}(\tau) \tag{11.6.23}$$

$$B = \frac{i}{\hbar}\int_{t'}^{t}d\tau\, e^{-i\omega(t'-\tau)}\left[F_+^{(2)}(\tau) - F_-^{(2)}(\tau)\right]. \tag{11.6.24}$$

Equation (11.6.22) is to be compared with the one obtained from a completeness unitarity expansion

$$\langle 0t''|0t''\rangle^{F\pm} = \sum_{N,M=0}^{\infty} \langle 0t''|Nt'\rangle^{F_-^{(1)}}\langle Nt'|Mt'\rangle^{F_\pm^{(2)}}\langle Mt'|0t''\rangle^{F_+^{(1)}} \tag{11.6.25}$$

where, for example, $\langle Mt'|0t''\rangle^{F_+^{(1)}}$ may be written down directly from (11.5.15) with an obvious change of notation, and is given by

$$\langle Mt'|0t''\rangle^{F_+^{(1)}} = \frac{(A_+)^M}{\sqrt{M!}}\langle 0t'|0t''\rangle^{F_+^{(1)}} \tag{11.6.26}$$

and similarly,

$$\langle 0t''|Nt'\rangle^{F_-^{(1)}} = \langle 0t''|0t'\rangle^{F_-^{(1)}}\frac{(A_-^*)^N}{\sqrt{N!}}. \tag{11.6.27}$$

To obtain the expression for $\langle nt'|nt'\rangle_0^{F_\pm^{(2)}}$ for arbitrary n, we expand the exponentials $\exp(A_-^* A_+)$, $\exp(A_-^* B)$, $\exp(-B^* A_+)$ in (11.6.22).

To the above end,

$$e^{A_-^* A_+}\, e^{A_-^* B}\, e^{-B^* A_+}$$

$$= \sum_{a,b,c=0}^{\infty} \frac{(A_-^*)^{a+b}}{\sqrt{(a+b)!}}\frac{(A_+)^{a+c}}{\sqrt{(a+c)!}}\frac{\sqrt{(a+b)!\,(a+c)!}}{a!}\frac{(B)^b}{b!}\frac{(-B^*)^c}{c!}.$$

$$\tag{11.6.28}$$

Upon setting $a + b = N$, $a + c = M$, selecting the diagonal element $N = M \equiv n$ in (11.6.25) we obtain by comparing (11.6.22) and (11.6.25) and using in the process (11.6.26)–(11.6.28), that[11]

$$\langle nt'|nt'\rangle^{F_\pm} = \langle 0t'|0t'\rangle \sum_{a=0}^{n} \frac{n!}{\left[(n-a)!\right]^2 a!}(-|B|^2)^{n-a} \quad (11.6.29)$$

now for arbitrary sources F_+, F_- in operation within the interval from t' to t and t back to t', respectively,

$$|B|^2 = \frac{1}{\hbar^2}\left|\int_{t'}^{t}d\tau\, e^{-i\omega(t'-\tau)}(F_+(\tau) - F_-(\tau))\right|^2 \quad (11.6.30)$$

and we have set $\langle 0t'|0t'\rangle^{F_\pm} \equiv \langle 0t'|0t'\rangle$ as in (11.6.6).

It is physically interesting to consider the initial state to be a thermal mixture of energy states. This is obtained by multiplying $\langle nt'|nt'\rangle$ in (11.6.29) by the Boltzmann factor $\exp(-\hbar\omega n/kT)$ (see also §6.3) and summing over all n,

$$\left(1 - e^{-\hbar\omega/kT}\right)\sum_{n=0}^{\infty} e^{-\hbar\omega n/kT} \langle nt'|nt'\rangle^{F_\pm} \equiv \langle t'|t'\rangle^T \quad (11.6.31)$$

where $\left(1 - e^{-\hbar\omega/kT}\right)$ is the normalization constant for the Boltzmann factor.

The summation in (11.6.31) is carried out by following the steps in (6.3.5)–(6.3.8) to obtain for a thermal mixture at temperature T

$$\langle t'|t'\rangle^T = \langle 0t'|0t'\rangle \exp\left[-\frac{1}{\hbar^2}\frac{1}{(e^{\hbar\omega/kT}-1)}\right.$$

$$\left.\times\left|\int_{t'}^{t}d\tau\, e^{-i\omega(t'-\tau)}[F_+(\tau) - F_-(\tau)]\right|^2\right] \quad (11.6.32)$$

where $\langle 0t'|0t'\rangle$ is given in (11.6.6). Equation (11.6.32) generalizes the expression for $\langle 0t'|0t'\rangle$ in (11.6.6) to finite temperatures with the initial state of the system being a thermal mixture.

$\langle t'|t'\rangle^T$ may be rewritten in a compact form as in (11.6.8). Applications of $\langle t'|t'\rangle^T$ such as in (11.6.18), (11.6.19) are given in Problem 11.18.

Interactions between bosons (and bosons/fermions) may be considered by the addition of interaction Hamiltonians to the free oscillators, in the presence of external sources, and consider the dynamics in the positive sense of time $t' \to t$ and then in the negative sense from t back to t' followed by an application of the quantum dynamical principle in the two segments. We will not, however, go into these details here.

---

[11] Similar methods are used in field theory, see: Manoukian (1991).

Problems

11.1. Work out the details specified below (11.2.19), involving the functional differentiations $(\delta/\delta S(\tau))^2$, to obtain the equality in (11.2.22).

11.2. Carry out the p-integration leading to the result in (11.2.25) and show its equivalence to one in (9.4.21).

11.3. Obtain an expression for $\langle qt|pt'\rangle^{(0)}$ corresponding to the Hamiltonian $H'_{(0)} = p^2/2m - qF(\tau) + pS(\tau)$, for $S(\tau) \neq 0$, generalizing the one in (11.2.22).

11.4. Use (11.2.22) to determine $\langle pt|p't'\rangle^{(0)}$ for the Hamiltonian in (11.2.24).

11.5. Generalize the expression in (11.2.57) for the forced harmonic oscillator by adding a term $pS(\tau)$ to the Hamiltonian in (11.2.32).

11.6. Establish the identity in (11.2.28).

11.7. Verify that $K(\tau,\tau')$ in (11.2.42) satisfies the conditions (i)–(iii) in (11.2.37)–(11.2.41).

11.8. Use the Fourier transforms defined in (11.3.4), (11.3.5) to derive (11.3.13).

11.9. Show that the double Fourier integral in (11.3.18) leads from (11.3.17) to (11.3.19).

11.10. Verify that (11.5.12) is also consistent with (11.5.10) for the creation operator $a^\dagger(\tau)$ by taking the functional differentiation $-i\hbar\delta/\delta F(\tau)$ of the former equation.

11.11. By a method similar to the one used to obtain $\langle nt|0t'\rangle$ in (11.5.15), derive an expression for $\langle nt|n't'\rangle$, and compare your result with the one in (6.2.30).

11.12. Interactions of bosons and fermions were considered in (6.5.18), (6.5.30) with the latter being supersymmetric. Investigate the nature of the transformation functions for such interactions by adding couplings to external sources in the Hamiltonians. What is the significance of the state $\langle 0\tau;\lambda| = \langle 0,0|U(\tau,\lambda)$ for τ large, given that $|0,0\rangle$ denotes the ground-state of the free Bose-Fermi oscillator in (6.5.8)?

11.13. Show that the functional differentiations in (11.6.12), as translation operators, lead to (11.6.13).

11.14. Find $\langle 0|p(\tau)|0\rangle\big|_F$ and $\langle 0|\delta(p-p(\tau))|0\rangle\big|_F$ in analogy to the results in (11.6.18), (11.6.19), respectively, for the momentum description of the oscillator.

11.15. Spell out the details leading from (11.6.28) to (11.6.29) for the transformation function $\langle nt'|nt'\rangle^{F\pm}$.

11.16. Follow the steps through (6.3.5)–(6.3.8) to obtain the expression for a thermal mixture in (11.6.32) starting from (11.6.29).

11.17. Rewrite $\langle t'|t'\rangle^T$ for a thermal mixture in a compact form as in (11.6.8).

11.18. Find the average displacement $\langle t'|q(\tau)|t'\rangle^T$ and the probability density $\langle t'|\delta(q-q(\tau))|t'\rangle^T$, generalizing the results in (11.6.18),

(11.6.19), for an initial state as a thermal mixture with transformation function given in (11.6.32), at finite temperature T.

12
Approximating Quantum Systems

Several approximations were already carried out in earlier chapters. In this respect, for example in Chapter 7, we have considered relativistic corrections to the energy levels of the hydrogen atom, also the non-relativistic Lamb shift, and treated the atom in external electromagnetic fields. In Chapter 8, we have discussed the validity of the exponential law in a two-level system (§8.1) and related approximations involved. In the same chapter, radiation loss in spin precession was studied (§8.4), a computation of the anomalous magnetic moment of the electron was made (§8.5), the problem of quantum decoherence by the environment was considered (§8.7, §8.9) and the so-called geometric phase in the adiabatic approximating regime was investigated (§8.13). Some approximations are also given in Chapter 15 on quantum scattering, just to mention a few of the applications of approximation methods. The present chapter supplements these studies by investigating the nature of several approximation procedures, some of which are related to the above applications in other chapters. Accordingly, the latter material may be read in conjunction with the present one.

Sections 12.1, 12.2 deal with conventional time-independent perturbation theories, followed by one on variational methods. High-order perturbations and related divergent series as applied to an anharmonic oscillator potential is the subject matter of §12.4. In §12.5, we study the so-called semi-classical WKB approximation. Time-dependent perturbation theory is treated in §12.5 dealing, in particular, with the sudden and adiabatic approximations in which a Hamiltonian may change rapidly in time in a very short time interval and in the other extreme a Hamiltonian may change very slowly during a long time span, respectively. In the last section, we study, in the density operator formalism, the response of a system, into consideration, to another system, such as the environment. In this section, we derive the master equation describing the dynamics of the reduced density operator of the system of interest after having traced the density operator of the combined two systems over the variables of the other one.

12.1 Non-Degenerate Perturbation Theory

Consider a Hamiltonian H^0 which, as part of its spectrum, has a discrete non-degenerate one with eigenvalue equation

$$H^0 \left|n\right\rangle_0 = E_n^0 \left|n\right\rangle_0 \tag{12.1.1}$$

where the E_n^0 are non-degenerate and the eigenvectors $\left|n\right\rangle_0$ are orthonormal. We add to H^0 a term H_1, referred to as a perturbation, which is in some sense small in comparison to H^0, thus introducing the Hamiltonian

$$H = H^0 + H_1. \tag{12.1.2}$$

We suppose that for the new system, we have an eigenvalue equation

$$H \left|n\right\rangle = E_n \left|n\right\rangle \tag{12.1.3}$$

and that E_n is near E_n^0 for a given quantum number n. The shift in energy due to the addition of the perturbation H_1 to H^0 is defined by

$$\Delta E_n = E_n - E_n^0 \tag{12.1.4}$$

and we set

$$\left|n\right\rangle = \left|n\right\rangle_0 + \left|n\right\rangle'. \tag{12.1.5}$$

Upon multiplying (12.1.3) from the left by ${}_0\langle n|$ and using (12.1.1)–(12.1.5), we obtain for the energy shift the expression

$$\Delta E_n = \frac{{}_0\langle n|H_1|n\rangle}{{}_0\langle n|n\rangle}. \tag{12.1.6}$$

This is invariant under phase transformations

$$\left|n\right\rangle \rightarrow \left|n\right\rangle e^{i\gamma} \tag{12.1.7}$$

as the phase factor cancels out from the numerator and the denominator in (12.1.6).

To first order in H_1, we may replace $\left|n\right\rangle$ by $\left|n\right\rangle_0$ in (12.1.6), to get for the energy shift

$$\Delta E_n^1 = {}_0\langle n|H_1|n\rangle_0 \tag{12.1.8}$$

or

$$E_n \simeq E_n^0 + E_n^1 \tag{12.1.9}$$

with

$$E_n^1 = {}_0\langle n|H_1|n\rangle_0. \tag{12.1.10}$$

To first order, we write

$$\left|n\right\rangle \simeq \left|n\right\rangle_0 + \left|n\right\rangle_1 \tag{12.1.11}$$

12.1 Non-Degenerate Perturbation Theory

and use the normalizability of $|n\rangle$, $|n\rangle_0$

$$1 = \langle n|n\rangle \simeq 1 + {}_0\langle n|n\rangle_1 + {}_1\langle n|n\rangle_0 \tag{12.1.12}$$

to infer that

$$ {}_0\langle n|n\rangle_1 = -{}_0\langle n|n\rangle_1^* . \tag{12.1.13}$$

That is, ${}_0\langle n|n\rangle_1$, if not zero, is pure imaginary. Now we invoke the invariance of ΔE_n in (12.1.6) under phase transformations as in (12.1.7) to carry out the transformation

$$|n\rangle \to |\bar{n}\rangle = |n\rangle\, e^{i\gamma} \simeq |n\rangle_0 \, (1 + i\gamma) + |n\rangle_1 \tag{12.1.14}$$

and choose

$$\gamma = i\,{}_0\langle n|n\rangle_1 \tag{12.1.15}$$

to note from (12.1.14) that

$${}_0\langle n|\bar{n}\rangle \simeq 1 + i\gamma + {}_0\langle n|n\rangle_1 = 1. \tag{12.1.16}$$

The latter means that if we write

$$|\bar{n}\rangle \simeq |n\rangle_0 + |\bar{n}\rangle_1 \tag{12.1.17}$$

then

$${}_0\langle n|\bar{n}\rangle_1 = 0. \tag{12.1.18}$$

Hence from now on, we may choose $|n\rangle_1$ such that

$${}_0\langle n|n\rangle_1 = 0 \tag{12.1.19}$$

using the same notation as in (12.1.11).

A non-homogeneous equation satisfied by $|n\rangle_1$ is readily obtained from (12.1.3) by using (12.1.9), (12.1.11) together with the eigenvalue equation (12.1.1) for the unperturbed system giving

$$\left(H^0 - E_n^0\right)|n\rangle_1 = \left(E_n^1 - H_1\right)|n\rangle_0 . \tag{12.1.20}$$

Since E_n^1 is determined from (12.1.10) and $|n\rangle_0$ is supposed to be known, this equation may be used to obtain $|n\rangle$ of the perturbed system to first order.

If the spectrum of H^0 consists only of a discrete non-degenerate one, then we may rewrite the right-hand of (12.1.20), by using in the precess of the completeness of the eigenvector of H^0, as

$$\left(E_n^1 - H_1\right)|n\rangle_0 = E_n^1|n\rangle_0 - \sum_m |m\rangle_0\, {}_0\langle m|H_1|n\rangle_0$$

$$= E_n^1|n\rangle_0 - E_n^1|n\rangle_0 - \sum_{m \neq n} |m\rangle_0\, {}_0\langle m|H_1|n\rangle_0$$

$$= -\sum_{m\neq n} |m\rangle_0 \, _0\langle m|H_1|n\rangle_0 \tag{12.1.21}$$

where in writing the second equality we have used (12.1.10). From (12.1.20), (12.1.21) gives

$$|n\rangle_1 = a|n\rangle_0 + \sum_{m\neq n} |m\rangle_0 \frac{1}{(E_n^0 - E_m^0)} \, _0\langle m|H_1|n\rangle_0 \tag{12.1.22}$$

where a is a constant. In general, the vector $a|n\rangle_0$ is introduced since for a Hamiltonian with a discrete non-degenerate spectrum, $|n\rangle_0$, up to a multiplicative constant, is the only vector annihilated by $H^0 - E_n^0$ in (12.1.20). Upon multiplying (12.1.22) by $_0\langle n|$ and using (12.1.19), we may infer that $a = 0$.

To first order, we then have from (12.1.11), (12.1.22)

$$|n\rangle = |n\rangle_0 + \sum_{m\neq n} |m\rangle_0 \frac{1}{(E_n^0 - E_m^0)} \, _0\langle m|H_1|n\rangle_0. \tag{12.1.23}$$

If $_0\langle m|H_1|n\rangle_0$ in (12.1.10) is zero, one may go to second order in H_1. To this end, we note from (12.1.6), (12.1.5), that the expression for the energy shift may be rewritten as

$$\Delta E_n = \frac{_0\langle n|H_1|n\rangle_0}{1 + _0\langle n|n\rangle'} + \frac{_0\langle n|H_1|n\rangle'}{1 + _0\langle n|n\rangle'}. \tag{12.1.24}$$

With the orthogonality relation in (12.1.19), ΔE_n, up to *second* order in H_1, is then given by

$$\Delta E_n \simeq {}_0\langle n|H_1|n\rangle_0 + {}_0\langle n|H_1|n\rangle_1 \tag{12.1.25}$$

or

$$E_n \simeq E_n^0 + {}_0\langle n|H_1|n\rangle_0 + {}_0\langle n|H_1|n\rangle_1. \tag{12.1.26}$$

Using (12.1.23) this may be rewritten as

$$E_n = E_n^0 + {}_0\langle n|H_1|n\rangle_0 + \sum_{m\neq n} \frac{|{}_0\langle m|H_1|n\rangle_0|^2}{E_n^0 - E_m^0} + \ldots \tag{12.1.27}$$

As an application, though a formal one,[1] consider the Hamiltonian

$$H = \frac{p^2}{2m} + \frac{m\omega^2}{2}x^2 + \lambda x^3 \tag{12.1.28}$$

with the interaction term $H_1 = \lambda x^3$ added to the harmonic oscillator Hamiltonian §6.1 in one dimension.

---

[1] Here it is sufficient to note that for $\lambda < 0$, $x \to \infty$ or for $\lambda > 0$, $x \to -\infty$, the interaction term $\to -\infty$ being unbounded from below in either cases.

12.1 Non-Degenerate Perturbation Theory

Since H_1 is odd in x, and the harmonic oscillator wavefunctions $\psi_n(x)$ (§6.1, (6.1.39), (6.1.41), (6.1.42)) have definite parities, i.e., $|\psi_n(x)|^2$ is even, we obtain from (12.1.10)

$$E_n^1 = 0. \tag{12.1.29}$$

To second order, we have to evaluate

$$E_n^2 = \sum_{m \neq n} \frac{|_0\langle m|H_1|n\rangle_0|^2}{E_n^0 - E_m^0} \tag{12.1.30}$$

where $E_n^0 = \hbar\omega(n + 1/2)$. By taking advantage of the completeness relations

$$_0\langle m|x^3|n\rangle_0 = \sum_{m'} {}_0\langle m|x|m'\rangle_0 \, _0\langle m'|x^2|n\rangle_0 \tag{12.1.31}$$

and so on, and using, in the process, such matrix elements as in (6.1.45), we obtain the selection rules $m = n \pm 3$, $m = n \pm 1$ for the non-vanishing of $_0\langle m|x^3|n\rangle_0$, leading to

$$_0\langle n+3|x^3|n\rangle_0 = \left(\frac{\hbar}{2m\omega}\right)^{3/2} \sqrt{(n+1)(n+2)(n+3)} \tag{12.1.32}$$

$$_0\langle n-3|x^3|n\rangle_0 = \left(\frac{\hbar}{2m\omega}\right)^{3/2} \sqrt{n(n-1)(n-2)} \tag{12.1.33}$$

$$_0\langle n+1|x^3|n\rangle_0 = 3\left(\frac{\hbar}{2m\omega}\right)^{3/2} (n+1)\sqrt{n+1} \tag{12.1.34}$$

$$_0\langle n-1|x^3|n\rangle_0 = 3\left(\frac{\hbar}{2m\omega}\right)^{3/2} n\sqrt{n}. \tag{12.1.35}$$

From (12.1.30), (12.1.27), one readily obtains

$$E_n \simeq \hbar\omega\left(n + \frac{1}{2}\right) - \left(\frac{\hbar}{2m\omega}\right)^3 \frac{(30n^2 + 30n + 11)}{\hbar\omega}\lambda^2 \tag{12.1.36}$$

to second order in λ.

For the eigenvector $|n\rangle$ to first order in λ, we have from (12.1.23) and (12.1.32)–(12.1.35),

$$|n\rangle \simeq |n\rangle_0 - \frac{\lambda}{3\hbar\omega}\left(\frac{\hbar}{2m\omega}\right)^{3/2}\sqrt{(n+1)(n+2)(n+3)}\,|n+3\rangle_0$$

$$+ \frac{\lambda}{3\hbar\omega}\left(\frac{\hbar}{2m\omega}\right)^{3/2}\sqrt{n(n-1)(n-2)}\,|n-3\rangle_0$$

$$- 3\frac{\lambda}{\hbar\omega}\left(\frac{\hbar}{2m\omega}\right)^{3/2}(n+1)\sqrt{n+1}\,|n+1\rangle_0$$

$$+3\frac{\lambda}{\hbar\omega}\left(\frac{\lambda}{2m\omega}\right)^{3/2}n\sqrt{n}\,|n-1\rangle_0 \qquad (12.1.37)$$

and we note that $_0\langle n|n\rangle_1 = 0$ where $|n\rangle_1$ consists of the last four terms in this equation.

The series in (12.1.27) may be extended to arbitrary orders for a Hamiltonian with a discrete non-degenerate spectrum. Such expansions are referred to as Rayleigh-Schrödinger series.

12.2 Degenerate Perturbation Theory

Suppose that E_n^0 in the discrete spectrum of a Hamiltonian H^0 is $k(n)$-fold degenerate,

$$H^0\,|n,\nu(n)\rangle_0 = E_n^0\,|n,\nu(n)\rangle_0 \qquad (12.2.1)$$

where the eigenvector $|n,\nu(n)\rangle_0$, $\nu(n) = 1,\ldots,k(n)$, may be chosen to be orthonormal. Now we add a perturbation H_1 to H^0, defining a Hamiltonian H, and consider the eigenvalue problem

$$H\,|n\rangle = E_n\,|n\rangle. \qquad (12.2.2)$$

When the perturbation H_1 is let to go to zero, $|n\rangle$ may not necessarily go to one of the particular eigenvectors $|n,\nu(n)\rangle_0$ but, in general, to some linear combination of them. Accordingly, we introduce an eigenvector of H^0 of the form

$$|n\rangle_0 = \sum_{\nu(n)=1}^{k(n)} a_{\nu(n)}\,|n,\nu(n)\rangle_0 \qquad (12.2.3)$$

with some expansion coefficients $a_\nu(n)$, $\nu = 1,\ldots,k(n)$.

Upon multiplying (12.2.2) from the left by $_0\langle n,\nu(n)|$ and using (12.2.1), we obtain

$$\Delta E_n \,_0\langle n,\nu(n)|n\rangle = \,_0\langle n,\nu(n)|H_1|n\rangle \qquad (12.2.4)$$

where $\Delta E_n = E_n - E_n^0$. To lowest order in H_1, we take $|n\rangle$ to coincide with $|n\rangle_0$, and use (12.2.3) to rewrite (12.2.4) as

$$\sum_{\nu'(n)}\left[E_n^1\delta_{\nu\nu'} - \,_0\langle n,\nu(n)|H_1|n,\nu'(n)\rangle_0\right]a_{\nu'(n)} = 0. \qquad (12.2.5)$$

where $\Delta E_n = E_n^1$ to first order.

To have a non-trivial solution for the $a_{\nu(n)}$ it is necessary that

$$\det\left|E_n^1\delta_{\nu\nu'} - \,_0\langle n,\nu(n)|H_1|n,\nu'(n)\rangle_0\right| = 0 \qquad (12.2.6)$$

otherwise one may be able to invert the matrix

$$[M_{\nu\nu'}] = [E_n^1 \delta_{\nu\nu'} - {}_0\langle n, \nu(n)|H_1|n, \nu'(n)\rangle_0]$$

and get zero for the $a_{\nu(n)}$. Equation (12.2.6) is of a $k(n)^{\text{th}}$ order one in term of the matrix elements ${}_0\langle n, \nu(n)|H_1|n, \nu'(n)\rangle_0$ and is called the *secular equation*.

From (12.2.6), we may infer that if, *a priori*, the eigenvectors $|n, \nu'(n)\rangle_0$ are properly chosen, by considering suitable linear combinations of eigenvectors of H^0 with eigenvalue E_n^0, such that $[{}_0\langle n, \nu(n)|H_1|n, \nu'(n)\rangle_0]$ are diagonal, them the matrix $[M_{\nu\nu'}]$ would be also diagonal and the analysis leading to (12.2.6) would simply give the solutions $E_{n\nu}^1 = \langle n, \nu(n)|H_1|n, \nu(n)\rangle$.

As an application, consider the Hamiltonian of the hydrogen atom in a sufficiently strong uniform magnetic field such that the fine-structure contribution V_F may be neglected as given in (7.9.13), and we treat the $e^2 \mathbf{A}_{ext}^2/2Mc^2$ term as a perturbation. That is, we set

$$H = H^0 + H_1 \quad (12.2.7)$$

where

$$H^0 = H_C - \frac{eB}{2Mc}(L_z + 2S_z) \quad (12.2.8)$$

$$H_1 = \frac{e^2 B^2}{8Mc^2}(x^2 + y^2) \quad (12.2.9)$$

(see (7.9.5)), where $\mathbf{B} = (0, 0, B)$, $\mathbf{A} = B(-y, x, 0)/2$, and H_C is the hydrogen atom Coulomb Hamiltonian. The eigenvalue of H^0 were given in (7.9.15)

$$E^0(n, m, m_s) = E_n^C + \eta(m + 2m_s), \quad \eta = -\frac{e\hbar B}{2Mc} > 0. \quad (12.2.10)$$

$m_s = \pm 1/2$, and the eigenstates are given in (7.9.14). Here $E_n^C = -\text{Ry}/n^2$.

Consider $n = 2$. In particular, we note that the eigenvalue E^0 in (12.2.10) is degenerate for $(\ell = 1, m = 0, m_s = 1/2)$ and $(\ell = 0, m = 0, m_s = 1/2)$, corresponding to eigenstates

$$R_{21}(r)Y_{10}(\Omega)\begin{pmatrix}1\\0\end{pmatrix} \equiv \Phi_1, \quad R_{20}(r)Y_{00}(\Omega)\begin{pmatrix}1\\0\end{pmatrix} \equiv \Phi_2 \quad (12.2.11)$$

with eigenvalue $E^0 = E_2^C + \eta$. We note that $(x^2 + y^2) = r^2 \sin^2 \theta$, and from the identity (5.8.41), used twice, that $\cos^2 \theta \, Y_{\ell m}$ is a linear combination of $Y_{\ell \pm 2, m}$ and $Y_{\ell m}$. Accordingly, from this, or by direct computation, we may infer that $\langle Y_{\ell' 0}|(1 - \cos^2 \theta)|Y_{\ell 0}\rangle$ is diagonal for $\ell, \ell' = 0, 1$, and from (5.8.43) or by direct evaluation

$$\langle Y_{00}|(1 - \cos^2 \theta)|Y_{00}\rangle = \frac{2}{3}, \quad \langle Y_{10}|(1 - \cos^2 \theta)|Y_{10}\rangle = \frac{2}{5}. \quad (12.2.12)$$

On the other hand from (7.3.35),

$$\langle R_{20}|r^2|R_{20}\rangle = 42 a_0^2, \qquad \langle R_{21}|r^2|R_{21}\rangle = 30\, a_0^2 \qquad (12.2.13)$$

where a_0 is the Bohr radius.

The matrix M in (12.2.5), (12.2.6) is then in diagonal form leading from (12.2.12), (12.2.13) to the splitting of the level $E_2^C + \eta$ to

$$E_2^C + \eta + 28\left(\frac{e^2 B^2}{8Mc^2}\right) a_0^2, \qquad E_2^C + \eta + 12\left(\frac{e^2 B^2}{8Mc^2}\right) a_0^2. \qquad (12.2.14)$$

12.3 Variational Methods

Variational methods will be used in Chapter 13 in minimizing a functional of the electron density in the so-called Thomas-Fermi atom (§13.1) as a first step in obtaining an expression for the ground-state energy of atoms as a function of the atomic number Z, and in Chapter 14 in the investigation of a "no-binding" theorem in the process of establishing the stability of matter. The variational method considered in the present section in its simplest form is the following one.

Given a Hamiltonian H, the problem considered is to choose a trial wavefunction Ψ, depending, in general, on one or more parameters, to minimize the following expectation value

$$F[\Psi] = \frac{\langle \Psi|H|\Psi\rangle}{\langle \Psi|\Psi\rangle} \qquad (12.3.1)$$

considered as a functional of Ψ. In practice, such a minimizing is achieved by optimizing $F[\Psi]$ over the parameter(s) on which *a priori* chosen Ψ may depend.

From the very definition of a ground-state energy, or just the lowest point of the spectrum of a Hamiltonian, such a procedure cannot provide a lower bound to it. Thus given a trial wavefunction, it will provide, in general, an *upper bound*[2] to the exact ground-state energy E_0 (Chapter 4). The assessment of the accuracy of such a bound *as an estimate* of E_0 is, in general, not always an easy task. On the other hand, if one also derives a lower bound to E_0 and it turns out that both bounds are close to each other, then such a procedure would provide an excellent way of estimating E_0.

As an example, consider first the inequality in (3.1.8) rewritten, for the simplicity of the notation, in terms of a variable z in such units that in one dimension

$$-\frac{1}{4}\langle z^2 g^2(z)\rangle \leq \left\langle -\frac{d^2}{dz^2} - \frac{1}{2}\frac{d}{dz}(zg(z))\right\rangle. \qquad (12.3.2)$$

For example, for

---

[2] That this method leads to an *upper* bound for the ground-state energy, in general, is not sufficiently emphasized in the literature.

12.3 Variational Methods

$$g(z) = 4 + 2z^2 \qquad (12.3.3)$$

the inequality (12.3.2) leads to

$$2 \leqslant \left\langle -\frac{d^2}{dz^2} + V(z) \right\rangle \qquad (12.3.4)$$

generating an anharmonic potential with fixed couplings (see also Problem 12.8),

$$V(z) = z^2 + 4z^4 + z^6. \qquad (12.3.5)$$

As a trial wavefunction for the ground-state of the Hamiltonian

$$H = -\frac{d^2}{dz^2} + V(z) \qquad (12.3.6)$$

we may choose

$$\Psi(z) = \frac{1}{(2\pi)^{1/4}\sigma^{1/2}} \exp\left(-\frac{z^2}{4\sigma^2}\right) \qquad (12.3.7)$$

where the parameter $\sigma > 0$ will be chosen optimally. Upon substituting (12.3.7) in (12.3.1), (12.3.4), we obtain

$$2 \leqslant E_0 \leqslant F[\Psi] = \frac{1}{4\sigma^2} + \sigma^2 + 12\sigma^4 + 15\sigma^6. \qquad (12.3.8)$$

The expression on the extreme right-hand side of (12.3.8) is minimized for σ^2 about .188, leading to the satisfactory bounds

$$2 \leqslant E_0 \leqslant 2.0416. \qquad (12.3.9)$$

The above interesting method was used in obtaining the *exact* ground-state energies of the harmonic oscillator (§6.1), of the hydrogen atom (§7.1) and for the Dirac delta potential (§4.2), by appropriately choosing, in the process, the trial wavefunctions in such a way that the resulting upper bounds *coincide* with the respective lower bounds, and this procedure may be applied to other cases as well (see also (3.3.6), (3.3.7), (4.2.83), (4.2.84)).

It is easy to apply the above method to generate, a priori, potentials with variable couplings. For example, consider the function

$$g(z) = \frac{1}{z}\frac{\sqrt{8\lambda}}{c} + c + \sqrt{2\lambda}\, z \qquad (12.3.10)$$

in (12.3.2), where $\lambda > 0, c > 0$ are constants. This leads to

$$\frac{1}{4}(zg(z))^2 - \frac{1}{2}(zg(z))' = -\frac{1}{2}\left(c - \frac{4\lambda}{c^2}\right) + \left(\frac{c^2}{4} + \frac{2\lambda}{c}\right)z^2$$

$$+ \sqrt{\frac{\lambda}{2}}\, cz^3 + \frac{\lambda}{2}z^4. \qquad (12.3.11)$$

Using the elementary bound

$$\sqrt{\frac{\lambda}{2}}\, cz^3 = 2\left(\sqrt{\frac{\lambda}{2}}\, z^2\right)\left(\frac{cz}{2}\right) \leqslant \frac{\lambda}{2}z^4 + \frac{c^2}{4}z^2 \qquad (12.3.12)$$

we have

$$\frac{1}{4}(zg(z))^2 - \frac{1}{2}(zg(z))' \leqslant -\frac{1}{2}\left(c - \frac{4\lambda}{c^2}\right) + \frac{1}{2}\left(c^2 + \frac{4\lambda}{c}\right)z^2 + \lambda z^4 \quad (12.3.13)$$

thus obtaining

$$\frac{1}{2}\left(c - \frac{4\lambda}{c^2}\right) \leqslant \left\langle -\frac{d^2}{dz^2} + \frac{1}{2}\left(c^2 + \frac{4\lambda}{c}\right)z^2 + \lambda z^4\right\rangle \qquad (12.3.14)$$

generating an anharmonic potential $V(z) = \lambda_0 z^2 + \lambda z^4$. The usefulness and limitations of this inequality in deriving bounds on the ground-state energy is the subject of Problem 12.9.

As another example, consider the hydrogen molecule in which one of its electrons has been stripped off forming an ion of net charge $+|e|$. To generate the potential energy for the latter, we introduce a vector field (see also (3.4.22)–(3.4.24))

$$\mathbf{F}(\mathbf{x}) = \frac{e^2}{2}\left(\frac{\mathbf{x} - \mathbf{R}/2}{|\mathbf{x} - \mathbf{R}/2|} + \frac{\mathbf{x} + \mathbf{R}/2}{|\mathbf{x} + \mathbf{R}/2|}\right) \qquad (12.3.15)$$

where $\mathbf{x}$ denotes the coordinate of the electron and $\pm \mathbf{R}/2$ denote position vectors of the two protons, $\mathbf{R} = (0, 0, R)$. Now invoking positivity

$$\left\|\left(\frac{\hbar \nabla}{\sqrt{2m}} + \frac{\sqrt{2m}}{\hbar}\mathbf{F}\right)\chi\right\|^2 \geqslant 0 \qquad (12.3.16)$$

we obtain the simple bound

$$-\frac{2m}{\hbar^2}\langle \mathbf{F}^2\rangle \leqslant \left\langle \frac{\mathbf{p}^2}{2m} - \nabla \cdot \mathbf{F}\right\rangle \qquad (12.3.17)$$

and the explicit equality

$$-\nabla \cdot \mathbf{F} = -\frac{e^2}{|\mathbf{x} - \mathbf{R}/2|} - \frac{e^2}{|\mathbf{x} + \mathbf{R}/2|}. \qquad (12.3.18)$$

On the other hand, $\mathbf{F}$ in (12.3.15) being equal to $e^2/2$ times the sum of two unit vectors, we have

$$\mathbf{F}^2 \leqslant e^4. \qquad (12.3.19)$$

From (12.3.17)–(12.3.19), we obtain

$$-\frac{2me^4}{\hbar^2} + \frac{e^2}{R} \leqslant \left\langle \frac{\mathbf{p}^2}{2m} + V(\mathbf{x}) \right\rangle \qquad (12.3.20)$$

with $R = |\mathbf{R}|$,

$$V(\mathbf{x}) = -\frac{e^2}{|\mathbf{x} - \mathbf{R}/2|} - \frac{e^2}{|\mathbf{x} + \mathbf{R}/2|} + \frac{e^2}{R} \qquad (12.3.21)$$

where in writing (12.3.20), we have added the term e^2/R on both sides of (12.3.17). $V(\mathbf{x})$ denotes the potential energy of the one-electron (ionized) hydrogen molecule, and $\mathbf{p}^2/2m + V(\mathbf{x})$ denotes the corresponding Hamiltonian taking into account the fact that the mass of the proton $\gg$ mass m of the electron.

As a trial wavefunction for the *electron*, we choose the normalized function

$$\Psi(\mathbf{x}) = \frac{\beta^{3/2}}{\sqrt{2\pi}N}\left[\exp\left(-\beta\left|\mathbf{x} - \frac{\mathbf{R}}{2}\right|\right) + \exp\left(-\beta\left|\mathbf{x} + \frac{\mathbf{R}}{2}\right|\right)\right] \qquad (12.3.22)$$

where

$$N = \sqrt{1 + \left(1 + \lambda + \frac{\lambda^2}{3}\right)\exp-\lambda}, \qquad \lambda \equiv \beta R \qquad (12.3.23)$$

and the parameters β, R will be chosen optimally.

If we set

$$\phi_1 = \frac{\beta^{3/2}}{\sqrt{\pi}}\exp\left(-\beta\left|\mathbf{x} - \frac{\mathbf{R}}{2}\right|\right), \qquad \phi_2 = \frac{\beta^{3/2}}{\sqrt{\pi}}\exp\left(-\beta\left|\mathbf{x} + \frac{\mathbf{R}}{2}\right|\right) \qquad (12.3.24)$$

then from symmetry

$$\langle\Psi|H|\Psi\rangle = \frac{1}{N^2}\left[\langle\phi_1|H|\phi_1\rangle + \langle\phi_2|H|\phi_1\rangle\right]. \qquad (12.3.25)$$

The computation of the above matrix elements is straightforward (see Problem 12.10) but tedious and are explicitly given by

$$\langle\phi_1|H|\phi_1\rangle = \frac{e^2}{a_0 N^2}\left[\frac{\xi^2}{2} - \xi + \xi\left(1 + \frac{1}{\lambda}\right)e^{-2\lambda}\right] \qquad (12.3.26)$$

$$\langle\phi_2|H|\phi_1\rangle = \frac{e^2}{a_0 N^2}\left[\frac{\xi^2}{2}\left(1 + \lambda - \frac{\lambda^2}{3}\right)e^{-\lambda} - 2\xi(1 + \lambda)e^{-\lambda} + \frac{(N^2-1)}{\lambda}\xi\right] \qquad (12.3.27)$$

where $a_0 = \hbar^2/me^2$, and we have introduced the parameter,

$$\xi = \beta a_0. \qquad (12.3.28)$$

From (12.3.25)–(12.3.28), we then have the following upper bound for the ground-state energy of the molecule

where

$$E_0 \leqslant \frac{e^2}{a_0} \frac{1}{N^2(\lambda)} \left[\frac{\xi^2}{2} A(\lambda) - \xi B(\lambda) \right] \quad (12.3.29)$$

where

$$A(\lambda) = 1 + \left(1 + \lambda - \frac{\lambda^2}{3}\right) e^{-\lambda} \quad (12.3.30)$$

$$B(\lambda) = 1 - \left(1 + \frac{1}{\lambda}\right) e^{-2\lambda} + \left(\frac{5\lambda}{3} + 1 - \frac{1}{\lambda}\right) e^{-\lambda} \quad (12.3.31)$$

and $N = N(\lambda)$ is defined in (12.3.23).

Optimizing the right-hand side of (12.3.29) over ξ gives

$$\xi = B(\lambda)/A(\lambda). \quad (12.3.32)$$

Upon substituting (12.3.32) in (12.3.29), we obtain

$$E_0 \leqslant -\frac{e^2}{2a_0} \frac{B^2(\lambda)}{N^2(\lambda) A(\lambda)}. \quad (12.3.33)$$

Optimizing the right-hand side of (12.3.33) over λ yields $\lambda \simeq 2.48$ leading to

$$E_0 \leqslant -\frac{e^2}{2a_0} 1.173 \quad (12.3.34)$$

and $\xi = 1.238$. From $\xi = \beta a_0$, $\lambda = \beta R$, these give $R = 2.003\, a_0$. All told, (12.3.20), (12.3.34) lead to the bounds

$$-\frac{me^4}{2\hbar^2} 3 \leqslant E_0 \leqslant -\frac{me^4}{2\hbar^2} 1.173. \quad (12.3.35)$$

By introducing a more complicated trial wavefunction depending on more parameters, one expects that the upper bound in (12.3.35) may be further reduced. The control of the lower bound, however, is more difficult.

Needless to say, the variational procedure may be carried out, optimizing the expectation value in (12.3.1) over the parameters that a trial wavefunction may depend, even if no lower bound to the exact ground-state energy E_0 is available. When both upper and lower bounds to E_0 are derived one may, or may not, be able to assess this estimation depending on how close, or not close, these two bounds are to each other, respectively. The first excited energy level to a given Hamiltonian may be also estimated by the variational procedure by optimizing the expectation value in (12.3.1) by choosing a trial wavefunction which is orthogonal to the one chosen for the ground-state, and extend further the analysis in a similar fashion, for other excited energy states.

12.4 High-Order Perturbations, Divergent Series; Padé Approximants

In this section, we investigate the nature of the high-order terms in the Rayleigh-Schrödinger perturbation series for the so-called anharmonic oscillator involving a x^4 term in the potential energy in addition to the harmonic oscillator x^2 one. This interaction turns out to be important as the perturbation series diverges and provides a prototype for other divergent series and has also been useful in clarifying related aspects in field theory. We first derive a dispersion relation that relates the ground-state energy for positive coupling, for which the theory is defined, in terms of an integral restricted to negative coupling. This integral leads to a formal expression for the perturbation coefficients of the ground-state energy to any order. We then carry out a qualitative study of these coefficients, using path integrals, which clearly shows that they grow factorially with the order for high orders implying the divergence of the perturbation series. Finally we comment on a procedure which re-arranges the perturbation terms, referred to as the method of Padé approximants, that converges, in the limit, to the actual energy value.

Consider an analytic function $f(\xi')$ of a complex variable ξ' in the complex cut-plane shown in Figure 12.1. Cauchy's Theorem implies that

$$f(\xi) = \frac{1}{2\pi i} \oint_C d\xi' \frac{f(\xi')}{(\xi' - \xi)} \tag{12.4.1}$$

where ξ lies within the contour away from the cut axis. Assuming that there is no contribution to the integral from the circles of radii $R \to \infty$, $\delta \to 0$, we obtain

$$f(\xi) = \frac{1}{2\pi i} \int_{-\infty}^{0} d\xi' \frac{\text{Disc } f(\xi')}{(\xi' - \xi)} \tag{12.4.2}$$

where $\text{Disc } f(\xi')$ is the discontinuity of $f(\xi')$ across the cut, i.e.,

$$\text{Disc } f(\xi') = f(\xi' + i\varepsilon) - f(\xi' - i\varepsilon), \qquad \varepsilon \to +0. \tag{12.4.3}$$

The variable ξ will be chosen to be real and positive for which $f(\xi)$ is real. Equation (12.4.2) is referred to as a dispersion relation relating $f(\xi)$ to the discontinuity of $f(\xi')$ across the cut.

One may subtract from $f(\xi)$ its value at $\xi \to +0$, obtaining the once subtracted dispersion relation

$$f(\xi) - f(0) = \frac{1}{2\pi i} \int_{-\infty}^{0} \xi \, d\xi' \frac{\text{Disc } f(\xi')}{\xi'(\xi' - \xi)}. \tag{12.4.4}$$

A formal series expansion

$$f(\xi) - f(0) = \sum_{K \geqslant 1} f_K (\xi)^K \tag{12.4.5}$$

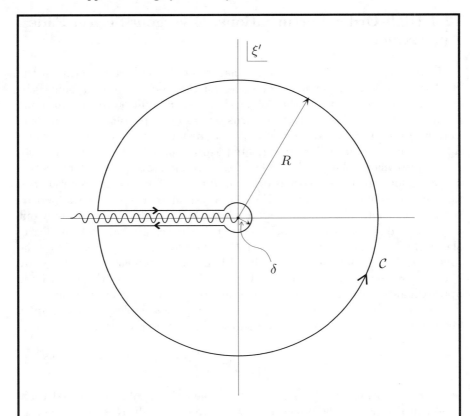

Fig. 12.1. Contour of integration $\mathcal{C}$ in the complex cut-plane ξ' used in (12.4.1). The radii of the circles are taken to be $R \to \infty$, $\delta \to 0$.

then leads to the following expression for the expansion coefficients

$$f_K = \frac{1}{2\pi i} \int_{-\infty}^{0} \frac{\mathrm{d}\xi'}{(\xi')^{K+1}} \operatorname{Disc} f(\xi') \tag{12.4.6}$$

where we have used the expansion

$$\frac{\xi}{(\xi' - \xi)} = \sum_{K \geqslant 1} \left(\frac{\xi}{\xi'}\right)^K \tag{12.4.7}$$

in (12.4.4).

In the notation of (12.3.6), we consider the Hamiltonian

$$H = -\frac{\mathrm{d}^2}{\mathrm{d}z^2} + z^2 + \lambda z^4, \qquad \lambda > 0. \tag{12.4.8}$$

12.4 High-Order Perturbations, Divergent Series; Padé Approximants

The nature of the potential $V(z) = z^2 + \lambda z^4$ is quite different for $\lambda > 0$ and $\lambda < 0$ (see Figure 12.2). In the latter case, the potential develops unstable states with energies involving imaginary parts and decay through the potential barriers of finite widths. Such an imaginary part for the ground-state energy $E(\lambda)$, when λ is continued to $|\lambda|\exp(\pm i\pi)$ from $\lambda > 0$, is what is needed to obtain the expansion coefficients E_K in a formal perturbation expansion ($\lambda > 0$),

$$E(\lambda) - E(0_+) = \sum_{K \geqslant 1} E_K(\lambda)^K \tag{12.4.9}$$

using an integral as in (12.4.6). In the present notation used $E(0_+) = 1$, corresponding to the harmonic oscillator potential.

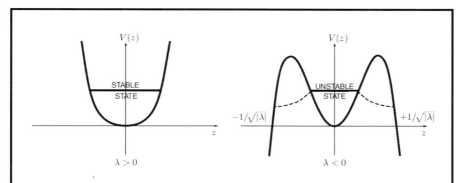

Fig. 12.2. Shapes of the potential energy for positive and negative λ. For $\lambda < 0$ the potential develops unstable decaying states with energies involving imaginary parts. The potential energy in the latter case is also unbounded from below. The figures are not based on actual numerical values.

For $\lambda < 0$, the potential is *unbounded* from below, with the properties of the system quite different from that with $\lambda > 0$, and the theory cannot just be extended from $\lambda > 0$ to $\lambda < 0$. Technically, this corresponds to the fact in the complex λ-plane, $E(\lambda)$ may be analytically continued at most only to a cut-plane with the cut along the negative real axis of λ. This, in particular, leads to a discontinuity of $E(\lambda)$ along the negative real axis, i.e., $\bigl[E(\lambda + i0) - E(\lambda - i0)\bigr] \neq 0$ for $\lambda < 0$, and gives rise to an imaginary part for the energy $E(\lambda)$ as discussed above.

Assuming the analyticity property of $E(\lambda)$ in the cut-plane and that no contribution arises from integrations over the circles of radii $R \to \infty$, $\delta \to 0$ in Figure 12.1 in the complex λ-plane,

$$E(\lambda + i0) - E(\lambda - i0) = 2i\,\text{Im}\,E(\lambda + i0) \tag{12.4.10}$$

for the discontinuity across the cut, we may infer from (12.4.6) that

$$E_K = \frac{1}{\pi} \int_{-\infty}^{0} d\lambda \, \frac{\operatorname{Im} E(\lambda)}{\lambda^{K+1}} \qquad (12.4.11)$$

for the coefficients E_K in (12.4.9), where $\operatorname{Im} E(\lambda) = \operatorname{Im} E(\lambda + i0)$ approaching from above the negative axis. We will not, however, go into the proof of the analyticity property of $E(\lambda)^3$ for which (12.4.10), (12.4.11) are assumed to hold.

The perturbation series for $E(\lambda)$ in (12.4.9) also diverges. We will provide a qualitative analysis of this fact by using path integrals. Before doing this, however, we will show how the ground-state energy may be formally extracted from a time evolution analysis if time t is replaced by $-i\hbar\beta$ and the limit $\beta \to \infty$ is taken.

To the above end, suppose that $E_1 > E_0$ denotes the energy level of a given Hamiltonian H just above the ground-state energy E_0, then

$$\langle \psi | e^{-\beta H} | \psi \rangle = \int_{E_0}^{\infty} d\|P_H(\lambda)\psi\|^2 \, e^{-\beta\lambda}$$

$$= e^{-\beta E_0} \left\{ \|P_H(E_0)\psi\|^2 + \int_{E_1}^{\infty} d\|P_H(\lambda)\psi\|^2 \, e^{-\beta(\lambda - E_0)} \right\}$$

$$(12.4.12)$$

where $P_H(\lambda)$ is the spectral measure of the given Hamiltonian (§1.8). Upon taking the limit $\beta \to \infty$, we obtain

$$\langle \psi | e^{-\beta H} | \psi \rangle \longrightarrow e^{-\beta E_0} \|P_H(E_0)\psi\|^2 \qquad (12.4.13)$$

for $P_H(E_0)|\psi\rangle$ not the zero vector.

Consider also, for example, the Green function of the harmonic oscillator given in (9.2.23). For $x' = x$, $T \to -i\hbar\beta$, one has explicitly,

$$G(xT; x0)\big|_{T=-i\hbar\beta} = \left(\frac{m\omega}{2\pi i\hbar \sin\omega T}\right)^{1/2}$$

$$\times \exp\left(\frac{im\omega x^2}{\hbar}\left[\cot\omega T - \csc\omega T\right]\right)\bigg|_{T=-i\hbar\beta}$$

$$\longrightarrow e^{-\beta(\hbar\omega/2)} \left(\frac{m\omega}{\pi\hbar}\right)^{1/2} \exp\left(-\frac{m\omega}{\hbar} x^2\right) \qquad (12.4.14)$$

for $\beta \to \infty$, where, again, we recognize the coefficient of β, in the first exponential, denoting the ground-state energy.

To obtain the imaginary part of $E(\lambda)$ for $\lambda < 0$ in (12.4.11), (with λ taken to approach the negative axis from above), we consider the imaginary part of the trace of the operator $\exp(-\beta H)$ for β large, with the trace normalized

---

[3] For the relevant details see, Loeffel et al. (1969), and references therein.

12.4 High-Order Perturbations, Divergent Series; Padé Approximants

with respect to the corresponding one for the harmonic oscillator one. For $K \to \infty$, we may also consider $\lambda \sim 0_-$. That is, we investigate the nature of the function

$$F(\lambda, \beta) = \operatorname{Im}\left(\frac{1}{C}\operatorname{Tr}\left[e^{-\beta H}\right]\right), \qquad \lambda < 0 \tag{12.4.15}$$

where

$$C = \operatorname{Tr}\left[e^{-\beta H^0}\right] \tag{12.4.16}$$

$$H^0 = -\frac{d^2}{dz^2} + z^2 \tag{12.4.17}$$

in the above mentioned limits.

For β large and $\lambda \sim 0_-$,

$$F(\lambda, \beta) \sim \operatorname{Im}\left(\frac{e^{-\beta[1+\mathcal{O}(\lambda)+i\Delta E]}}{e^{-\beta}}\right) \tag{12.4.18}$$

where $\mathcal{O}(\lambda)$ denotes the perturbative real correction to the harmonic oscillator ground-state energy equal to 1, in the units used in (12.4.8), (12.4.17), and ΔE is the imaginary part of the corresponding unstable energy (see Figure 12.2) arising for $\lambda \lesssim 0$. We will see that, self consistently, ΔE is exponentially small for $\lambda \sim 0_-$, and hence we may write

$$F(\lambda, \beta) \sim -\beta \Delta E. \tag{12.4.19}$$

By a path integral representation of $F(\lambda, \beta)$, we will learn that for β large, $\lambda \sim 0_-$, $F(\lambda, \beta)$ leads to

$$\Delta E \sim c \frac{e^{2/3\lambda}}{\sqrt{-\lambda}} \tag{12.4.20}$$

for some constant c of order 1 independent of λ (and, of course, independent of β) in the above limits.

Upon substituting the expression in (12.4.20) for $\operatorname{Im}\Delta E$ in (12.4.11) and explicitly carrying out the λ-integration, we obtain ($K \to \infty$), up to an overall numerical factor of order 1,[4]

$$E_K \sim (-1)^{K+1}\left(\frac{3}{2}\right)^K \Gamma(K+\tfrac{1}{2}), \tag{12.4.21}$$

where $\Gamma(K+1/2)$ is the gamma function with integral representation

$$\Gamma(K+\tfrac{1}{2}) = \int_0^\infty du\, e^{-u}(u)^{K-1/2}. \tag{12.4.22}$$

---

[4] This numerical factor is given by $2\sqrt{6/\pi^3} \sim 0.88$. This factor is not essential in showing the growth of E_K factorially in the order K. The explicit behavior in (12.4.21) was first derived by Bender and Wu (1971).

12 Approximating Quantum Systems

To carry out the integral (12.4.11), first make the substitution $\lambda \to -\lambda$, and then introduce the integration variable $u = 2/3\lambda$. The coefficients E_K grow factorially for $K \to \infty$ leading to a diverging Rayleigh-Schrödinger perturbation series.

To derive (12.4.20), we first write the path integral expression for $F(\lambda, \beta)$ in (12.4.15) with $\lambda < 0$,[5]

$$F(\lambda, \beta) = \operatorname{Im}\left(\frac{1}{C}\operatorname{Tr}\left[e^{-\left(\frac{\beta}{2}-\left(-\frac{\beta}{2}\right)\right)H}\right]\right)$$

$$= \operatorname{Im}\left(\frac{1}{C}\operatorname{Tr}\int \mathscr{D}(z(\cdot))\exp\left[-\int_{-\beta/2}^{\beta/2}d\tau\left(\frac{\dot{z}^2(\tau)}{4} + z^2(\tau) + \lambda z^4(\tau)\right)\right]\right)$$
(12.4.23)

with the boundary condition $z(-\beta/2) = z(\beta/2)$, and we consider the limit $\beta \to \infty$, where formally (see (10.1.4))

$$\mathscr{D}(z(\cdot)) = \lim_{N\to\infty}\left(\frac{N}{2\pi\hbar^2\beta}\right)^{N/2}\left(\prod_{i=1}^{N-1}dz_i\right).$$
(12.4.24)

Let $z(t) = \eta(t) + \delta z(t)$, where $\eta(t)$ defines classical path(s) and $\delta z(t)$ the deviations of $z(t)$ from $\eta(t)$. $\eta(t)$ satisfies the differential equation[6]

$$\frac{\ddot{\eta}(t)}{2} - 2\eta(t) - 4\lambda\eta^3(t) = 0$$
(12.4.25)

with boundary condition $\eta(-\beta/2) = \eta(\beta/2)$, $\beta \to \infty$. By direct substitution in (12.4.25), the solutions, for $\beta \to \infty$, are readily verified to be given by

$$\eta(t) = \frac{\pm 1}{\sqrt{-\lambda}\,\cosh\left[2(t-t_0)\right]}$$
(12.4.26)

where $\eta(-\beta/2) \to 0$, $\eta(\beta/2) \to 0$ for $\beta \to \infty$, t_0 arbitrary.

Now for $\lambda \sim 0_-$, we consider the contribution due to derivations about the classical paths in (12.4.26). To this end, we first note that

$$\lim_{\beta\to\infty}\exp\left(-\int_{-\beta/2}^{\beta/2}d\tau\left[\frac{\dot{\eta}^2(\tau)}{4} + \eta^2(\tau) + \lambda\eta^4(\tau)\right]\right) = \exp\left(\frac{2}{3\lambda}\right) \quad (12.4.27)$$

and for further reference that

---
[5] We note that classically the Lagrangian corresponding to the Hamiltonian $p^2 + z^2 + \lambda z^4$ is given by $(\dot{z}^2/4) - z^2 - \lambda z^4$ which upon the substitution $t \to -it$, leads to the exponential factor in (12.4.23).
[6] We use a method developed in: Zinn-Justin (1981).

12.4 High-Order Perturbations, Divergent Series; Padé Approximants

$$\|\dot\eta\| \equiv \left(\int_{-\infty}^{\infty} dt\,\dot\eta^2(t)\right)^{1/2} = \frac{2}{\sqrt{-3\lambda}} \qquad (12.4.28)$$

both of which follow directly from (12.4.26).

In terms of the deviations $\delta z(t)$ about $\eta(t)$, with boundary conditions $\delta z(\pm\beta/2) = 0$,

$$\int_{-\beta/2}^{\beta/2} d\tau \left(\frac{\dot z^2(\tau)}{4} + z^2(\tau) + \lambda z^4(\tau)\right) \simeq -\frac{2}{3\lambda} + \mathcal{A}(\delta z) \qquad (12.4.29)$$

where

$$\mathcal{A}(\delta z) = \int_{-\beta/2}^{\beta/2} dt\, \delta z \left(-\frac{1}{4}\frac{d^2}{dt^2} + 1 - \frac{6}{\cosh^2[2(t-t_0)]}\right)\delta z. \qquad (12.4.30)$$

Here we have used (12.4.25), (12.4.26), integrated by parts over t using the boundary conditions on $\delta z(t)$ mentioned above, and noted that the coefficients of the terms involving $(\delta z)^3$, $(\delta z)^4$ vanish for $\lambda \to 0_-$. We note that $f_0(t) = \dot\eta(t)/\|\dot\eta\|$ is a normalized eigenvector of the operator $M = [-d^2/4dt^2 + 1 - 6\cosh^{-2}(2(t-t_0))]$ with eigenvalue 0, i.e., $M f_0(t) = 0$.

The arbitrariness of the parameter t_0 in (12.4.26), corresponding to multiple solutions of (12.4.25), gives rise to an associated degree of freedom in the problem. Instead of working with the parameter t_0, we may consider the scaled one $-\|\dot\eta\| t_0$ for a reason which will be clear below. We carry out a change of the path-integration variables in (12.4.23) as follows. We supplement the normalized function $f_0(t) = \dot\eta(t)/\|\dot\eta\|$ by an infinite set of functions $f_1(t), f_2(t), \ldots$ such that we may generate a mutually *orthonormal* set of functions $f_0(t), f_1(t), f_2(t), \ldots$, and expand $\eta(t) + \delta z(t)$ as

$$z(t) = \eta(t) + \sum_{i\geq 1} \delta z_i f_i(t). \qquad (12.4.31)$$

In particular, we note from (12.4.26), (12.4.31) that

$$\frac{\partial z(t)}{\partial(-\|\dot\eta\|t_0)} = -\frac{1}{\|\dot\eta\|}\left(\frac{\partial \eta(t)}{\partial t_0}\right) = \frac{\dot\eta(t)}{\|\dot\eta\|} \equiv f_0(t) \qquad (12.4.32)$$

$$\frac{\partial z(t)}{\partial(\delta z_j)} = f_j(t), \qquad j \geq 1 \qquad (12.4.33)$$

and we have thus carried out a transformation :

$$\bigl(z(t_1), z(t_2), \ldots\bigr) \longrightarrow \bigl(-\|\dot\eta\|t_0, \delta z_1, \delta z_2, \ldots\bigr). \qquad (12.4.34)$$

Since $f_0(t), f_1(t), \ldots$ are chosen to be orthonormal, the Jacobian of the transformation is one.

From (12.4.23), (12.4.28)–(12.4.30), (12.4.34), we then have for $\beta \to \infty$, $\lambda \sim 0_-$

$$\operatorname{Im}\left(\frac{1}{C}\operatorname{Tr}\left[e^{-\beta H}\right]\right) \sim -\frac{e^{2/3\lambda}}{\sqrt{-\lambda}}\operatorname{Im}\left(\frac{1}{C}\operatorname{Tr}\int_{-\beta/2}^{\beta/2}dt_0\int\mathscr{D}\left(\delta z(\cdot)\right)\exp\left[-\mathcal{A}(\delta z)\right]\right)$$
(12.4.35)

up to a multiplicative numerical factor independent of λ and β for $\lambda \sim 0_-$, $\beta \to \infty$.

The purpose of the exercise in the transformation carried out in (12.4.34) was two-fold: (1) It was to show that the degree of freedom associated with the parameter t_0, corresponding to multiple classical solutions $\eta(t)$, lead to a multiplicative factor $\|\dot\eta\| \sim 1/\sqrt{-\lambda}$ as an *extra* factor to $\exp(2/3\lambda)$, coming from (12.4.27), as spelled out in (12.4.35). (2) The action in (12.4.35) being quadratic, we may infer, for example, from a generalization of the Gaussian integral in (9.8.20), that the path integral in (12.4.35) $\sim (\det M)^{-1/2}$, and $\det M$ does not involve the zero eigenvalue corresponding to f_0, which has been already extracted in the infinite t_0-integral for $\beta \to \infty$, and is independent of t_0 in this limit.

The t_0-integral in (12.4.35) may be then explicitly carried out giving an overall β factor in (12.4.35) for $\beta \to \infty$. Since for $\lambda < 0$, the system is unstable (see Figure 12.2) $(\det M)^{-1/2}$ must have an imaginary part and, otherwise, all the coefficients in (12.4.11) will be zero. Technically this happens[7] because M has also a negative eigenvalue making $\det M$ negative.

Because of the normalization factor $1/C$, the leading net β, λ-dependent multiplicative factor for $\beta \to \infty$, $\lambda \sim 0_-$, in $F(\lambda,\beta)$ in (12.4.35) is then just $-\beta\exp(2/3\lambda)/\sqrt{-\lambda}$, simply up to a numerical factor of order 1, independent of λ, β in the above mentioned limits. Upon comparison of the behavior $F(\lambda,\beta) \sim -\beta\exp(2/3\lambda)/\sqrt{-\lambda}$ with (12.4.19) gives (12.4.20) thus obtaining the result stated in (12.4.21). Although the "action" in (12.4.35) is quadratic and hence the path integral may be explicitly carried out, the actual magnitude and sign of this overall λ-independent factor, for $\lambda \sim 0_-$, are clearly not essential to establish the factorial growth in K for $K \to \infty$ as given in (12.4.21) (see also Problem 12.12).

Over years, several investigations were carried out in re-summation methods and/or re-groupings of various terms in perturbation expansions of divergent series, as of the type studied above, leading in some cases to convergent results. One of these methods is that of the Padé approximants one discussed below which leads to a convergent result for $E(\lambda)$.

In the Padé approximants method, one formally replaces a series $\sum_K a_K \lambda^K$, representing a function $F(\lambda)$, by a double-sequence of ratios of two polynomials

---

[7] See Zinn-Justin (1981); see also Auberson *et al.* (1978).

$$P^{[N,M]}(\lambda) = \frac{\sum_{n=0}^{N} A_n \lambda^n}{\sum_{n=0}^{M} B_n \lambda^n} \ , \qquad (12.4.36)$$

$B_0 = 1$, and chooses the $(N + M + 1)$ coefficients $A_0, A_1, \ldots, A_N, B_1, \ldots, B_M$ in such a manner that the first $(N + M + 1)$ terms in the Taylor series of $P^{[N,M]}(\lambda)$ coincide, term by term, with the first $(N + M + 1)$ terms of the power series $\sum_K a_K \lambda^K$. $P^{[N,M]}(\lambda)$ is referred to as a *Padé approximant* associated with the original formal power series in question.

The interesting situation arises if $P^{[N,M]}(\lambda)$ converges for $N, M \to \infty$ to the actual function $F(\lambda)$, even if $\sum_K a_K \lambda^K$ diverges.

As an example of a Padé approximant, consider the Rayleigh-Schrödinger series for the ground-state energy of the anharmonic oscillator with Hamiltonian given in (12.4.8). To second order in λ (see Problem 12.1),

$$E(\lambda) = 1 + \frac{3}{4}\lambda - \frac{21}{6}\lambda^2 + \ldots \qquad (12.4.37)$$

and the Padé approximant $P^{[1,1]}(\lambda)$ is given by

$$P^{[1,1]}(\lambda) = \frac{1 + \frac{5}{2}\lambda}{1 + \frac{7}{4}\lambda} \qquad (12.4.38)$$

as is easily verified (see also Problem 12.13).

The above method has been successfully used in the literature[8] showing that the Padé approximants $P^{[N,N]}(\lambda)$ converge for $N \to \infty$ to the actual value for $E(\lambda)$. For example, $P^{[10,10]}(0.1) = 1.065\,285\,509\,535$, $P^{[20,20]}(0.1) = 1.065\,285\,509\,543$, while the actual value $E(0.1) = 1.062\,285\,5 \pm 0.000\,000\,5$. And $P^{[10,10]}(1) = 1.392\,102\,495\,074$, $P^{[20,20]}(1) = 1.392\,337\,481\,861$, while the actual value $E(1) = 1.392\,751 \pm 0.000\,620$.

As mentioned above, there have been many other re-summation procedures introduced recently and the reader may wish to consult the relevant journals for related details.

12.5 WKB Approximation

12.5.1 General Theory

Consider the Schrödinger equation for a stationary state of energy E, in one dimension

---

[8] For the relevant details see, Loeffel et al. (1969), and references therein for values of $E(\lambda)$.

$$\left(\frac{d^2}{dx^2} + \frac{p^2(x)}{\hbar^2}\right)\psi(x) = 0 \qquad (12.5.1)$$

where

$$p^2(x) = 2m(E - V(x)). \qquad (12.5.2)$$

Within an interval, when $V(x)$ is constant, the formal solutions of (12.5.1) are of the form $\exp(\pm i\, xp/\hbar)$. More generally, one may formally introduce, locally, a scale $\lambda(x) = \hbar/|p(x)|$ — referred to by some as a reduced local de Broglie wavelength of the particle in question. We consider potentials $V(x)$ which vary slowly over distances of the order $\lambda(x)$ in the neighborhood of x. By setting $\Delta V = \Delta x V'(x)$, then for a change Δx of the order of $\lambda(x)$, for a slowly varying potential over $\lambda(x)$, we take, as a rule of thumb, $|\Delta V| \simeq |\lambda(x)V'(x)| \ll \hbar^2/(m\,\lambda^2(x))$. The latter condition may be equivalently rewritten as $|\Delta \lambda| \simeq |\lambda(x)\lambda'(x)| \ll |\lambda(x)|$ signifying that the approximation sought corresponds to cases where $\lambda(x)$ varies slowly over a distance of the order $\lambda(x)$ itself. This constraint in turn, may be rewritten formally in terms of the local classical "momentum" $|p(x)|$ as $|p'(x)/p(x)| \ll |p(x)|/\hbar$.[9] Such a restriction will be derived more precisely below.

To find the solutions of (12.5.1) under the above stated condition, we set

$$\psi(x) = \frac{1}{\sqrt{S'(x)}} \exp S(x) \qquad (12.5.3)$$

where S, and its derivative, are unknown. The specific way of writing $\psi(x)$ in this form turns out to be convenient and simplifies the analysis to some extent. In any case $S(x)$ is unknown and will be, self consistently, determined by substituting the expression for $\psi(x)$ in (12.5.1). We explicitly have

$$\psi'' = \left[\frac{1}{4}\left(\frac{S''}{S'}\right)^2 - \frac{1}{2}\left(\frac{S''}{S'}\right)' + (S')^2\right]\psi \qquad (12.5.4)$$

leading from (12.5.1) to

$$(S')^2 + \frac{p^2(x)}{\hbar^2} + \frac{1}{4}\left(\frac{S''}{S'}\right)^2 - \frac{1}{2}\left(\frac{S''}{S'}\right)' = 0. \qquad (12.5.5)$$

For $p(x)$ a constant, we have seen above that $S(x)$ is linear in x, and hence $S'(x)$ is a constant. More generally for slowly varying S', we neglect the last two terms in (12.5.5) and, in turn, investigate the nature of this approximation. To this end, we obtain

$$S'(x) \simeq \pm i\,\frac{p(x)}{\hbar}. \qquad (12.5.6)$$

---

[9] For $p(x) \neq 0$, the condition in question, written in this last form, corresponds formally to a $\hbar \sim 0$ analysis.

12.5 WKB Approximation

Upon substitution of (12.5.6) in (12.5.5), we may infer that this approximation may be carried out provided

$$\left| \frac{1}{4}\left(\frac{p'}{p}\right)^2 - \frac{1}{2}\left(\frac{p'}{p}\right)' \right| \ll \frac{|p(x)|^2}{\hbar^2} \qquad (12.5.7)$$

and obtain the asymptotic solution

$$\psi(x) \simeq \frac{C_1}{\sqrt{p(x)}} \exp\left(\frac{i}{\hbar} \int^x dx\, p(x)\right) + \frac{C_2}{\sqrt{p(x)}} \exp\left(-\frac{i}{\hbar} \int^x dx\, p(x)\right) \qquad (12.5.8)$$

where C_1, C_2 are some constants. This approximation method is referred to as the WKB approximation.[10]

Clearly, the approximation breaks down at such points, called turning point,[11] where $p(x) = 0$. A turning point is shown in Figure 12.3, where $V'(a) > 0$.

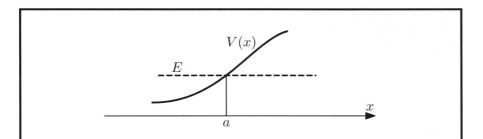

Fig. 12.3. The figure shows a turning point at $x = a$, where $V(a) = E$, and hence $p(a) = 0$. At this point, the approximation in (12.5.8) certainly breaks down. Here $V'(a) > 0$.

To find out how close can x come to the point a before the approximation breaks down, we carry out the expansion

$$p^2(x) \simeq -2m\left[V'(a)(x-a) + V''(a)\frac{(x-a)^2}{2}\right] \qquad (12.5.9)$$

where, here, $V'(a) > 0$.

For

---

[10] WKB stands for three of several contributors to this method, namely, G. Wentzel, H. A. Kramers and L. Brillouin.
[11] This terminology is taken from classical mechanics, where $p(x) = 0$ means that, at such a point, the kinetic energy of a particle is zero as the particle reverses its motion.

$$|V''(a)(x-a)| \ll 2|V'(a)| \qquad (12.5.10)$$

we have

$$p^2(x) \simeq 2m\,V'(a)(a-x). \qquad (12.5.11)$$

In the neighborhood of the point a for which (12.5.10) is satisfied, we may use (12.5.9) to rewrite (12.5.1) as

$$\left(\frac{d^2}{dx^2} - \frac{2mV'(a)(x-a)}{\hbar^2}\right)\chi = 0 \qquad (12.5.12)$$

where we have denoted the wavefunction in this region by χ. Upon making a change of variable x to

$$\kappa = (2mV'(a)/\hbar^2)^{1/3}(x-a), \qquad (12.5.13)$$

we obtain for (12.5.12)

$$\left(\frac{d^2}{d\kappa^2} - \kappa\right)\chi = 0. \qquad (12.5.14)$$

The special functions satisfying this differential equation are called Airy functions[12] and a pair of linearly independent solutions are denoted by $\mathrm{Ai}(\kappa)$, $\mathrm{Bi}(\kappa)$ having, in particular, the following properties:

$$\mathrm{Ai}(0) = \frac{\mathrm{Bi}(0)}{\sqrt{3}} = 3^{-2/3}\Gamma\!\left(\tfrac{2}{3}\right) \qquad (12.5.15)$$

and for $\kappa > 0$,

$$\mathrm{Ai}(\kappa) \xrightarrow[\kappa\to\infty]{} \frac{1}{2\sqrt{\pi}}\left(\frac{1}{\kappa}\right)^{1/4}\exp\!\left(-\frac{2}{3}\kappa^{3/2}\right) \qquad (12.5.16)$$

$$\mathrm{Bi}(\kappa) \xrightarrow[\kappa\to\infty]{} \frac{1}{\sqrt{\pi}}\left(\frac{1}{\kappa}\right)^{1/4}\exp\!\left(\frac{2}{3}\kappa^{3/2}\right). \qquad (12.5.17)$$

On the other hand for $\kappa < 0$,

$$\mathrm{Ai}(\kappa) \xrightarrow[|\kappa|\to\infty]{} \frac{1}{\sqrt{\pi}}\left(\frac{1}{|\kappa|}\right)^{1/4}\sin\!\left(\frac{2}{3}|\kappa|^{3/2} + \frac{\pi}{4}\right) \qquad (12.5.18)$$

$$\mathrm{Bi}(\kappa) \xrightarrow[|\kappa|\to\infty]{} \frac{1}{\sqrt{\pi}}\left(\frac{1}{|\kappa|}\right)^{1/4}\cos\!\left(\frac{2}{3}|\kappa|^{3/2} + \frac{\pi}{4}\right). \qquad (12.5.19)$$

The general solution of (12.5.14) is

$$\chi(x) = \alpha\,\mathrm{Ai}\!\left(\left(\frac{2mV'(a)}{\hbar^2}\right)^{1/3}(x-a)\right) + \beta\,\mathrm{Bi}\!\left(\left(\frac{2mV'(a)}{\hbar^2}\right)^{1/3}(x-a)\right)$$
$$(12.5.20)$$

---

[12] Cf. Abramowitz and Stegun (1972), p. 446.

with α, β some constants. We recall that in (12.5.14), x satisfies (12.5.10).

Bi(κ), unlike Ai(κ), grows exponentially in (12.5.17) for $x > a$ in the region $E < V$, and on physical grounds (see Figure 12.3) it is necessary to choose $\beta = 0$ in (12.5.20). Hence

$$\chi(x) = \alpha \, \text{Ai}\left(\left(\frac{2mV'(a)}{\hbar^2}\right)^{1/3}(x-a)\right). \tag{12.5.21}$$

In particular, we learn from the property in (12.5.15) that the solution is *finite* at the turning point a.

Now we have to find a *common region* to the *left* of the point a in which the solutions in (12.5.21) and (12.5.8) are valid. Hence in the region, we are seeking, both solutions must coincide.

To the above end, we note that the condition in (12.5.7) must be satisfied. This leads to

$$\left(\frac{5}{16}\right)^{1/3} < 1 \ll \left(\frac{2m\,V'(a)}{\hbar^2}\right)^{1/3}(a-x) \equiv |\kappa|. \tag{12.5.22}$$

In turn (12.5.10) requires from (12.5.22) that

$$|\kappa| \ll \frac{2}{|V''(a)|}\left(\frac{2m(V'(a))^4}{\hbar^2}\right)^{1/3} \tag{12.5.23}$$

for $V''(a) \neq 0$, for the justification of the neglect of the second term in (12.5.9) in comparison to the first one in a Taylor expression.

To the left of the point a, i.e., for $x < a$, conditions (12.5.10), (12.5.22), (12.5.23), then give $\kappa > 0$,

$$1 \ll |\kappa| \ll \frac{2}{|V''(a)|}\left(\frac{2m(V'(a))^4}{\hbar^2}\right)^{1/3} \tag{12.5.24}$$

for $V''(a) \neq 0$. From (12.5.18), we then write $\chi(\kappa)$ in (12.5.21), with $\kappa < 0$, $|\kappa|$ satisfying (12.5.24),

$$\chi(x) \simeq \frac{\alpha}{\sqrt{\pi}}\left[\left(\frac{2mV'(a)}{\hbar^2}\right)^{1/3}(a-x)\right]^{-1/4}$$

$$\times \sin\left(\frac{2}{3}\frac{\sqrt{2mV'(a)}}{\hbar}(a-x)^{3/2} + \frac{\pi}{4}\right). \tag{12.5.25}$$

Now we compare this solution to the one (12.5.8) for $x < a$, subject to the expansion of $p^2(x)$ in (12.5.9) and the condition (12.5.10). To do this, we further approximate the solution in (12.5.8) by using the expansion in (12.5.9), (12.5.10) and note that

$$\frac{1}{\hbar}\int_a^x dx\, p(x) \simeq -\frac{2}{3}\frac{\sqrt{2mV'(a)}}{\hbar}(a-x)^{3/2} + \frac{1}{10}\sqrt{\frac{2m}{V'(a)}}\frac{V''(a)(a-x)^{5/2}}{\hbar} \tag{12.5.26}$$

which may be rewritten as

$$\frac{1}{\hbar}\int_a^x dx\, p(x) \simeq -\frac{2}{3}|\kappa|^{3/2}\left\{1 - \frac{3}{20}\frac{V''(a)}{V'(a)}(a-x)\right\}. \tag{12.5.27}$$

From the condition (12.5.10), the second term in the curly brackets is very small in comparison to one and we may take

$$\exp\pm\frac{i}{\hbar}\int_a^x dx\, p(x) \simeq \exp\left(\mp\left(\frac{i}{\hbar}\right)\left(\frac{2}{3}\right)\sqrt{2mV'(a)}(a-x)^{3/2}\right). \tag{12.5.28}$$

Hence from (12.5.7) and (12.5.10), we have for $\kappa < 0$, $V''(a) \neq 0$,

$$1 \ll |\kappa| \ll \frac{2}{|V''(a)|}\left(\frac{2m(V'(a))^4}{\hbar^2}\right)^{1/3} \tag{12.5.29}$$

and the solution $\psi(x)$ in (12.5.8) takes the form

$$\psi(x) \simeq C_1\left(2mV'(a)(a-x)\right)^{-1/4}\exp\left(-\left(\frac{i}{\hbar}\right)\left(\frac{2}{3}\right)\sqrt{2mV'(a)}(a-x)^{3/2}\right)$$
$$+ C_2\left(2mV'(a)(a-x)\right)^{-1/4}\exp\left(+\left(\frac{i}{\hbar}\right)\left(\frac{2}{3}\right)\sqrt{2mV'(a)}(a-x)^{3/2}\right). \tag{12.5.30}$$

In writing (12.5.30), we have finally used (12.5.9), (12.5.10) again, and have used a lower limit a of integration for the integral in (12.5.8) and adjusted, accordingly, the coefficients C_1, C_2. Obviously for the common region in (12.5.24), (12.5.29) both solutions in (12.5.25), (12.5.30) should coincide, i.e., we must have

$$2iC_1 e^{i\pi/4} = -\frac{\alpha}{\sqrt{\pi}}(2mV'(a)\hbar)^{1/6} \equiv -c \tag{12.5.31}$$

$$2iC_2 e^{-i\pi/4} = \frac{\alpha}{\sqrt{\pi}}(2mV'(a)\hbar)^{1/6} \equiv c. \tag{12.5.32}$$

These conditions on C_1, C_2 may be now used in (12.5.8).

In summary, we have found the following approximations

$$\psi(x) = \frac{c}{\sqrt{p(x)}}\cos\left(\frac{1}{\hbar}\int_a^x dx\, p(x) + \frac{\pi}{4}\right) \tag{12.5.33}$$

for[13] $x < a$, $(a-x) \gg (\hbar^2/2mV'(a))^{1/3}$,

---

[13] Note that in writing the expression in (12.5.33) we have used the fact that $\sin(z - \pi/4) = -\cos(z + \pi/4)$.

$$\chi(x) = \alpha \operatorname{Ai}\left(\left(\frac{2mV'(a)}{\hbar^2}\right)^{1/3}(x-a)\right) \qquad (12.5.34)$$

for $|x-a| \ll 2|V'(a)|/|V''(a)|$, and the coefficients α, c are related through (12.5.31)/(12.5.32).

Now we consider the situation that when we move to the right of point a, we encounter a second turning point, say, b, where $V'(b) < 0$. This is illustrated in Figure 12.4, where for $a < x < b$, $V(x) > E$, i.e., $p(x)$ is imaginary, and beyond the point b, $V(x) < E$. This will allow us next to study the problem of tunneling through the potential barrier.

12.5.2 Barrier Penetration

The expression in (12.5.33), may be rewritten as

$$\psi(x) \equiv \psi_1(x) \simeq \frac{c\, e^{i\pi/4}}{2\sqrt{p(x)}} \exp\left(\frac{i}{\hbar}\int_a^x dx\, p(x)\right)$$

$$+ \frac{c\, e^{-i\pi/4}}{2\sqrt{p(x)}} \exp\left(-\frac{i}{\hbar}\int_a^x dx\, p(x)\right). \qquad (12.5.35)$$

Hence we recognize the first and second terms in (12.5.35) as corresponding to amplitudes of incidence on and reflection off the potential barrier.

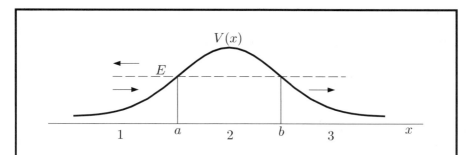

Fig. 12.4. The figure shows two turning points at $x = a$ and $x = b$, at which $V'(a) > 0$ and $V'(b) < 0$, respectively. For $a < x < b$, $V(x) > E$.

Repeating the analysis leading to the approximation in (12.5.8), we may infer that for $(x-a) \gg (\hbar^2/2mV'(a))^{1/3}$, in the region 2, the approximate solution corresponding to that in (12.5.8) is given by

$$\psi_2(x) = \frac{\gamma}{\sqrt{|p(x)|}} \exp\left(-\frac{1}{\hbar}\int_a^x dx |p(x)|\right) \qquad (12.5.36)$$

where we have used the fact that in region 2, $p(x)$ is imaginary and we have selected the damped solution required by the physics of the problem. $\psi_2(x)$ is not valid at turning point b (and also at a). Below we will follow the procedure leading to (12.5.20) to investigate the nature of the solution near the point b. Before doing this, we investigate the approximate solution in the region 3.

In the region 3, $p(x)$ is real and only an amplitude of transmission may arise. Accordingly for $|x - b| \gg (\hbar^2/2m|V'(b)|)^{1/3}$,

$$\psi_3(x) = \frac{c'}{\sqrt{p(x)}} \exp\left(\frac{i}{\hbar} \int_b^x dx\, p(x)\right). \tag{12.5.37}$$

In the neighborhood of the point b, i.e., for $|x - b| \ll 2|V'(b)|/|V''(b)|$, for $V''(b) \neq 0$, the approximate solution of the Schrödinger equation (12.5.1) is obtained in a similar way as the one (12.5.20), except now $V'(b) < 0$, giving

$$\chi(x) = \alpha'\, \text{Ai}\left(-\left(\frac{2mV'(b)}{\hbar^2}\right)^{1/3}(x-b)\right) + \beta'\, \text{Bi}\left(-\left(\frac{2mV'(b)}{\hbar^2}\right)^{1/3}(x-b)\right) \tag{12.5.38}$$

and for $2|V'(b)|/|V''(b)| \gg x - b \gg (\hbar^2/2m|V'(b)|)^{1/3}$, $x > b$, we have according to (12.5.18), (12.5.19),

$$\chi(x) \to \left(\frac{1}{\sqrt{\pi}}\right)\left(\frac{1}{|\kappa|}\right)^{1/4}\left[\alpha' \sin\left(\frac{2}{3}|\kappa|^{3/2} + \frac{\pi}{4}\right) + \beta' \cos\left(\frac{2}{3}|\kappa|^{3/2} + \frac{\pi}{4}\right)\right] \tag{12.5.39}$$

where now $|\kappa| = (2m|V'(b)|/\hbar^2)^{1/3}(x-b)$.

By expanding $p(x)$ in (12.5.37) about the point $x = b$, and finding a common region of the validity of the resulting solution with one in (12.5.39), carried out in a similar way as before, we obtain

$$\alpha' = \frac{\sqrt{\pi} e^{i\pi/4} c'}{(2m\hbar|V'(b)|)^{1/6}} \tag{12.5.40}$$

$$\beta' = \frac{\sqrt{\pi} e^{-i\pi/4} c'}{(2m\hbar|V'(b)|)^{1/6}}. \tag{12.5.41}$$

On the other hand, for $x < b$, $2|V'(b)|/|V''(b)| \gg |x - b| \gg (\hbar^2/2m|V'(b)|)^{1/3}$, we have from (12.5.16), (12.5.17) the leading contribution to $\chi(x)$

$$\chi(x) \to \left(\frac{\beta'}{\sqrt{\pi}}\right)\left(\frac{1}{|\kappa|}\right)^{1/4} \exp\left(\frac{2}{3}|\kappa|^{3/2}\right) \tag{12.5.42}$$

where $|\kappa|$ is defined below (12.5.39). The solution in the region 2 damps out as we move from a to the right on the way to b and, of course, the opposite arises as we move from b to a, as indicated in (12.5.42).

To compare the solution in (12.5.42) with the one in (12.5.36) in a common region of their validity, we first rewrite (12.5.36) as

$$\psi_2(x) = \exp\left(-\frac{1}{\hbar}\int_a^b \mathrm{d}x\,|p(x)|\right)\frac{\gamma}{\sqrt{|p(x)|}}\exp\left(\frac{1}{\hbar}\int_x^b \mathrm{d}x\,|p(x)|\right). \quad (12.5.43)$$

As a function of x, in the neighborhood of b ($x < b$),

$$\left(\frac{1}{\sqrt{|p(x)|}}\right)\exp\left(\frac{1}{\hbar}\int_x^b \mathrm{d}x\,|p(x)|\right)$$

$$\simeq \frac{1}{(2m|V'(b)||x-b|)^{1/4}}\exp\left(\frac{2}{3\hbar}\sqrt{2m|V'(b)|}|x-b|^{3/2}\right). \quad (12.5.44)$$

Upon comparison of (12.5.43)/(12.5.44) with that in (12.5.42), in their common range of validity, we obtain

$$\exp\left(-\frac{1}{\hbar}\int_a^b \mathrm{d}x\,|p(x)|\right)\gamma = \frac{(2m\hbar|V'(b)|)^{1/6}}{\sqrt{\pi}}\beta' = \mathrm{e}^{-\mathrm{i}\pi/4}c' \quad (12.5.45)$$

where we have also used (12.5.41).

Finally, the comparison of (12.5.21) with that in (12.5.36) in the neighborhood of point $x = a$, in their common region of validity, by using in the process of (12.5.16) and expanding $|p(x)|$ in (12.5.36) about the point a, gives

$$\gamma = (2m\hbar V'(a))^{1/6}\frac{\alpha}{2\sqrt{\pi}} = \frac{c}{2} \quad (12.5.46)$$

where we have also used (12.5.32).

Upon setting $c/2 = 1$ in (12.5.35), and taking advantage of the equalities (12.5.41), (12.5.45) and (12.5.46), we may infer that

$$c' = \mathrm{e}^{\mathrm{i}\pi/4}\exp\left(-\frac{1}{\hbar}\int_a^b \mathrm{d}x\,|p(x)|\right). \quad (12.5.47)$$

Finally, from (12.5.35), (12.5.37) for the expression of the wavefunctions in regions 1 and 3, and (12.5.47) we obtain for the transmission and reflection probabilities the leading expressions

$$T \simeq \exp\left(-\frac{2}{\hbar}\int_a^b \mathrm{d}x\,\sqrt{2m(V(x)-E)}\right) \quad (12.5.48)$$

$$R \simeq 1 \quad (12.5.49)$$

respectively. Obviously, these are valid if the exponential term in (12.5.48) is *small*. The fact that R turns out to be equal to one is of no violation of the conservation of probability ($R + T = 1$) as both results for R and T are *leading* contributions to these probabilities and the value "1" for R dominates over any small correction of the order T as given in (12.5.48).

Next we investigate the nature of quantization rules set up by the WKB approximation and make contact with the so-called old quantum theory. Before doing this we note that the WKB approximations of the wavefunctions $\psi_1(x)$, $\psi_2(x)$, $\psi_3(x)$ obtained in regions 1, 2, 3 as given, respectively, in (12.5.35), (12.5.36), (12.5.37), by making use of the constraints on their coefficients in (12.5.45), (12.5.46), (12.5.47), may be rewritten, away from the turning points a and b, as

$$\psi_1(x) = \frac{c}{\sqrt{p(x)}} \cos\left(\frac{1}{\hbar}\int_a^x dx\, p(x) + \frac{\pi}{4}\right) \qquad (12.5.50)$$

$$\psi_2(x) = \frac{c}{2\sqrt{|p(x)|}} \exp\left(-\frac{1}{\hbar}\int_a^x dx\, |p(x)|\right) \qquad (12.5.51)$$

$$\psi_3(x) = \frac{c}{2\sqrt{p(x)}} \exp\left(-\frac{1}{\hbar}\int_a^b dx\, |p(x)|\right) \exp i\left(\frac{1}{\hbar}\int_b^x dx\, p(x) + \frac{\pi}{4}\right). \qquad (12.5.52)$$

We have seen, how by examining the solution of (12.5.1) near the turning point $x = a$, we were able to find the connection between the coefficients C_1, C_2 in (12.5.8) and γ in (12.5.36), as given in (12.5.31), (12.5.32), (12.5.46), going *from* the damped solution (12.5.36), in region 2, *to* the oscillatory one in (12.5.8), in region 1. The connection between $\psi_1(x)$ and $\psi_2(x)$ on the left and right of point $x = a$, respectively, as indicated by the *direction* of the arrow in.[14]

$$\frac{c}{2\sqrt{|p(x)|}} \exp\left(-\frac{1}{\hbar}\int_a^x dx\, |p(x)|\right) \mapsto \frac{c}{\sqrt{p(x)}} \cos\left(\frac{1}{\hbar}\int_a^x dx\, p(x) + \frac{\pi}{4}\right) \qquad (12.5.53)$$

is called a *connection formula* and shows the correspondence between regions 2 and 1, remembering that in the physics of the problem we have chosen incidence on the potential barrier from left to right. Upon rewriting

$$\exp\left(-\frac{1}{\hbar}\int_a^x dx\, |p(x)|\right) = \exp\left(-\frac{1}{\hbar}\int_a^b dx\, |p(x)|\right) \exp\left(-\frac{i}{\hbar}\int_x^b dx\, p(x)\right) \qquad (12.5.54)$$

in (12.5.51), where we have used the fact that $p(x) = i|p(x)|$ in region 2, we see an obvious correspondence rule between the WKB wavefunctions in (12.5.51), (12.5.52) in region 2 and 3.

12.5.3 WKB Quantization Rules

Consider the potential energies depicted in Figure 12.5. Referring to part (A) of the figure, we may obtain the WKB approximation in the region

---
[14] Note that in order to avoid confusion with the notation of a limit, denoted by $\to$, we have used the notation $\mapsto$ in (12.5.53).

12.5 WKB Approximation 713

$a < x < b$ by two methods. One is to match its oscillatory solution with the exponentially decreasing one on the left-hand side of point a, and another one by matching it with the exponentially decreasing one as one moves away from the point b. By requiring that the solutions obtained by these two methods to be the same, leads to a quantization rule as follows.

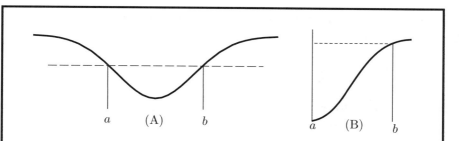

Fig. 12.5. (A) A potential energy with two turning points at a and b. (B) A potential energy with a turning point at b, and the potential energy is taken to go to infinity at a. This is equivalent in working in half space.

To the left of the turning point b, in part (A) of Figure 12.5, we may use the solution in (12.5.50) with a in it simply replaced by b leading to the WKB solution for $a < x < b$ given by

$$\phi(x) = \left(\frac{c}{\sqrt{p(x)}}\right) \cos\left(\frac{1}{\hbar}\int_b^x dx\, p(x) + \frac{\pi}{4}\right) \qquad (12.5.55)$$

for some constant c. A solution in $a < x < b$ obtained by matching it with the exponentially decreasing one on the left-hand side of point a, may be directly read from the one in (12.5.55) by merely replacing $\int_b^x dx\, p(x)$ by $\int_x^a dx\, p(x)$ in it, and replacing the coefficient c, in general, by some other constant, say, d. This second method then leads to

$$\phi(x) = \left(\frac{d}{\sqrt{p(x)}}\right) \cos\left(\frac{1}{\hbar}\int_x^a dx\, p(x) + \frac{\pi}{4}\right). \qquad (12.5.56)$$

We first note that for the cosine function in (12.5.56), we may write

$$\cos\left(\frac{1}{\hbar}\int_x^a dx\, p(x) + \frac{\pi}{4}\right)$$

$$= \cos\left[\left(\frac{1}{\hbar}\int_b^x dx\, p(x) + \frac{\pi}{4}\right) + \left(\frac{1}{\hbar}\int_a^b dx\, p(x) - \frac{\pi}{2}\right)\right]$$

$$= \cos\left(\frac{1}{\hbar}\int_b^x dx\, p(x) + \frac{\pi}{4}\right)\cos\left(\frac{1}{\hbar}\int_a^b dx\, p(x) - \frac{\pi}{2}\right)$$

$$-\sin\left(\frac{1}{\hbar}\int_b^x dx\, p(x) + \frac{\pi}{4}\right)\sin\left(\frac{1}{\hbar}\int_a^b dx\, p(x) - \frac{\pi}{2}\right). \quad (12.5.57)$$

Requiring that the solution (12.5.55), (12.5.56) obtained by the two methods coincide leads from (12.5.57) to

$$\frac{1}{\hbar}\int_b^a dx\, p(x) - \frac{\pi}{2} = n\pi, \qquad d = (-1)^n c \quad (12.5.58)$$

giving the quantization rule

$$\frac{1}{\hbar}\int_b^a dx\, p(x) = \pi\left(n + \frac{1}{2}\right) \quad (12.5.59)$$

where the n are non-negative integers. Since in the range $a < x < b$, $p(x)$ is real and positive (see Figure 12.5 (A)), n cannot take on negative integers. The quantization rule in (12.5.59) brings us into contact with the old quantum theory of Bohr and Sommerfeld. We recall from (12.5.48), that the validity of the WKB approximation, requires, in particular, that such exponential terms as in (12.5.48) be small. The latter implies that (12.5.59) is strictly valid for large positive integers. The expression in (12.5.59) is, nevertheless, useful in predicting the discrete spectrum for some Hamiltonians (see Problem 12.16). By considering the wavefunctions near the turning points a and b, using (12.5.59), and formally writing the cosine function in (12.5.56), for $x \to b - 0$, as $\cos[(\pi/4) + n\pi]$, n would correspond to the number of nodes of the wavefunction in $a < x < b$, since each time, $\pi/4$ is translated by π, the wavefunction on its way to the new value cuts the x-axis.

In part (B) of Figure 12.5, the wavefunction is to have a node as $x \to a+\varepsilon$, for $\varepsilon \to +0$, at which point the potential is defined to be infinite, and the WKB solution in (12.5.56) is to be replaced by

$$\phi(x) = \left(\frac{d'}{\sqrt{p(x)}}\right)\sin\left(\frac{1}{\hbar}\int_{a+0}^x dx\, p(x)\right), \qquad \varepsilon \to +0 \quad (12.5.60)$$

for $a < x < b$. Upon comparison of (12.5.55) with (12.5.60), we arrive at the quantization rule

$$\frac{1}{\hbar}\int_{a+0}^b dx\, p(x) = \pi\left(n + \frac{3}{4}\right) \quad (12.5.61)$$

for non-negative integers.

It is interesting to point out that for the elementary problem of a particle in a box with impenetrable walls at a and b, the WKB solutions are to have nodes at $b-0$, $a+0$ and the solutions in (12.5.55), (12.5.56) are to be replaced, respectively, by

$$\left(\frac{c'}{\sqrt{p(x)}}\right)\sin\left(\frac{1}{\hbar}\int_x^{b-0}\mathrm{d}x\,p(x)\right),\quad \left(\frac{d'}{\sqrt{p(x)}}\right)\sin\left(\frac{1}{\hbar}\int_{a+0}^x\mathrm{d}x\,p(x)\right) \tag{12.5.62}$$

which upon their comparison leads to the quantization rule

$$\frac{1}{\hbar}\int_{a+0}^{b-0}\mathrm{d}x\,p(x)=\pi(n+1). \tag{12.5.63}$$

[With $p(x)=\sqrt{2mE}$, (12.5.63) leads to the familiar energy levels $E_n=\hbar^2\pi^2(n+1)^2/2m(b-a)^2$.]

12.5.4 The Radial Equation

For a spherically symmetric potential $V(r)$, the radial equation is given by

$$\left[-\frac{\hbar^2}{2m}\left(\frac{1}{r^2}\frac{\mathrm{d}}{\mathrm{d}r}r^2\frac{\mathrm{d}}{\mathrm{d}r}\right)+\frac{\hbar^2}{2mr^2}\ell(\ell+1)+V(r)-E\right]R(r)=0, \tag{12.5.64}$$

where for simplicity of the notation, we suppressed the dependence of $R(r)$ on $\ell=0,1,2,...$, and other quantum numbers. We may take advantage of the analysis carried above to reduce (12.5.64) to the form (12.5.1). To do this, we set

$$R(r)=\frac{f(r)}{\sqrt{r}} \tag{12.5.65}$$

to rewrite (12.5.64) as

$$\left[r^2\frac{\mathrm{d}^2}{\mathrm{d}r^2}+r\frac{\mathrm{d}}{\mathrm{d}r}+\frac{2m}{\hbar^2}r^2(E-V)-(\ell+\tfrac{1}{2})^2\right]f=0. \tag{12.5.66}$$

Finally, let a be a conveniently chosen scale, and introduce the variable $\eta:-\infty<\eta<\infty$, by

$$r=ae^{\eta/a}. \tag{12.5.67}$$

In terms of this new variable, (12.5.66) becomes

$$\frac{\mathrm{d}^2}{\mathrm{d}\eta^2}f+\frac{p^2(\eta)}{\hbar^2}f=0,\quad -\infty<\eta<\infty \tag{12.5.68}$$

where

$$p^2(\eta)=2m\left(E-V\left(ae^{\eta/a}\right)\right)e^{2\eta/a}-\frac{\hbar^2}{a^2}(\ell+\tfrac{1}{2})^2. \tag{12.5.69}$$

Thus provided, the WKB approximation may be satisfied, we may infer, for example, from (12.5.59) that

$$\frac{1}{\hbar} \int_{\eta_a}^{\eta_b} d\eta \, p(\eta) = \pi \left(n + \frac{1}{2}\right) \tag{12.5.70}$$

and when the integral is rewritten in terms of the variable r, we have

$$\frac{1}{\hbar} \int_{r_a}^{r_b} dr \sqrt{2m\left(E - V(r)\right) - \frac{\hbar^2}{r^2}\left(\ell + \frac{1}{2}\right)^2} = \pi \left(n + \frac{1}{2}\right) \tag{12.5.71}$$

where r_a, r_b are the zeros of the integrand.

We may thus conclude that to apply the WKB procedure in the one-dimensional radial equation in the form.

$$\left[\frac{d^2}{dr^2} + \frac{1}{\hbar^2}\left(2m(E - V(r)) - \frac{\hbar^2}{r^2}\ell(\ell+1)\right)\right] u(r) = 0 \tag{12.5.72}$$

where

$$R(r) = u(r)/r, \tag{12.5.73}$$

it is necessary to replace $\ell(\ell+1)$ in it by $(\ell+1/2)^2$. This is true even for $\ell = 0$. The reader is asked to investigate conditions for the justification of the WKB asymptotic procedure for the attractive Coulomb potential in Problem 12.17 and investigate the eigenvalues resulting from (12.5.71).

12.6 Time-Dependence; Sudden Approximation and the Adiabatic Theorem

The time evolution of quantum systems for time-independent and time-dependent Hamiltonians was investigated quite generally in §2.5, and the general theory of quantum decay and its related time-dependence in §3.5. Exact treatments of transitions between harmonic oscillator states, driven by external time-dependent sources at zero and finite temperatures, were give in §6.2–§6.4. Chapter 8 was almost entirely devoted to time-dependent studies of spin 1/2 and two-level systems. A detailed account of Green functions and their time-dependence was given in Chapter 9. Path integrals and the quantum dynamical principle, as time-dependent problems, were, respectively, the subjects of Chapters 10 and 11. Scattering theory in a time-dependent setting will be carried out in Chapter 15, and relativistic dynamical aspects in Chapter 16. In the present section, we consider some time-dependence aspects which call for valid approximations as discussed below. The above mentioned material may be also consulted for completeness since some of the topics are related to the present ones as will be specified below.

12.6 Time-Dependence; Sudden Approximation and the Adiabatic Theorem

On a time scale set up by the physics of a problem, such as one given, for example, by a typical quantum mechanical period of oscillations, we investigate the dynamics corresponding to two extreme cases of Hamiltonians. For one case, suppose that a Hamiltonian varies from one form to a final form during a very short time τ in comparison to a time scale set up in a problem. The approximation corresponding to such a sudden change ($\tau \to 0$) of a Hamiltonian is called a *sudden* approximation. For the other extreme case, consider a Hamiltonian which varies very slowly in an interval $[0, T]$ in comparison to a time scale set up in a problem, such that one may effectively consider the limit $T \to \infty$. The corresponding approximation is called the *adiabatic* approximation (see also §8.13). In both cases, the underlying potential energies need *not* be weak.

Before going into the above details, we first treat another important situation, where a *weak* perturbation is added to a given Hamiltonian during a given interval of time $(0, \tau)$, and we investigate the nature of transitions that may occur between different states of the given Hamiltonian after the perturbation is switched off.

12.6.1 Weak Perturbations

Suppose we are given a time-independent Hamiltonian H^0, and during a time interval $(0, \tau)$, we switch on a *weak* perturbation $H^1(t)$, generating the Hamiltonian[15]

$$H(t) = H^0 + H^1(t) \qquad (12.6.1)$$

where $H^1(t) = 0$ for $t \geqslant \tau$, $t < 0$.

We are interested in investigating transitions that may occur between states of H^0 caused by the perturbation $H^1(t)$ after the latter is switched off. A state $|\psi(t)\rangle$, at time t, satisfies the Schrödinger equation

$$i\hbar \frac{d}{dt}|\psi(t)\rangle = H(t)|\psi(t)\rangle \qquad (12.6.2)$$

and just prior to the switching on of the perturbation, we specify an initial condition $|\psi(t)\rangle|_{t=0} = |\psi(0)\rangle$.

Let $P_{\mathbf{A}}(\boldsymbol{\lambda})$, $\boldsymbol{\lambda} = (\lambda_0, \lambda_1, \ldots)$, be the spectral measure associated with commuting set of spectral measures of operators $\mathbf{A} = (H^0, A_1, \ldots)$, thus including the Hamiltonian H^0, in question, to specify a state at $t \geqslant \tau$.

For measurements carried out on the system at time $t \geqslant \tau$, after the perturbation is switched off,

$$\int_\triangle \langle\psi(t)|dP_{\mathbf{A}}(\boldsymbol{\lambda})|\psi(t)\rangle \equiv \langle\psi(t)|P_{\mathbf{A}}(\triangle)|\psi(t)\rangle \qquad (12.6.3)$$

---

[15] The switching on and off of the perturbation in time may be carried out in a smooth manner as done in (8.7.21), (8.7.23), (12.6.33).

represents the probability of the measurements of the observables in $\mathbf{A} = (H^0, A_1, \ldots)$ to have values in the set $\triangle$, i.e., for which $\boldsymbol{\lambda} \in \triangle$, with $|\psi(0)\rangle$ denoting the initial state of the system. Here $P_{\mathbf{A}}(\triangle)$ denotes a projection operator.

To consider actual transitions of the system under the action of the perturbation, we choose $\triangle$ such that

$$\int_{\triangle} dP_{\mathbf{A}}(\boldsymbol{\lambda}') |\psi(0)\rangle = P_{\mathbf{A}}(\triangle) |\psi(0)\rangle = 0. \tag{12.6.4}$$

The physical significance of the condition in (12.6.4) is clear. It states that the values obtained in the measurements of the observables in $\mathbf{A} = (H^0, H_1, \ldots)$, specified by the condition $\boldsymbol{\lambda}' \notin \triangle$, at least one of them is different from its initial value as a result of the application of the perturbation thus the latter causing the system to make a transition. This obviously does *not* necessarily exclude transitions which conserve energy (i.e., for which $\lambda_0 = \lambda_0'$) since not all of the components of $\boldsymbol{\lambda}'$ need to be different from those of $\boldsymbol{\lambda}$ so that $\boldsymbol{\lambda}' \notin \triangle$.

To obtain the expression of the transition probability in question, we first set

$$|\psi(t)\rangle = e^{-itH^0/\hbar} |\phi(t)\rangle \tag{12.6.5}$$

and

$$\hat{H}^1(t) = e^{itH^0/\hbar} H^1(t) e^{-itH^0/\hbar}. \tag{12.6.6}$$

These lead to

$$|\phi(t)\rangle = |\phi(0)\rangle - \frac{i}{\hbar} \int_0^t dt' \hat{H}^1(t') |\phi(t')\rangle \tag{12.6.7}$$

$$|\psi(0)\rangle = |\phi(0)\rangle. \tag{12.6.8}$$

From (12.6.3)–(12.6.8), we then have for a *weak* perturbation, i.e., to the leading order in $H^1(t)$,

$$\langle \psi(t) | P_{\mathbf{A}}(\triangle) | \psi(t) \rangle = \langle \psi(\tau) | P_{\mathbf{A}}(\triangle) | \psi(\tau) \rangle$$

$$= \frac{1}{\hbar^2} \int_0^\tau dt' \int_0^\tau dt'' \left\langle \psi(0) \left| \hat{H}^1(t') P_{\mathbf{A}}(\triangle) \hat{H}^1(t'') \right| \psi(0) \right\rangle \tag{12.6.9}$$

where we have made explicit use of (12.6.4).

For the sequel, we consider $H^1(t)$, for $0 < t < \tau$ in (12.6.1), to be time-independent.

Suppose that initially,

$$|\psi(0)\rangle = |\Psi(\lambda_0', \lambda_1', \ldots)\rangle \equiv |\Psi(\boldsymbol{\lambda}')\rangle \tag{12.6.10}$$

12.6 Time-Dependence; Sudden Approximation and the Adiabatic Theorem

for which, according to (12.6.4), $\boldsymbol{\lambda}' \notin \triangle$. One may then write the double integral in (12.6.9) as

$$\int_\triangle \langle \psi(0) | H^1 \mathrm{d}P_{\mathbf{A}}(\boldsymbol{\lambda}) H^1 | \psi(0) \rangle \int_0^\tau \mathrm{d}t' \int_0^\tau \mathrm{d}t'' \mathrm{e}^{\mathrm{i}(\lambda_0' - \lambda_0)(t' - t'')/\hbar}$$

$$= \tau^2 \int_\triangle \langle H^1 \psi(0) | \mathrm{d}P_{\mathbf{A}}(\boldsymbol{\lambda}) | H^1 \psi(0) \rangle \frac{\sin^2\left[(\lambda_0 - \lambda_0')\tau/2\hbar\right]}{\tau^2\left[(\lambda_0 - \lambda_0')/2\hbar\right]^2} \quad (12.6.11)$$

where we have multiplied and divided the resulting integral by τ^2.

We formally introduce a density of states $\rho(\boldsymbol{\lambda})$, for values of the observables in $\mathbf{A} = (H^0, A_1, \ldots)$ lying in the range $(\boldsymbol{\lambda}, \boldsymbol{\lambda} + \mathrm{d}\boldsymbol{\lambda})$ in $\triangle$, and write

$$\langle H^1 \psi(0) | \mathrm{d}P_{\mathbf{A}}(\boldsymbol{\lambda}) | H^1 \psi(0) \rangle = \left| \langle \Psi(\boldsymbol{\lambda}) | H^1 | \Psi(\boldsymbol{\lambda}') \rangle \right|^2 \mathrm{d}^n \boldsymbol{\lambda}\, \rho(\boldsymbol{\lambda}) \quad (12.6.12)$$

where n specifies the number of observables in $\mathbf{A}$, and we have used (12.6.10). Suppose that λ_0, in $\triangle$, varies within a range $I(\varepsilon) = (\bar{\lambda}_0 - \varepsilon, \bar{\lambda}_0 + \varepsilon)$, and $\triangle = I \cup \triangle_1$, where $\triangle_1$ corresponds the range of values for $\lambda_1, \lambda_2, \ldots$.

The function

$$\frac{\sin^2\left[(\lambda_0 - \lambda_0')\tau/2\hbar\right]}{\tau^2\left[(\lambda_0 - \lambda_0')/2\hbar\right]^2} \equiv \operatorname{sinc}^2\left[(\lambda_0 - \lambda_0')\tau/2\hbar\right] \quad (12.6.13)$$

in (12.6.11) peaks at $\lambda_0 = \lambda_0'$ of width $\sim 2\pi$. For $\lambda_0' \in I(\varepsilon)$, this peak falls within the λ_0 interval of integration in (12.6.11). We consider $(\bar{\lambda}_0 - \lambda_0' - \varepsilon)\tau/2\hbar \ll -\pi$, $(\bar{\lambda}_0 - \lambda_0' + \varepsilon)\tau/2\hbar \gg \pi$, which, in particular, implies that $\varepsilon\tau/\hbar \gg 2\pi$. For such an ε, but the latter, nevertheless, sufficiently small, the integrand $\left|\langle \Psi(\boldsymbol{\lambda}) | H^1 | \psi(\boldsymbol{\lambda}') \rangle\right|^2 \rho(\boldsymbol{\lambda})$, for the λ_0-integral, may be evaluated at the central point $\lambda_0 = \lambda_0'$ as the main contribution to the λ_0-integral comes from this point.

By using the value of the integral

$$\frac{\tau}{2\hbar} \int_{I(\varepsilon)} \mathrm{d}\lambda_0 \operatorname{sinc}^2 \frac{(\lambda_0 - \lambda_0')\tau}{2\hbar} \simeq \pi \quad (12.6.14)$$

we obtain from (12.6.9)–(12.6.13), the approximation

$$\langle \psi(\tau) | P_{\mathbf{A}}(\triangle) | \psi(\tau) \rangle$$

$$= \frac{2\pi\tau}{\hbar} \int_{\triangle_1} \mathrm{d}\lambda_1 \ldots \left| \langle \Psi(\lambda_0', \lambda_1, \ldots) | H^1 | \Psi(\lambda_0', \lambda_1', \ldots) \rangle \right|^2 \rho(\lambda_0', \lambda_1, \ldots). \quad (12.6.15)$$

The approximation in (12.6.9) will be adequate if the integral in (12.6.15) is much smaller than $\hbar/2\pi\tau$.

Since the factor multiplying τ on the right-hand side of (12.6.15) is independent of τ, we may introduce the transition probability per unit time by

$$\frac{\langle \psi(\tau)|P_{\mathbf{A}}(\triangle)|\psi(\tau)\rangle}{\tau}$$

$$= \frac{2\pi}{\hbar}\int_{\triangle_1} d\lambda_1 \ldots \left|\langle \Psi(\lambda'_0,\lambda_1,\ldots)|H^1|\Psi(\lambda'_0,\lambda'_1,\ldots)\rangle\right|^2 \rho(\lambda'_0,\lambda_1,\ldots). \tag{12.6.16}$$

This expression emerging from work of Dirac is called the *Golden Rule*.[16]

The treatment of the case *when* $\lambda'_0 \notin I(\varepsilon)$ is left as an exercise to the reader (see Problem 12.19).

Of particular interest in the application of (12.6.16) is to the (elastic) scattering of a particle of initial momentum $\mathbf{p}'$ to momentum $\mathbf{p}$ via a potential $H^1 = V(\mathbf{x})$ conserving energy. Then by setting $\mathbf{p}'^2/2m = \mathbf{p}^2/2m = E$, $|\mathbf{p}|/m = v$ and

$$\frac{d^3\mathbf{p}}{(2\pi\hbar)^3} = \frac{m^3}{(2\pi\hbar)^3}\, v\, dE\, d\Omega \equiv \rho(E,\Omega)\, dE\, d\Omega \tag{12.6.17}$$

(see also (9.7.35)), where $d\Omega$ is the element of the solid angle about the vector $\mathbf{p}$, with $\langle \Psi(\lambda_0,\lambda_1)|H^1|\Psi(\lambda_0,\lambda'_1)\rangle$ identified with $\langle \mathbf{p}|V|\mathbf{p}'\rangle$, we obtain for the transition probability per unit time from (12.6.16) the expression[17] $((\theta,\phi)\in\triangle_1)$

$$\frac{2\pi}{\hbar}\int_{\triangle_1} d\Omega\, |\langle \mathbf{p}|V|\mathbf{p}'\rangle|^2\, \rho(E,\Omega) = \frac{m^2}{4\pi^2\hbar^4}\, v\int_{\triangle_1} d\Omega\, |V(\mathbf{p}-\mathbf{p}')|^2 \tag{12.6.18}$$

where $V(\mathbf{p}-\mathbf{p}')$ is the Fourier transform of $V(\mathbf{x})$:

$$V(\mathbf{p}-\mathbf{p}') = \int d^3\mathbf{x}\, e^{-i\mathbf{x}\cdot(\mathbf{p}-\mathbf{p}')/\hbar}\, V(\mathbf{x}). \tag{12.6.19}$$

For additional and related details to equations such as (12.6.16), the reader is referred to Chapter 15 (see, for example, (15.8.39)).

12.6.2 Sudden Approximation

Consider a change occurring in a Hamiltonian during a very short time τ,

$$H = \begin{cases} H_1,\, t \leqslant 0 \\ H_2,\, t > 0 \end{cases} \tag{12.6.20}$$

This may be equivalently rewritten as

---

[16] This name was coined by E. Fermi.
[17] Compare this with the one in (15.3.13) for later reference in Chapter 15. The approximation leading to the one in (12.6.18) is referred to as the *Born approximation* which will be discussed in detail in Chapter 15.

12.6 Time-Dependence; Sudden Approximation and the Adiabatic Theorem

$$H(t) = H_2 + V(t) \tag{12.6.21}$$

where

$$V(t) = \begin{cases} H_1 - H_2, & t \leqslant 0 \\ 0, & t > 0. \end{cases} \tag{12.6.22}$$

Here $V(t)$ need not be a weak perturbation to H_2.

At time $t = 0$, just prior to the "sudden" change in the Hamiltonian described above, when the Hamiltonian of the system is still given by H_1, we prepare the system in a state, say, $|\psi_1\rangle$. For $t > 0$, the latter state develops in time via the Hamiltonian $H(t)$ thus satisfying the Schrödinger equation

$$i\hbar \frac{d}{dt} |\psi_1(t)\rangle = H(t) |\psi_1(t)\rangle \tag{12.6.23}$$

with $H(t)$ given in (12.6.21) and $|\psi_1(t)\rangle|_{t=0} = |\psi_1\rangle$.

After a time $t \geqslant \tau$, when the Hamiltonian is given by H_2, a state $|\psi_2\rangle$ develops in time via H_2, i.e.,

$$|\psi_2(t)\rangle = e^{-itH_2/\hbar} |\psi_2\rangle. \tag{12.6.24}$$

We are interested in investigating the nature of the time-dependence of the transition probability $|\langle \psi_2(t) | \psi_1(t) \rangle|^2$ under a sudden change of the Hamiltonian as given in (12.6.20) during a very short time τ.

To the above end, let

$$|\psi_1(t)\rangle = e^{-itH_2/\hbar} |\phi_1(t)\rangle \tag{12.6.25}$$

$$\hat{V}(t) = e^{itH_2/\hbar} V(t) e^{-itH_2/\hbar} \tag{12.6.26}$$

in a similar approach as done in (12.6.5), (12.6.6). Then

$$|\phi_1(t)\rangle = |\phi_1\rangle - \frac{i}{\hbar} \int_0^t dt' \, \hat{V}(t') |\phi_1(t')\rangle \tag{12.6.27}$$

$$|\psi_1\rangle = |\phi_1\rangle. \tag{12.6.28}$$

These equations lead to the transition amplitude

$$\langle \psi_2(t) | \psi_1(t) \rangle = \langle \psi_2 | \psi_1 \rangle - \frac{i}{\hbar} \int_0^t dt' \, \langle \psi_2(t') | V(t') | \psi_1(t') \rangle. \tag{12.6.29}$$

We expect that if the Hamiltonian changes during a very short time τ, then the second term in (12.6.29) will be vanishingly small for $\tau \to 0$ since $V(t') \to 0$ for $t' > \tau$, and obtain that

$$\langle \psi_2(t) | \psi_1(t) \rangle \simeq \langle \psi_2 | \psi_1 \rangle. \tag{12.6.30}$$

That is, the transition amplitude does not significantly change for a rapid change of the Hamiltonian. To compute this amplitude, one may then evaluate the corresponding expression on the right-hand side of (12.6.30) which is time-independent. Note that $|\psi_2\rangle$, $|\psi_1\rangle$ develop in time with different Hamiltonians and (12.6.30), although approximate, does not simply follow from unitarity.

Clearly for

$$\frac{1}{\hbar}\left|\int_0^t \mathrm{d}t'\, \langle\psi_2(t')|V(t')|\psi_1(t')\rangle\right| \leqslant \frac{1}{\hbar}\int_0^t \mathrm{d}t'\, \|V(t')\psi_2(t')\| \ll 1 \qquad (12.6.31)$$

and $|\langle\psi_2|\psi_1\rangle|$ not too small, one may neglect the second term on the right-hand side of (12.6.29) in computing the corresponding transition probability. The corresponding approximation is referred to as the *sudden approximation*.

Intuitively, if $|V|$, as an order of magnitude of the effective potential operating only during a short period of time τ, then from (12.6.31), we ought to have $\tau|V|/\hbar \ll 1$, as a rule of thumb, for the validity of the sudden approximation.[18]

Ideally, one may rewrite the Hamiltonian $H(t)$ in (12.6.21) as

$$H(t) = H_2 + (H_1 - H_2)\,\Theta(-t) \qquad (12.6.32)$$

where $\Theta(-t)$ is the step function of negative argument corresponding to a change of the Hamiltonian in time $\tau \to 0$. More realistically and for a rigorous analysis, however, one may replace $\Theta(-t)$ by a smooth "function of negative argument":[19]

$$\Theta_c(-t) = \begin{cases} 1, & t < 0, \\ \left(1 - \dfrac{\tau}{t}\,\mathrm{e}\,\mathrm{e}^{-\tau/t}\right), & 0 \leqslant t < \tau, \\ 0, & \tau \leqslant t. \end{cases} \qquad (12.6.33)$$

This remarkable function is not only *continuous* but has a continuous derivative as well, and is non-vanishing for $t \leqslant \tau$ only. $\Theta_c(-t)$ and its derivative are plotted in Figure 12.6.

Typically, if $|\psi_2\rangle$ is an eigenvector of H_2, then

$$\frac{1}{\hbar}\int_0^t \mathrm{d}t'\, \|V(t')\psi_2(t')\| = \frac{1}{\hbar}\int_0^t \mathrm{d}t'\, \|V(t')\psi_2\|. \qquad (12.6.34)$$

For the change in the Hamiltonian occurring smoothly during a non-vanishing time interval $(0,\tau]$ with $V(t)$ taken now as $(H_1 - H_2)\,\Theta_c(-t)$, the above is bounded above by $\tau\|(H_1 - H_2)\psi_2\|/\hbar$, where we have used the fact that

---

[18] For example, for a charged particle moving at high speeds during its interaction with an atom of atomic radius $\sim a$, we may roughly take $\tau \sim a/c$, $V \sim e^2/a$, thus $\tau V/\hbar \sim e^2/\hbar c = \alpha$ which is of the order of the fine-structure constant.

[19] See also (8.7.21).

12.6 Time-Dependence; Sudden Approximation and the Adiabatic Theorem   723

$$\int_0^\tau dt\, \Theta_c(-t) < \tau \qquad (12.6.35)$$

(see also (8.7.21), (8.7.22)).

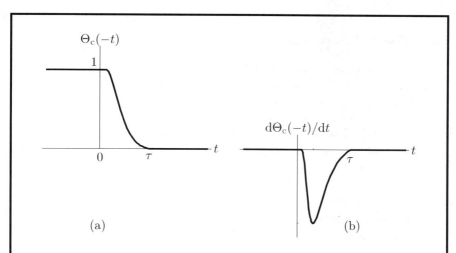

Fig. 12.6. (a) Plot of the function $\Theta_c(-t)$ and in (b) its derivative given by:
$$\frac{d\Theta_c(-t)}{dt} = \begin{cases} 0, & t < 0 \\ \frac{\tau}{t^2}\left(1 - \frac{\tau}{t}\right) e\, e^{-\tau/t}, & 0 \leqslant t < \tau \\ 0, & \tau \leqslant t \end{cases}$$
Both functions are continuous and rigorously vanish for $t > \tau$.

As an elementary example, suppose that a uniform electric field $\mathscr{E}$ is suddenly applied to a charged particle in a harmonic oscillator potential, in one dimension, being initially in the ground-state. In this case,

$$H_1 = \frac{p^2}{2m} + \frac{m\omega^2 x^2}{2} \qquad (12.6.36)$$

$$H_2 = \frac{p^2}{2m} + \frac{m\omega^2 x^2}{2} - e\mathscr{E}x. \qquad (12.6.37)$$

The latter Hamiltonian may be rewritten as

$$H_2 = \frac{p^2}{2m} + \frac{m\omega^2}{2}\left(x - \frac{e\mathscr{E}}{m\omega^2}\right)^2 - \frac{1}{2}\frac{e^2\mathscr{E}^2}{m\omega^2}. \qquad (12.6.38)$$

The transition probability, for example, that the harmonic oscillator will be excited to the first state is according to (12.6.30), (see (6.1.37), (6.1.38))

$$\left| \left(\frac{m\omega}{\pi\hbar}\right)^{1/2} \int_{-\infty}^{\infty} \mathrm{d}x \, \mathrm{e}^{-m\omega(x-a)^2/2\hbar} \sqrt{2} \sqrt{\frac{m\omega}{\hbar}} \, (x-a) \, \mathrm{e}^{-m\omega x^2/2\hbar} \right|^2 \quad (12.6.39)$$

where $a = e\mathscr{E}/m\omega^2$. The latter is easily integrated to yield the probability $a^2(m\omega/2\hbar)\exp(-m\omega a^2/2\hbar)$.

12.6.3 The Adiabatic Theorem

Consider a time-dependent Hamiltonian $H(t)$ which varies slowly in time t in an interval $[0,T]$, measured by a large value of the length T of the time interval (i.e., $T \to \infty$). The slow variation is made in comparison to a time scale set up in a problem. Such a time scale may be a typical quantum mechanical oscillation period $\sim \omega^{-1}$ (see, for example, (8.13.27)).

For each fixed t, let $|\eta_n(t)\rangle$ be a normalized eigenvector of $H(t)$ with eigenvalue $E_n(t)$,

$$H(t)|\eta_n(t)\rangle = E_n(t)|\eta_n(t)\rangle. \quad (12.6.40)$$

The spectrum of $H(t)$ need not to consist only of (discrete) eigenvalues. Given an eigenvector $|\eta_n(t)\rangle$, we may, conveniently, introduce the equivalent eigenvector defined by

$$|\phi_n(t)\rangle = |\eta_n(t)\rangle \exp\left(\mathrm{i}\int_0^t \mathrm{d}t' \, \langle \eta_n(t')|\mathrm{i}\dot\eta_n(t')\rangle\right) \quad (12.6.41)$$

(see also §8.13, (8.13.5)). The eigenvector $|\phi_n\rangle$ resulting from $|\eta_n(t)\rangle$ by a phase transformation ($\langle \eta_n|\mathrm{i}\dot\eta_n\rangle$ is real), has the advantage over $|\eta_n(t)\rangle$ in that

$$\left\langle \phi_n(t) \middle| \dot\phi_n(t) \right\rangle = 0 \quad (12.6.42)$$

as is easily verified.

We will assume that no crossing takes place between $E_n(t)$ and other eigenvalues for all $0 \leqslant t \leqslant T$.

Let $|\psi(t)\rangle$ be any state of the dynamical system satisfying the Schrödinger equation

$$\mathrm{i}\hbar \frac{\mathrm{d}}{\mathrm{d}t}|\psi(t)\rangle = H(t)|\psi(t)\rangle. \quad (12.6.43)$$

Upon integrating the elementary equation

$$\frac{\mathrm{d}}{\mathrm{d}t}\left[\langle\phi_n(t)|\psi(t)\rangle \, \mathrm{e}^{\frac{\mathrm{i}}{\hbar}\int_0^t \mathrm{d}t' E_n(t')}\right] = \left\langle \dot\phi_n(t) \middle| \psi(t) \right\rangle \exp\left(\frac{\mathrm{i}}{\hbar}\int_0^t \mathrm{d}t' E_n(t')\right) \quad (12.6.44)$$

where we have used the fact that

$$\left\langle \phi_n(t) \middle| \frac{\mathrm{d}}{\mathrm{d}t}\psi(t) \, \mathrm{e}^{\mathrm{i}\int_0^t \mathrm{d}t' E_n(t')/\hbar} \right\rangle$$

12.6 Time-Dependence; Sudden Approximation and the Adiabatic Theorem

$$= -\frac{i}{\hbar} \langle \phi_n(t) | (H(t) - E_n(t)) | \psi(t) \rangle \, e^{i \int_0^t dt' E_n(t')/\hbar}$$

$$= 0, \tag{12.6.45}$$

we obtain

$$\langle \eta_n(t) | \psi(t) \rangle \exp\left(-i \int_0^t dt' \, \langle \eta_n(t') | i\dot\eta_n(t') \rangle \right) \exp\frac{i}{\hbar} \int_0^t dt' E_n(t')$$

$$= \langle \eta_n(0) | \psi(0) \rangle + \int_0^t dt' \left\langle \dot\phi_n(t') \middle| \psi(t') \right\rangle \exp\left(\frac{i}{\hbar} \int_0^{t'} dt'' E_n(t'') \right). \tag{12.6.46}$$

In writing the expression on the right-hand side of (12.6.46) we have used (12.6.41). Equation (12.6.46) is exact.

Provided that for $T \to \infty$, the second term on the right-hand side of (12.6.46) becomes negligible, for $\langle \eta_n(0) | \psi(0) \rangle \neq 0$, we obtain

$$\langle \eta_n(t) | \psi(t) \rangle \exp\left(-i \int_0^t dt' \, \langle \eta_n(t') | i\dot\eta_n(t') \rangle \right) \exp\left(\frac{i}{\hbar} \int_0^t dt' E_n(t')\right)$$

$$\simeq \langle \eta_n(0) | \psi(0) \rangle \tag{12.6.47}$$

which is the content of the Adiabatic Theorem. Equation (12.6.47) states the fact that in the adiabatic regime of a very slow change of the Hamiltonian $H(t)$ in the interval $[0, T]$ ($T \to \infty$), the scalar product $\langle \eta_n(t) | \psi(t) \rangle$, with the appropriate phase factors as given on the left-hand side of (12.6.47), essentially remains *invariant* in time in the above mentioned interval.

To investigate the nature of the second term on the right-hand side of (12.6.46) suppose for simplicity that the spectrum of $H(t)$ consists only of a discrete non-degenerate spectrum for all $0 \leq t \leq T$. Quite generally, one may expand $|\psi(t)\rangle = \sum_m |\eta_m(t)\rangle \langle \eta_m(t) | \psi(t) \rangle$. For $|\psi(0)\rangle = |\eta_n(0)\rangle$, for a given n, then (12.6.47) implies that $\langle \eta_n(t) | \psi(t) \rangle$ is approximately one times the complex conjugate of the phases multiplying it, and otherwise $\langle \eta_m(t) | \psi(t) \rangle \simeq 0$ for $m \neq n$, since $\langle \eta_m(0) | \eta_n(0) \rangle = 0$ for $m \neq n$, and the second term on the right-hand side of (12.6.46) is expected to vanish for $T \to \infty$. Hence,

$$|\psi(t)\rangle = |\eta_n(t)\rangle \exp\left(i \int_0^t dt' \, \langle \eta_n(t') | i\dot\eta_n(t') \rangle \right)$$

$$\times \exp\left(-\frac{i}{\hbar} \int_0^t dt' E_n(t') \right) + \ldots \tag{12.6.48}$$

plus terms which are expected to vanish for $T \to \infty$, given that

$$|\psi(0)\rangle = |\eta_n(0)\rangle. \tag{12.6.49}$$

12 Approximating Quantum Systems

That is, the state $|\psi(t)\rangle$ evolves together with the Hamiltonian satisfying approximately the eigenvalue equation (12.6.40) for all $0 \leqslant t \leqslant T$.

We note that for any state $|\psi(t)\rangle$, of the dynamical system, the second term in (12.6.46), with the above discrete non-degenerate assumption of the spectrum of the $H(t)$, may be rewritten as

$$\sum_{m \neq n} \int_0^t dt' \left\langle \dot{\phi}_n(t') \middle| \phi_m(t') \right\rangle \langle \phi_m(t') | \psi(t') \rangle \exp\left(\frac{i}{\hbar} \int_0^{t'} dt'' E_n(t'') \right) \tag{12.6.50}$$

where we have used (12.6.42).

Now upon taking the time derivative of

$$\langle \phi_n(t) | H(t) = E_n(t) \langle \phi_n(t) | \tag{12.6.51}$$

and multiplying the resulting equation by $|\phi_m(t)\rangle$, for $m \neq n$, we obtain

$$\left\langle \dot{\phi}_n(t) \middle| \phi_m(t) \right\rangle = -\frac{\langle \phi_n(t) | \frac{\partial}{\partial t} H(t) | \phi_m(t) \rangle}{(E_m(t) - E_n(t))} \tag{12.6.52}$$

expressed in terms of the change of the Hamiltonian.

Also note that as in (12.6.45),

$$\frac{d}{dt}\left(|\psi(t)\rangle e^{\frac{i}{\hbar} \int_0^t dt' E_n(t')} \right) = -\frac{i}{\hbar} (H(t) - E_n(t)) |\psi(t)\rangle e^{\frac{i}{\hbar} \int_0^t dt' E_n(t')} \tag{12.6.53}$$

from which

$$|\psi(t)\rangle e^{\frac{i}{\hbar} \int_0^t dt' E_n(t')} = i\hbar \frac{1}{(H(t) - E_n(t))} \frac{d}{dt}\left(|\psi(t)\rangle e^{\frac{i}{\hbar} \int_0^t dt' E_n(t')} \right)$$

$$\equiv |\chi(t)\rangle. \tag{12.6.54}$$

That is, we may write

$$\left\langle \dot{\phi}_n \middle| \phi_m \right\rangle \langle \phi_m | \psi \rangle \exp\left(\frac{i}{\hbar} \int_0^t dt' E_n(t') \right)$$

$$= -i\hbar \frac{\langle \phi_n | \frac{\partial H}{\partial t} | \phi_m \rangle}{(E_m - E_n)^2} \left\langle \phi_m \middle| \frac{d}{dt} \chi(t) \right\rangle \tag{12.6.55}$$

for the integrand in (12.6.50) with $t \to t'$.

Upon substitution of (12.6.55) in (12.6.50) and integrating by parts over t', the latter equation becomes

$$\exp\left(\frac{i}{\hbar} \int_0^t dt' E_n(t') \right) \langle \xi(t) | \psi(t) \rangle \bigg|_0^t$$

$$-\int_0^t dt' \left\langle \frac{d}{dt'} \xi(t') \middle| \psi(t') \right\rangle \exp\left(\frac{i}{\hbar} \int_0^{t'} dt'' E_n(t'')\right) \qquad (12.6.56)$$

where

$$\langle \xi(t) | = -i\hbar \sum_{m \neq n} \frac{\langle \phi_n(t) | \partial H(t)/\partial t | \phi_m(t) \rangle}{(E_m(t) - E_n(t))^2} \langle \phi_m(t) |. \qquad (12.6.57)$$

The reason for integrating by parts is that if we make the change of variable $t = sT$ in (12.6.56) where $0 \leqslant s \leqslant 1$, and introduce the notation $|\xi(t)\rangle \to |\xi(s)\rangle$, $|\psi(t)\rangle \to |\psi(s)\rangle$, $E_n(t) \to E_n(s)$, we obtain a multiplicative factor $1/T$ in (12.6.56):

$$\frac{1}{T}\left[\exp\left(\frac{i}{\hbar} T \int_0^s ds' E_n(s')\right) \langle \xi(s) | \psi(s) \rangle \bigg|_0^s \right.$$
$$\left. - \int_0^s ds' \left\langle \frac{d}{ds'} \xi(s') \middle| \psi(s') \right\rangle \exp\left(\frac{i}{\hbar} T \int_0^{s'} ds'' E_n(s'')\right) \right] \qquad (12.6.58)$$

as is easily checked. This is unlike the expression in (12.6.50) in its original form as the latter involves the t'-integration variable and one time derivative in $\langle \dot{\phi}_n(t') | \phi_m(t') \rangle$, and the substitution $t = sT$ is not helpful in that form.

Therefore provided the expression within the square brackets in (12.6.58) remains bounded for $T \to \infty$, (12.6.47) holds up to a correction of the order $1/T$.

For a detailed application of the Adiabatic Theorem, see (8.13.19)–(8.13.27) where the correction to the adiabatic limit is given by the ratio of two frequencies.

The derivation of a sufficiency condition for the Theorem as obtained in (12.6.58) with a $\mathcal{O}(1/T)$ correction may be also carried out even if the spectrum of $H(t)$ does not consist only of a discrete non-degenerate one by a very similar procedure as given above and further generalizations are possible.[20]

12.7 Master Equation; Exponential Law, Coupling to the Environment

Suppose we have two interacting systems described by a Hamiltonian

$$H = H_{01} + H_{02} + H_I \qquad (12.7.1)$$

---

[20] See Kato (1951b). There has been much interest in this Theorem recently with various degrees of rigor, cf., Avron *et al.* (1990); Wu and Yang (2004), and references therein.

where H_{02} is the free Hamiltonian of system 2, and H_{01} is the Hamiltonian of system 1, which may include interactions terms between its sub-systems. H_I is an interaction term between system 1 and 2. The Hamiltonians H_{01}, H_{02} are assumed to commute. We carry out an investigation of the dynamical process described by the Hamiltonian in the density operator formalism. Our interest is in the dynamics and aspects of system 1 only in response to system 2. Accordingly, we trace the total density operator $\rho_T(t)$ over variables of system 2, thus introducing the *reduced* density operator of system 1, defined by

$$\rho(t) = \underset{2}{\mathrm{Tr}}[\rho_T(t)] \tag{12.7.2}$$

where

$$\rho_T(t) = \mathrm{e}^{-itH/\hbar}\rho_T(0)\mathrm{e}^{itH/\hbar}. \tag{12.7.3}$$

To study the nature of the reduced density operator in (12.7.2), we first introduce the operator

$$\eta_I(t) = \mathrm{e}^{it(H_{01}+H_{02})/\hbar}\,\rho_T(t)\mathrm{e}^{-it(H_{01}+H_{02})/\hbar} \tag{12.7.4}$$

(the so-called density operator in the interaction picture), to obtain, the equation

$$\frac{\mathrm{d}}{\mathrm{d}t}\eta_I(t) = -\frac{\mathrm{i}}{\hbar}\left[H_I(t), \eta_I(t)\right] \tag{12.7.5}$$

and

$$H_I(t) = \mathrm{e}^{it(H_{01}+H_{02})/\hbar}H_I\mathrm{e}^{-it(H_{01}+H_{02})/\hbar}. \tag{12.7.6}$$

In terms of $\eta_I(t)$, the reduced density operator $\rho(t)$ is then clearly given by

$$\rho(t) = \mathrm{e}^{-itH_{01}/\hbar}\eta(t)\mathrm{e}^{itH_{01}/\hbar} \tag{12.7.7}$$

where

$$\eta(t) = \underset{2}{\mathrm{Tr}}\left[\eta_I(t)\right]. \tag{12.7.8}$$

We first develop a differential equation for $\eta(t)$ based on some approximations. The resulting equation derived is referred to as the master equation. This is followed by applications to the exponential law presented in the analysis carried out after (8.1.67), and also to the concept of quantum decoherence as a result of coupling to the environment elaborated upon at the end of §8.7, notably in (8.7.47).

12.7.1 Master Equation

We may integrate (12.7.5) to obtain

$$\eta_I(t) = \eta_I(0) - \frac{\mathrm{i}}{\hbar}\int_0^t \mathrm{d}t'\,[H_I(t'), \eta_I(t')] \tag{12.7.9}$$

where $\eta_I(0) = \rho_T(0)$.

12.7 Master Equation; Exponential Law, Coupling to the Environment

A formal perturbation expansion in $H_I(t)$, gives

$$\eta_I(t) = \eta_I(0) + \sum_{n \geqslant 1} \left(\frac{-i}{\hbar}\right)^n \int_0^t dt_n \int_0^{t_n} dt_{n-1} \cdots$$
$$\times \int_0^{t_2} dt_1 \Big[H_I(t_n), \ldots, [H_I(t_1), \eta_I(0)] \ldots \Big]. \quad (12.7.10)$$

We set
$$\eta_I(0) = \eta_T(0) = \rho_1(0)\rho_2(0) \quad (12.7.11)$$

where the initial density operators of the two systems are taken to be uncorrelated. This suggests to introduce the operator

$$Q(t) = \mathbf{1} + \sum_{n \geqslant 1} \left(\frac{-i}{\hbar}\right)^n \int_0^t dt_n \int_0^{t_n} dt_{n-1} \cdots$$
$$\times \int_0^{t_2} dt_1 \operatorname*{Tr}_2 \Big[H_I(t_n), \ldots, [H_I(t_1), \rho_2(0)\bullet] \ldots \Big] \quad (12.7.12)$$

to write $\eta(t)$ in (12.7.8) as

$$\eta(t) = Q(t)\rho_2(0) \quad (12.7.13)$$

giving
$$\dot{\eta}(t) = \dot{Q}(t)\rho_2(0). \quad (12.7.14)$$

The latter may be conveniently rewritten as

$$\dot{\eta}(t) = \dot{Q}(t)Q^{-1}(t)\eta(t). \quad (12.7.15)$$

The operator $\dot{Q}(t)Q^{-1}(t)$ may be referred to as a time development generator.

We carry out a weak coupling expansion of the generator $\dot{Q}(t)Q^{-1}(t)$ in H_I. To this end, to second order

$$\dot{Q}(t) = -\frac{i}{\hbar}\operatorname*{Tr}_2\Big[H_I(t), \rho_2(0)\bullet\Big] - \frac{1}{\hbar^2}\int_0^t dt' \operatorname*{Tr}_2\Big[H_I(t), [H_I(t'), \rho_2(0)\bullet]\Big]. \quad (12.7.16)$$

In the sequel, we consider, interaction Hamiltonians such that

$$\operatorname*{Tr}_2 [H_I(t)\, \rho_2(0)] = 0 \quad (12.7.17)$$

and this property will be explicitly verified below. Therefore to go up to second order for the generator $\dot{Q}(t)Q^{-1}(t)$, we may take $Q^{-1}(t)$ simply to be the identity operator $\mathbf{1}$. That is, to a second order expansion of the generator, we obtain from (12.7.15), (12.7.8) the equation

$$\dot{\eta}(t) = -\frac{1}{\hbar^2}\int_0^t dt'\ \mathop{\mathrm{Tr}}_2\Big[H_{\mathrm{I}}(t),\ [H_{\mathrm{I}}(t'),\ \rho_2(0)\eta(t)]\Big]. \qquad (12.7.18)$$

The types of interaction Hamiltonians $H_{\mathrm{I}}(t)$ to be investigated are of the form

$$H_{\mathrm{I}}(t) = A^\dagger e^{i\omega t}\sum_k \lambda_k b_k e^{-i\omega_k t} + A e^{-i\omega t}\sum_k \lambda_k^* b_k^\dagger e^{i\omega_k t} \qquad (12.7.19)$$

where $b_k^\dagger$, b_k are creation, annihilation operators of excitation energy $\hbar\omega_k$ associated with system 2, with the free Hamiltonian H_{02} of the latter system, given by

$$H_{02} = \sum_k \hbar\omega_k\, b_k^\dagger b_k \qquad (12.7.20)$$

corresponding to an infinite set of independent harmonic oscillators, omitting the zero point energies.

$A^\dagger$, A are creation, annihilation operators of excitation energy $\hbar\omega$ pertaining to system 1.

We take the system 2 to be initially in the ground-state, i.e.,

$$\rho_2(0) = |0\rangle_{2\,2}\langle 0| \qquad (12.7.21)$$

hence

$$_2\langle 0|b_k|0\rangle_2 = 0 = {}_2\!\left\langle 0\left|b_k^\dagger\right|0\right\rangle_2. \qquad (12.7.22)$$

We also recall the following basic relations (see §6.1, (6.1.30)),

$$b_k^\dagger |n_1,\ldots,n_k,\ldots\rangle = \sqrt{n_k+1}\,|n_1\ldots,n_k+1,\ldots\rangle \qquad (12.7.23)$$

$$b_k |n_1,\ldots,n_k,\ldots\rangle = \sqrt{n_k}\,|n_1\ldots,n_k-1,\ldots\rangle \qquad (12.7.24)$$

and the orthogonality property of the different states $|n_1,n_2,\ldots\rangle$ and, in particular, the orthogonality of $|0\rangle$ and the excited states

$$\langle n_1,n_2,\ldots |0\rangle = 0. \qquad (12.7.25)$$

It is straightforward to evaluate the trace $\mathop{\mathrm{Tr}}_2$ in (12.7.18) (see Problem 12.23) giving

$$\dot{\eta}(t) = -\Big\{A^\dagger A\eta(t)I(t) + \eta(t)A^\dagger A I^*(t) - A\eta(t)A^\dagger\big(I(t)+I^*(t)\big)\Big\} \qquad (12.7.26)$$

where

$$I(t) = \frac{1}{\hbar^2}\int_0^t dt'\ {}_2\langle 0|H_{\mathrm{I}}(t)\,H_{\mathrm{I}}(t')|0\rangle_2$$

12.7 Master Equation; Exponential Law, Coupling to the Environment

$$= \frac{1}{\hbar^2} \sum_k |\lambda_k|^2 \int_0^t dt' e^{i(\omega-\omega_k)(t-t')} \qquad (12.7.27)$$

and $_2\langle 0|H_I(t)|0\rangle_2 = 0$ satisfying (12.7.17).

As we are summing over an infinite number of modes, we will replace the sum over k in (12.7.27) by an integral over the frequency $\omega_k \to \omega'$, and introduce, in turn a frequency density $n(\omega')$, to rewrite $I(t)$ as

$$I(t) = \frac{1}{\hbar^2} \int_0^\infty d\omega' \, |\lambda(\omega')|^2 \, n(\omega') \int_0^t dt' \, e^{i(\omega-\omega')(t-t')}. \qquad (12.7.28)$$

We will refer to (12.7.26) as the *master equation*,[21] and

$$_2\langle 0|H_I(t) H_I(t')|0\rangle_2 = \int_0^\infty d\omega' \, |\lambda(\omega')|^2 \, n(\omega') \, e^{i(\omega-\omega')(t-t')} \qquad (12.7.29)$$

depending on the time difference $(t-t')$, as the *correlation function* of system 2, to which we will return shortly.

After carrying out the time integral in (12.7.28), we obtain

$$\text{Re}\{I(t)\} = \frac{1}{\hbar^2} \int_0^\infty d\omega' |\lambda(\omega')|^2 \, n(\omega') \frac{\sin(\omega'-\omega)t}{(\omega'-\omega)} \qquad (12.7.30)$$

$$\text{Im}\{I(t)\} = -\frac{1}{\hbar^2} \int_0^\infty d\omega' |\lambda(\omega')|^2 n(\omega') \frac{\sin^2(\omega'-\omega)t/2}{(\omega'-\omega)/2}. \qquad (12.7.31)$$

In the applications to follow, the real part of $I(t)$ will be of central importance. To evaluate the latter, we note that it may be rewritten as

$$\text{Re}\{I(t)\} = \frac{1}{\hbar^2} \int_{-\omega t}^\infty dx \left|\lambda\left(\omega(1+\frac{x}{\omega t})\right)\right|^2 n\left(\omega(1+\frac{x}{\omega t})\right) \frac{\sin x}{x} \qquad (12.7.32)$$

(see also (8.1.90)). The function $\sin x/x$ peaks at the origin, and is concentrated mainly in the region $|x| \leqslant \pi$. We make the Markov approximation by assuming that $|\lambda(\omega')|^2 \, n(\omega')$ is slowly varying around the point $\omega' = \omega$, and hence for $\omega t \geqslant \pi$, it may be taken outside the integral evaluated at $\omega' = \omega$, i.e., at $x = 0$, with increasing accuracy for $\omega t \gg \pi$. This gives

$$\text{Re}\{I(t)\} = \frac{|\lambda(\omega)|^2 \, n(\omega)}{\hbar^2} \int_{-\omega t}^\infty dx \, \frac{\sin x}{x} \qquad (12.7.33)$$

which for $\omega t \gg \pi$,

$$\text{Re}\{I(t)\} = \frac{\pi |\lambda(\omega)|^2 \, n(\omega)}{\hbar^2}$$

---

[21] This name is usually given to (12.7.26) after some simplifications are made in the expression for $I(t)$.

$$\equiv \frac{\gamma}{2} \tag{12.7.34}$$

using the notation in (8.1.93).

Actually, the equality in (12.7.34) is a familiar one as following directly from (12.7.30), which for sufficiently large t, $\sin(\omega' - \omega)t/(\omega' - \omega)$ acts like a delta function with support at ω. We may also rewrite the correlation function in (12.7.29) as

$$_2\langle 0|H_{\rm I}(t)\,H_{\rm I}(t')|0\rangle_2 = \int_{-1}^{\infty} {\rm d}x\,|\lambda(\omega + \omega x)|^2\,n(\omega + \omega x)\,\cos(\omega\tau x)$$
$$- {\rm i}\int_{-1}^{\infty} {\rm d}x\,|\lambda(\omega + \omega x)|^2\,n(\omega + \omega x)\,\sin(\omega\tau x) \tag{12.7.35}$$

where $\tau = t - t'$. Assuming the integrability of $|\lambda(\omega + \omega x)|^2\,n(\omega + \omega x)$ over x, then for $\omega\tau$ sufficiently large, the so-called Riemann-Lebesgue Lemma implies that both integrals in (12.7.35) become small.

That is, for $\tau \gg 1/\omega$, the correlation function $_2\langle 0|H_{\rm I}(t)\,H_{\rm I}(t')|0\rangle_2$ becomes small, and the latter contributes mainly to the time integral in (12.7.27)/(12.7.28) for τ not large. As a matter of fact, if we formally replace the correlation function by the highly localized expression in τ: $\gamma/2\,\delta(\tau)$, then this gives from (12.7.27) the result that $I(t) = \gamma/2$ consistent with (12.7.34). This is the content of the Markov approximation. The operator $H_{\rm I}(t')$, in $_2\langle 0|H_{\rm I}(t)\,H_{\rm I}(t')|0\rangle_2$, creates single excitations from the vacuum (ground-state) at time t' which are annihilated by $H_{\rm I}(t)$ within a short interval of time of length $\tau = t - t'$, ending up again by the vacuum. Such fluctuations occur in short periods of time of the order $1/\omega$, referred to as the correlation time. On the other hand, in (12.7.34), we have taken $t \gg 1/\omega$, and the corresponding approximation requires that t is long compared to the correlation time of system 2. Finally, for the validity of the second order perturbation theory in (12.7.18), we may infer from (12.7.26), (12.7.34) that t should be short compared to γ^{-1}. As shown below γ is a constant associated with the decay of system 1 in response to system 2.

What is the significance of the small imaginary part in (12.7.31)? This, in general, contributes to an energy shift (as in the Lamb shift) for the system 1. This is easily seen from the dynamical equation for $\rho(t)$ in (12.7.2), which from (12.7.7), (12.7.26) is given by

$$\dot{\rho} = -\frac{\rm i}{\hbar}\left[{\rm e}^{-{\rm i}tH_{01}/\hbar}(H_{01} + \hbar I_i\,A^\dagger A){\rm e}^{{\rm i}tH_{01}/\hbar},\,\rho\right]$$
$$- \left\{\mathcal{A}^\dagger\mathcal{A}\rho + \rho\mathcal{A}^\dagger\mathcal{A} - \mathcal{A}\rho\mathcal{A}^\dagger\right\}I_r \tag{12.7.36}$$

where

$$\mathcal{A}^\sharp = {\rm e}^{-{\rm i}tH_{01}/\hbar}A^\sharp{\rm e}^{{\rm i}tH_{01}/\hbar} \tag{12.7.37}$$

12.7 Master Equation; Exponential Law, Coupling to the Environment

$$I_r = \text{Re}\{I\}, \quad I_i = \text{Im}\{I\}. \tag{12.7.38}$$

In (12.7.36), note that the Hamiltonian H_{01} of system is displaced by the quadratic term $\hbar I_i A^\dagger A$, involving the energy correction $\hbar I_i$, as a result of the interaction of systems 1 with that of 2.

For further reference, we note that

$$I(t) + I^*(t) = 2I_r = \gamma \tag{12.7.39}$$

as follows from (12.7.34).

12.7.2 Exponential Law

Consider the two-level system in (8.1.67) with interaction Hamiltonian as given in (8.1.68)–(8.1.70),

$$H_{01} = \frac{E_0 + E_1}{2} + \frac{E_1 - E_0}{2}\sigma_3. \tag{12.7.40}$$

Suppose the system is initially in the state $(1\ \ 0)^T$, i.e.,

$$\eta(0) = \rho(0) = \begin{pmatrix} 1 & 0 \\ 0 & 0 \end{pmatrix} \tag{12.7.41}$$

and let

$$\eta(t) = \begin{pmatrix} a & b \\ c & d \end{pmatrix}. \tag{12.7.42}$$

Then from (12.7.26), (8.1.69), (8.1.70)

$$\begin{pmatrix} \dot a & \dot b \\ \dot c & \dot d \end{pmatrix} = -\begin{pmatrix} a(I + I^*) & bI \\ cI^* & -a(I + I^*) \end{pmatrix} \tag{12.7.43}$$

which from (12.7.41) leads to

$$a(t) = \exp\left(-\int_0^t dt'\,(I(t') + I^*(t'))\right) \tag{12.7.44}$$

$$b(t) = 0 = c(t) \tag{12.7.45}$$

$$d(t) = 1 - \exp\left(-\int_0^t dt'\,(I(t') + I^*(t'))\right). \tag{12.7.46}$$

From (12.7.7), (12.7.40), we get

$$\rho(t) = \begin{pmatrix} a(t) & 0 \\ 0 & d(t) \end{pmatrix}. \tag{12.7.47}$$

This gives for the survival probability

$$P(t) = \begin{pmatrix} 1 & 0 \end{pmatrix} \begin{pmatrix} a(t) & 0 \\ 0 & d(t) \end{pmatrix} \begin{pmatrix} 1 \\ 0 \end{pmatrix}$$

$$= a(t) = e^{-\gamma t} \tag{12.7.48}$$

which coincides with (8.1.94) as expected. It is valid for $t \gg \omega^{-1}$ but for t short in comparison to γ^{-1}, where γ is the decay constant.

The integration in

$$\int_0^t dt' (I(t') + I^*(t')) = \gamma t \tag{12.7.49}$$

requires justification. To this end, note that from (12.7.33) that

$$I(t) + I^*(t) = \frac{\gamma}{\pi} \int_{-\omega t}^{\infty} dx \frac{\sin x}{x}. \tag{12.7.50}$$

Let $\omega t = z$. Then

$$\int_{-z}^{\infty} dx \frac{\sin x}{x} = \pi - \int_{z}^{\infty} dx \frac{\sin x}{x}$$

$$= \pi - \frac{\cos z}{z} - \frac{\sin z}{z^2} - \frac{2}{z^2} \int_1^{\infty} dx \frac{\sin(xz)}{x^3} \tag{12.7.51}$$

useful for $z \geq 1$ where in the last step we have integrated by parts twice.

Similarly,

$$\int_{-z}^{\infty} dx \frac{\sin x}{x} = \frac{\pi}{2} + \int_0^z dx \frac{\sin x}{x} \tag{12.7.52}$$

and for $0 < z \leqslant \pi$, (12.7.52) implies that

$$\left| \int_{-z}^{\infty} dx \frac{\sin x}{x} \right| \leqslant \frac{\pi}{2} + z. \tag{12.7.53}$$

From (12.7.50)–(12.7.53), we then have

$$\int_0^t dt' (I(t) + I^*(t')) = \gamma t \left(1 + \mathcal{O}\left(\frac{\ln(\omega t)}{\omega t}\right)\right) \tag{12.7.54}$$

justifying the expression given in (12.7.49) for $t \gg \omega^{-1}$.

12.7.3 Coupling to the Environment

Consider a spin 1/2 object interacting with an apparatus described by a harmonic oscillator as given in (8.7.24), with the initial state of the system of spin 1/2 and the apparatus as given in (8.7.25)

12.7 Master Equation; Exponential Law, Coupling to the Environment

$$|\Phi_0\rangle = \begin{pmatrix} c_+ \\ c_- \end{pmatrix} |-i\alpha_0\rangle \tag{12.7.55}$$

where $|-i\alpha_0\rangle$ is a coherent state describing the initial state of the apparatus, α_0 is real and positive. Here H_{01} is taken to be the expression in (8.7.24).

In the notation, (12.7.8), we have from (8.7.31)

$$\eta(0) = |c_+|^2 \begin{pmatrix} 1 & 0 \\ 0 & 0 \end{pmatrix} |\alpha\rangle\langle\alpha| + |c_-|^2 \begin{pmatrix} 0 & 0 \\ 0 & 1 \end{pmatrix} |-\alpha\rangle\langle-\alpha|$$

$$+ c_+ c_-^* \begin{pmatrix} 0 & 1 \\ 0 & 0 \end{pmatrix} |\alpha\rangle\langle-\alpha| + c_+^* c_- \begin{pmatrix} 0 & 0 \\ 1 & 0 \end{pmatrix} |-\alpha\rangle\langle\alpha| \tag{12.7.56}$$

where $\alpha = \alpha_0 \exp(-i\omega T)$, $T = \hbar\pi/2\lambda$ (see (8.7.29)).

For the interaction of the apparatus with the environment (see §8.7), we choose the one given in (8.7.44), where $a^\dagger$, a are the creation, annihilation operators associated with the harmonic oscillator describing the apparatus.

To solve (12.7.26), with $A^\dagger$, A now identified with $a^\dagger$, a, we introduce[22] the generating function

$$F[K^*, K; t] = \text{Tr}\left[e^{K^* a} \eta(t) e^{K a^\dagger}\right] \tag{12.7.57}$$

where the Tr is taken over the apparatus variables, and K, K^* are parameters.

Using the elementary properties

$$a\, e^{K a^\dagger} = e^{K a^\dagger}(a + K) \tag{12.7.58}$$

$$e^{K^* a} a^\dagger = (a^\dagger + K^*) e^{K^* a} \tag{12.7.59}$$

we obtain

$$\text{Tr}\left[e^{K^* a} a^\dagger a\, \eta\, e^{K a^\dagger}\right] = K^* \text{Tr}\left[e^{K^* a} a\eta\, e^{K a^\dagger}\right]$$

$$+ \text{Tr}\left[e^{K^* a} a\eta a^\dagger\, e^{K a^\dagger}\right] \tag{12.7.60}$$

$$\text{Tr}\left[e^{K^* a} \eta a^\dagger a\, e^{K a^\dagger}\right] = \text{Tr}\left[e^{K^* a} \eta a^\dagger\, e^{K a^\dagger}\right] K$$

$$+ \text{Tr}\left[e^{K^* a} a\eta a^\dagger\, e^{K a^\dagger}\right]. \tag{12.7.61}$$

Therefore, from (12.7.26),

$$\text{Tr}\left[e^{K^* a} \dot{\eta}\, e^{K a^\dagger}\right] = -K^* I\, \text{Tr}\left[e^{K^* a} a\eta e^{K a}\right] - KI^* \text{Tr}\left[e^{K^* a} \eta a^\dagger e^{K a^\dagger}\right]. \tag{12.7.62}$$

---

[22] We use a method described in: Walls and Milburn (1985).

In terms of the generating function in (12.7.57), this leads to

$$\frac{d}{dt}F[K^*, K; t] = -\left[K^*I\frac{d}{dK^*} + KI\frac{d}{dK}\right]F[K^*, K; t]. \quad (12.7.63)$$

This is readily integrated over t to give

$$\text{Tr}\left[e^{K^*a}\eta(t)e^{Ka^\dagger}\right] = \text{Tr}\left[\exp\left(K^*\,a\,e^{-\int_0^t I}\right)\eta(0)\exp\left(Ka^\dagger\,e^{-\int_0^t I^*}\right)\right]. \quad (12.7.64)$$

Using the properties

$$e^{ca}|\alpha\rangle = e^{c\alpha}|\alpha\rangle \quad (12.7.65)$$

(see (6.6.26)), $\langle\alpha|\alpha\rangle = 1$, $\langle\alpha|-\alpha\rangle = \exp\left(|\alpha|^2/2\right) \equiv \exp(\alpha_0^2/2)$ (see (6.6.38)), it is not difficult to see from (12.7.56), and (12.7.64), that

$$\eta(t) = |c_+|^2 \begin{pmatrix} 1 & 0 \\ 0 & 0 \end{pmatrix}\left|\alpha\,e^{-\int_0^t I}\right\rangle\left\langle\alpha\,e^{-\int_0^t I}\right|$$

$$+ |c_-|^2 \begin{pmatrix} 0 & 0 \\ 0 & 1 \end{pmatrix}\left|-\alpha\,e^{-\int_0^t I}\right\rangle\left\langle-\alpha\,e^{-\int_0^t I}\right|$$

$$+ c_+c_-^* \begin{pmatrix} 0 & 1 \\ 0 & 0 \end{pmatrix}\left|\alpha\,e^{-\int_0^t I}\right\rangle\left\langle-\alpha\,e^{-\int_0^t I}\right|\exp(-2\alpha_0^2(1-e^{-\gamma t}))$$

$$+ c_+^*c_- \begin{pmatrix} 0 & 0 \\ 1 & 0 \end{pmatrix}\left|-\alpha\,e^{-\int_0^t I}\right\rangle\left\langle\alpha\,e^{-\int_0^t I}\right|\exp(-2\alpha_0^2(1-e^{-\gamma t})). \quad (12.7.66)$$

Now it is not necessary to carry out the operation in $\exp\left(-i\frac{tH_{01}}{\hbar}\right)\eta(t)\exp\left(i\frac{tH_{01}}{\hbar}\right)$, defining $\rho(t)$ in (12.7.7), to note the following. Although γt is small, i.e., $\exp[-2\alpha_0^2(1-e^{-\gamma t})] \simeq \exp[-2\alpha_0^2\gamma t]$, then for a macroscopic value $\alpha_0^2 \gg 1/2\gamma t$, the non-diagonal part of (12.7.66) will be washed away relative to the diagonal one, with decoherence setting in exponentially on a decoherence time scale $\sim 1/\gamma\alpha_0^2$ as discussed at the end of §8.7. The operation in $\exp(-itH_{01}/\hbar)\eta(t)\exp(itH_{01}/\hbar)$ just multiplies the α's in $|\pm\alpha\rangle$ by phase factors without obviously destroying the Hermiticity of $\rho(t)$ (see Problem 12.24).

Problems

12.1. For the Hamiltonian of the anharmonic oscillator potential in (12.4.8), show that the Rayleigh-Schrödinger series for the ground-state energy up to third order in λ is given by

$$E(\lambda) = 1 + (3/4)\lambda - (21/16)\lambda^2 + (333/64)\lambda^3 + \ldots.$$

Also find the ground-state to first order.

12.2. Treating the electron-electron interaction term in the helium atom as a perturbation, find approximately, to the leading order in this interaction, the ground-state energy of the atom.

12.3. For a two-level system, let $|1\rangle, |2\rangle$ be two orthonormal vectors. Consider linear combinations $[a_1 |1\rangle + a_2 |2\rangle]$ as eigenstates of a given Hamiltonian H to find the exact energy levels and eigenvectors of the system in terms of the matrix elements $\langle i|H|j\rangle$. If $|1\rangle, |2\rangle$ are taken as eigenstates of a Hamiltonian H^0 with eigenvalues E_1, E_2, compare the expressions obtained with those of perturbation theory discussed in §12.1, §12.2 for the cases $E_1 \neq E_2, E_1 = E_2$ by treating $H - H^0$ as a perturbation to H^0.

12.4. If the degeneracy of a given energy level E_n^0 is not removed to fist order, then show how energy corrections may be determined to second order.

12.5. Compute the splitting of the ground-state energy of the 3D harmonic oscillator to the leading order if a perturbation $-Fz$ is added to the Hamiltonian by working in Cartesian coordinates.

12.6. Investigate the splitting of the energy level $E_2 - \eta$, in (12.2.10), corresponding to the states specified by $(\ell = 1, m = 0, m_s = -1/2), (\ell = 0, m = 0, m_s = -1/2)$ due to the perturbation in (12.2.9).

12.7. To the Hamiltonian of a plane rotator $H^0 = -(\hbar^2/2I)(\partial^2/\partial\phi^2)$, where I is a moment of inertia, a perturbation $-\mu E \cos\phi$ is added, where μ is an electric dipole moment of the rotator and E is the magnitude of an applied electric field. Study the leading order perturbations contributions to this coupled system. [Ref. Johnston and Sposito (1976).]

12.8. Scale the variable z in the Hamiltonian in (12.3.4) to generate one with variable couplings and re-express the bounds in (12.3.9) in terms of the coupling parameters.

12.9. Use the variational method with trial wavefunction given in (12.3.7) to obtain an upper-bound-estimate for the ground-state energy of the anharmonic potential with Hamiltonian given in (12.3.14). Study carefully the usefulness and limitations of the inequality in (12.3.14) in providing a lower bound to this positive Hamiltonian in question. Can you improve the lower bound to obtain a bound closer to the upper one?

12.10. Derive the expressions for the matrix elements, $\langle \phi_1|H|\phi_1\rangle, \langle \phi_2|H|\phi_1\rangle$ in (12.3.26), (12.3.27) to finally obtain the bound in (12.3.29).

12.11. Derive the behavior in (12.4.21) of the expansion coefficients E_K of the ground-state energy of the anharmonic oscillator potential with Hamiltonian in (12.4.8), from a direct *perturbative* analysis of the path integral expression (12.4.23) without using the dispersion relation expression leading to the integral in (12.4.11).

12.12. The numerical factor multiplying the expression for E_K in (12.4.21), independent of K, for $K \to \infty$, may be obtained by investigation of the determinant of the matrix $M = -(1/4)\mathrm{d}^2/\mathrm{d}t^2 + 1 - 6\cosh^{-2}[2(t-t_0)]$

738   12 Approximating Quantum Systems

in (12.4.30), in comparison to the matrix $M_0 = -(1/4)d^2/dt^2 + 1$, since the action in the path integral in (12.4.35) is quadratic. Using this fact, together with the λ-independent numerical factors such as in (12.4.28) omitted in obtaining (12.4.21), evaluate the overall numerical factor in question. [Ref. Zinn-Justin (1981).]

12.13. Determine the Padé approximant $P^{[2,1]}(\lambda)$ consistent with the expansion for the ground-state energy to third order in λ given in Problem 12.1 for the anharmonic oscillator potential with Hamiltonian in (12.4.8).

12.14. How can one modify the WKB approximation if $V'(a)$ is zero in (12.5.9)?

12.15. Reformulate the WKB approximation in the momentum description.

12.16. (i) In reference to part (A) of Figure 12.5, study the nature of the eigenvalues of the harmonic oscillator as following from (12.5.59).
(ii) For a Hamiltonian with a linear potential $-Fx$ defined for $x > 0$, study the nature of the eigenvalues according to (12.5.61) in reference to part (B) of Figure 12.5 with $a = 0$.

12.17. Justify the use of (12.5.71) for the WKB approximation as applied to the attractive Coulomb potential and find the corresponding eigenvalues according to this formula.

12.18. Attempt to extend the WKB approximation to 3D deriving in the process a Hamilton-Jacobi-like equation encountered in classical dynamics.

12.19. Derive the expression corresponding to (12.6.16) in the case when $\lambda'_0 \notin I(\varepsilon)$.

12.20. For the sudden approximation concerning the Hamiltonian H_1, H_2 in (12.6.36), (12.6.37), find the transition probability from the ground-state energy to *any* excited state.

12.21. Find the transition probability for the transition from the ground-state to an arbitrary excited state when the direction of the electric field in the Hamiltonian $H'_1 \equiv H_2$, where H_2 is defined in (12.6.37), is suddenly changed.

12.22. If the equilibrium point of a harmonic oscillator potential is changed adiabatically (i.e., slowly) with a uniform speed, find the transition probability from the ground-state to the fist excited state.

12.23. Evaluated the trace $\underset{2}{\mathrm{Tr}}$ in (12.7.18) to show that it is given by (12.7.26).

12.24. Carry out the action of the operator $\exp(-itH_{01}/\hbar)$ in $[\exp(-itH_{01}/\hbar)\eta(t)\exp(itH_{01}/\hbar)] = \rho(t)$, where H_{01} is taken to be the expression in (8.7.24), and $\eta(t)$ is given in (12.7.66), to show that this operation simply multiplies the α's in $|\pm\alpha\rangle$, in (12.7.66) by phase factors (without obviously destroying the Hermiticity of $\rho(t)$).

13

Multi-Electron Atoms: Beyond the Thomas-Fermi Atom

The purpose of this chapter is to determine, as an estimate, an *explicit* expression for the ground-state energy $E(Z)$ for multi-electron neutral atoms as a function of the atomic number Z.

The Hamiltonian of a neutral atom consisting of Z electrons and a heavy nucleus of charge $Z|e|$ is taken to be

$$H = \sum_{i=1}^{Z} \left(\frac{\mathbf{p}_i^2}{2m} - \frac{Ze^2}{r_i} \right) + \sum_{i<j}^{Z} \frac{e^2}{|\mathbf{x}_i - \mathbf{x}_j|} \qquad (13.1)$$

where $r_i = |\mathbf{x}_i|$, and m is the mass of an electron. Since the last term, responsible for the electron-electron interaction, is positive a lower bound to the ground-state energy of multi-electron atoms is readily obtained from (13.1) by using the bound

$$H \geqslant \sum_{i=1}^{Z} \left(\frac{\mathbf{p}_i^2}{2m} - \frac{Ze^2}{r_i} \right) \qquad (13.2)$$

with the right-hand side consisting of the sum of the Hamiltonians of Z "non-interacting" hydrogenic atoms each of ground-state energy $-mZ^2e^4/2\hbar^2$. This leads (§3.4) to the following conservative lower bound for the ground-state energy of atoms

$$E(Z) \geqslant -\frac{mZ^3e^4}{2\hbar^2}. \qquad (13.3)$$

which is sufficient theoretically in establishing the stability of atoms, and improvements to the lower bound in (13.3) may be certainly made (see Problem 13.14).

Our *starting* point for obtaining an explicit expression for $E(Z)$ is the so-called Thomas-Fermi (TF) model for the atom. The TF atom has captivated the hearts of physicists since its birth over a three quarter of a century ago in 1927, when quantum physics was still in its infancy, and will continue to do so due to its extreme simplicity. In simplest terms, the interaction that an electron in the TF atom at **x** experiences is described by an effective spherically symmetric potential $V(r)$, $|\mathbf{x}| = r$, and uses the TF semi-classical approximation (see (9.8.1)–(9.8.5)). The key point is that complicated expressions for the interactions and the multi-particle kinetic energy are replaced by simple functions of the electron density determined in the TF semi-classical approximation (see (9.8.4), (9.8.5), and also (9.7.31), (4.5.15) for the kinetic energy). In Appendix A to §13.1, on the TF atom, it is shown formally that the ground-state energy $E_{TF}(Z)$ computed according to this model actually gives the leading contribution to the *exact* one $E(Z)$ for large Z. Therefore, *this* model provides the *correct* starting point for the determination of an estimate for $E(Z)$. Corrections to $E_{TF}(Z)$ are then systematically investigated. The corrections turn out to be, however, not just mere perturbation expansion terms in power of $1/Z$ and the analysis is more involved than that. These corrections, based on physical grounds, are worked out in the subsequent sections (§13.2–§13.4) and the resulting final expression, as an estimate, for $E(Z)$ is given in §13.5 by adding up these contributions.

We urge the reader to review the contents of §9.7, §9.8, on Green functions, before reading this chapter. Here one will witness the power and the relative simplicity of using Green functions in dealing with endless situations encountered in quantum physics.

13.1 The Thomas-Fermi Atom

In the Thomas-Fermi atom, the interaction that an electron at **x** experiences is described by an effective spherically symmetric potential $V(\mathbf{x}) = V(r)$, $|\mathbf{x}| = r$, also that the TF semi-classical (see (9.8.1)–(9.8.5)) approximation is valid and that the complicated expressions for the multi-particle kinetic energy and the interactions may be self consistently replaced by simple functions of the electron density determined in the TF semi-classical approximation (see (9.8.4), (9.8.5), (9.7.31)), as will be now described. This is indeed correct for large Z values as is formally shown in Appendix A to this section. Deviations from large Z values will be then investigated in the next three sections.

From (9.8.5), we may infer that the electron density $n(r)$, allowing for spin degeneracy, is given by

$$n(\mathbf{x}) = \frac{1}{3\pi^2}\left(\frac{2m(\xi - V(r))}{\hbar^2}\right)^{3/2} = n(r) \qquad (13.1.1)$$

where

$$\int d^3\mathbf{x}\, n(\mathbf{x}) = Z. \qquad (13.1.2)$$

Here, according to (9.7.22), (9.8.4), the parameter ξ provides an upper bound to the maximum energy of a bound electron. The parameter ξ also determines

13.1 The Thomas-Fermi Atom

the boundary of the TF atom defined by $r = r_B$. Since the electron density $n(r) = 0$ for $r \geqslant r_B$, we have

$$V(r_B) = \xi \tag{13.1.3}$$

and due to the spherical symmetry of $V(r)$ (and $n(r)$) and the assumption of neutrality of the atom, with neutrality arising by taking the charge of the nucleus into account as well, we may infer that $V(r_B) = 0$ (see Problem 13.1), i.e., $\xi = 0$. [Although they are related, a vanishing property of both $n(\mathbf{x})$, $V(\mathbf{x})$ for $|\mathbf{x}| \to \infty$, would imply that $\xi = 0$.]

The Green function part $G_{\sigma\sigma'}(\mathbf{x}t; \mathbf{x}'0)$ (see (9.7.15), see also (9.7.34)), with spin indices σ, σ', satisfies the Schrödinger equation with potential $V(r)$, and in the TF semi-classical approximation, for coincident space points $\mathbf{x} = \mathbf{x}'$, is given by (see (9.8.3)), ($\tau = t/\hbar$)

$$G_{\sigma\sigma}(\mathbf{x}t; \mathbf{x}0) = \int \frac{d^3\mathbf{p}}{(2\pi\hbar)^3} \exp\left[-i\left(\frac{\mathbf{p}^2}{2m} + V(r)\right)\tau\right]. \tag{13.1.4}$$

For the sum of the kinetic energies of the electrons we obtain from (9.7.31), (13.1.4) (see also below (4.5.15)) the expression

$$\sum_\sigma \int d^3\mathbf{x} \frac{1}{2\pi i} \int_{-\infty}^{\infty} \frac{d\tau}{\tau - i\varepsilon} \left[i\frac{\partial}{\partial \tau} - V\right] G_{\sigma\sigma}(\mathbf{x}t; \mathbf{x}0)$$

$$\equiv \frac{(3\pi^2)^{5/3} \hbar^2}{10\pi^2 m} \int d^3\mathbf{x} \, (n(\mathbf{x}))^{5/3} \equiv T[n] \tag{13.1.5}$$

as followed by an elementary integration, where we have used (13.1.1), $\xi = 0$, and allowed for spin.

From electrostatics, one may define the interaction of the electron-nucleus system in terms of the electron density, and add to it the kinetic energy term (13.1.5). We then obtain the *energy* functional $F[n]$, dependent on the density n, defined by

$$F[n] = \frac{(3\pi^2)^{5/3} \hbar^2}{10\pi^2 m} \int d^3\mathbf{x} \, (n(\mathbf{x}))^{5/3} - Ze^2 \int \frac{d^3\mathbf{x}}{|\mathbf{x}|} n(\mathbf{x})$$

$$+ \frac{e^2}{2} \int d^3\mathbf{x}\, d^3\mathbf{x}' \, n(\mathbf{x}) \frac{1}{|\mathbf{x} - \mathbf{x}'|} n(\mathbf{x}') \tag{13.1.6}$$

with the ground-state energy $E_{TF}(Z)$ of the TF atom obtained by minimizing $F[n]$ over n. The second and third terms in (13.1.6) take into account, respectively, of the interactions of the electrons with the nucleus and the electron-electron interactions in this model.

By setting the variational (functional) derivative of (13.1.6), with respect to $n(\mathbf{x})$, equal to zero, as one proceeds in Lagrangian mechanics, gives

$$\frac{(3\pi^2)^{5/3}\hbar^2}{10\pi^2 m}\frac{5}{3}(n(\mathbf{x}))^{2/3} = \frac{Ze^2}{|\mathbf{x}|} - e^2\int\frac{d^3\mathbf{x}'}{|\mathbf{x}-\mathbf{x}'|}n(\mathbf{x}') \tag{13.1.7}$$

or by using (13.1.1) this leads to[1]

$$V(r) = -\frac{Ze^2}{r} + e^2\int\frac{d^3\mathbf{x}'}{|\mathbf{x}-\mathbf{x}'|}n(\mathbf{x}') \tag{13.1.8}$$

with $\xi = 0$, for the effective potential felt by an electron at $\mathbf{x}$ in the TF atom.

In Appendix B to this section, we show that the solution $n(\mathbf{x}) = n_{\text{TF}}(\mathbf{x})$ of (13.1.7) actually gives the *smallest* value for the energy density functional (13.1.6).

Let

$$V(r) = -\frac{Ze^2}{r}F(r) \tag{13.1.9}$$

and set

$$r = ax, \qquad F(r) \equiv f(x) \tag{13.1.10}$$

where

$$a = \left(\frac{3\pi}{4}\right)^{2/3}\frac{\hbar^2}{2me^2}\frac{1}{Z^{1/3}} \cong 0.8853 a_0 \frac{1}{Z^{1/3}} \tag{13.1.11}$$

and a_0 is the Bohr radius $\hbar^2/me^2$.

A straightforward integration over the angles in (13.1.8) gives

$$f(x) = 1 - x\int_0^\infty dx'\,\frac{\sqrt{x'}}{x_>}[f(x')]^{3/2} \tag{13.1.12}$$

satisfying the boundary condition (B.C.)

$$f(0) = 1. \tag{13.1.13}$$

The latter B.C. corresponds to $V(r) \sim -Ze^2/r$ for $r \to 0$, i.e., the Coulomb potential of the nucleus being dominant in this limit. Here $x_> = \max(x, x')$.

The integral equation in (13.1.12) may be reduced to a differential equation by differentiating twice with respect to x giving

$$\frac{d^2}{dx^2}f(x) = \frac{(f(x))^{3/2}}{x^{1/2}}, \qquad f(0) = 1. \tag{13.1.14}$$

The TF electron density and the effective potential may be written in terms of the dimensionless function $f(x)$ as follows

---

[1] Note that the right-hand side of (13.1.8) may be formally rewritten as

$$e^2\int\frac{d^3\mathbf{x}'}{|\mathbf{x}-\mathbf{x}'|}\left[-\frac{Z\delta(r')}{4\pi r'^2} + n(\mathbf{x}')\right], \qquad r' = |\mathbf{x}'|.$$

13.1 The Thomas-Fermi Atom

$$n(r) = \frac{32}{9\pi^3}\left(\frac{me^2}{\hbar^2}\right)^3 Z^2 \left(\frac{f(x)}{x}\right)^{3/2} \equiv n_{\mathrm{TF}}(r) \qquad (13.1.15)$$

$$V(r) = -Z^{4/3}\frac{2me^4}{\hbar^2}\left(\frac{4}{3\pi}\right)^{2/3}\frac{f(x)}{x}. \qquad (13.1.16)$$

The normalization condition (13.1.2), together with (13.1.14), give

$$Z = Z\int_0^{x_{\mathrm{B}}} dx\, x f''(x) = Z\left[x_{\mathrm{B}} f'(x_{\mathrm{B}}) - f(x_{\mathrm{B}}) + 1\right] \qquad (13.1.17)$$

where $x_{\mathrm{B}} = r_{\mathrm{B}}/a$ (see (13.1.10)) with r_{B} specifying the boundary of the TF atom. Thus from (13.1.3), with $\xi = 0$, and (13.1.9), (13.1.10), (13.1.16), (13.1.17), we obtain

$$x_{\mathrm{B}} f'(x_{\mathrm{B}}) = 0. \qquad (13.1.18)$$

To find the asymptotic behavior for $x \to \infty$, we write $f(x) = cx^{-\gamma}$ and substitute the latter in (13.1.14) to obtain $c = 144$, $\gamma = 3$, i.e.,

$$f(x) \sim \frac{144}{x^3}, \qquad (13.1.19)$$

for $x \to \infty$. [Although (13.1.19) exactly satisfies the differential equation in (13.1.14), it is not a solution for all x since it does not satisfy the normalization condition at $x = 0$.] This gives the asymptotic limits

$$V(r) \longrightarrow -144\, a^3 \frac{Ze^2}{r^4} \qquad (13.1.20)$$

$$n(r) \longrightarrow \frac{1}{3\pi^2}\left(\frac{2mZe^2}{\hbar^2} 144\, a^3\right)^{3/2} \frac{1}{r^6} \qquad (13.1.21)$$

for $r \to \infty$, with the latter giving a far slow decrease for $n(r)$ in comparison to an exponential decrease typical of bound states.

Equation (13.1.20) shows that $f(x)$ vanishes at *infinity* and that $x_{\mathrm{B}} = \infty$. Equation (13.1.18) then also implies that $f'(x) \to 0$ for $x \to \infty$. Accordingly, upon integrating (13.1.14) from 0 to ∞, we may infer that for the slope of $f(x)$ at the origin we have

$$f'(0) = -\int_0^\infty dx\, \frac{(f(x))^{3/2}}{x^{1/2}} < 0 \qquad (13.1.22)$$

i.e., it is strictly negative.

Actually the function $f(x)$ vanishes only at infinity. This is easily seen by integrating the differential equation (13.1.14), over x, between two points $x_1 < x_2$, to obtain

$$f'(x_2) - f'(x_1) = \int_{x_1}^{x_2} dx \, \frac{(f(x))^{3/2}}{x^{1/2}} \qquad (13.1.23)$$

and hence conclude that with $f'(0) < 0$, that

$$f'(0) \leqslant \cdots \leqslant f'(x_1) \leqslant \cdots \leqslant f'(x_2) \leqslant \cdots \leqslant 0 \qquad (13.1.24)$$

for $0 < \cdots < x_1 < \cdots < x_2 < \cdots < \infty$. That is, $f(x)$, starting at $f(0) = 1$, is monotonically non-increasing, approaching zero for $x \to \infty$. The function $f(x)$ cannot vanish for finite x and then increase again as this will be in contradiction with (13.1.24). Also note that the differential equation (13.1.14) implies that

$$f''(x) \xrightarrow[x \to \infty]{} 0. \qquad (13.1.25)$$

The function $f(x)$ and its derivative $f'(x)$ may be then determined numerically from the differential equation in (13.1.14) with the boundary conditions $f(0) = 1$, $f(x) \to 0$ for $x \to \infty$. In particular, $f'(0) \cong -1.58807$. For the integral in (13.1.22), we have numerically

$$\int_0^\infty dx \, \frac{(f(x))^{3/2}}{x^{1/2}} = -f'(0) \cong 1.58807. \qquad (13.1.26)$$

The numerical value of the following integral[2]

$$\int_0^\infty dx \, (f(x))^2 \cong 0.6154 \qquad (13.1.27)$$

will be also needed later on.

The explicit analytical expressions for the following integrals are easily established:

$$\int_x^\infty dy \, \frac{(f(y))^{3/2}}{y^{1/2}} = -f'(x) \qquad (13.1.28)$$

$$\int_0^\infty dx \, x f''(x) f'(x) = -\frac{1}{2} \int_0^\infty dx \, (f'(x))^2 \qquad (13.1.29)$$

$$\int_0^\infty dx \, \frac{(f(x))^{5/2}}{x^{1/2}} = -f'(0) - \int_0^\infty dx \, (f'(x))^2 \qquad (13.1.30)$$

$$\int_0^\infty dx \, \frac{(f(x))^{5/2}}{x^{1/2}} = \frac{5}{2} \int_0^\infty dx \, (f'(x))^2. \qquad (13.1.31)$$

From (13.1.29)–(13.1.31), we also have

---

[2] Note that, in particular, this means that the TF potential $V(r)$, as given in (13.1.16), is square-integrable with respect to the Euclidean measure $d^3\mathbf{x}$.

13.1 The Thomas-Fermi Atom

$$\int_0^\infty dx \, \frac{(f(x))^{5/2}}{x^{1/2}} = -\frac{5}{7} f'(0) \qquad (13.1.32)$$

$$\int_0^\infty dx \, (f'(x))^2 = -\frac{2}{7} f'(0). \qquad (13.1.33)$$

The behavior of $f(x)$ near $x = 0$, may be inferred by writing

$$f(x) = 1 + x f'(0) + a x^b + \ldots \qquad (13.1.34)$$

and by substituting the latter in the differential equation in (13.1.14) to obtain $a = 4/3$, $b = 3/2$, i.e.,

$$f(x) = 1 + x f'(0) + \frac{4}{3} x^{3/2} + \ldots \qquad (13.1.35)$$

for $x \to 0$. [Note that (13.1.14) does not represent a Taylor expansion about $x = 0$, since, from (13.1.14), $f''(x)$, for $x \to 0$, does not exist.]

Equation (13.1.16) leads to

$$V(r) \longrightarrow -\frac{Ze^2}{r} \left[1 + \frac{f'(0)}{a} r \right] \qquad (13.1.36)$$

and (13.1.15) to

$$n(r) \longrightarrow \frac{1}{3\pi^2} \left(\frac{2mZe^2}{\hbar^2} \right)^{3/2} \frac{1}{r^{3/2}} \left[1 + \frac{3}{2} \frac{f'(0)}{a} r \right] \qquad (13.1.37)$$

for $r \to 0$.

Computation of $E_{\text{TF}}(Z)$

For the kinetic energy term $T[n]$ in (13.1.5), with $n = n_{\text{TF}}$, given in (13.1.15), we have

$$T[n] = \frac{3}{5} \frac{e^2 Z^2}{a} \int_0^\infty dx \, \frac{(f(x))^{5/2}}{x^{1/2}} \qquad (13.1.38)$$

and from (13.1.32), this gives

$$T[n] = -\frac{3}{7} \frac{e^2 Z^2}{a} f'(0) \qquad (13.1.39)$$

where $f'(0) < 0$ is given in (13.1.26).

For the electrons-nucleus interaction part

$$-Ze^2 \int \frac{d^3\mathbf{x}}{r} n(r) = -\frac{Z^2 e^2}{a} \int_0^\infty dx \, \frac{(f(x))^{3/2}}{x^{1/2}}$$

$$= \frac{Z^2 e^2}{a} f'(0) \tag{13.1.40}$$

where we have used (13.1.26).

Finally, for the electron-electron interaction part

$$\frac{e^2}{2} \int d^3x\, d^3x'\, n(r) \frac{1}{|\mathbf{x}-\mathbf{x}'|} n(r')$$

$$= \frac{Z^2 e^2}{2a} \int_0^\infty dx \int_0^\infty dx' \frac{\sqrt{xx'}}{x_>} \left(f(x)\right)^{3/2} \left(f(x')\right)^{3/2}$$

$$= -\frac{Z^2 e^2}{2a} \left[f'(0) + \int_0^\infty dx \frac{(f(x))^{5/2}}{x^{1/2}} \right]$$

$$= -\frac{Z^2 e^2}{7a} f'(0) \tag{13.1.41}$$

where we have used (13.1.32).

Adding the contributions (13.1.39)–(13.1.41), for the minimizing density (13.1.15) of the energy functional (13.1.6), we obtain for the ground-state energy $E_{\mathrm{TF}}(Z)$ of the TF atom:

$$E_{\mathrm{TF}}(Z) = \frac{3}{7} \frac{e^2 Z^2}{a} f'(0)$$

$$= \frac{6}{7} \left(\frac{4}{3\pi}\right)^{2/3} f'(0) \left(\frac{me^4}{\hbar^2}\right) Z^{7/3} \tag{13.1.42}$$

which upon using the numerical value in (13.1.26) gives

$$E_{\mathrm{TF}}(Z) \cong -1.5375 \left(\frac{me^4}{2\hbar^2}\right) Z^{7/3}. \tag{13.1.43}$$

The power law $Z^{7/3}$ is to be noted.

For future reference, we rewrite (13.1.42) as

$$E_{\mathrm{TF}}(Z) = E_{\mathrm{TF}}(1) Z^{7/3} \tag{13.1.44}$$

where

$$E_{\mathrm{TF}}(1) = \frac{6}{7} \left(\frac{4}{3\pi}\right)^{2/3} f'(0) \left(\frac{me^4}{\hbar^2}\right). \tag{13.1.45}$$

Appendix A To §13.1: The TF Energy Gives the Leading Contribution to $E(Z)$ for Large Z

Let $E(Z)$ denote the exact ground-state energy of the multi-electron atom Hamiltonian with atomic number Z:

13.1 The Thomas-Fermi Atom

$$H = \sum_{i=1}^{Z}\left(\frac{\mathbf{p}_i^2}{2m} - \frac{Ze^2}{r_i}\right) + \sum_{i<j}^{Z}\frac{e^2}{|\mathbf{x}_i - \mathbf{x}_j|}. \qquad (A\text{-}13.1.1)$$

We show rather formally that[3]

$$\lim_{Z\to\infty} Z^{-7/3} E(Z) = E_{\mathrm{TF}}(1) \qquad (A\text{-}13.1.2)$$

where $E_{\mathrm{TF}}(1)$ is the coefficient of $Z^{7/3}$ in the TF energy $E_{\mathrm{TF}}(Z)$ in (13.1.44). Due to the complexity of the demonstration, the details provided in this appendix may be omitted at a first reading. The main result (A-13.1.2) is, however, of central importance for the multi-electron atom problem.

To establish (A-13.1.2), we derive formally upper and lower bounds to the left-hand side of (A-13.1.2) and show that the limits of both bounds coincide with $E_{\mathrm{TF}}(1)$ thus obtaining the result in question.

The Upper Bound

We consider first the seemingly unrelated problem of a one-body potential with Hamiltonian[4]

$$h = \frac{\mathbf{p}^2}{2m} + V(r) \qquad (A\text{-}13.1.3)$$

where $V(r)$ is the TF potential (13.1.8), (13.1.9). We set $\mathbf{x} = \mathbf{R}/Z^{1/3}$, and define

$$n(r) \equiv Z^2 \rho_{\mathrm{TF}}(R) \qquad (A\text{-}13.1.4)$$

where $n(r)$ ($\equiv n_{\mathrm{TF}}(r)$) is the TF density in (13.1.7), (13.1.15) satisfying (13.1.2). Thus from (13.1.16) we may write

$$V(r) = Z^{4/3} v(R)$$

$$= -\frac{\hbar^2}{2m}(3\pi^2)^{2/3} Z^{4/3} (\rho_{\mathrm{TF}}(R))^{2/3}. \qquad (A\text{-}13.1.5)$$

The Green function $G(\mathbf{x}t;\mathbf{x}'0;V)$, for the potential V, may be written as (see (9.7.56))

$$G(\mathbf{x}t;\mathbf{x}'0;V) = \int \frac{d^3\mathbf{k}}{(2\pi)^3} e^{i\mathbf{k}\cdot(\mathbf{x}-\mathbf{x}')} \exp\left[-i\left(\frac{\hbar^2 \mathbf{k}^2}{2m}\tau + U(\mathbf{x},\tau,\mathbf{k})\right)\right] \qquad (A\text{-}13.1.6)$$

where $\tau = t/\hbar$, with U satisfying the equation (see (9.7.59)),

---

[3] The content of this result is a Theorem due to Lieb and Simon (1973). The derivation given here is based on: Manoukian and Osaklung (2000).

[4] Note that this Hamiltonian with the given potential $V(r)$ in (13.1.9), (13.1.16) satisfies the conditions of Theorem 4.1.1, hence its spectrum is bounded from below and its negative part of the spectrum is discrete.

748   13 Multi-Electron Atoms: Beyond the Thomas-Fermi Atom

$$-\frac{\partial}{\partial \tau}U + V - \frac{\hbar^2 \mathbf{k}}{m}\cdot \nabla U + \frac{\hbar^2}{2m}(\nabla U)^2 + \frac{i\hbar^2}{2m}\nabla^2 U = 0 \qquad \text{(A-13.1.7)}$$

with the boundary condition $U\big|_{\tau=0} = 0$.

We are particularly interested in the integral

$$\int d^3\mathbf{x}\, G(\mathbf{x}t; \mathbf{x}0; V) \qquad \text{(A-13.1.8)}$$

for coincident space points. The $\exp[i\mathbf{k}\cdot(\mathbf{x}-\mathbf{x}')]$ in (A-13.1.6) then becomes simply replaced by one. Since the $\mathbf{x}$ in (A-13.1.8) and the $\mathbf{k}$ in (A-13.1.6) are merely integration variables, we may make *any* convenient change of these variables of integrations. In particular, we consider the change of variables of integrations: $\mathbf{x}\to\mathbf{R}$, $\mathbf{k}\to\mathbf{K}$, where (see (A-13.1.4)) $\mathbf{x} = \mathbf{R}/Z^{1/3}$, $\mathbf{k} = Z^{2/3}\mathbf{K}$ and also carry out the following scaling substitutions: $V = Z^{4/3}v$, $\tau = T/Z^{4/3}$. We note that with the τ, $\mathbf{k}$ scalings, just defined, the product $k^2\tau = K^2 T$ in the exponential in (A-13.1.6) remains invariant. With these new variables, (A-13.1.7) reduces to

$$-\frac{\partial}{\partial T}U + v - \frac{\hbar^2 \mathbf{K}\cdot\nabla_R U}{mZ^{1/3}} + \frac{\hbar^2}{2m}\frac{(\nabla_R U)^2}{Z^{2/3}} + \frac{i\hbar^2}{2m}\frac{\nabla_R^2 U}{Z^{2/3}} = 0. \qquad \text{(A-13.1.9)}$$

Let $\lim_{Z\to\infty} U = U_\infty$. Then (A-13.1.9) collapses to $-\partial U_\infty/\partial T + v = 0$, where we note that v is independent of Z (see (A-13.1.5)). Thus we obtain $U_\infty = vT$.

Accordingly, we have the following limits for large Z, as is readily verified upon substitution of vT for U, $Z\to\infty$

$$\int d^3\mathbf{x}\, \frac{2}{2\pi i}\int_{-\infty}^{\infty}\frac{d\tau}{\tau - i\varepsilon}\, G(\mathbf{x}t;\mathbf{x}0; V) \longrightarrow Z\int d^3\mathbf{R}\, \rho_{\text{TF}}(R) \equiv Z \qquad \text{(A-13.1.10)}$$

and

$$Z^{-7/3}\int d^3\mathbf{x}\, \frac{2}{2\pi i}\int_{-\infty}^{\infty}\frac{d\tau}{\tau - i\varepsilon}\, i\frac{\partial}{\partial \tau}G(\mathbf{x}t;\mathbf{x}0; V)$$

$$\longrightarrow 2\int d^3\mathbf{R}\int \frac{d^3\mathbf{K}}{(2\pi)^3}\left[\frac{\hbar^2 \mathbf{K}^2}{2m} + v(R)\right]\Theta\left(\sqrt{\frac{-2mv(R)}{\hbar^2}} - |\mathbf{K}|\right)$$

$$= (3\pi^2)^{5/3}\frac{\hbar^2}{10\pi^2 m}\int d^3\mathbf{R}\,(\rho_{\text{TF}}(R))^{5/3} - e^2\int \frac{d^3\mathbf{R}}{R}\rho_{\text{TF}}(R)$$

$$+ e^2\int d^3\mathbf{R}\, d^3\mathbf{R}'\, \rho_{\text{TF}}(R)\frac{1}{|\mathbf{R}-\mathbf{R}'|}\rho_{\text{TF}}(R') \qquad \text{(A-13.1.11)}$$

where the factor 2 multiplying the τ-integral is to account for spin. The τ-integrals project out the negative spectrum of h in (A-13.1.3).

Equation (A-13.1.10), in particular, is of fundamental importance. It states (see (9.7.24)) that for large Z, the Hamiltonian h, allowing for spin, has Z (orthonormal) eigenvectors corresponding to its *negative* spectrum. Let $g_1(\mathbf{x},\sigma),\ldots,g_Z(\mathbf{x},\sigma)$ denote these eigenvectors for large Z. Define the determinantal (anti-symmetric) function

$$\Phi_Z(\mathbf{x}_1\sigma_1,\ldots,\mathbf{x}_Z\sigma_Z) = \frac{1}{\sqrt{Z!}}\det[g_\alpha(\mathbf{x}_\beta,\sigma_\beta)]. \tag{A-13.1.12}$$

Since such an anti-symmetric function does not necessarily coincide with the ground-state function of the Hamiltonian H in (A-13.1.1), the expectation value $\langle\Phi_Z|H|\Phi_Z\rangle$, with respect to Φ_Z in (A-13.1.12) can only *over* estimate the exact ground-state energy $E(Z)$ of H or at best be equal to it.

We rewrite the Hamiltonian H in (A-13.1.1) equivalently as

$$H = \sum_{i=1}^{Z} h_i + \left(\sum_{i<j}^{Z} \frac{e^2}{|\mathbf{x}_i-\mathbf{x}_j|} - e^2\sum_{i=1}^{Z}\int\frac{d^3\mathbf{x}}{|\mathbf{x}-\mathbf{x}_i|}n(r)\right) \tag{A-13.1.13}$$

where h_i is defined in (A-13.1.3) with corresponding variables $\mathbf{x}_i$, $\mathbf{p}_i$.

Accordingly,

$$\lim_{Z\to\infty} Z^{-7/3} E(Z) \leqslant \lim_{Z\to\infty} Z^{-7/3}\langle\Phi_Z|H|\Phi_Z\rangle$$

$$= \lim_{Z\to\infty} Z^{-7/3}\sum_{i=1}^{Z}\langle g_i|h_i|g_i\rangle + \lim_{Z\to\infty} Z^{-7/3} F_Z \tag{A-13.1.14}$$

where

$$F_Z = -e^2\sum_\sigma\int\frac{d^3\mathbf{x}\,d^3\mathbf{x}'}{|\mathbf{x}-\mathbf{x}'|}n_Z(\mathbf{x}\sigma,\mathbf{x}\sigma)\,n(r')$$

$$+ \frac{e^2}{2}\sum_{\sigma,\sigma'}\int\frac{d^3\mathbf{x}\,d^3\mathbf{x}'}{|\mathbf{x}-\mathbf{x}'|}\left[n_Z(\mathbf{x}\sigma,\mathbf{x}\sigma)\,n_Z(\mathbf{x}'\sigma',\mathbf{x}'\sigma') - |n_Z(\mathbf{x}\sigma,\mathbf{x}'\sigma')|^2\right]$$

$$\tag{A-13.1.15}$$

$$n_Z(\mathbf{x}\sigma,\mathbf{x}'\sigma') = \sum_{i=1}^{Z} g_i(\mathbf{x},\sigma)\,g_i^*(\mathbf{x}',\sigma'). \tag{A-13.1.16}$$

Now using the fact that the second term in the square brackets in the second term in (A-13.1.15) is negative and

$$\lim_{Z\to\infty} Z^{-2}\sum_\sigma n_Z(\mathbf{x}\sigma,\mathbf{x}\sigma) = \lim_{Z\to\infty} Z^{-2}\frac{2}{2\pi\mathrm{i}}\int_{-\infty}^{\infty}\frac{d\tau}{\tau-\mathrm{i}\varepsilon}G(\mathbf{x}t;\mathbf{x}0;V)$$

$$\equiv \rho_{\mathrm{TF}}(R) \tag{A-13.1.17}$$

we obtain

$$\lim_{Z\to\infty} Z^{-7/3} F_Z \leqslant -e^2 \lim_{Z\to\infty} Z^{-7/3} \int \frac{d^3x\, d^3x'}{|\mathbf{x}-\mathbf{x}'|} \left[n(r') \left(\sum_\sigma n_Z(\mathbf{x}\sigma, \mathbf{x}\sigma) \right) \right.$$

$$\left. - \frac{1}{2} \left(\sum_\sigma n_Z(\mathbf{x}\sigma, \mathbf{x}\sigma) \right) \left(\sum_{\sigma'} n_Z(\mathbf{x}'\sigma', \mathbf{x}'\sigma') \right) \right]$$

$$= -\frac{e^2}{2} \int \frac{d^3\mathbf{R}\, d^3\mathbf{R}'}{|\mathbf{R}-\mathbf{R}'|} \rho_{\text{TF}}(R)\, \rho_{\text{TF}}(R'). \qquad \text{(A-13.1.18)}$$

Finally, we have

$$\lim_{Z\to\infty} Z^{-7/3} \sum_{i=1}^Z \langle g_i | h_i | g_i \rangle = \lim_{Z\to\infty} \left(Z^{-7/3} 2 \sum_{\lambda<0} \lambda \right)$$

$$= \lim_{Z\to\infty} Z^{-7/3} \int d^3\mathbf{x}\, \frac{2}{2\pi i} \int_{-\infty}^\infty \frac{d\tau}{\tau - i\varepsilon}\, i\frac{\partial}{\partial\tau} G(\mathbf{x}t; \mathbf{x}0; V)$$

$$\text{(A-13.1.19)}$$

where $\sum_{\lambda<0} \lambda$ in $2\sum_{\lambda<0} \lambda$ is a sum over all the negative eigenvalues of h in (A-13.1.3) (see (9.7.32)) allowing for multiplicity but not spin degeneracy. The factor 2 takes the latter into account.

From (A-13.1.14)–(A-13.1.19) *and* (A-13.1.11) we obtain

$$\lim_{Z\to\infty} Z^{-7/3} E(Z) \leqslant \frac{(3\pi^2)^{5/3} \hbar^2}{10\pi^2 m} \int d^3\mathbf{R}\, (\rho_{\text{TF}}(R))^{5/3} - e^2 \int \frac{d^3\mathbf{R}}{R} \rho_{\text{TF}}(R)$$

$$+ \frac{e^2}{2} \int \frac{d^3\mathbf{R}\, d^3\mathbf{R}'}{|\mathbf{R}-\mathbf{R}'|} \rho_{\text{TF}}(R)\, \rho_{\text{TF}}(R') \qquad \text{(A-13.1.20)}$$

where we recognize, from (13.1.6), (13.1.15), (13.1.16) and (13.1.44), that the right-hand side of this inequality coincides with the coefficient $E_{\text{TF}}(1)$ of $Z^{7/3}$ in $E_{\text{TF}}(Z)$, as given in (13.1.44), (13.1.45). That is,

$$\lim_{Z\to\infty} Z^{-7/3} E(Z) \leqslant E_{\text{TF}}(1). \qquad \text{(A-13.1.21)}$$

The Lower Bound

We use a special case of the lower bound of the repulsive electron-electron potential to be established in (14.1.3) given by

$$\sum_{i<j}^Z \frac{e^2}{|\mathbf{x}_i - \mathbf{x}_j|} \geqslant \sum_{i=1}^Z e^2 \int \frac{d^3\mathbf{x}}{|\mathbf{x}-\mathbf{x}_i|} n(r) - \frac{e^2}{2} \int d^3\mathbf{x}\, d^3\mathbf{x}'\, n(r) \frac{1}{|\mathbf{x}-\mathbf{x}'|} n(r')$$

$$-\frac{(3\pi^2)^{5/3} \hbar^2}{60\pi^2 mZ} \int d^3\mathbf{x} \left(n(r)\right)^{5/3} + 6Z^2 E_{\text{TF}}(1) \quad \text{(A-13.1.22)}$$

where $n(r)$ is taken to be the TF density in (A-13.1.4), (13.1.7), (13.1.15) and we have chosen $\beta = 6Z$, with $N = Z$.

Let Ψ be a normalized anti-symmetric function in $(\mathbf{x}_1 \sigma_1, \ldots, \mathbf{x}_Z \sigma_Z)$. Then from (A-13.1.4), (A-13.1.13) and (A-13.1.22),

$$\langle \Psi | H | \Psi \rangle \geqslant \langle \Psi | \sum_i h_i | \Psi \rangle - \frac{e^2}{2} Z^{7/3} \int d^3\mathbf{R} \, d^3\mathbf{R}' \, \rho_{\text{TF}}(R) \frac{1}{|\mathbf{R} - \mathbf{R}'|} \rho_{\text{TF}}(R')$$

$$- \frac{(3\pi^2)^{5/3} \hbar^2}{60\pi^2 mZ} Z^{7/3} \int d^3\mathbf{R} \left(\rho_{\text{TF}}(R)\right)^{5/3} + 6Z^2 E_{\text{TF}}(1).$$
(A-13.1.23)

Consider the lowest energy E of the Hamiltonian $\sum_i h_i$. The latter Hamiltonian describes Z "non-interacting" electrons, but each interacting with the external potential V. According to Pauli's exclusion principle, these Z electrons can be put in the lowest energy levels of $\sum_i h_i$ (allowing for spin degeneracy) if $Z \leqslant$ the number of such available levels, or else if Z is larger, then the remaining free electrons should have arbitrary small ($\to 0$) kinetic energies to define the lowest energy of $\sum_i h_i$. In either cases, $E \geqslant 2 \sum_{\lambda < 0} \lambda$, where $\sum_{\lambda < 0} \lambda$ is defined as before below (A-13.1.19).

Accordingly, from (A-13.1.23) and the last equality in (A-13.1.19), we have

$$\lim_{Z \to \infty} Z^{-7/3} \langle \Psi | H | \Psi \rangle \geqslant \lim_{Z \to \infty} Z^{-7/3} \int d^3\mathbf{x} \, \frac{2}{2\pi i} \int_{-\infty}^{\infty} \frac{d\tau}{\tau - i\varepsilon} \, i\frac{\partial}{\partial \tau} G(\mathbf{x}\tau; \mathbf{x}0; V)$$

$$- \frac{e^2}{2} \int d^3\mathbf{R} \, d^3\mathbf{R}' \, \rho_{\text{TF}}(R) \frac{1}{|\mathbf{R} - \mathbf{R}'|} \rho_{\text{TF}}(R')$$
(A-13.1.24)

involving reals, and noted that the last two terms on the right-hand side of (A-13.1.23) go to zero, when multiplied by $Z^{-7/3}$ in the limit $Z \to \infty$. From (A-13.1.11), (A-13.1.24) we then obtain

$$\lim_{Z \to \infty} Z^{-7/3} \langle \Psi | H | \Psi \rangle \geqslant \frac{(3\pi^2)^{5/3} \hbar^2}{10\pi^2 m} \int d^3\mathbf{R} \left(\rho_{\text{TF}}(R)\right)^{5/3} - e^2 \int \frac{d^3\mathbf{R}}{R} \rho_{\text{TF}}(R)$$

$$+ \frac{e^2}{2} \int \frac{d^3\mathbf{R} \, d^3\mathbf{R}'}{|\mathbf{R} - \mathbf{R}'|} \rho_{\text{TF}}(R) \rho_{\text{TF}}(R')$$

$$\equiv E_{\text{TF}}(1). \tag{A-13.1.25}$$

Since $|\Psi\rangle$ was arbitrary and can only overestimate $E(Z)$ or at best would lead to a value equal to it, (A-13.1.25) is also true for the (unknown) ground-state wavefunction, that is

$$\lim_{Z\to\infty} Z^{-7/3} E(Z) \geqslant E_{\text{TF}}(1). \tag{A-13.1.26}$$

From (A-13.1.21), (A-13.1.26), the statement in (A-13.1.2) then follows.

Appendix B to §13.1: The TF Density Actually Gives the Smallest Value for the Energy Density Functional in (13.1.6)

In this appendix, we show that the TF density $n_{\text{TF}}(\mathbf{x})$ satisfying (13.1.7) actually gives the smallest possible value for the energy functional (13.1.6) defined a priori for an arbitrary density $\rho(\mathbf{x}) \geqslant 0$ by

$$F[\rho] = A \int d^3\mathbf{x}\, (\rho(\mathbf{x}))^{5/3} - Ze^2 \int \frac{d^3\mathbf{x}}{|\mathbf{x}|} \rho(\mathbf{x})$$

$$+ \frac{e^2}{2} \int d^3\mathbf{x}\, d^3\mathbf{x}'\, \rho(\mathbf{x}) \frac{1}{|\mathbf{x}-\mathbf{x}'|} \rho(\mathbf{x}') \tag{B-13.1.1}$$

where

$$A = \frac{(3\pi^2)^{5/3} \hbar^2}{10\pi^2 m}. \tag{B-13.1.2}$$

Let $\rho(\mathbf{x}) = t\rho_1(\mathbf{x}) + (1-t)\rho_2(\mathbf{x}) \equiv t\rho_1 + (1-t)\rho_2$, $\rho(\mathbf{x}') = t\rho_1(\mathbf{x}') + (1-t)\rho_2(\mathbf{x}') \equiv t\rho_1' + (1-t)\rho_2'$, $0 \leqslant t \leqslant 1$, and where $\rho_1, \rho_2 \geqslant 0$. From Appendix II on convexity, we have the elementary inequality

$$(t\rho_1 + (1-t)\rho_2)^{5/3} \leqslant t(\rho_1)^{5/3} + (1-t)(\rho_2)^{5/3}. \tag{B-13.1.3}$$

Also

$$[t\rho_1 + (1-t)\rho_2][t\rho_1' + (1-t)\rho_2']$$
$$= t\rho_1\rho_1' + (1-t)\rho_2\rho_2' - t(1-t)(\rho_1 - \rho_2)(\rho_1' - \rho_2'). \tag{B-13.1.4}$$

A Fourier transform, for example, shows that

$$\int d^3\mathbf{x}\, d^3\mathbf{x}' [\rho_1(\mathbf{x}) - \rho_2(\mathbf{x})] \frac{1}{|\mathbf{x}-\mathbf{x}'|} [\rho_1(\mathbf{x}') - \rho_2(\mathbf{x}')] \geqslant 0. \tag{B-13.1.5}$$

to be used in conjunction with the last term in (B-13.1.4). Hence we may conclude form (B-13.1.1)–(B-13.1.5) that

$$F[t\rho_1 + (1-t)\rho_2] \leq tF[\rho_1] + (1-t)F[\rho_2]. \tag{B-13.1.6}$$

Also

$$\frac{d}{dt}F[t\rho_1 + (1-t)\rho_2] = \frac{5}{3}A\int d^3x \left[t\rho_1 + (1-t)\rho_2\right]^{3/2}(\rho_1 - \rho_2)$$

$$-Ze^2 \int \frac{d^3x}{|\mathbf{x}|}(\rho_1 - \rho_2)$$

$$+ e^2 \int \frac{d^3x\, d^3x'}{|\mathbf{x}-\mathbf{x}'|}\left[t\rho_1' + (1-t)\rho_2'\right](\rho_1 - \rho_2) \tag{B-13.1.7}$$

and

$$\left.\frac{d}{dt}F[t\rho_1 + (1-t)\rho_2]\right|_{t=0}$$

$$= \int d^3x\, (\rho_1 - \rho_2)\left[\frac{5}{3}A\rho_2^{2/3} - \frac{Ze^2}{|\mathbf{x}|} + e^2 \int \frac{d^3x'}{|\mathbf{x}-\mathbf{x}'|}\rho_2'\right]. \tag{B-13.1.8}$$

By choosing $\rho_2 = n_{\mathrm{TF}}$, and $\rho_1 = \sigma \geq 0$ arbitrary, we conclude from (13.1.7) that the expression within the square brackets in (B-13.1.8) is zero, thus

$$\left.\frac{d}{dt}F[t\sigma + (1-t)n_{\mathrm{TF}}]\right|_{t=0} = 0. \tag{B-13.1.9}$$

Also (B-13.1.6) leads to the bound

$$F[\sigma] - F[n_{\mathrm{TF}}] \geq \frac{F[t\sigma + (1-t)n_{\mathrm{TF}}] - F[n_{\mathrm{TF}}]}{t}. \tag{B-13.1.10}$$

Since the left-hand side of (B-13.1.10) is independent of t, we may take the limit $t \to 0$ and use (B-13.1.9) to conclude that

$$F[\sigma] \geq F[n_{\mathrm{TF}}] \tag{B-13.1.11}$$

with the TF density n_{TF} providing the smallest possible value for the energy functional in (B-13.1.1).

13.2 Correction due to Electrons Bound Near the Nucleus

According to (13.1.36), an electron near the nucleus, i.e., for $r \to 0$, feels the potential

13 Multi-Electron Atoms: Beyond the Thomas-Fermi Atom

$$V_0(r) = -\frac{Ze^2}{r} - \frac{Ze^2}{a} f'(0) \tag{13.2.1}$$

where $a \propto Z^{-1/3}$ is defined in (13.1.11), and $f'(0)$ is given in (13.1.26). The first term $-Ze^2/r$ is the familiar potential due to the nucleus, while

$$-\frac{Ze^2}{a} f'(0) = \frac{Ze^2}{a} |f'(0)| = e^2 \int \frac{d^3\mathbf{x}}{|\mathbf{x}|} n(\mathbf{x}) \tag{13.2.2}$$

(see (13.1.7)–(13.1.10), (13.1.22), (13.1.1)) is a background constant potential due to the electrons felt at the origin.

The potential $V_0(r)$ may be treated explicitly without recourse to a semi-classical approximation. From the theory of the hydrogen atom, we may place the electrons bound near the nucleus up to some energy level specified, say, by a principal quantum number n'. For a given n', the maximum energy ξ attained by an electron bound near the nucleus in the potential $V_0(r)$ is then given by

$$-\frac{Z^2 e^4 m}{2\hbar^2 n'^2} - \frac{Ze^2}{a} f'(0). \tag{13.2.3}$$

Accordingly, to obtain the correction to the TF semi-classical approximation $E_{\mathrm{TF}}(Z)$, due to electrons bound near the nucleus, we *replace* the TF semi-classical contribution,[5] in the potential $V_0(r)$, up to the maximum energy in (13.2.3), by the exact contribution due to this potential with electrons placed in energy levels, having at most the energy in (13.2.3), according to Pauli's exclusion principle.

Hence the correction to $E_{\mathrm{TF}}(Z)$ sought in this section is given by[6]

$$E_{\mathrm{Sc}} = \sum_{n=1}^{n'} (2n^2) \left[-\frac{mZ^2 e^4}{2\hbar^2 n^2} + \frac{Ze^2}{a} f'(0) \right]$$

$$- \frac{2}{2\pi \mathrm{i}} \int d^3\mathbf{x} \int_{-\infty}^{\infty} \frac{d\tau}{\tau - \mathrm{i}\varepsilon} e^{\mathrm{i}\xi\tau} \, \mathrm{i} \frac{\partial}{\partial \tau} G_{\mathrm{Sc}}(\mathbf{x}t; \mathbf{x}0; V_0) \tag{13.2.4}$$

where

$$G_{\mathrm{Sc}}(\mathbf{x}t; \mathbf{x}0; V_0) = \int \frac{d^3\mathbf{k}}{(2\pi)^3} \exp\left[-\mathrm{i}\left(\frac{\hbar^2 \mathbf{k}^2}{2m} + V_0(r) \right) \tau \right] \tag{13.2.5}$$

with the second term in (13.2.4) giving the sum of the eigenvalues, of a Hamiltonian with potential V_0, *less than* ξ (see (9.7.32)). Accordingly, we may introduce the parameter

---

[5] In reference to this contribution, as given in the second term in (13.2.4), see (9.7.32), (9.8.3) by finally incorporating spin.

[6] Sc in E_{Sc} refers to Scott (1952) who first gave the correction in (13.2.14) — see also Englert and Schwinger (1984).

13.2 Correction due to Electrons Bound Near the Nucleus

$$\eta = \frac{2\hbar^2}{mZ^2 e^4}\left(\xi + \frac{Ze^2}{a}f'(0)\right) \tag{13.2.6}$$

and take $(1/-\eta)^{1/2}$ (see (13.2.3)) as the average of the principle quantum numbers n' and $n'+1$, with the latter corresponding to the next energy level greater than ξ. That is,

$$\left(\frac{1}{-\eta}\right)^{1/2} = \frac{n' + (n'+1)}{2} = n' + \frac{1}{2}. \tag{13.2.7}$$

We recall that the $2n^2$ factor multiplying the summand in the first term in (13.2.4) denotes the number of electrons that may be put in the energy level specified by the principal quantum number n.

Upon summation over n, the first term in (13.2.4) is given by

$$-\frac{mZ^2 e^4}{\hbar^2} n' + \frac{Ze^2}{3a} n'(n'+1)(2n'+1) f'(0) \tag{13.2.8}$$

and by keeping track of the power of Z, we note that the second term in (13.2.8) is $\propto Z^{4/3}$ in contrast to Z^2 as appearing in the first term.

The second term in (13.2.4) is explicitly given by

$$-2\int d^3 x \int \frac{d^3 k}{(2\pi)^3}\left(\frac{\hbar^2 k^2}{2m} - \frac{Ze^2}{r}\right)\Theta\left(\sqrt{\frac{2m}{\hbar^2}\left(\xi' + \frac{Ze^2}{r}\right)} - |\mathbf{k}|\right)$$

$$= \frac{\hbar^2}{3m\pi^2}\left(\frac{2m(-\xi')}{\hbar^2}\right)^{5/2} \int_{(r \leqslant Ze^2/-\xi')} d^3 x \left\{\frac{1}{5}\left(\frac{Ze^2}{-\xi' r}\right)^{5/2}\left[1 - \left(\frac{-\xi' r}{Ze^2}\right)\right]^{5/2}\right.$$

$$\left. + \frac{1}{2}\left(\frac{Ze^2}{-\xi' r}\right)^{3/2}\left[1 - \left(\frac{-\xi' r}{Ze^2}\right)\right]^{3/2}\right\} \tag{13.2.9}$$

where

$$\xi' = \xi + \frac{Ze^2}{a} f'(0). \tag{13.2.10}$$

Upon setting

$$-\frac{\xi' r}{Ze^2} = y \tag{13.2.11}$$

and using the integral

$$\int_0^1 y^2 \, dy \left[\frac{1}{5}\frac{(1-y)^{5/2}}{y^{5/2}} + \frac{1}{2}\frac{(1-y)^{3/2}}{y^{3/2}}\right] = \frac{3\pi}{32} \tag{13.2.12}$$

we obtain for (13.2.9) the expression

$$m\frac{Z^2e^4}{\hbar^2}\left(\frac{1}{-\eta}\right)^{1/2} = m\frac{Z^2e^4}{\hbar^2}\left(n' + \frac{1}{2}\right) \qquad (13.2.13)$$

where $\eta = 2\hbar^2\xi'/mZ^2e^4$ was introduced in (13.2.6), and we have used (13.2.7), (13.2.10).

All told, we obtain from (13.2.8), (13.1.13) and (13.2.4), to the leading order in Z

$$E_{\text{Sc}} = \frac{mZ^2e^4}{2\hbar^2} \qquad (13.2.14)$$

where the first term in (13.2.8) cancels out, and the principal quantum number n' now appears only in the coefficient of $Z^{4/3}$ in the second term in (13.2.8). This term, however, may not be retained as a further correction to the Z^2 one in (13.2.14) since the corrections considered in the next sections are proportional to $Z^{5/3}$. Thus we learn, in particular, that for the correction due to electrons bound near the nucleus, the contribution of the background constant potential due to all of the electrons in (13.2.2) may be neglected and we may effectively take V_0 to be just the Coulomb potential

$$V_{\text{C}}(r) = -\frac{Ze^2}{r} \qquad (13.2.15)$$

due to the nucleus.

13.3 The Exchange Term

In considering the electron-electron interaction term

$$\frac{e^2}{2}\int d^3x\, d^3x'\, n(\mathbf{x})\frac{1}{|\mathbf{x} - \mathbf{x}'|}n(\mathbf{x}') \qquad (13.3.1)$$

in the TF theory in (13.1.6), we have not taken Pauli's exclusion principle into account that no two electrons may occupy the same state. In the present section we remedy this situation.

To the above end, we consider, the non-local density given in (9.7.26) with $\xi = 0$

$$n_{\sigma\sigma'}(\mathbf{x},\mathbf{x}') = \frac{1}{2\pi i}\int_{-\infty}^{\infty}\frac{d\tau}{\tau - i\varepsilon}\, G_{\sigma\sigma'}(\mathbf{x}t;\mathbf{x}'0). \qquad (13.3.2)$$

Since we are considering non-spin interactions,

$$G_{\sigma\sigma'}(\mathbf{x}t;\mathbf{x}'0) = \delta_{\sigma\sigma'}G(\mathbf{x}t;\mathbf{x}'0) \qquad (13.3.3)$$

$$n_{\sigma\sigma'}(\mathbf{x},\mathbf{x}') \equiv \delta_{\sigma\sigma'}N(\mathbf{x},\mathbf{x}') \qquad (13.3.4)$$

and

$$\sum_\sigma n_{\sigma\sigma}(\mathbf{x},\mathbf{x}') = 2N(\mathbf{x},\mathbf{x}') \equiv n(\mathbf{x},\mathbf{x}') \qquad (13.3.5)$$

13.3 The Exchange Term

where $N(\mathbf{x}, \mathbf{x}')$ does not take spin into account. We also note that

$$n(\mathbf{x}, \mathbf{x}) = n(\mathbf{x}) \tag{13.3.6}$$

in the earlier notation in (13.3.1), and we may rewrite the integrand in (13.3.1), not involving the $1/|\mathbf{x} - \mathbf{x}'|$ factor, rather trivially as

$$n(\mathbf{x}, \mathbf{x})\, n(\mathbf{x}', \mathbf{x}') = \sum_{\sigma, \sigma'} N(\mathbf{x}, \mathbf{x})\, N(\mathbf{x}', \mathbf{x}')$$

$$= \sum_{\sigma} N(\mathbf{x}, \mathbf{x})\, N(\mathbf{x}', \mathbf{x}') + \sum_{\sigma \neq \sigma'} N(\mathbf{x}, \mathbf{x})\, N(\mathbf{x}', \mathbf{x}'). \tag{13.3.7}$$

To satisfy Pauli's exclusion principle, we replace the first term in (13.3.7) by

$$\sum_{\sigma} \left[N(\mathbf{x}, \mathbf{x})\, N(\mathbf{x}', \mathbf{x}') - N(\mathbf{x}, \mathbf{x}')\, N(\mathbf{x}', \mathbf{x}) \right] \tag{13.3.8}$$

which when combined with the second term in (13.3.7) provides the substitution rule

$$n(\mathbf{x}, \mathbf{x})\, n(\mathbf{x}', \mathbf{x}') \longrightarrow \sum_{\sigma, \sigma'} \left[N(\mathbf{x}, \mathbf{x})\, N(\mathbf{x}', \mathbf{x}') - \frac{1}{2} N(\mathbf{x}, \mathbf{x}')\, N(\mathbf{x}', \mathbf{x}) \right]$$

$$\equiv n(\mathbf{x}, \mathbf{x})\, n(\mathbf{x}', \mathbf{x}') - \frac{1}{2} n(\mathbf{x}, \mathbf{x}')\, n(\mathbf{x}', \mathbf{x}) \tag{13.3.9}$$

as is easily checked by trivially summing over σ and σ'. We note the important complex conjugation property in (9.7.27):

$$n(\mathbf{x}', \mathbf{x}) = \big(n(\mathbf{x}, \mathbf{x}')\big)^*. \tag{13.3.10}$$

Hence to obtain the "exchange correction" to E_{TF}, in the TF semi-classical approximation, we have, from (13.3.1), (13.3.9), (13.3.10), to compute the integral

$$E_{\text{exc}} = -\frac{e^2}{2} \int d^3\mathbf{x}\, d^3\mathbf{x}'\, n(\mathbf{x}, \mathbf{x}') \frac{1}{2|\mathbf{x} - \mathbf{x}'|} \big(n(\mathbf{x}, \mathbf{x}')\big)^* \tag{13.3.11}$$

where we note the additional $1/2$ factor multiplying $1/|\mathbf{x} - \mathbf{x}'|$ in (13.3.11).
To the above end, by setting

$$\sqrt{-\frac{2mV(r)}{\hbar^2}} = k_{\text{F}}(r), \qquad \mathbf{x} - \mathbf{x}' = \boldsymbol{\zeta}, \qquad |\boldsymbol{\zeta}| = \zeta \tag{13.3.12}$$

where $V(r)$ is the TF potential (13.1.8), (13.1.16), we have (see (13.3.2), (9.8.3), (9.7.26))

13 Multi-Electron Atoms: Beyond the Thomas-Fermi Atom

$$n(\mathbf{x}, \mathbf{x}') = \frac{2}{2\pi i} \int_{-\infty}^{\infty} \frac{d\tau}{\tau - i\varepsilon} \int \frac{d^3 k}{(2\pi)^3} e^{i\mathbf{k}\cdot(\mathbf{x}-\mathbf{x}')} \exp\left[-i\left(\frac{\hbar^2 k^2}{2m} + V(r)\right)\tau\right]$$

$$= \frac{4\pi}{i\zeta} \int_0^{k_F(r)} \frac{k\,dk}{(2\pi)^3} \left(e^{ik\zeta} - e^{-ik\zeta}\right)$$

$$= -\frac{1}{\pi^2 \zeta} \frac{\partial}{\partial \zeta} \frac{1}{\zeta} \int_0^{\zeta k_F(r)} dx \cos(x) \tag{13.3.13}$$

or

$$n(\mathbf{x}, \mathbf{x}') = \frac{1}{\pi^2 \zeta^3} \Big[\sin\big(\zeta k_F(r)\big) - \zeta k_F(r) \cos\big(\zeta k_F(r)\big)\Big]. \tag{13.3.14}$$

Let

$$\zeta k_F(r) = y, \qquad y = |\mathbf{y}| \tag{13.3.15}$$

to obtain

$$n(\mathbf{x}, \mathbf{x}') = \frac{k_F^3(r)}{\pi^2 y^3} \big[\sin y - y \cos y\big]. \tag{13.3.16}$$

For $\mathbf{x}' \to \mathbf{x}$, it is easily checked that $n(\mathbf{x}, \mathbf{x}')$ goes over to $n(\mathbf{x})$ in (13.1.1), with $\xi = 0$, (see Problem 13.9).

Accordingly, for the exchange effect E_{exc} in (13.3.11) we obtain

$$E_{\text{exc}} = -\frac{e^2}{2} \int \frac{d^3\mathbf{x}\, d^3\mathbf{x}'\, k_F^6(r)}{\pi^4 2 |\mathbf{x} - \mathbf{x}'|} \left(\frac{\sin y - y \cos y}{y^3}\right)^2. \tag{13.3.17}$$

Upon making a change of the variable of integration $\mathbf{x}'$ to $\mathbf{y}$, this leads to

$$E_{\text{exc}} = -\frac{e^2}{4\pi^4} \int d^3\mathbf{x}\, k_F^4(r) \int \frac{d^3\mathbf{y}}{y} \left(\frac{\sin y - y \cos y}{y^3}\right)^2. \tag{13.3.18}$$

Finally using the integral

$$\int_0^\infty \frac{dy}{y^5} \left(\sin y - y \cos y\right)^2 = \frac{1}{4} \tag{13.3.19}$$

and the definition of $k_F(r)$ in (13.3.12), rewritten in terms of the density $n(r)$, we obtain

$$E_{\text{exc}} = -\frac{e^2}{4\pi^3} (3\pi^2)^{4/3} \int d^3\mathbf{x}\, \big(n(r)\big)^{4/3}. \tag{13.3.20}$$

This may be simply rewritten in terms of the TF function $f(x)$ in (13.1.15) as

$$E_{\text{exc}} = -\left(\frac{9}{2\pi^4}\right)^{1/3} Z^{5/3} \left(\frac{me^4}{\hbar^2}\right) \int_0^\infty dx\, \big(f(x)\big)^2. \tag{13.3.21}$$

The numerical value of the integral in (13.1.27) gives

$$E_{\text{exc}} \cong -0.4416 \left(\frac{me^4}{2\hbar^2}\right) Z^{5/3}. \tag{13.3.22}$$

13.4 Quantum Correction

The deviation of the (space coincident) Green function $\delta G(\mathbf{x}t; \mathbf{x}0; V)$ part, for a given potential, from the Thomas-Fermi semi-classical approximation in (9.8.3) has been worked out in §9.8, (9.8.23) and is given by ($\tau = t/\hbar$)

$$\delta G(\mathbf{x}\,\hbar\tau; \mathbf{x}\,0; V) = \frac{\hbar^2\tau^2}{12m}\left[(\boldsymbol{\nabla}^2 V) - \frac{\mathrm{i}\tau}{2}(\boldsymbol{\nabla}V)^2\right]$$

$$\times \int \frac{\mathrm{d}^3\mathbf{k}}{(2\pi)^3}\,\exp\left[-\mathrm{i}\left(\frac{\hbar^2\mathbf{k}^2}{2m} + V\right)\tau\right] \quad (13.4.1)$$

using the mass m for μ, and emphasizing its dependence on V.

The (quantum) correction due to electrons near the nucleus has been already taken into account in §13.2 where we found that the effective potential felt by such an electron may be taken to be simply the Coulomb potential of the nucleus alone

$$V_\mathrm{C}(r) = -\frac{Ze^2}{r}. \quad (13.4.2)$$

Accordingly, from (13.4.1) and (9.7.32), we may introduce a further quantum correction to the ground-state energy by[7]

$$E_\mathrm{qua} = \int \mathrm{d}^3\mathbf{x}\,\frac{2}{2\pi\mathrm{i}}\int_{-\infty}^{\infty}\frac{\mathrm{d}\tau}{\tau - \mathrm{i}\varepsilon}\left[\mathrm{i}\frac{\partial}{\partial\tau}\delta G(\mathbf{x}\hbar\tau; \mathbf{x}0; V_\mathrm{TF}) - \mathrm{i}\frac{\partial}{\partial\tau}\delta G(\mathbf{x}\hbar\tau; \mathbf{x}0; V_\mathrm{C})\right]$$

$$(13.4.3)$$

where V_TF is the TF potential (13.1.9), (13.1.10), (13.1.16) and V_C is the Coulomb potential in (13.4.2) due to the nucleus, and the τ-integral projects out the respective negative spectra (see (9.7.32)).

Integrating by parts over τ, we have for the τ-integral in the following

$$I[V(\mathbf{x})] = \frac{2}{2\pi\mathrm{i}}\int_{-\infty}^{\infty}\frac{\mathrm{d}\tau}{\tau - \mathrm{i}\varepsilon}\,\mathrm{i}\frac{\partial}{\partial\tau}\delta G(\mathbf{x}\,\hbar\tau; \mathbf{x}\,0; V)$$

$$= \frac{1}{\pi}\int_{-\infty}^{\infty}\frac{\mathrm{d}\tau}{\tau^2}\,\delta G(\mathbf{x}\,\hbar\tau; \mathbf{x}\,0; V)$$

$$= \frac{\hbar^2}{12\pi m}\int_{-\infty}^{\infty}\mathrm{d}\tau\left[\boldsymbol{\nabla}^2 V - \frac{\mathrm{i}\tau}{2}(\boldsymbol{\nabla}V)^2\right]$$

$$\times \int \frac{\mathrm{d}^3\mathbf{k}}{(2\pi)^3}\,\exp\left[-\mathrm{i}\left(\frac{\hbar^2\mathbf{k}^2}{2m} + V\right)\tau\right]$$

---

[7] This section is based on: Manoukian and Bantitadawit (1999).

$$= \frac{\hbar^2}{12\pi m}\left[\nabla^2 V + \frac{(\nabla V)^2}{2}\frac{\mathrm{d}}{\mathrm{d}V}\right]\int\frac{\mathrm{d}^3\mathbf{k}}{(2\pi)^3}$$

$$\times \int_{-\infty}^{\infty}\mathrm{d}\tau\,\exp\left[-\mathrm{i}\left(\frac{\hbar^2\mathbf{k}^2}{2m}+V\right)\tau\right] \quad (13.4.4)$$

or

$$I[V(\mathbf{x})] = \frac{\hbar^2}{6m}\left[\nabla^2 V + \frac{(\nabla V)^2}{2}\frac{\mathrm{d}}{\mathrm{d}V}\right]\int\frac{\mathrm{d}^3\mathbf{k}}{(2\pi)^3}\,\delta\!\left(\frac{\hbar^2\mathbf{k}^2}{2m}+V(r)\right). \quad (13.4.5)$$

Upon integration over $\mathbf{k}$, and taking the derivative with respect to V, we obtain

$$I[V(\mathbf{x})] = \frac{\sqrt{2m}}{12\pi^2\hbar}\left[\sqrt{-V}\,\nabla^2 V - \frac{1}{4}\frac{(\nabla V)^2}{\sqrt{-V}}\right]. \quad (13.4.6)$$

Now we use the identity

$$\nabla\cdot\left[\nabla(g)^{3/2}\right] = \frac{3}{2}\sqrt{g}\,\nabla^2 g + \frac{3}{4}\frac{(\nabla g)^2}{\sqrt{g}} \quad (13.4.7)$$

or

$$\frac{1}{4}\frac{(\nabla V)^2}{\sqrt{-V}} = \frac{1}{3}\nabla^2(-V)^{3/2} + \frac{1}{2}\sqrt{-V}\,\nabla^2 V \quad (13.4.8)$$

to get

$$I[V(\mathbf{x})] = \frac{\sqrt{2m}}{24\pi^2\hbar}\left[\sqrt{-V}\,\nabla^2 V - \frac{2}{3}\nabla^2(-V)^{3/2}\right]. \quad (13.4.9)$$

The expression for E_{qua} in (13.4.3) then becomes

$$E_{\mathrm{qua}} = \int\mathrm{d}^3\mathbf{x}\left(I[V_{\mathrm{TF}}(\mathbf{x})] - I[V_{\mathrm{C}}(\mathbf{x})]\right). \quad (13.4.10)$$

Consider the contribution of the second term in (13.4.9) to the $\mathbf{x}$-integral in (13.4.10). To this end, using (13.1.9), (13.1.10), (13.4.2), the integral

$$I_2 = \int\mathrm{d}^3\mathbf{x}\,\nabla^2\!\left[(-V_{\mathrm{TF}})^{3/2} - (-V_{\mathrm{C}})^{3/2}\right]$$

$$= \frac{(Ze^2)^{3/2}(4\pi)}{a^{1/2}}\int_{-\infty}^{\infty}x\,\mathrm{d}x\,\frac{\mathrm{d}^2}{\mathrm{d}x^2}x\left[\frac{(f(x))^{3/2}}{x^{3/2}}-\frac{1}{x^{3/2}}\right] \quad (13.4.11)$$

integrates to

13.4 Quantum Correction

$$I_2 = \frac{(Ze^2)^{3/2}(4\pi)}{a^{1/2}} \frac{3}{2} \left[x^{1/2}(f(x))^{1/2} f'(x) - \frac{\left[(f(x))^{3/2} - 1\right]}{x^{1/2}} \right]_{x \to 0}^{x \to \infty}. \tag{13.4.12}$$

From (13.1.19), (13.1.35), it is readily checked that the expression within the brackets in (13.4.12) goes to zero for $x \to \infty$ and $x \to 0$. That is, the second term in (13.4.9) gives no contribution to (13.4.10).

Using the Poisson equations (see (13.1.8), (13.1.1), with $\xi = 0$)

$$\nabla^2 V_{\text{TF}} = 4\pi Z e^2 \delta^3(\mathbf{x}) - 4\pi e^2 n_{\text{TF}}(r) \tag{13.4.13}$$

$$\nabla^2 V_{\text{C}} = 4\pi Z e^2 \delta^3(\mathbf{x}). \tag{13.4.14}$$

We then obtain for (13.4.10)

$$E_{\text{qua}} = \frac{\sqrt{2m}\, Ze^2}{6\pi\hbar} \int d^3\mathbf{x} \left[(-V_{\text{TF}})^{1/2} - (-V_{\text{C}})^{1/2} \right] \delta^3(\mathbf{x})$$

$$- \frac{e^2}{6\pi\hbar} \int d^3\mathbf{x}\, (-2m V_{\text{TF}})^{1/2} n_{\text{TF}}(r). \tag{13.4.15}$$

The first integral is proportional to

$$\int_0^\infty dx\, \frac{\delta(x)}{x^{1/2}} \left[(f(x))^{1/2} - 1 \right] \tag{13.4.16}$$

which, from (13.1.35), integrates to zero.

All told, we have from (13.4.15), (13.1.1),

$$E_{\text{qua}} = -\frac{e^2}{18\pi^3} (3\pi^2)^{4/3} \int d^3\mathbf{x}\, (n(\mathbf{x}))^{4/3} \tag{13.4.17}$$

where we have used the notation $n(\mathbf{x})$ for $n_{\text{TF}}(r)$.

Upon comparison of (13.4.17) with E_{exc} in (13.3.20) we conclude that[8]

$$E_{\text{qua}} = \frac{2}{9} E_{\text{exc}}$$

$$= -\left(\frac{4}{81\pi^4}\right)^{1/3} Z^{5/3} \left(\frac{me^4}{\hbar^2}\right) \int_0^\infty dx\, (f(x))^2. \tag{13.4.18}$$

From the numerical value of the integral in (13.1.27) we then obtain

$$E_{\text{qua}} \cong -0.09814 \left(\frac{me^4}{2\hbar^2}\right) Z^{5/3}. \tag{13.4.19}$$

---
[8] The total contribution $E_{\text{exc}} + E_{\text{qua}} = (11/9) E_{\text{exc}}$ was evaluated by Schwinger (1981) by modelling his analysis after the harmonic oscillator potential.

13.5 Adding Up the Various Contributions: Estimation of $E(Z)$

The various contributions to the estimation of the ground-state energy $E(Z)$ of neutral atoms evaluated in the previous sections are as follows:

- the TF energy E_{TF}, as the leading contribution, ((13.1.42)–(13.1.45)) given by

$$E_{TF} = \frac{6}{7}\left(\frac{4}{3\pi}\right)^{2/3} f'(0) \left(\frac{me^4}{\hbar^2}\right) Z^{7/3} \qquad (13.5.1)$$

with $f'(0)$ given in (13.1.26),

- correction due to electrons bound near the nucleus (see (13.2.14)) given by

$$E_{Sc} = \left(\frac{me^4}{\hbar^2}\right) \frac{Z^2}{2} \qquad (13.5.2)$$

- exchange effect (see (13.3.21)) given by

$$E_{exc} = -\left(\frac{9}{2\pi^4}\right)^{1/3} \left(\frac{me^4}{\hbar^2}\right) Z^{5/3} \int_0^\infty dx \, (f(x))^2 \qquad (13.5.3)$$

- a quantum correction (see (13.4.18)) given by

$$E_{qua} = -\left(\frac{4}{81\pi^4}\right)^{1/3} \left(\frac{me^4}{\hbar^2}\right) Z^{5/3} \int_0^\infty dx \, (f(x))^2$$

$$= \frac{2}{9} E_{exc} \qquad (13.5.4)$$

where the numerical value of the integral in (13.5.3), (13.5.4) is given in (13.1.27).

Adding up the various contributions (13.5.1)–(13.5.4), we obtain, as an estimate for $E(Z)$, the following functional dependence on Z:

$$E(Z) \cong \left[-1.5375 Z^{7/3} + Z^2 - 0.5397 Z^{5/3}\right] \left(\frac{me^4}{2\hbar^2}\right). \qquad (13.5.5)$$

This expression turns out to be remarkably reliable and we refer the reader to the general survey: Morgan III (1996), Chapter 20, p. 233, for the assessment of its accuracy, for further developments and for other methods.

Problems

13.1. For a potential

$$V(\mathbf{x}) = \int d^3 \mathbf{x}' \, \frac{\rho(\mathbf{x}')}{|\mathbf{x} - \mathbf{x}'|}, \qquad \int d^3 \mathbf{x} \, \rho(\mathbf{x}) = 0$$

corresponding to a neutral system, where $\rho(\mathbf{x}) = \rho(|\mathbf{x}|)$ is a spherically symmetric density such that $\rho(|\mathbf{x}|) = 0$ for $|\mathbf{x}| \geqslant r_B$, show that $V(\mathbf{x}) = 0$ for $|\mathbf{x}| \geqslant r_B$. The radius r_B may be finite or infinite. [This is a special case of Newton's classic result.]

13.2. Derive the expression for the kinetic energy in (13.1.5) by using (13.1.4) and (13.1.1) with $\xi = 0$.

13.3. Show that the variational (functional) derivative of (13.1.6) gives (13.1.7), (13.1.8).

13.4. Starting from the expression of $V(r)$ in (13.1.8), (13.1.9) derive the equations (13.1.12), (13.1.14)–(13.1.16).

13.5. Establish the asymptotic limits in (13.1.20), (13.1.21) for $r \to \infty$.

13.6. Prove the explicit analytical expressions for the integrals in (13.1.28)–(13.1.33).

13.7. Establish the limits of the integrals in (A-13.1.10), (A-13.1.11).

13.8. Verify in detail the steps leading from (13.2.9) to (13.2.13).

13.9. Show that $n(\mathbf{x}, \mathbf{x}')$ in (13.3.16) goes over to the local density $n(r)$ in (13.1.1), with $\xi = 0$, for $\mathbf{x}' \to \mathbf{x}$.

13.10. Show that the expression within the square brackets in (13.4.12) goes to zero for $x \to \infty$ and $x \to 0$.

13.11. Investigate the nature of the TF theory described in §13.1 for an ion with N electrons, $N < Z$, and, in particular, the properties of the corresponding TF function $f(x)$.

13.12. Show that the probability $\text{Prob}\left[r < r_0\right]$ of finding an electron within a sphere of radius r_0 about the nucleus for the TF density (13.1.15) is given by

$$\text{Prob}\left[r < r_0\right] = 1 - f\left(\frac{r_0}{a}\right) + \frac{r_0}{a} f'\left(\frac{r_0}{a}\right)$$

where $a \propto Z^{-1/3} a_0$ is defined in (13.1.11). Make a careful study of this probability for various r_0 by noting, in the process, its dependence on the atomic number Z.

13.13. Reconsider the TF theory in §13.1 by investigating the nature of the next to the leading term $\mathbf{p}^2/2m$ in the relativistic expression for the kinetic energy $\sqrt{\mathbf{p}^2 c^2 + m^2 c^4} - mc^2$ for an electron, where c is the speed of light. When would such a term be important in your analysis?

13.14. In obtaining the lower bound in (13.3) from the lower bound Hamiltonian in (13.2), no use was made of Pauli's exclusion principle. Make use of the latter to obtain an improvement to the bound in (13.3) at least for sufficiently large Z.

14

Quantum Physics and the Stability of Matter

If one is asked to prepare a short list of most significant problems in physics of theoretical nature and are critical for our existence and that of the universe, the subject matter treated in this chapter would undoubtedly be on such a list.

God forbid the Pauli exclusion principle becomes abolished making the electron in matter to behave as a boson, and converts matter to a "bosonic one", then such matter would collapse and our world will cease to exist. This is what quantum physics predicts. Here we see this monumental theory at its best. The Pauli exclusion principle is not only sufficient for the stability of our matter but is also necessary. This result alone promotes the Pauli exclusion principle, or more generally the spin and statistics connection, as probably one of the most important results in physics, and in the sciences, in general.

In regard to such "bosonic matter", F. J. Dyson writes:[1] *"[Bosonic] matter in bulk would collapse into a condensed high-density phase. The assembly of any two macroscopic objects would release energy comparable to that of an atomic bomb.... Matter without the exclusion principle is unstable."* E. H. Lieb writes:[2] *"Such "matter" would be very unpleasant stuff to have lying around the house."*

The drastic difference between matter (with the exclusion principle) and "bosonic matter", for systems considered with Coulombic interactions, with N negative and N positive charges, is that the ground-state energy has the power law N^α, where $\alpha = 1$ for matter,[3] while $\alpha > 1$ for "bosonic matter". Such a power law behavior *with* $\alpha > 1$, for the ground-state energy, implies

---

[1] Dyson (1967).

[2] E. H. Lieb, in: Thirring (1991), p. 23. This volume and subsequent editions consisting of many of his publications with various collaborators, such as W. E. Thirring and others, contain a wealth of information on the subject.

[3] Lenard and Dyson (1968); Lieb and Thirring (1975).

the collapse of "bosonic matter",[4,5] since the formation of a single system consisting of $(2N+2N)$ particles is favored over two separate systems brought into contact, each consisting of $(N+N)$ particles, and the energy released upon collapse of two separate systems into a single system, being proportional to $[(2N)^\alpha - 2(N)^\alpha]$, will be overwhelmingly large for realistic large N, e.g., $N \sim 10^{23}$.

The Hamiltonian under consideration for the stability matter is taken to be the N-electron one

$$H = \sum_{i=1}^{N} \frac{\mathbf{p}_i^2}{2m} + \sum_{i<j}^{N} \frac{e^2}{|\mathbf{x}_i - \mathbf{x}_j|} - \sum_{i=1}^{N}\sum_{j=1}^{k} \frac{Z_j\, e^2}{|\mathbf{x}_i - \mathbf{R}_j|} + \sum_{i<j}^{k} \frac{Z_i Z_j\, e^2}{|\mathbf{R}_i - \mathbf{R}_j|} \quad (14.1)$$

where m denotes the mass of the electron and the $\mathbf{x}_i$, $\mathbf{R}_j$ correspond, respectively, to positions of the electrons and nuclei. Also we consider neutral matter, i.e.,

$$\sum_{i=1}^{k} Z_i = N. \quad (14.2)$$

The Hamiltonian in (14.1) is a typical one in that it corresponds to motionless (i.e., infinitely massive and hence with arbitrary large rest mass energies) fixed point-like nuclei. This is non-academic. By doing so, one does not dwell on the fate and the dynamics of the positive background, and one is looking at, and monitoring the fate of, the electrons through the "eye" of the former system.

The key result in the problem of the stability of matter, with the exclusion principle, is the single power law behavior $E_N \sim -N$ of the ground-state energy, and the physically expected result that the ground-state energy per electron $|E_N/N|$ remains bounded for all N unlike "bosonic matter" for which the latter becomes larger and larger as N increases. What we will actually learn in §14.3 is that, for a non-vanishing probability of having the electrons within a sphere of radius R, the volume v_R in which the electrons are confined grows not any slower than the first power of N for $N \to \infty$. That is, *necessarily, the radius R of spatial extension of matter grows not any slower than $N^{1/3}$ for $N \to \infty$*. No wonder why matter occupies so large a volume! Here it is worth recalling the words addressed by Paul Ehrenfest to Wolfgang Pauli in 1931 on the occasion of the Lorentz medal[6] to this effect: *"We take a piece of metal, or a stone. When we think about it, we are astonished that this quantity of matter should occupy so large a volume"*. He went on

---

[4] Dyson and Lenard (1967); Lieb (1979, 1976); Manoukian and Muthaporn (2003b). The corresponding law here is $N^{5/3}$ with motionless fixed (i.e., infinitely massive) positive charges.

[5] Dyson (1967); Manoukian and Muthaporn (2002); Conlon et al. (1988). The corresponding law here is $N^{7/5}$, where the positive charges are treated dynamically as well with finite masses restricted to Coulombic interactions.

[6] See Ehrenfest (1959), p. 617, as quoted in Dyson (1967).

14.1 Lower Bound to the Multi-Particle Repulsive Coulomb Potential ...

by stating that the Pauli exclusion principle is the reason: *"Answer: only the Pauli principle, no two electrons in the same state"*.

It is important to emphasize that the collapse of "bosonic matter" occurs even if the positive charges are treated dynamically with finite masses with Coulomb interactions.[7] Also that the collapse of "bosonic matter" is not a characteristic of the dimensionality of space,[8] and that such matter does not change, for example, from an "implosive" to a "stable" or to an "explosive" phase with change of dimensionality.

In §14.1, we obtain a lower bound to the electron-electron repulsive Coulomb potential energy, which when combined with the lower bound in (4.6.24) derived for the expectation value of the kinetic energy of the electrons provides a lower bound to the ground-state energy E_N in §14.2. In the latter section, an upper bound for E_N is also derived consistent with the single power of N obtained for E_N in the lower bound. The high density limit of matter is investigated in §14.3. The final section §14.4, deals with the collapse of "bosonic matter".

To make this work accessible to a wider audience we have relegated rather some technical aspects of the analyses to the appendices of the relevant sections.

14.1 Lower Bound to the Multi-Particle Repulsive Coulomb Potential Energy

In the appendix to this section, we derive the following inequality which follows from (A-14.1.28) and reads in detail,

$$(3\pi^2)^{5/3} \frac{\hbar^2}{10\pi^2 m \beta} \int d^3\mathbf{x}\, \rho^{5/3}(\mathbf{x}) - \sum_{j=1}^{k} Z_j\, e^2 \int d^3\mathbf{x}\, \frac{\rho(\mathbf{x})}{|\mathbf{x} - \mathbf{R}_j|}$$

$$+ \frac{e^2}{2} \int d^3\mathbf{x}\, d^3\mathbf{x}'\, \rho(\mathbf{x}) \frac{1}{|\mathbf{x} - \mathbf{x}'|} \rho(\mathbf{x}') + \sum_{i<j}^{k} \frac{Z_i Z_j\, e^2}{|\mathbf{R}_i - \mathbf{R}_j|}$$

$$\geqslant \beta E_{\text{TF}}(1) \sum_{i=1}^{k} Z_i^{7/3} \qquad (14.1.1)$$

where $\beta > 0$ is an arbitrary dimensionless parameter, $\rho(\mathbf{x})$ is an arbitrary positive function, and $E_{\text{TF}}(1) \cong -1.5375\,(me^4/2\hbar^2)$ is the coefficient of the TF ground-state energy defined in (13.1.43), (13.1.45).

The energy density functional, expressed in terms of the density $\rho(\mathbf{x})$ on the left-hand side of (14.1.1) is in the spirit of the TF energy functional

---
[7] Dyson (1967); Manoukian and Muthaporn (2002).
[8] Manoukian and Muthaporn (2003a); Muthaporn and Manoukian (2004a,b).

considered earlier in (13.1.6) in the TF theory, with the mass m of the electron replaced by $m\beta$, and with the further generalization of including k nuclei, with the last term, involving '$Z_i Z_j e^2$', describing their interactions.

The inequality in (14.1.1) gives rise to a lower bound to the (repulsive) Coulomb potential energy of k particles of charges $Z_1|e|, \ldots, Z_k|e|$, or charges $-Z_1|e|, \ldots, -Z_k|e|$, i.e., for charges of the same signs as follows:

$$\sum_{i<j}^{k} \frac{Z_i Z_j e^2}{|\mathbf{R}_i - \mathbf{R}_j|} \geqslant \sum_{j=1}^{k} Z_j e^2 \int d^3\mathbf{x}\, \frac{\rho(\mathbf{x})}{|\mathbf{x} - \mathbf{R}_j|}$$

$$- \frac{e^2}{2} \int d^3\mathbf{x}\, d^3\mathbf{x}'\, \rho(\mathbf{x}) \frac{1}{|\mathbf{x} - \mathbf{x}'|} \rho(\mathbf{x}')$$

$$- \left(3\pi^2\right)^{5/3} \frac{\hbar^2}{10\pi^2 m\beta} \int d^3\mathbf{x}\, \rho^{5/3}(\mathbf{x}) + \beta E_{\mathrm{TF}}(1) \sum_{i=1}^{k} Z_i^{7/3}.$$

(14.1.2)

In particular for the interaction of N electrons we have, with substitutions $k \to N$, $Z_j \to 1$, $\mathbf{R}_j \to \mathbf{x}_j$ for $j = 1, \ldots, N$:

$$\sum_{i<j}^{N} \frac{e^2}{|\mathbf{x}_i - \mathbf{x}_j|} \geqslant \sum_{j=1}^{N} e^2 \int d^3\mathbf{x}\, \frac{\rho(\mathbf{x})}{|\mathbf{x} - \mathbf{x}_j|}$$

$$- \frac{e^2}{2} \int d^3\mathbf{x}\, d^3\mathbf{x}'\, \rho(\mathbf{x}) \frac{1}{|\mathbf{x} - \mathbf{x}'|} \rho(\mathbf{x}')$$

$$- \left(3\pi^2\right)^{5/3} \frac{\hbar^2}{10\pi^2 m\beta} \int d^3\mathbf{x}\, \rho^{5/3}(\mathbf{x}) + \beta N E_{\mathrm{TF}}(1). \quad (14.1.3)$$

The two inequalities (14.1.1), (14.1.3) combined with the lower bound of the expectation of the kinetic energy given in (4.6.24) for $s = 1/2$, will be used in the next section to derive a lower bound for the exact ground-state energy of matter with Coulomb interaction by appropriately choosing $\rho(\mathbf{x})$ in (14.1.1), (14.1.3) to coincide with the particle number density defined in (4.6.16).[9]

---

[9] Note that the positive function $\rho(x)$ in (14.1.1)–(14.1.3) being arbitrary may be chosen to be the same in all of these three inequalities.

14.1 Lower Bound to the Multi-Particle Repulsive Coulomb Potential ...

Appendix to §14.1: A Thomas-Fermi-Like Energy Functional and No Binding

Due to the technical nature of this appendix, its content may be omitted at a first reading. The main result obtained (inequality (A-14.1.28)), however, is important in obtaining a lower bound to the repulsive Coulomb potential energy for the e^--e^- interaction as given in (14.1.3).

We introduce the functional of a positive function $\rho(\mathbf{x})$ defined by

$$F[\rho; Z_1, \ldots, Z_k, \mathbf{R}_1, \ldots, \mathbf{R}_k]$$

$$= (3\pi^2)^{5/3} \frac{\hbar^2}{10\pi^2 m\beta} \int d^3\mathbf{x}\, \rho^{5/3}(\mathbf{x}) - \sum_{j=1}^{k} Z_j\, e^2 \int d^3\mathbf{x}\, \frac{\rho(\mathbf{x})}{|\mathbf{x} - \mathbf{R}_j|}$$

$$+ \frac{e^2}{2} \int d^3\mathbf{x}\, d^3\mathbf{x}'\, \rho(\mathbf{x}) \frac{1}{|\mathbf{x} - \mathbf{x}'|} \rho(\mathbf{x}') + \sum_{i<j}^{k} \frac{Z_i Z_j\, e^2}{|\mathbf{R}_i - \mathbf{R}_j|} \qquad \text{(A-14.1.1)}$$

(compare with (13.1.6)), depending on positive parameters $Z_1, \ldots, Z_k$ and vectors $\mathbf{R}_1, \ldots, \mathbf{R}_k$. Here $\beta > 0$ is an arbitrary dimensionless parameter. [In particular, for $k = 1$, the last term in (A-14.1.1) is absent, and by setting $\mathbf{R}_i = \mathbf{0}$, $\beta = 1$, we obtain the energy functional in (13.1.6), (B-13.1.1).]

The main result (A-14.1.24)/(A-14.1.28) established in this appendix was used in this section to obtain a lower bound for the (repulsive) Coulomb potential for many particles having charges of the same signs as given in (14.1.2), (14.1.3).

Let $\rho_0(\mathbf{x}; k)$ satisfy the equation (see also (13.1.7))

$$(3\pi^2)^{2/3} \frac{\hbar^2}{2m\beta} \rho_0^{2/3}(\mathbf{x}; k) = \sum_{i=1}^{k} \frac{Z_i\, e^2}{|\mathbf{x} - \mathbf{R}_i|} - e^2 \int d^3\mathbf{x}'\, \frac{1}{|\mathbf{x} - \mathbf{x}'|} \rho_0(\mathbf{x}'; k)$$

(A-14.1.2)

as obtained by functional differentiation of (A-14.1.1) with respect to $\rho(\mathbf{x})$ and by setting the result equal to zero as done in Lagrangian mechanics.

Following the proof given in Appendix B to §13.1, which shows that the TF density satisfying (13.1.7) actually provides the smallest value (see (B-13.1.11)), for the energy density functional (B-13.1.1), we conclude (see Problem 14.8) that $\rho_0(\mathbf{x}; k)$ satisfying (A-14.1.2) provides the smallest value for the functional $F[\rho; Z_1, \ldots, Z_k, \mathbf{R}_1, \ldots, \mathbf{R}_k]$ in (A-14.1.1), with the normalization condition

$$\int d^3\mathbf{x}\, \rho_0(\mathbf{x}; k) = \sum_{i=1}^{k} Z_i \qquad \text{(A-14.1.3)}$$

satisfied. That is

$$F[\rho; Z_1, \ldots, Z_k, \mathbf{R}_1, \ldots, \mathbf{R}_k] \geqslant F[\rho_0; Z_1, \ldots, Z_k, \mathbf{R}_1, \ldots, \mathbf{R}_k]. \qquad \text{(A-14.1.4)}$$

We introduce the functionals

$$F[\rho; \lambda Z_1, \ldots, \lambda Z_l, Z_{l+1}, \ldots, Z_k, \mathbf{R}_1, \ldots, \mathbf{R}_k] \tag{A-14.1.5}$$

and

$$F[\rho; \lambda Z_1, \ldots, \lambda Z_l, \mathbf{R}_1, \ldots, \mathbf{R}_l] \tag{A-14.1.6}$$

where $l < k$, and $\lambda > 0$ is an arbitrary parameter.

Let $\rho_1(\mathbf{x})$, $\rho_2(\mathbf{x})$ be the corresponding solutions to (A-14.1.2) for the functionals in (A-14.1.5), (A-14.1.6), respectively:

$$(3\pi^2)^{2/3} \frac{\hbar^2}{2m\beta} \rho_1^{2/3}(\mathbf{x}) = \sum_{j=1}^{l} \frac{\lambda Z_j\, e^2}{|\mathbf{x} - \mathbf{R}_j|} + \sum_{j=l+1}^{k} \frac{Z_j\, e^2}{|\mathbf{x} - \mathbf{R}_j|}$$

$$- e^2 \int d^3\mathbf{x}' \frac{1}{|\mathbf{x} - \mathbf{x}'|} \rho_1(\mathbf{x}') \tag{A-14.1.7}$$

$$(3\pi^2)^{2/3} \frac{\hbar^2}{2m\beta} \rho_2^{2/3}(\mathbf{x}) = \sum_{j=1}^{l} \frac{\lambda Z_j\, e^2}{|\mathbf{x} - \mathbf{R}_j|} - e^2 \int d^3\mathbf{x}' \frac{1}{|\mathbf{x} - \mathbf{x}'|} \rho_2(\mathbf{x}'). \tag{A-14.1.8}$$

For simplicity of the notation only, we have suppressed the dependence of ρ_1, ρ_2 on λ, k, l.

By setting

$$(3\pi^2)^{2/3} \frac{\hbar^2}{2m\beta} \rho_j^{2/3}(\mathbf{x}) \equiv Q_j(\mathbf{x}), \qquad j = 1, 2 \tag{A-14.1.9}$$

we obtain from (A-14.1.7), (A-14.1.8), upon subtraction,

$$Q_1(\mathbf{x}) - Q_2(\mathbf{x}) = \sum_{j=l+1}^{k} \frac{Z_j\, e^2}{|\mathbf{x} - \mathbf{R}_j|} - e^2 \int d^3\mathbf{x}' \frac{1}{|\mathbf{x} - \mathbf{x}'|} \left[\rho_1(\mathbf{x}') - \rho_2(\mathbf{x}')\right]$$

$$= \sum_{j=l+1}^{k} \frac{Z_j\, e^2}{|\mathbf{x} - \mathbf{R}_j|}$$

$$- \frac{1}{3\pi^2} \left(\frac{2m\beta}{\hbar^2}\right)^{3/2} e^2 \int d^3\mathbf{x}' \frac{1}{|\mathbf{x} - \mathbf{x}'|} \left[Q_1^{3/2}(\mathbf{x}') - Q_2^{3/2}(\mathbf{x}')\right].$$

$$\tag{A-14.1.10}$$

Since the sum over j in (A-14.1.10) is non-negative, $[Q_1(\mathbf{x}) - Q_2(\mathbf{x})]$ cannot be strictly negative for all $\mathbf{x}$ otherwise this will be in contradiction with the equation (A-14.1.10) itself.

We introduce the set

$$S = \left\{ \mathbf{x} \,\middle|\, Q_1(\mathbf{x}) - Q_2(\mathbf{x}) < 0 \right\} \tag{A-14.1.11}$$

14.1 Lower Bound to the Multi-Particle Repulsive Coulomb Potential ...

and we will show that this set is empty, thus concluding that $Q_1(\mathbf{x}) - Q_2(\mathbf{x}) \geqslant 0$.

We assume that S is non-empty and then run into a contradiction. As we move away from the boundary Ω of S, $[Q_1(\mathbf{x}) - Q_2(\mathbf{x})]$ changes sign or vanishes, by definition of S, and we then have

$$\hat{\mathbf{n}} \cdot \boldsymbol{\nabla}[Q_1(\mathbf{x}) - Q_2(\mathbf{x})] \geqslant 0 \qquad \text{(A-14.1.12)}$$

otherwise, we would run into a region beyond S where $[Q_1(\mathbf{x}) - Q_2(\mathbf{x})]$ is still strictly negative. [If S is of infinite extension the non-negativity of $\hat{\mathbf{n}} \cdot \boldsymbol{\nabla}[Q_1(\mathbf{x}) - Q_2(\mathbf{x})]$ on the boundary still holds.]

The application of the Laplacian to (A-14.1.10) gives

$$\nabla^2 [Q_1(\mathbf{x}) - Q_2(\mathbf{x})] = -4\pi \sum_{j=l+1}^{k} Z_j\, e^2 \delta^3(\mathbf{x} - \mathbf{R_j})$$

$$+ 4\pi e^2 \left(\frac{2m\beta}{\hbar^2 (3\pi^2)^{2/3}} \right)^{3/2} \left[Q_1^{3/2}(\mathbf{x}) - Q_2^{3/2}(\mathbf{x}) \right]$$

(A-14.1.13)

and for $\mathbf{x}$ in the set S, the expression on the right-hand side of this equation is strictly negative since $\left[Q_1^{3/2}(\mathbf{x}) - Q_2^{3/2}(\mathbf{x}) \right] < 0$ for such $\mathbf{x}$ by hypothesis.

Accordingly,

$$0 > \int_S d^3\mathbf{x}\, \nabla^2 [Q_1(\mathbf{x}) - Q_2(\mathbf{x})] = \int_\Omega d\Omega\, \hat{\mathbf{n}} \cdot \boldsymbol{\nabla}[Q_1(\mathbf{x}) - Q_2(\mathbf{x})] \qquad \text{(A-14.1.14)}$$

in contradiction with (A-14.1.12), hence S is empty and

$$Q_1(\mathbf{x}) - Q_2(\mathbf{x}) \geqslant 0 \qquad \text{(A-14.1.15)}$$

as a function of $\mathbf{x}$.

In reference to the functional

$$F[\rho; Z_{l+1}, \ldots, Z_k, \mathbf{R}_{l+1}, \ldots, \mathbf{R}_k] \qquad \text{(A-14.1.16)}$$

let $\rho_3(\mathbf{x})$ satisfy

$$(3\pi^2)^{2/3} \frac{\hbar^2}{2m\beta} \rho_3^{2/3}(\mathbf{x}) = \sum_{j=l+1}^{k} \frac{Z_j\, e^2}{|\mathbf{x} - \mathbf{R}_j|} - e^2 \int d^3\mathbf{x}'\, \frac{1}{|\mathbf{x} - \mathbf{x}'|} \rho_3(\mathbf{x}')$$

(A-14.1.17)

in analogy to (A-14.1.7), (A-14.1.8).

We define

$$g(\lambda) = F[\rho_1; \lambda Z_1, \ldots, \lambda Z_l, Z_{l+1}, \ldots, Z_k, \mathbf{R}_1, \ldots, \mathbf{R}_k]$$

$$- F[\rho_2; \lambda Z_1, \ldots, \lambda Z_l, \mathbf{R}_1, \ldots, \mathbf{R}_l]$$
$$- F[\rho_3; Z_{l+1}, \ldots, Z_k, \mathbf{R}_{l+1}, \ldots, \mathbf{R}_k] \quad \text{(A-14.1.18)}$$

with $l < k$ and the ρ_i non-negative. Since for $\lambda = 0$, ρ_1 and ρ_3 denote the same density, and ρ_2, in (A-14.1.8) is obviously equal to zero for $\lambda = 0$, as the left-hand side of (A-14.1.8) is non-negative while its right-hand side is non-positive for $\lambda = 0$, we may infer that

$$g(0) = 0. \quad \text{(A-14.1.19)}$$

We will show that
$$g(1) \geqslant 0. \quad \text{(A-14.1.20)}$$

From (A-14.1.19), we may write

$$g(1) = \int_0^1 d\lambda\, g'(\lambda) \quad \text{(A-14.1.21)}$$

and hence to establish (A-14.1.20) it is sufficient to show that $g'(\lambda) \geqslant 0$ for $0 \leqslant \lambda \leqslant 1$.

To the above end, we note from (A-14.1.1) with $Z_1 \to \lambda Z_1, \ldots, Z_l \to \lambda Z_l$, $\rho \to \rho_1$ that

$$\frac{\partial}{\partial \lambda} F[\rho_1; \lambda Z_1, \ldots, \lambda Z_l, Z_{l+1}, \ldots, Z_k, \mathbf{R}_1, \ldots, \mathbf{R}_k]$$

$$= \int d^3\mathbf{x} \left[(3\pi^2)^{2/3} \frac{\hbar^2}{2m\beta} \rho_1^{2/3}(\mathbf{x}) - e^2 \sum_{j=1}^{l} \frac{\lambda Z_j}{|\mathbf{x} - \mathbf{R}_j|} \right.$$

$$\left. - e^2 \sum_{j=l+1}^{k} \frac{Z_j}{|\mathbf{x} - \mathbf{R}_j|} + e^2 \int d^3\mathbf{x}' \frac{1}{|\mathbf{x} - \mathbf{x}'|} \rho_1(\mathbf{x}') \right] \frac{\partial}{\partial \lambda} \rho_1(\mathbf{x})$$

$$- \sum_{j=1}^{l} Z_j\, e^2 \int d^3\mathbf{x} \frac{1}{|\mathbf{x} - \mathbf{R}_j|} \rho_1(\mathbf{x})$$

$$+ e^2 \left(2\lambda \sum_{i=1}^{l-1} \sum_{j=i+1}^{l} \frac{Z_i Z_j}{|\mathbf{R}_i - \mathbf{R}_j|} + \sum_{i=1}^{l} Z_i \sum_{j=l+1}^{k} \frac{Z_j}{|\mathbf{R}_i - \mathbf{R}_j|} \right).$$
$$\text{(A-14.1.22)}$$

On account of (A-14.1.7), the expression within the square brackets of the $\mathbf{x}$-integral in the first term on the right-hand side of (A-14.1.22) is zero. An expression similar to the one in (A-14.1.22) for

14.1 Lower Bound to the Multi-Particle Repulsive Coulomb Potential ...   773

$$\frac{\partial}{\partial \lambda} F[\rho_2; \lambda Z_1, \ldots, \lambda Z_l, \mathbf{R}_1, \ldots, \mathbf{R}_l]$$

may be also readily derived. Hence from (A-14.1.18)

$$\frac{\partial}{\partial \lambda} g(\lambda) = \sum_{i=1}^{l} Z_i \left(\sum_{j=l+1}^{k} \frac{Z_j e^2}{|\mathbf{R}_i - \mathbf{R}_j|} - e^2 \int d^3\mathbf{x} \frac{[\rho_1(\mathbf{x}) - \rho_2(\mathbf{x})]}{|\mathbf{x} - \mathbf{R}_i|} \right)$$

$$\equiv \sum_{i=1}^{l} Z_i [Q_1(\mathbf{R}_i) - Q_2(\mathbf{R}_i)] \geqslant 0 \qquad (\text{A-14.1.23})$$

where we have used (A-14.1.10) and (A-14.1.22), thus establishing (A-14.1.20). Here we note that the summation over j in the first term in (A-14.1.10) is from $(l+1)$ to k, while the one on the extreme right-hand side of (A-14.1.23) is over i from 1 to l, and there are no ambiguities in the expression in (A-14.1.23).

Accordingly, from (A-14.1.18), (A-14.1.20) we obtain

$$F[\rho_1; Z_1, \ldots, Z_k, \mathbf{R}_1, \ldots, \mathbf{R}_k] \geqslant F[\rho_2; Z_1, \ldots, Z_l, \mathbf{R}_1, \ldots, \mathbf{R}_l]$$
$$+ F[\rho_3; Z_{l+1}, \ldots, Z_k, \mathbf{R}_{l+1}, \ldots, \mathbf{R}_k] \qquad (\text{A-14.1.24})$$

for any $1 \leqslant l < k$, where ρ_1, ρ_2, ρ_3 are the densities which provide the smallest values for the corresponding functionals, respectively.

Since l, k (with $l < k$) are arbitrary natural numbers, (A-14.1.24) implies that

$$F[\rho_0; Z_1, \ldots, Z_k, \mathbf{R}_1, \ldots, \mathbf{R}_k] \geqslant \sum_{i=1}^{k} F[\rho_{\text{TF}}^i; Z_i, \mathbf{R}_i] \qquad (\text{A-14.1.25})$$

where each $F[\rho_{\text{TF}}^i; Z_i, \mathbf{R}_i]$ is the TF functional (13.1.7), evaluated with the TF density ρ_{TF}^i with nuclear charge $Z_i|e|$, situated at $\mathbf{R}_i$, and the mass m of each negatively charged particle simply scaled by β. That is,

$$(3\pi^2)^{2/3} \frac{\hbar^2}{2m\beta} (\rho_{\text{TF}}^i(\mathbf{x}))^{2/3} = \frac{Z_i e^2}{|\mathbf{x} - \mathbf{R}_i|} - e^2 \int d^3\mathbf{x}' \frac{1}{|\mathbf{x} - \mathbf{x}'|} \rho_{\text{TF}}^i(\mathbf{x}').$$
(A-14.1.26)

Upon replacing $\mathbf{x}$ by $\mathbf{x} + \mathbf{R}_i$ and setting

$$\rho_{\text{TF}}^i(\mathbf{x} + \mathbf{R}_i) = n_{\text{TF}}(\mathbf{x}) \Big|_{\substack{m \to m\beta \\ Z \to Z_i}} \qquad (\text{A-14.1.27})$$

where $n_{\text{TF}}(\mathbf{x})$ is the TF density (13.1.7) of §13.1, we obtain from (A-14.1.25) and (13.1.42)–(13.1.45), (A-14.1.4),

774   14 Quantum Physics and the Stability of Matter

$$F[\rho; Z_1, \ldots, Z_k, \mathbf{R}_1, \ldots, \mathbf{R}_k] \geq \beta E_{\mathrm{TF}}(1) \sum_{i=1}^{k} Z_i^{7/3} \qquad \text{(A-14.1.28)}$$

for arbitrary positive $\rho(\mathbf{x})$, where $E_{\mathrm{TF}}(1) \cong -1.5375 \left(me^4/2\hbar^2\right)$ corresponding to particles of masses m.

The basic inequality in (A-14.1.24), shows that a system identified by the parameters $[Z_1, \ldots, Z_k, \mathbf{R}_1, \ldots, \mathbf{R}_k]$ cannot have an (optimized) energy functional (A-14.1.1) less than the sum of the (optimized) energy functional of any two subsystems identified by parameters $[Z_1, \ldots, Z_l, \mathbf{R}_1, \ldots, \mathbf{R}_l]$, $[Z_{l+1}, \ldots, Z_k, \mathbf{R}_{l+1}, \ldots, \mathbf{R}_k]$, $l < k$. Because of this last property, the Theorem embodied in the inequalities (A-14.1.24), (A-14.1.25) is referred to as a "No Binding Theorem".[10]

14.2 Lower and Upper Bounds for the Ground-State Energy and the Stability of Matter

14.2.1 A Lower Bound

For anti-symmetric normalized functions $\Psi(\mathbf{x}_1\sigma_1, \ldots, \mathbf{x}_N\sigma_N)$ of N electrons, we have for the expectation value of the Hamiltonian H in (14.1)

$$\langle \Psi | H | \Psi \rangle = \sum_{i=1}^{N} \left\langle \Psi \left| \frac{\mathbf{p}_i^2}{2m} \right| \Psi \right\rangle - \sum_{i=1}^{N} \sum_{j=1}^{k} Z_j\, e^2 \left\langle \Psi \left| \frac{1}{|\mathbf{x}_i - \mathbf{R}_j|} \right| \Psi \right\rangle$$

$$+ \sum_{i<j}^{N} e^2 \left\langle \Psi \left| \frac{1}{|\mathbf{x}_i - \mathbf{x}_j|} \right| \Psi \right\rangle + \sum_{i<j}^{k} e^2 \frac{Z_i Z_j}{|\mathbf{R}_i - \mathbf{R}_j|}. \qquad (14.2.1)$$

To derive a lower bound to this expectation value, we recall the definition of electron density

$$\rho(\mathbf{x}) = N \sum_{\sigma_1, \ldots, \sigma_N} \int \mathrm{d}^3\mathbf{x}_2 \ldots \mathrm{d}^3\mathbf{x}_N\, |\Psi(\mathbf{x}\sigma_1, \mathbf{x}_2\sigma_2, \ldots, \mathbf{x}_N\sigma_N)|^2 \qquad (14.2.2)$$

normalized to

$$\int \mathrm{d}^3\mathbf{x}\, \rho(\mathbf{x}) = N \qquad (14.2.3)$$

which we will use in (A-14.1.28), and the lower bound (4.6.24) to the expectation value of the kinetic energy for $s = 1/2$:

---

[10] This important result was discovered by Teller (1962) and was established rigorously by Lieb and Simon (1973); Lieb (1976) where the existence of a consistent positive solution of (A-14.1.2) for the density is also studied.

14.2 Lower and Upper Bounds for the Ground-State Energy and ...

$$\sum_{i=1}^{N}\left\langle\Psi\left|\frac{\mathbf{p}_i^2}{2m}\right|\Psi\right\rangle \geq \frac{3}{5}\left(\frac{3\pi}{4}\right)^{2/3}\frac{\hbar^2}{2m}\int d^3\mathbf{x}\,\rho^{5/3}(\mathbf{x}). \tag{14.2.4}$$

For the second term on the right-hand side of (14.2.1), we have the explicit equality

$$\sum_{i=1}^{N}\sum_{j=1}^{k} Z_j\, e^2\left\langle\Psi\left|\frac{1}{|\mathbf{x}_i-\mathbf{R}_j|}\right|\Psi\right\rangle = \sum_{j=1}^{k} Z_j e^2 \int d^3\mathbf{x}\,\frac{1}{|\mathbf{x}-\mathbf{R}_j|}\rho(\mathbf{x}). \tag{14.2.5}$$

We derive a lower bound to the third term in (14.2.1) by using, in the process, the lower bound in (14.1.3) for the Coulomb potential energy of repulsion of the electrons. To this end, we first note that

$$\sum_{i=1}^{N} e^2\int d^3\mathbf{x}\,\rho(\mathbf{x})\left\langle\Psi\left|\frac{1}{|\mathbf{x}-\mathbf{x}_i|}\right|\Psi\right\rangle = e^2\int d^3\mathbf{x}\,d^3\mathbf{x}'\,\rho(\mathbf{x})\frac{1}{|\mathbf{x}-\mathbf{x}'|}\rho(\mathbf{x}') \tag{14.2.6}$$

and hence from (14.1.3)

$$\sum_{i<j}^{N} e^2\left\langle\Psi\left|\frac{1}{|\mathbf{x}_i-\mathbf{x}_j|}\right|\Psi\right\rangle \geq \frac{e^2}{2}\int d^3\mathbf{x}\,d^3\mathbf{x}'\,\rho(\mathbf{x})\frac{1}{|\mathbf{x}-\mathbf{x}'|}\rho(\mathbf{x}')$$

$$-(3\pi^2)^{5/3}\frac{\hbar^2}{10\pi^2 m\beta}\int d^3\mathbf{x}\,\rho^{5/3}(\mathbf{x}) + \beta N E_{\mathrm{TF}}(1). \tag{14.2.7}$$

From (14.2.4)–(14.2.7), we then obtain the following lower bound for (14.2.1)

$$\langle\Psi|H|\Psi\rangle \geq (3\pi^2)^{5/3}\frac{\hbar^2}{10\pi^2 m\beta'}\int d^3\mathbf{x}\,\rho^{5/3}(\mathbf{x}) - \sum_{j=1}^{k} Z_j e^2\int d^3\mathbf{x}\,\frac{\rho(\mathbf{x})}{|\mathbf{x}-\mathbf{R}_j|}$$

$$+\frac{e^2}{2}\int d^3\mathbf{x}\,d^3\mathbf{x}'\,\rho(\mathbf{x})\frac{1}{|\mathbf{x}-\mathbf{x}'|}\rho(\mathbf{x}') + \sum_{i<j}^{k}\frac{Z_i Z_j e^2}{|\mathbf{R}_i-\mathbf{R}_j|} + \beta N E_{\mathrm{TF}}(1) \tag{14.2.8}$$

where we have set

$$\frac{\dfrac{3}{5}\left(\dfrac{3\pi}{4}\right)^{2/3}-\dfrac{(3\pi^2)^{5/3}}{5\pi^2}\dfrac{1}{\beta}}{\dfrac{(3\pi^2)^{5/3}}{5\pi^2}} = \left(\frac{1}{4\pi}\right)^{2/3} - \frac{1}{\beta} = \frac{1}{\beta'}. \tag{14.2.9}$$

For a positive β' we must choose $\beta > (4\pi)^{2/3}$.

The sum of the first four terms on the right-hand side of the inequality in (14.2.8) coincide with the expression on the left-hand side of the inequality in (14.1.1) with β in the latter replaced by β'. Hence

$$\langle \Psi | H | \Psi \rangle \geqslant \beta' E_{\text{TF}}(1) \sum_{i=1}^{k} Z_i^{7/3} + \beta N E_{\text{TF}}(1) \qquad (14.2.10)$$

or

$$\langle \Psi | H | \Psi \rangle \geqslant E_{\text{TF}}(1) \left[\beta N + \frac{\sum_{i=1}^{k} Z_i^{7/3}}{\left(\frac{1}{4\pi}\right)^{2/3} - \frac{1}{\beta}} \right]. \qquad (14.2.11)$$

Optimizing over β, we obtain

$$\beta = (4\pi)^{2/3} \left[1 + \left(\frac{\sum_{i=1}^{k} Z_i^{7/3}}{N} \right)^{1/2} \right] \qquad (14.2.12)$$

giving finally the Lieb-Thirring bound[11]

$$\langle \Psi | H | \Psi \rangle \geqslant E_{\text{TF}}(1)(4\pi)^{2/3} N \left[1 + \left(\sum_{i=1}^{k} \frac{Z_i^{7/3}}{N} \right)^{1/2} \right]^2 \qquad (14.2.13)$$

where

$$E_{\text{TF}}(1) = -1.5375 \left(\frac{me^4}{2\hbar^2} \right). \qquad (14.2.14)$$

If Z corresponds to the nucleus with the maximum charge, in units of $|e|$, then

$$\sum_{i=1}^{k} Z_i^{7/3} \leqslant Z^{4/3} \sum_{i=1}^{k} Z_i = N Z^{4/3} \qquad (14.2.15)$$

giving for the ground-state energy E_N the lower bound

$$E_N \geqslant -8.3104 \left(\frac{me^4}{2\hbar^2} \right) N \left[1 + Z^{2/3} \right]^2 \qquad (14.2.16)$$

where we have used the fact that Ψ is arbitrary and hence (14.2.13) is true for the ground-state as well. The numerical coefficient 8.3104 may be further reduced but we will not attempt to do so here.

---

[11] Lieb and Thirring (1975).

14.2.2 Upper Bounds

A quick and rather conservative upper bound for E_N may be derived by considering the following trial determinantal function

$$\Phi(\mathbf{x}_1\sigma_1,\ldots,\mathbf{x}_N\sigma_N) = \frac{1}{\sqrt{N!}} \det[\phi_j(\mathbf{x}_k,\sigma_k)] \qquad (14.2.17)$$

$(j,k = 1,\ldots,N)$, where

$$\phi_j(\mathbf{x},\sigma) = \phi\left(\mathbf{x} - \mathbf{L}^{(j)}\right) \chi_j(\sigma) \qquad (14.2.18)$$

with normalized spin functions $\chi_j(\sigma)$, which for simplicity make be taken to be all the same, and

$$\phi(\mathbf{x}) = \prod_i \left(\frac{1}{\sqrt{L}} \cos\left(\frac{\pi x_i}{2L}\right)\right), \quad |x_i| \leqslant L \qquad (14.2.19)$$

$i = 1,2,3$, and is zero otherwise, $\mathbf{x} = (x_1,x_2,x_3)$. We choose the vectors $\mathbf{L}^{(1)},\ldots,\mathbf{L}^{(N)}$ as follows

$$\mathbf{L}^{(j)} = jD(1,1,1), \quad j = 1,\ldots,N \qquad (14.2.20)$$

and we may choose

$$4L < D. \qquad (14.2.21)$$

It is easy to see that the intervals: $\{jD - L \leqslant x_i \leqslant jD + L\}$, for $j = 1,\ldots,N$, are disjoint, for each $i = 1,2,3$, and the functions $\phi(\mathbf{x} - \mathbf{L}^{(j)})$ are then non-overlapping, and orthogonal with respect to *each* of the components x_i of $\mathbf{x}$.

We choose

$$\mathbf{R}_j = \mathbf{L}^{(j)}, \quad j = 1,\ldots,k. \qquad (14.2.22)$$

The above construction consists of conveniently placing the k nuclei at $\mathbf{L}^{(1)},\ldots,\mathbf{L}^{(k)}$ and one electron in each one of the k boxes with centers at $\mathbf{L}^{(1)},\ldots,\mathbf{L}^{(k)}$. One electron is also placed in each of the remaining $(N-k)$ nuclei-free boxes with centers at $\mathbf{L}^{(k+1)},\ldots,\mathbf{L}^{(N)}$. The Coulomb potential being of long range, interactions occur between particles in the different boxes as well.

Due to the localizations of the functions $\phi_j(\mathbf{x},\sigma)$, as described above, the electrons are well separated, and we may write

$$|\mathbf{x}_i - \mathbf{x}_j| \geqslant D/\sqrt{2}, \quad i \neq j \qquad (14.2.23)$$

and bound the repulsive e–e interaction term as

$$\sum_{i<j}^{N} \frac{e^2}{|\mathbf{x}_i - \mathbf{x}_j|} \leqslant \frac{\sqrt{2}}{D} e^2 \sum_{i<j}^{N} (1). \qquad (14.2.24)$$

From (14.2.22), (14.2.20), $|\mathbf{R}_i - \mathbf{R}_j| > D$ for $i \neq j$, and we have the inequality

$$\sum_{i<j}^{k} \frac{Z_i Z_j e^2}{|\mathbf{R}_i - \mathbf{R}_j|} < \frac{e^2}{D} \sum_{i<j}^{k} Z_i Z_j. \tag{14.2.25}$$

Finally, we use the conservative bound

$$-\sum_{i=1}^{N}\sum_{j=1}^{k} \frac{Z_j e^2}{|\mathbf{x}_i - \mathbf{R}_j|} \leqslant -\sum_{i=1}^{k}\sum_{j=1}^{k} \frac{Z_j e^2}{|\mathbf{x}_i - \mathbf{R}_j|}$$

$$\leqslant -\sum_{i=1}^{k} \frac{Z_i e^2}{|\mathbf{x}_i - \mathbf{R}_i|}$$

$$= -\sum_{i=1}^{k} \frac{Z_i e^2}{|\mathbf{x}_i - \mathbf{L}^{(i)}|} \tag{14.2.26}$$

to obtain

$$\langle \Psi|H|\Psi\rangle \leqslant \left\langle \Phi \left| \sum_{i=1}^{N} \frac{\mathbf{p}_i^2}{2m} \right| \Phi \right\rangle - \sum_{i=1}^{k} Z_i e^2 \int \frac{d^3\mathbf{x}}{|\mathbf{x} - \mathbf{L}^{(i)}|} \phi_i^2(\mathbf{x})$$

$$+ \frac{e^2}{D}\left[\sqrt{2}\sum_{i<j}^{N}(1) + \sum_{i<j}^{k} Z_i Z_j\right]. \tag{14.2.27}$$

The kinetic energy part is explicitly given by

$$\left\langle \Phi \left| \sum_{i=1}^{N} \frac{\mathbf{p}_i^2}{2m} \right| \Phi \right\rangle = N \frac{3\hbar^2}{2m} \left(\frac{\pi}{2L}\right)^2 \tag{14.2.28}$$

and

$$\int \frac{d^3\mathbf{x}}{|\mathbf{x} - \mathbf{L}^{(i)}|} \phi_i^2(\mathbf{x}) = \int \frac{d^3\mathbf{x}}{|\mathbf{x}|} \phi^2(\mathbf{x}) \geqslant \frac{1}{\sqrt{3}L} \tag{14.2.29}$$

since $|\mathbf{x}| \leqslant \sqrt{3}L$ in the latter integral.

All told, we obtain

$$\langle \Psi|H|\Psi\rangle \leqslant \frac{3\hbar^2\pi^2}{8m}\frac{N}{L^2} - \frac{e^2}{\sqrt{3}L}N + \frac{e^2}{D}\left[\sqrt{2}\sum_{i<j}^{N}(1) + \sum_{i<j}^{k} Z_i Z_j\right]. \tag{14.2.30}$$

Optimization over L gives

$$L = \frac{3\sqrt{3}}{4}\pi^2\left(\frac{\hbar^2}{me^2}\right) \tag{14.2.31}$$

14.2 Lower and Upper Bounds for the Ground-State Energy and ...

and leads to the bound

$$\langle \Psi | H | \Psi \rangle \leqslant -\frac{4}{9\pi^2}\left(\frac{me^4}{2\hbar^2}\right)N + \frac{e^2}{D}\left[\sqrt{2}\sum_{i<j}^{N}(1) + \sum_{i<j}^{k} Z_i Z_j\right]. \quad (14.2.32)$$

We may choose D large enough to make the second term as small as we please in comparison to the first one (e.g., make it equal to $0.00031(me^2/2\hbar^2)N$) to obtain

$$\langle \Psi | H | \Psi \rangle \leqslant -0.0450 \left(\frac{me^4}{2\hbar^2}\right) N. \quad (14.2.33)$$

Since Φ does not necessarily coincide with the ground-state wavefunction, and the configuration positions of the nuclei does not necessarily correspond to the lowest possible energy, (14.2.33) leads to an upper bound for E_N:

$$E_N \leqslant -0.0450 \left(\frac{me^4}{2\hbar^2}\right) N. \quad (14.2.34)$$

with the upper bound having the same power of N as the lower bound.

As an estimation, the coefficient 0.0450 in (14.2.34) may be significantly increased. For example, and quite formally, however, we may consider the following infinitely separated N clusters: k ions (atoms), each in the ground-state, of nuclear charges $Z_1|e|, \ldots, Z_k|e|$ having each one electron, and $(N-k)$ free electrons with vanishingly small kinetic energies. Formally, the ground-state of such a system is $-\sum_{i=1}^{k} Z_i^2 me^4/2\hbar^2$, and since $Z_i^2 \geqslant Z_i$, we obtain

$$E_N \leqslant -\left(\frac{me^4}{2\hbar^2}\right) N. \quad (14.2.35)$$

thus increasing the above coefficient to one.

Note that the lower bound in (14.2.13), and the upper bound $-\sum_{i=1}^{k} Z_i^2 me^4/2\hbar^2$ given above (14.2.35) imply the following interesting conclusion. Suppose for some q, $2 \leqslant q \ll N$, $Z_1 = \ldots = Z_q, Z_{q+1} = \ldots = Z_k = 0$, then the ground-state energy will grow not slower than $(-1)N^2$. Stability then implies that as more and more matter is put together, thus increasing the number N of electrons, the number k of nuclei in such matter, as separate clusters, would necessarily increase and not arbitrarily fuse together and their individual charges remain bounded. That is, technically, as $N \to \infty$, then stability implies that $k \to \infty$ as well, and no nuclei may be found in matter that would carry arbitrary large portions of the total positive charge available. With such bounded positive charges, (14.2.16), (14.2.35) provide bounds linear in N for large N. [Of course Z is bounded in nature.]

14.3 Investigation of the High-Density Limit for Matter and Its Stability

This section addresses the important physical problem of investigating rigorously the high-density limit of matter and of its spatial extension, as a function of N, as the amount of matter is increased in the light of our analysis in §3.1 (see (3.1.27)–(3.1.37)). In reference to this analysis, we recall the words of Paul Ehrenfest to Wolfgang Pauli, quoted in the introduction to this chapter, regarding as to why matter occupies such a large volume of space. Our strategy of attack is the following. For a non-vanishing probability of having the electrons within a sphere of radius R, we prove rigorously that the radius of this spatial extension grows *necessarily* not slower than $N^{1/3}$ for large N. (Manoukian and Sirininlakul (2005)). An explicit quantitative statement will be also made regarding this inflation of matter. In passing, the result embodied in this investigation immediately provides also a lower bound to the average spatial extension of matter as a function of N.

14.3.1 Upper Bound of the Average Kinetic Energy of Electrons in Matter

Let $|\Psi(m)\rangle$ denote a strictly negative-energy state of matter, not necessarily the ground-state,

$$-\varepsilon_N[m] \leqslant \langle \Psi(m)|H|\Psi(m)\rangle < 0 \tag{14.3.1}$$

where $-\varepsilon_N[m] = E_N < 0$ is the ground-state energy, and we have emphasized its dependence on the mass m of the electron.

By definition of the ground-state energy, the state $|\Psi(m/2)\rangle$ cannot lead for $\langle \Psi(m/2)|H|\Psi(m/2)\rangle$ a numerical value lower than $-\varepsilon_N[m]$. That is,

$$-\varepsilon_N[m] \leqslant \langle \Psi(m/2)|H|\Psi(m/2)\rangle \tag{14.3.2}$$

where we note that the interaction part V of the Hamiltonian H in (14.1) is not explicitly dependent on m:

$$V = -\sum_{i=1}^{N}\sum_{j=1}^{k}\frac{Z_j e^2}{|\mathbf{x}_i - \mathbf{R}_j|} + \sum_{i<j}^{N}\frac{e^2}{|\mathbf{x}_i - \mathbf{x}_j|} + \sum_{i<j}^{k}\frac{Z_i Z_j e^2}{|\mathbf{R}_i - \mathbf{R}_j|}. \tag{14.3.3}$$

Accordingly (14.3.2) implies that

$$-\varepsilon_N[2m] \leqslant \left\langle \Psi(m)\left|\left(\sum_{i=1}^{N}\frac{\mathbf{p}_i^2}{4m} + V\right)\right|\Psi(m)\right\rangle. \tag{14.3.4}$$

Upon writing, trivially,

14.3 Investigation of the High-Density Limit for Matter and Its Stability

$$\sum_{i=1}^{N} \frac{\mathbf{p}_i^2}{2m} + V = \sum_{i=1}^{N} \frac{\mathbf{p}_i^2}{4m} + \left(\sum_{i=1}^{N} \frac{\mathbf{p}_i^2}{4m} + V \right) \quad (14.3.5)$$

the extreme right-hand of the inequality (14.3.1) then leads to

$$\left\langle \Psi(m) \left| \sum_{i=1}^{N} \frac{\mathbf{p}_i^2}{4m} \right| \Psi(m) \right\rangle < - \left\langle \Psi(m) \left| \left(\sum_{i=1}^{N} \frac{\mathbf{p}_i^2}{4m} + V \right) \right| \Psi(m) \right\rangle \quad (14.3.6)$$

which upon multiplying by two, (14.3.4) gives

$$\left\langle \Psi(m) \left| \sum_{i=1}^{N} \frac{\mathbf{p}_i^2}{2m} \right| \Psi(m) \right\rangle < 2\varepsilon_N[2m] \quad (14.3.7)$$

for all states $|\Psi(m)\rangle$ such that (14.3.1) is true including the ground-state.

Thus from (14.3.7), (14.2.15), (14.2.4), we have the following bounds for the expectation value T of the total kinetic energy of all the electrons in such states

$$\frac{3}{5} \left(\frac{3\pi}{4} \right)^{2/3} \frac{\hbar^2}{2m} \int d^3\mathbf{x}\, \rho^{5/3}(\mathbf{x}) \leqslant T < 16.6208 \left(\frac{me^4}{\hbar^2} \right) N \left[1 + Z^{2/3} \right]^2. \quad (14.3.8)$$

Now we are ready for the main investigation of this section.

14.3.2 Inflation of Matter

Let $\mathbf{x}$ denote the position of an electron relative, for example, to the center of mass of the nuclei. We define the set function

$$\chi_R(\mathbf{x}) = \begin{cases} 1, & \text{if } \mathbf{x} \text{ lies within a sphere of radius } R \\ 0, & \text{otherwise.} \end{cases} \quad (14.3.9)$$

We are interested in the expression

$$\sum_{\sigma_1,\ldots,\sigma_N} \int \left(\prod_{i=1}^{N} d^3\mathbf{x}_i\, \chi_R(\mathbf{x}_i) \right) |\Psi(\mathbf{x}_1\sigma_1,\ldots,\mathbf{x}_N\sigma_N)|^2$$

$$= \text{Prob}\left[|\mathbf{x}_1| \leqslant R, \ldots, |\mathbf{x}_N| \leqslant R \right] \quad (14.3.10)$$

which gives the probability of finding all the electrons within the sphere of radius R.

Clearly,

$$\text{Prob}\left[|\mathbf{x}_1| \leqslant R, \ldots, |\mathbf{x}_N| \leqslant R \right] \leqslant \text{Prob}\left[|\mathbf{x}_1| \leqslant R, \ldots, |\mathbf{x}_j| \leqslant R \right]$$

$$\leqslant \ldots \leqslant \text{Prob}\left[|\mathbf{x}_1| \leqslant R \right] = \frac{1}{N} \int d^3\mathbf{x}\, \chi_R(\mathbf{x}) \rho(\mathbf{x}) \quad (14.3.11)$$

for $j < N$, with $\rho(\mathbf{x})$ given in (14.2.2).

By Hölder's inequality in Appendix II,

$$\int d^3\mathbf{x}\, \chi_R(\mathbf{x})\rho(\mathbf{x}) \leqslant \left(\int d^3\mathbf{x}\, \rho^{5/3}(\mathbf{x})\right)^{3/5} \left(\int d^3\mathbf{x}\, \chi_R(\mathbf{x})\right)^{2/5} \qquad (14.3.12)$$

where $\chi_R^{5/2}(\mathbf{x}) = \chi_R(\mathbf{x})$,

$$\int d^3\mathbf{x}\, \chi_R(\mathbf{x}) = v_R \qquad (14.3.13)$$

denotes the volume in which the electrons are confined.

Hence, in particular, (14.3.11) gives

$$\text{Prob}\,[|\mathbf{x}_1| \leqslant R, \ldots, |\mathbf{x}_N| \leqslant R] \leqslant \text{Prob}\,[|\mathbf{x}_1| \leqslant R]$$

$$\leqslant \frac{(v_R)^{2/5}}{N} \left(\int d^3\mathbf{x}\, \rho^{5/3}(\mathbf{x})\right)^{3/5} \qquad (14.3.14)$$

and (14.3.8) finally leads to the simple bound

$$\text{Prob}\,[|\mathbf{x}_1| \leqslant R, \ldots, |\mathbf{x}_N| \leqslant R] \left(\frac{N}{v_R}\right)^{2/5} < \left(\frac{1}{a_0^3}\right)^{2/5} 10\left[1 + Z^{2/3}\right]^{6/5} \qquad (14.3.15)$$

where $a_0 = \hbar^2/me^2$ is the Bohr radius and Z is the maximum of the nuclear charges.

We immediately infer from (14.3.15) the inescapable fact that *necessarily*, for a non-vanishing probability of having the electrons within a sphere of radius R, the corresponding volume v_R grows not any slower than the first power of N for $N \to \infty$, since otherwise the left-hand side of (14.3.15) would go to infinity in this limit and would be in contradiction with the finite upper bound in (14.3.15). That is, *necessarily, the radius R of spatial extension of matter grows not any slower than $N^{1/3}$ for $N \to \infty$*. (Manoukian and Sirininlakul (2005)). No wonder why matter occupies so large a volume!

In turn, one may infer from (14.3.15), that the infinite density limit $N/v_R \to \infty$, i.e., of the system collapsing onto itself, does not occur, as the probability on the left-hand side of (14.3.15) goes to zero in such a limit upon multiplying the latter equation first by $(v_R/N)^{2/5}$.

From (14.3.14), (14.3.8), we may also write

$$\frac{1}{N}\int d^3\mathbf{x}\, \chi_R(\mathbf{x})\,\rho(\mathbf{x}) = \text{Prob}\,[|\mathbf{x}| \leqslant R] < \left(\frac{v_R}{N}\right)^{2/5} \left(\frac{1}{a_0^3}\right)^{2/5} 10\left[1 + Z^{2/3}\right]^{6/5}. \qquad (14.3.16)$$

This immediately leads a lower bound to the expectation value

$$\left\langle \sum_{i=1}^{N} \frac{|\mathbf{x}_i|}{N} \right\rangle = \sum_{\sigma_1,\ldots,\sigma_N} \int d^3\mathbf{x}_1 \ldots d^3\mathbf{x}_N \left(\sum_{i=1}^{N} \frac{|\mathbf{x}_i|}{N}\right) |\Psi(\mathbf{x}_1\sigma_1,\ldots,\mathbf{x}_N\sigma_N)|^2$$

$$= \frac{1}{N}\int d^3\mathbf{x}\,|\mathbf{x}|\rho(\mathbf{x}) \tag{14.3.17}$$

for a measure of the extension of matter. Using the facts that

$$\frac{1}{N}\int d^3\mathbf{x}\,|\mathbf{x}|\rho(\mathbf{x}) \geqslant \frac{1}{N}\int_{|\mathbf{x}|>R} d^3\mathbf{x}\,|\mathbf{x}|\rho(\mathbf{x}) \geqslant \frac{R}{N}\int_{|\mathbf{x}|>R} d^3\mathbf{x}\,\rho(\mathbf{x})$$

$$= R\,\text{Prob}\,[|\mathbf{x}|>R] \tag{14.3.18}$$

$$\text{Prob}\,[|\mathbf{x}|>R] = 1 - \text{Prob}\,[|\mathbf{x}|\leqslant R] \tag{14.3.19}$$

$v_R = 4\pi R^3/3$, and (14.3.16) we obtain

$$\left\langle \sum_{i=1}^{N}\frac{|\mathbf{x}_i|}{N}\right\rangle > R\left[1 - \left(\frac{R^3}{Na_0^3}\right)^{2/5}\left(\frac{4\pi}{3}\right)^{2/5}10\left[1+Z^{2/3}\right]^{6/5}\right]. \tag{14.3.20}$$

Upon optimizing the right-hand side of the above inequality over R, this gives

$$R = \left(\frac{1}{22}\right)^{5/6}\left(\frac{3}{4\pi}\right)^{1/3}a_0 N^{1/3}\frac{1}{[1+Z^{2/3}]} \tag{14.3.21}$$

leading for (14.3.20) the explicit bound

$$\left\langle \sum_{i=1}^{N}\frac{|\mathbf{x}_i|}{N}\right\rangle > 0.02575\,a_0\frac{N^{1/3}}{[1+Z^{2/3}]}. \tag{14.3.22}$$

14.4 The Collapse of "Bosonic Matter"

We derive lower and upper bounds for the ground-state energy E_N^B for "bosonic matter" consisting of N negatively charged (spin 0) bosons and N positively charged motionless bosons of charges $-|e|, +|e|$ respectively, with Coulombic interactions. That is, we consider the Hamiltonian

$$H = \sum_{i=1}^{N}\frac{\mathbf{p}_i^2}{2m_i} + V(\mathbf{x}_1,\ldots,\mathbf{x}_N;\mathbf{R}_1,\ldots,\mathbf{R}_N) \tag{14.4.1}$$

where

$$V(\mathbf{x}_1,\ldots,\mathbf{x}_N;\mathbf{R}_1,\ldots,\mathbf{R}_N)$$
$$= \sum_{i<j}^{N}\frac{e^2}{|\mathbf{x}_i-\mathbf{x}_j|} - \sum_{i=1}^{N}\sum_{j=1}^{N}\frac{e^2}{|\mathbf{x}_i-\mathbf{R}_j|} + \sum_{i<j}^{N}\frac{e^2}{|\mathbf{R}_i-\mathbf{R}_j|} \tag{14.4.2}$$

the masses $m_1, \ldots, m_N$ of the negatively charged particles are not necessarily taken to be equal. Here the $\mathbf{x}_i$, $\mathbf{R}_j$ respectively, refer to the negatively, positively charged particles.

Since $\sum_{i=1}^{N} \mathbf{p}_i^2/2m_i$ is a positive operator, we may consider the bounds

$$\sum_{i=1}^{N} \frac{\mathbf{p}_i^2}{2m} \leqslant \sum_{i=1}^{N} \frac{\mathbf{p}_i^2}{2m_i} \leqslant \sum_{i=1}^{N} \frac{\mathbf{p}_i^2}{2\underline{m}} \tag{14.4.3}$$

where $m = \max_i(m_i)$, $\underline{m} = \min_i(m_i)$, and it is sufficient to consider, respectively, the Hamiltonians

$$H_{\mathrm{L}} = \sum_{i=1}^{N} \frac{\mathbf{p}_i^2}{2m} + V(\mathbf{x}_1, \ldots, \mathbf{x}_N; \mathbf{R}_1, \ldots, \mathbf{R}_N) \; (\leqslant H) \tag{14.4.4}$$

and

$$H_{\mathrm{U}} = \sum_{i=1}^{N} \frac{\mathbf{p}_i^2}{2\underline{m}} + V(\mathbf{x}_1, \ldots, \mathbf{x}_N; \mathbf{R}_1, \ldots, \mathbf{R}_N) \; (\geqslant H) \tag{14.4.5}$$

in determining the lower and upper bounds for the ground-state energy E_N^{B}. The analyses will give rise to an $N^{5/3}$ law[12] for such systems.

14.4.1 A Lower Bound

We first reconsider the TF theory in §13.1 when the electrons are replaced by (spin 0) bosons.

In this case, the density in (13.1.1) becomes replaced by (see also (9.8.5))

$$n_{\mathrm{B}}(\mathbf{x}) = \frac{1}{6\pi^2} \left(\frac{-2mV(r)}{\hbar^2}\right)^{3/2} \tag{14.4.6}$$

with $\xi = 0$. The TF energy functional (13.1.6) becomes then simply replaced by

$$F^{\mathrm{B}}[n_{\mathrm{B}}] = (3\pi^2)^{5/3} \frac{\hbar^2}{10\pi^2} \left(\frac{2^{2/3}}{m}\right) \int d^3\mathbf{x} \, (n_{\mathrm{B}}(\mathbf{x}))^{5/3} - Ze^2 \int \frac{d^3\mathbf{x}}{|\mathbf{x}|} n_{\mathrm{B}}(\mathbf{x})$$

$$+ \frac{e^2}{2} \int d^3\mathbf{x} \, d^3\mathbf{x}' \, n_{\mathrm{B}}(\mathbf{x}) \frac{1}{|\mathbf{x} - \mathbf{x}'|} n_{\mathrm{B}}(\mathbf{x}'). \tag{14.4.7}$$

That is, in the TF theory for bosons we simply have to replace m by $m/2^{2/3}$ in the TF theory for electrons. Hence, in particular, $E_{\mathrm{TF}}^{\mathrm{B}}(Z)$ for boson, we have from (13.1.43):

---
[12] Dyson and Lenard (1967); Lenard and Dyson (1968); Lieb (1976, 1979); Manoukian and Muthaporn (2003b).

14.4 The Collapse of "Bosonic Matter"

$$E^{\text{B}}_{\text{TF}}(Z) = \frac{1}{2^{2/3}} E_{\text{TF}}(Z). \tag{14.4.8}$$

In the TF-like energy functional in (14.1.1), likewise, we replace m by $m/2^{2/3}$.

For the expectation value T of the total kinetic energy for bosons, we may use, for example, the bound in (4.6.26)

$$\frac{3}{5}\left(\frac{3\pi}{2N}\right)^{2/3} \frac{\hbar^2}{2m} \int d^3 x \, \rho^{5/3}(\mathbf{x}) \leqslant T \tag{14.4.9}$$

where

$$\rho(\mathbf{x}) = N \int d^3 x_2 \ldots d^3 x_N \, |\Psi(\mathbf{x}, \mathbf{x}_2, \ldots, \mathbf{x}_N)|^2. \tag{14.4.10}$$

All told, we then have from (14.4.10), (14.4.8), for the basic inequality corresponding to the present case,

$$\langle \Psi | H_{\text{L}} | \Psi \rangle \geqslant \frac{3}{5}\left(\frac{3\pi}{2N}\right)^{2/3} \frac{\hbar^2}{2m} \int d^3 x \, \rho^{5/3}(\mathbf{x})$$

$$- (3\pi^2)^{5/3} \frac{\hbar^2}{10\pi^2 \beta} \left(\frac{2^{2/3}}{m}\right) \int d^3 x \, \rho^{5/3}(\mathbf{x}) - \sum_{j=1}^{k} Z_j e^2 \int d^3 x \, \frac{\rho(\mathbf{x})}{|\mathbf{x} - \mathbf{R}_j|}$$

$$+ \frac{e^2}{2} \int d^3 x \, d^3 x' \, \rho(\mathbf{x}) \frac{1}{|\mathbf{x} - \mathbf{x}'|} \rho(\mathbf{x}') + \sum_{i<j}^{k} \frac{Z_i Z_j e^2}{|\mathbf{R}_i - \mathbf{R}_j|} + \frac{\beta N}{2^{2/3}} E_{\text{TF}}(1)$$
$$\tag{14.4.11}$$

where now we have to make the replacements $Z_i \to 1$, $k \to N$, but we will do that later for greater generality.

Upon setting

$$\frac{\frac{3}{5}\left(\frac{3\pi}{2N}\right)^{2/3} - \frac{(3\pi^2)^{5/3}}{5\pi^2 \beta} 2^{2/3}}{\frac{(3\pi^2)^{5/3}}{5\pi^2} 2^{2/3}} = \frac{1}{\beta''} \tag{14.4.12}$$

we obtain instead of (14.2.10), (14.2.11) in the present case

$$\langle \Psi | H_{\text{L}} | \Psi \rangle \geqslant \frac{\beta''}{2^{2/3}} E_{\text{TF}}(1) \sum_{i=1}^{k} Z_i^{7/3} + \frac{\beta N}{2^{2/3}} E_{\text{TF}}(1) \tag{14.4.13}$$

or

$$\langle \Psi | H_{\text{L}} | \Psi \rangle \geqslant \frac{1}{2^{2/3}} E_{\text{TF}}(1) \left[\beta N + \frac{\sum_{i=1}^{k} Z_i^{7/3}}{\left(\left(\frac{1}{4\pi N}\right)^{2/3} - \frac{1}{\beta}\right)} \right]. \tag{14.4.14}$$

Optimization over β, this gives

$$\beta = (4\pi N)^{2/3} \left[1 + \left(\frac{\sum_{i=1}^{k} Z_i^{7/3}}{N} \right)^{1/2} \right] \quad (14.4.15)$$

leading to the bound

$$\langle \Psi | H | \Psi \rangle \geqslant E_{\mathrm{TF}}(1)(2\pi)^{2/3} N^{5/3} \left[1 + \left(\sum_{i=1}^{k} \frac{Z_i^{7/3}}{N} \right)^{1/2} \right]^2 \quad (14.4.16)$$

or from (13.1.43), (13.1.45)

$$\langle \Psi | H_{\mathrm{L}} | \Psi \rangle \geqslant -5.2352 \left(\frac{me^4}{2\hbar^2} \right) N^{5/3} \left[1 + \left(\frac{\sum_{i=1}^{k} Z_i^{7/3}}{N} \right)^{1/2} \right]^2. \quad (14.4.17)$$

[The numerical factor 5.2352 may be decreased[13] further but we will not attempt to do so here.]

Accordingly, for the bosonic system at hand we have ($Z_i \to 1$, $k \to N$)

$$E_N^{\mathrm{B}} = -20.941 \left(\frac{me^4}{2\hbar^2} \right) N^{5/3}. \quad (14.4.18)$$

14.4.2 An Upper Bound

We consider arbitrary $N \geqslant 8$. Since $(N/8)^{1/3}$ is some real number, it may be written as

$$(N/8)^{1/3} = n + \varepsilon \quad (14.4.19)$$

where n is a strictly positive integer and $0 \leqslant \varepsilon < 1$. Let

$$8n^3 \equiv \kappa \quad (14.4.20)$$

then we have the useful bounds

$$\frac{2\kappa\varepsilon}{n} \left(1 + \frac{\varepsilon}{n} \right) \leqslant (N - \kappa) \leqslant \frac{4\kappa\varepsilon}{n} \left(1 + \frac{\varepsilon}{n} \right) \quad (14.4.21)$$

$$1 \leqslant \left(1 + \frac{\varepsilon}{n} \right) < 2. \quad (14.4.22)$$

---

[13] See Manoukian and Sirininlakul (2004).

14.4 The Collapse of "Bosonic Matter"

We introduce the trial wavefunction $\Psi(\mathbf{x}_1,\ldots,\mathbf{x}_N)$ by considering the following construction. We put[14] κ of the negatively charged particles and κ of the positively charged particles in a box of sides $2L$, $2L$, $2L$ centered at the origin of the coordinate system. The remaining particles are placed as follows. We consider $(N-\kappa)$ non-overlapping boxes, each of sides $2L_0$, $2L_0$, $2L_0$ which also do not overlap with the box at the origin. The centers of these $(N-\kappa)$ boxes are defined by the tip of the vectors $\mathbf{L}^{(1)},\ldots,\mathbf{L}^{(N-\kappa)}$, (see Figure 14.1), where

$$\mathbf{L}^{(j)} = jD(1,1,1), \quad j=1,\ldots,N-\kappa. \tag{14.4.23}$$

To ensure that all the $(N-\kappa)+1$ boxes are non-overlapping, it is sufficient, but not necessary, to choose

$$6L \leqslant 6L_0 \leqslant D. \tag{14.4.24}$$

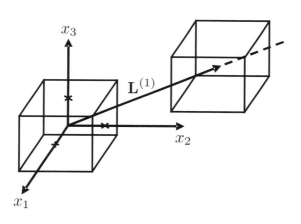

Fig. 14.1. The figure shows the regions (non-overlapping boxes) where particles are localized. The centers of the boxes are situated at $\mathbf{0}, \mathbf{L}^{(1)},\ldots,\mathbf{L}^{(N-\kappa)}$ with the latter $(N-\kappa)$ vectors being along the vector $(1,1,1)$. The sides of the box at the origin are equal to $2L, 2L, 2L$, while the ones of the other $(N-\kappa)$ boxes are $2L_0, 2L_0, 2L_0$, where $L_0 \geqslant L$. The Coulomb potential being of long range, there are non-trivial interactions between particles in different boxes as well.

The numerical factor 6 is chosen for convenience to simplify the algebra.

---

[14] The following analysis is based on: Manoukian and Muthaporn (2003b) deriving the upper bound for all $N \geqslant 8$ and is an extension of the classic work of Lieb (1979) for N restricted to $N = 8, 64, 216, \ldots$.

Now we put $(N-\kappa)$ of the positively charged particles at the *centers* of the $(N-\kappa)$ boxes introduced above by choosing

$$\mathbf{R}_{\kappa+1} = \mathbf{L}^{(1)}, \ldots, \mathbf{R}_N = \mathbf{L}^{(N-\kappa)} \qquad (14.4.25)$$

and we put one negatively charged particle in each of the $(N-\kappa)$ boxes. $\mathbf{R}_1, \ldots, \mathbf{R}_\kappa$ are chosen to lie in the first box, with the latter centered at the origin, as will be discussed later.

To localize the negatively charged particles in the $(N-\kappa)+1$ boxes as discussed above we introduce the following normalized trial wavefunction:

$$\Psi(\mathbf{x}_1, \ldots, \mathbf{x}_N) = \frac{1}{\sqrt{N!\kappa!}} \sum_\pi \phi(\mathbf{x}(\pi_1)) \ldots \phi(\mathbf{x}(\pi_\kappa))$$

$$\times \psi_1(\mathbf{x}(\pi_{\kappa+1})) \ldots \psi_{N-\kappa}(\mathbf{x}(\pi_N)) \qquad (14.4.26)$$

where the sum is over all permutations $\{\pi_1, \ldots, \pi_N\}$ of $\{1, \ldots, N\}$ such that

$$\int d^3\mathbf{x}\, \psi_i^*(\mathbf{x})\psi_j(\mathbf{x}) = \delta_{ij}, \quad \int d^3\mathbf{x}\, \phi^*(\mathbf{x})\psi_i(\mathbf{x}) = 0, \quad \int d^3\mathbf{x}\, |\phi(\mathbf{x})|^2 = 1. \qquad (14.4.27)$$

Since Ψ does not necessarily coincide with the ground-state wavefunction, we have for the ground-state energy E_N^B

$$E_N^\mathrm{B} \leqslant \langle \Psi | H_\mathrm{U} | \Psi \rangle. \qquad (14.4.28)$$

We choose the following localized single-particle trial wavefunctions consistent with the above construction by placing the negatively charged particles in the $(N-\kappa)+1$ boxes:

$$\phi(\mathbf{x}) = \prod_i \left(\frac{1}{\sqrt{L}} \cos\left(\frac{\pi x_i}{2L}\right) \right) \equiv \phi_L(\mathbf{x}), \quad |x_i| \leqslant L \qquad (14.4.29)$$

$i = 1, 2, 3$ and is zero otherwise, and for $j = 1, \ldots, N-\kappa$,

$$\psi_j(\mathbf{x}) = \prod_i \left(\frac{1}{\sqrt{L_0}} \cos\left(\frac{\pi\left(x_i - L_i^{(j)}\right)}{2L_0}\right) \right) \equiv \phi_{L_0}(\mathbf{x} - \mathbf{L}^{(j)}), \quad |x_i - L_i^{(j)}| \leqslant L_0 \qquad (14.4.30)$$

and are zero otherwise, $i = 1, 2, 3$.

Since the intervals $\{-L \leqslant x_i \leqslant L\}$, $\{jD - L_0 \leqslant x_i \leqslant jD + L_0\}$ for $j = 1, \ldots, N-\kappa$, $(i = 1, 2, 3)$ are all disjoint, the wavefunctions $\phi(\mathbf{x})$, $\psi_j(\mathbf{x})$ are non-overlapping and automatically satisfy (14.4.27), with orthogonality relations holding with respect to *each* component x_i of $\mathbf{x}$.

The single-particle average kinetic energies are given by

$$T^0 = \frac{\hbar^2}{2m} \int d^3\mathbf{x} |\boldsymbol{\nabla}\phi(\mathbf{x})|^2 = \frac{3\hbar^2}{2m}\left(\frac{\pi}{2L}\right)^2 \qquad (14.4.31)$$

$$T_j = \frac{\hbar^2}{2m}\int d^3x |\nabla\psi_j(\mathbf{x})|^2 = \frac{3\hbar^2}{2m}\left(\frac{\pi}{2L_0}\right)^2 \equiv T^1 \qquad (14.4.32)$$

and for the multi-particle state

$$\sum_{j=1}^{N}\frac{\hbar^2}{2m}\int d^3x_1\ldots d^3x_N |\nabla_j\Psi(\mathbf{x}_1,\ldots\mathbf{x}_N)|^2 = [\kappa T^0 + (N-\kappa)T^1] \qquad (14.4.33)$$

as is easily checked since the functions in each of the products in (14.4.26) are orthogonal with respect to each of the components of $\mathbf{x}$.

A detailed calculation gives

$$\langle\Psi|H_U|\Psi\rangle = [\kappa T^0 + (N-\kappa)T^1] + \langle V_1\rangle + \langle V_2\rangle + \sum_{i<j}^{N}\frac{e^2}{|\mathbf{R}_i-\mathbf{R}_j|} \qquad (14.4.34)$$

where

$$\langle V_1\rangle = -e^2\sum_{j=1}^{N}\int d^3x\left[\frac{\kappa}{|\mathbf{x}-\mathbf{R}_j|}\phi_L^2(\mathbf{x}) + \left(\frac{1}{|\mathbf{x}+\mathbf{L}^{(1)}-\mathbf{R}_j|}\right.\right.$$
$$\left.\left.+\ldots+\frac{1}{|\mathbf{x}+\mathbf{L}^{(N-\kappa)}-\mathbf{R}_j|}\right)\phi_{L_0}^2(\mathbf{x})\right] \qquad (14.4.35)$$

$$\langle V_2\rangle = \frac{e^2}{2}\kappa(\kappa-1)\int d^3x\, d^3x'\, \phi_L^2(\mathbf{x})\frac{1}{|\mathbf{x}-\mathbf{x}'|}\phi_L^2(\mathbf{x}')$$
$$+ e^2\kappa\sum_{j=1}^{N-\kappa}\int d^3x\, d^3x'\, \phi_L^2(\mathbf{x})\frac{1}{|\mathbf{x}-\mathbf{x}'-\mathbf{L}^{(j)}|}\phi_{L_0}^2(\mathbf{x}')$$
$$+ e^2\kappa\sum_{i<j}^{N-\kappa}\int d^3x\, d^3x'\, \phi_{L_0}^2(\mathbf{x})\frac{1}{|\mathbf{x}-\mathbf{x}'+\mathbf{L}^{(i)}-\mathbf{L}^{(j)}|}\phi_{L_0}^2(\mathbf{x}'). \qquad (14.4.36)$$

By noting the overall *negative* sign of $\langle V_1\rangle$, we may bound the latter as

$$\langle V_1\rangle \leqslant -e^2\sum_{j=1}^{\kappa}\int d^3x\frac{\kappa}{|\mathbf{x}-\mathbf{R}_j|}\phi_L^2(\mathbf{x}) - e^2(N-\kappa)\int d^3x\frac{1}{|\mathbf{x}|}\phi_{L_0}^2(\mathbf{x}) \qquad (14.4.37)$$

where we have conveniently chosen an upper bound with the summation going up to κ, and in writing the second term, we have chosen only the terms with $j = \kappa+1,\ldots,j=N$, respectively, for the $(N-\kappa)$ terms multiplying $\phi_{L_0}^2$ and used, in the process, (14.4.25).

Since $|\mathbf{x}| \leqslant \sqrt{3}L_0$ in the last integral in (14.4.37), we obtain

$$\langle V_1 \rangle \leqslant -e^2 \sum_{j=1}^{\kappa} \int d^3\mathbf{x} \frac{\kappa}{|\mathbf{x} - \mathbf{R}_j|} \phi_L^2(\mathbf{x}) - \frac{e^2(N-\kappa)}{\sqrt{3}L_0}. \qquad (14.4.38)$$

To derives an upper bound for $\langle V_2 \rangle$, we note that

$$\left|\mathbf{L}^{(j)}\right| \geqslant \sqrt{3}D, \quad \left|\mathbf{L}^{(i)} - \mathbf{L}^{(j)}\right| \geqslant \sqrt{3}D \qquad (14.4.39)$$

for $i \neq j$, and because of the vanishing properties of $\phi_L^2(\mathbf{x})$, $\phi_{L_0}^2(\mathbf{x})$ outside the corresponding intervals in questions, we obtain in reference to the second and third set of integrals

$$\left|\mathbf{x} - \mathbf{x}' - \mathbf{L}^{(j)}\right| \geqslant \left|\mathbf{L}^{(j)}\right| \left(1 - \frac{4\sqrt{3}L_0}{\left|\mathbf{L}^{(j)}\right|}\right)^{1/2} \geqslant D \qquad (14.4.40)$$

where we have used (14.4.24), and similarly

$$\left|\mathbf{x} - \mathbf{x}' + \mathbf{L}^{(i)} - \mathbf{L}^{(j)}\right| \geqslant D. \qquad (14.4.41)$$

Accordingly, the second and third set of integrals on the right-hand of (14.4.36), combined, may be bounded above by

$$\frac{e^2}{D}\left[\kappa(N-\kappa) + \frac{(N-\kappa)(N-\kappa-1)}{2}\right] \qquad (14.4.42)$$

thus obtaining

$$\langle V_2 \rangle \leqslant \frac{e^2}{2}\kappa(\kappa-1)\int d^3\mathbf{x}\, d^3\mathbf{x}'\, \phi_L^2(\mathbf{x})\frac{1}{|\mathbf{x}-\mathbf{x}'|}\phi_L^2(\mathbf{x}') + \frac{e^2(N-\kappa)(N+\kappa-1)}{2D}. \qquad (14.4.43)$$

Finally we use the bounds

$$|\mathbf{R}_i - \mathbf{R}_j| \geqslant D \qquad (14.4.44)$$

for $j = \kappa+1, \ldots, N$ and all i such that $1 \leqslant i < j$. This follows since for $i = \kappa+1, \ldots, N$ (such that $i < j$) we may use the equalities in (14.4.25), while for $i = 1, \ldots, \kappa$, $|\mathbf{R}_i| \leqslant \sqrt{3}L$ and (14.4.44) follows. The decomposition

$$\sum_{i<j}^{N} \frac{1}{|\mathbf{R}_i - \mathbf{R}_j|} = \sum_{i<j}^{\kappa} \frac{1}{|\mathbf{R}_i - \mathbf{R}_j|} + \sum_{j=\kappa+1}^{N}\sum_{i=1}^{j-1} \frac{1}{|\mathbf{R}_i - \mathbf{R}_j|} \qquad (14.4.45)$$

then leads to

$$\langle \Psi | H_\mathrm{U} | \Psi \rangle \leqslant \kappa T^0 + \langle H_1 \rangle + (N-\kappa)\left[T^1 + \frac{e^2(N+\kappa-1)}{D} - \frac{e^2}{\sqrt{3}L_0}\right] \qquad (14.4.46)$$

where

$$\langle H_1 \rangle = -\kappa e^2 \sum_{j=1}^{\kappa} \int \frac{d^3\mathbf{x}}{|\mathbf{x} - \mathbf{R}_j|} \phi_L^2(\mathbf{x})$$

$$+ \frac{e^2}{2} \kappa(\kappa - 1) \int d^3\mathbf{x}\, d^3\mathbf{x}'\, \phi_L^2(\mathbf{x}) \frac{1}{|\mathbf{x} - \mathbf{x}'|} \phi_L^2(\mathbf{x}') + e^2 \sum_{i<j}^{\kappa} \frac{1}{|\mathbf{R}_i - \mathbf{R}_j|}. \quad (14.4.47)$$

In the appendix to this section, the following upper bound for $\langle H_1 \rangle$ is obtained

$$\langle H_1 \rangle \leqslant -\frac{e^2}{6L} \kappa^{4/3} \quad (14.4.48)$$

by appropriately fixing the positions of the κ positive charges. Again this latter configuration does not necessarily correspond to the lowest possible energy, and hence we obtain an upper bound to E_N^{B} given by

$$E_N^{\mathrm{B}} \leqslant \frac{3\hbar^2}{8\underline{m}} \frac{\pi^2}{L^2} \kappa - \frac{e^2}{6L} \kappa^{4/3}$$

$$+ (N - \kappa) \left[\frac{3\hbar^2}{8\underline{m}} \frac{\pi^2}{L_0^2} + \frac{e^2(N + \kappa - 1)}{xL_0} - \frac{e^2}{\sqrt{3}L_0} \right] \quad (14.4.49)$$

where we have set $D = xL_0$, with $x \geqslant 6$, which will be conveniently and consistently chosen.

Optimization over L and L_0, gives

$$L = \frac{9\pi^2 \hbar^2}{2\underline{m} e^2} \frac{1}{\kappa^{1/3}}, \quad L_0 = \frac{3\sqrt{3}\pi^2 \hbar^2}{2\underline{m} e^2} \frac{1}{\left[2 - 2\sqrt{3} \frac{(N + \kappa - 1)}{x}\right]} \quad (14.4.50)$$

with

$$0 < \left[2 - 2\sqrt{3} \frac{(N + \kappa - 1)}{x}\right] \leqslant \frac{\sqrt{3}}{3} \kappa^{1/2} \quad (14.4.51)$$

and with $\kappa^{1/3} \geqslant 2$, we may choose $x = 2\sqrt{3}(N + \kappa - 1)$, which is obviously larger than 6, giving

$$L_0 = \frac{3\sqrt{3}\pi^2 \hbar^2}{2\underline{m} e^2} > L. \quad (14.4.52)$$

The last term on the right-hand side of (14.4.49), involving the $(N-\kappa)$ factor, then leads to a strict negative contribution proportional to N, that is with a power of N less than that of the sum of the first two terms with L as given in (14.4.50). Hence we may further bound the right-hand side of (14.4.49) from above by the sum of the first two terms only.

Accordingly, we obtain the strict upper bound

$$E_N^{\mathrm{B}} < -\frac{me^4}{2\hbar^2}\frac{N^{5/3}}{27\pi^2}\frac{1}{(1+\varepsilon/n)^5} \qquad (14.4.53)$$

for all $N \geqslant 8$, where we have used the fact that (see (14.4.19), (14.4.20))

$$\kappa = N\left[1+\frac{\varepsilon}{n}\right]^{-3} = 8n^3. \qquad (14.4.54)$$

From (14.4.22), a conservative bound is $[1+\varepsilon/n]^5 < 2^5$ for this factor in (14.4.53). For the cases where $\varepsilon = 0$,

$$E_N^{\mathrm{B}} < -\frac{1}{27\pi^2}\left(\frac{me^4}{2\hbar^2}\right)N^{5/3}. \qquad (14.4.55)$$

More generally, for all $0 \leqslant \varepsilon < 1$, and for large bosonic systems, e.g., with $n \geqslant 4$, i.e., with $N \geqslant 512$,

$$E_N^{\mathrm{B}} < -\frac{1}{83\pi^2}\left(\frac{me^4}{2\hbar^2}\right)N^{5/3} \qquad (14.4.56)$$

and more interestingly for larger systems, e.g., with $n \geqslant 50$, i.e., for $N \geqslant 10^6$,

$$E_N^{\mathrm{B}} < -\frac{1}{30\pi^2}\left(\frac{me^4}{2\hbar^2}\right)N^{5/3}. \qquad (14.4.57)$$

A larger numerical coefficient than $1/(30\pi^2)$ of $(me^4/2\hbar^2)N^{5/3}$, may be obtained but this will not be attempted here.

We note the presence of the same power $N^{5/3}$ for the lower (14.4.18) and upper bounds (14.4.53)–(14.4.57) for E_N^{B}.

We have thus established the main objective of this section by showing the power law behavior $N^{5/3}$ for bosons, implying the collapse of such systems. It is interesting to point out that the collapse is not a characteristic of the dimensionality of (Euclidean) space and persists in arbitrary dimensions[15] (see also Problem 14.7) with $1/r$ interactions. The $N^{5/3}$ law for bosons just established corresponds to fixed positively charged particles (i.e., infinitely massive motionless point particles). As mentioned in the introductory section to this chapter this is non-academic. The problem with the positively charged particles treated dynamically with finite masses as well restricted, however, to the Coulombic interaction, gives rise to a $N^{7/5}$ power law implies again the instability of such systems and we refer the reader to the literature[16] for the derivation of this law.

---

[15] Muthaporn and Manoukian (2004a).
[16] Dyson (1967); Manoukian and Muthaporn (2002); Conlon et al. (1988). For the corresponding law in other dimensions see, Manoukian and Muthaporn (2003a).

Appendix to §14.4: Upper Bounds for $\langle H_1 \rangle$ in (14.4.47)

Consider the following integrals written in terms of dimensionless variables

$$I_1(\kappa; \mathbf{X}_1, \ldots, \mathbf{X}_\kappa) = -\kappa \sum_{j=1}^{\kappa} \int \frac{\mathrm{d}^3 \mathbf{x}}{|\mathbf{x} - \mathbf{X}_j|} f^2(\mathbf{x}) \qquad \text{(A-14.4.1)}$$

$$I_2(\kappa) = \frac{\kappa^2}{2} \int \mathrm{d}^3 \mathbf{x}\, \mathrm{d}^3 \mathbf{x}'\, f^2(\mathbf{x}) \frac{1}{|\mathbf{x} - \mathbf{x}'|} f^2(\mathbf{x}') \qquad \text{(A-14.4.2)}$$

$$I_3(\kappa; \mathbf{X}_1, \ldots, \mathbf{X}_\kappa) = \sum_{i<j}^{\kappa} \frac{1}{|\mathbf{X}_i - \mathbf{X}_j|} \qquad \text{(A-14.4.3)}$$

where $f(\mathbf{x})$ is a real function defined by

$$f(\mathbf{x}) = g(x_1) g(x_2) g(x_3) \qquad \text{(A-14.4.4)}$$

$g(x_i) \neq 0$ for $-1 \leqslant x_i \leqslant 1$, and $g(x_i) = 0$ otherwise, $g(x) = g(-x)$,

$$\int_{-1}^{1} g^2(x)\, \mathrm{d}x = 1. \qquad \text{(A-14.4.5)}$$

By choosing, in particular,

$$g(x) = \cos \frac{\pi x}{2} \qquad \text{(A-14.4.6)}$$

setting $\mathbf{X}_j = \mathbf{R}_j / L$, and noting that $\kappa(\kappa - 1) < \kappa^2$, for the coefficient of the second term in (14.4.47), we conclude that

$$\langle H_1 \rangle \leqslant \frac{e^2}{L} I(\kappa; \mathbf{X}_1, \ldots, \mathbf{X}_\kappa) \qquad \text{(A-14.4.7)}$$

where

$$I(\kappa; \mathbf{X}_1, \ldots, \mathbf{X}_\kappa) = I_1(\kappa; \mathbf{X}_1, \ldots, \mathbf{X}_\kappa) + I_2(\kappa) + I_3(\kappa; \mathbf{X}_1, \ldots, \mathbf{X}_\kappa). \qquad \text{(A-14.4.8)}$$

Therefore, to derive an upper bound for $\langle H_1 \rangle$ we may work directly with the expressions in $I(\kappa; \mathbf{X}_1, \ldots, \mathbf{X}_\kappa)$.

We partition[17] the unit interval $[0, 1]$ into n subintervals: $0 = a_0 < a_1 < \ldots a_n = 1$ such that

$$\int_{a_{j-1}}^{a_j} \mathrm{d}x\, g^2(x) = \frac{1}{2n} \qquad \text{(A-14.4.9)}$$

for $j = 1, \ldots, \kappa$. By doing so, we divide the box of sides $2, 2, 2$ into $(2n)^3 = 8n^3 \equiv \kappa$ smaller boxes.

---

[17] We follow the construction of Lieb (1979).

794 14 Quantum Physics and the Stability of Matter

Let $\alpha_j = a_j - a_{j-1}$, then we note that

$$\sum_{j=1}^{n} \alpha_j = 1. \qquad \text{(A-14.4.10)}$$

A box of sides α_i, α_j, α_l may be denoted by B_{ijl}. We label the smaller boxes thus generated arbitrarily by $B_1, \ldots, B_\kappa$, and note from (A-14.4.4), (14.4.5) that

$$\int_{B_i} d^3\mathbf{x} \, f^2(\mathbf{x}) = \frac{1}{8n^3} = \frac{1}{\kappa} \qquad \text{(A-14.4.11)}$$

for $i = 1, \ldots, \kappa$.

The integrals in (A-14.4.1), (A-14.4.2) may be rewritten as sums of integrals over such boxes as follows:

$$I(\kappa; \mathbf{X}_1, \ldots, \mathbf{X}_\kappa) = -\kappa \sum_{i=1}^{\kappa} \sum_{j=1}^{\kappa} \int_{B_i} \frac{d^3\mathbf{x}}{|\mathbf{x} - \mathbf{X}_j|} f^2(\mathbf{x}) \qquad \text{(A-14.4.12)}$$

$$I_2(\kappa) = \frac{\kappa^2}{2} \sum_{i=1}^{\kappa} \sum_{j=1}^{\kappa} \int_{B_i} d^3\mathbf{x} \int_{B_j} d^3\mathbf{x}' \, f^2(\mathbf{x}) \frac{1}{|\mathbf{x} - \mathbf{x}'|} f^2(\mathbf{x}'). \qquad \text{(A-14.4.13)}$$

Now we place $\mathbf{X}_1$ in box B_1, $\mathbf{X}_2$ in box $B_2, \ldots, \mathbf{X}_\kappa$ in box B_κ and then average the expression of $I(\kappa; \mathbf{X}_1, \ldots, \mathbf{X}_\kappa)$, over $\mathbf{X}_1, \ldots, \mathbf{X}_\kappa$ by multiplying it by the normalized density

$$\frac{f^2(\mathbf{X}_1)}{\int_{B_1} d^3\mathbf{X} f^2(\mathbf{X})} \cdots \frac{f^2(\mathbf{X}_\kappa)}{\int_{B_\kappa} d^3\mathbf{X} f^2(\mathbf{X})} = \prod_{i=1}^{\kappa} \left(\kappa f^2(\mathbf{X}_i)\right) \qquad \text{(A-14.4.14)}$$

and integrating over $\mathbf{X}_1, \ldots, \mathbf{X}_\kappa$, respectively, over the boxes $B_1, \ldots, B_\kappa$ to obtain for this average the expression

$$\langle I(\kappa; \mathbf{X}_1, \ldots, \mathbf{X}_\kappa) \rangle = -\kappa^2 \sum_{i=1}^{\kappa} \sum_{j=1}^{\kappa} \int_{B_i} d^3\mathbf{x} \int_{B_j} d^3\mathbf{x}' f^2(\mathbf{x}) \frac{1}{|\mathbf{x} - \mathbf{x}'|} f^2(\mathbf{x}')$$

$$+ \frac{\kappa^2}{2} \sum_{i=1}^{\kappa} \sum_{j=1}^{\kappa} \int_{B_i} d^3\mathbf{x} \int_{B_j} d^3\mathbf{x}' f^2(\mathbf{x}) \frac{1}{|\mathbf{x} - \mathbf{x}'|} f^2(\mathbf{x}')$$

$$+ \kappa^2 \sum_{i<j}^{\kappa} \int_{B_i} d^3\mathbf{x} \int_{B_j} d^3\mathbf{x}' f^2(\mathbf{x}) \frac{1}{|\mathbf{x} - \mathbf{x}'|} f^2(\mathbf{x}')$$

$$\equiv -\frac{\kappa^2}{2} \sum_{i=1}^{\kappa} \int_{B_i} d^3\mathbf{x} \int_{B_i} d^3\mathbf{x}' f^2(\mathbf{x}) \frac{1}{|\mathbf{x} - \mathbf{x}'|} f^2(\mathbf{x}').$$

$$\text{(A-14.4.15)}$$

14.4 The Collapse of "Bosonic Matter"

By the definition of an *average*, there must exist at least one set of $\{\mathbf{X}_1, \ldots, \mathbf{X}_\kappa\}$ with $\mathbf{X}_1$ in box $B_1, \ldots, \mathbf{X}_\kappa$ in box B_κ such that

$$I(\kappa; \mathbf{X}_1, \ldots, \mathbf{X}_\kappa) \leqslant \langle I(\kappa; \mathbf{X}_1, \ldots, \mathbf{X}_\kappa) \rangle \qquad \text{(A-14.4.16)}$$

That is, *with such* a set $\{\mathbf{X}_1, \ldots, \mathbf{X}_\kappa\}$,

$$I(\kappa; \mathbf{X}_1, \ldots, \mathbf{X}_\kappa) \leqslant -\frac{\kappa^2}{2} \sum_{i=1}^{\kappa} \int_{B_i} d^3\mathbf{x} \int_{B_i} d^3\mathbf{x}' \, f^2(\mathbf{x}) \frac{1}{|\mathbf{x}-\mathbf{x}'|} f^2(\mathbf{x}'). \qquad \text{(A-14.4.17)}$$

A box B_{ijl} of sides $\alpha_i, \alpha_j, \alpha_l$ may be inserted in a sphere of radius $\frac{1}{2}\sqrt{\alpha_i^2 + \alpha_j^2 + \alpha_l^2}$ and from the normalization condition (A-14.4.11) we have (see Problem 14.6)

$$\int_{B_{ijl}} d^3\mathbf{x} \int_{B_{ijl}} d^3\mathbf{x}' \, f^2(\mathbf{x}) \frac{1}{|\mathbf{x}-\mathbf{x}'|} f^2(\mathbf{x}') \geqslant \frac{2}{\sqrt{\alpha_i^2 + \alpha_j^2 + \alpha_l^2}} \left(\frac{1}{\kappa}\right)^2. \qquad \text{(A-14.4.18)}$$

Accordingly, we may bound $I(\kappa; \mathbf{X}_1, \ldots, \mathbf{X}_\kappa)$ as follows

$$I(\kappa; \mathbf{X}_1, \ldots, \mathbf{X}_\kappa) \leqslant -8 \sum_{i,j,l=1}^{n} \frac{1}{\sqrt{\alpha_i^2 + \alpha_j^2 + \alpha_l^2}} \qquad \text{(A-14.4.19)}$$

where, by symmetry, the factor 8 takes into account all of the boxes, since the summation over i, j, l, accounts only for n^3 boxes corresponding to $0 \leqslant x_1 \leqslant 1, 0 \leqslant x_2 \leqslant 1, 0 \leqslant x_3 \leqslant 1$.

By noting that for $a_i > 0$,

$$\sum_{i=1}^{K} \sqrt{a_i} \frac{1}{\sqrt{a_i}} = K \qquad \text{(A-14.4.20)}$$

an elementary application of the Cauchy-Schwarz inequality then yields

$$\left(\sum_{i=1}^{K} a_i\right) \left(\sum_{i=1}^{K} \frac{1}{a_i}\right) \geqslant K^2 \qquad \text{(A-14.4.21)}$$

or

$$\left(\sum_{i=1}^{K} \frac{1}{a_i}\right) \geqslant \frac{K^2}{\left(\sum_{i=1}^{K} a_i\right)}. \qquad \text{(A-14.4.22)}$$

Applying this elementary inequality to (A-14.4.19) gives

$$I(\kappa; \mathbf{X}_1, \ldots, \mathbf{X}_\kappa) \leqslant -8 \left(n^3\right)^2 \frac{1}{\sum_{i,j,l=1}^{n} \sqrt{\alpha_i^2 + \alpha_j^2 + \alpha_l^2}}. \qquad \text{(A-14.4.23)}$$

On the other hand,

$$\sum_{i,j,l=1}^{n} \sqrt{\alpha_i^2 + \alpha_j^2 + \alpha_l^2} \leqslant \sum_{i,j,l=1}^{n} (\alpha_i + \alpha_j + \alpha_l) = 3n^2 \qquad \text{(A-14.4.24)}$$

where we have used (A-14.4.10), which from (A-14.4.23) leads to

$$I(\kappa; \mathbf{X}_1, \ldots, \mathbf{X}_\kappa) \leqslant -\frac{8}{3}n^4 = -\frac{\kappa^{4/3}}{6} \qquad \text{(A-14.4.25)}$$

since $\kappa = 8n^3$. The improvement of the bound in (A-14.4.25) is not ruled out. From (A-14.4.7), the inequality in (14.4.48) then follows.

Problems

14.1. Establish the equality in (14.2.6).
14.2. Verify the details leading from (14.4.11) to (14.4.17).
14.3. Derive the expressions for the average kinetic energies given in (14.4.31)–(14.4.33).
14.4. Show that the expectation value $\langle \Psi | H_\mathrm{U} | \Psi \rangle$ is as given in (14.4.34)–(14.4.36).
14.5. Repeat the derivation of the upper bound for E_N^B given after (14.4.19) by considering the following construction: Rewrite (14.4.19) (if $\varepsilon_1 \neq 0$)

$$\left(\frac{N}{8}\right)^{113} = n_1 + \varepsilon_1, \quad 0 < \varepsilon_1 < 1$$

where the natural number $n_1 \geqslant 1$. Let $k_1 = 8n_1^3$, and continuing in this manner, we have (if $\varepsilon_2 \neq 0$)

$$\left(\frac{(N-k_1)}{8}\right)^{113} = n_2 + \varepsilon_2, \quad 0 < \varepsilon_2 < 1$$

and so on, by defining in turn $k_2 = 8n_2^3, \ldots, k_b = 8n_b^3$, you will reach a natural number b such that

$$\left(\frac{[N - (k_1 + \ldots + k_b)]}{8}\right)^{113} = \varepsilon_{b+1}, \quad 0 < \varepsilon_{b+1} < 1.$$

For example, for $N = 58410$, $b = 5$. For $b > 1$, place $k_1, \ldots, k_b$, pairs of negatively and positively charged particles in b non-overlapping boxes, and the remaining $2[N - (k_1 + \ldots + k_b)]$ particles in other $[N - (k_1 + \ldots + k_b)]$ non-overlapping boxes with each containing one positive and one negative charge. By such a construction, can you increase the numerical factor multiplying $-N^{5/3}\left(me^4/2\hbar^2\right)$ in (14.4.53)?

14.6. Let $\rho(\mathbf{x})$ be a charge density vanishing outside, a region $\mathcal{R}$ corresponding to a total charge Q, i.e.,

$$\int_{\mathcal{R}} \mathrm{d}^3\mathbf{x}\, \rho(\mathbf{x}) = Q$$

such that $|\mathbf{x}| \leqslant R$ for $\mathbf{x}$ in $\mathcal{R}$ for some R. Show that

$$I[\rho] \equiv \int_{\mathcal{R}} \mathrm{d}^3\mathbf{x} \int_{\mathcal{R}} \mathrm{d}^3\mathbf{x}'\, \rho(\mathbf{x}) \frac{1}{|\mathbf{x} - \mathbf{x}'|} \rho(\mathbf{x}') \geqslant \frac{Q^2}{R}.$$

14.7. Repeat the derivation of the upper bound for E_N^B in §14.4 in arbitrary dimensions ν [with $1/r$ potentials] to show that $E_N^\mathrm{B} \leqslant -\left(\frac{me^4}{2\hbar^2}\right) c_\mathrm{B}(\nu) N^{(2+\nu)/\nu}$ and obtain an explicit expression for such a positive constant $c_\mathrm{B}(\nu)$ depending on ν. [Ref: Muthaporn and Manoukian (2004a).]

14.8. Follow the proof given in Appendix B to §13.1, which shows that the TF density satisfying (13.1.7) actually gives the smallest value for the functional (B-13.1.1), to prove simply that $\rho_0(\mathbf{x}; k)$ in (A-14.1.2) provides the smallest value for the functional (A-14.1.1), i.e., (A-14.1.4) holds true.

14.9. Derive the equality in (A-14.1.22), and the corresponding expression for

$$\frac{\partial}{\partial \lambda} F[\rho_2; \lambda Z_1, \ldots, \lambda Z_l; \mathbf{R}_1, \ldots, \mathbf{R}_l]$$

to finally obtain the result given in (A-14.1.23).

14.10. Derive the counterpart of the expression for the probability in (14.3.15) and the counterpart of the lower bound in (14.3.22) for "bosonic matter". Do these bounds provide useful information on the nature of such "matter" as they do for matter with the exclusion principle? Explain.

15

Quantum Scattering

Significant progress has been made over the years in physics through scattering experiments. The discovery of the atomic nucleus, the visible tracks of particles observed in cloud chambers, the determination of the structure of matter, the emergence of the endless variety of particles by accelerators and the extraction of information on their interactions, are just a few of the examples concerned with the analyses of scattering processes.

In the present chapter, dealing with the theoretical development of quantum scattering, we approach the basic problem of scattering as a time evolution process involving, in general, three stages: the preparatory stage occurring in the remote past to the interacting stage, followed finally by the detection stage in the distant future. As a time evolution process, asymptotic boundary conditions arise, dictated by the physical situation under study, linking the three stages. These aspects are investigated in §15.1, §15.2. Special emphasis is put in the latter section on the connection between the momentum a particle has acquired in a collision and its emergence spatially within a cone on its way to a detector. Differential cross sections are studied in §15.3. In terms of number of particles, we here recall the definition of a differential cross section as the number of particles scattered per unit time into a solid angle, about the scattering angle, divided by the flux, with the latter being the incident number of particles per unit area per unit time. In a scattering process, the detection of a scattered particle is carried out away from the forward direction to avoid or minimize interference effects between incident and scattering components of the interacting state. The analysis of the scattering process in the forward direction leads to the so-called optical theorem which together its physical interpretation is the subject matter of §15.4. In this section, a phase shift analysis is also carried out dealing with the expansion of scattering amplitudes in terms of angular momentum states. §15.5 provides a detailed treatment of Coulomb scattering which requires special attention due to the long range nature of the underlying interaction. In §15.6, §15.7, we will see how the elegant functional formulation in Chapter 11 may be used in scattering theory, and an application is then carried out to scattering at small deflection angles at high energies. §15.1 through §15.7 deal with elastic scattering theory where the initial and final particles in the process are the same and their is no change in the internal energies of the particles involved so that the total kinetic energy of the system remains conserved. Inelastic processes and the underlying theory are treated in §15.8 which deal with cluster of particles as introduced in §2.5. Here we also elaborate on some subtleties of systems involving three particles. In the final section (§15.9), we consider the energy loss of a charged particle moving through a medium (hydrogen). In the same section, this is followed by a treatment of neutron interferometry dealing with the splitting and recombination of a neutron beam with an investigation of the interference effect, resulting upon recombination, in the Earth gravitational field.

15.1 Interacting States and Asymptotic Boundary Conditions

In a scattering process, one would initially prepare a particle in some state, say, $|\Phi_{in}(t)\rangle$, let it eventually interact with another system, assumed widely separated in the beginning of the experiment, and then finally study the outcome of the process. In the preparatory stage, in the remote past, before a particle participates in the scattering process, its interaction with the other system, if it is of short range,[1] is negligible and the state $|\Phi_{in}(t)\rangle$, for $t \to -\infty$, develops in time via the free Hamiltonian $H_0 = -\hbar^2 \nabla^2/2m$, where m is the mass of the particle. In time, as the particle approaches the other system in question and its interaction with the latter becomes non-negligible, the particle would be described by a state $|\psi_-(t)\rangle$, satisfying the time-dependent Schrödinger equation

$$\left(i\hbar \frac{\partial}{\partial t} - H \right) |\psi_-(t)\rangle = 0 \qquad (15.1.1)$$

where $H \neq H_0$ is the total Hamiltonian, and $|\Phi_{in}(t)\rangle$ would provide the asymptotic boundary condition for $|\psi_-(t)\rangle$ in the limit $t \to -\infty$.

Similarly, a sufficiently long time after scattering, as a particle emerges from such a process, and if its interaction, assumed of short range, with the remaining system becomes negligible, one, in a statistical sense, may investigate its localizability in a detection region and enquire as well of the momentum it has acquired that has led it to such a region. In this detection stage, one may then assign a state, say, $|\Phi_{out}(t)\rangle$ for such a statistical study

---

[1] By an interaction of short range, it is meant an interaction which vanishes faster than the Coulomb potential at large distances of separations of the particle in question from the rest of the system, such as $(distance)^{-2}$ for distance $\to \infty$, or faster.

15.1 Interacting States and Asymptotic Boundary Conditions

which, for $t \to +\infty$, develops in time via a free Hamiltonian. Before the interaction becomes negligible, however, the particle would be described by some state $|\psi_+(t)\rangle$, satisfying the time-dependent Schrödinger equation

$$\left(i\hbar \frac{\partial}{\partial t} - H\right)|\psi_+(t)\rangle = 0 \tag{15.1.2}$$

with $|\Phi_{\text{out}}(t)\rangle$ taken as the asymptotic time limit $t \to +\infty$ of $|\psi_+(t)\rangle$.

The purpose of this section is to study the nature of the fully *interacting* states $|\psi_\pm(t)\rangle$ and their asymptotic boundary conditions: $|\psi_\pm(t)\rangle \to |\Phi_{\text{out/in}}(t)\rangle$ for $t \to \pm\infty$. The scattering of a particle off a Coulomb potential, as of a long range interaction, will be treated in detail in §15.5.

Consider the retarded/advanced free Green functions $G_\pm^0(x, x')$, introduced in §9.1 (see (9.1.12), (9.1.35)), where we have used the convenient notation[2] $(t, \mathbf{x}) = x$, given by

$$G_\pm^0(x, x') = \pm i\hbar \int \frac{(dp)}{(2\pi\hbar)^4} \frac{e^{i(x-x')p/\hbar}}{\left(p^0 - \frac{\mathbf{p}^2}{2m} \pm i\varepsilon\right)}, \qquad \varepsilon \to +0 \tag{15.1.3}$$

$$(x - x')p = (\mathbf{x} - \mathbf{x}') \cdot \mathbf{p} - (t - t')p^0 \tag{15.1.4}$$

$$(dp) = dp^0 d^3\mathbf{p} \tag{15.1.5}$$

with boundary conditions (see (9.1.9), (9.1.33))

$$G_\pm^0(x, x') = 0 \qquad \text{for} \qquad t - t' \lessgtr 0 \tag{15.1.6}$$

satisfying the differential equations

$$\left(i\hbar \frac{\partial}{\partial t} - H_0\right) G_\pm^0(x, x') = \pm i\hbar \, \delta^4(x - x') \tag{15.1.7}$$

where $H_0 = -\hbar^2 \nabla^2/2m$.

We introduce the integral equation

$$A_\pm(x, x') = \delta^4(x - x') \pm \frac{i}{\hbar} \int (dx'') \, G_\mp^0(x, x'') \, V(x'') \, A_\pm(x'', x') \tag{15.1.8}$$

where $V(x'')$ is a given potential which may be formally considered to be time-dependent, in general, and

$$(dx'') = dt'' d^3\mathbf{x}''. \tag{15.1.9}$$

It is then easily verified from (15.1.7), (15.1.8), that

---

[2] Needless to say, this is just a convenient notation having nothing to do with a relativistic notation.

$$\psi_{\pm}(x) = \int (\mathrm{d}x') \, A_{\pm}(x, x') \, \Phi_{\text{out/in}}(x'), \tag{15.1.10}$$

upon applying the operator $[i\hbar \, \partial/\partial t - H_0]$ to the latter, that they satisfy the interacting Schrödinger equations (15.1.1), (15.1.2), where $H = H_0 + V$, and use has been made of the facts that $[i\hbar \, \partial/\partial t - H_0]\Phi_{\text{out/in}}(x) = 0$. That is, the compact expressions in (15.1.10) provide *solutions of the fully interacting systems*. We will eventually consider only time-independent potentials.

From (15.1.8), (15.1.10), we note that $\psi_{\pm}(x)$ may be rewritten as[3]

$$\psi_{\pm}(x) = \Phi_{\text{out/in}}(x) \pm \frac{i}{\hbar} \int (\mathrm{d}x') \, (\mathrm{d}x'') \, G_{\mp}^0(x, x'') \, V(x'')$$
$$\times A_{\pm}(x'', x') \, \Phi_{\text{out/in}}(x') \tag{15.1.11}$$

We will see that the solutions $\psi_{\pm}(x)$ formally satisfy the asymptotic boundary conditions[4] $\psi_{\pm}(x) \to \Phi_{\text{out/in}}(x)$ for $t \to \pm\infty$.

To the above end, we introduce the Fourier transforms

$$V(x) = \int \frac{(\mathrm{d}p)}{(2\pi\hbar)^4} \, e^{ipx/\hbar} \, V(p) \tag{15.1.12}$$

$p = (p^0, \mathbf{p})$,

$$A_{\pm}(x'', x') = \int \frac{(\mathrm{d}p'')(\mathrm{d}p')}{(2\pi\hbar)^4} A_{\pm}(p'', p') \, e^{ip''x''/\hbar} \, e^{-ip'x'/\hbar} \tag{15.1.13}$$

$$\Phi_{\text{out/in}}(x') = \int \frac{(\mathrm{d}p)}{(2\pi\hbar)^4} \left[2\pi\hbar \, \delta(p^0 - \mathbf{p}^2/2m)\right] e^{ipx'/\hbar} \, \tilde{\Phi}_{\text{out/in}}(p) \tag{15.1.14}$$

and we may rewrite the second term on the right-hand side of (15.1.11) as

$$\int \frac{(\mathrm{d}p)}{(2\pi\hbar)^4} \frac{e^{-i[p^0 - E(p)]t/\hbar}}{[p^0 - E(p) \mp i\varepsilon]} e^{i[\mathbf{p}\cdot\mathbf{x} - E(\mathbf{p})t]/\hbar} F_{\pm}(p) \tag{15.1.15}$$

where

$$F_{\pm}(p) = \int \frac{(\mathrm{d}p'')(\mathrm{d}p')}{(2\pi\hbar)^4} V(p - p'') A_{\pm}(p'', p') \phi_{\text{out/in}}(\mathbf{p}') \, 2\pi\hbar \, \delta(p'^0 - E(\mathbf{p}')) \tag{15.1.16}$$

$$E(\mathbf{p}) = \frac{\mathbf{p}^2}{2m}. \tag{15.1.17}$$

In reference to the p^0-integral in (15.1.15) we note that

---

[3] Equations such as (15.1.11), (15.1.21), (15.1.33) are usually referred to as Lippmann-Schwinger equations.

[4] The nature of the $t \to \pm\infty$ limits will be further discussed below.

15.1 Interacting States and Asymptotic Boundary Conditions

$$\lim_{t\to+\infty} \frac{e^{-i[p^0-E(\mathbf{p})]t/\hbar}}{[p^0 - E(\mathbf{p}) \mp i\varepsilon]} = \begin{cases} 0 \\ -2\pi i\, \delta(p^0 - E(\mathbf{p})) \end{cases} \tag{15.1.18}$$

as obtained by closing the contour of integration, in the complex p^0-plane, from below with $\operatorname{Im}(p^0) < 0$ on the infinite semi-circle part of the contour. For the $-i\varepsilon$ term the contour does not enclose the pole $p^0 = E(\mathbf{p})+i\varepsilon$, which lies in the upper half p^0-plane, thus giving the value 0. For the $+i\varepsilon$ term, the pole $p^0 = E(\mathbf{p}) - i\varepsilon$ is in the lower half p^0-plane giving $-2\pi i\,\delta(p^0 - E(\mathbf{p}))$ by the residue theorem with the minus sign arising since the direction of the contour is in the clockwise direction. Similarly, by closing the p^0-contour from above, we have

$$\lim_{t\to-\infty} \frac{e^{-i[p^0-E(\mathbf{p})]t/\hbar}}{[p^0 - E(\mathbf{p}) \mp i\varepsilon]} = \begin{cases} +2\pi i\,\delta(p^0 - E(\mathbf{p})) \\ 0. \end{cases} \tag{15.1.19}$$

That is, in particular,

$$\lim_{t\to\pm\infty} \frac{e^{-i[p^0-E(\mathbf{p})]t/\hbar}}{[p^0 - E(\mathbf{p}) \mp i\varepsilon]} = 0 \tag{15.1.20}$$

which from (15.1.15), (15.1.11) formally establish the asymptotic boundary conditions $\psi_\pm(x) \to \Phi_{\text{out/in}}(x)$ for $t \to \pm\infty$.

The following representation for $\psi_\pm(x)$ is easily obtained from (15.1.11), (15.1.15)

$$\psi_\pm(x) = \Phi_{\text{out/in}}(x) + \int \frac{(dp)}{(2\pi\hbar)^4} \frac{e^{-i[p^0-E(\mathbf{p})]t/\hbar}}{[p^0 - E(\mathbf{p}) \mp i\varepsilon]} e^{i[\mathbf{x}\cdot\mathbf{p}-E(\mathbf{p})t]/\hbar} F_\pm(p). \tag{15.1.21}$$

Also we note that by using the integral representation

$$\delta^4(x - x') = \int \frac{(dp)(dp')}{(2\pi\hbar)^4} \delta^4(p - p')\, e^{ipx/\hbar} e^{-ip'x'/\hbar} \tag{15.1.22}$$

one obtains from (15.1.8), (15.1.13) the following integral equations for $A_\pm(p,p')$:

$$A_\pm(p,p') = \delta^4(p-p') + \frac{1}{[p^0 - E(\mathbf{p}) \mp i\varepsilon]} \int \frac{(dp'')}{(2\pi\hbar)^4} V(p-p'')\, A_\pm(p'',p'). \tag{15.1.23}$$

For a time-independent potential $V(\mathbf{x})$,

$$V(p) = \int (dx)\, e^{-ip\,x/\hbar}\, V(\mathbf{x})$$

$$= (2\pi\hbar)\,\delta(p^0)\, V(\mathbf{p}) \tag{15.1.24}$$

and we may write

$$A_\pm (p, p') = \delta (p^0 - p'^0) A_\pm (\mathbf{p}, \mathbf{p}'; p^0). \tag{15.1.25}$$

Upon substitution of (15.1.24), (15.1.25) in (15.1.23), and integrating over p'^0, the equations in (15.1.23) reduce to

$$A_\pm (\mathbf{p}, \mathbf{p}'; p^0) = \delta^3 (\mathbf{p} - \mathbf{p}') + \frac{1}{[p^0 - E(\mathbf{p}) \mp i\varepsilon]}$$
$$\times \int \frac{d^3 \mathbf{p}''}{(2\pi\hbar)^3} V (\mathbf{p} - \mathbf{p}'') A_\pm (\mathbf{p}'', \mathbf{p}'; p^0). \tag{15.1.26}$$

As we will see in §15.3, the following object defined by

$$f(\mathbf{p}, \mathbf{p}') = -4\pi^2 \hbar \, m \left([p^0 - E(p)] A_- (\mathbf{p}, \mathbf{p}'; p^0) \right) \Big|_{p^0 = E(p) = E(p')} \tag{15.1.27}$$

turns out to be important in scattering theory, and for reasons discussed there is referred to as the scattering amplitude. The restriction with a bar, on the right-hand side of (15.1.27) indicating to set $p^0 = E(p) = E(p')$, is referred to as the energy shell restriction.

For a time-independent potential, $\psi_\pm (x)$ in (15.1.21) becomes

$$\psi_\pm (x) = \Phi_{\text{out/in}} (x) + \int \frac{d^3 \mathbf{p}}{(2\pi\hbar)^3} \frac{d^3 \mathbf{p}'}{(2\pi\hbar)^3} \frac{e^{-i[E(\mathbf{p}') - E(\mathbf{p})]t/\hbar}}{[E(\mathbf{p}') - E(\mathbf{p}) \mp i\varepsilon]}$$
$$\times e^{i[\mathbf{x}\cdot\mathbf{p} - E(\mathbf{p})t]/\hbar} F_\pm (\mathbf{p}, \mathbf{p}'; E(\mathbf{p}')) \tilde{\Phi}_{\text{out/in}} (\mathbf{p}') \tag{15.1.28}$$

where

$$F_\pm (\mathbf{p}, \mathbf{p}'; E(\mathbf{p}')) = \int d^3 \mathbf{p}'' \, V (\mathbf{p} - \mathbf{p}'') A_\pm (\mathbf{p}'', \mathbf{p}'; E(\mathbf{p}')). \tag{15.1.29}$$

$A_\pm (p, p')$ are related to the full Green functions $G_\pm (x, x')$ in the theory which satisfy the equations

$$\left[i\hbar \frac{\partial}{\partial t} - H \right] G_\pm (x, x') = \pm i\hbar \, \delta^4 (x - x'). \tag{15.1.30}$$

It is readily verified, that the solutions of (15.1.30) are given by

$$G_\pm (x, x') = G_\pm^0 (x, x') \mp \frac{i}{\hbar} \int (dx'') \, G_\pm^0 (x, x'') \, V(x'') \, G_\pm (x'', x') \tag{15.1.31}$$

by the application, in the process, of the operator $[i\hbar \, \partial/\partial t - H_0]$ to the latter and using (15.1.7).

15.1 Interacting States and Asymptotic Boundary Conditions

Upon carrying out the Fourier transforms

$$G_{\pm}(x,x') = \pm i\hbar \int \frac{(dp)(dp')}{(2\pi\hbar)^4} e^{ipx/\hbar} e^{-ip'x'/\hbar} G_{\pm}(p,p') \qquad (15.1.32)$$

we note from (15.1.31), that the $G_{\pm}(p,p')$ satisfy the integral equations

$$G_{\pm}(p,p') = \frac{\delta^4(p-p')}{[p'^0 - E(\mathbf{p}') \pm i\varepsilon]}$$

$$+ \frac{1}{[p^0 - E(\mathbf{p}) \pm i\varepsilon]} \int \frac{(dp'')}{(2\pi\hbar)^4} V\left(\frac{p-p''}{\hbar}\right) G_{\pm}(p'',p'). \qquad (15.1.33)$$

By comparing (15.1.33) with (15.1.23), we may infer that

$$A_{\pm}(p,p') = G_{\mp}(p,p')[p'^0 - E(\mathbf{p}')]. \qquad (15.1.34)$$

For a time-independent potential,

$$G_{\mp}(p,p') = G_{\mp}(\mathbf{p},\mathbf{p}';p^0)\,\delta(p^0 - p'^0) \qquad (15.1.35)$$

and we have

$$A_{\pm}(\mathbf{p},\mathbf{p}';p^0) = G_{\mp}(\mathbf{p},\mathbf{p}';p^0)[p^0 - E(\mathbf{p}')] \qquad (15.1.36)$$

which will have an important application in determining transition amplitudes in scattering processes in §15.3.

We close this section, by elaborating rigorously on the nature of the limits $|\psi_{\pm}(t)\rangle \to |\phi_{\text{out/in}}(t)\rangle$ for $t \to \pm\infty$, under some sufficiency conditions. To this end, consider a time-independent square-integrable potential $V(\mathbf{x})$. We treat the time $t \to +\infty$ limit only. The $t \to -\infty$ limit may be treated in the same manner.

Suppose that we are given a normalized state $|\Phi_{\text{out}}\rangle$, which in the **x**-description satisfies the condition

$$\int d^3\mathbf{x}\,|\Phi_{\text{out}}(\mathbf{x})| \leqslant C < \infty, \qquad (15.1.37)$$

and develops in time as $|\Phi_{\text{out}}(t)\rangle = \exp(-itH_0/\hbar)|\Phi_{\text{out}}\rangle$.

The solution of (15.1.2) is given by

$$|\psi_+(t)\rangle = \exp(-itH/\hbar)|\psi_+\rangle \qquad (15.1.38)$$

and the asymptotic boundary condition $|\psi_+(t)\rangle \to |\Phi_{\text{out}}(t)\rangle$ may be defined[5] by

---

[5] Such a limit is referred to a strong one which in turn implies the weak limit $\langle \chi|(\psi_+(t) - \Phi_{\text{out}}(t))\rangle \to 0$, for $t \to +\infty$, as follows by an elementary application of the Cauchy-Schwarz inequality in conjunction with (15.1.39) for any normalizable state $|\chi\rangle$.

$$\lim_{t \to +\infty} \|\psi_+(t) - \Phi_{\text{out}}(t)\| = 0. \tag{15.1.39}$$

The unitarity of the operator $\exp(-itH/\hbar)$ implies from (15.1.39) that

$$\lim_{t \to +\infty} \|\psi_+ - \Omega(t)\Phi_{\text{out}}\| = 0 \tag{15.1.40}$$

where

$$\Omega(t) = \exp(itH/\hbar)\exp(-itH_0/\hbar). \tag{15.1.41}$$

That the limit $t \to +\infty$ in (15.1.40) exists follows by noting that

$$(\Omega(t_1) - \Omega(t_2))|\Phi_{\text{out}}\rangle = \int_{t_2}^{t_1} d\tau \left(\frac{d}{d\tau}\Omega(\tau)\right)|\Phi_{\text{out}}\rangle$$

$$= \frac{i}{\hbar}\int_{t_2}^{t_1} d\tau \, e^{i\tau H/\hbar} V e^{-i\tau H_0/\hbar}|\Phi_{\text{out}}\rangle \tag{15.1.42}$$

leads to

$$\|(\Omega(t_1) - \Omega(t_2))\phi_{\text{out}}\| \leq \frac{1}{\hbar}\left|\int_{t_2}^{t_1} d\tau \|V\Phi_{\text{out}}(\tau)\|\right|. \tag{15.1.43}$$

On the other hand from (9.1.4), (9.1.8),

$$\Phi_{\text{out}}(\mathbf{x}, \tau) = \left(\frac{m}{2\pi i\hbar\tau}\right)^{3/2} \int d^3\mathbf{x}' \exp\left(\frac{im|\mathbf{x}-\mathbf{x}'|^2}{2\hbar\tau}\right) \Phi_{\text{out}}(\mathbf{x}') \tag{15.1.44}$$

which from (15.1.37) gives

$$|\Phi_{\text{out}}(\mathbf{x}, \tau)| \leq \left(\frac{m}{2\pi\hbar}\right)^{3/2} C\Big/|\tau|^{3/2} \tag{15.1.45}$$

and

$$\|V\Phi_{\text{out}}(\tau)\| \leq \left(\frac{m}{2\pi\hbar}\right)^{3/2} C\|V\|\Big/|\tau|^{3/2}. \tag{15.1.46}$$

From this inequality, we may conclude that the left-hand of (15.1.43) vanishes for $|t_1|, |t_2| \to \infty$. That is, $\{\Omega(t)|\Phi_{\text{out}}\rangle\}$ forms a Cauchy sequence[6] whose limit (in the strong sense), denoted by $|\psi_+\rangle$, exists for $t \to +\infty$, thus establishing (15.1.39)/(15.1.40).[7]

A similar analysis may be carried out for $|\psi_-(t)\rangle$ corresponding to a state $|\Phi_{\text{in}}(t)\rangle$ in the limit $t \to -\infty$.

The operators[8] $\Omega_\pm$, $\Omega_\pm^\dagger$ defined as the (strong) limits of $\Omega(t)$, $\Omega^\dagger(t)$:

---

[6] See property (iv) in the definition of a Hilbert space in §1.7.
[7] The condition in (15.1.37) may be relaxed but we will not go into these details here.
[8] Some authors interchange the $\pm$ signs in $\Omega_\pm$ with $\mp$ corresponding to the limits $t \to \pm\infty$ in their notation.

$$\Omega_\pm = \lim_{t \to \pm\infty} e^{itH/\hbar} \, e^{-itH_0/\hbar} \qquad (15.1.47)$$

$$\Omega_\pm^\dagger = \lim_{t \to \pm\infty} e^{itH_0/\hbar} \, e^{-itH/\hbar} \qquad (15.1.48)$$

are referred to as Møller wave operators. In terms of $\Omega_\pm$, we may write

$$|\psi_\pm\rangle = \Omega_\pm \left|\Phi_{\text{out/in}}\right\rangle. \qquad (15.1.49)$$

15.2 Particle Detection and Connection between Configuration and Momentum Spaces in Scattering

Consider the scattering of a particle off a given potential of short range.[9] In the remote past, before any interaction occurs, the particle is prepared in some state $|\Phi_{\text{in}}(t)\rangle$, for $t \to -\infty$, and develops in time via the free Hamiltonian H_0. Eventually in time, the particle would feel the presence of the potential and would be described by an interacting state $|\psi_-(t)\rangle$, satisfying the Schrödinger equation in (15.1.1), consistent with the boundary condition (§15.1): $|\psi_-(t)\rangle \to |\Phi_{\text{in}}(t)\rangle$ for $t \to -\infty$.

Typically, one would then enquire about the probability of finding the emerging particle from the scattering process beyond an arbitrary large radial distance within a cone,[10] with apex at the scattering center (the origin of the coordinate system), on its way to the detection region (see Figure 15.1). This probability is given by

$$\text{Prob}[\mathbf{x} \in C_D] = \int_{C_D} d^3\mathbf{x} \, |\psi_-(\mathbf{x}, t)|^2 \qquad (15.2.1)$$

for t positive and large, $\psi_-(\mathbf{x}, t) \equiv \psi_-(x)$ is given in (15.1.11), and

$$C_D = C_0 \cap \{|\mathbf{x}| > D\} \qquad (15.2.2)$$

where D will be taken to be large, and for a given unit vector $\mathbf{N}$,

$$C_0 = \{\mathbf{x} : |\mathbf{x}| \geq \mathbf{x} \cdot \mathbf{N} \geq \alpha |\mathbf{x}|, \text{ for some } \alpha \in (0, 1]\}. \qquad (15.2.3)$$

In the preparatory stage, a wavepacket is prepared which propagates with some average momentum, say, $\mathbf{p}'$ whose direction may be taken to define the

---

[9] The scattering off a Coulomb potential is given in §15.5, where it will be seen, in particular, that the time development of the preparatory state does not develop via the free Hamiltonian H_0 due to the slow decrease of the Coulomb potential at large distances.

[10] Such a point is emphasized to a large extent in: Dollard (1969); Amrein et al. (1970); Manoukian and Prugovečki (1971); Prugovečki (1971); Amrein (1981), and others.

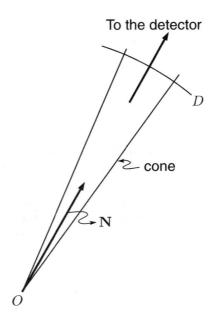

Fig. 15.1. After a particle emerges from a scattering process, a non vanishing probability of finding the particle within a cone, beyond an arbitrary large radial distance D, implies that the momentum it has acquired is directed within the same cone in conformity with one's perception of scattering.

z-axis of the coordinate system chosen. The packet will be assumed to have a given lateral width, i.e., in a direction perpendicular to $\mathbf{p}'$, denoted by $2a$.

We first set the detector in (15.2.1), at a large distance $|\mathbf{x}| = \sqrt{\mathbf{x}_\parallel^2 + z^2} > D$ from the scattering center such that $|\mathbf{x}_\parallel| \gg a$, so that no appreciable interference occurs between the incident part $\Phi_{\text{in}}(x)$ and the scattered part, represented by the integral on the right-hand side of the equation for $\psi_-(x)$ in (15.1.11). We will consider the scattering process for $|\mathbf{x}_\parallel| \leqslant a$, i.e., in the forward direction, later in §15.4.

So what we need is a wavepacket which at time $t > 0$, corresponding to the detection time, of the form[11]

$$\Phi_{\text{in}}(\mathbf{x}, t) = \int d^2 x'_\parallel \, G^0_+ \left(\mathbf{x}_\parallel, \mathbf{x}'_\parallel; t \right) \chi\left(\mathbf{x}'_\parallel \right) \xi(z, t) \qquad (15.2.4)$$

---

[11] A particular choice of a wavepacket is not necessary, but the form given in (15.2.4) clarifies many aspects of the theory.

where $\chi\left(\mathbf{x}'_{||}\right)$ vanishes for $\left|\mathbf{x}'_{||}\right| > a$. For a uniform lateral distribution this may be taken simply to be a step function[11,12]

$$\chi\left(\mathbf{x}'_{||}\right) = \frac{1}{\sqrt{\pi a^2}} \begin{cases} 1 & \left|\mathbf{x}'_{||}\right| \leqslant a \\ 0 & \left|\mathbf{x}'_{||}\right| > a. \end{cases} \tag{15.2.5}$$

On the other hand, $\xi(z)$ is so chosen that its Fourier transform $\tilde{\xi}(q) \equiv \tilde{\eta}(q-p')$ has a pronounced peak at $q = p'$ and the only appreciable value of $|\tilde{\eta}(q-p')|^2$ occurs in the neighborhood of this point. With $E(q) = q^2/2m$,

$$\xi(z,t) = \int_{-\infty}^{\infty} \frac{dq}{2\pi\hbar}\, e^{i[qz - E(q)t]/\hbar}\, \tilde{\eta}(q-p'). \tag{15.2.6}$$

By a change of the integration variable from q to Q,

$$Q = q - p' \tag{15.2.7}$$

and

$$q^2 \simeq p'^2 + 2p'Q \tag{15.2.8}$$

since only a small Q^2 is important, we may rewrite

$$\xi(z,t) = e^{i[p'z - E(p')t]/\hbar} \int_{-\infty}^{\infty} \frac{dQ}{2\pi\hbar}\, e^{iQ\left(z - \frac{p'}{m}t\right)/\hbar}\, \tilde{\eta}(Q) \tag{15.2.9}$$

or

$$\xi(z,t) = e^{i[p'z - E(p')t]/\hbar}\, \eta\left(z - \frac{p'}{m}t\right). \tag{15.2.10}$$

The integral multiplying $\xi(z,t)$ in (15.2.4) will be denoted by $\chi(\mathbf{x}_{||},t)$

$$\chi\left(\mathbf{x}_{||},t\right) = \frac{1}{\sqrt{\pi a^2}} \int_{|\mathbf{x}'_{||}| \leqslant a} d^2\mathbf{x}'_{||}\, G^0_+\left(\mathbf{x}_{||},\mathbf{x}'_{||};t\right) \tag{15.2.11}$$

which from (9.1.38) is given by

$$\chi\left(\mathbf{x}_{||},t\right) = \frac{m}{2\pi i \hbar t}\, \frac{1}{\sqrt{\pi a^2}} \int_{|\mathbf{x}'_{||}| \leqslant a} d^2\mathbf{x}'_{||}\, \exp\frac{im}{2\hbar t}\left(\mathbf{x}_{||} - \mathbf{x}'_{||}\right)^2. \tag{15.2.12}$$

This integral has many interesting properties.[13] It may be expressed as

---

[12] More precisely, $\chi(\mathbf{x}'_{||})$ may be taken to vanish smoothly at $|\mathbf{x}''| = a$, in a continuous manner, starting at $1/\sqrt{\pi a^2}$ for $|\mathbf{x}''| = 0$, to avoid technical problems arising near the boundaries of the wavepacket. This, however, complicates the analysis to some extent.

[13] For the relevant details of $\chi(\mathbf{x}_{||},t)$ in (15.2.12), (15.2.13) see the appendix to this section.

$$\chi\left(\mathbf{x}_{\|}, t\right) = \frac{1}{2\pi i} \frac{u}{\sqrt{\pi a^2}} e^{iv^2/2u} F(u, v) \tag{15.2.13}$$

with

$$u = \frac{ma^2}{\hbar t}, \qquad v = \frac{|\mathbf{x}_{\|}|}{a} u \tag{15.2.14}$$

$$F(u, v) = 2\pi \int_0^1 \rho \, d\rho \, J_0(\rho v) e^{i(u/2)\rho^2} \tag{15.2.15}$$

where $J_0(\rho v)$ is the zeroth order Bessel function, and we have the normalization condition

$$\frac{1}{2\pi^2} \int_0^\infty v \, dv \, |F(u, v)|^2 = 1 \tag{15.2.16}$$

that is, $\|\chi(\cdot, t)\| = 1$.

At $t = 0$, $\chi(\mathbf{x}_{\|}, 0)$ vanishes for $|\mathbf{x}_{\|}| > a$. On the other hand for $t \neq 0$, and $|\mathbf{x}_{\|}| \gg a$, corresponding to the position of the detector set, mentioned earlier,

$$\chi(\mathbf{x}_{\|}, t) = \mathcal{O}\left((a/|\mathbf{x}_{\|}|)^{3/2}\right). \tag{15.2.17}$$

In the sequel, we consider the limits[14]

$$|\mathbf{x}| \gg a \gg \hbar/p' \tag{15.2.18}$$

and

$$a^2 p'/\hbar |\mathbf{x}| \gg 2\pi. \tag{15.2.19}$$

In (A-15.2.11), we will see by noting that $m|\mathbf{x}|/t \sim p'$, and this implies from (15.2.19), that

$$\int_{|\mathbf{x}_{\|}| \leq a} d^2 \mathbf{x}_{\|} \, |\chi(\mathbf{x}_{\|}, t)|^2 \simeq 1 \tag{15.2.20}$$

and the lateral width of the wavepacket does not spread significantly.

To compute the probability in (15.2.1), we use the representation

$$G_+^0(x, x'') = \frac{m}{2\pi i \hbar} \frac{1}{|\mathbf{x} - \mathbf{x}''|} \int_{-\infty}^\infty \frac{dp^0}{2\pi \hbar} e^{-ip^0(t-t'')/\hbar}$$

$$\times \exp\left(\frac{i}{\hbar} \sqrt{2m p^0} \, |\mathbf{x} - \mathbf{x}''|\right) \tag{15.2.21}$$

in the integral equation for $\psi_-(x)$ in (15.1.11). For $|\mathbf{x}| > D$, D large

---

[14] For example, for $a \simeq 10^{-3}$ meters, $|\mathbf{x}| \sim 1$ meter, $a^2 p'/\hbar |\mathbf{x}| \sim p' 10^{28}$/kg.meter.s^{-1}, where $\hbar/p'$ denotes the wavelength of the packet. For $\hbar/p' \sim 10^{-10}$ meters, $a^2 p'/\hbar |\mathbf{x}| \sim 10^4$.

15.2 Particle Detection...

$$G_+^0(x, x'') \to \frac{m}{2\pi i \hbar |\mathbf{x}|} \int_{-\infty}^{\infty} \frac{dp^0}{2\pi \hbar} \, e^{-ip^0(t-t'')/\hbar} \exp\left(\frac{i}{\hbar}\sqrt{2mp^0}\,|\mathbf{x}|\right)$$

$$\times \exp\left(-\frac{i}{\hbar}\sqrt{2mp^0}\,\mathbf{n}\cdot\mathbf{x}''\right) \qquad (15.2.22)$$

where
$$\mathbf{n} = \mathbf{x}/|\mathbf{x}|. \qquad (15.2.23)$$

For $|\mathbf{x}_\parallel| \gg a$, $|\mathbf{x}| > D$, D large, we see from (15.2.17), (15.2.18), (15.2.22), that $\Phi_{\text{in}}(\mathbf{x},t)$ may be neglected on the right-hand side of (15.1.11). That is, no interference[15] between the incident and scattered components of $\psi_-(x)$ needs to be considered.

Using the Fourier transforms of $V(x)$, $A_-(x'',x')$, in (15.1.12)/(15.1.24), (15.1.13)/(15.1.25), respectively, we obtain from (15.1.11), (15.2.22), (15.2.17), the asymptotic equality

$$\psi_-(\mathbf{x},t) = -\frac{m}{2\pi\hbar^2 |\mathbf{x}|} \int \frac{d^3q}{(2\pi\hbar)^3}\, e^{i[|\mathbf{q}||\mathbf{x}| - E(\mathbf{q})t]/\hbar}$$

$$\times F_-(|\mathbf{q}|\,\mathbf{n}, \mathbf{q}; E(\mathbf{q}))\, \tilde{\Phi}_{\text{in}}(\mathbf{q}) \qquad (15.2.24)$$

for $|\mathbf{x}_\parallel|^{3/2} \gg a^{1/2} |\mathbf{x}|$, in particular, and hence for $|\mathbf{x}_\parallel| \gg a$. Here

$$\tilde{\Phi}_{\text{in}}(\mathbf{q}) = \int d^3x\, e^{-i\mathbf{q}\cdot\mathbf{x}/\hbar}\, \Phi_{\text{in}}(\mathbf{x},0) \qquad (15.2.25)$$

or

$$\tilde{\Phi}_{\text{in}}(\mathbf{q}) = \frac{1}{\sqrt{\pi a^2}} \int_{|\mathbf{x}'_\parallel| \leq a} d^2x'_\parallel\, e^{-i\mathbf{q}_\parallel \cdot \mathbf{x}'_\parallel/\hbar}\, \tilde{\eta}(q_\perp - p'). \qquad (15.2.26)$$

Upon writing $\psi_-(\mathbf{x},t) \equiv \psi_-(|\mathbf{x}|, \mathbf{n}; t)$, we already see from (15.2.24), that a particle initially prepared with momentum $\mathbf{q}$, with distribution provided by $\left|\tilde{\Phi}_{\text{in}}(\mathbf{q})\right|^2$, the particle emerges from the scattering process with momentum $|\mathbf{q}|\,\mathbf{n}$ in the direction $\mathbf{n}$. In particular, for a particle emerging, in configuration space, within a cone C_0 in (15.2.3), its momentum would be directed within the same cone. This is in conformity with one's perception of a scattering process.

In (15.2.24), (15.2.26) we make a change of the variables of integrations $\mathbf{q}_\parallel$, $\mathbf{x}'_\parallel$ to $\mathbf{Q}$, $\boldsymbol{\rho}'$

$$\boldsymbol{\rho}' = a\,\mathbf{x}'_\parallel \sqrt{\frac{p'}{\hbar |\mathbf{x}|}}, \qquad \mathbf{Q} = \mathbf{q}_\parallel \sqrt{\frac{\hbar |\mathbf{x}|}{p'}}\, \frac{1}{a} \qquad (15.2.27)$$

---

[15] We will look into the forward direction, that is in the effective direction of propagation of the incident packet, in §15.4.

and note the invariance of the product measures

$$d^2 q_\| \, d^2 x''_\| = d^2 Q \, d^2 \rho'. \tag{15.2.28}$$

From the limit in (15.2.19), the integral over ρ' corresponding to the $\mathbf{x}''$-integral in (15.2.26) becomes effectively replaced by $(2\pi\hbar)^2 \delta^3(\mathbf{Q})$. On the other hand since $\tilde{\eta}(q_\perp - p')$ has a pronounced peak at $q_\perp = p'$ and is appreciably non-vanishing in the neighborhood of this point, we can carry out the expansion $q_\perp^2 \simeq p'^2 + 2p'(q_\perp - p')$ as in (15.2.8) for $\exp i [q_\perp |\mathbf{x}| - E(q_\perp) t]/\hbar$ in (15.2.24), and we may effectively evaluate $F_-(q_\perp \mathbf{n}, \mathbf{q}_\perp; E(q_\perp))$ at $\mathbf{q}_\perp = (\mathbf{0}, p')$.

Accordingly, we obtain asymptotically for $\psi_-(\mathbf{x}, t)$ in (15.2.24),

$$\psi_-(\mathbf{x}, t) = -\frac{m}{2\pi\hbar^2 |\mathbf{x}|} F_-(p'\mathbf{n}, \mathbf{p}'; E(\mathbf{p}')) e^{\frac{i}{\hbar}[|\mathbf{p}'||\mathbf{x}| - E(\mathbf{p}')t]} \frac{\eta\left(|\mathbf{x}| - \frac{p'}{m} t\right)}{\sqrt{\pi a^2}} \tag{15.2.29}$$

where we have used the fact that

$$\int_{-\infty}^{\infty} \frac{dq'}{2\pi\hbar} \exp\left(i\frac{q'}{\hbar}\left(|\mathbf{x}| - \frac{p'}{m} t\right)\right) \tilde{\eta}(q') = \eta\left(|\mathbf{x}| - \frac{p'}{m} t\right) \tag{15.2.30}$$

and $q' = q_\perp - p'$.

For D large, we then have from (15.2.1), (15.2.30),

$$\int_{C_D} d^3\mathbf{x} \, |\psi_-(\mathbf{x}, t)|^2 = \int_{C_0(\mathbf{n})} d\Omega \left(\frac{m}{2\pi\hbar^2}\right)^2 |F_-(p'\mathbf{n}, \mathbf{p}'; E(\mathbf{p}'))|^2$$

$$\times \int_D^\infty d|\mathbf{x}| \, \frac{\left|\eta\left(|\mathbf{x}| - \frac{p'}{m} t\right)\right|^2}{\pi a^2} \tag{15.2.31}$$

where

$$C_0(\mathbf{n}) = \{\mathbf{n} : 1 \geqslant \mathbf{n} \cdot \mathbf{N} \geqslant \alpha, \text{for some } \alpha \in (0, 1]\} \tag{15.2.32}$$

and the unit vector $\mathbf{N}$, and α are chosen such that $|\mathbf{x}_\|| \gg a$.

In the next section we will use (15.2.31) to obtain an expression for the differential cross section of the process.

Appendix to §15.2: Some Properties of $F(u, v)$

The function[16] $F(u, v)$ defined in (15.2.15), of the variables u, and v in (15.2.14), may be expressed in terms of so-called Lommel functions. For $u/v < 1$,

---

[16] For more related details, see: Manoukian (1989); Wolf (1951); Born and Wolf (1975).

$$F(u,v) = \frac{2\pi}{u} e^{iu/2} [U_1(u,v) - i\, U_2(u,v)] \tag{A-15.2.1}$$

and for $u/v > 1$,

$$F(u,v) = \frac{2\pi}{u} \left[ie^{-iv^2/2u} - ie^{iu/2} V_0(u,v) - e^{iu/2} V_1(u,v) \right] \tag{A-15.2.2}$$

where $U_n(u,v)$, $V_n(u,v)$ are Lommel functions defined by

$$U_n(u,v) = \sum_{j=0}^{\infty} (-1)^j \left(\frac{u}{v}\right)^{2j+n} J_{2j+n}(v) \tag{A-15.2.3}$$

$$V_n(u,v) = \sum_{j=0}^{\infty} (-1)^j \left(\frac{v}{u}\right)^{2j+n} J_{2j+n+1}(v) \tag{A-15.2.4}$$

and $J_m(v)$ is a Bessel function of the first kind of order m.
In particular

$$U_1(u,u) = \frac{\sin u}{2} \tag{A-15.2.5}$$

$$U_2(u,u) = \frac{J_0(u) - \cos u}{2} \tag{A-15.2.6}$$

and

$$\int_u^\infty v\, dv\, [U_1^2(u,v) + U_2^2(u,v)] = \frac{u^2}{2} [J_0(u)\cos u + J_1(u)\sin u] \tag{A-15.2.7}$$

$F(u,v)$ satisfies the normalization condition in (15.2.16).
For $u/v < 1$,

$$\left| F(u,v) - \frac{2\pi}{v} e^{iu/2} J_1(v) \right| = \frac{1}{v} \mathcal{O}\left(\frac{u}{v}\right) \tag{A-15.2.8}$$

giving rise to the behavior in (15.2.17).
For $u/v > 1$,

$$\left| F(u,v) - \frac{2\pi i}{u} e^{-iv^2/2u} + \frac{2\pi i}{u} e^{iu/2} J_0(v) \right| = \frac{1}{u} \mathcal{O}\left(\frac{v}{u}\right). \tag{A-15.2.9}$$

Finally we note from (A-15.2.7), (15.2.13) that

$$\int_{|\mathbf{x}_\parallel| \leq a} d^2\mathbf{x}_\parallel\, |\chi(\mathbf{x}_\parallel, t)|^2 = 1 - (J_0(u)\cos u + J_1(u)\sin u) \tag{A-15.2.10}$$

and for $u \gg 1$,

$$\int_{|\mathbf{x}_{||}|\leqslant a} d^2\mathbf{x}_{||}\, |\chi(\mathbf{x}_{||},t)|^2 \simeq 1. \qquad (A\text{-}15.2.11)$$

For $u = 50$, the right-hand side of (A-15.2.10) attains the value 0.92. The condition (15.2.19) implies that $u \gg \frac{m|\mathbf{x}|}{t}\frac{2\pi}{p'}$ and for $p' \sim m|\mathbf{x}|/t$, this leads to the condition $u \gg 2\pi$. That is, under this limit the lateral width of the wavepacket does not spread significantly.

15.3 Differential Cross Sections

15.3.1 Expression for the Differential Cross Section

In reference to (15.2.4)–(15.2.6),

$$\int_D^\infty \frac{d|\mathbf{x}|\,\left|\eta\left(|\mathbf{x}| - \frac{p'}{m}t\right)\right|^2}{\pi a^2} \simeq \int_{|\mathbf{x}_{||}|\leqslant a} d^2\mathbf{x}_{||} \int_D^\infty dz\, \frac{|\Phi_{\text{in}}(\mathbf{x}_{||},z,t)|^2}{\pi a^2} \qquad (15.3.1)$$

denotes the probability, per unit area, that the incident particle flows through the plane $z = D$ at time t, for both large, in the absence of a potential.[17] On the other hand, the left-hand side of (15.2.31) represents the probability that an incident particle is scattered within a cone, and is found beyond a radial distance D at time t. Accordingly for a sufficiently narrow cone, i.e., for which α in (15.2.32) is close to one, $\mathbf{n} \to \mathbf{N}$, we obtain, by dividing (15.2.31) by the probability per unit area in (15.3.1), the expression for the differential cross section

$$\frac{d\sigma}{d\Omega} = \left(\frac{m}{2\pi\hbar^2}\right)^2 |F_-(p'\mathbf{n},\mathbf{p}';E(\mathbf{p}'))|^2 \qquad (15.3.2)$$

where the conservation of energy is evident.

One may define the scattering amplitude

$$f(\mathbf{p},\mathbf{p}') = -\frac{m}{2\pi\hbar^2}\int d^3\mathbf{p}''\, V(\mathbf{p}-\mathbf{p}'')\, A_-(\mathbf{p}'',\mathbf{p}';E(\mathbf{p}')) \qquad (15.3.3)$$

with

$$\mathbf{p} = |\mathbf{p}'|\mathbf{n} \qquad (15.3.4)$$

to rewrite simply

$$\frac{d\sigma}{d\Omega} = |f(\mathbf{p},\mathbf{p}')|^2 \qquad (15.3.5)$$

where we have used the definition of $F_-(\mathbf{p},\mathbf{p}';E(\mathbf{p}'))$ in (15.1.29). The reason for choosing the multiplicative (-1) factor in (15.3.3) will be seen in the next section when studying the so-called optical theorem.

---

[17] In the presence of a potential, the probability of observing the particle in the forward direction is altered due to the interference of the incident wavepacket and the scattered amplitude in that direction as will be investigated in the next section.

15.3 Differential Cross Sections

From (15.1.27), (15.1.36), one may also write the scattering amplitude as

$$f(\mathbf{p}, \mathbf{p}') = -4\pi^2 \hbar\, m \bigg([p^0 - E(\mathbf{p})]$$
$$\times G_+(\mathbf{p}, \mathbf{p}'; p^0) [p^0 - E(\mathbf{p}')] \bigg) \bigg|_{p^0 = E(\mathbf{p}) = E(\mathbf{p}')} \qquad (15.3.6)$$

involving the full Green function $G_+(\mathbf{p}, \mathbf{p}'; p^0)$ (see (15.1.33), (15.1.35)) with a restriction set in (15.3.6) on the energy shell as indicated.

The cross section is defined by

$$\sigma = \int d\Omega\, |f(\mathbf{p}, \mathbf{p}')|^2 \qquad (15.3.7)$$

and a further important relation between σ and the scattering amplitude will be derived in the next section referred to as the optical theorem.

For a systematic study of the scattering amplitude $f(\mathbf{p}, \mathbf{p}')$ we derive an important integral equation which follows from (15.1.26). For arbitrary p^0,

$$A_-(\mathbf{q}, \mathbf{p}'; p^0) = \delta^3(\mathbf{q} - \mathbf{p}') + \frac{1}{[p^0 - E(\mathbf{q}) + i\varepsilon]}$$
$$\times \int \frac{d^3 q'}{(2\pi\hbar)^3} V(\mathbf{q} - \mathbf{q}')\, A_-(\mathbf{q}', \mathbf{p}'; p^0). \qquad (15.3.8)$$

Upon multiplying this equation by $V(\mathbf{p} - \mathbf{q})$, integrating over $\mathbf{q}$, and setting, in general[18]

$$\int d^3 q'\, V(\mathbf{q} - \mathbf{q}')\, A_-(\mathbf{q}', \mathbf{p}'; p^0) = T(\mathbf{q}, \mathbf{p}'; p^0) \qquad (15.3.9)$$

off the energy shell, gives

$$T(\mathbf{p}, \mathbf{p}'; p^0) = V(\mathbf{p} - \mathbf{p}') + \int \frac{d^3 q}{(2\pi\hbar)^3} \frac{V(\mathbf{p} - \mathbf{q})}{[p^0 - E(\mathbf{q}) + i\varepsilon]} T(\mathbf{q}, \mathbf{p}'; p^0) \qquad (15.3.10)$$

referred to as the T-matrix. In particular, from (15.1.26), (15.1.27) and (15.3.10), the scattering amplitude may be written as

$$f(\mathbf{p}, \mathbf{p}') = -\frac{m}{2\pi\hbar^2} T(\mathbf{p}, \mathbf{p}'; p^0) \bigg|_{p^0 = E(\mathbf{p}) = E(\mathbf{p}')}. \qquad (15.3.11)$$

To first order in V, i.e., for a weak potential,

$$T(\mathbf{p}, \mathbf{p}'; p^0) \simeq V(\mathbf{p} - \mathbf{p}') \qquad (15.3.12)$$

---

[18] Note that for $p^0 = E(\mathbf{p}')$, $T(\mathbf{q}, \mathbf{p}'; E(\mathbf{p}'))$ coincides with $F_-(\mathbf{q}, \mathbf{p}'; E(\mathbf{p}'))$ — see (15.1.29).

referred to as the first Born approximation, the differential cross section takes the simple form

$$\frac{d\sigma}{d\Omega} \simeq \frac{m^2}{4\pi^2\hbar^4} |V(\mathbf{p} - \mathbf{p}')|^2 \qquad (15.3.13)$$

with $\mathbf{p} = |\mathbf{p}'|\mathbf{n}$. For some applications of the Born approximation see Problems 15.1, 15.2.

15.3.2 Sufficiency Conditions for the Validity of the Born Expansion

We carry out a rigorous study of sufficiency conditions for the validity of the Born approximation in (15.3.12) and to arbitrary orders in the potential energy $V(\mathbf{x})$ starting from the integral equation for the T-matrix given in (15.3.10). To this end, with m replaced by the reduced mass μ of a two-particle system and for $V(\mathbf{x}) = V(r)$, $r = |\mathbf{x}|$ as a function of the distance between the two particles, suppose that the following sufficiency conditions are satisfied:

$$\int_0^\infty r^2 dr\, |V(r)| < \infty \quad \text{and} \quad \int_0^\infty r\, dr\, |V(r)| < \infty. \qquad (15.3.14)$$

The first condition refers to the absolute integrability of $V(\mathbf{x})$.

We set $p^0 = E(\mathbf{p}')$ in the integral equation (15.3.10) and denote

$$T(\mathbf{p}, \mathbf{p}'; E(\mathbf{p}')) \equiv T(\mathbf{p}, \mathbf{p}'). \qquad (15.3.15)$$

To first order in the potential,

$$T^{(1)}(\mathbf{p}, \mathbf{p}') = \int d^3\mathbf{x}\, e^{-i(\mathbf{p}-\mathbf{p}')\cdot\mathbf{x}/\hbar}\, V(\mathbf{x}) \qquad (15.3.16)$$

and more generally, we carry out an expansion

$$T(\mathbf{p}, \mathbf{p}') = \sum_{n \geq 1} T^{(n)}(\mathbf{p}, \mathbf{p}') \qquad (15.3.17)$$

where $T^{(n)}(\mathbf{p}, \mathbf{p}')$ is of nth order in the potential. Using the Fourier transform in (15.3.16), and the integral (see (9.1.25), (9.1.31)) $(E(\mathbf{q}) = \mathbf{q}^2/2\mu)$

$$\int \frac{d^3\mathbf{q}}{(2\pi\hbar)^3} \frac{e^{i\mathbf{q}\cdot\mathbf{x}/\hbar}}{[p^0 - E(\mathbf{q}) + i\varepsilon]} = -\frac{\mu}{2\pi\hbar^2} \frac{e^{i\sqrt{2mp^0}\,|\mathbf{x}|}}{|\mathbf{x}|} \qquad (15.3.18)$$

one obtains the explicit expression $(n \geq 2)$

$$T^{(n)}(\mathbf{p}, \mathbf{p}') = \left(-\frac{\mu}{2\pi\hbar^2}\right)^{n-1} \int d^3\mathbf{x}_1 \ldots d^3\mathbf{x}_n\, e^{-i\mathbf{x}_1\cdot\mathbf{p}/\hbar}\, V(\mathbf{x}_1) \frac{e^{ip|\mathbf{x}_1-\mathbf{x}_2|/\hbar}}{|\mathbf{x}_1 - \mathbf{x}_2|}$$

$$\times V(\mathbf{x}_2) \frac{e^{ip|\mathbf{x}_2-\mathbf{x}_3|/\hbar}}{|\mathbf{x}_2-\mathbf{x}_3|} \ldots V(\mathbf{x}_{n-1}) \frac{e^{ip|\mathbf{x}_{n-1}-\mathbf{x}_n|/\hbar}}{|\mathbf{x}_{n-1}-\mathbf{x}_n|} V(\mathbf{x}_n) e^{i\mathbf{x}_n \cdot \mathbf{p}'/\hbar}. \tag{15.3.19}$$

We may then bound $T^{(n)}(\mathbf{p},\mathbf{p}')$ in absolute value as

$$\left|T^{(n)}(\mathbf{p},\mathbf{p}')\right| \leqslant \left(\frac{\mu}{2\pi\hbar^2}\right)^{n-1} \int d^3\mathbf{x}_n |V(\mathbf{x}_n)| \left|I^{(n-1)}(\mathbf{x}_n)\right| \tag{15.3.20}$$

where

$$I^{(n-1)}(\mathbf{x}_n) = \int d^3\mathbf{x}_1 \ldots d^3\mathbf{x}_{n-1}\, e^{-i\mathbf{x}_1\cdot\mathbf{p}/\hbar}$$

$$\times V(\mathbf{x}_1) \frac{e^{ip|\mathbf{x}_1-\mathbf{x}_2|/\hbar}}{|\mathbf{x}_1-\mathbf{x}_2|} \ldots V(\mathbf{x}_{n-1}) \frac{e^{ip|\mathbf{x}_{n-1}-\mathbf{x}_n|/\hbar}}{|\mathbf{x}_{n-1}-\mathbf{x}_n|}. \tag{15.3.21}$$

Using the inequality

$$\int \frac{d\Omega_1}{|\mathbf{x}_1-\mathbf{x}_2|} \leqslant \frac{4\pi}{r_1} \tag{15.3.22}$$

shown later, we have the following $\mathbf{x}_2$-independent bound

$$\left|\int d^3\mathbf{x}_1 \frac{V(\mathbf{x}_1)}{|\mathbf{x}_1-\mathbf{x}_2|}\right| \leqslant 4\pi \int_0^\infty r_1\, dr_1 |V(r_1)| \tag{15.3.23}$$

which by hypothesis (see (15.3.14)) exists.

From (15.3.23), we then have

$$\left|I^{(n-1)}(\mathbf{x}_n)\right| \leqslant \left(4\pi \int_0^\infty r\, dr |V(r)|\right)^{n-1} \tag{15.3.24}$$

and, for all $n \geqslant 1$,

$$\left|T^{(n)}(\mathbf{p},\mathbf{p}')\right| \leqslant \left(\int d^3\mathbf{x} |V(\mathbf{x})|\right) \left(\frac{2\mu}{\hbar^2} \int_0^\infty r\, dr |V(r)|\right)^{n-1} \tag{15.3.25}$$

where both factors on the right-hand side exist by hypothesis.

The series in (15.3.17) is then absolutely convergent for

$$\frac{2\mu}{\hbar^2} \int_0^\infty r\, dr |V(r)| < 1 \tag{15.3.26}$$

giving rise to a sufficiency condition for the validity of the Born series in (15.3.17).

For a particle of mass m in a bounded potential $|V(r)| \leqslant V_0$ of finite range $V(r) = 0$ for $r > R$, the condition in (15.3.26) is satisfied for

$$\frac{m}{\hbar^2} V_0 R^2 < 1. \tag{15.3.27}$$

To establish the inequality in (15.3.22), we use the expansion

$$\frac{1}{|\mathbf{x}_1 - \mathbf{x}_2|} = \sum_{\ell=0}^{\infty} \left(\frac{r_<}{r_>}\right)^\ell \frac{1}{r_>} P_\ell(\cos\theta)$$

$$= 4\pi \sum_{\ell=0}^{\infty} \sum_{m=-\ell}^{\ell} \frac{1}{(2\ell+1)} \left(\frac{r_<}{r_>}\right)^\ell \frac{1}{r_>} Y_{\ell\,m}(\hat{x}_1) Y^*_{\ell\,m}(\hat{x}_2) \quad (15.3.28)$$

where $r_<$ and $r_>$ denote, respectively, the smaller and the larger of r_1, r_2. Hence

$$\int \frac{d\Omega_1}{|\mathbf{x}_1 - \mathbf{x}_2|} = 4\pi \left[\frac{\Theta(r_1 - r_2)}{r_1} + \frac{\Theta(r_2 - r_1)}{r_2} \right] \quad (15.3.29)$$

and since $\Theta(r_2 - r_1)/r_2 < \Theta(r_2 - r_1)/r_1$, and $\Theta(r_1 - r_2) + \Theta(r_2 - r_1) = 1$, the inequality in (15.3.22) follows. Note that an equality in (15.3.22) holds if $r_1 \geqslant r_2$.

15.3.3 Two-Particle Scattering

Needless to say, the scattering theory developed above and the expression for the differential cross section obtained in (15.3.5) hold for the interaction of a particle of mass $m = m_1$ with a much heavier (approximately motionless) particle of mass m_2. If m_2 is not large, and the particles are not identical and of spin 0, all of the above formulae are still valid in the center of mass of the two particles if one simply replaces $m = m_1$ in them by the reduced mass $\mu = m_1 m_2/(m_1 + m_2)$. The transformation to the laboratory system is the same as in classical dynamics.[19] For definiteness, consider the particle of mass m_2 (the target particle) at rest, and a spherically symmetric potential so that the scattering, in the center of mass, is specified by the angle θ only. The scattering angles ϑ_1 and ϑ_2 of the particles of masses m_1 and m_2, respectively, in the laboratory system are given by the well know formulae

$$\tan\vartheta_1 = \sin\theta \Big/ \left(\frac{m_1}{m_2} + \cos\theta\right) \quad (15.3.30)$$

$$\vartheta_2 = (\pi - \theta)/2 \quad (15.3.31)$$

with the latter giving the recoil angle of the target particle. The differential cross section in the laboratory system is given by

$$\left.\frac{d\sigma}{d\Omega_1}(\vartheta_1)\right|_{\text{LAB}} = \frac{d\sigma}{d\Omega}(\theta) \frac{d(\cos\theta)}{d(\cos\vartheta_1)} \quad (15.3.32)$$

---

[19] Cf. Marion and Thornton (1988), p. 310, p. 326.

where $d\sigma(\theta)/d\Omega$ is the one obtained in the center of mass system.

For identical (i.e., indistinguishable) particles of mass $m_1 = m_2 = m$,

$$\vartheta_1 = \frac{\theta}{2}, \tag{15.3.33}$$

and we have to use the proper symmetrization of the scattering amplitudes, based on the statistics of the particles, since a detector cannot distinguish between them. In the center of mass system two possible routes of scattering are shown in Figure 15.2.

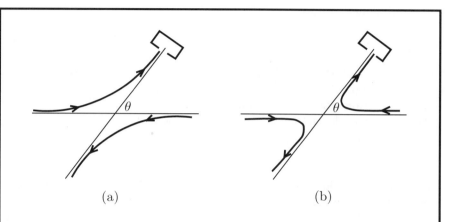

Fig. 15.2. For identical particles, the detector (set at angle θ-shown in the center of mass frame) cannot distinguish between the two possible routes (a) and (b) of scattering. The necessary symmetrization, based on the statistics of the particles, of the scattering amplitudes leads to interference effects which are observed experimentally.

Consider two indistinguishable particles of spin 0. In this case we have to add to $f(\theta)$ the amplitude $f(\pi - \theta)$ corresponding to the route of scattering shown in Figure 15.2 (b) before defining the differential cross section. The amplitude $f(\theta)$ is referred to as the direct scattering amplitude, while $f(\pi - \theta)$ as the exchange one. The differential cross section (in the center of mass) is then given by

$$\begin{aligned}\frac{d\sigma}{d\Omega} &= |f(\theta) + f(\pi - \theta)|^2 \\ &= |f(\theta)|^2 + |f(\pi - \theta)|^2 + 2\,\mathrm{Re}\,(f^*(\theta)f(\pi - \theta)).\end{aligned} \tag{15.3.34}$$

For a detector set at $\theta = \pi/2$, $d\sigma/d\Omega$ is given by $4|f(\theta)|^2$ rather than $2|f(\theta)|^2$ as the latter would be expected by a naive classical argument. Such a doubling is observed experimentally and is due to the interference term in (15.3.34) as a result of the correlation between the particles arising from the symmetry of the states under the interchange of the two particles.

The situation with fermions turns out to be quite interesting. For definiteness consider the scattering of two indistinguishable spin 1/2 particles interacting via a spin-independent interaction as above. Let $|\chi\rangle$, $|\chi'\rangle$ denote the normalized spin states of the particles. The effective scattering amplitude would then be given by

$$f(\theta)|\chi_1\rangle|\chi_2'\rangle - f(\pi-\theta)|\chi_1'\rangle|\chi_2\rangle \qquad (15.3.35)$$

and the differential cross section is

$$\frac{d\sigma}{d\Omega} = |f(\theta)|^2 + |f(\pi-\theta)|^2 - 2\,\text{Re}\,(f^*(\theta)f(\pi-\theta))\,|\langle\chi|\chi'\rangle|^2 \qquad (15.3.36)$$

Therefore an interference term arises only if the spin states are non-orthogonal. In particular if the spins of the particles are in opposite directions of a quantization axis no interference term arises. On the other hand, if the spins of the particles are along the same direction, then

$$\frac{d\sigma}{d\Omega} = |f(\theta) - f(\pi-\theta)|^2 \qquad (15.3.37)$$

This expression is interesting in the sense that at $\theta = \pi/2$, the differential cross vanishes, unlike for the bosonic case, and is verified experimentally.

Now consider the scattering of unpolarized spin 1/2 particles. We know from §2.8, §5.5, from the addition of spin 1/2's, we have the triplet state, corresponding to a spin 1, and the singlet state, corresponding to a spin 0. The Fermi character of the spin 1/2 particles, requires from (15.3.36) that

$$\frac{d\sigma}{d\Omega}\uparrow\uparrow = |f(\theta) - f(\pi-\theta)|^2$$

$$\frac{d\sigma}{d\Omega}\uparrow\downarrow = |f(\theta)|^2 + |f(\pi-\theta)|^2$$

$$\frac{d\sigma}{d\Omega}\downarrow\uparrow = |f(\theta)|^2 + |f(\pi-\theta)|^2 \qquad (15.3.38)$$

$$\frac{d\sigma}{d\Omega}\downarrow\downarrow = |f(\theta) - f(\pi-\theta)|^2$$

with weight factors 1/4 for unpolarized particles, giving

$$\frac{d\sigma_u}{d\Omega} = |f(\theta)|^2 + |f(\pi-\theta)|^2 - \text{Re}\,(f^*(\theta)f(\pi-\theta)) \qquad (15.3.39)$$

where u stands for unpolarized, and unlike (15.3.36) note that there is no factor of 2 in the interference term in (15.3.39).

15.4 The Optical Theorem and Its Interpretation; Phase Shifts   821

The result in (15.3.39) may be also obtained by noting the symmetry properties of the triplet and singlet states. To this end, note from (5.6.21) that for spin states of spins s_1, s_2,

$$\langle s_1, m_1; s_2, m_2 | s, m \rangle = (-1)^{s_1+s_2-s} \langle s_2, m_2; s_1, m_1 | s, m \rangle \qquad (15.3.40)$$

where in the above case $s_1 = s_2 = 1/2$. Therefore the triplet state ($s = 1$) is symmetric in spin space, while the singlet ($s = 0$) is anti-symmetric. The Pauli exclusion principle then requires that

$$\frac{d\sigma_u}{d\Omega} = \frac{3}{4}|f(\theta) - f(\pi - \theta)|^2 + \frac{1}{4}|f(\theta) + f(\pi - \theta)|^2 \qquad (15.3.41)$$

which coincides with the expression in (15.3.39). The 3 to 1 factors above correspond to the number of allowed values for m for $s = 1$, and 0, respectively.

For spin-dependent interactions of spin 1/2 particles, we may use the very general analysis carried out in §8.6 for the scattering of a spin 1/2 particle off a spin 0 target as well as the one for the scattering off a spin 1/2 target. Given an M-matrix relating an initial and final state, the differential cross section of the scattering process is given by

$$\frac{d\sigma}{d\Omega} = \frac{\text{Tr}\left[M \rho^{(i)} M^\dagger\right]}{\text{Tr}\left[\rho^{(i)}\right]} \qquad (15.3.42)$$

where $\rho^{(i)}$ is the initial density operator, and may be normalized as $\text{Tr}[\rho^{(i)}] = 1$, and $M \rho^{(i)} M^\dagger$ is the final density operator. General structures of the M matrices were determined in §8.6 and polarized as well as unpolarized incident beams were considered. For details we refer the reader to that section.

15.4 The Optical Theorem and Its Interpretation; Phase Shifts

We derive a relation between the cross section and the scattering amplitude in the forward direction, which is referred to as the optical theorem. The physical meaning embodied in this result will be also emphasized. Finally, we decompose the scattering amplitude into partial amplitudes specified by angular momentum states, referred to as a phase shift analysis, by making use, in the precess, of the optical theorem.

15.4.1 The Optical Theorem

To obtain the above mentioned relation, we multiply the equation for $A_-(\mathbf{q}, \mathbf{p}'; p^0)$ in (15.3.8) by $T^*(\mathbf{q}, \mathbf{p}; p^0)$ integrate over $\mathbf{q}$, and use the definition in (15.3.9) to obtain after re-arrangements of terms

$$T^*\left(\mathbf{p}', \mathbf{p}; p^0\right) = \int d^3\mathbf{q}\, d^3\mathbf{q}'\, A_-^*\left(\mathbf{q}', \mathbf{p}; p^0\right) V(\mathbf{q}' - \mathbf{q})\, A_-\left(\mathbf{q}, \mathbf{p}'; p^0\right)$$

$$- \int \frac{d^3\mathbf{q}}{(2\pi\hbar)^3}\, T^*\left(\mathbf{q}, \mathbf{p}; p^0\right) \frac{1}{[p^0 - E(\mathbf{q}) + i\varepsilon]}\, T\left(\mathbf{q}, \mathbf{p}'; p^0\right) \quad (15.4.1)$$

where we have used, in the process, the reality of the potential $V(\mathbf{x})$ to infer that $V^*(\mathbf{q}) = V(-\mathbf{q})$.

Upon taking the complex conjugate of (15.4.1) and *interchanging* the momenta $\mathbf{p} \leftrightarrow \mathbf{p}'$, we also have

$$T(\mathbf{p}, \mathbf{p}'; p^0) = \int d^3\mathbf{q}\, d^3\mathbf{q}'\, A_-^*(\mathbf{q}', \mathbf{p}; p^0) V(\mathbf{q}' - \mathbf{q})\, A_-(\mathbf{q}, \mathbf{p}'; p^0)$$

$$- \int \frac{d^3\mathbf{q}}{(2\pi\hbar)^3}\, T^*(\mathbf{q}, \mathbf{p}; p^0) \frac{1}{[p^0 - E(\mathbf{q}) - i\varepsilon]}\, T(\mathbf{q}, \mathbf{p}'; p^0) \quad (15.4.2)$$

where in writing the first integral we have used again the reality of the potential and conveniently relabelled the integration variables $\mathbf{q} \leftrightarrow \mathbf{q}'$. The first integral in (15.4.2) is the same as the first integral in (15.4.1).

We set $p^0 = E(\mathbf{p}') = E(\mathbf{p})$, and use the relation

$$[E(\mathbf{p}') - E(\mathbf{q}) - i\varepsilon]^{-1} - [E(\mathbf{p}') - E(\mathbf{q}) + i\varepsilon]^{-1} = 2\pi i\, \delta(E(\mathbf{p}') - E(\mathbf{q})) \quad (15.4.3)$$

to obtain by subtracting (15.4.1) from (15.4.2)

$$T(\mathbf{p}, \mathbf{p}'; E(\mathbf{p}')) - T^*(\mathbf{p}', \mathbf{p}; E(\mathbf{p}'))$$

$$= -\frac{2\pi i\, m|\mathbf{p}'|}{(2\pi\hbar)^3} \int d\Omega''\, T^*\left(|\mathbf{p}'|\mathbf{n}'', \mathbf{p}; E(\mathbf{p}')\right) T\left(|\mathbf{p}'|\mathbf{n}'', \mathbf{p}'; E(\mathbf{p}')\right) \quad (15.4.4)$$

where we note the $(-)$ sign multiplying the integral on the right-hand side. Finally using the definitions in (15.3.3)/(15.1.29), (15.3.9), (15.4.4) may be rewritten in terms of the scattering amplitude,

$$f(\mathbf{p}, \mathbf{p}') - f^*(\mathbf{p}', \mathbf{p}) = \frac{i|\mathbf{p}'|}{2\pi\hbar} \int d\Omega''\, f^*(|\mathbf{p}'|\mathbf{n}'', \mathbf{p})\, f(|\mathbf{p}'|\mathbf{n}'', \mathbf{p}') \quad (15.4.5)$$

In particular for $\mathbf{p}' = \mathbf{p}$, we have from the expression of the cross section in (15.3.7), the following equality referred to as the *optical theorem*

$$\sigma = \frac{4\pi\hbar}{|\mathbf{p}|} \operatorname{Im} f(\mathbf{p}, \mathbf{p}) \quad (15.4.6)$$

relating the cross section to the imaginary part of the scattering amplitude in the *forward* direction. The choice of the minus sign (-1) factor (a phase)

15.4 The Optical Theorem and Its Interpretation; Phase Shifts

in (15.3.3) adjusts the sign of Im $f(\mathbf{p}, \mathbf{p})$ to be positive matching consistently the positivity of σ in (15.3.7), (15.4.6).

To understand why σ is related to the scattering amplitude in the forward direction, we will investigate the scattering process in this direction and see what quantum physics has to say about this.

In the forward direction, i.e., *for* $|\mathbf{x}_{||}| \lesssim a$, the incident component $\Phi_{in}(x)$ of $\psi_-(x)$ in (15.1.11) cannot be neglected in comparison to the integral on the right-hand side of (15.1.11) as we have done in (15.2.24) for the case $|\mathbf{x}_{||}| \gg a$, using the vanishing property of $\chi(\mathbf{x}_{||}, t)$. In the present case for $|\mathbf{x}_{||}| \lesssim a, |\mathbf{x}| > D$, D large, (15.2.24) becomes simply replaced by

$$\psi_-(x) \to \Phi_{in}(x) - \frac{m}{2\pi\hbar^2 |\mathbf{x}|} \int \frac{d^3\mathbf{q}}{(2\pi\hbar)^3} e^{i[|\mathbf{q}||\mathbf{x}|-E(\mathbf{q})t]/\hbar}$$
$$\times F_-\left(|\mathbf{q}|\mathbf{n}, \mathbf{q}; E(\mathbf{q})\right) \tilde{\Phi}_{in}(\mathbf{q}) \quad (15.4.7)$$

leading to the asymptotic equality

$$|\psi_-(\mathbf{x}, t)|^2 = |\Phi_{in}(\mathbf{x}, t)|^2 - \frac{m}{\pi\hbar^2 |\mathbf{x}|} \text{Re}\{M(\mathbf{x}, t)\} \quad (15.4.8)$$

where Re$\{\cdot\}$ denotes the real part, and

$$M(\mathbf{x}, t) = \int \frac{d^3\mathbf{q}}{(2\pi\hbar)^3} e^{i[|\mathbf{q}||\mathbf{x}|-E(\mathbf{q})t]/\hbar} F_-\left(|\mathbf{q}|\mathbf{n}, \mathbf{q}; E(\mathbf{q})\right) \tilde{\Phi}_{in}(\mathbf{q})\Phi_{in}^*(\mathbf{x}, t)$$
$$(15.4.9)$$

The second term on the right-hand side of (15.4.8) provides an *interference* term between the incident and the scattered components in the forward direction.

Now we look at the intensity in the forward direction, as obtained from (15.4.8), on a screen, set parallel to the x-y plane, at a distance $z = D$, by integrating $|\psi_-(\mathbf{x}, t)|^2$ over an area $\lesssim \pi a^2$.

Accordingly, we are led to consider the integral

$$\int_{|\mathbf{x}_{||}|\lesssim a} d^2\mathbf{x}_{||} \, M(\mathbf{x}, t) \quad (15.4.10)$$

and, as in (15.2.18), (15.2.19), we consider the limits ($p' = |\mathbf{p}'|$)

$$z \gg a \gg \hbar/p' \quad (15.4.11)$$

and simultaneously,[20]

$$\frac{p'}{\hbar} a^2/z \gg 2\pi. \quad (15.4.12)$$

For $|\mathbf{x}_{||}| \leqslant a$, $z = D \gg a$, we have

---
[20] A similar method was used by van de Hulst (1949). See also Newton (1976).

$$|\mathbf{x}| \simeq z + \frac{\mathbf{x}_\|^2}{2z} \tag{15.4.13}$$

and we may effectively set $\mathbf{n} \simeq \mathbf{n}_0 = (0,0,1)$ in F_- since $\theta \simeq 0$ for $D \gg a$. By making a change of the variables integrations $\mathbf{q}_\|, \mathbf{x}_\|, \mathbf{x}'_\|$ in (15.4.9), (15.4.10), (15.2.26), respectively to $\mathbf{Q}', \boldsymbol{\rho}, \boldsymbol{\rho}'$, with the latter two dimensionless,

$$\boldsymbol{\rho} = \mathbf{x}_\| \sqrt{p'/z\hbar}, \qquad \boldsymbol{\rho}' = \mathbf{x}'_\| \sqrt{p'/z\hbar} \tag{15.4.14}$$

$$\mathbf{Q}' = \mathbf{q}_\| \sqrt{z\hbar/p'} \tag{15.4.15}$$

the corresponding product measure changes as follows,

$$d^2\mathbf{q}_\| \, d^2\mathbf{x}_\| \, d^2\mathbf{x}'_\| \rightarrow \frac{z\hbar}{p'} \, d^2\mathbf{Q}' \, d^2\boldsymbol{\rho} \, d^2\boldsymbol{\rho}' \tag{15.4.16}$$

and the z factor in the latter cancels out the multiplicative $1/|\mathbf{x}|$ factor in the second term on the right-hand side of (15.4.8) for $z \gg a \gtrsim |\mathbf{x}_\||$.

From (A-15.2.9) and (15.2.12) (see also (15.2.20)), one may effectively replace $\chi(\mathbf{x}_\|, t)$ in (15.4.9)/(15.4.10) by $1/\sqrt{\pi a^2}$. One may now repeat a similar analysis as the one carried out through (15.2.25)–(15.2.29). We also note by writing in the exponent in (15.4.9) (see also (15.2.10), (15.2.25), (15.2.26))

$$\frac{p'|\mathbf{x}|}{\hbar} \simeq \frac{p'z}{\hbar} + \frac{p'\mathbf{x}_\|^2}{2\hbar z}$$

$$= \frac{p'z}{\hbar} + \frac{\rho^2}{2} \tag{15.4.17}$$

with $|\boldsymbol{\rho}| \leqslant a\sqrt{p'/z\hbar}$, we are led to the evaluation of the simple Gaussian integral in (15.4.10)

$$\int_{|\boldsymbol{\rho}| \leqslant a\sqrt{p'/z\hbar}} d^2\boldsymbol{\rho} \, \exp(i\rho^2/2) \simeq 2\pi i \tag{15.4.18}$$

which from (15.4.12), we have extended the limit of integration to infinity.

All told, we have from (15.2.20), (15.2.10) and (15.4.8)

$$\int_{|\mathbf{x}_\||<a} d^2\mathbf{x}_\| \, |\psi_-(\mathbf{x},t)|^2 = \frac{\left|\eta(z - \frac{p'}{m}t)\right|^2}{\pi a^2} \left[\pi a^2 - \frac{4\pi\hbar}{p'} \operatorname{Im} f(\mathbf{p}', \mathbf{p}')\right] \tag{15.4.19}$$

where we have used the fact that $\operatorname{Re}(i\ldots) = -\operatorname{Im}(\ldots)$, the definitions in (15.3.3)/(15.1.29), and recall that $\mathbf{p}' = (0,0,p') \equiv p'\mathbf{n}_0$.

The result in (15.4.19) is remarkable. In the absence of a potential the expression within the square brackets would be just πa^2. In the presence of an

15.4 The Optical Theorem and Its Interpretation; Phase Shifts

interaction, the interference between the incident and scattered components of $\psi_-(\mathbf{x},t)$, in the forward direction, reduces the intensity observed in that direction. The cross sectional area on the screen is thus reduced from πa^2 by just the correct amount (see (15.4.6) with $p' = |\mathbf{p}'| = |\mathbf{p}|$) to compensate for the probability loss when a particle is scattered into all angles. It is a consequence of the conservation of probability. The optical theorem will be also studied in the context of inelastic processes in §15.8.

There are many authors associated with the optical theorem in physics, in general, such as Bohr, Peierls, Placzek (in quantum physics), and several others, dating back to Lord Rayleigh in his studies of the color and polarization of the sky.[21]

The first Born approximation in (15.3.12), (15.3.16) obviously violates the optical theorem since $V(\mathbf{p}-\mathbf{p}')$ for $\mathbf{p} = \mathbf{p}'$ is real. From (15.4.6),(15.3.7),

$$\frac{4\pi\hbar}{|\mathbf{p}|} \operatorname{Im} f(\mathbf{p}',\mathbf{p}') = \int d\Omega |f(\mathbf{p},\mathbf{p}')|^2 \qquad (15.4.20)$$

where $\mathbf{p} = |\mathbf{p}'|\mathbf{n}$ and thus the result embodied in the optical theorem relates, consistently, different orders in the potential (see Problem 15.2).

15.4.2 Phase Shifts Analysis

Consider the elastic scattering of a particle off a target particle initially at rest both of spin 0. Let $\mathbf{p}'$, $\mathbf{p}$ denote the initial, final relative momenta, respectively, and μ the reduced mass. For the interaction between the particles given by a potential energy which depends solely on the magnitude of the relative position vectors between the particles, we have a rotational invariant theory.

For spin 0 particles, the scattering matrix (15.3.10)/(15.3.11) may be expanded in terms of the relative angular momentum states, in the center of mass frame, as follows ($|\mathbf{p}| = |\mathbf{p}'| \equiv p$)

$$\langle \mathbf{p}|T|\mathbf{p}'\rangle = \sum_{\ell,m}\sum_{\ell',m'} Y_{\ell m}(\hat{\mathbf{p}}) Y^*_{\ell' m'}(\hat{\mathbf{p}}') \langle p,\ell,m|T|p,\ell',m'\rangle \qquad (15.4.21)$$

where we have used the expansion of the states $\langle \mathbf{p}|$ in (5.10.95) in terms of angular momentum states.

T must be diagonal in ℓ, m and also be independent of m due to the independence of the theory on the orientation of the system. Accordingly, we write

$$\langle p,\ell,m|T|p',\ell',m'\rangle = -\frac{[\rho_\ell(p)-1]}{2\mathrm{i}}\frac{8\pi^2\hbar^3}{\mu p}\delta_{\ell\ell'}\delta_{mm'} \qquad (15.4.22)$$

where the numerical coefficient was chosen for convenience.

---
[21] For the fascinating history of the optical theorem, in general, see Newton (1976).

By using the addition theorem of the spherical harmonics in (5.3.67), (15.4.22) and (15.4.21) give

$$\langle \mathbf{p}|T|\mathbf{p}'\rangle = -\frac{2\pi\hbar^3}{\mu p}\sum_{\ell=0}^{\infty}(2\ell+1)\frac{[\rho_\ell(p)-1]}{2i}P_\ell(\cos\theta) \qquad (15.4.23)$$

where θ is the angle between the momenta, i.e., $\hat{\mathbf{p}}'\cdot\hat{\mathbf{p}} = \cos\theta$.

From (15.3.11), we then have the following expansion for the scattering amplitude $f(\mathbf{p},\mathbf{p}') \equiv f(p,\theta)$:

$$f(p,\theta) = \frac{\hbar}{2ip}\sum_{\ell=0}^{\infty}(2\ell+1)\left[\rho_\ell(p)-1\right]P_\ell(\cos\theta) \qquad (15.4.24)$$

where the coefficient $[\rho_\ell(p)-1]$ is to be determined. For $V=0$, $\rho_\ell \to 1$.

By using the orthogonality relation of the Legendre polynomials

$$\int_0^\pi \sin\theta\, d\theta\, P_\ell(\cos\theta)\, P_{\ell'}(\cos\theta) = \frac{2}{2\ell+1}\delta_{\ell\ell'} \qquad (15.4.25)$$

we obtain from (15.3.7)/(15.4.20), i.e., from the optical theorem,

$$\frac{\pi\hbar^2}{p^2}\sum_{\ell=0}^{\infty}(2\ell+1)|\rho_\ell-1|^2 = \frac{2\pi\hbar^2}{p^2}\sum_{\ell=0}^{\infty}(2\ell+1)\,\text{Im}\left(\frac{\rho_\ell-1}{i}\right) \qquad (15.4.26)$$

suppressing the p-dependence of ρ_ℓ for simplicity of the notation. Using $\text{Im}\left[(\rho_\ell-1)/i\right] = 1 - \text{Re}\,\rho_\ell$, we may infer that

$$|\rho_\ell-1|^2 = 2 - 2\,\text{Re}\,\rho_\ell \qquad (15.4.27)$$

or that $|\rho_\ell|^2 = 1$. That is, ρ_ℓ is a phase factor which we may write as

$$\rho_\ell = e^{2i\delta_\ell} \qquad (15.4.28)$$

with δ_ℓ real, and is referred to as the phase shift, which depends on p and ℓ.

The scattering amplitude in (15.4.24) takes the form of a sum of partial scattering amplitudes

$$f(p,\theta) = \frac{\hbar}{2ip}\sum_{\ell=0}^{\infty}(2\ell+1)\left(e^{2i\delta_\ell}-1\right)P_\ell(\cos\theta) \qquad (15.4.29)$$

To obtain information on the phase shift δ_ℓ, we note from (15.4.27), in the process, that with $\tilde{\Phi}_{\text{in}}(\mathbf{q}) = (2\pi\hbar)^3\delta^3(\mathbf{q}-\mathbf{p}')$, $\psi_-(\mathbf{x}) = \psi_-(\mathbf{x},\mathbf{p})\exp(iE(\mathbf{p})t/\hbar)$, we have, for $|\mathbf{x}| \to \infty$,

$$\psi_-(\mathbf{x},\mathbf{p}) \to e^{i\mathbf{x}\cdot\mathbf{p}'/\hbar} + f(p,\theta)\frac{e^{i|\mathbf{x}||\mathbf{p}'|/\hbar}}{|\mathbf{x}|} \qquad (15.4.30)$$

15.4 The Optical Theorem and Its Interpretation; Phase Shifts

where we have used the definitions (15.3.3)/(15.3.9) and (15.1.21) (see also (15.4.7)).

Taking the advantage of the expansion of $\exp i\mathbf{p}\cdot\mathbf{x}/\hbar$ in terms[22] of the Legendre polynomials $P_\ell(\cos\theta)$ and the spherical Bessel functions $j_\ell(pr/\hbar)$ in (5.10.36), we have from (15.4.29),(15.4.30), with $|\mathbf{x}|\equiv r$,

$$\psi_-(\mathbf{x},\mathbf{p}) \longrightarrow \frac{\hbar}{pr}\sum_{\ell=0}^{\infty}(2\ell+1)i^\ell e^{i\delta_\ell}\cos\delta_\ell\left[\sin\left(\frac{pr}{\hbar}-\frac{\ell\pi}{2}\right)\right.$$
$$\left. + \tan\delta_\ell\,\cos\left(\frac{pr}{\hbar}-\frac{\ell\pi}{2}\right)\right]P_\ell(\cos\theta) \quad (15.4.31)$$

where we have used the asymptotic behavior

$$j_\ell\!\left(\frac{pr}{\hbar}\right)\xrightarrow[r\to\infty]{}\frac{\hbar}{pr}\sin\left(\frac{pr}{\hbar}-\frac{\ell\pi}{2}\right) \quad (15.4.32)$$

for the spherical Bessel function.

For a spherically symmetric potential $V(r)$, we may write

$$\psi_-(\mathbf{x},\mathbf{p}) = \frac{\hbar}{p}\sum_{\ell=0}^{\infty}(2\ell+1)\,i^\ell\,e^{i\delta_\ell}\cos\delta_\ell\,\frac{g_\ell(r)}{r}P_\ell(\cos\theta) \quad (15.4.33)$$

and from (15.4.31), $g_\ell(r)$ satisfies the boundary condition,

$$g_\ell(r)\xrightarrow[r\to\infty]{}\sin\left(\frac{pr}{\hbar}-\frac{\ell\pi}{2}\right) + \tan\delta_\ell\,\cos\left(\frac{pr}{\hbar}-\frac{\ell\pi}{2}\right) \quad (15.4.34)$$

More generally, for all r, it is easy to show that $g_\ell(r)$ satisfies the one dimensional differential equation

$$\left[\frac{d^2}{dr^2}-\frac{\ell(\ell+1)}{r^2}-\frac{2\mu}{\hbar^2}V(r)+\frac{p^2}{\hbar^2}\right]g_\ell(r)=0 \quad (15.4.35)$$

as obtained from the time-independent Schrödinger equation.

Two real solutions of (15.4.35) for $V(r)=0$ are given by

$$a_\ell(pr/\hbar)=\frac{pr}{\hbar}j_\ell\!\left(\frac{pr}{\hbar}\right),\quad b_\ell\!\left(\frac{pr}{\hbar}\right)=-\frac{pr}{\hbar}n_\ell\!\left(\frac{pr}{\hbar}\right) \quad (15.4.36)$$

where $n_\ell(x)$ is the so-closed spherical Neumann function,[23] behaving asymptotically for $r\to\infty$, as

$$a_\ell(pr/\hbar)\xrightarrow[r\to\infty]{}\sin\left(\frac{pr}{\hbar}-\frac{\ell\pi}{2}\right) \quad (15.4.37)$$

---

[22] Note that for $\tilde{\Phi}_{in}(\mathbf{q})=(2\pi\hbar)^3\delta^3(\mathbf{q}-\mathbf{p}')$, $\mathbf{x}$ points in the same direction as $\mathbf{p}$ (see below (15.2.26)).

[23] Cf. Arfken and Weber (1995), p. 678, p. 682.

$$b_\ell(pr/\hbar) \xrightarrow[r \to \infty]{} \cos\left(\frac{pr}{\hbar} - \frac{\ell\pi}{2}\right) \qquad (15.4.38)$$

Upon defining the Green function, associated with (15.4.35),

$$G_\ell(r, r') = -\frac{\hbar}{p}\left[a_\ell\left(\frac{pr}{\hbar}\right) b_\ell\left(\frac{pr'}{\hbar}\right) \Theta(r' - r) + a_\ell\left(\frac{pr'}{\hbar}\right) b_\ell\left(\frac{pr}{\hbar}\right) \Theta(r - r')\right] \qquad (15.4.39)$$

satisfying (see Problem 15.4),

$$\left[\frac{d^2}{dr^2} - \frac{\ell(\ell+1)}{r^2} + \frac{\hbar^2}{p^2}\right] G_\ell(r, r') = \delta(r - r') \qquad (15.4.40)$$

as verified by using, in the process that, $d\Theta(r' - r)/dr = -\delta(r - r')$, $d\Theta(r - r')/dr = \delta(r - r')$, and the property

$$\left(\frac{d}{dr} a_\ell\left(\frac{pr}{\hbar}\right)\right) b_\ell\left(\frac{pr}{\hbar}\right) - a_\ell\left(\frac{pr}{\hbar}\right)\left(\frac{d}{dr} b_\ell\left(\frac{pr}{\hbar}\right)\right) = \frac{p}{\hbar}, \qquad (15.4.41)$$

the solution of (15.4.35) may be given in the form of an integral equation

$$g_\ell(r) = a_\ell\left(\frac{pr}{\hbar}\right) + \int_0^\infty dr' \, G_\ell(r, r') \, u(r') \, g_\ell(r') \qquad (15.4.42)$$

where

$$u(r) = \frac{2\mu}{\hbar^2} V(r). \qquad (15.4.43)$$

Form the boundary condition (15.4.34), we may then use (15.4.42), and the asymptotic behavior of $G_\ell(r, r')$ as obtained from (15.4.36)–(15.4.38), for $r \to \infty$, to get

$$\tan \delta_\ell = -\frac{\hbar}{p} \int_0^\infty dr \, a_\ell\left(\frac{pr}{\hbar}\right) u(r) \, g_\ell(r). \qquad (15.4.44)$$

In the Born approximation, we have from (15.4.42)

$$g_\ell(r) \simeq a_\ell(r)$$

giving

$$\tan \delta_\ell \simeq \delta_\ell \simeq -\frac{p}{\hbar} \int_0^\infty r^2 dr \left(j_\ell\left(\frac{pr}{\hbar}\right)\right)^2 u(r) \qquad (15.4.45)$$

for small δ_ℓ, assuming, of course, the convergence of the latter integral.

For small r,

$$a_\ell\left(\frac{pr}{\hbar}\right) \xrightarrow[r \to 0]{} \left(\frac{pr}{\hbar}\right)^{\ell+1} \bigg/ (2\ell + 1)!! \qquad (15.4.46)$$

showing that $g_\ell(r) \to 0$ for $r \to 0$. From (15.4.45) one may also formally infer that

15.4 The Optical Theorem and Its Interpretation; Phase Shifts

$$\delta_\ell \propto \mathcal{O}\left(\left(\frac{p}{\hbar}\right)^{2\ell+1}\right) \quad (15.4.47)$$

for $p \to 0$ under suitable conditions imposed on $u(r)$ as, for example, will be the case if $u(r)$ is of finite range.

A Born series may be formally developed for $g_\ell(r)$ from (15.4.42) given by

$$g_\ell(r) = a_\ell\left(\frac{pr}{\hbar}\right) + \sum_{n \geqslant 1} \int_0^\infty dr'_1 \ldots \int_0^\infty dr'_n \; G_\ell(r, r'_1) \; u(r'_1)$$

$$\times G_\ell(r'_1, r'_2) \; u(r'_2) \ldots G_\ell(r'_{n-1}, r'_n) \; u(r'_n) \; a_\ell\left(\frac{pr'_n}{\hbar}\right) \quad (15.4.48)$$

and is real.

For a given potential $V(r)$, information may be obtained on the sign of δ_ℓ as follows. We scale the potential energy $V(r)$ by two parameters $V(r) \to \lambda_1 V(r), V(r) \to \lambda_2 V(r)$. The corresponding solutions $g_\ell(r)$ and phase shifts δ_ℓ will be denoted by $g_\ell(\lambda_i, r)$ and $\delta_\ell(\lambda_i), i = 1, 2$. Upon multiplying the differential equation satisfied by $g_\ell(\lambda_1, r)$ from the left by $g_\ell(\lambda_2, r)$, and similarly upon multiplying the differential equation satisfied by $g_\ell(\lambda_2, r)$ from the left by $g_\ell(\lambda_1, r)$ and subtracting the latter equation from the former, we obtain

$$g_\ell(\lambda_2, r)\frac{d}{dr}g_\ell(\lambda_1, r) - g_\ell(\lambda_1, r)\frac{d}{dr}g_\ell(\lambda_2, r)$$

$$= (\lambda_1 - \lambda_2)\int_0^r dr' u(r') g_\ell(\lambda_1, r') g_\ell(\lambda_2, r'). \quad (15.4.49)$$

Taking the limit $r \to \infty$, and using (15.4.34), this gives

$$\tan \delta_\ell(\lambda_1) - \tan \delta_\ell(\lambda_2) = -\frac{\hbar}{p}(\lambda_1 - \lambda_2)\int_0^\infty dr' u(r') g_\ell(\lambda_1, r') g_\ell(\lambda_2, r') \quad (15.4.50)$$

Setting $\lambda_1 = \lambda + \Delta\lambda, \lambda_2 = \lambda$, for $\Delta\lambda \to 0$, we get

$$\frac{d}{d\lambda}\delta_\ell(\lambda) = -\frac{\hbar}{p}\int_0^\infty dr \; u(r) \left(g_\ell(\lambda, r) \cos \delta_\ell(\lambda)\right)^2 \quad (15.4.51)$$

which upon integrating over λ from 0 to 1, leads to

$$\delta_\ell = -\frac{\hbar}{p}\int_0^\infty dr \; u(r) \left[\int_0^1 d\lambda \left(g_\ell(\lambda, r) \cos \delta_\ell(\lambda)\right)^2\right] \quad (15.4.52)$$

showing a correlation between the sign of δ_ℓ and the potential $u(r)$. The phase shift is positive for $u(r) < 0$, and negative for $u(r) > 0$, in general.

Intuitively, for a potential energy of finite range, say, a, a particle impinging on a target with "impact parameter" $\sim \hbar\ell/p$, with ℓ such that $\hbar\ell/p > a$ is expected to be unimportant in scattering and one may formally cut-off the sum in (15.4.29) at $\bar{\ell} = [pa/\hbar]$, where $[x]$ is largest positive integer $\leqslant x$. In particular, for low energy, one may take

$$f(p,\theta) \simeq \frac{\hbar}{2ip}(e^{2i\delta_0} - 1) = f_0(p,\theta) \qquad (15.4.53)$$

for the scattering amplitude.

The phase shift analysis carried out at the end of this section will be generalized in §15.8 to include inelastic processes and see how the counterpart of formula (15.4.29) is modified in such cases. For some applications see the problems section.

15.5 Coulomb Scattering

This section deals with a systematic treatment of Coulomb scattering. We first extend the result of Problem 15.3 to the Coulomb interaction to extract the asymptotic "free" propagator corresponding to the propagation of a particle at large distances from the scattering center. We then show that the asymptotic time development of a charged particle is not given by the free Hamiltonian and, in turn, obtain the corresponding modified asymptotic time evolution operator. The modifications arise because of the long range nature of the Coulomb potential and a particle feels its presence no matter how far it is from the scattering center. This is followed by investigating the behavior of the full Coulomb Green function G_+ near the energy shell. Finally the Coulomb scattering amplitude is obtained from a time-dependent setting. The Coulomb scattering of two identical particles will be also discussed.

15.5.1 Asymptotically "Free" Coulomb Green Functions

We extend the result of Problem 15.3 to study the nature of the integrals

$$\int d^3\mathbf{p}\, e^{i\mathbf{p}\cdot\mathbf{x}/\hbar}\, G_\pm(\mathbf{p},\mathbf{p}';p^0) \qquad \text{for } |\mathbf{x}| \to \infty \qquad (15.5.1)$$

for p^0 near $E(\mathbf{p}') = \mathbf{p}'^2/2\mu$ for a two-particle system of reduced mass μ, for the full Coulomb Green function in (9.9.54).

By changing the variable of integration in this latter equation from ξ to z, where

$$z = \frac{\xi 4p^0}{(p^0 - E(\mathbf{p}') \pm i\varepsilon)} \qquad (15.5.2)$$

we have (see (9.9.54))

$$G_\pm(\mathbf{p},\mathbf{p}';p^0) = \mp \frac{\mathrm{i}}{\pi^2}\sqrt{\frac{\mu}{8p^0}}\left(\frac{4p^0}{p^0 - E(\mathbf{p}') \pm \mathrm{i}\varepsilon}\right)\left(\frac{p^0 - E(\mathbf{p}') \pm \mathrm{i}\varepsilon}{4p^0}\right)^{\pm \mathrm{i}\gamma}$$

$$\times \int_0^\infty \mathrm{d}z\, z^{\pm \mathrm{i}\gamma}\frac{\mathrm{d}}{\mathrm{d}z}\frac{z}{[z(\mathbf{p}-\mathbf{p}')^2 - 2\mu(p^0 - E(\mathbf{p}) \pm \mathrm{i}\varepsilon)]^2} \quad (15.5.3)$$

for $p^0 \simeq E(\mathbf{p}')$,

$$\gamma = \frac{\lambda \mu/\hbar}{\sqrt{2\mu p^0}}, \qquad V(\mathbf{x}) = \frac{\lambda}{|\mathbf{x}|} \quad (15.5.4)$$

with $V(\mathbf{x})$ denoting a Coulomb potential. The denominator in the integrand in (15.5.3) in the square brackets may be rewritten as

$$\left[(z+1)\left(\mathbf{p} - \frac{\mathbf{p}'z}{z+1}\right)^2 - \frac{\mathbf{p}'^2}{(z+1)} \mp \mathrm{i}\varepsilon\right] \quad (15.5.5)$$

By changing the integration variable in (15.5.1) from $\mathbf{p}$ to $\mathbf{Q} = \mathbf{p} - \mathbf{p}'z/(z+1)$, and using the integral

$$\int \mathrm{d}^3\mathbf{Q}\, \frac{\mathrm{e}^{\mathrm{i}\mathbf{Q}\cdot\mathbf{x}/\hbar}}{\left[\mathbf{Q}^2 - \frac{\mathbf{p}'^2}{(z+1)^2} \mp \mathrm{i}\varepsilon\right]^2} = \pm \mathrm{i}\pi^2\frac{(z+1)}{|\mathbf{p}'|}\exp\left(\pm\frac{\mathrm{i}|\mathbf{p}'||\mathbf{x}|}{(z+1)\hbar}\right) \quad (15.5.6)$$

we obtain for the integral in (15.5.1), for $p^0 \simeq E(\mathbf{p}')$, the expression

$$\frac{1}{p'}\sqrt{\frac{\mu}{8p^0}}\, \mathrm{e}^{\mathrm{i}\mathbf{p}'\cdot\mathbf{x}/\hbar}\left(\frac{4p^0}{p^0 - E(\mathbf{p}') \pm \mathrm{i}\varepsilon}\right)\left(\frac{p^0 - E(\mathbf{p}') \pm \mathrm{i}\varepsilon}{4p^0}\right)^{\pm \mathrm{i}\gamma}$$

$$\times \int_0^\infty \mathrm{d}z\, z^{\pm \mathrm{i}\gamma}\frac{\mathrm{d}}{\mathrm{d}z}\left[\frac{z}{(z+1)}\, \mathrm{e}^{\pm \mathrm{i}a_\pm/(z+1)}\right] \quad (15.5.7)$$

where

$$a_\pm = \frac{1}{\hbar}\left[|\mathbf{p}'||\mathbf{x}| - \mathbf{p}'\cdot\mathbf{x}\right] \quad (15.5.8)$$

and we note that we have multiplied and divided the integral in (15.5.1) by $\exp(\mathrm{i}\,\mathbf{p}'\cdot\mathbf{x}/\hbar)$.

We are interested in the limit $a_\pm \to \infty$, where we note the positivity of $a_\pm$, in general. To this end, we change the variable of integration in (15.5.7) from z to ρ

$$z = \frac{a_\pm}{\rho}. \quad (15.5.9)$$

Then it readily follows from (15.5.7) that for $a_\pm \to \infty$,

$$\int \mathrm{d}^3\mathbf{p}\, \mathrm{e}^{\mathrm{i}\mathbf{p}\cdot\mathbf{x}/\hbar}\, G_\pm(\mathbf{p},\mathbf{p}';p^0) \longrightarrow \mathrm{e}^{\mathrm{i}\mathbf{p}'\cdot\mathbf{x}/\hbar}\, \mathrm{e}^{\pm \mathrm{i}\gamma \ln\left(\frac{|\mathbf{p}'||\mathbf{x}| - \mathbf{p}'\cdot\mathbf{x}}{\hbar}\right)} G_\pm^{0C}(\mathbf{p}')$$

$$(15.5.10)$$

where the **x**-independent part

$$G_{\pm}^{0C}(\mathbf{p}') = \frac{1}{[p^0 - E(\mathbf{p}') \pm i\varepsilon]^{1\mp i\gamma}} \exp\left[\mp i\gamma \ln\left(\frac{2\mathbf{p}^2}{\mu}\right)\right] e^{\pi\gamma/2} \Gamma(1\mp i\gamma) \tag{15.5.11}$$

with $\Gamma(1\mp i\gamma)$ denoting the gamma function, defines, for $p^0 \simeq E(\mathbf{p})$, the asymptotic "free" propagators. We note the change in the nature of the singularity at $p^0 \simeq E(\mathbf{p}')$ from a simple pole with a non-trivial dependence on the Coulomb coupling λ. Also the exponential factor $\exp(i\,\mathbf{p}' \cdot \mathbf{x}/\hbar)$ is changed from the asymptotically free one, corresponding to a short range potential, to one with a distortion given by the additional **x**-dependent exponential factor in (15.5.10).

15.5.2 Asymptotic Time Development of a Charged Particle State

Consider the integral

$$\int_{-\infty}^{\infty} dx\, \frac{e^{-iax}}{(\beta \mp ix)^\nu} = \Theta(\pm a) 2\pi \frac{|a|^{\nu-1}}{\Gamma(\nu)} e^{\mp \beta a} \tag{15.5.12}$$

for a real, $\operatorname{Re}\beta > 0$, $\operatorname{Re}\nu > 0$ and $\Theta(\pm a)$ is the step function. In reference to the asymptotic propagator in (15.5.11), let (with $\mathbf{p}' \to \mathbf{p}$)

$$\mp i(\pm i\varepsilon - E(\mathbf{p})) = \beta \tag{15.5.13}$$

$$1 \mp i\gamma = \nu \tag{15.5.14}$$

$$t/\hbar = a \tag{15.5.15}$$

to infer from (15.5.12) that for $\varepsilon \to +0$,

$$(\pm i\hbar) \int_{-\infty}^{\infty} \frac{dp^0}{2\pi\hbar}\, e^{-ip^0 t/\hbar}\, G_{\pm}^{0C}(\mathbf{p})$$

$$= \Theta(\pm t) \exp -\frac{i}{\hbar}\left[E(\mathbf{p})t + (\operatorname{sgn} t)\frac{\lambda\mu}{|\mathbf{p}|} \ln\left(\frac{2\mathbf{p}^2 |t|}{\mu\hbar}\right)\right] \tag{15.5.16}$$

where we have cancelled $\Gamma(1\mp i\gamma)$ in (15.5.11) with the one in (15.5.12) and used the property

$$(\mp i)^{\mp i\gamma} = \left(e^{\mp i\frac{\pi}{2}}\right)^{\mp i\gamma} = e^{-\gamma\pi/2} \tag{15.5.17}$$

to cancel out the $\exp(\pi\gamma/2)$ factor as well.

From (15.5.16) we see that the asymptotic "free" time evolution operator for $t \to \pm\infty$, is modified, for the Coulomb case, from the free one $\exp(-iH_0 t/\hbar)$ to $\exp(-iH_{0C}(t)/\hbar)$, where[24]

---

[24] This expression was discovered by Dollard (1964) by a completely different approach.

$$H_{0C}(t) = H_0 t + (\text{sgn } t)\frac{\lambda\mu}{\sqrt{2\mu H_0}} \ln\left(\frac{4H_0|t|}{\hbar}\right) \qquad (15.5.18)$$

and reflects the fact that a charged particle always feels the presence of the Coulomb force due to its long range effect.

An intuitive argument[25] leading to (15.5.16)/(15.5.18) nay be readily given. For $t \to \pm\infty$, the momentum and the position of a particle are related classically by $|\mathbf{p}| \sim m|\mathbf{x}|/|t|$. Accordingly, replacing $|\mathbf{x}|$ by $|t||\mathbf{p}|/m$ in $V(\mathbf{x})$ and integrating over t, to generate the asymptotic "free" time evolution operator we obtain, in addition to $t\mathbf{p}^2/2m$, for a particle of mass m, the correction $(\lambda\mu/|\mathbf{p}|)\ln|t|$ which has the correct time dependence as in (15.5.18) with $|\mathbf{p}| \to \sqrt{2mH_0}$. Incidentally, for a potential which vanishes faster than the Coulomb potential, the above substitution $|\mathbf{x}| \to |t||\mathbf{p}|/m$ in the potential gives a vanishing contribution to $\int V(|t||\mathbf{p}|/m) \, dt$ for $|t| \to \infty$ and hence no modification to the term $H_0 t$ in (15.5.18) arises.

15.5.3 The Full Green Function G_+ Near the Energy Shell

We make a change of the variable of the integration variable ξ in the Coulomb Green function $G_+(\mathbf{p},\mathbf{p}';p^0)$, relevant to the problem at hand, in (9.9.54) to $z = \rho_+ z$, where ρ_+ is defined in (9.9.55). Near the energy shell $p^0 \simeq E(\mathbf{p}) = E(\mathbf{p}')$, i.e., for $\rho_+ \to \infty$, the expression for $G_+(\mathbf{p},\mathbf{p}';p^0)$ in (9.9.54) takes the simple form

$$G_+(\mathbf{p},\mathbf{p}';p^0) = -\frac{i}{\pi^2}\sqrt{\frac{\mu}{8p^0}}\frac{(\rho_+)^{1-i\gamma}}{[(\mathbf{p}-\mathbf{p}')^2]^2}\int_0^\infty dz\, z^{i\gamma}\frac{d}{dz}\frac{z}{(z-1)^2}$$

$$= -\frac{\gamma}{\pi^2}\sqrt{\frac{\mu}{8p^0}}\frac{(\rho_+)^{1-i\gamma}}{[(\mathbf{p}-\mathbf{p}')^2]^2}\int_0^\infty dz\,\frac{z^{+i\gamma}}{(z-1)^2} \qquad (15.5.19)$$

where in the last equality we have integrated by parts.

We use the integral

$$\int_0^\infty dz\, \frac{z^{+i\gamma}}{(z-1)^2} = -e^{\pi\gamma}\,\Gamma(1+i\gamma)\Gamma(1-i\gamma)$$

$$= -e^{\pi\gamma}\left(\frac{\Gamma^*(1-i\gamma)}{\Gamma(1-i\gamma)}\right)(\Gamma(1-i\gamma))^2 \qquad (15.5.20)$$

and the definitions of ρ_+ in (9.9.55), γ in (15.5.4), to obtain near the energy shell

$$G_+(\mathbf{p},\mathbf{p}';p^0) \to G_+^{0C}(\mathbf{p})\left[\frac{\lambda}{2\hbar\pi^2}\frac{1}{(\mathbf{p}-\mathbf{p}')^2}\exp\left(-i\frac{\lambda\mu}{\hbar|\mathbf{p}|}\ln\left[\frac{(\mathbf{p}-\mathbf{p}')^2}{4\mathbf{p}^2}\right]\right)\right.$$

---

[25] Dollard and Vello (1966).

$$\times \frac{\Gamma^*(1-i\gamma)}{\Gamma(1-i\gamma)} \right] G_+^{0C}(\mathbf{p}') \qquad (15.5.21)$$

where $G_+^{0C}(\mathbf{p})$ is the asymptotic "free" propagator in (15.5.11).

Upon comparison of (15.5.21) with (15.3.6), with $G_+^0(\mathbf{p})$ now replaced by $G_+^{0C}(\mathbf{p})$ for the Coulomb case, one is tempted to identify the Coulomb scattering amplitude by

$$f_C(\mathbf{p}, \mathbf{p}') = -\frac{2\mu\lambda}{(\mathbf{p}-\mathbf{p}')^2} \exp\left(\frac{-i\lambda\mu}{\hbar|\mathbf{p}|} \ln\left[\frac{(\mathbf{p}-\mathbf{p}')^2}{4\mathbf{p}^2}\right]\right) \frac{\Gamma^*(1-i\gamma)}{\Gamma(1-i\gamma)} \qquad (15.5.22)$$

where $\Gamma^*(1-i\gamma)/\Gamma(1-i\gamma)$ is a phase factor. That the expression for the Coulomb scattering problem is indeed given by (15.5.22) will be now shown by a direct time-dependent treatment via the evolution operators.

For a decomposition of $f_C(\mathbf{p}, \mathbf{p}') \equiv f_C(p, \theta)$ into partial scattering amplitudes see Problem 15.10.

15.5.4 The Scattering Amplitude via Evolution Operators

With the asymptotic "free" time evolution operator given by $\exp\left(-iH_{0C}(t)/\hbar\right)$, with $H_{0C}(t)$ defined in (15.5.18), the transition amplitude for the scattering off a Coulomb potential with initial momentum $\mathbf{p}'$ to a final momentum $\mathbf{p}$ is given by

$$\mathcal{A}(\mathbf{p}, \mathbf{p}') = \langle \mathbf{p} | \, e^{iH_{0C}(t)/\hbar} \, e^{-itH/\hbar} \, e^{i\tau H/\hbar} \, e^{-iH_{0C}(\tau)/\hbar} \, | \mathbf{p}' \rangle \qquad (15.5.23)$$

with $\tau \to -\infty, t \to +\infty$. The operator $\exp\left(i\tau H/\hbar\right) \exp\left(-iH_{0C}(\tau)/\hbar\right)$ for $\tau \to -\infty$, for example, replaces the conventional Møller wave operator Ω_- in (15.1.47) (see also Problem 15.17).

Using the representation

$$\int_{-\infty}^{\infty} dp^0 \, \frac{e^{-i(t-\tau)p^0/\hbar}}{p^0 - H + i\varepsilon} = -2\pi i e^{-i(t-\tau)H/\hbar} \qquad (15.5.24)$$

for $t > \tau$, and the definition (see (9.7.1), (9.9.4))

$$\langle \mathbf{p} | \frac{1}{p^0 - H + i\varepsilon} | \mathbf{p}' \rangle = i\hbar \, (2\pi\hbar)^3 \, G_+(\mathbf{p}, \mathbf{p}'; p^0) \qquad (15.5.25)$$

we may rewrite the amplitude $\mathcal{A}(\mathbf{p}, \mathbf{p}')$ as

$$\mathcal{A}(\mathbf{p}, \mathbf{p}') = -\frac{(2\pi\hbar)^3 \hbar}{\pi} \int_{-\infty}^{\infty} dp^0 \, e^{-i[p^0 - E(\mathbf{p})]t/\hbar} \, e^{i[p^0 - E(\mathbf{p}')]\tau/\hbar}$$

$$\times \exp\left[\frac{i\lambda\mu}{\hbar|\mathbf{p}|} \ln\left(\frac{2\mathbf{p}^2|t|}{\mu\hbar}\right)\right] \exp\left[\frac{i\lambda\mu}{\hbar|\mathbf{p}'|} \ln\left(\frac{2\mathbf{p}'^2|\tau|}{\mu\hbar}\right)\right] G_+(\mathbf{p}, \mathbf{p}'; p^0)$$

$$(15.5.26)$$

for $t \to +\infty, \tau \to -\infty$.

The integrals in (15.5.12) alow us to use the representation ($\varepsilon \to 0$)

$$\Theta(t)\left(\frac{|t|}{\hbar}\right)^{i\gamma} = \frac{i\Gamma(1+i\gamma)}{2\pi} e^{-\pi\gamma/2} \int_{-\infty}^{\infty} dq^0 \frac{e^{-iq^0 t/\hbar}}{(q^0 + i\varepsilon)^{1+i\gamma}} \quad (15.5.27)$$

Replacing this in (15.2.26), thus introducing an integral over q^0 and p^0, multiplying and dividing the resulting integrand by $[p^0 - E(\mathbf{p}) + i\varepsilon]^{1-i\gamma}$, and making a change of the variable of integration p^0 to $Q^0 = p^0 + q^0$, we have

$$\mathcal{A}(\mathbf{p}, \mathbf{p}') = \int_{-\infty}^{\infty} dQ^0 \int_{-\infty}^{\infty} dq^0 \frac{e^{-i(Q^0 - E(\mathbf{p}))t/\hbar}}{(Q^0 - E(\mathbf{p}) - q^0 + i\varepsilon)^{1-i\gamma}}$$

$$\times \frac{1}{(q^0 + i\varepsilon)^{1+i\gamma}} K(\mathbf{p}, \mathbf{p}', \tau; Q^0 - q^0) \quad (15.5.28)$$

where we have set

$$-\frac{(2\pi\hbar)^2 \hbar}{\pi} \cdot \frac{\Gamma(1+i\gamma)}{2\pi} e^{-\pi\gamma/2} e^{i[p^0 - E(\mathbf{p}')]\tau/\hbar} \left(\exp\frac{i\lambda\mu}{\hbar|\mathbf{p}|}\ln\left(\frac{2\mathbf{p}^2}{\mu}\right)\right)$$

$$\times \left(\exp\frac{i\lambda\mu}{\hbar|\mathbf{p}'|}\ln\left(\frac{2\mathbf{p}'^2|\tau|}{\hbar}\right)\right) [p^0 - E(\mathbf{p}) + i\varepsilon]^{1-i\gamma} G_+(\mathbf{p}, \mathbf{p}'; p^0)$$

$$= K(\mathbf{p}, \mathbf{p}', \tau; p^0). \quad (15.5.29)$$

For $t \to +\infty$ (15.5.28) reduces formally to

$$\mathcal{A}(\mathbf{p}, \mathbf{p}') = -2\pi i \int_{-\infty}^{\infty} dQ^0 \, \delta(Q^0 - E(\mathbf{p}))$$

$$\times \lim_{a \to 0} \int_{-\infty}^{\infty} dx \frac{a^2}{(a(1-x) + i\varepsilon)^{1-i\gamma}(ax + i\varepsilon)^{1+i\gamma}} K(\mathbf{p}, \mathbf{p}', \tau; Q^0 - ax) \quad (15.5.30)$$

by formally making a change of variable of integration from q^0 to $x = q^0/(Q^0 - E(\mathbf{p}))$.[26] In the x-integrand in (15.5.30), as the product of two factors, we first replace a by zero in K, and the first factor for $a \gtrless 0$, $\varepsilon \to +0$ is independent of a. Using the integral

$$\int_{-\infty}^{\infty} dx \frac{1}{((1-x) + i\varepsilon)^{1-i\gamma}(x + i\varepsilon)^{1+i\gamma}} = \frac{2\pi}{i} \frac{1}{\Gamma(1-i\gamma)\Gamma(1+i\gamma)} \quad (15.5.31)$$

for $\varepsilon \to 0$, we get

---

[26] Similar limits appear in the work of Papanicolaou (1976).

$$\mathcal{A}(\mathbf{p},\mathbf{p}') = \frac{2\pi}{i} \int_{-\infty}^{\infty} dQ^0 \, \delta(Q^0 - E(\mathbf{p})) \frac{K(\mathbf{p},\mathbf{p}',\tau;Q^0)}{\Gamma(1-i\gamma)\,\Gamma(1+i\gamma)} \quad (15.5.32)$$

Repeating the same procedure for the $\tau \to -\infty$ limit, by using, in the process, the representation (see (15.5.12))

$$\Theta(-\tau)\left(\frac{|\tau|}{\hbar}\right)^{i\gamma} = \frac{i\Gamma(1+i\gamma)}{2\pi} e^{-\pi\gamma/2} \int_{-\infty}^{\infty} dq^0 \frac{e^{iq^0\tau/\hbar}}{(q^0+i\varepsilon)^{1+i\gamma}} \quad (15.5.33)$$

in the expression for $K(\mathbf{p},\mathbf{p}',\tau;Q^0)$ (with $p^0 \to Q^0$), multiplying and dividing the resulting (Q^0, q^0)-integrand in (15.5.32) by $[Q^0 - E(\mathbf{p}') + i\varepsilon]^{1-i\gamma}$ we obtain for $\tau \to -\infty$,

$$\mathcal{A}(\mathbf{p},\mathbf{p}') = -\frac{(2\pi\hbar)^2}{\mu} f_C(\mathbf{p},\mathbf{p}') \, \delta(E(\mathbf{p}) - E(\mathbf{p}')) \quad (15.5.34)$$

where in integrating over Q^0, with the energy shell constraints provided by $\delta(Q^0 - E(\mathbf{p}))\delta(Q^0 - E(\mathbf{p}'))$, we have used (15.5.21), (15.5.22).

To see that $f_C(\mathbf{p},\mathbf{p}')$ actually represents the scattering amplitude for the Coulomb potential, we note that the transition probability for the process is given from (15.5.34) to be

$$\mathcal{P}(\mathbf{p},\mathbf{p}') = \frac{(2\pi\hbar)^4}{\mu^2} |f_C(\mathbf{p},\mathbf{p}')|^2$$

$$\times \left(\int_{-T/2}^{T/2} \frac{dt}{2\pi\hbar} e^{-i[E(\mathbf{p})-E(\mathbf{p}')]t/\hbar}\right) \delta(E(\mathbf{p}) - E(\mathbf{p}')) \quad (15.5.35)$$

where we have used the integral representation of the delta distribution for $T \to \infty$. Equation (15.5.35) is formally interpreted as giving for the transition probability per unit time the expression

$$\frac{\mathcal{P}(\mathbf{p},\mathbf{p}')}{T} = \frac{(2\pi\hbar)^4}{\mu^2} |f_C(\mathbf{p},\mathbf{p}')|^2 \frac{\delta(E(\mathbf{p}) - E(\mathbf{p}'))}{2\pi\hbar} \quad (15.5.36)$$

Upon integrating (15.5.36) over $|\mathbf{p}|^2 \, d|\mathbf{p}|/(2\pi\hbar)^3$, for a fixed solid angle about a unit vector $\mathbf{n}$, we obtain

$$\int_0^\infty \frac{|\mathbf{p}|^2 d|\mathbf{p}|}{(2\pi\hbar)^3} \frac{\mathcal{P}(\mathbf{p},\mathbf{p}')}{T} = \frac{p'}{\mu} |f_C(|\mathbf{p}'|\mathbf{n},\mathbf{p}')|^2 \quad (15.5.37)$$

Finally we note that for a scattering process with an initial sharp momentum $\mathbf{p}'$ with normalization of the state carried out within a unit volume of space, p'/μ represents the probability that an incident particle crosses a unit cross sectional area, perpendicular to $\mathbf{p}'$, per unit time, i.e., it denotes the flux. Thus we obtain the differential cross section by dividing the right-hand side of (15.5.37) by p'/μ,

15.5 Coulomb Scattering

$$\frac{d\sigma}{d\Omega} = |f_C(\mathbf{p},\mathbf{p}')|^2 \qquad (15.5.38)$$

$\mathbf{p} = |\mathbf{p}'|\mathbf{n}$, where $f_C(\mathbf{p},\mathbf{p}')$ is given in (15.5.22).
Defining the angle θ by $\mathbf{p}\cdot\mathbf{p}' = |\mathbf{p}'|^2 \cos\theta$, in (15.5.22), (15.5.38) gives

$$\frac{d\sigma}{d\Omega} = \frac{\mu^2 \lambda^2}{4|\mathbf{p}'|^4 \sin^4(\theta/2)}. \qquad (15.5.39)$$

It is remarkable that the quantum mechanical answer in (15.5.39) for Coulomb scattering agrees with the classical one.[27] Due to the long range nature of the potential, the cross section does not exist since it is not integrable at $\theta \simeq 0$. In practice, the Coulomb potential, however, is usually screened at large distances and this divergence problem does not arise.

For the scattering particles of two indistinguishable particles, the solution of amplitude $f_C(\mathbf{p},\mathbf{p}') \equiv f_C(\theta)$ in (15.5.22) allows us to determine the explicit expression of the interference term (the exchange effect) of the differential cross section discussed at the end of §15.3.

In terms of the scattering angle θ in the center of mass, (15.5.22) may be rewritten as

$$f_C(\theta) = -\frac{e^2}{mv^2 \sin^2\theta/2} \exp\left(-i\frac{e^2}{\hbar v}\ln[\sin^2\theta/2]\right) \frac{\Gamma^*(1-i\gamma)}{\Gamma(1-i\gamma)} \qquad (15.5.40)$$

for particles of charge $\pm e$, and we have used the fact that $\mu = m/2$. Also the scattering angle of the incident particle is given by $\vartheta = \theta/2$, $d(\cos\theta)/d\cos\vartheta = 4\cos\vartheta$, and v denotes the speed of the incident particle.

According to the symmetrization to be carried out based on the statistics of the particles, we have from the discussion at the end of §15.3 the following result for the differential cross section for the scattering of two indistinguishable charged particles of charges $\pm e$ and of spin 0 or spin 1/2, in the laboratory frame with one of the particles initially at rest,[28]

$$\left.\frac{d\sigma}{d\Omega}\right|_{\text{LAB}} = \frac{4e^4 \cos\vartheta}{m^2 v^4}\left[\frac{1}{\sin^4\vartheta} + \frac{1}{\cos^4\vartheta} + 2\varepsilon\frac{\cos[(e^2/\hbar v)\ln\tan^2\vartheta]}{\sin^2\vartheta \cos^2\vartheta}\right] \qquad (15.5.41)$$

Here $\varepsilon = 1$ for spin 0 particles, $\varepsilon = -1$ for spin 1/2 particles polarized in the same direction, $\varepsilon = 0$ for spin 1/2 polarized in opposite directions, $\varepsilon = -1/2$ for completely unpolarized spin 1/2 particles.

For $\hbar \to 0$, the rapid oscillations of the last term in (15.5.41) gives rise to a vanishing contribution when averaged over a small angular breadth thus recovering the classical result. This same conclusion of the vanishing of the exchange term in (15.5.41) also follows for small scattering angles and for

---
[27] In quantum electrodynamics, the differential cross section is modified.
[28] Such scattering processes are usually called *Mott scatterings* for the pioneering work of N. F. Mott on the Coulomb scattering of identical charged particles.

small speeds. As discussed in §15.3, the doubling of the differential cross section, over the classical one, for spin 0 particles and its vanishing for polarized spin 1/2 particles, in the same direction, at $\vartheta = \pi/4$ (i.e., $\theta = \pi/2$ in the center of mass system) should be noted.

15.6 Functional Treatment of Scattering Theory

The functional treatment of the transformation function $\langle \mathbf{x}t | \mathbf{p}'t' \rangle$ based on the quantum dynamical principle studied in §11.2 will be used to write down an explicit expression for the scattering amplitude involving functional differentiations with respect to external sources.[29] The functional approach turns to be quite useful for studying, in particular, scattering at small deflection angles at high energies. It is also useful for the determination of the asymptotic "free" propagators for long range interactions such as the Coulomb one in a straightforward manner without using the explicit solution of full Green function. These applications of the general functional treatment provided here will be given in the next section.

For recasting our functional solution in (15.6.6), (15.6.20) derived below into a path integral form see Problem 15.13.[29]

From the analysis carried out in §11.2, the transformation function $\langle \mathbf{x}t | \mathbf{p}'t' \rangle$, for a particle in a given potential $V(\mathbf{x})$ is explicitly given form (11.2.22), (11.2.4), (11.2.27), to be

$$\langle \mathbf{x}t | \mathbf{p}'t' \rangle = \exp\left(-\frac{i}{\hbar}\int_{t'}^{t} d\tau\, V\left(-i\hbar\frac{\delta}{\delta \mathbf{F}(\tau)}\right)\right) \langle \mathbf{x}t | \mathbf{p}'t' \rangle^{(0)}\bigg|_{\mathbf{F}=0} \quad (15.6.1)$$

where ($E(\mathbf{p}) = \mathbf{p}^2/2\mu$, with μ denoting a reduced mass)

$$\langle \mathbf{x}t | \mathbf{p}'t' \rangle^{(0)} = \exp\frac{i}{\hbar}\left[\mathbf{x}\cdot\mathbf{p}' - E(\mathbf{p}')(t-t')\right]$$

$$\times \exp\left(\frac{i}{\hbar}\int_{t'}^{t} d\tau\, \mathbf{F}(\tau)\cdot\left[\mathbf{x} - \frac{\mathbf{p}'}{\mu}(t-\tau)\right]\right)$$

$$\times \exp\left(-\frac{i}{2\mu\hbar}\int_{t'}^{t} d\tau \int_{t'}^{t} d\tau'\, [t-\tau_>]\mathbf{F}(\tau)\cdot\mathbf{F}(\tau')\right) \quad (15.6.2)$$

$$\tau_> = \max(\tau, \tau') \quad (15.6.3)$$

and $\mathbf{F}(\tau)$ is an external (vector) source coupled linearly to $\mathbf{x}$ in the Hamiltonian

$$H = \frac{\mathbf{p}^2}{2\mu} + V(\mathbf{x}) - \mathbf{x}\cdot\mathbf{F}(\tau) \quad (15.6.4)$$

---

[29] See also Manoukian (1985, 1987a).

15.6 Functional Treatment of Scattering Theory

By using the identity in (11.2.28) and the translational operation property of the functional differential operator

$$\exp \int_{t'}^{t} d\tau \left[\mathbf{x} - \frac{\mathbf{p}'}{\mu}(t-\tau)\right] \cdot \frac{\delta}{\delta \mathbf{F}(\tau)} \quad (15.6.5)$$

we may rewrite (15.6.1), (15.6.2) as

$$\langle \mathbf{x} t | \mathbf{p}' t' \rangle = \exp \frac{i}{\hbar} [\mathbf{x} \cdot \mathbf{p}' - E(\mathbf{p}')(t-t')]$$

$$\times \exp \left(\frac{i\hbar}{2\mu} \int_{t'}^{t} d\tau \int_{t'}^{t} d\tau' [t - \tau_{>}] \frac{\delta}{\delta \mathbf{F}(\tau)} \cdot \frac{\delta}{\delta \mathbf{F}(\tau')}\right)$$

$$\times \exp \left(-\frac{i}{\hbar} \int_{t'}^{t} d\tau \, V\left(\mathbf{x} - \frac{\mathbf{p}'}{m}(t-\tau) + \mathbf{F}(\tau)\right)\right)\bigg|_{\mathbf{F}=0} \quad (15.6.6)$$

in analogy to (11.2.31). An interesting application of (15.6.6) to scattering at small deflection angles will be given in the next section.

Since in (15.6.6), we finally set $\mathbf{F} = 0$, the theory becomes translational invariant in time and $\langle \mathbf{x} t | \mathbf{p}' t' \rangle$ is a function of $t - t'$.

For $t > t'$,

$$\langle \mathbf{x} t | \mathbf{x}' t' \rangle = G_{+}(\mathbf{x} t, \mathbf{x}' t') \quad (15.6.7)$$

with $G_{+}(\mathbf{x} t, \mathbf{x}' t') = 0$ for $t < t'$ (see (9.7.13)) and

$$\langle \mathbf{x} t | \mathbf{p}' t' \rangle = G_{+}(\mathbf{x} t, \mathbf{p}' t')$$

$$= \int d^3 \mathbf{x}' e^{i \mathbf{p}' \cdot \mathbf{x}'/\hbar} G_{+}(\mathbf{x} t, \mathbf{x}' t') \quad (15.6.8)$$

From the Fourier transform of $G_{+}(\mathbf{x} t, \mathbf{x}' t')$ in (15.1.32), (15.1.35), and (15.6.8), we then have

$$G_{+}(\mathbf{p}, \mathbf{p}'; p^0) = -\frac{i}{\hbar} \frac{1}{(2\pi\hbar)^3} \int_0^{\infty} dT \, e^{i(p^0 + i\varepsilon)T/\hbar} \int d^3 \mathbf{x} \, e^{-i\mathbf{p} \cdot \mathbf{x}/\hbar} \langle \mathbf{x} T | \mathbf{p}' 0 \rangle \quad (15.6.9)$$

for $\varepsilon \to +0$, with the $i\varepsilon$ prescription involving the boundary condition in (9.7.12).

Upon writing $(T = t - t')$

$$\langle \mathbf{x} T | \mathbf{p}' 0 \rangle = \exp \left(\frac{i}{\hbar}[\mathbf{x} \cdot \mathbf{p}' - E(\mathbf{p}')T]\right) K(\mathbf{x}, \mathbf{p}'; T) \quad (15.6.10)$$

thus defining $K(\mathbf{x}, \mathbf{p}'; T)$ from (15.6.6) to be

$$K(\mathbf{x}, \mathbf{p}'; T) = \exp \left(\frac{i\hbar}{2\mu} \int_{t'}^{t} d\tau \int_{t'}^{t} d\tau' [t - \tau_{>}] \frac{\delta}{\delta \mathbf{F}(\tau)} \cdot \frac{\delta}{\delta \mathbf{F}(\tau')}\right)$$

$$\times \exp\left(-\frac{i}{\hbar}\int_{t'}^{t} d\tau\, V(\mathbf{x} - \frac{\mathbf{p}'}{\mu}(t-\tau) + \mathbf{F}(\tau))\right)\bigg|_{\mathbf{F}=0} \quad (15.6.11)$$

We then obtain

$$G_+(\mathbf{p}, \mathbf{p}'; p^0) = -\frac{i}{\hbar}\frac{1}{(2\pi\hbar)^3}\int_0^\infty d\alpha\, e^{i[p^0 - E(\mathbf{p}') + i\varepsilon]\alpha/\hbar}$$

$$\times \int d^3\mathbf{x}\, e^{-i\mathbf{x}\cdot(\mathbf{p}-\mathbf{p}')/\hbar} K(\mathbf{x}, \mathbf{p}'; \alpha) \quad (15.6.12)$$

where α plays the role of time and this notation for it is introduced quite often in the literature.[30]

In the integrand in (15.6.12), we recognize $[p^0 - E(\mathbf{p}') + i\varepsilon]$ as the inverse of the free Green function in the energy-momentum description.

To use (15.6.12) in deriving an expression for the scattering amplitude $f(\mathbf{p}, \mathbf{p}')$, we note from (15.3.11), (15.3.9), (15.1.29), (15.1.36)

$$f(\mathbf{p}, \mathbf{p}') = -\frac{\mu}{2\pi\hbar^2}\int d^3\mathbf{p}''\, V(\mathbf{p} - \mathbf{p}'')G_+(\mathbf{p}'', \mathbf{p}'; p^0)[p^0 - E(\mathbf{p}')]\bigg|_{p^0=E(\mathbf{p}')}$$

$$(15.6.13)$$

This suggests to multiply $G_+(\mathbf{p}, \mathbf{p}'; p^0)$ in (15.6.12) by $[p^0 - E(\mathbf{p}')]$ thus giving

$$G_+(\mathbf{p}, \mathbf{p}'; p^0)[p^0 - E(\mathbf{p}')] = -\frac{1}{(2\pi\hbar)^3}\int_0^\infty d\alpha\left(\frac{\partial}{\partial\alpha} e^{i\alpha[p^0 - E(\mathbf{p}') + i\varepsilon]/\hbar}\right)$$

$$\times \int d^3\mathbf{x}\, e^{-i\mathbf{x}\cdot(\mathbf{p}-\mathbf{p}')/\hbar} K(\mathbf{x}, \mathbf{p}'; \alpha) \quad (15.6.14)$$

where we have used the fact that $\partial/\partial\alpha$ of the first exponential generates the factor $i[p^0 - E(\mathbf{p}')]/\hbar$ in the integrand (for $\varepsilon \to +0$).

Now from (15.6.10), and the fact that $\langle \mathbf{x}|\mathbf{p}'\rangle = \exp i\mathbf{x}\cdot\mathbf{p}'/\hbar$, we have

$$K(\mathbf{x}, \mathbf{p}'; 0) = 1 \quad (15.6.15)$$

We now consider the cases for which

$$\lim_{\alpha\to\infty}\int d^3\mathbf{x}\, e^{-i\mathbf{x}\cdot(\mathbf{p}-\mathbf{p}')/\hbar} K(\mathbf{x}, \mathbf{p}'; \alpha) \quad (15.6.16)$$

exits. This, in particular, implies that

$$\lim_{\alpha\to\infty} e^{-\varepsilon\alpha}\int d^3\mathbf{x}\, e^{-i\mathbf{x}\cdot(\mathbf{p}-\mathbf{p}')/\hbar} K(\mathbf{x}, \mathbf{p}'; \alpha) = 0. \quad (15.6.17)$$

---

[30] This is especially the case in field theory.

15.6 Functional Treatment of Scattering Theory

We may then integrate (15.6.14) over α by parts to obtain on the *energy shell* $p^0 = E(\mathbf{p}')$, $\varepsilon \to +0$,

$$(2\pi\hbar)^3 G_+(\mathbf{p}, \mathbf{p}'; p^0)[p^0 - E(\mathbf{p}')]\Big|_{p^0 = E(\mathbf{p}')}$$

$$= -[(2\pi\hbar)^3 \delta^3(\mathbf{p} - \mathbf{p}') - \int_0^\infty d\alpha \int d^3x \, e^{-i\mathbf{x}\cdot(\mathbf{p}-\mathbf{p}')/\hbar} \frac{\partial}{\partial \alpha} K(\mathbf{x}, \mathbf{p}'; \alpha)] \quad (15.6.18)$$

or

$$G_+(\mathbf{p}, \mathbf{p}'; p^0)[p^0 - E(\mathbf{p}')]\Big|_{p^0 = E(\mathbf{p}')}$$

$$= \lim_{\alpha \to \infty} \frac{1}{(2\pi\hbar)^3} \int d^3x \, e^{-i\mathbf{x}\cdot(\mathbf{p}-\mathbf{p}')/\hbar} K(\mathbf{x}, \mathbf{p}'; \alpha) \quad (15.6.19)$$

From this the scattering amplitude $f(\mathbf{p}, \mathbf{p}')$ in (15.6.13) may be then obtained from the expression for $K(\mathbf{x}, \mathbf{p}'; \alpha)$ to be

$$f(\mathbf{p}, \mathbf{p}') = -\frac{\mu}{2\pi\hbar^2} \lim_{\alpha \to \infty} \int d^3x \, e^{-i\mathbf{x}\cdot(\mathbf{p}-\mathbf{p}')/\hbar} V(\mathbf{x}) K(\mathbf{x}, \mathbf{p}'; \alpha) \quad (15.6.20)$$

where $K(\mathbf{x}, \mathbf{p}'; t - t')$ is given in (15.6.11), and we have integrated over $\mathbf{p}''$ in (15.6.13). The presence of the $K(\mathbf{x}, \mathbf{p}'; \alpha)$ factor in the integrand in (15.6.20) gives an obvious modification to the Born approximation in (15.3.11)/(15.3.12) in the exact theory.

In case that the limit in (15.6.16) does not exist, as we will encounter for Coulomb scattering, (15.6.14) cannot be integrated parts. To deal with such situations, one may study the behavior of $G_+(\mathbf{p}, \mathbf{p}'; p^0)$ near the energy shell directly from (15.6.12). To this end, define the integration variable

$$z = \frac{\alpha}{\hbar}[p^0 - E(\mathbf{p}')] \quad (15.6.21)$$

to obtain

$$G_+(\mathbf{p}, \mathbf{p}'; p^0)[p^0 - E(\mathbf{p}')] = -\frac{i}{(2\pi\hbar)^3} \int_0^\infty dz \, e^{iz(1+i\varepsilon)}$$

$$\times \int d^3x \, e^{-i\mathbf{x}\cdot(\mathbf{p}-\mathbf{p}')/\hbar} K\left(\mathbf{x}, \mathbf{p}'; \frac{z\hbar}{p^0 - E(\mathbf{p}')}\right) \quad (15.6.22)$$

for $p^0 - E(\mathbf{p}') \gtrsim 0$, $\varepsilon \to +0$.

Applications of (15.6.20) and (15.6.22) will be given in the next section.

The scattering amplitude for the Coulomb interaction was already derived and explicitly given in §15.5. A functional treatment of scattering amplitudes for *long range* interactions, in general, including the Coulomb one

842   15 Quantum Scattering

may be given by a modification of transformation functions, such as the one in (15.6.1), based on an extension of the quantum dynamical principle, which takes into account the unescapable fact that asymptotically in time (see, e.g., (15.5.18)) a particle still feels the presence of the interaction and the corresponding asymptotic "free" states depend on the coupling parameter(s) of the interaction.[31]

For a path integral treatment of scattering theory see Problem 15.13, and the reference below.[31]

15.7 Scattering at Small Deflection Angles at High Energies: Eikonal Approximation

As mentioned in the previous section, the functional approach turns out to be quite suited for studying scattering at small deflection angles at high energies. Such an analysis will be carried and the resulting approximate expression for the scattering amplitude will be obtained and is shown to satisfy the optical theorem at high energies as an important consistency check of the conservation of probability. This approximation is referred to as the eikonal approximation.[32] An application will be also carried out to determine the asymptotic "free" Green function for the Coulomb potential without the use of the explicit solution of the Coulomb Green function.

15.7.1 Eikonal Approximation

We have seen in the previous section that the scattering amplitude $f(\mathbf{p}, \mathbf{p}')$ for a scattering process may be rewritten as in (15.6.20), where the factor $K(\mathbf{x}, \mathbf{p}'; \alpha)$ in the integrand, defined in (15.6.11) involving functional differentiations, may be conveniently re-expressed as

$$K(\mathbf{x}, \mathbf{p}'; T) = \exp\left(\frac{i}{2\mu}\int_{t'}^{t} d\tau \int_{t'}^{t} d\tau' \, [t - \tau_>] \frac{\delta}{\delta \mathbf{Q}(\tau)} \cdot \frac{\delta}{\delta \mathbf{Q}(\tau')}\right)$$

$$\times \exp\left(-\frac{i}{\hbar}\int_{t'}^{t} d\tau \, V\left(\mathbf{x} - \frac{\mathbf{p}'}{\mu}(t - \tau) + \sqrt{\hbar}\, \mathbf{Q}(\tau)\right)\right)\bigg|_{\mathbf{Q}=0}.$$

(15.7.1)

Here as in (11.2.31), we have made the substitution $\mathbf{F}(\tau) \to \sqrt{\hbar}\, \mathbf{Q}(\tau)$. The latter, in the argument of V takes into account the deviation of the dynamics from a *straight line* $\left[\mathbf{x} - \frac{\mathbf{p}'}{\mu}(t - \tau)\right]$. The effect of this deviation is obtained

---

[31] See Manoukian (1985).
[32] The word *eikonal* (from Greek) means *image* and this term is borrowed from optics where some similar formulae as in here appear.

15.7 Scattering at Small Deflection Angles at High Energies

by the functional differentiations provided by the exponential factor involving the functional differential operators $(\delta/\delta \mathbf{Q}(\tau)) \cdot (\delta/\delta \mathbf{Q}(\tau'))$.

For scattering at small deflection angles, we may as a first approximation, neglect the contribution of the functional differential operators arising in (15.7.1) and set $\mathbf{Q} = 0$. Corrections to this approximation may be then systematically taken into account by carrying out the functional differential operations spelled out in (15.7.1).

Accordingly from (15.6.20), the scattering amplitude takes the simple form

$$f(\mathbf{p}, \mathbf{p}') = -\frac{\mu}{2\pi\hbar^2} \int d^3\mathbf{x}\, e^{-i\mathbf{x}\cdot(\mathbf{p}-\mathbf{p}')/\hbar}\, V(\mathbf{x})$$

$$\times \exp\left(-\frac{i}{\hbar}\int_0^\infty d\beta\, V\left(\mathbf{x} - \frac{\mathbf{p}'}{\mu}\beta\right)\right) \quad (15.7.2)$$

where we have taken the limit $\alpha \to \infty$ in (15.6.20) and introduced a new integration variable β.

We note that (15.7.2) modifies the Born approximation in (15.3.12) / (15.3.11) by the presence of a phase factor in the integrand, depending on the potential, accumulated during the scattering process.

It is convenient to write the integration variable $\mathbf{x}$ as

$$\mathbf{x} = \mathbf{b} + \xi \frac{\mathbf{p}'}{\mu} \quad (15.7.3)$$

where $\mathbf{b}$ is orthogonal to $\mathbf{p}'$, $-\infty < \xi < \infty$, and

$$d^3\mathbf{x} = \left|\frac{\mathbf{p}'}{\mu}\right| d\xi\, d^2\mathbf{b}. \quad (15.7.4)$$

Also we note that

$$(\mathbf{p} - \mathbf{p}') \cdot \mathbf{p}' = -\frac{1}{2}(\mathbf{p}-\mathbf{p}')^2 + \frac{1}{2}(\mathbf{p}^2 - \mathbf{p}'^2)$$

$$= -\frac{1}{2}(\mathbf{p}-\mathbf{p}')^2 \quad (15.7.5)$$

since $\mathbf{p}^2 = \mathbf{p}'^2$, and hence for small deflection angles, we may replace $\mathbf{x} \cdot (\mathbf{p} - \mathbf{p}')$ by $\mathbf{b} \cdot (\mathbf{p} - \mathbf{p}')$ in the first exponential in (15.7.2). By making a change of the variable integration β to $\eta = \beta - \xi$, we may then rewrite (15.7.2) as

$$f(\mathbf{p}, \mathbf{p}') = -\frac{|\mathbf{p}'|}{2\pi\hbar^2} \int_{-\infty}^\infty d\xi \int d^2\mathbf{b}\, V\left(\mathbf{b} + \xi\frac{\mathbf{p}'}{\mu}\right) e^{-i\mathbf{b}\cdot(\mathbf{p}-\mathbf{p}')/\hbar}$$

$$\times \exp\left(\frac{-i}{\hbar}\int_{-\xi}^\infty d\eta\, V\left(\mathbf{b} - \frac{\mathbf{p}'}{\mu}\eta\right)\right)$$

$$= -\frac{i|\mathbf{p}'|}{2\pi\hbar} \int_{-\infty}^{\infty} d\xi \int d^2\mathbf{b}\, e^{-i\mathbf{b}\cdot(\mathbf{p}-\mathbf{p}')/\hbar}$$

$$\times \frac{\partial}{\partial \xi} \exp\left(-\frac{i}{\hbar} \int_{-\xi}^{\infty} d\eta\, V\left(\mathbf{b} - \frac{\mathbf{p}'}{\mu}\eta\right)\right). \tag{15.7.6}$$

Carrying out the ξ-integration, this gives

$$f(\mathbf{p},\mathbf{p}') = \frac{i|\mathbf{p}'|}{2\pi\hbar} \int d^2\mathbf{b}\, e^{-i\mathbf{b}\cdot(\mathbf{p}-\mathbf{p}')/\hbar}$$

$$\times \left[1 - \exp\left(\frac{-i\mu}{\hbar|\mathbf{p}'|}\int_{-\infty}^{\infty} d\rho\, V(\mathbf{b},\rho)\right)\right] \tag{15.7.7}$$

where we have finally introduced the integration variable ρ. The expression in (15.7.7) is usually referred to as the *impact parameter* representation of the scattering amplitude, where $|\mathbf{b}|$ plays the role of the impact parameter. If for some R, $|\mathbf{b}| > R$, $V(\mathbf{b},\rho) = 0$, for all ρ, then the **b**-integral in (15.7.7) will be restricted for $|\mathbf{b}| \leqslant R$.

For $V(\mathbf{x})$, depending on the distance $|\mathbf{x}| = r$ between two particles, the angular part of the integral in (15.7.7) may be explicitly carried out giving

$$f(\mathbf{p},\mathbf{p}') = \frac{i|\mathbf{p}'|}{\hbar} \int_0^\infty b\, db\, J_0\left(\frac{2|\mathbf{p}'|}{\hbar} b \sin\frac{\theta}{2}\right) [1 - \exp i\chi(b,|\mathbf{p}'|)] \tag{15.7.8}$$

where J_0 is the zeroth order Bessel function and

$$\chi(b,|\mathbf{p}'|) = -\frac{\mu}{\hbar|\mathbf{p}'|} \int_{-\infty}^{\infty} d\rho\, V\left(\sqrt{b^2 + \rho^2}\right) \tag{15.7.9}$$

is called the eikonal phase function. The integrability of $V\left(\sqrt{b^2+\rho^2}\right)$ on ρ restricts the class of potentials for the validity of the treatment.

It is remarkable that the eikonal approximation in (15.7.8) satisfies the optical theorem at high energies. To this end, in the forward direction, i.e., for $\theta = 0$, $J_0(0) = 1$,

$$\frac{4\pi\hbar}{|\mathbf{p}'|} \operatorname{Im} f(\mathbf{p}',\mathbf{p}') = 4\pi \int_0^\infty b\, db\, [1 - \cos\chi(b,|\mathbf{p}'|)]. \tag{15.7.10}$$

On the other hand,

$$d\Omega = 2\pi \sin\theta\, d\theta = 8\pi \sin\frac{\theta}{2}\, d\left(\sin\frac{\theta}{2}\right) \tag{15.7.11}$$

for the element of the solid angle, and with $\kappa = 2|\mathbf{p}'|\sin(\theta/2)/\hbar$, we have

$$\int d\Omega\, |f(\mathbf{p},\mathbf{p}')|^2 = 2\pi \int_0^\infty b\, db \int_0^\infty b'\, db' \left[1 - e^{-i\chi(b,|\mathbf{p}'|)}\right]\left[1 - e^{i\chi(b',|\mathbf{p}'|)}\right]$$

15.7 Scattering at Small Deflection Angles at High Energies 845

$$\times \int_0^{2|\mathbf{p}'|/\hbar} \kappa \, d\kappa \, J_0(\kappa b) J_0(\kappa b'). \tag{15.7.12}$$

At *high energies*, i.e., for $|\mathbf{p}'|$ large,

$$\int_0^{2|\mathbf{p}'|/\hbar} \kappa \, d\kappa \, J_0(\kappa b) J_0(\kappa b') \longrightarrow \frac{\delta(b-b')}{b} \tag{15.7.13}$$

from which

$$\int d\Omega \, |f(\mathbf{p},\mathbf{p}')|^2 = 4\pi \int_0^\infty b \, db \, [1 - \cos\chi(b,|\mathbf{p}'|)]. \tag{15.7.14}$$

This, unlike the Born approximation expression, satisfies the equality in (15.4.20) of the optical theorem as a consistency check. The Born approximation, however, is not restricted to small deflection scattering angles.

For an eikonal approximation for the Coulomb potential see Problem 15.15.

15.7.2 Determination of Asymptotic "Free" Green Function of the Coulomb Interaction

The starting point for the determination of the asymptotic "free" propagators is provided by examining (15.6.22) near the energy shell.

Upon substituting

$$K\left(\mathbf{x},\mathbf{p}'; z\hbar/(p^0 - E(\mathbf{p}'))\right)$$

$$\simeq \exp{-\frac{i}{\hbar} \int_0^{z\hbar/(p^0-E(\mathbf{p}'))} d\beta \, V\left(\mathbf{x} - \frac{\mathbf{p}'}{\mu}\beta\right)} \tag{15.7.15}$$

in (15.6.22), we obtain

$$\int d^3\mathbf{p} \, e^{i\mathbf{p}\cdot\mathbf{x}/\hbar} G_+(\mathbf{p},\mathbf{p}';p^0) \simeq \frac{-i e^{i\mathbf{x}\cdot\mathbf{p}'/\hbar}}{[p^0 - E(\mathbf{p}') + i\varepsilon]} \int_0^\infty dz \, e^{iz(1+i\varepsilon)}$$

$$\times \exp{-\frac{i}{\hbar} \int_0^{z\hbar/(p^0-E(\mathbf{p}'))} d\beta \, V\left(\mathbf{x} - \frac{\mathbf{p}'}{\mu}\beta\right)} \tag{15.7.16}$$

for $p^0 - E(\mathbf{p}') \gtrsim 0$, $\varepsilon \to +0$.

For the Coulomb potential $V(\mathbf{x}) = \lambda/|\mathbf{x}|$,

$$\int_0^{z\hbar/(p^0-E(\mathbf{p}'))} d\beta \, V\left(\mathbf{x} - \frac{\mathbf{p}'}{\mu}\beta\right) \simeq \frac{\lambda\mu}{|\mathbf{p}'|} \ln\left(\frac{2|\mathbf{p}'|z\hbar}{\mu(p^0-E(\mathbf{p}))|\mathbf{x}|(1-\cos\theta)}\right) \tag{15.7.17}$$

846   15 Quantum Scattering

for $(p^0 - E(\mathbf{p}')) \simeq 0$. Hence

$$\exp -\frac{i}{\hbar}\int_0^{z\hbar/(p^0-E(\mathbf{p}'))} d\beta\, V\left(\mathbf{x}-\frac{\mathbf{p}'}{\mu}\beta\right)$$

$$\simeq \frac{1}{[p^0-E(\mathbf{p}')+i\varepsilon]^{-i\gamma}} \exp -i\gamma\ln\left(\frac{2\mathbf{p}'^2 z\hbar}{\mu(|\mathbf{p}'|\,|\mathbf{x}|-\mathbf{p}'\cdot\mathbf{x})}\right) \quad (15.7.18)$$

where $\gamma = \lambda\mu/\hbar|\mathbf{p}'|$.

Finally using the integral $(\varepsilon \to +0)$

$$\int_0^\infty dz\, e^{iz(1+i\varepsilon)}(z)^{-i\gamma} = ie^{\pi\gamma/2}\,\Gamma(1-i\gamma) \quad (15.7.19)$$

this gives from (15.7.18), (15.7.16),

$$\int d^3\mathbf{p}\, e^{i\mathbf{p}\cdot\mathbf{x}/\hbar}\, G_+(\mathbf{p},\mathbf{p}';p^0)$$

$$\simeq e^{i\mathbf{x}\cdot\mathbf{p}'/\hbar}\, \frac{e^{-i\gamma\ln(2\mathbf{p}'^2/\mu)}}{[p^0-E(\mathbf{p}')+i\varepsilon]^{1-i\gamma}}\, e^{i\gamma\ln\left(\frac{|\mathbf{p}'|\,|\mathbf{x}|-\mathbf{p}'\cdot\mathbf{x}}{\hbar}\right)} e^{\pi\gamma/2}\Gamma(1-i\gamma)$$

(15.7.20)

near the energy shell, which is to be compared with (15.5.10), (15.5.11), with the latter obtained from the explicit solution of G_+. Equation (15.7.20) gives the asymptotic "free" Coulomb Green function $G_+^{0C}(\mathbf{p}')$ in (15.5.11).

For additional applications of the eikonal approximation see Problem 15.14, 15.15. See also Problem 15.16.

15.8 Multi-Channel Scatterings of Clusters and Bound Systems

In the present section, we extend our earlier analysis in this chapter to describe the scattering of bound particles and clusters (§2.5), in general, as well, where the initial and final particles in a process are not necessarily the same. The internal states of the particles are also allowed to change, in the analysis, upon scattering. For example, a hydrogen atom may be excited from its ground state or be completely ionized, with its electron being stripped from the proton, upon the impact of the atom by an electron. To carry out such analyses, we first introduce the concept of channels and channel Hamiltonians. This will allow us to discuss various possible processes, such as the ones mentioned above, and define the Hamiltonians corresponding to the asymptotic time evolutions immediately after the preparatory stage and just before the detection stage in scattering. This is followed by finding the solutions of

15.8 Multi-Channel Scatterings of Clusters and Bound Systems

the fully interacting theories, corresponding to the preparatory states, obtaining the expressions for the differential cross sections and generalizing the optical theorem of §15.4 to accommodate the above cases. We then study some subtleties occurring in carrying out the Born approximation for three-particle systems and develop equations for handling such a problem. Finally, we extend the phase shift analysis carried out in §15.4 to extract information on inelastic processes.

15.8.1 Channels and Channel Hamiltonians

We consider a system of particles interacting via two-body potentials. To describe the general scattering processes intended in this section, we follow the analysis carried out in §2.5 and group the particles in consideration into clusters. The Hamiltonian of the system may be then written quite generally in the form in (2.5.48), where V_A, V_E in (2.5.49), (2.5.50) are the *intra*-clusteral and *inter*-clusteral interaction potentials.

For example consider a system of three particles forming two clusters, with one consisting of one particle and the other one consisting of two particles. Such a grouping may, for example, correspond to studying a scattering process involving a positron, defining the first cluster, and a hydrogen atom, defining the second cluster, composed of an electron and a proton.

The Hamiltonian of the system of the three particles, interacting with two-body potentials, may be written as

$$H = \sum_{i=1}^{3} \mathbf{p}_i^2/2m_i + V_{12}(\mathbf{x}_1 - \mathbf{x}_2) + V_{13}(\mathbf{x}_1 - \mathbf{x}_3) + V_{23}(\mathbf{x}_2 - \mathbf{x}_3) \quad (15.8.1)$$

For the two cluster system, the kinetic energy term of H in (15.8.1) may be *rewritten* (see §2.5, in particular, (2.5.48)), consistently with Galilean invariance, as

$$H_0(\{1,(2,3)\}) = \frac{\mathbf{P}^2}{2M} + \frac{\left(\mathbf{p}_1 - \frac{m_1}{M}\mathbf{P}\right)^2}{2m_1} + \frac{\left(\mathbf{p}_2 + \mathbf{p}_3 - \frac{(m_2+m_3)}{M}\mathbf{P}\right)^2}{2(m_2+m_3)}$$

$$+ \frac{\left(\mathbf{p}_2 - \frac{m_2}{m_2+m_3}(\mathbf{p}_2 + \mathbf{p}_3)\right)^2}{2m_2} + \frac{\left(\mathbf{p}_3 - \frac{m_3}{m_2+m_3}(\mathbf{p}_2 + \mathbf{p}_3)\right)^2}{2m_3}$$

$$(15.8.2)$$

where

$$\mathbf{P} = \mathbf{p}_1 + \mathbf{p}_2 + \mathbf{p}_3, \qquad M = m_1 + m_2 + m_3 \quad (15.8.3)$$

and $\mathbf{p_1}$ denotes the momentum of the cluster involving the one particle. The physical interpretation of the terms in (15.8.2) is as follows. $\mathbf{P}^2/2M$ represents the kinetic energy associated with the center of mass motion. The second and third term denote, respectively, the kinetic energies of the two clusters with the center of mass motion removed. The fourth and fifth terms denote the kinetic energies of the two particles making up the second cluster with their center of mass motion removed.

We note that the second and the third terms in (15.8.2) may be combined to yield

$$\frac{((m_2 + m_3)\mathbf{p}_1 - m_1(\mathbf{p}_2 + \mathbf{p}_3))^2}{2M^2\mu_1} \tag{15.8.4}$$

where

$$\mu_1 = m_1(m_2 + m_3)/M \tag{15.8.5}$$

Similarly the fourth and the fifth terms in (15.8.2) combine to yield

$$\frac{(m_3\mathbf{p}_2 - m_2\mathbf{p}_3)^2}{2(m_2 + m_3)^2\mu_{23}} \tag{15.8.6}$$

where

$$\mu_{23} = m_2 m_3/(m_2 + m_3) \tag{15.8.7}$$

Also note that

$$\mathbf{q}_{23} = (m_3\mathbf{p}_2 - m_2\mathbf{p}_3)/(m_2 + m_3) \tag{15.8.8}$$

denotes the relative momentum of particle 2 with respect to particle 3, and

$$\mathbf{q}_1 = [(m_2 + m_3)\mathbf{p}_1 - m_1(\mathbf{p}_2 + \mathbf{p}_3)]/M \tag{15.8.9}$$

is the momentum of particle 1 relative to the center of mass of the cluster consisting of particles 2 and 3.

To introduce the concept of a channel and the corresponding channel Hamiltonian, we introduce the variables

$$\mathbf{r} = \mathbf{x}_2 - \mathbf{x}_3 \tag{15.8.10}$$

$$\mathbf{R}_2 = (m_2\mathbf{x}_2 + m_3\mathbf{x}_3)/(m_2 + m_3) \tag{15.8.11}$$

$$\mathbf{R} = (m_1\mathbf{x}_1 + m_2\mathbf{x}_2 + m_3\mathbf{x}_3)/M \tag{15.8.12}$$

$$\boldsymbol{\eta} = \mathbf{x}_1 - \mathbf{R}_2 \tag{15.8.13}$$

where $\mathbf{r}, \boldsymbol{\eta}$ are conjugate to $\mathbf{q}_{23}, \mathbf{q}_1$, respectively, to rewrite the Hamiltonian in (15.8.1)/(15.8.2), as (see (2.5.58))

$$H = -\frac{\hbar^2}{2M}\nabla_{\mathbf{R}}^2 - \frac{\hbar^2}{2\mu_1}\nabla_{\boldsymbol{\eta}}^2 - \frac{\hbar^2}{2\mu_{23}}\nabla_{\mathbf{r}}^2 + V_{12}\left(\boldsymbol{\eta} - \frac{m_3}{m_2 + m_3}\mathbf{r}\right)$$

$$+ V_{13}\left(\boldsymbol{\eta} + \frac{m_2}{m_2 + m_3}\mathbf{r}\right) + V_{23}(\mathbf{r}) \tag{15.8.14}$$

15.8 Multi-Channel Scatterings of Clusters and Bound Systems

and μ_1, μ_{23} are the reduced masses defined in (15.8.5), (15.8.7). Here V_{12}, V_{13} are referred to as inter-clusteral interaction potentials describing the interaction of the first cluster, consisting of one particle, with the two particles making up the second cluster. On the other hand, $V_{23}(\mathbf{r})$ is referred to as the intra-clusteral interaction potential responsible for the interactions between the two particles within the second cluster.

The concept of a channel and the corresponding channel Hamiltonian naturally arises from (15.8.14) as follows. If for $|\boldsymbol{\eta}| \to \infty$, V_{12}, $V_{13} \to 0$ in (15.8.14), then we may define a *channel* Hamiltonian H_1,

$$H_1 = -\frac{\hbar^2}{2M}\nabla_\mathbf{R}^2 - \frac{\hbar^2}{2\mu_1}\nabla_{\boldsymbol{\eta}}^2 - \frac{\hbar^2}{2\mu_{23}}\nabla_\mathbf{r}^2 + V_{23}(\mathbf{r}) \qquad (15.8.15)$$

with one cluster remotely separated from the other. Here there is no break up of the second cluster made up of particles 2 and 3 interacting via the potential $V_{23}(\mathbf{r})$. This Hamiltonian may be then used to describe the initial (or final) stage of a scattering process when the two clusters are widely separated. The grouping of the three particles into two groups (clusters) defines a possible arrangement channel. This latter channel may be denoted by $\{(1),(2,3)\}$ consisting of two clusters, one involving particle 1, and the other involving particles 2 and 3. By interchanging particle 1 and 2, we may introduce the channel $\{(2),(1,3)\}$ and, in turn introduce the corresponding channel Hamiltonian H_2. Another possible channel is $\{(1),(2),(3)\}$ consisting of three clusters each made up of one particle and so on.

A multi-channel scattering from channel $\{(1),(2,3)\}$ to channel $\{(2),(1,3)\}$ is depicted in Figure 15.3.

Most importantly, we note that the total Hamiltonian H in (15.8.1), may be equivalently rewritten in terms of the channel Hamiltonians H_1, H_2 introduced above as

$$H = H_1 + V_1 = H_2 + V_2 \qquad (15.8.16)$$

where

$$H_1 = H_0\left(\{1,(2,3)\}\right) + V_{23}(\mathbf{x}_2 - \mathbf{x}_3) \qquad (15.8.17)$$

$$V_1 = V_{12}(\mathbf{x}_1 - \mathbf{x}_2) + V_{13}(\mathbf{x}_1 - \mathbf{x}_3) \qquad (15.8.18)$$

$$H_2 = H_0\left(\{2,(1,3)\}\right) + V_{13}(\mathbf{x}_1 - \mathbf{x}_3) \qquad (15.8.19)$$

$$V_2 = V_{12}(\mathbf{x}_1 - \mathbf{x}_2) + V_{23}(\mathbf{x}_2 - \mathbf{x}_3) \qquad (15.8.20)$$

where $H_0\left(\{1,(2,3)\}\right)$ is defined in (15.8.2). Here V_1, defines the inter-clusteral interaction between the clusters $(1),(2,3)$, and similarly, V_2 defines the inter-clusteral interaction between the clusters $(2),(1,3)$.

For the collision of two clusters containing, respectively, n_1, n_2 particles, the Hamiltonian, consistent with Galilean invariance, may be written as

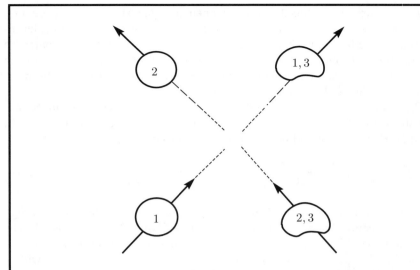

Fig. 15.3. The figure depicts the scattering of two clusters from channel $\{(1),(2,3)\}$ to channel $\{(2),(1,3)\}$, with channel Hamiltonians H_1, H_2, respectively, as described in the text.

$$H = \frac{\mathbf{P}^2}{2M} + \sum_{\beta=1}^{n_1} \frac{\left(\mathbf{P}_{1\beta} - \frac{m_{1\beta}}{M_1}\mathbf{P}_{1\cdot}\right)^2}{2m_{1\beta}} + \sum_{\beta=1}^{n_2} \frac{\left(\mathbf{P}_{2\beta} - \frac{m_{2\beta}}{M_2}\mathbf{P}_{2\cdot}\right)^2}{2m_{2\beta}}$$

$$+ \frac{\left(\mathbf{P}_{1\cdot} - \frac{M_1}{M}\mathbf{P}\right)^2}{2M_1} + \frac{\left(\mathbf{P}_{2\cdot} - \frac{M_2}{M}\mathbf{P}\right)^2}{2M_2} + \sum_{\alpha<\beta}^{n_1} V_{1\alpha\beta}\left(\mathbf{x}_{1\alpha} - \mathbf{x}_{1\beta}\right)$$

$$+ \sum_{\alpha<\beta}^{n_2} V_{2\alpha\beta}\left(\mathbf{x}_{2\alpha} - \mathbf{x}_{2\beta}\right) + \sum_{\alpha=1}^{n_2}\sum_{\beta=1}^{n_1} V_{3\alpha\beta}\left(\mathbf{x}_{1\beta} - \mathbf{x}_{2\alpha}\right) \quad (15.8.21)$$

in the notation of (2.5.48), where, in particular, the last term is responsible for the inter-clusteral interaction. The fourth and fifth kinetic energy terms may combined to

$$\left(\frac{M_2\mathbf{P}_{1\cdot} - M_1\mathbf{P}_{2\cdot}}{M}\right)^2 \frac{1}{2\mu_{12}} \quad (15.8.22)$$

where

$$\mu_{12} = \frac{M_1 M_2}{M_1 + M_2} \quad (15.8.23)$$

is the reduced mass of the two clusters. $(M_2\mathbf{P}_{1\cdot} - M_1\mathbf{P}_{2\cdot})/M$ denotes the relative momentum of the first cluster with respect to the second one.

15.8 Multi-Channel Scatterings of Clusters and Bound Systems

On the other hand, the five kinetic energy terms in (15.8.21) may be also simply combined (see, for example (2.5.21)) to

$$\sum_{\alpha=1}^{2}\sum_{\beta=1}^{n_\alpha} \mathbf{P}_{\alpha\beta}^2/2m_{\alpha\beta} \qquad (15.8.24)$$

which allows us to re-group together the momenta, as done in the kinetic energy terms in (15.8.21), in different manners to define new clusters. Finally, we note that we may also re-group the potential energy terms in (15.8.21) consistently with the emerging newly chosen set of clusters and, in turn, extract the inter-clusteral part of the potential energies as done for the example corresponding to the one depicted in Figure 15.3, and carried out in (15.8.16)–(15.8.20).

15.8.2 Interacting States Corresponding to Preparatory Channels

Consider a system of non-interacting clusters with total energy E_a as the sum of their binding energies and their kinetic energies. Let $|\Phi_{\text{in}}^a(E_a)\rangle \exp(-iE_a t/\hbar)$ denotes the state describing the system, where $|\Phi_{\text{in}}^a(E_a)\rangle$ satisfies the time-independent Schrödinger equation

$$(H_a - E_a)|\Phi_{\text{in}}^a(E_a)\rangle = 0 \qquad (15.8.25)$$

where H_a is the channel Hamiltonian for the system of the non-interacting clusters. Note that, H_a includes the intra-clusteral interaction potentials responsible for keeping (binding) the particles within the clusters together. Let V_a denote the inter-cluster potential responsible for the interaction of clusters when they eventually merge together in a scattering process.

The total Hamiltonian of the interacting system is, by definition, given by

$$H = H_a + V_a \qquad (15.8.26)$$

The interacting state corresponding to the initial system of non-interacting clusters may be written as $|\psi_-^a(E_a)\rangle \exp(-iE_a t/\hbar)$, where

$$|\psi_-^a(E_a)\rangle = |\Phi_{\text{in}}^a(E_a)\rangle + \frac{1}{(E_a - H_a + i\varepsilon)} V_a |\psi_-^a(E_a)\rangle \qquad (15.8.27)$$

as seen below.

Upon multiplying this equation by $(E_a - H_a + i\varepsilon)$ and by re-arrangement of terms we get

$$(E_a - H + i\varepsilon)|\psi_-^a(E_a)\rangle = (E_a - H_a + i\varepsilon)|\Phi_{\text{in}}^a(E_a)\rangle \qquad (15.8.28)$$

where we have used (15.8.26). By taking the limit $\varepsilon \to +0$ and using (15.8.25), we verify that $|\psi_-^a(E_a)\rangle$ satisfies the time-dependent Schrödinger equation

$$(H - E_a)\left|\psi_-^a(E_a)\right\rangle = 0 \qquad (15.8.29)$$

involving the fully interacting Hamiltonian H.

Formally, if we replace E_a in (15.8.27) by p^0, multiply the resulting equation by $\chi(p^0 - E_a)\exp{-ip^0 t/\hbar}$ and integrate over p^0, for some function $\chi(p^0 - E_a)$ which is appreciably non-zero for $p^0 \simeq E_a$, we obtain

$$\int_{-\infty}^{\infty} \frac{dp^0}{2\pi\hbar}\left[\left|\psi_-^a(p^0)\right\rangle - \left|\Phi_{\text{in}}^a(p^0)\right\rangle\right]\chi(p^0 - E_a)\exp(-ip^0 t/\hbar)$$

$$= \int_{-\infty}^{\infty}\frac{dp^0}{2\pi\hbar}\,\chi(p^0 - E_a)\,\frac{e^{-itp^0/\hbar}}{(p^0 - H_a + i\varepsilon)}\,V_a\left|\psi_-^a(p^0)\right\rangle \qquad (15.8.30)$$

For $t < 0$ we may, in the complex p^0-plane, close the contour of integration from above, where $\operatorname{Im} p^0 > 0$, thus avoiding any pole that may arise from the Hamiltonian H_a for $\varepsilon \to +0$. We may thus formally infer that for $t \to -\infty$, the right-hand side of (15.8.30) would be zero. This gives the correct asymptotic boundary condition for the interacting state. In the sequel we consider the limit where $\chi(p^0 - E_a)$ becomes $(2\pi\hbar)\,\delta(p^0 - E_a)$ as done in (15.8.27).

Eventually the initial clusters merge together and the inter-cluster interaction V_a becomes non-negligible. After their interactions, some new clusters may emerge in the process with a corresponding new channel Hamiltonian H_b. By definition the same Hamiltonian H in (15.8.26) may be rewritten as

$$H = H_b + V_b \qquad (15.8.31)$$

where V_b is the inter-clusteral interaction potential of the new clusters.

To obtain the amplitude for transition from the a channel to the b one, we first multiply and divide the right-hand side of (15.8.27) by $[E_a - H_b + i\varepsilon]$, to get

$$\left|\psi_-^a(E_a)\right\rangle = \frac{1}{[E_a - H_b + i\varepsilon]}\left[(E_a - H_a + i\varepsilon) + (H_a - H_b)\right]\left|\Phi_{\text{in}}^a(E_a)\right\rangle$$

$$+ \frac{1}{[E_a - H_b + i\varepsilon]}\left[(E_a - H_a + i\varepsilon) + (H_a - H_b)\right]\frac{1}{E_a - H_a + i\varepsilon}\,V_a\left|\psi_-^a(E_a)\right\rangle$$
$$(15.8.32)$$

From (15.8.26), (15.8.31), $H_a - H_b = V_b - V_a$. Accordingly, by using (15.8.29), and the defining (15.8.27) all over again on the right-hand side of (15.8.32), we obtain from the latter

$$\left|\psi_-^a(E_a)\right\rangle = \frac{1}{[E_a - H_b + i\varepsilon]}\left[i\varepsilon\left|\Phi_{\text{in}}^a(E_a)\right\rangle + V_b\left|\psi_-^a(E_a)\right\rangle\right] \qquad (15.8.33)$$

and note the presence of the $i\varepsilon$ factor multiplying $\left|\Phi_{\text{in}}^a(E_a)\right\rangle$ on the right-hand side of the equation.

15.8.3 Transition Probabilities and the Optical Theorem

The transition amplitude from the a channel to the b channel, with the latter described by an asymptotic state $|\Phi^b_{\text{out}}(E_b)\rangle \exp(-itE_b/\hbar)$, is then obtained by considering the limit $t \to +\infty$ of

$$\langle \Phi^b_{\text{out}}(t) | \psi^a_-(t) \rangle = \delta(a,b) + \frac{e^{-i(E_a - E_b)t/\hbar}}{[E_a - E_b + i\varepsilon]} \langle \Phi^b_{\text{out}}(E_b) | V_b \, \psi^a_-(E_a) \rangle \tag{15.8.34}$$

where we have multiplied the equation by $\exp -i(E_a - E_b)t/\hbar$ to define the corresponding time-dependent states, and have used the equation

$$\langle \Phi^b_{\text{out}}(E_b) | (E_a - H_b + i\varepsilon) = (E_a - E_b + i\varepsilon) \langle \Phi^b_{\text{out}}(E_b) | \tag{15.8.35}$$

Note the cancellation of the $i\varepsilon$ factor multiplying $|\Phi^a_{\text{in}}(E_a)\rangle$ in (15.8.33) on account of the formal orthogonality relation

$$\langle \Phi^b_{\text{out}}(E_b) | \Phi^a_{\text{in}}(E_a) \rangle = \delta(a,b) \tag{15.8.36}$$

where $\delta(a,b)$ is symbolic standing for all the variables defining the state. Even for elastic scattering, where the initial and final clusters are the same, if there is a change in the momentum of a cluster in the process then $\delta(a,b) = 0$.

The transition probability rate (probability per unit time) from the a channel to the b channel is then given from (15.8.34) to be $(t \to +\infty)$

$$\frac{d}{dt} |\langle \Phi^b_{\text{out}}(t) | \psi^a_-(t) \rangle|^2 = 2 \operatorname{Im} \left\{ T_{aa}\, \delta(a,b) + \frac{1}{\hbar} |T_{ba}|^2 \, \frac{1}{[E_a - E_b - i\varepsilon]} \right\}$$

$$= 2\delta(a,b) \operatorname{Im}(T_{aa}) + \frac{2\pi}{\hbar} \delta(E_a - E_b) |T_{ba}|^2 \tag{15.8.37}$$

where

$$T_{ba} = \langle \Phi^b_{\text{out}}(E_b) | V_b \, \psi^a_-(E_a) \rangle. \tag{15.8.38}$$

For $\delta(a,b) = 0$, corresponding to any changes occurring in the variables of the process by the collision of the clusters, $(t \to +\infty)$

$$\frac{d}{dt} |\langle \Phi^b_{\text{out}}(t) | \psi^a_-(t) \rangle|^2 = \frac{2\pi}{\hbar} \delta(E_a - E_b) |T_{ba}|^2 \tag{15.8.39}$$

where the conservation of the total energy is evident.

For an application, consider the collision of a cluster of momentum $\mathbf{p}'$, with wavefunction normalized within a unit volume of space, with a cluster at rest. In the center of mass frame, if μ_a denotes the reduced mass of the two cluster system, then p'/μ_a denotes the incident flux. Upon summing (integrating) $\sum^{f(b)}$ over the group of final states observed in the experiment pertaining to the b channel, we obtain from (15.8.39)

854   15 Quantum Scattering

$$\sigma_{ba} = \frac{2\pi\mu_a}{|\mathbf{p}'|\hbar} \sum^{f(b)} \delta(E_b - E_a) |T_{ba}|^2 \qquad (15.8.40)$$

for the cross section for the transition from the a channel to the b channel.

On the other hand, if we sum the left-hand side of (15.8.37) over *all* possible channels with all configurations that may make transitions to from the a channel, we have to obtain zero on account of the conservation probability and the presence of the time derivative d/dt. Accordingly, (15.8.37) leads from (15.8.40) to the equality,

$$\sigma = \sum_b \sigma_{ba} = \frac{2\mu_a}{|\mathbf{p}'|} \sum_b \sum^{f(b)} \delta(a,b) \operatorname{Im}(-T'_{aa}) \qquad (15.8.41)$$

where σ is the *total* cross section for transition from the a channel, and T'_{aa} corresponds to the transition for the *elastic* scattering where *no* changes occur in the variables of the scattering process. Equation (15.8.41) is the generalization of the optical theorem in (15.4.6).

15.8.4 Basic Processes

As an application of (15.8.40), we first consider the scattering of a particle off a bound state. More specifically we treat the problem of the scattering of a positron (e^+) off a hydrogen atom. Let m denote the mass of e^+ or e^-, and m_0 the mass of the proton, taking into account that $m_0 \gg m$. Labelling e^+, e^- to be particles 1, 2, we have for the Hamiltonian in (15.8.14),

$$H \longrightarrow \frac{\mathbf{p}_1^2}{2m} \frac{2m+m_0}{(m+m_0)} + \frac{\mathbf{p}_2^2}{2m} \frac{(m+m_0)}{m_0} - \frac{e^2}{\left|\boldsymbol{\eta} - \frac{m_0}{m_0+m}\mathbf{r}\right|}$$

$$+ \frac{e^2}{\left|\boldsymbol{\eta} + \frac{m}{m+m_0}\mathbf{r}\right|} - \frac{e^2}{|\mathbf{r}|}$$

$$\longrightarrow \frac{\mathbf{p}_1^2}{2m} + \frac{\mathbf{p}_2^2}{2m} - \frac{e^2}{|\boldsymbol{\eta}-\mathbf{r}|} + \frac{e^2}{|\boldsymbol{\eta}|} - \frac{e^2}{|\mathbf{r}|} \qquad (15.8.42)$$

where $\boldsymbol{\eta}$, $\mathbf{r}$ denote the position vectors of e^+, e^- from the origin where the proton is situated.

We consider the elastic scattering of e^+ off the hydrogen atom with the latter remaining in its initial ground state. Then

$$V_a = -\frac{e^2}{|\boldsymbol{\eta}-\mathbf{r}|} + \frac{e^2}{|\boldsymbol{\eta}|} = V_b \qquad (15.8.43)$$

signalling the fact that the electron remains bound to the proton in the scattering process.

15.8 Multi-Channel Scatterings of Clusters and Bound Systems

Let $\mathbf{p}'$ denotes the initial momentum of e^+, and (see (7.1.4)), $(m_0 \gg m)$

$$\Phi_0(\mathbf{x}) = \frac{1}{\sqrt{\pi a_0^3}} \exp(-|\mathbf{x}|/a_0), \qquad a_0 = \hbar^2/me^2 \qquad (15.8.44)$$

is the ground-state wavefunction of the hydrogen atom, with E_0, below, denoting the ground-state energy.

As an approximation to (15.8.40), (15.8.38) we may take[33]

$$\frac{d\sigma}{d\Omega} = \frac{2\pi m}{\hbar |\mathbf{p}'|} \int_0^\infty \frac{p^2 dp}{(2\pi\hbar)^3} \delta\left(E_0 + \frac{\mathbf{p}'^2}{2m} - E_0 - \frac{\mathbf{p}^2}{2m}\right) |I(\mathbf{p}, \mathbf{p}')|^2 \qquad (15.8.45)$$

with $(\mathbf{q} = \mathbf{p} - \mathbf{p}')$

$$I(\mathbf{p}, \mathbf{p}') = \int d^3\boldsymbol{\eta}\, d^3\mathbf{r}\, e^{-i\mathbf{q}\cdot\boldsymbol{\eta}/\hbar} |\Phi_0(\mathbf{r})|^2 \left[-\frac{e^2}{|\boldsymbol{\eta} - \mathbf{r}|} + \frac{e^2}{|\boldsymbol{\eta}|}\right] \qquad (15.8.46)$$

Using the integrals

$$\int d^3\boldsymbol{\eta}\, \frac{e^{-i\mathbf{q}\cdot\boldsymbol{\eta}/\hbar}}{|\boldsymbol{\eta} - \mathbf{r}|} = \frac{4\pi\hbar^2}{q^2} e^{-i\mathbf{q}\cdot\mathbf{r}/\hbar} \qquad (15.8.47)$$

$$\int d^3\mathbf{r}\, e^{-i\mathbf{q}\cdot\mathbf{r}/\hbar} |\Phi_0(\mathbf{r})|^2 = \left[1 + \frac{1}{4}\frac{q^2 a_0^2}{\hbar^2}\right]^{-2}$$

$$\equiv F(|\mathbf{q}|) \qquad (15.8.48)$$

we obtain from (15.8.45)

$$\frac{d\sigma}{d\Omega} = \frac{m^2 e^4}{4(\mathbf{p}')^4 \sin^4(\theta/2)} [1 - F(|\mathbf{q}|)]^2 \qquad (15.8.49)$$

$$|\mathbf{q}| = 2|\mathbf{p}'|\sin\theta/2 \qquad (15.8.50)$$

with θ denoting the scattering angle of e^+.

$F(|\mathbf{q}|)$ is called the atomic form factor. For $|\mathbf{p}'|\sin\theta/2 \ll \hbar/a_0$, $F(|\mathbf{q}|) \simeq 1$, and the hydrogen atom appears neutral to e^+, $d\sigma/d\Omega \simeq 0$ and no scattering occurs. For $|\mathbf{p}'|\sin\theta/2 \gg \hbar/a_0$, $F(|\mathbf{q}|) \simeq 0$, and the differential cross section in (15.8.49) approaches the ordinary Coulomb scattering one in (15.5.39) with e^+ experiencing the charge of the proton. For intermediate values of $|\mathbf{p}'|\sin\theta/2$, e^+ sees the atom with an effective charge

$$e_{\text{eff}} = e\sqrt{\mathcal{Z}(|\mathbf{q}|)} \qquad (15.8.51)$$

---

[33] Here we have also neglected logarithmic distortions as occurring in (15.5.10) accompanying charged particles.

with
$$\mathcal{Z}(|\mathbf{q}|) = [1 - F(|\mathbf{q}|)]^2 \qquad (15.8.52)$$
as an effective charge "renormalization" factor, with total screening of the charge of the proton $\mathcal{Z}(|\mathbf{q}|) \to 0$ at zero momentum transfer $|\mathbf{q}| \to 0$.

As another application of (15.8.40), we treat the problem of the ionization of the hydrogen atom upon the impact of an electron, with the atom initially in its ground-state.

The Hamiltonian is given by the expression $(m_0 \gg m)$
$$H = \frac{\mathbf{p}_1^2}{2m} + \frac{\mathbf{p}_2^2}{2m} + \frac{e^2}{|\boldsymbol{\eta} - \mathbf{r}|} - \frac{e^2}{|\boldsymbol{\eta}|} - \frac{e^2}{|\mathbf{r}|} \qquad (15.8.53)$$
and
$$H_a = \frac{\mathbf{p}_1^2}{2m} + \frac{\mathbf{p}_2^2}{2m} - \frac{e^2}{|\mathbf{r}|}, \qquad V_a = \frac{e^2}{|\boldsymbol{\eta} - \mathbf{r}|} - \frac{e^2}{|\boldsymbol{\eta}|} \qquad (15.8.54)$$
with $\mathbf{r}$ denoting the position vector of the initially bound electron.

Let $\Phi_0(\mathbf{x})$, as in (15.8.44), denote the ground state of the hydrogen atom. We take $H_b = H_a$, with $\Phi(\mathbf{x}, \mathbf{p}_2)$ denoting an eigenfunction corresponding to the continuous spectrum of the hydrogen atom. We know from (1.8.51), that
$$\langle \Phi(\cdot, \mathbf{p}_2) | \Phi_0(\cdot) \rangle = 0. \qquad (15.8.55)$$

As a first approximation to (15.8.38), we take the amplitude
$$A(\mathbf{p}_1, \mathbf{p}_1', \mathbf{p}_2) = \int d^3\boldsymbol{\rho}\, d^3\mathbf{r}\, e^{-i\mathbf{p}_1 \cdot \boldsymbol{\eta}/\hbar}\, \Phi^*(\mathbf{r}, \mathbf{p}_2) \left[\frac{e^2}{|\boldsymbol{\eta} - \mathbf{r}|} - \frac{e^2}{|\boldsymbol{\eta}|} \right]$$
$$\times \Phi_0(\mathbf{r})\, e^{i\mathbf{p}_1' \cdot \boldsymbol{\eta}/\hbar} \qquad (15.8.56)$$

For the exchange amplitude, for the two-electron system, we may then define
$$A_{\text{exc}}(\mathbf{p}_1, \mathbf{p}_1', \mathbf{p}_2) = \int d^3\boldsymbol{\rho}\, d^3\mathbf{r}\, \Phi^*(\boldsymbol{\eta}, \mathbf{p}_2)\, e^{-i\mathbf{p}_1 \cdot \mathbf{r}/\hbar} \left[\frac{e^2}{|\boldsymbol{\eta} - \mathbf{r}|} - \frac{e^2}{|\boldsymbol{\eta}|} \right]$$
$$\times \Phi_0(\mathbf{r})\, e^{i\mathbf{p}_1' \cdot \boldsymbol{\eta}/\hbar} \qquad (15.8.57)$$
by interchanging $\boldsymbol{\eta} \leftrightarrow \mathbf{r}$ in the final states in (15.8.56).

Due to the orthogonality relation in (15.8.55), the second term $-e^2/|\boldsymbol{\eta}|$ in the square brackets in (15.8.56) vanishes upon integration over $\mathbf{r}$. Using the integral (15.8.47), we obtain
$$A(\mathbf{p}_1, \mathbf{p}_1', \mathbf{p}_2) = \frac{4\pi e^2 \hbar^2}{\mathbf{q}^2}\, f(\mathbf{q}, \mathbf{p}_2) \qquad (15.8.58)$$
where

15.8 Multi-Channel Scatterings of Clusters and Bound Systems

$$f(\mathbf{q}, \mathbf{p}_2) = \int d^3\mathbf{r} \, e^{-i\mathbf{q}\cdot\mathbf{r}/\hbar} \, \Phi^*(\mathbf{r}, \mathbf{p}_2) \, \Phi_0(\mathbf{r}) \tag{15.8.59}$$

$$\mathbf{q} = \mathbf{p}_1 - \mathbf{p}'_1. \tag{15.8.60}$$

In (15.8.57), we cannot dismiss with the $-e^2/|\boldsymbol{\eta}|$ term in the square brackets since $\Phi^*(\boldsymbol{\eta}, \mathbf{p}_2)$ and $\Phi_0(\mathbf{r})$ depend on different variables. Using, (15.8.47), and the integral

$$\int d^3\mathbf{r} \, e^{-i\mathbf{Q}\cdot\mathbf{r}/\hbar} \, \Phi_0(\mathbf{r}) = 8\sqrt{\pi a_0^3} \left[1 + \mathbf{Q}^2 a_0^2/\hbar^2\right]^{-2}$$

$$\equiv g(|\mathbf{Q}|) \tag{15.8.61}$$

we may rewrite

$$A_{\text{exc}}(\mathbf{p}_1, \mathbf{p}'_1, \mathbf{p}_2) = \int \frac{d^3\mathbf{q}'}{(2\pi\hbar)^3} \frac{4\pi e^2 \hbar^2}{\mathbf{q}'^2} \int d^3\boldsymbol{\eta} \, e^{-i(\mathbf{q}' - \mathbf{p}'_1)\cdot\boldsymbol{\eta}/\hbar} \, \Phi^*(\boldsymbol{\eta}, \mathbf{p}_2)$$

$$\times \left[g(|\mathbf{q}' - \mathbf{p}_1|) - g(|\mathbf{p}_1|)\right]. \tag{15.8.62}$$

As a second approximation, we replace $\Phi^*(\mathbf{x}, \mathbf{p}_2)$ in (15.8.59), (15.8.62) by $\exp(-i\mathbf{p}_2 \cdot \mathbf{x}/\hbar)$. This gives

$$A(\mathbf{p}_1, \mathbf{p}'_1; \mathbf{p}_2) = \frac{4\pi e^2 \hbar^2}{\mathbf{q}^2} \, g(|\mathbf{q} + \mathbf{p}_2|). \tag{15.8.63}$$

Conservation of energy, implies that

$$|\mathbf{p}_1| = \sqrt{\mathbf{p}'^2 - \mathbf{p}_2^2 + 2mE_0} \tag{15.8.64}$$

Hence for an initial incident electron of sufficiently high energy $\mathbf{p}'^2 a_0^2/\hbar^2 \gg 1$, $\mathbf{p}_2^2$ not large, $g(|\mathbf{p}_1|) \simeq 0$, and we obtain from (15.8.62)

$$A_{\text{exc}}(\mathbf{p}_1, \mathbf{p}'_1, \mathbf{p}_2) = \frac{4\pi e^2 \hbar^2}{|\mathbf{p}_2 - \mathbf{p}'|^2} \, g(|\mathbf{q} + \mathbf{p}_2|) \tag{15.8.65}$$

or

$$A_{\text{exc}}(\mathbf{p}_1, \mathbf{p}'_1, \mathbf{p}_2) = \frac{\mathbf{q}^2}{|\mathbf{p}_2 - \mathbf{p}'|^2} \, A(\mathbf{p}_1, \mathbf{p}'_1, \mathbf{p}_2) \tag{15.8.66}$$

where $\mathbf{q}$ is given in (15.8.60) and $A(\mathbf{p}_1, \mathbf{p}'_1, \mathbf{p}_2)$ in (15.8.63).

For an unpolarized beam of electrons, we have to take the following combination of A and A_{exc} (see (15.3.39))

$$|A|^2 + |A_{\text{exc}}|^2 - \text{Re}(A^* A_{\text{exc}}) = \mathcal{P} \tag{15.8.67}$$

in computing $d\sigma$ for the process.

In detail, (15.8.63), (15.8.66) give

$$\mathcal{P} = \left(\frac{4\pi e^2 \hbar^2}{\mathbf{q}^2}\right)^2 g^2 \left(|\mathbf{q}+\mathbf{p}_2|\right) \left[1 + \frac{\mathbf{q}^4}{|\mathbf{p}_2 - \mathbf{p}'|^4} - \frac{\mathbf{q}^2}{|\mathbf{p}_2 - \mathbf{p}'|^2}\right]. \quad (15.8.68)$$

From (15.8.40), (15.8.68),

$$\sigma = \frac{2\pi m}{\hbar |\mathbf{p}'_1|} \int \frac{d^3\mathbf{p}_1}{(2\pi\hbar)^3} \frac{d^3\mathbf{p}_2}{(2\pi\hbar)^3} \, \delta\left(\frac{{\mathbf{p}'_1}^2}{2m} + E_0 - \frac{\mathbf{p}_1^2}{2m} - \frac{\mathbf{p}_2^2}{2m}\right) \mathcal{P} \quad (15.8.69)$$

and by setting

$$\varepsilon_2 = \mathbf{p}_2^2/2m \quad (15.8.70)$$

we obtain, by integrating over $|\mathbf{p}_1|$ in the process over the delta distribution in (15.8.69),

$$\frac{d\sigma}{d\varepsilon_2} = \frac{2\pi m^3}{\hbar (2\pi\hbar)^6} \frac{|\mathbf{p}_1||\mathbf{p}_2|}{|\mathbf{p}'_1|} \int d\Omega_1 d\Omega_2 \, \mathcal{P} \quad (15.8.71)$$

where $|\mathbf{p}_1|$ is given in (15.8.64), $\mathcal{P}$ in (15.8.68), (15.8.61). For carrying out the angular integrals under a simplifying assumption made on the correction due to exchange in $\mathcal{P}$, as given in (15.8.68), see Problem 15.19.

15.8.5 Born Approximation, Connectedness and Faddeev Equations

Consider a *two*-particle system with a two-body potential V_{12}. The operator form of the Green function[34] $G = \left[p^0 - H + i\varepsilon\right]^{-1}$, up to a multiplicative constant, may be written as

$$G = G^0 + G^0 V_{12} G \quad (15.8.72)$$

or formally in a perturbative expansion as

$$G = \sum_{n \geq 0} \left(G^0 V_{12}\right)^n G^0 \quad (15.8.73)$$

where $G^0 = \left[p^0 - H_0 + i\varepsilon\right]^{-1}$ and H_0 is the free two-particle Hamiltonian.

In the center of mass of the two-particle system, the matrix element $\langle \mathbf{p}_1, \mathbf{p}_2 | G^0 V_{12} | \mathbf{p}'_1, \mathbf{p}'_2 \rangle$, in detail, is given by

$$\langle \mathbf{p}_1, \mathbf{p}_2 | G^0 V_{12} G^0 | \mathbf{p}'_1, \mathbf{p}'_2 \rangle$$

---

[34] To simplify the notation further, we omit the $+$ sign in G_+, G_+^0, corresponding to $\varepsilon \to +0$, in the remaining part of this section. G is related to the resolvent of H.

$$= \frac{\delta^3\left(\mathbf{p}_1+\mathbf{p}_2-\mathbf{p}'_1-\mathbf{p}'_2\right)}{\left[p^0-\frac{\mathbf{p}_1^2}{2m_1}-\frac{\mathbf{p}_2^2}{2m_2}+i\varepsilon\right]}\langle\mathbf{q}_{12}|V_{12}|\mathbf{q}'_{12}\rangle\frac{1}{\left[p^0-\frac{\mathbf{p}'^2_1}{2m_1}-\frac{\mathbf{p}'^2_2}{2m_2}+i\varepsilon\right]}$$
(15.8.74)

up to a finite multiplicative constant, and where, in the notation of (15.8.8),

$$\mathbf{q}_{12}=(m_2\mathbf{p}_1-m_1\mathbf{p}_2)/(m_1+m_2) \qquad (15.8.75)$$

denotes the relative momentum of particle 1 with respect to 2, and note that

$$\mathbf{q}_{12}-\mathbf{q}'_{12}=\mathbf{p}_1-\mathbf{p}'_1 \qquad (15.8.76)$$

denotes the momentum transfer to particle 1. The $\delta^3\left(\mathbf{p}_1+\mathbf{p}_2-\mathbf{p}'_1-\mathbf{p}'_2\right)$ factor arises as a consequence of momentum conservation.

The expression on the right-hand side of (15.8.74) is depicted diagrammatically in Figure 15.4 (a), where the horizontal lines represent the two particles with initial and final momenta $\mathbf{p}'_1, \mathbf{p}'_2$ and $\mathbf{p}_1, \mathbf{p}_2$, respectively. They are connected by the wiggly line, representing the interaction potential, giving rise to a net momentum transfer $\mathbf{p}_1-\mathbf{p}'_1=-(\mathbf{p}_2-\mathbf{p}'_2)$. The right and left of the interaction lines are multiplied by G^0, with H_0 replaced by the sum of the kinetic energies of the particles with momenta $\mathbf{p}'_1, \mathbf{p}'_2$ and $\mathbf{p}_1, \mathbf{p}_2$, respectively. An overall $\delta^3\left(\mathbf{p}_1+\mathbf{p}_2-\mathbf{p}'_1-\mathbf{p}'_2\right)$ arises in $G^0 V G^0 V G^0$, and for the higher orders as well, as a consequence of momentum conservation.

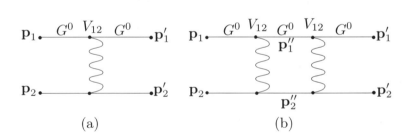

Fig. 15.4. (a) This figure depicts the expression on the right-hand side of (15.8.74). The wiggly line represents the potential energy V_{12}, connecting the two particles, with H_0 in the G^0, on the right and left of V_{12}, evaluated, respectively, at the sum of the kinetic energies of the two particles with momenta $\mathbf{p}'_1, \mathbf{p}'_2$ and $\mathbf{p}_1, \mathbf{p}_2$. The figure in (b) represents the product $G^0 V_{12} G^0 V_{12} G^0$, with $\mathbf{p}''_1=\mathbf{p}_1+\mathbf{p}_2-\mathbf{p}''_2$, as a net integration variable, and H_0 in G_0 in the middle, is evaluated at $\mathbf{p}''^2_1/2m_1+\mathbf{p}''^2_2/2m_2$.

Equation (15.8.72) may be depicted diagrammatically as shown in Figure 15.5, where the two horizontal unconnected lines represent two free particles. The remaining part of figure represents the *connected* part of G where the two particles are necessarily interacting connected by at least one wiggly line as seen in Figure 15.4.

Fig. 15.5. A diagrammatic representation of G in (15.8.72), where the first two horizontal lines on the right-hand side show two non-interacting particles, i.e., not connected with wiggly lines as in Figure 15.4. The second term gives the *connected* part of G, represented by graphs of the type shown in Figure 15.4, and others of arbitrary orders, where the two lines, representing the particles, are connected by at least one wiggly line.

From (15.8.72) and Figure 15.4, the connected part C of G is given by

$$C = \frac{1}{(p^0 - H + i\varepsilon)} V_{12} \frac{1}{(p^0 - H_0 + i\varepsilon)} \tag{15.8.77}$$

Now let us investigate the situation corresponding to an expansion as in (15.8.73) for a *three*-particle system interacting with two-body potentials. As in (15.8.8), (15.8.9), we define the momenta

$$\mathbf{q}_{ij} = (m_j \mathbf{p}_i - m_i \mathbf{p}_j) / (m_i + m_j) \tag{15.8.78}$$

$$\mathbf{q}_i = [(m_j + m_k) \mathbf{p}_i - m_i (\mathbf{p}_j + \mathbf{p}_k)] / M \tag{15.8.79}$$

with $i, j, k = 1, 2, 3, \; j \neq i \neq k$.

The free three-particle Hamiltonian may be written in various forms (see (15.8.2)),

$$\begin{aligned} H_0 &= \frac{\mathbf{p}_1^2}{2m_1} + \frac{\mathbf{p}_2^2}{2m_2} + \frac{\mathbf{p}_3^2}{2m_3} = \frac{\mathbf{q}_{12}^2}{2\mu_{12}} + \frac{\mathbf{q}_3^2}{2\mu_3} \\ &= \frac{\mathbf{q}_{23}^2}{2\mu_{23}} + \frac{\mathbf{q}_1^2}{2\mu_1} \\ &= \frac{\mathbf{q}_{13}^2}{2\mu_{13}} + \frac{\mathbf{q}_2^2}{2\mu_2} \end{aligned} \tag{15.8.80}$$

where

$$\mu_{ij} = m_i m_j / (m_i + m_j) \tag{15.8.81}$$

15.8 Multi-Channel Scatterings of Clusters and Bound Systems

$$\mu_i = m_i \left(m_j + m_k \right)/M \tag{15.8.82}$$

with the indices defined as above.

The total potential energy V may be defined as in (15.8.1) consisting of three terms V_{12}, V_{13}, V_{23}. We may also introduce four channel Hamiltonians

$$\begin{aligned} H_0 &= H_0 \\ H_1 &= H_0 + V_{23} \\ H_2 &= H_0 + V_{13} \\ H_3 &= H_0 + V_{12} \end{aligned} \tag{15.8.83}$$

corresponding, respectively, to all the particles are free, particle 1 is free, particle 2 is free, and particle 3 is free. The inter-clusteral potentials V_a, $a = 0, 1, 2, 3$, such that $H_a + V_a = H$, with the latter denoting the total Hamiltonian, are then given by

$$\begin{aligned} V_0 &= V_{12} + V_{13} + V_{23} \equiv V \\ V_1 &= V_{12} + V_{13} \\ V_2 &= V_{12} + V_{23} \\ V_3 &= V_{13} + V_{23} \end{aligned} \tag{15.8.84}$$

From (15.8.27), (15.8.38), the transition operator, whose matrix elements are given by T_{ba} in (15.8.38), from the a channel to the b channel, may be written as

$$U_{ba} = V_b + V_b \frac{1}{p^0 - H + i\varepsilon} V_a \tag{15.8.85}$$

off the energy shell.

As before

$$G = G^0 + G^0 \, V \, G \tag{15.8.86}$$

where now V consists of the three terms in (15.8.84), and $G^0 = \left(p^0 - H_0 + i\varepsilon\right)^{-1}$ for a free three-particle system.

A perturbation expansion for G as in (15.8.73) is not useful as will be seen below. For example, up to a finite multiplicative constant,

$$\langle \mathbf{p}_1, \mathbf{p}_2, \mathbf{p}_3 | G^0 V \, G^0 | \mathbf{p}'_1, \mathbf{p}'_2, \mathbf{p}'_3 \rangle = \frac{\delta^3 \left(\mathbf{p}_1 + \mathbf{p}_2 + \mathbf{p}_3 - \mathbf{p}'_1 - \mathbf{p}'_2 - \mathbf{p}'_3 \right)}{\left(p^0 - \sum_{i=1}^{3} \mathbf{p}_i^2 / 2m_i + i\varepsilon \right)}$$

$$\times \left[\langle \mathbf{q}_{12} | V_{12} | \mathbf{q}'_{12} \rangle \, \delta^3 \left(\mathbf{p}_3 - \mathbf{p}'_3 \right) + \langle \mathbf{q}_{13} | V_{13} | \mathbf{q}'_{13} \rangle \, \delta^3 \left(\mathbf{p}_2 - \mathbf{p}'_2 \right) \right.$$

$$\left. + \langle \mathbf{q}_{23} | V_{23} | \mathbf{q}'_{23} \rangle \, \delta^3 \left(\mathbf{p}_1 - \mathbf{p}'_1 \right) \right] \frac{1}{\left[p^0 - \sum_{i=1}^{3} \mathbf{p}'^2_i / 2m_i + i\varepsilon \right]} \tag{15.8.87}$$

where the first term within the square brackets indicates that particle 3 is just a "spectator" and does not participate in the interaction, while particles

1 and 2 interact via the potential V_{12}. A similar interpretation is given to the other two terms involving $\delta^3(\mathbf{p}_2 - \mathbf{p}'_2)$ and $\delta^3(\mathbf{p}_1 - \mathbf{p}'_1)$ as factors. The expression on the right-hand side of (15.8.87) is represented diagrammatically in Figure 15.6 (a). The deltas *within* the square brackets corresponding to "spectator" particle will be referred to as dangerous deltas for reasons discussed below.

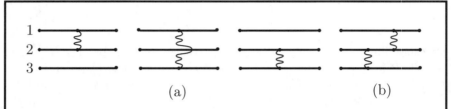

Fig. 15.6. (a) The first graph shows that particle 3, a "spectator", does not participate in the interaction, while particles 1 and 2 interact via the potential V_{12}, depicted by the vertical wiggly line connecting lines 1 and 2. A similar interpretation is also given for the other two graphs. All these graphs are *disconnected* including dangerous deltas associated with the "spectator" particles. (b) A connected graph, where all the particles interact, representing the term $G^0 V_{23} G^0 V_{12} G^0$, involving no dangerous deltas, coming from a higher order expansion of G.

The problem with (15.8.87) and similar terms of higher orders is as follows. In computing transition amplitudes, one would take the absolute values squared of terms such as in (15.8.87). The overall delta $\delta^3(\mathbf{p}_1 + \mathbf{p}_2 + \mathbf{p}_3 - \mathbf{p}'_1 - \mathbf{p}'_2 - \mathbf{p}'_3)$ is harmless in the sense that in computing a physical quantity, such as a transition amplitude, *per unit volume* v, as $v \to \infty$, as done in going from (15.5.37) to (15.5.38), and also in going from (15.8.39) to (15.8.40), in obtaining the differential cross sections, the squaring of $\delta^3(\mathbf{p}_1 + \mathbf{p}_2 + \mathbf{p}_3 - \mathbf{p}'_1 - \mathbf{p}'_2 - \mathbf{p}'_3)$ gives $\delta^3(0)\delta^3(\mathbf{p}_1 + \mathbf{p}_2 + \mathbf{p}_3 - \mathbf{p}'_1 - \mathbf{p}'_2 - \mathbf{p}'_3) = (v/(2\pi\hbar)^3)\delta^3(\mathbf{p}_1 + \mathbf{p}_2 + \mathbf{p}_3 - \mathbf{p}'_1 - \mathbf{p}'_2 - \mathbf{p}'_3)$, $v \to \infty$, and the volume v does not contribute in such a computation. This is unlike the squaring of the deltas *within* the square brackets in (15.8.87) and any manipulation involving them becomes meaningless.

It is thus clear that an expansion such as in (15.8.73), for the present case, is not useful, and we have to obtain the *connected* part of G, as done for the two-particle case in (15.8.77). We also have to find an iterative summation procedure in evaluating the connected part of G involving no dangerous deltas of the types mentioned above.

The connected part of G may be found as follows. We collect all the graphs where particle 3 is a spectator, and denote the resultant graph by

15.8 Multi-Channel Scatterings of Clusters and Bound Systems

Λ_3. Similarly define Λ_2 and Λ_1. Finally, we note that G^0 represents the three particles as free. The connected part C of G is then given by inspection to be

$$C = G - \Lambda_3 - \Lambda_2 - \Lambda_1 - G^0 \qquad (15.8.88)$$

From (15.8.77), we note that

$$\Lambda_3 = G_{12}V_{12}G^0 \qquad (15.8.89)$$

$$\Lambda_2 = G_{13}V_{13}G^0 \qquad (15.8.90)$$

$$\Lambda_1 = G_{23}V_{23}G^0 \qquad (15.8.91)$$

where *now* H_0, of course, is the free three-particle Hamiltonian,

$$G_{ij} = \left[p^0 - H_0 - V_{ij}\right]^{-1} \qquad (15.8.92)$$

The connected part C of G is depicted in Figure 15.7.

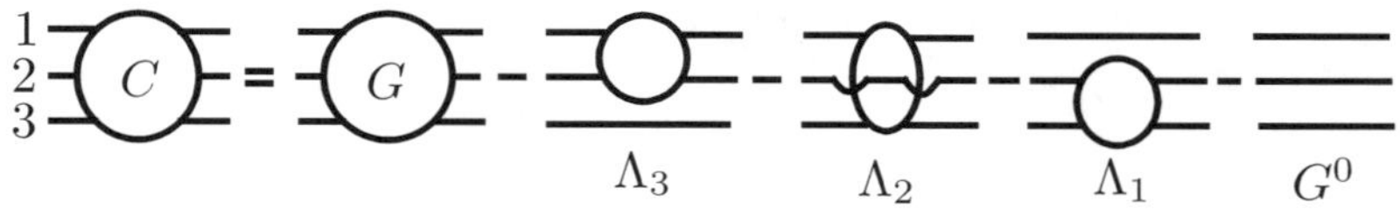

Fig. 15.7. The connected part of G, where $\Lambda_3, \Lambda_2, \Lambda_1$ are defined in (15.8.89), (15.8.90), (15.8.91), respectively.

The explicit demonstration of (15.8.88) is straightforward. To this end, we note that the transition operator from the channel 0 to channel 0, U_{00} in (15.8.85), i.e., for which the three particles are initially and finally *free*, is given by

$$U_{00} = V + V\, G^0 U_{00} = V + U_{00} G^0 V$$

$$= V + V\, G\, V$$

$$\equiv U \qquad (15.8.93)$$

In particular,
$$V\, G = U\, G^0 \qquad (15.8.94)$$

which, in turn implies that

$$G = G^0 + G^0 U\, G^0 \qquad (15.8.95)$$

Define two-particle amplitudes

$$U_i = V_{jk} + V_{jk} G_{jk} V_{jk} \qquad (15.8.96)$$

with cyclic interchange of i, j, k, $j < k$, $i, j, k = 1, 2, 3$, with G_{jk} defined in (15.8.92), involving the *three*-particle free Hamiltonian H_0, and the potential V_{jk}.

Also set

$$V_{jk} + V_{jk} G^0 U = U^{(i)} \qquad (15.8.97)$$

and note that

$$U^{(1)} + U^{(2)} + U^{(3)} = U \qquad (15.8.98)$$

as implied by the first equality in (15.8.93), and the fact that $V_{12} + V_{13} + V_{23} = V$.

Upon substituting

$$G^0 = V_{jk} - G_{jk} V_{jk} G^0 \qquad (15.8.99)$$

in (15.8.97), using the latter equation all over again in the resulting expression, and (15.8.98), we obtain

$$U^{(i)} = V_{jk} + V_{jk} G_{jk} \left(V_{jk} + U^{(j)} + U^{(k)} \right)$$

$$= U_i + V_{jk} G_{jk} \left(U^{(j)} + U^{(k)} \right) \qquad (15.8.100)$$

where we have, in the process, used (15.8.94). Finally, since

$$V_{jk} G_{jk} = U_i G^0 \qquad (15.8.101)$$

as in (15.8.94), we have

$$U^{(i)} = U_i + U_i G^0 \left(U^{(j)} + U^{(k)} \right) \qquad (15.8.102)$$

These equations are referred to as the *Faddeev equations*.[35] We note, in particular, that each $U^{(i)}$ is coupled to two different $U^{(j)}$'s *with $j \neq i$*, only.

Equations (15.8.102) provide an iterative procedure to determine the $U^{(i)}$ in terms of two-particle amplitudes U_1, U_2, U_3 defined in (15.8.96).

From (15.8.95), (15.8.97), (15.8.98) and (15.8.102), we then have

$$G = G^0 + G^0 \left(U_1 + U_2 + U_3 \right) G^0 + G^0 U_1 G^0 \left(U^{(2)} + U^{(3)} \right) G^0$$

$$+ G^0 U_2 G^0 \left(U^{(1)} + U^{(3)} \right) G^0 + G^0 U_3 G^0 \left(U^{(1)} + U^{(2)} \right) G^0 \qquad (15.8.103)$$

Now it is easy to see that

---

[35] Faddeev (1960, 1961, 1962).

$$G^0 \left(U_1 + U_2 + U_3\right) G^0 = \Lambda_1 + \Lambda_2 + \Lambda_3 \tag{15.8.104}$$

with Λ_1, Λ_2, Λ_3 defined in (15.8.89)–(15.8.91), from which we obtain

$$C = G^0 U_1 G^0 \left(U^{(2)} + U^{(3)}\right) G^0 + G^0 U_2 G^0 \left(U^{(1)} + U^{(3)}\right) G^0$$
$$+ G^0 U_3 \left(U^{(1)} + U^{(2)}\right) G^0 \tag{15.8.105}$$

for the connected part of G. That C is indeed connected follows by noting that from iterating the Faddeev equations in determining $U^{(1)}$, $U^{(2)}$, $U^{(3)}$, in terms of U_1, U_2, U_3, we always obtain for C the sum of terms of the form

$$\cdots \left(G^0 U_i\right) \left(G^0 U_j\right) G^0 \cdots \tag{15.8.106}$$

with $i \neq j$, for any two consecutive terms in a product in the sum. The fact that $i \neq j$ in (15.8.106), as just discussed, implies that such terms are necessarily connected and thus do not involve the dangerous one particle deltas mentioned earlier. Equations (15.8.105), (15.8.102) allow thus the determination of the connected part C of G.[36]

15.8.6 Phase Shifts Analysis

Consider a two-cluster system in the center of mass frame. The total angular momentum in this frame consists (see (2.7.39)/(2.7.40)) of the orbital angular momentum of relative motion and the internal angular momenta of each of the two clusters. For simplicity of presentation, we study the collision of a cluster with a target cluster both of zero internal angular momenta. A generalization including a target with non zero internal angular momentum is given in Problem 15.20 using the angular momentum decomposition in (5.10.99), (5.10.103).

Let T_{aa} denote the elastic T-scattering matrix, in the center of mass, of the collision of the two clusters with zero internal angular momenta. In this case the angular momentum will be just the orbital angular momentum of the relative motion. Assuming rotational invariance, the same analysis leading to (15.4.23) gives

$$T_{aa} = -\frac{2\pi \hbar^3}{\mu_a p'_a} \sum_{\ell=0}^{\infty} (2\ell + 1) \frac{[\eta_\ell^a(p) - 1]}{2i} P_\ell(\cos\theta) \tag{15.8.107}$$

where μ_a denotes the reduced mass of the two-cluster system, and, as we will see below, when other channels are available for the scattering process, $\eta_\ell^a(p)$ is *not* just a phase for all ℓ.

---

[36] For analyses corresponding to more than three particles, cf., Weinberg (1964); Yakubovsky (1967); Grinyuk (1980).

The elastic cross section σ_{aa} is given from (15.8.40) to be

$$\sigma_{aa} = \frac{2\pi\mu_a}{p'_a \hbar} \int \frac{d^3\mathbf{p}_a}{(2\pi\hbar)^3} \, \delta\left(\frac{\mathbf{p}_a^2}{2\mu_a} + E_a - \frac{\mathbf{p}'^2_a}{2\mu_a} - E_a\right) |T_{aa}|^2 \qquad (15.8.108)$$

where E_a is the sum of internal energies of the two clusters. The integration over p_a gives

$$\sigma_{aa} = \frac{\mu_a^2}{4\pi^2 \hbar^4} \int d\Omega_a \, |T_{aa}|^2 \qquad (15.8.109)$$

which allows us to define the elastic scattering amplitude ($p_a = |\mathbf{p}_a| = |\mathbf{p}'_a|$)

$$f_{aa}(p_a, \theta) = -\frac{\mu_a}{2\pi\hbar^2} T_{aa} \qquad (15.8.110)$$

as in (15.3.11), with

$$f_{aa}(p_a, \theta) = \frac{\hbar}{2ip_a} \sum_{\ell=0}^{\infty} (2\ell+1) [\eta_\ell^a(p) - 1] P_\ell(\cos\theta) \qquad (15.8.111)$$

From (15.8.109) the elastic cross section σ_{aa} may be rewritten as (see (15.4.25))

$$\sigma_{aa} = \frac{\pi\hbar^2}{p_a^2} \sum_{\ell=0}^{\infty} (2\ell+1) |\eta_\ell^a(p) - 1|^2 \qquad (15.8.112)$$

On the other hand, the optical theorem in (15.8.41) relates Im $f_{aa}(p_a, 0)$ to the *total* cross section σ. The latter may, in general, be different from σ_{aa} when other channels are available for the scattering process to go into from the a channel. One may then introduce the concept of the reactive cross section σ_{re}, to account for this, defined by

$$\sigma_{re} = \sigma - \sigma_{el} \qquad (15.8.113)$$

where $\sigma_{el} = \sigma_{aa}$ is the scattering cross section for the a channel in question.

From (15.8.41), the total cross section is given by

$$\sigma = \frac{4\pi\hbar}{p_a} \, \text{Im} \, f_{aa}(p_a, 0) \qquad (15.8.114)$$

or from (15.8.111)

$$\sigma = \frac{2\pi\hbar^2}{p_a^2} \sum_{\ell=0}^{\infty} (2\ell+1) \, \text{Re}\,(1 - \eta_\ell^a(p)) \qquad (15.8.115)$$

For the reactive cross section we then have

$$\sigma_{re} = \frac{\pi\hbar^2}{p_a^2} \sum_{\ell=0}^{\infty} (2\ell+1) \left[1 - |\eta_\ell^a(p)|^2\right] \qquad (15.8.116)$$

as is readily checked from (15.8.113), (15.8.115). Therefore, when there are other channels available to go into from the a channel, we must have

$$|\eta_\ell^a(p)|^2 < 1 \tag{15.8.117}$$

That is, we may write

$$\eta_\ell^a(p) = |\eta_\ell^a(p)| \, e^{2i\delta_\ell^a(p)} \tag{15.8.118}$$

with $|\eta_\ell^a(p)| < 1$ in this more general case. It is interesting to note that if inelastic processes are permissible, i.e., $|\eta_\ell^a(p)| < 1$, then from (15.8.111) this implies that the "elastic channel" a is available as well for the collision process as $f_{aa}(p,\theta)$ would be non-zero. Note also that σ_{re} attains its maximum, for $\sigma_{re} = \sigma_{el}$.

For further generalizations, allowing non-zero internal angular momenta see Problem 15.20.

15.9 Passage of Particles through Media; Neutron Interferometer

An application of great practical importance connected with the passage of charged particles through matter is that of their energy loss by collisions with the underlying medium. We here consider the one associated with the passage of charged particles in hydrogen as a direct application of the formalism developed in the previous section. The analysis may, however, be applied to other media as well. Finally, we study the problem of neutron interferometry which has been quite important in recent years in quantum physics and is expected to have further applications in the future. Here we are interested in the splitting and re-combination of a beam of neutrons in the Earth's gravitational field and in the determination of the gravity induced quantum mechanical phase shift arising from the interference of the combined beams. The phase shift depends, in particular, on the gravitational and the Planck constants.

15.9.1 Passage of Charged Particles through Hydrogen

Consider the collision of a charged particle of mass M and charge e_0 with a hydrogen atom. The amplitude for the excitation of the atom from the ground state Φ_0 to an excited state Φ_n is given, approximately from (15.8.58) to be

$$A(\mathbf{p}, \mathbf{p}') = \frac{4\pi e_0 e \hbar^2}{\mathbf{q}^2} \left\langle \Phi_n \left| e^{-i\mathbf{q}\cdot\mathbf{x}/\hbar} \right| \Phi_0 \right\rangle \tag{15.9.1}$$

where $\mathbf{q} = \mathbf{p}_n - \mathbf{p}'$ is the momentum transfer to the charged particle. The cross section of excitation is then (see, for example, (15.8.69))

$$\sigma_{n0} = \frac{32\pi^3 e_0^2 e^2 \hbar^4 \mu}{\hbar |\mathbf{p}'|} \int \frac{d^3 \mathbf{p}_n}{(2\pi\hbar)^3} \delta\left(\frac{\mathbf{p}'^2}{2\mu} + E_0 - \frac{\mathbf{p}_n^2}{2\mu} - E_n\right)$$

$$\times \frac{\left|\langle \Phi_n | e^{-i\mathbf{q}\cdot\mathbf{x}/\hbar} | \Phi_0 \rangle\right|^2}{|\mathbf{q}|^4} \quad (15.9.2)$$

where

$$\mu = \frac{Mm}{M+m} \quad (15.9.3)$$

is the reduced mass of the system, with the reduced mass of the atom taken to be the mass of the electron. Here E_0, E_n denote the energies associated with the states Φ_0, Φ_n, respectively.

The $|\mathbf{p}_n|$-integration may be readily carried out by using the property of the delta function to set

$$|\mathbf{p}_n| = \sqrt{\mathbf{p}'^2 + 2\mu(E_0 - E_n)}. \quad (15.9.4)$$

To carry out the angular integration, we note that for given $|\mathbf{p}'|, |\mathbf{p}_n|$, with the latter defined in (15.9.4),

$$\mathbf{q}^2 = \mathbf{p}_n^2 - 2|\mathbf{p}_n||\mathbf{p}'|\cos\theta + \mathbf{p}'^2 \equiv q^2 \quad (15.9.5)$$

we have

$$2\pi \sin\theta d\theta = \frac{2\pi q dq}{|\mathbf{p}_n||\mathbf{p}'|} \quad (15.9.6)$$

and

$$q_{max} = |\mathbf{p}'| + |\mathbf{p}_n|, \quad q_{min} = |\mathbf{p}'| - |\mathbf{p}_n|. \quad (15.9.7)$$

Hence,

$$\sigma_{n0} = \frac{8\pi e_0^2 e^2 \mu^2}{|\mathbf{p}'|^2} \int_{q_{min}}^{q_{max}} \frac{dq}{q^3} F_{n0}(q) \quad (15.9.8)$$

where

$$F_{n0}(q) = \left|\langle \Phi_n | e^{-iqz/\hbar} | \Phi_0 \rangle\right|^2 \quad (15.9.9)$$

and we have arbitrarily chosen the direction of $\mathbf{q}$ to be along the z-axis in computing $F_{n0}(q)$.

Consider a system of atoms with $\mathcal{N}$ denoting the number of atoms per unit volume. We may then introduce the energy loss per unit path length of the charged particle in passing through the medium by

$$-\frac{d\mathcal{E}}{dL} = \sum_n{}' (E_n - E_0) \mathcal{N} \sigma_{n0} \quad (15.9.10)$$

where the summation (integration) goes over all the accessible states for the discrete as well as the continuous energy levels.

We may rewrite (15.9.10) in an equivalent form by setting

15.9 Passage of Particles Through Media; Neutron Interferometer

$$\frac{2m}{q^2}(E_n - E_0)F_{n0}(q) = f_{n0}(q) \qquad (15.9.11)$$

to obtain

$$-\frac{d\mathscr{E}}{dL} = \frac{4\pi e_0^2 e^2 \mu^2 \mathscr{N}}{m\,|\mathbf{p}'|^2} \sum_n{}' \int_{q_{\min}}^{q_{\max}} \frac{dq}{q} f_{n0}(q) \qquad (15.9.12)$$

where we recall that $q_{\min/\max}$ depend on n.

The function $f_{n0}(q)$ has the following important completeness property

$$\sum_n{}' f_{n0}(q) = 1 \qquad (15.9.13)$$

for all q to be shown later below.

Because of the singular nature of $F_{n0}(q)/q^3$ for $q \to 0$, and its rapid damping for large q, because of the $1/q^3$ factor and the matrix element of the exponential $\exp -iqz/\hbar$ in $F_{n0}(q)$ (see, for example, (15.8.48)), the main region of integration in (15.9.8) comes form small q. The latter corresponds to a small momentum transfer to the charged particle resulting in an almost undeflected straight path of the particle.

We consider the high-energy limit $|\mathbf{p}'| \gg |\mathbf{p}'| - |\mathbf{p}_n|$, from which

$$q_{\max} = |\mathbf{p}'| + |\mathbf{p}_n| = 2\,|\mathbf{p}'| - (|\mathbf{p}'| - |\mathbf{p}_n|)$$

$$\simeq 2\,|\mathbf{p}'| \equiv q'_{\max} \qquad (15.9.14)$$

and from (15.9.4),

$$q_{\min} = |\mathbf{p}'| - |\mathbf{p}_n| = \frac{2\mu(E_n - E_0)}{q_{\max}} \simeq \frac{\mu(E_n - E_0)}{|\mathbf{p}'|} \equiv q'_{\min}. \qquad (15.9.15)$$

Since the main contribution to the integral in (15.9.8) comes from small q, we may effectively carry out the so-called dipole approximation to $F_{n0}(q)$ and replace the latter by $q^2\,|\langle\Phi_n|z|\Phi_0\rangle|^2/\hbar^2$ by expanding the exponential $\exp(-iqz/\hbar) \simeq 1 - iqz/\hbar$, and using the orthogonality of $|\Phi_n\rangle$ and $|\Phi_0\rangle$. This gives the q-independent quantity for $f_{n0}(q)$

$$\frac{2m}{\hbar^2}(E_n - E_0)\,|\langle\Phi_n|z|\Phi_0\rangle|^2 \equiv f_{n0}^d \qquad (15.9.16)$$

referred to as the oscillator strength, while $f_{n0}(q)$ in (15.9.11) as the generalized oscillator strength, with d in f_{n0}^d corresponding to the dipole approximation.

Accordingly, we may rewrite (15.9.12) as

$$-\frac{d\mathscr{E}}{dL} = \frac{4\pi e_0^2 e^2 \mu^2 \mathscr{N}}{m\,|\mathbf{p}'|^2}$$

$$\times \left[\ln(q'_{\max}) - \frac{2m}{\hbar^2} {\sum_n}' (E_n - E_0) |\langle \Phi_n | z | \Phi_0 \rangle|^2 \ln(q'_{\min}) \right] \quad (15.9.17)$$

where in writing the first term we have used the completeness summation formula in (15.9.13) and the fact that $q'_{\max}$ is independent of n.

Upon defining an "average ionization potential" I through

$$\frac{2m}{\hbar^2} {\sum_n}' (E_n - E_0) |\langle \Phi_n | z | \Phi_0 \rangle|^2 \ln(E_n - E_0) = \ln I \quad (15.9.18)$$

we have from (15.9.13), (15.9.14), (15.9.15), (15.9.17)

$$-\frac{d\mathscr{E}}{dL} = \frac{4\pi e_0^2 e^2 \mathscr{N}}{mv^2} \ln\left(\frac{2\mu v^2}{I}\right) \quad (15.9.19)$$

where $\mathbf{p}' = \mu \mathbf{v}$, $\mathbf{v}$ is the initial velocity of the incident charged particle relative to an atom, M in (15.9.3) denotes its mass and m denotes the mass of the *electron* in the atom.

It is interesting to note that for $M \gg m$, (15.9.19) reduces to

$$-\frac{d\mathscr{E}}{dL} = \frac{4\pi e_0^2 e^2 \mathscr{N}}{mv^2} \ln\left(\frac{2mv^2}{I}\right) \quad (15.9.20)$$

which is *independent* of the mass of the incident charged particle.

On the other hand for an incident electron $\mu = m/2$, and (neglecting the exchange effect),

$$-\frac{d\mathscr{E}}{dL} = \frac{4\pi e_0^2 e^2 \mathscr{N}}{mv^2} \ln\left(\frac{mv^2}{I}\right). \quad (15.9.21)$$

It remains to establish the completeness summation formula in (15.9.13). To this end, we note that from the Schrödinger equation,

$$(E_n - E_0) \int d^3\mathbf{x}\, \Phi_n^* e^{-iqz/\hbar} \Phi_0 = -\frac{\hbar^2}{2m} \int d^3\mathbf{x}\, e^{-iqz/\hbar} \nabla \cdot (\nabla \Phi_n^* \, \Phi_0 - \Phi_n^* \nabla \Phi_0)$$

$$= -\frac{i\hbar q}{2m} \int d^3\mathbf{x}\, e^{-iqz/\hbar} \left(\frac{\partial}{\partial z} \Phi_n^* \, \Phi_0 - \Phi_n^* \frac{\partial}{\partial z} \Phi_0 \right)$$

$$= \frac{\hbar^2}{2m} \left[\frac{2iq}{\hbar} \left\langle \Phi_n \left| e^{-iqz/\hbar} \frac{\partial}{\partial z} \right| \Phi_0 \right\rangle + \frac{q^2}{\hbar^2} \left\langle \Phi_n \left| e^{-iqz/\hbar} \right| \Phi_0 \right\rangle \right]. \quad (15.9.22)$$

Upon multiplying the above equation by $\left\langle \Phi_0 \left| e^{iqz'/\hbar} \right| \Phi_n \right\rangle 2m/q^2$ and summing (integrating) over n, we obtain from (15.9.11)

$${\sum_n}' f_{n0}(q) = \frac{\hbar^2}{q^2} \left[\frac{2iq}{\hbar} {\sum_n}' \left\langle \Phi_0 \left| e^{iqz'/\hbar} \right| \Phi_n \right\rangle \left\langle \Phi_n \left| e^{-iqz/\hbar} \frac{\partial}{\partial z} \right| \Phi_0 \right\rangle \right.$$

$$+ \frac{q^2}{\hbar^2} \sum_n{}' \langle \Phi_0 | e^{iqz'/\hbar} | \Phi_n \rangle \langle \Phi_n | e^{-iqz/\hbar} | \Phi_0 \rangle \Big]$$

$$= \frac{\hbar^2}{q^2} \left[\frac{2iq}{\hbar} \left\langle \Phi_0 \left| \frac{\partial}{\partial z} \right| \Phi_0 \right\rangle + \frac{q^2}{\hbar^2} \langle \Phi_0 | \Phi_0 \rangle \right]$$

$$= 1 \qquad (15.9.23)$$

where we have used the fact that $\langle \Phi_0 | \partial \Phi_0 / \partial z \rangle = 0$, $\langle \Phi_0 | \Phi_0 \rangle = 1$.

A similar analysis as above may be carried out for multi-electron atoms, where one of the changes to be made in the analysis is to replace $\exp -i\mathbf{q} \cdot \mathbf{x}/\hbar$ by $\sum_j \exp(-i\mathbf{q} \cdot \mathbf{x}_j/\hbar)$, with a summation going over the electrons in the atom (see Problem 15.21). For the hydrogen case, $I \simeq 15$ eV. For other atoms, I is proportional to the atomic number Z with a proportionality factor roughly of the order of 10 eV.

Needless to say, the charged particle would also emit radiation but we will not go into this here.

15.9.2 Neutron Interferometer

Consider a neutron beam which strikes the surface of a (cubic) crystal at an incident angle θ as shown in Figure 15.8, with wave vector $\mathbf{k} = \mathbf{p}/\hbar$. For the so-called Laue scattering,[37] the transmitted beam wave vector $\mathbf{k}_t$ is the same as the incident one, while the one of the reflected beam $\mathbf{k}_r$ differs from $\mathbf{k}_t$ by a reciprocal lattice vector $\mathbf{Q}$, with the latter parallel to the face of the crystal. The initial incident state may be written as

$$\psi_i = e^{i\mathbf{k} \cdot \mathbf{x}} = e^{ik_1 x_1}(\chi_1 + i\chi_2) \qquad (15.9.24)$$

where χ_1, χ_2 are standing waves given by

$$\chi_1 = \cos k_2 x_2, \qquad \chi_2 = \sin k_2 x_2 \qquad (15.9.25)$$

and

$$|k_2| = \frac{|\mathbf{Q}|}{2} = \frac{\pi}{a_2} \qquad (15.9.26)$$

with atomic locations given by

$$\mathbf{a_n} = (n_1 a_1, n_2 a_2, n_3 a_3) \qquad (15.9.27)$$

where n_1, n_2, n_3 are integers. We note that[38]

---

[37] For additional details of Laue scattering from crystals see Greenberger and Overhauser (1979); Olariu and Iovitzu Popescu (1985); Rauch and Petrascheck (1978).

[38] Without loss of generality it is assumed that there is an atom located at $x_2 = 0$ as indicted in (15.9.27).

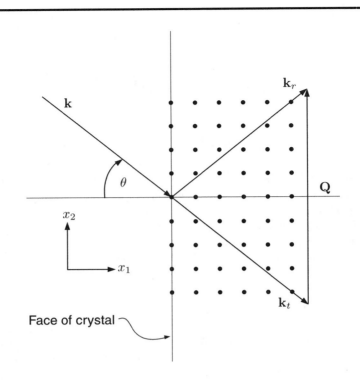

Fig. 15.8. The figure depicts the so-called Laue scattering of a beam of neutrons from a (cubic) crystal where the associated transmitted wave vector $\mathbf{k}_t$ is the same as that of the incident one, while the wave vector $\mathbf{k}_r$ of the reflected beam differs from $\mathbf{k}_t$ by a reciprocal lattice vector $\mathbf{Q}$. The incident angle θ is referred to as the Bragg angle.

$$|\chi_1|^2 = \cos^2 \frac{\pi x_2}{a_2} \tag{15.9.28}$$

takes its maximum value at atomic sites, while

$$|\chi_2|^2 = \sin^2 \frac{\pi x_2}{a_2} \tag{15.9.29}$$

takes rather its minimum value of zero at atomic sites. That is, $i(\exp ik_1 x_1)\chi_2$ essentially goes through the crystal undisturbed (see Figure 15.9), while $(\exp ik_1 x_1)\chi_1$ interacts relatively stronger with the crystal. Accordingly, where the beam emerges from the opposite face of the crystal, we may write

$$\psi_f = e^{ik_1 x_1}(\eta \chi_1 + i\chi_2) \tag{15.9.30}$$

where η is a complex number

15.9 Passage of Particles Through Media; Neutron Interferometer 873

$$\eta = \alpha e^{i\beta} \tag{15.9.31}$$

with α, β real, are characteristics of the crystal and its thickness, and where α is the absorption coefficient. For crystals with negligible absorption of neutrons, such as silicon of several millimeters thickness, one may take $\alpha = 1$.

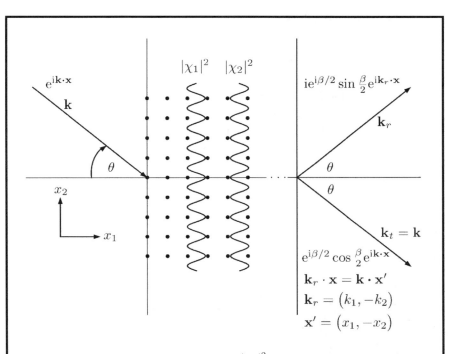

Fig. 15.9. For the standing wave χ_1, $|\chi_1|^2$ takes its maximum values at the atomic sites, and hence $(\exp i k_1 x_1)\chi_1$ interacts rather strongly with the crystal. On the other hand, for the standing wave χ_2, $|\chi_2|^2$ takes its minimum value of zero at the atomic sites, and hence $i(\exp i k_1 x_1)\chi_2$ propagates essentially undisturbed in the crystal. For a crystal of negligible absorption of neutrons, when the beam reaches the opposite face of the crystal, it emerges from it with a reflection coefficient $i(\exp i\beta/2)\sin\beta/2$ and transmission coefficient $(\exp i\beta/2)\cos\beta/2$, where β is real and is a characteristic of the crystal and its thickness.

With $\alpha = 1$, we may rewrite ψ_f as

$$\psi_f = \rho\, e^{i\mathbf{k}\cdot\mathbf{x}} + \kappa\, e^{i\mathbf{k}\cdot\mathbf{x}'} \tag{15.9.32}$$

where

$$\rho = e^{i\beta/2}\cos\frac{\beta}{2}, \qquad \kappa = i\, e^{i\beta/2}\sin\frac{\beta}{2} \tag{15.9.33}$$

and

$$\mathbf{x}' = (x_1, -x_2) \tag{15.9.34}$$

$$\mathbf{k} \cdot \mathbf{x}' = \mathbf{k}_r \cdot \mathbf{x} \tag{15.9.35}$$

$$\mathbf{k}_r = (k_1, -k_2). \tag{15.9.36}$$

One may then introduce the transmission and reflection probabilities

$$T = |\rho|^2 = \cos^2 \frac{\beta}{2}, \qquad R = |\kappa|^2 = \sin^2 \frac{\beta}{2}. \tag{15.9.37}$$

Note the presence of the i multiplicative factor in κ given in (15.9.33).

Referring to the experimental situation of a neutron interferometer, depicted in Figure 15.10, we note that since the amplitude to go from A to B, for example, outside the crystals is, up to an unimportant multiplicative factor, just the phase factor $\exp\left(\mathrm{i}M(\mathbf{x}_B - \mathbf{x}_A)\right)^2/2\hbar t_{BA}$, where M is the mass of the neutron, the intensities I_1, I_2 at the detection sites D_1, D_2 are given from (15.9.32)/(15.9.33), up to the same overall unimportant factor

$$I_1 = \left|\kappa^3 + \rho\kappa\rho\right|^2 = R(T-R)^2 \tag{15.9.38}$$

$$I_2 = \left|\kappa^2 \rho + \rho\kappa^2\right|^2 = 4TR^2 \tag{15.9.39}$$

using the symmetry between the segments BC, AD and similarly of the segments AB, DC for propagation outside the crystals.

Now we rotate the whole system in Figure 15.10 by an angle ϕ, raising the segment BC at a height H above the segment AD, both in horizontal direction, in the Earth's gravitational field as shown in Figure 15.11.

For the amplitude of propagation between two points in a given potential in almost a straight line, we may take the corresponding approximation from the expression already derived in (11.2.31),

$$\langle \mathbf{x}t | \mathbf{x}'t' \rangle = \langle \mathbf{x}t | \mathbf{x}'t' \rangle_0 \exp -\frac{\mathrm{i}}{\hbar} \int_{t'}^{t} \mathrm{d}\tau\, V\left(\mathbf{x}' + (\mathbf{x} - \mathbf{x}')\frac{\tau - t'}{(t - t')}\right) \tag{15.9.40}$$

where $\langle \mathbf{x}t | \mathbf{x}'t' \rangle_0$ is the free Green function.

For the subsequent analysis we consider the thickness of the slabs $a \ll d$. We choose the potential energy on the segment AD to be zero as a reference point, and take

$$\frac{1}{\hbar}\int_{t_B}^{t_C} \mathrm{d}\tau\, V\left(\mathbf{x}_B + (\mathbf{x}_C - \mathbf{x}_B)\frac{\tau - t_B}{(t_C - t_B)}\right) \simeq (t_C - t_B)\frac{MgH}{\hbar}$$

$$= \left(\frac{d}{v\cos\theta}\right)\frac{2Mgd\sin\theta\sin\phi}{\hbar}$$

$$\equiv \Delta \tag{15.9.41}$$

15.9 Passage of Particles Through Media; Neutron Interferometer   875

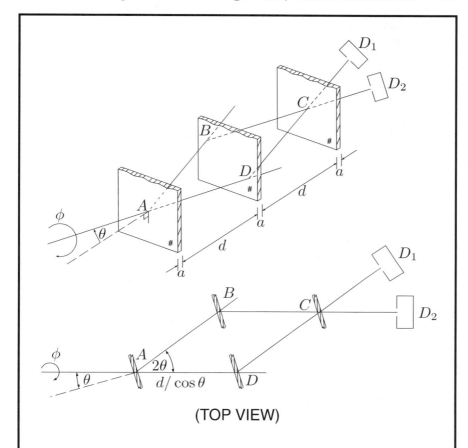

Fig. 15.10. (Neutron Interferometer) A beam of neutrons strikes a slab of crystal in a system of three parallel slabs equally spaced at distances d apart and each of thickness a. The beams BC, DC interfere at C and the intensities of neutrons emerging from C are measured by the detectors D_1, D_2. Eventually the whole system will be rotated about the direction of the incident beam by an angle ϕ raising the beam BC above the beam AD, both moving horizontally, in the Earth's gravitational field (see also Figure 15.11).

corresponding to the line segment BC for neutrons travelling at a height H as given in Figure 15.11. outside the crystals, where v is the speed of a neutron.

By symmetry, $\langle \mathbf{x}_C t_C | \mathbf{x}_B t_B \rangle_0 = \langle \mathbf{x}_D t_D | x_A t_A \rangle_0$ for the free parts of the Green function, while $\langle \mathbf{x}_B t_B | \mathbf{x}_A t_A \rangle = \langle \mathbf{x}_C t_C | \mathbf{x}_D t_D \rangle$, for propagation outside the crystals, which according to (15.9.40), are up to unimportant numerical factors, just phase factors which as easily seen below may be factored out in computing the amplitudes corresponding to the intensities I_1, I_2 determined at the detection sites D_1, D_2, and hence are unimportant.

From Figure 15.11, (15.9.40), (15.9.41) and (15.9.33) we have, *up to an unimportant common multiplicative factor*, in analogy to (15.9.38), (15.9.39),

$$I_1 = \left|\kappa^3 e^{-i\Delta} + \rho^2 \kappa\right|^2$$

$$= R\left[1 - 2TR(1 + \cos \Delta)\right] \tag{15.9.42}$$

and

$$I_2 = \left|\kappa^2 e^{-i\Delta}\rho + \rho\kappa^2\right|^2$$

$$= 2TR^2(1 + \cos \Delta). \tag{15.9.43}$$

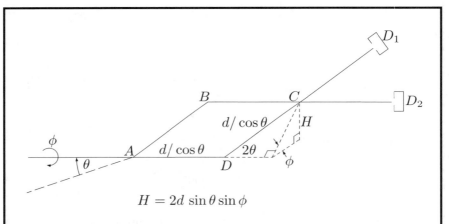

$$H = 2d \sin \theta \sin \phi$$

Fig. 15.11. By a rotation of the whole system in Figure 15.10 about the direction of the incident beam by an angle ϕ, the beam BC is raised above the beam AD, both in a horizontal direction, by a height H, as indicated, in the Earth's gravitational field. For $\theta \neq 0$, the difference $I_2 - I_1$ of the intensities measured by the detectors D_2 and D_1, oscillates as a function of ϕ with a phase depending on the gravitational constant g and the Planck constant h.

We note that $I_1 + I_2$ is independent of the phase, while

$$I_2 - I_1 = R(4TR - 1) + 4TR^2 \cos \Delta \tag{15.9.44}$$

has an oscillatory character in ϕ, and may be recast in the form

$$\frac{(I_2 - I_1)(\phi)}{(I_2 - I_1)(0)} = \alpha_1 + \alpha_2 \cos\left(\frac{2Mgd^2 \tan\theta \sin\phi}{\hbar v}\right) \tag{15.9.45}$$

and is a function of a gravity induced *quantum* mechanical phase depending on the gravitational constant g and the Planck constant h.

An experiment of the sort described above has been carried out[39] and the induced phase shift has been observed. The parameters taken were $a = 0.2$ cm, $d = 3.5$ cm, $\theta = 22.1°$, $h/Mv = 1.445$ Å. If a is not neglected in the above analysis, then Δ in (15.9.44) is to be multiplied[40] by the corrective factor $(1 + a/d)$.

Problems

15.1. Find the scattering amplitude and the cross section in the Born approximation for the Yukawa potential $\lambda(\exp -\mu |\mathbf{x}|)/|\mathbf{x}|$, and the exponential potential $\lambda'(\exp -\mu |\mathbf{x}|)$, $\mu > 0$. Use the sufficiency condition (15.3.26) to set bounds on the parameters in question for the absolute convergence of the Born series.

15.2. Form (15.4.20) you have seen that the optical theorem relates different orders in the potential. As a matter of fact, the Born approximation violates the optical theorem, as seen, for example, from the previous problem, since $\operatorname{Im} f_B = 0$, while $\sigma_B \neq 0$. Accordingly, it is necessary to go to second order in the potential in determining $\operatorname{Im} f(\mathbf{p}', \mathbf{p}')$ to verify the optical theorem. Use (15.3.10)/(15.3.11) to *second* order in V to verify the optical theorem in (15.4.6) with σ determined in the Born approximation σ_B.

15.3. For a potential of short range, i.e., for one which vanishes faster than the Coulomb potential, show that for $|\mathbf{x}| \to \infty$ $(p^0 \simeq E(\mathbf{p}'))$

$$\int d^3 \mathbf{p}\, e^{i\mathbf{p}\cdot\mathbf{x}/\hbar}\, G_\pm(\mathbf{p}, \mathbf{p}'; p^0) \longrightarrow \frac{e^{i\mathbf{p}'\cdot\mathbf{x}/\hbar}}{(p^0 - E(\mathbf{p}') \pm i\varepsilon)}.$$

15.4. Show that the radial partial Green function $G_\ell(r, r')$ in (15.4.39) satisfies the differential equation (15.4.40) and hence (15.4.42) provides a solution for $g_\ell(r)$.

15.5. Restrict the sum in (15.4.29) for the scattering amplitude $f(p, \theta)$ to S and P waves, i.e., for $\ell = 0, \ell = 1$, to write the differential cross section as

$$\frac{d\sigma}{d\Omega} = \frac{\hbar^2}{p^2}(A + B\cos\theta + C\cos^2\theta).$$

Find bounds on the coefficients A, B, C. Also express these coefficients in terms of the phase shifts δ_0, δ_1 and find the general structure for σ.

15.6. For the Born approximation to the phase shift δ_ℓ in (15.4.45), for small δ_ℓ such that $\tan \delta_\ell \simeq \delta_\ell$, $(\exp 2i\delta_\ell) - 1 \simeq 2i\delta_\ell$, use the expansion formula

---
[39] Colella et al. (1975). See also Greenberger and Overhauser (1979).
[40] Greenberger and Overhauser (1979).

$$\sum_{\ell=0}^{\infty}(2\ell+1)\left(j_\ell\left(\frac{pr}{\hbar}\right)\right)^2 P_\ell(\cos\theta) = \frac{\sin Qr}{Qr}$$

where $Q = 2(p/\hbar)\sin(\theta/2)$, to obtain an expression for the scattering amplitude and compare it with the one given in (15.3.12)/(15.3.11) for a spherically symmetric potential $V(r)$.

15.7. For the $\ell = 0$ contribution in (15.4.29), set $g_0(r) = g(k,r)$ in (15.4.33) with $p/\hbar = k$. Use the Schrödinger equation to show that for R large

$$Q(k,R) = Q(0,R) - k^2 \frac{\int_0^R dr\, g(k,r)g(0,r)}{g(k,R)g(0,R)}$$

where $Q(k,R) = (dg(k,R)/dR)/g(k,R)$, and $g(0,R)$ may be taken to be $(R-a)$ for some constant a. Use this equation, together with the asymptotic form for $g(k,R)$ in (15.4.34) to obtain the so-called *effective range expansion*

$$k \cot \delta_0 = -\frac{1}{a} + \frac{r_0}{2}k^2 + \cdots$$

at low energies $k^2 \simeq 0$, where

$$r_0 = \frac{2}{a^2} \int_0^R dr\, [(r-a)^2 - g^2(0,r)],$$

$g''(0,r) - u(r)g(0,r) = 0$. The parameters a, r_0 are respectively called the scattering length and the effective range. Interpret the significance of these parameters, and obtain the $\ell = 0$ contribution to the cross section at low energies.

15.8. Consider a particle in the spherically symmetric potential

$$V(r) = \begin{cases} V_0, & r < R \\ 0, & r \geqslant R \end{cases}$$

at low energies.
(i) For $V_0 < 0$, determine the scattering length a and effective r_0 introduced in the previous problem.
(ii) Carry out a qualitative analysis of the sign of a versus the sign and magnitude of V_0 in a plot of $g_0(r)$ (i.e., for $\ell = 0$) versus r.
(iii) At low energies, determine the $\ell = 0$ contribution of the cross section. Under what condition the latter and the phase shift δ_0 vanish? The vanishing property of the $\ell = 0$ contribution to the cross section is referred to as the *Ramsauer effect*.

15.9. For the spherically symmetric potential in the previous problem with $V_0 < 0$, consider the low energy $\ell = 0$ scattering amplitude $f_0(p,\theta)$ in (15.4.53). Investigate the nature of the poles that $f_0(p,\theta)$ may have on the positive imaginary axis in the complex p-plane and the bound states. Investigate also the analyticity property of $f_0(p,\theta)$ in the complex energy $E = p^2/2m$ plane.

15.10. Expand the scattering amplitude $f_C(p,\theta)$ for the Coulomb potential in (15.5.22) in terms of the Legendre polynomials $P_\ell(\cos\theta)$, where $|\mathbf{p} - \mathbf{p}'| = 2p\sin\theta/2$, and show that up to an overall phase factor

$$f_C(p,\theta) = \frac{\hbar}{2ip} \sum_{\ell=0}^{\infty} (2\ell+1) \frac{\Gamma(\ell+1+i\gamma)}{\Gamma(\ell+1-i\gamma)} P_\ell(\cos\theta)$$

where $\gamma = \lambda\mu/\hbar p$. For an attractive potential $\gamma = -|\lambda|\mu/\hbar p$, and $\Gamma(\ell+1+i|\lambda|\mu/\hbar)$ develops poles at non-positive integers for $p = i\kappa$, $\kappa > 0$. Use this fact to show that the corresponding energy values $p^2/2\mu$ coincide with the discrete energy levels of the hydrogen atom.

15.11. For the potential $V(r) = \infty$ for $r < R$ and zero otherwise, compute the differential cross section. Investigate the high energy behavior of the cross section and show that it does not coincide with the one obtained from a naïve classical computation. Interpret this result.

15.12. Use the intuitive argument given below (15.5.18) to obtain the asymptotic "free" time evolution operator for a potential $V(r)$ which vanishes at infinity like $r^{-\delta}$ for some $0 < \delta < 1$ in analogy to the Coulomb case $\exp(-iH_{0C}(t)/\hbar)$ with $H_{0C}(t)$ given in (15.5.18).
[Ref.: Manoukian and Prugovečki (1971). See also this reference for related technical details and physical considerations.]

15.13. Use the analysis carried out in §11.4 to recast the expression for the scattering amplitude as obtained from (15.6.6) in a path integral formalism.

15.14. Develop the eikonal approximation in (15.7.7) for the Yukawa potential $V(\mathbf{x}) = V(\mathbf{b}, \rho) = \lambda \exp\left(-\mu_0\sqrt{\mathbf{b}^2 + \rho^2}\right)/\sqrt{\mathbf{b}^2 + \rho^2}, \mu_0 > 0$.
[Hint: You may express $\chi(b, |\mathbf{p}'|)$ in (15.7.9) in terms of the Bessel function $K_0(\mu_0|\mathbf{b}|)$.]

15.15. Use the asymptotic behavior of the Bessel function $K_0(\mu_0|\mathbf{b}|)$ for $\mu_0 \to +0$ to obtain an eikonal approximation for the Coulomb potential up to an overall unimportant (infinite) phase factor by the application of the previous problem.

15.16. Use the method developed through (15.7.15)–(15.7.20) to derive an expression for the asymptotic "free" Green function corresponding to potential $V(r) = \lambda r^{-\delta}$, for some $0 < \delta < 1$. See also Problem 15.12.

15.17. Repeat the procedure worked out through (15.1.41)–(15.1.46), to prove the existence of the modified Møller wave operators $\Omega_{\pm C}$ for the Coulomb potential as the (strong) limits $t \to \pm\infty$ of $\Omega_C(t) = e^{itH/\hbar}e^{-iH_{0C}(t)/\hbar}$ where $H_{0C}(t)$ is given in (15.5.18).
[Ref.: Dollard (1964).]

15.18. The scattering S operator is so defined that $\langle\Phi_{\text{out}}^b|S|\Phi_{\text{in}}^a\rangle = \langle\psi_+^b|\psi_-^a\rangle$, where $|\psi_-^a\rangle = \Omega_-^a|\Phi_{\text{in}}^a\rangle$, $\psi_+^b = \Omega_+^b|\Phi_{\text{out}}^b\rangle$, and $\Omega_\pm^c$ are given by the (strong) limits $t \to \pm\infty$ of $\exp(itH/\hbar)\exp(-itH_c/\hbar)$, where H_c is the corresponding c-channel Hamiltonian. Relate this operator to the transition matrix T_{ba} in (15.8.38) for an incoming $|\Phi_{\text{in}}^a\rangle$ state and an outgoing $|\Phi_{\text{out}}^b\rangle$ state. Formulate and spell out precisely the conditions for which this operator is unitary.

15.19. To obtain a better agreement with experiments, one may replace* the factor $\left[1 + \dfrac{\mathbf{q}^4}{|\mathbf{p}_2 - \mathbf{p}'|^4} - \dfrac{\mathbf{q}^2}{|\mathbf{p}_2 - \mathbf{p}'|^2}\right]$ in (15.8.68), taking into account of the exchange effect, by $\left[1 + \left(\dfrac{\mathbf{q}^2}{\mathbf{p}'^2 - \mathbf{p}_2^2}\right)^2 - \dfrac{\mathbf{q}^2}{\mathbf{p}'^2 - \mathbf{p}_2^2}\right]$. Make such a replacement in (15.8.71) and carry out the angular integrations. A change of an angular integration variable as in (15.9.6) is useful.
[*Ochkur (1963, 1964).]

15.20. Carry out the decomposition of T_{aa} in (15.8.107) for the scattering of a particle of spin zero off a cluster whose internal angular momentum s in not zero.
[Hint. Follow an analysis as in (15.4.21) and use expansions as in (5.10.99)/(5.10.103) in terms of Clebsch-Gordan coefficients. See also §5.9.]
Can you extend the analysis to the case for the scattering of one cluster off another if the internal angular momenta of both clusters are non-zero?

15.21. Recapitulate the analysis carried out in §15.9 for the energy loss per unit path length by a charged particle passing through a medium with multi-electron atoms. Simply replace $\exp(-i\mathbf{q}\cdot\mathbf{x}/\hbar)$ in (15.9.9) by $\sum_j \exp(-i\mathbf{q}\cdot\mathbf{x}_j/\hbar)$ as a sum over all the electrons in an atom. The corresponding completeness summation formula to (15.9.13) would be equal to the atomic number Z. The "average ionization potential" I will then involve $\left|\left\langle \Phi_n \left| \sum_j \mathbf{x}_j \right| \Phi_0 \right\rangle\right|^2$ instead, and is defined by dividing the resulting new expression corresponding to the left-hand side of (15.9.18) by Z. You should then obtain the same expression as in (15.9.19) with $\mathcal{N}$ simply replaced by $\mathcal{N}Z$, where the latter denotes the number of electrons per unit volume present in the medium.

16
Quantum Description of Relativistic Particles

This chapter is involved with the method of providing a quantum description of relativistic particles. One is confronted with the requirement of developing such a formalism, as imposed by nature, when the energy and momentum of a particle are large enough so that the Schrödinger equation, with a non-relativistic kinetic energy, becomes inapplicable. A relativistic theory, as a result of the exchange that takes place between energy and matter, allows the creation of an unlimited number of particles and the number of particles in a physical process need not be conserved. An appropriate description of such physical processes for which a variable number of particles may be created or destroyed is provided by the very rich concept of a quantum field. For example, photon emissions and annihilations are explained by the introduction of the electromagnetic quantum field. The theory which emerges from extending quantum physics to the relativistic regime is called *Relativistic Quantum Field Theory* or just *Quantum Field Theory*. Quantum Electrodynamics is an example of a quantum field theory and is the most precise theory devised by man when confronted with experiments. The essence of special relativity is that all inertial frames are completely equivalent in explaining a physical theory as one inertial frame cannot be distinguished from another. This invariance property of physical theories in all inertial frames as required by the special theory of relativity is also readily implemented in quantum field theory.

This chapter is not involved with the intricacies of quantum field theory, except for the moderate exception in §16.9, but with the details of the precursor of such a theory. It provides the bridge between quantum physics developed so far in the text and the theory of quantum fields. As a first step, we have to find replacements for Schrödinger's equation when the energies and momenta of the particles to be described are too high for a non-relativistic treatment. This will lead us to the Klein-Gordon equation for spin 0 particles and the Dirac equation for spin 1/2 particles, and later on (§16.8) relativistic wave equations will be developed for higher spins as well. These equations are Lorentz covariant in the sense that they have the same form in all inertial frames. In our presentation, massive as well as massless particles will be considered.

We have already dealt with the Dirac equation in §7.4, §7.5 in describing relativistic corrections to the hydrogen atom spectrum. The relativistic wave equations predict the existence of negative energy states with negative mass, with energies going down to $-\infty$ implying the instability of the corresponding systems. Dirac, in his spin 1/2 theory, was led to postulate that the negative energy states are completely filled by electrons, according to the Pauli exclusion principle, so that no transitions to these states are possible and stability would be achieved. This vacuum state thickly populated by electrons has been called the Dirac sea. In 1930, he has interpreted the absence of an electron in the sea (a "hole") as an anti-particle — the positron. The physics community found it difficult to accept his prediction until C. D. Anderson discovered the positron in 1932 who apparently was not aware of Dirac's prediction[1] at the time of his discovery. It is also interesting to note that G. Gamow referred[1] to the electrons in the negative energy states, postulated by Dirac, as "*donkey electrons*" because they would move in the opposite direction of the applied force. It is important to realize that a theory which started with the hope of providing a quantum description of a single relativistic particle led eventually to a multi-particle theory.

Historically, the Klein-Gordon equation was introduced before Dirac introduced his. Because of the emergence of negative energy states and the difficulty in defining a positive definite probability density, the Klein-Gordon equation was ignored until Dirac developed his formalism and led to his prediction of the anti-particle (the positron), and its discovery, at which time the importance of the Klein-Gordon equation was recognized and consistent field theories were formulated.

Dirac's theory provided tremendous insight into the nature of a combined theory consisting of quantum physics and relativity leading eventually to the birth of quantum field theory. As far as the achievements of the Dirac theory were, we may summarize by saying that besides being a Lorentz covariant theory and describes spin 1/2 particles, it also predicted approximately the gyromagnetic ratio $g = 2$ of the electron,[2] it gave the correct fine-structure of the hydrogen atom and, of course, led to the discovery of anti-matter. These achievements were so great, that apparently *Dirac himself remarked*[3] *in one of his talks that his equation was more intelligent than its author.*

The first six sections of this chapter deal with the Dirac equation. Although some knowledge of the basics of special relativity is assumed of the reader, the intricacies of the so-called Lorentz transformations are spelled out in §16.2. Special emphasis is put on the concept of helicity in this latter

---

[1] cf, Weisskopf (1980).
[2] A correction to the $g = 2$ value was derived in §8.5.
[3] Weisskopf (1980).

reference, where one learns that helicity as the spin component taken parallel to a particle's momentum, as the latter is determined in each of the corresponding inertial frames, has a Lorentz (relativistic) invariant meaning and is well defined for massive as well as for massless particles. §16.6 provides a detailed physical interpretation of the Dirac theory and, in particular, to the origin of the relativistic corrections to the hydrogen atom worked out in §7.4, §7.5. The exact bound Coulomb problem is solved in the Appendix to §16.6. The Klein-Gordon equation and some equations which follow from it are the subject of §16.7. Relativistic wave equations for any mass $M \geqslant 0$ and higher spins are developed in §16.8. The last section, §16.9 deals with the spin and statistics connection whose importance can never be overemphasized. We use the ingenious approach due to Schwinger in examining typical Lagrangian densities in relativistic quantum field theories to investigate the nature of the spin and statistics connection. Here familiarity with Lagrangian dynamics is essential. In the appendix to this section a detailed treatment of the so-called action integral is given, however, for systems admitting Grassmann variables as well.

The Spin and Statistics Theorem, in its simplest form, states that no two identical particles of half-odd integer spins (fermions) can occupy the same state, while any number of identical particles of integer spins (bosons) may do so without limitation. The practical effect of this theorem prevails over the whole of science and provides the basis for explaining the periodic table of elements from which our world and we are made of. Matter is stable because of the spin and statistics connection (the so-called Pauli exclusion principle) as applied to electrons as particles of spin 1/2. It also explains as to why matter occupies such a large volume of space.[4] We have seen in Chapter 14, §14.4 on *The Collapse of "Bosonic Matter"* as to what happens to matter if the spin and statistics connection (the Pauli exclusion principle) were abolished. In regard to matter without the spin and statistics connection, and before carrying out the analyses intended in this chapter, it is worth recalling[5] F. Dyson's words: "*Matter in bulk would collapse into a condensed high density phase. The assembly of any two macroscopic objects would release energy comparable to that of an atomic bomb. . . . Matter without the exclusion principle is unstable.*", and also quote from the translator's Preface of the classic book[6] by S.-I. Tomonaga on The Story of Spin: *The existence of spin, and the statistics associated with it, is the most subtle and ingenious design of Nature — without it the whole universe would collapse.*

---

[4] See §14.4 and the Introduction to Chapter 14.
[5] See the Introduction to Chapter 14.
[6] T. Oka in the Translator's Preface of the book by Tomonaga (1997).

16.1 The Dirac Equation and Pauli's Fundamental Theorem

Let us go over the method used in §7.4 to obtain the Dirac equation of spin $1/2$. Using the constraint between the energy E, and the momentum $\mathbf{p}$, of a free particle of mass M

$$\sqrt{\mathbf{p}^2 c^2 + M^2 c^4} - E = 0, \qquad (16.1.1)$$

where c denotes the speed of light, we formally obtain, upon the substitutions $E \to i\hbar \partial/\partial t$, $\mathbf{p} \to -i\hbar \boldsymbol{\nabla}$, the equation

$$i\hbar \frac{\partial}{\partial t} \psi(\mathbf{x}, t) = \left(\sqrt{-\hbar^2 c^2 \boldsymbol{\nabla}^2 + M^2 c^4} \right) \psi(\mathbf{x}, t) \qquad (16.1.2)$$

where for spin $1/2$, $\psi(\mathbf{x}, t)$ is a two-component spinor. The positive operator $\sqrt{-\hbar^2 c^2 \boldsymbol{\nabla}^2 + M^2 c^4}$ is a well defined operator,[7] by Fourier transform, for example, and by using the relations

$$\sqrt{-\hbar^2 c^2 \boldsymbol{\nabla}^2 + M^2 c^4}\, \delta^3(\mathbf{x} - \mathbf{x}') = \int \frac{d^3 \mathbf{p}}{(2\pi\hbar)^3} \sqrt{\hbar^2 c^2 \mathbf{p}^2 + M^2 c^4}\, e^{i\mathbf{p}\cdot(\mathbf{x}-\mathbf{x}')/\hbar}$$

$$\equiv K(\mathbf{x} - \mathbf{x}') \qquad (16.1.3)$$

$$\psi(\mathbf{x}, t) = \int d^3 \mathbf{x}'\, \delta^3(\mathbf{x} - \mathbf{x}')\, \psi(\mathbf{x}', t), \qquad (16.1.4)$$

(16.1.2) leads to

$$i\hbar \frac{\partial}{\partial t} \psi(\mathbf{x}, t) = \int d^3 \mathbf{x}'\, K(\mathbf{x} - \mathbf{x}')\, \psi(\mathbf{x}', t). \qquad (16.1.5)$$

This equation is non-local in $\mathbf{x}$. It is also not easily handed, in general, in the presence of interactions unless one is dealing with very weak interactions which would allow one to carry out an expansion under the square root. Accordingly, to obtain a more manageable equation, one may go back to (16.1.1), multiply the latter by $\left[\sqrt{\mathbf{p}^2 c^2 + M^2 c^4} + E\right]$ to get

$$\left(\mathbf{p}^2 c^2 + M^2 c^4 - E^2\right) \psi = 0 \qquad (16.1.6)$$

with $E \to i\hbar \partial/\partial t$, $\mathbf{p} \to -i\hbar \boldsymbol{\nabla}$. Using the identity in (7.4.4), involving the Pauli matrices, and the defining coupled equations (7.4.6), (7.4.7), we obtain

---

[7] That the square root of such an operator is well defined is not sufficiently emphasized in the literature. For handling such a square-root operator in a mathematical rigorous way, see, for example, Daubechies and Lieb (1983).

16.1 The Dirac Equation and Pauli's Fundamental Theorem

$$\left(\boldsymbol{\gamma} \cdot \frac{\hbar \boldsymbol{\nabla}}{i} + \gamma^0 \frac{\hbar}{ic} \frac{\partial}{\partial t} + Mc\right)\Psi = 0 \quad (16.1.7)$$

where Ψ is a four-component spinor, and it is understood that Mc is multiplied by the 4×4 unit matrix diag$[1, 1, 1, 1]$. The matrices γ^0, $\boldsymbol{\gamma} = (\gamma^1, \gamma^2, \gamma^3)$ are given by (see also the Appendix to §7.4)

$$\gamma^0 = \begin{pmatrix} I & 0 \\ 0 & -I \end{pmatrix}, \quad \boldsymbol{\gamma} = \begin{pmatrix} 0 & \boldsymbol{\sigma} \\ -\boldsymbol{\sigma} & 0 \end{pmatrix} \quad (16.1.8)$$

where I in here is the 2×2 unit matrix, $\boldsymbol{\sigma} = (\sigma_1, \sigma_2, \sigma_3)$ are the Pauli matrices, and the gamma matrices have the properties ($\mu = 0, 1, 2, 3$)

$$\text{Tr}(\gamma^\mu) = 0, \quad (\gamma^0)^\dagger = \gamma^0, \quad (\boldsymbol{\gamma})^\dagger = -\boldsymbol{\gamma}, \quad (\gamma^\mu)^\dagger = \gamma^0 \gamma^\mu \gamma^0, \quad (16.1.9)$$

$$(\gamma^0)^* = \gamma^0, \quad (\gamma^1)^* = \gamma^1, \quad (\gamma^2)^* = -\gamma^2, \quad (\gamma^3)^* = \gamma^3, \quad (16.1.10)$$

$$(\gamma^0)^T = \gamma^0, \quad (\gamma^1)^T = -\gamma^1, \quad (\gamma^2)^T = \gamma^2, \quad (\gamma^3)^T = -\gamma^3, \quad (16.1.11)$$

$$(\gamma^0)^2 = I, \quad (\gamma^1)^2 = (\gamma^2)^2 = (\gamma^3)^2 = -I, \quad (16.1.12)$$

where I in (16.1.12) now is the 4×4 unit matrix.

Using the notation,

$$\frac{\partial}{\partial x^i} = \partial_i, \quad i = 1, 2, 3, \quad \frac{\partial}{c \partial t} = \partial_0 \quad (16.1.13)$$

we may rewrite (16.1.7) in the form

$$\left(\frac{\gamma^\mu \partial_\mu}{i} + \frac{Mc}{\hbar}\right)\Psi = 0 \quad (16.1.14)$$

with a summation over upper and lower repeated indices $\mu = 0, 1, 2, 3$, understood, or simply as

$$\left(\frac{\gamma \partial}{i} + \frac{Mc}{\hbar}\right)\Psi = 0 \quad (16.1.15)$$

where $\gamma \partial = \gamma^\mu \partial_\mu$. Equation (16.1.15) is the celebrated Dirac equation.

Here we emphasize the positions of the indices μ in γ^μ and ∂_μ. This point will be clear when we study the relativistic invariance of the theory in the next section.

To obtain the Dirac equation for a charged particle interacting with an extend electromagnetic field specified by ϕ, $\mathbf{A}$, we carry out the minimal substitutions $i\hbar \partial/\partial t \to i\hbar \partial/\partial t - e\phi$, $-i\hbar \boldsymbol{\nabla} \to -i\hbar \boldsymbol{\nabla} - (e/c)\mathbf{A}$, where e is the charge of the particle in question. That is, we make the substitutions

$$\frac{\partial_\mu}{i} \longrightarrow \frac{\partial_\mu}{i} - \frac{e}{\hbar c} A_\mu \quad (16.1.16)$$

where[8]
$$A_0 = -\phi, \qquad \mathbf{A} = (A_1, A_2, A_3) \qquad (16.1.17)$$

to obtain from (16.1.14)

$$\left[\gamma^\mu \left(\frac{\partial_\mu}{i} - \frac{e}{\hbar c} A_\mu\right) + \frac{Mc}{\hbar}\right] \Psi = 0 \qquad (16.1.18)$$

where again we emphasize the positions of the indices $\mu = 0, 1, 2, 3$ in (16.1.18).

The matrices γ^μ in (16.1.8) satisfy the key anti-commutation relations

$$\{\gamma^\mu, \gamma^\nu\} = -2g^{\mu\nu} \qquad (16.1.19)$$

where[9]

$$[g^{\mu\nu}] = \begin{pmatrix} -1 & 0 & 0 & 0 \\ 0 & 1 & 0 & 0 \\ 0 & 0 & 1 & 0 \\ 0 & 0 & 0 & 1 \end{pmatrix} = [g_{\mu\nu}] \qquad (16.1.20)$$

is the so-called Minkowski metric on which more will be said in the next section.

The Dirac equations (16.1.14) may be equivalently rewritten in terms of other sets of gamma matrices γ'^μ that satisfy the *same* anti-commutation relations (16.1.19) as the original ones in (16.1.8), in the form

$$\left(\gamma'^\mu \frac{\partial_\mu}{i} + \frac{Mc}{\hbar}\right) \Psi' = 0 \qquad (16.1.21)$$

for some new spinors Ψ' related simply to Ψ, thus showing the representation independence of the Dirac equation. That is, one may appropriately choose, in general, different sets of gamma matrices γ'^μ satisfying the anti-commutation relations in (16.1.19) in setting up the Dirac equation as well. This statement is the content of a theorem called *Pauli's Fundamental Theorem* which is proved in the appendix to this section. It states that there exists a 4×4 non-singular matrix G (i.e., $\det G \neq 0$) such that

$$\gamma'^\mu = G \gamma^\mu G^{-1} \qquad (16.1.22)$$

$$\Psi' = G\Psi \qquad (16.1.23)$$

---

[8] We will see in the next section that, consistently, $A^1 = A_1$, $A^2 = A_2$, $A^3 = A_3$, $A^0 = -A_0$.

[9] Several authors, such as J. Schwinger, S. Weinberg, W. Pauli,..., have adopted the definition in (16.1.20) for the metric. Equivalently, some other authors prefer to use their $g^{\mu\nu}$ as the negative of the one in (16.1.20), with their right-hand side of (16.1.19) being replaced by $2g^{\mu\nu}$. It is interesting to note that Dirac has used both (signatures) definitions, see [Dirac (1959, 2001)].

16.1 The Dirac Equation and Pauli's Fundamental Theorem

relating (γ'^μ, Ψ') to (γ^μ, Ψ), which are consistent with (16.1.14), (16.1.21) with the latter obtained by multiplying the former from the left by G and inserting the identity $G^{-1}G = I$ just to the left of Ψ in (16.1.14).

The matrices γ^μ in (16.1.8) are said to provide the Dirac-Pauli representation of the gamma matrices satisfying (16.1.19). Given the metric $g^{\mu\nu}$ as defined in (16.1.20), and the anti-commutation relation in (16.1.19), some relations involving gamma matrices which are representation independent are given in Table 16.1. In the Dirac-Pauli representation γ^5, defined in the Table, is given by

$$\gamma^5 = \begin{pmatrix} 0 & I \\ I & 0 \end{pmatrix}. \tag{16.1.24}$$

Table 16.1. Given the definition of the metric $g^{\mu\nu}$ in (16.1.20), and the anti-commutation relations of gamma matrices in (16.1.19), the Table gives some relations involving them which are representation independent. The Greek indices go over $0, 1, 2, 3$.

$$[g^{\mu\nu}] = \operatorname{diag}[-1, 1, 1, 1], \qquad \{\gamma^\mu, \gamma^\nu\} = -2g^{\mu\nu},$$

$$\gamma^\mu \gamma^\nu = -g^{\mu\nu} I + \tfrac{1}{2}[\gamma^\mu, \gamma^\nu],$$

$$(\gamma^0)^2 = I, \qquad (\gamma^i)^2 = -I, \quad i = 1, 2, 3.$$

$$g_{\mu\nu} \gamma^\mu \gamma^\nu = -4I,$$

$$g_{\mu\nu} \gamma^\mu (\gamma^\sigma) \gamma^\nu = 2\gamma^\sigma,$$

$$g_{\mu\nu} \gamma^\mu (\gamma^\sigma \gamma^\lambda) \gamma^\nu = 4g^{\sigma\lambda},$$

$$g_{\mu\nu} \gamma^\mu (\gamma^\sigma \gamma^\lambda \gamma^\rho) \gamma^\nu = 2\gamma^\rho \gamma^\lambda \gamma^\sigma,$$

$$\big[\gamma^\mu, [\gamma^\sigma, \gamma^\rho]\big] = 4(\gamma^\sigma g^{\mu\rho} - \gamma^\rho g^{\mu\sigma}),$$

$$\operatorname{Tr}[\gamma^\mu \gamma^\nu] = -4g^{\mu\nu},$$

$$\operatorname{Tr}[\gamma^\alpha \gamma^\beta \gamma^\mu \gamma^\nu] = 4(g^{\alpha\beta} g^{\mu\nu} - g^{\alpha\mu} g^{\beta\nu} + g^{\alpha\nu} g^{\beta\mu}),$$

$$\operatorname{Tr}[\text{odd number of } \gamma\text{'s}] = 0.$$

$$\text{For } \gamma^5 = i\gamma^0 \gamma^1 \gamma^2 \gamma^3, \quad (\gamma^5)^2 = I, \quad \{\gamma^5, \gamma^\mu\} = 0,$$

$$(\gamma^\mu a_\mu)^2 = -I[\mathbf{a}^2 - (a^0)^2], \qquad (\boldsymbol{\gamma} \cdot \mathbf{a})^2 = -I\mathbf{a}^2,$$

$$\mathbf{a} = (a_1, a_2, a_3), \qquad a_0 = -a^0, \qquad a_i = a^i, \quad i = 1, 2, 3.$$

As an example of a G transformation which will find important applications to problems with zero mass particles is the unitary matrix

$$G = \frac{1}{\sqrt{2}} \begin{pmatrix} I & I \\ -I & I \end{pmatrix}, \qquad G^{-1} = G^\dagger = \frac{1}{\sqrt{2}} \begin{pmatrix} I & -I \\ I & I \end{pmatrix} \qquad (16.1.25)$$

leading to the so-called chiral representation in which γ^5 is diagonal,

$$\gamma'^0 = \begin{pmatrix} 0 & -I \\ -I & 0 \end{pmatrix}, \qquad \boldsymbol{\gamma}' = \begin{pmatrix} 0 & \boldsymbol{\sigma} \\ -\boldsymbol{\sigma} & 0 \end{pmatrix}, \qquad \gamma'^5 = \begin{pmatrix} I & 0 \\ 0 & -I \end{pmatrix} \qquad (16.1.26)$$

where note that $\boldsymbol{\gamma}'$ coincides with the one in the Dirac-Pauli representation.

Another useful representation is the so-called Majorana representation provided by the unitary matrix

$$G = G^{-1} = G^\dagger = \frac{1}{\sqrt{2}} \begin{pmatrix} I & \sigma_2 \\ \sigma_2 & -I \end{pmatrix} \qquad (16.1.27)$$

leading to

$$\gamma''^0 = \begin{pmatrix} 0 & \sigma_2 \\ \sigma_2 & 0 \end{pmatrix}, \qquad \gamma''^5 = \begin{pmatrix} \sigma_2 & 0 \\ 0 & -\sigma_2 \end{pmatrix} \qquad (16.1.28)$$

$$\gamma''^1 = \begin{pmatrix} i\sigma_3 & 0 \\ 0 & i\sigma_3 \end{pmatrix}, \qquad \gamma''^2 = \begin{pmatrix} 0 & -\sigma_2 \\ \sigma_2 & 0 \end{pmatrix}, \qquad \gamma''^3 = \begin{pmatrix} -i\sigma_1 & 0 \\ 0 & -i\sigma_1 \end{pmatrix} \qquad (16.1.29)$$

in which γ''^5 is diagonal, and $(\gamma'''^\mu/i)^* = (\gamma'''^\mu/i)$ which makes the Dirac operator $(\gamma''\partial/i + Mc/\hbar)$ real.

As a final example, we introduce the unitary matrix of the form

$$G = I \cos\theta + \boldsymbol{\gamma} \cdot \mathbf{n} \sin\theta \qquad (16.1.30)$$

where $\mathbf{n}$ is a unit 3-vector, and θ is an arbitrary angle. Since $\boldsymbol{\gamma}^\dagger = -\boldsymbol{\gamma}$, $(\boldsymbol{\gamma} \cdot \mathbf{n})^2 = -I$, we have

$$G^\dagger = G^{-1} = I \cos\theta - \boldsymbol{\gamma} \cdot \mathbf{n} \sin\theta. \qquad (16.1.31)$$

In Problem 16.1, the reader is asked to find the new representation of the gamma matrices for fixed $\mathbf{n}$ and θ.

The matrix G in (16.1.30) will be also applied in §16.6 in a slightly different context in studying, in the process, the physical content of the Dirac equation, where we use the fact that for any dimensionless non-zero three-vector $\mathbf{a}$ such that

$$\mathbf{n} = \frac{\mathbf{a}}{|\mathbf{a}|} \qquad (16.1.32)$$

$$\cos\theta = \left[\frac{\sqrt{\mathbf{a}^2 + 1} + 1}{2\sqrt{\mathbf{a}^2 + 1}} \right]^{1/2} \qquad (16.1.33)$$

16.1 The Dirac Equation and Pauli's Fundamental Theorem

$$\sin\theta = |\mathbf{a}|\left[\frac{1}{2\sqrt{\mathbf{a}^2+1}\left(\sqrt{\mathbf{a}^2+1}+1\right)}\right]^{1/2} \tag{16.1.34}$$

we have (see Problem 16.2)

$$G\gamma^0(\boldsymbol{\gamma}\cdot\mathbf{a}+1)G^{-1} = \gamma^0\sqrt{\mathbf{a}^2+1}. \tag{16.1.35}$$

Appendix to §16.1: Pauli's Fundamental Theorem

Consider the Dirac-Pauli representation of the gamma matrices γ^μ in (16.1.8) satisfying the anti-commutation relations in (16.1.19). In the vector space generated by all 4×4 matrices, we may choose as bases the following set of 16 matrices constructed out of the γ^μ:

$$\{\Gamma_A\} = \{I, \gamma^0, i\gamma^0\gamma^1\gamma^2\gamma^3, \gamma^1\gamma^2\gamma^3, i\gamma^1, i\gamma^2, i\gamma^3, \gamma^0\gamma^1, \gamma^0\gamma^2, \gamma^0\gamma^3,$$

$$i\gamma^2\gamma^3, i\gamma^3\gamma^1, i\gamma^1\gamma^2, i\gamma^0\gamma^2\gamma^3, i\gamma^0\gamma^3\gamma^1, i\gamma^0\gamma^1\gamma^2\}$$

$$\equiv \left\{I, \gamma^0, \gamma^5, \gamma^1\gamma^2\gamma^3, i\gamma^j, \gamma^0\gamma^j, \frac{i}{2}\varepsilon^{jkm}\gamma^k\gamma^m, \frac{i}{2}\gamma^0\varepsilon^{jkm}\gamma^k\gamma^m\right\}$$

$$\tag{A-16.1.1}$$

$j, k, m = 1, 2, 3$, $A = 1, 2, \ldots, 16$, where γ^5 is defined in (16.1.24). We note that for all $A = 1, \ldots, 16$

$$\Gamma_A^\dagger = \Gamma_A, \qquad (\Gamma_A)^2 = \Gamma_A. \tag{A-16.1.2}$$

It is interesting to actually spell out the explicit forms of the ordered set of elements in $\{\Gamma_A\}$:

$$\{\Gamma_A\} = \left\{\begin{pmatrix}I & 0 \\ 0 & I\end{pmatrix}, \begin{pmatrix}I & 0 \\ 0 & -I\end{pmatrix}, \begin{pmatrix}0 & I \\ I & 0\end{pmatrix}, i\begin{pmatrix}0 & -I \\ I & 0\end{pmatrix},\right.$$

$$\left.i\begin{pmatrix}0 & \sigma \\ -\sigma & 0\end{pmatrix}, \begin{pmatrix}0 & \sigma \\ \sigma & 0\end{pmatrix}, \begin{pmatrix}\sigma & 0 \\ 0 & \sigma\end{pmatrix}, \begin{pmatrix}\sigma & 0 \\ 0 & -\sigma\end{pmatrix}\right\}. \tag{A-16.1.3}$$

The simplicity of the structure of the matrices in the set $\{\Gamma_A\}$ involving first the unit matrix followed by matrices involving the Pauli ones, explains the *ordering* of the elements originally taken in (A-16.1.1).

In particular, we note that for the product of any two elements Γ_A, Γ_B in $\{\Gamma_A\}$, we have

$$\Gamma_A \Gamma_B = \xi_{AB}\Gamma_C \tag{A-16.1.4}$$

for some numbers $\xi_{AB} = \pm 1, \pm i$, Γ_C is in $\{\Gamma_A\}$. Since every matrix Γ_A is its own inverse (see (A-16.1.2)), we also have

$$\Gamma_B \Gamma_A = \Gamma_B^{-1} \Gamma_A^{-1} = (\Gamma_A \Gamma_B)^{-1} = \frac{1}{\xi_{AB}} \Gamma_C. \qquad \text{(A-16.1.5)}$$

Now for matrices γ'^μ, which satisfy the anti-commutation relations

$$\{\gamma'^\mu, \gamma'^\nu\} = -2g^{\mu\nu}. \qquad \text{(A-16.1.6)}$$

We may, in one-to-one correspondence to the elements Γ_A in $\{\Gamma_A\}$, defined in (A-16.1.1), introduce the ordered set

$$\{\Gamma'_A\} = \left\{ I, \gamma'^0, \gamma'^5, \ldots, \frac{i}{2} \gamma'^0 \varepsilon^{jkm} \gamma'^k \gamma'^m \right\}. \qquad \text{(A-16.1.7)}$$

Because of the representation independent properties of the gamma matrices in Table 16.1, we have

$$\Gamma'_A \Gamma'_B = \xi_{AB} \Gamma'_C, \qquad \Gamma'_B \Gamma'_A = \frac{1}{\xi_{AB}} \Gamma'_C \qquad \text{(A-16.1.8)}$$

where the ξ_{AB} are the same numbers appearing in (A-16.1.4), (A-16.1.5) for the corresponding elements $\Gamma_A, \Gamma_B, \Gamma_C$.

The matrices in $\{\Gamma_A\}$, and similarly in $\{\Gamma'_A\}$, are linearly independent. Given a 4×4 matrix U, we may introduce a matrix

$$\widetilde{U} = \sum_A \Gamma'_A U \Gamma_A. \qquad \text{(A-16.1.9)}$$

Upon multiplying the latter from the left by Γ'_B, and from the right by Γ_B, we obtain by using (A-16.1.4), (A-16.1.8)

$$\Gamma'_B \widetilde{U} \Gamma_B = \sum_C \Gamma'_C U \Gamma_C = \widetilde{U} \qquad \text{(A-16.1.10)}$$

where we have noted that the sum over A is equivalent to a sum over C corresponding to the generated matrices Γ'_C, Γ_C. That is,

$$\Gamma'_B \widetilde{U} = \widetilde{U} \Gamma_B. \qquad \text{(A-16.1.11)}$$

It is easy to see that the linear independence of the matrices Γ_A implies that the matrix U may be chosen in (A-16.1.9) so that $\widetilde{U}$ is not the zero matrix. Accordingly, suppose that some matrix element $(\widetilde{U})_{a_0 b_0}$ of $\widetilde{U}$ is not zero, i.e., $(\widetilde{U})_{a_0 b_0} = \alpha \neq 0$ for some pair (a_0, b_0). Now introduce a matrix V such that all of its matrix elements $(V)_{ab}$ are zero except the one element $(V)_{b_0 a_0} = 1/4\alpha$.

Given the above constructed matrix V, we may introduce, in turn, the matrix

$$\widetilde{V}' = \sum_A \Gamma_A V \Gamma'_A. \qquad \text{(A-16.1.12)}$$

16.1 The Dirac Equation and Pauli's Fundamental Theorem

Almost an identical analysis as the one carried out in going from (A-16.1.9) to (A-16.1.11) then shows that

$$\Gamma_B \widetilde{V}' = \widetilde{V}' \Gamma'_B. \qquad \text{(A-16.1.13)}$$

Upon multiplying the latter from the right by $\widetilde{U}$ an using (A-16.1.11) give

$$\Gamma_B \widetilde{V}' \widetilde{U} = \widetilde{V}' \widetilde{U} \Gamma_B \qquad \text{(A-16.1.14)}$$

for all $B = 1, \ldots, 16$. That is, $\widetilde{V}' \widetilde{U}$ commutes with *all* the elements of the set $\{\Gamma_A\}$ in (A-16.1.1), which forms a complete set of 4×4 matrices, and hence $\widetilde{V}' \widetilde{U}$ is, necessarily, some multiple of the identity matrix I.[10] This is the content of Schur's Lemma, and we may write

$$\widetilde{V}' \widetilde{U} = \beta I \qquad \text{(A-16.1.15)}$$

for some number β to be determined.

Taking the trace of $\widetilde{V}' \widetilde{U}$ in (A-16.1.15) and using the definition of $\widetilde{V}'$ in (A-16.1.12) give

$$4\beta = \text{Tr}\left[\widetilde{V}' \widetilde{U}\right] = \sum_A \text{Tr}\left[\Gamma_A V \Gamma'_A \widetilde{U}\right]$$

$$= \sum_A \text{Tr}\left[\Gamma_A V \widetilde{U} \Gamma_A\right]$$

$$= \sum_A \text{Tr}\left[V \widetilde{U} (\Gamma_A)^2\right]$$

$$= 16 (V)_{b_0 a_0} (\widetilde{U})_{a_0 b_0}$$

$$= 4 \qquad \text{(A-16.1.16)}$$

where in writing the third equality we have, in the process, used (A-16.1.11). Hence $\beta = 1$, and from (A-16.1.15), we may take $G = \widetilde{U}$, $G^{-1} = \widetilde{V}'$ in the notation of (16.1.22). We multiply (A-16.1.13) from the left by G and choose $\Gamma_B = \gamma^\mu$, $\Gamma'_B = \gamma'^\mu$, to obtain

$$G \gamma^\mu G^{-1} = \gamma'^\mu \qquad \text{(A-16.1.17)}$$

thus establishing the theorem. We note that G may be multiplied by any non-zero constant for the validity of (A-16.1.17). That is, G is defined up to an arbitrary finite multiplicative constant.

---

[10] Actually, it is sufficient that $\widetilde{V}' \widetilde{U}$ commutes with all the γ^μ for $\mu = 0, 1, 2, 3$ since this will imply that $\widetilde{V}' \widetilde{U}$ commutes with all the matrices Γ_A in $\{\Gamma_A\}$ in (A-16.1.1).

16.2 Lorentz Covariance, Boosts and Spatial Rotations

A basic physical requirement which goes into the heart of the theory of relativity is that physical laws should be the same in all inertial frames. This means, in particular, that dynamical equations should not change as one goes from one inertial frame to another up to a mere *relabelling* of the variables of the underlying theories. Because of such relabellings of the variables, the equations are said to transform covariantly from one inertial frame to another and the corresponding rules of transformations are called, in general, Lorentz transformations. This will be spelled out for the Dirac equation below. In particular, we will develop the rules for the transformation of a Dirac spinor under pure Lorentz transformations (relativistic boosts) and under spatial rotations.

16.2.1 Lorentz Transformations

If an event is labelled by $(t, \mathbf{x})$, in one inertial frame, then this *same* event will be labelled, say, by $(t', \mathbf{x}')$ in another inertial frame. The transformation rules which connect the different labellings $(t, \mathbf{x})$ to $(t', \mathbf{x}')$, for the same event, for two inertial frames are called *Lorentz Transformations*. One inertial frame F' may move with a uniform velocity $\mathbf{v}$ with respect to another frame F, as determined in F, with a possible orientation of the Cartesian space coordinate axes of F' as also determined in F at time $t = 0$. If a $(t = 0, \mathbf{x} = 0)$ reading in F corresponds to a $(t' = 0, \mathbf{x}' = 0)$ one in F', then the underlying Lorentz Transformations are called homogeneous ones, otherwise they are called inhomogeneous. In the former case, the origins of the coordinate axes set up by F and F' coincide at the time readings $t = 0$, $t' = 0$ by observers located at the corresponding origins of the respective coordinate systems of F and F'.

We use the notation $x^0 = ct$, $\mathbf{x} = (x^1, x^2, x^3)$, $(x^\mu) = (x^0, \mathbf{x}) \equiv x$, $\mu = 0, 1, 2, 3$. In Euclidean space, the distance squared between two points labelled by $\mathbf{x}$ and $\mathbf{y}$ in a given coordinate system remains invariant under rotation of the coordinate system for which the labelling of the points $\mathbf{x}$, $\mathbf{y}$ change to, say, $\mathbf{x}'$, $\mathbf{y}'$, i.e., $(\mathbf{x} - \mathbf{y})^2 = (\mathbf{x}' - \mathbf{y}')^2$. Similarly, for Lorentz transformations $x, y \to x', y'$, the following quadratic form remains invariant

$$(\mathbf{x} - \mathbf{y})^2 - (x^0 - y^0)^2 = (\mathbf{x}' - \mathbf{y}')^2 - (x'^0 - y'^0)^2 \tag{16.2.1}$$

which one may conveniently rewrite as

$$(x^\mu - y^\mu)g_{\mu\nu}(x^\nu - y^\nu) = \text{invariant} \tag{16.2.2}$$

where $g_{\mu\nu} = g_{\nu\mu}$ is referred to as the Minkowski metric and the matrix $[g_{\mu\nu}]$, with matrix elements $g_{\mu\nu} = g^{\mu\nu}$, $\mu, \nu = 0, 1, 2, 3$, is defined in (16.1.20). One may set

$$g_{\mu\nu} x^\nu = x_\mu \tag{16.2.3}$$

16.2 Lorentz Covariance, Boosts and Spatial Rotations

and rewrite the left-hand side of (16.2.2) simply as $(x^\mu - y^\mu)(x_\mu - y_\mu)$. That is $g_{\mu\nu}$ may be used to lower the index ν in x^ν. We note, in particular, that (16.2.3) implies that

$$x_0 = -x^0, \quad x_i = x^i, \quad i = 1, 2, 3 \tag{16.2.4}$$

One may raise the index μ in x_μ by multiplying it by $g^{\mu\nu}$, i.e., $x^\nu = x_\mu g^{\mu\nu}$, where $g^{\mu\nu}$ is numerically equal to $g_{\mu\nu}$. We also note that

$$g_{\rho\nu} g^{\nu\mu} = \delta_\rho{}^\mu \tag{16.2.5}$$

where $[\delta_\rho{}^\mu] = \text{diag}[1,1,1,1]$ is the identity 4×4 matrix.

Under a homogeneous Lorentz transformation $\mathbf{x} \to \mathbf{x}'$, we have

$$x'^\mu = \Lambda^\mu{}_\nu x^\nu \tag{16.2.6}$$

For a pure boost along the z-axis, for example, $x'^0 = \eta(x^0 - x^3 v/c)$, $x'^1 = x^1$, $x'^2 = x^2$, $x'^3 = \eta(x^3 - vt)$, and hence $(\eta = (1 - v^2/c^2)^{-1/2})$

$$\Lambda^i{}_j = \delta^i{}_j + (\eta - 1)\delta^{i3}\delta_{j3} \tag{16.2.7}$$

$$\Lambda^0{}_0 = \eta, \quad \Lambda^i{}_0 = -\delta^{i3}\eta v/c, \quad \Lambda^0{}_i = -\delta_{i3}\eta v/c \tag{16.2.8}$$

More generally, the elements $\Lambda^\mu{}_\nu$, depending on an arbitrary uniform relative velocity $\mathbf{v}$ of F' relative to F, as determined in F, and an arbitrary angle of rotation φ about an arbitrary unit three-vector $\mathbf{n}$, are given in Table 16.2.

From (16.2.6),

$$\frac{\partial x'^\mu}{\partial x^\nu} = \Lambda^\mu{}_\nu \tag{16.2.9}$$

since $\Lambda^\mu{}_\nu$ is independent of x, and from the chain rule

$$\frac{\partial}{\partial x^\nu} = \frac{\partial x'^\mu}{\partial x^\nu}\frac{\partial}{\partial x'^\mu} \tag{16.2.10}$$

we obtain the simple rule, to be used below, that

$$\partial_\nu = \Lambda^\mu{}_\nu \partial'_\mu \tag{16.2.11}$$

where $\partial_\nu = \partial/\partial x^\nu$ in the notation in (16.1.13).

From the inverse transformation $x' \to x$ to the one in (16.2.6), it is readily shown that

$$\Lambda^{\nu\mu} = (\Lambda^{-1})^{\mu\nu} \tag{16.2.12}$$

and hence

$$\Lambda^\rho{}_\lambda \Lambda^{\nu\lambda} = g^{\rho\nu}. \tag{16.2.13}$$

Table 16.2. Explicit expressions for the elements $\Lambda^\mu{}_\nu$ (including the infinitesimal ones for small $\delta\mathbf{v}$, $\delta\varphi$) of the homogeneous Lorentz transformations in (16.2.6). The rotation matrix elements R^{ij} are given in (2.1.4). Note that $\Lambda^{ij} = \Lambda^i{}_j$ (see (16.2.4)), $i,j = 1,2,3$; $\mu,\nu = 0,1,2,3$.

$\Lambda^{\nu\mu} = (\Lambda^{-1})^{\mu\nu}$	$\Lambda^\mu{}_\nu \simeq \delta^\mu{}_\nu + \delta\omega^\mu{}_\nu$
$\Lambda^\rho{}_\lambda \Lambda^{\nu\lambda} = g^{\rho\nu}$	$\delta\omega^{\mu\nu} = -\delta\omega^{\nu\mu}$
$\Lambda^{ij} = R^{ij} + (\eta - 1)R^{ik}\dfrac{v^k v^j}{\mathbf{v}^2}$	$\delta\omega^{ij} = \varepsilon^{ijk} n^k \delta\varphi$
$\Lambda^0{}_0 = \eta \equiv \left(1 - \dfrac{\mathbf{v}^2}{c^2}\right)^{-1/2}$	$\delta\omega^0{}_0 = 0$
$\Lambda^0{}_i = -\eta\, \dfrac{v^i}{c}$	$\delta\omega^0{}_i = -\dfrac{\delta v^i}{c}$
$\Lambda^i{}_0 = -\eta\, R^{ij}\dfrac{v^j}{c}$	$\delta\omega^i{}_0 = -\dfrac{\delta v^i}{c}$

16.2.2 Lorentz Covariance, Boosts and Spatial Rotations

From (16.2.11), we may replace ∂_μ by $\Lambda^\nu{}_\mu \partial'_\nu$ in the Dirac equation (16.1.14) to obtain

$$0 = \left(\frac{\gamma^\mu \partial_\mu}{i} + \frac{Mc}{\hbar}\right)\Psi = \left((\Lambda^\nu{}_\mu \gamma^\mu)\frac{\partial'_\nu}{i} + \frac{Mc}{\hbar}\right)\Psi \qquad (16.2.14)$$

Let

$$\Lambda^\nu{}_\mu \gamma^\mu = \gamma'^\nu \qquad (16.2.15)$$

and on account of (16.2.13),

$$\{\gamma'^\nu, \gamma'^\rho\} = -2\Lambda^{\nu\lambda}\Lambda^\rho{}_\lambda$$

$$= -2g^{\nu\rho} \qquad (16.2.16)$$

Hence from Pauli's Fundamental Theorem in the last section, there exists a non-singular matrix U such that

$$\gamma'^\nu = U^{-1}\gamma^\nu U \qquad (16.2.17)$$

Here we find it more convenient to use U^{-1} for G in our previous notation in §16.1.

16.2 Lorentz Covariance, Boosts and Spatial Rotations

Upon multiplying (16.2.14) from the left by U, using (16.2.15), (16.2.17), and setting
$$U\Psi(x) = \Psi'(x'), \qquad (16.2.18)$$
where x' is defined in terms of x in (16.2.6), we obtain
$$\left(\frac{\gamma^\nu \partial'_\nu}{i} + \frac{Mc}{\hbar}\right)\Psi'(x') = 0 \qquad (16.2.19)$$

This amounts into a mere relabelling of x and Ψ in the original Dirac equation in (16.1.14). Pauli's Fundamental Theorem guarantees the existence of a non-singular matrix U satisfying (16.2.17) such that (16.2.19) holds true.

For example, for transformations implemented by $\Lambda^\nu{}_\mu$ corresponding to pure boosts ($R^{ij} \to \delta^{ij}$), (16.2.15), (16.2.17) lead from the expression for $\Lambda^\nu{}_\mu$ in Table 16.2, ($i = 1, 2, 3$)

$$\left(\gamma^i + (\eta - 1)\frac{\beta^i \boldsymbol{\gamma}\cdot\boldsymbol{\beta}}{\beta^2}\right) - \eta\beta^i\gamma^0 = U^{-1}\gamma^i U \qquad (16.2.20)$$

$$\eta(\gamma^0 - \boldsymbol{\gamma}\cdot\boldsymbol{\beta}) = U^{-1}\gamma^0 U \qquad (16.2.21)$$

where we have set $v^i/c = \beta^i$, $|\mathbf{v}|/c = \beta$.[11] Multiplying (16.2.20) by $N^i = \beta^i/\beta$ gives

$$\eta\,\boldsymbol{\gamma}\cdot\mathbf{N} - \eta\beta\gamma^0 = U^{-1}\boldsymbol{\gamma}\cdot\mathbf{N}\, U \qquad (16.2.22)$$

We will see that (16.2.21), (16.2.22) are solved for U by setting

$$U = I\cosh\frac{\vartheta}{2} - \gamma^0\boldsymbol{\gamma}\cdot\mathbf{N}\sinh\frac{\vartheta}{2}, \quad U^{-1} = I\cosh\frac{\vartheta}{2} + \gamma^0\boldsymbol{\gamma}\cdot\mathbf{N}\sinh\frac{\vartheta}{2} \qquad (16.2.23)$$

where ϑ is to be determined, and (16.2.23) is consistent with the identity to be shown later
$$U^\dagger = \gamma^0 U^{-1}\gamma^0 \qquad (16.2.24)$$

Upon substitution of the expression for U in (16.2.20)–(16.2.21) gives

$$\cosh^2\frac{\vartheta}{2} + \sinh^2\frac{\vartheta}{2} = \cosh\vartheta = \eta \qquad (16.2.25)$$

$$2\sinh\frac{\vartheta}{2}\cosh\frac{\vartheta}{2} = \sinh\vartheta = \eta\beta \qquad (16.2.26)$$

or

$$\vartheta = \ln(\eta\beta + \eta)$$

$$= \ln\left(2\eta\beta + \sqrt{\frac{1-\beta}{1+\beta}}\right) \qquad (16.2.27)$$

---
[11] Note that β is a standard relativistic notation for $|\mathbf{v}|/c$ and should not be confused with the notation for γ^0 sometimes used in the literature.

Using the facts that $(\gamma^0\gamma^3)^2 = I$, and that for an arbitrary matrix A, $\cosh A - \sinh A = \exp(-A)$, we obtain from (16.2.23)

$$U_B = \exp\left(-\frac{\vartheta}{2}\gamma^0 \frac{\boldsymbol{\gamma}\cdot\boldsymbol{\beta}}{\beta}\right) \tag{16.2.28}$$

where B corresponds to boosts.

In (16.2.20)–(16.2.28), $\mathbf{v}$ denotes the velocity of the frame F' relative to the frame F as determined in F.

On the other hand for a pure spatial rotation by an angle φ about a unit three-vector $\mathbf{n}$, $\Lambda^{ij} = R^{ij}$ (see (2.1.4)), $\Lambda^0{}_0 = 1$, $\Lambda^0{}_i = 0$, $\Lambda^i{}_0 = 0$, and almost an identical analysis as given above for pure boosts, with hyperbolic functions replaced, in the process, by trigonometric ones gives

$$U_R = \exp\left(\frac{i}{2}\varphi \mathbf{n}\cdot\boldsymbol{\Sigma}\right) \tag{16.2.29}$$

where

$$\boldsymbol{\Sigma} = \begin{pmatrix} \boldsymbol{\sigma} & 0 \\ 0 & \boldsymbol{\sigma} \end{pmatrix} = \gamma^5\gamma^0\boldsymbol{\gamma}, \qquad \Sigma^j = \frac{i}{4}\varepsilon^{jk\ell}[\gamma^k,\gamma^\ell] \tag{16.2.30}$$

which should be compared with the rotations of spinors in (2.8.1) discussed there in a non-relativistic context. Since

$$(\mathbf{n}\cdot\boldsymbol{\Sigma})^2 = I, \qquad (\mathbf{n}\cdot\boldsymbol{\Sigma})^3 = \mathbf{n}\cdot\boldsymbol{\Sigma} \tag{16.2.31}$$

we note again from (16.2.29) the *double-valuedness* of the spinor under rotation with 4π rather than 2π radians to return the spinor to its original state.

Since $\boldsymbol{\Sigma}$ is Hermitian, U_R is clearly *unitary*, this is *unlike* U_B for a pure boost given in (16.2.28) where for the latter $U_B^\dagger \neq U_B^{-1}$. U_B, however, satisfies the relation in (16.2.24) (see also (16.2.53)). Since $\gamma^0\boldsymbol{\Sigma}\gamma^0 = \boldsymbol{\Sigma}$, U_R, apart from being unitary, also satisfies the relation in (16.2.24) with U in it replaced by U_R. We will use this fact in (16.2.61) to formulate quite generally the invariance of the scalar product of spinors under homogeneous Lorentz transformations, where γ^0 is introduced and so designed to make up for the non-unitary character of U_B without spoiling the unitary property of U_R.

We will encounter the matrix $\boldsymbol{\Sigma}$ again in (16.2.46) and in sections to follow as well, and, as expected from its expression in (16.2.30), it is associated with spin.

For more general transformations given in Table 16.1, but for infinitesimal ones we have

$$\Lambda^\nu{}_\mu \simeq \delta^\nu{}_\mu + \delta\omega^\nu{}_\mu, \qquad \delta\omega^{\nu\mu} = -\delta\omega^{\mu\nu} \tag{16.2.32}$$

where $\delta\omega^\nu{}_\mu \to 0$ for $\mathbf{v} \to 0$, $\varphi \to 0$. Since in this limit, $U \to I$, we may also set

16.2 Lorentz Covariance, Boosts and Spatial Rotations   897

$$U \simeq I + \frac{i}{2\hbar}\delta\omega^{\mu\nu}S_{\mu\nu} \tag{16.2.33}$$

$$U^{-1} \simeq I - \frac{i}{2\hbar}\delta\omega^{\mu\nu}S_{\mu\nu} \tag{16.2.34}$$

where $S_{\mu\nu}$ is to be determined.

By equating (16.2.15), (16.2.17) and using (16.2.32)–(16.2.34), we are led to

$$\delta\omega^{\nu\mu}\gamma_\mu = -\frac{i}{2\hbar}\delta\omega^{\lambda\mu}[S_{\lambda\mu},\gamma^\nu] \tag{16.2.35}$$

Since $\delta\omega^\nu{}_\mu$ is anti-symmetric, the left-hand side of (16.2.35) may rewritten as

$$\frac{1}{2}\left(\delta^\nu{}_\lambda \delta\omega^{\lambda\mu}\gamma_\mu + \delta^\nu{}_\mu \delta\omega^{\mu\lambda}\gamma_\lambda\right) = \frac{\delta\omega^{\lambda\mu}}{2}\left(\delta^\nu{}_\lambda \gamma_\mu - \delta^\nu{}_\mu \gamma_\lambda\right) \tag{16.2.36}$$

which upon comparison with the right-hand side of (16.2.35) gives

$$[S^{\lambda\mu},\gamma^\nu] = i\hbar\left(g^{\lambda\nu}\gamma^\mu - g^{\mu\nu}\gamma^\lambda\right) \tag{16.2.37}$$

The solution is readily verified to be given by

$$S^{\lambda\mu} = \frac{i\hbar}{4}\left[\gamma^\lambda,\gamma^\mu\right]. \tag{16.2.38}$$

To find the interpretation of this matrix we first note from (16.2.6), (16.2.32)

$$x'^\mu \simeq x^\mu + \delta\omega^\mu{}_\nu x^\nu \tag{16.2.39}$$

Hence from (16.2.18), we have

$$\Psi'\left(x^\sigma + \delta\omega^\sigma{}_\lambda x^\lambda\right) \simeq \left(I + \frac{i}{2\hbar}\delta\omega^{\mu\nu}S_{\mu\nu}\right)\Psi(x^\sigma) \tag{16.2.40}$$

or

$$\Psi'(x^\sigma) \simeq \left(I + \frac{i}{2\hbar}\delta\omega^{\mu\nu}S_{\mu\nu}\right)\Psi\left(x^\sigma - \delta\omega^\sigma{}_\lambda x^\lambda\right) \tag{16.2.41}$$

Now we may carry out a Taylor expansion of Ψ in $\delta\omega$ on the right-hand side of (16.2.41) to get

$$\Psi'(x) \simeq \Psi(x) + \frac{i}{2\hbar}\delta\omega^{\mu\nu}J_{\mu\nu}\Psi(x) \tag{16.2.42}$$

where

$$J_{\mu\nu} = S_{\mu\nu} + \left(x_\mu \frac{\hbar}{i}\partial_\nu - x_\nu \frac{\hbar}{i}\partial_\mu\right) \tag{16.2.43}$$

In particular for $i,j,k = 1,2,3$,

$$J_i = \frac{1}{2}\varepsilon_{ijk}J_{jk}$$

$$= S_i + L_i \tag{16.2.44}$$

where **L** is the orbital angular momentum and

$$S_i = \frac{1}{2}\varepsilon_{ijk} S_{jk}$$

$$= \frac{i\hbar}{8}\varepsilon_{ijk}\left[\gamma_i, \gamma_k\right] \tag{16.2.45}$$

is identified with the spin. The remaining part of $J_{\mu\nu}$, that is J_{i0}, is usually referred to as the "booster".

In the notation of (16.2.30), we may rewrite the spin **S** as

$$\mathbf{S} = \frac{\hbar}{2}\mathbf{\Sigma}, \qquad \mathbf{S}^2 = \hbar^2 \frac{1}{2}\left(\frac{1}{2}+1\right) I \tag{16.2.46}$$

Upon setting $P^\mu = \hbar \partial^\mu / i$, the following commutations relations are readily derived

$$[P^\mu, P^\nu] = 0 \tag{16.2.47}$$

$$\left[P^\mu, J^{\nu\lambda}\right] = i\hbar \left(g^{\mu\lambda} P^\nu - g^{\mu\nu} P^\lambda\right) \tag{16.2.48}$$

$$\left[J^{\lambda\beta}, J^{\alpha\sigma}\right] = i\hbar \left(g^{\lambda\alpha} J^{\beta\sigma} - g^{\beta\alpha} J^{\lambda\sigma} + g^{\beta\sigma} J^{\lambda\alpha} - g^{\lambda\sigma} J^{\beta\alpha}\right) \tag{16.2.49}$$

establishing the algebra of the generators of the inhomogeneous Lorentz transformations, where the commutation relation of any two of the generators gives rise to a linear combination of the generators.

16.2.3 Lorentz Invariant Scalar Products of Spinors, Lorentz Scalars and Lorentz Vectors

We will establish the identity in (16.2.24) and consider some of its consequences.

Using the reality of $\Lambda^\nu{}_\mu$ in (16.2.15), and hence the identity

$$\gamma'^\nu = \gamma^0 (\gamma'^\nu)^\dagger \gamma^0 \tag{16.2.50}$$

as obtained from (16.1.9), we have from (16.2.50)

$$U^{-1} \gamma^\nu U = \gamma^0 \left(U^{-1} \gamma^\nu U\right)^\dagger \gamma^0$$

$$= \gamma^0 U^\dagger \gamma^0 \gamma^\nu \gamma^0 \left(\gamma^0 U^\dagger\right)^{-1} \tag{16.2.51}$$

where we have used (16.1.9), (16.1.12), or

$$\gamma^\nu = \left(U \gamma^0 U^\dagger \gamma^0\right) \gamma^\nu \left(U \gamma^0 U^\dagger \gamma^0\right)^{-1} \tag{16.2.52}$$

16.2 Lorentz Covariance, Boosts and Spatial Rotations

This in turn implies that

$$\gamma^\nu \left(U\gamma^0 U^\dagger \gamma^0\right) = \left(U\gamma^0 U^\dagger \gamma^0\right) \gamma^\nu \qquad (16.2.53)$$

That is, $U\gamma^0 U^\dagger \gamma^0$ commutes with all the γ^ν and hence the former is some multiple of the unit matrix (see also Appendix to §16.1)

$$\left(U\gamma^0 U^\dagger \gamma^0\right) = aI \qquad (16.2.54)$$

From (16.2.54), we may also write

$$U^\dagger = a\gamma^0 U^{-1} \gamma^0 \qquad (16.2.55)$$

Upon substitution of the expressions for

$$U^\dagger \simeq I - \frac{i}{2\hbar}\delta\omega^{\mu\nu} S^\dagger_{\mu\nu} \qquad (16.2.56)$$

and U^{-1}, given in (16.2.34), for infinitesimal transformations, where we recall that $\delta\omega^{\mu\nu}$ is real, in (16.2.55), and using the property

$$S^\dagger_{\mu\nu} = \gamma^0 S_{\mu\nu} \gamma^0 \qquad (16.2.57)$$

as follows from (16.2.38), give $a = 1$. Thus from (16.2.55) we have established (16.2.24).

Now we may formulate the *Lorentz invariant scalar product* of two spinors $\Psi(x)$, $\chi(x)$ under homogeneous Lorentz transformations in analogy to the invariance of the quadratic form

$$x^\mu g_{\mu\nu} y^\nu = \text{invariant} \qquad (16.2.58)$$

in Minkowski spacetime. To this end, we note from (16.2.18) that under a homogeneous Lorentz transformation $\Psi(x) \to \Psi'(x')$, $\chi(x) \to \chi'(x')$, there exists a non-singular matrix U such that

$$\Psi'(x') = U\Psi(x), \qquad \chi'(x') = U\chi(x) \qquad (16.2.59)$$

and from (16.2.24) that $U^\dagger = \gamma^0 U^{-1} \gamma^0$. The later implies tat

$$U^\dagger \gamma^0 U = \gamma^0. \qquad (16.2.60)$$

Hence for the combination $\Psi'^\dagger(x')\gamma^0 \chi'(x')$, we have

$$\Psi'^\dagger(x')\gamma^0 \chi'(x') = \Psi^\dagger(x) U^\dagger \gamma^0 U \chi(x) = \Psi^\dagger(x)\gamma^0 \chi(x). \qquad (16.2.61)$$

That is,

$$\Psi^\dagger(x)\gamma^0 \chi(x) = \text{invariant} \qquad (16.2.62)$$

The presence of γ^0 in (16.2.62), which is in analogy to the metric in (16.2.58), stems from the fact that although the corresponding transformation for pure spatial rotations U_R, given in (16.2.29), is *unitary* $U_R^\dagger = U_R^{-1}$, but nevertheless satisfies the relation $U_R^\dagger = \gamma^0 U_R^{-1} \gamma^0$ in (16.2.24), the one involving a pure homogeneous Lorentz transformation (a boost) U_B, given in (16.2.28), is *not* unitary $U_B^\dagger \neq U_B^{-1}$ but satisfies the relation $U_B^\dagger = \gamma^0 U_B^{-1} \gamma^0$ in (16.2.24). This explains the essential presence of γ^0 in (16.2.62) to make up for this distinction between the two transformations and, in turn, ensures the invariance property stated in (16.2.62). Because of the importance of the combination $\Psi^\dagger \gamma^0$, one introduces a special notation for it

$$\Psi^\dagger(x)\gamma^0 = \overline{\Psi} \qquad (16.2.63)$$

In particular, the invariant combination $\overline{\Psi}(x)\Psi(x)$ is referred to as a *Lorentz scalar*.

Similarly, by using in the process (16.2.15), (16.2.17), we obtain

$$\left(\overline{\Psi}'(x')\gamma^\mu \Psi'(x')\right) = \Lambda^\mu{}_\nu \left(\overline{\Psi}(x)\gamma^\nu \Psi(x)\right) \qquad (16.2.64)$$

(see Problem 16.5). That is, $\overline{\Psi}(x)\gamma^\nu \Psi(x)$ transforms as x^ν in (16.2.6) and is referred to as a *Lorentz vector*.

16.3 Spin, Helicity and $\mathcal{P}$, $\mathcal{C}$, $\mathcal{T}$ Transformations

In the next section, we will see that neither the spin **S** nor the orbital angular momentum **L** introduced in (16.2.44) are conserved but the total angular momentum **J**, however, is. The spin component parallel to the momentum of a particle is referred to as the *helicity* (see also §5.10). The helicity as the component of the transformed spin taken parallel to the momentum of the particle in question, with the momentum as determined in each corresponding inertial frame, has a Lorentz invariant meaning. The concept of helicity is of utmost importance when dealing with massless particles as spin measurement along arbitrary directions has no meaning for such particles. The application of helicity to massive Dirac particles will be given in the next section and to massless ones in §16.5. The investigation of helicity for massless particles of higher spins turns out to be quite interesting and will be discussed in §16.8. In the present section, we also study the nature of the Dirac equation under parity $\mathcal{P}$, time reversal $\mathcal{T}$ (see also §2.6), and under the so-called charge conjugation $\mathcal{C}$ whose physical meaning will emerge later in §16.6.

16.3.1 Spin & Helicity

The spin matrix **S** was introduced in (16.2.44)–(16.2.46), and from (16.2.30) is given by

$$\mathbf{S} = \frac{\hbar}{2}\begin{pmatrix} \boldsymbol{\sigma} & 0 \\ 0 & \boldsymbol{\sigma} \end{pmatrix} = \frac{\hbar}{2}\gamma^5 \gamma^0 \boldsymbol{\gamma}. \qquad (16.3.1)$$

This may be taken as the spin operator for a massive particle in its rest frame.

In a momentum description (§16.4), we may define the spin in the frame in which observations are made (the "laboratory") and in which the particle in question is moving with momentum $\mathbf{p}$ by applying the reversed boost operation given by the transformation matrix U in (16.2.23), (16.2.17), as reformulated in the present context, where now $\mathbf{v}$ in it represents the particle's velocity. That is, we have to find the matrix $USU^{-1} = \mathbf{S}'$ rather than $U^{-1}SU$ and the reason for applying the former operation than the latter is, as just mentioned, we are going from the particle's frame to the observer's frame.

To the above end

$$\mathbf{S}' = \left(I\cosh\frac{\vartheta}{2} - \gamma^0\boldsymbol{\gamma}\cdot\mathbf{N}\sinh\frac{\vartheta}{2}\right)\frac{\hbar}{2}\gamma^5\gamma^0\boldsymbol{\gamma}\left(I\cosh\frac{\vartheta}{2} + \gamma^0\boldsymbol{\gamma}\cdot\mathbf{N}\sinh\frac{\vartheta}{2}\right) \quad (16.3.2)$$

(see (16.2.23)), where $\mathbf{p} = M\eta\mathbf{v}$, $p^0 = M\eta c$, $\eta = (1-v^2/c^2)^{-1/2}$

$$\mathbf{N} = \frac{\mathbf{p}}{|\mathbf{p}|}, \qquad p^0 \equiv \sqrt{\mathbf{p}^2 + M^2c^2} \quad (16.3.3)$$

and from (16.2.25), (16.2.26),

$$\cosh\vartheta = \frac{p^0}{Mc}, \qquad \sinh\vartheta = \frac{|\mathbf{p}|}{Mc} \quad (16.3.4)$$

$$\sinh\frac{\vartheta}{2} = \sqrt{\frac{p^0 - Mc}{2Mc}}, \quad \cosh\frac{\vartheta}{2} = \sqrt{\frac{p^0 + Mc}{2Mc}}, \quad \tanh\frac{\vartheta}{2} = \frac{|\mathbf{p}|}{p^0 + Mc}. \quad (16.3.5)$$

The expression on the right-hand side of (16.3.2) simplifies to

$$\mathbf{S}' = \frac{p^0}{Mc}\mathbf{S} - \frac{\mathbf{p}(\mathbf{S}\cdot\mathbf{p})}{Mc(p^0 + Mc)} - \frac{\hbar|\mathbf{p}|}{2Mc}\gamma^5\left[\boldsymbol{\gamma}(\boldsymbol{\gamma}\cdot\mathbf{N}) + \mathbf{N}\right]. \quad (16.3.6)$$

The last term on the right-hand side of this equation may be expressed in terms of S^{0i} defined in (16.2.38) (see Problem 16.6).

We consider the components of $\mathbf{S}' = (S'_\parallel, \mathbf{S}'_\perp)$ parallel and perpendicular to $\mathbf{p}$. For $S'_\parallel$, we have

$$S'_\parallel = S_\parallel\left(\frac{p^0}{Mc} - \frac{\mathbf{p}^2}{Mc(p^0 + Mc)}\right) \quad (16.3.7)$$

and the coefficient of $S_\parallel$ is one, giving

$$S'_\parallel = S_\parallel. \quad (16.3.8)$$

This result is remarkable. It states that the component of spin parallel to $\mathbf{p}$ has a Lorentz invariant meaning and quite importantly it is *independent* of the mass of the particle and hence exists rigorously for $M \to 0$ as well. It is also Hermitian. In any other inertial frame, say, F'', if the momentum of the particle in question is determined to be $\mathbf{p}''$, then the component of the transformed spin $\mathbf{S}''$ parallel now to $\mathbf{p}''$ will also satisfy $S''_{\parallel} = S_{\parallel}$.

For the component perpendicular to $\mathbf{p}$,

$$\mathbf{S}'_{\perp} = \frac{p^0}{Mc}\mathbf{S}_{\perp} - \frac{\hbar}{2}\frac{|\mathbf{p}|}{Mc}\gamma^5\left[\boldsymbol{\gamma}(\boldsymbol{\gamma}\cdot\mathbf{N}) + \mathbf{N}\right]$$

$$= \frac{p^0}{Mc}\mathbf{S}_{\perp} - i\frac{\hbar}{2}\frac{|\mathbf{p}|}{Mc}\gamma^5(\boldsymbol{\Sigma}\times\mathbf{N}) \qquad (16.3.9)$$

and $\mathbf{S}'_{\perp} \neq \mathbf{S}_{\perp}$, that is, it is frame dependent. It has no zero mass limit, and, as is easily verified, the second term is not Hermitian.

Summarizing then, we have found the privileged direction for spin measurement, to be given by the direction of momentum of the particle itself, with the momentum as determined in each of the corresponding inertial frames, and has a Lorentz invariant meaning, and is applicable both for massive as well as for massless particles and the corresponding operator is Hermitian. Applications of this important result will be given in the forthcoming sections.

16.3.2 $\mathcal{P}$, $\mathcal{C}$, $\mathcal{T}$ Transformations

Let $x' = (x^0, -\mathbf{x})$, then we may rewrite the Dirac equation (16.1.14) as

$$\left(-\frac{\boldsymbol{\gamma}\cdot\partial'}{i} - \frac{\gamma^0\partial'^0}{i} + \frac{Mc}{\hbar}\right)\Psi(x) = 0. \qquad (16.3.10)$$

Multiplying the latter from the left by γ^0 gives

$$\left(\frac{\gamma\partial'}{i} + \frac{Mc}{\hbar}\right)\gamma^0\Psi(x) = 0. \qquad (16.3.11)$$

Hence we may set

$$\Psi'(x') = \eta_\mathcal{P}\,\gamma^0\Psi(x) \qquad (16.3.12)$$

and define the parity transformation by

$$\mathcal{P}:\ \Psi(x) \longrightarrow \eta_\mathcal{P}\,\gamma^0\Psi(x') \qquad (16.3.13)$$

where $\eta_\mathcal{P}$ is a phase factor.

The products $\overline{\Psi}(x)\gamma^5\Psi(x)$ and $\overline{\Psi}(x)\gamma^5\gamma^\mu\Psi(x)$ transform, respectively, as a pseudo-scalar and a pseudo-vector.

We consider next time reversal and this analysis will be followed by charge conjugation. Let $x' = (-x^0, \mathbf{x})$, then

$$\left(\frac{\boldsymbol{\gamma}\cdot\partial'}{i} + \frac{\gamma^0\partial'^0}{i} + \frac{Mc}{\hbar}\right)\Psi(x) = 0. \qquad (16.3.14)$$

Upon multiplying the latter from the left by $\gamma^1\gamma^3 K$, where K denotes the operation of complex conjugation, and making use of the fact that in the Dirac-Pauli representation $\gamma^0, \gamma^1, \gamma^3$ are real and $(\gamma^2)^* = -\gamma^2$ (see (16.1.10)), it is easy to verify that

$$\left(\frac{\gamma'\partial'}{i} + \frac{Mc}{\hbar}\right)\gamma^1\gamma^3 K\Psi(x) = 0 \qquad (16.3.15)$$

and we may set

$$\Psi'(x') = \eta_\mathcal{T}\,\gamma^1\gamma^3\Psi^*(x) \qquad (16.3.16)$$

and define the time reversal transformation by

$$\mathcal{T}:\ \Psi(x) \longrightarrow \eta_\mathcal{T}\,\gamma^1\gamma^3\Psi^*(x') \qquad (16.3.17)$$

up to a phase factor $\eta_\mathcal{T}$.

For the so-called charge conjugation transformation, we consider the Dirac equation in (16.1.18) in a given external electromagnetic field $A_\mu(x)$ given in (16.1.17). We first take the complex conjugate of the equation (16.1.18) to obtain from (16.1.10),

$$\left[-\gamma^\lambda\left(\frac{\partial_\lambda}{i} + \frac{e}{\hbar c}A_\lambda\right) + \gamma^2\left(\frac{\partial_2}{i} + \frac{e}{\hbar c}A_2\right) + \frac{Mc}{\hbar}\right]\Psi^*(x) = 0 \qquad (16.3.18)$$

where here $\lambda = 0, 1, 3$. Multiplying this equation from the left by $i\gamma^2$ gives in the Dirac-Pauli representation[12]

$$\left[\gamma^\mu\left(\frac{\partial_\mu}{i} + \frac{e}{\hbar c}A_\mu\right) + \frac{Mc}{\hbar}\right]\left(i\gamma^2\gamma^0\right)\overline{\Psi}^\top(x) = 0. \qquad (16.3.19)$$

We may then define charge conjugation by

$$\mathcal{C}:\ \Psi(x) \longrightarrow \Psi_C(x) = \eta_C\left(i\gamma^2\gamma^0\right)\overline{\Psi}^\top(x), \qquad (16.3.20)$$

where η_C is a phase factor, and we have used the fact that $\gamma^0\overline{\Psi}^\top = \gamma^0\left(\Psi^\dagger\gamma^0\right)^\top = \Psi^*$. That is, for a given external electromagnetic field, $\Psi_C(x)$ satisfies the same equation as $\Psi(x)$ with the sign of the charge e in it simply reversed $e \to -e$. The physical significance of this will be discussed in §16.6.

16.4 General Solution of the Dirac Equation

We consider the general solution of the Dirac equation

$$\left(\frac{\gamma\partial}{i} + \frac{Mc}{\hbar}\right)\Psi(x) = 0 \qquad (16.4.1)$$

---

[12] The i factor in $i\gamma^2$ is chosen for convenience to make $(i\sigma_2)$ a real matrix.

$x = (x^0, \mathbf{x})$, $x^0 = ct$. We carry out a four dimensional Fourier transform

$$\Psi(x) = \int \frac{(\mathrm{d}p)}{(2\pi\hbar)^4} \, \mathrm{e}^{\mathrm{i}px/\hbar} \, \Psi(p) \tag{16.4.2}$$

with the Lorentz invariant measure

$$(\mathrm{d}p) = \mathrm{d}p^0 \mathrm{d}p^1 \mathrm{d}p^2 \mathrm{d}p^3 \tag{16.4.3}$$

and the Lorentz scalar

$$px = \mathbf{p} \cdot \mathbf{x} - p^0 x^0, \tag{16.4.4}$$

$p = (p^0, \mathbf{p})$. From (16.4.1), (16.4.2), we have

$$(\gamma p + Mc)\, \Psi(p) = 0 \tag{16.4.5}$$

where $\gamma p = \gamma^\mu p_\mu = \boldsymbol{\gamma} \cdot \mathbf{p} + \gamma^0 p_0 = \boldsymbol{\gamma} \cdot \mathbf{p} - \gamma^0 p^0$, and we note that in the Dirac-Pauli representation given in (16.1.8), no i factors appear in (16.4.5) in the p-description multiplying γ^μ

Upon multiplying (16.4.5) from the left by $(-\gamma p + Mc)$ gives

$$\left(p^2 + M^2 c^2\right) \Psi(p) = 0 \tag{16.4.6}$$

and the solution is of the form

$$\Psi(p) = \delta\!\left(p^2 + M^2 c^2\right) \Phi(p) \tag{16.4.7}$$

where $\delta(p^2 + M^2 c^2)$ is the Dirac delta distribution, and we have noted that $af(a) = 0$ implies that $f(a) = g(a)\delta(a) = g(0)\delta(a)$. Since $p^2 + M^2 c^2 = \mathbf{p}^2 + M^2 c^2 - {p^0}^2$, we may use the well known property

$$\delta\!\left(p^2 + M^2 c^2\right) = \frac{\delta\!\left(p^0 - \sqrt{\mathbf{p}^2 + M^2 c^2}\right) + \delta\!\left(p^0 + \sqrt{\mathbf{p}^2 + M^2 c^2}\right)}{2\sqrt{\mathbf{p}^2 + M^2 c^2}} \tag{16.4.8}$$

leading to two parts contributing to $\Psi(p)$, one for $p^0 > 0$ and one for $p^0 < 0$. Equivalently, (16.4.5) has a non-trivial solution only if $\gamma p + Mc$ has *no inverse*, i.e., $\det(\gamma p + Mc) = M^2 c^2 - {p^0}^2 + \mathbf{p}^2 = 0$.

To obtain a more symmetrical expression for $\psi(p)$, we make the transformation $\mathbf{p} \to -\mathbf{p}$ *corresponding to* $p^0 < 0$, and integrate over p^0 in (16.4.2)) using the constraints imposed by the two deltas in (16.4.8). This gives the general structure

$$\Psi(x) = \int 2Mc \, \mathrm{d}\omega_p \left[\mathrm{e}^{\mathrm{i}px/\hbar} \Phi_+(\mathbf{p}) + \mathrm{e}^{-\mathrm{i}px/\hbar} \Phi_-(\mathbf{p}) \right] \tag{16.4.9}$$

using a rather standard notation for the Lorentz invariant measure

$$\mathrm{d}\omega_p = \frac{\mathrm{d}^3 \mathbf{p}}{(2\pi\hbar)^3} \frac{1}{2p^0}, \quad p^0 > 0 \tag{16.4.10}$$

16.4 General Solution of the Dirac Equation

as obtained from the Lorentz invariant measure (dp) and the Lorentz invariant delta distribution $\delta(p^2 + M^2c^2)$, where *now*, and from now on,

$$p^0 = +\sqrt{\mathbf{p}^2 + M^2c^2}, \qquad (16.4.11)$$

px is defined in (16.4.4), and $p^0 = -p_0$ as always. The $2Mc$ factor multiplying $d\omega_p$ in (16.4.9) is introduced for convenience. The restriction $p^2 + M^2c^2 = 0$, i.e.,

$$p^2 = -M^2c^2 \qquad (16.4.12)$$

as provided by $\delta(p^2 + M^2c^2)$ in (16.4.8) is a restriction on the mass shell relating p^0 and $\mathbf{p}$. Also, $\Phi_+(\mathbf{p})$, $\Phi_-(\mathbf{p})$ satisfy the equations

$$(\gamma p + Mc)\Phi_+(\mathbf{p}) = 0 \qquad (16.4.13)$$

$$(-\gamma p + Mc)\Phi_-(\mathbf{p}) = 0. \qquad (16.4.14)$$

Now it is straightforward to solve these two equations. We consider (16.4.13) first.

For[13] $\mathbf{p} = 0$, $p^0 = Mc$, and (16.4.13) gives

$$(I - \gamma^0)\Phi_+(\mathbf{0}) = 0. \qquad (16.4.15)$$

From the expression for γ^0 in (16.1.8), we see that $I - \gamma^0 = \text{diag}[0, 0, 2, 2]$, and hence the last two entries of $\Phi_+(\mathbf{0})$ must be zero, i.e.,

$$\Phi_+(\mathbf{0}, \sigma) = \begin{pmatrix} \xi_\sigma \\ 0 \end{pmatrix} \qquad (16.4.16)$$

which has only two rows and two independent (orthogonal) solutions may be chosen

$$\xi_\sigma^\dagger \xi_{\sigma'} = \delta_{\sigma\sigma'} \qquad (16.4.17)$$

such as $(1\ 0)^\dagger$, $(0\ 1)^\dagger$.

Clearly, the solution of (16.4.13) is then

$$\Phi_+(\mathbf{p}, \sigma) = (-\gamma p + Mc)\Phi_+(\mathbf{0}, \sigma), \qquad (16.4.18)$$

with $\Phi(\mathbf{0}, \sigma)$ given in (16.4.16), since $(\gamma p + Mc)(-\gamma p + Mc) = (p^2 + M^2c^2) = 0$.

Therefore, up to an arbitrary (one component) function $a(\mathbf{p}, \sigma)$ of $(\mathbf{p}, \sigma)$, we have from (16.4.18)

$$\Phi_+(\mathbf{p}, \sigma) = u(\mathbf{p}, \sigma)\, a(\mathbf{p}, \sigma) \qquad (16.4.19)$$

---

[13] Here we assume $M \neq 0$. The mass zero case will be considered in detail in the next section.

where $u(\mathbf{p}, \sigma)$ is taken the property normalized four component object

$$u(\mathbf{p}, \sigma) = \sqrt{\frac{p^0 + Mc}{2Mc}} \begin{pmatrix} \xi_\sigma \\ \dfrac{\boldsymbol{\sigma} \cdot \mathbf{p}}{p^0 + Mc} \xi_\sigma \end{pmatrix} \qquad (16.4.20)$$

corresponding to two solutions in (16.4.16). The normalization condition adopted is given by

$$u^\dagger(\mathbf{p}, \sigma) \gamma^0 u(\mathbf{p}, \sigma') \equiv \overline{u}(\mathbf{p}, \sigma) u(\mathbf{p}, \sigma') = \delta_{\sigma\sigma'} \qquad (16.4.21)$$

where

$$\overline{u}(\mathbf{p}, \sigma) = u^\dagger(\mathbf{p}, \sigma) \gamma^0. \qquad (16.4.22)$$

With such a normalization given in (16.4.21),

$$u^\dagger(\mathbf{p}, \sigma) u(\mathbf{p}, \sigma') = \frac{p^0}{Mc} \delta_{\sigma\sigma'} \qquad (16.4.23)$$

and obviously

$$(\gamma p + Mc) u(\mathbf{p}, \sigma) = 0. \qquad (16.4.24)$$

Also

$$\overline{u}(\mathbf{p}, \sigma)(\gamma p + Mc) = 0 \qquad (16.4.25)$$

A more interesting way of deriving (16.4.20) is to apply the booster operation U^{-1} with $\mathbf{N} = \mathbf{p}/|\mathbf{p}|$ to $u(\mathbf{0}, \sigma) = (\xi_\sigma\ 0)^\dagger$ to obtain $u(\mathbf{p}, \sigma)$. As mentioned in the previous section, the reason why we must apply U^{-1} rather than U is that the latter gives rise to the motion of the observation frame (the "laboratory") with respect to the rest frame of the particle in question, while we are interested in carrying out observations on the particle itself (in the "laboratory" frame).

To the above end, with ($\boldsymbol{\beta} = \mathbf{v}/c$)

$$\mathbf{N} = \frac{\boldsymbol{\beta}}{\beta} = \frac{\mathbf{p}}{|\mathbf{p}|} \qquad (16.4.26)$$

$$\mathbf{p} = M\eta\mathbf{v} \qquad (16.4.27)$$

$$\eta = \left(1 - \frac{\mathbf{v}^2}{c^2}\right)^{-1/2} \qquad (16.4.28)$$

and with the expressions for functions of ϑ in (16.3.4), (16.3.5),

$$U^{-1} \begin{pmatrix} \xi_\sigma \\ 0 \end{pmatrix} = \cosh \frac{\vartheta}{2} \begin{pmatrix} \xi_\sigma \\ \dfrac{\boldsymbol{\sigma} \cdot \mathbf{p}}{p^0 + Mc} \xi_\sigma \end{pmatrix} \qquad (16.4.29)$$

which upon normalization as in (16.4.21) gives (16.4.20). It is important to emphasize such that such a matrix U may be defined afresh in a momentum description where the two frames under consideration are the particle's (for $M \neq 0$), and the observation frame. Massless particles will be described in the next section.

Given the solution $u(\mathbf{p}, \sigma)$ in (16.4.20) of (16.4.24), we may introduce its charge conjugate transform obtained from (16.3.20) given by

$$v(\mathbf{p}, \sigma) = i\gamma^2 \gamma^0 \overline{u}^\top(\mathbf{p}, \sigma) \qquad (16.4.30)$$

up to a phase factor, which upon using the identity

$$\sigma_2 \boldsymbol{\sigma}^\top = -\boldsymbol{\sigma} \sigma_2 \qquad (16.4.31)$$

gives

$$v(\mathbf{p}, \sigma) = \sqrt{\frac{p^0 + Mc}{2Mc}} \begin{pmatrix} \dfrac{\boldsymbol{\sigma} \cdot \mathbf{p}}{p^0 + Mc} \xi'_\sigma \\ \xi'_\sigma \end{pmatrix} \qquad (16.4.32)$$

where

$$\xi'_\sigma = (-i\sigma_2)\xi^*_\sigma \qquad (16.4.33)$$

for the corresponding ξ_σ in (16.4.20), and

$$\xi'^\dagger_\sigma \xi'_{\sigma'} = \delta_{\sigma\sigma'}, \qquad (16.4.34)$$

$v(\mathbf{p}, \sigma)$ satisfies the equation (16.4.14), i.e.,[14]

$$(-\gamma p + Mc) v(\mathbf{p}, \sigma) = 0, \qquad \overline{v}(\mathbf{p}, \sigma)(-\gamma p + Mc) = 0 \qquad (16.4.35)$$

and $\overline{v}(\mathbf{p}, \sigma) = v^\dagger(\mathbf{p}, \sigma) \gamma^0$. It satisfies the normalization condition

$$\overline{v}(\mathbf{p}, \sigma) v(\mathbf{p}, \sigma') = -\delta_{\sigma\sigma'} \qquad (16.4.36)$$

and

$$v^\dagger(\mathbf{p}, \sigma) v(\mathbf{p}, \sigma') = \frac{p^0}{Mc} \delta_{\sigma\sigma'}. \qquad (16.4.37)$$

Note the minus sign in (16.4.36).

We also have the orthogonality conditions

$$\overline{u}(\mathbf{p}, \sigma) v(\mathbf{p}, \sigma') = 0 \qquad (16.4.38)$$

for both values of σ, σ' adopted. From (16.4.20), (16.1.8), we note that $\overline{u}(\mathbf{p}, \sigma) = u^\dagger(-\mathbf{p}, \sigma)$, and we may rewrite (16.4.38) as

---

[14] Note that for $\mathbf{p} = 0$, $(I + \gamma^0) v(0, \sigma) = 0$ and hence the first two entries of $v(0, \sigma)$ must vanish and is consistent with the expression in (16.4.32) as expected.

$$u^\dagger(-\mathbf{p},\sigma)v(\mathbf{p},\sigma') = 0 \tag{16.4.39}$$

The action of $(-i\sigma_2)$ on ξ_σ^* in (16.4.33) is clear. If the ξ_σ in (16.4.20) are given by

$$\xi_{+\mathbf{N}} = \begin{pmatrix} \cos\dfrac{\theta}{2}\,e^{-i\phi/2} \\ \sin\dfrac{\theta}{2}\,e^{i\phi/2} \end{pmatrix}, \quad \xi_{-\mathbf{N}} = \begin{pmatrix} -\sin\dfrac{\theta}{2}\,e^{-i\phi/2} \\ \cos\dfrac{\theta}{2}\,e^{i\phi/2} \end{pmatrix} \tag{16.4.40}$$

(see (8.1.16)) with a unit three-vector

$$\mathbf{N} = \bigl(\sin\theta\cos\phi,\ \sin\theta\sin\phi,\ \cos\theta\bigr) \tag{16.4.41}$$

then

$$\xi'_{\pm\mathbf{N}} = (-i\sigma_2)\xi^*_{\pm\mathbf{N}} = \pm\xi_{\mp\mathbf{N}}. \tag{16.4.42}$$

That is, in particular, $(-i\sigma_2)$ acting on $\xi^*_{\pm\mathbf{N}}$ reverses the direction of spin.

The solutions $\Phi_-(\mathbf{p},\sigma)$ of (16.4.14) are given by

$$\Phi_-(\mathbf{p},\sigma) = v(\mathbf{p},\sigma)\,b^*(\mathbf{p},\sigma) \tag{16.4.43}$$

with $b^*(\mathbf{p},\sigma)$ an arbitrary function of $(\mathbf{p},\sigma)$.

We have thus obtained four orthogonal spinors $u(\mathbf{p},\sigma)$, $v(\mathbf{p},\sigma)$ two for each value taken by σ in the sense of (16.4.21), (16.4.36), (16.4.38).

The general solution of (16.4.1) may be then written from (16.4.2), (16.4.9), (16.4.19), (16.4.43), (16.4.20), (16.4.32) as

$$\Psi(x) = \sum_\sigma \int 2Mc\,d\omega_p \left[e^{ipx/\hbar}u(\mathbf{p},\sigma)\,a(\mathbf{p},\sigma) + e^{-ipx/\hbar}v(\mathbf{p},\sigma)\,b^*(\mathbf{p},\sigma)\right] \tag{16.4.44}$$

For the moment, if we adopt the normalization condition[15]

$$\frac{1}{\hbar c}\int d^3\mathbf{x}\,\Psi^\dagger(x)\,\Psi(x) = 1 \tag{16.4.45}$$

thus providing, in the process, specific units for Ψ, then (16.4.23), (16.4.37), (16.4.39) together with (16.4.45) gives the following restriction on the coefficients $a(\mathbf{p},\sigma)$, $b^*(\mathbf{p},\sigma)$

$$\frac{1}{\hbar c}\sum_\sigma\int 2Mc\,d\omega_p\left[|a(\mathbf{p},\sigma)|^2 + |b(\mathbf{p},\sigma)|^2\right] = 1 \tag{16.4.46}$$

---

[15] The alert reader might wonder why we have not considered the Lorentz scalar $\Psi^\dagger(x)\gamma^0\Psi(x)$ (see (16.2.62)) instead of $\Psi^\dagger(x)\Psi(x)$ in (16.4.45). The reason is that neither $\Psi^\dagger(x)\Psi(x)$ nor $d^3\mathbf{x}$ are Lorentz invariant but their product is. More will be said about the normalization condition adopted in (16.4.45) in §16.6.

16.4 General Solution of the Dirac Equation

where because of the normalization factor p^0/Mc in (16.4.23), (16.4.37), the $2Mc$ factor multiplying $d\omega_p$ appears again in (16.4.46).

The Hamiltonian may be obtained from (16.4.1) by multiplying the latter by $-\hbar c \gamma^0$ giving

$$i\hbar \frac{\partial}{\partial t} \Psi(x) = \gamma^0 \left(c\hbar \frac{\boldsymbol{\gamma} \cdot \boldsymbol{\nabla}}{i} + Mc^2 \right) \Psi(x) \equiv H\Psi(x). \qquad (16.4.47)$$

We may rewrite (16.4.44) as

$$\Psi(x) = \sum_\sigma \int 2Mc \, d\omega_p \, e^{i\mathbf{p}\cdot\mathbf{x}/\hbar} \left[e^{-ip^0 x^0/\hbar} u(\mathbf{p}, \sigma) \, a(\mathbf{p}, \sigma) \right.$$

$$\left. + e^{ip^0 x^0/\hbar} v(-\mathbf{p}, \sigma) \, b^*(-\mathbf{p}, \sigma) \right] \qquad (16.4.48)$$

to obtain the two equations

$$Hu(\mathbf{p}, \sigma) = p^0 c \, u(\mathbf{p}, \sigma) \qquad (16.4.49)$$

$$Hv(-\mathbf{p}, \sigma) = -p^0 c \, v(-\mathbf{p}, \sigma) \qquad (16.4.50)$$

where the Hamiltonian in the **p**-description is given by

$$H = \gamma^0 (c\boldsymbol{\gamma} \cdot \mathbf{p} + Mc^2) = \begin{pmatrix} Mc^2 & c\boldsymbol{\sigma} \cdot \mathbf{p} \\ c\boldsymbol{\sigma} \cdot \mathbf{p} & -Mc^2 \end{pmatrix} \qquad (16.4.51)$$

with $u(\mathbf{p}, \sigma)$ and $v(-\mathbf{p}, \sigma)$ corresponding, respectively, to positive and negative energies $E_\pm = \pm p^0 c$. The significance of a negative energy solution will be discussed in §16.6. Equations (16.4.49), (16.4.50) also follows directly from (16.4.24), (16.4.35) if one uses the definitions of H in (16.4.51).

We consider the parity operation in (16.3.13) as applied to $u(\mathbf{p}, \sigma)$, $v(\mathbf{p}, \sigma)$, giving

$$\mathcal{P} u(\mathbf{p}, \sigma) = \eta_\mathcal{P} \, u(-\mathbf{p}, \sigma) \qquad (16.4.52)$$

$$\mathcal{P} v(\mathbf{p}, \sigma) = -\eta_\mathcal{P} \, v(-\mathbf{p}, \sigma) \qquad (16.4.53)$$

Without loss of generality we may set $\eta_\mathcal{P} = +1$ or -1, and the numerical factor $\eta_\mathcal{P}$ chosen is referred to as the relative, intrinsic parity of a particle. Equations (16.4.52), (16.4.53) show that the intrinsic parities associated with the negative energy solution is *opposite* to the one adopted for the positive energy. The interpretation of this will be also discussed in §16.6 when studying the particle content of the theory.

For the spin related matrix $\boldsymbol{\Sigma}$ in (16.2.30), (16.2.46) we note that

$$\boldsymbol{\Sigma} \cdot \mathbf{N} \, u(\mathbf{0}, \sigma) = \pm u(\mathbf{0}, \sigma) \qquad (16.4.54)$$

for ξ_σ corresponding to $\xi_{\pm \mathbf{N}}$, respectively, in (16.4.40), while

$$\mathbf{\Sigma} \cdot \mathbf{N}\, v(\mathbf{0}, \sigma) = \mp v(\mathbf{0}, \sigma). \tag{16.4.55}$$

On the other hand for $\mathbf{N}$ taken along $\mathbf{p}$, i.e., $\mathbf{N} = \mathbf{p}/|\mathbf{p}|$,

$$u(\mathbf{p}, \sigma) = \sqrt{\frac{p^0 + Mc}{2Mc}} \begin{pmatrix} \xi_{\pm \mathbf{N}} \\ \pm \dfrac{|\mathbf{p}|}{p^0 + Mc}\, \xi_{\pm \mathbf{N}} \end{pmatrix} \tag{16.4.56}$$

with ξ_σ corresponding, respectively, to $\xi_{\pm \mathbf{N}}$, and

$$v(\mathbf{p}, \sigma) = \sqrt{\frac{p^0 + Mc}{2Mc}} \begin{pmatrix} -\dfrac{|\mathbf{p}|}{p^0 + Mc}\, \xi_{\mp \mathbf{N}} \\ \pm \xi_{\mp \mathbf{N}} \end{pmatrix}. \tag{16.4.57}$$

For the helicity $S_\parallel$ (see (16.3.8)), with $S_\parallel = \hbar \Sigma_\parallel/2$, and with $u(\mathbf{p}, \sigma)$, $v(\mathbf{p}, \sigma)$ given above in (16.4.56), (16.4.57),

$$\Sigma_\parallel u(\mathbf{p}, \sigma) = \pm u(\mathbf{p}, \sigma) \tag{16.4.58}$$

$$\Sigma_\parallel v(\mathbf{p}, \sigma) = \mp v(\mathbf{p}, \sigma). \tag{16.4.59}$$

Unlike the non-relativistic case

$$[H, \mathbf{S}] \neq 0, \qquad [H, \mathbf{L}] \neq 0 \tag{16.4.60}$$

where the spin $\mathbf{S}$ and orbital angular momentum $\mathbf{L}$ are defined in (16.2.44)–(16.2.46), but the total angular momentum $\mathbf{J} = \mathbf{L} + \mathbf{S}$ is conserved, i.e., (see Problem 16.9)

$$[H, \mathbf{J}] = 0. \tag{16.4.61}$$

The spin component $S_\parallel$ parallel to $\mathbf{p}$, i.e., the helicity (§16.3, (16.3.8)) has the distinction that it commutes with H:

$$[H, S_\parallel] = 0, \tag{16.4.62}$$

as is readily verified.

We introduce the following matrices

$$\mathbb{P}_+(\mathbf{p}) = \sum_\sigma u(\mathbf{p}, \sigma)\, \overline{u}(\mathbf{p}, \sigma) \tag{16.4.63}$$

$$\mathbb{P}_-(\mathbf{p}) = -\sum_\sigma v(\mathbf{p}, \sigma)\, \overline{v}(\mathbf{p}, \sigma). \tag{16.4.64}$$

They are orthogonal projection operations, i.e., they satisfy the following equations

16.4 General Solution of the Dirac Equation

$$\mathbb{P}_+(\mathbf{p})\,\mathbb{P}_+(\mathbf{p}) = \mathbb{P}_+(\mathbf{p}), \qquad \mathbb{P}_-(\mathbf{p})\,\mathbb{P}_-(\mathbf{p}) = \mathbb{P}_-(\mathbf{p}) \tag{16.4.65}$$

$$\mathbb{P}_+(\mathbf{p})\,\mathbb{P}_-(\mathbf{p}) = 0 = \mathbb{P}_-(\mathbf{p})\,\mathbb{P}_+(\mathbf{p}). \tag{16.4.66}$$

They project out, respectively, positive and negative energy solutions

$$\mathbb{P}_+(\mathbf{p})\,u(\mathbf{p},\sigma) = u(\mathbf{p},\sigma), \qquad \mathbb{P}_+(\mathbf{p})\,v(\mathbf{p},\sigma) = 0 \tag{16.4.67}$$

$$\mathbb{P}_-(\mathbf{p})\,v(\mathbf{p},\sigma) = v(\mathbf{p},\sigma), \qquad \mathbb{P}_-(\mathbf{p})\,u(\mathbf{p},\sigma) = 0. \tag{16.4.68}$$

The explicit expressions for these operators may be worked out directly from their definitions in (16.4.63), (16.4.64) and are given by

$$\mathbb{P}_+(\mathbf{p}) = \frac{(-\gamma p + Mc)}{2Mc} \tag{16.4.69}$$

$$\mathbb{P}_-(\mathbf{p}) = \frac{(\gamma p + Mc)}{2Mc} \tag{16.4.70}$$

and satisfy the completeness relation

$$\mathbb{P}_+(\mathbf{p}) + \mathbb{P}_-(\mathbf{p}) = I. \tag{16.4.71}$$

We close this section by noting that $u(\mathbf{p},\sigma)$ in (16.4.20) may be rewritten, in general, as

$$u(\mathbf{p},\sigma) = \sqrt{\frac{p^0 + Mc}{2Mc}} \begin{pmatrix} \xi_\sigma \\ \dfrac{|\mathbf{p}|}{p^0 + Mc}\,\boldsymbol{\sigma}\cdot\widehat{\mathbf{p}}\,\xi_\sigma \end{pmatrix} \tag{16.4.72}$$

where $\widehat{\mathbf{p}} = \mathbf{p}/|\mathbf{p}|$. At low energies, i.e., for $|\mathbf{p}|/Mc \ll 1$, the upper component is large in comparison to the lower one with the tater suppressed by a factor of the order $|\mathbf{p}|/Mc$, and exactly the opposite happens for $v(\mathbf{p},\sigma)$, giving

$$u(\mathbf{p},\sigma) \sim \begin{pmatrix} \xi_\sigma \\ 0 \end{pmatrix}, \qquad v(\mathbf{p},\sigma) \sim \begin{pmatrix} 0 \\ \xi'_\sigma \end{pmatrix}. \tag{16.4.73}$$

For later reference, we note that by using the elementary property

$$(H)^2 = I {p^0}^2 c^2 \tag{16.4.74}$$

for the Hamiltonian in (16.4.51), the Dirac time evolution operator takes the simple form

$$e^{-itH/\hbar} = I \cos\left(\frac{p^0 ct}{\hbar}\right) - i\frac{H}{p^0 c}\sin\left(\frac{p^0 ct}{\hbar}\right) \tag{16.4.75}$$

where $p^0 = +\sqrt{\mathbf{p}^2 + M^2 c^2}$.

In the next section, we consider massless particles in the light of the Dirac equation.

16.5 Massless Dirac Particles

To investigate the nature of Dirac massless particles, it turns out to be more suitable to work in the chiral representation of the γ matrices in (16.1.26) obtained via the unitary operator G given in (16.1.25). As we have seen in §16.3, it is essential to deal with the helicity, i.e., with the component of spin parallel to the direction of the momentum $\mathbf{p}$ of the particle.

To treat massless particles it is most instructive to consider first the zero mass limit $M \to 0$ of the solution of the Dirac equation and then compare the solution obtained with the one derived by studying the Dirac equation with the Mass M set equal to zero at the outset.

In the chiral representation, we have explicitly from (16.1.25), (16.4.20),

$u'(\mathbf{p}, \sigma) = G u(\mathbf{p}, \sigma)$

$$= \frac{1}{\sqrt{2}} \begin{pmatrix} \left[\sqrt{\dfrac{p^0 + Mc}{2Mc}} + \dfrac{\boldsymbol{\sigma} \cdot \mathbf{p}}{\sqrt{2Mc(p^0 + Mc)}}\right] \xi_\sigma \\ -\left[\sqrt{\dfrac{p^0 + Mc}{2Mc}} - \dfrac{\boldsymbol{\sigma} \cdot \mathbf{p}}{\sqrt{2Mc(p^0 + Mc)}}\right] \xi_\sigma \end{pmatrix}. \qquad (16.5.1)$$

Using the fact that

$$\frac{1}{\sqrt{p^0 + Mc}} = \frac{\sqrt{p^0 - Mc}}{|\mathbf{p}|} \qquad (16.5.2)$$

and the expression for $\cosh \vartheta/2$ and $\sinh \vartheta/2$ in (16.3.5) leads, using in the process (16.2.28), to

$$u'(\mathbf{p}, \sigma) = \frac{1}{\sqrt{2}} \begin{pmatrix} \exp\left(+\dfrac{\vartheta}{2} \boldsymbol{\sigma} \cdot \mathbf{N}\right) \xi_\sigma \\ -\exp\left(-\dfrac{\vartheta}{2} \boldsymbol{\sigma} \cdot \mathbf{N}\right) \xi_\sigma \end{pmatrix} \qquad (16.5.3)$$

where $\mathbf{N} = \mathbf{p}/|\mathbf{p}|$, and from (16.2.27), (16.4.26), (16.4.27),

$$\vartheta = \ln\left(\frac{2|\mathbf{p}|}{Mc} + \sqrt{\frac{1 - |\mathbf{p}|/p^0}{1 + |\mathbf{p}|/p^0}}\right) \qquad (16.5.4)$$

where $p^0 = +\sqrt{\mathbf{p}^2 + M^2 c^2}$. $u'(\mathbf{p}, \sigma)$ satisfies the normalization condition

$$\overline{u}'(\mathbf{p}, \sigma) u'(\mathbf{p}, \sigma') = \delta_{\sigma \sigma'} \qquad (16.5.5)$$

with $\overline{u}' = u'^\dagger \gamma'^0$, as expected.

For $\xi_\sigma \equiv \xi_{-\mathbf{N}}$ (see (16.4.40))

16.5 Massless Dirac Particles

$$u'(\mathbf{p}, \sigma) = \frac{1}{\sqrt{2}} \begin{pmatrix} e^{-\vartheta/2} \xi_{-\mathbf{N}} \\ -e^{\vartheta/2} \xi_{-\mathbf{N}} \end{pmatrix} \tag{16.5.6}$$

and from (16.5.4), we note that for $M \to 0$

$$u'(\mathbf{p}, \sigma) \longrightarrow -\frac{1}{\sqrt{2}} \exp\left[\frac{1}{2} \ln\left(\frac{2|\mathbf{p}|}{M}\right)\right] \begin{pmatrix} 0 \\ \xi_{-\mathbf{N}} \end{pmatrix} \tag{16.5.7}$$

for which using now the *normalization* condition

$$u^\dagger(\mathbf{p}, \sigma) u'(\mathbf{p}, \sigma') = \delta_{\sigma\sigma'} \tag{16.5.8}$$

gives rigorously for $M \to 0$,

$$u'(\mathbf{p}, \sigma) = \begin{pmatrix} 0 \\ \xi_{-\mathbf{N}} \end{pmatrix}. \tag{16.5.9}$$

For the charge conjugate spinor, we have directly from (16.4.32) in the chiral representation

$$v'(\mathbf{p}, \sigma) = \frac{1}{\sqrt{2}} \begin{pmatrix} \exp\left(+\frac{\vartheta}{2} \boldsymbol{\sigma} \cdot \mathbf{N}\right) \xi'_\sigma \\ \exp\left(-\frac{\vartheta}{2} \boldsymbol{\sigma} \cdot \mathbf{N}\right) \xi'_\sigma \end{pmatrix} \tag{16.5.10}$$

where $\xi'_{\pm\mathbf{N}} = \pm\xi_{\mp\mathbf{N}}$ (see (16.4.42)). Hence for $\xi_\sigma \equiv \xi_{-\mathbf{N}}$,

$$v'(\mathbf{p}, \sigma) \longrightarrow -\frac{e^{\vartheta/2}}{\sqrt{2}} \begin{pmatrix} \xi_{+\mathbf{N}} \\ 0 \end{pmatrix} \tag{16.5.11}$$

for $M \to 0$, and now with the normalization condition

$$v'^\dagger(\mathbf{p}, \sigma) v'(\mathbf{p}, \sigma') = \delta_{\sigma\sigma'} \tag{16.5.12}$$

we obtain

$$v'(\mathbf{p}, \sigma) = \begin{pmatrix} \xi_{+\mathbf{N}} \\ 0 \end{pmatrix}. \tag{16.5.13}$$

It is interesting to note from (16.5.9), (16.5.13), that the solution of positive and negative energies decouple.

Similarly, for $\xi_\sigma = \xi_{+\mathbf{N}}$, we have for $M \to 0$,

$$u'(\mathbf{p}, \sigma) = \begin{pmatrix} \xi_{+\mathbf{N}} \\ 0 \end{pmatrix}, \quad v'(\mathbf{p}, \sigma) = \begin{pmatrix} 0 \\ \xi_{-\mathbf{N}} \end{pmatrix}. \tag{16.5.14}$$

When the mass M is set rigorously equal to zero at the outset in the Dirac equation, the Hamiltonian in the chiral representation is given by

$$H' = \gamma'^0 \gamma' \cdot \mathbf{p} = \begin{pmatrix} c\boldsymbol{\sigma} \cdot \mathbf{p} & 0 \\ 0 & -c\boldsymbol{\sigma} \cdot \mathbf{p} \end{pmatrix} \tag{16.5.15}$$

and

$$\boldsymbol{\Sigma}' = \boldsymbol{\Sigma} = \begin{pmatrix} \boldsymbol{\sigma} & 0 \\ 0 & \boldsymbol{\sigma} \end{pmatrix}, \quad \gamma'^5 = \begin{pmatrix} I & 0 \\ 0 & -I \end{pmatrix}. \tag{16.5.16}$$

Now it is an easy matter to show (see Problem 16.11) that the only matrices in the complete set of the 16 matrices in the ordered set[16] $\{\Gamma'_A\}$ in the chiral representation that commute with H', apart from the identity, are γ'^5 and the linear combinations ($\mathbf{N} = \mathbf{p}/|\mathbf{p}|$)

$$a\mathbf{N} \cdot \begin{pmatrix} \boldsymbol{\sigma} & 0 \\ 0 & -\boldsymbol{\sigma} \end{pmatrix} + b\mathbf{N} \cdot \begin{pmatrix} \boldsymbol{\sigma} & 0 \\ 0 & \boldsymbol{\sigma} \end{pmatrix} \tag{16.5.17}$$

for arbitrary numerical factors a and b. The first matrix in (16.5.17) is nothing but H', up to a multiplicative numerical factor, while the second is $\boldsymbol{\Sigma} \cdot \mathbf{N}$ also up to a numerical factor in the momentum description. Also note that $[\gamma'^5, \boldsymbol{\Sigma} \cdot \mathbf{N}] = 0$.

That is, we have the commuting set $\{H', \mathbf{S} \cdot \mathbf{N}, \gamma'^5\}$ of operators:[17]

$$[H', \mathbf{S} \cdot \mathbf{N}] = 0, \quad [H', \gamma'^5] = 0, \quad [\gamma'^5, \mathbf{S} \cdot \mathbf{N}] = 0 \tag{16.5.18}$$

where $\mathbf{S} = \hbar\boldsymbol{\Sigma}/2$ (see (16.2.44), (16.2.46)), and we must find the simultaneous eigenstates of these operators to specify the state of a particle. For positive energy

$$H' u'(\mathbf{p}, \sigma) = |\mathbf{p}|c\, u'(\mathbf{p}, \sigma) \tag{16.5.19}$$

$$\mathbf{S} \cdot \mathbf{N}\, u'(\mathbf{p}, \sigma) = \hbar\lambda\, u'(\mathbf{p}, \sigma) \tag{16.5.20}$$

$$\gamma'^5 u'(\mathbf{p}, \sigma) = \zeta\, u'(\mathbf{p}, \sigma). \tag{16.5.21}$$

From (16.5.15), (16.5.16), it is easy to see that (16.5.19)–(16.5.21) are compatible only if $u'(\mathbf{p}, \sigma)$ has *either* an upper component *or* a lower one. It is straightforward to show that the solutions for $u'(\mathbf{p}, \sigma)$ are given by

$$u'(\mathbf{p}, -1) = \begin{pmatrix} 0 \\ \xi_{-\mathbf{N}} \end{pmatrix}, \quad u'(\mathbf{p}, +1) = \begin{pmatrix} \xi_{+\mathbf{N}} \\ 0 \end{pmatrix} \tag{16.5.22}$$

as before, with

---
[16] See the appendix to §16.1.
[17] The role of parity will be discussed below.

16.5 Massless Dirac Particles

$$\mathbf{S} \cdot \mathbf{N} u'(\mathbf{p}, \sigma) = -\frac{\hbar}{2} u'(\mathbf{p}, \sigma) \qquad (16.5.23)$$

$$\gamma'^5 u'(\mathbf{p}, \sigma) = -u'(\mathbf{p}, \sigma) \qquad (16.5.24)$$

and

$$\mathbf{S} \cdot \mathbf{N} u'(\mathbf{p}, \sigma) = +\frac{\hbar}{2} u'(\mathbf{p}, \sigma) \qquad (16.5.25)$$

$$\gamma'^5 u'(\mathbf{p}, \sigma) = +u'(\mathbf{p}, \sigma) \qquad (16.5.26)$$

respectively.

γ'^5 is called the chirality operator, and we see from (16.5.23)–(16.5.26), that for a positive energy solution we always have

$$\mathbf{\Sigma} \cdot \mathbf{N} u'(\mathbf{p}, \sigma) = \gamma'^5 u'(\mathbf{p}, \sigma). \qquad (16.5.27)$$

That is, for positive energy massless Dirac particles, helicity and chirality have the same sign.

For the charge conjugate spinor

$$v'(\mathbf{p}, \sigma) = i\gamma'^2 \gamma'^0 u'^\top(\mathbf{p}, \sigma) \qquad (16.5.28)$$

the solutions are

$$\begin{pmatrix} \xi_{+\mathbf{N}} \\ 0 \end{pmatrix}, \quad \begin{pmatrix} 0 \\ \xi_{-\mathbf{N}} \end{pmatrix} \qquad (16.5.29)$$

for $\xi_\sigma = \xi_{-\mathbf{N}}, \xi_{+\mathbf{N}}$, respectively.

The corresponding equations to (16.5.19)–(16.5.26) are then

$$H' v'(-\mathbf{p}, \sigma) = -|\mathbf{p}| v'(-\mathbf{p}, \sigma) \qquad (16.5.30)$$

$$\mathbf{S} \cdot \mathbf{N} v'(-\mathbf{p}, \sigma) = -\frac{\hbar}{2} v'(-\mathbf{p}, \sigma) \qquad (16.5.31)$$

$$\gamma'^5 v'(-\mathbf{p}, \sigma) = +v'(-\mathbf{p}, \sigma) \qquad (16.5.32)$$

and

$$\mathbf{S} \cdot \mathbf{N} v'(-\mathbf{p}, \sigma) = +\frac{\hbar}{2} v'(-\mathbf{p}, \sigma) \qquad (16.5.33)$$

$$\gamma'^5 v'(-\mathbf{p}, \sigma) = -v'(\mathbf{p}, \sigma) \qquad (16.5.34)$$

respectively, and we always have

$$\mathbf{\Sigma} \cdot \mathbf{N} v'(-\mathbf{p}, \sigma) = -\gamma'^5 v'(-\mathbf{p}, \sigma) \qquad (16.5.35)$$

with chirality and helicity of opposite signs in this case.

916   16 Quantum Description of Relativistic Particles

We note that the solutions in (16.5.22) obtained for $u'(\mathbf{p}, \sigma)$ are connected by the parity operation

$$\gamma'^0 \begin{pmatrix} 0 \\ \xi_{-\mathbf{N}} \end{pmatrix} \xrightarrow{\mathbf{N} \to -\mathbf{N}} \begin{pmatrix} \xi_{+\mathbf{N}} \\ 0 \end{pmatrix} \qquad (16.5.36)$$

up to a phase, where γ'^0 is given in (16.1.26). Massless particles that are produced through processes which do not conserve *parity*, so that a massless particle with only one of the helicities, say, $-\hbar/2$ in (16.5.23) is observed, may be described by the spinor

$$\widetilde{u}(\mathbf{p}, \sigma) = \frac{1}{2} \left(I - \gamma'^5 \right) u'(\mathbf{p}, \sigma) \qquad (16.5.37)$$

which is *non-zero* for $u'(\mathbf{p}, \sigma) = \begin{pmatrix} 0 & \xi_{-\mathbf{N}} \end{pmatrix}^\top$, and that

$$\gamma'^5 \widetilde{u}(\mathbf{p}, \sigma) = -\widetilde{u}(\mathbf{p}, \sigma), \qquad (16.5.38)$$

$$\mathbf{S} \cdot \mathbf{N}\, \widetilde{u}(\mathbf{p}, \sigma) = -\frac{\hbar}{2} \widetilde{u}(\mathbf{p}, \sigma). \qquad (16.5.39)$$

A particle with negative helicity is said to be left-handed, while one with positive helicity is said to be right-handed.[18]

The charge conjugate spinor $\widetilde{v}(\mathbf{p}, \sigma)$ corresponding to $\widetilde{u}(\mathbf{p}, \sigma)$ would be then defined as in (16.5.31), satisfying

$$\mathbf{S} \cdot (-\mathbf{N})\, \widetilde{v}(-\mathbf{p}, \sigma) = +\frac{\hbar}{2} \widetilde{v}(-\mathbf{p}, \sigma). \qquad (16.5.40)$$

The fact that the positive and energy solutions in (16.5.22), (16.5.29) decouple, a two-component formulation for massless particles may be set up, but we will not go into it here.

For massless particles of higher spin s, we will see in §16.8 that the only possible values for helicities are $\pm s$ and no intermediate $2s - 1$ values appear.

16.6 Physical Interpretation, Localization and Particle Content

In the previous sections of this chapter, we have developed the Dirac equation to provide a quantum description of relativistic spin 1/2 particles in a Lorentz covariant manner. We have witnessed the existence of negative energy states. This, in particular, implies the unboundedness of the spectrum of

---
[18] The word chirality is derived from a Greek word referring to "hand" or in this context to "handedness".

16.6 Physical Interpretation, Localization and Particle Content

the corresponding Hamiltonian from below as the kinetic energies associated with such states become more and more negative. One cannot also simply exclude these states, in general, from the underlying Hilbert space due to the lack of its completeness without them for a correct statistical interpretation of the theory. As a matter of fact, quantum transition probabilities of occurrence of basic fundamental relativistic processes turn out to be incorrect, i.e., in contradiction with experiments, if such states are not included in their calculations. The purpose of this section, is to study the nature of the negative energy states, their non-trivial consequences, the idea of localization of a particle and that of vacuum fluctuations. As mentioned in the Introduction to this chapter, these concepts provide the first steps in the development of quantum field theory.

In the light of the Dirac equation, we first derive expressions for the probability density and probability current as follow from this equation and we consider the initial value problem. One will soon realize that a single-particle interpretation with such a probability density turns out to be not complete, and one is led to a multi-particle theory. This is due to the fact that relativity allows the creation of an unlimited number of pairs of particles, and we must take them into account even when discussing the motion of a single particle. We then study the concept of a position of a particle, its localization and its role in the nature of relativistic corrections of the spectrum of the hydrogen atom, together, with the proper interpretation of the negative energy solutions of negative mass.

16.6.1 Probability, Probability Current and the Initial Value Problem

Consider the Dirac equation in an external electromagnetic field

$$\left[\gamma^\mu \left(\frac{\partial_\mu}{i} - \frac{e}{c\hbar} A_\mu\right) + \frac{Mc}{\hbar}\right] \Psi = 0 \qquad (16.6.1)$$

and of its adjoint multiplied by γ^0: $\overline{\Psi} = \Psi^\dagger \gamma^0$,

$$\overline{\Psi}\left[\gamma^\mu \left(\frac{\overleftarrow{\partial}_\mu}{i} + \frac{e}{c\hbar} A_\mu\right) - \frac{Mc}{\hbar}\right] = 0. \qquad (16.6.2)$$

Multiplying (16.6.1) from the left by $\overline{\Psi}$, and (16.6.2) from the right by Ψ, and subtracting one equation from the other, we obtain

$$\partial_\mu \left(\overline{\Psi}(x)\gamma^\mu \Psi(x)\right) = 0 \qquad (16.6.3)$$

or

$$\frac{\partial}{\partial t}\Psi^\dagger(x)\Psi(x) + \nabla \cdot \left(c\overline{\Psi}(x)\boldsymbol{\gamma}\Psi(x)\right) = 0 \qquad (16.6.4)$$

providing a conservation law for the probability four-current density $J^\mu(x) = \overline{\Psi}(x)\gamma^\mu \Psi(x)/\hbar$, in the unit adopted for $\Psi(x)$ in (16.4.45). The latter implies that the decrease of probability within a given volume of space is compensated by the flow of the probability three-current density $\mathbf{J}(x) = \overline{\Psi}(x)\boldsymbol{\gamma}\Psi(x)/\hbar$ out of the volume in question. The product $\Psi^\dagger(x)\Psi(x)/\hbar c = J^0(x)/c$ allows one formally to adopt the normalization condition in (16.4.45). We will see below at the end of this section, however, that a single-particle probability density interpretation of this in the relativistic domain turns out to be not complete.

The total probability three-current $\mathbf{J}_{\text{Tot}}(t)$ at any given time t, may be determined from (16.4.44), with $u(\mathbf{p},\sigma)$ and $v(\mathbf{p},\sigma)$ satisfying the normalization conditions in (16.4.23), (16.4.37), and the easily derived properties

$$\overline{u}(\mathbf{p},\sigma)\boldsymbol{\gamma} u(\mathbf{p},\sigma') = \frac{\mathbf{p}}{Mc}\delta_{\sigma\sigma'} \qquad (16.6.5)$$

$$\overline{v}(\mathbf{p},\sigma)\boldsymbol{\gamma} v(\mathbf{p},\sigma') = \frac{\mathbf{p}}{Mc}\delta_{\sigma\sigma'}, \qquad (16.6.6)$$

to be given by

$$\mathbf{J}_{\text{Tot}}(t) = \frac{1}{\hbar}\int d^3\mathbf{x}\, \overline{\Psi}(x)\boldsymbol{\gamma}\Psi(x)$$

$$= \frac{1}{\hbar c}\sum_\sigma \int 2Mc\,d\omega_p \left(\frac{\mathbf{p}}{M}\right)\left[|a(\mathbf{p},\sigma)|^2 + |b(\mathbf{p},\sigma)|^2\right]$$

$$+ \frac{1}{\hbar c}\sum_{\sigma,\sigma'}\int 2Mc\,d\omega_p \left[\overline{u}(-\mathbf{p},\sigma)c\boldsymbol{\gamma} v(\mathbf{p},\sigma')\right.$$

$$\left.\times e^{2ip^0 ct/\hbar} a^*(-\mathbf{p},\sigma)\, b^*(\mathbf{p},\sigma') + \text{c.c.}\right]. \qquad (16.6.7)$$

Due to the interference between the negative and positive energy solutions, the current $\mathbf{J}_{\text{Tot}}(t)$ oscillates rapidly with angular frequencies $\omega = 2\sqrt{\mathbf{p}^2 c^2 + M^2 c^4}/\hbar \geqslant 2Mc^2/\hbar \sim 10^{21}$ sec^{-1}. Such oscillations were referred to as "Zitterbewegung" by Schrödinger. We will say more about them below.

Given an initial condition $\Psi(\mathbf{x},0)$, describing, for example, the localization of a particle, we will see how the negative energy solution contributes to the solution for $t > 0$. *Given an initial condition, the Dirac equation provides in* (16.4.44) *the solution* of the problem for $t > 0$. By using, in the process, the normalization properties in (16.4.23), (16.4.37), (16.4.39) we obtain

$$a(\mathbf{p},\sigma) = u^\dagger(\mathbf{p},\sigma)\int d^3\mathbf{x}\, e^{-i\mathbf{p}\cdot\mathbf{x}/\hbar}\,\Psi(\mathbf{x},0) \qquad (16.6.8)$$

$$b^*(\mathbf{p},\sigma) = v^\dagger(\mathbf{p},\sigma)\int d^3\mathbf{x}\, e^{i\mathbf{p}\cdot\mathbf{x}/\hbar}\,\Psi(\mathbf{x},0). \qquad (16.6.9)$$

16.6 Physical Interpretation, Localization and Particle Content

For example, if the electron is initially prepared with spin, say, along the z-axis, and

$$\Psi(\mathbf{x}, 0) = F(\mathbf{x}) \begin{pmatrix} 1 \\ 0 \\ 0 \\ 0 \end{pmatrix} \qquad (16.6.10)$$

then

$$a(\mathbf{p}, \sigma) = \sqrt{\frac{p^0 + Mc}{2Mc}}\, \xi_\sigma^\dagger \begin{pmatrix} 1 \\ 0 \end{pmatrix} F(\mathbf{p}) \qquad (16.6.11)$$

$$b^*(\mathbf{p}, \sigma) = \sqrt{\frac{p^0 + Mc}{2Mc}}\, \frac{|\mathbf{p}|}{p^0 + Mc}\, \xi_\sigma^{\prime\dagger} \begin{pmatrix} \cos\theta' \\ \sin\theta'\, e^{i\phi'} \end{pmatrix} F(-\mathbf{p}) \qquad (16.6.12)$$

(see (16.4.20), (16.4.32)), where $\mathbf{p} = |\mathbf{p}|(\sin\theta'\cos\phi', \sin\theta'\sin\phi', \cos\theta')$ and $F(\mathbf{p})$ is the Fourier transform of $F(\mathbf{x})$ in (16.6.10). That is, given an initial condition, the coefficients $b^*(\mathbf{p}, \sigma)$, corresponding to the negative energy solution, would, in general, contribute to $\Psi(\mathbf{x}, t)$ for $t > 0$ in (16.4.44). They are suppressed by a factor $|\mathbf{p}|/(p^0 + Mc)$ relative to the positive energy coefficients at low energies. On the other hand for higher energies at which $|\mathbf{p}|$ is, say, comparable to Mc, $|\mathbf{p}|/(p^0 + Mc)$ is of the order one and $b^*(\mathbf{p}, \sigma)$ would be equally important as $a(\mathbf{p}, \sigma)$. Accordingly, if the wavepacket $F(\mathbf{x})$ is such that the particle is localized within a volume of extension R, an hence typically $|\mathbf{p}| \sim \hbar/R$, then energies for which $|\mathbf{p}| \sim Mc$ will contribute to the integral in (16.4.44) and make $b^*(\mathbf{p}, \sigma)$ of the same order as $a(\mathbf{p}, \sigma)$. Roughly, as soon as a particle is localized within a radius of the order of its Compton wavelength or less, the negative energy contribution to $\Psi(x)$ cannot be neglected.

16.6.2 Diagonalization of the Hamiltonian and Definitions of Position Operators

We carry out a diagonalization of the Hamiltonian in (16.4.51) of a Dirac particle by using, in the process, a transformation of the type given in (16.1.30)–(16.1.35), which provides a representation of the Dirac equation in which the positive and negative energy states may be separately represented by two-component spinors. This new representation of the Dirac theory is referred to as the Foldy-Wouthuysen-Tani[19] representation, and provides insight into the problem of the position of a relativistic particle and into the physics of the relativistic corrections of the hydrogen atom studied at length in §7.4.

---

[19] Foldy and Wouthuysen (1950); Tani (1951).

Let $\mathbf{a} = \mathbf{p}/Mc$ in (16.1.32)–(16.1.34), then the unitary transformation matrix G in (16.1.30) takes the form

$$G = \frac{1}{\sqrt{2p^0(p^0 + Mc)}} \left[p^0 + Mc + \boldsymbol{\gamma} \cdot \mathbf{p} \right] \tag{16.6.13}$$

and

$$G^\dagger = G^{-1} = \frac{1}{\sqrt{2p^0(p^0 + Mc)}} \left[p^0 + Mc - \boldsymbol{\gamma} \cdot \mathbf{p} \right]. \tag{16.6.14}$$

It is understood that $(p^0 + Mc)$ in the square brackets in (16.6.13), (16.6.14) is multiplied by the identity matrix.

From the expression of the Hamiltonian $H = \gamma^0 \left(c\boldsymbol{\gamma} \cdot \mathbf{p} + Mc^2 \right)$ in the usual Dirac-Pauli representation, and the identity in (16.1.35), we obtain the following expression for the Hamiltonian in the new representation

$$G H G^{-1} = H' = \gamma^0 \sqrt{\mathbf{p}^2 c^2 + M^2 c^2} \equiv \gamma^0 p^0 c. \tag{16.6.15}$$

The Dirac general solution $\Psi(x)$ in (16.4.44) then transforms to

$$\Psi'(x) = G\Psi(x)$$

$$= \sum_\sigma \int \frac{2Mc\, d\omega_p}{\sqrt{2p^0(p^0+Mc)}} \, e^{i\mathbf{p}\cdot\mathbf{x}/\hbar} \left[e^{-ip^0 x^0/\hbar} p^0 \left(I + \gamma^0 \right) u(\mathbf{p},\sigma) a(\mathbf{p},\sigma) \right.$$

$$\left. - e^{ip^0 x^0/\hbar} p^0 \left(I - \gamma^0 \right) v(-\mathbf{p},\sigma) b^*(-\mathbf{p},\sigma) \right]$$

$$\equiv \begin{pmatrix} \psi_+(x) \\ 0 \end{pmatrix} + \begin{pmatrix} 0 \\ \psi_-(x) \end{pmatrix} \tag{16.6.16}$$

where we have used (16.4.49), (16.4.50) after having multiplied these two equations by γ^0 from the left. Here we have identified $\psi_\pm(x)$ with the two component spinors as extracted, respectively, from the projection operators $(I \pm \gamma^0)/2$ onto upper/lower components as occurring within the square brackets in (16.6.16). From (16.6.15), (16.6.16), we then have the two-component equations

$$i\hbar \frac{\partial}{\partial t} \psi_\pm(x) = \pm \sqrt{-\hbar^2 c^2 \boldsymbol{\nabla}^2 + M^2 c^4} \, \psi_\pm(x) \tag{16.6.17}$$

where we recall the definition of γ^0 in (16.1.8).

On the other hand, the Dirac position variable $\mathbf{x}$ in the new representation becomes

$$\mathbf{x}' = G \mathbf{x} G^{-1}$$

16.6 Physical Interpretation, Localization and Particle Content   921

$$= \mathbf{x} - \frac{i\hbar \boldsymbol{\gamma}}{2p^0} + \frac{\hbar}{2(p^0)^2(p^0+Mc)}\left\{i\mathbf{p}\,\boldsymbol{\gamma}\cdot\mathbf{p} - p^0 \boldsymbol{\Sigma}\times\mathbf{p}\right\}$$

$$\equiv \mathbf{x} + \mathbf{R} \qquad (16.6.18)$$

as is easily verified, where we have used the fact that in the **p**-description $\mathbf{x} = i\hbar \nabla_{\mathbf{p}}$, and that the momentum **p** is the *same* in both representations since G in (16.6.13) commutes with **p**. In writing the last expression within the curly brackets in (16.6.18) we have also used the identity

$$i\left(\mathbf{p} + \boldsymbol{\gamma}\cdot\mathbf{p}\boldsymbol{\gamma}\right) = \boldsymbol{\Sigma}\times\mathbf{p}. \qquad (16.6.19)$$

The first term **x** on the right-hand side of (16.6.18) in the new representation at time t is given by

$$\mathbf{x}(t) = e^{itH'/\hbar}\,\mathbf{x}\,e^{-itH'/\hbar} = \mathbf{x} + t\,\frac{\mathbf{p}}{(p^0)^2}\,H' \qquad (16.6.20)$$

and for the associated velocity

$$\frac{d\mathbf{x}(t)}{dt} = \frac{\mathbf{p}}{(p^0)^2}\,H'. \qquad (16.6.21)$$

The latter is uniform in t, and is closely analogous to the classical concept of velocity, and gives the expression $\dfrac{\mathbf{p}}{p^0}\,c$ when applied to a positive energy solution.

The position operator $\mathbf{x}' \equiv \mathbf{x}'_{\text{op}}$, however, is far more complex. For a given well behaved spinor $\Psi'(x)$, we may write

$$\mathbf{x}'_{\text{op}}\Psi'(x) = \int d^3x'\,\kappa(\mathbf{x},\mathbf{x}')\,\Psi'(x^0,\mathbf{x}') \qquad (16.6.22)$$

and formally define, in the sense of distributions, the kernel,

$$\kappa(\mathbf{x},\mathbf{x}') = \mathbf{x}\,\delta^3(\mathbf{x}-\mathbf{x}') + \frac{i\boldsymbol{\gamma}\nabla^2}{2\pi^2}\,I_1(a) - \frac{i}{4\pi^2}\boldsymbol{\gamma}\cdot\nabla\,\nabla I_2(a)$$

$$+ \frac{i}{4\pi^2}\frac{\hbar}{Mc}\boldsymbol{\Sigma}\times\nabla I_3(a) \qquad (16.6.23)$$

where

$$I_1(a) = \frac{1}{a}\int_0^\infty \frac{dz}{z}\,\frac{\sin(za)}{\sqrt{z^2+1}} \qquad (16.6.24)$$

$$I_2(a) = \frac{1}{a}\int_0^\infty \frac{z\,dz\,\sin(za)}{(z^2+1)\left(\sqrt{z^2+1}+1\right)} \qquad (16.6.25)$$

$$I_3(a) = \frac{1}{a}\int_0^\infty \frac{z\,dz\,\sin(za)}{(z^2+1+\sqrt{z^2+1})} \qquad (16.6.26)$$

922 16 Quantum Description of Relativistic Particles

$$a \equiv \frac{Mc}{\hbar} |\mathbf{x} - \mathbf{x}'|. \tag{16.6.27}$$

The kernel $\kappa(\mathbf{x}, \mathbf{x}')$ does not vanish for $\mathbf{x} \neq \mathbf{x}'$, i.e., it is non-local. The functions $I_1(a)$, $I_2(a)$, $I_3(a)$ are plotted in Figure 16.1. They become small for $a \gg 1$, i.e., for $|\mathbf{x} - \mathbf{x}'| \gg \hbar/Mc$. That is, $\mathbf{x}'$ defines a non-local operator with a non-locality spread roughly of the order of the Compton wavelength $\hbar/Mc$ of the particle.

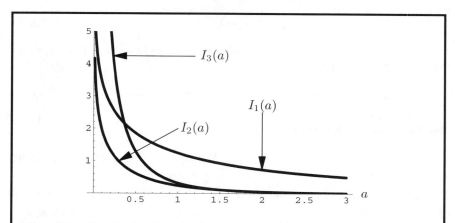

Fig. 16.1. Graphs showing the behavior of $I_1(a)$, $I_2(a)$, $I_3(a)$ in (16.6.24), (16.6.25), (16.6.26) on their dependence on $a > 0$ in (16.6.27). Their rapid vanishing properties for $|\mathbf{x} - \mathbf{x}'| \gg \hbar/Mc$ are evident.

The time evolution operator in the new representation may be explicitly written as

$$e^{-itH'/\hbar} = \cos\left(\frac{tp^0 c}{\hbar}\right) - i\gamma^0 \sin\left(\frac{tp^0 c}{\hbar}\right) \tag{16.6.28}$$

where we have used the fact that $(\gamma^0)^2 = I$, $(\gamma^0)^3 = \gamma^0$. Accordingly, the time evolution of $\mathbf{x}'$ is given by

$$\mathbf{x}'(t) = e^{itH'/\hbar} \mathbf{x}' e^{-itH'/\hbar} \tag{16.6.29}$$

and from (16.6.18), (16.6.28), this works out to be

$$\mathbf{x}'(t) = \mathbf{x} - \hbar \frac{\mathbf{\Sigma} \times \mathbf{p}}{2p^0(p^0 + Mc)} + t \frac{\mathbf{p}}{p^{0 2}} H'$$

$$- \frac{i\hbar}{2p^0} \left(\boldsymbol{\gamma} - \frac{\mathbf{p}\, \boldsymbol{\gamma} \cdot \mathbf{p}}{p^0(p^0 + Mc)}\right) \left[\cos\left(\frac{2p^0 tc}{\hbar}\right) - i\gamma^0 \sin\left(\frac{2p^0 tc}{\hbar}\right)\right] \tag{16.6.30}$$

and coincides with the expression in (16.6.18) for $t = 0$, as expected.

Equation (16.6.30) also gives

$$\frac{d\mathbf{x}'(t)}{dt} = \frac{\mathbf{p}}{p^{0^2}} H' + c\left(i\boldsymbol{\gamma} - \frac{i\mathbf{p}\,\boldsymbol{\gamma}\cdot\mathbf{p}}{p^0(p^0 + Mc)}\right)\left[\sin\left(\frac{2p^0 tc}{\hbar}\right) + i\gamma^0 \cos\left(\frac{2p^0 tc}{\hbar}\right)\right]. \tag{16.6.31}$$

From (16.6.18), (16.6.21), (16.6.31) we may write

$$\frac{d}{dt}\mathbf{R}(t) = c\left(i\boldsymbol{\gamma} - \frac{i\mathbf{p}\,\boldsymbol{\gamma}\cdot\mathbf{p}}{p^0(p^0 + Mc)}\right)\left[\sin\left(\frac{2p^0 tc}{\hbar}\right) + i\gamma^0 \cos\left(\frac{2p^0 tc}{\hbar}\right)\right] \tag{16.6.32}$$

which gives the deviation of the velocity $d\mathbf{x}'/dt$ from $d\mathbf{x}/dt$.

Therefore in the new representation in which the Hamiltonian is diagonal, one may, in general, define two position operators $\mathbf{x}(t)$, $\mathbf{x}'(t)$. The former is closely analogous to the classical concept of position corresponding to a uniform velocity. The latter, however, executes a complex motion, involving a rapidly oscillating one with frequencies $2p^0 c/\hbar \geqslant 2Mc^2/\hbar \sim 10^{21}$ sec^{-1} (the so-called Schrödinger's "Zitterbewegung") about $\mathbf{x}(t)$ and is responsible for the non-locality and the associated spread of a particle over distances roughly of the order of its Compton wavelength. Because of the reasons just mentioned, $\mathbf{x}$ is usually referred to as a *"mean" position operator*, while $\mathbf{x}'$, with its inherited complex motion in time, is responsible for the "jittery" behavior of the particle.

The orbital angular momentum in the new representation reads

$$\mathbf{L}' = \mathbf{x}' \times \mathbf{p}$$

$$= \mathbf{x} \times \mathbf{p} + \mathbf{R} \times \mathbf{p}$$

$$= \mathbf{L} + \mathbf{R} \times \mathbf{p} \tag{16.6.33}$$

and for the spin

$$\mathbf{S}' = \frac{\hbar}{2} G \boldsymbol{\Sigma} G^{-1}$$

$$= \frac{\hbar}{2}\boldsymbol{\Sigma} + \frac{i\hbar}{2p^0}\boldsymbol{\gamma} \times \mathbf{p} + \frac{\hbar}{2p^0(p^0 + Mc)}(\boldsymbol{\Sigma} \times \mathbf{p}) \times \mathbf{p}$$

$$= \mathbf{S} - \mathbf{R} \times \mathbf{p} \tag{16.6.34}$$

where now $\mathbf{x} \times \mathbf{p}$ and $\hbar\boldsymbol{\Sigma}/2$ define "mean" orbital angular momentum and spin, and both commute with H',

$$[\mathbf{x} \times \mathbf{p}, H'] = 0, \quad \frac{\hbar}{2}[\boldsymbol{\Sigma}, H'] = 0 \tag{16.6.35}$$

(see (16.6.15)), i.e., they are separately *conserved*. It is interesting to note that the total angular momentum remains invariant under the transformation

16 Quantum Description of Relativistic Particles

$$\mathbf{L} + \mathbf{S} = \mathbf{L}' + \mathbf{S}'. \tag{16.6.36}$$

The derivations $\mathbf{L}' - \mathbf{L} = \mathbf{R} \times \mathbf{p}$, $\mathbf{S}' - \mathbf{S} = -\mathbf{R} \times \mathbf{p}$ from the "mean" operators are expressed in terms of the deviation of the position operator $\mathbf{x}'$ from the "mean" one $\mathbf{R} = \mathbf{x}' - \mathbf{x}$, and are of orbital angular momentum of origin. With the fluctuating character of $\mathbf{R}(t)$, with velocity as given in (16.6.32), one may formally interpret the above spin deviation as an additional orbital angular momentum associated with the complex motion of the particle over extended regions as discussed above.

A similar analysis as above through (16.6.18)–(16.6.34) may be also carried in the Dirac representation. In particular, the operator $\mathbf{X}$ in the Dirac representation whose transformation in the new representation is $\mathbf{x}$ is clearly given by

$$\mathbf{X} = G^{-1} \mathbf{x} G. \tag{16.6.37}$$

In the **p**-description, $\mathbf{x} = i\hbar \boldsymbol{\nabla}_\mathbf{p}$, and a straightforward application of (16.6.37) gives

$$\mathbf{X} = \mathbf{x} + \frac{i\hbar \boldsymbol{\gamma}}{2p^0} - \frac{\hbar}{2(p^0)^2(p^0 + Mc)} \left\{ i\mathbf{p}\, \boldsymbol{\gamma} \cdot \mathbf{p} + p^0 \boldsymbol{\Sigma} \times \mathbf{p} \right\}$$

$$\equiv \mathbf{x} - \boldsymbol{\eta}. \tag{16.6.38}$$

This position operator is referred to as the Pryce-Newton-Wigner[20] operator. It satisfies the commutation relations

$$[X_i, X_j] = 0, \quad [X_i, p_j] = i\hbar \delta_{ij} \tag{16.6.39}$$

and with $\mathbf{X}(t) = \exp\left(itH/\hbar\right) \mathbf{X} \exp\left(-itH/\hbar\right)$, with H given in (16.4.51),

$$\frac{d}{dt} \mathbf{X}(t) = \frac{\mathbf{p}}{p^0} \frac{H}{p^0} = \frac{\mathbf{p}}{p^{0\,2}} \gamma^0 \left(\boldsymbol{\gamma} \cdot \mathbf{p} c + Mc^2\right) \tag{16.6.40}$$

This velocity operator is given by a satisfactory expression. It is uniform and gives the expected expression $\mathbf{p}c/p^0$ when applied to a positive energy solution. The operator $\mathbf{x}(t)$, however, is far more complex as discussed below.

By using the fact that $\mathbf{x}(t) = \exp\left(itH/\hbar\right) \mathbf{x} \exp\left(-itH/\hbar\right)$, and the explicit expression for the Dirac time-evolution operator given in (16.4.75), we obtain

$$\mathbf{x}(t) = \mathbf{x} - \frac{i\hbar}{2p^{0\,2}c} \left(\gamma^0 \boldsymbol{\gamma} H - \mathbf{p}c\right) + \frac{tH\mathbf{p}}{p^{0\,2}}$$

$$+ \frac{i\hbar}{2p^{0\,2}c} \left(\gamma^0 \boldsymbol{\gamma} H - \mathbf{p}c\right) \left[\cos\left(\frac{2p^0 tc}{\hbar}\right) - \frac{iH}{p^0 c} \sin\left(\frac{2p^0 tc}{\hbar}\right)\right] \tag{16.6.41}$$

---

[20] Pryce (1948); Newton and Wigner (1949).

16.6 Physical Interpretation, Localization and Particle Content

where the identity

$$(\gamma^0\boldsymbol{\gamma} H - c\mathbf{p}) = -(H\gamma^0\boldsymbol{\gamma} - c\mathbf{p}) \qquad (16.6.42)$$

should be noted.

For the deviation $\boldsymbol{\eta}(t) = \mathbf{x}(t) - \mathbf{X}(t)$, we also have

$$\frac{d\boldsymbol{\eta}(t)}{dt} = -\frac{i}{p^0}(\gamma^0\boldsymbol{\gamma} H - \mathbf{p}c)\left[\sin\left(\frac{2p^0 tc}{\hbar}\right) + \frac{iH}{p^0 c}\cos\left(\frac{2p^0 tc}{\hbar}\right)\right] \qquad (16.6.43)$$

which has the oscillatory characteristic behavior with frequencies $2p^0 c/\hbar \geqslant 2Mc^2/\hbar$.

Therefore in the Dirac representation, one may also, in general, define two position operators $\mathbf{X}(t)$, $\mathbf{x}(t)$. The velocity corresponding to the former is analogous to the classical one, while the Dirac position operator executes a complicated motion involving a rapidly fluctuating one with high frequencies (the "Zitterbewegung") about $\mathbf{X}(t)$ and is, again, responsible for the non-locality and the associated spread of the particle. These fluctuations were also encountered when studying the total probability current in (16.6.7) arising from the interference of positive and negative energy solutions. We note that due to the fact that $\mathbf{X}(t)$ is not in "diagonal" form,[21] it mixes the positive and negative states, i.e., it does not commute with the sign of the energy operator. Also note that $d\mathbf{x}/dt|_{t=0} = c\gamma^0\boldsymbol{\gamma}$, and $(c\gamma^0\boldsymbol{\gamma})^2 = 3c^2$ for its square. This is unlike the velocity operator $d\mathbf{X}/dt$ in (16.6.40). The operator $\mathbf{X}(t)$ is usually referred to as a "mean" position operator, while $\mathbf{x}$, with its inherited complex motion, simply as the Dirac position operator.

The "mean" orbital angular momentum in the Dirac representation is defined by

$$\mathbb{L} = \mathbf{X} \times \mathbf{p}$$

$$= \mathbf{x} \times \mathbf{p} - \boldsymbol{\eta} \times \mathbf{p} \qquad (16.6.44)$$

and the "mean" spin operator by

$$\mathbb{S} = G^{-1}\frac{\hbar}{2}\boldsymbol{\Sigma} G$$

$$= \mathbf{S} - \frac{i\hbar}{2p^0}\boldsymbol{\gamma} \times \mathbf{p} + \frac{\hbar}{2p^0(p^0 + Mc)}(\boldsymbol{\Sigma} \times \mathbf{p}) \times \mathbf{p}$$

---

[21] By a matrix in "diagonal" form, in this context, it is meant a matrix of the form $\begin{pmatrix} A' & 0 \\ 0 & B' \end{pmatrix}$ where A' and B' are 2×2 matrices. Such matrices are also referred to as *even* matrices, while a matrix of the form $\begin{pmatrix} 0 & A' \\ B' & 0 \end{pmatrix}$ is referred to as an *odd* matrix.

$$= \mathbf{S} + \boldsymbol{\eta} \times \mathbf{p}. \tag{16.6.45}$$

These "mean" angular momentum operators are separately conserved. Also the deviation $\mathbf{S}-\$ = -\boldsymbol{\eta}\times\mathbf{p}$ may be formally interpreted as an additional orbital angular momentum contributing to spin due to the complicated motion of the particle in an extended region as before.

16.6.3 Origin of Relativistic Corrections in the Hydrogen Atom

Consider the Dirac Hamiltonian in the presence of an electrostatic potential $e\phi(\mathbf{x})$:

$$H_1 = \gamma^0 \left(c\boldsymbol{\gamma}\cdot\mathbf{p} + Mc^2\right) + e\phi(\mathbf{x})$$

$$= Mc^2 \left[\gamma^0 + \frac{\gamma^0\boldsymbol{\gamma}\cdot\mathbf{p}}{Mc} + \frac{e}{Mc^2}\phi(\mathbf{x})\right]$$

$$= Mc^2 \widetilde{H}_1 \tag{16.6.46}$$

We wish to *diagonalize* this Hamiltonian. We will eventually consider corrections up to the order $1/c^2$ only.

We first apply the unitary G transformation given in (16.6.13) to $\widetilde{H}_1$ to obtain

$$\widetilde{H}_1' = G\widetilde{H}_1 G^{-1} = \gamma^0 \sqrt{1 + \frac{\mathbf{p}^2}{M^2c^2}} + \frac{e}{Mc^2}\phi(G\mathbf{x}G^{-1}) \tag{16.6.47}$$

where we have used (16.6.15), and that

$$G\phi(\mathbf{x})G^{-1} = \phi(G\mathbf{x}G^{-1})$$

$$= \phi(\mathbf{x}') \tag{16.6.48}$$

Here $\mathbf{x}'$ is given explicitly in (16.6.18), where $\mathbf{x}$ is the "mean" position and $\mathbf{x}'(t) - \mathbf{x}(t)$ has the complex motion elaborated upon earlier associated with "Zitterbewegung".

We are interested in finding the corrections in H_1 up to order $1/c^2$. That is, we have to find corrections in $\widetilde{H}_1$ in (16.6.47) up to $1/c^4$. The ϕ term is already multiplied by $1/c^2$, hence we may solve for $\mathbf{x}'$ in (16.6.18) up to the order $1/c^2$. This is given by

$$\mathbf{x}' \simeq \mathbf{x} - \frac{i\hbar}{2Mc}\boldsymbol{\gamma} - \frac{\hbar}{4M^2c^2}\boldsymbol{\Sigma}\times\mathbf{p} \tag{16.6.49}$$

where we note that $(1/p^0 - 1/Mc)$ is of the order $(Mc)^{-3}$.

16.6 Physical Interpretation, Localization and Particle Content

Up to $1/c^4$, we then have[22]

$$\frac{1}{c^2}\phi(\mathbf{x}') \simeq \frac{1}{c^2}\phi\left(\mathbf{x} - \frac{i\hbar}{2Mc}\boldsymbol{\gamma} - \frac{\hbar}{4M^2c^2}\boldsymbol{\Sigma}\times\mathbf{p}\right)$$

$$\simeq \frac{1}{c^2}\phi(\mathbf{x}) - \frac{i\hbar}{2Mc^3}\boldsymbol{\gamma}\cdot\boldsymbol{\nabla}\phi(\mathbf{x}) + \frac{\hbar^2}{8M^2c^4}\boldsymbol{\nabla}^2\phi(\mathbf{x})$$

$$- \frac{\hbar}{4M^2c^4}\boldsymbol{\nabla}\phi(\mathbf{x})\cdot(\boldsymbol{\Sigma}\times\mathbf{p}) \qquad (16.6.50)$$

expanded about the "mean" position in the Foldy-Wouthuysen-Tani representation.

Accordingly, up to $1/c^4$, $\widetilde{H}'_1$ is given by

$$\widetilde{H}'_1 \simeq \gamma^0\left(1 + \frac{\mathbf{p}^2}{2M^2c^2} - \frac{\mathbf{p}^4}{8M^4c^4}\right) + \frac{e}{Mc^2}\phi(\mathbf{x})$$

$$- \frac{i\hbar e}{2M^2c^3}\boldsymbol{\gamma}\cdot\boldsymbol{\nabla}\phi(\mathbf{x}) + \frac{\hbar^2 e}{8M^3c^4}\boldsymbol{\nabla}^2\phi(\mathbf{x})$$

$$- \frac{\hbar e}{4M^3c^4}\boldsymbol{\nabla}\phi(\mathbf{x})\cdot(\boldsymbol{\Sigma}\times\mathbf{p}). \qquad (16.6.51)$$

This is almost in diagonal form except for the $\boldsymbol{\gamma}\cdot\boldsymbol{\nabla}\phi(\mathbf{x})$ term which is in "non-diagonal" form and sticks out as a sore thumb, and we have to carry out a further transformation to diagonalize $\widetilde{H}'_1$.

To the above end, we use the following key equation. Given any *odd* matrix, as defined before, i.e., a matrix $\widetilde{O}$ of the form

$$\widetilde{O} = \begin{pmatrix} 0 & A' \\ B' & 0 \end{pmatrix}$$

where A', B' are any 2×2 matrices, then

$$\left[\frac{1}{2}\gamma^0\widetilde{O}, \gamma^0\right] = -\widetilde{O} \qquad (16.6.52)$$

and note that minus sign on the right-hand of this commutation relation.

Therefore to diagonalize $\widetilde{H}'_1$ and cancel out the $\boldsymbol{\gamma}\cdot\boldsymbol{\nabla}\phi$ term in (16.6.51), which is of the order $1/c^3$, we choose $\widetilde{O} = \left(-i\hbar e/2M^2c^3\right)\boldsymbol{\gamma}\cdot\boldsymbol{\nabla}\phi$, and introduce the unitary transformation

---

[22] For any three-vector $\mathbf{k}$ which commutes with $\mathbf{x}$, $\boldsymbol{\Sigma}$ and $\mathbf{p}$, we note that $\mathbf{k}\cdot\mathbf{x}$ and $\mathbf{k}\cdot(\boldsymbol{\Sigma}\times\mathbf{p})$ *commute*. This allows us to carry out the Taylor expansion on the right-hand side of (16.6.50). $\boldsymbol{\Sigma}\times\mathbf{p}$ and $\boldsymbol{\gamma}$ do not commute, however, but keeping track of their commutators is not important since this will lead to corrections of the order $1/c^5$. Also note that $(\boldsymbol{\gamma}\cdot\boldsymbol{\nabla})^2 = -\boldsymbol{\nabla}^2$.

$$G' = I + \frac{1}{2}\gamma^0 \widetilde{O} \qquad (16.6.53)$$

correct up to order $1/c^3$. The term $\gamma^0 \widetilde{O}/2$ can act only on the γ^0 term, with coefficient one in (16.6.51), as the product of $\gamma^0 \widetilde{O}/2$ with any of the other terms in (16.6.51) are of the order $1/c^5$ or higher. Hence

$$\widetilde{H}_1'' \simeq G' \widetilde{H}_1' G'^{-1}$$

$$\simeq \gamma^0 \left(1 + \frac{\mathbf{p}^2}{2M^2 c^2} - \frac{\mathbf{p}^4}{8M^4 c^4}\right) + \frac{e}{Mc^2}\phi(\mathbf{x})$$

$$+ \frac{\hbar^2 e}{8M^3 c^4}\boldsymbol{\nabla}^2 \phi(\mathbf{x}) - \frac{\hbar e}{4M^3 c^4}\boldsymbol{\nabla}\phi(\mathbf{x})\cdot(\boldsymbol{\Sigma}\times\mathbf{p}) \qquad (16.6.54)$$

where we have used (16.6.52) to cancel out the $\widetilde{O}$ term, proportional to $\boldsymbol{\gamma}\cdot\boldsymbol{\nabla}\phi$, identified above.

Setting $e\phi(\mathbf{x}) = U(\mathbf{x})$, and using the identity $\boldsymbol{\nabla}\phi(\mathbf{x})\cdot(\boldsymbol{\Sigma}\times\mathbf{p}) = -\boldsymbol{\Sigma}\cdot(\boldsymbol{\nabla}\phi(\mathbf{x})\times\mathbf{p})$, the diagonalized form of H_1 in (16.6.46), up to the order $1/c^2$, is given by

$$H_1'' = \gamma^0 \left(Mc^2 + \frac{\mathbf{p}^2}{2M} - \frac{\mathbf{p}^4}{8M^3 c^2}\right) + U(\mathbf{x}) + \frac{\hbar^2}{8M^2 c^2}\boldsymbol{\nabla}^2 U(\mathbf{x})$$

$$+ \frac{1}{2M^2 c^2}\mathbf{S}\cdot(\boldsymbol{\nabla}U(\mathbf{x})\times\mathbf{p}) \qquad (16.6.55)$$

which is obviously in diagonal form. When restricted to the upper component of the Dirac spinor, it should be compared with (7.4.32) (see also (7.4.38)), where we have generated the Darwin and the spin-orbit coupling terms, as well as the leading relativistic correction to the kinetic energy. H_1'' is expressed in terms of the "mean" position operator $\mathbf{x}$, in the Foldy-Wouthuysen-Tani representation, with the corrections arising from the deviation of the position operator in this representation about the "mean" one associated, in particular, with "Zitterbewegung". This provides a beautiful explanation of the Darwin term arising, as shown in (16.6.50), from the fluctuations of the position of the electron which encounters a smeared-out Coulomb potential leading to the $\boldsymbol{\nabla}^2\phi$ term in (16.6.55). An exact treatment of the bound Coulomb potential is given in the appendix to this section.

Now we generalize the above analysis leading to (16.6.55) by including a time-independent three-vector potential $\mathbf{A}$ as well. That is, we consider the Dirac equation given in (16.1.18) with time-independent A_μ. The Hamiltonian associated with such a system is then given by

$$H_2 = \gamma^0 \left(c\boldsymbol{\gamma}\cdot\mathbf{p} + Mc^2\right) - e\gamma^0\,\boldsymbol{\gamma}\cdot\mathbf{A}(\mathbf{x}) - eA_0(\mathbf{x}) \qquad (16.6.56)$$

where $A_0(\mathbf{x}) = -\phi(\mathbf{x})$.

16.6 Physical Interpretation, Localization and Particle Content

As in (16.6.46), we rewrite (16.6.56) in the form

$$H_2 = Mc^2 \left[\gamma^0 + \frac{\gamma^0 \boldsymbol{\gamma} \cdot \mathbf{p}}{Mc} + \frac{e\phi(\mathbf{x})}{Mc^2} - \frac{e}{Mc^2}\gamma^0 \boldsymbol{\gamma} \cdot \mathbf{A}(\mathbf{x})\right]$$

$$\equiv Mc^2 \widetilde{H}_2 \qquad (16.6.57)$$

We will diagonalize $\widetilde{H}_2$ up to the order $1/c^4$ as before.

The odd matrix in the square brackets in (16.6.57) is

$$\frac{1}{c}\widetilde{O}' = \frac{\gamma^0 \boldsymbol{\gamma}}{Mc} \cdot \left(\mathbf{p} - \frac{e}{c}\mathbf{A}\right) \qquad (16.6.58)$$

The commutation relation in (16.6.52) suggests to consider the matrix

$$\frac{1}{c}D' = \frac{1}{2c}\gamma^0 \widetilde{O}' = \frac{\boldsymbol{\gamma}}{2Mc} \cdot \left(\mathbf{p} - \frac{e}{c}\mathbf{A}\right) \qquad (16.6.59)$$

We introduce the transformation

$$_1G' = \exp\left(\frac{1}{c}D'\right) \qquad (16.6.60)$$

as a first step to diagonalize $\widetilde{H}_2$.

Since we are interested in going up to the order $1/c^4$, we may use the expansion

$$e^{D'/c} T e^{-D'/c} \simeq T + \frac{1}{c}[D', T] + \frac{1}{2!c^2}\Big[D', [D', T]\Big]$$

$$+ \frac{1}{3!c^3}\Big[D', \big[D', [D', T]\big]\Big]$$

$$+ \frac{1}{4!c^4}\Big[D', \Big[D', \big[D', [D', T]\big]\Big]\Big] \qquad (16.6.61)$$

for any term T in $\widetilde{H}_2$. For the terms proportional to $1/c^2$ in $\widetilde{H}_2$, we may restrict the right-hand side of (16.6.61) to the first three terms only and so on.

The first transformation is readily worked out (see Problem 16.15), and is given by

$$\widetilde{H}_2' = e^{D'/c} \widetilde{H}_2 e^{-D'/c}$$

$$= \gamma^0 + \frac{e}{Mc^2}\phi(\mathbf{x}) + \frac{1}{2M^2c^2}\gamma^0 \left[\gamma^0 \boldsymbol{\gamma} \cdot \left(\mathbf{p} - \frac{e}{c}\mathbf{A}\right)\right]^2$$

$$- \frac{1}{8M^4c^4}\gamma^0 [\gamma^0 \boldsymbol{\gamma} \cdot \mathbf{p}]^4 - \frac{e}{8M^3c^4}\Big[\gamma^0 \boldsymbol{\gamma} \cdot \mathbf{p}, [\gamma^0 \boldsymbol{\gamma} \cdot \mathbf{p}, \phi]\Big]$$

$$+ \frac{1}{c^3}\widetilde{O}'',\tag{16.6.62}$$

where $\widetilde{O}''$ consists only of *odd* matrices, and all the other terms in (16.6.62) consist of even matrices. Is the explicit expression of $\widetilde{O}''$ in (16.6.62) important? The answer is no, since it already involves the factor $1/c^3$, and we may just use the commutation relation in (16.6.52) and define the transformation

$$_2G'' \simeq I + \frac{\gamma^0}{2c^3}\widetilde{O}''\tag{16.6.63}$$

correct up to the order $1/c^3$, to *cancel out* the $\widetilde{O}''/c^3$ term in (16.6.62). The term $\gamma^0\widetilde{O}''/2c^3$ in (16.6.63) is allowed to act only on the γ^0 term in (16.6.62), with coefficient one, as all the other terms in (16.6.62) are of the order $1/c^2$ and of higher orders, and would otherwise lead to corrections of the order $1/c^5$ and higher.

Finally, note that ($\boldsymbol{\nabla}\cdot\mathbf{A}=0$)

$$\left[\gamma^0\boldsymbol{\gamma}\cdot\left(\mathbf{p}-\frac{e}{c}\mathbf{A}\right)\right]^2 = \left(\mathbf{p}-\frac{e}{c}\mathbf{A}\right)^2 - \frac{e\hbar}{c}\boldsymbol{\Sigma}\cdot\mathbf{B}\tag{16.6.64}$$

(see also (7.4.31)), where $\mathbf{B}=\boldsymbol{\nabla}\times\mathbf{A}$,

$$\left[\gamma^0\boldsymbol{\gamma}\cdot\mathbf{p}\right]^4 = \left(\mathbf{p}^2\right)^2\tag{16.6.65}$$

and

$$\left[\gamma^0\boldsymbol{\gamma}\cdot\mathbf{p},\left[\gamma^0\boldsymbol{\gamma}\cdot\mathbf{p},\phi\right]\right] = -\hbar^2\boldsymbol{\nabla}^2\phi - 2\hbar\boldsymbol{\Sigma}\cdot(\boldsymbol{\nabla}\phi\times\mathbf{p})\tag{16.6.66}$$

All told, we obtain by restricting to the upper components, denoted by $|_+$, the diagonalized Hamiltonian

$$H_2''|_+ \simeq {}_2G''\left(Mc^2\widetilde{H}_2'\right){}_2G''^{-1}\Big|_+$$

$$\simeq Mc^2 + \frac{1}{2M}\left(\mathbf{p}-\frac{e}{c}\mathbf{A}\right)^2 - \frac{\mathbf{p}^4}{8M^3c^2} + U(\mathbf{x})$$

$$+ \frac{\hbar^2}{8M^2c^2}\boldsymbol{\nabla}^2 U + \frac{1}{2M^2c^2}\mathbf{S}\cdot(\boldsymbol{\nabla}U\times\mathbf{p})$$

$$- \frac{e}{Mc}\mathbf{S}\cdot\mathbf{B}\tag{16.6.67}$$

up to the order $1/c^2$. If $\mathbf{A}$ has a built in $1/c$ factor one may linearize $(\mathbf{p}-e\mathbf{A}/c)^2$ in $\mathbf{A}$ thus obtaining the Hamiltonian in (7.4.32) up to the rest energy Mc^2. As discussed in §7.4, upon rewriting the last term in (16.6.67) as $-(2e/2Mc)\mathbf{S}\cdot\mathbf{B}$, the theory provides, the approximate g-factor equal to 2 for the electron. The correction to g has been investigated in §8.5. In (16.6.67), $\mathbf{S}=\hbar\boldsymbol{\sigma}/2$, $U=e\phi$.

16.6.4 The Positron and Emergence of a Many-Particle Theory

The fact that the Dirac theory predicts negative energy states with energies going down to $-\infty$ would imply the instability of such a system. For example, an electron in the ground-state energy of an atom would spontaneously decay to such lower and lower negative energy states emitting radiation of arbitrary large energies leading eventually to the collapse of the atom with a release of an infinite amount of energy. To resolve such a dilemma, Dirac assumed that *a priori* all the negative energy states are filled with electrons, giving rise to the so-called Dirac sea or the Dirac vacuum, in accord to the Pauli exclusion principle so that no transitions to such states are possible, thus ensuring the stability of the atom.

The consequences of the above assumption of a completely filled vacuum with negative energy electrons are many. A negative energy electron in the Dirac sea, may absorb radiation of sufficient energy so as to overcome an energy gap such as from $-Mc^2$ to Mc^2, or the corresponding one to the problem at hand, thus making a jump to a positive energy state leaving behind a surplus of positive energy and a surplus of positive charge $+|e|$ *relative to* the Dirac sea. This has led Dirac in 1930 to interpret the "hole" left behind by the transition of the negative energy electron to a positive energy state, or the absence of the negative electron, as a particle that has the same mass as the electron but of opposite charge.[23] This particle called the positron (e^+) was discovered shortly after in 1932 by C. D. Anderson who apparently, as mentioned before, was not aware of Dirac's prediction at the time of his discovery. Incidentally the above argument has also provided an explanation of the so-called *pair production* $\gamma \to e^+e^-$ by a photon (in the vicinity of a nucleus[24]). The "donkey" electrons, as G. Gamow named them, led to the birth of the positrons with opposite intrinsic parity to that of electrons. Conversely, if a "hole" is created in the vacuum, then a positive energy electron may make a transition to such a state releasing radiation giving rise to the phenomenon of *pair annihilation*.

In the field of a nucleus, a pair e^+e^- may be created, violating conservation of energy for a short time $\Delta t \sim \hbar/2Mc^2$. An electron in orbit around the nucleus may jump into the "hole" thus created, while the electron created would travel a distance of the order $c\Delta t \sim \hbar/2Mc$ replacing the initial electron in orbit during this time. This process may be repeated providing a simple explanation of *"Zitterbewegung"* as an exchange process between a "primary" electron in an atom and a "secondary" electron ejected from the Dirac sea in the field of the nucleus occurring within distances of the order of the Compton wavelength $\hbar/Mc$ of the electron. Pairs created in the vicinity of the nucleus would lead to a particle screening of the charge of the nucleus.

---

[23] Dirac initially assumed that the particle is the proton since there were no positrons at that time. Apart from the large mass difference between the proton and the electron, there were other inconsistencies with this assumption.

[24] The presence of the nucleus is to conserve energy and momentum.

An electron, in the atom, at large distances from the nucleus would then see a smaller charge on the nucleus than an electron nearby (such as one in s-state). This leads to the concept of *vacuum polarization*, by the field of the nucleus, and also to the concept of charge *renormalization* as a result of the charge screening mentioned above.

Thus by putting relativity with quantum physics one has encountered negative energy states, going down to $-\infty$. By invoking stability of the atom one was led to the discovery of anti-matter which finally led to a multi-particle theory. The wavefunction of a single-particle quantum particle in the relativistic regime turned out to be not complete, and a formalism which would naturally describe creation and destruction of particles became necessary. The so-called "hole" theory, with a filled Dirac sea of negative energy electrons, although it gave insight into the nature of fundamental processes involving relativistic quantum particles and of vacuum fluctuations, turned out to be also incomplete. For example, in the "hole" theory the number of electrons minus the number of positrons, created, is conserved by the simultaneous creation of a "hole" for every electrons ejected from the Dirac sea. In nature, there are processes, where just an electron or just a positron is created. Examples of such process are β^- decay or muon decay: $\mu^- \to e^- + \tilde{\nu}_e + \nu_\mu$, and β^+ decay: $p \to n + e^+ + \nu_e$, for a bound proton in a nucleus, where ν_μ, ν_e are neutrinos, and $\tilde{\nu}_e$ is an anti-neutrino, associated with the respective particles. Also we have seen in (16.3.19) that the Dirac equation, in a *given* external potential A_μ, may be transformed by charge conjugation to describe *a priori* a particle of positive charge $+|e|$. That is, if we were set initially to develop a relativistic theory for the positron and go through a "hole" theory with a sea filled with negative energy positrons would we have discovered the electron?

The Dirac theory with its tremendous accomplishments led to the development of modern (relativistic) quantum field theory, where fields describe the creation and annihilation of particles and the number of particles need not be conserved. The concept of a potential, with its inherited inconsistencies, in a relativistic setting, in such a multi-particle relativistic theory with particles created and annihilated, was abandoned in describing the interaction of elementary particles. The interactions between the particles were thus described by the exchange of particles rather than by potentials. Lorentz invariance turned out to be readily implemented in quantum field theory[25] and vacuum fluctuations were found to be a natural consequence of the theory.

---

[25] See, for example, the expression of the Lagrangian density in (16.9.32).

Appendix to §16.6: Exact Treatment of the Dirac Equation in the Bound Coulomb Problem

In §7.4–§7.7, we have given a detailed treatment of relativistic corrections to the hydrogen atom spectrum and further physical insight into the problem was gained through the analysis worked out in (16.6.46)–(16.6.55), in the problem of approximately diagonalizing the Dirac Hamiltonian with the Coulomb potential and in the light of the concept of "Zitterbewegung" of the electron. In this appendix, we give an exact treatment of the Coulomb potential problem in the Dirac equation. The exact expression derived for the bound energy spectrum, when expanded in powers of the fine-structure constant, agrees, as expected, with the corresponding one obtained in §7.4, §7.5 to the order kept there. As we have already seen in §7.6, §7.7, however, there are additional physical effects that contribute to the hydrogen spectrum. The retention of higher order terns, to arbitrary orders, in the expansion of the exact energy expression obtained below cannot then be justified as there are other competing contributions to the spectrum not accounted for by an analysis of the Dirac equation in this potential alone.

Upon writing the Dirac spinor Ψ as $\begin{pmatrix} \psi_1 & \psi_2 \end{pmatrix}^\top$, then the eigenvalue equation as obtained from the Hamiltonian H_1 in (16.6.46) gives the following two two-component coupled equations

$$c\boldsymbol{\sigma}\cdot\mathbf{p}\,\psi_1 + \left(U - Mc^2 - E\right)\psi_2 = 0 \qquad \text{(A-16.6.1)}$$

$$c\boldsymbol{\sigma}\cdot\mathbf{p}\,\psi_2 + \left(U + Mc^2 - E\right)\psi_1 = 0 \qquad \text{(A-16.6.2)}$$

where $U = e\phi$.

We use the identity

$$(\boldsymbol{\sigma}\cdot\mathbf{x})(\boldsymbol{\sigma}\cdot\mathbf{p}) = \mathbf{x}\cdot\mathbf{p} + i\boldsymbol{\sigma}\cdot(\mathbf{x}\times\mathbf{p}) \qquad \text{(A-16.6.3)}$$

which upon multiplying from the left by $\boldsymbol{\sigma}\cdot\mathbf{x}$ leads to

$$(\boldsymbol{\sigma}\cdot\mathbf{p}) = \boldsymbol{\sigma}\cdot\hat{\mathbf{x}}\left(\hat{\mathbf{x}}\cdot\mathbf{p} + \frac{i}{r}\boldsymbol{\sigma}\cdot\mathbf{L}\right) \qquad \text{(A-16.6.4)}$$

where $r = |\mathbf{x}|$, $\hat{\mathbf{x}}$ is a unit vector along $\mathbf{x}$.

The Hamiltonian H_1 commutes with $\mathbf{J}^2$, J_3, $\mathbf{L}^2$, and upon using

$$\mathbf{J} = \mathbf{L} + \frac{\hbar}{2}\boldsymbol{\Sigma} \qquad \text{(A-16.6.5)}$$

we may write

$$\boldsymbol{\Sigma}\cdot\mathbf{L} = \frac{1}{\hbar^2}\left(\mathbf{J}^2 - \mathbf{L}^2 - \frac{3}{4}\hbar^2\right) \qquad \text{(A-16.6.6)}$$

and also

$$\boldsymbol{\sigma}\cdot\mathbf{L} = \frac{1}{\hbar^2}\left(\mathbf{J}^2 - \mathbf{L}^2 - \frac{3}{4}\hbar^2\right) \qquad \text{(A-16.6.7)}$$

where it is understood that the right-hand sides of the latter two equations are multiplied by unit matrices.

In Appendix to §7.5, we have combined the orbital angular momentum and spin, and given the expressions for $\langle\theta,\phi|j,\ell = j \mp 1/2, m_j\rangle$ in (A-7.5.4), (A-7.5.5) as *two*-component objects. It is easily seen that

$$\boldsymbol{\sigma}\cdot\mathbf{L}\,\langle\theta,\phi|j,\ell = j \mp 1/2, m_j\rangle = [\pm\hbar\,(j+1/2) - \hbar]\,\langle\theta,\phi|j,\ell = j \mp 1/2, m_j\rangle \qquad \text{(A-16.6.8)}$$

or

$$(\boldsymbol{\sigma}\cdot\mathbf{L} + \hbar)\,\langle\theta,\phi|j,\ell = j \mp 1/2, m_j\rangle = \pm\hbar\,(j+1/2)\,\langle\theta,\phi|j,\ell = j \mp 1/2, m_j\rangle \qquad \text{(A-16.6.9)}$$

We also use the identity in (A-7.5.7),

$$\boldsymbol{\sigma}\cdot\hat{\mathbf{x}}\,\langle\theta,\phi|j,\ell = j \mp 1/2, m_j\rangle = -\,\langle\theta,\phi|j,\ell = j \pm 1/2, m_j\rangle \qquad \text{(A-16.6.10)}$$

The eigenvalue equation in (A-16.6.9) suggests to rewrite $\boldsymbol{\sigma}\cdot\mathbf{p}$ in (A-16.6.4) as

$$\boldsymbol{\sigma}\cdot\mathbf{p} = \boldsymbol{\sigma}\cdot\hat{\mathbf{x}}\left(p_r + \frac{\mathrm{i}}{r}(\boldsymbol{\sigma}\cdot\mathbf{L} + \hbar)\right) \qquad \text{(A-16.6.11)}$$

where p_r is defined in (7.2.2).

Clearly, if we write

$$\psi_1(\mathbf{x}) = (-\mathrm{i})F(r)\,\langle\theta,\phi|j,\ell = j \mp 1/2, m_j\rangle \qquad \text{(A-16.6.12)}$$

where the $(-\mathrm{i})$ factor is chosen for convenience, then

$$\boldsymbol{\sigma}\cdot\mathbf{p}\,\psi_1(\mathbf{x}) = \mathrm{i}\left[p_r \pm \frac{\mathrm{i}\hbar}{r}\left(j + \frac{1}{2}\right)\right]F(r)\,\langle\theta,\phi|j,\ell = j \pm 1/2, m_j\rangle \qquad \text{(A-16.6.13)}$$

where we have used (A-16.6.9), (A-16.6.10) and noted that $\boldsymbol{\sigma}\cdot\hat{\mathbf{x}}$ is independent of r.

Accordingly, we have succeeded in obtaining a separation of the variables r, (θ,ϕ) in (A-16.6.1), (A-16.6.2), by using in the process (A-16.6.13), by writing

$$\psi_2(\mathbf{x}) = G(r)\,\langle\theta,\phi|j,\ell = j \pm 1/2, m_j\rangle \qquad \text{(A-16.6.14)}$$

to obtain the radial equations

$$\left(\frac{1}{r}\frac{\mathrm{d}}{\mathrm{d}r}r - \frac{\kappa}{r}\right)F(r) + \frac{(U - E - Mc^2)}{\hbar c}G(r) = 0 \qquad \text{(A-16.6.15)}$$

$$\left(\frac{1}{r}\frac{\mathrm{d}}{\mathrm{d}r}r + \frac{\kappa}{r}\right)G(r) - \frac{(U - E + Mc^2)}{\hbar c}F(r) = 0 \qquad \text{(A-16.6.16)}$$

16.6 Physical Interpretation, Localization and Particle Content

where $\kappa = \mp(j + 1/2)$, and in obtaining (A-16.6.16), we have multiplied (A-16.6.2) by $(-i)$.

We apply the operator $\dfrac{1}{r}\dfrac{d}{dr}r$ to both of the equations in (A-16.6.15), (A-16.6.16), and rewrite the resulting equations in matrix form to obtain

$$\left[\frac{d^2}{dr^2} + \frac{2}{r}\frac{d}{dr}\right]\begin{pmatrix}F\\G\end{pmatrix} = \left[\frac{\kappa^2}{r^2} + \frac{M^2c^4 - (E-U)^2}{\hbar^2 c^2}\right]\begin{pmatrix}F\\G\end{pmatrix}$$

$$+ \frac{1}{r^2}\begin{pmatrix}-\kappa & -Z\alpha\\ Z\alpha & \kappa\end{pmatrix}\begin{pmatrix}F\\G\end{pmatrix}. \qquad \text{(A-16.6.17)}$$

Here $\alpha = e^2/\hbar c$ is the fine-structure constant, and we have used in turn (A-16.6.15), (A-16.6.16) all over again to obtain the final form in (A-16.6.17).

We set

$$\gamma = \sqrt{\kappa^2 - Z^2\alpha^2} \qquad \text{(A-16.6.18)}$$

and we may diagonalize the last matrix on the right-hand side of (A-16.6.17) via the matrix[26]

$$Q = \frac{1}{\sqrt{2\kappa(\kappa+\gamma)}}\begin{pmatrix}Z\alpha & -(\kappa+\gamma)\\ -(\kappa+\gamma) & Z\alpha\end{pmatrix} \qquad \text{(A-16.6.19)}$$

whose inverse is

$$Q^{-1} = -\frac{(\kappa/\gamma)}{\sqrt{2\kappa(\kappa+\gamma)}}\begin{pmatrix}Z\alpha & \kappa+\gamma\\ \kappa+\gamma & Z\alpha\end{pmatrix}. \qquad \text{(A-16.6.20)}$$

In reference to the last 2×2 matrix on the right-hand side of (A-16.6.17), the Q matrix gives

$$Q^{-1}\begin{pmatrix}-\kappa & -Z\alpha\\ Z\alpha & \kappa\end{pmatrix}Q = \begin{pmatrix}\gamma & 0\\ 0 & -\gamma\end{pmatrix} \qquad \text{(A-16.6.21)}$$

Hence upon setting

$$Q^{-1}\begin{pmatrix}F\\G\end{pmatrix} = \begin{pmatrix}u_+\\ u_-\end{pmatrix} \qquad \text{(A-16.6.22)}$$

$$\frac{\sqrt{M^2c^4 - E^2}}{\hbar c} = \lambda, \quad \rho = 2\lambda r, \qquad \text{(A-16.6.23)}$$

with $|E| < Mc^2$, and multiplying (A-16.6.17) by Q^{-1} and inserting the identity QQ^{-1} between $(F\ G)^{\mathsf{T}}$ and the last 2×2 matrix in the latter equation, we obtain

$$\left[\frac{d^2}{d\rho^2} + \frac{2}{\rho}\frac{d}{d\rho} + \left(-\frac{1}{4} + \frac{E}{\hbar c\lambda}\frac{Z\alpha}{\rho} - \frac{\gamma(\gamma\pm 1)}{\rho^2}\right)\right]u_\pm = 0 \qquad \text{(A-16.6.24)}$$

---
[26] We use a method of Goodman and Ignjatović (1997).

We carry out the substitutions

$$u_+ = \rho^\gamma \, e^{-\rho/2} \, y_+, \quad u_- = \rho^{\gamma-1} \, e^{-\rho/2} \, y_- \quad (A\text{-}16.6.25)$$

and set

$$n' = \frac{E}{\hbar c \lambda} Z\alpha - \gamma \quad (A\text{-}16.6.26)$$

to obtain from (A-16.6.24)

$$\rho \frac{d^2}{d\rho^2} y_+ + [2(\gamma+1) - \rho] \frac{d}{d\rho} y_+ - (1-n') y_+ = 0 \quad (A\text{-}16.6.27)$$

$$\rho \frac{d^2}{d\rho^2} y_- + [2\gamma - \rho] \frac{d}{d\rho} y_- + n' y_- = 0. \quad (A\text{-}16.6.28)$$

These equations are particular cases of Kummer's equation[27]

$$\rho \frac{d^2}{d\rho^2} w + (b - \rho) \frac{d}{d\rho} w - a w = 0 \quad (A\text{-}16.6.29)$$

where solution, regular at $\rho = 0$, is the confluent hypergeometric function

$$F(a, b; \rho) = 1 + \frac{a}{1!}\rho + \frac{a(a+1)}{b(b+1)} \frac{\rho^2}{2!} + \cdots \quad (A\text{-}16.6.30)$$

and has the asymptotic behavior

$$F(a, b; \rho) \sim \frac{\Gamma(b)}{\Gamma(a)} \, e^\rho \rho^{a-b} \left[1 + \mathcal{O}\left(\tfrac{1}{\rho}\right)\right] \quad (A\text{-}16.6.31)$$

for $\rho \to \infty$. It becomes a polynomial, however, for a a non-positive integer.

From (A-16.6.25), (A-16.6.27), (A-16.6.28) we may infer that

$$y_+ = c_+ \, F(1 - n', 2(\gamma + 1); \rho) \quad (A\text{-}16.6.32)$$

$$y_- = c_- \, F(-n', 2\gamma; \rho) \quad (A\text{-}16.6.33)$$

The asymptotic behavior in (A-16.6.31), in conjunction with the definitions of $u_\pm$ in (A-16.6.25), necessitate that $y_\pm$ are polynomials. For (A-16.6.32) this means that the possible values for n' are $1, 2, \ldots$, while for (A-16.6.33) that $n' = 0, 1, 2, \ldots$. For $n' = 0$, however, $F(1, 2(\gamma + 1); \rho)$ diverges as e^ρ for $\rho \to \infty$ and hence c_+ must be taken to be zero for $n' = 0$.

All told, we may solve for E from (A-16.6.26) giving

$$E = Mc^2 \left[1 + \frac{Z^2 \alpha^2}{(n' + \gamma)^2}\right]^{-1/2} \quad (A\text{-}16.6.34)$$

---

[27] Cf., Abramowitz and Stegun (1972), p. 504.

One may introduce the principal quantum number $n = n' + (j + 1/2)$, to finally rewrite (A-16.6.34) as

$$E = Mc^2 \left[1 + \frac{Z^2\alpha^2}{\left[n - (j+1/2) + \sqrt{(j+1/2)^2 - Z^2\alpha^2}\right]^2} \right]^{-1/2} \qquad \text{(A-16.6.35)}$$

where we have used the fact that $|\kappa| = j + 1/2$ (see below (A-16.6.16)). The reader may which to carry out an expansion of (A-16.6.35) in powers of $Z^2\alpha^2$ to make a comparison with the results in §7.5.

The radial functions $F(r)$, $G(r)$ may be now determined from (A-16.6.22), (A-16.6.25), (A-16.6.32), (A-16.6.33), with $c_+ = 0$ for $n' = 0$, and may be then property normalized.[28]

16.7 The Klein-Gordon Equation

In this section we consider the quantum description of relativistic spin 0 particles. As in the spin 1/2 case, we encounter the expected negative energy states. For bosons, however, an argument based on a completely filled sea of bosons of negative energies to ensure stability breaks down as an arbitrary number of bosons may be put in a given energy level. Accordingly transitions to such negative energy levels would be possible and the argument collapses. A quantum field theory treatment, however, overcomes such a difficulty. To some extent the analysis given here follows the one of the Dirac theory in the previous sections.

16.7.1 Setting Up Spin 0 Equations

Under a homogeneous Lorentz transformation $x \to x'$, as given in (16.2.6), a Lorentz scalar $\Phi(x)$, by definition, remains invariant, i.e.,

$$\Phi'(x') = \Phi(x). \qquad (16.7.1)$$

For infinitesimal transformation given in (16.2.39), this becomes

$$\Phi'(x^\sigma + \delta\omega^\sigma{}_\lambda x^\lambda) \simeq \Phi(x) \qquad (16.7.2)$$

or

$$\Phi'(x) \simeq \Phi(x^\sigma - \delta\omega^\sigma{}_\lambda x^\lambda) \simeq \Phi(x) + \frac{i}{2\hbar}\delta\omega^{\mu\nu} L_{\mu\nu}\Phi(x) \qquad (16.7.3)$$

where

---

[28] See Goodman and Ignjatović (1997).

$$L_{\mu\nu} = \left(x_\mu \frac{\hbar}{i}\partial_\nu - x_\nu \frac{\hbar}{i}\partial_\mu\right) \tag{16.7.4}$$

which upon comparison with (16.2.43), for example, one may infer the spin 0 content of a Lorentz scalar (also pseudo-scalar).

From the energy-momentum constraint in (16.1.6), followed by the so-called minimal substitution in (16.1.16) in the presence of an external electromagnetic field, the wave equation of a spin 0 follows and is given by[29]

$$\left[\left(p^\mu - \frac{e}{c}A^\mu\right)\left(p_\mu - \frac{e}{c}A_\mu\right) + M^2 c^2\right]\Phi(x) = 0 \tag{16.7.5}$$

and is referred to as the Klein-Gordon equation of a spin 0 charged particle. Here we identify p_μ with $\hbar \partial_\mu / i$.

From one's experience with the Dirac equation, the first thing that might come into one's mind, regarding the above equation, is how to rewrite it in first order in space and time derivatives. This is easily achieved.

To the above end, set

$$\frac{1}{Mc}\left(p^\mu - \frac{e}{c}A^\mu\right)\Phi = \frac{1}{i}\chi^\mu \tag{16.7.6}$$

from which, we may rewrite (16.7.5) as

$$\frac{1}{i}\left(p_\mu - \frac{e}{c}A_\mu\right)\chi^\mu + Mc\Phi = 0 \tag{16.7.7}$$

We may recast this equation elegantly in a form similar to the Dirac equation in (16.1.18) as

$$\left[\beta^\mu\left(p_\mu - \frac{e}{c}A_\mu\right) + Mc\right]\Xi(x) = 0 \tag{16.7.8}$$

where

$$\Xi = \begin{pmatrix} \chi^0 \\ \chi^1 \\ \chi^2 \\ \chi^3 \\ \Phi \end{pmatrix} \tag{16.7.9}$$

and the 5×5 matrices β^μ are pure imaginary and are given by

$$\beta^0 = i\begin{pmatrix} 0 & 0 & 0 & 0 & 1 \\ 0 & 0 & 0 & 0 & 0 \\ 0 & 0 & 0 & 0 & 0 \\ 0 & 0 & 0 & 0 & 0 \\ -1 & 0 & 0 & 0 & 0 \end{pmatrix}, \quad \beta^1 = i\begin{pmatrix} 0 & 0 & 0 & 0 & 0 \\ 0 & 0 & 0 & 0 & -1 \\ 0 & 0 & 0 & 0 & 0 \\ 0 & 0 & 0 & 0 & 0 \\ 0 & -1 & 0 & 0 & 0 \end{pmatrix} \tag{16.7.10}$$

---

[29] Recall the definition of our metric in (16.1.20), and the property of raising or lowering a four-vector index 'μ' as given in (16.2.4). Note also we have here divided the energy-momentum constraint equation in (16.1.6) by c^2.

$$\beta^2 = i \begin{pmatrix} 0 & 0 & 0 & 0 & 0 \\ 0 & 0 & 0 & 0 & 0 \\ 0 & 0 & 0 & 0 & -1 \\ 0 & 0 & 0 & 0 & 0 \\ 0 & 0 & -1 & 0 & 0 \end{pmatrix}, \quad \beta^3 = i \begin{pmatrix} 0 & 0 & 0 & 0 & 0 \\ 0 & 0 & 0 & 0 & 0 \\ 0 & 0 & 0 & 0 & 0 \\ 0 & 0 & 0 & 0 & -1 \\ 0 & 0 & 0 & -1 & 0 \end{pmatrix} \quad (16.7.11)$$

as is easily verified. It is understood that Mc in (16.7.8) is multiplied by the identity element.

The wave equation in (16.7.8) is referred to as the *Duffin-Kemmer-Petiau* equation. It is also readily shown, that the matrices β^μ satisfy the relations

$$\beta^\mu \beta^\nu \beta^\rho + \beta^\rho \beta^\nu \beta^\mu = -\left(\beta^\mu g^{\nu\rho} + \beta^\rho g^{\nu\mu}\right). \quad (16.7.12)$$

The matrix β^0, in particular, has no inverse. This is problematic in recasting (16.7.8) in the form of the Dirac equation given in (16.4.47) with the time derivative simply appearing on its left-hand side with no β^0 multiplying it. This is due to the fact of the redundancy of some of the components in (16.7.9) as seen through the definition in (16.7.6).

We may also rewrite the Klein-Gordon equation in another form which is of first order in the time derivative where no (singular) matrix, such as β^0 in (16.7.8), multiplies it on its left-hand side, and readily allows the determination of the Hamiltonian theory. This, however, turns out to sacrifice the relativistic appearance of the resulting equation. It is a two-component equation and is usually known as the *Feshbach-Villars*[30] equation which we study next.

To obtain the above mentioned equation, we rewrite (16.7.5) as

$$\left(1 + \frac{\pi^0}{Mc}\right)\left(1 - \frac{\pi^0}{Mc}\right)\Phi + \frac{\boldsymbol{\pi} \cdot \boldsymbol{\pi}}{M^2 c^2}\Phi = 0 \quad (16.7.13)$$

where

$$\pi^\mu = \left(p^\mu - \frac{e}{c}A^\mu\right) \quad (16.7.14)$$

This suggests to set

$$\left(1 + \frac{\pi^0}{Mc}\right)\Phi = \psi_1, \quad \left(1 - \frac{\pi^0}{Mc}\right)\Phi = \psi_2 \quad (16.7.15)$$

obtaining the following two equations from (16.7.13),

$$\frac{\pi^0}{Mc}\psi_1 = \psi_1 + \frac{\boldsymbol{\pi} \cdot \boldsymbol{\pi}}{M^2 c^2}(\psi_1 + \psi_2) \quad (16.7.16)$$

$$\frac{\pi^0}{Mc}\psi_2 = -\psi_2 - \frac{\boldsymbol{\pi} \cdot \boldsymbol{\pi}}{M^2 c^2}(\psi_1 + \psi_2) \quad (16.7.17)$$

By using the properties of the Pauli matrices in (2.8.2), and setting

---
[30] Feshbach and Villars (1958).

940   16 Quantum Description of Relativistic Particles

$$\psi = \begin{pmatrix} \psi_1 \\ \psi_2 \end{pmatrix}, \tag{16.7.18}$$

we may combine (16.7.16), (16.7.17) into the two-component equation

$$i\hbar \frac{\partial}{\partial t}\psi = \left(eA^0 + Mc^2\sigma_3\right)\psi + (\sigma_3 + i\sigma_2)\frac{\pi^2}{2M}\psi \tag{16.7.19}$$

which looks like a non-relativistic equation.

By taking the complex conjugate of the (16.7.19) and multiplying it from the left by $-\sigma_1$, we get

$$i\hbar \frac{\partial}{\partial t}\begin{pmatrix} \psi_2^* \\ \psi_1^* \end{pmatrix} = \left(-eA^0 + Mc^2\sigma_3\right)\begin{pmatrix} \psi_2^* \\ \psi_1^* \end{pmatrix} + (\sigma_3 + i\sigma_2)\frac{\left(\mathbf{p} + \frac{e}{c}\mathbf{A}\right)^2}{2M}\begin{pmatrix} \psi_2^* \\ \psi_1^* \end{pmatrix} \tag{16.7.20}$$

thus obtaining the charge conjugate wavefunction

$$\psi_c = \begin{pmatrix} \psi_2^* \\ \psi_1^* \end{pmatrix} \tag{16.7.21}$$

That is, given an external field A_μ, ψ_c satisfies the same equation in (16.7.19) with the charge $e \to -e$.

16.7.2 A Continuity Equation

We obtain a continuity equation in analogy to the Dirac case in (16.6.4). We will see below that the definition of the adjoint operation is to be modified here and is to be taken as

$$C_{\text{adj}} = \sigma_3 \left(C^*\right)^\top \sigma_3 \tag{16.7.22}$$

where $(C^*)^\top$ is the familiar one.

Multiplying (16.7.19) from the left by $(\Psi^*)^\top \sigma_3$, and by following, in the process, a procedure similar to the one in deriving (16.6.4), we obtain

$$\frac{\partial}{\partial t}\left[\psi^{*\top}\sigma_3\psi\right] + \boldsymbol{\nabla}\cdot\mathbf{J} = 0 \tag{16.7.23}$$

where

$$\mathbf{J} = \frac{\hbar}{2iM}\left(\psi^{*\top}(\sigma_3 + i\sigma_2)\boldsymbol{\nabla}\psi - \left(\boldsymbol{\nabla}\psi^{*\top}\right)(\sigma_3 + i\sigma_2)\psi\right)$$

$$- \frac{e}{Mc}\mathbf{A}\psi^{*\top}(\sigma_3 + i\sigma_2)\psi \tag{16.7.24}$$

The quantity $\psi^{*\top}\sigma_3\psi$ is given by

$$\psi^{*\top}\sigma_3\psi = |\psi_1|^2 - |\psi_2|^2 \tag{16.7.25}$$

and is the difference between two positive definite densities. Because of this property, $\psi^{*\top}\sigma_3\psi$ is interpreted as a charge density when multiplied by e, rather than a probability density and is consistent with a theory describing particles of both signs of the charge.

In reference to (16.7.25), we note that if we adopt the normalization

$$\int d^3x\, \psi^{*\top}\sigma_3\psi = q, \qquad (16.7.26)$$

where q is some constant, then this implies that for the charge conjugate ψ_c in (16.7.12)

$$\int d^3x\, \psi_c^{*\top}\sigma_3\psi_c = -q \qquad (16.7.27)$$

as expected.

The emergence of $\psi^{*\top}\sigma_3\psi$ rather than of the more familiar expression $\psi^{*\top}\psi$ has led to define expectation value of an operator O as[31]

$$\langle O \rangle = \int d^3x\, \psi^{*\top}\sigma_3 O\psi \qquad (16.7.28)$$

and define the adjoint of O as in (16.7.22). This is important because we may infer from the expression of the Hamiltonian

$$H = \left(eA^0 + Mc^2\sigma_3\right) + (\sigma_3 + i\sigma_2)\frac{\pi^2}{2M}, \qquad (16.7.29)$$

as obtained from (16.7.19), that it is Hermitian with the definition given in (16.7.22), where we note that

$$\sigma_3(i\sigma_2)^{*\top}\sigma_3 = i\sigma_2 \qquad (16.7.30)$$

16.7.3 General Solution of the Free Feshbach-Villars Equation

Following the procedure developed for the Dirac equation in §16.4, we write the solution of (16.7.19) for $A_\mu = 0$ as

$$\psi(x) = \int 2Mc\, d\omega_p \left[e^{ipx/\hbar} u(\mathbf{p})a(\mathbf{p}) + e^{-ipx/\hbar} v(\mathbf{p})b^*(\mathbf{p})\right] \qquad (16.7.31)$$

where $u(\mathbf{p})$ is the positive-energy solution satisfying

$$\begin{pmatrix} \left[p^0 - Mc - \frac{\mathbf{p}^2}{2Mc}\right] & -\frac{\mathbf{p}^2}{2Mc} \\ \frac{\mathbf{p}^2}{2Mc} & \left[p^0 + Mc + \frac{\mathbf{p}^2}{2Mc}\right] \end{pmatrix} u(\mathbf{p}) = 0 \qquad (16.7.32)$$

---

[31] For additional details on the definition of the expectation value see Feshbach and Villars (1958).

with $p^0 = +\sqrt{\mathbf{p}^2 + M^2c^2}$. The solution of (16.7.32) is elementary and with the normalization condition

$$u^{*\,\mathrm{T}} \sigma_3 u = \frac{p^0}{Mc} \qquad (16.7.33)$$

(see (16.7.25)), is given by

$$u(\mathbf{p}) = \frac{1}{2Mc} \begin{pmatrix} Mc + p^0 \\ Mc - p^0 \end{pmatrix} \qquad (16.7.34)$$

Similarly, for $v(\mathbf{p})$ we have from (16.7.32) with $p^0 \to -p^0$, the solution

$$v(\mathbf{p}) = \frac{1}{2Mc} \begin{pmatrix} Mc - p^0 \\ Mc + p^0 \end{pmatrix} \qquad (16.7.35)$$

where $p^0 = +\sqrt{\mathbf{p}^2 + M^2c^2}$, with the normalization condition

$$v^{*\,\mathrm{T}} \sigma_3 v = -\frac{p^0}{Mc} \qquad (16.7.36)$$

and

$$u^{*\,\mathrm{T}} \sigma_3 v = 0 \qquad (16.7.37)$$

The normalization condition

$$\int \mathrm{d}^3\mathbf{x}\, \psi^{*\,\mathrm{T}}(x)\, \sigma_3\, \psi(x) = q \qquad (16.7.38)$$

for a given constant C, then leads to the constraint

$$\int 2Mc\,\mathrm{d}\omega_p\, \left[|a(\mathbf{p})|^2 - |b(\mathbf{p})|^2\right] = q \qquad (16.7.39)$$

on the expansion coefficients a, b^* in (16.7.31). The integral in (16.7.39) is, in general, not positive definite. This has led to interpret e times the integrand as a charge density as mentioned before.

16.7.4 Diagonalization of the Hamiltonian and Definition of Position Operators

The Hamiltonian

$$H = Mc^2 \sigma_3 + (\sigma_3 + i\sigma_2) \frac{\mathbf{p}^2}{2M} \qquad (16.7.40)$$

for the free system may be diagonalized by the transformation matrix

$$G = \frac{1}{2\sqrt{Mcp^0}} \left[p^0 + Mc + (p^0 - Mc)\sigma_1\right] \qquad (16.7.41)$$

16.7 The Klein-Gordon Equation

whose inverse is

$$G^{-1} = \frac{1}{2\sqrt{Mcp^0}} \left[p^0 + Mc - (p^0 - Mc)\sigma_1 \right] \qquad (16.7.42)$$

The transformed Hamiltonian works out to be

$$H' = GHG^{-1}$$
$$= \sigma_3 p^0 c \qquad (16.7.43)$$

and the position variable $\mathbf{x}$ becomes

$$\mathbf{x}' = G\mathbf{x}G^{-1}$$
$$= \mathbf{x} - \frac{i\hbar}{2p^{0^2}} \sigma_1 \mathbf{p}$$
$$\equiv \mathbf{x} + \mathbf{R} \qquad (16.7.44)$$

The first term $\mathbf{x}$ in the new representation develops in time as

$$\mathbf{x}(t) = e^{itH'/\hbar} \mathbf{x} e^{-itH'/\hbar}$$
$$= \mathbf{x} + t \frac{\mathbf{p}}{p^{0^2}} H' \qquad (16.7.45)$$

with associated velocity

$$\frac{d\mathbf{x}(t)}{dt} = \frac{\mathbf{p}}{p^{0^2}} H' \qquad (16.7.46)$$

and is in close analogy to the classical concept of velocity.

As in the Dirac case, $\mathbf{x}' \equiv \mathbf{x}'_{op}$, however, defines a non-local operator, and for a given well behaved function $\psi'(x)$,

$$\mathbf{x}'_{op} \psi'(x) = \int d^3 x' \, \kappa(\mathbf{x}, \mathbf{x}') \, \psi'(x^0, \mathbf{x}') \qquad (16.7.47)$$

where

$$\kappa(\mathbf{x}, \mathbf{x}') = \mathbf{x}\, \delta^3(\mathbf{x} - \mathbf{x}') - \frac{\sigma_1 \hbar^2}{2} \nabla \int \frac{d^3 p}{(2\pi\hbar)^3} \frac{e^{i\mathbf{p}\cdot(\mathbf{x}-\mathbf{x}')/\hbar}}{[\mathbf{p}^2 + M^2 c^2]}$$

$$= \mathbf{x}\, \delta^3(\mathbf{x} - \mathbf{x}') - \frac{\sigma_1}{2} \nabla \left[\frac{e^{-Mc|\mathbf{x}-\mathbf{x}'|/\hbar}}{4\pi |\mathbf{x} - \mathbf{x}'|} \right]. \qquad (16.7.48)$$

This shows a clear non-locality, associated with the position operator $\mathbf{x}'$, with a spread roughly of the order of the Compton wavelength $\hbar/Mc$ of the particle.

To investigate the time development of the operator $\mathbf{x}'$, we note that the time evolution operator in the new representation is given explicitly by

$$e^{-itH'/\hbar} = \cos\left(\frac{p^0 ct}{\hbar}\right) - i\sigma_3 \sin\left(\frac{p^0 ct}{\hbar}\right) \quad (16.7.49)$$

This leads to

$$\mathbf{x}'(t) = e^{itH'/\hbar} \mathbf{x}' e^{-itH'/\hbar}$$

$$= \mathbf{x} + t\frac{\mathbf{P}}{p^{0\,2}} H' - \frac{i\hbar \mathbf{p}\sigma_1}{2p^{0\,2}}\left[\cos\left(\frac{2p^0 ct}{\hbar}\right) - i\sigma_3 \sin\left(\frac{2p^0 ct}{\hbar}\right)\right] \quad (16.7.50)$$

and for the velocity associated with the deviation $\mathbf{R}(t)$ in (16.7.44), we have

$$\frac{d\mathbf{R}(t)}{dt} = \frac{ic\mathbf{p}\sigma_1}{p^0}\left[\sin\left(\frac{2p^0 ct}{\hbar}\right) + i\sigma_3 \cos\left(\frac{2p^0 ct}{\hbar}\right)\right] \quad (16.7.51)$$

As in the Dirac theory, $\mathbf{R}(t)$ has a complex "jittery" motion, the "Zitterbewegung", with high frequencies $2p^0 c/\hbar \geqslant 2Mc^2/\hbar$, where now M is the mass of the spin 0 particle. In analogy to the spin 1/2 case, the position operator $\mathbf{x}(t)$, in (16.7.45), is referred as a "mean" position operator with associated velocity which is uniform and is closely analogous to the classical concept of velocity, with $\mathbf{x}'(t)$ executing a complex motion about it. A similar analysis for corresponding position operators may be also carried in the original representation with the Hamiltonian given in (16.7.40) and is left as an exercise to the reader (see Problem 16.20).

16.7.5 The External Field Problem

We consider the Hamiltonian in (16.7.29) in the presence of an external time-independent weak electromagnetic field $A_\mu(\mathbf{x})$, and rewrite first H as

$$H = Mc^2 \left\{ \sigma_3 \left[1 + \frac{\pi^2}{2M^2 c^2}\right] + \frac{U}{Mc^2} + (i\sigma_2)\frac{\pi^2}{2M^2 c^2} \right\}$$

$$\equiv Mc^2 \widetilde{H} \quad (16.7.52)$$

where $U = eA^0 = -eA_0$.

We diagonalize the Hamiltonian H by considering relativistic corrections in it up to the order $1/c^4$, and hence up to $1/c^6$ in $\widetilde{H}$.

The only non-diagonal term in (16.7.52) is the one involving σ_2. The other Pauli matrix which is non-diagonal is σ_1. To diagonalize H, we use a key commutation relation given below in (16.7.54).

Consider the non-diagonal 2×2 matrix $\widetilde{O}$ of the form

16.7 The Klein-Gordon Equation

$$\widetilde{O} = i\sigma_2 a + b\sigma_1 \tag{16.7.53}$$

where a and b are any numbers, then from the anti-commutativity of σ_3 with σ_1, σ_2, we have

$$[\frac{1}{2}\sigma_3\widetilde{O}, \sigma_3] = -\widetilde{O} \tag{16.7.54}$$

where note the minus sign on the right-hand side of this equality.

The commutation relation in (16.7.54) suggests to choose, as a first step of the diagonalization process, the transformation

$$G_1 = \exp\left(\frac{\sigma_3}{2c^2}\widetilde{O}_1\right) \tag{16.7.55}$$

where

$$\frac{1}{2c^2}\sigma_3\widetilde{O}_1 = \frac{1}{2}\sigma_3\left(i\sigma_2\frac{\pi^2}{2M^2c^2}\right)$$

$$= \frac{\sigma_1}{4M^2c^2}\pi^2 \equiv \frac{1}{c^2}D_1 \tag{16.7.56}$$

From the expansion

$$\widetilde{H}' = G_1\widetilde{H}G_1^{-1} \simeq \widetilde{H} + \frac{1}{c^2}[D_1, H] + \frac{1}{2!c^4}\Big[D_1, [D_1, H]\Big]$$

$$+ \frac{1}{3!c^6}\bigg[D_1, \Big[D_1, [D_1, H]\Big]\bigg] \tag{16.7.57}$$

valid up to $1/c^6$, we obtain the expression

$$\widetilde{H}' \simeq \sigma_3\left[1 + \frac{\pi^2}{2M^2c^2} - \frac{\pi^4}{8M^4c^4} + \frac{\pi^6}{16M^6c^6}\right] + \frac{U}{Mc^2} + \frac{[\pi^2, [\pi^2, U]]}{32M^5c^6}$$

$$+ \frac{1}{c^4}\widetilde{O}_2 \tag{16.7.58}$$

where

$$\frac{1}{c^4}\widetilde{O}_2 = -i\sigma_2\left[\frac{\pi^4}{4M^4c^4} - \frac{\pi^6}{24M^6c^6}\right] + \sigma_1\frac{[\pi^2, U]}{4M^3c^4} \tag{16.7.59}$$

The matrix $\widetilde{O}_2$ is the only non-diagonal one in (16.7.58) and has the structure given in (16.7.53).

To cancel the $\widetilde{O}_2/c^4$ term in (16.7.58), we introduce the transformation $G_2 = \exp\left(D_2/c^4\right)$ such that the commutator of D_2/c^4 with the σ_3 term with coefficient one in (16.7.58) cancels the term in question. This is clearly given by (see (16.7.54))

$$G_2 \simeq 1 + \frac{\sigma_3}{2}\frac{1}{c^4}\widetilde{O}_2$$

$$= 1 - \sigma_1 \frac{\pi^4}{8M^4c^4} + i\sigma_2 \frac{[\pi^2, U]}{8M^3c^4} + \sigma_1 \frac{\pi^6}{48M^6c^6} \quad (16.7.60)$$

The above transformation gives

$$\widetilde{H}'' = G_2 \widetilde{H}' G_2^{-1}$$

$$\simeq \sigma_3 \left[1 + \frac{\pi^2}{2M^2c^2} - \frac{\pi^4}{8M^4c^4} + \frac{\pi^6}{16M^6c^6}\right] + \frac{U}{Mc^2} + \frac{\left[\pi^2, [\pi^2, U]\right]}{32M^5c^6}$$

$$+ \frac{1}{c^6}\widetilde{O}_3 \quad (16.7.61)$$

where $\widetilde{O}_3$ is not-diagonal in the form given in (16.7.53). Its explicit expression, however, is not important since by choosing a transformation $G_3 \simeq 1 + \sigma_3 \widetilde{O}_3/2c^6$ we may cancel out the $\widetilde{O}_3/c^6$ term in (16.7.61) by applying to (16.7.54) one more time. Here it is important to note that the only term that $\sigma_3 \widetilde{O}_3/2c^6$ can operate on is the first term involving σ_3 with coefficient one in (16.7.61). All the other terms would, otherwise, lead to corrections of the order $1/c^8$ or higher.

All told, the diagonalized Hamiltonian corresponding to the one in (16.7.52), up to the order $1/c^4$, is given by

$$H'' = \sigma_3 \left[Mc^2 + \frac{\pi^2}{2M} - \frac{\pi^4}{8M^3c^2} + \frac{\pi^6}{16M^5c^4}\right] + U$$

$$+ \frac{\left[\pi^2, [\pi^2, U]\right]}{32M^4c^4} \quad (16.7.62)$$

For a Hamiltonian with an electric field only which is weak, (16.7.62) simplifies to

$$H'' = \sigma_3 \left[Mc^2 + \frac{\mathbf{p}^2}{2M} - \frac{\mathbf{p}^4}{8M^3c^2} + \frac{\mathbf{p}^6}{16M^5c^4}\right] + U$$

$$+ \frac{\left[\mathbf{p}^2, [\mathbf{p}^2, U]\right]}{32M^4c^4} \quad (16.7.63)$$

Up to the order $1/c^2$, corrections arise only in the kinetic energy term, and the next order correction is $1/c^4$ rather than $1/c^3$. On the other hand for a weak magnetic field a correction of the order $1/c^3$, linear in $\mathbf{A}$, comes from the $-\pi^4/8M^3c^2$ term within the square brackets multiplying σ_3 in (16.7.62).

This concludes our treatment of the spin 0 case. As mentioned in the introduction to this section, an argument based on a filled sea of bosons of negative energies to ensure stability breaks down. The quantum field theory treatment, however, overcomes such a difficulty. The spin 0 case with equation in (16.7.8) will be considered again in §16.9 in discussing the spin and statistics connection.

In the next section, we develop relativistic wave equations for arbitrary spins.

16.8 Relativistic Wave Equations for Any Mass and Any Spin

We generalize the Dirac equation for $M \geqslant 0$ to any spin $s = 1, 3/2, 2, \ldots$, with $s = 0$ studied in the previous section and spin $s = 1/2$ studied in §16.4, §16.5. We consider in turn the cases $M > 0$ and $M = 0$.

16.8.1 $M > 0$:

We introduce a totally *symmetric* spinor $\Psi^{\alpha_1\cdots\alpha_k}$, with $k = 2s$, where each of the indices α_j goes over from 1 to 4, satisfying the equations

$$\left(\frac{\gamma\partial}{i} + \frac{Mc}{\hbar}\right)^{\alpha_1\beta_1} \Psi^{\beta_1\alpha_2\cdots\alpha_k}(x) = 0$$

$$\vdots \qquad (16.8.1)$$

$$\left(\frac{\gamma\partial}{i} + \frac{Mc}{\hbar}\right)^{\alpha_k\beta_k} \Psi^{\alpha_1\cdots\alpha_{k-1}\beta_k}(x) = 0$$

That is, $\Psi^{\alpha_1\cdots\alpha_k}(x)$ satisfies a Dirac equation with respect to each of its k indices. From the complete symmetry of $\Psi^{\alpha_1\cdots\alpha_k}$ in the α_i's, an equation in (16.8.1) satisfied with respect to one of its indices implies the validity of the equations with respect to the other indices by permutations. The set of equations in (16.8.1) is called the *Bargmann-Wigner* set of equations.

Lorentz covariance of the above equations implies, by invoking, in the process, Pauli's Fundamental Theorem as in §16.1, that there exists a non-singular matrix, say, G such that under homogeneous Lorentz transformations $\Psi^{\beta_1\cdots\beta_k}(x) \to \Psi'^{\alpha_1\cdots\alpha_k}(x')$, with

$$\Psi'^{\alpha_1\cdots\alpha_k}(x') = (G)^{\alpha_1\beta_1} \cdots (G)^{\alpha_k\beta_k} \Psi^{\beta_1\cdots\beta_k}(x) \qquad (16.8.2)$$

Fourier transforming $\Psi^{\alpha_1\cdots\alpha_k}(x)$, as in (16.4.2), and multiplying the resulting equation corresponding to the first one in (16.8.1) by $(-\gamma p + Mc)^{\alpha_1'\alpha_1}$, or the second one by $(-\gamma p + Mc)^{\alpha_2'\alpha_2}$, and so on, we get

$$\left(p^2 + M^2 c^2\right) \Psi^{\alpha_1 \dots \alpha_k}(p) = 0 \qquad (16.8.3)$$

Therefore repeating the analysis given through (16.4.5)–(16.4.14), we have

$$\Psi^{\alpha_1 \dots \alpha_k}(x) = \int 2Mc \, d\omega_p \left[e^{ipx/\hbar} \Phi_+{}^{\alpha_1 \dots \alpha_k}(\mathbf{p}) + e^{-ipx/\hbar} \Phi_-{}^{\alpha_1 \dots \alpha_k}(\mathbf{p}) \right] \qquad (16.8.4)$$

with $p^0 = +\sqrt{\mathbf{p}^2 + M^2 c^2}$,

$$(\gamma p + Mc)^{\alpha_1 \beta_1} \Phi_+{}^{\beta_1 \alpha_2 \dots \alpha_k}(\mathbf{p}) = 0$$
$$\vdots \qquad (16.8.5)$$
$$(\gamma p + Mc)^{\alpha_k \beta_k} \Phi_+{}^{\alpha_1 \dots \alpha_{k-1} \beta_k}(\mathbf{p}) = 0$$

and

$$(-\gamma p + Mc)^{\alpha_1 \beta_1} \Phi_-{}^{\beta_1 \alpha_2 \dots \alpha_k}(\mathbf{p}) = 0$$
$$\vdots \qquad (16.8.6)$$
$$(-\gamma p + Mc)^{\alpha_k \beta_k} \Phi_-{}^{\alpha_1 \dots \alpha_{k-1} \beta_k}(\mathbf{p}) = 0$$

Due to the symmetry of $\Phi_+{}^{\alpha_1 \dots \alpha_k}(\mathbf{p})$ in the indices $\alpha_1, \dots, \alpha_k$, it may be written as a symmetrized sum of the positive energy Dirac spinors $u(\mathbf{p}, \sigma)$ in (16.4.56) to satisfy the set of equations in (16.8.5). We may then write

$$\Phi_+{}^{\alpha_1 \dots \alpha_k}(\mathbf{p}, \sigma) = \sum_{\sigma_1, \dots, \sigma_k = \pm 1} C(\sigma_1, \dots, \sigma_k; \sigma) \left[u^{\alpha_1}(\mathbf{p}, \sigma_1) \cdots u^{\alpha_k}(\mathbf{p}, \sigma_k) \right] a(\mathbf{p}, \sigma)$$

$$\equiv u^{\alpha_1 \dots \alpha_k}(\mathbf{p}, \sigma) a(\mathbf{p}, \sigma) \qquad (16.8.7)$$

with the ξ_σ taken to be the $\xi_{\pm \mathbf{N}}$ given in (16.4.40). Here $C(\sigma_1, \dots, \sigma_k; \sigma)$ is a symmetrization operation over the indices $\sigma_1, \dots, \sigma_k$ thus leading to a symmetric $\Phi_+{}^{\alpha_1 \dots \alpha_k}$ in $(\alpha_1, \dots, \alpha_k)$, and the σ_j take on the values ± 1 corresponding to $\xi_{\pm \mathbf{N}}$. In (16.8.7), σ denotes the number of $\xi_{+\mathbf{N}}$ occurring in $[u^{\alpha_1}(\mathbf{p}, \sigma_1) \cdots u^{\alpha_k}(\mathbf{p}, \sigma_k)]$, i.e., to the number of $\sigma_1, \dots, \sigma_k$ taking the value $+1$. The $a(\mathbf{p}, \sigma)$ are arbitrary (one component) coefficients. Clearly, σ takes on $(k+1)$ values.

An explicit expansion for the symmetrization operation $C(\sigma_1, \dots, \sigma_k; \sigma)$ may be given, in the spirit of the one given in (2.8.18), by

$$C(\sigma_1, \dots, \sigma_k; \sigma) = \frac{1}{\sqrt{k! \sigma! (k-\sigma)!}} \left[\frac{\partial}{\partial g_{\sigma_1}} \cdots \frac{\partial}{\partial g_{\sigma_k}} (g_{+1})^\sigma (g_{-1})^{k-\sigma} \right] \qquad (16.8.8)$$

where g_+, g_- are two independent variables. We note that for any given fixed value assigned to σ, $C(\sigma_1, \dots, \sigma_k; \sigma)$, after carrying out the differentiations with respect to $g_{\sigma_1}, \dots, g_{\sigma_k}$, with the simple rule

$$\frac{\partial}{\partial g_{\sigma_j}} (g_{\pm 1}) = \delta_{\sigma_j, \pm 1}, \qquad (16.8.9)$$

16.8 Relativistic Wave Equations for Any Mass and Any Spin

is independent of g_+, g_-. For example, for $k = 2$,

$$C(\sigma_1, \sigma_2; 0) = \delta_{\sigma_1, -1} \delta_{\sigma_2, -1} \tag{16.8.10}$$

$$C(\sigma_1, \sigma_2; 1) = \frac{1}{\sqrt{2}} \left(\delta_{\sigma_1, +1} \delta_{\sigma_2, -1} + \delta_{\sigma_1, -1} \delta_{\sigma_2, +1} \right) \tag{16.8.11}$$

$$C(\sigma_1, \sigma_2; 2) = \delta_{\sigma_1, +1} \delta_{\sigma_2, +1} \tag{16.8.12}$$

(see also Problem 16.21).

The multiplicative nature of the transformation rule via the product $(G)^{\alpha_1 \beta_1} \cdots (G)^{\alpha_k \beta_k}$ in (16.8.2), implies, from (16.2.29), that the spin operator has the additive nature

$$\mathbf{S}^{\alpha_1 \beta_1, \ldots, \alpha_k \beta_k} = (\mathbf{S})^{\alpha_1 \beta_1} \delta^{\alpha_2 \beta_2} \cdots \delta^{\alpha_k \beta_k} + \cdots + \delta^{\alpha_1 \beta_1} \cdots \delta^{\alpha_{k-1} \beta_{k-1}} (\mathbf{S})^{\alpha_k \beta_k} \tag{16.8.13}$$

where $(\mathbf{S})^{\alpha \beta} = \dfrac{\hbar}{2} (\mathbf{\Sigma})^{\alpha \beta}$ is defined in (16.2.46).

With $\mathbf{N} = \mathbf{p}/|\mathbf{p}|$, in the $\mathbf{p}$-description, we may rewrite (16.4.58), in the present notation, as

$$\mathbf{\Sigma} \cdot \mathbf{N}\, u(\mathbf{p}, \pm 1) = \pm u(\mathbf{p}, \pm 1) \tag{16.8.14}$$

Accordingly, from (16.8.7), (16.8.13), (16.8.14), we have

$$(\mathbf{S} \cdot \mathbf{N})^{\alpha_1 \beta_1, \ldots, \alpha_k \beta_k} \Phi_+^{\beta_1 \ldots \beta_k}(\mathbf{p}, \sigma) = \frac{\hbar}{2} [\sigma - (k - \sigma)] \Phi_+^{\alpha_1 \ldots \alpha_k}(\mathbf{p}, \sigma) \tag{16.8.15}$$

where $(k - \sigma)$ denotes the number of $\sigma_1, \ldots, \sigma_k$ taking the value -1. At this stage, we write $k = 2s$, and set

$$\sigma = m_s + s \tag{16.8.16}$$

Since $\sigma = 0, 1, 2, \ldots, 2s$, we may infer that

$$m_s = -s, -s+1, \ldots, s-1, s \tag{16.8.17}$$

and obtain from (16.8.15),

$$(\mathbf{S} \cdot \mathbf{N})^{\alpha_1 \beta_1, \ldots, \alpha_{2s} \beta_{2s}} \Phi_+^{\beta_1 \ldots \beta_{2s}}(\mathbf{p}, m_s + s) = \hbar m_s \Phi_+^{\alpha_1 \ldots \alpha_{2s}}(\mathbf{p}, m_s + s) \tag{16.8.18}$$

thus establishing the spin s character of $\Phi_+^{\alpha_1 \ldots \alpha_{2s}}(\mathbf{p}, m_s + s)$. Needless to say, the same conclusion is reached if we go to the rest frame of the particle, choose the quantization along, say, the z-axis, and apply the spin component S_3 to $\Phi_+^{\alpha_1 \ldots \alpha_{2s}}(\mathbf{0}, m_s + s)$.

Similarly, we may carry out the expansion

$$\Phi_-^{\alpha_1 \ldots \alpha_k}(\mathbf{p}, \sigma) = \sum_{\sigma_1, \ldots, \sigma_k = \pm 1} C(\sigma_1, \ldots \sigma_k; \sigma) [v(\mathbf{p}, \sigma_1) \ldots v(\mathbf{p}, \sigma_k)] b^*(\mathbf{p}, \sigma) \tag{16.8.19}$$

in terms of the charge conjugate spinors $v(\mathbf{p}, \sigma)$ in (16.4.57), and repeat the analysis given above now applied to $\Phi_-^{\alpha_1 \ldots \alpha_{2s}}(\mathbf{p}, \sigma)$. The relevant details are left as an exercise to the reader.

16.8.2 $M = 0$:

For $M = 0$, we work in the chiral representation and with the measure $2Mc\,d\omega_p$ in (16.8.4) simply now replaced by $d\omega_p$. The equations (16.8.5), (16.8.6) take the form

$$(\gamma'p)^{\alpha_1\beta_1}\,\Phi_{\pm 0}^{\beta_1\alpha_2\ldots\alpha_k}(\mathbf{p}) = 0$$
$$\vdots \qquad (16.8.20)$$
$$(\gamma'p)^{\alpha_k\beta_k}\,\Phi_{\pm 0}^{\alpha_1\ldots\alpha_{k-1}\beta_k}(\mathbf{p}) = 0$$

where the 0 in the subscript corresponds to mass zero.

For example, for $\Phi_{+0}^{\alpha_1\ldots\alpha_{2s}}(\mathbf{p})$, we may rewrite the above, more conveniently, as

$$\left(\gamma'^0\boldsymbol{\gamma}'\cdot\mathbf{N}\right)^{\alpha_1\beta_1}\Phi_{+0}^{\beta_1\alpha_2\ldots\alpha_k}(\mathbf{p}) = \Phi_{+0}^{\alpha_1\ldots\alpha_k}(\mathbf{p})$$
$$\vdots \qquad (16.8.21)$$
$$\left(\gamma'^0\boldsymbol{\gamma}'\cdot\mathbf{N}\right)^{\alpha_k\beta_k}\Phi_{+0}^{\alpha_1\ldots\alpha_{k-1}\beta_k}(\mathbf{p}) = \Phi_{+0}^{\alpha_1\ldots\alpha_k}(\mathbf{p})$$

where we have used the fact that $p^0 = +|\mathbf{p}|$, and set $\mathbf{N} = \mathbf{p}/|\mathbf{p}|$.

As a direct generalization of the analysis carried out in §16.5, we consider the set of operators which commute with the operators occurring on the left-hand sides of (16.8.21):

$$\left(\gamma'^0\,\boldsymbol{\gamma}'\cdot\mathbf{N}\right)^{\alpha_1\beta_1}\delta^{\alpha_2\beta_2}\ldots\delta^{\alpha_k\beta_k}$$
$$\delta^{\alpha_1\beta_1}\left(\gamma'^0\,\boldsymbol{\gamma}'\cdot\mathbf{N}\right)^{\alpha_2\beta_2}\delta^{\alpha_3\beta_3}\ldots\delta^{\alpha_k\beta_k}$$
$$\vdots \qquad (16.8.22)$$
$$\delta^{\alpha_1\beta_1}\ldots\delta^{\alpha_{k-1}\beta_{k-1}}\left(\gamma'^0\,\boldsymbol{\gamma}'\cdot\mathbf{N}\right)^{\alpha_k\beta_k}$$

The analysis in §16.5 shows that the operators commuting with the above operators are of the form

$$\left(\gamma'^5\right)^{\alpha_1\beta_1}\delta^{\alpha_2\beta_2}\ldots\delta^{\alpha_k\beta_k}$$
$$\vdots \qquad (16.8.23)$$
$$\delta^{\alpha_1\beta_1}\ldots\delta^{\alpha_{k-1}\beta_{k-1}}\left(\gamma'^5\right)^{\alpha_k\beta_k}$$

or ones involving two, or three, or...or k γ'^5 matrices. Such an operator is

$$\left(\gamma'^5\right)^{\alpha_1\beta_1}\delta^{\alpha_2\beta_2}\ldots\delta^{\alpha_{k-1}\beta_{k-1}}\left(\gamma'^5\right)^{\alpha_k\beta_k} \qquad (16.8.24)$$

involving two γ'^5 matrices. Also other operators commuting with the ones in (16.8.22) are

$$(\mathbf{S}\cdot\mathbf{N})^{\alpha_1\beta_1}\delta^{\alpha_2\beta_2}\ldots\delta^{\alpha_k\beta_k}$$
$$\vdots \qquad (16.8.25)$$
$$\delta^{\alpha_1\beta_1}\ldots\delta^{\alpha_{k-1}\beta_{k-1}}(\mathbf{S}\cdot\mathbf{N})^{\alpha_k\beta_k}$$

16.8 Relativistic Wave Equations for Any Mass and Any Spin 951

or ones involving two, or three, or...or k $(\mathbf{S} \cdot \mathbf{N})$ matrices, also operators of the form in (16.8.22) involving two, or three, or...or k $(\gamma'^0 \gamma' \cdot \mathbf{N})$ matrices.

To find simultaneous eigenstates of these commuting set of matrices is straightforward. To this end, we carry out the expansion

$$\Phi_{+0}^{\alpha_1 \ldots \alpha_k}(\mathbf{p}, \sigma) = \sum_{\sigma_1, \ldots \sigma_k = \pm 1} C(\sigma_1, \sigma_2, \ldots \sigma_k; \sigma)$$

$$\times [u'^{\alpha_1}(\mathbf{p}, \sigma_1) \ldots u'^{\alpha_k}(\mathbf{p}, \sigma_k)] a(\mathbf{p}, \sigma) \quad (16.8.26)$$

as in (16.8.7), where in the present notation

$$u'(\mathbf{p}, +1) = \begin{pmatrix} \xi_{+\mathbf{N}} \\ 0 \end{pmatrix}, \quad u'(\mathbf{p}, -1) = \begin{pmatrix} 0 \\ \xi_{-\mathbf{N}} \end{pmatrix} \quad (16.8.27)$$

(see (16.5.22)).

From the definition of $C(\sigma_1, \ldots, \sigma_k; \sigma)$ in (16.8.8), the following equalities are readily established (see Problem 16.22)

$$C(+1, \sigma_2, \ldots \sigma_k; \sigma) = \sqrt{\frac{\sigma}{k}} C(\sigma_2, \ldots \sigma_k; \sigma - 1) \quad (16.8.28)$$

$$C(-1, \sigma_2, \ldots \sigma_k; \sigma) = \sqrt{\frac{k - \sigma}{k}} C(\sigma_2, \ldots \sigma_k; \sigma) \quad (16.8.29)$$

which in turn vanish for $\sigma = 0$, $\sigma = k$, respectively.

The equalities in (16.8.28), (16.8.29) allow us to rewrite (16.8.26) as

$$\Phi_{+0}^{\alpha_1 \ldots \alpha_k}(\mathbf{p}, \sigma) = \sum_{\sigma_1, \ldots \sigma_{k-2} = \pm 1} \{ \cdot \} \, u'^{\alpha_3}(\mathbf{p}, \sigma_1) \ldots u'^{\alpha_k}(\mathbf{p}, \sigma_{k-2}) a(\mathbf{p}, \sigma)$$

$$(16.8.30)$$

where

$$\{ \cdot \} = \sqrt{\frac{\sigma(\sigma - 1)}{k(k - 1)}} \, u'^{\alpha_1}(\mathbf{p}, +1) u'^{\alpha_2}(\mathbf{p}, +1) C(\sigma_1, \ldots \sigma_{k-2}; \sigma - 2)$$

$$+ \sqrt{\frac{\sigma(k - \sigma)}{k(k - 1)}} \left(u'^{\alpha_1}(\mathbf{p}, +1) u'^{\alpha_2}(\mathbf{p}, -1) + u'^{\alpha_1}(\mathbf{p}, -1) u'^{\alpha_2}(\mathbf{p}, +1) \right)$$

$$\times C(\sigma_1, \ldots \sigma_{k-2}; \sigma - 1)$$

$$+ \sqrt{\frac{(k - \sigma)(k - \sigma - 1)}{k(k - 1)}} u'^{\alpha_1}(\mathbf{p}, -1) u'^{\alpha_2}(\mathbf{p}, -1) C(\sigma_1, \ldots \sigma_{k-2}; \sigma)$$

$$(16.8.31)$$

Now consider the application of the first operator in (16.8.23) to $\Phi_{+0}^{\alpha_1 \ldots \alpha_k}(\mathbf{p}, \sigma)$ given in (16.8.30). This leads from (16.5.24), (16.5.26) to

$$\left(\gamma'^5\right)^{\alpha_1\beta_1} \Phi_{+0}^{\beta_1\alpha_2\ldots\alpha_k}(\mathbf{p},\sigma)$$

$$= \sum_{\sigma_1,\ldots\sigma_{k-2}=\pm 1} \{\cdot\}' u'^{\alpha_3}(\mathbf{p},\sigma_1)\ldots u'^{\alpha_k}(\mathbf{p},\sigma_{k-2})a(\mathbf{p},\sigma) \quad (16.8.32)$$

where

$$\{\cdot\}' = \sqrt{\frac{\sigma(\sigma-1)}{k(k-1)}}\, u'^{\alpha_1}(\mathbf{p},+1)u'^{\alpha_2}(\mathbf{p},+1)C(\sigma_1,\ldots\sigma_{k-2};\sigma-2)$$

$$+ \sqrt{\frac{\sigma(k-\sigma)}{k(k-1)}}\, [u'^{\alpha_1}(\mathbf{p},+1)u'^{\alpha_2}(\mathbf{p},-1) - u'^{\alpha_1}(\mathbf{p},-1)u'^{\alpha_2}(\mathbf{p},+1)]$$

$$\times C(\sigma_1,\ldots\sigma_{k-2};\sigma-1)$$

$$- \sqrt{\frac{(k-\sigma)(k-\sigma-1)}{k(k-1)}}\, u'^{\alpha_1}(\mathbf{p},-1)u'^{\alpha_2}(\mathbf{p},-1)C(\sigma_1,\ldots\sigma_{k-2};\sigma).$$

$$(16.8.33)$$

Note, in particular, the minus sign within the square brackets in the second term in (16.8.33).

Clearly, in order $\Phi_{+0}^{\beta_1\ldots\alpha_k}(\mathbf{p},\sigma)$ be an eigenstate of $(\gamma'^5)^{\alpha_1\beta_1}$, either $\sigma=0$, or $\sigma=k$, so that the second term in (16.8.33), in particular, vanishes. For $\sigma=0$, the first term in (16.8.33) also vanishes, while for $\sigma=k$, the third term vanishes. For the latter we have, respectively,

$$\left(\gamma'^5\right)^{\alpha_1\beta_1} \Phi_{+0}^{\beta_1\alpha_2\ldots\alpha_k}(\mathbf{p},0) = -\Phi_{+0}^{\alpha_1\alpha_2\ldots\alpha_k}(\mathbf{p},0), \quad \sigma=0,$$

where

$$\Phi_{+0}^{\alpha_1\alpha_2\ldots\alpha_k}(\mathbf{p},0) = u'^{\alpha_1}(\mathbf{p},-1)\ldots u'^{\alpha_k}(\mathbf{p},-1)a(\mathbf{p},0) \quad (16.8.34)$$

or

$$\left(\gamma'^5\right)^{\alpha_1\beta_1} \Phi_{+0}^{\beta_1\alpha_2\ldots\alpha_k}(\mathbf{p},k) = +\Phi_{+0}^{\alpha_1\alpha_2\ldots\alpha_k}(\mathbf{p},k), \quad \sigma=k \quad (16.8.35)$$

where

$$\Phi_{+0}^{\alpha_1\alpha_2\ldots\alpha_k}(\mathbf{p},k) = u'^{\alpha_1}(\mathbf{p},+1)\ldots u'^{\alpha_k}(\mathbf{p},+1)a(\mathbf{p},k) \quad (16.8.36)$$

Now it is obvious that to have $\Phi_{+0}^{\alpha_1\ldots\alpha_k}(\mathbf{p},\sigma)$ as a simultaneous eigenstate of all the commuting operators enumerated above, we must have either $\sigma=0$, or $\sigma=k$, i.e., $a(\mathbf{p},\sigma)\equiv 0$, for $\sigma=1,\ldots,k-1$.

Upon writing $k=2s$, and setting $\sigma=m_s+s$, we then obtain

$$\Phi_{+0}^{\alpha_1\ldots\alpha_{2s}}(\mathbf{p},m_s+s) = \delta(m_s,s)u'^{\alpha_1}(\mathbf{p},+1)\ldots u'^{\alpha_{2s}}(\mathbf{p},+1)a(\mathbf{p},2s)$$

$$+ \delta(m_s, -s) u'^{\alpha_1}(\mathbf{p}, -1) \ldots u'^{\alpha_{2s}}(\mathbf{p}, -1) a(\mathbf{p}, 0) \tag{16.8.37}$$

written in terms of Kronecker deltas.

For the helicity operator, as obtained from (16.8.13), (16.8.37) gives

$$(\mathbf{S} \cdot \mathbf{N})^{\alpha_1 \beta_1, \ldots, \alpha_{2s} \beta_{2s}} \Phi_{+0}^{\beta_1 \ldots \beta_{2s}}(\mathbf{p}, m_s + s)$$

$$= [\hbar s \delta(m_s, s) - \hbar s \delta(m_s, -s)] \Phi_{+0}^{\alpha_1 \ldots \alpha_{2s}}(\mathbf{p}, m_s + s) \tag{16.8.38}$$

yielding only two possible values for the helicity of a particle.

For a massless particle of definite helicity, as discussed in §16.5, say of helicity $\varepsilon \hbar s$, with $\varepsilon = -1$ (or $+1$), we may set

$$\widetilde{\Phi}_{+,\varepsilon}^{\alpha_1 \ldots \alpha_{2s}}(\mathbf{p}) = \frac{1}{2}(I + \varepsilon \gamma'^5)^{\alpha_1 \beta_1} \ldots \frac{1}{2}(I + \varepsilon \gamma'^5)^{\alpha_{2s} \beta_{2s}} \Phi_{+0}^{\beta_1 \ldots \beta_{2s}}(\mathbf{p}, (1+\varepsilon)s) \tag{16.8.39}$$

where we recall that $\gamma'^5 u'(\mathbf{p}, \varepsilon) = \varepsilon u'(\mathbf{p}, \varepsilon)$, and $\widetilde{\Phi}_{+,\varepsilon}^{\alpha_1 \ldots \alpha_{2s}}(\mathbf{p})$ satisfies the set of equations in (16.8.21), and

$$(\mathbf{S} \cdot \mathbf{N})^{\alpha_1 \beta_1, \ldots, \alpha_{2s} \beta_{2s}} \widetilde{\Phi}_{+,\varepsilon}^{\beta_1 \ldots \beta_{2s}}(\mathbf{p}) = \varepsilon \hbar s \widetilde{\Phi}_{+,\varepsilon}^{\alpha_1 \ldots \alpha_{2s}}(\mathbf{p}) \tag{16.8.40}$$

One may treat the charge conjugate solutions in the same manner.

16.9 Spin & Statistics

This section is involved with the spin and statistics connection. We consider only spin 0 and spin 1/2. Familiarity with Lagrangian dynamics is essential. In the appendix to this section we provide, however, a fairly detailed treatment of the action integral including for systems admitting Grassmann variables. The method developed there is extended within this section to systems with infinite degrees of freedom. Accordingly the reader is advised to consult this appendix while reading this section.

First we examine the solutions of the free Klein-Gordon and Dirac equations and interpret their expansion coefficients as operators for creation and annihilation of particles and anti-particles thus generating quantum fields. We set up field equations for both systems and corresponding Lagrangian densities which are of first order in ∂_μ, and are expressed in terms of Hermitian fields. We then introduce Schwinger's ingenious constructive approach[32] which treats such Lagrangians in a unified manner and leads naturally to the spin and statistics connection. Finally we use the results stemming from this analysis to establish the spin and statistics connection by examining the Lagrangian densities set up earlier. In the present section we do not dwell on the Spin and Statistics Theorem in arbitrary dimensions of space.

---

[32] Schwinger (1953); see also Schwinger (1951b, 1958a,b, 1961b).

For obvious reasons there has been much interest in the Spin and Statistics Theorem over the years since the early work of Pauli on the so-called exclusion principle named after him. I am pleased to see that the activities on this problem are still going on and even escalating in the recent literature.

16.9.1 Quantum Fields

The solution of the Klein-Gordon equation in (16.7.5) for free spin 0 particles, i.e., for $A^\mu = 0$, may be quite generally written as

$$\Phi(x) = \sqrt{\hbar c^2} \int \frac{d^3\mathbf{p}}{(2\pi\hbar)^3 2p^0} \left[A(\mathbf{p}) e^{ipx/\hbar} + B^*(\mathbf{p}) e^{-ipx/\hbar} \right] \qquad (16.9.1)$$

where the numerical factor $\sqrt{\hbar c^2}$ is inserted for convenience, $px = \mathbf{p}\cdot\mathbf{x} - p^0 x^0$, $p^0 = +\sqrt{\mathbf{p}^2 + M^2 c^2}$, $x^0 = ct$, and, of course, the mass M does *not* have to be the same as for the spin 1/2 case.

Similarly for spin 1/2, the general solution in (16.4.44) of the free Dirac equation in (16.4.1) is given by

$$\Psi(x) = \sum_\sigma \int 2Mc \, \frac{d^3\mathbf{p}}{(2\pi\hbar)^3 2p^0} \left[u(\mathbf{p},\sigma) a(\mathbf{p},\sigma) e^{ipx/\hbar} + v(\mathbf{p},\sigma) b^*(\mathbf{p},\sigma) e^{-ipx/\hbar} \right] \qquad (16.9.2)$$

The expansion coefficients $A(\mathbf{p})$, $a(\mathbf{p},\sigma)$ correspond to positive energy solutions, while $B^*(\mathbf{p})$, $b^*(\mathbf{p},\sigma)$ correspond to the negative energy ones.

To overcome the dilemma of negative energy solutions and associated instability problems, $A(\mathbf{p})$, $a(\mathbf{p},\sigma)$, in quantum field theory, are interpreted as annihilation operators of the corresponding particles of spin 0 and spin 1/2, while their adjoints $A^\dagger(\mathbf{p})$, $a^\dagger(\mathbf{p},\sigma)$ as creation of the particles in question acting on the vacuum state. Similarly, $B(\mathbf{p})$, $b(\mathbf{p},\sigma)$ are interpreted as annihilation operators of the associated anti-particles, and their adjoints $B^\dagger(\mathbf{p},\sigma)$, $b^\dagger(\mathbf{p},\sigma)$, $(B^*(\mathbf{p},\sigma) \to B^\dagger(\mathbf{p},\sigma),\ b^*(\mathbf{p},\sigma) \to b^\dagger(\mathbf{p},\sigma))$ as creation operators of the anti-particles, all with *positive* energies, acting on the vacuum state.

The operators, $A(\mathbf{p})$, $A^\dagger(\mathbf{p})$, $B(\mathbf{p})$, $B^\dagger(\mathbf{p})$ may be written in terms the field Φ, now as an operator, *integrated over all space* and evaluated at any given time t by the method of Problem 16.23 as follows.

We use the facts that

$$i\hbar \int d^3\mathbf{x}\, e^{-ipx/\hbar} \overleftrightarrow{\partial}_0 e^{ip'x/\hbar} = 2p^0 (2\pi\hbar^3) \delta^3(\mathbf{p} - \mathbf{p}') \qquad (16.9.3)$$

and

$$i\hbar \int d^3\mathbf{x}\, e^{ipx/\hbar} \overleftrightarrow{\partial}_0 e^{ip'x/\hbar} = 0 \qquad (16.9.4)$$

where

$$\overleftrightarrow{\partial}_0 = \overrightarrow{\partial}_0 - \overleftarrow{\partial}_0 \qquad (16.9.5)$$

to obtain

$$A(\mathbf{p}) = \frac{i\hbar}{\sqrt{\hbar c^2}} \int d^3x \, e^{-ipx/\hbar} \overleftrightarrow{\partial}_0 \Phi(x) \tag{16.9.6}$$

$$B^\dagger(\mathbf{p}) = \frac{i\hbar}{\sqrt{\hbar c^2}} \int d^3x \, e^{ipx/\hbar} \overleftrightarrow{\partial}_0 \Phi(x) \tag{16.9.7}$$

and similarly for their adjoints.

The operators $a(\mathbf{p}, \sigma)$, $b^\dagger(\mathbf{p}, \sigma)$ may be also extracted in analogous relations as in (16.6.8), (16.6.9), but evaluated at any time t, as follows

$$a(\mathbf{p}, \sigma) = \int d^3x \, e^{-ipx/\hbar} \, u^\dagger(\mathbf{p}, \sigma) \Psi(x) \tag{16.9.8}$$

$$b^\dagger(\mathbf{p}, \sigma) = \int d^3x \, e^{ipx/\hbar} \, v^\dagger(\mathbf{p}, \sigma) \Psi(x) \tag{16.9.9}$$

in terms of the operator $\Psi(x)$. Similar relations as in (16.9.8), (16.9.9) may be also written for $a^\dagger(\mathbf{p}, \sigma)$, $b(\mathbf{p}, \sigma)$.

For physical interpretations and applications, commutativity properties of the above creation and annihilation operators have to be consistently established.

To establish commutativity properties of $A(\mathbf{p})$, $A^\dagger(\mathbf{p}')$, $B(\mathbf{p}'')$, $B^\dagger(\mathbf{p}''')$, we note from (16.9.6), (16.9.7) and the equations for their adjoints, that commutativity properties of the quantum field $\Phi(x)$, as well as of its adjoint $\Phi^\dagger(x)$, have to be established at all points in space including at *different* ones, but may be taken all at the same time. A similar statement holds for the operators $a(\mathbf{p}, \sigma)$, $a^\dagger(\mathbf{p}', \sigma')$, $b(\mathbf{p}'', \sigma'')$, $b^\dagger(\mathbf{p}''', \sigma''')$ in terms of the quantum field $\Psi(x)$, as well as of its adjoint $\Psi^\dagger(x)$, taken at different points in space also.

The burden of this section is to establish commutativity properties of $\Phi(x)$, $\Phi^\dagger(x)$ as well as of $\Psi(x)$, $\Psi^\dagger(x)$, and in turn establish the spin and statistics connection for particles of spin 0 and of spin 1/2.

We next set up Lagrangian densities for the above two systems.

16.9.2 Lagrangian for Spin 0 Particles

We consider the wave equation in (16.7.8) for spin 0 free particles

$$\left(\frac{\beta^\mu}{i} \partial_\mu + \frac{Mc}{\hbar} \right) \Xi = 0 \tag{16.9.10}$$

where Ξ is defined in (16.7.9), and the β^μ matrices are given in (16.7.10), (16.7.11). The matrices β^μ/i are real,

$$(\beta^0)^\dagger = \beta^0, \quad (\boldsymbol{\beta})^\dagger = -\boldsymbol{\beta} \tag{16.9.11}$$

To obtain the equation satisfied by the adjoint of Ξ, we introduce the matrix

$$\Lambda = 2\left(\beta^0\right)^2 - I = \begin{pmatrix} 1 & 0 & 0 & 0 & 0 \\ 0 & -1 & 0 & 0 & 0 \\ 0 & 0 & -1 & 0 & 0 \\ 0 & 0 & 0 & -1 & 0 \\ 0 & 0 & 0 & 0 & 1 \end{pmatrix} \quad (16.9.12)$$

which satisfies the relations

$$\left[\Lambda, \beta^0\right] = 0, \quad \{\Lambda, \boldsymbol{\beta}\} = \mathbf{0} \quad (16.9.13)$$

Accordingly, by taking the adjoint of (16.9.10), multiplying the resulting equation from the right by Λ, and using (16.9.11), (16.9.13) we get

$$\Xi^\dagger \Lambda \left(-\frac{\beta^\mu}{i}\overleftarrow{\partial}_\mu + \frac{Mc}{\hbar}\right) = 0 \quad (16.9.14)$$

We may rewrite (16.9.10) in terms of real components by defining, in the process,

$$\Phi = \frac{1}{\sqrt{2}}\left(\Phi_1 + i\Phi_2\right), \quad \chi^\mu = \frac{1}{\sqrt{2}}\left(\chi_1^\mu + i\chi_2^\mu\right) \quad (16.9.15)$$

where the χ^μ are given in terms of Φ in (16.7.6), now for $A^\mu = 0$, i.e.,

$$\chi^\mu = \frac{\hbar}{Mc}\partial^\mu \Phi \quad (16.9.16)$$

Introducing the ten-component entity

$$\phi = (\phi_a) = \sqrt{\frac{Mc}{\hbar}} \left(\chi_1^0 \; \chi_1^1 \; \chi_1^2 \; \chi_1^3 \; \Phi_1 \; \chi_2^0 \; \chi_2^1 \; \chi_2^2 \; \chi_2^3 \; \Phi_2\right)^{\mathsf{T}} \quad (16.9.17)$$

with $a = 1, 2, \ldots, 10$, and the 10×10 matrices

$$B^\mu = \begin{pmatrix} \dfrac{\beta^\mu}{i} & 0 \\ 0 & \dfrac{\beta^\mu}{i} \end{pmatrix}, \quad \Gamma = \begin{pmatrix} \Lambda & 0 \\ 0 & \Lambda \end{pmatrix} \quad (16.9.18)$$

we may rewrite (16.9.10), (16.9.14) as

$$\left(B^\mu \partial_\mu + \frac{Mc}{\hbar}\right)\phi = 0 \quad (16.9.19)$$

$$\phi \Gamma \left(-B^\mu \overleftarrow{\partial}_\mu + \frac{Mc}{\hbar}\right) = 0 \quad (16.9.20)$$

written as matrix multiplication. The advantage of these equations over the earlier ones is that they are written completely in terms of *real* objects.

The following two matrices B^μ, Γ are of central importance in our spin 0 case. In particular, they have the following properties

$$\Gamma^2 = I, \qquad \Gamma B^0 = B^0, \qquad (\Gamma B^\mu)^\dagger = -\Gamma B^\mu, \qquad \Gamma^\dagger = \Gamma \qquad (16.9.21)$$

and

$$(\Gamma B^\mu)^\top = -(\Gamma B^\mu), \qquad \Gamma^\top = \Gamma. \qquad (16.9.22)$$

We introduce the Lagrangian density for spin 0 particles in which the derivative ∂_μ appears linearly

$$\mathscr{L} = -\frac{1}{4}[\phi \Gamma B^\mu \partial_\mu \phi - \partial_\mu \phi \Gamma B^\mu \phi] - \frac{Mc}{2\hbar} \phi \Gamma \phi \qquad (16.9.23)$$

and the properties of ΓB^μ, Γ in (16.9.21) ensure the Hermiticity of the Lagrangian density.

We will see below, while studying Schwinger's Theorem, how the Lagrangian density in (16.9.23) leads to the equations (16.9.19), (16.9.20) satisfied by ψ and investigate the nature of the spin and statistics connection.

We next set up the Lagrangian density for free spin 1/2 particles.

16.9.3 Lagrangian for Spin 1/2 Particles

We work in the Majorana representation, with the gamma matrices given in (16.1.28), (16.1.29). To simplify the notation, we will still denote these corresponding gamma matrices by γ^μ, and recall that in this representation the matrices γ^μ/i are *real*.

In the Majorana representation, defining the 8×8 matrices

$$\Gamma^\mu = \begin{pmatrix} \gamma^\mu & 0 \\ 0 & \gamma^\mu \end{pmatrix} \qquad (16.9.24)$$

and writing the Dirac spinor Ψ in terms of real and imaginary parts

$$\Psi = \frac{1}{\sqrt{2}} (\Psi_1 + i\Psi_2) \qquad (16.9.25)$$

we may rewrite the Dirac equation in this representation as

$$\left(\frac{\Gamma^\mu}{i} \partial_\mu + \frac{Mc}{\hbar}\right) \psi = 0 \qquad (16.9.26)$$

where

$$\psi = (\psi_a) = \begin{pmatrix} \Psi_1^1 & \Psi_1^2 & \Psi_1^3 & \Psi_1^4 & \Psi_2^1 & \Psi_2^2 & \Psi_2^3 & \Psi_2^4 \end{pmatrix}^\top. \qquad (16.9.27)$$

For the adjoint of the equation (16.9.26), we have

$$\psi \, \Gamma^0 \left(-\frac{\Gamma^\mu}{i} \overleftarrow{\partial}_\mu + \frac{Mc}{\hbar} \right) = 0. \tag{16.9.28}$$

The matrices of central importance for spin 1/2 here are the matrices Γ^μ. In particular

$$(\Gamma^0 \Gamma^\mu)^\dagger = \Gamma^0 \Gamma^\mu, \qquad (\Gamma^0)^\dagger = \Gamma^0, \qquad (\Gamma^0)^2 = I \tag{16.9.29}$$

$$(\Gamma^0 \Gamma^\mu)^\top = \Gamma^0 \Gamma^\mu, \qquad {\Gamma^0}^\top = -\Gamma^0 \tag{16.9.30}$$

The Lagrangian density of spin 1/2 is defined by

$$\mathscr{L} = -\frac{1}{4}\left(\psi \frac{\Gamma^0 \Gamma^\mu}{i} \partial_\mu \psi - \partial_\mu \psi \frac{\Gamma^0 \Gamma^\mu}{i} \psi \right) - \frac{Mc}{2\hbar} \psi \Gamma^0 \psi \tag{16.9.31}$$

The Hermiticity of the latter follows from the Hermiticity of $\Gamma^0 \Gamma^\mu$, Γ^0.

As for the spin 0 case, we will show how the Lagrangian density in (16.9.31) leads to the Dirac equations in (16.9.26), (16.9.28) and investigate the spin and statistics connection after having developed Schwinger's constructive approach.

16.9.4 Schwinger's Constructive Approach

The Lagrangian densities in (16.9.23), (16.9.31) are of the form[33]

$$\mathscr{L}(x) = \frac{1}{4}\left[\eta_a(x) Q^\mu_{ab} \partial_\mu \eta_b(x) - \partial_\mu \eta_a(x) Q^\mu_{ab} \eta_b(x) \right] + \frac{1}{2} \eta_a(x) Q_{ab} \eta_b(x) \tag{16.9.32}$$

where the numerical constant matrices Q^μ, Q are such that

$$(Q^\mu)^\dagger = -Q^\mu, \qquad Q^\dagger = Q \tag{16.9.33}$$

(see (16.9.21), (16.9.29)), thus guaranteeing that $\mathscr{L}$ is Hermitian, and *either*

$$(Q^\mu)^\top = -Q^\mu, \qquad Q^\top = Q \tag{16.9.34}$$

as in the spin 0 case in (16.9.22), *or*

$$(Q^\mu)^\top = Q^\mu, \qquad Q^\top = -Q \tag{16.9.35}$$

as in the spin 1/2 case in (16.9.30).

Here it is worth noting that any square matrix C may be written as the sum of a symmetric matrix S and an anti-symmetric one A,

$$C_{ab} = \frac{1}{2}(C_{ab} + C_{ba}) + \frac{1}{2}(C_{ab} - C_{ba})$$

---
[33] See also the appendix to this section.

$$\equiv S_{ab} + A_{ab} \qquad (16.9.36)$$

$A^\top = -A$, $S^\top = S$. For a matrix C such that $C^\dagger = -C$, this means that $A^\dagger = -A$, $S^\dagger = -S$ which correspond to the cases encountered above. It is important to note that in the Lagrangian density in (16.9.32), a symmetric matrix Q goes with the anti-symmetric ones Q^μ, while an anti-symmetric matrix Q goes with the symmetric ones Q^μ as shown, respectively, in (16.9.34), (16.9.35).

The Lagrangian density in (16.9.32) is local, i.e., all the terms in it are evaluated at the same spacetime point. At a given time $t = x^0/c$, locality means that we may add up the Lagrangian densities $\mathscr{L}(ct, \mathbf{x})$ evaluated at different points to obtain a Lagrangian $L(x^0) = \int d^3 \mathbf{x}\, \mathscr{L}(ct, \mathbf{x})$ since relativity implies that different regions in space at the same time are dynamically independent as no signal can propagate between them. Here the position variable $\mathbf{x}$ specifies the infinitely uncountable number of degrees of freedom in the fields $\eta(x^0, \mathbf{x}) = \eta_\mathbf{x}(ct)$, with the latter as dynamical variables, in the same way that the index i in $q_i(t)$ specifies the various degrees of freedom in particle mechanics. One may then define the action associated with the Lagrangian density $\mathscr{L}(x)$ in (16.9.32) by

$$A = \frac{1}{c}\int dx^0 \int d^3\mathbf{x}\, \mathscr{L}(x^0, \mathbf{x}) = \frac{1}{c}\int (dx)\, \mathscr{L}(x) \qquad (16.9.37)$$

where we have integrated over all time. The action in (16.9.37) with Lagrangian density in (16.9.32) should be compared with the one in (A-16.9.9). In the present case we are, however, dealing with a system of infinite degrees of freedom.

To emphasize the functional dependence of the action on the field η, we write $\mathscr{L}(x) = \mathscr{L}[\eta(x)]$. We consider the variation of the action in response to variation of the field $\eta(x) \to \eta(x) - \delta\eta(x)$ to first order in $\delta\eta(x)$. The variation of the Lagrangian density is then given by

$$\delta\mathscr{L}[\eta(x)] = \mathscr{L}[\eta(x)] - \mathscr{L}[\eta(x) - \delta\eta(x)] \qquad (16.9.38)$$

paying special attention to the order in which $\delta\eta(x)$ appears relative to $\eta(x)$ and $\partial_\mu \eta(x)$ in (16.9.38).

Following Schwinger, we consider variations of the action as arising from c-number variations $\delta\eta(x)$ of the field. We will then encounter that such variations we were set to achieve with $\delta\eta(x)$ commuting with the field itself is possible only for the system satisfying the conditions in (16.9.34), while $\delta\eta(x)$, as a c-number Grassmann variable, anti-commuting with the field, is possible only for the system satisfying the conditions in (16.9.35).

The variation of the action, multiplied by c, corresponding to the variation of the Lagrangian density in (16.9.38) is explicitly given from (16.9.32) to be

$$\int (dx)\, \delta\mathscr{L}[\eta(x)] = \int (dx)\, \left(\frac{1}{4}[\delta\eta_a(x) Q^\mu_{ab} \partial_\mu \eta_b(x) + \eta_a(x) Q^\mu_{ab} \partial_\mu \delta\eta_b(x) \right.$$

$$-\partial_\mu \delta\eta_a(x) Q^\mu_{ab} \eta_b(x) - \partial_\mu \eta_a(x) Q^\mu_{ab} \delta\eta_b(x)]$$

$$+ \frac{1}{2} [\delta\eta_a(x) Q_{ab} \eta_b(x) + \eta_a(x) Q_{ab} \delta\eta_b(x)] \Big)$$

$$= \frac{1}{2} \int (\mathrm{d}x)\, \partial_\mu \left(\frac{1}{2} [\eta_a(x) Q^\mu_{ab} \delta\eta_b(x) - \delta\eta_a(x) Q^\mu_{ab} \eta_b(x)] \right)$$

$$+ \frac{1}{2} \int (\mathrm{d}x)\, [\delta\eta_a(x) Q^\mu_{ab} \partial_\mu \eta_b(x) - \partial_\mu \eta_a(x) Q^\mu_{ab} \delta\eta_b(x)$$

$$+ \delta\eta_a(x) Q_{ab} \eta_b(x) + \eta_a(x) Q_{ab} \delta\eta_b(x)]$$
(16.9.39)

with the order in which $\delta\eta(x)$ appears kept intact.

To treat both cases in (16.9.34), (16.9.35) simultaneously, we may write

$$Q^\mu_{ab} = \varepsilon Q^\mu_{ba}, \qquad Q_{ab} = -\varepsilon Q_{ba} \qquad (16.9.40)$$

where $\varepsilon = -1$ corresponds to the case in (16.9.34), while $\varepsilon = +1$ corresponds to the case in (16.9.35).

To simplify the expressions within the second pair of square brackets on the extreme right-hand side of (16.9.39) we use the relation

$$\partial_\mu \eta_b(x) = \partial_\mu \int (\mathrm{d}x')\, \delta^4(x'-x) \eta_b(x')$$

$$= \int (\mathrm{d}x')\, \eta_b(x') \partial_\mu \delta^4(x'-x)$$

$$= -\int (\mathrm{d}x')\, \eta_b(x') \partial'_\mu \delta^4(x'-x) \qquad (16.9.41)$$

to write

$$\int (\mathrm{d}x)\, Q^\mu_{ab} \delta\eta_a(x) \partial_\mu \eta_b(x) = -\int (\mathrm{d}x)(\mathrm{d}x')\, \delta\eta_a(x) \eta_b(x') Q^\mu_{ab} \partial'_\mu \delta^4(x'-x) \qquad (16.9.42)$$

Similarly, we have

$$\int (\mathrm{d}x)\, Q^\mu_{ab} \partial_\mu \eta_a(x) \delta\eta_b(x) = -\int (\mathrm{d}x)(\mathrm{d}x')\, \eta_a(x') \delta\eta_b(x) Q^\mu_{ab} \partial'_\mu \delta^4(x'-x)$$

$$= -\varepsilon \int (\mathrm{d}x)(\mathrm{d}x')\, \eta_a(x') \delta\eta_b(x) Q^\mu_{ba} \partial'_\mu \delta^4(x'-x)$$

$$= -\varepsilon \int (\mathrm{d}x)(\mathrm{d}x')\, \eta_b(x') \delta\eta_a(x) Q^\mu_{ab} \partial'_\mu \delta^4(x'-x) \qquad (16.9.43)$$

Thus the *second* integral on the extreme right-hand side of (16.9.39) (which is not the integral of a total differential) may be rewritten as

$$-\frac{1}{2}\int(\mathrm{d}x)(\mathrm{d}x')\,[\delta\eta_a(x)\eta_b(x') - \varepsilon\eta_b(x')\delta\eta_a(x)]$$

$$\times [Q^\mu_{ab}\partial'_\mu \delta^4(x'-x) - Q_{ab}\delta^4(x'-x)] \quad (16.9.44)$$

where we have used (16.9.40). It is remarkable that one is able to combine the terms depending on Q^μ_{ab} with the terms depending on Q_{ab} in (16.9.44) precisely *because* of the *opposite* symmetry properties of Q^μ and Q in (16.9.40).

From the expression in the first pair of square brackets in (16.9.44) we learn that a c-number variation with $\delta\eta_a(x)$ *commuting* with the field $\eta_b(x')$ is possible only for $\varepsilon = -1$ satisfying the conditions in (16.9.34), i.e., for spin 0, while a variation $\delta\eta_a(x)$ as a c-number Grassmann variable, *anti-commuting* with the field $\eta_b(x')$ is possible only for $\varepsilon = +1$ satisfying the conditions in (16.9.35), i.e., for spin 1/2, otherwise the expression in the first pair of square brackets is identically equal to zero. Therefore

$$[\delta\eta_a(x), \eta_b(x')] = 0 \quad (16.9.45)$$

for $\varepsilon = -1$, while

$$\{\delta\eta_a(x), \eta_b(x')\} = 0 \quad (16.9.46)$$

for $\varepsilon = +1$, i.e.,

$$\delta\eta_a(x)\,\eta_b(x') = -\varepsilon\eta_b(x')\,\delta\eta_a(x). \quad (16.9.47)$$

In *both* cases, (16.9.44), (16.9.39), (16.9.40), (16.9.47) then give

$$\int(\mathrm{d}x)\,\delta\mathscr{L}[\eta(x)] = \int(\mathrm{d}x)\,\partial_\mu \left[\frac{\eta_a(x)\,Q^\mu_{ab}}{2}\delta\eta_b(x)\right]$$

$$+ \int(\mathrm{d}x)\,\delta\eta_a(x)\,[Q^\mu_{ab}\partial_\mu\eta_b(x) + Q_{ab}\eta_b(x)] \quad (16.9.48)$$

For $\delta\eta_a(x)$ arbitrary, the principle of stationary action gives the field equation

$$Q^\mu \partial_\mu \eta(x) + Q\eta(x) = 0 \quad (16.9.49)$$

As for the system in (A-16.9.22)–(A-16.9.32), the variation of the action, multiplied by c, in (16.9.48) now extended to a system with an infinite (uncountable) numbers of degrees of freedom introduces the generator ($\partial_0 = \partial/\partial(ct)$),

$$G = \frac{1}{c}\int \mathrm{d}^3\mathbf{x}\,\frac{\eta(x)Q^0}{2}\delta\eta(x) \quad (16.9.50)$$

which brings about the change of the dynamical variables $\eta(x)$ by $\delta\eta(x)/2$ with

$$\frac{\delta\eta(x)}{2} = \frac{1}{i\hbar}[\eta(x), G] \qquad (16.9.51)$$

In detail, (16.9.51), with $x'^0 = x^0$, reads

$$\frac{\delta\eta_e(x)}{2} = \frac{1}{i\hbar c}Q^0_{ab}\int d^3x' \left[\eta_e(x), \eta_a(x')\frac{\delta\eta_b(x')}{2}\right]$$

$$= \frac{1}{i\hbar c}Q^0_{ab}\int d^3x' \left(\eta_e(x)\eta_a(x') + \varepsilon\eta_a(x')\eta_e(x)\right)\frac{\delta\eta_b(x')}{2} \qquad (16.9.52)$$

or

$$\left[\eta_e(x)\eta_a(x') + \varepsilon\eta_a(x')\eta_e(x)\right]Q^0_{ab} = i\hbar c\,\delta_{be}\,\delta^3(\mathbf{x}-\mathbf{x}'). \qquad (16.9.53)$$

We apply the above results to spin 0 and spin 1/2.

16.9.5 The Spin and Statistics Connection

Spin 0

By comparing (16.9.23) with (16.9.32) we see that

$$Q^\mu = -\Gamma B^\mu, \quad Q = -\frac{Mc}{\hbar}\Gamma \qquad (16.9.54)$$

where Γ, B^μ are given in (16.9.18). From (16.9.49), we obtain

$$\Gamma\left(B^\mu\partial_\mu + \frac{Mc}{\hbar}\right)\phi = 0 \qquad (16.9.55)$$

which upon multiplying from the left by Γ gives the field equation in (16.9.19) and hence finally to the Klein-Gordon equation.

From (16.9.21), we note $Q^0 = -\Gamma B^0 = -B^0$, with B^0 given in (16.9.18) and β^0 in (16.7.10). That is,

$$Q^0_{ab} = -\delta(a,1)\delta(b,5) + \delta(a,5)\delta(b,1)$$

$$- \delta(a,6)\delta(b,10) + \delta(a,10)\delta(b,6) \qquad (16.9.56)$$

written in term of Kronecker deltas.

From (16.9.53), (16.9.56), (16.9.17), with $x^0 = x'^0$ and $\varepsilon = -1$ (see (16.9.33), (16.9.40)), we obtain

$$-\left[\phi_e(x), \phi_1(x')\right]\delta(b,5) + \left[\phi_e(x), \phi_5(x')\right]\delta(b,1)$$

$$-\left[\phi_e(x), \phi_6(x')\right]\delta(b,10) + \left[\phi_e(x), \phi_{10}(x')\right]\delta(b,6) = i\hbar c\,\delta_{be}\,\delta^3(\mathbf{x}-\mathbf{x}'). \qquad (16.9.57)$$

That is, the only non-vanishing commutators are

$$[\phi_1(x), \phi_5(x')] = i\hbar c\, \delta^3(\mathbf{x} - \mathbf{x}') \tag{16.9.58}$$

$$[\phi_6(x), \phi_{10}(x')] = i\hbar c\, \delta^3(\mathbf{x} - \mathbf{x}') \tag{16.9.59}$$

From (16.9.17), (16.9.16), these give

$$[\Phi_1(x), \dot{\Phi}_1(x')] = i\hbar c^2 \delta^3(\mathbf{x} - \mathbf{x}') \tag{16.9.60}$$

$$[\Phi_2(x), \dot{\Phi}_2(x')] = i\hbar c^2 \delta^3(\mathbf{x} - \mathbf{x}') \tag{16.9.61}$$

For the Klein-Gordon field $\Phi(x)$ in (16.9.15), we then have, in particular, $(x^0 = x'^0)$,

$$[\Phi(x), \dot{\Phi}^\dagger(x')] = i\hbar c^2 \delta^3(\mathbf{x} - \mathbf{x}') \tag{16.9.62}$$

$$[\Phi(x), \Phi(x')] = 0, \quad [\Phi(x), \Phi^\dagger(x')] = 0 \tag{16.9.63}$$

$$[\dot{\Phi}(x), \dot{\Phi}^\dagger(x')] = 0 \tag{16.9.64}$$

From (16.9.62)–(16.9.64), and the expression for the creation and annihilation operators $A^\dagger(\mathbf{p})$, $A(\mathbf{p})$, as obtained from (16.9.6), we have

$$[A(\mathbf{p}), A^\dagger(\mathbf{p}')] = \frac{\hbar}{c} 2p^0 (2\pi\hbar)^3 \delta^3(\mathbf{p} - \mathbf{p}') \tag{16.9.65}$$

$$[A(\mathbf{p}), A(\mathbf{p}')] = 0, \quad [A^\dagger(\mathbf{p}), A^\dagger(\mathbf{p}')] = 0 \tag{16.9.66}$$

If we denote the vacuum state by $|\text{vac}\rangle$, that is, in particular, $A(\mathbf{p})|\text{vac}\rangle = 0$, then a two-particle state involving spin 0 (bosons) of momenta $\mathbf{p}, \mathbf{p}'$ may be constructed as follows

$$A^\dagger(\mathbf{p}) A^\dagger(\mathbf{p}') |\text{vac}\rangle = |\mathbf{p}, \mathbf{p}'\rangle \tag{16.9.67}$$

The second commutator in (16.9.66) then implies that

$$|\mathbf{p}, \mathbf{p}'\rangle = |\mathbf{p}', \mathbf{p}\rangle \tag{16.9.68}$$

which is the spin and statistics connection for bosons in its simplest form. Commutation relations involving the anti-particle operators are similarly obtained.

Spin 1/2

Upon comparing (16.9.32) and (16.9.31) we obtain

$$Q^\mu = -\frac{\Gamma^0 \Gamma^\mu}{i}, \quad Q = -\frac{Mc}{\hbar} \Gamma^0 \tag{16.9.69}$$

where Γ^μ is defined in (16.9.24). Equation (16.9.49) gives the equation

$$\Gamma^0 \left(\Gamma^\mu \frac{\partial_\mu}{i} + \frac{Mc}{\hbar} \right) \psi = 0 \tag{16.9.70}$$

which after multiplying it from the left by Γ^0 gives the field equation in (16.9.26).

The matrix Q^0, relevant to the relation in (16.9.53) takes particularly the simple form iI, where I is the identity matrix, as follows from the first relation in (16.9.69). Hence (16.9.53) with $x^0 = x'^0$ gives

$$\{\psi_e(x), \psi_a(x')\} = \hbar c\, \delta_{ea}\, \delta^3(\mathbf{x} - \mathbf{x}') \qquad (16.9.71)$$

where we have used the fact that $\varepsilon = +1$ (see (16.9.35), (16.9.40)).

From (16.9.27) and (16.9.25), we conclude that for the Dirac field $\Psi(x)$ with $x^0 = x'^0$:

$$\{\Psi_a(x), \Psi_b^\dagger(x')\} = \hbar c\, \delta_{ab}\, \delta^3(\mathbf{x} - \mathbf{x}') \qquad (16.9.72)$$

$$\{\Psi_a(x), \Psi_b(x')\} = 0, \qquad \{\Psi_a^\dagger(x), \Psi_b^\dagger(x')\} = 0 \qquad (16.9.73)$$

where now $a, b = 1, 2, 3, 4$.

The expressions for the creation and annihilation operators $a^\dagger(\mathbf{p}, \sigma)$, $a(\mathbf{p}', \sigma)$ as obtained from (16.9.8), then imply the following anti-commutation relations:

$$\{a(\mathbf{p}, \sigma), a^\dagger(\mathbf{p}', \sigma')\} = \hbar c \frac{p^0}{Mc} \delta_{\sigma\sigma'} (2\pi\hbar)^3 \delta^3(\mathbf{p} - \mathbf{p}') \qquad (16.9.74)$$

$$\{a(\mathbf{p}, \sigma), a(\mathbf{p}', \sigma')\} = 0, \qquad \{a^\dagger(\mathbf{p}, \sigma), a^\dagger(\mathbf{p}', \sigma')\} = 0 \qquad (16.9.75)$$

Again if we denote the vacuum state by $|\text{vac}\rangle$, that is, in particular, $a(\mathbf{p}, \sigma)|\text{vac}\rangle = 0$, then a two-particle state involving two spin 1/2 (fermions) particles of momenta $\mathbf{p}, \mathbf{p}'$ and spin indices σ, σ' may be constructed as follows

$$a^\dagger(\mathbf{p}, \sigma) a^\dagger(\mathbf{p}', \sigma') |\text{vac}\rangle = |\mathbf{p}, \sigma; \mathbf{p}', \sigma'\rangle \qquad (16.9.76)$$

The second anti-commutation relation in (16.9.75) then implies that

$$|\mathbf{p}, \sigma; \mathbf{p}', \sigma'\rangle = -|\mathbf{p}', \sigma'; \mathbf{p}, \sigma\rangle \qquad (16.9.77)$$

which is the spin and statistics connection for fermions in its simplest form. Anti-commutation relations involving the anti-particle operators are similarly obtained.

Similar methods may be applied to spin 1 particles. There are some complications, however, associated with higher spins and we will not go into such details here. In theoretical descriptions of higher spin particles the latter are often considered as composite particles and not as fundamental as the lower spin ones.

Appendix to §16.9: The Action Integral

As in Chapters 10 and 11, we use the $(\mathbf{q}, \mathbf{p})$ language and to simplify the notation, we often suppress the indices in (q_i, p_i) corresponding to various degrees of freedom. For example, we may often write $\sum_i p_i \dot{q}_i$ simply as $p\dot{q}$.

The action integral is defined by

$$A = \int_{T_1}^{T_2} \left[\frac{1}{2} (p\,\mathrm{d}q + \mathrm{d}q\,p) - H(q, p, t)\mathrm{d}t \right] \tag{A-16.9.1}$$

written in a symmetrized form in the product $p\mathrm{d}q$ to ensure Hermiticity, where $H(q, p, t)$ is the Hamiltonian and should be Hermitian.

We consider variations of the action that arise by the variation of the dynamical variables only:

$$q \to \bar{q} = q - \delta q \tag{A-16.9.2}$$

$$p \to \bar{p} = p - \delta p \tag{A-16.9.3}$$

Since $\delta(\mathrm{d}q) = \delta\left(q(t+\mathrm{d}t) - q(t)\right) = \mathrm{d}\delta q$, and so on, the variations in (A-16.9.2), (A-16.9.3) lead to

$$\delta A = \int_{T_1}^{T_2} \mathrm{d}\left[\frac{1}{2}(p\,\delta q + \delta q\,p)\right]$$

$$+ \int_{T_1}^{T_2} \left[\frac{1}{2}(\delta p\,\mathrm{d}q + \mathrm{d}q\,\delta p - \mathrm{d}p\,\delta q - \delta q\,\mathrm{d}p) - \delta H\,\mathrm{d}t\right] \tag{A-16.9.4}$$

where

$$\delta H(q, p, t) = H(q, p, t) - H(q - \delta q, p - \delta p, t) \tag{A-16.9.5}$$

For infinitesimal numerical changes δq, δp in (A-16.9.2), (A-16.9.3) which commute with the dynamical variables, we may write

$$\delta A = \int_{T_1}^{T_2} \mathrm{d}\,[p\,\delta q] + \int_{T_1}^{T_2} [\delta p\,\mathrm{d}q - \mathrm{d}p\,\delta q - \delta H\,\mathrm{d}t] \tag{A-16.9.6}$$

Here we recognize the $p\delta q$ term in the first integral (see (2.3.13)) associated with the boundary terms, as the generator for the infinitesimal transformation in (A-16.9.2). To obtain the generator for both transformations in (A-16.9.2), (A-16.9.3) in a unified manner, one rewrites the action integral in (A-16.9.1) in a more symmetrical way in q and p as follows by omitting, in the process the integral

$$\frac{1}{4}\int_{T_1}^{T_2} \mathrm{d}(qp + pq)$$

of a total differential, to obtain

$$A = \int_{T_1}^{T_2} \left[\frac{1}{4}(p\,\mathrm{d}q + \mathrm{d}q\,p - \mathrm{d}p\,q - q\,\mathrm{d}p) - H(q, p, t)\,\mathrm{d}t\right]. \tag{A-16.9.7}$$

It is more convenient to set

$$z = (p, q), \quad Q = \begin{pmatrix} 0 & I \\ -I & 0 \end{pmatrix}, \quad Q^\mathsf{T} = \begin{pmatrix} 0 & -I \\ I & 0 \end{pmatrix} \tag{A-16.9.8}$$

and rewrite A as

$$A = \int_{T_1}^{T_2} \mathrm{d}t \left[\frac{1}{4} (z_i Q_{ij} \dot{z}_j - \dot{z}_i Q_{ij} z_j) - H(z, t) \right] \tag{A-16.9.9}$$

with the indices enumerating the various degrees of freedom.

For infinitesimal variations

$$z \to \bar{z} = z - \delta z \tag{A-16.9.10}$$

with $\delta H(z, t) = H(z, t) - H(z - \delta z, t)$, we have for the corresponding variation in A

$$\delta A = \int_{T_1}^{T_2} \mathrm{d} \left[\frac{1}{2} \left(\frac{z_i Q_{ij}}{2} \delta z_j - \delta z_i \frac{Q_{ij} z_j}{2} \right) \right]$$

$$+ \int_{T_1}^{T_2} \left[\frac{1}{2} \left(\delta z_i Q_{ij} \, \mathrm{d}z_j - \mathrm{d}z_i Q_{ij} \, \delta z_j \right) - \delta H \, \mathrm{d}t \right] \tag{A-16.9.11}$$

with the order in which the variations δz_i appear kept intact.

We now consider *numerical* variations which commute with the dynamical variables, i.e., $[z_i(t), \delta z_j(t')] = 0$. From the property of Q in (A-16.9.8), the expression for δA simplifies to

$$\delta A = \int_{T_1}^{T_2} \mathrm{d} \left[\frac{z_i Q_{ij}}{2} \delta z_j \right] + \int_{T_1}^{T_2} \left[\delta z_i Q_{ij} \, \mathrm{d}z_j - \delta H \, \mathrm{d}t \right] \tag{A-16.9.12}$$

The principle of stationary action gives

$$\delta H = \delta z_i(t) Q_{ij} \dot{z}_j(t) = -\dot{z}_i(t) Q_{ij} \delta z_j(t) \tag{A-16.9.13}$$

and the variation induces a generator

$$G = \frac{z(t) Q}{2} \delta z(t) \tag{A-16.9.14}$$

which is Hermitian, and generates a unitary operator $U = 1 + iG/\hbar$ for infinitesimal δz, changing the dynamical variables z by $\delta z/2$, i.e.,

$$\frac{\delta z_j}{2} = \frac{1}{i\hbar} [z_j, G] \tag{A-16.9.15}$$

The change brought about in an operator $B(z(t), t)$ by the change of the dynamical variables $z(t)$, as generated by the generator G, is linear in

16.9 Spin & Statistics

$\delta z(t)$, and should also coincide with the transformation given in (A-16.9.15) for $B(z(t), t) = z(t)$. Accordingly quite generally,

$$\frac{\delta B}{2} = \frac{1}{i\hbar}[B, G] \qquad \text{(A-16.9.16)}$$

In particular, (A-16.9.15) implies that

$$\frac{\delta z_j(t)}{2} = \frac{1}{i\hbar}\left[z_j(t), z_k(t)\frac{\delta z_\ell}{2}\right] Q_{k\ell} \qquad \text{(A-16.9.17)}$$

or

$$[z_j(t), z_k(t)] Q_{k\ell} = i\hbar\, \delta_{j\ell}. \qquad \text{(A-16.9.18)}$$

From the expression for Q in (A-16.9.8) this gives

$$[p_i(t), p_j(t)] = 0, \qquad [q_i(t), q_j(t)] = 0, \qquad [q_j(t), p_k(t)] = i\hbar\, \delta_{jk} \qquad \text{(A-16.9.19)}$$

as expected. On the other hand, from (A-16.9.13), (A-16.9.16) we have

$$\frac{\delta H}{2} = \frac{1}{i\hbar}[H, z_k(t)] Q_{k\ell} \frac{\delta z_\ell(t)}{2} = -\dot{z}_j(t) Q_{j\ell} \frac{\delta z_\ell(t)}{2} \qquad \text{(A-16.9.20)}$$

leading to

$$i\hbar \dot{z}_j(t) = [z_j(t), H] \qquad \text{(A-16.9.21)}$$

again as expected, verifying the consistency of the analysis.

We note that Q in (A-16.9.8) is anti-Hermitian ($Q^\dagger = -Q$) which ensures the Hermiticity of the Lagrangian (the integrand) in (A-16.9.9). It also satisfies the restrictive relation $Q^\top = -Q$ as it is real.

Now any square matrix may be written as the sum of a symmetric matrix and an anti-symmetric one (see (16.9.36)). If the square matrix is anti-Hermitian then so are its symmetric and anti-symmetric parts. This allows us to consider, more generally, two classes of Lagrangians:

$$L = \frac{1}{4}(\eta_i Q_{ij} \dot{\eta}_j - \dot{\eta}_i Q_{ij} \eta_j) - H(\eta, t) \qquad \text{(A-16.9.22)}$$

with η real, as before, where ($Q = [Q_{ij}]$)

$$Q^\dagger = -Q \qquad \text{(A-16.9.23)}$$

and *either*

$$Q^\top = Q \qquad \text{(A-16.9.24)}$$

or

$$Q^\top = -Q \qquad \text{(A-16.9.25)}$$

The system satisfying condition (A-16.9.25) was already considered above for a specific matrix Q. The same analysis as above then gives for an arbitrary matrix Q satisfying (A-16.9.23), (A-16.9.25),

$$G = \frac{\eta(t)\, Q}{2}\, \delta\eta(t) \tag{A-16.9.26}$$

and, in particular,

$$[\eta_j(t), \eta_k(t)]\, Q_{k\ell} = i\hbar\, \delta_{j\ell}, \qquad Q^\top = -Q. \tag{A-16.9.27}$$

We will see that c-number variations $\delta\eta_i(t)$ commuting with the variables $\eta_j(t')$ is possible for Q satisfying (A-16.9.25) only, while c-number variations anti-commuting with the variables, so called Grassmann variables already encountered in Chapters 6 and 10, are possible only for Q in (A-16.9.23) satisfying (A-16.9.24).

To reach the above conclusion we may refer to (A-16.9.11) where now Q ($Q^\dagger = -Q$) is arbitrary but satisfies either (A-16.9.24), (A-16.9.25). For Q satisfying (A-16.9.25), δz_j cannot anti-commute with z_i otherwise both expressions within the round brackets in the two integrals in (A-16.9.11) will vanish. On the other hand, for Q satisfying (A-16.9.24), δz_j cannot commute with z_i otherwise both expressions in the round brackets just mentioned will vanish again. Therefore, we may state that

$$[\delta\eta_i(t), \eta_j(t')] = 0, \quad Q^\top = -Q \tag{A-16.9.28}$$

$$\{\delta\eta_i(t), \eta_j(t')\} = 0, \quad Q^\top = Q \tag{A-16.9.29}$$

are *admissible for the corresponding matrix Q but not vice versa*.

In both cases, we have, by referring to (A-16.9.11), (A-16.9.14), that

$$G = \eta Q\, \frac{\delta\eta}{2} \tag{A-16.9.30}$$

and, in particular, the *commutator* in (A-16.9.15), together *with* (A-16.9.28), (A-16.9.29) give ($Q = [Q_{k\ell}]$)

$$[\eta_j(t), \eta_k(t)]\, Q_{k\ell} = i\hbar\, \delta_{j\ell}, \qquad Q^\top = -Q \tag{A-16.9.31}$$

and

$$\{\eta_j(t), \eta_k(t)\}\, Q_{k\ell} = i\hbar\, \delta_{j\ell}, \qquad Q^\top = Q. \tag{A-16.9.32}$$

The above results are extended to systems with an infinite degrees of freedom for actual physical systems which *exist* in the just given section in studying the spin and statistics connection. Here the matrix Q corresponds to Q^0/c in (16.9.32). Further generalizations as arising, for example, from more complicated variations are possible but we will not go into these here.

Problems

16.1. Find the expressions of the gamma matrices as obtained from the Dirac-Pauli representation via the transformation G given in (16.1.30) for fixed **n** and θ.

16.2. Establish the transformation of the matrix $\gamma^0(\boldsymbol{\gamma}\cdot\mathbf{a}+1)$ given in (16.1.35) via the transformation G defined in (16.1.30).

16.3. Verify the expressions (16.2.25)–(16.2.27) as obtained from (16.2.20)–(16.2.23) to obtained the explicit expression of the transformation U_B of a spinor for pure boost in (16.2.28). Show that U_B is not unitary but it satisfies the identity in (16.2.24). Show, as expected, that the transformation U_R for pure rotations in (16.2.29) is unitary.

16.4. In analogy to the analysis carried out through (16.2.20)–(16.2.28) for pure boosts, carry out the corresponding one leading to the expression for the transformation of pure rotations U_R given in (16.2.29).

16.5. Establish the Lorentz vector character of $\overline{\Psi}(x)\gamma^\mu\Psi(x)$ in (16.2.64).

16.6. Derive the expression for the transformed spin $\mathbf{S}'$ in (16.3.6) from (16.3.2), and prove the second equality in (16.3.9).

16.7. Obtain the $\mathcal{PCT}$ product transformation rule of a Dirac spinor.

16.8. Derive the normalization conditions in (16.4.23), (16.4.37), (16.4.39), for $u(\mathbf{p},\sigma)$, $v(\mathbf{p},\sigma')$.

16.9. Show that for the Dirac Hamiltonian in (16.4.51), $[H,\mathbf{S}]\neq\mathbf{0}$, $[H,\mathbf{L}]\neq\mathbf{0}$ but for the total angular momentum $\mathbf{J}=\mathbf{L}+\mathbf{S}$, $[H,\mathbf{J}]=\mathbf{0}$, i.e., total angular momentum is conserved.

16.10. Show that the time evolution operator $\exp(-itH/\hbar)$ via the Dirac Hamiltonian in (16.4.51) takes the simple form given in (16.4.75).

16.11. Show that the only matrices in the complete set of 16 matrices in the ordered set $\{\Gamma'_A\}$ (see (A-16.1.1), (A-16.1.7)) in the chiral representation that commute with H' in (16.5.15), apart from the identity are γ'^5 and linear combinations of the matrices in (16.5.17).

16.12. Carry out the details in the application of $G\mathbf{x}G^{-1}$ leading to the expression in (16.6.18) for the Foldy-Wouthuysen-Tani representation $\mathbf{x}'$ of $\mathbf{x}$.

16.13. Derive the time-dependent expression for the position operator $\mathbf{x}'(t)$ in (16.6.30) by using the time evolution operator given in (16.6.28).

16.14. Derive the Pryce-Newton-Wigner position operator $\mathbf{X}$ in (16.6.38) as obtained from the transformation given (16.6.37), verify the commutation rules in (16.6.39) and obtain the expression for corresponding velocity given in (16.6.40). Finally derive the time-dependence of the Dirac position operator in (16.6.41).

16.15. Carry out the transformation $e^{D'/c}\widetilde{H}_2 e^{-D'/c}$ to obtain, as a first step of the diagonalization process of the Hamiltonian H_2 in (16.6.56), and obtain the expression given in (16.6.62). What is the explicit expression for $\widetilde{O}''$ in the latter equation?.

16.16. Carry out an expansion of E in (A-16.6.35) in powers of $Z^2\alpha^2$ and compare with the results obtained in §7.5.

16.17. (*The Klein-Paradox*) Solve the Dirac equation in a potential which varies only along the direction x given by $V(x)=0$ for $x<0$, $V(x)=V_0$ for $x>0$. For a given energy $E=p^0c$, with $0<E<V_0$ determine

970   16 Quantum Description of Relativistic Particles

the reflected and transmitted current through this potential step. The paradoxical result encountered in your physical interpretation is called the Klein-paradox.

16.18. Verify that with the definitions in (16.7.6), (16.7.9), the Klein-Gordon equation may be rewritten in the form in (16.7.8) which is referred to as the Duffin-Kemmer-Petiau equation with the β^μ matrices as given in (16.7.10), (16.7.11). Establish also the validity of (16.7.12) as satisfied by the β^μ matrices.

16.19. Show that the Hamiltonian of spin 0 particles in the Feshbach-Villars form in (16.7.40) is diagonalized to the form in (16.7.43) via the transformation matrix G in (16.7.41) and obtain the transformed position operator in (16.7.44). Derive also the time-dependence of the latter as given in (16.7.50).

16.20. Find the position operator of a spin 0 particle in the original Feshbach-Villars representation whose Foldy-Wouthuysen-Tani representation is **x**. Interpret your result.

16.21. In analogy to the symmetrization operation $C(\sigma_1, \sigma_2; \sigma)$, for $\sigma = 0, 1, 2$, in (16.8.10)–(16.8.12), find the explicit expressions for the symmetrization operation $C(\sigma_1, \sigma_2, \sigma_3, \sigma_4; \sigma)$ for $\sigma = 0, 1, 2, 3, 4$.

16.22. Prove the properties of the symmetrization operation $C(\sigma_1, \ldots, \sigma_k; \sigma)$ given in (16.8.28), (16.8.29).

16.23. Prove the equations in (16.9.3), (16.9.4), and show that $A(\mathbf{p})$, $B^\dagger(\mathbf{p})$ are given by (16.9.6), (16.9.7) by integrating $\Phi(\mathbf{x})$ over all space and evaluated at any given time. Write the integral expressions for $A^\dagger(\mathbf{p})$, $B(\mathbf{p})$ as well.

16.24. Show that the second equality in (16.9.39) follows from the first one.

16.25. Verify all the commutators involving the Klein-Gordon field $\Phi(x)$ as obtained from the system in (16.9.57). Work out also all the commutators involving the operators A, $A^\dagger$, B, $B^\dagger$ as obtained from (16.9.6), (16.9.7).

Mathematical Appendices

I

Variations of the Baker-Campbell-Hausdorff Formula

The purpose of this appendix is to derive expressions for the product of two exponentials $\exp(A)\exp(B)$ of two non-commuting operators A and B, as well as of taking the derivative of $\exp(A(\tau))$, with respect to a variable τ, of operator-valued functions $A(\tau)$ which are not necessarily linear in τ. Several formulae will be given.

1. Integral Expression for the Product of the Exponentials of Operators

By carrying out the derivative with respect to the parameter λ in the following

$$\frac{d}{d\lambda}\left[e^{\lambda A} e^{-\lambda B}\right] = e^{\lambda A}(A-B)e^{-\lambda B} \tag{I.1.1}$$

gives upon integration over λ from 0 to 1, an integral expression for a product

$$e^A e^{-B} = 1 + \int_0^1 d\lambda\, e^{\lambda A}(A-B)e^{-\lambda B}. \tag{I.1.2}$$

2. Derivative of the Exponential of Operator-Valued Functions

To take the derivative of $\exp(A(\tau))$, with respect to τ, where $A(\tau)$ is an operator valued function, it is wrong to say that it is given by $A'(\tau)\exp(A(\tau))$ or by $\exp(A(\tau))A'(\tau)$. This is not true even for

$$A(\tau) = A_1 + \tau A_2 \tag{I.2.1}$$

where $[A_1, A_2] \neq 0$. It is true, however, for $A(\tau) = \tau A$. More generally, (I.1.2) gives

I Variations of the Baker-Campbell-Hausdorff Formula

$$e^{A(\tau+\Delta\tau)}e^{-A(\tau)} = 1 + \int_0^1 d\lambda\, e^{\lambda A(\tau+\Delta\tau)}\left[A(\tau+\Delta\tau) - A(\tau)\right]e^{-\lambda A(\tau)}. \quad (\text{I.2.2})$$

If we multiply the latter from the right by $\exp(A(\tau))$, this gives for $\Delta\tau \to 0$

$$\left[e^{A(\tau+\Delta\tau)} - e^{A(\tau)}\right] = \Delta\tau\left[\int_0^1 d\lambda\, e^{\lambda A(\tau)}A'(\tau)\,e^{-\lambda A(\tau)}\right]e^{A(\tau)} \quad (\text{I.2.3})$$

or

$$\frac{d}{d\tau}e^{A(\tau)} = \left[\int_0^1 d\lambda\, e^{\lambda A(\tau)}A'(\tau)\,e^{-\lambda A(\tau)}\right]e^{A(\tau)}. \quad (\text{I.2.4})$$

Similarly, one may derive that

$$\frac{d}{d\tau}e^{A(\tau)} = e^{A(\tau)}\left[\int_0^1 d\lambda\, e^{-\lambda A(\tau)}A'(\tau)\,e^{\lambda A(\tau)}\right]. \quad (\text{I.2.5})$$

One may, of course, carry out a formal Taylor expansion in λ and integrate term by term over λ in (I.2.4) to obtain

$$\frac{d}{d\tau}e^{A(\tau)} = \sum_{n=0}^{\infty}\left(\frac{\left[A(\tau),\ldots,\left[A(\tau),A'(\tau)\right]\ldots\right]}{(n+1)!}\right)e^{A(\tau)} \quad (\text{I.2.6})$$

where n denotes the number of $A(\tau)$'s that appear in each term in the summand.

As an example, consider the seemingly simple expression of the form in (I.2.1)

$$A(\tau) = aip + bx\tau \quad (\text{I.2.7})$$

where a, b are some real constants, and p is the momentum operator $p = -i\hbar d/dx$. Then $A'(\tau) = bx$, $[A(\tau), A'(\tau)] = aib\,[p,x] = ab\hbar$, $\left[A(\tau),[A(\tau),A'(\tau)]\right] = 0$. Accordingly, (I.2.6) gives

$$\frac{d}{d\tau}\exp(aip + bx\tau) = \left(bx + \frac{ab\hbar}{2}\right)\exp(aip + bx\tau) \quad (\text{I.2.8})$$

and is neither equal to $A'(\tau)\exp(A(\tau))$ nor to $\exp(A(\tau))A'(\tau)$. Note that (I.2.5) also gives

$$\frac{d}{d\tau}e^{A(\tau)} = e^{A(\tau)}\sum_{n=0}^{\infty}\frac{\left[[\ldots[A'(\tau),A(\tau)],\ldots,A(\tau)\right]}{(n+1)!}. \quad (\text{I.2.9})$$

From which we also obtain

$$\frac{d}{d\tau}\exp(aip + bx\tau) = \exp(aip + bx\tau)\left(bx - \frac{ab\hbar}{2}\right). \quad (\text{I.2.10})$$

3. The Classic Baker-Campbell-Hausdorff Formula

To obtain the formula in question, let $G(\tau) = \tau(A+B) + C(\tau)$, such that $C(0) = 0$, $C'(0) = 0$. Then (I.2.6) or (I.2.9) lead to

$$e^{G(\tau)} = 1 + (A+B)\tau + \left[C''(0) + (A+B)^2\right]\frac{\tau^2}{2!}$$

$$+ \left[C'''(0) + \frac{3}{2}\left[C''(0)(A+B) + (A+B)C''(0)\right] + (A+B)^3\right]\frac{\tau^3}{3!}$$

$$+ \ldots \quad \text{(I.3.1)}$$

On the other hand

$$e^{\tau A} e^{\tau B} = 1 + (A+B)\tau + \left(B^2 + 2AB + A^2\right)\frac{\tau^2}{2!}$$

$$+ \left(B^3 + 3AB^2 + 3A^2B + A^3\right)\frac{\tau^3}{3!} + \ldots \quad \text{(I.3.2)}$$

Upon *equating* (I.3.1) and (I.3.2), and comparing the coefficients of τ^2 and τ^3 we obtain

$$C''(0) = [A, B] \quad \text{(I.3.3)}$$

$$C'''(0) = \frac{1}{2}\left[A, [A, B]\right] + \frac{1}{2}\left[B, [B, A]\right]. \quad \text{(I.3.4)}$$

We may also write

$$\exp G(\tau) = \exp\left[(A+B)\tau + C''(0)\frac{\tau^2}{2!} + C'''(0)\frac{\tau^3}{3!} + \ldots\right] \quad \text{(I.3.5)}$$

which from (I.3.3), (I.3.4), and the fact that we have equated the left-hand sides of (I.3.2), (I.3.5) give

$$e^A e^B = e^{(A+B+C)} \quad \text{(I.3.6)}$$

where

$$C = \frac{1}{2}[A, B] + \frac{1}{12}\left[A, [A, B]\right] + \frac{1}{12}\left[B, [B, A]\right] + \ldots. \quad \text{(I.3.7)}$$

4. A Modification of the Baker-Campbell-Hausdorff Formula

A modification of the formula in (I.3.6) may be obtained as follows. We may use the latter to write

I Variations of the Baker-Campbell-Hausdorff Formula

$$\exp(A+B)\exp(-A)\exp(-B)$$

$$= \exp(A+B)\exp\bigg(-(A+B) + \frac{1}{2}[A,B]$$

$$-\frac{1}{12}[A,[A,B]] - \frac{1}{12}[B,[B,A]] + \ldots\bigg)$$

$$= \exp\bigg(\frac{1}{2}[A,B] + \frac{1}{6}[A,[A,B]] - \frac{1}{3}[B,[B,A]] + \ldots\bigg) \quad \text{(I.4.1)}$$

where in writing the last equality we have used (I.3.6)/(I.3.7) all over again to the product of the two exponentials occurring on its immediate left-hand side. This gives

$$e^{(A+B)} e^{-A} e^{-B} = e^D \quad \text{(I.4.2)}$$

where

$$D = \frac{1}{2}[A,B] + \frac{1}{6}[A,[A,B]] - \frac{1}{3}[B,[B,A]] + \ldots \quad \text{(I.4.3)}$$

II

Convexity and Basic Inequalities

In this appendix, we study the concept of convexity and some of the inequalities which follow from such a property. The inequalities derived, include the Minkowski, the Hölder and Young's inequalities of integrals.

1. General Convexity Theorem

Let $h(x)$ be a real function of a real variable x such that its derivative $h'(x)$ is *non-decreasing*, i.e., $h'(x_1) \leqslant h'(x_2)$ for $x_1 < x_2$. Then for any real numbers $\alpha_1 > 0$, $\alpha_2 > 0$ such that $\alpha_1 + \alpha_2 = 1$

$$h(\alpha_1 a + \alpha_2 b) \leqslant \alpha_1 h(a) + \alpha_2 h(b). \tag{II.1.1}$$

For $a = b$, this trivially holds true with an equality sign. Therefore, without loss of generality, we may take $b > a$.

To prove this, we note that by the Mean-Value Theorem, we may find numbers c and d $(b > a)$:

$$\text{(figure: number line with points } a, c, (\alpha_1 a + \alpha_2 b), d, b\text{)}$$

such that

$$h'(c) = \frac{h(\alpha_1 a + \alpha_2 b) - h(a)}{(\alpha_1 a + \alpha_2 b) - a} = \frac{h(\alpha_1 a + \alpha_2 b) - h(a)}{\alpha_2 (b - a)} \tag{II.1.2}$$

$$h'(d) = \frac{h(b) - h(\alpha_1 a + \alpha_2 b)}{b - (\alpha_1 a + \alpha_2 b)} = \frac{h(b) - h(\alpha_1 a + \alpha_2 b)}{\alpha_1 (b - a)}. \tag{II.1.3}$$

But, by the property of $h'(x)$, $h'(c) \leqslant h'(d)$, hence

$$\frac{h(\alpha_1 a + \alpha_2 b) - h(a)}{\alpha_2(b-a)} \leqslant \frac{h(b) - h(\alpha_1 a + \alpha_2 b)}{\alpha_1(b-a)} \tag{II.1.4}$$

which by rearrangement of terms and recalling the property $\alpha_1 + \alpha_2 = 1$, leads to the inequality in (II.1.1).

As an example, consider the function $h(x) = x^p$, for $x > 0$, $p \geqslant 1$. Then $h'(x) = px^{p-1}$ and is non-decreasing in x, and

$$(\alpha_1 a + \alpha_2 b)^p \leqslant \alpha_1(a)^p + \alpha_2(b)^p \tag{II.1.5}$$

for a, $b > 0$.

As another example, consider the function $h(x) = -\ln x$, $x > 0$. Then $h'(x) = -1/x$, and is increasing in x. That is,

$$-\ln(\alpha_1 a + \alpha_2 b) \leqslant -\alpha_1 \ln a - \alpha_2 \ln b \tag{II.1.6}$$

or

$$\ln(a^{\alpha_1} b^{\alpha_2}) \leqslant \ln(\alpha_1 a + \alpha_2 b) \tag{II.1.7}$$

and

$$a^{\alpha_1} b^{\alpha_2} \leqslant \alpha_1 a + \alpha_2 b \tag{II.1.8}$$

$a, b > 0$.

2. Minkowski's Inequality for Integrals

For $1 \leqslant p$, let

$$\|f\|_p = \left(\int_{\mathbb{R}^\nu} d^\nu \mathbf{x}\, |f(\mathbf{x})|^p\right)^{1/p} > 0 \tag{II.2.1}$$

where $f(\mathbf{x})$ may, in general, be complex. Then the inequality in question reads

$$\|f + g\|_p \leqslant \|f\|_p + \|g\|_p. \tag{II.2.2}$$

To prove this, we note that

$$\frac{|f+g|}{\|f\|_p + \|g\|_p} \leqslant \frac{|f|}{\|f\|_p + \|g\|_p} + \frac{|g|}{\|f\|_p + \|g\|_p} = \alpha_1 \frac{|f|}{\|f\|_p} + \alpha_2 \frac{|g|}{\|g\|_p} \tag{II.2.3}$$

where

$$\alpha_1 = \frac{\|f\|_p}{\|f\|_p + \|g\|_p}, \qquad \alpha_2 = \frac{\|g\|_p}{\|f\|_p + \|g\|_p} \tag{II.2.4}$$

and with $a = |f|/\|f\|_p$, $b = |g|/\|g\|_p$, (II.1.5), dealing with real numbers, gives

$$|f+g|^p \leqslant \left(\|f\|_p + \|g\|_p\right)^p \left(\alpha_1 \frac{|f|^p}{(\|f\|_p)^p} + \alpha_2 \frac{|g|^p}{(\|g\|_p)^p}\right) \qquad \text{(II.2.5)}$$

where we recall that

$$\left(\|f\|_p\right)^p = \int_{\mathbb{R}^\nu} d^\nu \mathbf{x}\, |f(\mathbf{x})|^p. \qquad \text{(II.2.6)}$$

Therefore upon integrating (II.2.5) over $\mathbf{x}$, and taking the $1/p$ root leads to the inequality in (II.2.2) finally by using the fact that $\alpha_1 + \alpha_2 = 1$.

3. Hölder's Inequality for Integrals

Let

$$\langle f | g \rangle = \int_{\mathbb{R}^\nu} d^\nu \mathbf{x}\, f^*(\mathbf{x})\, g(\mathbf{x}) \qquad \text{(II.3.1)}$$

and $\|f\|_p$ as defined in (II.2.1), then inequality in question reads

$$|\langle f | g \rangle| \leqslant \|f\|_p \|g\|_q \qquad \text{(II.3.2)}$$

where $p, q > 1$, such that

$$\frac{1}{p} + \frac{1}{q} = 1. \qquad \text{(II.3.3)}$$

To prove the inequality in (II.3.2), we use (II.1.8) to write

$$\left(\frac{|f(\mathbf{x})|^p}{(\|f\|_p)^p}\right)^{1/p} \left(\frac{|g(\mathbf{x})|^q}{(\|g\|_q)^q}\right)^{1/q} \leqslant \frac{1}{p}\frac{|f(\mathbf{x})|^p}{(\|f\|_p)^p} + \frac{1}{q}\frac{|g(\mathbf{x})|^q}{(\|g\|_q)^q} \qquad \text{(II.3.4)}$$

which upon integration over $\mathbf{x}$ gives

$$\int_{\mathbb{R}^\nu} d^\nu \mathbf{x}\, |f(\mathbf{x})|\, |g(\mathbf{x})| \leqslant \|f\|_p \|g\|_q \left(\frac{1}{p} + \frac{1}{q}\right) = \|f\|_p \|g\|_q. \qquad \text{(II.3.5)}$$

But

$$|\langle f | g \rangle| \leqslant \int_{\mathbb{R}^\nu} d^\nu \mathbf{x}\, |f(\mathbf{x})|\, |g(\mathbf{x})| \qquad \text{(II.3.6)}$$

from which the inequality in (II.3.2) follows.

For $p = q = 2$, (II.3.2), gives the *Cauchy-Schwarz inequality*

$$\left|\int_{\mathbb{R}^\nu} d^\nu \mathbf{x}\, f^*(\mathbf{x})\, g(\mathbf{x})\right| \leqslant \left(\int_{\mathbb{R}^\nu} d^\nu \mathbf{x}\, |f(\mathbf{x})|^2\right)^{1/2} \left(\int_{\mathbb{R}^\nu} d^\nu \mathbf{x}\, |g(\mathbf{x})|^2\right)^{1/2}. \qquad \text{(II.3.7)}$$

4. Young's Inequality for Integrals

We provide one of Young's inequalities and it is given not in the most general cases but in a form which is sufficient for the applications considered in the text as stated:

$$\left| \int_{\mathbb{R}^\nu} d^\nu \mathbf{x} \int_{\mathbb{R}^\nu} d^\nu \mathbf{y}\, f^*(\mathbf{x})\, G(\mathbf{x}-\mathbf{y})\, g(\mathbf{y}) \right| \leqslant \|f\|_p \|G\|_1 \|g\|_q \qquad \text{(II.4.1)}$$

where $p, q > 1$ and satisfy (II.3.3).

To prove (II.4.1), we note that its left-hand side of (II.4.1) is bounded above by

$$\int_{\mathbb{R}^\nu} d^\nu \mathbf{x} \int_{\mathbb{R}^\nu} d^\nu \mathbf{y}\, \left(|f(\mathbf{x})||G(\mathbf{x}-\mathbf{y})|^{1/p}\right)\left(|g(\mathbf{y})||G(\mathbf{x}-\mathbf{y})|^{1/q}\right) \qquad \text{(II.4.2)}$$

which from Hölder's inequality (II.3.2) in 2ν dimensional space, is bounded above by

$$\left(\int_{\mathbb{R}^\nu} d^\nu \mathbf{x} \int_{\mathbb{R}^\nu} d^\nu \mathbf{y}\, |f(\mathbf{x})|^p |G(\mathbf{x}-\mathbf{y})|\right)^{1/p} \left(\int_{\mathbb{R}^\nu} d^\nu \mathbf{x} \int_{\mathbb{R}^\nu} d^\nu \mathbf{y}\, |g(\mathbf{y})|^q |G(\mathbf{x}-\mathbf{y})|\right)^{1/q}$$

$$= \|f\|_p \|G\|_1 \|g\|_q \qquad \text{(II.4.3)}$$

where we have used (II.3.3) in writing $\|G\|_1$,

$$\|G\|_1 = \int_{\mathbb{R}^\nu} d^\nu \mathbf{x}\, |G(\mathbf{x})|. \qquad \text{(II.4.4)}$$

III

The Poisson Equation in 4D

In four dimensional Euclidean space, a vector
$$\mathbf{r} = \left(x^1, x^2, x^3, x^4\right) \tag{III.1}$$
in spherical coordinates may be written with components as
$$\begin{aligned} x^1 &= r \sin\theta \cos\phi \sin\chi \\ x^2 &= r \sin\theta \sin\phi \sin\chi \\ x^3 &= r \cos\theta \sin\chi \\ x^4 &= r \cos\chi \end{aligned} \tag{III.2}$$
when $r = |\mathbf{r}|$, and
$$0 \leqslant r < \infty, \quad 0 \leqslant \theta \leqslant \pi, \quad 0 \leqslant \phi \leqslant 2\pi, \quad 0 \leqslant \chi \leqslant \pi. \tag{III.3}$$

For the Jacobian of the transformation $(x^1, x^2, x^3, x^4) \to (r, \theta, \phi, \chi)$, we have
$$|J| = \left| \frac{\partial\left(x^1, x^2, x^3, x^4\right)}{\partial(r, \theta, \phi, \chi)} \right| = r^3 \sin\theta \sin^2\chi \tag{III.4}$$
and the volume element is given by
$$d^4\mathbf{r} = r^3 dr \, \sin\theta \, d\theta \, d\phi \, \sin^2\chi \, d\chi. \tag{III.5}$$

The volume $V(R)$ of a sphere of radius R in 4D is then equal to
$$V(R) = \int_0^R r^3 dr \int_0^\pi \sin\theta \, d\theta \int_0^{2\pi} d\phi \int_0^\pi \sin^2\chi \, d\chi = \frac{\pi^2 R^4}{2}. \tag{III.6}$$

To obtain the surface area $S(R)$ of the sphere of radius R in 4D, we replace, in the process, $r^3 dr$ by R^3 in (III.5), giving

III The Poisson Equation in 4D

$$S(R) = 2\pi^2 R^3 \tag{III.7}$$

with the solid angle element

$$d\Omega = \sin\theta \, d\theta \, d\phi \, \sin^2\chi \, d\chi. \tag{III.8}$$

The Dirac delta distributions are given by

$$\delta^4(\mathbf{r} - \mathbf{r}') = \frac{\delta(r - r')}{r^3} \frac{\delta(\theta - \theta')\,\delta(\phi - \phi')\,\delta(\chi - \chi')}{\sin\theta \, \sin^2\chi} \tag{III.9}$$

and

$$\delta(\Omega - \Omega') = \frac{\delta(\theta - \theta')\,\delta(\phi - \phi')\,\delta(\chi - \chi')}{\sin\theta \, \sin^2\chi}. \tag{III.10}$$

1. The Poisson Equation

The Poisson equation in $4D$ is the content of the following result:

$$\nabla^2 \frac{1}{|\mathbf{r} - \mathbf{r}'|^2} = -4\pi^2 \delta^4(\mathbf{r} - \mathbf{r}'). \tag{III.1.1}$$

To establish this, let

$$\nabla^2 G(\mathbf{r}, \mathbf{r}') = -\delta^4(\mathbf{r} - \mathbf{r}') \tag{III.1.2}$$

and upon writing

$$\delta^4(\mathbf{r} - \mathbf{r}') = \int \frac{d^4k}{(2\pi)^4} e^{i\mathbf{k}\cdot(\mathbf{r} - \mathbf{r}')} \tag{III.1.3}$$

we obtain

$$G(\mathbf{r}, \mathbf{r}') = \int \frac{d^4k}{(2\pi)^4} \frac{e^{i\mathbf{k}\cdot(\mathbf{r} - \mathbf{r}')}}{k^2}. \tag{III.1.4}$$

The expression on the right-hand side may be rewritten as: $(\rho = |\mathbf{r} - \mathbf{r}'|)$

$$\frac{(2\pi)}{(2\pi)^4} \frac{2}{\rho} \int_0^\infty dk \int_0^\pi d\chi \, i \left(\frac{d}{d\chi} e^{ik\rho\cos\chi} \right) \sin\chi$$

$$= -\frac{i}{4\pi^3 \rho} \int_0^\infty dk \int_0^\pi d\chi \, \cos\chi \, e^{ik\rho\cos\chi}$$

$$= -\frac{1}{4\pi^3 \rho^2} \int_0^\infty dk \, \frac{d}{dk} \int_0^\pi d\chi \, e^{ik\rho\cos\chi}. \tag{III.1.5}$$

Using the properties,

$$\int_0^\pi d\chi \, e^{ik\rho\cos\chi} = \pi J_0(k\rho), \qquad J_0(0) = 1, \qquad J_0(\infty) = 0 \tag{III.1.6}$$

where $J_0(z)$ is the Bessel function of order zero, in (III.1.5) establishes the result stated in (III.1.1).

2. Generating Function

In 3D, $|\mathbf{r}-\mathbf{r}'|^{-1}$ may be expanded in terms of Legendre polynomials with well known orthogonality properties. We extend such a procedure to 4D.

To the above end, we introduce the special functions referred to as Chebyshev's polynomials $U_n(x)$ of type II. The generating function of these special functions is given by

$$\frac{1}{(1+t^2-2tx)} = \sum_{n=0}^{\infty} t^n U_n(x), \qquad |x|<1, \quad |t|<1. \tag{III.2.1}$$

The following properties of the $U_n(x)$ are to be noted:

$$U_0(x) = 1, \quad U_1(x) = 2x, \quad U_2(x) = 4x^2 - 1,$$

$$U_3(x) = 4x(2x^2-1), \quad U_4(x) = 16x^4 - 12x^2 + 1, \ldots \tag{III.2.2}$$

$$U_n(-x) = (-1)^n U_n(x) \tag{III.2.3}$$

$$U_{n+1}(x) = 2x U_n(x) - U_{n-1}(x). \tag{III.2.4}$$

They satisfy the differential equation

$$(1-x^2)U_n''(x) - 3x U_n'(x) + n(n+2) U_n(x) = 0, \tag{III.2.5}$$

the orthogonality relation

$$\int_{-1}^{1} dx \sqrt{1-x^2}\, U_n(x) U_m(x) = \frac{\pi}{2} \delta_{nm}, \tag{III.2.6}$$

and the following integral

$$\int_{-1}^{1} dz\, U_n\!\left(xy + \sqrt{1-x^2}\sqrt{1-y^2}\, z\right) = 2\frac{U_n(x) U_n(y)}{(n+1)}. \tag{III.2.7}$$

In particular, since $U_0(x) = 1$, (III.2.6) gives

$$\int_{-1}^{1} dx \sqrt{1-x^2}\, U_n(x) = \frac{\pi}{2} \delta_{n0}. \tag{III.2.8}$$

The following explicit expression is also useful

$$U_n(x) = \frac{\sin\bigl((n+1)\arccos x\bigr)}{\sin(\arccos x)}. \tag{III.2.9}$$

For $x \to 1$, the latter gives

$$U_n(1) = (n+1). \tag{III.2.10}$$

For $x = 0$, we also have

$$U_{2n}(0) = (-1)^n \tag{III.2.11}$$

$$U_{2n+1}(0) = 0. \tag{III.2.12}$$

3. Expansion Theorem

Let

$$r_< = \min(r, r'), \quad r_> = \max(r, r'), \quad x = \frac{\mathbf{r} \cdot \mathbf{r}'}{rr'}, \tag{III.3.1}$$

then the expansion theorem (Theorem III.1) directly follows from (III.2.1).

Theorem III.1

$$\frac{1}{|\mathbf{r} - \mathbf{r}'|^2} = \sum_{n=0}^{\infty} \frac{(r_<)^n}{(r_>)^{2+n}} U_n\left(\frac{\mathbf{r} \cdot \mathbf{r}'}{rr'}\right). \tag{III.3.2}$$

To establish a result given below, we note that the above expression may be conveniently rewritten as

$$\frac{1}{|\mathbf{r} - \mathbf{r}'|^2} = \sum_{n=0}^{\infty} \left\{ \Theta(r - r') \, (r')^n \left[r^{-2-n} U_n(x) \right] \right.$$

$$\left. + \Theta(r' - r) \, (r')^{-2-n} \left[r^n U_n(x) \right] \right\} \tag{III.3.3}$$

where x is defined in (III.3.1).

By using the properties

$$\frac{\partial}{\partial x^i} \Theta(r - r') = \frac{x^i}{r} \delta(r - r') \tag{III.3.4}$$

$$\nabla^2 x = -\frac{3x}{r^2} \tag{III.3.5}$$

$$\nabla^2 r = \frac{3}{r} \tag{III.3.6}$$

$$\sum_{i=1}^{4} \left(\frac{\partial r}{\partial x^i}\right)^2 = 1 \tag{III.3.7}$$

$$\sum_{i=1}^{4} \left(\frac{\partial x}{\partial x^i}\right)^2 = \frac{(1 - x^2)}{r^2} \tag{III.3.8}$$

$$\sum_{i=1}^{4} \left(\frac{\partial r}{\partial x^i}\right)\left(\frac{\partial x}{\partial x^i}\right) = 0 \tag{III.3.9}$$

where x is defined in (III.3.1), and the differential equation (III.2.5), we obtain from (III.3.3) that

$$\nabla^2 \frac{1}{|\mathbf{r} - \mathbf{r}'|^2} = -\frac{\delta(r - r')}{r^3} \sum_{n=0}^{\infty} 2(n+1) U_n(x). \qquad \text{(III.3.10)}$$

Upon comparison of (III.3.10) with (III.1.1) and using (III.9), (III.10), we may infer the following result:

$$\sum_{n=0}^{\infty} \frac{(n+1)}{2\pi^2} U_n(x) = \delta(\Omega - \Omega') \qquad \text{(III.3.11)}$$

where

$$x = \mathbf{N} \cdot \mathbf{N}' \qquad \text{(III.3.12)}$$

and $\mathbf{N}, \mathbf{N}'$ are unit vectors in $4D$.

Now we prove the following indispensable orthogonality/completeness relation over angles.

4. Generalized Orthogonality Relation

Theorem IV.1

$$\int d\Omega'' \, U_n(\mathbf{N} \cdot \mathbf{N}'') \, U_m(\mathbf{N}'' \cdot \mathbf{N}') = \frac{2\pi^2}{(n+1)} \delta_{nm} \, U_n(\mathbf{N} \cdot \mathbf{N}') \qquad \text{(III.4.1)}$$

where $\mathbf{N}, \mathbf{N}', \mathbf{N}''$ are unit vectors in $4D$.

Without loss of generality we may choose the x^4 axis along $\mathbf{N}'$, and set

$$\mathbf{N} \cdot \mathbf{N}' = \cos \chi, \qquad \mathbf{N}'' \cdot \mathbf{N}' = \cos \chi''. \qquad \text{(III.4.2)}$$

We may then write

$$\mathbf{N} \cdot \mathbf{N}'' = \cos \chi \cos \chi'' + \sin \chi \sin \chi'' \cos \alpha \qquad \text{(III.4.3)}$$

where

$$\cos \alpha = \cos \theta \cos \theta'' + \sin \theta \sin \theta'' \cos (\phi - \phi''). \qquad \text{(III.4.4)}$$

We may expand $U_n(\mathbf{N} \cdot \mathbf{N}'')$ in terms of Legendre polynomials in the angle α, i.e.,

$$U_n(\mathbf{N} \cdot \mathbf{N}'') = \sum_{k=0}^{n} A_{nk}(\chi, \chi'') \, P_k(\cos \alpha) \qquad \text{(III.4.5)}$$

where, in detail,

$$P_k(\cos \alpha) = \frac{4\pi}{(2k+1)} \sum_{m'=-k}^{k} Y_{km'}(\theta, \phi) \, Y_{km'}^*(\theta'', \phi''). \qquad \text{(III.4.6)}$$

Upon integration of (III.4.5) over $(\cos\alpha)$: $-1 \leq \cos\alpha \leq 1$, using the integral (III.2.7), with the argument of U_n in the integrand given in (III.4.3), and the fact that

$$\int_{-1}^{1} d(\cos\alpha)\, P_k(\cos\alpha) = 2\delta_{k0} \tag{III.4.7}$$

we obtain for the coefficient A_{n0}:

$$A_{n0}(\chi,\chi'') = \frac{U_n(\cos\chi)\, U_n(\cos\chi'')}{(n+1)}. \tag{III.4.8}$$

For the integral on the left-hand side of (III.4.1), the latter is explicitly given by

$$\sum_{k=0}^{n} \sum_{m'=-k}^{k} \frac{4\pi}{(2k+1)} Y_{km'}(\theta,\phi) \int d\Omega''\, A_{nk}(\chi,\chi'')\, Y^*_{km'}(\theta'',\phi'')\, U_m(\cos\chi''). \tag{III.4.9}$$

From the definition of the solid angle element $d\Omega''$ (see (III.8)), and the integral

$$\int_0^{2\pi} d\phi'' \int_0^{\pi} d\theta''\, \sin\theta''\, Y_{km'}(\theta'',\phi'') = \sqrt{4\pi}\, \delta_{k0}\, \delta_{m'0} \tag{III.4.10}$$

(III.4.9) reduces to the expression

$$4\pi \frac{1}{\sqrt{4\pi}} \sqrt{4\pi} \int_0^{\pi} d\chi''\, A_{n0}(\chi,\chi'')\, U_m(\cos\chi'') \sin^2\chi''$$

$$= \frac{4\pi}{(n+1)} U_n(\cos\chi) \int_0^{\pi} d\chi''\, U_n(\cos\chi'')\, U_m(\cos\chi'') \sin^2\chi'' \tag{III.4.11}$$

where in writing the last step, we have used (III.4.8). The integral in (III.4.11) satisfies the orthogonality relation in (III.2.6). Hence by using the fact that $\cos\chi = \mathbf{N} \cdot \mathbf{N}'$, the result stated in Theorem IV.1 immediately follows.

References

Abramowitz, M. and Stegun, I. A. (Editors) (1972). *Handbook of Mathematical Functions*. Dover, New York.

Aharonov, Y. and Anandan, J. (1987). "Phase Change during a Cyclic Quantum Evolution". *Phys. Rev. Lett.* **58** (16), pp. 1593–1596. Reprinted in Shapere and Wilczek (1989) pp. 145–148.

Aitchison, I. J. R. (1987). "Berry Phases, Magnetic Monopoles, and Wess-Zumino Terms of How the Skyrmion Got Its Spin". *Acta Phys. Pol. B* **18** (3), pp. 207–235. Reprinted in Shapere and Wilczek (1989) pp. 380–408.

Amrein, W. O. (1981). *Non-Relativistic Quantum Dynamics*. D. Reidel Publishing Co., Dordrecht.

Amrein, W. O., Martin, P. A. and Misra, B. (1970). "On the Asymptotic Condition of Scattering Theory". *Helv. Phys. Acta* **43**, pp. 313–344.

Arfken, G. B. and Weber, H. J. (1995). *Mathematical Methods for Physicists*. Academic Press, San Diego, 4th edn.

Arunasalam, V. (1969). "The Electromagnetic Shift of Landau Levels and the Magnetic Moment of the Electron". *Am. J. Phys.* **37** (9), pp. 877–881.

Aspect, A., Grangier, P. and Roger, G. (1982). "Experimental Realization of Einstein-Podolsky-Rosen-Bohm Gedankenexperiment: A New Violation of Bell's Inequalities". *Phys. Rev. Lett.* **49** (2), pp. 91–94.

Au, C.-K. and Feinberg, G. (1974). "Effects of Retardation on Electromagnetic Self-Energy of Atomic States". *Phys. Rev. A* **9** (5), pp. 1794–1800. [Erratum: *ibid.* **12**, 1733 (1975). Addendum: *ibid.* **12**, 1722 (1975).].

Auberson, G., Mahoux, G. and Mennessier, G. (1978). "On the Perturbation Theory of the Anharmonic Oscillator at Large Orders". *Nuovo Cimento* **48 A**, pp. 1–23.

Avron, J. E., Howland, J. S. and Simon, B. (1990). "Adiabatic Theorems for Dense Point Spectra". *Commun. Math. Phys.* **128** (3), pp. 497–507.

Bargmann, V. (1952). "On the Number of Bound States in a Central Field of Force". *Proc. Nat. Acad. Sci. USA* **38** (11), pp. 961–966.

Batelaan, H., Gay, T. J. and Schwendiman, J. J. (1997). "Stern-Gerlach Effect for Electron Beams". *Phys. Rev. Lett.* **79** (23), pp. 4517–4521.

Bell, J. S. (1989). *Speakable and Unspeakable in Quantum Mechanics*. Cambridge University Press, Cambridge.

Bender, C. M. and Wu, T. T. (1971). "Large-Order Behavior of Perturbation Theory". *Phys. Rev. Lett.* **27** (7), pp. 461–465.

Bennett, C. H., Brassard, G. and Ekert, A. K. (1992). "Quantum Cryptography". *Scientific American* **267** (10), pp. 50–57.

Bennett, C. H. et al. (1993). "Teleporting an Unknown Quantum State via Dual Classical and Einstein-Podolsky-Rosen Channels". *Phys. Rev. Lett.* **70** (13), pp. 1895–1899.

Berry, M. V. (1984). "Quantal Phase Factors Accompanying Adiabatic Changes". *Proc. R. Soc. Lond. A* **392** (1802), pp. 45–57. Reprinted in Shapere and Wilczek (1989) pp. 124–136.

Berry, M. V. (1987). "The Adiabatic Phase and Pancharatnam's Phase for Polarized Light". *J. Mod. Optics* **34** (11), pp. 1401–1407. Reprinted in Shapere and Wilczek (1989) pp. 67–73.

Bethe, H. A. (1947). "The Electromagnetic Shift of Energy Levels". *Phys. Rev.* **72** (4), pp. 339–341.

Bethe, H. A., Brown, L. M. and Stehn, J. R. (1950). "Numerical Value of the Lamb Shift". *Phys. Rev.* **77** (3), pp. 370–374.

Bethe, H. A. and Salpeter, E. E. (1977). *Quantum Mechanics of One- and Two-Electron Atoms*. Plenum Publishing, New York.

Bitter, T. and Dubbers, D. (1987). "Manifestation of Berry's Topological Phase in Neutron Spin Rotation". *Phys. Rev. Lett.* **59** (3), pp. 251–254.

Born, M. and Wolf, E. (1975). *Principles of Optics*. Pergamon Press, Oxford, 5th edn.

Bouwmeester, D. et al. (1997). "Experimental Quantum Teleportation". *Nature* **390**, pp. 575–579.

Brillouin, L. (1928). "Is it Possible to Test by a Direct Experiment the Hypothesis of the Spinning Electron?" *Proc. Nat. Acad. Sci. USA* **14** (10), pp. 755–763.

Brune, M. et al. (1992). "Manipulation of Photons in a Cavity by Dispersive Atom-Field Coupling: Quantum-Nondemolition Measurements and Generation of "Schrödinger Cat" States". *Phys. Rev. A* **45** (7), pp. 5193–5214.

Brune, M. et al. (1996). "Observing the Progressive Decoherence of the "Meter" in a Quantum Measurement". *Phys. Rev. Lett.* **77** (24), pp. 4887–4890.

Byrne, J. (1978). "Young's Double Beam Interference Experiment with Spinor and Vector Waves". *Nature* **275** (5677), pp. 188–191.

Chiao, R. Y., Kwiat, P. G. and Steinberg, A. M. (1994). "Optical Tests of Quantum Mechanicss". In Bederson, B. and Walther, H. (Editors), "Advances in Atomic, Molecular, and Optical Physics, Vol. 34", Academic Press, New York, p. 35.

Clauser, J. F. and Horne, M. A. (1974). "Experimental Consequences of Objective Local Theories". *Phys. Rev. D* **10** (2), pp. 526–535.

Clauser, J. F. and Shimony, A. (1978). "Bell's Theorem. Experimental Tests and Implications". *Rep. Prog. Phys.* **41** (12), pp. 1881–1927.

Clauser, J. F. *et al.* (1969). "Proposed Experiment to Test Local Hidden-Variable Theories". *Phys. Rev. Lett.* **23** (15), pp. 880–884. [Erratum: *ibid.* **24** (10), 549 (1970).].

Colella, R., Overhauser, A. W. and Werner, S. A. (1975). "Observation of Gravitationally Induced Quantum Interference". *Phys. Rev. Lett.* **34** (23), pp. 1472–1474.

Conlon, J. G., Lieb, E. H. and Yau, H.-T. (1988). "The $N^{7/5}$ Law for Charged Bosons". *Commun. Math. Phys.* **116** (3), pp. 417–448. Reprinted in Thirring (1991) pp. 435–466.

Conte, M., Penzo, A. and Pusterla, M. (1995). "Spin Splitting due to Longitudinal Stern-Gerlach Kicks". *Nuovo Cimento* **108 A** (1), pp. 127–136.

Cooper, F., Khare, A. and Sukhatme, U. (1995). "Supersymmetry and Quantum Mechanics". *Phys. Rep.* **251** (5–6), pp. 267–385. [ArXiv: hep-th/9405029].

Daubechies, I. and Lieb, E. H. (1983). "One-Electron Relativistic Molecules with Coulomb Interaction". *Commun. Math. Phys.* **90** (4), pp. 497–510. Reprinted in Thirring (1991) pp. 467–480.

de Crombrugghe, M. and Rittenberg, V. (1983). "Supersymmetric Quantum Mechanics". *Ann. Phys. (N.Y.)* **151** (1), pp. 99–126.

Degasperis, A., Fonda, L. and Ghirardi, G. C. (1973). "Does the Lifetime of an Unstable System Depend on the Measuring Apparatus?" *Nuovo Cimento* **21 A**, pp. 471–492.

Dehmelt, H. (1990). "Experiments on the Structure of an Individual Elementary Particle". *Science* **247** (4942), pp. 539–545.

Dirac, P. A. M. (1931). "Quantised Singularities in the Electromagnetic Field". *Proc. R. Soc. Lond. A* **133**, pp. 60–72.

Dirac, P. A. M. (1959). "Fixation of Coordinates in the Hamiltonian Theory of Gravitation". *Phys. Rev.* **114** (3), pp. 924–930.

Dirac, P. A. M. (2001). *Lectures on Quantum Mechanics*. Dover, Mineola, New York.

Dollard, J. D. (1964). "Asymptotic Convergence and the Coulomb Interaction". *J. Math. Phys.* **5** (6), pp. 729–738.

Dollard, J. D. (1969). "Scattering into Cones I: Potential Scattering". *Commun. Math. Phys.* **12** (3), pp. 193–203.

Dollard, J. D. and Vello, G. (1966). "Asymptotic Behaviour of a Dirac Particle in a Coulomb Potential". *Nuovo Cimento* **45**, pp. 801–812.

Dyson, F. J. (1967). "Ground-State Energy of a Finite System of Charged Particles". *J. Math. Phys.* **8** (8), pp. 1538–1545.

Dyson, F. J. and Lenard, A. (1967). "Stability of Matter. I". *J. Math. Phys.* **8** (3), pp. 423–434.

Ehrenfest, P. (1959). "Ansprache zur Verleihung der Lorentzmedaille an Professor Wolfgang Pauli am 31 Oktober 1931. (Address on award of Lorentz medal to Professor Wolfgang Pauli on 31 October 1931)". In Klein, M. J. (Editor), "Paul Ehrenfest: Collected scientific papers", North-Holland, Amsterdam, p. 617. [The address appeared originally in P. Ehrenfest (1931). *Versl. Akad. Amsterdam* **40**, pp. 121–126.].

Einstein, A., Podolsky, B. and Rosen, N. (1935). "Can Quantum-Mechanical Description of Physical Reality Be Considered Complete?" *Phys. Rev.* **47** (10), pp. 777–780.

Englert, B.-G. and Schwinger, J. (1984). "Statistical Atom: Handling the Strongly Bound Electrons". *Phys. Rev. A* **29** (5), pp. 2331–2338.

Faddeev, L. D. (1960). "Scattering Theory for a System of Three Particles". *Zh. Eksp. Teor. Fiz.* **39**, pp. 1459–1467. [English translation: *Sov. Phys. JETP* **12**, 1014–1019 (1961).].

Faddeev, L. D. (1961). "The Resolvent of the Schrödinger Operator for a System of Three Particles Interacting in Pairs". *Dokl. Akad. Nauk SSSR* **138**, pp. 565–567. [English translation: *Sov. Phys. Dokl.* **6**, 384–386 (1961).].

Faddeev, L. D. (1962). "The Construction of the Resolvent of the Schrödinger Operator for a Three-Particle System, and the Scattering Problem". *Dokl. Akad. Nauk SSSR* **145**, pp. 301–304. [English translation: *Sov. Phys. Dokl.* **7**, 600–602 (1963).].

Faddeev, L. D. (1969). "Feynman Integral for Singular Lagrangians". *Teor. Mat. Fiz.* **1** (1), pp. 3–18. [English translation: *Theor. Math. Phys.* **1** (1), 1–13 (1969).].

Fernow, R. C. (1976). "Expansions of Spin-1/2 Expectation Values". *Am. J. Phys.* **44** (6), pp. 560–563.

Feshbach, H. and Villars, F. (1958). "Elementary Relativistic Wave Mechanics of Spin 0 and Spin 1/2 Particles". *Rev. Mod. Phys.* **30** (1), pp. 24–45.

Feynman, R. P. (1948). "Space-Time Approach to Non-Relativistic Quantum Mechanics". *Rev. Mod. Phys.* **20** (2), pp. 367–387.

Feynman, R. P. (1985). *QED: The Strange Theory of Light and Matter*. Princeton University Press, Princeton, New Jersey.

Feynman, R. P. and Hibbs, A. R. (1965). *Quantum Mechanics and Path Integrals*. McGraw-Hill, New York.

Feynman, R. P., Vernon Jr., F. L. and Hellwarth, R. W. (1957). "Geometrical Representation of the Schrödinger Equation for Solving Maser Problems". *J. Appl. Phys.* **28** (1), pp. 49–52.

Foldy, L. L. and Wouthuysen, S. A. (1950). "On the Dirac Theory of Spin 1/2 Particles and Its Non-Relativistic Limit". *Phys. Rev.* **78** (1), pp. 29–36.

Fonda, L., Ghirardi, G. C. and Rimini, A. (1978). "Decay Theory of Unstable Quantum Systems". *Rep. Prog. Phys.* **41** (4), pp. 587–631.

Gallup, G. A., Batelaan, H. and Gay, T. J. (2001). "Quantum-Mechanical Analysis of a Longitudinal Stern-Gerlach Effect". *Phys. Rev. Lett.* **86** (20), pp. 4508–4511.

Gendenshteïn, L. E. (1983). "Derivation of Exact Spectra of the Schrödinger Equation by Means of Supersymmetry". *JETP Lett.* **38** (6), pp. 356–359.

Gerry, C. C. and Knight, P. L. (1997). "Quantum Superpositions and Schrödinger Cat States in Quantum Optics". *Am. J. Phys.* **65** (10), pp. 964–974.

Gerry, C. C. and Singh, V. A. (1983). "Remarks on the Effects of Topology in the Aharonov-Bohm Effect". *Nuovo Cimento* **73 B**, pp. 161–170.

Goldberger, M. L. and Watson, K. M. (1964). *Collision Theory.* John Wiley & Sons, New York.

Goodman, B. and Ignjatović, S. R. (1997). "A Simpler Solution of the Dirac Equation in a Coulomb Potential". *Am. J. Phys.* **65** (3), pp. 214–221.

Gottfried, K. (1989). *Quantum Mechanics.* Addison-Wesley, Reading, Massachusetts.

Gradshteyn, I. S. and Ryzhik, I. M. (1965). *Table of Integrals, Series and Products.* Academic Press, San Diego.

Greenberger, D. M. and Overhauser, A. W. (1979). "Coherence Effects in Neutron Diffraction and Gravity Experiments". *Rev. Mod. Phys.* **51** (1), pp. 43–78.

Grinyuk, B. E. (1980). "New Variants of Many-Particle Faddeev Equations". *Teor. Mat. Fiz.* **43** (3), pp. 386–400. [English translation: *Theor. Math. Phys.* **43** (3), 532–542 (1980).].

Grotch, H. (1981). "Lamb Shift in Nonrelativistic Quantum Electrodynamics". *Am. J. Phys.* **49** (1), pp. 48–51. [Erratum: *ibid.* **49** (7), 699 (1981).].

Grotch, H. and Kazes, E. (1977). "Nonrelativistic Quantum Mechanics and the Anomalous Part of the Electron g Factor". *Am. J. Phys.* **45** (7), pp. 618–623.

Hannay, J. H. (1985). "Angle Variable Holonomy in Adiabatic Excursion of an Integrable Hamiltonian". *J. Phys. A: Math. Gen.* **18** (2), pp. 221–230. Reprinted in Shapere and Wilczek (1989) pp. 426–435.

Itano, W. M. *et al.* (1990). "Quantum Zeno Effect". *Phys. Rev. A* **41** (5), pp. 2295–2300.

Johnston, G. L. and Sposito, G. (1976). "On the Stark Effect of the Plane Rotator". *Am. J. Phys.* **44** (8), pp. 723–728.

Kaempffer, F. A. (1965). *Concepts in Quantum Mechanics.* Academic Press, New York.

Kato, T. (1951a). "Fundamental Properties of Hamiltonian Operator of the Schrödinger Type". *Trans. Amer. Math. Soc.* **70**, pp. 196–211.

Kato, T. (1951b). "On the Adiabatic Theorem of Quantum Mechanics". *J. Phys. Soc. Jap.* **5**, pp. 435–439.

Kato, T. (1966). *Perturbation Theory for Linear Operators.* Springer-Verlag, Berlin.

Kato, T. (1967). "Some Mathematical Problems in Quantum Mechanics". *Prog. Theor. Phys. Suppl.* **40**, pp. 3–19.

Khalfin, L. A. (1957). "Contribution to the Decay Theory of a Quasi-Stationary State". *Zh. Eksp. Teor. Fiz.* **33**, p. 1371. [English translation: *Sov. Phys. JETP* **6**, 1053 (1958).].

Khalfin, L. A. (1990). *Usp. Fiz. Nauk* **160**, p. 185.

Khare, A. and Maharana, J. (1984). "Supersymmetric Quantum Mechanics in One, Two and Three Dimensions". *Nucl. Phys. B* **244** (2), pp. 409–420.

Klein, A. G. and Opat, G. I. (1976). "Observation of 2π Rotations by Fresnel Diffraction of Neutrons". *Phys. Rev. Lett.* **37** (5), pp. 238–240.

Klempt, E. (1976). "Observability of the Sign of Wave Functions". *Phys. Rev. D* **13** (11), pp. 3125–3129.

Koshino, K. and Shimizu, A. (2004). "Quantum Zeno Effect for Exponentially Decaying Systems". *Phys. Rev. Lett.* **92** (3), p. 030401.

Lam, C. S. (1965). "Feynman Rules and Feynman Integrals for System with Higher-Spin Fields". *Nuovo Cimento* **38** (4), pp. 1755–1765.

Lamb, W. E. and Retherford, R. C. (1947). "Fine Structure of the Hydrogen Atom by a Microwave Method". *Ann. Phys. (N.Y.)* **72** (3), pp. 241–243.

Lenard, A. and Dyson, F. J. (1968). "Stability of Matter. II". *J. Math. Phys.* **9** (5), pp. 698–711.

Lieb, E. H. (1976). "The Stability of Matter". *Rev. Mod. Phys.* **48** (4), pp. 553–569. Reprinted in Thirring (1991) pp. 483–499.

Lieb, E. H. (1979). "The $N^{5/3}$ Law for Bosons". *Phys. Lett. A* **70** (2), pp. 71–73. Reprinted in Thirring (1991) pp. 327–329.

Lieb, E. H. (2000). "Lieb-Thirring Inequalities". In "Kluwer Encyclopedia of Mathematics, Supplement Vol. II", Kluwer, Dordrecht, pp. 311–313. [ArXiv: math-ph/0003039].

Lieb, E. H. and Simon, B. (1973). "Thomas-Fermi Theory Revisited". *Phys. Rev. Lett.* **31** (11), pp. 681–683.

Lieb, E. H. and Thirring, W. E. (1975). "Bound for the Kinetic Energy of Fermions Which Proves the Stability of Matter". *Phys. Rev. Lett.* **35** (11), pp. 687–689. [Erratum: *ibid.* **35** (16), 1116 (1975).]. Reprinted in Thirring (1991) pp. 323–326.

Lieb, E. H. and Thirring, W. E. (1976). "Inequalities for the Moments of the Eigenvalues of the Schrödinger Hamiltonian and Their Relation to Sobolev Inequalities". In Lieb, E. H., Simon, B. and Wightman, A. S. (Editors), "Studies in Mathematical Physics: Essays in Honor of Valentine Bargmann", Princeton University Press, Princeton, New Jersey, pp. 269–303. Reprinted in Thirring (1991) pp. 135–169.

Lin, Q.-G. (2002). "Geometric Phases for Neutral and Charged Particles in a Time-Dependent Magnetic Field". *J. Phys. A: Math. Gen.* **35** (2), pp. 377–391.

Loeffel, J. J. *et al.* (1969). "Padé Approximants and the Anharmonic Oscillator". *Phys. Lett. B* **30** (9), pp. 656–658.

Manoukian, E. B. (1983). *Renormalization*. Academic Press, New York.

Manoukian, E. B. (1985). "Quantum Action Principle and Path Integrals for Long-Range Interactions". *Nuovo Cimento* **90 A** (11), pp. 295–307.

Manoukian, E. B. (1986a). "Action Principle and Quantization of Gauge Fields". *Phys. Rev. D* **34** (12), pp. 3739–3749.

Manoukian, E. B. (1986b). "Exact Solutions for Stimulated Emissions by External Source". *Int. J. Theor. Phys.* **25** (2), pp. 147–158.

Manoukian, E. B. (1986c). *Modern Concepts and Theorems of Mathematical Statistics*. Springer-Verlag, New York.

Manoukian, E. B. (1987a). "Functional Approach to Scattering in Quantum Mechanics". *Int. J. Theor. Phys.* **26** (10), pp. 981–989.

Manoukian, E. B. (1987b). "Functional Differential Equations for Gauge Theories". *Phys. Rev. D* **35** (6), pp. 2047–2048.

Manoukian, E. B. (1987c). "Reflection Off a Reflecting Surface in Quantum Mechanics: Where do the Reflections Actually Occur? II. Law of Reflection". *Nuovo Cimento* **100 B** (2), pp. 185–194.

Manoukian, E. B. (1989). "Theoretical Intricacies of the Single-Slit, the Double-Slit and Related Experiments in Quantum Mechanics". *Found. Phys.* **19** (5), pp. 479–504.

Manoukian, E. B. (1990). "Temperature-Dependent Particle Production and Stimulated Emissions by External Sources". *Int. J. Theor. Phys.* **29** (12), pp. 1313–1325.

Manoukian, E. B. (1991). "Derivation of the Closed-Time Path of Quantum Field Theory at Finite Temperatures". *J. Phys. G: Nucl. Part. Phys.* **17** (11), pp. L173–L175.

Manoukian, E. B. and Bantitadawit, P. (1999). "Direct Derivation of the Schwinger Quantum Correction to the Thomas-Fermi Atom". *Int. J. Theor. Phys.* **38** (3), pp. 897–899.

Manoukian, E. B. and Muthaporn, C. (2002). "The Collapse of "Bosonic Matter"". *Prog. Theor. Phys.* **107** (5), pp. 927–939.

Manoukian, E. B. and Muthaporn, C. (2003a). "Is 'Bosonic Matter' Unstable in 2D?" *J. Phys. A: Math. Gen.* **36** (3), pp. 653–663.

Manoukian, E. B. and Muthaporn, C. (2003b). "$N^{5/3}$ Law for Bosons for Arbitrary Large N". *Prog. Theor. Phys.* **110** (2), pp. 385–391.

Manoukian, E. B. and Osaklung, J. (2000). "A Derivation of the $Z \to \infty$ Limit for Atoms". *Prog. Theor. Phys.* **103** (4), pp. 697–702.

Manoukian, E. B. and Prugovečki, E. (1971). "On the Existence of Wave Operators in Time-Dependent Potential Scattering for Long-Range Potentials". *Can. J. Phys.* **49** (1), pp. 102–107.

Manoukian, E. B. and Rotjanakusol, A. (2003). "Quantum Dynamics of the Stern-Gerlach (S-G) Effect". *Eur. Phys. J. D* **25** (3), pp. 253–259.

Manoukian, E. B. and Siranan, S. (2005). "Action Principle and Algebraic Approach to Gauge Transformations in Gauge Theories". *Int. J. Theor. Phys.* **44** (1), pp. 53–62.

Manoukian, E. B. and Sirininlakul, S. (2004). "Rigorous Lower Bounds for the Ground State Energy of Matter". *Phys. Lett. A* **332** (1–2), pp. 54–59. [Erratum: *ibid.* **337** (4–6), p. 496].

Manoukian, E. B. and Sirininlakul, S. (2005). "High-Density Limit and Inflation of Matter". *Phys. Rev. Lett.* **95** 190402, pp. 1–3.

Manoukian, E. B. and Yongram, N. (2002). "Field Theory Methods in Classical Dynamics". *Int. J. Theor. Phys.* **41** (7), pp. 1327–1337. [ArXiv: hep-th/0411273].

Manoukian, E. B. and Yongram, N. (2004). "Speed Dependent Polarization Correlations in QED and Entanglement". *Eur. Phys. J. D* **31** (1), pp. 137–143. [ArXiv: quant-ph/0411079].

Marc, G. and McMillan, W. G. (1985). "The Virial Theorem". In Prigogine, I. and Rice, S. A. (Editors), "Advances in Chemical Physics, Vol. 58", Wiley, New York, pp. 209–361.

Marion, J. B. and Thornton, S. T. (1988). *Classical Dynamics of Particles and Systems*. Harcourt Brace Jovanovich Publishers, San Diego, 3rd edn.

Miranowicz, A. and Tamaki, K. (2002). "An Introduction to Quantum Teleportation". *Math. Sciences (Suri-Kagaku)* **473**, pp. 28–34. [ArXiv: quant-ph/0302114].

Misner, C. W., Thorne, K. S. and Wheeler, J. A. (1973). *Gravitation.* W. H. Freeman & Co., San Francisco.

Misra, B. and Sudarshan, E. C. G. (1977). "The Zeno's Paradox in Quantum Theory". *J. Math. Phys.* **18** (4), pp. 756–763.

Monroe, C. *et al.* (1996). "A "Schrödinger Cat" Superposition State of an Atom". *Science* **272** (5265), pp. 1131–1136.

Morgan III, J. D. (1996). In Drake, G. W. F. (Editor), "Atomic, Molecular and Optical Physics Handbook", American Institute of Physics, Woodbury, New York.

Morse, P. M. and Feshbach, H. (1953). *Methods of Theoretical Physics, Part I*. McGraw-Hill, New York.

Muthaporn, C. and Manoukian, E. B. (2004a). "Instability of "Bosonic Matter" in All Dimensions". *Phys. Lett. A* **321** (3), pp. 152–154.

Muthaporn, C. and Manoukian, E. B. (2004b). "N^2 Law for Bosons in $2D$". *Rep. Math. Phys.* **53** (3), pp. 415–424.

Nakazato, H. *et al.* (1995). "On the Quantum Zeno Effect". *Phys. Lett. A* **199** (1–2), pp. 27–32. [ArXiv: quant-ph/9605008].

Newton, R. G. (1976). "Optical Theorem and Beyond". *Am. J. Phys.* **44** (7), pp. 639–642.

Newton, T. D. and Wigner, E. P. (1949). "Localized States for Elementary Systems". *Rev. Mod. Phys.* **21** (3), pp. 400–406.

Nielsen, M. A., Knill, E. and Laflamme, R. (1998). "Complete Quantum Teleportation Using Nuclear Magnetic Resonance". *Nature* **396**, pp. 52–55. [ArXiv: quant-ph/9811020].

Ochkur, V. I. (1963). *Zh. Eksp. Teor. Fiz.* **45**, pp. 734–741. [English translation: *Sov. Phys. JETP* **18**, 503–508 (1964).].

Ochkur, V. I. (1964). "Ionization of the Hydrogen Atom by Electron Impact with Allowance for Exchange". *Zh. Eksp. Teor. Fiz.* **47** (5), pp. 1746–1750. [English translation: *Sov. Phys. JETP* **20**, 1175–1178 (1965).].

Olariu, S. and Iovitzu Popescu, I. (1985). "The Quantum Effects of Electromagnetic Fluxes". *Rev. Mod. Phys.* **57** (2), pp. 339–436.

Paley, R. E. A. C. and Wiener, N. (1934). *Fourier Transforms in the Complex Domain, $\mathcal{AMS}$ Colloquium Puplications, Vol. 19*. American Mathematical Society, Providence, RI.

Pancharatnam, S. (1956). "Generalized Theory of Interference, and Its Applications. I: Coherent Pencils". *Proc. Ind. Acad. Sci. A* **44** (5), pp. 247–262.

Papanicolaou, N. (1976). "Infrared Problems in Quantum Electrodynamics". *Phys. Rep.* **24** (4), pp. 229–313.

Pauli, W. (1964). In Kronig, R. and Weisskopf, V. F. (Editors), "Wolfgang Pauli, Collected Scientific Papers, Vol. 2", John Wiley & Sons, New York, pp. 544–552.

Perès, A. (1986). "When is a Quantum Measurement?" *Am. J. Phys.* **54** (8), pp. 688–692.

Provost, J. P. and Vallee, C. (1980). "Riemannian Structure on Manifolds of Quantum States". *Commun. Math. Phys.* **76** (3), pp. 289–301.

Prugovečki, E. (1971). "On Time-Dependent Scattering Theory for Long-Range Interactions". *Nuovo Cimento* **4 B**, pp. 105–123.

Pryce, M. H. L. (1948). "The Mass-Centre in the Restricted Theory of Relativity and Its Connexion with the Quantum Theory of Elementary Particles". *Proc. R. Soc. Lond. A* **195** (1040), pp. 62–81.

Rabi, I. I. (1988). "Otto Stern and the Discovery of Space Quantization, I. I. Rabi as told to John S. Rigden". *Z. Phys. D* **10**, pp. 119–120.

Ramsey, N. F. (1950). "A Molecular Beam Resonance Method with Separated Oscillating Fields". *Phys. Rev.* **78** (6), pp. 695–699.

Ramsey, N. F. (1990). "Experiments with Separated Oscillatory Fields and Hydrogen Masers". *Rev. Mod. Phys.* **62** (3), pp. 541–552.

Ramsey, N. F. and Silsbee, H. B. (1951). "Phase Shifts in the Molecular Beam Method of Separated Oscillating Fields". *Phys. Rev.* **84** (3), pp. 506–507.

Rauch, H. and Petrascheck, D. (1978). "Neutron Diffraction". In Dachs, H. (Editor), "Topics in Current Physics, Vol. 6", Springer-Verlag, Berlin, pp. 303–351.

Rauch, H. et al. (1975). "Verification of Coherent Spinor Rotation of Fermions". *Phys. Lett. A* **54** (6), pp. 425–427.

Reed, M. C. and Simon, B. (1978). *Methods of Modern Mathematical Physics, Vol. IV: Analysis of Operators*. Academic Press, New York.

Rudin, W. (1966). *Real and Complex Analysis*. McGraw-Hill, New York.

Sakurai, J. J. (1994). *Modern Quantum Mechanics*. Addison-Wesley, Reading, Massachusetts, 2nd edn. Edited by Tuan, S.-F.

Samuel, J. and Bhandari, R. (1988). "General Setting for Berry's Phase". *Phys. Rev. Lett.* **60** (23), pp. 2339–2342. Reprinted in Shapere and Wilczek (1989) pp. 149–152.

Schulman, L. S. (1971). "Approximate Topologies". *J. Math. Phys.* **12** (2), pp. 304–308.

Schwartz, C. and Tiemann, J. J. (1959). "New Calculation of the Numerical Value of the Lamb Shift". *Ann. Phys. (N.Y.)* **6** (2), pp. 178–187.

Schwinger, J. (1951a). "On the Green's Functions of Quantized Fields. I". *Proc. Nat. Acad. Sci. USA* **37** (7), pp. 452–455.

Schwinger, J. (1951b). "The Theory of Quantized Fields. I". *Phys. Rev.* **82** (6), pp. 914–927.

Schwinger, J. (1953). "The Theory of Quantized Fields. II". *Phys. Rev.* **91** (3), pp. 713–728.

Schwinger, J. (1958a). "Addendum to Spin, Statistics, and the TCP Theorem". *Proc. Nat. Acad. Sci. USA* **44** (6), pp. 617–619.

Schwinger, J. (1958b). "On the Euclidean Structure of Relativistic Field Theory". *Proc. Nat. Acad. Sci. USA* **44** (9), pp. 956–965.

Schwinger, J. (1960a). "Quantum Variables and the Action Principle". *Proc. Nat. Acad. Sci. USA* **47** (7), pp. 1075–1083.

Schwinger, J. (1960b). "The Special Canonical Group". *Proc. Nat. Acad. Sci. USA* **46** (10), pp. 1401–1415.

Schwinger, J. (1960c). "Unitary Transformations and the Action Principle". *Proc. Nat. Acad. Sci. USA* **46** (6), pp. 883–897.

Schwinger, J. (1961a). "Brownian Motion of a Quantum Oscillator". *J. Math. Phys.* **2** (3), pp. 407–432.

Schwinger, J. (1961b). "On the Bound States of a Given Potential". *Proc. Nat. Acad. Sci. USA* **47** (1), pp. 122–129.

Schwinger, J. (1962). "Exterior Algebra and the Action Principle I". *Proc. Nat. Acad. Sci. USA* **48** (4), pp. 603–611.

Schwinger, J. (1964). "Coulomb Green's Function". *J. Math. Phys* **5** (11), pp. 1606–1608.

Schwinger, J. (1970). *Particles, Sources and Fields, Vol. I.* Addison-Wesley, Reading, MA.

Schwinger, J. (1981). "Thomas-Fermi Model: The Second Correction". *Phys. Rev. A* **24** (5), pp. 2353–2361.

Schwinger, J. (1991). *Quantum Kinematics and Dynamics.* Addison-Wesley, Redwood City.

Schwinger, J. (2001). *Quantum Mechanics: Symbolism of Atomic Measurements.* Springer-Verlag, Berlin. Edited by Englert, B.-G.

Scott, J. M. C. (1952). "The Binding Energy of the Thomas-Fermi Atom". *Phil. Mag.* **43** (343), pp. 859–867.

Senjanovic, P. (1976). "Path Integral Quantization of Field Theories with Second-Class Constraints". *Ann. Phys. (N.Y.)* **100** (1–2), pp. 227–261.

Shapere, A. and Wilczek, F. (Editors) (1989). *Geometric Phases in Physics.* World Scientific, Singapore.

Simon, B. (1983). "Holonomy, the Quantum Adiabatic Theorem, and Berry's Phase". *Phys. Rev. Lett.* **51** (24), pp. 2167–2170. Reprinted in Shapere and Wilczek (1989) pp. 137–140.

Stoll, M. E., Vega, A. J. and Vaughan, R. (1977). "Explicit Demonstration of Spinor Character for a Spin-1/2 Nucleus via NMR Interferometry". *Phys. Rev. A* **16** (4), pp. 1521–1524.

Stump, D. R. and Pollack, G. L. (1998). "Radiation by a Neutron in a Magnetic Field". *Eur. J. Phys.* **19** (1), pp. 59–67.

Sudarshan, E. C. G. and Mukunda, N. (1974). *Classical Dynamics: A Modern Perspective.* John Wiley & Sons, New York.

Suter, D., Mueller, K. T. and Pines, A. (1988). "Study of the Aharonov-Anandan Quantum Phase by NMR Interferometry". *Phys. Rev. Lett.* **60** (13), pp. 1218–1220. Reprinted in Shapere and Wilczek (1989) pp. 221–223.

Tani, S. (1951). "Connection Between Particle Models and Field Theory. I". *Prog. Theor. Phys.* **6**, p. 267.

Teller, E. (1962). "On the Stability of Molecules in the Thomas-Fermi Theory". *Rev. Mod. Phys.* **34** (4), pp. 627–631.

Thirring, W. E. (Editor) (1991). *The Stability of Matter: From Atoms to Stars: Selecta of Elliott H. Lieb.* Springer-Verlag, Berlin.

Titchmarsh, E. C. (1937). *Introduction to the Theory of Fourier Integrals.* Clarendon Press, Oxford, England.

Tittel, W., Ribordy, G. and Gisin, N. (1998). "Quantum Cryptography". *Physics World* **11** (3), pp. 41–45.

Tomonaga, S.-I. (1997). *The Story of Spin.* University of Chicago Press, Chicago. English Translated by Oka, T.

Tycko, R. (1987). "Adiabatic Rotational Splittings and Berry's Phase in Nuclear Quadrupole Resonance". *Phys. Rev. Lett.* **58** (22), pp. 2281–2284. Reprinted in Shapere and Wilczek (1989) pp. 217–220.

van de Hulst, H. C. (1949). "On the Attenuation of Plane Waves by Obstacles of Arbitrary Size and Form". *Physica* **15** (8–9), pp. 740–746.

Wagh, A. G. and Rakhecha, V. C. (1993). "Exact Fixed-Angle Spinor Evolutions via the Rotating-Frame Formalism". *Phys. Rev. A* **48** (3), pp. R1729–R1732.

Walls, D. F. and Milburn, G. J. (1985). "Effect of Dissipation on Quantum Coherence". *Phys. Rev. A* **31** (4), pp. 2403–2408.

Watson, G. N. (1966). *A Treatise on the Theory of Bessel Functions.* Cambridge University Press, Cambridge, 2nd edn.

Weinberg, S. (1964). "Systematic Solution of Multiparticle Scattering Problems". *Phys. Rev.* **133** (1B), pp. B232–B256.

Weinberg, S. (1995). *The Quantum Theory of Fields, Vol. 1: Foundations.* Cambridge University Press, Cambridge.

Weisskopf, V. F. (1980). "Growning Up with Field Theory, The Development of Quantum Electrodynamics in Half a Century, ...". In Weisskopf, V. F. (Editor), "The 1979 Bernard Gregory Lectures", CERN, Geneva, pp. 1–53.

Werner, S. A. *et al.* (1975). "Observation of the Phase Shift of a Neutron Due to Precession in a Magnetic Field". *Phys. Rev. Lett.* **35** (16), pp. 1053–1055.

Wheeler, J. A. and Zurek, W. H. (Editors) (1983). *Quantum Theory and Measurement*. Princeton University Press, Princeton, New Jersey.

Wilczek, F. and Zee, A. (1984). "Appearance of Gauge Structure in Simple Dynamical Systems". *Phys. Rev. Lett.* **52** (24), pp. 2111–2114. Reprinted in Shapere and Wilczek (1989) pp. 141–144.

Wilkinson, D. T. and Crane, H. R. (1963). "Precision Measurement of the g Factor of the Free Electron". *Phys. Rev.* **130** (3), pp. 852–863.

Wolf, E. (1951). "Light Distribution Near Focus in an Error-Free Diffraction Image". *Proc. R. Soc. Lond. A* **204** (1079), pp. 533–548.

Wu, T. T. and Yang, C. N. (1975). "Concept of Nonintegrable Phase Factors and Global Formulation of Gauge Fields". *Phys. Rev. D* **12** (12), pp. 3845–3857.

Wu, Z. and Yang, H. (2004). "Validity of Quantum Adiabatic Theorem". [ArXiv: quant-ph/0410118].

Yakubovsky, O. A. (1967). "On the Integral Equations in the Theory of N Particle Scattering". *Yad. Fiz.* **5**, pp. 1312–1320. [English translation: *Sov. J. Nucl. Phys.* **5**, 937–942 (1967).].

Yongram, N. and Manoukian, E. B. (2003). "Joint Probabilities of Photon Polarization Correlations in e^+e^- Annihilation". *Int. J. Theor. Phys.* **42** (8), pp. 1755–1764. [ArXiv: quant-ph/0411072 with Errata].

Yurke, B. and Stoler, D. (1986). "Generating quantum mechanical superpositions of macroscopically distinguishable states via amplitude dispersion". *Phys. Rev. Lett.* **57** (1), pp. 13–16.

Zhu, S.-L., Wang, Z. D. and Zhang, Y.-D. (2000). "Nonadiabatic Noncyclic Geometric Phase of a Spin-1/2 Particle Subject to an Arbitrary Magnetic Field". *Phys. Rev. B* **61** (2), pp. 1142–1148.

Zinn-Justin, J. (1981). "Perturbation Series at Large Orders in Quantum Mechanics and Field Theories: Application to the Problem of Resummation". *Phys. Rep.* **70** (2), pp. 109–167.

Zukowski, M. *et al.* (1993). ""Event-Ready-Detectors" Bell Experiment via Entanglement Swapping". *Phys. Rev. Lett.* **71** (26), pp. 4287–4290.

Zurek, W. H. (1991). "Decoherence and the Transition from Quantum to Classical". *Physics Today* **44** (10), pp. 36–44.

Index

A-description, 13
A-measurement, 6
abelian transformations, 516
absorption coefficient, 873
absorption of neutrons, 873
absorption by a machine, 19
action integral, 601, 606, 959, 965
addition of states, 17
adiabatic approximation, 717
adiabatic hypothesis, 514
adiabatic process, 514
adiabatic regime, 513
adiabatic theorem, 724, 725, 727
adjoint, 10, 23
advanced Green function, 554
Aharonov-Anandan (AA) phase, 520
Aharonov-Bohm effect, 572, 576, 579
Airy functions, 706, 710
  properties of, 706
algebra of the generators of the inhomogeneous Lorentz transformations, 898
amplitude, 9, 548, 602
angular integration, 400
angular momentum, 116, 249
  addition of angular momenta, 275
  eigenvalue problem, 251
angular momentum states
  arbitrary spins, 312
  single particle states, 307
  two particle states, 317
  two particles of arbitrary spins, 319

anharmonic potentials, 683, 691, 692, 696
anholonomy, 524
annihilation operators, 471, 954
anomalous magnetic moment
  expression for, 452
anomalous magnetic moment of the electron, 444
  computation of, 446–453
  observational aspect, 445, 446
anti-commutation relations, 136, 343, 886, 964, 968
anti-commuting c-number variables, 136, 225, 226, 633, 637, 674
anti-linear functional $|x\rangle$, 38
anti-linear functionals, 37
anti-linear operator, 57
anti-matter, 882, 932
anti-particles, 882, 953
anti-unitary, 57, 89
anti-unitary operator, 112
associated Laguerre polynomials, 367, 413, 439
asymptotic "free" propagators, 832
asymptotic boundary conditions, 800, 803
asymptotic time development of charged particle states, 832
asymptotically "free" Coulomb Green functions, 830
atom in external magnetic field, 406
atomic bomb, 765, 883
average ionization potential, 870

1000   Index

B-filter, 7
$|b\rangle\langle a|$-type operation, 4
Baker-Campbell-Hausdorff formula, 354, 605, 975
Bargmann-Wigner set of equations, 947
Bell state measurement, 502
Bell states, 502
Bell's Test, 486
Berry phase, 513, 515
Bessel functions, 220, 810, 844
    modified Bessel function, 220
Bessel's inequality, 35, 51
β^- decay, 932
β^+ decay, 932
Bethe's non-relativistic approximation, 400
bits, 504
blocked spin components, 13
boosts, 91, 892, 894
Born approximation, 720, 816, 858
    validity of, 816
Bose excitations, 628
Bose/Fermi excitations, 672
Bose-Einstein factor, 342
Bose-Fermi oscillator, 136, 346
    supersymmetric transformations, 346
bosonic matter, 765
bosonic matter in bulk, 765, 792
bosons, 37, 783, 955, 963
bound-states, 187, 192–199
boundary conditions, 554, 800, 801, 806, 807
boundedness of Hamiltonians from below, 102, 143, 152–163, 183, 219, 329, 360, 776, 786
box diagram, 88, 137
bra, 23
Breit-Weisskopf-Wigner formula, 172

canonical transformations, 615, 617, 624, 625
Cartesian components in terms of spherical harmonics, 265
Cauchy sequence, 34, 806
Cauchy's Theorem, 177, 695
Cauchy-Schwarz inequality, 24, 34, 979
center of mass, 107, 108, 819, 847
change of integration variables of Grassmann type, 637

channel Hamiltonians, 847–849
channels, 847
charge conjugation, 903, 907, 940, 941
charge distribution of proton, 417
charge renormalization, 856, 932
Chebyshev's polynomials of type II, 983
    generalized orthogonality relation, 985
    orthogonality relation, 983
    properties of, 983
chirality, 915
chronological time anti-ordering, 111
chronological time ordering, 110
classical Lagrangian, 562, 564, 599
Clauser-Horne (C-H) inequality, 499, 500
Clebsch-Gordan coefficients, 275, 281, 288
    explicit (Racah) expression, 287
clone, 503
closed curve in parameter space, 522
closed path, 86, 87, 91, 136, 138
closed-time path and expectation-value formalism, 675, 676
clusters and bound systems, 846
clusters of particles, 106
    inter-clusteral interactions, 106, 109
    intra-clusteral interactions, 106, 109
coherent state of harmonic oscillator, 349
coherent states, 354
collapse, 765
collapse of "bosonic matter", 783
commutation relations, 85, 92, 93, 103, 139, 144, 249, 330, 963, 968
commutator, 92
complete set of observables, 17
completeness, 34, 80
completeness relation, 911
completeness relation for Green functions, 547
completeness unitarity expansion, 679
complex conjugation, 9, 10
complex Grassmann variables, 637
complex quantities, 15, 16, 18
Compton wavelength, 922, 943
conditional amplitude, 567

conditional probability, 68, 69, 72, 73, 78, 80, 494, 498, 506
Condon and Shortley convention, 286
confluent hypergeometric function, 936
connected part, 860, 863, 865
connectedness and Faddeev equations, 858
connection between configuration and momentum spaces in scattering, 807
connection between the quantum dynamical principle and the path integral formalisms, 672
connection formula, 712
conservation of angular momentum, 97
conservation of momentum, 97
constrained dynamics, 614
constrained path integrals, 623
construction of Hamiltonians, 102–109
construction of supersymmetric Hamiltonians, 226–230, 247
contact term, 449
continuity condition, 354
continuity equation, 917, 940
continuity property of transformations approaching the identity, 89
continuous spectrum, 183
conversion factor, 90
convexity theorem, 977
correction due to electrons bound near the nucleus, 753
correlation between apparatus and physical system, 66, 71, 462
correlation between dynamical variables, 535
correlation function, 731
correlations, 531
Coulomb gauge, 533
Coulomb Green function, 590, 831, 833
  negative spectrum, 594
  positive spectrum, 596
Coulomb potential, 362, 363, 372, 392, 406, 412, 837, 933
Coulomb scattering, 830
counter-terms, 398, 401, 402
counting number of eigenvalues, 203, 206
coupling to the environment, 727, 734

creation operators, 471, 955
cross sections, 815, 854, 866
current density, 496
curved nature of parameter space, 525
curved space, 527
cyclotron frequency, 436

damping factor, 403
dangerous deltas, 862
Darwin term, 375, 928
decay constant, 404, 405, 443, 444
decay of excited states, 397, 403
decay of quantum systems, 168
decay width, 405
decoherence time scale, 736
decomposition into clusters, 119
definite state, 72
deflection angles, 843
degeneracy, 53, 518
degenerate perturbation theory, 688
degree of degeneracy, 25, 584, 594
density of states, 204, 582, 585
density operator, 26, 31, 468, 472, 534, 728, 730
derivative of exponential of operator-valued functions, 973
destroying quantum superpositions, 472
destructive interference, 74
detection of intruder, 507, 508
deuteron, 244, 325
deviation from the Thomas-Fermi approximation, 590
differential cross sections, 814, 818–821, 837, 858
differential operator, 635
dimensionality, 16
dipole-dipole interaction of two spin 1/2 particles, 326
Dirac bracket, 627
Dirac delta distribution, 38, 571
Dirac delta for Grassmann variables, 635
Dirac delta potential, 190, 558
Dirac equation, 371, 885
Dirac equation in the bound Coulomb problem, 933
Dirac position variable **x**, 920
Dirac sea or Dirac vacuum, 931
Dirac string, 528

Dirac time evolution operator, 911
disappearance of interference, 77, 478
discontinuity across a cut, 697
discrete eigenvalue, 43
discrete set, 65
discrete spectrum, 183, 200
dispersion relation, 695
distinguishable states of a physical system, 67
disturbing transition of a system by a measurement, 71
divergent series, 695
domain of A, 41
double-slit experiment, 74, 78, 551, 552, 599
double-valuedness of spinors under rotation, 129, 896
dual vector space, 23
Duffin-Kemmer-Petiau equation, 939, 970
dynamical phase factor, 515

effective charge "renormalization", 856
effective range, 878
effective range expansion, 878
eigenvalue equation, 43
eigenvalue problem, 24
eigenvalue problem and supersymmetry, 224
eigenvalues, 53, 183
eigenvectors, 43
eikonal approximation, 842
eikonal phase function, 844
elastic scattering, 800
elastic scattering amplitude, 866
electric dipole moment, 404
electromagnetic potential, 371
energy "width", 170
energy functional, 741
energy loss per unit path length of charged particle, 868
energy shell, 804
energy-time uncertainty principle, 170
entangled ancillary pair of particles, 501
entangled states, 72, 499, 500, 503
entangled swapping, 503
EPR paradox, 487
essential spectrum, 47, 48, 53

Euclidean space, $4D$, 592, 981
Euler angles, 255, 257
even permutation, 83
event, 892
exchange term, 756
existence of bound-states for $\nu = 1$, 192
existence of bound-states for $\nu = 2$, 194
existence of bound-states for $\nu = 3$, 195
expansion theorem, 984
expectation values, 10, 92, 246, 675
explicit expression for the Clebsch-Gordan coefficients, 284
exponential decay, 172, 427
exponential law, 432, 727, 728, 733
exponential representation of rotation matrix, 84
expression for g-factor, 440
expression for the ground-state energy of atoms, 762
expressions for transformations functions, 656–661, 664
external field problem, 926, 930, 933, 944
external force, 335
external sources, 335, 674

Faddeev equations, 864
"fall" of a particle into another, 143
Fermi excitations, 640
Fermi oscillator, 343
  annihilation, creation operators, 343
  ground-state persistence amplitude, 344
Fermi-Dirac factor, 345
fermions, 36
Feshbach-Villars equation, 939
field equation, 531
filter, 19
fine-structure, 689
fine-structure constant, 372, 375, 393
fine-structure correction, 382, 383
fine-structure of the H-atom, 377, 379
  $\ell = 0$ states, 381
  $\ell \neq 0$ states, 382
finite degeneracy, 183
first class constraints, 618
fixing a gauge, 621
fluctuations of the position of the electron, 375

flux, 572, 573
Foldy-Wouthuysen-Tani representation, 919, 927, 928, 969, 970
forced harmonic oscillator, 661
Fourier series, 573
Fourier transform, 99
Fourier transform in the complex domain, 174
"free" time evolution operator for the Coulomb case, 832
functional, 654
functional derivative, 653, 654

g-factor, 375
g-factor for the electron, 930
g-factor of the proton, 376
Galilean frame, 81
Galilean invariant theory, 101
Galilean scalar, 89
Galilean scalar (field), 117
Galilean space-time coordinate transformations, 83
Galilean transformation of operator, 92
Galilean transformations, 81, 82, 86
Galilean vector (field), 117
gamma matrices, 885–888
  chiral representation, 888
  Dirac-Pauli representation, 885
  Majorana representation, 888
gauge fixing parameter, 621
gauge freedom, 621
gauge transformation, 529
Gaussian distribution, 467
Gaussian integrals, 543, 613, 659, 702
general solution of Dirac equation, 908
generalized orthogonality relation, 985
generalized oscillator strength, 869
generating function, 570, 735, 983
generation of arbitrary spins, 121
generation of states, 15
  $\langle a|$ and $|b\rangle$ symbols, 16
  two-stage process, 15
    annihilation, 16
    production, 16
generators of the Galilean transformations, 91
geodesic, 529–531
geodesic equation, 529
geometric phases, 513, 515, 519, 520

Golden rule, 720
Grassmann variables, 137, 344, 633, 635, 674
Green function near the energy shell, 833
Green functions, 547, 580, 740
  matrix notation, 580
group properties, 55, 87, 88
  associativity, 87
  inverse, 87
grouping of particles into clusters, 107

Hölder's inequality, 148, 222, 782, 979
Hamilton-Jacobi-like equation, 738
Hamiltonian operator, 55, 94
harmonic oscillator, 329
  annihilation and creation operators, 335
  finite temperature, 340
  Gaussian probability density, 355
  ground-state persistence amplitude, 336
  persistence probability at high temperatures, 342
  spectrum, 331
  stimulated excitations, 340
  time-dependent disturbance, 335
  transition to and between excited states, 335
  wavefunctions in the x-description, 333
harmonic oscillator wavefunctions, 238
$\hbar$ (Planck's constant divided by 2π), 90, 91
Heisenberg representation, 652
Heisenberg's equations of motion, 97
Heisenberg-like picture, 92
helicity, 313, 322, 900, 915, 953
helicity states, 313, 322
helicity versus standard spin states, 315
  two particle systems, 324
Hermitian operator, 24, 41
Hermiticity, 40
high-density limit for matter, 780
high-order perturbations, 695
Hilbert space, 33
"hole" theory, 932
hydrogen atom, 370
  eigenstates, 366

eigenvalue problem, 363
orthogonality relation, 368, 369
stability of, 363
hydrogen molecule, 692
hyperfine-structure of the H-atom, 377, 384
$\ell = 0$, 387
$\ell \neq 0$, 391

ideal apparatus, 71, 73, 465, 477, 480
identity operation, 4, 12
identity operator, 66
impact parameter, 830
impact parameter representation of scattering amplitude, 844
indistinguishable particles, 36, 819
indistinguishable spin 1/2 particles, 820
inelastic processes, 800, 867
inertial frame, 881, 892
infimum, 51
infinitely many degrees of freedom, 486
infinitesimal Lorentz transformations, 896
infinitesimal rotation, 83, 894
infinitesimal transformations, 83, 91, 92, 894
inflation of matter, 781
initial value problem for Dirac equation, 917
initially prepared state, 16
inner product, 23
inner product space, 23, 34
intensity distribution, 535
intensity of light, 33
inter-clusteral interactions, 108, 109, 847
interacting states, 800, 801, 851
interaction picture, 429
interaction time, 471
interaction with a harmonic oscillator, 467
interaction with an apparatus, 463
interactions of short range, 800
interference effect, 524
interference terms, 74–77, 469, 477
internal angular momentum, 119, 120
internal energy, 104
intra-clusteral interaction potential, 108, 847

intrinsic parities, 496, 909, 931
intrinsic parity of a photon, 496
invariance, 89, 97
invariant, 8
inverse operator, 45
irreducibility of constraints, 618
isolated point, 43

Jacobian, 118
Jacobian of transformation for Grassmann variables, 638
Jensen's formula, 177
joint probability of spins correlations, 494

ket, 23
key (cryptography), 504, 506
kittens, 419
Klein-Gordon equation, 937, 938, 970
Kramers degeneracy, 132
Kummer's equation, 936

L-H polarizations, 496
labelling of space-time points, 82
ladder operators, 250, 254
Lagrange multipliers, 617, 621
Lagrangian for spin 0 particles, 955
Lagrangian for spin 1/2 particles, 957
Lamb shift, 391
computation of, 394–401
Lamb shift and renormalization, 398
Landé g-factor, 294
Landau levels, 436
Landau-Larmor energy, 437
Laplacian in parabolic coordinate system, 417
law of reflection, 565, 570
left-hand derivative, 634
left-handed, 916
left-handed (L-H) polarization, 495
Lieb-Thirring bound, 217, 776
Lieb-Thirring kinetic energy inequality, 221, 224
lifetime, 170
limit of potential well, 190
line of singularities, 528
linear and quadratic potentials, 555
linear combination, 17

linear combination of selective measurement symbols, 11
linear operator, 57
linear superposition, 17
"linearity" condition, 37
Lippmann-Schwinger equations, 802
Local Hidden Variables (LHV) theories, 488
locally square-integrable, 182
Lommel functions, 812
long wavelength approximation, 406
longitudinal component of magnetic field, 534, 537
Lorentz covariance, 892, 894
Lorentz force, 531, 538
Lorentz invariant scalar product, 899
Lorentz scalar, 900
Lorentz transformations, 892, 893
Lorentz vector, 900
lower and upper bounds for the ground-state energy for matter, 774
lower bound for the expectation value of the separation distance between two particles in negative energy, 148
lower bounds to the expectation value of the kinetic energy, 220
  multi-particle states: bosons, 224
  multi-particle states: fermions, 222
  one-particle systems, 220

Møller wave operators, 807, 879
machine, 14–16
machine operating in reverse, 15
machine represented by 2×2 matrix, 20
magnetic dipole moment, 372
magnetic dipole moment of the proton, 373
magnetic field gradient, 537, 538
magnetic monopole, 528
Malus formula, 31
Markov approximation, 431, 731, 732
mass renormalization, 401
massless Dirac particles, 912
master equation, 727, 728, 731
matrix elements, 12
matrix elements of $\exp(i\varphi \mathbf{n} \cdot \mathbf{S}/\hbar)$, 127

matrix elements of finite rotations, 254
$D_{mm'}^{(j)}(\alpha, \beta, \gamma)$ functions, 257, 258
$d_{mm'}^{(j)}(\beta)$ functions, 258
orthogonality properties, 284
matrix multiplication, 12
matter without the exclusion principle, 765
mean lifetime, 404, 444
mean lifetime of the state $2P_{1/2}$, 404
"mean" orbital angular momentum, 923, 925
"mean" position, 926
"mean" position operator, 923, 925
"mean" spin, 923
"mean" spin operator, 925
mean square radius of proton, 417
mesoscopic states, 484
metric, 524, 525
metric on the unit sphere, 527
minimal electromagnetic coupling, 378
Minkowski metric, 886
Minkowski's inequality, 978
mixed-description, 12
mixtures, 25
  mixture, 26, 31
  random mixture, 27, 29
modified Bessel functions, 220, 570
momentum operator, 96
monitoring observables, 79
monitoring spin, 478
motion restricted to the surface of unit sphere, 527
Mott scatterings, 837
multi-channel scattering, 846, 849
multi-component object, 89
multi-electron atoms, 105, 739
multi-particle Hamiltonians, 104, 167, 739, 766
multi-particle systems, 104, 106, 148, 163–168, 739, 766, 850
multi-particle systems with Coulomb interactions, 167, 739, 766
muon decay, 932
Møller wave operators for Coulomb potential, 879

$N^{5/3}$ law for bosons, 784
natural units, 90
"needle" registering-value, 65

neutral state, 71
neutron interferometer, 867, 871, 875
9-j symbol, 304, 306
    orthogonality property, 307
    relation to 6-j symbols, 306
No-Binding Theorems, 197, 774
non-abelian gauge theories of fundamental interactions, 519
non-abelian transformations, 519
non-commutativity of measurements, 31, 423
non-degenerate perturbation theory, 684
non-empty subspace, 43
non-flip of spin, 75, 476
non-holonomic system, 524
non-local density of states, 204, 583
non-relativistic electron, 403
non-zero spins, 89
normalization condition, 9, 18, 24, 100
normalization condition for momenta, 100
normalization factor, 70
normalization of probability, 492
number of eigenvalues, 210, 583, 668
    Bargmann inequality for number of eigenvalues, 215
    number of eigenvalues in one-dimensional case, 213
    number of eigenvalues in three-dimensional case, 211
    number of eigenvalues in two-dimensional case, 212
    Schwinger inequality for number of eigenvalues, 211
numerical factor $\langle b|a \rangle$, 4
numerical factors, 7, 8, 10

observable, 2
    compatible observables, 2
    complete set of compatible observables, 2
    continuous set of values, 2
    finite number of discrete values, 2
    incompatible observables, 2
    infinite number of discrete values, 2
odd permutation, 83
old quantum theory, 712
one-time pad scheme, 504

optical theorem, 822, 853, 854
optical theorem and its interpretation, 821
orbital angular momentum, 117, 258
    half-odd integral values?, 259
order of measurements, 505
orthogonal, 51
orthogonal entangled states, 502
orthogonal projection operations, 910
orthogonality conditions, 66, 79
orthogonality relation, 45
orthogonality relation of photon states, 416
orthonormal basis, 58
orthonormality property, 122
oscillator strength, 869
oscillatory magnetic field, 13

p-description, 100
Padé approximant, 738
Padé approximants method, 702
pair annihilation, 931
pair production, 931
Paley-Wiener Theorem, 172, 174
Pancharatnam definition, 524, 546
parabolic coordinate system, 417
parallel transport, 524
parallel transport of a vector, 527
parameter space, 517
parity, 112, 114
parity transformation, 130, 902
particle detection, 807
Paschen-Back effect, 408
passage of charged particles through hydrogen, 867
passage of particles through media, 867
path integrals
    and velocity dependent potentials, 608
    completeness relation, 601
    constrained dynamics, 614
    for a given potential, 604
    particle in external electromagnetic fields, 608
Pauli equation, 433
Pauli exclusion principle, 765
Pauli Hamiltonian, 229, 432
Pauli matrices, 421, 885
Pauli's fundamental theorem, 886, 889

$\mathcal{P}, \mathcal{C}, \mathcal{T}$ transformations, 902
$\mathcal{PCT}$ transformation, 969
permutations, 123
phase factor, 61, 522
phase shifts analysis, 825, 865
photon, 391, 442, 447
physical system transitions, 69
physically observed mass, 398, 403
plane rotator, 737
Poincaré sphere, 526
Poisson equation, 415
Poisson equation in $4D$, 982
Poisson probability mass function, 352
Poisson sum formula, 571
polarization of light, 29
    ϑ-polarizer, 32
    x-polarized state, 30
    x-polarizer, 30
    y-polarized state, 30
    unpolarized light, 32
polarization vector, 391
polarized along the z-axis, 457
polarized beam, 457
position operators, 93, 96, 919, 942
positivity, 536
positivity constraint, 537
positron, 109, 393, 494, 882, 932
possible outcomes of spin measurements, 492
potential well, 187
power of test of detection of the intruder, 507
preparatory channels, 851
presence of radiation, 403
primary constraints, 617
probabilities of counts, 488
probability, 9, 18, 68
probability current, 917
probability density, 29, 536
probability distribution of the kinetic energy, 414
probability for the "fall" of one particle into the other, 148
probability mass function, 488
probability of decay, 480, 481
product of the exponentials of operators, 973
projection operators, 42, 80

Pryce-Newton-Wigner operator, 924, 969
pseudo-scalar, 902
pseudo-vector, 902
pure ensembles, 25

quadrupole moment
    reduced matrix element, 327
quantization of the electric charge, 528
quantized flux, 575, 576
quantum anti-Zeno effect, 482
quantum correction, 759
quantum cryptography, 501, 503
quantum decoherence, 79, 463, 469, 472, 482, 728
quantum dynamical principle, 649, 650, 652, 663
quantum dynamical principle to path integrals, 669
quantum dynamics, 100
quantum electrodynamics, 375, 490, 881
quantum field theory, 881
quantum fields, 954
quantum mechanical counterpart of Lorentz-force, 534
quantum mechanical phase depending on the gravitational constant and the Planck constant, 877
quantum superpositions, 79
quantum teleportation, 501
quantum Zeno effect, 482

R-H polarizations, 496
radial functions, 404
radiation field, 391
Ramsauer effect, 878
Ramsey apparatus, 13, 14, 473, 478
Ramsey oscillatory fields method, 473
Ramsey zones, 477, 478, 509
N Ramsey zones, 480
Ramsey-like method, 508
Rayleigh-Schrödinger series, 688, 700, 703, 736
rays, 37
re-summation methods, 702
reactive cross section, 866
real Grassmann variables, 633
reduced density operator, 468, 472, 728

reduced mass effect, 377
reduced masses, 849
reduced matrix element, 293, 296
reflection coefficient, 561
reflection probability, 711, 874
registering apparatus, 66
relative momentum, 848
relative phases, 56
relativistic correction to the kinetic energy, 375
relativistic quantum field theory, 881
relativistic wave equations for any mass and any spin, 947
relativistically invariant, 392
relativity, 392
renormalization
  contact term, 449
  counter term, 398, 402
  effective charge, 855, 932
  of mass, 402, 403
  physical significance and stability of the electron, 398, 402, 403, 449
  Z_3: wavefunction renormalization constant, 393
renormalized mass, 398, 402, 403
representation of simple machines, 14
representations of the generators **P**, **X**, 96
repulsive Coulomb potential energy lower bound for multi-particles, 767
"resistance" of Fermi particles to increase in density, 149
resolution of the identity, 42, 353
resolution of the identity in the x-description, 100
resolvent, 46, 152
resolvent set, 47
resonance, 481
resonance condition, 509
rest energy, 371
retarded Green function, 549
retarted/advanced Green functions, 801
reversal of phase, 513
Riemann-Lebesgue Lemma, 732
Rigged Hilbert space, 38, 39
right-hand derivative, 634
right-handed, 916
right-handed (R-H) polarization, 495

role of the environment, 469
rotation matrix, 82
rotation of a coordinate system, 84
rotation of a spinor, 508, 511, 512
rotation of a spinor by 2π radians, 129, 511, 512
Rydberg, 372, 400

6-j symbol, 304
  orthogonality relation, 305
  relation to 3-j symbols, 304
  sum rule, 305
Samuel-Bhandari (SB) phase, 529
scale transformations, 8, 146
scattering S operator, 879
scattering amplitudes, 804, 815, 826, 834
scattering at small deflection angles at high energies, 842
scattering length, 878
scattering of indistinguishable charged particles, 837
scattering of spin 1/2 particle off a spin 0 target, 454
scattering of spin 1/2 particles off a spin 1/2 target, 459
scattering theory
  functional treatment, 838
Schrödinger equation, 101
Schrödinger's cat, 482
Schwinger's constructive approach, 958
scrambled message, 504
second class constraints, 618
second rank anti-symmetric spinor, 129
secondary constraints, 618
secular equation, 689
selection rule, 266
selective measurement symbols, 2, 14, 30
selective measurements, 2, 3, 18, 486
  polarizer, 3
    polarized light, 3, 30, 495
    unpolarized light, 3, 32
  prism, 3
  Stern-Gerlach apparatus, 3, 4, 13
self-adjoint, 41
self-adjoint operators, 39
separability, 34, 80
separated oscillatory fields zones, 473

shape invariant partner potentials, 233
silicon, 873
smooth function, 722
solenoid, 572
solid angle, 517, 982
solution of the free Feshbach-Villars equation, 941
solutions of the fully interacting systems, 802
space reflection, 112
space translations, 91
spatial extension of bound-state systems, 148, 150, 151
spatial extension of matter, 782, 783
spatial extension of the fermionic systems, 150
spatial rotations, 892, 894
spectra of self-adjoint operators, 41, 45
spectral decomposition, 42
speed of light, 370, 399, 892
spherical Bessel functions, 827
spherical harmonics, 262
    addition theorem, 267
    integral involving three spherical harmonics, 301
    reduced matrix element, 303
    special values, 265
    tensor operator Y_M^L, 299
spherically symmetric potential, 214
spin, 269, 898, 900, 947
    arbitrary spins, 274, 947
    density operators, 453
spin s out of $2s$ spin 1/2's, 129
spin and relativistic corrections, 370
spin and statistics connection, 962
spin component in the $+z$ direction, 14
spin of the proton, 385
spin precession, 441
spin versus helicity states, 313
spin-flip, 73, 74, 76, 131, 469, 475
spin-orbit (SL) coupling, 376, 928
spinors, 121
spin 0, 117, 783, 819, 820, 937, 955, 963
spin 1, 6, 118, 129, 272
spin 1/2, 5, 7, 74, 77, 270
    general aspects, 420
spin 1/2 in external magnetic fields, 423, 425

splitting of beam along quantization axis, 538
spontaneously broken theories, 226
square-integrable, 172
square-integrable functions, 36, 44, 228, 229, 233, 242, 243
stability, 145
stability and multi-particle systems, 148
stability of matter, 774
stability of the hydrogen atom, 360
stable system, 150
Stark effect, 412
state, 16
state of combined (correlated) system, 66, 468
state of the apparatus, 67
state vector, 18
states of a physical system, 37
statistical (density) operator, 26, 31, 468, 472, 534, 728, 730
step-function, 41, 42
Stern-Gerlach, 2, 11
    filtering machine, 2
    filtering processes, 2, 6
Stern-Gerlach effect, 531
Stokes's theorem, 516
strong limits, 354, 806
strong magnetic field, 408, 411
subsidiary constraint, 622
successive Galilean transformations, 86, 91
successive Lorentz transformations, 376
successive operations, 3
successive operations of two machines, 15
successive selective measurements, 31
sudden approximation, 717, 720, 722
sum of the negative eigenvalues, 216
    one-dimensional case, 218
    three-dimensional case, 216
    two-dimensional case, 217
superposition principle, 17, 419, 482
superpotential, 231
superselection rules, 37, 484
supersymmetric Hamiltonians in higher dimensions, 247

supersymmetric partner Hamiltonians, 231
supersymmetric partner potentials, 231
supersymmetric transformations, 138
supersymmetry, 136, 224, 434
supersymmetry generators, 136, 225, 226, 346, 348
supremum, 51
survival probability, 169, 432
"switching on" of apparatus, 466
symbol $\Lambda(a) = |a\rangle\langle a|$, 2
symmetric spinors of rank k, 122
symmetrization operation, 948, 970
symmetry operation, 132
symmetry transformation, 61
systems of n particles, 104, 106, 118, 148, 163, 166, 167, 739, 766, 850

21.1 cm wavelength, 388
3-j symbol
  orthogonality property, 289
  relations to Clebsch-Gordan coefficients, 289
T-scattering matrix, 865
tensor operators, 296–303
TF electron density and the effective potential, 740, 742
thermal average, 340, 680
thermal mixture, 340, 342, 345, 680
Thomas factor, 376
Thomas-Fermi (TF) approximation, 587, 588, 740
Thomas-Fermi atom, 739, 740
  computation of ground-state energy, 745, 746
Thomas-Fermi-like energy functional, 769
time reversal, 112, 130, 903
time-dependent forced dynamics, 561, 564, 631, 643, 656, 660
time-dependent Hamiltonians, 111
time-evolution, 101, 109, 111, 665
trace, 12, 86
trace functionals, 665
trace operation, 10
transformation function $\langle \mathbf{x} | \mathbf{p} \rangle$, 98, 99
transformation function from the B- to A-descriptions, $\langle a | b \rangle$, 19

transformation functions, 22, 649, 652, 653
transformation law under coordinate rotation, 95, 120, 121, 249, 258
transformation law under time reversal, 131, 903
transformation rule of a spinor of rank one under a coordinate rotation, 133
transition amplitude, 337
transition probabilities, 719, 722, 853
translational independent contribution to $\mathbf{J}$, 117
transmission coefficient, 561
transmission probability, 711, 874
transpose, 86
trial wavefunctions, 690, 777, 788
triangular inequality, 24, 35
"tuning" condition, 468
turning point, 705, 707
two-dimensional Green function, 570
two-level atom, 428, 484
two-level systems, 427, 737
two-particle scattering, 818–821, 837
two-particle states of arbitrary spins versus helicity states, 322
two-particle systems and relative motion, 104

ultraviolet cut-off, 396
uncertainties, 144, 145, 362
uniform time-independent magnetic field, 13
unitarity, 67
unitary operator, 55, 57, 78, 89, 112
unpolarized beam, 455, 536, 538
upper bound cut-off, 399
upper bound for the expectation value of the kinetic energy operator for negative energies, 147, 150, 781

vacuum fluctuations, 932
variational form, 652
variational methods, 690
vector operators, 290, 298
vector space, 16, 34
vectors, 16, 33
velocity dependent potentials, 608, 612
Vernam cipher, 503

Index   1011

virial theorem, 245
volume of sphere in $4D$, 981

wavefunction in different descriptions, 18
Z_3: wavefunction renormalization constant, 393
wavefunctions of hydrogen atom, 368
wavepacket, 533, 809, 919
weak limit, 354, 805
weak magnetic field, 411
weak perturbations, 717, 718
Wigner's Theorem on symmetry transformations, 57, 81, 89
Wigner-Eckart theorem, 293, 297, 326, 389
winding number, 571
WKB approximation, 703, 705
  barrier penetration, 709
  quantization rules, 712, 714, 715
  radial equation, 715
WKB procedure in the one-dimensional radial equation, 716

x-description, 96, 100

Young's inequality, 211, 213, 980
Yukawa potential, 157, 198, 244
Yukawa term, 347

Zeeman effect, 406, 412
zero point energy, 334, 343, 350
0 vector, 17
Zitterbewegung, 375, 918, 923, 925, 926, 928, 931, 944